Craftsman 2008
NATIONAL
REPAIR & REMODELING
ESTIMATOR

by Albert S. Paxton
Edited by J. A. O'Grady

Includes inside the back cover:

- An estimating CD with all the costs in this book, plus,
- An estimating program that makes it easy to use these costs,
- An interactive video guide to the National Estimator program,
- A program that converts your estimates into invoices,
- A program that exports your estimates to QuickBooks Pro.

Monthly price updates on the Web are free and automatic all during 2008. You'll be prompted when it's time to collect the next update. A connection to the Web is required.

Download all of Craftsman's most popular costbooks for one low price with the Craftsman Site License. http://www.craftsmansitelicense.com

Craftsman Book Company
6058 Corte del Cedro / P.O. Box 6500 / Carlsbad, CA 92018

Preface

The author has corresponded with manufacturers and wholesalers of building material supplies and surveyed retail pricing services. From these sources, he has developed Average Material Unit Costs which should apply in most parts of the country.

Wherever possible, the author has listed Average Labor Unit Costs which are derived from the Average Manhours per Unit, the Crew Size, and the Wage Rates used in this book. Please read How to Use This Book for a more in-depth explanation of the arithmetic.

If you prefer, you can develop your own local labor unit costs. You can do this by simply multiplying the Average Manhours per Unit by your local crew wage rates per hour. Using your actual local labor wage rates for the trades will make your estimate more accurate.

What is a realistic labor unit cost to one reader may well be low or high to another reader, because of variations in labor efficiency. The Average Manhours per Unit figures were developed by time studies at job sites around the country. To determine the daily production rate for the crew, divide the total crew manhours per day by the Average Manhours per Unit.

The subject topics in this book are arranged in alphabetical order, A to Z. To help you find specific construction items, there is a complete alphabetical index at the end of the book, and a main subject index at the beginning of the book.

This manual shows crew, manhours, material, labor and equipment cost estimates based on Large or Small Volume work, then a total cost and a total including overhead and profit. No single price fits all repair and remodeling jobs. Generally, work done on smaller jobs costs more per unit installed and work on larger jobs costs less. The estimates in this book reflect that simple fact. The two estimates you find for each work item show the author's opinion of the likely range of costs for most contractors and for most jobs. So, which cost do you use, High Volume or Low Volume?

The only right price is the one that gets the job and earns a reasonable profit. Finding that price always requires estimating judgment. Use Small Volume cost estimates when some or most of the following conditions are likely:

- The crews won't work more than a few days on site.
- Better quality work is required.
- Productivity will probably be below average.
- Volume discounts on materials aren't available.
- Bidding is less competitive.
- Your overhead is higher than most contractors.

When few or none of those conditions apply, use Large Volume cost estimates.

Credits and Acknowledgments

This book has over 12,000 cost estimates for 2008. To develop these estimates, the author and editors relied on information supplied by hundreds of construction cost authorities. We offer our sincere thanks to the contractors, engineers, design professionals, construction estimators, material suppliers and manufacturers who, in the spirit of cooperation, have assisted in the preparation of this 31st edition of the *National Repair & Remodeling Estimator*. Our appreciation is extended to those listed below.

American Standard Products
DAP Products
Outwater Plastic Industries
Con-Rock Concrete
Georgia Pacific Products

Kohler Products
Wood Mode Cabinets
Transit Mixed Concrete
U.S. Gypsum Products
Henry Roofing Products

Special thanks to: Dal-Tile Corporation, 1713 Stewart, Santa Monica, California

About the Author

Albert Paxton is president of Professional Construction Analysts, Inc., located at 5823 Filaree Heights Ave., Malibu, CA 90265-3944; alpaxton@pca-group.net; fax (310) 589-0400. Mr. Paxton is a California licensed General Contractor (B1-425946). PCA's staff is comprised of estimators, engineers and project managers who are also expert witnesses, building appraisers and arbitrators operating throughout the United States.

PCA clients include property insurance carriers, financial institutions, self-insureds, and private individuals. The expertise of **PCA** is in both new and repair-remodel construction, both commercial and residential structures. In addition to individual structures, **PCA** assignments have included natural disasters such as Hurricanes Hugo, Andrew and Iniki, the Northridge earthquake in California, Hurricanes Charley, Frances, Ivan and Jeanne striking Florida and the Southeastern states, and the catastrophic Hurricane Katrina, whose destruction in the Gulf Coast will be felt in the building and repair industry for years to come.

©2007 Craftsman Book Company ISBN 978-1-57218-197-7

Cover design by Bill Grote

Cover photos by Ron South Photography & Ed Kessler Studios

Main Subject Index

How to Use This Book

	1		2	3	4	5	6	7	8	9	10	11	12
	Description		Oper	Unit	Vol	Crew Size	Man-hours per Unit	Crew Output per Day	Avg Mat'l Unit Cost	Avg Labor Unit Cost	Avg Equip Unit Cost	Avg Total Unit Cost	Avg Price Incl O&P

The descriptions and cost data in this book are arranged in a series of columns, which are described below. The cost data is divided into two categories: Costs Based On Large Volume and Costs Based On Small Volume. These two categories provide the estimator with a pricing range for each construction topic.

The Description column (1) contains the pertinent, specific information necessary to make the pricing information relevant and accurate.

The Operation column (2) contains a description of the construction repair or remodeling operation being performed. Generally the operations are Demolition, Install, and Reset.

The Unit column (3) contains the unit of measurement or quantity which applies to the item described.

The Volume column (4) breaks jobs into Large and Small categories. Based on the information given regarding volume (on page 2), select your job size.

The Crew Size column (5) contains a description of the trade that usually installs or labors on the specified item. It includes information on the labor trade that installs the material and the typical crew size. Letters and numbers are used in the abbreviations in the crew size column. Full descriptions of these abbreviations are in the Crew Compositions and Wage Rates table, beginning on page 15.

The Manhours per Unit column (6) is for the listed operation and listed crew.

The units per day in this book don't take into consideration unusually large or small quantities. But items such as travel, accessibility to work, experience of workers, and protection of undamaged property, which can favorably or adversely affect productivity, have been considered in developing Average Manhours per Unit. For further information about labor, see "Notes — Labor" in the Notes Section of some specific items.

Crew Output per Day (7) is based on how many units, on average, a crew can install or demo in one 8-hour day.

Crew Output per Day and Average Material Unit (8) Cost should assist the estimator in:

1. Checking prices quoted by others.

2. Developing local prices.

The Average Material Unit Cost column contains an average material cost for products (including, in many cases, the by-products used in installing the products) for both large and small volume. It doesn't include an allowance for sales tax, delivery charges, overhead and profit. Percentages for waste, shrinkage, or coverage have been taken into consideration unless indicated. For other information, see "Dimensions" or "Installation" in the Notes Section.

If the item described has many or very unusual by-products which are essential to determining the Average Material Unit Cost, the author has provided examples of material pricing. These examples are placed throughout the book in the Notes Section.

You should verify labor rates and material prices locally. Though the prices in this book are average material prices, prices vary from locality to locality. A local hourly wage rate should normally include taxes, benefits, and insurance. Some contractors may also include overhead and profit in the hourly rate.

The Average Labor Unit Cost column (9) contains an average labor cost based on the Average Manhours per Unit and the Crew Compositions and Wage Rates table. The average labor unit cost equals the Average Manhours per Unit multiplied by the Average Crew Rate per hour. The rates include fringe benefits, taxes, and insurance. Examples that show how to determine the average labor unit cost are provided in the Notes Section.

The Average Equipment Unit Cost column (10) contains an average equipment cost, based on both the average daily rental and the cost per day if owned and depreciated. The costs of daily maintenance and the operator are included.

The Average Total Unit Cost column (11) includes the sum of the Material, Equipment, and Labor Cost columns. It doesn't include an allowance for overhead and profit.

The Average (Total) Price Including Overhead and Profit column (12) results from adding an overhead and profit allowance to Total Cost. This allowance reflects the author's interpretation of average fixed and variable overhead expenses and the labor intensiveness of the operation vs. the costs of materials for the operation. This allowance factor varies throughout the book, depending on the operation. Each contractor interprets O&P differently. The range can be from 15 percent to 80 percent of the Average Total Unit Cost.

Estimating Techniques

Estimating Repair/Remodeling Jobs: The unforeseen, unpredictable, or unexpected can ruin you.

Each year, the residential repair and remodeling industry grows. It's currently outpacing residential new construction due to increases in land costs, labor wage rates, interest rates, material costs, and economic uncertainty. When people can't afford a new home, they tend to remodel their old one. And there are always houses that need repair, from natural disasters or accidents like fire. The professional repair and remodeling contractor is moving to the forefront of the industry.

Repair and remodeling spawns three occupations: the contractor and his workers, the insurance company property claims adjuster, and the property damage appraiser. Each of these professionals shares common functions, including estimating the cost of the repair or remodeling work.

Estimating isn't an exact science. Yet the estimate determines the profit or loss for the contractor, the fairness of the claim payout by the adjuster, and the amount of grant or loan by the appraiser. Quality estimating must be uppermost in the mind of each of these professionals. And accurate estimates are possible only when you know exactly what materials are needed and the number of manhours required for demolition, removal, and installation. Remember, profits follow the professional. To be profitable you must control costs — and cost control is directly related to accurate, professional estimates.

There are four general types of estimates, each with a different purpose and a corresponding degree of accuracy:

- The guess method: "All bathrooms cost $5,000." or "It looks like an $8,000 job to me."

- The per measure method: (I like to call it the surprise package.) "Remodeling costs $60 per SF, the job is 500 SF, so the price is $30,000."

These two methods are the least accurate and accomplish little for the adjuster or the appraiser. The contractor might use the methods for qualifying customers (e.g., "I thought a bathroom would only cost $2,000."), but never as the basis for bidding or negotiating a price.

- The *piece estimate* or *stick-by-stick* method.

- The *unit cost estimate* method.

These two methods yield a detailed estimate itemizing all of the material quantities and costs, the labor manhours and wage rates, the subcontract costs, and the allowance for overhead and profit.

Though time-consuming, the detailed estimate is the most accurate and predictable. It's a very satisfactory tool for negotiating either the contract price or the adjustment of a building loss. The piece estimate and the unit cost estimate rely on historical data, such as manhours per specific job operation and recent material costs. The successful repair and remodeling contractor, or insurance/appraisal company, maintains records of previous jobs detailing allocation of crew manhours per day and materials expended.

While new estimators don't have historical data records, they can rely on reference books, magazines, and newsletters to estimate manhours and material costs. It is important to remember that **the reference must pertain to repair and remodeling**. This book is designed *specifically* to meet this requirement.

The reference material must specialize in repair and remodeling work because there's a large cost difference between new construction and repair and remodeling. Material and labor construction costs vary radically with the size of the job or project. Economies of scale come into play. The larger the quantity of materials, the better the purchase price should be. The larger the number of units to be installed, the greater the labor efficiency.

Repair and remodeling work, compared to new construction, is more expensive due to a normally smaller volume of work. Typical repair work involves only two or three rooms of a house, or one roof. In new construction, the job size may be three to five complete homes or an entire development. And there's another factor: a lot of repair and remodeling is done with the house occupied, forcing the crew to work around the normal, daily activities of the occupants. In new construction, the approach is systematic and logical — work proceeds from the ground up to the roof and to the inside of the structure.

Since the jobs are small, the repair and remodeling contractor doesn't employ trade specialists. Repairers employ the "jack-of-all-trades" who is less specialized and therefore less efficient. This isn't to say the repairer is less professional than the trade specialist. On the contrary, the repairer must know about many more facets of construction: not just framing, but painting, finish carpentry, roofing, and electrical as well. But because the repairer has to spread his expertise over a greater area, he will be less efficient than the specialist who repeats the same operation all day long.

Another factor reducing worker efficiency is poor access to the work area. With new construction, where building is an orderly "from the ground up" approach, workers have easy access to the work area for any given operation. The workers can spread out as much as needed, which facilitates efficiency and minimizes the manhours required to perform a given operation.

The opposite situation exists with repair and remodeling construction. Consider an example where the work area involves fire damage on the second floor. Materials either go up through the interior stairs or through a second story window. Neither is easy when the exterior and interior walls have a finished covering such as siding and drywall. That results in greater labor costs with repair and remodeling because it takes more manhours to perform many of the same tasks.

If, as a professional estimator, you want to start collecting historical data, the place to begin is with daily worker time sheets that detail:

1. total hours worked by each worker per day

2. what specific operations each worker performed that day

3. how many hours (to the nearest tenth) each worker used in each operation performed that day.

Second, you must catalog all material invoices daily, being sure that quantities and unit costs per item are clearly indicated.

Third, maintain a record of overhead expenses attributable to the particular project. Then, after a number of jobs, you'll be able to calculate an average percentage of the job's gross amount that's attributable to overhead. Many contractors add 45% for overhead and profit to their total direct costs (direct labor, direct material, and direct subcontract costs). But that figure may not be right for your jobs.

Finally, each week you should reconcile in a job summary file the actual costs versus the estimated costs, and determine why there is any difference. This information can't immediately help you on this job since the contract has been signed, but it will be invaluable to you on your next job.

Up to now I've been talking about general estimating theory. Now let's be more specific. On page 8 is a Building Repair Estimate form. Each line is keyed to an explanation. A filled-out copy of the form is also provided, and on page 10, a blank, full-size copy that you can reproduce for your own use.

You can adapt the Building Repair Estimate form, whether you're a contractor, adjuster, or appraiser. Use of the form will yield a detailed estimate that will identify:

- The room or area involved, including sizes, dimensions and measurements.

- The kind and quality of material to be used.

- The quantities of materials to be used and verification of their prices.

- The type of work to be performed (demolish, remove, install, remove and reset) by what type of crew.

- The crew manhours per job operation and verification of the hourly wage scale.

- All arithmetical calculations that can be verified.

- Areas of difference between your estimate and others.

- Areas that will be a basis for negotiation and discussion of details.

Each job estimate begins with a visual inspection of the work site. If it's a repair job, you've got to see the damage. Without a visual inspection, you can't select a method of repair and you can't properly evaluate the opinions of others regarding repair or replacement. With either repair or remodeling work, the visual inspection is essential to uncover the "hiders" — the unpredictable, unforeseen, and unexpected problems that can turn profit into loss, or simplicity into nightmare. You're looking for the many variables and unknowns that exist behind an exterior or interior wall covering.

Along with the Building Repair Estimate form, use this checklist to make sure you're not forgetting anything.

Checklist

- Site accessibility: Will you store materials and tools in the garage? Is it secure? You can save a half hour to an hour each day by storing tools in the garage. Will the landscaping prevent trucks from reaching the work site? Are wheelbarrows or concrete pumpers going to be required?

- Soil: What type and how much water content? Will the soil change your excavation estimate?

- Utility lines: What's under the soil and where? Should you schedule the utilities to stake their lines?

- Soundness of the structure: If you're going to remodel, repair or add on, how sound is that portion of the house that you're going to have to work around? Where are the load-bearing walls? Are you going to remove and reset any walls? Do the floor joists sag?

- Roof strength: Can the roof support the weight of another layer of shingles. (Is four layers of composition shingles already too much?)

- Electrical: Is another breaker box required for the additional load?

This checklist is by no means complete, but it is a start. Take pictures! A digital camera will quickly pay for itself. When you're back at the office, the picture helps reconstruct the scene. Before and after pictures are also a sales tool representing your professional expertise.

During the visual inspection always be asking yourself "what if" this or that happened. Be looking for potential problem areas that would be extremely labor intensive or expensive in material to repair or replace.

Also spend some time getting to know your clients and their attitudes. Most of repair and remodeling work occurs while the house is occupied. If the work will be messy, let the homeowners know in advance. Their satisfaction is your ultimate goal — and their satisfaction will provide you a pleasant working atmosphere. You're there to communicate with them. At the end of an estimate and visual inspection, the homeowner should have a clear idea of what you can or can't do, how it will be done, and approximately how long it will take. Don't discuss costs now! Save the estimating for your quiet office with a print-out calculator and your cost files or reference books.

What you create on your estimate form during a visual inspection is a set of rough notes and diagrams that make the estimate speak. To avoid duplications and omissions, estimate in a *systematic sequence of inspection*. There are two questions to consider. First, where do you start the estimate? Second, In what order will you list the damaged or replaced items? It's customary to start in the room having either the most damage or requiring the most extensive remodeling. The sequence of listing is important. Start with either the floor or the ceiling. When starting with the floor, you might list items in the following sequence: Joists, subfloor, finish floor, base — listing from bottom to top. When starting with the ceiling, you reverse, and list from top to bottom. The important thing is to be *consistent* as you go from room to room! It's a good idea to figure the roof and foundation separately, instead of by the room.

After completing your visual inspection, go back to your office to cost out the items. Talk to your material supply houses and get unit costs for the quantity involved. Consult your job files or reference books and assign crew manhours to the different job operations.

There's one more reason for creating detailed estimates. Besides an estimate, what else have your notes given you? A material take-off sheet, a lumber list, a plan and specification sheet — the basis for writing a job summary for comparing estimated costs and profit versus actual costs and profit — and a project schedule that minimizes down time.

Here's the last step: Enter an amount for overhead and profit. No matter how small or large your work volume is, be realistic — everyone has overhead. An office,

even in your home, costs money to operate. If family members help out, pay them. Everyone's time is valuable!

Don't forget to charge for performing your estimate. A professional expects to be paid. You'll render a better product if you know you're being paid for your time. If you want to soften the blow to the owner, say the first hour is free or that the cost of the estimate will be deducted from the job price if you get the job.

In conclusion, whether you're a contractor, adjuster, or appraiser, you're selling your personal service, your ideas, and your reputation. To be successful you must:

- Know yourself and your capabilities.
- Know what the job will require by ferreting out the "hiders."
- Know your products and your work crew.
- Know your productivity and be able to deliver in a reasonable manner and within a reasonable time frame.
- Know your client and make it clear that all change orders, no matter how large or small, will cost money.

National Estimator 08 Inside the back cover of this book you'll find an envelope with a compact disk. The disk has National Estimator, an easy-to-use estimating program with all the cost estimates in this book. Insert the CD in your computer and wait a few seconds. Installation should begin automatically. (If not, click Start, Settings, Control Panel, double-click Add/Remove Programs and Install.) Select ShowMe from the installation menu and Dan will show you how to use National Estimator. When ShowMe is complete, select Install Program. When the National Estimator program has been installed, click Help on the menu bar, click Contents, click Print all Topics, click File and click Print Topic to print a 28-page instruction manual for National Estimator.

Building Repair Estimate (completed example)

Date _____

Insured	John Q. Smith
Loss Address	123 A. Main St.
City	Anywhere, Anystate 00010
Bldg. R.C.V.	100,000 Bldg. A.C.V. 80,000

Claim or Policy No. DP 0029   Page 1 of 2
Cause of Loss Fire
Home Ph. 555-1241
Bus. Ph. 555-1438   Other Ins. Y Ⓝ
Insurance Amount $100,000

Insurance Required R.C.V.(80%) A.C.V.(80%)

Description of Item	Unit Cost or Material Price Only			Labor Price Only		
	Unit	Unit Price	Total (Col. A)	Hours	Rate	Total Col. B
Install 1/2" sheetrock (standard,) on walls, including tape and finish 400 (page 149)	400	0.43	172.00	9.6	49.91	479.14
Paint walls, roller, smooth finish						
1 coat sealer 600 (page 322)	600	0.09	54.00	4.2	49.64	208.49
2 coats latex flat 600 (page 325)	600	0.19	114.00	7.2	49.64	357.41
Totals			340.00			1045.04

Total Column A   340.00

6% Tax   83.10
   1385.04

10% Overhead   146.81
   1468.14

10% Profit   146.81

Grand Total   1761.76

This is not an order to Repair

The undersigned agrees to complete and guarantee repairs at a total of $

Repairer ABC Construction
Street 316 E. 2nd Street
City Anywhere   Phone
By Jack Williams
Adjuster Stan Jones   Date of A/P N/A
Adj. License No. (If Any) 561-84
Service Office Name Phoenix

Note: This form does not replace the need for field notes, sketches and measurements.

Building Repair Estimate (blank template with callouts)

Date ㉝

Insured ①	
Loss Address ④	
City	
Bldg. R.C.V. ⑦ Bldg. A.C.V. ⑧	

Claim or Policy No. ②   Page of ③
Cause of Loss ⑤
Home Ph.
Bus. Ph.   Other Ins. Y N ⑥
Insurance Amount ⑨

Insurance Required R.C.V. (⑩ %) A.C.V.(⑪)%

Description of Item	Unit Cost or Material Price Only			Labor Price Only		
	Unit	Unit Price	Total (Col. A)	Hours	Rate	Total Col. B
⑳	⑬	⑭	⑮	⑰	⑱	⑲
Totals		㉑				㉒

Total Column A ㉙

㉓
㉔

㉕
㉖

㉗

Grand Total ㉘

This is not an order to Repair

The undersigned agrees to complete and guarantee repairs at a total of $ ㉚

Repairer ㉚
Street
City ㉛   Phone
By ㉜
Adjuster ㉞   Date of A/P ㉟
Adj. License No. (If Any)
Service Office Name ㊱

Note: This form does not replace the need for field notes, sketches and measurements.

Keyed Explanations of the Building Repair Estimate Form

1. Insert name of insured(s).

2. Insert claim number or, if claim number is not available, insert policy number or binder number.

3. Insert the page number and the total number of pages.

4. Insert street address, city and state where loss or damage occurred.

5. Insert type of loss (wind, hail, fire, water, etc.)

6. Check YES if there is other insurance, whether collectible or not. Check NO if there's only one insurer.

7. Insert the present replacement cost of the building. What would it cost to build the structure today?

8. Insert present actual cash value of the building.

9. Insert the amount of insurance applicable. If there is more than one insurer, insert the total amount of applicable insurance provided by all insurers.

10. If the amount of insurance required is based on replacement cost value, circle RCV and insert the percent required by the policy, if any.

11. If the amount of insurance required is based on actual cash value, circle ACV and insert the percent required by the policy, if any.

   Note: (regarding 10 and 11) if there is a non-concurrency, i.e., one insurer requires insurance to 90% of value while another requires insurance to 80% of value, make a note here. Comment on the non-concurrency in the settlement report.

12. The installed price and/or material price only, as expressed in columns 13 through 15, may include any of the following (expressed in units and unit prices):

   Material only (no labor)

   Material and labor to replace

   Material and labor to remove and replace

   Unit Cost is determined by dividing dollar cost by quantity. The term cost, as used in unit cost, is not intended to include any allowance, percentage or otherwise, for overhead or profit. Usually, overhead and profit are expressed as a percentage of cost. Cost must be determined first. Insert a line or dash in a space(s) in columns 13, 14, 15, 17, 18 or 19 if the space is not to be used.

13. The *units* column includes both the quantity and the unit of measure, i.e., 100 SF, 100 BF, 200 CF, 100 CY, 20 ea., etc.

14. The *unit price* may be expressed in dollars, cents or both. If the units column has 100 SF and if the unit price column has $.10, this would indicate the price to be $.10 per SF.

15. The *total* column is merely the dollar product of the quantity (in column 13) times the price per unit measure (in column 14).

16-19. These columns are normally used to express labor as follows: hours times rate per hour. However, it is possible to express labor as a unit price, i.e., 100 SF in column 13, a dash in column 17, $.05 in column 18 and $5.00 in column 19.

20. Under *description of item*, the following may be included:

   Description of item to be repaired or replaced (studs 2" x 4" 8'0" #2 Fir, Sheetrock 1/2", etc.)

   Quantities or dimensions (20 pcs., 8'0" x 14'0", etc.)

   Location within a room or area (north wall, ceiling, etc.)

   Method of correcting damage (paint - 1 coat; sand, fill and finish; R&R; remove only; replace; resize; etc.)

21-22. Dollar totals of columns A and B respectively.

23-27. Spaces provided for items not included in the body of the estimate (subtotals, overhead, profit, sales tax, etc.)

28. Total cost of repair.

29. Insert the agreed amount here. The agreement may be between the claim representative and the insured or between the claim rep and the repairer. If the agreed price is different from the grand total, the reason(s) for the difference should be itemized on the estimate sheet. If there is no room, attach an additional estimate sheet.

30. PRINT the name of the insured or the repairer so that it is legible.

31. PRINT the address of the insured or repairer legibly. Include phone number.

32. The insured or a representative of the repairer should sign here indicating agreement with the claim rep's estimate.

33. Insured or representative of the repairer should insert date here.

34. Claim rep should sign here.

35. Claim rep should insert date here.

36. Insert name of service office here.

Building Repair Estimate

Date

| Insured | Claim or Policy No. | | Page | of |

| Loss Address | Home Ph. | Cause of Loss |

| City | Bus. Ph. | Other Ins. Y N |

| Building. R.C.V. | Bldg. A.C.V. | Insurance Amount |

| Insurance Required R.C.V.(%) A.C.V.(%) | Unit Cost or Material Price Only | | | Labor Price Only | | |

Description of Item	Unit	Unit Price	Total (Col. A)	Hours	Rate	Total Col. B)

THIS IS NOT AN ORDER TO REPAIR   **TOTALS**

The undersigned agrees to complete and guarantee repairs at a total of $	**Total Column A**	
Repairer		
Street		
City Phone		
By		
Adjuster Date of A/P		
Adj. License No. (If Any)	**Grand Total**	
Service Office Name		

Note: This form does not replace the need for field notes, sketches and measurements.

Wage Rates Used in This Book

Wage rates listed here and used in this book were compiled in the fall of 2007 and projected to mid-2008. Wage rates are in dollars per hour.

"Base Wage Per Hour" (Col. 1) includes items such as vacation pay and sick leave which are normally taxed as wages. Nationally, these benefits average 5.15% of the Base Wage Per Hour. This amount is paid by the Employer in addition to the Base Wage Per Hour.

"Liability Insurance and Employer Taxes" (Cols. 3 & 4) include national averages for state unemployment insurance (4.00%), federal unemployment insurance (0.80%), Social Security and Medicare tax (7.65%), lia-

bility insurance (2.29%), and Workers' Compensation Insurance which varies by trade. This total percentage (Col. 3) is applied to the sum of Base Wage Per Hour and Taxable Fringe Benefits (Col. 1 + Col. 2) and is listed in Dollars (Col. 4). This amount is paid by the Employer in addition to the Base Wage Per Hour and the Taxable Fringe Benefits.

"Non-Taxable Fringe Benefits" (Col. 5) include employer-sponsored medical insurance and other benefits, which nationally average 4.55% of the Base Wage Per Hour.

"Total Hourly Cost Used In This Book" is the sum of Columns 1, 2, 4, & 5.

	1	2	3	4	5	6
Trade	Base Wage Per Hour	Taxable Fringe Benefits (5.15% of Base Wage)	Liability Insurance & Employer Taxes %	Liability Insurance & Employer Taxes $	Non-Taxable Fringe Benefits (4.55% of Base Wage)	Total Hourly Cost Used in This Book
Air Tool Operator	$21.86	$1.13	32.93%	$7.57	$0.99	$31.55
Bricklayer or Stone Mason	$24.70	$1.27	25.54%	$6.63	$1.12	$33.72
Bricktender	$18.81	$0.97	25.54%	$5.05	$0.86	$25.69
Carpenter	$23.23	$1.20	31.83%	$7.78	$1.06	$33.27
Cement Mason	$23.49	$1.21	23.35%	$5.77	$1.07	$31.54
Electrician, Journeyman Wireman	$27.25	$1.40	20.04%	$5.74	$1.24	$35.63
Equipment Operator	$27.32	$1.41	25.42%	$7.30	$1.24	$37.27
Fence Erector	$24.77	$1.28	26.21%	$6.83	$1.13	$34.01
Floorlayer: Carpet, Linoleum, Soft Tile	$22.67	$1.17	24.02%	$5.73	$1.03	$30.60
Floorlayer: Hardwood	$23.80	$1.23	24.02%	$6.01	$1.08	$32.12
Glazier	$22.51	$1.16	25.98%	$6.15	$1.02	$30.84
Laborer, General Construction	$19.01	$0.98	32.93%	$6.58	$0.86	$27.43
Lather	$24.42	$1.26	21.48%	$5.52	$1.11	$32.31
Marble and Terrazzo Setter	$21.82	$1.12	21.56%	$4.95	$0.99	$28.88
Painter, Brush	$24.65	$1.27	25.08%	$6.50	$1.12	$33.54
Painter, Spray-Gun	$25.39	$1.31	25.08%	$6.70	$1.16	$34.56
Paperhanger	$25.88	$1.33	25.08%	$6.82	$1.18	$35.21
Plasterer	$24.08	$1.24	28.78%	$7.29	$1.10	$33.71
Plumber	$27.81	$1.43	24.47%	$7.16	$1.27	$37.67
Reinforcing Ironworker	$22.86	$1.18	28.81%	$6.93	$1.04	$32.01
Roofer, Foreman	$26.07	$1.34	44.34%	$12.15	$1.19	$40.75
Roofer, Journeyman	$23.70	$1.22	44.34%	$11.05	$1.08	$37.05
Roofer, Hot Mop Pitch	$24.41	$1.26	44.34%	$11.38	$1.11	$38.16
Roofer, Wood Shingles	$24.89	$1.28	44.34%	$11.60	$1.13	$38.90
Sheet Metal Worker	$26.92	$1.39	26.21%	$7.42	$1.22	$36.95
Tile Setter	$24.10	$1.24	21.56%	$5.46	$1.10	$31.90
Tile Setter Helper	$18.44	$0.95	21.56%	$4.18	$0.84	$24.41
Truck Driver	$19.98	$1.03	26.42%	$5.55	$0.91	$27.47

Area Modification Factors

Construction costs are higher in some areas than in other areas. Add or deduct the percentages shown on the following pages to adapt the costs in this book to your job site. Adjust your cost estimate by the appropriate percentages in this table to find the estimated cost for the site selected. Where 0% is shown, it means no modification is required.

Modification factors are listed alphabetically by state and province. Areas within each state are listed by the first three digits of the postal zip code. For convenience, one representative city is identified in each three-digit zip or range of zips. Percentages are based on the average of all data points in the table.

Factors listed for each state and province are the average of all data points in that state or province. Figures for three-digit zips are the average of all five-digit zips in that area, and are the weighted average of factors for labor, material and equipment.

The National Estimator program will apply an area modification factor for any five-digit zip you select. Click Utilities. Click Options. Then select the Area Modification Factors tab.

These percentages are composites of many costs and will not necessarily be accurate when estimating the cost of any particular part of a building. But when used to modify costs for an entire structure, they should improve the accuracy of your estimates.

Alabama		-8%
Anniston	362	-12%
Auburn	368	-10%
Bellamy	369	-12%
Birmingham	350-352	-1%
Dothan	363	-9%
Evergreen	364	-14%
Gadsden	359	-11%
Huntsville	358	-4%
Jasper	355	-15%
Mobile	365-366	4%
Montgomery	360-361	-4%
Scottsboro	357	-7%
Selma	367	-10%
Sheffield	356	-5%
Tuscaloosa	354	-5%
Alaska		**29%**
Anchorage	995	38%
Fairbanks	997	32%
Juneau	998	34%
Ketchikan	999	10%
King Salmon	996	30%
Arizona		**-7%**
Chambers	865	-13%
Douglas	855	-12%
Flagstaff	860	-10%
Kingman	864	-10%
Mesa	852-853	3%
Phoenix	850	3%
Prescott	863	-7%
Show Low	859	-10%
Tucson	856-857	-6%
Arkansas		**-10%**
Batesville	725	-15%
Camden	717	-6%
Fayetteville	727	6%
Fort Smith	729	-8%
Harrison	726	-17%
Hope	718	-13%
Hot Springs	719	-15%
Jonesboro	724	-12%
Little Rock	720-722	-4%
Pine Bluff	716	-7%
Russellville	728	-9%
West Memphis	723	-11%
California		**10%**
Alhambra	917-918	11%
Bakersfield	932-933	1%
El Centro	922	1%
Eureka	955	-2%
Fresno	936-938	0%
Herlong	961	4%
Inglewood	902-905	11%
Irvine	926-927	16%
Lompoc	934	4%
Long Beach	907-908	12%
Los Angeles	900-901	10%
Marysville	959	2%
Modesto	953	1%
Mojave	935	7%
Novato	949	18%
Oakland	945-947	20%
Orange	928	14%
Oxnard	930	5%
Pasadena	910-912	12%
Rancho Cordova	956-957	9%
Redding	960	1%
Richmond	948	22%
Riverside	925	4%
Sacramento	958	9%
Salinas	939	7%
San Bernardino	923-924	4%
San Diego	919-921	9%
San Francisco	941	30%
San Jose	950-951	21%
San Mateo	943-944	25%
Santa Barbara	931	10%
Santa Rosa	954	11%
Stockton	952	7%
Sunnyvale	940	24%
Van Nuys	913-916	10%
Whittier	906	11%
Colorado		**0%**
Aurora	800-801	8%
Boulder	803-804	4%
Colorado Springs	808-809	1%
Denver	802	8%
Durango	813	-5%
Fort Morgan	807	-5%
Glenwood Springs	816	6%
Grand Junction	814-815	-2%
Greeley	806	5%
Longmont	805	3%
Pagosa Springs	811	-6%
Pueblo	810	-6%
Salida	812	-7%
Connecticut		**12%**
Bridgeport	066	13%
Bristol	060	15%
Fairfield	064	15%
Hartford	061	13%
New Haven	065	12%
Norwich	063	7%
Stamford	068-069	17%
Waterbury	067	12%
West Hartford	062	5%
Delaware		**6%**
Dover	199	-2%
Newark	197	10%
Wilmington	198	9%
District of Columbia		**11%**
Washington	200-205	11%
Florida		**-3%**
Altamonte Springs	327	-0%
Bradenton	342	-2%
Brooksville	346	-5%
Daytona Beach	321	-9%
Fort Lauderdale	333	5%
Fort Myers	339	0%
Fort Pierce	349	-7%
Gainesville	326	-8%
Jacksonville	322	-2%
Lakeland	338	-7%
Melbourne	329	-3%
Miami	330-332	4%
Naples	341	3%
Ocala	344	-11%
Orlando	328	5%
Panama City	324	-9%
Pensacola	325	-7%
Saint Augustine	320	-3%
Saint Cloud	347	-1%
St Petersburg	337	-3%
Tallahassee	323	-6%
Tampa	335-336	-1%
W. Palm Beach	334	2%
Georgia		**-0%**
Albany	317	-12%
Athens	306	-6%
Atlanta	303	15%
Augusta	308-309	-7%
Buford	305	0%
Calhoun	307	-9%
Columbus	318-319	-11%
Dublin/Fort Valley	310	-13%
Hinesville	313	-7%
Kings Bay	315	-12%
Macon	312	-4%
Marietta	300-302	5%
Savannah	314	-4%
Statesboro	304	-12%
Valdosta	316	-10%
Hawaii		**29%**
Aliamanu	968	30%
Ewa	967	28%
Halawa Heights	967	28%
Hilo	967	28%
Honolulu	968	30%
Kailua	968	30%
Lualualei	967	28%
Mililani Town	967	28%
Pearl City	967	28%
Wahiawa	967	28%
Waianae	967	28%
Wailuku (Maui)	967	28%
Idaho		**-8%**
Boise	837	1%
Coeur d'Alene	838	-8%
Idaho Falls	834	-12%
Lewiston	835	-9%
Meridian	836	-7%
Pocatello	832	-13%
Sun Valley	833	-7%
Illinois		**6%**
Arlington Heights	600	23%
Aurora	605	22%
Belleville	622	-2%
Bloomington	617	1%
Carbondale	629	-8%
Carol Stream	601	22%
Centralia	628	-9%
Champaign	618	0%
Chicago	606-608	22%
Decatur	623	-8%
Galesburg	614	-8%
Granite City	620	0%
Green River	612	0%
Joliet	604	19%
Kankakee	609	1%
Lawrenceville	624	-8%
Oak Park	603	25%
Peoria	615-616	4%
Peru	613	5%
Quincy	602	23%
Rockford	610-611	9%
Springfield	625-627	-1%
Urbana	619	-4%
Indiana		**-3%**
Aurora	470	-8%
Bloomington	474	-3%
Columbus	472	-4%
Elkhart	465	0%
Evansville	476-477	1%

City	Code	%
Fort Wayne	467-468	-1%
Gary	463-464	6%
Indianapolis	460-462	6%
Jasper	475	-8%
Jeffersonville	471	-5%
Kokomo	469	-6%
Lafayette	479	-5%
Muncie	473	-7%
South Bend	466	1%
Terre Haute	478	-7%
Iowa		**-4%**
Burlington	526	-5%
Carroll	514	-14%
Cedar Falls	506	-6%
Cedar Rapids	522-524	2%
Cherokee	510	-6%
Council Bluffs	515	23%
Creston	508	-9%
Davenport	527-528	0%
Decorah	521	-2%
Des Moines	500-503	2%
Dubuque	520	-3%
Fort Dodge	505	-8%
Mason City	504	-8%
Ottumwa	525	-9%
Sheldon	512	-9%
Shenandoah	516	-11%
Sioux City	511	-3%
Spencer	513	-9%
Waterloo	507	-8%
Kansas		**-7%**
Colby	677	-1%
Concordia	669	-15%
Dodge City	678	-6%
Emporia	668	-11%
Fort Scott	667	-7%
Hays	676	-16%
Hutchinson	675	-8%
Independence	673	-1%
Kansas City	660-662	7%
Liberal	679	-20%
Salina	674	-6%
Topeka	664-666	-5%
Wichita	670-672	-5%
Kentucky		**-6%**
Ashland	411-412	-8%
Bowling Green	421	-6%
Campton	413-414	-8%
Covington	410	2%
Elizabethtown	427	-10%
Frankfort	406	-8%
Hazard	417-418	-11%
Hopkinsville	422	-8%
Lexington	403-405	-1%
London	407-409	-7%
Louisville	400-402	1%
Owensboro	423	-10%
Paducah	420	-6%
Pikeville	415-416	-6%
Somerset	425-426	-13%
White Plains	424	-1%
Louisiana		**3%**
Alexandria	713-714	-10%
Baton Rouge	707-708	8%
Houma	703	16%
Lafayette	705	7%
Lake Charles	706	0%
Mandeville	704	7%
Minden	710	-7%
Monroe	712	-7%
New Orleans	700-701	22%
Shreveport	711	-5%

City	Code	%
Maine		**-7%**
Auburn	042	-7%
Augusta	043	-8%
Bangor	044	-8%
Bath	045	-7%
Brunswick	039-040	-1%
Camden	048	-9%
Cutler	046	-10%
Dexter	049	-7%
Northern Area	047	-12%
Portland	041	3%
Maryland		**3%**
Annapolis	214	11%
Baltimore	210-212	7%
Bethesda	208-209	16%
Church Hill	216	0%
Cumberland	215	-10%
Elkton	219	-4%
Frederick	217	4%
Laurel	206-207	10%
Salisbury	218	-6%
Massachusetts		**14%**
Ayer	015-016	10%
Bedford	017	20%
Boston	021-022	33%
Brockton	023-024	23%
Cape Cod	026	7%
Chicopee	010	9%
Dedham	019	17%
Fitchburg	014	15%
Hingham	020	24%
Lawrence	018	18%
Nantucket	025	13%
New Bedford	027	8%
Northfield	013	2%
Pittsfield	012	3%
Springfield	011	11%
Michigan		**2%**
Battle Creek	490-491	2%
Detroit	481-482	12%
Flint	484-485	0%
Grand Rapids	493-495	2%
Grayling	497	-7%
Jackson	492	1%
Lansing	488-489	4%
Marquette	498-499	-3%
Pontiac	483	14%
Royal Oak	480	10%
Saginaw	486-487	-3%
Traverse City	496	-3%
Minnesota		**1%**
Bemidji	566	-1%
Brainerd	564	0%
Duluth	556-558	2%
Fergus Falls	565	-8%
Magnolia	561	-8%
Mankato	560	0%
Minneapolis	553-555	16%
Rochester	559	0%
St Cloud	563	2%
St Paul	550-551	13%
Thief River Falls	567	-3%
Willmar	562	0%
Mississippi		**-9%**
Clarksdale	386	-9%
Columbus	397	-9%
Greenville	387	-18%
Greenwood	389	-12%
Gulfport	395	5%
Jackson	390-392	-6%
Laurel	394	-5%
McComb	396	-13%

City	Code	%
Meridian	393	-10%
Tupelo	388	-9%
Missouri		**-5%**
Cape Girardeau	637	-8%
Caruthersville	638	-12%
Chillicothe	646	-10%
Columbia	652	-4%
East Lynne	647	-5%
Farmington	636	-11%
Hannibal	634	6%
Independence	640	7%
Jefferson City	650-651	-5%
Joplin	648	-12%
Kansas City	641	9%
Kirksville	635	-18%
Knob Noster	653	-8%
Lebanon	654-655	-15%
Poplar Bluff	639	-9%
Saint Charles	633	4%
Saint Joseph	644-645	-1%
Springfield	656-658	-9%
St Louis	630-631	11%
Montana		**-5%**
Billings	590-591	-2%
Butte	597	-4%
Fairview	592	-10%
Great Falls	594	-2%
Havre	595	-11%
Helena	596	-4%
Kalispell	599	-6%
Miles City	593	1%
Missoula	598	-6%
Nebraska		-10%
Alliance	693	-13%
Columbus	686	-3%
Grand Island	688	-8%
Hastings	689	-12%
Lincoln	683-685	-4%
McCook	690	-14%
Norfolk	687	-13%
North Platte	691	-11%
Omaha	680-681	-1%
Valentine	692	-17%
Nevada		**6%**
Carson City	897	-1%
Elko	898	5%
Ely	893	10%
Fallon	894	5%
Las Vegas	889-891	9%
Reno	895	6%
New Hampshire		**1%**
Charlestown	036	0%
Concord	034	3%
Dover	038	3%
Lebanon	037	-1%
Littleton	035	-6%
Manchester	032-033	3%
New Boston	030-031	8%
New Jersey		**16%**
Atlantic City	080-084	9%
Brick	087	10%
Dover	078	17%
Edison	088-089	17%
Hackensack	076	15%
Monmouth	077	20%
Newark	071-073	16%
Passaic	070	17%
Paterson	074-075	15%
Princeton	085	15%
Summit	079	23%
Trenton	086	13%

City	Code	%
New Mexico		**-10%**
Alamogordo	883	-14%
Albuquerque	870-871	-2%
Clovis	881	-12%
Farmington	874	-5%
Fort Sumner	882	-8%
Gallup	873	-6%
Holman	877	-12%
Las Cruces	880	-14%
Santa Fe	875	-8%
Socorro	878	-15%
Truth or Consequences	879	-17%
Tucumcari	884	-11%
New York		**9%**
Albany	120-123	7%
Amityville	117	14%
Batavia	140	2%
Binghamton	137-139	1%
Bronx	104	14%
Brooklyn	112	11%
Buffalo	142	2%
Elmira	149	-1%
Flushing	113	23%
Garden City	115	20%
Hicksville	118	19%
Ithaca	148	-1%
Jamaica	114	21%
Jamestown	147	-6%
Kingston	124	-2%
Long Island	111	38%
Montauk	119	12%
New York (Manhattan)	100-102	37%
New York City	100-102	37%
Newcomb	128	-1%
Niagara Falls	143	-5%
Plattsburgh	129	-3%
Poughkeepsie	125-126	6%
Queens	110	23%
Rochester	144-146	2%
Rockaway	116	15%
Rome	133-134	-5%
Staten Island	103	15%
Stewart	127	-1%
Syracuse	130-132	2%
Tonawanda	141	0%
Utica	135	-5%
Watertown	136	-3%
West Point	109	12%
White Plains	105-108	20%
North Carolina		**-3%**
Asheville	287-289	-6%
Charlotte	280-282	5%
Durham	277	2%
Elizabeth City	279	-5%
Fayetteville	283	-9%
Goldsboro	275	0%
Greensboro	274	0%
Hickory	286	-8%
Kinston	285	-11%
Raleigh	276	5%
Rocky Mount	278	-8%
Wilmington	284	-5%
Winston-Salem	270-273	-3%
North Dakota		**-3%**
Bismarck	585	-4%
Dickinson	586	0%
Fargo	580-581	-2%
Grand Forks	582	-2%
Jamestown	584	-7%
Minot	587	-3%

Nekoma	583	-11%
Williston	588	7%
Ohio		**-1%**
Akron	442-443	2%
Canton	446-447	-3%
Chillicothe	456	-4%
Cincinnati	450-452	4%
Cleveland	440-441	5%
Columbus	432	6%
Dayton	453-455	-1%
Lima	458	-6%
Marietta	457	-6%
Marion	433	-7%
Newark	430-431	4%
Sandusky	448-449	-2%
Steubenville	439	-9%
Toledo	434-436	8%
Warren	444	-5%
Youngstown	445	-2%
Zanesville	437-438	-3%
Oklahoma		**-8%**
Adams	739	-10%
Ardmore	734	-10%
Clinton	736	-13%
Durant	747	7%
Enid	737	-11%
Lawton	735	-14%
McAlester	745	-14%
Muskogee	744	-12%
Norman	730	-6%
Oklahoma City	731	-4%
Ponca City	746	-7%
Poteau	749	-9%
Pryor	743	-14%
Shawnee	748	-15%
Tulsa	740-741	-3%
Woodward	738	1%
Oregon		**-3%**
Adrian	979	-15%
Bend	977	-2%
Eugene	974	0%
Grants Pass	975	-4%
Klamath Falls	976	-8%
Pendleton	978	-6%
Portland	970-972	10%
Salem	973	0%
Pennsylvania		**-1%**
Allentown	181	9%
Altoona	166	-8%
Beaver Springs	178	-3%
Bethlehem	180	9%
Bradford	167	-9%
Butler	160	1%
Chambersburg	172	-4%
Clearfield	168	-2%
DuBois	158	-10%
East Stroudsburg	183	-1%
Erie	164-165	-6%
Genesee	169	-11%
Greensburg	156	-4%
Harrisburg	170-171	5%
Hazleton	182	-5%
Johnstown	159	-10%
Kittanning	162	-8%
Lancaster	175-176	8%
Meadville	163	-12%
Montrose	188	-9%
New Castle	161	-3%
Philadelphia	190-191	16%
Pittsburgh	152	5%
Pottsville	179	-7%
Punxsutawney	157	-7%
Reading	195-196	4%
Scranton	184-185	-2%
Somerset	155	-9%
Southeastern	193	11%
Uniontown	154	-6%
Valley Forge	194	17%
Warminster	189	16%
Warrendale	150-151	5%
Washington	153	8%
Wilkes Barre	186-187	-3%
Williamsport	177	-4%
York	173-174	3%
Rhode Island		**8%**
Bristol	028	7%
Coventry	028	7%
Cranston	029	9%
Davisville	028	7%
Narragansett	028	7%
Newport	028	7%
Providence	029	9%
Warwick	028	7%
South Carolina		**-3%**
Aiken	298	5%
Beaufort	299	-2%
Charleston	294	-2%
Columbia	290-292	-5%
Greenville	296	-3%
Myrtle Beach	295	-7%
Rock Hill	297	-5%
Spartanburg	293	-5%
South Dakota		**-10%**
Aberdeen	574	-5%
Mitchell	573	-11%
Mobridge	576	-15%
Pierre	575	-15%
Rapid City	577	-9%
Sioux Falls	570-571	-3%
Watertown	572	-9%
Tennessee		**-3%**
Chattanooga	374	0%
Clarksville	370	3%
Cleveland	373	-2%
Columbia	384	-10%
Cookeville	385	-9%
Jackson	383	-7%
Kingsport	376	-5%
Knoxville	377-379	-3%
McKenzie	382	-10%
Memphis	380-381	1%
Nashville	371-372	6%
Texas		**-3%**
Abilene	795-796	-8%
Amarillo	790-791	-6%
Arlington	760	3%
Austin	786-787	6%
Bay City	774	18%
Beaumont	776-777	7%
Brownwood	768	-9%
Bryan	778	-8%
Childress	792	-22%
Corpus Christi	783-784	-1%
Dallas	751-753	6%
Del Rio	788	-16%
El Paso	798-799	-12%
Fort Worth	761-762	3%
Galveston	775	8%
Giddings	789	-4%
Greenville	754	5%
Houston	770-772	9%
Huntsville	773	6%
Longview	756	-4%
Lubbock	793-794	-10%
Lufkin	759	-10%
McAllen	785	-16%
Midland	797	-4%
Palestine	758	-7%
Plano	750	7%
San Angelo	769	-9%
San Antonio	780-782	-2%
Texarkana	755	-10%
Tyler	757	-8%
Victoria	779	1%
Waco	765-767	-7%
Wichita Falls	763	-8%
Woodson	764	-10%
Utah		**-5%**
Clearfield	840	-2%
Green River	845	-3%
Ogden	843-844	-11%
Provo	846-847	-8%
Salt Lake City	841	0%
Vermont		**-4%**
Albany	058	-6%
Battleboro	053	-3%
Beecher Falls	059	-8%
Bennington	052	-7%
Burlington	054	3%
Montpelier	056	-4%
Rutland	057	-7%
Springfield	051	-4%
White River Junction	050	-4%
Virginia		**-6%**
Abingdon	242	-12%
Alexandria	220-223	12%
Charlottesville	229	-6%
Chesapeake	233	-3%
Culpeper	227	-1%
Farmville	239	-17%
Fredericksburg	224-225	-4%
Galax	243	-12%
Harrisonburg	228	-7%
Lynchburg	245	-12%
Norfolk	235-237	-4%
Petersburg	238	-5%
Radford	241	-13%
Richmond	232	7%
Roanoke	240	-10%
Staunton	244	-11%
Tazewell	246	-11%
Virginia Beach	234	-4%
Williamsburg	230-231	-1%
Winchester	226	-4%
Washington		**1%**
Clarkston	994	0%
Everett	982	5%
Olympia	985	5%
Pasco	993	-2%
Seattle	980-981	13%
Spokane	990-992	-2%
Tacoma	983-984	5%
Vancouver	986	1%
Wenatchee	988	-7%
Yakima	989	-5%
West Virginia		**-8%**
Beckley	258-259	-7%
Bluefield	247-248	-9%
Charleston	250-253	4%
Clarksburg	263-264	-10%
Fairmont	266	-9%
Huntington	255-257	-3%
Lewisburg	249	-13%
Martinsburg	254	-7%
Morgantown	265	-10%
New Martinsville	262	-11%
Parkersburg	261	1%
Romney	267	-11%
Sugar Grove	268	-13%
Wheeling	260	-9%
Wisconsin		**2%**
Amery	540	2%
Beloit	535	7%
Clam Lake	545	-4%
Eau Claire	547	0%
Green Bay	541-543	3%
La Crosse	546	-2%
Ladysmith	548	-6%
Madison	537	10%
Milwaukee	530-534	10%
Oshkosh	549	3%
Portage	539	2%
Prairie du Chien	538	1%
Wausau	544	1%
Wyoming		**-2%**
Casper	826	-2%
Cheyenne/ Laramie	820	-6%
Gillette	827	4%
Powell	824	-3%
Rawlins	823	9%
Riverton	825	-8%
Rock Springs	829-831	5%
Sheridan	828	-7%
Wheatland	822	-9%

CANADA

Alberta		**39.8%**
Calgary	39.4	37.0%
Edmonton	40.9	40.6%
Ft. McMurray	41.6	41.8%
British Columbia		**39.6%**
Fraser Valley	39.4	38.1%
Okanagan	34.9	34.2%
Vancouver	43.9	46.6%
Manitoba		**43.0%**
North Manitoba	46.1	50.9%
South Manitoba	42.2	42.8%
Selkirk	39.4	37.3%
Winnipeg	40.9	41.0%
New Brunswick		**36.9%**
Moncton	36.4	36.9%
Nova Scotia		**43.5%**
Amherst	41.6	42.8%
Nova Scotia	40.2	39.6%
Sydney	43.0	48.1%
Newfoundland/ Labrador		**41.6%**
Ontario		**42.0%**
London	42.4	44.5%
Thunder Bay	37.9	36.3%
Toronto	43.9	45.3%
Quebec		**44.7%**
Montreal	42.4	47.8%
Quebec City	40.2	41.7%
Saskatchewan		**44.1%**
La Ronge	42.4	46.9%
Prince Albert	40.9	43.4%
Saskatoon	40.2	41.9%

Crew Compositions & Wage Rates

Crew Code	Manhours per day	Total costs $/Hr.	$/Day	Average Crew Rate (ACR) per Hour	Average Crew Rate (ACROP) per Hour w/O&P
AB					
1 Air tool operator	8.00	$31.55	$252.41		
1 Laborer	8.00	$27.43	$219.44		
TOTAL	**16.00**		**$471.85**	**$29.49**	**$43.65**
AD					
2 Air tool operators	16.00	$31.55	$504.82		
1 Laborer	8.00	$27.43	$219.44		
TOTAL	**24.00**		**$724.26**	**$30.18**	**$44.66**
BD					
3 Bricklayers	24.00	$33.72	$809.28		
2 Bricktenders	16.00	$25.69	$411.04		
TOTAL	**40.00**		**$1,220.32**	**$30.51**	**$45.46**
BK					
1 Bricklayer	8.00	$33.72	$269.76		
1 Bricktender	8.00	$25.69	$205.52		
TOTAL	**16.00**		**$475.28**	**$29.71**	**$44.26**
BO					
2 Bricklayers	16.00	$33.72	$539.52		
2 Bricktenders	16.00	$25.69	$411.04		
TOTAL	**32.00**		**$950.56**	**$29.71**	**$44.26**
2C					
2 Carpenters	16.00	$33.27	$532.32	$33.27	$49.91
CA					
1 Carpenter	8.00	$33.27	$266.16	$33.27	$49.91
CH					
1 Carpenter	8.00	$33.27	$266.16		
½ Laborer	4.00	$27.43	$109.72		
TOTAL	**12.00**		**$375.88**	**$31.32**	**$46.99**
CJ					
1 Carpenter	8.00	$33.27	$266.16		
1 Laborer	8.00	$27.43	$219.44		
TOTAL	**16.00**		**$485.60**	**$30.35**	**$45.53**
CN					
2 Carpenters	16.00	$33.27	$532.32		
½ Laborer	4.00	$27.43	$109.72		
TOTAL	**20.00**		**$642.04**	**$32.10**	**$48.15**
CS					
2 Carpenters	16.00	$33.27	$532.32		
1 Laborer	8.00	$27.43	$219.44		
TOTAL	**24.00**		**$751.76**	**$31.32**	**$46.99**
CU					
4 Carpenters	32.00	$33.27	$1,064.64		
1 Laborer	8.00	$27.43	$219.44		
TOTAL	**40.00**		**$1,284.08**	**$32.10**	**$48.15**
CW					
2 Carpenters	16.00	$33.27	$532.32		
2 Laborers	16.00	$27.43	$438.88		
TOTAL	**32.00**		**$971.20**	**$30.35**	**$45.53**

Crew Code	Manhours per day	Total costs $/Hr.	$/Day	Average Crew Rate (ACR) per Hour	Average Crew Rate (ACROP) per Hour w/O&P
CX					
4 Carpenters	32.00	$33.27	$1,064.64	$33.27	$49.91
CY					
3 Carpenters	24.00	$33.27	$798.48		
2 Laborers	16.00	$27.43	$438.88		
1 Equipment operator	8.00	$37.27	$298.16		
1 Laborer	8.00	$27.43	$219.44		
TOTAL	**56.00**		**$1,754.96**	**$31.34**	**$47.01**
CZ					
4 Carpenters	32.00	$33.27	$1,064.64		
3 Laborers	24.00	$27.43	$658.32		
1 Equipment operator	8.00	$37.27	$298.16		
1 Laborer	8.00	$27.43	$219.44		
TOTAL	**72.00**		**$2,240.56**	**$31.12**	**$46.68**
DD					
2 Cement masons	16.00	$31.54	$504.64		
1 Laborer	8.00	$27.43	$219.44		
TOTAL	**24.00**		**$724.08**	**$30.17**	**$44.65**
DF					
3 Cement masons	24.00	$31.54	$756.96		
5 Laborers	40.00	$27.43	$1,097.20		
TOTAL	**64.00**		**$1,854.16**	**$28.97**	**$42.88**
EA					
1 Electrician	8.00	$35.63	$285.04	$35.63	$52.02
EB					
2 Electricians	16.00	$35.63	$570.08	$35.63	$52.02
ED					
1 Electrician	8.00	$35.63	$285.04		
1 Carpenter	8.00	$33.27	$266.16		
TOTAL	**16.00**		**$551.20**	**$34.45**	**$50.30**
FA					
1 Floorlayer	8.00	$30.60	$244.80	$30.60	$44.98
FB					
2 Floorlayers	16.00	$30.60	$489.60		
1/4 Laborer	2.00	$27.43	$54.86		
TOTAL	**18.00**		**$544.46**	**$30.25**	**$44.46**
FC					
1 Floorlayer (hardwood)	8.00	$32.12	$256.99	$32.12	$47.22
FD					
2 Floorlayers (hardwood)	16.00	$32.12	$513.98		
1/4 Laborer	2.00	$27.43	$54.86		
TOTAL	**18.00**		**$568.84**	**$31.60**	**$46.45**
GA					
1 Glazier	8.00	$30.84	$246.72	$30.84	$45.64
GB					
2 Glaziers	16.00	$30.84	$493.44	$30.84	$45.64
GC					
3 Glaziers	24.00	$30.84	$740.16	$30.84	$45.64
HA					
1 Fence erector	8.00	$34.01	$272.05	$34.01	$50.67
HB					
2 Fence erectors	16.00	$34.01	$544.10		
1 Laborer	8.00	$27.43	$219.44		
TOTAL	**24.00**		**$763.54**	**$31.81**	**$47.40**

Crew Code	Manhours per day	Total costs $/Hr.	$/Day	Average Crew Rate (ACR) per Hour	Average Crew Rate (ACROP) per Hour w/O&P
1L					
1 Laborer	8.00	$27.43	$219.44	$27.43	$40.87
LB					
2 Laborers	16.00	$27.43	$438.88	$27.43	$40.87
LC					
2 Laborers	16.00	$27.43	$438.88		
1 Carpenter	8.00	$33.27	$266.16		
TOTAL	**24.00**		**$705.04**	**$29.38**	**$43.77**
LD					
2 Laborers	16.00	$27.43	$438.88		
2 Carpenters	16.00	$33.27	$532.32		
TOTAL	**32.00**		**$971.20**	**$30.35**	**$45.22**
LG					
5 Laborers	40.00	$27.43	$1,097.20		
1 Carpenter	8.00	$33.27	$266.16		
TOTAL	**48.00**		**$1,363.36**	**$28.40**	**$42.32**
LH					
3 Laborers	24.00	$27.43	$658.32	$27.43	$40.87
LJ					
4 Laborers	32.00	$27.43	$877.76	$27.43	$40.87
LK					
2 Laborers	16.00	$27.43	$438.88		
2 Carpenters	16.00	$33.27	$532.32		
1 Equipment operator	8.00	$37.27	$298.16		
1 Laborer	8.00	$27.43	$219.44		
TOTAL	**48.00**		**$1,488.80**	**$31.02**	**$46.21**
LL					
3 Laborers	24.00	$27.43	$658.32		
1 Carpenter	8.00	$33.27	$266.16		
1 Equipment operator	8.00	$37.27	$298.16		
1 Laborer	8.00	$27.43	$219.44		
TOTAL	**48.00**		**$1,442.08**	**$30.04**	**$44.76**
LR					
1 Lather	8.00	$32.31	$258.48	$32.31	$48.14
ML					
2 Bricklayers	16.00	$33.72	$539.52		
1 Bricktender	8.00	$25.69	$205.52		
TOTAL	**24.00**		**$745.04**	**$31.04**	**$46.25**
NA					
1 Painter (brush)	8.00	$33.54	$268.32	$33.54	$49.64
NC					
1 Painter (spray)	8.00	$34.56	$276.48	$34.56	$51.15
P3					
2 Plasterers	16.00	$33.71	$539.36		
1 Laborer	8.00	$27.43	$219.44		
TOTAL	**24.00**		**$758.80**	**$31.62**	**$47.11**
PE					
3 Plasterers	24.00	$33.71	$809.04		
2 Laborers	16.00	$27.43	$438.88		
TOTAL	**40.00**		**$1,247.92**	**$31.20**	**$46.49**

Crew Code	Manhours per day	Total costs $/Hr.	$/Day	Average Crew Rate (ACR) per Hour	Average Crew Rate (ACROP) per Hour w/O&P
QA					
1 Paperhanger	8.00	$35.21	$281.70	$35.21	$52.11
2R					
2 Roofers (composition)	16.00	$37.05	$592.80	$37.05	$57.43
RG					
2 Roofers (composition)	16.00	$37.05	$592.80		
1 Laborer	8.00	$27.43	$219.44		
TOTAL	**24.00**		**$812.24**	**$33.84**	**$52.46**
RJ					
2 Roofers (wood shingles)	16.00	$38.90	$622.32	$38.90	$60.29
RL					
2 Roofers (composition)	16.00	$37.05	$592.80		
1/2 Laborer	4.00	$27.43	$109.72		
TOTAL	**20.00**		**$702.52**	**$35.13**	**$54.45**
RM					
2 Roofers (wood shingles)	16.00	$38.90	$622.32		
1/2 Laborer	4.00	$27.43	$109.72		
TOTAL	**20.00**		**$732.04**	**$36.60**	**$56.73**
RQ					
2 Roofers (wood shingles)	16.00	$38.90	$622.32		
7/8 Laborer	7.00	$27.43	$192.01		
TOTAL	**23.00**		**$814.33**	**$35.41**	**$54.88**
RS					
1 Roofer (foreman)	8.00	$40.75	$326.00		
3 Roofers (pitch)	24.00	$38.16	$915.86		
2 Laborers	16.00	$27.43	$438.88		
TOTAL	**48.00**		**$1,680.74**	**$35.02**	**$54.27**
RT					
2 Roofers (pitch)	16.00	$38.16	$610.58		
1 Laborer	8.00	$27.43	$219.44		
TOTAL	**24.00**		**$830.02**	**$34.58**	**$53.61**
SA					
1 Plumber	8.00	$37.67	$301.36	$37.67	$55.37
SB					
1 Plumber	8.00	$37.67	$301.36		
1 Laborer	8.00	$27.43	$219.44		
TOTAL	**16.00**		**$520.80**	**$32.55**	**$47.85**
SC					
1 Plumber	8.00	$37.67	$301.36		
1 Electrician	8.00	$35.63	$285.04		
TOTAL	**16.00**		**$586.40**	**$36.65**	**$53.88**
SD					
1 Plumber	8.00	$37.67	$301.36		
1 Laborer	8.00	$27.43	$219.44		
1 Electrician	8.00	$35.63	$285.04		
TOTAL	**24.00**		**$805.84**	**$33.58**	**$49.36**
SE					
2 Plumbers	16.00	$37.67	$602.72		
1 Laborer	8.00	$27.43	$219.44		
1 Electrician	8.00	$35.63	$285.04		
TOTAL	**32.00**		**$1,107.20**	**$34.60**	**$50.86**

Crew Code	Manhours per day	Total costs $/Hr.	$/Day	Average Crew Rate (ACR) per Hour	Average Crew Rate (ACROP) per Hour w/O&P
SF					
2 Plumbers	16.00	$37.67	$602.72		
1 Laborer	8.00	$27.43	$219.44		
TOTAL	**24.00**		**$822.16**	**$34.26**	**$50.36**
SG					
3 Plumbers	24.00	$37.67	$904.08		
1 Laborer	8.00	$27.43	$219.44		
TOTAL	**32.00**		**$1,123.52**	**$35.11**	**$51.61**
TB					
1 Tile setter (ceramic)	8.00	$31.90	$255.20		
1 Tile setter's helper (ceramic)	8.00	$24.41	$195.28		
TOTAL	**16.00**		**$450.48**	**$28.15**	**$41.39**
UA					
1 Sheet metal worker	8.00	$36.95	$295.60	$36.95	$54.69
UB					
2 Sheet metal workers	16.00	$36.95	$591.20	$36.95	$54.69
UC					
2 Sheet metal workers	16.00	$36.95	$591.20		
2 Laborers	16.00	$27.43	$438.88		
TOTAL	**32.00**		**$1,030.08**	**$32.19**	**$47.64**
UD					
1 Sheet metal worker	8.00	$36.95	$295.60		
1 Laborer	8.00	$27.43	$219.44		
TOTAL	**16.00**		**$515.04**	**$32.19**	**$47.64**
UE					
1 Sheet metal worker	8.00	$36.95	$295.60		
1 Laborer	8.00	$27.43	$219.44		
½ Electrician	4.00	$35.63	$142.52		
TOTAL	**20.00**		**$657.56**	**$32.88**	**$48.66**
UF					
2 Sheet metal workers	16.00	$36.95	$591.20		
1 Laborer	8.00	$27.43	$219.44		
TOTAL	**24.00**		**$810.64**	**$33.78**	**$49.99**
VB					
1 Equipment operator	8.00	$37.27	$298.16		
1 Laborer	8.00	$27.43	$219.44		
TOTAL	**16.00**		**$517.60**	**$32.35**	**$47.88**

Abbreviations Used in This Book

ABS	acrylonitrile butadiene styrene
ACR	average crew rate
AGA	American Gas Association
AMP	ampere
Approx.	approximately
ASME	American Society of Mechanical Engineers
auto.	automatic
Avg.	average
Bdle.	bundle
BTU	British thermal unit
BTUH	British thermal unit per hour
C	100
cc	center to center or cubic centimeter
CF	cubic foot
CFM	cubic foot per minute
CLF	100 linear feet
Const.	construction
Corr.	corrugated
CSF	100 square feet
CSY	100 square yards
Ctn	carton
CWT	100 pounds
CY	cubic yard
Cu.	cubic
d	penny
D	deep or depth
Demo	demolish
dia.	diameter
D.S.A.	double strength, A grade
D.S.B.	double strength, B grade
Ea	each
e.g.	for example
etc.	et cetera
exp.	exposure
FAS	First and Select grade

F.H.A.	Federal Housing Administration
fl. oz.	fluid ounce
flt	flight
ft.	foot
ga.	gauge
gal	gallon
galv.	galvanized
GFI	ground fault interrupter
GPH	gallons per hour
GPM	gallons per minute
H	height or high
HP, hp	horsepower
Hr.	hour
HVAC	heating, ventilating, air conditioning
i.d.	inside diameter
i.e.	that is
Inst	install
I.P.S.	iron pipe size
KD	knocked down
KW, kw	kilowatts
L	length or long
lb, lbs.	pound(s)
LF	linear feet
LS	lump sum
M	1000
Mat'l	material
Max.	maximum
MBF	1000 board feet
MBHP	1000 boiler horsepower
Mi	miles
Min.	minimum
MSF	1000 square feet
O.B.	opposed blade
oc	on center
o.d.	outside dimension
oz.	ounce
pcs.	pieces
pkg.	package

PSI	per square inch
PVC	polyvinyl chloride
Qt.	quart
R.E.	rounded edge
R/L	random length
RS	rapid start (lamps)
R/W/L	random width and length
S4S	surfaced-four-sides
SF	square foot
SL	slimline (lamps)
Sq.	1 square or 100 square foot
S.S.B.	single strength, B quality
std.	standard
SY	square yard
T	thick
T&G	tongue and groove
U	thermal conductivity
U.I.	united inches
UL	Underwriters Laboratories
U.S.G.	United States Gypsum
VLF	vertical linear feet
W	width or wide
yr.	year

Symbols

/	per
%	percent
"	inches
'	foot or feet
x	by
o	degree
#	number or pounds
$	dollar
+/-	plus or minus

For crew abbreviations, please see Crew Compositions & Wage Rates chart, pages 15 to 19.

Acoustical and insulating tile

1. **Dimensions**

   a. Acoustical tile. $\frac{1}{2}$" thick x 12" x 12", 24".

   b. Insulating tile, decorative. $\frac{1}{2}$" thick x 12" x 12", 24"; $\frac{1}{2}$" thick x 16" x 16", 32".

2. **Installation.** Tile may be applied to existing plaster (if joist spacing is suitable) or to wood furring strips. Tile may have a square edge or flange. Depending on the type and shape of the tile, you can use adhesive, staples, nails or clips to attach the tile.

3. **Estimating Technique.** Determine area and add 5% to 10% for waste.

4. **Notes on Material Pricing.** A material price of $19.00 a gallon for adhesive was used to compile the Average Material Cost/Unit on the following pages. Here are the coverage rates:

12" x 12"	1.25 Gal/CSF
12" x 24"	0.95 Gal/CSF
16" x 16"	0.75 Gal/CSF
16" x 32"	0.55 Gal/CSF

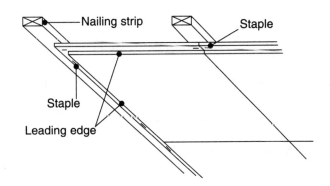

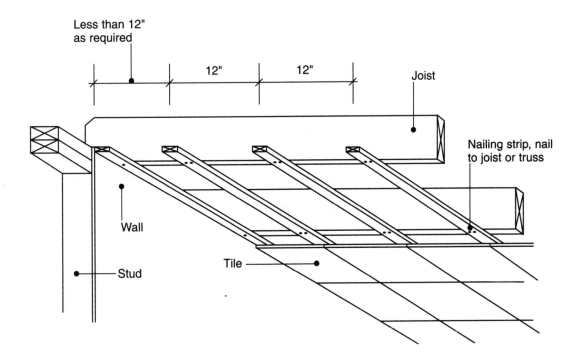

Description	Oper	Unit	Vol	Crew Size	Man-hours per Unit	Crew Output per Day	Avg Mat'l Unit Cost	Avg Labor Unit Cost	Avg Equip Unit Cost	Avg Total Unit Cost	Avg Price Incl O&P

Acoustical treatment
See also Suspended ceiling systems, page 424

Ceiling and wall tile
Adhesive set

Tile only, no grid system

| | Demo | SF | Lg | LB | .012 | 1300 | --- | .33 | --- | .33 | **.49** |
| | Demo | SF | Sm | LB | .018 | 910.0 | --- | .49 | --- | .49 | **.74** |

Tile on furring strips

| | Demo | SF | Lg | LB | .009 | 1710 | --- | .25 | --- | .25 | **.37** |
| | Demo | SF | Sm | LB | .013 | 1197 | --- | .36 | --- | .36 | **.53** |

Mineral fiber, vinyl coated, tile only
Applied in square pattern by adhesive to solid backing; 5% tile waste

1/2" thick, 12" x 12" or 12" x 24"

Description	Oper	Unit	Vol	Crew Size	Man-hours per Unit	Crew Output per Day	Avg Mat'l Unit Cost	Avg Labor Unit Cost	Avg Equip Unit Cost	Avg Total Unit Cost	Avg Price Incl O&P
Economy, mini perforated	Inst	SF	Lg	2C	.018	880.0	1.03	.60	---	1.63	**2.08**
	Inst	SF	Sm	2C	.026	616.0	1.12	.87	---	1.99	**2.59**
Standard, random perforated	Inst	SF	Lg	2C	.018	880.0	1.44	.60	---	2.04	**2.55**
	Inst	SF	Sm	2C	.026	616.0	1.56	.87	---	2.43	**3.09**
Designer, swirl perforation	Inst	SF	Lg	2C	.018	880.0	2.32	.60	---	2.92	**3.57**
	Inst	SF	Sm	2C	.026	616.0	2.51	.87	---	3.38	**4.18**
Deluxe, sculptured face	Inst	SF	Lg	2C	.018	880.0	2.64	.60	---	3.24	**3.93**
	Inst	SF	Sm	2C	.026	616.0	2.86	.87	---	3.73	**4.59**

5/8" thick, 12" x 12" or 12" x 24"

Description	Oper	Unit	Vol	Crew Size	Man-hours per Unit	Crew Output per Day	Avg Mat'l Unit Cost	Avg Labor Unit Cost	Avg Equip Unit Cost	Avg Total Unit Cost	Avg Price Incl O&P
Economy, mini perforated	Inst	SF	Lg	2C	.018	880.0	1.15	.60	---	1.75	**2.22**
	Inst	SF	Sm	2C	.026	616.0	1.25	.87	---	2.12	**2.74**
Standard, random perforated	Inst	SF	Lg	2C	.018	880.0	1.62	.60	---	2.22	**2.76**
	Inst	SF	Sm	2C	.026	616.0	1.76	.87	---	2.63	**3.32**
Designer, swirl perforation	Inst	SF	Lg	2C	.018	880.0	2.63	.60	---	3.23	**3.92**
	Inst	SF	Sm	2C	.026	616.0	2.85	.87	---	3.72	**4.58**
Deluxe, sculptured face	Inst	SF	Lg	2C	.018	880.0	3.00	.60	---	3.60	**4.35**
	Inst	SF	Sm	2C	.026	616.0	3.25	.87	---	4.12	**5.04**

3/4" thick, 12" x 12" or 12" x 24"

Description	Oper	Unit	Vol	Crew Size	Man-hours per Unit	Crew Output per Day	Avg Mat'l Unit Cost	Avg Labor Unit Cost	Avg Equip Unit Cost	Avg Total Unit Cost	Avg Price Incl O&P
Economy, mini perforated	Inst	SF	Lg	2C	.018	880.0	1.28	.60	---	1.88	**2.37**
	Inst	SF	Sm	2C	.026	616.0	1.39	.87	---	2.26	**2.90**
Standard, random perforated	Inst	SF	Lg	2C	.018	880.0	1.82	.60	---	2.42	**2.99**
	Inst	SF	Sm	2C	.026	616.0	1.98	.87	---	2.85	**3.57**
Designer, swirl perforation	Inst	SF	Lg	2C	.018	880.0	2.98	.60	---	3.58	**4.33**
	Inst	SF	Sm	2C	.026	616.0	3.22	.87	---	4.09	**5.00**
Deluxe, sculptured face	Inst	SF	Lg	2C	.018	880.0	3.41	.60	---	4.01	**4.82**
	Inst	SF	Sm	2C	.026	616.0	3.69	.87	---	4.56	**5.54**

Applied by adhesive to furring strips ADD

| | Inst | SF | Lg | 2C | .002 | --- | --- | .07 | --- | .07 | **.10** |
| | Inst | SF | Sm | 2C | .002 | --- | --- | .07 | --- | .07 | **.10** |

Description	Oper	Unit	Vol	Crew Size	Man-hours per Unit	Crew Output per Day	Avg Mat'l Unit Cost	Avg Labor Unit Cost	Avg Equip Unit Cost	Avg Total Unit Cost	Avg Price Incl O&P

Stapled

Tile only, no grid system	Demo	SF	Lg	LB	.014	1170	---	.38	---	.38	.57
	Demo	SF	Sm	LB	.020	819.0	---	.55	---	.55	.82
Tile on furring strips	Demo	SF	Lg	LB	.010	1540	---	.27	---	.27	.41
	Demo	SF	Sm	LB	.015	1078	---	.41	---	.41	.61

Mineral fiber, vinyl coated, tile only
Applied in square pattern by staples, nails, or clips; 5% tile waste
1/2" thick, 12" x 12" or 12" x 24"

Economy, mini perforated	Inst	SF	Lg	2C	.017	960.0	.89	.57	---	1.46	1.87
	Inst	SF	Sm	2C	.024	672.0	.96	.80	---	1.76	2.30
Standard, random perforated	Inst	SF	Lg	2C	.017	960.0	1.32	.57	---	1.89	2.37
	Inst	SF	Sm	2C	.024	672.0	1.43	.80	---	2.23	2.84
Designer, swirl perforation	Inst	SF	Lg	2C	.017	960.0	2.11	.57	---	2.68	3.27
	Inst	SF	Sm	2C	.024	672.0	2.29	.80	---	3.09	3.83
Deluxe, sculptured face	Inst	SF	Lg	2C	.017	960.0	2.40	.57	---	2.97	3.61
	Inst	SF	Sm	2C	.024	672.0	2.60	.80	---	3.40	4.19

5/8" thick, 12" x 12" or 12" x 24"

Economy, mini perforated	Inst	SF	Lg	2C	.017	960.0	1.06	.57	---	1.63	2.07
	Inst	SF	Sm	2C	.024	672.0	1.14	.80	---	1.94	2.51
Standard, random perforated	Inst	SF	Lg	2C	.017	960.0	1.48	.57	---	2.05	2.55
	Inst	SF	Sm	2C	.024	672.0	1.60	.80	---	2.40	3.04
Designer, swirl perforation	Inst	SF	Lg	2C	.017	960.0	2.40	.57	---	2.97	3.61
	Inst	SF	Sm	2C	.024	672.0	2.60	.80	---	3.40	4.19
Deluxe, sculptured face	Inst	SF	Lg	2C	.017	960.0	2.72	.57	---	3.29	3.98
	Inst	SF	Sm	2C	.024	672.0	2.95	.80	---	3.75	4.59

3/4" thick, 12" x 12" or 12" x 24"

Economy, mini perforated	Inst	SF	Lg	2C	.017	960.0	1.18	.57	---	1.75	2.21
	Inst	SF	Sm	2C	.024	672.0	1.27	.80	---	2.07	2.66
Standard, random perforated	Inst	SF	Lg	2C	.017	960.0	1.66	.57	---	2.23	2.76
	Inst	SF	Sm	2C	.024	672.0	1.79	.80	---	2.59	3.26
Designer, swirl perforation	Inst	SF	Lg	2C	.017	960.0	2.70	.57	---	3.27	3.95
	Inst	SF	Sm	2C	.024	672.0	2.93	.80	---	3.73	4.57
Deluxe, sculptured face	Inst	SF	Lg	2C	.017	960.0	3.10	.57	---	3.67	4.41
	Inst	SF	Sm	2C	.024	672.0	3.35	.80	---	4.15	5.05

Applied by staples, nails or clips to furring strips ADD

| | Inst | SF | Lg | 2C | .017 | 960.0 | --- | .57 | --- | .57 | .85 |
| | Inst | SF | Sm | 2C | .024 | 672.0 | --- | .80 | --- | .80 | 1.20 |

Tile patterns, effect on labor

Herringbone, Increase manhours

	Inst	%	Lg	2C	25.0	25.00	---	---	---	---	---
	Inst	%	Sm	2C	25.0	18.00	---	---	---	---	---
Diagonal, Increase manhours	Inst	%	Lg	2C	20.0	20.00	---	---	---	---	---
	Inst	%	Sm	2C	20.0	14.00	---	---	---	---	---
Ashlar, Increase manhours	Inst	%	Lg	2C	30.0	30.00	---	---	---	---	---
	Inst	%	Sm	2C	30.0	21.00	---	---	---	---	---

Description	Oper	Unit	Vol	Crew Size	Man-hours per Unit	Crew Output per Day	Avg Mat'l Unit Cost	Avg Labor Unit Cost	Avg Equip Unit Cost	Avg Total Unit Cost	Avg Price Incl O&P

Furring strips, 8% waste included

Over wood

Description	Oper	Unit	Vol	Crew Size	Man-hours	Output	Mat'l	Labor	Equip	Total	Price
1" x 4", 12" oc	Inst	SF	Lg	2C	.010	1600	.28	.33	---	.61	.82
	Inst	SF	Sm	2C	.014	1120	.30	.47	---	.77	1.04
1" x 4", 16" oc	Inst	SF	Lg	2C	.008	1920	.24	.27	---	.51	.68
	Inst	SF	Sm	2C	.012	1344	.26	.40	---	.66	.90

Over plaster

Description	Oper	Unit	Vol	Crew Size	Man-hours	Output	Mat'l	Labor	Equip	Total	Price
1" x 4", 12" oc	Inst	SF	Lg	2C	.013	1280	.28	.43	---	.71	.97
	Inst	SF	Sm	2C	.018	896.0	.30	.60	---	.90	1.24
1" x 4", 16" oc	Inst	SF	Lg	2C	.010	1600	.24	.33	---	.57	.78
	Inst	SF	Sm	2C	.014	1120	.26	.47	---	.73	1.00

Adhesives

Better quality, gun-applied in continuous bead to wood or metal framing or furring members.
Per 100 SF of surface area including 6% waste.

Panel adhesives

Subfloor adhesive, on floors

12" oc members

Description	Oper	Unit	Vol	Crew Size	Man-hours	Output	Mat'l	Labor	Equip	Total	Price
1/8" diameter (11 fl.oz./CSF)	Inst	CSF	Lg	CA	.075	106.0	2.31	2.50	---	4.81	6.40
	Inst	CSF	Sm	CA	.101	79.50	2.70	3.36	---	6.06	8.15
1/4" diameter (43 fl.oz./CSF)	Inst	CSF	Lg	CA	.075	106.0	9.21	2.50	---	11.71	14.30
	Inst	CSF	Sm	CA	.101	79.50	10.80	3.36	---	14.16	17.40
3/8" diameter (97 fl.oz./CSF)	Inst	CSF	Lg	CA	.075	106.0	20.80	2.50	---	23.30	27.60
	Inst	CSF	Sm	CA	.101	79.50	24.30	3.36	---	27.66	33.00
1/2" diameter (172 fl.oz./CSF)	Inst	CSF	Lg	CA	.075	106.0	36.80	2.50	---	39.30	46.00
	Inst	CSF	Sm	CA	.101	79.50	43.00	3.36	---	46.36	54.50

16" oc members

Description	Oper	Unit	Vol	Crew Size	Man-hours	Output	Mat'l	Labor	Equip	Total	Price
1/8" diameter (9 fl.oz./CSF)	Inst	CSF	Lg	CA	.056	143.0	1.85	1.86	---	3.71	4.92
	Inst	CSF	Sm	CA	.075	107.3	2.16	2.50	---	4.66	6.23
1/4" diameter (34 fl.oz./CSF)	Inst	CSF	Lg	CA	.056	143.0	7.37	1.86	---	9.23	11.30
	Inst	CSF	Sm	CA	.075	107.3	8.61	2.50	---	11.11	13.60
3/8" diameter (78 fl.oz./CSF)	Inst	CSF	Lg	CA	.056	143.0	16.60	1.86	---	18.40	21.90
	Inst	CSF	Sm	CA	.075	107.3	19.40	2.50	---	21.90	26.10
1/2" diameter (137 fl.oz./CSF)	Inst	CSF	Lg	CA	.056	143.0	29.40	1.86	---	31.26	36.60
	Inst	CSF	Sm	CA	.075	107.3	34.40	2.50	---	36.90	43.30

24" oc members

Description	Oper	Unit	Vol	Crew Size	Man-hours	Output	Mat'l	Labor	Equip	Total	Price
1/8" diameter (7 fl.oz./CSF)	Inst	CSF	Lg	CA	.052	154.0	1.39	1.73	---	3.12	4.19
	Inst	CSF	Sm	CA	.069	115.5	1.62	2.30	---	3.92	5.31
1/4" diameter (26 fl.oz./CSF)	Inst	CSF	Lg	CA	.052	154.0	5.53	1.73	---	7.26	8.95
	Inst	CSF	Sm	CA	.069	115.5	6.46	2.30	---	8.76	10.90
3/8" diameter (58 fl.oz./CSF)	Inst	CSF	Lg	CA	.052	154.0	12.50	1.73	---	14.23	16.90
	Inst	CSF	Sm	CA	.069	115.5	14.60	2.30	---	16.90	20.20
1/2" diameter (103 fl.oz./CSF)	Inst	CSF	Lg	CA	.052	154.0	22.10	1.73	---	23.83	28.00
	Inst	CSF	Sm	CA	.069	115.5	25.80	2.30	---	28.10	33.10

Description	Oper	Unit	Vol	Crew Size	Man-hours per Unit	Crew Output per Day	Avg Mat'l Unit Cost	Avg Labor Unit Cost	Avg Equip Unit Cost	Avg Total Unit Cost	Avg Price Incl O&P

Wall sheathing or shear panel wall adhesive on walls, floors or ceilings

12" oc members

Description	Oper	Unit	Vol	Crew Size	Man-hours per Unit	Crew Output per Day	Avg Mat'l Unit Cost	Avg Labor Unit Cost	Avg Equip Unit Cost	Avg Total Unit Cost	Avg Price Incl O&P
1/8" diameter (11 fl.oz./CSF)	Inst	CSF	Lg	CA	.100	80.00	2.97	3.33	---	6.30	8.41
	Inst	CSF	Sm	CA	.133	60.00	3.47	4.42	---	7.89	10.60
1/4" diameter (43 fl.oz./CSF)	Inst	CSF	Lg	CA	.100	80.00	11.90	3.33	---	15.23	18.60
	Inst	CSF	Sm	CA	.133	60.00	13.90	4.42	---	18.32	22.60
3/8" diameter (97 fl.oz./CSF)	Inst	CSF	Lg	CA	.100	80.00	26.70	3.33	---	30.03	35.70
	Inst	CSF	Sm	CA	.133	60.00	31.20	4.42	---	35.62	42.60
1/2" diameter (172 fl.oz./CSF)	Inst	CSF	Lg	CA	.100	80.00	47.30	3.33	---	50.63	59.40
	Inst	CSF	Sm	CA	.133	60.00	55.30	4.42	---	59.72	70.20

16" oc members

Description	Oper	Unit	Vol	Crew Size	Man-hours per Unit	Crew Output per Day	Avg Mat'l Unit Cost	Avg Labor Unit Cost	Avg Equip Unit Cost	Avg Total Unit Cost	Avg Price Incl O&P
1/8" diameter (9 fl.oz./CSF)	Inst	CSF	Lg	CA	.091	88.00	2.38	3.03	---	5.41	7.28
	Inst	CSF	Sm	CA	.121	66.00	2.78	4.03	---	6.81	9.24
1/4" diameter (34 fl.oz./CSF)	Inst	CSF	Lg	CA	.091	88.00	9.48	3.03	---	12.51	15.40
	Inst	CSF	Sm	CA	.121	66.00	11.10	4.03	---	15.13	18.80
3/8" diameter (78 fl.oz./CSF)	Inst	CSF	Lg	CA	.091	88.00	21.40	3.03	---	24.43	29.10
	Inst	CSF	Sm	CA	.121	66.00	25.00	4.03	---	29.03	34.80
1/2" diameter (137 fl.oz./CSF)	Inst	CSF	Lg	CA	.091	88.00	37.80	3.03	---	40.83	48.10
	Inst	CSF	Sm	CA	.121	66.00	44.20	4.03	---	48.23	56.90

24" oc members

Description	Oper	Unit	Vol	Crew Size	Man-hours per Unit	Crew Output per Day	Avg Mat'l Unit Cost	Avg Labor Unit Cost	Avg Equip Unit Cost	Avg Total Unit Cost	Avg Price Incl O&P
1/8" diameter (7 fl.oz./CSF)	Inst	CSF	Lg	CA	.084	95.00	1.79	2.79	---	4.58	6.25
	Inst	CSF	Sm	CA	.112	71.25	2.09	3.73	---	5.82	7.99
1/4" diameter (26 fl.oz./CSF)	Inst	CSF	Lg	CA	.084	95.00	7.11	2.79	---	9.90	12.40
	Inst	CSF	Sm	CA	.112	71.25	8.31	3.73	---	12.04	15.20
3/8" diameter (58 fl.oz./CSF)	Inst	CSF	Lg	CA	.084	95.00	16.00	2.79	---	18.79	22.60
	Inst	CSF	Sm	CA	.112	71.25	18.70	3.73	---	22.43	27.10
1/2" diameter (103 fl.oz./CSF)	Inst	CSF	Lg	CA	.084	95.00	28.40	2.79	---	31.19	36.80
	Inst	CSF	Sm	CA	.112	71.25	33.20	3.73	---	36.93	43.70

Polystyrene or polyurethane foam panel adhesive, on walls

12" oc members

Description	Oper	Unit	Vol	Crew Size	Man-hours per Unit	Crew Output per Day	Avg Mat'l Unit Cost	Avg Labor Unit Cost	Avg Equip Unit Cost	Avg Total Unit Cost	Avg Price Incl O&P
1/8" diameter (11 fl.oz./CSF)	Inst	CSF	Lg	CA	.100	80.00	3.64	3.33	---	6.97	9.18
	Inst	CSF	Sm	CA	.133	60.00	4.25	4.42	---	8.67	11.50
1/4" diameter (43 fl.oz./CSF)	Inst	CSF	Lg	CA	.100	80.00	14.50	3.33	---	17.83	21.70
	Inst	CSF	Sm	CA	.133	60.00	16.90	4.42	---	21.32	26.10
3/8" diameter (97 fl.oz./CSF)	Inst	CSF	Lg	CA	.100	80.00	32.70	3.33	---	36.03	42.60
	Inst	CSF	Sm	CA	.133	60.00	38.20	4.42	---	42.62	50.60
1/2" diameter (172 fl.oz./CSF)	Inst	CSF	Lg	CA	.100	80.00	57.80	3.33	---	61.13	71.50
	Inst	CSF	Sm	CA	.133	60.00	67.60	4.42	---	72.02	84.40

16" oc members

Description	Oper	Unit	Vol	Crew Size	Man-hours per Unit	Crew Output per Day	Avg Mat'l Unit Cost	Avg Labor Unit Cost	Avg Equip Unit Cost	Avg Total Unit Cost	Avg Price Incl O&P
1/8" diameter (9 fl.oz./CSF)	Inst	CSF	Lg	CA	.091	88.00	2.91	3.03	---	5.94	7.89
	Inst	CSF	Sm	CA	.121	66.00	3.40	4.03	---	7.43	9.95
1/4" diameter (34 fl.oz./CSF)	Inst	CSF	Lg	CA	.091	88.00	11.60	3.03	---	14.63	17.90
	Inst	CSF	Sm	CA	.121	66.00	13.50	4.03	---	17.53	21.60
3/8" diameter (78 fl.oz./CSF)	Inst	CSF	Lg	CA	.091	88.00	26.20	3.03	---	29.23	34.60
	Inst	CSF	Sm	CA	.121	66.00	30.60	4.03	---	34.63	41.20
1/2" diameter (137 fl.oz./CSF)	Inst	CSF	Lg	CA	.091	88.00	46.30	3.03	---	49.33	57.80
	Inst	CSF	Sm	CA	.121	66.00	54.10	4.03	---	58.13	68.20

Description	Oper	Unit	Vol	Crew Size	Man-hours per Unit	Crew Output per Day	Avg Mat'l Unit Cost	Avg Labor Unit Cost	Avg Equip Unit Cost	Avg Total Unit Cost	Avg Price Incl O&P
24" oc members											
1/8" diameter (7 fl.oz./CSF)	Inst	CSF	Lg	CA	.084	95.00	2.18	2.79	---	4.97	**6.70**
	Inst	CSF	Sm	CA	.129	62.00	2.55	4.29	---	6.84	**9.37**
1/4" diameter (26 fl.oz./CSF)	Inst	CSF	Lg	CA	.084	95.00	8.70	2.79	---	11.49	**14.20**
	Inst	CSF	Sm	CA	.129	62.00	10.20	4.29	---	14.49	**18.10**
3/8" diameter (58 fl.oz./CSF)	Inst	CSF	Lg	CA	.084	95.00	19.60	2.79	---	22.39	**26.80**
	Inst	CSF	Sm	CA	.129	62.00	22.90	4.29	---	27.19	**32.80**
1/2" diameter (103 fl.oz./CSF)	Inst	CSF	Lg	CA	.084	95.00	34.70	2.79	---	37.49	**44.10**
	Inst	CSF	Sm	CA	.129	62.00	40.60	4.29	---	44.89	**53.10**

Gypsum drywall adhesive, on ceilings

Description	Oper	Unit	Vol	Crew Size	Man-hours per Unit	Crew Output per Day	Avg Mat'l Unit Cost	Avg Labor Unit Cost	Avg Equip Unit Cost	Avg Total Unit Cost	Avg Price Incl O&P
12" oc members											
1/8" diameter (11 fl.oz./CSF)	Inst	CSF	Lg	CA	.100	80.00	3.31	3.33	---	6.64	**8.80**
	Inst	CSF	Sm	CA	.133	60.00	3.87	4.42	---	8.29	**11.10**
1/4" diameter (43 fl.oz./CSF)	Inst	CSF	Lg	CA	.100	80.00	13.20	3.33	---	16.53	**20.20**
	Inst	CSF	Sm	CA	.133	60.00	15.40	4.42	---	19.82	**24.40**
3/8" diameter (97 fl.oz./CSF)	Inst	CSF	Lg	CA	.100	80.00	29.80	3.33	---	33.13	**39.20**
	Inst	CSF	Sm	CA	.133	60.00	34.80	4.42	---	39.22	**46.60**
1/2" diameter (172 fl.oz./CSF)	Inst	CSF	Lg	CA	.100	80.00	52.70	3.33	---	56.03	**65.50**
	Inst	CSF	Sm	CA	.133	60.00	61.50	4.42	---	65.92	**77.40**
16" oc members											
1/8" diameter (9 fl.oz./CSF)	Inst	CSF	Lg	CA	.091	88.00	2.64	3.03	---	5.67	**7.58**
	Inst	CSF	Sm	CA	.121	66.00	3.09	4.03	---	7.12	**9.59**
1/4" diameter (34 fl.oz./CSF)	Inst	CSF	Lg	CA	.091	88.00	10.60	3.03	---	13.63	**16.70**
	Inst	CSF	Sm	CA	.121	66.00	12.30	4.03	---	16.33	**20.20**
3/8" diameter (78 fl.oz./CSF)	Inst	CSF	Lg	CA	.091	88.00	23.80	3.03	---	26.83	**31.90**
	Inst	CSF	Sm	CA	.121	66.00	27.80	4.03	---	31.83	**38.00**
1/2" diameter (137 fl.oz./CSF)	Inst	CSF	Lg	CA	.091	88.00	42.10	3.03	---	45.13	**53.00**
	Inst	CSF	Sm	CA	.121	66.00	49.20	4.03	---	53.23	**62.60**
24" oc members											
1/8" diameter (7 fl.oz./CSF)	Inst	CSF	Lg	CA	.084	95.00	1.98	2.79	---	4.77	**6.47**
	Inst	CSF	Sm	CA	.112	71.25	2.32	3.73	---	6.05	**8.26**
1/4" diameter (26 fl.oz./CSF)	Inst	CSF	Lg	CA	.084	95.00	7.91	2.79	---	10.70	**13.30**
	Inst	CSF	Sm	CA	.112	71.25	9.25	3.73	---	12.98	**16.20**
3/8" diameter (58 fl.oz./CSF)	Inst	CSF	Lg	CA	.084	95.00	17.90	2.79	---	20.69	**24.70**
	Inst	CSF	Sm	CA	.112	71.25	20.90	3.73	---	24.63	**29.60**
1/2" diameter (103 fl.oz./CSF)	Inct	CCF	Lg	CA	.084	95.00	31.60	2.79	---	34.39	**40.50**
	Inst	CSF	Sm	CA	.112	71.25	36.90	3.73	---	40.63	**48.00**

Gypsum drywall adhesive, on walls

Description	Oper	Unit	Vol	Crew Size	Man-hours per Unit	Crew Output per Day	Avg Mat'l Unit Cost	Avg Labor Unit Cost	Avg Equip Unit Cost	Avg Total Unit Cost	Avg Price Incl O&P
12" oc members											
1/8" diameter (11 fl.oz./CSF)	Inst	CSF	Lg	CA	.100	80.00	3.31	3.33	---	6.64	**8.80**
	Inst	CSF	Sm	CA	.133	60.00	3.87	4.42	---	8.29	**11.10**
1/4" diameter (43 fl.oz./CSF)	Inst	CSF	Lg	CA	.100	80.00	13.20	3.33	---	16.53	**20.20**
	Inst	CSF	Sm	CA	.133	60.00	15.40	4.42	---	19.82	**24.40**
3/8" diameter (97 fl.oz./CSF)	Inst	CSF	Lg	CA	.100	80.00	29.80	3.33	---	33.13	**39.20**
	Inst	CSF	Sm	CA	.133	60.00	34.80	4.42	---	39.22	**46.60**
1/2" diameter (172 fl.oz./CSF)	Inst	CSF	Lg	CA	.100	80.00	52.70	3.33	---	56.03	**65.50**
	Inst	CSF	Sm	CA	.133	60.00	61.50	4.42	---	65.92	**77.40**

Description	Oper	Unit	Vol	Crew Size	Man-hours per Unit	Crew Output per Day	Avg Mat'l Unit Cost	Avg Labor Unit Cost	Avg Equip Unit Cost	Avg Total Unit Cost	Avg Price Incl O&P
16" oc members											
1/8" diameter (9 fl.oz./CSF)	Inst	CSF	Lg	CA	.091	88.00	2.64	3.03	---	5.67	**7.58**
	Inst	CSF	Sm	CA	.121	66.00	3.09	4.03	---	7.12	**9.59**
1/4" diameter (34 fl.oz./CSF)	Inst	CSF	Lg	CA	.091	88.00	10.60	3.03	---	13.63	**16.70**
	Inst	CSF	Sm	CA	.121	66.00	12.30	4.03	---	16.33	**20.20**
3/8" diameter (78 fl.oz./CSF)	Inst	CSF	Lg	CA	.091	88.00	23.80	3.03	---	26.83	**31.90**
	Inst	CSF	Sm	CA	.121	66.00	27.80	4.03	---	31.83	**38.00**
1/2" diameter (137 fl.oz./CSF)	Inst	CSF	Lg	CA	.091	88.00	42.10	3.03	---	45.13	**53.00**
	Inst	CSF	Sm	CA	.121	66.00	49.20	4.03	---	53.23	**62.60**
24" oc members											
1/8" diameter (7 fl.oz./CSF)	Inst	CSF	Lg	CA	.084	95.00	1.98	2.79	---	4.77	**6.47**
	Inst	CSF	Sm	CA	.112	71.25	2.32	3.73	---	6.05	**8.26**
1/4" diameter (26 fl.oz./CSF)	Inst	CSF	Lg	CA	.084	95.00	7.91	2.79	---	10.70	**13.30**
	Inst	CSF	Sm	CA	.112	71.25	9.25	3.73	---	12.98	**16.20**
3/8" diameter (58 fl.oz./CSF)	Inst	CSF	Lg	CA	.084	95.00	17.90	2.79	---	20.69	**24.70**
	Inst	CSF	Sm	CA	.112	71.25	20.90	3.73	---	24.63	**29.60**
1/2" diameter (103 fl.oz./CSF)	Inst	CSF	Lg	CA	.084	95.00	31.60	2.79	---	34.39	**40.50**
	Inst	CSF	Sm	CA	.112	71.25	36.90	3.73	---	40.63	**48.00**

Hardboard or plastic panel adhesive, on walls

Description	Oper	Unit	Vol	Crew Size	Man-hours per Unit	Crew Output per Day	Avg Mat'l Unit Cost	Avg Labor Unit Cost	Avg Equip Unit Cost	Avg Total Unit Cost	Avg Price Incl O&P
12" oc members											
1/8" diameter (11 fl.oz./CSF)	Inst	CSF	Lg	CA	.100	80.00	3.31	3.33	---	6.64	**8.80**
	Inst	CSF	Sm	CA	.133	60.00	3.87	4.42	---	8.29	**11.10**
1/4" diameter (43 fl.oz./CSF)	Inst	CSF	Lg	CA	.100	80.00	13.20	3.33	---	16.53	**20.20**
	Inst	CSF	Sm	CA	.133	60.00	15.40	4.42	---	19.82	**24.40**
3/8" diameter (97 fl.oz./CSF)	Inst	CSF	Lg	CA	.100	80.00	29.80	3.33	---	33.13	**39.20**
	Inst	CSF	Sm	CA	.133	60.00	34.80	4.42	---	39.22	**46.60**
1/2" diameter (172 fl.oz./CSF)	Inst	CSF	Lg	CA	.100	80.00	52.70	3.33	---	56.03	**65.50**
	Inst	CSF	Sm	CA	.133	60.00	61.50	4.42	---	65.92	**77.40**
16" oc members											
1/8" diameter (9 fl.oz./CSF)	Inst	CSF	Lg	CA	.091	88.00	2.64	3.03	---	5.67	**7.58**
	Inst	CSF	Sm	CA	.121	66.00	3.09	4.03	---	7.12	**9.59**
1/4" diameter (34 fl.oz./CSF)	Inst	CSF	Lg	CA	.091	88.00	10.60	3.03	---	13.63	**16.70**
	Inst	CSF	Sm	CA	.121	66.00	12.30	4.03	---	16.33	**20.20**
3/8" diameter (78 fl.oz./CSF)	Inst	CSF	Lg	CA	.091	88.00	23.80	3.03	---	26.83	**31.90**
	Inst	CSF	Sm	CA	.121	66.00	27.80	4.03	---	31.83	**38.00**
1/2" diameter (137 fl.oz./CSF)	Inst	CSF	Lg	CA	.091	88.00	42.10	3.03	---	45.13	**53.00**
	Inst	CSF	Sm	CA	.121	66.00	49.20	4.03	---	53.23	**62.60**
24" oc members											
1/8" diameter (7 fl.oz./CSF)	Inst	CSF	Lg	CA	.084	95.00	1.98	2.79	---	4.77	**6.47**
	Inst	CSF	Sm	CA	.112	71.25	2.32	3.73	---	6.05	**8.26**
1/4" diameter (26 fl.oz./CSF)	Inst	CSF	Lg	CA	.084	95.00	7.91	2.79	---	10.70	**13.30**
	Inst	CSF	Sm	CA	.112	71.25	9.25	3.73	---	12.98	**16.20**
3/8" diameter (58 fl.oz./CSF)	Inst	CSF	Lg	CA	.084	95.00	17.90	2.79	---	20.69	**24.70**
	Inst	CSF	Sm	CA	.112	71.25	20.90	3.73	---	24.63	**29.60**
1/2" diameter (103 fl.oz./CSF)	Inst	CSF	Lg	CA	.084	95.00	31.60	2.79	---	34.39	**40.50**
	Inst	CSF	Sm	CA	.112	71.25	36.90	3.73	---	40.63	**48.00**

Description	Oper	Unit	Vol	Crew Size	Man-hours per Unit	Crew Output per Day	Avg Mat'l Unit Cost	Avg Labor Unit Cost	Avg Equip Unit Cost	Avg Total Unit Cost	Avg Price Incl O&P

Air conditioning and ventilating systems

System components

Condensing units

Air cooled, compressor, standard controls

Description	Oper	Unit	Vol	Crew Size	Man-hours per Unit	Crew Output per Day	Avg Mat'l Unit Cost	Avg Labor Unit Cost	Avg Equip Unit Cost	Avg Total Unit Cost	Avg Price Incl O&P
1.0 ton	Inst	Ea	Lg	SB	8.00	2.00	749.00	260.00	---	1009.00	**1240.00**
	Inst	Ea	Sm	SB	10.0	1.60	845.00	326.00	---	1171.00	**1320.00**
1.5 ton	Inst	Ea	Lg	SB	9.14	1.75	918.00	298.00	---	1216.00	**1490.00**
	Inst	Ea	Sm	SB	11.4	1.40	1040.00	371.00	---	1411.00	**1580.00**
2.0 ton	Inst	Ea	Lg	SB	10.7	1.50	1050.00	348.00	---	1398.00	**1720.00**
	Inst	Ea	Sm	SB	13.3	1.20	1190.00	433.00	---	1623.00	**1820.00**
2.5 ton	Inst	Ea	Lg	SB	12.8	1.25	1260.00	417.00	---	1677.00	**2060.00**
	Inst	Ea	Sm	SB	16.0	1.00	1420.00	521.00	---	1941.00	**2180.00**
3.0 ton	Inst	Ea	Lg	SB	16.0	1.00	1490.00	521.00	---	2011.00	**2480.00**
	Inst	Ea	Sm	SB	20.0	0.80	1680.00	651.00	---	2331.00	**2640.00**
4.0 ton	Inst	Ea	Lg	SB	21.3	0.75	1960.00	693.00	---	2653.00	**3270.00**
	Inst	Ea	Sm	SB	26.7	0.60	2210.00	869.00	---	3079.00	**3490.00**
5.0 ton	Inst	Ea	Lg	SB	32.0	0.50	2400.00	1040.00	---	3440.00	**4290.00**
	Inst	Ea	Sm	SB	40.0	0.40	2710.00	1300.00	---	4010.00	**4620.00**
Minimum Job Charge											
	Inst	Job	Lg	SB	5.33	3.00	---	173.00	---	173.00	**255.00**
	Inst	Job	Sm	SB	6.67	2.40	---	217.00	---	217.00	**319.00**

Diffusers

Aluminum, opposed blade damper, unless otherwise noted

Ceiling or sidewall, linear

Description	Oper	Unit	Vol	Crew Size	Man-hours per Unit	Crew Output per Day	Avg Mat'l Unit Cost	Avg Labor Unit Cost	Avg Equip Unit Cost	Avg Total Unit Cost	Avg Price Incl O&P
2" wide	Inst	LF	Lg	UA	.267	30.00	27.50	9.87	---	37.37	**46.20**
	Inst	LF	Sm	UA	.333	24.00	31.00	12.30	---	43.30	**49.20**
4" wide	Inst	LF	Lg	UA	.296	27.00	36.90	10.90	---	47.80	**58.60**
	Inst	LF	Sm	UA	.370	21.60	41.60	13.70	---	55.30	**61.80**
6" wide	Inst	LF	Lg	UA	.333	24.00	45.60	12.30	---	57.90	**70.70**
	Inst	LF	Sm	UA	.417	19.20	51.50	15.40	---	66.90	**74.30**
8" wide	Inst	LF	Lg	UA	.400	20.00	53.20	14.80	---	68.00	**83.10**
	Inst	LF	Sm	UA	.500	16.00	60.10	18.50	---	78.60	**87.40**

Perforated, 24" x 24" panel size

Description	Oper	Unit	Vol	Crew Size	Man-hours per Unit	Crew Output per Day	Avg Mat'l Unit Cost	Avg Labor Unit Cost	Avg Equip Unit Cost	Avg Total Unit Cost	Avg Price Incl O&P
6" x 6"	Inst	Ea	Lg	UA	.500	16.00	74.90	18.50	---	93.40	**113.00**
	Inst	Ea	Sm	UA	.625	12.80	84.50	23.10	---	107.60	**119.00**
8" x 8"	Inst	Ea	Lg	UA	.571	14.00	77.20	21.10	---	98.30	**120.00**
	Inst	Ea	Sm	UA	.714	11.20	87.10	26.40	---	113.50	**126.00**
10" x 10"	Inst	Ea	Lg	UA	.667	12.00	78.40	24.70	---	103.10	**127.00**
	Inst	Ea	Sm	UA	.833	9.60	88.40	30.80	---	119.20	**134.00**
12" x 12"	Inst	Ea	Lg	UA	.800	10.00	81.30	29.60	---	110.90	**137.00**
	Inst	Ea	Sm	UA	1.00	8.00	91.70	37.00	---	128.70	**146.00**
18" x 18"	Inst	Ea	Lg	UA	1.00	8.00	108.00	37.00	---	145.00	**179.00**
	Inst	Ea	Sm	UA	1.25	6.40	122.00	46.20	---	168.20	**190.00**

Description	Oper	Unit	Vol	Crew Size	Man-hours per Unit	Crew Output per Day	Avg Mat'l Unit Cost	Avg Labor Unit Cost	Avg Equip Unit Cost	Avg Total Unit Cost	Avg Price Incl O&P
Rectangular, one- to four-way blow											
6" x 6"	Inst	Ea	Lg	UA	.500	16.00	43.90	18.50	---	62.40	**77.80**
	Inst	Ea	Sm	UA	.625	12.80	49.50	23.10	---	72.60	**83.70**
12" x 6"	Inst	Ea	Lg	UA	.571	14.00	56.20	21.10	---	77.30	**95.80**
	Inst	Ea	Sm	UA	.714	11.20	63.40	26.40	---	89.80	**102.00**
12" x 9"	Inst	Ea	Lg	UA	.667	12.00	66.10	24.70	---	90.80	**113.00**
	Inst	Ea	Sm	UA	.833	9.60	74.60	30.80	---	105.40	**120.00**
12" x 12"	Inst	Ea	Lg	UA	.800	10.00	77.20	29.60	---	106.80	**133.00**
	Inst	Ea	Sm	UA	1.00	8.00	87.10	37.00	---	124.10	**142.00**
24" x 12"	Inst	Ea	Lg	UA	1.00	8.00	131.00	37.00	---	168.00	**205.00**
	Inst	Ea	Sm	UA	1.25	6.40	148.00	46.20	---	194.20	**216.00**
T-bar mounting, 24" x 24" lay in frame											
6" x 6"	Inst	Ea	Lg	UA	.500	16.00	70.80	18.50	---	89.30	**109.00**
	Inst	Ea	Sm	UA	.625	12.80	79.90	23.10	---	103.00	**114.00**
9" x 9"	Inst	Ea	Lg	UA	.571	14.00	79.00	21.10	---	100.10	**122.00**
	Inst	Ea	Sm	UA	.714	11.20	89.10	26.40	---	115.50	**128.00**
12" x 12"	Inst	Ea	Lg	UA	.667	12.00	104.00	24.70	---	128.70	**156.00**
	Inst	Ea	Sm	UA	.833	9.60	117.00	30.80	---	147.80	**162.00**
15" x 15"	Inst	Ea	Lg	UA	.800	10.00	108.00	29.60	---	137.60	**168.00**
	Inst	Ea	Sm	UA	1.00	8.00	121.00	37.00	---	158.00	**176.00**
18" x 18"	Inst	Ea	Lg	UA	1.00	8.00	142.00	37.00	---	179.00	**218.00**
	Inst	Ea	Sm	UA	1.25	6.40	160.00	46.20	---	206.20	**228.00**
Minimum Job Charge											
	Inst	Job	Lg	UA	2.29	3.50	---	84.60	---	84.60	**125.00**
	Inst	Job	Sm	UA	2.86	2.80	---	106.00	---	106.00	**156.00**

Ductwork

Fabricated rectangular, includes fittings, joints, supports
Aluminum alloy

Description	Oper	Unit	Vol	Crew Size	Man-hours per Unit	Crew Output per Day	Avg Mat'l Unit Cost	Avg Labor Unit Cost	Avg Equip Unit Cost	Avg Total Unit Cost	Avg Price Incl O&P
Under 100 lbs	Inst	Lb	Lg	UF	.343	70.00	3.62	11.60	---	15.22	**21.30**
	Inst	Lb	Sm	UF	.429	56.00	4.08	14.50	---	18.58	**25.50**
100 to 500 lbs	Inst	Lb	Lg	UF	.267	90.00	2.70	9.02	---	11.72	**16.50**
	Inst	Lb	Sm	UF	.333	72.00	3.05	11.30	---	14.35	**19.70**
500 to 1,000 lbs	Inst	Lb	Lg	UF	.218	110.0	1.86	7.36	---	9.22	**13.00**
	Inst	Lb	Sm	UF	.273	88.00	2.10	9.22	---	11.32	**15.80**
Over 1,000 lbs	Inst	Lb	Lg	UF	.185	130.0	1.42	6.25	---	7.67	**10.90**
	Inst	Lb	Sm	UF	.231	104.0	1.60	7.80	---	9.40	**13.20**
Galvanized steel											
Under 400 lbs	Inst	Lb	Lg	UF	.120	200.0	.76	4.05	---	4.81	**6.87**
	Inst	Lb	Sm	UF	.150	160.0	.86	5.07	---	5.93	**8.36**
400 to 1,000 lbs	Inst	Lb	Lg	UF	.112	215.0	.64	3.78	---	4.42	**6.33**
	Inst	Lb	Sm	UF	.140	172.0	.73	4.73	---	5.46	**7.73**
1,000 to 2,000 lbs	Inst	Lb	Lg	UF	.104	230.0	.55	3.51	---	4.06	**5.83**
	Inst	Lb	Sm	UF	.130	184.0	.62	4.39	---	5.01	**7.12**
2,000 to 5,000 lbs	Inst	Lb	Lg	UF	.100	240.0	.47	3.38	---	3.85	**5.54**
	Inst	Lb	Sm	UF	.125	192.0	.53	4.22	---	4.75	**6.78**
Over 10,000 lbs	Inst	Lb	Lg	UF	.096	250.0	.41	3.24	---	3.65	**5.27**
	Inst	Lb	Sm	UF	.120	200.0	.46	4.05	---	4.51	**6.46**

Description	Oper	Unit	Vol	Crew Size	Man-hours per Unit	Crew Output per Day	Avg Mat'l Unit Cost	Avg Labor Unit Cost	Avg Equip Unit Cost	Avg Total Unit Cost	Avg Price Incl O&P
Flexible, coated fabric on spring steel, aluminum, or corrosion-resistant metal											
Non-insulated											
3" diameter	Inst	LF	Lg	UD	.058	275.0	.91	1.87	---	2.78	**3.81**
	Inst	LF	Sm	UD	.073	220.0	1.03	2.35	---	3.38	**4.51**
5" diameter	Inst	LF	Lg	UD	.071	225.0	1.06	2.29	---	3.35	**4.60**
	Inst	LF	Sm	UD	.089	180.0	1.20	2.86	---	4.06	**5.44**
6" diameter	Inst	LF	Lg	UD	.080	200.0	1.19	2.58	---	3.77	**5.18**
	Inst	LF	Sm	UD	.100	160.0	1.35	3.22	---	4.57	**6.11**
7" diameter	Inst	LF	Lg	UD	.091	175.0	1.57	2.93	---	4.50	**6.14**
	Inst	LF	Sm	UD	.114	140.0	1.77	3.67	---	5.44	**7.20**
8" diameter	Inst	LF	Lg	UD	.107	150.0	1.63	3.44	---	5.07	**6.97**
	Inst	LF	Sm	UD	.133	120.0	1.83	4.28	---	6.11	**8.17**
10" diameter	Inst	LF	Lg	UD	.128	125.0	2.08	4.12	---	6.20	**8.49**
	Inst	LF	Sm	UD	.160	100.0	2.35	5.15	---	7.50	**9.97**
12" diameter	Inst	LF	Lg	UD	.160	100.0	2.47	5.15	---	7.62	**10.50**
	Inst	LF	Sm	UD	.200	80.00	2.79	6.44	---	9.23	**12.30**
Insulated											
3" diameter	Inst	LF	Lg	UD	.064	250.0	1.57	2.06	---	3.63	**4.85**
	Inst	LF	Sm	UD	.080	200.0	1.77	2.58	---	4.35	**5.58**
4" diameter	Inst	LF	Lg	UD	.071	225.0	1.57	2.29	---	3.86	**5.19**
	Inst	LF	Sm	UD	.089	180.0	1.77	2.86	---	4.63	**6.01**
5" diameter	Inst	LF	Lg	UD	.080	200.0	1.84	2.58	---	4.42	**5.93**
	Inst	LF	Sm	UD	.100	160.0	2.07	3.22	---	5.29	**6.83**
6" diameter	Inst	LF	Lg	UD	.091	175.0	2.02	2.93	---	4.95	**6.66**
	Inst	LF	Sm	UD	.114	140.0	2.28	3.67	---	5.95	**7.71**
7" diameter	Inst	LF	Lg	UD	.107	150.0	2.47	3.44	---	5.91	**7.94**
	Inst	LF	Sm	UD	.133	120.0	2.79	4.28	---	7.07	**9.13**
8" diameter	Inst	LF	Lg	UD	.128	125.0	2.53	4.12	---	6.65	**9.01**
	Inst	LF	Sm	UD	.160	100.0	2.85	5.15	---	8.00	**10.50**
10" diameter	Inst	LF	Lg	UD	.160	100.0	3.29	5.15	---	8.44	**11.40**
	Inst	LF	Sm	UD	.200	80.00	3.71	6.44	---	10.15	**13.20**
12" diameter	Inst	LF	Lg	UD	.213	75.00	3.83	6.86	---	10.69	**14.60**
	Inst	LF	Sm	UD	.267	60.00	4.32	8.59	---	12.91	**17.00**
14" diameter	Inst	LF	Lg	UD	.320	50.00	4.63	10.30	---	14.93	**20.60**
	Inst	LF	Sm	UD	.400	40.00	5.23	12.90	---	18.13	**24.30**

Duct accessories

Air extractors

Description	Oper	Unit	Vol	Crew Size	Man-hours per Unit	Crew Output per Day	Avg Mat'l Unit Cost	Avg Labor Unit Cost	Avg Equip Unit Cost	Avg Total Unit Cost	Avg Price Incl O&P
12" x 4"	Inst	Ea	Lg	UA	.333	24.00	14.30	12.30	---	26.60	**34.70**
	Inst	Ea	Sm	UA	.417	19.20	16.20	15.40	---	31.60	**39.00**
8" x 6"	Inst	Ea	Lg	UA	.400	20.00	14.30	14.80	---	29.10	**38.40**
	Inst	Ea	Sm	UA	.500	16.00	16.20	18.50	---	34.70	**43.50**
20" x 8"	Inst	Ea	Lg	UA	.500	16.00	32.20	18.50	---	50.70	**64.40**
	Inst	Ea	Sm	UA	.625	12.80	36.30	23.10	---	59.40	**70.50**
18" x 10"	Inst	Ea	Lg	UA	.571	14.00	31.60	21.10	---	52.70	**67.60**
	Inst	Ea	Sm	UA	.714	11.20	35.60	26.40	---	62.00	**74.70**
24" x 12"	Inst	Ea	Lg	UA	.800	10.00	43.90	29.60	---	73.50	**94.20**
	Inst	Ea	Sm	UA	1.00	8.00	49.50	37.00	---	86.50	**104.00**

Description	Oper	Unit	Vol	Crew Size	Man-hours per Unit	Crew Output per Day	Avg Mat'l Unit Cost	Avg Labor Unit Cost	Avg Equip Unit Cost	Avg Total Unit Cost	Avg Price Incl O&P
Dampers, fire, curtain-type, vertical											
8" x 4"	Inst	Ea	Lg	UA	.333	24.00	19.50	12.30	---	31.80	**40.60**
	Inst	Ea	Sm	UA	.417	19.20	22.00	15.40	---	37.40	**44.80**
12" x 4"	Inst	Ea	Lg	UA	.364	22.00	19.50	13.50	---	33.00	**42.30**
	Inst	Ea	Sm	UA	.455	17.60	22.00	16.80	---	38.80	**46.90**
16" x 14"	Inst	Ea	Lg	UA	.667	12.00	31.00	24.70	---	55.70	**72.10**
	Inst	Ea	Sm	UA	.833	9.60	35.00	30.80	---	65.80	**80.50**
24" x 20"	Inst	Ea	Lg	UA	1.00	8.00	39.20	37.00	---	76.20	**99.80**
	Inst	Ea	Sm	UA	1.25	6.40	44.20	46.20	---	90.40	**113.00**
Dampers, multi-blade, opposed blade											
12" x 12"	Inst	Ea	Lg	UA	.500	16.00	28.70	18.50	---	47.20	**60.30**
	Inst	Ea	Sm	UA	.625	12.80	32.30	23.10	---	55.40	**66.50**
12" x 18"	Inst	Ea	Lg	UA	.667	12.00	38.00	24.70	---	62.70	**80.20**
	Inst	Ea	Sm	UA	.833	9.60	42.90	30.80	---	73.70	**88.50**
24" x 24"	Inst	Ea	Lg	UA	1.00	8.00	80.70	37.00	---	117.70	**148.00**
	Inst	Ea	Sm	UA	1.25	6.40	91.10	46.20	---	137.30	**159.00**
48" x 36"	Inst	Ea	Lg	UD	2.67	6.00	245.00	86.00	---	331.00	**408.00**
	Inst	Ea	Sm	UD	3.33	4.80	276.00	107.00	---	383.00	**435.00**
Dampers, variable volume modulating, motorized											
12" x 12"	Inst	Ea	Lg	UA	0.80	10.00	147.00	29.60	---	176.60	**213.00**
	Inst	Ea	Sm	UA	1.00	8.00	166.00	37.00	---	203.00	**221.00**
24" x 12"	Inst	Ea	Lg	UA	1.33	6.00	177.00	49.10	---	226.10	**276.00**
	Inst	Ea	Sm	UA	1.67	4.80	199.00	61.70	---	260.70	**291.00**
30" x 18"	Inst	Ea	Lg	UA	2.00	4.00	328.00	73.90	---	401.90	**486.00**
	Inst	Ea	Sm	UA	2.50	3.20	370.00	92.40	---	462.40	**506.00**
Thermostat, ADD	Inst	Ea	Lg	UA	1.33	6.00	35.10	49.10	---	84.20	**113.00**
	Inst	Ea	Sm	UA	1.67	4.80	39.60	61.70	---	101.30	**131.00**
Dampers, multi-blade, parallel blade											
8" x 8"	Inst	Ea	Lg	UA	.400	20.00	59.70	14.80	---	74.50	**90.50**
	Inst	Ea	Sm	UA	.500	16.00	67.30	18.50	---	85.80	**94.70**
16" x 10"	Inst	Ea	Lg	UA	.500	16.00	62.60	18.50	---	81.10	**99.30**
	Inst	Ea	Sm	UA	.625	12.80	70.60	23.10	---	93.70	**105.00**
18" x 12"	Inst	Ea	Lg	UA	.667	12.00	64.40	24.70	---	89.10	**110.00**
	Inst	Ea	Sm	UA	.833	9.60	72.60	30.80	---	103.40	**118.00**
28" x 16"	Inst	Ea	Lg	UA	1.00	8.00	87.20	37.00	---	124.20	**155.00**
	Inst	Ea	Sm	UA	1.25	6.40	98.30	46.20	---	144.50	**167.00**
Mixing box, with electric or pneumatic motor, with silencer											
150 to 250 CFM	Inst	Ea	Lg	UD	1.60	10.00	532.00	51.50	---	583.50	**688.00**
	Inst	Ea	Sm	UD	2.00	8.00	601.00	64.40	---	665.40	**696.00**
270 to 600 CFM	Inst	Ea	Lg	UD	2.00	8.00	550.00	64.40	---	614.40	**728.00**
	Inst	Ea	Sm	UD	2.50	6.40	620.00	80.50	---	700.50	**740.00**

Description	Oper	Unit	Vol	Crew Size	Man-hours per Unit	Crew Output per Day	Avg Mat'l Unit Cost	Avg Labor Unit Cost	Avg Equip Unit Cost	Avg Total Unit Cost	Avg Price Incl O&P

Fans

Air conditioning or processed air handling

Axial flow, compact, low sound 2.5" self-propelled

Description	Oper	Unit	Vol	Crew Size	Man-hours per Unit	Crew Output per Day	Avg Mat'l Unit Cost	Avg Labor Unit Cost	Avg Equip Unit Cost	Avg Total Unit Cost	Avg Price Incl O&P
3,800 CFM, 5 HP	Inst	Ea	Lg	UE	6.67	3.00	3890.00	219.00	---	4109.00	**4800.00**
	Inst	Ea	Sm	UE	8.33	2.40	4390.00	274.00	---	4664.00	**4790.00**
15,600 CFM, 10 HP	Inst	Ea	Lg	UE	13.3	1.50	6790.00	437.00	---	7227.00	**8450.00**
	Inst	Ea	Sm	UE	16.7	1.20	7660.00	549.00	---	8209.00	**8470.00**

In-line centrifugal, supply/exhaust booster, aluminum wheel/hub, disconnect switch, 1/4" self-propelled

Description	Oper	Unit	Vol	Crew Size	Man-hours per Unit	Crew Output per Day	Avg Mat'l Unit Cost	Avg Labor Unit Cost	Avg Equip Unit Cost	Avg Total Unit Cost	Avg Price Incl O&P
500 CFM, 10" dia. connect	Inst	Ea	Lg	UE	6.67	3.00	819.00	219.00	---	1038.00	**1270.00**
	Inst	Ea	Sm	UE	8.33	2.40	924.00	274.00	---	1198.00	**1330.00**
1,520 CFM, 16" dia. connect	Inst	Ea	Lg	UE	10.0	2.00	1070.00	329.00	---	1399.00	**1720.00**
	Inst	Ea	Sm	UE	12.5	1.60	1210.00	411.00	---	1621.00	**1820.00**
3,480 CFM, 20" dia. connect	Inst	Ea	Lg	UE	20.0	1.00	1400.00	658.00	---	2058.00	**2590.00**
	Inst	Ea	Sm	UE	25.0	0.80	1580.00	822.00	---	2402.00	**2800.00**

Ceiling fan, right angle, extra quiet 0.10" self-propelled

Description	Oper	Unit	Vol	Crew Size	Man-hours per Unit	Crew Output per Day	Avg Mat'l Unit Cost	Avg Labor Unit Cost	Avg Equip Unit Cost	Avg Total Unit Cost	Avg Price Incl O&P
200 CFM	Inst	Ea	Lg	UE	1.25	16.00	174.00	41.10	---	215.10	**261.00**
	Inst	Ea	Sm	UE	1.56	12.80	197.00	51.30	---	248.30	**273.00**
900 CFM	Inst	Ea	Lg	UE	1.67	12.00	415.00	54.90	---	469.90	**559.00**
	Inst	Ea	Sm	UE	2.08	9.60	469.00	68.40	---	537.40	**570.00**
3,000 CFM	Inst	Ea	Lg	UE	2.50	8.00	790.00	82.20	---	872.20	**1030.00**
	Inst	Ea	Sm	UE	3.13	6.40	891.00	103.00	---	994.00	**1040.00**
Exterior wall or roof cap	Inst	Ea	Lg	UA	.667	12.00	115.00	24.70	---	139.70	**169.00**
	Inst	Ea	Sm	UA	.833	9.60	130.00	30.80	---	160.80	**176.00**

Roof ventilators, corrosive fume resistant, plastic blades

Roof ventilator, centrifugal V belt drive motor, 1/4" self-propelled

Description	Oper	Unit	Vol	Crew Size	Man-hours per Unit	Crew Output per Day	Avg Mat'l Unit Cost	Avg Labor Unit Cost	Avg Equip Unit Cost	Avg Total Unit Cost	Avg Price Incl O&P
250 CFM, 1/4 HP	Inst	Ea	Lg	UE	5.00	4.00	2430.00	164.00	---	2594.00	**3040.00**
	Inst	Ea	Sm	UE	6.25	3.20	2740.00	206.00	---	2946.00	**3040.00**
900 CFM, 1/3 HP	Inst	Ea	Lg	UE	6.67	3.00	2630.00	219.00	---	2849.00	**3350.00**
	Inst	Ea	Sm	UE	8.33	2.40	2970.00	274.00	---	3244.00	**3380.00**
1,650 CFM, 1/2 HP	Inst	Ea	Lg	UE	8.00	2.50	3130.00	263.00	---	3393.00	**3990.00**
	Inst	Ea	Sm	UE	10.0	2.00	3530.00	329.00	---	3859.00	**4020.00**
2,250 CFM, 1 HP	Inst	Ea	Lg	UE	10.0	2.00	3280.00	329.00	---	3609.00	**4250.00**
	Inst	Ea	Sm	UE	12.5	1.60	3700.00	411.00	---	4111.00	**4300.00**

Utility set centrifugal V belt drive motor, 1/4" self-propelled

Description	Oper	Unit	Vol	Crew Size	Man-hours per Unit	Crew Output per Day	Avg Mat'l Unit Cost	Avg Labor Unit Cost	Avg Equip Unit Cost	Avg Total Unit Cost	Avg Price Incl O&P
1,200 CFM, 1/4 HP	Inst	Ea	Lg	UE	5.00	4.00	2430.00	164.00	---	2594.00	**3040.00**
	Inst	Ea	Sm	UE	6.25	3.20	2740.00	206.00	---	2946.00	**3040.00**
1,500 CFM, 1/3 HP	Inst	Ea	Lg	UE	6.67	3.00	2460.00	219.00	---	2679.00	**3150.00**
	Inst	Ea	Sm	UE	8.33	2.40	2770.00	274.00	---	3044.00	**3180.00**
1,850 CFM, 1/2 HP	Inst	Ea	Lg	UE	8.00	2.50	2460.00	263.00	---	2723.00	**3210.00**
	Inst	Ea	Sm	UE	10.0	2.00	2770.00	329.00	---	3099.00	**3260.00**
2,200 CFM, 3/4 HP	Inst	Ea	Lg	UE	10.0	2.00	2490.00	329.00	---	2819.00	**3350.00**
	Inst	Ea	Sm	UE	12.5	1.60	2810.00	411.00	---	3221.00	**3410.00**

Direct drive

Description	Oper	Unit	Vol	Crew Size	Man-hours per Unit	Crew Output per Day	Avg Mat'l Unit Cost	Avg Labor Unit Cost	Avg Equip Unit Cost	Avg Total Unit Cost	Avg Price Incl O&P
320 CFM, 11" x 11" damper	Inst	Ea	Lg	UE	5.00	4.00	294.00	164.00	---	458.00	**581.00**
	Inst	Ea	Sm	UE	6.25	3.20	331.00	206.00	---	537.00	**635.00**
600 CFM, 11" x 11" damper	Inst	Ea	Lg	UE	5.00	4.00	323.00	164.00	---	487.00	**615.00**
	Inst	Ea	Sm	UE	6.25	3.20	364.00	206.00	---	570.00	**668.00**

Description	Oper	Unit	Vol	Crew Size	Man-hours per Unit	Crew Output per Day	Avg Mat'l Unit Cost	Avg Labor Unit Cost	Avg Equip Unit Cost	Avg Total Unit Cost	Avg Price Incl O&P

Ventilation, residential

Attic

Roof-type ventilators

Aluminum dome, damper and curb

Description	Oper	Unit	Vol	Crew Size	Man-hours per Unit	Crew Output per Day	Avg Mat'l Unit Cost	Avg Labor Unit Cost	Avg Equip Unit Cost	Avg Total Unit Cost	Avg Price Incl O&P
6" diameter, 300 CFM	Inst	Ea	Lg	EA	.667	12.00	246.00	23.80	---	269.80	**317.00**
	Inst	Ea	Sm	EA	.833	9.60	277.00	29.70	---	306.70	**321.00**
7" diameter, 450 CFM	Inst	Ea	Lg	EA	.727	11.00	268.00	25.90	---	293.90	**346.00**
	Inst	Ea	Sm	EA	.909	8.80	302.00	32.40	---	334.40	**350.00**
9" diameter, 900 CFM	Inst	Ea	Lg	EA	.800	10.00	433.00	28.50	---	461.50	**539.00**
	Inst	Ea	Sm	EA	1.00	8.00	488.00	35.60	---	523.60	**540.00**
12" diameter, 1,000 CFM	Inst	Ea	Lg	EA	1.00	8.00	266.00	35.60	---	301.60	**357.00**
	Inst	Ea	Sm	EA	1.25	6.40	300.00	44.50	---	344.50	**365.00**
16" diameter, 1,500 CFM	Inst	Ea	Lg	EA	1.14	7.00	321.00	40.60	---	361.60	**428.00**
	Inst	Ea	Sm	EA	1.43	5.60	362.00	51.00	---	413.00	**436.00**
20" diameter, 2,500 CFM	Inst	Ea	Lg	EA	1.33	6.00	392.00	47.40	---	439.40	**520.00**
	Inst	Ea	Sm	EA	1.67	4.80	442.00	59.50	---	501.50	**529.00**
26" diameter, 4,000 CFM	Inst	Ea	Lg	EA	1.60	5.00	474.00	57.00	---	531.00	**628.00**
	Inst	Ea	Sm	EA	2.00	4.00	535.00	71.30	---	606.30	**639.00**
32" diameter, 6,500 CFM	Inst	Ea	Lg	EA	2.00	4.00	655.00	71.30	---	726.30	**858.00**
	Inst	Ea	Sm	EA	2.50	3.20	739.00	89.10	---	828.10	**869.00**
38" diameter, 8,000 CFM	Inst	Ea	Lg	EA	2.67	3.00	971.00	95.10	---	1066.10	**1260.00**
	Inst	Ea	Sm	EA	3.33	2.40	1100.00	119.00	---	1219.00	**1270.00**
50" diameter, 13,000 CFM	Inst	Ea	Lg	EA	4.00	2.00	1400.00	143.00	---	1543.00	**1820.00**
	Inst	Ea	Sm	EA	5.00	1.60	1580.00	178.00	---	1758.00	**1840.00**
Plastic ABS dome											
900 CFM	Inst	Ea	Lg	EA	.667	12.00	78.40	23.80	---	102.20	**125.00**
	Inst	Ea	Sm	EA	.833	9.60	88.40	29.70	---	118.10	**132.00**
1,600 CFM	Inst	Ea	Lg	EA	.800	10.00	117.00	28.50	---	145.50	**176.00**
	Inst	Ea	Sm	EA	1.00	8.00	132.00	35.60	---	167.60	**184.00**

Wall-type ventilators, one speed, with shutter

Description	Oper	Unit	Vol	Crew Size	Man-hours per Unit	Crew Output per Day	Avg Mat'l Unit Cost	Avg Labor Unit Cost	Avg Equip Unit Cost	Avg Total Unit Cost	Avg Price Incl O&P
12" diameter, 1,000 CFM	Inst	Ea	Lg	EA	.667	12.00	145.00	23.80	---	168.80	**202.00**
	Inst	Ea	Sm	EA	.833	9.60	164.00	29.70	---	193.70	**207.00**
14" diameter, 1,500 CFM	Inst	Ea	Lg	EA	.800	10.00	177.00	28.50	---	205.50	**245.00**
	Inst	Ea	Sm	EA	1.00	8.00	199.00	35.60	---	234.60	**251.00**
16" diameter, 2,000 CFM	Inst	Ea	Lg	EA	1.00	8.00	266.00	35.60	---	301.60	**357.00**
	Inst	Ea	Sm	EA	1.25	6.40	300.00	44.50	---	344.50	**365.00**

Entire structure, wall-type, one speed, with shutter

Description	Oper	Unit	Vol	Crew Size	Man-hours per Unit	Crew Output per Day	Avg Mat'l Unit Cost	Avg Labor Unit Cost	Avg Equip Unit Cost	Avg Total Unit Cost	Avg Price Incl O&P
30" diameter, 4,800 CFM	Inst	Ea	Lg	EA	1.60	5.00	456.00	57.00	---	513.00	**608.00**
	Inst	Ea	Sm	EA	2.00	4.00	515.00	71.30	---	586.30	**619.00**
36" diameter, 7,000 CFM	Inst	Ea	Lg	EA	2.00	4.00	497.00	71.30	---	568.30	**676.00**
	Inst	Ea	Sm	EA	2.50	3.20	561.00	89.10	---	650.10	**691.00**
42" diameter, 10,000 CFM	Inst	Ea	Lg	EA	2.67	3.00	608.00	95.10	---	703.10	**839.00**
	Inst	Ea	Sm	EA	3.33	2.40	686.00	119.00	---	805.00	**860.00**
48" diameter, 16,000 CFM	Inst	Ea	Lg	EA	4.00	2.00	796.00	143.00	---	939.00	**1120.00**
	Inst	Ea	Sm	EA	5.00	1.60	898.00	178.00	---	1076.00	**1160.00**
Two speeds, ADD	Inst	Ea	Lg	---	---	---	56.80	---	---	56.80	**65.30**
	Inst	Ea	Sm	---	---	---	64.00	---	---	64.00	**73.60**

Description	Oper	Unit	Vol	Crew Size	Man-hours per Unit	Crew Output per Day	Avg Mat'l Unit Cost	Avg Labor Unit Cost	Avg Equip Unit Cost	Avg Total Unit Cost	Avg Price Incl O&P
Entire structure, lay-down type, one speed, with shutter											
30" diameter, 4,500 CFM	Inst	Ea	Lg	EA	1.33	6.00	497.00	47.40	---	544.40	**641.00**
	Inst	Ea	Sm	EA	1.67	4.80	561.00	59.50	---	620.50	**648.00**
36" diameter, 6,500 CFM	Inst	Ea	Lg	EA	1.60	5.00	544.00	57.00	---	601.00	**709.00**
	Inst	Ea	Sm	EA	2.00	4.00	614.00	71.30	---	685.30	**718.00**
42" diameter, 9,000 CFM	Inst	Ea	Lg	EA	2.00	4.00	644.00	71.30	---	715.30	**844.00**
	Inst	Ea	Sm	EA	2.50	3.20	726.00	89.10	---	815.10	**856.00**
48" diameter, 12,000 CFM	Inst	Ea	Lg	EA	2.67	3.00	848.00	95.10	---	943.10	**1110.00**
	Inst	Ea	Sm	EA	3.33	2.40	957.00	119.00	---	1076.00	**1130.00**
Two speeds, ADD	Inst	Ea	Lg	---	---	---	13.90	---	---	13.90	**16.00**
	Inst	Ea	Sm	---	---	---	15.70	---	---	15.70	**18.10**
12-hour timer, ADD	Inst	Ea	Lg	EA	.500	16.00	25.20	17.80	---	43.00	**54.90**
	Inst	Ea	Sm	EA	.625	12.80	28.40	22.30	---	50.70	**60.90**
Minimum Job Charge											
	Inst	Job	Lg	EA	2.29	3.50	---	81.60	---	81.60	**119.00**
	Inst	Job	Sm	EA	2.86	2.80	---	102.00	---	102.00	**149.00**

Grilles

Aluminum
Air return

Description	Oper	Unit	Vol	Crew Size	Man-hours per Unit	Crew Output per Day	Avg Mat'l Unit Cost	Avg Labor Unit Cost	Avg Equip Unit Cost	Avg Total Unit Cost	Avg Price Incl O&P
6" x 6"	Inst	Ea	Lg	UA	.333	24.00	11.60	12.30	---	23.90	**31.50**
	Inst	Ea	Sm	UA	.417	19.20	13.10	15.40	---	28.50	**35.90**
10" x 6"	Inst	Ea	Lg	UA	.364	22.00	14.20	13.50	---	27.70	**36.20**
	Inst	Ea	Sm	UA	.455	17.60	16.00	16.80	---	32.80	**40.90**
16" x 8"	Inst	Ea	Lg	UA	.400	20.00	20.60	14.80	---	35.40	**45.60**
	Inst	Ea	Sm	UA	.500	16.00	23.20	18.50	---	41.70	**50.60**
12" x 12"	Inst	Ea	Lg	UA	.500	16.00	20.60	18.50	---	39.10	**51.00**
	Inst	Ea	Sm	UA	.625	12.80	23.20	23.10	---	46.30	**57.40**
24" x 12"	Inst	Ea	Lg	UA	.571	14.00	36.90	21.10	---	58.00	**73.60**
	Inst	Ea	Sm	UA	.714	11.20	41.60	26.40	---	68.00	**80.60**
48" x 24"	Inst	Ea	Lg	UA	.800	10.00	142.00	29.60	---	171.60	**207.00**
	Inst	Ea	Sm	UA	1.00	8.00	160.00	37.00	---	197.00	**214.00**
Minimum Job Charge											
	Inst	Job	Lg	UA	2.29	3.50	---	84.60	---	84.60	**125.00**
	Inst	Job	Sm	UA	2.86	2.80	---	106.00	---	106.00	**156.00**

Registers, air supply

Ceiling/wall, O.B. damper, anodized aluminum
One- or two-way deflection, adjustable curved face bars

Description	Oper	Unit	Vol	Crew Size	Man-hours per Unit	Crew Output per Day	Avg Mat'l Unit Cost	Avg Labor Unit Cost	Avg Equip Unit Cost	Avg Total Unit Cost	Avg Price Incl O&P
14" x 8"	Inst	Ea	Lg	UA	.500	16.00	24.60	18.50	---	43.10	**55.60**
	Inst	Ea	Sm	UA	.625	12.80	27.70	23.10	---	50.80	**61.90**

Description	Oper	Unit	Vol	Crew Size	Man-hours per Unit	Crew Output per Day	Avg Mat'l Unit Cost	Avg Labor Unit Cost	Avg Equip Unit Cost	Avg Total Unit Cost	Avg Price Incl O&P
Baseboard, adjustable damper, enameled steel											
10" x 6"	Inst	Ea	Lg	UA	.333	24.00	10.70	12.30	---	23.00	**30.50**
	Inst	Ea	Sm	UA	.417	19.20	12.00	15.40	---	27.40	**34.80**
12" x 5"	Inst	Ea	Lg	UA	.364	22.00	11.60	13.50	---	25.10	**33.20**
	Inst	Ea	Sm	UA	.455	17.60	13.10	16.80	---	29.90	**38.00**
12" x 6"	Inst	Ea	Lg	UA	.364	22.00	10.70	13.50	---	24.20	**32.20**
	Inst	Ea	Sm	UA	.455	17.60	12.00	16.80	---	28.80	**36.90**
12" x 8"	Inst	Ea	Lg	UA	.400	20.00	15.50	14.80	---	30.30	**39.70**
	Inst	Ea	Sm	UA	.500	16.00	17.50	18.50	---	36.00	**44.80**
14" x 6"	Inst	Ea	Lg	UA	.444	18.00	11.60	16.40	---	28.00	**37.60**
	Inst	Ea	Sm	UA	.556	14.40	13.10	20.50	---	33.60	**43.50**
Minimum Job Charge											
	Inst	Job	Lg	UA	2.29	3.50	---	84.60	---	84.60	**125.00**
	Inst	Job	Sm	UA	2.86	2.80	---	106.00	---	106.00	**156.00**

Ventilators

Base, damper, screen; 8" neck diameter

Description	Oper	Unit	Vol	Crew Size	Man-hours per Unit	Crew Output per Day	Avg Mat'l Unit Cost	Avg Labor Unit Cost	Avg Equip Unit Cost	Avg Total Unit Cost	Avg Price Incl O&P
215 CFM @ 5 MPH wind	Inst	Ea	Lg	UD	2.67	6.00	64.90	86.00	---	150.90	**202.00**
	Inst	Ea	Sm	UD	3.33	4.80	73.30	107.00	---	180.30	**232.00**

System units complete
Fan coil air conditioning
Cabinet mounted, with filters
Chilled water

Description	Oper	Unit	Vol	Crew Size	Man-hours per Unit	Crew Output per Day	Avg Mat'l Unit Cost	Avg Labor Unit Cost	Avg Equip Unit Cost	Avg Total Unit Cost	Avg Price Incl O&P
0.5 ton cooling	Inst	Ea	Lg	SB	4.00	4.00	842.00	130.00	---	972.00	**1160.00**
	Inst	Ea	Sm	SB	5.00	3.20	950.00	163.00	---	1113.00	**1190.00**
1 ton cooling	Inst	Ea	Lg	SB	5.33	3.00	954.00	173.00	---	1127.00	**1350.00**
	Inst	Ea	Sm	SB	6.67	2.40	1080.00	217.00	---	1297.00	**1390.00**
2.5 ton cooling	Inst	Ea	Lg	SB	8.00	2.00	1730.00	260.00	---	1990.00	**2370.00**
	Inst	Ea	Sm	SB	10.0	1.60	1950.00	326.00	---	2276.00	**2430.00**
3 ton cooling	Inst	Ea	Lg	SB	16.0	1.00	1900.00	521.00	---	2421.00	**2950.00**
	Inst	Ea	Sm	SB	20.0	0.80	2150.00	651.00	---	2801.00	**3100.00**
10 ton cooling	Inst	Ea	Lg	SF	12.0	2.00	2750.00	411.00	---	3161.00	**3770.00**
	Inst	Ea	Sm	SF	15.0	1.60	3100.00	514.00	---	3614.00	**3860.00**
15 ton cooling	Inst	Ea	Lg	SF	16.0	1.50	3830.00	548.00	---	4378.00	**5210.00**
	Inst	Ea	Sm	SF	20.0	1.20	4320.00	685.00	---	5005.00	**5330.00**
20 ton cooling	Inst	Ea	Lg	SF	32.0	0.75	4910.00	1100.00	---	6010.00	**7260.00**
	Inst	Ea	Sm	SF	40.0	0.60	5540.00	1370.00	---	6910.00	**7560.00**
30 ton cooling	Inst	Ea	Lg	SF	48.0	0.50	7280.00	1640.00	---	8920.00	**10800.00**
	Inst	Ea	Sm	SF	60.0	0.40	8220.00	2060.00	---	10280.00	**11200.00**
Minimum Job Charge											
	Inst	Job	Lg	SB	5.33	3.00	---	173.00	---	173.00	**255.00**
	Inst	Job	Sm	SB	6.67	2.40	---	217.00	---	217.00	**319.00**

Description	Oper	Unit	Vol	Crew Size	Man-hours per Unit	Crew Output per Day	Avg Mat'l Unit Cost	Avg Labor Unit Cost	Avg Equip Unit Cost	Avg Total Unit Cost	Avg Price Incl O&P

Heat pumps

Air to air, split system, not including curbs or pads

Description	Oper	Unit	Vol	Crew Size	Man-hours per Unit	Crew Output per Day	Avg Mat'l Unit Cost	Avg Labor Unit Cost	Avg Equip Unit Cost	Avg Total Unit Cost	Avg Price Incl O&P
2 ton cool, 8.5 MBHP heat	Inst	Ea	Lg	SB	16.0	1.00	2400.00	521.00	---	2921.00	**3520.00**
	Inst	Ea	Sm	SB	20.0	0.80	2710.00	651.00	---	3361.00	**3660.00**
3 ton cool, 13 MBHP heat	Inst	Ea	Lg	SB	32.0	0.50	3100.00	1040.00	---	4140.00	**5100.00**
	Inst	Ea	Sm	SB	40.0	0.40	3500.00	1300.00	---	4800.00	**5410.00**
7.5 ton cool, 33 MBHP heat	Inst	Ea	Lg	SB	48.5	0.33	7520.00	1580.00	---	9100.00	**11000.00**
	Inst	Ea	Sm	SB	61.5	0.26	8480.00	2000.00	---	10480.00	**11400.00**
15 ton cool, 64 MBHP heat	Inst	Ea	Lg	SB	64.0	0.25	13500.00	2080.00	---	15580.00	**18500.00**
	Inst	Ea	Sm	SB	80.0	0.20	15200.00	2600.00	---	17800.00	**19000.00**

Air to air, single package, not including curbs, pads, or plenums

Description	Oper	Unit	Vol	Crew Size	Man-hours per Unit	Crew Output per Day	Avg Mat'l Unit Cost	Avg Labor Unit Cost	Avg Equip Unit Cost	Avg Total Unit Cost	Avg Price Incl O&P
2 ton cool, 6.5 MBHP heat	Inst	Ea	Lg	SB	12.8	1.25	2900.00	417.00	---	3317.00	**3940.00**
	Inst	Ea	Sm	SB	16.0	1.00	3270.00	521.00	---	3791.00	**4030.00**
3 ton cool, 10 MBHP heat	Inst	Ea	Lg	SB	21.3	0.75	3480.00	693.00	---	4173.00	**5020.00**
	Inst	Ea	Sm	SB	26.7	0.60	3930.00	869.00	---	4799.00	**5200.00**

Water source to air, single package

Description	Oper	Unit	Vol	Crew Size	Man-hours per Unit	Crew Output per Day	Avg Mat'l Unit Cost	Avg Labor Unit Cost	Avg Equip Unit Cost	Avg Total Unit Cost	Avg Price Incl O&P
1 ton cool, 13 MBHP heat	Inst	Ea	Lg	SB	9.14	1.75	1110.00	298.00	---	1408.00	**1720.00**
	Inst	Ea	Sm	SB	11.4	1.40	1250.00	371.00	---	1621.00	**1800.00**
2 ton cool, 19 MBHP heat	Inst	Ea	Lg	SB	10.7	1.50	1400.00	348.00	---	1748.00	**2130.00**
	Inst	Ea	Sm	SB	13.3	1.20	1580.00	433.00	---	2013.00	**2220.00**
5 ton cool, 29 MBHP heat	Inst	Ea	Lg	SB	21.3	0.75	2460.00	693.00	---	3153.00	**3840.00**
	Inst	Ea	Sm	SB	26.7	0.60	2770.00	869.00	---	3639.00	**4050.00**

Minimum Job Charge

Description	Oper	Unit	Vol	Crew Size	Man-hours per Unit	Crew Output per Day	Avg Mat'l Unit Cost	Avg Labor Unit Cost	Avg Equip Unit Cost	Avg Total Unit Cost	Avg Price Incl O&P
	Inst	Job	Lg	SB	8.00	2.00	---	260.00	---	260.00	**383.00**
	Inst	Job	Sm	SB	10.0	1.60	---	326.00	---	326.00	**479.00**

Roof top air conditioners

Standard controls, curb, energy economizer

Single zone, electric-fired cooling, gas-fired heating

Description	Oper	Unit	Vol	Crew Size	Man-hours per Unit	Crew Output per Day	Avg Mat'l Unit Cost	Avg Labor Unit Cost	Avg Equip Unit Cost	Avg Total Unit Cost	Avg Price Incl O&P
3 ton cool, 60 MBHP heat	Inst	Ea	Lg	SB	13.3	1.20	4650.00	433.00	---	5083.00	**5980.00**
	Inst	Ea	Sm	SB	16.7	0.96	5250.00	544.00	---	5794.00	**6050.00**
4 ton cool, 95 MBHP heat	Inst	Ea	Lg	SB	16.0	1.00	4830.00	521.00	---	5351.00	**6320.00**
	Inst	Ea	Sm	SB	20.0	0.80	5450.00	651.00	---	6101.00	**6400.00**
10 ton cool, 200 MBHP heat	Inst	Ea	Lg	SF	60.0	0.40	10000.00	2060.00	---	12960.00	**15500.00**
	Inst	Ea	Sm	SF	75.0	0.32	12300.00	2570.00	---	14870.00	**16100.00**
30 ton cool, 540 MBHP heat	Inst	Ea	Lg	SG	160	0.20	33500.00	5620.00	---	39120.00	**46700.00**
	Inst	Ea	Sm	SG	200	0.16	37800.00	7020.00	---	44820.00	**48100.00**
40 ton cool, 675 MBHP heat	Inst	Ea	Lg	SG	200	0.16	43500.00	7020.00	---	50520.00	**60400.00**
	Inst	Ea	Sm	SG	246	0.13	49100.00	8640.00	---	57740.00	**61800.00**
50 ton cool, 810 MBHP heat	Inst	Ea	Lg	SG	267	0.12	52800.00	9370.00	---	62170.00	**74500.00**
	Inst	Ea	Sm	SG	320	0.10	59500.00	11200.00	---	70700.00	**76000.00**

Single zone, gas-fired cooling and heating

Description	Oper	Unit	Vol	Crew Size	Man-hours per Unit	Crew Output per Day	Avg Mat'l Unit Cost	Avg Labor Unit Cost	Avg Equip Unit Cost	Avg Total Unit Cost	Avg Price Incl O&P
3 ton cool, 90 MBHP heat	Inst	Ea	Lg	SB	16.0	1.00	5790.00	521.00	---	6311.00	**7430.00**
	Inst	Ea	Sm	SB	20.0	0.80	6530.00	651.00	---	7181.00	**7490.00**

Description	Oper	Unit	Vol	Crew Size	Man-hours per Unit	Crew Output per Day	Avg Mat'l Unit Cost	Avg Labor Unit Cost	Avg Equip Unit Cost	Avg Total Unit Cost	Avg Price Incl O&P
Multizone, electric-fired cooling, gas-fired cooling											
20 ton cool, 360 MBHP heat	Inst	Ea	Lg	SG	160	0.20	52300.00	5620.00	---	57920.00	**68400.00**
	Inst	Ea	Sm	SG	200	0.16	59000.00	7020.00	---	66020.00	**69300.00**
30 ton cool, 540 MBHP heat	Inst	Ea	Lg	SG	229	0.14	65500.00	8040.00	---	73540.00	**87200.00**
	Inst	Ea	Sm	SG	286	0.11	73900.00	10000.00	---	83900.00	**88700.00**
70 ton cool, 1,500 MBHP heat	Inst	Ea	Lg	SG	400	0.08	123000.00	14000.00	---	137000.00	**162000.00**
	Inst	Ea	Sm	SG	500	0.06	139000.00	17600.00	---	156600.00	**164000.00**
80 ton cool, 1,500 MBHP heat	Inst	Ea	Lg	SG	457	0.07	140000.00	16000.00	---	156000.00	**185000.00**
	Inst	Ea	Sm	SG	571	0.06	158000.00	20000.00	---	178000.00	**188000.00**
105 ton cool, 1,500 MBHP heat	Inst	Ea	Lg	SG	640	0.05	184000.00	22500.00	---	206500.00	**245000.00**
	Inst	Ea	Sm	SG	800	0.04	208000.00	28100.00	---	236100.00	**249000.00**
Minimum Job Charge											
	Inst	Job	Lg	SB	5.00	3.20	---	163.00	---	163.00	**239.00**
	Inst	Job	Sm	SB	6.25	2.56	---	203.00	---	203.00	**299.00**

Window unit air conditioners

Semi-permanent installation, 3-speed fan, 125-volt GFI receptacle, energy-efficient models

Description	Oper	Unit	Vol	Crew Size	Man-hours per Unit	Crew Output per Day	Avg Mat'l Unit Cost	Avg Labor Unit Cost	Avg Equip Unit Cost	Avg Total Unit Cost	Avg Price Incl O&P
6,000 BTUH (0.5 ton cool)	Inst	Ea	Lg	EB	4.00	4.00	562.00	143.00	---	705.00	**854.00**
	Inst	Ea	Sm	EB	5.00	3.20	634.00	178.00	---	812.00	**894.00**
9,000 BTUH (0.75 ton cool)	Inst	Ea	Lg	EB	4.00	4.00	620.00	143.00	---	763.00	**921.00**
	Inst	Ea	Sm	EB	5.00	3.20	700.00	178.00	---	878.00	**960.00**
12,000 BTUH (1.0 ton cool)	Inst	Ea	Lg	EB	4.00	4.00	661.00	143.00	---	804.00	**968.00**
	Inst	Ea	Sm	EB	5.00	3.20	746.00	178.00	---	924.00	**1010.00**
18,000 BTUH (1.5 ton cool)	Inst	Ea	Lg	EB	5.33	3.00	755.00	190.00	---	945.00	**1150.00**
	Inst	Ea	Sm	EB	6.67	2.40	851.00	238.00	---	1089.00	**1200.00**
24,000 BTUH (2.0 ton cool)	Inst	Ea	Lg	EB	5.33	3.00	825.00	190.00	---	1015.00	**1230.00**
	Inst	Ea	Sm	EB	6.67	2.40	931.00	238.00	---	1169.00	**1280.00**
36,000 BTUH (3.0 ton cool)	Inst	Ea	Lg	EB	5.33	3.00	1060.00	190.00	---	1250.00	**1500.00**
	Inst	Ea	Sm	EB	6.67	2.40	1200.00	238.00	---	1438.00	**1550.00**
Minimum Job Charge											
	Inst	Job	Lg	EA	2.67	3.00	---	95.10	---	95.10	**139.00**
	Inst	Job	Sm	EA	3.33	2.40	---	119.00	---	119.00	**173.00**

Awnings. See Canopies, page 76
Backfill. See Excavation, page 156

Bath accessories.
For plumbing fixtures, see individual items.

The material cost of an item includes the fixture, water supply, and trim (includes fittings and faucets). The labor cost of an item includes installation of the fixture and connection of water supply and/or electricity, but no demolition or clean-up is included. Average rough-in of pipe, waste, and vent is an "add" item shown below each major grouping of fixtures, unless noted otherwise.

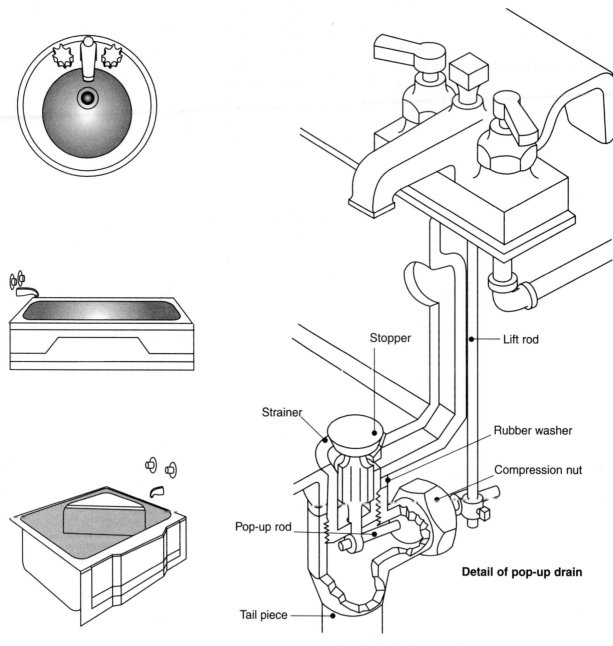

Detail of pop-up drain

From: *Basic Plumbing with Illustrations* Craftsman Book Company

Bath accessories

Description	Oper	Unit	Vol	Crew Size	Man-hours per Unit	Crew Output per Day	Avg Mat'l Unit Cost	Avg Labor Unit Cost	Avg Equip Unit Cost	Avg Total Unit Cost	Avg Price Incl O&P
Average quality											
Cup holder, surface mounted											
Chrome	Inst	Ea	Lg	CA	.333	24.00	5.22	11.10	---	16.32	**22.60**
	Inst	Ea	Sm	CA	.444	18.00	6.29	14.80	---	21.09	**29.40**
Cup and toothbrush holder											
Brass	Inst	Ea	Lg	CA	.333	24.00	7.30	11.10	---	18.40	**25.00**
	Inst	Ea	Sm	CA	.444	18.00	8.80	14.80	---	23.60	**32.30**
Chrome	Inst	Ea	Lg	CA	.333	24.00	5.84	11.10	---	16.94	**23.30**
	Inst	Ea	Sm	CA	.444	18.00	7.04	14.80	---	21.84	**30.30**
Cup, toothbrush and soapholder, recessed											
Chrome	Inst	Ea	Lg	CA	.500	16.00	11.80	16.60	---	28.40	**38.50**
	Inst	Ea	Sm	CA	.667	12.00	14.20	22.20	---	36.40	**49.60**
Glass shelf, chrome											
18" long	Inst	Ea	Lg	CA	.400	20.00	35.00	13.30	---	48.30	**60.20**
	Inst	Ea	Sm	CA	.533	15.00	42.20	17.70	---	59.90	**75.10**
24" long	Inst	Ea	Lg	CA	.400	20.00	36.50	13.30	---	49.80	**61.90**
	Inst	Ea	Sm	CA	.533	15.00	44.00	17.70	---	61.70	**77.20**
Grab bars, concealed mounting											
Satin stainless steel											
12" long	Inst	Ea	Lg	CA	.400	20.00	19.90	13.30	---	33.20	**42.90**
	Inst	Ea	Sm	CA	.533	15.00	24.00	17.70	---	41.70	**54.20**
18" long	Inst	Ea	Lg	CA	.400	20.00	19.90	13.30	---	33.20	**42.90**
	Inst	Ea	Sm	CA	.533	15.00	24.00	17.70	---	41.70	**54.20**
24" long	Inst	Ea	Lg	CA	.400	20.00	19.90	13.30	---	33.20	**42.90**
	Inst	Ea	Sm	CA	.533	15.00	24.00	17.70	---	41.70	**54.20**
30" long	Inst	Ea	Lg	CA	.400	20.00	21.50	13.30	---	34.80	**44.60**
	Inst	Ea	Sm	CA	.533	15.00	25.90	17.70	---	43.60	**56.40**
36" long	Inst	Ea	Lg	CA	.400	20.00	21.70	13.30	---	35.00	**44.90**
	Inst	Ea	Sm	CA	.533	15.00	26.10	17.70	---	43.80	**56.70**
42" long	Inst	Ea	Lg	CA	.400	20.00	23.40	13.30	---	36.70	**46.90**
	Inst	Ea	Sm	CA	.533	15.00	28.30	17.70	---	46.00	**59.10**
48" long	Inst	Ea	Lg	CA	.400	20.00	25.60	13.30	---	38.90	**49.40**
	Inst	Ea	Sm	CA	.533	15.00	30.90	17.70	---	48.60	**62.10**
Angle bar, 16" L x 32" H	Inst	Ea	Lg	CA	.400	20.00	47.50	13.30	---	60.80	**74.50**
	Inst	Ea	Sm	CA	.533	15.00	57.20	17.70	---	74.90	**92.40**
Tub-shower bar, 36" L x 54" H	Inst	Ea	Lg	CA	.400	20.00	74.50	13.30	---	87.80	**106.00**
	Inst	Ea	Sm	CA	.533	15.00	89.80	17.70	---	107.50	**130.00**
Robe hooks, single or double	Inst	Ea	Lg	CA	.250	32.00	3.39	8.32	---	11.71	**16.40**
	Inst	Ea	Sm	CA	.333	24.00	4.09	11.10	---	15.19	**21.30**
Shower curtain rod, 1" dia x 5' 5" L with adjacent rod holder											
	Inst	Ea	Lg	CA	.250	32.00	10.90	8.32	---	19.22	**25.00**
	Inst	Ea	Sm	CA	.333	24.00	13.10	11.10	---	24.20	**31.70**
Soap dish, chrome											
Surface	Inst	Ea	Lg	CA	.250	32.00	5.22	8.32	---	13.54	**18.50**
	Inst	Ea	Sm	CA	.333	24.00	6.29	11.10	---	17.39	**23.90**
Recessed	Inst	Ea	Lg	CA	.250	32.00	9.89	8.32	---	18.21	**23.90**
	Inst	Ea	Sm	CA	.333	24.00	11.90	11.10	---	23.00	**30.30**
Soap holder with bar											
Surface	Inst	Ea	Lg	CA	.250	32.00	5.84	8.32	---	14.16	**19.20**
	Inst	Ea	Sm	CA	.333	24.00	7.04	11.10	---	18.14	**24.70**
Recessed	Inst	Ea	Lg	CA	.250	32.00	13.70	8.32	---	22.02	**28.30**
	Inst	Ea	Sm	CA	.333	24.00	16.50	11.10	---	27.60	**35.60**

Description	Oper	Unit	Vol	Crew Size	Man-hours per Unit	Crew Output per Day	Avg Mat'l Unit Cost	Avg Labor Unit Cost	Avg Equip Unit Cost	Avg Total Unit Cost	Avg Price Incl O&P
Toilet roll holder											
Surface	Inst	Ea	Lg	CA	.250	32.00	8.76	8.32	---	17.08	**22.60**
	Inst	Ea	Sm	CA	.333	24.00	10.60	11.10	---	21.70	**28.80**
Recessed	Inst	Ea	Lg	CA	.250	32.00	17.20	8.32	---	25.52	**32.30**
	Inst	Ea	Sm	CA	.333	24.00	20.70	11.10	---	31.80	**40.50**
Toilet roll holder with hood, recessed											
Surface	Inst	Ea	Lg	CA	.333	24.00	18.30	11.10	---	29.40	**37.60**
	Inst	Ea	Sm	CA	.444	18.00	22.00	14.80	---	36.80	**47.50**
Recessed	Inst	Ea	Lg	CA	.333	24.00	29.20	11.10	---	40.30	**50.20**
	Inst	Ea	Sm	CA	.444	18.00	35.20	14.80	---	50.00	**62.60**
Towel bars, square or round											
Stainless steel											
18"	Inst	Ea	Lg	CA	.250	32.00	8.80	8.32	---	17.12	**22.60**
	Inst	Ea	Sm	CA	.333	24.00	10.60	11.10	---	21.70	**28.80**
24"	Inst	Ea	Lg	CA	.250	32.00	9.64	8.32	---	17.96	**23.60**
	Inst	Ea	Sm	CA	.333	24.00	11.60	11.10	---	22.70	**30.00**
30"	Inst	Ea	Lg	CA	.250	32.00	10.40	8.32	---	18.72	**24.50**
	Inst	Ea	Sm	CA	.333	24.00	12.60	11.10	---	23.70	**31.10**
36"	Inst	Ea	Lg	CA	.250	32.00	11.30	8.32	---	19.62	**25.50**
	Inst	Ea	Sm	CA	.333	24.00	13.60	11.10	---	24.70	**32.30**
Aluminum											
18"	Inst	Ea	Lg	CA	.250	32.00	8.07	8.32	---	16.39	**21.80**
	Inst	Ea	Sm	CA	.333	24.00	9.72	11.10	---	20.82	**27.80**
24"	Inst	Ea	Lg	CA	.250	32.00	8.69	8.32	---	17.01	**22.50**
	Inst	Ea	Sm	CA	.333	24.00	10.50	11.10	---	21.60	**28.70**
30"	Inst	Ea	Lg	CA	.250	32.00	9.31	8.32	---	17.63	**23.20**
	Inst	Ea	Sm	CA	.333	24.00	11.20	11.10	---	22.30	**29.50**
36"	Inst	Ea	Lg	CA	.250	32.00	10.00	8.32	---	18.32	**24.00**
	Inst	Ea	Sm	CA	.333	24.00	12.10	11.10	---	23.20	**30.50**
Towel pin											
Chrome	Inst	Ea	Lg	CA	.222	36.00	5.22	7.39	---	12.61	**17.10**
	Inst	Ea	Sm	CA	.296	27.00	6.29	9.85	---	16.14	**22.00**
Towel ring											
Chrome	Inst	Ea	Lg	CA	.222	36.00	7.15	7.39	---	14.54	**19.30**
	Inst	Ea	Sm	CA	.296	27.00	8.62	9.85	---	18.47	**24.70**
Antique brass	Inst	Ea	Lg	CA	.222	36.00	7.48	7.39	---	14.87	**19.70**
	Inst	Ea	Sm	CA	.296	27.00	9.02	9.85	---	18.87	**25.20**
Towel ladder, 4 arms											
Antique brass	Inst	Ea	Lg	CA	.333	24.00	43.10	11.10	---	54.20	**66.20**
	Inst	Ea	Sm	CA	.444	18.00	51.90	14.80	---	66.70	**81.90**
Polished chrome	Inst	Ea	Lg	CA	.333	24.00	34.60	11.10	---	45.70	**56.40**
	Inst	Ea	Sm	CA	.444	18.00	41.70	14.80	---	56.50	**70.10**
Towel supply shelf, with towel bar											
18" long	Inst	Ea	Lg	CA	.250	32.00	58.80	8.32	---	67.12	**80.00**
	Inst	Ea	Sm	CA	.333	24.00	70.80	11.10	---	81.90	**98.10**
24" long	Inst	Ea	Lg	CA	.250	32.00	105.00	8.32	---	113.32	**133.00**
	Inst	Ea	Sm	CA	.333	24.00	126.00	11.10	---	137.10	**162.00**
Wall-to-floor angle bar with flange, bolt, washer and screws											
	Inst	Ea	Lg	CA	.444	18.00	84.00	14.80	---	98.80	**119.00**
	Inst	Ea	Sm	CA	.593	13.50	101.00	19.70	---	120.70	**146.00**

Description	Oper	Unit	Vol	Crew Size	Man-hours per Unit	Crew Output per Day	Avg Mat'l Unit Cost	Avg Labor Unit Cost	Avg Equip Unit Cost	Avg Total Unit Cost	Avg Price Incl O&P

Medicine cabinets
No electrical work included; for wall outlet cost, see Electrical, page 151

Surface mounting, no wall opening

Swing door cabinets with reversible mirror door

Description	Oper	Unit	Vol	Crew Size	Man-hours per Unit	Crew Output per Day	Avg Mat'l Unit Cost	Avg Labor Unit Cost	Avg Equip Unit Cost	Avg Total Unit Cost	Avg Price Incl O&P
16" x 22"	Inst	Ea	Lg	CA	.800	10.00	43.40	26.60	---	70.00	**89.90**
	Inst	Ea	Sm	CA	1.07	7.50	52.40	35.60	---	88.00	**114.00**
16" x 26"	Inst	Ea	Lg	CA	.800	10.00	58.10	26.60	---	84.70	**107.00**
	Inst	Ea	Sm	CA	1.07	7.50	70.00	35.60	---	105.60	**134.00**

Swing door, corner cabinets with reversible mirror door

Description	Oper	Unit	Vol	Crew Size	Man-hours per Unit	Crew Output per Day	Avg Mat'l Unit Cost	Avg Labor Unit Cost	Avg Equip Unit Cost	Avg Total Unit Cost	Avg Price Incl O&P
16" x 36"	Inst	Ea	Lg	CA	.800	10.00	76.30	26.60	---	102.90	**128.00**
	Inst	Ea	Sm	CA	1.07	7.50	91.90	35.60	---	127.50	**159.00**

Sliding door cabinets, bypassing mirror doors
Toplighted, stainless steel

Description	Oper	Unit	Vol	Crew Size	Man-hours per Unit	Crew Output per Day	Avg Mat'l Unit Cost	Avg Labor Unit Cost	Avg Equip Unit Cost	Avg Total Unit Cost	Avg Price Incl O&P
20" x 20"	Inst	Ea	Lg	CA	.800	10.00	91.30	26.60	---	117.90	**145.00**
	Inst	Ea	Sm	CA	1.07	7.50	110.00	35.60	---	145.60	**180.00**
24" x 20"	Inst	Ea	Lg	CA	.800	10.00	110.00	26.60	---	136.60	**166.00**
	Inst	Ea	Sm	CA	1.07	7.50	132.00	35.60	---	167.60	**205.00**
28" x 20"	Inst	Ea	Lg	CA	.800	10.00	120.00	26.60	---	146.60	**178.00**
	Inst	Ea	Sm	CA	1.07	7.50	145.00	35.60	---	180.60	**220.00**

Cosmetic box with framed mirror stainless steel, Builders series

Description	Oper	Unit	Vol	Crew Size	Man-hours per Unit	Crew Output per Day	Avg Mat'l Unit Cost	Avg Labor Unit Cost	Avg Equip Unit Cost	Avg Total Unit Cost	Avg Price Incl O&P
18" x 26"	Inst	Ea	Lg	CA	.800	10.00	60.40	26.60	---	87.00	**109.00**
	Inst	Ea	Sm	CA	1.07	7.50	72.80	35.60	---	108.40	**137.00**
24" x 32"	Inst	Ea	Lg	CA	.800	10.00	63.80	26.60	---	90.40	**113.00**
	Inst	Ea	Sm	CA	1.07	7.50	77.00	35.60	---	112.60	**142.00**
30" x 32"	Inst	Ea	Lg	CA	.800	10.00	69.70	26.60	---	96.30	**120.00**
	Inst	Ea	Sm	CA	1.07	7.50	84.00	35.60	---	119.60	**150.00**
36" x 32"	Inst	Ea	Lg	CA	.800	10.00	73.80	26.60	---	100.40	**125.00**
	Inst	Ea	Sm	CA	1.07	7.50	89.00	35.60	---	124.60	**156.00**
48" x 32"	Inst	Ea	Lg	CA	.800	10.00	84.00	26.60	---	110.60	**136.00**
	Inst	Ea	Sm	CA	1.07	7.50	101.00	35.60	---	136.60	**170.00**

3-way mirror, tri-view

30" x 30"

Description	Oper	Unit	Vol	Crew Size	Man-hours per Unit	Crew Output per Day	Avg Mat'l Unit Cost	Avg Labor Unit Cost	Avg Equip Unit Cost	Avg Total Unit Cost	Avg Price Incl O&P
Frameless, beveled mirror	Inst	Ea	Lg	CA	.889	9.00	193.00	29.60	---	222.60	**266.00**
	Inst	Ea	Sm	CA	1.18	6.80	232.00	39.30	---	271.30	**326.00**
Natural oak	Inst	Ea	Lg	CA	.889	9.00	222.00	29.60	---	251.60	**300.00**
	Inst	Ea	Sm	CA	1.18	6.80	268.00	39.30	---	307.30	**367.00**
White finish	Inst	Ea	Lg	CA	.889	9.00	207.00	29.60	---	236.60	**283.00**
	Inst	Ea	Sm	CA	1.18	6.80	250.00	39.30	---	289.30	**346.00**

36" x 30"

Description	Oper	Unit	Vol	Crew Size	Man-hours per Unit	Crew Output per Day	Avg Mat'l Unit Cost	Avg Labor Unit Cost	Avg Equip Unit Cost	Avg Total Unit Cost	Avg Price Incl O&P
Frameless, beveled mirror	Inst	Ea	Lg	CA	.889	9.00	219.00	29.60	---	248.60	**297.00**
	Inst	Ea	Sm	CA	1.18	6.80	265.00	39.30	---	304.30	**363.00**
Natural oak	Inst	Ea	Lg	CA	.889	9.00	248.00	29.60	---	277.60	**330.00**
	Inst	Ea	Sm	CA	1.18	6.80	299.00	39.30	---	338.30	**403.00**
White finish	Inst	Ea	Lg	CA	.889	9.00	234.00	29.60	---	263.60	**313.00**
	Inst	Ea	Sm	CA	1.18	6.80	282.00	39.30	---	321.30	**383.00**

Description	Oper	Unit	Vol	Crew Size	Man-hours per Unit	Crew Output per Day	Avg Mat'l Unit Cost	Avg Labor Unit Cost	Avg Equip Unit Cost	Avg Total Unit Cost	Avg Price Incl O&P
48" x 30"											
Frameless, beveled mirror	Inst	Ea	Lg	CA	1.00	8.00	284.00	33.30	---	317.30	**376.00**
	Inst	Ea	Sm	CA	1.33	6.00	342.00	44.30	---	386.30	**460.00**
Natural oak	Inst	Ea	Lg	CA	1.00	8.00	305.00	33.30	---	338.30	**401.00**
	Inst	Ea	Sm	CA	1.33	6.00	368.00	44.30	---	412.30	**489.00**
White finish	Inst	Ea	Lg	CA	1.00	8.00	298.00	33.30	---	331.30	**392.00**
	Inst	Ea	Sm	CA	1.33	6.00	359.00	44.30	---	403.30	**479.00**
With matching light fixture											
with 2 lights	Inst	Ea	Lg	EA	2.00	4.00	51.10	71.30	---	122.40	**163.00**
	Inst	Ea	Sm	EA	2.67	3.00	61.60	95.10	---	156.70	**210.00**
with 3 lights	Inst	Ea	Lg	EA	2.00	4.00	76.70	71.30	---	148.00	**192.00**
	Inst	Ea	Sm	EA	2.67	3.00	92.40	95.10	---	187.50	**245.00**
with 4 lights	Inst	Ea	Lg	EA	2.00	4.00	84.00	71.30	---	155.30	**201.00**
	Inst	Ea	Sm	EA	2.67	3.00	101.00	95.10	---	196.10	**255.00**
with 6 lights	Inst	Ea	Lg	EA	2.00	4.00	102.00	71.30	---	173.30	**222.00**
	Inst	Ea	Sm	EA	2.67	3.00	123.00	95.10	---	218.10	**281.00**
With matching light fixture for beveled mirror cabinets only											
with 2 lights	Inst	Ea	Lg	EA	2.00	4.00	69.40	71.30	---	140.70	**184.00**
	Inst	Ea	Sm	EA	2.67	3.00	83.60	95.10	---	178.70	**235.00**
with 3 lights	Inst	Ea	Lg	EA	2.00	4.00	84.00	71.30	---	155.30	**201.00**
	Inst	Ea	Sm	EA	2.67	3.00	101.00	95.10	---	196.10	**255.00**
with 4 lights	Inst	Ea	Lg	EA	2.00	4.00	110.00	71.30	---	181.30	**230.00**
	Inst	Ea	Sm	EA	2.67	3.00	132.00	95.10	---	227.10	**291.00**
with 6 lights	Inst	Ea	Lg	EA	2.00	4.00	128.00	71.30	---	199.30	**251.00**
	Inst	Ea	Sm	EA	2.67	3.00	154.00	95.10	---	249.10	**316.00**

Recessed mounting, overall sizes

Swing door with mirror

Builders series

Description	Oper	Unit	Vol	Crew Size	Man-hours per Unit	Crew Output per Day	Avg Mat'l Unit Cost	Avg Labor Unit Cost	Avg Equip Unit Cost	Avg Total Unit Cost	Avg Price Incl O&P
14" x 18"											
Polished brass strip	Inst	Ea	Lg	CA	1.00	8.00	44.50	33.30	---	77.80	**101.00**
	Inst	Ea	Sm	CA	1.33	6.00	53.70	44.30	---	98.00	**128.00**
Polished edge strip	Inst	Ea	Lg	CA	1.00	8.00	63.50	33.30	---	96.80	**123.00**
	Inst	Ea	Sm	CA	1.33	6.00	76.60	44.30	---	120.90	**154.00**
Stainless steel strip	Inst	Ea	Lg	CA	1.00	8.00	53.30	33.30	---	86.60	**111.00**
	Inst	Ea	Sm	CA	1.33	6.00	64.20	44.30	---	108.50	**140.00**
14" x 24"											
Polished brass strip	Inst	Ea	Lg	CA	1.00	8.00	48.90	33.30	---	82.20	**106.00**
	Inst	Ea	Sm	CA	1.33	6.00	59.00	44.30	---	103.30	**134.00**
Polished edge strip	Inst	Ea	Lg	CA	1.00	8.00	67.90	33.30	---	101.20	**128.00**
	Inst	Ea	Sm	CA	1.33	6.00	81.80	44.30	---	126.10	**161.00**
Stainless steel strip	Inst	Ea	Lg	CA	1.00	8.00	58.40	33.30	---	91.70	**117.00**
	Inst	Ea	Sm	CA	1.33	6.00	70.40	44.30	---	114.70	**147.00**

Description	Oper	Unit	Vol	Crew Size	Man-hours per Unit	Crew Output per Day	Avg Mat'l Unit Cost	Avg Labor Unit Cost	Avg Equip Unit Cost	Avg Total Unit Cost	Avg Price Incl O&P
Decorator series											
14" x 18"											
Frameless bevel mirror	Inst	Ea	Lg	CA	1.00	8.00	72.30	33.30	---	105.60	**133.00**
	Inst	Ea	Sm	CA	1.33	6.00	87.10	44.30	---	131.40	**167.00**
Natural oak	Inst	Ea	Lg	CA	1.00	8.00	97.80	33.30	---	131.10	**162.00**
	Inst	Ea	Sm	CA	1.33	6.00	118.00	44.30	---	162.30	**202.00**
White birch	Inst	Ea	Lg	CA	1.00	8.00	94.90	33.30	---	128.20	**159.00**
	Inst	Ea	Sm	CA	1.33	6.00	114.00	44.30	---	158.30	**198.00**
White finish	Inst	Ea	Lg	CA	1.00	8.00	117.00	33.30	---	150.30	**184.00**
	Inst	Ea	Sm	CA	1.33	6.00	141.00	44.30	---	185.30	**228.00**
14" x 24"											
Frameless bevel mirror	Inst	Ea	Lg	CA	1.00	8.00	73.00	33.30	---	106.30	**134.00**
	Inst	Ea	Sm	CA	1.33	6.00	88.00	44.30	---	132.30	**168.00**
Natural oak	Inst	Ea	Lg	CA	1.00	8.00	98.60	33.30	---	131.90	**163.00**
	Inst	Ea	Sm	CA	1.33	6.00	119.00	44.30	---	163.30	**203.00**
White birch	Inst	Ea	Lg	CA	1.00	8.00	95.60	33.30	---	128.90	**160.00**
	Inst	Ea	Sm	CA	1.33	6.00	115.00	44.30	---	159.30	**199.00**
White finish	Inst	Ea	Lg	CA	1.00	8.00	120.00	33.30	---	153.30	**188.00**
	Inst	Ea	Sm	CA	1.33	6.00	145.00	44.30	---	189.30	**233.00**

Oak framed cabinet with oval mirror

Description	Oper	Unit	Vol	Crew Size	Man-hours per Unit	Crew Output per Day	Avg Mat'l Unit Cost	Avg Labor Unit Cost	Avg Equip Unit Cost	Avg Total Unit Cost	Avg Price Incl O&P
20" x 36"											
	Inst	Ea	Lg	CA	1.00	8.00	146.00	33.30	---	179.30	**218.00**
	Inst	Ea	Sm	CA	1.33	6.00	176.00	44.30	---	220.30	**269.00**

Mirrors
Decorator oval mirrors, antique gold

Description	Oper	Unit	Vol	Crew Size	Man-hours per Unit	Crew Output per Day	Avg Mat'l Unit Cost	Avg Labor Unit Cost	Avg Equip Unit Cost	Avg Total Unit Cost	Avg Price Incl O&P
16" x 24"	Inst	Ea	Lg	CA	.160	50.00	47.50	5.32	---	52.82	**62.60**
	Inst	Ea	Sm	CA	.213	37.50	57.20	7.09	---	64.29	**76.40**
16" x 32"	Inst	Ea	Lg	CA	.160	50.00	58.40	5.32	---	63.72	**75.20**
	Inst	Ea	Sm	CA	.213	37.50	70.40	7.09	---	77.49	**91.60**

Shower equipment. See Shower stalls, page 375
Vanity cabinets. See Cabinets, page 72

Description	Oper	Unit	Vol	Crew Size	Man-hours per Unit	Crew Output per Day	Avg Mat'l Unit Cost	Avg Labor Unit Cost	Avg Equip Unit Cost	Avg Total Unit Cost	Avg Price Inc O&P

Bathtubs (includes whirlpools)

Plumbing fixtures with good quality supply fittings and faucets included in material cost. Labor cost for installation only of fixture, supply fittings, and faucets. For rough-in, see Adjustments in this bathtub section

Frequently encountered applications

Detach & reset operations

Description	Oper	Unit	Vol	Crew	MH	Output	Mat'l	Labor	Equip	Total	Price
Free standing tub	Reset	Ea	Lg	SB	3.20	5.00	38.50	104.00	---	142.50	197.00
	Reset	Ea	Sm	SB	4.57	3.50	46.80	149.00	---	195.80	272.00
Recessed tub	Reset	Ea	Lg	SB	4.00	4.00	38.50	130.00	---	168.50	236.00
	Reset	Ea	Sm	SB	5.71	2.80	46.80	186.00	---	232.80	327.00
Sunken tub	Reset	Ea	Lg	SB	5.33	3.00	38.50	173.00	---	211.50	299.00
	Reset	Ea	Sm	SB	7.62	2.10	46.80	248.00	---	294.80	418.00

Remove operations

Free standing tub	Demo	Ea	Lg	SB	1.33	12.00	10.50	43.30	---	53.80	75.70
	Demo	Ea	Sm	SB	1.90	8.40	12.80	61.90	---	74.70	106.00
Recessed tub	Demo	Ea	Lg	SB	2.00	8.00	21.00	65.10	---	86.10	120.00
	Demo	Ea	Sm	SB	2.86	5.60	25.50	93.10	---	118.60	166.00
Sunken tub	Demo	Ea	Lg	SB	2.67	6.00	31.50	86.90	---	118.40	164.00
	Demo	Ea	Sm	SB	3.81	4.20	38.30	124.00	---	162.30	226.00

Replace operations

Recessed into wall (above floor installation)
Enameled cast iron, colors
60" L x 30" W x 14" H

	Inst	Ea	Lg	SB	6.67	2.40	717.00	217.00	---	934.00	1140.00
	Inst	Ea	Sm	SB	9.5	1.68	870.00	309.00	---	1179.00	1460.00

66" L x 32" W x 16-1/4" H

	Inst	Ea	Lg	SB	6.67	2.40	1580.00	217.00	---	1797.00	2140.00
	Inst	Ea	Sm	SB	9.5	1.68	1920.00	309.00	---	2229.00	2670.00

Enameled steel, colors
60" L x 30" W x 17-1/2" H

	Inst	Ea	Lg	SB	6.67	2.40	347.00	217.00	---	564.00	718.00
	Inst	Ea	Sm	SB	9.5	1.68	422.00	309.00	---	731.00	939.00

Sunken (below floor installation) or full frame support with full sidewall finish
Enameled cast iron, colors
60" L x 32" W x 18-1/4" H

	Inst	Ea	Lg	SB	6.67	2.40	1770.00	217.00	---	1987.00	2350.00
	Inst	Ea	Sm	SB	9.5	1.68	2150.00	309.00	---	2459.00	2920.00
Bathtub with whirlpool											
Plumbing installation	Inst	Ea	Lg	SB	13.3	1.20	3060.00	433.00	---	3493.00	4160.00
	Inst	Ea	Sm	SB	19.0	0.84	3720.00	618.00	---	4338.00	5190.00
Electrical installation	Inst	Ea	Lg	EA	6.15	1.30	123.00	219.00	---	342.00	461.00
	Inst	Ea	Sm	EA	8.79	0.91	149.00	313.00	---	462.00	628.00

Description	Oper	Unit	Vol	Crew Size	Man-hours per Unit	Crew Output per Day	Avg Mat'l Unit Cost	Avg Labor Unit Cost	Avg Equip Unit Cost	Avg Total Unit Cost	Avg Price Incl O&P
Frequently encountered applications (continued)											
71-3/4" L x 36" W x 20-7/8" H											
	Inst	Ea	Lg	SB	6.67	2.40	3060.00	217.00	---	3277.00	**3840.00**
	Inst	Ea	Sm	SB	9.5	1.68	3720.00	309.00	---	4029.00	**4730.00**
Bathtub with whirlpool											
Plumbing installation	Inst	Ea	Lg	SB	13.3	1.20	5110.00	433.00	---	5543.00	**6520.00**
	Inst	Ea	Sm	SB	19.0	0.84	6210.00	618.00	---	6828.00	**8050.00**
Electrical installation	Inst	Ea	Lg	EA	6.15	1.30	123.00	219.00	---	342.00	**461.00**
	Inst	Ea	Sm	EA	8.79	0.91	149.00	313.00	---	462.00	**628.00**
Fiberglass, colors											
60" L x 42" W x 20" H											
	Inst	Ea	Lg	SB	6.67	2.40	1030.00	217.00	---	1247.00	**1500.00**
	Inst	Ea	Sm	SB	9.5	1.68	1250.00	309.00	---	1559.00	**1890.00**
Bathtub with whirlpool											
Plumbing installation	Inst	Ea	Lg	SB	13.3	1.20	1880.00	433.00	---	2313.00	**2800.00**
	Inst	Ea	Sm	SB	19.0	0.84	2280.00	618.00	---	2898.00	**3530.00**
Electrical installation	Inst	Ea	Lg	EA	6.15	1.30	123.00	219.00	---	342.00	**461.00**
	Inst	Ea	Sm	EA	8.79	0.91	149.00	313.00	---	462.00	**628.00**
72" L x 42" W x 20" H											
	Inst	Ea	Lg	SB	6.67	2.40	1070.00	217.00	---	1287.00	**1550.00**
	Inst	Ea	Sm	SB	9.5	1.68	1300.00	309.00	---	1609.00	**1950.00**
Bathtub with whirlpool											
Plumbing installation	Inst	Ea	Lg	SB	13.3	1.20	1980.00	433.00	---	2413.00	**2920.00**
	Inst	Ea	Sm	SB	19.0	0.84	2410.00	618.00	---	3028.00	**3680.00**
Electrical installation	Inst	Ea	Lg	EA	6.15	1.30	123.00	219.00	---	342.00	**461.00**
	Inst	Ea	Sm	EA	8.79	0.91	149.00	313.00	---	462.00	**628.00**
60" L x 60" W x 21" H, corner											
	Inst	Ea	Lg	SB	6.67	2.40	1220.00	217.00	---	1437.00	**1720.00**
	Inst	Ea	Sm	SB	9.5	1.68	1480.00	309.00	---	1789.00	**2150.00**
Bathtub with whirlpool											
Plumbing installation	Inst	Ea	Lg	SB	13.3	1.20	2310.00	433.00	---	2743.00	**3300.00**
	Inst	Ea	Sm	SB	19.0	0.84	2810.00	618.00	---	3428.00	**4140.00**
Electrical installation	Inst	Ea	Lg	EA	6.15	1.30	123.00	219.00	---	342.00	**461.00**
	Inst	Ea	Sm	EA	8.79	0.91	149.00	313.00	---	462.00	**628.00**
Adjustments											
Install rough-In											
	Inst	Ea	Lg	SB	10.0	1.60	59.50	326.00	---	385.50	**547.00**
	Inst	Ea	Sm	SB	14.3	1.12	72.30	465.00	---	537.30	**767.00**
Install shower head with mixer valve over tub											
Open wall, ADD	Inst	Ea	Lg	SA	2.00	4.00	80.50	75.30	---	155.80	**203.00**
	Inst	Ea	Sm	SA	2.86	2.80	97.80	108.00	---	205.80	**271.00**
Closed wall, ADD	Inst	Ea	Lg	SA	4.00	2.00	126.00	151.00	---	277.00	**366.00**
	Inst	Ea	Sm	SA	5.71	1.40	153.00	215.00	---	368.00	**492.00**

Description	Oper	Unit	Vol	Crew Size	Man-hours per Unit	Crew Output per Day	Avg Mat'l Unit Cost	Avg Labor Unit Cost	Avg Equip Unit Cost	Avg Total Unit Cost	Avg Price Incl O&P

Free-standing

American Standard Products, enameled steel

Reminiscence Slipper Soaking Bathtub, white or black textured exterior
72-1/8" L x 37-1/2" W x 28-3/4" H

Description	Oper	Unit	Vol	Crew Size	Man-hours per Unit	Crew Output per Day	Avg Mat'l Unit Cost	Avg Labor Unit Cost	Avg Equip Unit Cost	Avg Total Unit Cost	Avg Price Incl O&P
White	Inst	Ea	Lg	SB	8.00	2.00	2400.00	260.00	---	2660.00	3150.00
	Inst	Ea	Sm	SB	11.4	1.40	2920.00	371.00	---	3291.00	3900.00
Colors	Inst	Ea	Lg	SB	8.00	2.00	2450.00	260.00	---	2710.00	3200.00
	Inst	Ea	Sm	SB	11.4	1.40	2980.00	371.00	---	3351.00	3970.00
Premium Colors	Inst	Ea	Lg	SB	8.00	2.00	2500.00	260.00	---	2760.00	3260.00
	Inst	Ea	Sm	SB	11.4	1.40	3030.00	371.00	---	3401.00	4030.00
Required accessories											
Ball and claw legs (four)	Inst	Set	Lg	SB	---	---	395.00	---	---	395.00	454.00
	Inst	Set	Sm	SB	---	---	479.00	---	---	479.00	551.00
Filler w/ Shower, brass	Inst	Ea	Lg	SB	---	---	615.00	---	---	615.00	708.00
	Inst	Ea	Sm	SB	---	---	747.00	---	---	747.00	859.00
Drain, chain & stopper	Inst	Set	Lg	SB	---	---	160.00	---	---	160.00	184.00
	Inst	Set	Sm	SB	---	---	195.00	---	---	195.00	224.00
Supply valves, brass	Inst	Set	Lg	SB	---	---	371.00	---	---	371.00	427.00
	Inst	Set	Sm	SB	---	---	451.00	---	---	451.00	518.00

Kohler Products, enameled cast iron

Birthday Bath
72" L x 37-1/2" W x 21-1/2" H

Description	Oper	Unit	Vol	Crew Size	Man-hours per Unit	Crew Output per Day	Avg Mat'l Unit Cost	Avg Labor Unit Cost	Avg Equip Unit Cost	Avg Total Unit Cost	Avg Price Incl O&P
White	Inst	Ea	Lg	SB	8.00	2.00	4880.00	260.00	---	5140.00	6000.00
	Inst	Ea	Sm	SB	11.4	1.40	5930.00	371.00	---	6301.00	7360.00
Colors	Inst	Ea	Lg	SB	8.00	2.00	5770.00	260.00	---	6030.00	7020.00
	Inst	Ea	Sm	SB	11.4	1.40	7000.00	371.00	---	7371.00	8600.00
Premium Colors	Inst	Ea	Lg	SB	8.00	2.00	6430.00	260.00	---	6690.00	7780.00
	Inst	Ea	Sm	SB	11.4	1.40	7810.00	371.00	---	8181.00	9530.00
Required accessories											
Ball-and-claw legs (four)	Inst	Set	Lg	SB	---	---	735.00	---	---	735.00	845.00
	Inst	Set	Sm	SB	---	---	893.00	---	---	893.00	1030.00
Filler w/ Shower, brass	Inst	Ea	Lg	SB	---	---	525.00	---	---	525.00	604.00
	Inst	Ea	Sm	SB	---	---	638.00	---	---	638.00	733.00
Drain, chain & stopper	Inst	Set	Lg	SB	---	---	263.00	---	---	263.00	302.00
	Inst	Set	Sm	SB	---	---	319.00	---	---	319.00	367.00
Supply valves, brass	Inst	Set	Lg	SB	---	---	87.50	---	---	87.50	101.00
	Inst	Set	Sm	SB	---	---	106.00	---	---	106.00	122.00

Description	Oper	Unit	Vol	Crew Size	Man-hours per Unit	Crew Output per Day	Avg Mat'l Unit Cost	Avg Labor Unit Cost	Avg Equip Unit Cost	Avg Total Unit Cost	Avg Price Incl O&P
Vintage Bath											
72" L x 42" W x 22" H											
White	Inst	Ea	Lg	SB	8.00	2.00	4650.00	260.00	---	4910.00	**5730.00**
	Inst	Ea	Sm	SB	11.4	1.40	5640.00	371.00	---	6011.00	**7030.00**
Colors	Inst	Ea	Lg	SB	8.00	2.00	5470.00	260.00	---	5730.00	**6680.00**
	Inst	Ea	Sm	SB	11.4	1.40	6650.00	371.00	---	7021.00	**8190.00**
Required accessories											
Wood base	Inst	Ea	Lg	SB	---	---	2000.00	---	---	2000.00	**2300.00**
	Inst	Ea	Sm	SB	---	---	2430.00	---	---	2430.00	**2790.00**
Ceramic base	Inst	Ea	Lg	SB	---	---	852.00	---	---	852.00	**980.00**
	Inst	Ea	Sm	SB	---	---	1030.00	---	---	1030.00	**1190.00**
Adjustable feet	Inst	Set	Lg	SB	---	---	46.30	---	---	46.30	**53.30**
	Inst	Set	Sm	SB	---	---	56.30	---	---	56.30	**64.70**

Recessed

American Standard Products, Americast

Description	Oper	Unit	Vol	Crew Size	Man-hours per Unit	Crew Output per Day	Avg Mat'l Unit Cost	Avg Labor Unit Cost	Avg Equip Unit Cost	Avg Total Unit Cost	Avg Price Incl O&P
60" L x 32" W x 17-3/4" H, w/ grab bar (Cambridge)											
White	Inst	Ea	Lg	SB	6.7	2.40	731.00	218.00	---	949.00	**1160.00**
	Inst	Ea	Sm	SB	9.5	1.68	887.00	309.00	---	1196.00	**1480.00**
Colors	Inst	Ea	Lg	SB	6.7	2.40	834.00	218.00	---	1052.00	**1280.00**
	Inst	Ea	Sm	SB	9.5	1.68	1060.00	309.00	---	1369.00	**1670.00**
Premium Colors	Inst	Ea	Lg	SB	6.7	2.40	907.00	218.00	---	1125.00	**1360.00**
	Inst	Ea	Sm	SB	9.5	1.68	1150.00	309.00	---	1459.00	**1780.00**
Bathtub with whirlpool (System-I)											
Plumbing installation											
White	Inst	Ea	Lg	SB	13.3	1.20	2270.00	433.00	---	2703.00	**3240.00**
	Inst	Ea	Sm	SB	19.0	0.84	2750.00	618.00	---	3368.00	**4070.00**
Colors	Inst	Ea	Lg	SB	13.3	1.20	2380.00	433.00	---	2813.00	**3370.00**
	Inst	Ea	Sm	SB	19.0	0.84	2890.00	618.00	---	3508.00	**4230.00**
Premium Colors	Inst	Ea	Lg	SB	13.3	1.20	2460.00	433.00	---	2893.00	**3470.00**
	Inst	Ea	Sm	SB	19.0	0.84	2990.00	618.00	---	3608.00	**4350.00**
Electrical installation											
	Inst	Ea	Lg	EA	6.15	1.30	123.00	219.00	---	342.00	**461.00**
	Inst	Ea	Sm	EA	8.79	0.91	149.00	313.00	---	462.00	**628.00**
60" L x 34" W x 17-1/2" H, w/ luxury ledge (Princeton)											
White	Inst	Ea	Lg	SB	6.7	2.40	782.00	218.00	---	1000.00	**1220.00**
	Inst	Ea	Sm	SB	9.5	1.68	949.00	309.00	---	1258.00	**1550.00**
Colors	Inst	Ea	Lg	SB	6.7	2.40	849.00	218.00	---	1067.00	**1300.00**
	Inst	Ea	Sm	SB	9.5	1.68	1030.00	309.00	---	1339.00	**1640.00**
Premium Colors	Inst	Ea	Lg	SB	6.7	2.40	911.00	218.00	---	1129.00	**1370.00**
	Inst	Ea	Sm	SB	9.5	1.68	1110.00	309.00	---	1419.00	**1730.00**

Description	Oper	Unit	Vol	Crew Size	Man-hours per Unit	Crew Output per Day	Avg Mat'l Unit Cost	Avg Labor Unit Cost	Avg Equip Unit Cost	Avg Total Unit Cost	Avg Price Incl O&P
66" L x 32" W x 20" H (Stratford)											
White	Inst	Ea	Lg	SB	6.7	2.40	1100.00	218.00	---	1318.00	**1580.00**
	Inst	Ea	Sm	SB	9.5	1.68	1330.00	309.00	---	1639.00	**1990.00**
Colors	Inst	Ea	Lg	SB	6.7	2.40	1270.00	218.00	---	1488.00	**1790.00**
	Inst	Ea	Sm	SB	9.5	1.68	1550.00	309.00	---	1859.00	**2230.00**
Bathtub with whirlpool											
Plumbing installation											
White	Inst	Ea	Lg	SB	13.3	1.20	2910.00	433.00	---	3343.00	**3980.00**
	Inst	Ea	Sm	SB	19.0	0.84	3530.00	618.00	---	4148.00	**4970.00**
Colors	Inst	Ea	Lg	SB	13.3	1.20	3090.00	433.00	---	3523.00	**4190.00**
	Inst	Ea	Sm	SB	19.0	0.84	3750.00	618.00	---	4368.00	**5220.00**
Electrical installation											
	Inst	Ea	Lg	EA	6.15	1.30	123.00	219.00	---	342.00	**461.00**
	Inst	Ea	Sm	EA	8.79	0.91	149.00	313.00	---	462.00	**628.00**
Options											
Grab bar kit											
Chrome	Inst	Ea	Lg	SB	---	---	127.00	---	---	127.00	**147.00**
	Inst	Ea	Sm	SB	---	---	155.00	---	---	155.00	**178.00**
Brass	Inst	Ea	Lg	SB	---	---	216.00	---	---	216.00	**249.00**
	Inst	Ea	Sm	SB	---	---	263.00	---	---	263.00	**302.00**
White	Inst	Ea	Lg	SB	---	---	216.00	---	---	216.00	**249.00**
	Inst	Ea	Sm	SB	---	---	263.00	---	---	263.00	**302.00**
Satin	Inst	Ea	Lg	SB	---	---	216.00	---	---	216.00	**249.00**
	Inst	Ea	Sm	SB	---	---	263.00	---	---	263.00	**302.00**
Drain kit											
Chrome	Inst	Ea	Lg	SB	---	---	87.50	---	---	87.50	**101.00**
	Inst	Ea	Sm	SB	---	---	106.00	---	---	106.00	**122.00**
Brass	Inst	Ea	Lg	SB		---	153.00	---	---	153.00	**175.00**
	Inst	Ea	Sm	SB	---	---	185.00	---	---	185.00	**213.00**
White	Inst	Ea	Lg	SB	---	---	174.00	---	---	174.00	**200.00**
	Inst	Ea	Sm	SB	---	---	211.00	---	---	211.00	**242.00**
Satin	Inst	Ea	Lg	SB	---	---	153.00	---	---	153.00	**175.00**
	Inst	Ea	Sm	SB	---	---	185.00	---	---	185.00	**213.00**

American Standard Products, enameled steel

Description	Oper	Unit	Vol	Crew Size	Man-hours per Unit	Crew Output per Day	Avg Mat'l Unit Cost	Avg Labor Unit Cost	Avg Equip Unit Cost	Avg Total Unit Cost	Avg Price Incl O&P
60" L x 30" W x 15" H (Salem)											
White, slip resistant	Inst	Ea	Lg	SB	6.67	2.40	321.00	217.00	---	538.00	**688.00**
	Inst	Ea	Sm	SB	9.5	1.68	389.00	309.00	---	698.00	**902.00**
Colors, slip resistant	Inst	Ea	Lg	SB	6.67	2.40	340.00	217.00	---	557.00	**710.00**
	Inst	Ea	Sm	SB	9.5	1.68	413.00	309.00	---	722.00	**930.00**
60" L x 30" W x 17-1/2" H (Solar)											
White, slip resistant	Inst	Ea	Lg	SB	6.67	2.40	325.00	217.00	---	542.00	**693.00**
	Inst	Ea	Sm	SB	9.5	1.68	394.00	309.00	---	703.00	**908.00**
Colors, slip resistant	Inst	Ea	Lg	SB	6.67	2.40	347.00	217.00	---	564.00	**718.00**
	Inst	Ea	Sm	SB	9.5	1.68	422.00	309.00	---	731.00	**939.00**

Description	Oper	Unit	Vol	Crew Size	Man-hours per Unit	Crew Output per Day	Avg Mat'l Unit Cost	Avg Labor Unit Cost	Avg Equip Unit Cost	Avg Total Unit Cost	Avg Price Incl O&P

Kohler Products, enameled cast iron

48" L x 44" W x 14" H, corner (Mayflower)

Description	Oper	Unit	Vol	Crew Size	Man-hours per Unit	Crew Output per Day	Avg Mat'l Unit Cost	Avg Labor Unit Cost	Avg Equip Unit Cost	Avg Total Unit Cost	Avg Price Incl O&P
White	Inst	Ea	Lg	SB	6.67	2.40	1900.00	217.00	---	2117.00	**2510.00**
	Inst	Ea	Sm	SB	9.5	1.68	2310.00	309.00	---	2619.00	**3110.00**
Colors	Inst	Ea	Lg	SB	6.67	2.40	2310.00	217.00	---	2527.00	**2980.00**
	Inst	Ea	Sm	SB	9.5	1.68	2800.00	309.00	---	3109.00	**3680.00**

54" L x 30-1/4" W x 14" H (Seaforth)

Description	Oper	Unit	Vol	Crew Size	Man-hours per Unit	Crew Output per Day	Avg Mat'l Unit Cost	Avg Labor Unit Cost	Avg Equip Unit Cost	Avg Total Unit Cost	Avg Price Incl O&P
White	Inst	Ea	Lg	SB	6.67	2.40	962.00	217.00	---	1179.00	**1430.00**
	Inst	Ea	Sm	SB	9.5	1.68	1170.00	309.00	---	1479.00	**1800.00**
Colors	Inst	Ea	Lg	SB	6.67	2.40	1150.00	217.00	---	1367.00	**1640.00**
	Inst	Ea	Sm	SB	9.5	1.68	1390.00	309.00	---	1699.00	**2060.00**

60" L x 32" W x 16-1/4" H (Mendota)

Description	Oper	Unit	Vol	Crew Size	Man-hours per Unit	Crew Output per Day	Avg Mat'l Unit Cost	Avg Labor Unit Cost	Avg Equip Unit Cost	Avg Total Unit Cost	Avg Price Incl O&P
White	Inst	Ea	Lg	SB	6.67	2.40	856.00	217.00	---	1073.00	**1300.00**
	Inst	Ea	Sm	SB	9.5	1.68	1040.00	309.00	---	1349.00	**1650.00**
Colors	Inst	Ea	Lg	SB	6.67	2.40	1050.00	217.00	---	1267.00	**1520.00**
	Inst	Ea	Sm	SB	9.5	1.68	1270.00	309.00	---	1579.00	**1920.00**
Premium Colors	Inst	Ea	Lg	SB	6.67	2.40	1170.00	217.00	---	1387.00	**1670.00**
	Inst	Ea	Sm	SB	9.5	1.68	1420.00	309.00	---	1729.00	**2090.00**

60" L x 34-1/4" W x 14" H (Villager)

Description	Oper	Unit	Vol	Crew Size	Man-hours per Unit	Crew Output per Day	Avg Mat'l Unit Cost	Avg Labor Unit Cost	Avg Equip Unit Cost	Avg Total Unit Cost	Avg Price Incl O&P
White	Inst	Ea	Lg	SB	6.67	2.40	602.00	217.00	---	819.00	**1010.00**
	Inst	Ea	Sm	SB	9.5	1.68	731.00	309.00	---	1040.00	**1300.00**
Colors	Inst	Ea	Lg	SB	6.67	2.40	717.00	217.00	---	934.00	**1140.00**
	Inst	Ea	Sm	SB	9.5	1.68	870.00	309.00	---	1179.00	**1460.00**

66" L x 32" W x 16-1/4" H (Dynametric)

Description	Oper	Unit	Vol	Crew Size	Man-hours per Unit	Crew Output per Day	Avg Mat'l Unit Cost	Avg Labor Unit Cost	Avg Equip Unit Cost	Avg Total Unit Cost	Avg Price Incl O&P
White	Inst	Ea	Lg	SB	6.67	2.40	1310.00	217.00	---	1527.00	**1830.00**
	Inst	Ea	Sm	SB	9.5	1.68	1590.00	309.00	---	1899.00	**2290.00**
Colors	Inst	Ea	Lg	SB	6.67	2.40	1580.00	217.00	---	1797.00	**2140.00**
	Inst	Ea	Sm	SB	9.5	1.68	1920.00	309.00	---	2229.00	**2670.00**

Sunken

American Standard Products, acrylic

58-1/4" L x 58-1/4" W x 22-1/2" H, corner, with grab bar drilling (Savona)

Bathtub with whirlpool (System-II)

Plumbing installation

Description	Oper	Unit	Vol	Crew Size	Man-hours per Unit	Crew Output per Day	Avg Mat'l Unit Cost	Avg Labor Unit Cost	Avg Equip Unit Cost	Avg Total Unit Cost	Avg Price Incl O&P
White	Inst	Ea	Lg	SB	13.3	1.20	2700.00	433.00	---	3133.00	**3740.00**
	Inst	Ea	Sm	SB	19.0	0.84	3280.00	618.00	---	3898.00	**4680.00**
Colors	Inst	Ea	Lg	SB	13.3	1.20	2740.00	433.00	---	3173.00	**3790.00**
	Inst	Ea	Sm	SB	19.0	0.84	3330.00	618.00	---	3948.00	**4740.00**
Premium Colors	Inst	Ea	Lg	SB	13.3	1.20	2790.00	433.00	---	3223.00	**3840.00**
	Inst	Ea	Sm	SB	19.0	0.84	3390.00	618.00	---	4008.00	**4800.00**

Electrical installation

Description	Oper	Unit	Vol	Crew Size	Man-hours per Unit	Crew Output per Day	Avg Mat'l Unit Cost	Avg Labor Unit Cost	Avg Equip Unit Cost	Avg Total Unit Cost	Avg Price Incl O&P
	Inst	Ea	Lg	EA	6.15	1.30	123.00	219.00	---	342.00	**461.00**
	Inst	Ea	Sm	EA	8.79	0.91	149.00	313.00	---	462.00	**628.00**

Description	Oper	Unit	Vol	Crew Size	Man-hours per Unit	Crew Output per Day	Avg Mat'l Unit Cost	Avg Labor Unit Cost	Avg Equip Unit Cost	Avg Total Unit Cost	Avg Price Incl O&P
Options											
Grab bar kit											
Chrome	Inst	Ea	Lg	SB	---	---	77.70	---	---	77.70	**89.40**
	Inst	Ea	Sm	SB	---	---	94.40	---	---	94.40	**109.00**
Brass	Inst	Ea	Lg	SB	---	---	109.00	---	---	109.00	**125.00**
	Inst	Ea	Sm	SB	---	---	132.00	---	---	132.00	**152.00**
White	Inst	Ea	Lg	SB	---	---	97.30	---	---	97.30	**112.00**
	Inst	Ea	Sm	SB	---	---	118.00	---	---	118.00	**136.00**
Satin	Inst	Ea	Lg	SB	---	---	109.00	---	---	109.00	**125.00**
	Inst	Ea	Sm	SB	---	---	132.00	---	---	132.00	**152.00**
Drain kit											
Chrome	Inst	Ea	Lg	SB	---	---	87.50	---	---	87.50	**101.00**
	Inst	Ea	Sm	SB	---	---	106.00	---	---	106.00	**122.00**
Brass	Inst	Ea	Lg	SB	---	---	153.00	---	---	153.00	**175.00**
	Inst	Ea	Sm	SB	---	---	185.00	---	---	185.00	**213.00**
White	Inst	Ea	Lg	SB	---	---	174.00	---	---	174.00	**200.00**
	Inst	Ea	Sm	SB	---	---	211.00	---	---	211.00	**242.00**
Satin	Inst	Ea	Lg	SB	---	---	153.00	---	---	153.00	**175.00**
	Inst	Ea	Sm	SB	---	---	185.00	---	---	185.00	**213.00**
59-1/2" L x 42-1/4" W x 21-1/4" H, oval (Savona)											
White	Inst	Ea	Lg	SB	6.67	2.40	834.00	217.00	---	1051.00	**1280.00**
	Inst	Ea	Sm	SB	9.5	1.68	1010.00	309.00	---	1319.00	**1620.00**
Colors	Inst	Ea	Lg	SB	6.67	2.40	874.00	217.00	---	1091.00	**1320.00**
	Inst	Ea	Sm	SB	9.5	1.68	1060.00	309.00	---	1369.00	**1670.00**
Premium Colors	Inst	Ea	Lg	SB	6.67	2.40	913.00	217.00	---	1130.00	**1370.00**
	Inst	Ea	Sm	SB	9.5	1.68	1110.00	309.00	---	1419.00	**1730.00**
Bathtub with whirlpool (System-II)											
Plumbing installation											
White	Inst	Ea	Lg	SB	13.3	1.20	1590.00	433.00	---	2023.00	**2470.00**
	Inst	Ea	Sm	SB	19.0	0.84	1930.00	618.00	---	2548.00	**3130.00**
Colors	Inst	Ea	Lg	SB	13.3	1.20	1640.00	433.00	---	2073.00	**2520.00**
	Inst	Ea	Sm	SB	19.0	0.84	1990.00	618.00	---	2608.00	**3200.00**
Premium Colors	Inst	Ea	Lg	SB	13.3	1.20	1680.00	433.00	---	2113.00	**2570.00**
	Inst	Ea	Sm	SB	19.0	0.84	2040.00	618.00	---	2658.00	**3250.00**
Electrical installation											
	Inst	Ea	Lg	EA	6.15	1.30	123.00	219.00	---	342.00	**461.00**
	Inst	Ea	Sm	EA	8.79	0.91	149.00	313.00	---	462.00	**628.00**
Drain kit											
Chrome	Inst	Ea	Lg	SB	---	---	87.50	---	---	87.50	**101.00**
	Inst	Ea	Sm	SB	---	---	106.00	---	---	106.00	**122.00**
Brass	Inst	Ea	Lg	SB	---	---	153.00	---	---	153.00	**175.00**
	Inst	Ea	Sm	SB	---	---	185.00	---	---	185.00	**213.00**
White	Inst	Ea	Lg	SB	---	---	174.00	---	---	174.00	**200.00**
	Inst	Ea	Sm	SB	---	---	211.00	---	---	211.00	**242.00**
Satin	Inst	Ea	Lg	SB	---	---	153.00	---	---	153.00	**175.00**
	Inst	Ea	Sm	SB	---	---	185.00	---	---	185.00	**213.00**

Description	Oper	Unit	Vol	Crew Size	Man-hours per Unit	Crew Output per Day	Avg Mat'l Unit Cost	Avg Labor Unit Cost	Avg Equip Unit Cost	Avg Total Unit Cost	Avg Price Incl O&P
60" L x 32" W x 21-1/2" H (Ellisse Built-In)											
Bathtub with whirlpool (System-I)											
Plumbing installation											
White	Inst	Ea	Lg	SB	13.3	1.20	2170.00	433.00	---	2603.00	**3130.00**
	Inst	Ea	Sm	SB	19.0	0.84	2630.00	618.00	---	3248.00	**3940.00**
Colors	Inst	Ea	Lg	SB	13.3	1.20	2210.00	433.00	---	2643.00	**3180.00**
	Inst	Ea	Sm	SB	19.0	0.84	2690.00	618.00	---	3308.00	**4000.00**
Electrical installation											
	Inst	Ea	Lg	EA	6.15	1.30	123.00	219.00	---	342.00	**461.00**
	Inst	Ea	Sm	EA	8.79	0.91	149.00	313.00	---	462.00	**628.00**
Bathtub with whirlpool (System-II)											
Plumbing installation											
White	Inst	Ea	Lg	SB	13.3	1.20	2340.00	433.00	---	2773.00	**3320.00**
	Inst	Ea	Sm	SB	19.0	0.84	2840.00	618.00	---	3458.00	**4170.00**
Colors	Inst	Ea	Lg	SB	13.3	1.20	2380.00	433.00	---	2813.00	**3370.00**
	Inst	Ea	Sm	SB	19.0	0.84	2890.00	618.00	---	3508.00	**4230.00**
Electrical installation											
	Inst	Ea	Lg	EA	6.15	1.30	123.00	219.00	---	342.00	**461.00**
	Inst	Ea	Sm	EA	8.79	0.91	149.00	313.00	---	462.00	**628.00**
Options											
Grab bar kit											
Chrome	Inst	Ea	Lg	SB	---	---	123.00	---	---	123.00	**142.00**
	Inst	Ea	Sm	SB	---	---	150.00	---	---	150.00	**172.00**
Brass	Inst	Ea	Lg	SB	---	---	206.00	---	---	206.00	**237.00**
	Inst	Ea	Sm	SB	---	---	250.00	---	---	250.00	**287.00**
White	Inst	Ea	Lg	SB	---	---	206.00	---	---	206.00	**237.00**
	Inst	Ea	Sm	SB	---	---	250.00	---	---	250.00	**287.00**
Drain kit											
Chrome	Inst	Ea	Lg	SB	---	---	87.50	---	---	87.50	**101.00**
	Inst	Ea	Sm	SB	---	---	106.00	---	---	106.00	**122.00**
Brass	Inst	Ea	Lg	SB	---	---	153.00	---	---	153.00	**175.00**
	Inst	Ea	Sm	SB	---	---	185.00	---	---	185.00	**213.00**
White	Inst	Ea	Lg	SB	---	---	174.00	---	---	174.00	**200.00**
	Inst	Ea	Sm	SB	---	---	211.00	---	---	211.00	**242.00**
69-1/4" L x 38-1/2" W x 19-3/4" H (Ellisse Oval)											
Bathtub with whirlpool (System-I)											
Plumbing installation											
White	Inst	Ea	Lg	SB	13.3	1.20	2300.00	433.00	---	2733.00	**3280.00**
	Inst	Ea	Sm	SB	19.0	0.84	2790.00	618.00	---	3408.00	**4120.00**
Colors	Inst	Ea	Lg	SB	13.3	1.20	2340.00	433.00	---	2773.00	**3330.00**
	Inst	Ea	Sm	SB	19.0	0.84	2840.00	618.00	---	3458.00	**4180.00**
Premium Colors	Inst	Ea	Lg	SB	13.3	1.20	2390.00	433.00	---	2823.00	**3380.00**
	Inst	Ea	Sm	SB	19.0	0.84	2900.00	618.00	---	3518.00	**4240.00**
Electrical installation											
	Inst	Ea	Lg	EA	6.15	1.30	123.00	219.00	---	342.00	**461.00**
	Inst	Ea	Sm	EA	8.79	0.91	149.00	313.00	---	462.00	**628.00**

Description	Oper	Unit	Vol	Crew Size	Man-hours per Unit	Crew Output per Day	Avg Mat'l Unit Cost	Avg Labor Unit Cost	Avg Equip Unit Cost	Avg Total Unit Cost	Avg Price Incl O&P
Options											
Grab bar kit											
Chrome	Inst	Ea	Lg	SB	---	---	123.00	---	---	123.00	**142.00**
	Inst	Ea	Sm	SB	---	---	150.00	---	---	150.00	**172.00**
Brass	Inst	Ea	Lg	SB	---	---	206.00	---	---	206.00	**237.00**
	Inst	Ea	Sm	SB	---	---	250.00	---	---	250.00	**287.00**
White	Inst	Ea	Lg	SB	---	---	206.00	---	---	206.00	**237.00**
	Inst	Ea	Sm	SB	---	---	250.00	---	---	250.00	**287.00**
Satin	Inst	Ea	Lg	SB	---	---	206.00	---	---	206.00	**237.00**
	Inst	Ea	Sm	SB	---	---	250.00	---	---	250.00	**287.00**
Drain kit											
Chrome	Inst	Ea	Lg	SB	---	---	87.50	---	---	87.50	**101.00**
	Inst	Ea	Sm	SB	---	---	106.00	---	---	106.00	**122.00**
Brass	Inst	Ea	Lg	SB	---	---	153.00	---	---	153.00	**175.00**
	Inst	Ea	Sm	SB	---	---	185.00	---	---	185.00	**213.00**
White	Inst	Ea	Lg	SB	---	---	174.00	---	---	174.00	**200.00**
	Inst	Ea	Sm	SB	---	---	211.00	---	---	211.00	**242.00**
Satin	Inst	Ea	Lg	SB	---	---	153.00	---	---	153.00	**175.00**
	Inst	Ea	Sm	SB	---	---	185.00	---	---	185.00	**213.00**
72" L x 48" W x 21-1/2" H (Ellisse)											
Bathtub with whirlpool											
Plumbing installation											
White	Inst	Ea	Lg	SB	13.3	1.20	2830.00	433.00	---	3263.00	**3900.00**
	Inst	Ea	Sm	SB	19.0	0.84	3440.00	618.00	---	4058.00	**4870.00**
Colors	Inst	Ea	Lg	SB	13.3	1.20	2880.00	433.00	---	3313.00	**3940.00**
	Inst	Ea	Sm	SB	19.0	0.84	3490.00	618.00	---	4108.00	**4930.00**
Premium Colors	Inst	Ea	Lg	SB	13.3	1.20	2920.00	433.00	---	3353.00	**4000.00**
	Inst	Ea	Sm	SB	19.0	0.84	3550.00	618.00	---	4168.00	**4990.00**
Electrical installation											
	Inst	Ea	Lg	EA	6.15	1.30	123.00	219.00	---	342.00	**461.00**
	Inst	Ea	Sm	EA	8.79	0.91	149.00	313.00	---	462.00	**628.00**
Drain kit											
Chrome	Inst	Ea	Lg	SB	---	---	87.50	---	---	87.50	**101.00**
	Inst	Ea	Sm	SB	---	---	106.00	---	---	106.00	**122.00**
Brass	Inst	Ea	Lg	SB	---	---	153.00	---	---	153.00	**175.00**
	Inst	Ea	Sm	SB	---	---	185.00	---	---	185.00	**213.00**
White	Inst	Ea	Lg	SB	---	---	174.00	---	---	174.00	**200.00**
	Inst	Ea	Sm	SB	---	---	211.00	---	---	211.00	**242.00**
Satin	Inst	Ea	Lg	SB	---	---	153.00	---	---	153.00	**175.00**
	Inst	Ea	Sm	SB	---	---	185.00	---	---	185.00	**213.00**

Kohler Products, enameled cast iron

60" L x 36" W x 20-3/8" H (Steeping Bath)

Description	Oper	Unit	Vol	Crew Size	Man-hours per Unit	Crew Output per Day	Avg Mat'l Unit Cost	Avg Labor Unit Cost	Avg Equip Unit Cost	Avg Total Unit Cost	Avg Price Incl O&P
White	Inst	Ea	Lg	SB	6.67	2.40	1970.00	217.00	---	2187.00	**2580.00**
	Inst	Ea	Sm	SB	9.5	1.68	2390.00	309.00	---	2699.00	**3200.00**
Colors	Inst	Ea	Lg	SB	6.67	2.40	2390.00	217.00	---	2607.00	**3060.00**
	Inst	Ea	Sm	SB	9.5	1.68	2900.00	309.00	---	3209.00	**3790.00**
Premium Colors	Inst	Ea	Lg	SB	6.67	2.40	2700.00	217.00	---	2917.00	**3430.00**
	Inst	Ea	Sm	SB	9.5	1.68	3280.00	309.00	---	3589.00	**4230.00**

Description	Oper	Unit	Vol	Crew Size	Man-hours per Unit	Crew Output per Day	Avg Mat'l Unit Cost	Avg Labor Unit Cost	Avg Equip Unit Cost	Avg Total Unit Cost	Avg Price Incl O&P
Bathtub with whirlpool (System-III)											
Plumbing installation											
White	Inst	Ea	Lg	SB	13.3	1.20	5090.00	433.00	---	5523.00	**6490.00**
	Inst	Ea	Sm	SB	19.0	0.84	6180.00	618.00	---	6798.00	**8020.00**
Colors	Inst	Ea	Lg	SB	13.3	1.20	5540.00	433.00	---	5973.00	**7010.00**
	Inst	Ea	Sm	SB	19.0	0.84	6730.00	618.00	---	7348.00	**8640.00**
Premium Colors	Inst	Ea	Lg	SB	13.3	1.20	5790.00	433.00	---	6223.00	**7290.00**
	Inst	Ea	Sm	SB	19.0	0.84	7030.00	618.00	---	7648.00	**8990.00**
Electrical installation											
	Inst	Ea	Lg	EA	6.15	1.30	123.00	219.00	---	342.00	**461.00**
	Inst	Ea	Sm	EA	8.79	0.91	149.00	313.00	---	462.00	**628.00**
60" L x 32" W x 18-1/4" H (Tea-for-Two)											
White	Inst	Ea	Lg	SB	6.67	2.40	1480.00	217.00	---	1697.00	**2030.00**
	Inst	Ea	Sm	SB	9.5	1.68	1800.00	309.00	---	2109.00	**2530.00**
Colors	Inst	Ea	Lg	SB	6.67	2.40	1770.00	217.00	---	1987.00	**2350.00**
	Inst	Ea	Sm	SB	9.5	1.68	2150.00	309.00	---	2459.00	**2920.00**
Premium Colors	Inst	Ea	Lg	SB	6.67	2.40	1980.00	217.00	---	2197.00	**2600.00**
	Inst	Ea	Sm	SB	9.5	1.68	2410.00	309.00	---	2719.00	**3220.00**
Bathtub with whirlpool (System-II)											
Plumbing installation											
White	Inst	Ea	Lg	SB	13.3	1.20	2840.00	433.00	---	3273.00	**3900.00**
	Inst	Ea	Sm	SB	19.0	0.84	3450.00	618.00	---	4068.00	**4880.00**
Colors	Inst	Ea	Lg	SB	13.3	1.20	3060.00	433.00	---	3493.00	**4160.00**
	Inst	Ea	Sm	SB	19.0	0.84	3720.00	618.00	---	4338.00	**5190.00**
Premium Colors	Inst	Ea	Lg	SB	13.3	1.20	3190.00	433.00	---	3623.00	**4300.00**
	Inst	Ea	Sm	SB	19.0	0.84	3870.00	618.00	---	4488.00	**5360.00**
Electrical installation											
	Inst	Ea	Lg	EA	6.15	1.30	123.00	219.00	---	342.00	**461.00**
	Inst	Ea	Sm	EA	8.79	0.91	149.00	313.00	---	462.00	**628.00**
66" L x 32" W x 18-1/4" H (Maestro)											
White	Inst	Ea	Lg	SB	6.67	2.40	1170.00	217.00	---	1387.00	**1660.00**
	Inst	Ea	Sm	SB	9.5	1.68	1420.00	309.00	---	1729.00	**2090.00**
Colors	Inst	Ea	Lg	SB	6.67	2.40	1380.00	217.00	---	1597.00	**1910.00**
	Inst	Ea	Sm	SB	9.5	1.68	1680.00	309.00	---	1989.00	**2380.00**
Premium Colors	Inst	Ea	Lg	SB	6.67	2.40	1540.00	217.00	---	1757.00	**2090.00**
	Inst	Ea	Sm	SB	9.5	1.68	1870.00	309.00	---	2179.00	**2600.00**
Bathtub with whirlpool (System-II)											
Plumbing installation											
White	Inst	Ea	Lg	SB	13.3	1.20	3570.00	433.00	---	4003.00	**4740.00**
	Inst	Ea	Sm	SB	19.0	0.84	4330.00	618.00	---	4948.00	**5890.00**
Colors	Inst	Ea	Lg	SB	13.3	1.20	3880.00	433.00	---	4313.00	**5090.00**
	Inst	Ea	Sm	SB	19.0	0.84	4710.00	618.00	---	5328.00	**6320.00**
Premium Colors	Inst	Ea	Lg	SB	13.3	1.20	4050.00	433.00	---	4483.00	**5290.00**
	Inst	Ea	Sm	SB	19.0	0.84	4910.00	618.00	---	5528.00	**6560.00**
Electrical installation											
	Inst	Ea	Lg	EA	6.15	1.30	123.00	219.00	---	342.00	**461.00**
	Inst	Ea	Sm	EA	8.79	0.91	149.00	313.00	---	462.00	**628.00**

Description	Oper	Unit	Vol	Crew Size	Man-hours per Unit	Crew Output per Day	Avg Mat'l Unit Cost	Avg Labor Unit Cost	Avg Equip Unit Cost	Avg Total Unit Cost	Avg Price Incl O&P
66" L x 32" W x 18" H (Tea-for-Two)											
White	Inst	Ea	Lg	SB	6.67	2.40	1570.00	217.00	---	1787.00	**2120.00**
	Inst	Ea	Sm	SB	9.5	1.68	1900.00	309.00	---	2209.00	**2640.00**
Colors	Inst	Ea	Lg	SB	6.67	2.40	1880.00	217.00	---	2097.00	**2480.00**
	Inst	Ea	Sm	SB	9.5	1.68	2280.00	309.00	---	2589.00	**3070.00**
Premium Colors	Inst	Ea	Lg	SB	6.67	2.40	2110.00	217.00	---	2327.00	**2740.00**
	Inst	Ea	Sm	SB	9.5	1.68	2560.00	309.00	---	2869.00	**3400.00**
Bathtub with whirlpool (System-II)											
Plumbing installation											
White	Inst	Ea	Lg	SB	13.3	1.20	3910.00	433.00	---	4343.00	**5130.00**
	Inst	Ea	Sm	SB	19.0	0.84	4750.00	618.00	---	5368.00	**6370.00**
Colors	Inst	Ea	Lg	SB	13.3	1.20	4220.00	433.00	---	4653.00	**5490.00**
	Inst	Ea	Sm	SB	19.0	0.84	5120.00	618.00	---	5738.00	**6800.00**
Premium Colors	Inst	Ea	Lg	SB	13.3	1.20	4450.00	433.00	---	4883.00	**5760.00**
	Inst	Ea	Sm	SB	19.0	0.84	5410.00	618.00	---	6028.00	**7130.00**
Electrical installation											
	Inst	Ea	Lg	EA	6.15	1.30	123.00	219.00	---	342.00	**461.00**
	Inst	Ea	Sm	EA	8.79	0.91	149.00	313.00	---	462.00	**628.00**
71-3/4" L x 36" W x 20-7/8" H (Tea-for-Two)											
White	Inst	Ea	Lg	SB	6.67	2.40	2520.00	217.00	---	2737.00	**3220.00**
	Inst	Ea	Sm	SB	9.5	1.68	3060.00	309.00	---	3369.00	**3970.00**
Colors	Inst	Ea	Lg	SB	6.67	2.40	3060.00	217.00	---	3277.00	**3840.00**
	Inst	Ea	Sm	SB	9.5	1.68	3720.00	309.00	---	4029.00	**4730.00**
Premium Colors	Inst	Ea	Lg	SB	6.67	2.40	3470.00	217.00	---	3687.00	**4310.00**
	Inst	Ea	Sm	SB	9.5	1.68	4210.00	309.00	---	4519.00	**5300.00**
Bathtub with whirlpool (System-III)											
Plumbing installation											
White	Inst	Ea	Lg	SB	13.3	1.20	4700.00	433.00	---	5133.00	**6050.00**
	Inst	Ea	Sm	SB	19.0	0.84	5710.00	618.00	---	6328.00	**7480.00**
Colors	Inst	Ea	Lg	SB	13.3	1.20	5110.00	433.00	---	5543.00	**6520.00**
	Inst	Ea	Sm	SB	19.0	0.84	6210.00	618.00	---	6828.00	**8050.00**
Premium Colors	Inst	Ea	Lg	SB	13.3	1.20	5340.00	433.00	---	5773.00	**6780.00**
	Inst	Ea	Sm	SB	19.0	0.84	6480.00	618.00	---	7098.00	**8360.00**
Electrical installation											
	Inst	Ea	Lg	EA	6.15	1.30	123.00	219.00	---	342.00	**461.00**
	Inst	Ea	Sm	EA	8.79	0.91	149.00	313.00	---	462.00	**628.00**
72" L x 36" W x 18-1/4" H (Caribbean)											
White	Inst	Ea	Lg	SB	6.67	2.40	2150.00	217.00	---	2367.00	**2790.00**
	Inst	Ea	Sm	SB	9.5	1.68	2610.00	309.00	---	2919.00	**3460.00**
Colors	Inst	Ea	Lg	SB	6.67	2.40	2570.00	217.00	---	2787.00	**3270.00**
	Inst	Ea	Sm	SB	9.5	1.68	3120.00	309.00	---	3429.00	**4040.00**
Premium Colors	Inst	Ea	Lg	SB	6.67	2.40	2890.00	217.00	---	3107.00	**3640.00**
	Inst	Ea	Sm	SB	9.5	1.68	3510.00	309.00	---	3819.00	**4490.00**

Description	Oper	Unit	Vol	Crew Size	Man-hours per Unit	Crew Output per Day	Avg Mat'l Unit Cost	Avg Labor Unit Cost	Avg Equip Unit Cost	Avg Total Unit Cost	Avg Price Incl O&P
Bathtub with whirlpool (System-III)											
Plumbing installation											
White	Inst	Ea	Lg	SB	13.3	1.20	5080.00	433.00	---	5513.00	**6480.00**
	Inst	Ea	Sm	SB	19.0	0.84	6170.00	618.00	---	6788.00	**8010.00**
Colors	Inst	Ea	Lg	SB	13.3	1.20	5530.00	433.00	---	5963.00	**7000.00**
	Inst	Ea	Sm	SB	19.0	0.84	6720.00	618.00	---	7338.00	**8640.00**
Premium Colors	Inst	Ea	Lg	SB	13.3	1.20	5780.00	433.00	---	6213.00	**7280.00**
	Inst	Ea	Sm	SB	19.0	0.84	7020.00	618.00	---	7638.00	**8980.00**
Electrical installation											
	Inst	Ea	Lg	EA	6.15	1.30	123.00	219.00	---	342.00	**461.00**
	Inst	Ea	Sm	EA	8.79	0.91	149.00	313.00	---	462.00	**628.00**
72" L x 42" W x 22" H (Seawall Bath) with personal shower system											
White	Inst	Ea	Lg	SB	6.67	2.40	5160.00	217.00	---	5377.00	**6260.00**
	Inst	Ea	Sm	SB	9.5	1.68	6270.00	309.00	---	6579.00	**7660.00**
Colors	Inst	Ea	Lg	SB	6.67	2.40	5960.00	217.00	---	6177.00	**7170.00**
	Inst	Ea	Sm	SB	9.5	1.68	7240.00	309.00	---	7549.00	**8780.00**
Premium Colors	Inst	Ea	Lg	SB	6.67	2.40	6560.00	217.00	---	6777.00	**7860.00**
	Inst	Ea	Sm	SB	9.5	1.68	7960.00	309.00	---	8269.00	**9610.00**
Bathtub with whirlpool (System-III)											
Plumbing installation											
White	Inst	Ea	Lg	SB	13.3	1.20	8930.00	433.00	---	9363.00	**10900.00**
	Inst	Ea	Sm	SB	19.0	0.84	10800.00	618.00	---	11418.00	**13400.00**
Colors	Inst	Ea	Lg	SB	13.3	1.20	9460.00	433.00	---	9893.00	**11500.00**
	Inst	Ea	Sm	SB	19.0	0.84	11500.00	618.00	---	12118.00	**14100.00**
Premium Colors	Inst	Ea	Lg	SB	13.3	1.20	9750.00	433.00	---	10183.00	**11900.00**
	Inst	Ea	Sm	SB	19.0	0.84	11800.00	618.00	---	12418.00	**14500.00**
Electrical installation											
	Inst	Ea	Lg	EA	6.15	1.30	123.00	219.00	---	342.00	**461.00**
	Inst	Ea	Sm	EA	8.79	0.91	149.00	313.00	---	462.00	**628.00**
Options											
Grab bar kit											
Chrome	Inst	Ea	Lg	SB	---	---	221.00	---	---	221.00	**255.00**
	Inst	Ea	Sm	SB	---	---	269.00	---	---	269.00	**309.00**
Brass	Inst	Ea	Lg	SB	---	---	299.00	---	---	299.00	**344.00**
	Inst	Ea	Sm	SB	---	---	363.00	---	---	363.00	**418.00**
Nickel	Inst	Ea	Lg	SB	---	---	374.00	---	---	374.00	**430.00**
	Inst	Ea	Sm	SB	---	---	454.00	---	---	454.00	**522.00**
Drain kit											
Chrome	Inst	Ea	Lg	SB	---	---	240.00	---	---	240.00	**276.00**
	Inst	Ea	Sm	SB	---	---	291.00	---	---	291.00	**335.00**
Brass	Inst	Ea	Lg	SB	---	---	251.00	---	---	251.00	**289.00**
	Inst	Ea	Sm	SB	---	---	305.00	---	---	305.00	**351.00**
Nickel	Inst	Ea	Lg	SB	---	---	317.00	---	---	317.00	**364.00**
	Inst	Ea	Sm	SB	---	---	385.00	---	---	385.00	**443.00**

Description	Oper	Unit	Vol	Crew Size	Man-hours per Unit	Crew Output per Day	Avg Mat'l Unit Cost	Avg Labor Unit Cost	Avg Equip Unit Cost	Avg Total Unit Cost	Avg Price Incl O&P

Kohler Products, fiberglass

60" L x 36" W x 20" H (Mariposa)

Description	Oper	Unit	Vol	Crew Size	Man-hours per Unit	Crew Output per Day	Avg Mat'l Unit Cost	Avg Labor Unit Cost	Avg Equip Unit Cost	Avg Total Unit Cost	Avg Price Incl O&P
White	Inst	Ea	Lg	SB	6.67	2.40	866.00	217.00	---	1083.00	**1310.00**
	Inst	Ea	Sm	SB	9.5	1.68	1050.00	309.00	---	1359.00	**1660.00**
Colors	Inst	Ea	Lg	SB	6.67	2.40	891.00	217.00	---	1108.00	**1340.00**
	Inst	Ea	Sm	SB	9.5	1.68	1080.00	309.00	---	1389.00	**1700.00**

Bathtub with whirlpool (System-I)
Plumbing installation

Description	Oper	Unit	Vol	Crew Size	Man-hours per Unit	Crew Output per Day	Avg Mat'l Unit Cost	Avg Labor Unit Cost	Avg Equip Unit Cost	Avg Total Unit Cost	Avg Price Incl O&P
White	Inst	Ea	Lg	SB	13.3	1.20	1820.00	433.00	---	2253.00	**2730.00**
	Inst	Ea	Sm	SB	19.0	0.84	2210.00	618.00	---	2828.00	**3450.00**
Colors	Inst	Ea	Lg	SB	13.3	1.20	1880.00	433.00	---	2313.00	**2800.00**
	Inst	Ea	Sm	SB	19.0	0.84	2280.00	618.00	---	2898.00	**3530.00**

Electrical installation

Description	Oper	Unit	Vol	Crew Size	Man-hours per Unit	Crew Output per Day	Avg Mat'l Unit Cost	Avg Labor Unit Cost	Avg Equip Unit Cost	Avg Total Unit Cost	Avg Price Incl O&P
	Inst	Ea	Lg	EA	6.15	1.30	123.00	219.00	---	342.00	**461.00**
	Inst	Ea	Sm	EA	8.79	0.91	149.00	313.00	---	462.00	**628.00**

60" L x 42" W x 20" H (Hourglass)

Description	Oper	Unit	Vol	Crew Size	Man-hours per Unit	Crew Output per Day	Avg Mat'l Unit Cost	Avg Labor Unit Cost	Avg Equip Unit Cost	Avg Total Unit Cost	Avg Price Incl O&P
White	Inst	Ea	Lg	SB	6.67	2.40	824.00	217.00		1041.00	**1270.00**
	Inst	Ea	Sm	SB	9.5	1.68	1000.00	309.00	---	1309.00	**1610.00**
Colors	Inst	Ea	Lg	SB	6.67	2.40	848.00	217.00	---	1065.00	**1290.00**
	Inst	Ea	Sm	SB	9.5	1.68	1030.00	309.00	---	1339.00	**1640.00**

Bathtub with whirlpool (System-I)
Plumbing installation

Description	Oper	Unit	Vol	Crew Size	Man-hours per Unit	Crew Output per Day	Avg Mat'l Unit Cost	Avg Labor Unit Cost	Avg Equip Unit Cost	Avg Total Unit Cost	Avg Price Incl O&P
White	Inst	Ea	Lg	SB	13.3	1.20	1700.00	433.00	---	2133.00	**2590.00**
	Inst	Ea	Sm	SB	19.0	0.84	2060.00	618.00	---	2678.00	**3280.00**
Colors	Inst	Ea	Lg	SB	13.3	1.20	1750.00	433.00	---	2183.00	**2650.00**
	Inst	Ea	Sm	SB	19.0	0.84	2120.00	618.00	---	2738.00	**3350.00**

Electrical installation

Description	Oper	Unit	Vol	Crew Size	Man-hours per Unit	Crew Output per Day	Avg Mat'l Unit Cost	Avg Labor Unit Cost	Avg Equip Unit Cost	Avg Total Unit Cost	Avg Price Incl O&P
	Inst	Ea	Lg	EA	6.15	1.30	123.00	219.00	---	342.00	**461.00**
	Inst	Ea	Sm	EA	8.79	0.91	149.00	313.00	---	462.00	**628.00**

60" L x 42" W x 20" H (Symbio)

Description	Oper	Unit	Vol	Crew Size	Man-hours per Unit	Crew Output per Day	Avg Mat'l Unit Cost	Avg Labor Unit Cost	Avg Equip Unit Cost	Avg Total Unit Cost	Avg Price Incl O&P
White	Inst	Ea	Lg	SB	6.67	2.40	1230.00	217.00	---	1447.00	**1730.00**
	Inst	Ea	Sm	SB	9.5	1.68	1490.00	309.00	---	1799.00	**2170.00**
Colors	Inst	Ea	Lg	SB	6.67	2.40	1270.00	217.00	---	1487.00	**1780.00**
	Inst	Ea	Sm	SB	9.5	1.68	1550.00	309.00	---	1859.00	**2230.00**

Bathtub with whirlpool (System-I)
Plumbing installation

Description	Oper	Unit	Vol	Crew Size	Man-hours per Unit	Crew Output per Day	Avg Mat'l Unit Cost	Avg Labor Unit Cost	Avg Equip Unit Cost	Avg Total Unit Cost	Avg Price Incl O&P
White	Inst	Ea	Lg	SB	13.3	1.20	2970.00	433.00	---	3403.00	**4050.00**
	Inst	Ea	Sm	SB	19.0	0.84	3610.00	618.00	---	4228.00	**5060.00**
Colors	Inst	Ea	Lg	SB	13.3	1.20	3090.00	433.00	---	3523.00	**4190.00**
	Inst	Ea	Sm	SB	19.0	0.84	3750.00	618.00	---	4368.00	**5220.00**

Electrical installation

Description	Oper	Unit	Vol	Crew Size	Man-hours per Unit	Crew Output per Day	Avg Mat'l Unit Cost	Avg Labor Unit Cost	Avg Equip Unit Cost	Avg Total Unit Cost	Avg Price Incl O&P
	Inst	Ea	Lg	EA	6.15	1.30	123.00	219.00	---	342.00	**461.00**
	Inst	Ea	Sm	EA	8.79	0.91	149.00	313.00	---	462.00	**628.00**

60" L x 42" W x 20" H (Synchrony)

Description	Oper	Unit	Vol	Crew Size	Man-hours per Unit	Crew Output per Day	Avg Mat'l Unit Cost	Avg Labor Unit Cost	Avg Equip Unit Cost	Avg Total Unit Cost	Avg Price Incl O&P
White	Inst	Ea	Lg	SB	6.67	2.40	994.00	217.00	---	1211.00	**1460.00**
	Inst	Ea	Sm	SB	9.5	1.68	1210.00	309.00	---	1519.00	**1840.00**
Colors	Inst	Ea	Lg	SB	6.67	2.40	1030.00	217.00	---	1247.00	**1500.00**
	Inst	Ea	Sm	SB	9.5	1.68	1250.00	309.00	---	1559.00	**1890.00**

Description	Oper	Unit	Vol	Crew Size	Man-hours per Unit	Crew Output per Day	Avg Mat'l Unit Cost	Avg Labor Unit Cost	Avg Equip Unit Cost	Avg Total Unit Cost	Avg Price Incl O&P
Bathtub with whirlpool (System-I)											
Plumbing installation											
White	Inst	Ea	Lg	SB	13.3	1.20	1830.00	433.00	---	2263.00	**2740.00**
	Inst	Ea	Sm	SB	19.0	0.84	2220.00	618.00	---	2838.00	**3470.00**
Colors	Inst	Ea	Lg	SB	13.3	1.20	1880.00	433.00	---	2313.00	**2800.00**
	Inst	Ea	Sm	SB	19.0	0.84	2280.00	618.00	---	2898.00	**3530.00**
Electrical installation											
	Inst	Ea	Lg	EA	6.15	1.30	123.00	219.00	---	342.00	**461.00**
	Inst	Ea	Sm	EA	8.79	0.91	149.00	313.00	---	462.00	**628.00**
60" L x 42-1/4" W x 18" H (Dockside)											
White	Inst	Ea	Lg	SB	6.67	2.40	986.00	217.00	---	1203.00	**1450.00**
	Inst	Ea	Sm	SB	9.5	1.68	1200.00	309.00	---	1509.00	**1830.00**
Colors	Inst	Ea	Lg	SB	6.67	2.40	1020.00	217.00	---	1237.00	**1490.00**
	Inst	Ea	Sm	SB	9.5	1.68	1240.00	309.00	---	1549.00	**1880.00**
Bathtub with whirlpool (System-I)											
Plumbing installation											
White	Inst	Ea	Lg	SB	13.3	1.20	1830.00	433.00	---	2263.00	**2740.00**
	Inst	Ea	Sm	SB	19.0	0.84	2220.00	618.00	---	2838.00	**3470.00**
Colors	Inst	Ea	Lg	SB	13.3	1.20	1890.00	433.00	---	2323.00	**2810.00**
	Inst	Ea	Sm	SB	19.0	0.84	2300.00	618.00	---	2918.00	**3550.00**
Electrical installation											
	Inst	Ea	Lg	EA	6.15	1.30	123.00	219.00	---	342.00	**461.00**
	Inst	Ea	Sm	EA	8.79	0.91	149.00	313.00	---	462.00	**628.00**
60" L x 60" W x 21" H, corner (Sojourn)											
White	Inst	Ea	Lg	SB	6.67	2.40	1180.00	217.00	---	1397.00	**1670.00**
	Inst	Ea	Sm	SB	9.5	1.68	1430.00	309.00	---	1739.00	**2100.00**
Colors	Inst	Ea	Lg	SB	6.67	2.40	1220.00	217.00	---	1437.00	**1720.00**
	Inst	Ea	Sm	SB	9.5	1.68	1480.00	309.00	---	1789.00	**2150.00**
Bathtub with whirlpool (System-I)											
Plumbing installation											
White	Inst	Ea	Lg	SB	13.3	1.20	2230.00	433.00	---	2663.00	**3200.00**
	Inst	Ea	Sm	SB	19.0	0.84	2710.00	618.00	---	3328.00	**4030.00**
Colors	Inst	Ea	Lg	SB	13.3	1.20	2310.00	433.00	---	2743.00	**3300.00**
	Inst	Ea	Sm	SB	19.0	0.84	2810.00	618.00	---	3428.00	**4140.00**
Electrical installation											
	Inst	Ea	Lg	EA	6.15	1.30	123.00	219.00	---	342.00	**461.00**
	Inst	Ea	Sm	EA	8.79	0.91	149.00	313.00	---	462.00	**628.00**
66" L x 35-7/8" W x 20" H (Mariposa)											
White	Inst	Ea	Lg	SB	6.67	2.40	870.00	217.00	---	1087.00	**1320.00**
	Inst	Ea	Sm	SB	9.5	1.68	1060.00	309.00	---	1369.00	**1670.00**
Colors	Inst	Ea	Lg	SB	6.67	2.40	896.00	217.00	---	1113.00	**1350.00**
	Inst	Ea	Sm	SB	9.5	1.68	1090.00	309.00	---	1399.00	**1710.00**
Bathtub with whirlpool (System-I)											
Plumbing installation											
White	Inst	Ea	Lg	SB	13.3	1.20	1900.00	433.00	---	2333.00	**2830.00**
	Inst	Ea	Sm	SB	19.0	0.84	2310.00	618.00	---	2928.00	**3570.00**
Colors	Inst	Ea	Lg	SB	13.3	1.20	1970.00	433.00	---	2403.00	**2900.00**
	Inst	Ea	Sm	SB	19.0	0.84	2390.00	618.00	---	3008.00	**3660.00**
Electrical installation											
	Inst	Ea	Lg	EA	6.15	1.30	123.00	219.00	---	342.00	**461.00**
	Inst	Ea	Sm	EA	8.79	0.91	149.00	313.00	---	462.00	**628.00**

Description	Oper	Unit	Vol	Crew Size	Man-hours per Unit	Crew Output per Day	Avg Mat'l Unit Cost	Avg Labor Unit Cost	Avg Equip Unit Cost	Avg Total Unit Cost	Avg Price Incl O&P
66" L x 42" W x 22-5/8" H (Folio)											
White	Inst	Ea	Lg	SB	6.67	2.40	999.00	217.00	---	1216.00	**1470.00**
	Inst	Ea	Sm	SB	9.5	1.68	1210.00	309.00	---	1519.00	**1850.00**
Colors	Inst	Ea	Lg	SB	6.67	2.40	1030.00	217.00	---	1247.00	**1500.00**
	Inst	Ea	Sm	SB	9.5	1.68	1250.00	309.00	---	1559.00	**1890.00**
Bathtub with whirlpool (System-I)											
Plumbing installation											
White	Inst	Ea	Lg	SB	13.3	1.20	1890.00	433.00	---	2323.00	**2810.00**
	Inst	Ea	Sm	SB	19.0	0.84	2290.00	618.00	---	2908.00	**3550.00**
Colors	Inst	Ea	Lg	SB	13.3	1.20	1950.00	433.00	---	2383.00	**2880.00**
	Inst	Ea	Sm	SB	19.0	0.84	2370.00	618.00	---	2988.00	**3640.00**
Electrical installation											
	Inst	Ea	Lg	EA	6.15	1.30	123.00	219.00	---	342.00	**461.00**
	Inst	Ea	Sm	EA	8.79	0.91	149.00	313.00	---	462.00	**628.00**
72" L x 36" W x 20" H (Mariposa)											
White	Inst	Ea	Lg	SB	6.67	2.40	874.00	217.00	---	1091.00	**1320.00**
	Inst	Ea	Sm	SB	9.5	1.68	1060.00	309.00	---	1369.00	**1680.00**
Colors	Inst	Ea	Lg	SB	6.67	2.40	901.00	217.00	---	1118.00	**1350.00**
	Inst	Ea	Sm	SB	9.5	1.68	1090.00	309.00	---	1399.00	**1710.00**
Bathtub with whirlpool (System-I)											
Plumbing installation											
White	Inst	Ea	Lg	SB	13.3	1.20	1910.00	433.00	---	2343.00	**2830.00**
	Inst	Ea	Sm	SB	19.0	0.84	2320.00	618.00	---	2938.00	**3580.00**
Colors	Inst	Ea	Lg	SB	13.3	1.20	1980.00	433.00	---	2413.00	**2910.00**
	Inst	Ea	Sm	SB	19.0	0.84	2400.00	618.00	---	3018.00	**3670.00**
Electrical installation											
	Inst	Ea	Lg	EA	6.15	1.30	123.00	219.00	---	342.00	**461.00**
	Inst	Ea	Sm	EA	8.79	0.91	149.00	313.00	---	462.00	**628.00**
72" L x 42" W x 20" H (Synchrony)											
White	Inst	Ea	Lg	SB	6.67	2.40	1040.00	217.00	---	1257.00	**1510.00**
	Inst	Ea	Sm	SB	9.5	1.68	1260.00	309.00	---	1569.00	**1900.00**
Colors	Inst	Ea	Lg	SB	6.67	2.40	1070.00	217.00	---	1287.00	**1550.00**
	Inst	Ea	Sm	SB	9.5	1.68	1300.00	309.00	---	1609.00	**1950.00**
Bathtub with whirlpool (System-I)											
Plumbing installation											
White	Inst	Ea	Lg	SB	13.3	1.20	1890.00	433.00	---	2323.00	**2810.00**
	Inst	Ea	Sm	SB	19.0	0.84	2300.00	618.00	---	2918.00	**3550.00**
Colors	Inst	Ea	Lg	SB	13.3	1.20	1980.00	433.00	---	2413.00	**2920.00**
	Inst	Ea	Sm	SB	19.0	0.84	2410.00	618.00	---	3028.00	**3680.00**
Electrical installation											
	Inst	Ea	Lg	EA	6.15	1.30	123.00	219.00	---	342.00	**461.00**
	Inst	Ea	Sm	EA	8.79	0.91	149.00	313.00	---	462.00	**628.00**

Description	Oper	Unit	Vol	Crew Size	Man-hours per Unit	Crew Output per Day	Avg Mat'l Unit Cost	Avg Labor Unit Cost	Avg Equip Unit Cost	Avg Total Unit Cost	Avg Price Incl O&P
Adjustments											
Detach & Reset Tub											
Free standing tub	Reset	Ea	Lg	SB	3.20	5.00	38.50	104.00	---	142.50	**197.00**
	Reset	Ea	Sm	SB	4.57	3.50	46.80	149.00	---	195.80	**272.00**
Recessed tub	Reset	Ea	Lg	SB	4.00	4.00	38.50	130.00	---	168.50	**236.00**
	Reset	Ea	Sm	SB	5.71	2.80	46.80	186.00	---	232.80	**327.00**
Sunken tub	Reset	Ea	Lg	SB	5.33	3.00	38.50	173.00	---	211.50	**299.00**
	Reset	Ea	Sm	SB	7.62	2.10	46.80	248.00	---	294.80	**418.00**
Remove Tub											
Free standing tub	Demo	Ea	Lg	SB	1.33	12.00	10.50	43.30	---	53.80	**75.70**
	Demo	Ea	Sm	SB	1.90	8.40	12.80	61.90	---	74.70	**106.00**
Recessed tub	Demo	Ea	Lg	SB	2.00	8.00	21.00	65.10	---	86.10	**120.00**
	Demo	Ea	Sm	SB	2.86	5.60	25.50	93.10	---	118.60	**166.00**
Sunken tub	Demo	Ea	Lg	SB	2.67	6.00	31.50	86.90	---	118.40	**164.00**
	Demo	Ea	Sm	SB	3.81	4.20	38.30	124.00	---	162.30	**226.00**
Install Rough-In											
	Inst	Ea	Lg	SB	10.0	1.60	59.50	326.00	---	385.50	**547.00**
	Inst	Ea	Sm	SB	14.3	1.12	72.30	465.00	---	537.30	**767.00**
Install shower head w/ mixer valve over tub											
Open wall, ADD	Inst	Ea	Lg	SA	2.00	4.00	80.50	75.30	---	155.80	**203.00**
	Inst	Ea	Sm	SA	2.86	2.80	97.80	108.00	---	205.80	**271.00**
Closed wall, ADD	Inst	Ea	Lg	SA	4.00	2.00	126.00	151.00	---	277.00	**366.00**
	Inst	Ea	Sm	SA	5.71	1.40	153.00	215.00	---	368.00	**492.00**

Block, concrete. See Masonry, page 282

Brick. See Masonry, page 276

Cabinets

Top quality cabinets are built with the structural stability of fine furniture. Framing stock is kiln dried and a full 1" thick. Cabinets have backs, usually 5-ply $3/16$"-thick plywood, with all backs and interiors finished. Frames should be constructed of hardwood with mortise and tenon joints; corner blocks should be used on all four corners of all base cabinets. Doors are usually of select $7/16$" thick solid core construction using semi-concealed hinges. End panels are $1/2$" thick and attached to frames with mortise and tenon joints, glued and pinned under pressure. Panels should also be dadoed to receive the tops and bottoms of wall cabinets. Shelves are adjustable with veneer faces and front edges. The hardware includes magnetic catches, heavy duty die cast pulls and hinges, and ball-bearing suspension system. The finish is scratch and stain resistant, including a first coat of hand-wiped stain, a sealer coat, and a synthetic varnish with plastic laminate characteristics.

Average quality cabinets feature hardwood frame construction with plywood backs and veneered plywood end panels. Joints are glued mortise and tenon. Doors are solid core attached with exposed self-closing hinges. Shelves are adjustable, and drawers ride on a ball-bearing side suspension glide system. The finish is usually three coats including stain, sealer, and a mar-resistant top coat for easy cleaning.

Economy quality cabinets feature pine construction with joints glued under pressure. Doors, drawer fronts, and side or end panels are constructed of either $1/2$"-thick wood composition board or $1/2$"-thick veneered pine. Face frames are $3/4$"-thick wood composition board or $3/4$"-thick pine. Features include adjustable shelves, hinge straps, and a three-point suspension system on drawers (using nylon rollers). The finish consists of a filler coat, base coat, and final polyester top coat.

Kitchen Cabinet Installation Procedure

To develop a layout plan, measure and write down the following:

1. Floor space.

2. Height and width of all walls.

3. Location of electrical outlets.

4. Size and position of doors, windows, and vents.

5. Location of any posts or pillars. Walls must be prepared if chair rails or baseboards are located where cabinets will be installed.

6. Common height and depth of base cabinets (including 1" for countertops) and wall cabinets.

When you plan the cabinet placement, consider this Rule of Thumb:

Allow between $4^1/2$' and $5^1/2$' of counter surface between the refrigerator and sink. Allow between 3' and 4' between the sink and range.

1. What do you have to fit into the available space?

2. Is there enough counter space on both sides of all appliances and sinks? The kitchen has three work centers, each with a major appliance as its hub, and each needing adequate counter space. They are:

    a. Fresh and frozen food center — Refrigerator-freezer

    b. Clean-up center — Sink with disposal-dishwasher

    c. Cooking center — Range-oven

3. Will the sink workspace fit neatly in front of a window?

4. The kitchen triangle is the most efficient kitchen design; it means placing each major center at approximately equidistant triangle points. The ideal triangle is 22 feet total. It should never be less than 13 feet or more than 25 feet.

5. Where are the centers located? A logical working and walking pattern is from refrigerator to sink to range. The refrigerator should be at a triangle point near a door, to minimize the distance to bring in groceries and reduce traffic that could interfere with food preparation. The range should be at a triangle point near the serving and dining area. The sink is located between the two. The refrigerator should be located far enough from the range so that the heat will not affect the refrigerator's cooling efficiency.

6. Does the plan allow for lighting the range and sink work centers and for ventilating the range center?

7. Make sure that open doors (such as cabinet doors or entrance/exit doors) won't interfere with access to an appliance. To clear appliances and cabinets, a door opening should not be less than 30" from the corner, since such equipment is 24" to 28" in depth. A clearance of 48" is necessary when a range is next to a door.

8. Next locate the wall studs with a stud finder or hammer, since all cabinets attach to walls with screws, never nails. Also remove chair rails or baseboards where they conflict with cabinets.

Wall Cabinets: From the highest point on the floor, measure up approximately 84" to determine the top of wall cabinets. Using two #10 x $2^1/2$" wood screws, drill through hanging strips built into the cabinet backs at top and bottom. Use a level to assure that cabinets and doors are aligned, then tighten the screws.

Base Cabinets: Start with a corner unit and a base unit on each side of the corner unit. Place this combination in the corner and work out from both sides. Use "C" clamps when connecting cabinets to draw adjoining units into alignment. With the front face plumb and the unit level from front to back and across the front edge, attach the unit to wall studs by screwing through the hanging strips. To attach adjoining cabinets, drill two holes in the vertical side of one cabinet, inside the door (near top and bottom), and just into the stile of the adjoining cabinet.

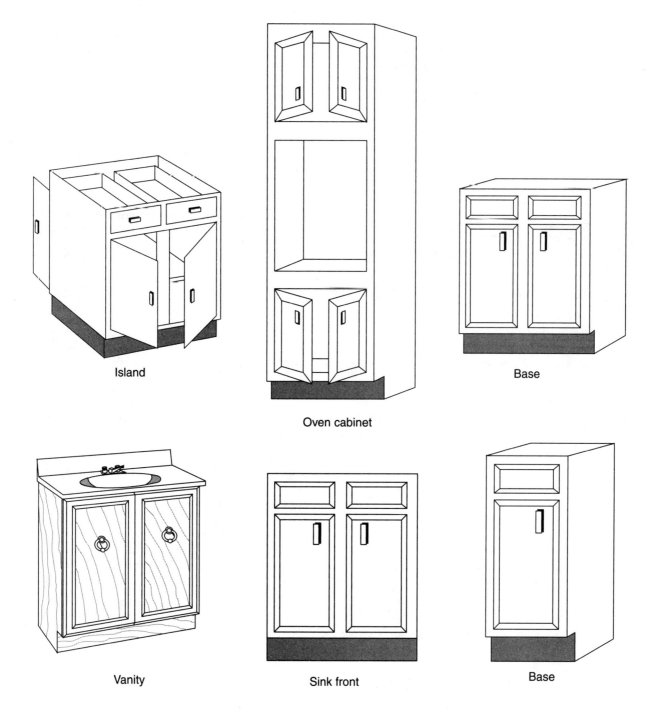

Island

Oven cabinet

Base

Vanity

Sink front

Base

Description	Oper	Unit	Vol	Crew Size	Man-hours per Unit	Crew Output per Day	Avg Mat'l Unit Cost	Avg Labor Unit Cost	Avg Equip Unit Cost	Avg Total Unit Cost	Avg Price Incl O&P

Cabinets

Labor costs include hanging and fitting of cabinets

Kitchen

3' x 4', wood; base, wall, or peninsula

Description	Oper	Unit	Vol	Crew Size	Man-hours per Unit	Crew Output per Day	Avg Mat'l Unit Cost	Avg Labor Unit Cost	Avg Equip Unit Cost	Avg Total Unit Cost	Avg Price Incl O&P
	Demo	Ea	Lg	LB	.640	25.00	---	17.60	---	17.60	**26.20**
	Demo	Ea	Sm	LB	1.067	15.00	---	29.30	---	29.30	**43.60**

Kitchen; all hardware included
Base cabinets, 35" H, 24" D; no tops

12" W, 1 door, 1 drawer

Description	Oper	Unit	Vol	Crew Size	Man-hours per Unit	Crew Output per Day	Avg Mat'l Unit Cost	Avg Labor Unit Cost	Avg Equip Unit Cost	Avg Total Unit Cost	Avg Price Incl O&P
High quality workmanship	Inst	Ea	Lg	CJ	1.00	16.00	171.00	30.40	---	201.40	**251.00**
	Inst	Ea	Sm	CJ	1.67	9.60	193.00	50.70	---	243.70	**308.00**
Good quality workmanship	Inst	Ea	Lg	CJ	.667	24.00	140.00	20.20	---	160.20	**198.00**
	Inst	Ea	Sm	CJ	1.11	14.40	158.00	33.70	---	191.70	**240.00**
Average quality workmanship	Inst	Ea	Lg	CJ	.667	24.00	107.00	20.20	---	127.20	**158.00**
	Inst	Ea	Sm	CJ	1.11	14.40	120.00	33.70	---	153.70	**195.00**

15" W, 1 door, 1 drawer

Description	Oper	Unit	Vol	Crew Size	Man-hours per Unit	Crew Output per Day	Avg Mat'l Unit Cost	Avg Labor Unit Cost	Avg Equip Unit Cost	Avg Total Unit Cost	Avg Price Incl O&P
High quality workmanship	Inst	Ea	Lg	CJ	1.00	16.00	181.00	30.40	---	211.40	**262.00**
	Inst	Ea	Sm	CJ	1.67	9.60	204.00	50.70	---	254.70	**320.00**
Good quality workmanship	Inst	Ea	Lg	CJ	.667	24.00	149.00	20.20	---	169.20	**209.00**
	Inst	Ea	Sm	CJ	1.11	14.40	168.00	33.70	---	201.70	**252.00**
Average quality workmanship	Inst	Ea	Lg	CJ	.667	24.00	113.00	20.20	---	133.20	**166.00**
	Inst	Ea	Sm	CJ	1.11	14.40	127.00	33.70	---	160.70	**203.00**

18" W, 1 door, 1 drawer

Description	Oper	Unit	Vol	Crew Size	Man-hours per Unit	Crew Output per Day	Avg Mat'l Unit Cost	Avg Labor Unit Cost	Avg Equip Unit Cost	Avg Total Unit Cost	Avg Price Incl O&P
High quality workmanship	Inst	Ea	Lg	CJ	1.00	16.00	187.00	30.40	---	217.40	**270.00**
	Inst	Ea	Sm	CJ	1.67	9.60	211.00	50.70	---	261.70	**329.00**
Good quality workmanship	Inst	Ea	Lg	CJ	.667	24.00	154.00	20.20	---	174.20	**215.00**
	Inst	Ea	Sm	CJ	1.11	14.40	173.00	33.70	---	206.70	**259.00**
Average quality workmanship	Inst	Ea	Lg	CJ	.667	24.00	118.00	20.20	---	138.20	**172.00**
	Inst	Ea	Sm	CJ	1.11	14.40	133.00	33.70	---	166.70	**210.00**

21" W, 1 door, 1 drawer

Description	Oper	Unit	Vol	Crew Size	Man-hours per Unit	Crew Output per Day	Avg Mat'l Unit Cost	Avg Labor Unit Cost	Avg Equip Unit Cost	Avg Total Unit Cost	Avg Price Incl O&P
High quality workmanship	Inst	Ea	Lg	CJ	1.16	13.80	207.00	35.20	---	242.20	**302.00**
	Inst	Ea	Sm	CJ	1.93	8.30	234.00	58.60	---	292.60	**368.00**
Good quality workmanship	Inst	Ea	Lg	CJ	.741	21.60	171.00	22.50	---	193.50	**239.00**
	Inst	Ea	Sm	CJ	1.23	13.00	193.00	37.30	---	230.30	**288.00**
Average quality workmanship	Inst	Ea	Lg	CJ	.741	21.60	132.00	22.50	---	154.50	**192.00**
	Inst	Ea	Sm	CJ	1.23	13.00	149.00	37.30	---	186.30	**234.00**

Description	Oper	Unit	Vol	Crew Size	Man-hours per Unit	Crew Output per Day	Avg Mat'l Unit Cost	Avg Labor Unit Cost	Avg Equip Unit Cost	Avg Total Unit Cost	Avg Price Incl O&P
24" W, 1 door, 1 drawer											
High quality workmanship	Inst	Ea	Lg	CJ	1.16	13.80	228.00	35.20	---	263.20	**326.00**
	Inst	Ea	Sm	CJ	1.93	8.30	257.00	58.60	---	315.60	**396.00**
Good quality workmanship	Inst	Ea	Lg	CJ	.741	21.60	187.00	22.50	---	209.50	**258.00**
	Inst	Ea	Sm	CJ	1.23	13.00	211.00	37.30	---	248.30	**309.00**
Average quality workmanship	Inst	Ea	Lg	CJ	.741	21.60	165.00	22.50	---	187.50	**232.00**
	Inst	Ea	Sm	CJ	1.23	13.00	186.00	37.30	---	223.30	**279.00**
30" W, 2 doors, 2 drawers											
High quality workmanship	Inst	Ea	Lg	CJ	1.54	10.40	290.00	46.70	---	336.70	**419.00**
	Inst	Ea	Sm	CJ	2.58	6.20	327.00	78.30	---	405.30	**510.00**
Good quality workmanship	Inst	Ea	Lg	CJ	1.04	15.40	251.00	31.60	---	282.60	**349.00**
	Inst	Ea	Sm	CJ	1.74	9.20	283.00	52.80	---	335.80	**419.00**
Average quality workmanship	Inst	Ea	Lg	CJ	1.04	15.40	190.00	31.60	---	221.60	**275.00**
	Inst	Ea	Sm	CJ	1.74	9.20	214.00	52.80	---	266.80	**336.00**
36" W, 2 doors, 2 drawers											
High quality workmanship	Inst	Ea	Lg	CJ	1.54	10.40	334.00	46.70	---	380.70	**471.00**
	Inst	Ea	Sm	CJ	2.58	6.20	377.00	78.30	---	455.30	**570.00**
Good quality workmanship	Inst	Ea	Lg	CJ	1.04	15.40	276.00	31.60	---	307.60	**379.00**
	Inst	Ea	Sm	CJ	1.74	9.20	312.00	52.80	---	364.80	**453.00**
Average quality workmanship	Inst	Ea	Lg	CJ	1.04	15.40	210.00	31.60	---	241.60	**300.00**
	Inst	Ea	Sm	CJ	1.74	9.20	237.00	52.80	---	289.80	**364.00**
42" W, 2 doors, 2 drawers											
High quality workmanship	Inst	Ea	Lg	CJ	2.00	8.00	386.00	60.70	---	446.70	**555.00**
	Inst	Ea	Sm	CJ	3.33	4.80	435.00	101.00	---	536.00	**674.00**
Good quality workmanship	Inst	Ea	Lg	CJ	1.33	12.00	320.00	40.40	---	360.40	**445.00**
	Inst	Ea	Sm	CJ	2.22	7.20	361.00	67.40	---	428.40	**534.00**
Average quality workmanship	Inst	Ea	Lg	CJ	1.33	12.00	243.00	40.40	---	283.40	**353.00**
	Inst	Ea	Sm	CJ	2.22	7.20	274.00	67.40	---	341.40	**430.00**
48" W, 4 doors, 2 drawers											
High quality workmanship	Inst	Ea	Lg	CJ	2.00	8.00	372.00	60.70	---	432.70	**538.00**
	Inst	Ea	Sm	CJ	3.33	4.80	419.00	101.00	---	520.00	**655.00**
Good quality workmanship	Inst	Ea	Lg	CJ	1.33	12.00	308.00	40.40	---	348.40	**430.00**
	Inst	Ea	Sm	CJ	2.22	7.20	347.00	67.40	---	414.40	**517.00**
Average quality workmanship	Inst	Ea	Lg	CJ	1.33	12.00	232.00	40.40	---	272.40	**339.00**
	Inst	Ea	Sm	CJ	2.22	7.20	262.00	67.40	---	329.40	**415.00**

Base corner cabinet, blind, 35" H, 24" D, no tops

Description	Oper	Unit	Vol	Crew Size	Man-hours per Unit	Crew Output per Day	Avg Mat'l Unit Cost	Avg Labor Unit Cost	Avg Equip Unit Cost	Avg Total Unit Cost	Avg Price Incl O&P
36" W, 1 door, 1 drawer											
High quality workmanship	Inst	Ea	Lg	CJ	1.43	11.20	243.00	43.40	---	286.40	**357.00**
	Inst	Ea	Sm	CJ	2.39	6.70	274.00	72.50	---	346.50	**438.00**
Good quality workmanship	Inst	Ea	Lg	CJ	.988	16.20	195.00	30.00	---	225.00	**279.00**
	Inst	Ea	Sm	CJ	1.65	9.70	219.00	50.10	---	269.10	**339.00**
Average quality workmanship	Inst	Ea	Lg	CJ	.988	16.20	155.00	30.00	---	185.00	**232.00**
	Inst	Ea	Sm	CJ	1.65	9.70	175.00	50.10	---	225.10	**285.00**

Description	Oper	Unit	Vol	Crew Size	Man-hours per Unit	Crew Output per Day	Avg Mat'l Unit Cost	Avg Labor Unit Cost	Avg Equip Unit Cost	Avg Total Unit Cost	Avg Price Incl O&P
42" W, 1 door, 1 drawer											
High quality workmanship	Inst	Ea	Lg	CJ	1.82	8.80	268.00	55.20	---	323.20	**405.00**
	Inst	Ea	Sm	CJ	3.02	5.30	303.00	91.70	---	394.70	**501.00**
Good quality workmanship	Inst	Ea	Lg	CJ	1.16	13.80	221.00	35.20	---	256.20	**318.00**
	Inst	Ea	Sm	CJ	1.93	8.30	250.00	58.60	---	308.60	**387.00**
Average quality workmanship	Inst	Ea	Lg	CJ	1.16	13.80	168.00	35.20	---	203.20	**254.00**
	Inst	Ea	Sm	CJ	1.93	8.30	189.00	58.60	---	247.60	**315.00**
36" x 36" (lazy Susan)											
High quality workmanship	Inst	Ea	Lg	CJ	1.43	11.20	363.00	43.40	---	406.40	**500.00**
	Inst	Ea	Sm	CJ	2.39	6.70	409.00	72.50	---	481.50	**599.00**
Good quality workmanship	Inst	Ea	Lg	CJ	.988	16.20	300.00	30.00	---	330.00	**405.00**
	Inst	Ea	Sm	CJ	1.65	9.70	338.00	50.10	---	388.10	**401.00**
Average quality workmanship	Inst	Ea	Lg	CJ	.988	16.20	229.00	30.00	---	259.00	**320.00**
	Inst	Ea	Sm	CJ	1.65	9.70	258.00	50.10	---	308.10	**385.00**

Utility closets, 81" to 85" H, 24" W, 24" D

Description	Oper	Unit	Vol	Crew Size	Man-hours per Unit	Crew Output per Day	Avg Mat'l Unit Cost	Avg Labor Unit Cost	Avg Equip Unit Cost	Avg Total Unit Cost	Avg Price Incl O&P
High quality workmanship	Inst	Ea	Lg	CJ	2.00	8.00	556.00	60.70	---	616.70	**758.00**
	Inst	Ea	Sm	CJ	3.33	4.80	627.00	101.00	---	728.00	**904.00**
Good quality workmanship	Inst	Ea	Lg	CJ	1.33	12.00	433.00	40.40	---	473.40	**581.00**
	Inst	Ea	Sm	CJ	2.22	7.20	489.00	67.40	---	556.40	**687.00**
Average quality workmanship	Inst	Ea	Lg	CJ	1.33	12.00	330.00	40.40	---	370.40	**456.00**
	Inst	Ea	Sm	CJ	2.22	7.20	372.00	67.40	---	439.40	**547.00**

Drawer base cabinets, 35" H, 24" D

4 drawers, no tops
18" W

Description	Oper	Unit	Vol	Crew Size	Man-hours per Unit	Crew Output per Day	Avg Mat'l Unit Cost	Avg Labor Unit Cost	Avg Equip Unit Cost	Avg Total Unit Cost	Avg Price Incl O&P
High quality workmanship	Inst	Ea	Lg	CJ	1.00	16.00	279.00	30.40	---	309.40	**380.00**
	Inst	Ea	Sm	CJ	1.67	9.60	314.00	50.70	---	364.70	**453.00**
Good quality workmanship	Inst	Ea	Lg	CJ	.667	24.00	239.00	20.20	---	259.20	**317.00**
	Inst	Ea	Sm	CJ	1.11	14.40	270.00	33.70	---	303.70	**374.00**
Average quality workmanship	Inst	Ea	Lg	CJ	.667	24.00	183.00	20.20	---	203.20	**250.00**
	Inst	Ea	Sm	CJ	1.11	14.40	206.00	33.70	---	239.70	**298.00**
24" W											
High quality workmanship	Inst	Ea	Lg	CJ	1.16	13.80	313.00	35.20	---	348.20	**428.00**
	Inst	Ea	Sm	CJ	1.93	8.30	353.00	58.60	---	411.60	**511.00**
Good quality workmanship	Inst	Ea	Lg	CJ	.741	21.60	258.00	22.50	---	280.50	**343.00**
	Inst	Ea	Sm	CJ	1.23	13.00	291.00	37.30	---	328.30	**405.00**
Average quality workmanship	Inst	Ea	Lg	CJ	.741	21.60	198.00	22.50	---	220.50	**271.00**
	Inst	Ea	Sm	CJ	1.23	13.00	223.00	37.30	---	260.30	**324.00**

Description	Oper	Unit	Vol	Crew Size	Man-hours per Unit	Crew Output per Day	Avg Mat'l Unit Cost	Avg Labor Unit Cost	Avg Equip Unit Cost	Avg Total Unit Cost	Avg Price Incl O&P
Island cabinets											
Island base cabinets, 35" H, 24" D											
24" W, 1 door both sides											
High quality workmanship	Inst	Ea	Lg	CJ	1.23	13.00	405.00	37.30	---	442.30	**542.00**
	Inst	Ea	Sm	CJ	2.05	7.80	457.00	62.20	---	519.20	**641.00**
Good quality workmanship	Inst	Ea	Lg	CJ	.821	19.50	333.00	24.90	---	357.90	**437.00**
	Inst	Ea	Sm	CJ	1.37	11.70	375.00	41.60	---	416.60	**513.00**
Average quality workmanship	Inst	Ea	Lg	CJ	.821	19.50	256.00	24.90	---	280.90	**344.00**
	Inst	Ea	Sm	CJ	1.37	11.70	289.00	41.60	---	330.60	**409.00**
30" W, 2 doors both sides											
High quality workmanship	Inst	Ea	Lg	CJ	1.60	10.00	429.00	48.60	---	477.60	**587.00**
	Inst	Ea	Sm	CJ	2.67	6.00	483.00	81.00	---	564.00	**701.00**
Good quality workmanship	Inst	Ea	Lg	CJ	1.07	15.00	353.00	32.50	---	385.50	**473.00**
	Inst	Ea	Sm	CJ	1.78	9.00	398.00	54.00	---	452.00	**559.00**
Average quality workmanship	Inst	Ea	Lg	CJ	1.07	15.00	268.00	32.50	---	300.50	**371.00**
	Inst	Ea	Sm	CJ	1.78	9.00	303.00	54.00	---	357.00	**444.00**
36" W, 2 doors both sides											
High quality workmanship	Inst	Ea	Lg	CJ	1.60	10.00	451.00	48.60	---	499.60	**614.00**
	Inst	Ea	Sm	CJ	2.67	6.00	508.00	81.00	---	589.00	**731.00**
Good quality workmanship	Inst	Ea	Lg	CJ	1.07	15.00	372.00	32.50	---	404.50	**495.00**
	Inst	Ea	Sm	CJ	1.78	9.00	419.00	54.00	---	473.00	**584.00**
Average quality workmanship	Inst	Ea	Lg	CJ	1.07	15.00	283.00	32.50	---	315.50	**388.00**
	Inst	Ea	Sm	CJ	1.78	9.00	319.00	54.00	---	373.00	**463.00**
48" W, 4 doors both sides											
High quality workmanship	Inst	Ea	Lg	CJ	2.13	7.50	484.00	64.70	---	548.70	**677.00**
	Inst	Ea	Sm	CJ	3.56	4.50	545.00	108.00	---	653.00	**816.00**
Good quality workmanship	Inst	Ea	Lg	CJ	1.39	11.50	399.00	42.20	---	441.20	**542.00**
	Inst	Ea	Sm	CJ	2.32	6.90	450.00	70.40	---	520.40	**645.00**
Average quality workmanship	Inst	Ea	Lg	CJ	1.39	11.50	303.00	42.20	---	345.20	**427.00**
	Inst	Ea	Sm	CJ	2.32	6.90	342.00	70.40	---	412.40	**516.00**
Corner island base cabinets, 35" H, 24" D											
42" W, 4 doors, 2 drawers											
High quality workmanship	Inst	Ea	Lg	CJ	2.13	7.50	429.00	64.70	---	493.70	**611.00**
	Inst	Ea	Sm	CJ	3.56	4.50	483.00	108.00	---	591.00	**742.00**
Good quality workmanship	Inst	Ea	Lg	CJ	1.39	11.50	350.00	42.20	---	392.20	**483.00**
	Inst	Ea	Sm	CJ	2.32	6.90	395.00	70.40	---	465.40	**579.00**
Average quality workmanship	Inst	Ea	Lg	CJ	1.39	11.50	264.00	42.20	---	306.20	**380.00**
	Inst	Ea	Sm	CJ	2.32	6.90	297.00	70.40	---	367.40	**462.00**
48" W, 6 doors, 2 drawers											
High quality workmanship	Inst	Ea	Lg	CJ	2.67	6.00	449.00	81.00	---	530.00	**660.00**
	Inst	Ea	Sm	CJ	4.44	3.60	506.00	135.00	---	641.00	**810.00**
Good quality workmanship	Inst	Ea	Lg	CJ	1.78	9.00	371.00	54.00	---	425.00	**526.00**
	Inst	Ea	Sm	CJ	2.96	5.40	418.00	89.80	---	507.80	**636.00**
Average quality workmanship	Inst	Ea	Lg	CJ	1.78	9.00	281.00	54.00	---	335.00	**418.00**
	Inst	Ea	Sm	CJ	2.96	5.40	317.00	89.80	---	406.80	**515.00**

Description	Oper	Unit	Vol	Crew Size	Man-hours per Unit	Crew Output per Day	Avg Mat'l Unit Cost	Avg Labor Unit Cost	Avg Equip Unit Cost	Avg Total Unit Cost	Avg Price Incl O&P
Hanging corner island wall cabinets, 18" H, 24" D											
18" W, 3 doors											
High quality workmanship	Inst	Ea	Lg	CJ	1.45	11.00	311.00	44.00	---	355.00	439.00
	Inst	Ea	Sm	CJ	2.42	6.60	350.00	73.50	---	423.50	531.00
Good quality workmanship	Inst	Ea	Lg	CJ	.970	16.50	228.00	29.40	---	257.40	317.00
	Inst	Ea	Sm	CJ	1.62	9.90	257.00	49.20	---	306.20	382.00
Average quality workmanship	Inst	Ea	Lg	CJ	.970	16.50	174.00	29.40	---	203.40	253.00
	Inst	Ea	Sm	CJ	1.62	9.90	196.00	49.20	---	245.20	310.00
24" W, 3 doors											
High quality workmanship	Inst	Ea	Lg	CJ	1.45	11.00	320.00	44.00	---	364.00	450.00
	Inst	Ea	Sm	CJ	2.42	6.60	361.00	73.50	---	434.50	543.00
Good quality workmanship	Inst	Ea	Lg	CJ	.970	16.50	237.00	29.40	---	266.40	329.00
	Inst	Ea	Sm	CJ	1.62	9.90	267.00	49.20	---	316.20	394.00
Average quality workmanship	Inst	Ea	Lg	CJ	.970	16.50	179.00	29.40	---	208.40	259.00
	Inst	Ea	Sm	CJ	1.62	9.90	202.00	49.20	---	251.20	316.00
30" W, 3 doors											
High quality workmanship	Inst	Ea	Lg	CJ	1.88	8.50	330.00	57.10	---	387.10	481.00
	Inst	Ea	Sm	CJ	3.14	5.10	372.00	95.30	---	467.30	589.00
Good quality workmanship	Inst	Ea	Lg	CJ	1.26	12.70	246.00	38.20	---	284.20	353.00
	Inst	Ea	Sm	CJ	2.11	7.60	278.00	64.00	---	342.00	430.00
Average quality workmanship	Inst	Ea	Lg	CJ	1.26	12.70	188.00	38.20	---	226.20	283.00
	Inst	Ea	Sm	CJ	2.11	7.60	212.00	64.00	---	276.00	351.00
Hanging island cabinets, 18" H, 12" D											
30" W, 2 doors both sides											
High quality workmanship	Inst	Ea	Lg	CJ	1.70	9.40	330.00	51.60	---	381.60	473.00
	Inst	Ea	Sm	CJ	2.86	5.60	372.00	86.80	---	458.80	576.00
Good quality workmanship	Inst	Ea	Lg	CJ	1.14	14.00	242.00	34.60	---	276.60	342.00
	Inst	Ea	Sm	CJ	1.90	8.40	273.00	57.70	---	330.70	414.00
Average quality workmanship	Inst	Ea	Lg	CJ	1.14	14.00	184.00	34.60	---	218.60	272.00
	Inst	Ea	Sm	CJ	1.90	8.40	207.00	57.70	---	264.70	335.00
36" W, 2 doors both sides											
High quality workmanship	Inst	Ea	Lg	CJ	1.70	9.40	339.00	51.60	---	390.60	484.00
	Inst	Ea	Sm	CJ	2.86	5.60	382.00	86.80	---	468.80	589.00
Good quality workmanship	Inst	Ea	Lg	CJ	1.14	14.00	250.00	34.60	---	284.60	351.00
	Inst	Ea	Sm	CJ	1.90	8.40	281.00	57.70	---	338.70	424.00
Average quality workmanship	Inst	Ea	Lg	CJ	1.14	14.00	190.00	34.60	---	224.60	280.00
	Inst	Ea	Sm	CJ	1.90	8.40	214.00	57.70	---	271.70	344.00
Hanging island cabinets, 24" H, 12" D											
24" W, 2 doors both sides											
High quality workmanship	Inst	Ea	Lg	CJ	1.45	11.00	311.00	44.00	---	355.00	439.00
	Inst	Ea	Sm	CJ	2.42	6.60	350.00	73.50	---	423.50	531.00
Good quality workmanship	Inst	Ea	Lg	CJ	.970	16.50	217.00	29.40	---	246.40	304.00
	Inst	Ea	Sm	CJ	1.62	9.90	244.00	49.20	---	293.20	367.00
Average quality workmanship	Inst	Ea	Lg	CJ	.970	16.50	160.00	29.40	---	189.40	236.00
	Inst	Ea	Sm	CJ	1.62	9.90	181.00	49.20	---	230.20	290.00

Description	Oper	Unit	Vol	Crew Size	Man-hours per Unit	Crew Output per Day	Avg Mat'l Unit Cost	Avg Labor Unit Cost	Avg Equip Unit Cost	Avg Total Unit Cost	Avg Price Incl O&P
30" W, 2 doors both sides											
High quality workmanship	Inst	Ea	Lg	CJ	1.88	8.50	323.00	57.10	---	380.10	**474.00**
	Inst	Ea	Sm	CJ	3.14	5.10	365.00	95.30	---	460.30	**581.00**
Good quality workmanship	Inst	Ea	Lg	CJ	1.26	12.70	234.00	38.20	---	272.20	**338.00**
	Inst	Ea	Sm	CJ	2.11	7.60	264.00	64.00	---	328.00	**413.00**
Average quality workmanship	Inst	Ea	Lg	CJ	1.26	12.70	173.00	38.20	---	211.20	**265.00**
	Inst	Ea	Sm	CJ	2.11	7.60	195.00	64.00	---	259.00	**330.00**
36" W, 2 doors both sides											
High quality workmanship	Inst	Ea	Lg	CJ	1.88	8.50	339.00	57.10	---	396.10	**493.00**
	Inst	Ea	Sm	CJ	3.14	5.10	382.00	95.30	---	477.30	**602.00**
Good quality workmanship	Inst	Ea	Lg	CJ	1.26	12.70	250.00	38.20	---	288.20	**357.00**
	Inst	Ea	Sm	CJ	2.11	7.60	281.00	64.00	---	345.00	**434.00**
Average quality workmanship	Inst	Ea	Lg	CJ	1.26	12.70	190.00	38.20	---	228.20	**285.00**
	Inst	Ea	Sm	CJ	2.11	7.60	214.00	64.00	---	278.00	**353.00**
42" W, 2 doors both sides											
High quality workmanship	Inst	Ea	Lg	CJ	2.46	6.50	410.00	74.70	---	484.70	**604.00**
	Inst	Ea	Sm	CJ	4.10	3.90	462.00	124.00	---	586.00	**741.00**
Good quality workmanship	Inst	Ea	Lg	CJ	1.63	9.80	301.00	49.50	---	350.50	**436.00**
	Inst	Ea	Sm	CJ	2.71	5.90	340.00	82.30	---	422.30	**531.00**
Average quality workmanship	Inst	Ea	Lg	CJ	1.63	9.80	231.00	49.50	---	280.50	**351.00**
	Inst	Ea	Sm	CJ	2.71	5.90	260.00	82.30	---	342.30	**436.00**
48" W, 4 doors both sides											
High quality workmanship	Inst	Ea	Lg	CJ	2.46	6.50	491.00	74.70	---	565.70	**702.00**
	Inst	Ea	Sm	CJ	4.10	3.90	554.00	124.00	---	678.00	**851.00**
Good quality workmanship	Inst	Ea	Lg	CJ	1.63	9.80	363.00	49.50	---	412.50	**509.00**
	Inst	Ea	Sm	CJ	2.71	5.90	409.00	82.30	---	491.30	**614.00**
Average quality workmanship	Inst	Ea	Lg	CJ	1.63	9.80	276.00	49.50	---	325.50	**406.00**
	Inst	Ea	Sm	CJ	2.71	5.90	312.00	82.30	---	394.30	**497.00**

Oven cabinets

81" to 85" H, 24" D, 27" W

Description	Oper	Unit	Vol	Crew Size	Man-hours per Unit	Crew Output per Day	Avg Mat'l Unit Cost	Avg Labor Unit Cost	Avg Equip Unit Cost	Avg Total Unit Cost	Avg Price Incl O&P
High quality workmanship	Inst	Ea	Lg	CJ	2.00	8.00	593.00	60.70	---	653.70	**803.00**
	Inst	Ea	Sm	CJ	3.33	4.80	669.00	101.00	---	770.00	**954.00**
Good quality workmanship	Inst	Ea	Lg	CJ	1.33	12.00	462.00	40.40	---	502.40	**614.00**
	Inst	Ea	Sm	CJ	2.22	7.20	520.00	67.40	---	587.40	**726.00**
Average quality workmanship	Inst	Ea	Lg	CJ	1.33	12.00	353.00	40.40	---	393.40	**484.00**
	Inst	Ea	Sm	CJ	2.22	7.20	398.00	67.40	---	465.40	**579.00**

Sink/range

Base cabinets, 35" H, 24" D, 2 doors, no tops included

30" W

Description	Oper	Unit	Vol	Crew Size	Man-hours per Unit	Crew Output per Day	Avg Mat'l Unit Cost	Avg Labor Unit Cost	Avg Equip Unit Cost	Avg Total Unit Cost	Avg Price Incl O&P
High quality workmanship	Inst	Ea	Lg	CJ	1.33	12.00	232.00	40.40	---	272.40	**339.00**
	Inst	Ea	Sm	CJ	2.22	7.20	261.00	67.40	---	328.40	**415.00**
Good quality workmanship	Inst	Ea	Lg	CJ	.889	18.00	190.00	27.00	---	217.00	**269.00**
	Inst	Ea	Sm	CJ	1.48	10.80	215.00	44.90	---	259.90	**325.00**
Average quality workmanship	Inst	Ea	Lg	CJ	.889	18.00	145.00	27.00	---	172.00	**215.00**
	Inst	Ea	Sm	CJ	1.48	10.80	164.00	44.90	---	208.90	**264.00**

Description	Oper	Unit	Vol	Crew Size	Man-hours per Unit	Crew Output per Day	Avg Mat'l Unit Cost	Avg Labor Unit Cost	Avg Equip Unit Cost	Avg Total Unit Cost	Avg Price Incl O&P
36" W											
High quality workmanship	Inst	Ea	Lg	CJ	1.33	12.00	296.00	40.40	---	336.40	**416.00**
	Inst	Ea	Sm	CJ	2.22	7.20	333.00	67.40	---	400.40	**501.00**
Good quality workmanship	Inst	Ea	Lg	CJ	.889	18.00	245.00	27.00	---	272.00	**334.00**
	Inst	Ea	Sm	CJ	1.48	10.80	276.00	44.90	---	320.90	**399.00**
Average quality workmanship	Inst	Ea	Lg	CJ	.889	18.00	187.00	27.00	---	214.00	**264.00**
	Inst	Ea	Sm	CJ	1.48	10.80	210.00	44.90	---	254.90	**320.00**
42" W											
High quality workmanship	Inst	Ea	Lg	CJ	1.74	9.20	328.00	52.80	---	380.80	**473.00**
	Inst	Ea	Sm	CJ	2.91	5.50	370.00	88.30	---	458.30	**576.00**
Good quality workmanship	Inst	Ea	Lg	CJ	1.16	13.80	271.00	35.20	---	306.20	**378.00**
	Inst	Ea	Sm	CJ	1.93	8.30	306.00	58.60	---	364.60	**455.00**
Average quality workmanship	Inst	Ea	Lg	CJ	1.16	13.80	205.00	35.20	---	240.20	**299.00**
	Inst	Ea	Sm	CJ	1.93	8.30	232.00	58.00	---	290.60	**366.00**
48" W											
High quality workmanship	Inst	Ea	Lg	CJ	1.74	9.20	371.00	52.80	---	423.80	**525.00**
	Inst	Ea	Sm	CJ	2.91	5.50	418.00	88.30	---	506.30	**635.00**
Good quality workmanship	Inst	Ea	Lg	CJ	1.16	13.80	305.00	35.20	---	340.20	**419.00**
	Inst	Ea	Sm	CJ	1.93	8.30	344.00	58.60	---	402.60	**501.00**
Average quality workmanship	Inst	Ea	Lg	CJ	1.16	13.80	234.00	35.20	---	269.20	**333.00**
	Inst	Ea	Sm	CJ	1.93	8.30	263.00	58.60	---	321.60	**404.00**

Sink/range front cabinets, 35" H, 2 doors

Description	Oper	Unit	Vol	Crew Size	Man-hours per Unit	Crew Output per Day	Avg Mat'l Unit Cost	Avg Labor Unit Cost	Avg Equip Unit Cost	Avg Total Unit Cost	Avg Price Incl O&P
30" W											
High quality workmanship	Inst	Ea	Lg	CJ	1.16	13.80	145.00	35.20	---	180.20	**227.00**
	Inst	Ea	Sm	CJ	1.93	8.30	164.00	58.60	---	222.60	**284.00**
Good quality workmanship	Inst	Ea	Lg	CJ	.800	20.00	121.00	24.30	---	145.30	**181.00**
	Inst	Ea	Sm	CJ	1.33	12.00	136.00	40.40	---	176.40	**224.00**
Average quality workmanship	Inst	Ea	Lg	CJ	.800	20.00	92.30	24.30	---	116.60	**147.00**
	Inst	Ea	Sm	CJ	1.33	12.00	104.00	40.40	---	144.40	**185.00**
36" W											
High quality workmanship	Inst	Ea	Lg	CJ	1.16	13.80	196.00	35.20	---	231.20	**288.00**
	Inst	Ea	Sm	CJ	1.93	8.30	221.00	58.60	---	279.60	**353.00**
Good quality workmanship	Inst	Ea	Lg	CJ	.800	20.00	162.00	24.30	---	186.30	**231.00**
	Inst	Ea	Sm	CJ	1.33	12.00	183.00	40.40	---	223.40	**280.00**
Average quality workmanship	Inst	Ea	Lg	CJ	.800	20.00	126.00	24.30	---	150.30	**188.00**
	Inst	Ea	Sm	CJ	1.33	12.00	142.00	40.40	---	182.40	**231.00**
42" W											
High quality workmanship	Inst	Ea	Lg	CJ	1.51	10.00	215.00	45.80	---	260.80	**326.00**
	Inst	Ea	Sm	CJ	2.50	6.40	242.00	75.90	---	317.90	**404.00**
Good quality workmanship	Inst	Ea	Lg	CJ	1.00	16.00	179.00	30.40	---	209.40	**260.00**
	Inst	Ea	Sm	CJ	1.67	9.60	202.00	50.70	---	252.70	**318.00**
Average quality workmanship	Inst	Ea	Lg	CJ	1.00	16.00	136.00	30.40	---	166.40	**208.00**
	Inst	Ea	Sm	CJ	1.67	9.60	153.00	50.70	---	203.70	**260.00**
48" W											
High quality workmanship	Inst	Ea	Lg	CJ	1.51	10.60	226.00	45.80	---	271.80	**340.00**
	Inst	Ea	Sm	CJ	2.50	6.40	255.00	75.90	---	330.90	**420.00**
Good quality workmanship	Inst	Ea	Lg	CJ	1.00	16.00	194.00	30.40	---	224.40	**278.00**
	Inst	Ea	Sm	CJ	1.67	9.60	219.00	50.70	---	269.70	**339.00**
Average quality workmanship	Inst	Ea	Lg	CJ	1.00	16.00	147.00	30.40	---	177.40	**222.00**
	Inst	Ea	Sm	CJ	1.67	9.60	166.00	50.70	---	216.70	**275.00**

Description	Oper	Unit	Vol	Crew Size	Man-hours per Unit	Crew Output per Day	Avg Mat'l Unit Cost	Avg Labor Unit Cost	Avg Equip Unit Cost	Avg Total Unit Cost	Avg Price Incl O&P

Wall cabinets
Refrigerator cabinets, 15" H, 12" D, 2 doors

30" W

Description	Oper	Unit	Vol	Crew Size	Man-hours	Output	Mat'l	Labor	Equip	Total	Price
High quality workmanship	Inst	Ea	Lg	CJ	1.54	10.40	163.00	46.70	---	209.70	**266.00**
	Inst	Ea	Sm	CJ	2.58	6.20	184.00	78.30	---	262.30	**338.00**
Good quality workmanship	Inst	Ea	Lg	CJ	1.03	15.60	121.00	31.30	---	152.30	**192.00**
	Inst	Ea	Sm	CJ	1.70	9.40	136.00	51.60	---	187.60	**241.00**
Average quality workmanship	Inst	Ea	Lg	CJ	1.03	15.60	92.60	31.30	---	123.90	**158.00**
	Inst	Ea	Sm	CJ	1.70	9.40	104.00	51.60	---	155.60	**203.00**

33" W

Description	Oper	Unit	Vol	Crew Size	Man-hours	Output	Mat'l	Labor	Equip	Total	Price
High quality workmanship	Inst	Ea	Lg	CJ	1.54	10.40	170.00	46.70	---	216.70	**274.00**
	Inst	Ea	Sm	CJ	2.58	6.20	191.00	78.30	---	269.30	**347.00**
Good quality workmanship	Inst	Ea	Lg	CJ	1.03	15.60	124.00	31.30	---	155.30	**196.00**
	Inst	Ea	Sm	CJ	1.70	9.40	140.00	51.60	---	191.60	**245.00**
Average quality workmanship	Inst	Ea	Lg	CJ	1.03	15.60	94.20	31.30	---	125.50	**160.00**
	Inst	Ea	Sm	CJ	1.70	9.40	106.00	51.60	---	157.60	**205.00**

36" W

Description	Oper	Unit	Vol	Crew Size	Man-hours	Output	Mat'l	Labor	Equip	Total	Price
High quality workmanship	Inst	Ea	Lg	CJ	1.54	10.40	179.00	46.70	---	225.70	**285.00**
	Inst	Ea	Sm	CJ	2.58	6.20	202.00	78.30	---	280.30	**360.00**
Good quality workmanship	Inst	Ea	Lg	CJ	1.03	15.60	132.00	31.30	---	163.30	**205.00**
	Inst	Ea	Sm	CJ	1.70	9.40	149.00	51.60	---	200.60	**256.00**
Average quality workmanship	Inst	Ea	Lg	CJ	1.03	15.60	100.00	31.30	---	131.30	**167.00**
	Inst	Ea	Sm	CJ	1.70	9.40	113.00	51.60	---	164.60	**213.00**

Range cabinets, 21" H, 12" D

24" W, 1 door

Description	Oper	Unit	Vol	Crew Size	Man-hours	Output	Mat'l	Labor	Equip	Total	Price
High quality workmanship	Inst	Ea	Lg	CJ	1.51	10.60	140.00	45.80	---	185.80	**236.00**
	Inst	Ea	Sm	CJ	2.50	6.40	158.00	75.90	---	233.90	**303.00**
Good quality workmanship	Inst	Ea	Lg	CJ	1.01	15.80	104.00	30.70	---	134.70	**170.00**
	Inst	Ea	Sm	CJ	1.68	9.50	117.00	51.00	---	168.00	**217.00**
Average quality workmanship	Inst	Ea	Lg	CJ	1.01	15.80	78.50	30.70	---	109.20	**140.00**
	Inst	Ea	Sm	CJ	1.68	9.50	88.50	51.00	---	139.50	**183.00**

30" W, 2 doors

Description	Oper	Unit	Vol	Crew Size	Man-hours	Output	Mat'l	Labor	Equip	Total	Price
High quality workmanship	Inst	Ea	Lg	CJ	1.51	10.60	155.00	45.80	---	200.80	**255.00**
	Inst	Ea	Sm	CJ	2.50	6.40	175.00	75.90	---	250.90	**324.00**
Good quality workmanship	Inst	Ea	Lg	CJ	1.01	15.80	115.00	30.70	---	145.70	**184.00**
	Inst	Ea	Sm	CJ	1.68	9.50	129.00	51.00	---	180.00	**232.00**
Average quality workmanship	Inst	Ea	Lg	CJ	1.01	15.80	87.90	30.70	---	118.60	**151.00**
	Inst	Ea	Sm	CJ	1.68	9.50	99.10	51.00	---	150.10	**195.00**

33" W, 2 doors

Description	Oper	Unit	Vol	Crew Size	Man-hours	Output	Mat'l	Labor	Equip	Total	Price
High quality workmanship	Inst	Ea	Lg	CJ	1.60	10.00	162.00	48.60	---	210.60	**267.00**
	Inst	Ea	Sm	CJ	2.67	6.00	182.00	81.00	---	263.00	**340.00**
Good quality workmanship	Inst	Ea	Lg	CJ	1.07	15.00	119.00	32.50	---	151.50	**192.00**
	Inst	Ea	Sm	CJ	1.78	9.00	135.00	54.00	---	189.00	**242.00**
Average quality workmanship	Inst	Ea	Lg	CJ	1.07	15.00	91.10	32.50	---	123.60	**158.00**
	Inst	Ea	Sm	CJ	1.78	9.00	103.00	54.00	---	157.00	**204.00**

Description	Oper	Unit	Vol	Crew Size	Man-hours per Unit	Crew Output per Day	Avg Mat'l Unit Cost	Avg Labor Unit Cost	Avg Equip Unit Cost	Avg Total Unit Cost	Avg Price Incl O&P
36" W, 2 doors											
High quality workmanship	Inst	Ea	Lg	CJ	1.60	10.00	173.00	48.60	---	221.60	**280.00**
	Inst	Ea	Sm	CJ	2.67	6.00	195.00	81.00	---	276.00	**355.00**
Good quality workmanship	Inst	Ea	Lg	CJ	1.07	15.00	126.00	32.50	---	158.50	**199.00**
	Inst	Ea	Sm	CJ	1.78	9.00	142.00	54.00	---	196.00	**251.00**
Average quality workmanship	Inst	Ea	Lg	CJ	1.07	15.00	95.80	32.50	---	128.30	**164.00**
	Inst	Ea	Sm	CJ	1.78	9.00	108.00	54.00	---	162.00	**211.00**
42" W, 2 doors											
High quality workmanship	Inst	Ea	Lg	CJ	1.60	10.00	188.00	48.60	---	236.60	**299.00**
	Inst	Ea	Sm	CJ	2.67	6.00	212.00	81.00	---	293.00	**376.00**
Good quality workmanship	Inst	Ea	Lg	CJ	1.07	15.00	140.00	32.50	---	172.50	**216.00**
	Inst	Ea	Sm	CJ	1.78	9.00	158.00	54.00	---	212.00	**270.00**
Average quality workmanship	Inst	Ea	Lg	CJ	1.07	15.00	107.00	32.50	---	139.50	**177.00**
	Inst	Ea	Sm	CJ	1.78	9.00	120.00	54.00	---	174.00	**225.00**
48" W, 2 doors											
High quality workmanship	Inst	Ea	Lg	CJ	2.00	8.00	214.00	60.70	---	274.70	**347.00**
	Inst	Ea	Sm	CJ	3.33	4.80	241.00	101.00	---	342.00	**440.00**
Good quality workmanship	Inst	Ea	Lg	CJ	1.33	12.00	159.00	40.40	---	199.40	**251.00**
	Inst	Ea	Sm	CJ	2.22	7.20	179.00	67.40	---	246.40	**316.00**
Average quality workmanship	Inst	Ea	Lg	CJ	1.33	12.00	121.00	40.40	---	161.40	**206.00**
	Inst	Ea	Sm	CJ	2.22	7.20	136.00	67.40	---	203.40	**265.00**

Range cabinets, 30" H, 12" D

Description	Oper	Unit	Vol	Crew Size	Man-hours per Unit	Crew Output per Day	Avg Mat'l Unit Cost	Avg Labor Unit Cost	Avg Equip Unit Cost	Avg Total Unit Cost	Avg Price Incl O&P
12" W, 1 door											
High quality workmanship	Inst	Ea	Lg	CJ	1.18	13.60	122.00	35.80	---	157.80	**201.00**
	Inst	Ea	Sm	CJ	1.95	8.20	138.00	59.20	---	197.20	**254.00**
Good quality workmanship	Inst	Ea	Lg	CJ	.780	20.50	89.50	23.70	---	113.20	**143.00**
	Inst	Ea	Sm	CJ	1.30	12.30	101.00	39.50	---	140.50	**180.00**
Average quality workmanship	Inst	Ea	Lg	CJ	.780	20.50	67.50	23.70	---	91.20	**117.00**
	Inst	Ea	Sm	CJ	1.30	12.30	76.10	39.50	---	115.60	**151.00**
15" W, 1 door											
High quality workmanship	Inst	Ea	Lg	CJ	1.18	13.60	135.00	35.80	---	170.80	**216.00**
	Inst	Ea	Sm	CJ	1.95	8.20	152.00	59.20	---	211.20	**271.00**
Good quality workmanship	Inst	Ea	Lg	CJ	.780	20.50	98.90	23.70	---	122.60	**154.00**
	Inst	Ea	Sm	CJ	1.30	12.30	112.00	39.50	---	151.50	**193.00**
Average quality workmanship	Inst	Ea	Lg	CJ	.780	20.50	75.40	23.70	---	99.10	**126.00**
	Inst	Ea	Sm	CJ	1.30	12.30	85.00	39.50	---	124.50	**161.00**
18" W, 1 door											
High quality workmanship	Inst	Ea	Lg	CJ	1.18	13.60	141.00	35.80	---	176.80	**223.00**
	Inst	Ea	Sm	CJ	1.95	8.20	159.00	59.20	---	218.20	**280.00**
Good quality workmanship	Inst	Ea	Lg	CJ	.780	20.50	104.00	23.70	---	127.70	**160.00**
	Inst	Ea	Sm	CJ	1.30	12.30	117.00	39.50	---	156.50	**199.00**
Average quality workmanship	Inst	Ea	Lg	CJ	.780	20.50	78.50	23.70	---	102.20	**130.00**
	Inst	Ea	Sm	CJ	1.30	12.30	88.50	39.50	---	128.00	**165.00**

Description	Oper	Unit	Vol	Crew Size	Man-hours per Unit	Crew Output per Day	Avg Mat'l Unit Cost	Avg Labor Unit Cost	Avg Equip Unit Cost	Avg Total Unit Cost	Avg Price Incl O&P
21" W, 1 door											
High quality workmanship	Inst	Ea	Lg	CJ	1.33	12.00	149.00	40.40	---	189.40	**240.00**
	Inst	Ea	Sm	CJ	2.22	7.20	168.00	67.40	---	235.40	**303.00**
Good quality workmanship	Inst	Ea	Lg	CJ	.889	18.00	110.00	27.00	---	137.00	**172.00**
	Inst	Ea	Sm	CJ	1.48	10.80	124.00	44.90	---	168.90	**216.00**
Average quality workmanship	Inst	Ea	Lg	CJ	.889	18.00	83.20	27.00	---	110.20	**140.00**
	Inst	Ea	Sm	CJ	1.48	10.80	93.80	44.90	---	138.70	**180.00**
24" W, 1 door											
High quality workmanship	Inst	Ea	Lg	CJ	1.33	12.00	166.00	40.40	---	206.40	**260.00**
	Inst	Ea	Sm	CJ	2.22	7.20	188.00	67.40	---	255.40	**326.00**
Good quality workmanship	Inst	Ea	Lg	CJ	.889	18.00	122.00	27.00	---	149.00	**187.00**
	Inst	Ea	Sm	CJ	1.48	10.80	138.00	44.90	---	182.90	**233.00**
Average quality workmanship	Inst	Ea	Lg	CJ	.889	18.00	92.60	27.00	---	119.60	**152.00**
	Inst	Ea	Sm	CJ	1.48	10.80	104.00	44.90	---	148.90	**193.00**
27" W, 2 doors											
High quality workmanship	Inst	Ea	Lg	CJ	1.33	12.00	184.00	40.40	---	224.40	**281.00**
	Inst	Ea	Sm	CJ	2.22	7.20	207.00	67.40	---	274.40	**350.00**
Good quality workmanship	Inst	Ea	Lg	CJ	.889	18.00	135.00	27.00	---	162.00	**203.00**
	Inst	Ea	Sm	CJ	1.48	10.80	152.00	44.90	---	196.90	**250.00**
Average quality workmanship	Inst	Ea	Lg	CJ	.889	18.00	104.00	27.00	---	131.00	**165.00**
	Inst	Ea	Sm	CJ	1.48	10.80	117.00	44.90	---	161.90	**208.00**
30" W, 2 doors											
High quality workmanship	Inst	Ea	Lg	CJ	1.78	9.00	195.00	54.00	---	249.00	**315.00**
	Inst	Ea	Sm	CJ	2.96	5.40	219.00	89.80	---	308.80	**398.00**
Good quality workmanship	Inst	Ea	Lg	CJ	1.19	13.50	144.00	36.10	---	180.10	**228.00**
	Inst	Ea	Sm	CJ	1.98	8.10	163.00	60.10	---	223.10	**286.00**
Average quality workmanship	Inst	Ea	Lg	CJ	1.19	13.50	108.00	36.10	---	144.10	**184.00**
	Inst	Ea	Sm	CJ	1.98	8.10	122.00	60.10	---	182.10	**237.00**
33" W, 2 doors											
High quality workmanship	Inst	Ea	Lg	CJ	1.78	9.00	206.00	54.00	---	260.00	**328.00**
	Inst	Ea	Sm	CJ	2.96	5.40	232.00	89.80	---	321.80	**413.00**
Good quality workmanship	Inst	Ea	Lg	CJ	1.19	13.50	152.00	36.10	---	188.10	**237.00**
	Inst	Ea	Sm	CJ	1.98	8.10	172.00	60.10	---	232.10	**296.00**
Average quality workmanship	Inst	Ea	Lg	CJ	1.19	13.50	116.00	36.10	---	152.10	**194.00**
	Inst	Ea	Sm	CJ	1.98	8.10	131.00	60.10	---	191.10	**247.00**
36" W, 2 doors											
High quality workmanship	Inst	Ea	Lg	CJ	1.78	9.00	214.00	54.00	---	268.00	**337.00**
	Inst	Ea	Sm	CJ	2.96	5.40	241.00	89.80	---	330.80	**424.00**
Good quality workmanship	Inst	Ea	Lg	CJ	1.19	13.50	159.00	36.10	---	195.10	**244.00**
	Inst	Ea	Sm	CJ	1.98	8.10	179.00	60.10	---	239.10	**305.00**
Average quality workmanship	Inst	Ea	Lg	CJ	1.19	13.50	121.00	36.10	---	157.10	**199.00**
	Inst	Ea	Sm	CJ	1.98	8.10	136.00	60.10	---	196.10	**254.00**

Description	Oper	Unit	Vol	Crew Size	Man-hours per Unit	Crew Output per Day	Avg Mat'l Unit Cost	Avg Labor Unit Cost	Avg Equip Unit Cost	Avg Total Unit Cost	Avg Price Incl O&P
42" W, 2 doors											
High quality workmanship	Inst	Ea	Lg	CJ	2.29	7.00	232.00	69.50	---	301.50	**383.00**
	Inst	Ea	Sm	CJ	3.81	4.20	262.00	116.00	---	378.00	**488.00**
Good quality workmanship	Inst	Ea	Lg	CJ	1.52	10.50	171.00	46.10	---	217.10	**275.00**
	Inst	Ea	Sm	CJ	2.54	6.30	193.00	77.10	---	270.10	**347.00**
Average quality workmanship	Inst	Ea	Lg	CJ	1.52	10.50	130.00	46.10	---	176.10	**226.00**
	Inst	Ea	Sm	CJ	2.54	6.30	147.00	77.10	---	224.10	**292.00**
48" W, 2 doors											
High quality workmanship	Inst	Ea	Lg	CJ	2.29	7.00	275.00	69.50	---	344.50	**434.00**
	Inst	Ea	Sm	CJ	3.81	4.20	310.00	116.00	---	426.00	**545.00**
Good quality workmanship	Inst	Ea	Lg	CJ	1.52	10.50	203.00	46.10	---	249.10	**312.00**
	Inst	Ea	Sm	CJ	2.54	6.30	228.00	77.10	---	305.10	**390.00**
Average quality workmanship	Inst	Ea	Lg	CJ	1.52	10.50	154.00	46.10	---	200.10	**254.00**
	Inst	Ea	Sm	CJ	2.54	6.30	173.00	77.10	---	250.10	**324.00**
24" W, blind corner unit, 1 door											
High quality workmanship	Inst	Ea	Lg	CJ	1.68	9.50	149.00	51.00	---	200.00	**255.00**
	Inst	Ea	Sm	CJ	2.81	5.70	168.00	85.30	---	253.30	**330.00**
Good quality workmanship	Inst	Ea	Lg	CJ	1.14	14.00	118.00	34.60	---	152.60	**193.00**
	Inst	Ea	Sm	CJ	1.90	8.40	133.00	57.70	---	190.70	**246.00**
Average quality workmanship	Inst	Ea	Lg	CJ	1.14	14.00	115.00	34.60	---	149.60	**189.00**
	Inst	Ea	Sm	CJ	1.90	8.40	129.00	57.70	---	186.70	**242.00**
24" x 24" angle corner unit, stationary											
High quality workmanship	Inst	Ea	Lg	CJ	1.68	9.50	214.00	51.00	---	265.00	**333.00**
	Inst	Ea	Sm	CJ	2.81	5.70	241.00	85.30	---	326.30	**417.00**
Good quality workmanship	Inst	Ea	Lg	CJ	1.14	14.00	157.00	34.60	---	191.60	**240.00**
	Inst	Ea	Sm	CJ	1.90	8.40	177.00	57.70	---	234.70	**299.00**
Average quality workmanship	Inst	Ea	Lg	CJ	1.14	14.00	122.00	34.60	---	156.60	**199.00**
	Inst	Ea	Sm	CJ	1.90	8.40	138.00	57.70	---	195.70	**252.00**
24" x 24" angle corner unit, lazy Susan											
High quality workmanship	Inst	Ea	Lg	CJ	1.68	9.50	317.00	51.00	---	368.00	**457.00**
	Inst	Ea	Sm	CJ	2.81	5.70	358.00	85.30	---	443.30	**557.00**
Good quality workmanship	Inst	Ea	Lg	CJ	1.14	14.00	257.00	34.60	---	291.60	**361.00**
	Inst	Ea	Sm	CJ	1.90	8.40	290.00	57.70	---	347.70	**435.00**
Average quality workmanship	Inst	Ea	Lg	CJ	1.14	14.00	176.00	34.60	---	210.60	**263.00**
	Inst	Ea	Sm	CJ	1.90	8.40	198.00	57.70	---	255.70	**324.00**

Vanity cabinets and sink top

Description	Oper	Unit	Vol	Crew Size	Man-hours per Unit	Crew Output per Day	Avg Mat'l Unit Cost	Avg Labor Unit Cost	Avg Equip Unit Cost	Avg Total Unit Cost	Avg Price Incl O&P
Disconnect plumbing and remove to dumpster											
	Demo	Ea	Lg	LB	1.00	16.00	---	27.40	---	27.40	**40.90**
	Demo	Ea	Sm	LB	1.67	9.60	---	45.80	---	45.80	**68.30**
Remove old unit, replace with new unit, reconnect plumbing											
	Reset	Ea	Lg	SB	2.29	7.00	---	74.50	---	74.50	**110.00**
	Reset	Ea	Sm	SB	3.81	4.20	---	124.00	---	124.00	**182.00**

Description	Oper	Unit	Vol	Crew Size	Man-hours per Unit	Crew Output per Day	Avg Mat'l Unit Cost	Avg Labor Unit Cost	Avg Equip Unit Cost	Avg Total Unit Cost	Avg Price Incl O&P

Vanity units, with marble tops, good quality fittings and faucets; hardware, deluxe, finished models

Stained ash and birch primed composition construction. Labor costs include fitting and hanging only of vanity units for rough-in costs, see Adjustments in this cabinets section, page 74

2-door units
20" x 16"

Description	Oper	Unit	Vol	Crew Size	Man-hours per Unit	Crew Output per Day	Avg Mat'l Unit Cost	Avg Labor Unit Cost	Avg Equip Unit Cost	Avg Total Unit Cost	Avg Price Incl O&P
Ash	Inst	Ea	Lg	SB	5.00	3.20	355.00	163.00	---	518.00	**665.00**
	Inst	Ea	Sm	SB	8.42	1.90	400.00	274.00	---	674.00	**883.00**
Birch	Inst	Ea	Lg	SB	5.00	3.20	323.00	163.00	---	486.00	**627.00**
	Inst	Ea	Sm	SB	8.42	1.90	365.00	274.00	---	639.00	**840.00**
Composition construction	Inst	Ea	Lg	SB	5.00	3.20	245.00	163.00	---	408.00	**533.00**
	Inst	Ea	Sm	SB	8.42	1.90	276.00	274.00	---	550.00	**734.00**

25" x 19"

Description	Oper	Unit	Vol	Crew Size	Man-hours per Unit	Crew Output per Day	Avg Mat'l Unit Cost	Avg Labor Unit Cost	Avg Equip Unit Cost	Avg Total Unit Cost	Avg Price Incl O&P
Ash	Inst	Ea	Lg	SB	5.00	3.20	413.00	163.00	---	576.00	**735.00**
	Inst	Ea	Sm	SB	8.42	1.90	466.00	274.00	---	740.00	**962.00**
Birch	Inst	Ea	Lg	SB	5.00	3.20	377.00	163.00	---	540.00	**691.00**
	Inst	Ea	Sm	SB	8.42	1.90	425.00	274.00	---	699.00	**913.00**
Composition construction	Inst	Ea	Lg	SB	5.00	3.20	268.00	163.00	---	431.00	**561.00**
	Inst	Ea	Sm	SB	8.42	1.90	303.00	274.00	---	577.00	**766.00**

31" x 19"

Description	Oper	Unit	Vol	Crew Size	Man-hours per Unit	Crew Output per Day	Avg Mat'l Unit Cost	Avg Labor Unit Cost	Avg Equip Unit Cost	Avg Total Unit Cost	Avg Price Incl O&P
Ash	Inst	Ea	Lg	SB	5.00	3.20	491.00	163.00	---	654.00	**829.00**
	Inst	Ea	Sm	SB	8.42	1.90	554.00	274.00	---	828.00	**1070.00**
Birch	Inst	Ea	Lg	SB	5.00	3.20	446.00	163.00	---	609.00	**774.00**
	Inst	Ea	Sm	SB	8.42	1.90	503.00	274.00	---	777.00	**1010.00**
Composition construction	Inst	Fa	Lg	SB	5.00	3.20	319.00	163.00	---	482.00	**622.00**
	Inst	Ea	Sm	SB	8.42	1.90	359.00	274.00	---	633.00	**834.00**

35" x 19"

Description	Oper	Unit	Vol	Crew Size	Man-hours per Unit	Crew Output per Day	Avg Mat'l Unit Cost	Avg Labor Unit Cost	Avg Equip Unit Cost	Avg Total Unit Cost	Avg Price Incl O&P
Ash	Inst	Ea	Lg	SB	5.33	3.00	531.00	173.00	---	704.00	**892.00**
	Inst	Ea	Sm	SB	8.89	1.80	598.00	289.00	---	887.00	**1140.00**
Birch	Inst	Ea	Lg	SB	5.33	3.00	480.00	173.00	---	653.00	**832.00**
	Inst	Ea	Sm	SB	8.89	1.80	542.00	289.00	---	831.00	**1080.00**
Composition construction	Inst	Ea	Lg	SB	5.33	3.00	347.00	173.00	---	520.00	**671.00**
	Inst	Ea	Sm	SB	8.89	1.80	391.00	289.00	---	680.00	**895.00**

For drawers in any above unit, ADD per drawer

Description	Oper	Unit	Vol	Crew Size	Man-hours per Unit	Crew Output per Day	Avg Mat'l Unit Cost	Avg Labor Unit Cost	Avg Equip Unit Cost	Avg Total Unit Cost	Avg Price Incl O&P
	Inst	Ea	Lg	SB	---	---	47.10	---	---	47.10	**56.50**
	Inst	Ea	Sm	SB	---	---	53.10	---	---	53.10	**63.70**

2-door cutback units with 3 drawers
36" x 19"

Description	Oper	Unit	Vol	Crew Size	Man-hours per Unit	Crew Output per Day	Avg Mat'l Unit Cost	Avg Labor Unit Cost	Avg Equip Unit Cost	Avg Total Unit Cost	Avg Price Incl O&P
Ash	Inst	Ea	Lg	SB	5.33	3.00	615.00	173.00	---	788.00	**994.00**
	Inst	Ea	Sm	SB	8.89	1.80	694.00	289.00	---	983.00	**1260.00**
Birch	Inst	Ea	Lg	SB	5.33	3.00	553.00	173.00	---	726.00	**918.00**
	Inst	Ea	Sm	SB	8.89	1.80	623.00	289.00	---	912.00	**1170.00**
Composition construction	Inst	Ea	Lg	SB	5.33	3.00	413.00	173.00	---	586.00	**751.00**
	Inst	Ea	Sm	SB	8.89	1.80	466.00	289.00	---	755.00	**984.00**

Description	Oper	Unit	Vol	Crew Size	Man- hours per Unit	Crew Output per Day	Avg Mat'l Unit Cost	Avg Labor Unit Cost	Avg Equip Unit Cost	Avg Total Unit Cost	Avg Price Incl O&P
49" x 19"											
Ash	Inst	Ea	Lg	SB	5.33	3.00	769.00	173.00	---	942.00	**1180.00**
	Inst	Ea	Sm	SB	8.89	1.80	867.00	289.00	---	1156.00	**1470.00**
Birch	Inst	Ea	Lg	SB	5.33	3.00	666.00	173.00	---	839.00	**1050.00**
	Inst	Ea	Sm	SB	8.89	1.80	750.00	289.00	---	1039.00	**1330.00**
Composition construction	Inst	Ea	Lg	SB	5.33	3.00	498.00	173.00	---	671.00	**852.00**
	Inst	Ea	Sm	SB	8.89	1.80	561.00	289.00	---	850.00	**1100.00**
60" x 19"											
Ash	Inst	Ea	Lg	SB	6.67	2.40	970.00	217.00	---	1187.00	**1480.00**
	Inst	Ea	Sm	SB	11.4	1.40	1090.00	371.00	---	1461.00	**1860.00**
Birch	Inst	Ea	Lg	SB	6.67	2.40	805.00	217.00	---	1022.00	**1290.00**
	Inst	Ea	Sm	SB	11.4	1.40	908.00	371.00	---	1279.00	**1640.00**
Composition construction	Inst	Ea	Lg	SB	6.67	2.40	626.00	217.00	---	843.00	**1070.00**
	Inst	Ea	Sm	SB	11.4	1.40	706.00	371.00	---	1077.00	**1390.00**

Corner unit, 1 door
22" x 22"

Description	Oper	Unit	Vol	Crew Size	Man- hours per Unit	Crew Output per Day	Avg Mat'l Unit Cost	Avg Labor Unit Cost	Avg Equip Unit Cost	Avg Total Unit Cost	Avg Price Incl O&P
Ash	Inst	Ea	Lg	SB	5.00	3.20	457.00	163.00	---	620.00	**787.00**
	Inst	Ea	Sm	SB	8.42	1.90	515.00	274.00	---	789.00	**1020.00**
Birch	Inst	Ea	Lg	SB	5.00	3.20	432.00	163.00	---	595.00	**757.00**
	Inst	Ea	Sm	SB	8.42	1.90	487.00	274.00	---	761.00	**987.00**
Composition construction	Inst	Ea	Lg	SB	5.00	3.20	352.00	163.00	---	515.00	**661.00**
	Inst	Ea	Sm	SB	8.42	1.90	396.00	274.00	---	670.00	**879.00**

Adjustments
To remove and reset

Description	Oper	Unit	Vol	Crew Size	Man- hours per Unit	Crew Output per Day	Avg Mat'l Unit Cost	Avg Labor Unit Cost	Avg Equip Unit Cost	Avg Total Unit Cost	Avg Price Incl O&P
Vanity units with tops	Reset	Ea	Lg	SA	3.33	2.40	---	125.00	---	125.00	**184.00**
	Reset	Ea	Sm	SA	5.71	1.40	---	215.00	---	215.00	**316.00**
Top only	Reset	Ea	Lg	SA	2.00	4.00	---	75.30	---	75.30	**111.00**
	Reset	Ea	Sm	SA	3.33	2.40	---	125.00	---	125.00	**184.00**

To install rough-in

Description	Oper	Unit	Vol	Crew Size	Man- hours per Unit	Crew Output per Day	Avg Mat'l Unit Cost	Avg Labor Unit Cost	Avg Equip Unit Cost	Avg Total Unit Cost	Avg Price Incl O&P
	Inst	Ea	Lg	SB	4.00	4.00	---	130.00	---	130.00	**191.00**
	Inst	Ea	Sm	SB	6.67	2.40	---	217.00	---	217.00	**319.00**

Description	Oper	Unit	Vol	Crew Size	Man-hours per Unit	Crew Output per Day	Avg Mat'l Unit Cost	Avg Labor Unit Cost	Avg Equip Unit Cost	Avg Total Unit Cost	Avg Price Incl O&P
Canopies											
Costs for awnings include all hardware											
Residential prefabricated											
Aluminum											
Carport, freestanding											
16' x 8'	Inst	Ea	Lg	2C	5.33	3.00	1140.00	177.00	---	1317.00	**1570.00**
	Inst	Ea	Sm	2C	7.11	2.25	1330.00	237.00	---	1567.00	**1880.00**
20' x 10'	Inst	Ea	Lg	2C	8.00	2.00	1200.00	266.00	---	1466.00	**1780.00**
	Inst	Ea	Sm	2C	10.7	1.50	1400.00	356.00	---	1756.00	**2140.00**
Door canopies, 36" projection											
4' wide	Inst	Ea	Lg	CA	1.00	8.00	189.00	33.30	---	222.30	**268.00**
	Inst	Ea	Sm	CA	1.33	6.00	221.00	44.30	---	265.30	**321.00**
5' wide	Inst	Ea	Lg	CA	1.33	6.00	212.00	44.30	---	256.30	**310.00**
	Inst	Ea	Sm	CA	1.78	4.50	248.00	59.20	---	307.20	**374.00**
6' wide	Inst	Ea	Lg	CA	2.00	4.00	229.00	66.50	---	295.50	**363.00**
	Inst	Ea	Sm	CA	2.67	3.00	268.00	88.80	---	356.80	**441.00**
8' wide	Inst	Ea	Lg	2C	3.20	5.00	313.00	106.00	---	419.00	**520.00**
	Inst	Ea	Sm	2C	4.27	3.75	366.00	142.00	---	508.00	**634.00**
10' wide	Inst	Ea	Lg	2C	5.33	3.00	359.00	177.00	---	536.00	**678.00**
	Inst	Ea	Sm	2C	7.11	2.25	419.00	237.00	---	656.00	**837.00**
12' wide	Inst	Ea	Lg	2C	8.00	2.00	426.00	266.00	---	692.00	**889.00**
	Inst	Ea	Sm	2C	10.7	1.50	498.00	356.00	---	854.00	**1110.00**
Door canopies, 42" projection											
4' wide	Inst	Ea	Lg	CA	1.00	8.00	217.00	33.30	---	250.30	**300.00**
	Inst	Ea	Sm	CA	1.33	6.00	254.00	44.30	---	298.30	**358.00**
5' wide	Inst	Ea	Lg	CA	1.33	6.00	237.00	44.30	---	281.30	**339.00**
	Inst	Ea	Sm	CA	1.78	4.50	277.00	59.20	---	336.20	**408.00**
6' wide	Inst	Ea	Lg	CA	2.00	4.00	300.00	66.50	---	366.50	**445.00**
	Inst	Ea	Sm	CA	2.67	3.00	350.00	88.80	---	438.80	**536.00**
8' wide	Inst	Ea	Lg	2C	3.20	5.00	350.00	106.00	---	456.00	**563.00**
	Inst	Ea	Sm	2C	4.27	3.75	409.00	142.00	---	551.00	**684.00**
10' wide	Inst	Ea	Lg	2C	5.33	3.00	406.00	177.00	---	583.00	**733.00**
	Inst	Ea	Sm	2C	7.11	2.25	474.00	237.00	---	711.00	**900.00**
12' wide	Inst	Ea	Lg	2C	8.00	2.00	477.00	266.00	---	743.00	**948.00**
	Inst	Ea	Sm	2C	10.7	1.50	557.00	356.00	---	913.00	**1170.00**
Door canopies, 48" projection											
4' wide	Inst	Ea	Lg	CA	1.00	8.00	237.00	33.30	---	270.30	**323.00**
	Inst	Ea	Sm	CA	1.33	6.00	277.00	44.30	---	321.30	**385.00**
5' wide	Inst	Ea	Lg	CA	1.33	6.00	278.00	44.30	---	322.30	**386.00**
	Inst	Ea	Sm	CA	1.78	4.50	325.00	59.20	---	384.20	**462.00**
6' wide	Inst	Ea	Lg	CA	2.00	4.00	315.00	66.50	---	381.50	**462.00**
	Inst	Ea	Sm	CA	2.67	3.00	368.00	88.80	---	456.80	**557.00**
8' wide	Inst	Ea	Lg	2C	3.20	5.00	429.00	106.00	---	535.00	**653.00**
	Inst	Ea	Sm	2C	4.27	3.75	502.00	142.00	---	644.00	**790.00**
10' wide	Inst	Ea	Lg	2C	5.33	3.00	480.00	177.00	---	657.00	**818.00**
	Inst	Ea	Sm	2C	7.11	2.25	561.00	237.00	---	798.00	**999.00**
12' wide	Inst	Ea	Lg	2C	8.00	2.00	540.00	266.00	---	806.00	**1020.00**
	Inst	Ea	Sm	2C	10.7	1.50	631.00	356.00	---	987.00	**1260.00**

Description	Oper	Unit	Vol	Crew Size	Man-hours per Unit	Crew Output per Day	Avg Mat'l Unit Cost	Avg Labor Unit Cost	Avg Equip Unit Cost	Avg Total Unit Cost	Avg Price Incl O&P
Door canopies, 54" projection											
4' wide	Inst	Ea	Lg	CA	1.00	8.00	273.00	33.30	---	306.30	**364.00**
	Inst	Ea	Sm	CA	1.33	6.00	319.00	44.30	---	363.30	**433.00**
5' wide	Inst	Ea	Lg	CA	1.33	6.00	290.00	44.30	---	334.30	**400.00**
	Inst	Ea	Sm	CA	1.78	4.50	339.00	59.20	---	398.20	**478.00**
6' wide	Inst	Ea	Lg	CA	2.00	4.00	361.00	66.50	---	427.50	**514.00**
	Inst	Ea	Sm	CA	2.67	3.00	421.00	88.80	---	509.80	**618.00**
8' wide	Inst	Ea	Lg	2C	3.20	5.00	492.00	106.00	---	598.00	**725.00**
	Inst	Ea	Sm	2C	4.27	3.75	575.00	142.00	---	717.00	**874.00**
10' wide	Inst	Ea	Lg	2C	5.33	3.00	556.00	177.00	---	733.00	**905.00**
	Inst	Ea	Sm	2C	7.11	2.25	649.00	237.00	---	886.00	**1100.00**
12' wide	Inst	Ea	Lg	2C	8.00	2.00	631.00	266.00	---	897.00	**1130.00**
	Inst	Ea	Sm	2C	10.7	1.50	738.00	356.00	---	1094.00	**1380.00**
Patio cover											
16' x 8'	Inst	Ea	Lg	2C	5.33	3.00	606.00	177.00	---	783.00	**963.00**
	Inst	Ea	Sm	2C	7.11	2.25	708.00	237.00	---	945.00	**1170.00**
20' x 10'	Inst	Ea	Lg	2C	8.00	2.00	820.00	266.00	---	1086.00	**1340.00**
	Inst	Ea	Sm	2C	10.7	1.50	958.00	356.00	---	1314.00	**1640.00**
Window awnings, 3' high											
4' wide	Inst	Ea	Lg	CA	1.14	7.00	63.10	37.90	---	101.00	**130.00**
	Inst	Ea	Sm	CA	1.52	5.25	73.80	50.60	---	124.40	**161.00**
6' wide	Inst	Ea	Lg	CA	1.60	5.00	76.30	53.20	---	129.50	**168.00**
	Inst	Ea	Sm	CA	2.13	3.75	89.10	70.90	---	160.00	**209.00**
9' wide	Inst	Ea	Lg	2C	2.67	6.00	105.00	88.80	---	193.80	**253.00**
	Inst	Ea	Sm	2C	3.56	4.50	122.00	118.00	---	240.00	**318.00**
12' wide	Inst	Ea	Lg	2C	4.00	4.00	131.00	133.00	---	264.00	**351.00**
	Inst	Ea	Sm	2C	5.33	3.00	153.00	177.00	---	330.00	**442.00**
Window awnings, 4' high											
4' wide	Inst	Ea	Lg	CA	1.14	7.00	72.70	37.90	---	110.60	**141.00**
	Inst	Ea	Sm	CA	1.52	5.25	85.00	50.60	---	135.60	**174.00**
6' wide	Inst	Ea	Lg	CA	1.60	5.00	96.00	53.20	---	149.20	**190.00**
	Inst	Ea	Sm	CA	2.13	3.75	112.00	70.90	---	182.90	**235.00**
9' wide	Inst	Ea	Lg	2C	2.67	6.00	131.00	88.80	---	219.80	**284.00**
	Inst	Ea	Sm	2C	3.56	4.50	153.00	118.00	---	271.00	**354.00**
12' wide	Inst	Ea	Lg	2C	4.00	4.00	172.00	133.00	---	305.00	**397.00**
	Inst	Ea	Sm	2C	5.33	3.00	201.00	177.00	---	378.00	**497.00**
Window awnings, 6' high											
4' wide	Inst	Ea	Lg	CA	1.14	7.00	119.00	37.90	---	156.90	**194.00**
	Inst	Ea	Sm	CA	1.52	5.25	139.00	50.60	---	189.60	**236.00**
6' wide	Inst	Ea	Lg	CA	1.60	5.00	162.00	53.20	---	215.20	**266.00**
	Inst	Ea	Sm	CA	2.13	3.75	189.00	70.90	---	259.90	**323.00**
9' wide	Inst	Ea	Lg	2C	2.67	6.00	202.00	88.80	---	290.80	**366.00**
	Inst	Ea	Sm	2C	3.56	4.50	236.00	118.00	---	354.00	**449.00**
12' wide	Inst	Ea	Lg	2C	4.00	4.00	227.00	133.00	---	360.00	**461.00**
	Inst	Ea	Sm	2C	5.33	3.00	266.00	177.00	---	443.00	**571.00**

Description	Oper	Unit	Vol	Crew Size	Man-hours per Unit	Crew Output per Day	Avg Mat'l Unit Cost	Avg Labor Unit Cost	Avg Equip Unit Cost	Avg Total Unit Cost	Avg Price Incl O&P
Roll-up											
3' wide	Inst	Ea	Lg	CA	1.00	8.00	49.50	33.30	---	82.80	**107.00**
	Inst	Ea	Sm	CA	1.33	6.00	57.80	44.30	---	102.10	**133.00**
4' wide	Inst	Ea	Lg	CA	1.14	7.00	63.60	37.90	---	101.50	**130.00**
	Inst	Ea	Sm	CA	1.52	5.25	74.30	50.60	---	124.90	**161.00**
6' wide	Inst	Ea	Lg	CA	1.60	5.00	78.80	53.20	---	132.00	**170.00**
	Inst	Ea	Sm	CA	2.13	3.75	92.00	70.90	---	162.90	**212.00**
9' wide	Inst	Ea	Lg	2C	2.67	6.00	114.00	88.80	---	202.80	**265.00**
	Inst	Ea	Sm	2C	3.56	4.50	133.00	118.00	---	251.00	**331.00**
12' wide	Inst	Ea	Lg	2C	4.00	4.00	139.00	133.00	---	272.00	**360.00**
	Inst	Ea	Sm	2C	5.33	3.00	163.00	177.00	---	340.00	**453.00**

Canvas

Traditional fabric awning, with waterproof, colorfast acrylic duck, double-stitched seams, tubular metal framing and all hardware Included

Description	Oper	Unit	Vol	Crew Size	Man-hours per Unit	Crew Output per Day	Avg Mat'l Unit Cost	Avg Labor Unit Cost	Avg Equip Unit Cost	Avg Total Unit Cost	Avg Price Incl O&P
Window awning, 24" drop											
3' wide	Inst	Ea	Lg	CA	1.00	8.00	162.00	33.30	---	195.30	**236.00**
	Inst	Ea	Sm	CA	1.33	6.00	189.00	44.30	---	233.30	**284.00**
4' wide	Inst	Ea	Lg	CA	1.14	7.00	198.00	37.90	---	235.90	**285.00**
	Inst	Ea	Sm	CA	1.52	5.25	231.00	50.60	---	281.60	**342.00**
Window awning, 30" drop											
3' wide	Inst	Ea	Lg	CA	1.00	8.00	185.00	33.30	---	218.30	**262.00**
	Inst	Ea	Sm	CA	1.33	6.00	216.00	44.30	---	260.30	**315.00**
4' wide	Inst	Ea	Lg	CA	1.14	7.00	222.00	37.90	---	259.90	**312.00**
	Inst	Ea	Sm	CA	1.52	5.25	260.00	50.60	---	310.60	**374.00**
5' wide	Inst	Ea	Lg	CA	1.33	6.00	253.00	44.30	---	297.30	**357.00**
	Inst	Ea	Sm	CA	1.78	4.50	295.00	59.20	---	354.20	**428.00**
6' wide	Inst	Ea	Lg	CA	1.60	5.00	283.00	53.20	---	336.20	**405.00**
	Inst	Ea	Sm	CA	2.13	3.75	330.00	70.90	---	400.90	**486.00**
8' wide	Inst	Ea	Lg	2C	3.20	5.00	315.00	106.00	---	421.00	**522.00**
	Inst	Ea	Sm	2C	4.27	3.75	368.00	142.00	---	510.00	**637.00**
10' wide	Inst	Ea	Lg	2C	4.00	4.00	359.00	133.00	---	492.00	**612.00**
	Inst	Ea	Sm	2C	5.33	3.00	419.00	177.00	---	596.00	**748.00**
Minimum Job Charge											
	Inst	Ea	Lg	CA	2.00	4.00	---	66.50	---	66.50	**99.80**
	Inst	Ea	Sm	CA	2.67	3.00	---	88.80	---	88.80	**133.00**

Description	Oper	Unit	Vol	Crew Size	Man-hours per Unit	Crew Output per Day	Avg Mat'l Unit Cost	Avg Labor Unit Cost	Avg Equip Unit Cost	Avg Total Unit Cost	Avg Price Incl O&P

Carpet

Detach & reset operations

Detach and relay

Existing carpet only	Reset	SY	Lg	FA	.095	84.00	---	2.91	.25	3.16	4.52
	Reset	SY	Sm	FA	.159	50.40	.20	4.87	.42	5.49	7.80
Existing carpet w/ new pad	Reset	SY	Lg	FA	.111	72.00	2.43	3.40	---	5.83	7.79
	Reset	SY	Sm	FA	.185	43.20	3.04	5.66	---	8.70	11.80

Remove operations

Carpet only, tacked	Demo	SY	Lg	FA	.035	228.0	---	1.07	---	1.07	1.57
	Demo	SY	Sm	FA	.058	136.8	---	1.77	---	1.77	2.61
Pad, minimal glue	Demo	SY	Lg	FA	.018	456.0	---	.55	---	.55	.81
	Demo	SY	Sm	FA	.029	273.6	---	.89	---	.89	1.30
Scrape up backing residue	Demo	SY	Lg	FA	.083	96.00	---	2.54	---	2.54	3.73
	Demo	SY	Sm	FA	.139	57.60	---	4.25	---	4.25	6.25

Replace operations

Price includes consultation, measurement, pad separate, carpet separate, installation using tack strips and hot melt tape on seams

Standard quality

Poly, thin pile density	Inst	SY	Lg	FA	.083	96.00	11.00	2.54	.25	13.79	16.70
	Inst	SY	Sm	FA	.139	57.60	13.80	4.25	.42	18.47	22.50
Loop pile, 20 oz.	Inst	SY	Lg	FA	.083	96.00	7.36	2.54	.25	10.15	12.50
	Inst	SY	Sm	FA	.139	57.60	9.20	4.25	.42	13.87	17.30
Pad	Inst	SY	Lg	FA	.018	456.0	1.98	.55	---	2.53	3.09
	Inst	SY	Sm	FA	.029	273.6	2.47	.89	---	3.36	4.14

Average quality

Poly, medium pile density	Inst	SY	Lg	FA	.083	96.00	14.70	2.54	.25	17.49	20.90
	Inst	SY	Sm	FA	.139	57.60	18.40	4.25	.42	23.07	27.80
Loop pile, 26 oz.	Inst	SY	Lg	FA	.083	96.00	10.10	2.54	.25	12.89	15.60
	Inst	SY	Sm	FA	.139	57.60	12.70	4.25	.42	17.37	21.20
Cut pile, 36 oz.	Inst	SY	Lg	FA	.083	96.00	12.00	2.54	.25	14.79	17.70
	Inst	SY	Sm	FA	.139	57.60	15.00	4.25	.42	19.67	23.90
Wool	Inst	SY	Lg	FA	.083	96.00	18.40	2.54	.25	21.19	25.10
	Inst	SY	Sm	FA	.139	57.60	23.00	4.25	.42	27.67	33.10
Pad	Inst	SY	Lg	FA	.018	456.0	2.43	.55	---	2.98	3.60
	Inst	SY	Sm	FA	.029	273.6	3.04	.89	---	3.93	4.80

Description	Oper	Unit	Vol	Crew Size	Man-hours per Unit	Crew Output per Day	Avg Mat'l Unit Cost	Avg Labor Unit Cost	Avg Equip Unit Cost	Avg Total Unit Cost	Avg Price Incl O&P
High quality											
Nylon, medium pile density	Inst	SY	Lg	FA	.083	96.00	22.10	2.54	.25	24.89	**29.40**
	Inst	SY	Sm	FA	.139	57.60	27.60	4.25	.42	32.27	**38.40**
Loop pile, 36 oz.	Inst	SY	Lg	FA	.083	96.00	12.00	2.54	.25	14.79	**17.70**
	Inst	SY	Sm	FA	.139	57.60	15.00	4.25	.42	19.67	**23.90**
Cut pile, 46 oz.	Inst	SY	Lg	FA	.083	96.00	16.60	2.54	.25	19.39	**23.00**
	Inst	SY	Sm	FA	.139	57.60	20.70	4.25	.42	25.37	**30.50**
Wool	Inst	SY	Lg	FA	.083	96.00	36.80	2.54	.25	39.59	**46.30**
	Inst	SY	Sm	FA	.139	57.60	46.00	4.25	.42	50.67	**59.60**
Pad	Inst	SY	Lg	FA	.018	456.0	5.11	.55	---	5.66	**6.69**
	Inst	SY	Sm	FA	.029	273.6	6.39	.89	---	7.28	**8.65**
Premium quality											
Nylon, thick pile density	Inst	SY	Lg	FA	.083	96.00	24.80	2.54	.25	27.59	**32.60**
	Inst	SY	Sm	FA	.139	57.60	31.10	4.25	.42	35.77	**42.40**
Cut pile, 54 oz.	Inst	SY	Lg	FA	.083	96.00	23.00	2.54	.25	25.79	**30.40**
	Inst	SY	Sm	FA	.139	57.60	28.80	4.25	.42	33.47	**39.70**
Wool	Inst	SY	Lg	FA	.083	96.00	73.60	2.54	.25	76.39	**88.60**
	Inst	SY	Sm	FA	.139	57.60	92.00	4.25	.42	96.67	**112.00**
Decorator, floral or design	Inst	SY	Lg	FA	.160	50.00	87.40	4.90	.25	92.55	**108.00**
	Inst	SY	Sm	FA	.267	30.00	109.00	8.17	.42	117.59	**138.00**
Pad	Inst	SY	Lg	FA	.018	456.0	6.51	.55	---	7.06	**8.30**
	Inst	SY	Sm	FA	.029	273.6	8.14	.89	---	9.03	**10.70**
Steps											
Waterfall (wrap over nose)	Inst	Ea	Lg	FA	.100	80.00	7.36	3.06	---	10.42	**13.00**
	Inst	Ea	Sm	FA	.167	48.00	9.20	5.11	---	14.31	**18.10**
Tucked (under tread nose)	Inst	Ea	Lg	FA	.160	50.00	7.36	4.90	---	12.26	**15.70**
	Inst	Ea	Sm	FA	.267	30.00	9.20	8.17	---	17.37	**22.60**
Cove or wall wrap											
4" high	Inst	LF	Lg	FA	.062	130.0	1.64	1.90	---	3.54	**4.67**
	Inst	LF	Sm	FA	.103	78.00	2.05	3.15	---	5.20	**6.99**
6" high	Inst	LF	Lg	FA	.067	120.0	2.46	2.05	---	4.51	**5.84**
	Inst	LF	Sm	FA	.111	72.00	3.07	3.40	---	6.47	**8.52**
8" high	Inst	LF	Lg	FA	.073	110.0	3.68	2.23	---	5.91	**7.52**
	Inst	LF	Sm	FA	.121	66.00	4.60	3.70	---	8.30	**10.70**

Description	Oper	Unit	Vol	Crew Size	Man-hours per Unit	Crew Output per Day	Avg Mat'l Unit Cost	Avg Labor Unit Cost	Avg Equip Unit Cost	Avg Total Unit Cost	Avg Price Inc O&P

Caulking

Material costs are typical costs for the listed bead diameters. Figures in parentheses, following bead diameter, indicate approximate coverage including 5% waste. Labor costs are per LF of bead length and assume good quality application on smooth to slightly irregular surfaces

Multi-purpose caulk, good quality

1/8" (11.6 LF/fluid oz.)	Inst	LF	Lg	CA	.018	445.0	.01	.60	---	.61	.91
	Inst	LF	Sm	CA	.024	333.8	.01	.80	---	.81	1.21
1/4" (2.91 LF/fluid oz.)	Inst	LF	Lg	CA	.025	320.0	.05	.83	---	.88	1.31
	Inst	LF	Sm	CA	.033	240.0	.06	1.10	---	1.16	1.72
3/8" (1.29 LF/fluid oz.)	Inst	LF	Lg	CA	.030	265.0	.11	1.00	---	1.11	1.62
	Inst	LF	Sm	CA	.040	198.8	.14	1.33	---	1.47	2.16
1/2" (.729 LF/fluid oz.)	Inst	LF	Lg	CA	.033	240.0	.20	1.10	---	1.30	1.88
	Inst	LF	Sm	CA	.044	180.0	.25	1.46	---	1.71	2.48

Butyl flex caulk, premium quality

1/8" (11.6 LF/fluid oz.)	Inst	LF	Lg	CA	.018	445.0	.02	.60	---	.62	.92
	Inst	LF	Sm	CA	.024	333.8	.02	.80	---	.82	1.22
1/4" (2.91 LF/fluid oz.)	Inst	LF	Lg	CA	.025	320.0	.08	.83	---	.91	1.34
	Inst	LF	Sm	CA	.033	240.0	.09	1.10	---	1.19	1.75
3/8" (1.29 LF/fluid oz.)	Inst	LF	Lg	CA	.030	265.0	.18	1.00	---	1.18	1.70
	Inst	LF	Sm	CA	.040	198.8	.22	1.33	---	1.55	2.25
1/2" (.729 LF/fluid oz.)	Inst	LF	Lg	CA	.033	240.0	.32	1.10	---	1.42	2.02
	Inst	LF	Sm	CA	.044	180.0	.40	1.46	---	1.86	2.66

Latex, premium quality

1/8" (11.6 LF/fluid oz.)	Inst	LF	Lg	CA	.018	445.0	.03	.60	---	.63	.93
	Inst	LF	Sm	CA	.024	333.8	.03	.80	---	.83	1.23
1/4" (2.91 LF/fluid oz.)	Inst	LF	Lg	CA	.025	320.0	.11	.83	---	.94	1.37
	Inst	LF	Sm	CA	.033	240.0	.14	1.10	---	1.24	1.81
3/8" (1.29 LF/fluid oz.)	Inst	LF	Lg	CA	.030	265.0	.24	1.00	---	1.24	1.77
	Inst	LF	Sm	CA	.040	198.8	.30	1.33	---	1.63	2.34
1/2" (.729 LF/fluid oz.)	Inst	LF	Lg	CA	.033	240.0	.44	1.10	---	1.54	2.15
	Inst	LF	Sm	CA	.044	180.0	.54	1.46	---	2.00	2.82

Latex caulk, good quality

1/8" (11.6 LF/fluid oz.)	Inst	LF	Lg	CA	.018	445.0	.02	.60	---	.62	.92
	Inst	LF	Sm	CA	.024	333.8	.02	.80	---	.82	1.22
1/4" (2.91 LF/fluid oz.)	Inst	LF	Lg	CA	.025	320.0	.08	.83	---	.91	1.34
	Inst	LF	Sm	CA	.033	240.0	.09	1.10	---	1.19	1.75
3/8" (1.29 LF/fluid oz.)	Inst	LF	Lg	CA	.030	265.0	.18	1.00	---	1.18	1.70
	Inst	LF	Sm	CA	.040	198.8	.22	1.33	---	1.55	2.25
1/2" (.729 LF/fluid oz.)	Inst	LF	Lg	CA	.033	240.0	.32	1.10	---	1.42	2.02
	Inst	LF	Sm	CA	.044	180.0	.40	1.46	---	1.86	2.66

Description	Oper	Unit	Vol	Crew Size	Man-hours per Unit	Crew Output per Day	Avg Mat'l Unit Cost	Avg Labor Unit Cost	Avg Equip Unit Cost	Avg Total Unit Cost	Avg Price Incl O&P
Oil base caulk, good quality											
1/8" (11.6 LF/fluid oz.)	Inst	LF	Lg	CA	.018	445.0	.03	.60	---	.63	**.93**
	Inst	LF	Sm	CA	.024	333.8	.03	.80	---	.83	**1.23**
1/4" (2.91 LF/fluid oz.)	Inst	LF	Lg	CA	.025	320.0	.10	.83	---	.93	**1.36**
	Inst	LF	Sm	CA	.033	240.0	.12	1.10	---	1.22	**1.79**
3/8" (1.29 LF/fluid oz.)	Inst	LF	Lg	CA	.030	265.0	.23	1.00	---	1.23	**1.76**
	Inst	LF	Sm	CA	.040	198.8	.28	1.33	---	1.61	**2.32**
1/2" (.729 LF/fluid oz.)	Inst	LF	Lg	CA	.033	240.0	.39	1.10	---	1.49	**2.10**
	Inst	LF	Sm	CA	.044	180.0	.49	1.46	---	1.95	**2.76**
Oil base caulk, economy quality											
1/8" (11.6 LF/fluid oz.)	Inst	LF	Lg	CA	.018	445.0	.01	.60	---	.61	**.91**
	Inst	LF	Sm	CA	.024	333.8	.01	.80	---	.81	**1.21**
1/4" (2.91 LF/fluid oz.)	Inst	LF	Lg	CA	.025	320.0	.05	.83	---	.88	**1.31**
	Inst	LF	Sm	CA	.033	240.0	.06	1.10	---	1.16	**1.72**
3/8" (1.29 LF/fluid oz.)	Inst	LF	Lg	CA	.030	265.0	.11	1.00	---	1.11	**1.62**
	Inst	LF	Sm	CA	.040	198.8	.14	1.33	---	1.47	**2.16**
1/2" (.729 LF/fluid oz.)	Inst	LF	Lg	CA	.033	240.0	.20	1.10	---	1.30	**1.88**
	Inst	LF	Sm	CA	.044	180.0	.25	1.46	---	1.71	**2.48**
Silicone caulk, good quality											
1/8" (11.6 LF/fluid oz.)	Inst	LF	Lg	CA	.018	445.0	.03	.60	---	.63	**.93**
	Inst	LF	Sm	CA	.024	333.8	.04	.80	---	.84	**1.24**
1/4" (2.91 LF/fluid oz.)	Inst	LF	Lg	CA	.025	320.0	.12	.83	---	.95	**1.39**
	Inst	LF	Sm	CA	.033	240.0	.15	1.10	---	1.25	**1.82**
3/8" (1.29 LF/fluid oz.)	Inst	LF	Lg	CA	.030	265.0	.27	1.00	---	1.27	**1.81**
	Inst	LF	Sm	CA	.040	198.8	.33	1.33	---	1.66	**2.38**
1/2" (.729 LF/fluid oz.)	Inst	LF	Lg	CA	.033	240.0	.48	1.10	---	1.58	**2.20**
	Inst	LF	Sm	CA	.044	180.0	.59	1.46	---	2.05	**2.87**
Silicone caulk, premium quality											
1/8" (11.6 LF/fluid oz.)	Inst	LF	Lg	CA	.018	445.0	.03	.60	---	.63	**.93**
	Inst	LF	Sm	CA	.024	333.8	.04	.80	---	.84	**1.24**
1/4" (2.91 LF/fluid oz.)	Inst	LF	Lg	CA	.025	320.0	.14	.83	---	.97	**1.41**
	Inst	LF	Sm	CA	.033	240.0	.18	1.10	---	1.28	**1.85**
3/8" (1.29 LF/fluid oz.)	Inst	LF	Lg	CA	.030	265.0	.33	1.00	---	1.33	**1.88**
	Inst	LF	Sm	CA	.040	198.8	.41	1.33	---	1.74	**2.47**
1/2" (.729 LF/fluid oz.)	Inst	LF	Lg	CA	.033	240.0	.58	1.10	---	1.68	**2.31**
	Inst	LF	Sm	CA	.044	180.0	.72	1.46	---	2.18	**3.02**
Tub caulk, white siliconized											
1/8" (11.6 LF/fluid oz.)	Inst	LF	Lg	CA	.018	445.0	.01	.60	---	.61	**.91**
	Inst	LF	Sm	CA	.024	333.8	.01	.80	---	.81	**1.21**
1/4" (2.91 LF/fluid oz.)	Inst	LF	Lg	CA	.025	320.0	.05	.83	---	.88	**1.31**
	Inst	LF	Sm	CA	.033	240.0	.06	1.10	---	1.16	**1.72**
3/8" (1.29 LF/fluid oz.)	Inst	LF	Lg	CA	.030	265.0	.11	1.00	---	1.11	**1.62**
	Inst	LF	Sm	CA	.040	198.8	.14	1.33	---	1.47	**2.16**
1/2" (.729 LF/fluid oz.)	Inst	LF	Lg	CA	.033	240.0	.20	1.10	---	1.30	**1.88**
	Inst	LF	Sm	CA	.044	180.0	.25	1.46	---	1.71	**2.48**

Description	Oper	Unit	Vol	Crew Size	Man-hours per Unit	Crew Output per Day	Avg Mat'l Unit Cost	Avg Labor Unit Cost	Avg Equip Unit Cost	Avg Total Unit Cost	Avg Price Incl O&P
Anti-algae and mildew-resistant tub caulk, premium quality white or clear silicone											
1/8" (11.6 LF/fluid oz.)	Inst	LF	Lg	CA	.018	445.0	.03	.60	---	.63	.93
	Inst	LF	Sm	CA	.024	333.8	.04	.80	---	.84	1.24
1/4" (2.91 LF/fluid oz.)	Inst	LF	Lg	CA	.025	320.0	.15	.83	---	.98	1.42
	Inst	LF	Sm	CA	.033	240.0	.19	1.10	---	1.29	1.87
3/8" (1.29 LF/fluid oz.)	Inst	LF	Lg	CA	.030	265.0	.34	1.00	---	1.34	1.89
	Inst	LF	Sm	CA	.040	198.8	.42	1.33	---	1.75	2.48
1/2" (.729 LF/fluid oz.)	Inst	LF	Lg	CA	.033	240.0	.60	1.10	---	1.70	2.34
	Inst	LF	Sm	CA	.044	180.0	.74	1.46	---	2.20	3.05
Elastomeric caulk, premium quality											
1/8" (11.6 LF/fluid oz.)	Inst	LF	Lg	CA	.018	445.0	.03	.60	---	.63	.93
	Inst	LF	Sm	CA	.024	333.8	.04	.80	---	.84	1.24
1/4" (2.91 LF/fluid oz.)	Inst	LF	Lg	CA	.025	320.0	.14	.83	---	.97	1.41
	Inst	LF	Sm	CA	.033	240.0	.18	1.10	---	1.28	1.85
3/8" (1.29 LF/fluid oz.)	Inst	LF	Lg	CA	.030	265.0	.31	1.00	---	1.31	1.85
	Inst	LF	Sm	CA	.040	198.8	.38	1.33	---	1.71	2.43
1/2" (.729 LF/fluid oz.)	Inst	LF	Lg	CA	.033	240.0	.55	1.10	---	1.65	2.28
	Inst	LF	Sm	CA	.044	180.0	.69	1.46	---	2.15	2.99
Add for irregular surfaces such as vertical masonry or lap siding											
	Inst	%	Lg	CA	---	---	5.0	---	---	---	---
	Inst	%	Sm	CA	---	---	5.0	---	---	---	---
Caulking gun, heavy duty, professional type											
	Inst	Ea	Lg	CA	---	---	25.20	---	---	25.20	29.00
	Inst	Ea	Sm	CA	---	---	31.20	---	---	31.20	35.90
Caulking gun, economy grade											
11-oz. cartridge	Inst	Ea	Lg	CA	---	---	4.77	---	---	4.77	5.49
	Inst	Ea	Sm	CA	---	---	5.91	---	---	5.91	6.80
29-oz. cartridge	Inst	Ea	Lg	CA	---	---	16.70	---	---	16.70	19.20
	Inst	Ea	Sm	CA	---	---	20.70	---	---	20.70	23.80

Ceramic tile

1. **Dimensions**. There are many sizes of ceramic tile. Only $4^1/_4$" x $4^1/_4$" and 1" x 1" will be discussed here.

   a. $4^1/_4$" x $4^1/_4$" tile is furnished both unmounted and back-mounted. Back-mounted tile are usually furnished in sheets of 12 tile.

   b. 1" x 1" mosaic tile is furnished face-mounted and back-mounted in sheets; normally, 2'-0" x 1'-0".

2. **Installation.** There are three methods:

   a. Conventional, which uses portland cement, sand and wet tile grout.

   b. Dry-set, which uses dry-set mix and dry tile grout mix.

   c. Organic adhesive, which uses adhesive and dry tile grout mix.

   The conventional method is the most expensive and is used less frequently than the other methods.

3. **Estimating Technique.** For tile, determine the area and add 5% to 10% for waste. For cove, base or trim, determine the length in linear feet and add 5% to 10% for waste.

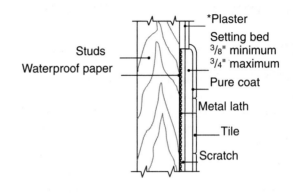

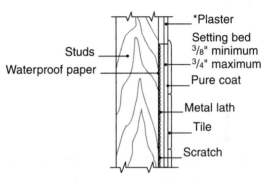

*This installation would be similar if wall finish above tile wainscot were of other material such as wallboard, plywood, etc.

Wood or steel construction with plaster above tile wainscot

Courtesy: *Ceramic Tile Institute of America* 700 N. Virgil Ave., Ste 300, Los Angeles, CA 90029

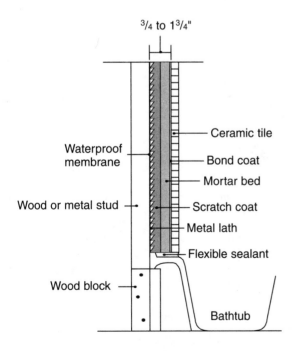

Cross section of bathtub wall using cement mortar

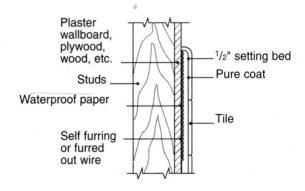

Wood or steel construction with solid covered backing

Description	Oper	Unit	Vol	Crew Size	Man-hours per Unit	Crew Output per Day	Avg Mat'l Unit Cost	Avg Labor Unit Cost	Avg Equip Unit Cost	Avg Total Unit Cost	Avg Price Incl O&P

Ceramic tile

Countertop/backsplash

Adhesive set with backmounted tile

Description	Oper	Unit	Vol	Crew Size	Man-hours per Unit	Crew Output per Day	Avg Mat'l Unit Cost	Avg Labor Unit Cost	Avg Equip Unit Cost	Avg Total Unit Cost	Avg Price Incl O&P
1" x 1"	Inst	SF	Lg	TB	.229	70.00	6.41	6.45	---	12.86	**16.90**
	Inst	SF	Sm	TB	.381	42.00	7.24	10.70	---	17.94	**24.10**
4-1/4" x 4-1/4"	Inst	SF	Lg	TB	.200	80.00	5.12	5.63	---	10.75	**14.20**
	Inst	SF	Sm	TB	.333	48.00	5.79	9.37	---	15.16	**20.40**

Cove/base

Adhesive set with unmounted tile

Description	Oper	Unit	Vol	Crew Size	Man-hours per Unit	Crew Output per Day	Avg Mat'l Unit Cost	Avg Labor Unit Cost	Avg Equip Unit Cost	Avg Total Unit Cost	Avg Price Incl O&P
4-1/4" x 4-1/4"	Inst	LF	Lg	TB	.200	80.00	4.92	5.63	---	10.55	**13.90**
	Inst	LF	Sm	TB	.333	48.00	5.56	9.37	---	14.93	**20.20**
6" x 4-1/4"	Inst	LF	Lg	TB	.168	95.00	4.81	4.73	---	9.54	**12.50**
	Inst	LF	Sm	TB	.281	57.00	5.43	7.91	---	13.34	**17.90**

Conventional mortar set with unmounted tile

Description	Oper	Unit	Vol	Crew Size	Man-hours per Unit	Crew Output per Day	Avg Mat'l Unit Cost	Avg Labor Unit Cost	Avg Equip Unit Cost	Avg Total Unit Cost	Avg Price Incl O&P
4-1/4" x 4-1/4"	Inst	LF	Lg	TB	.356	45.00	4.82	10.00	---	14.82	**20.30**
	Inst	LF	Sm	TB	.593	27.00	5.45	16.70	---	22.15	**30.80**
6" x 4-1/4"	Inst	LF	Lg	TB	.320	50.00	4.66	9.01	---	13.67	**18.60**
	Inst	LF	Sm	TB	.533	30.00	5.27	15.00	---	20.27	**28.10**

Dry-set mortar with unmounted tile

Description	Oper	Unit	Vol	Crew Size	Man-hours per Unit	Crew Output per Day	Avg Mat'l Unit Cost	Avg Labor Unit Cost	Avg Equip Unit Cost	Avg Total Unit Cost	Avg Price Incl O&P
4-1/4" x 4-1/4"	Inst	LF	Lg	TB	.246	65.00	4.84	6.92	---	11.76	**15.80**
	Inst	LF	Sm	TB	.410	39.00	5.47	11.50	---	16.97	**23.30**
6" x 4-1/4"	Inst	LF	Lg	TB	.213	75.00	4.69	6.00	---	10.69	**14.20**
	Inst	LF	Sm	TB	.356	45.00	5.30	10.00	---	15.30	**20.80**

Floors

1-1/2" x 1-1/2"

Description	Oper	Unit	Vol	Crew Size	Man-hours per Unit	Crew Output per Day	Avg Mat'l Unit Cost	Avg Labor Unit Cost	Avg Equip Unit Cost	Avg Total Unit Cost	Avg Price Incl O&P
Adhesive or dry-set base	Demo	SF	Lg	LB	.029	550.0	---	.80	---	.80	**1.19**
	Demo	SF	Sm	LB	.048	330.0	---	1.32	---	1.32	**1.96**
Conventional mortar base	Demo	SF	Lg	LB	.034	475.0	---	.93	---	.93	**1.39**
	Demo	SF	Sm	LB	.056	285.0	---	1.54	---	1.54	**2.29**

Adhesive set with backmounted tile

Description	Oper	Unit	Vol	Crew Size	Man-hours per Unit	Crew Output per Day	Avg Mat'l Unit Cost	Avg Labor Unit Cost	Avg Equip Unit Cost	Avg Total Unit Cost	Avg Price Incl O&P
1-1/2" x 1-1/2"	Inst	SF	Lg	TB	.133	120.0	6.41	3.74	---	10.15	**12.90**
	Inst	SF	Sm	TB	.222	72.00	7.24	6.25	---	13.49	**17.50**
4-1/4" x 4-1/4"	Inst	SF	Lg	TB	.123	130.0	5.12	3.46	---	8.58	**11.00**
	Inst	SF	Sm	TB	.205	78.00	5.79	5.77	---	11.56	**15.10**

Conventional mortar set with backmounted tile

Description	Oper	Unit	Vol	Crew Size	Man-hours per Unit	Crew Output per Day	Avg Mat'l Unit Cost	Avg Labor Unit Cost	Avg Equip Unit Cost	Avg Total Unit Cost	Avg Price Incl O&P
1-1/2" x 1-1/2"	Inst	SF	Lg	TB	.267	60.00	6.11	7.52	---	13.63	**18.10**
	Inst	SF	Sm	TB	.444	36.00	6.90	12.50	---	19.40	**26.30**
4-1/4" x 4-1/4"	Inst	SF	Lg	TB	.229	70.00	4.83	6.45	---	11.28	**15.00**
	Inst	SF	Sm	TB	.381	42.00	5.46	10.70	---	16.16	**22.10**

Dry-set mortar with backmounted tile

Description	Oper	Unit	Vol	Crew Size	Man-hours per Unit	Crew Output per Day	Avg Mat'l Unit Cost	Avg Labor Unit Cost	Avg Equip Unit Cost	Avg Total Unit Cost	Avg Price Incl O&P
1-1/2" x 1-1/2"	Inst	SF	Lg	TB	.168	95.00	6.18	4.73	---	10.91	**14.10**
	Inst	SF	Sm	TB	.281	57.00	6.98	7.91	---	14.89	**19.70**
4-1/4" x 4-1/4"	Inst	SF	Lg	TB	.152	105.0	4.89	4.28	---	9.17	**11.90**
	Inst	SF	Sm	TB	.254	63.00	5.53	7.15	---	12.68	**16.90**

Description	Oper	Unit	Vol	Crew Size	Man-hours per Unit	Crew Output per Day	Avg Mat'l Unit Cost	Avg Labor Unit Cost	Avg Equip Unit Cost	Avg Total Unit Cost	Avg Price Incl O&P
Wainscot cap											
Adhesive set with unmounted tile											
2" x 6"	Inst	LF	Lg	TB	.133	120.0	2.51	3.74	---	6.25	**8.39**
	Inst	LF	Sm	TB	.222	72.00	2.83	6.25	---	9.08	**12.40**
Conventional mortar set with unmounted tile											
2" x 6"	Inst	LF	Lg	TB	.213	75.00	2.46	6.00	---	8.46	**11.70**
	Inst	LF	Sm	TB	.356	45.00	2.78	10.00	---	12.78	**17.90**
Dry-set mortar with unmounted tile											
2" x 6"	Inst	LF	Lg	TB	.160	100.0	2.46	4.50	---	6.96	**9.45**
	Inst	LF	Sm	TB	.267	60.00	2.78	7.52	---	10.30	**14.30**
Walls, 1-1/2" x 1-1/2" or 4-1/4" x 4-1/4"											
Adhesive or dry-set base											
	Demo	SF	Lg	LB	.033	480.0	---	.91	---	.91	**1.35**
	Demo	SF	Sm	LB	.056	288.0	---	1.54	---	1.54	**2.29**
Conventional mortar base											
	Demo	SF	Lg	LB	.040	400.0	---	1.10	---	1.10	**1.63**
	Demo	SF	Sm	LB	.067	240.0	---	1.84	---	1.84	**2.74**
Adhesive set with backmounted tile											
1-1/2" x 1-1/2"	Inst	SF	Lg	TB	.160	100.0	6.41	4.50	---	10.91	**14.00**
	Inst	SF	Sm	TB	.267	60.00	7.24	7.52	---	14.76	**19.40**
4-1/4" x 4-1/4"	Inst	SF	Lg	TB	.145	110.0	5.12	4.08	---	9.20	**11.90**
	Inst	SF	Sm	TB	.242	66.00	5.79	6.81	---	12.60	**16.70**
Conventional mortar set with backmounted tile											
1-1/2" x 1-1/2"	Inst	SF	Lg	TB	.320	50.00	6.11	9.01	---	15.12	**20.30**
	Inst	SF	Sm	TB	.533	30.00	6.90	15.00	---	21.90	**30.00**
4-1/4" x 4-1/4"	Inst	SF	Lg	TB	.267	60.00	4.83	7.52	---	12.35	**16.60**
	Inst	SF	Sm	TB	.444	36.00	5.46	12.50	---	17.96	**24.70**
Dry-set mortar with backmounted tile											
1-1/2" x 1-1/2"	Inst	SF	Lg	TB	.200	80.00	6.18	5.63	---	11.81	**15.40**
	Inst	SF	Sm	TB	.333	48.00	6.98	9.37	---	16.35	**21.80**
4-1/4" x 4-1/4"	Inst	SF	Lg	TB	.178	90.00	4.89	5.01	---	9.90	**13.00**
	Inst	SF	Sm	TB	.296	54.00	5.53	8.33	---	13.86	**18.60**

Description	Oper	Unit	Vol	Crew Size	Man-hours per Unit	Crew Output per Day	Avg Mat'l Unit Cost	Avg Labor Unit Cost	Avg Equip Unit Cost	Avg Total Unit Cost	Avg Price Incl O&P

Closet door systems
Labor costs include hanging and fitting of doors and hardware

Bi-folding units
Includes hardware and pine fascia trim

Unfinished
Birch, flush face, 1-3/8" T, hollow core

Description	Oper	Unit	Vol	Crew	MH	Output	Mat'l	Labor	Equip	Total	Price
2'-0" x 6'-8", 2 doors	Inst	Set	Lg	2C	1.60	10.00	83.80	53.20	---	137.00	176.00
	Inst	Set	Sm	2C	2.46	6.50	98.10	81.80	---	179.90	236.00
2'-6" x 6'-8", 2 doors	Inst	Set	Lg	2C	1.60	10.00	101.00	53.20	---	154.20	196.00
	Inst	Set	Sm	2C	2.46	6.50	118.00	81.80	---	199.80	259.00
3'-0" x 6'-8", 2 doors	Inst	Set	Lg	2C	1.60	10.00	101.00	53.20	---	154.20	196.00
	Inst	Set	Sm	2C	2.46	6.50	118.00	81.80	---	199.80	259.00
4'-0" x 6'-8", 4 doors	Inst	Set	Lg	2C	2.00	8.00	144.00	66.50	---	210.50	265.00
	Inst	Set	Sm	2C	3.08	5.20	168.00	102.00	---	270.00	347.00
6'-0" x 6'-8", 4 doors	Inst	Set	Lg	2C	2.00	8.00	188.00	66.50	---	254.50	316.00
	Inst	Set	Sm	2C	3.08	5.20	220.00	102.00	---	322.00	407.00
8'-0" x 6'-8", 4 doors	Inst	Set	Lg	2C	2.00	8.00	258.00	66.50	---	324.50	397.00
	Inst	Set	Sm	2C	3.08	5.20	302.00	102.00	---	404.00	501.00
4'-0" x 8'-0", 4 doors	Inst	Set	Lg	2C	2.00	8.00	242.00	66.50	---	308.50	378.00
	Inst	Set	Sm	2C	3.08	5.20	283.00	102.00	---	385.00	479.00
6'-0" x 8'-0", 4 doors	Inst	Set	Lg	2C	2.00	8.00	306.00	66.50	---	372.50	452.00
	Inst	Set	Sm	2C	3.08	5.20	358.00	102.00	---	460.00	565.00
8'-0" x 8'-0", 4 doors	Inst	Set	Lg	2C	2.00	8.00	385.00	66.50	---	451.50	543.00
	Inst	Set	Sm	2C	3.08	5.20	450.00	102.00	---	552.00	672.00

Lauan, flush face, 1-3/8" T, hollow core

Description	Oper	Unit	Vol	Crew	MH	Output	Mat'l	Labor	Equip	Total	Price
2'-0" x 6'-8", 2 doors	Inst	Set	Lg	2C	1.60	10.00	64.40	53.20	---	117.60	154.00
	Inst	Set	Sm	2C	2.46	6.50	75.40	81.80	---	157.20	209.00
2'-6" x 6'-8", 2 doors	Inst	Set	Lg	2C	1.60	10.00	68.80	53.20	---	122.00	159.00
	Inst	Set	Sm	2C	2.46	6.50	80.50	81.80	---	162.30	215.00
3'-0" x 6'-8", 2 doors	Inst	Set	Lg	2C	1.60	10.00	75.70	53.20	---	128.90	167.00
	Inst	Set	Sm	2C	2.46	6.50	88.60	81.80	---	170.40	225.00
4'-0" x 6'-8", 4 doors	Inst	Set	Lg	2C	2.00	8.00	114.00	66.50	---	180.50	231.00
	Inst	Set	Sm	2C	3.08	5.20	133.00	102.00	---	235.00	307.00
6'-0" x 6'-8", 4 doors	Inst	Set	Lg	2C	2.00	8.00	138.00	66.50	---	204.50	259.00
	Inst	Set	Sm	2C	3.08	5.20	161.00	102.00	---	263.00	339.00
8'-0" x 6'-8", 4 doors	Inst	Set	Lg	2C	2.00	8.00	186.00	66.50	---	252.50	314.00
	Inst	Set	Sm	2C	3.08	5.20	218.00	102.00	---	320.00	404.00
4'-0" x 8'-0", 4 doors	Inst	Set	Lg	2C	2.00	8.00	180.00	66.50	---	246.50	307.00
	Inst	Set	Sm	2C	3.08	5.20	211.00	102.00	---	313.00	396.00
6'-0" x 8'-0", 4 doors	Inst	Set	Lg	2C	2.00	8.00	226.00	66.50	---	292.50	360.00
	Inst	Set	Sm	2C	3.08	5.20	264.00	102.00	---	366.00	458.00
8'-0" x 8'-0", 4 doors	Inst	Set	Lg	2C	2.00	8.00	283.00	66.50	---	349.50	425.00
	Inst	Set	Sm	2C	3.08	5.20	331.00	102.00	---	433.00	535.00

Description	Oper	Unit	Vol	Crew Size	Man-hours per Unit	Crew Output per Day	Avg Mat'l Unit Cost	Avg Labor Unit Cost	Avg Equip Unit Cost	Avg Total Unit Cost	Avg Price Incl O&P
Ponderosa pine, colonial raised panel (2/door), 1-3/8" T, solid core											
2'-0" x 6'-8", 2 doors	Inst	Set	Lg	2C	1.60	10.00	88.00	53.20	---	141.20	**181.00**
	Inst	Set	Sm	2C	2.46	6.50	103.00	81.80	---	184.80	**241.00**
4'-0" x 6'-8", 4 doors	Inst	Set	Lg	2C	2.00	8.00	177.00	66.50	---	243.50	**303.00**
	Inst	Set	Sm	2C	3.08	5.20	207.00	102.00	---	309.00	**392.00**
6'-0" x 6'-8", 4 doors	Inst	Set	Lg	2C	2.00	8.00	229.00	66.50	---	295.50	**363.00**
	Inst	Set	Sm	2C	3.08	5.20	268.00	102.00	---	370.00	**462.00**
8'-0" x 6'-8", 4 doors	Inst	Set	Lg	2C	2.00	8.00	285.00	66.50	---	351.50	**428.00**
	Inst	Set	Sm	2C	3.08	5.20	333.00	102.00	---	435.00	**537.00**
Ponderosa pine, raised panel louver, 1-3/8" T, solid core											
2'-0" x 6'-8", 2 doors	Inst	Set	Lg	2C	1.60	10.00	111.00	53.20	---	164.20	**208.00**
	Inst	Set	Sm	2C	2.46	6.50	130.00	81.80	---	211.80	**272.00**
4'-0" x 6'-8", 4 doors	Inst	Set	Lg	2C	2.00	8.00	217.00	66.50	---	283.50	**349.00**
	Inst	Set	Sm	2C	3.08	5.20	254.00	102.00	---	356.00	**446.00**
6'-0" x 6'-8", 4 doors	Inst	Set	Lg	2C	2.00	8.00	286.00	66.50	---	352.50	**429.00**
	Inst	Set	Sm	2C	3.08	5.20	335.00	102.00	---	437.00	**539.00**
Hardboard, flush face, 1-3/8" T, hollow core											
2'-0" x 6'-8", 2 doors	Inst	Set	Lg	2C	1.60	10.00	55.80	53.20	---	109.00	**144.00**
	Inst	Set	Sm	2C	2.46	6.50	65.30	81.80	---	147.10	**198.00**
2'-6" x 6'-8", 2 doors	Inst	Set	Lg	2C	1.60	10.00	59.50	53.20	---	112.70	**148.00**
	Inst	Set	Sm	2C	2.46	6.50	69.60	81.80	---	151.40	**203.00**
3'-0" x 6'-8", 2 doors	Inst	Set	Lg	2C	1.60	10.00	64.30	53.20	---	117.50	**154.00**
	Inst	Set	Sm	2C	2.46	6.50	75.20	81.80	---	157.00	**209.00**
4'-0" x 6'-8", 4 doors	Inst	Set	Lg	2C	2.00	8.00	106.00	66.50	---	172.50	**222.00**
	Inst	Set	Sm	2C	3.08	5.20	124.00	102.00	---	226.00	**296.00**
6'-0" x 6'-8", 4 doors	Inst	Set	Lg	2C	2.00	8.00	125.00	66.50	---	191.50	**244.00**
	Inst	Set	Sm	2C	3.08	5.20	146.00	102.00	---	248.00	**322.00**
8'-0" x 6'-8", 4 doors	Inst	Set	Lg	2C	2.00	8.00	188.00	66.50	---	254.50	**316.00**
	Inst	Set	Sm	2C	3.08	5.20	220.00	102.00	---	322.00	**407.00**
4'-0" x 8'-0", 4 doors	Inst	Set	Lg	2C	2.00	8.00	192.00	66.50	---	258.50	**321.00**
	Inst	Set	Sm	2C	3.08	5.20	225.00	102.00	---	327.00	**412.00**
6'-0" x 8'-0", 4 doors	Inst	Set	Lg	2C	2.00	8.00	231.00	66.50	---	297.50	**365.00**
	Inst	Set	Sm	2C	3.08	5.20	270.00	102.00	---	372.00	**465.00**
8'-0" x 8'-0", 4 doors	Inst	Set	Lg	2C	2.00	8.00	322.00	66.50	---	388.50	**470.00**
	Inst	Set	Sm	2C	3.08	5.20	377.00	102.00	---	479.00	**587.00**
Ash, flush face, 1-3/8" T, hollow core											
2'-0" x 6'-8", 2 doors	Inst	Set	Lg	2C	1.60	10.00	101.00	53.20	---	154.20	**196.00**
	Inst	Set	Sm	2C	2.46	6.50	118.00	81.80	---	199.80	**259.00**
2'-6" x 6'-8", 2 doors	Inst	Set	Lg	2C	1.60	10.00	112.00	53.20	---	165.20	**209.00**
	Inst	Set	Sm	2C	2.46	6.50	131.00	81.80	---	212.80	**273.00**
3'-0" x 6'-8", 2 doors	Inst	Set	Lg	2C	1.60	10.00	122.00	53.20	---	175.20	**220.00**
	Inst	Set	Sm	2C	2.46	6.50	143.00	81.80	---	224.80	**287.00**
4'-0" x 6'-8", 4 doors	Inst	Set	Lg	2C	2.00	8.00	175.00	66.50	---	241.50	**301.00**
	Inst	Set	Sm	2C	3.08	5.20	205.00	102.00	---	307.00	**389.00**

Description	Oper	Unit	Vol	Crew Size	Man-hours per Unit	Crew Output per Day	Avg Mat'l Unit Cost	Avg Labor Unit Cost	Avg Equip Unit Cost	Avg Total Unit Cost	Avg Price Incl O&P
6'-0" x 6'-8", 4 doors	Inst	Set	Lg	2C	2.00	8.00	215.00	66.50	---	281.50	**347.00**
	Inst	Set	Sm	2C	3.08	5.20	252.00	102.00	---	354.00	**443.00**
8'-0" x 6'-8", 4 doors	Inst	Set	Lg	2C	2.00	8.00	295.00	66.50	---	361.50	**439.00**
	Inst	Set	Sm	2C	3.08	5.20	345.00	102.00	---	447.00	**551.00**
4'-0" x 8'-0", 4 doors	Inst	Set	Lg	2C	2.00	8.00	138.00	66.50	---	204.50	**259.00**
	Inst	Set	Sm	2C	3.08	5.20	161.00	102.00	---	263.00	**339.00**
6'-0" x 8'-0", 4 doors	Inst	Set	Lg	2C	2.00	8.00	400.00	66.50	---	466.50	**560.00**
	Inst	Set	Sm	2C	3.08	5.20	468.00	102.00	---	570.00	**692.00**
8'-0" x 8'-0", 4 doors	Inst	Set	Lg	2C	2.00	8.00	570.00	66.50	---	636.50	**755.00**
	Inst	Set	Sm	2C	3.08	5.20	667.00	102.00	---	769.00	**921.00**

Prefinished

Walnut tone, mar-resistant finish
Embossed (distressed wood appearance) lauan, flush face, 1-3/8" T, hollow core

Description	Oper	Unit	Vol	Crew Size	Man-hours per Unit	Crew Output per Day	Avg Mat'l Unit Cost	Avg Labor Unit Cost	Avg Equip Unit Cost	Avg Total Unit Cost	Avg Price Incl O&P
2'-0" x 6'-8", 2 doors	Inst	Set	Lg	2C	1.60	10.00	59.00	53.20	---	112.20	**148.00**
	Inst	Set	Sm	2C	2.46	6.50	69.00	81.80	---	150.80	**202.00**
4'-0" x 6'-8", 4 doors	Inst	Set	Lg	2C	2.00	8.00	118.00	66.50	---	184.50	**236.00**
	Inst	Set	Sm	2C	3.08	5.20	138.00	102.00	---	240.00	**312.00**
6'-0" x 6'-8", 4 doors	Inst	Set	Lg	2C	2.00	8.00	155.00	66.50	---	221.50	**278.00**
	Inst	Set	Sm	2C	3.08	5.20	181.00	102.00	---	283.00	**362.00**

Lauan, flush face, 1-3/8" T, hollow core

Description	Oper	Unit	Vol	Crew Size	Man-hours per Unit	Crew Output per Day	Avg Mat'l Unit Cost	Avg Labor Unit Cost	Avg Equip Unit Cost	Avg Total Unit Cost	Avg Price Incl O&P
2'-0" x 6'-8", 2 doors	Inst	Set	Lg	2C	1.60	10.00	66.00	53.20	---	119.20	**156.00**
	Inst	Set	Sm	2C	2.46	6.50	77.20	81.80	---	159.00	**212.00**
4'-0" x 6'-8", 4 doors	Inst	Set	Lg	2C	2.00	8.00	131.00	66.50	---	197.50	**250.00**
	Inst	Set	Sm	2C	3.08	5.20	153.00	102.00	---	255.00	**330.00**
6'-0" x 6'-8", 4 doors	Inst	Set	Lg	2C	2.00	8.00	176.00	66.50	---	242.50	**302.00**
	Inst	Set	Sm	2C	3.08	5.20	206.00	102.00	---	308.00	**391.00**

Ponderosa pine, full louver, 1-3/8" T, hollow core ---

Description	Oper	Unit	Vol	Crew Size	Man-hours per Unit	Crew Output per Day	Avg Mat'l Unit Cost	Avg Labor Unit Cost	Avg Equip Unit Cost	Avg Total Unit Cost	Avg Price Incl O&P
2'-0" x 6'-8", 2 doors	Inst	Set	Lg	2C	1.60	10.00	137.00	53.20	---	190.20	**237.00**
	Inst	Set	Sm	2C	2.46	6.50	160.00	81.80	---	241.80	**307.00**
4'-0" x 6'-8", 4 doors	Inst	Set	Lg	2C	2.00	8.00	268.00	66.50	---	334.50	**408.00**
	Inst	Set	Sm	2C	3.08	5.20	314.00	102.00	---	416.00	**514.00**
6'-0" x 6'-8", 4 doors	Inst	Set	Lg	2C	2.00	8.00	341.00	66.50	---	407.50	**492.00**
	Inst	Set	Sm	2C	3.08	5.20	399.00	102.00	---	501.00	**613.00**

Ponderosa pine, raised louver, 1-3/8" T, hollow core ---

Description	Oper	Unit	Vol	Crew Size	Man-hours per Unit	Crew Output per Day	Avg Mat'l Unit Cost	Avg Labor Unit Cost	Avg Equip Unit Cost	Avg Total Unit Cost	Avg Price Incl O&P
2'-0" x 6'-8", 2 doors	Inst	Set	Lg	2C	1.60	10.00	145.00	53.20	---	198.20	**247.00**
	Inst	Set	Sm	2C	2.46	6.50	170.00	81.80	---	251.80	**318.00**
4'-0" x 6'-8", 4 doors	Inst	Set	Lg	2C	2.00	8.00	283.00	66.50	---	349.50	**425.00**
	Inst	Set	Sm	2C	3.08	5.20	331.00	102.00	---	433.00	**535.00**
6'-0" x 6'-8", 4 doors	Inst	Set	Lg	2C	2.00	8.00	341.00	66.50	---	407.50	**492.00**
	Inst	Set	Sm	2C	3.08	5.20	399.00	102.00	---	501.00	**613.00**

Closet door systems, sliding

Description	Oper	Unit	Vol	Crew Size	Man-hours per Unit	Crew Output per Day	Avg Mat'l Unit Cost	Avg Labor Unit Cost	Avg Equip Unit Cost	Avg Total Unit Cost	Avg Price Incl O&P

Sliding or bypassing units
Includes hardware, 4-5/8" jambs, header, and fascia
Wood inserts, 1-3/8" T, hollow core
Unfinished birch

Description	Oper	Unit	Vol	Crew Size	MH/Unit	Output/Day	Mat'l	Labor	Equip	Total	O&P
4'-0" x 6'-8", 2 doors	Inst	Set	Lg	2C	2.29	7.00	166.00	76.20	---	242.20	**305.00**
	Inst	Set	Sm	2C	3.48	4.60	194.00	116.00	---	310.00	**397.00**
6'-0" x 6'-8", 2 doors	Inst	Set	Lg	2C	2.29	7.00	204.00	76.20	---	280.20	**349.00**
	Inst	Set	Sm	2C	3.48	4.60	239.00	116.00	---	355.00	**448.00**
8'-0" x 6'-8", 2 doors	Inst	Set	Lg	2C	2.29	7.00	344.00	76.20	---	420.20	**510.00**
	Inst	Set	Sm	2C	3.48	4.60	402.00	116.00	---	518.00	**637.00**
4'-0" x 8'-0", 2 doors	Inst	Set	Lg	2C	2.29	7.00	288.00	76.20	---	364.20	**445.00**
	Inst	Set	Sm	2C	3.48	4.60	337.00	116.00	---	453.00	**561.00**
6'-0" x 8'-0", 2 doors	Inst	Set	Lg	2C	2.29	7.00	334.00	76.20	---	410.20	**498.00**
	Inst	Set	Sm	2C	3.48	4.60	391.00	116.00	---	507.00	**623.00**
8'-0" x 8'-0", 2 doors	Inst	Set	Lg	2C	2.29	7.00	462.00	76.20	---	538.20	**646.00**
	Inst	Set	Sm	2C	3.48	4.60	541.00	116.00	---	657.00	**795.00**
10'-0" x 6'-8", 3 doors	Inst	Set	Lg	2C	3.20	5.00	515.00	106.00	---	621.00	**752.00**
	Inst	Set	Sm	2C	4.85	3.30	603.00	161.00	---	764.00	**935.00**
12'-0" x 6'-8", 3 doors	Inst	Set	Lg	2C	3.20	5.00	555.00	106.00	---	661.00	**798.00**
	Inst	Set	Sm	2C	4.85	3.30	649.00	161.00	---	810.00	**989.00**
10'-0" x 8'-0", 3 doors	Inst	Set	Lg	2C	3.20	5.00	641.00	106.00	---	747.00	**897.00**
	Inst	Set	Sm	2C	4.85	3.30	750.00	161.00	---	911.00	**1100.00**
12'-0" x 8'-0", 3 doors	Inst	Set	Lg	2C	3.20	5.00	697.00	106.00	---	803.00	**961.00**
	Inst	Set	Sm	2C	4.85	3.30	815.00	161.00	---	976.00	**1180.00**

Unfinished hardboard

Description	Oper	Unit	Vol	Crew Size	MH/Unit	Output/Day	Mat'l	Labor	Equip	Total	O&P
4'-0" x 6'-8", 2 doors	Inst	Set	Lg	2C	2.29	7.00	87.00	76.20	---	163.20	**214.00**
	Inst	Set	Sm	2C	3.48	4.60	102.00	116.00	---	218.00	**291.00**
6'-0" x 6'-8", 2 doors	Inst	Set	Lg	2C	2.29	7.00	107.00	76.20	---	183.20	**237.00**
	Inst	Set	Sm	2C	3.48	4.60	125.00	116.00	---	241.00	**318.00**
8'-0" x 6'-8", 2 doors	Inst	Set	Lg	2C	2.29	7.00	196.00	76.20	---	272.20	**340.00**
	Inst	Set	Sm	2C	3.48	4.60	229.00	116.00	---	345.00	**437.00**
4'-0" x 8'-0", 2 doors	Inst	Set	Lg	2C	2.29	7.00	135.00	76.20	---	211.20	**270.00**
	Inst	Set	Sm	2C	3.48	4.60	158.00	116.00	---	274.00	**355.00**
6'-0" x 8'-0", 2 doors	Inst	Set	Lg	2C	2.29	7.00	155.00	76.20	---	231.20	**293.00**
	Inst	Set	Sm	2C	3.48	4.60	181.00	116.00	---	297.00	**382.00**
8'-0" x 8'-0", 2 doors	Inst	Set	Lg	2C	2.29	7.00	200.00	76.20	---	276.20	**344.00**
	Inst	Set	Sm	2C	3.48	4.60	234.00	116.00	---	350.00	**443.00**
10'-0" x 6'-8", 3 doors	Inst	Set	Lg	2C	3.20	5.00	242.00	106.00	---	348.00	**438.00**
	Inst	Set	Sm	2C	4.85	3.30	283.00	161.00	---	444.00	**568.00**
12'-0" x 6'-8", 3 doors	Inst	Set	Lg	2C	3.20	5.00	296.00	106.00	---	402.00	**500.00**
	Inst	Set	Sm	2C	4.85	3.30	346.00	161.00	---	507.00	**640.00**
10'-0" x 8'-0", 3 doors	Inst	Set	Lg	2C	3.20	5.00	330.00	106.00	---	436.00	**539.00**
	Inst	Set	Sm	2C	4.85	3.30	386.00	161.00	---	547.00	**686.00**
12'-0" x 8'-0", 3 doors	Inst	Set	Lg	2C	3.20	5.00	300.00	106.00	---	406.00	**505.00**
	Inst	Set	Sm	2C	4.85	3.30	351.00	161.00	---	512.00	**646.00**

89

Description	Oper	Unit	Vol	Crew Size	Man-hours per Unit	Crew Output per Day	Avg Mat'l Unit Cost	Avg Labor Unit Cost	Avg Equip Unit Cost	Avg Total Unit Cost	Avg Price Incl O&P
Unfinished lauan											
4'-0" x 6'-8", 2 doors	Inst	Set	Lg	2C	2.29	7.00	120.00	76.20	---	196.20	**252.00**
	Inst	Set	Sm	2C	3.48	4.60	140.00	116.00	---	256.00	**335.00**
6'-0" x 6'-8", 2 doors	Inst	Set	Lg	2C	2.29	7.00	141.00	76.20	---	217.20	**276.00**
	Inst	Set	Sm	2C	3.48	4.60	165.00	116.00	---	281.00	**363.00**
8'-0" x 6'-8", 2 doors	Inst	Set	Lg	2C	2.29	7.00	204.00	76.20	---	280.20	**349.00**
	Inst	Set	Sm	2C	3.48	4.60	239.00	116.00	---	355.00	**448.00**
4'-0" x 8'-0", 2 doors	Inst	Set	Lg	2C	2.29	7.00	195.00	76.20	---	271.20	**339.00**
	Inst	Set	Sm	2C	3.48	4.60	228.00	116.00	---	344.00	**436.00**
6'-0" x 8'-0", 2 doors	Inst	Set	Lg	2C	2.29	7.00	230.00	76.20	---	306.20	**379.00**
	Inst	Set	Sm	2C	3.48	4.60	269.00	116.00	---	385.00	**483.00**
8'-0" x 8'-0", 2 doors	Inst	Set	Lg	2C	2.29	7.00	310.00	76.20	---	386.20	**471.00**
	Inst	Set	Sm	2C	3.48	4.60	363.00	116.00	---	479.00	**591.00**
10'-0" x 6'-8", 3 doors	Inst	Set	Lg	2C	3.20	5.00	280.00	106.00	---	386.00	**482.00**
	Inst	Set	Sm	2C	4.85	3.30	328.00	161.00	---	489.00	**619.00**
12'-0" x 6'-8", 3 doors	Inst	Set	Lg	2C	3.20	5.00	301.00	106.00	---	407.00	**506.00**
	Inst	Set	Sm	2C	4.85	3.30	352.00	161.00	---	513.00	**647.00**
10'-0" x 8'-0", 3 doors	Inst	Set	Lg	2C	3.20	5.00	400.00	106.00	---	506.00	**620.00**
	Inst	Set	Sm	2C	4.85	3.30	468.00	161.00	---	629.00	**780.00**
12'-0" x 8'-0", 3 doors	Inst	Set	Lg	2C	3.20	5.00	438.00	106.00	---	544.00	**663.00**
	Inst	Set	Sm	2C	4.85	3.30	512.00	161.00	---	673.00	**831.00**
Unfinished red oak											
4'-0" x 6'-8", 2 doors	Inst	Set	Lg	2C	2.29	7.00	180.00	76.20	---	256.20	**321.00**
	Inst	Set	Sm	2C	3.48	4.60	211.00	116.00	---	327.00	**416.00**
6'-0" x 6'-8", 2 doors	Inst	Set	Lg	2C	2.29	7.00	221.00	76.20	---	297.20	**368.00**
	Inst	Set	Sm	2C	3.48	4.60	259.00	116.00	---	375.00	**471.00**
8'-0" x 6'-8", 2 doors	Inst	Set	Lg	2C	2.29	7.00	364.00	76.20	---	440.20	**533.00**
	Inst	Set	Sm	2C	3.48	4.60	426.00	116.00	---	542.00	**663.00**
4'-0" x 8'-0", 2 doors	Inst	Set	Lg	2C	2.29	7.00	310.00	76.20	---	386.20	**471.00**
	Inst	Set	Sm	2C	3.48	4.60	363.00	116.00	---	479.00	**591.00**
6'-0" x 8'-0", 2 doors	Inst	Set	Lg	2C	2.29	7.00	340.00	76.20	---	416.20	**505.00**
	Inst	Set	Sm	2C	3.48	4.60	398.00	116.00	---	514.00	**631.00**
8'-0" x 8'-0", 2 doors	Inst	Set	Lg	2C	2.29	7.00	467.00	76.20	---	543.20	**651.00**
	Inst	Set	Sm	2C	3.48	4.60	546.00	116.00	---	662.00	**802.00**
10'-0" x 6'-8", 3 doors	Inst	Set	Lg	2C	3.20	5.00	464.00	106.00	---	570.00	**693.00**
	Inst	Set	Sm	2C	4.85	3.30	543.00	161.00	---	704.00	**866.00**
12'-0" x 6'-8", 3 doors	Inst	Set	Lg	2C	3.20	5.00	468.00	106.00	---	574.00	**698.00**
	Inst	Set	Sm	2C	4.85	3.30	548.00	161.00	---	709.00	**872.00**
10'-0" x 8'-0", 3 doors	Inst	Set	Lg	2C	3.20	5.00	640.00	106.00	---	746.00	**896.00**
	Inst	Set	Sm	2C	4.85	3.30	749.00	161.00	---	910.00	**1100.00**
12'-0" x 8'-0", 3 doors	Inst	Set	Lg	2C	3.20	5.00	655.00	106.00	---	761.00	**913.00**
	Inst	Set	Sm	2C	4.85	3.30	766.00	161.00	---	927.00	**1120.00**

Description	Oper	Unit	Vol	Crew Size	Man-hours per Unit	Crew Output per Day	Avg Mat'l Unit Cost	Avg Labor Unit Cost	Avg Equip Unit Cost	Avg Total Unit Cost	Avg Price Incl O&P

Mirror bypass units

Frameless unit with 1/2" beveled mirror edges

Description	Oper	Unit	Vol	Crew	MH	Out	Mat'l	Labor	Equip	Total	O&P
4'-0" x 6'-8", 2 doors	Inst	Set	Lg	2C	2.67	6.00	212.00	88.80	---	300.80	**377.00**
	Inst	Set	Sm	2C	4.10	3.90	248.00	136.00	---	384.00	**490.00**
6'-0" x 6'-8", 2 doors	Inst	Set	Lg	2C	2.67	6.00	273.00	88.80	---	361.80	**447.00**
	Inst	Set	Sm	2C	4.10	3.90	319.00	136.00	---	455.00	**572.00**
8'-0" x 6'-8", 2 doors	Inst	Set	Lg	2C	2.67	6.00	379.00	88.80	---	467.80	**569.00**
	Inst	Set	Sm	2C	4.10	3.90	443.00	136.00	---	579.00	**715.00**
4'-0" x 8'-0", 2 doors	Inst	Set	Lg	2C	2.67	6.00	259.00	88.80	---	347.80	**431.00**
	Inst	Set	Sm	2C	4.10	3.90	303.00	136.00	---	439.00	**553.00**
6'-0" x 8'-0", 2 doors	Inst	Set	Lg	2C	2.67	6.00	307.00	88.80	---	395.80	**486.00**
	Inst	Set	Sm	2C	4.10	3.90	359.00	136.00	---	495.00	**618.00**
8'-0" x 8'-0", 2 doors	Inst	Set	Lg	2C	2.67	6.00	402.00	88.80	---	490.80	**596.00**
	Inst	Set	Sm	2C	4.10	3.90	470.00	136.00	---	606.00	**746.00**
10'-0" x 6'-8", 3 doors	Inst	Set	Lg	2C	3.20	5.00	510.00	106.00	---	616.00	**746.00**
	Inst	Set	Sm	2C	4.85	3.30	597.00	161.00	---	758.00	**928.00**
12'-0" x 6'-8", 3 doors	Inst	Set	Lg	2C	3.20	5.00	552.00	106.00	---	658.00	**795.00**
	Inst	Set	Sm	2C	4.85	3.30	646.00	161.00	---	807.00	**985.00**
10'-0" x 8'-0", 3 doors	Inst	Set	Lg	2C	3.20	5.00	582.00	106.00	---	688.00	**829.00**
	Inst	Set	Sm	2C	4.85	3.30	681.00	161.00	---	842.00	**1030.00**
12'-0" x 8'-0", 3 doors	Inst	Set	Lg	2C	3.20	5.00	637.00	106.00	---	743.00	**892.00**
	Inst	Set	Sm	2C	4.85	3.30	745.00	161.00	---	906.00	**1100.00**

Aluminum frame unit

Description	Oper	Unit	Vol	Crew	MH	Out	Mat'l	Labor	Equip	Total	O&P
4'-0" x 6'-8", 2 doors	Inst	Set	Lg	2C	2.67	6.00	217.00	88.80	---	305.80	**383.00**
	Inst	Set	Sm	2C	4.10	3.90	254.00	136.00	---	390.00	**497.00**
6'-0" x 6'-8", 2 doors	Inst	Set	Lg	2C	2.67	6.00	281.00	88.80	---	369.80	**456.00**
	Inst	Set	Sm	2C	4.10	3.90	329.00	136.00	---	465.00	**583.00**
8'-0" x 6'-8", 2 doors	Inst	Set	Lg	2C	2.67	6.00	340.00	88.80	---	428.80	**524.00**
	Inst	Set	Sm	2C	4.10	3.90	398.00	136.00	---	534.00	**662.00**
4'-0" x 8'-0", 2 doors	Inst	Set	Lg	2C	2.67	6.00	246.00	88.80	---	334.80	**416.00**
	Inst	Set	Sm	2C	4.10	3.90	288.00	136.00	---	424.00	**536.00**
6'-0" x 8'-0", 2 doors	Inst	Set	Lg	2C	2.67	6.00	318.00	88.80	---	406.80	**499.00**
	Inst	Set	Sm	2C	4.10	3.90	372.00	136.00	---	508.00	**633.00**
8'-0" x 8'-0", 2 doors	Inst	Set	Lg	2C	2.67	6.00	395.00	88.80	---	483.80	**588.00**
	Inst	Set	Sm	2C	4.10	3.90	462.00	136.00	---	598.00	**736.00**
10'-0" x 6'-8", 3 doors	Inst	Set	Lg	2C	3.20	5.00	480.00	106.00	---	586.00	**712.00**
	Inst	Set	Sm	2C	4.85	3.30	562.00	161.00	---	723.00	**888.00**
12'-0" x 6'-8", 3 doors	Inst	Set	Lg	2C	3.20	5.00	523.00	106.00	---	629.00	**761.00**
	Inst	Set	Sm	2C	4.85	3.30	612.00	161.00	---	773.00	**946.00**
10'-0" x 8'-0", 3 doors	Inst	Set	Lg	2C	3.20	5.00	551.00	106.00	---	657.00	**793.00**
	Inst	Set	Sm	2C	4.85	3.30	645.00	161.00	---	806.00	**983.00**
12'-0" x 8'-0", 3 doors	Inst	Set	Lg	2C	3.20	5.00	603.00	106.00	---	709.00	**853.00**
	Inst	Set	Sm	2C	4.85	3.30	706.00	161.00	---	867.00	**1050.00**

Description	Oper	Unit	Vol	Crew Size	Man-hours per Unit	Crew Output per Day	Avg Mat'l Unit Cost	Avg Labor Unit Cost	Avg Equip Unit Cost	Avg Total Unit Cost	Avg Price Incl O&P
Golden oak frame unit											
4'-0" x 6'-8", 2 doors	Inst	Set	Lg	2C	2.67	6.00	257.00	88.80	---	345.80	**429.00**
	Inst	Set	Sm	2C	4.10	3.90	301.00	136.00	---	437.00	**550.00**
6'-0" x 6'-8", 2 doors	Inst	Set	Lg	2C	2.67	6.00	334.00	88.80	---	422.80	**517.00**
	Inst	Set	Sm	2C	4.10	3.90	391.00	136.00	---	527.00	**654.00**
8'-0" x 6'-8", 2 doors	Inst	Set	Lg	2C	2.67	6.00	403.00	88.80	---	491.80	**597.00**
	Inst	Set	Sm	2C	4.10	3.90	472.00	136.00	---	608.00	**747.00**
4'-0" x 8'-0", 2 doors	Inst	Set	Lg	2C	2.67	6.00	291.00	88.80	---	379.80	**468.00**
	Inst	Set	Sm	2C	4.10	3.90	340.00	136.00	---	476.00	**596.00**
6'-0" x 8'-0", 2 doors	Inst	Set	Lg	2C	2.67	6.00	377.00	88.80	---	465.80	**567.00**
	Inst	Set	Sm	2C	4.10	3.90	441.00	136.00	---	577.00	**712.00**
8'-0" x 8'-0", 2 doors	Inst	Set	Lg	2C	2.67	6.00	468.00	88.80	---	556.80	**671.00**
	Inst	Set	Sm	2C	4.10	3.90	548.00	136.00	---	684.00	**834.00**
10'-0" x 6'-8", 3 doors	Inst	Set	Lg	2C	3.20	5.00	569.00	106.00	---	675.00	**814.00**
	Inst	Set	Sm	2C	4.85	3.30	666.00	161.00	---	827.00	**1010.00**
12'-0" x 6'-8", 3 doors	Inst	Set	Lg	2C	3.20	5.00	620.00	106.00	---	726.00	**873.00**
	Inst	Set	Sm	2C	4.85	3.30	725.00	161.00	---	886.00	**1080.00**
10'-0" x 8'-0", 3 doors	Inst	Set	Lg	2C	3.20	5.00	654.00	106.00	---	760.00	**912.00**
	Inst	Set	Sm	2C	4.85	3.30	765.00	161.00	---	926.00	**1120.00**
12'-0" x 8'-0", 3 doors	Inst	Set	Lg	2C	3.20	5.00	715.00	106.00	---	821.00	**982.00**
	Inst	Set	Sm	2C	4.85	3.30	837.00	161.00	---	998.00	**1200.00**
Steel frame unit											
4'-0" x 6'-8", 2 doors	Inst	Set	Lg	2C	2.67	6.00	152.00	88.80	---	240.80	**308.00**
	Inst	Set	Sm	2C	4.10	3.90	178.00	136.00	---	314.00	**409.00**
6'-0" x 6'-8", 2 doors	Inst	Set	Lg	2C	2.67	6.00	199.00	88.80	---	287.80	**362.00**
	Inst	Set	Sm	2C	4.10	3.90	233.00	136.00	---	369.00	**472.00**
8'-0" x 6'-8", 2 doors	Inst	Set	Lg	2C	2.67	6.00	244.00	88.80	---	332.80	**414.00**
	Inst	Set	Sm	2C	4.10	3.90	285.00	136.00	---	421.00	**533.00**
4'-0" x 8'-0", 2 doors	Inst	Set	Lg	2C	2.67	6.00	167.00	88.80	---	255.80	**325.00**
	Inst	Set	Sm	2C	4.10	3.90	195.00	136.00	---	331.00	**429.00**
6'-0" x 8'-0", 2 doors	Inst	Set	Lg	2C	2.67	6.00	226.00	88.80	---	314.80	**393.00**
	Inst	Set	Sm	2C	4.10	3.90	264.00	136.00	---	400.00	**509.00**
8'-0" x 8'-0", 2 doors	Inst	Set	Lg	2C	2.67	6.00	284.00	88.80	---	372.80	**460.00**
	Inst	Set	Sm	2C	4.10	3.90	332.00	136.00	---	468.00	**587.00**
10'-0" x 6'-8", 3 doors	Inst	Set	Lg	2C	3.20	5.00	358.00	106.00	---	464.00	**571.00**
	Inst	Set	Sm	2C	4.85	3.30	419.00	161.00	---	580.00	**724.00**
12'-0" x 6'-8", 3 doors	Inst	Set	Lg	2C	3.20	5.00	400.00	106.00	---	506.00	**620.00**
	Inst	Set	Sm	2C	4.85	3.30	468.00	161.00	---	629.00	**780.00**
10'-0" x 8'-0", 3 doors	Inst	Set	Lg	2C	3.20	5.00	407.00	106.00	---	513.00	**628.00**
	Inst	Set	Sm	2C	4.85	3.30	476.00	161.00	---	637.00	**790.00**
12'-0" x 8'-0", 3 doors	Inst	Set	Lg	2C	3.20	5.00	457.00	106.00	---	563.00	**685.00**
	Inst	Set	Sm	2C	4.85	3.30	535.00	161.00	---	696.00	**857.00**

Description	Oper	Unit	Vol	Crew Size	Man-hours per Unit	Crew Output per Day	Avg Mat'l Unit Cost	Avg Labor Unit Cost	Avg Equip Unit Cost	Avg Total Unit Cost	Avg Price Incl O&P
Accordion doors											
Custom prefinished woodgrain print											
2'-0" x 6'-8"	Inst	Set	Lg	2C	1.60	10.00	245.00	53.20	---	298.20	**362.00**
	Inst	Set	Sm	2C	2.46	6.50	287.00	81.80	---	368.80	**452.00**
3'-0" x 6'-8"	Inst	Set	Lg	2C	1.60	10.00	332.00	53.20	---	385.20	**462.00**
	Inst	Set	Sm	2C	2.46	6.50	388.00	81.80	---	469.80	**569.00**
4'-0" x 6'-8"	Inst	Set	Lg	2C	1.60	10.00	430.00	53.20	---	483.20	**574.00**
	Inst	Set	Sm	2C	2.46	6.50	503.00	81.80	---	584.80	**701.00**
5'-0" x 6'-8"	Inst	Set	Lg	2C	1.60	10.00	497.00	53.20	---	550.20	**651.00**
	Inst	Set	Sm	2C	2.46	6.50	581.00	81.80	---	662.80	**791.00**
6'-0" x 6'-8"	Inst	Set	Lg	2C	1.60	10.00	607.00	53.20	---	660.20	**778.00**
	Inst	Set	Sm	2C	2.46	6.50	710.00	81.80	---	791.80	**940.00**
7'-0" x 6'-8"	Inst	Set	Lg	2C	2.00	8.00	722.00	66.50	---	788.50	**930.00**
	Inst	Set	Sm	2C	3.08	5.20	845.00	102.00	---	947.00	**1130.00**
8'-0" x 6'-8"	Inst	Set	Lg	2C	2.00	8.00	787.00	66.50	---	853.50	**1000.00**
	Inst	Set	Sm	2C	3.08	5.20	921.00	102.00	---	1023.00	**1210.00**
9'-0" x 6'-8"	Inst	Set	Lg	2C	2.00	8.00	881.00	66.50	---	947.50	**1110.00**
	Inst	Set	Sm	2C	3.08	5.20	1030.00	102.00	---	1132.00	**1340.00**
10'-0" x 6'-8"	Inst	Set	Lg	2C	2.00	8.00	1010.00	66.50	---	1076.50	**1270.00**
	Inst	Set	Sm	2C	3.08	5.20	1190.00	102.00	---	1292.00	**1520.00**
For 8'-0"H, ADD	Inst	%	Lg	2C	---		12.0	---	---	---	**---**
	Inst	%	Sm	2C	---		12.0	---	---	---	**---**
Heritage prefinished real wood veneer											
2'-0" x 6'-8"	Inst	Set	Lg	2C	1.60	10.00	478.00	53.20	---	531.20	**630.00**
	Inst	Set	Sm	2C	2.46	6.50	559.00	81.80	---	640.80	**766.00**
3'-0" x 6'-8"	Inst	Set	Lg	2C	1.60	10.00	672.00	53.20	---	725.20	**853.00**
	Inst	Set	Sm	2C	2.46	6.50	786.00	81.80	---	867.80	**1030.00**
4'-0" x 6'-8"	Inst	Set	Lg	2C	1.60	10.00	871.00	53.20	---	924.20	**1080.00**
	Inst	Set	Sm	2C	2.46	6.50	1020.00	81.80	---	1101.80	**1290.00**
5'-0" x 6'-8"	Inst	Set	Lg	2C	1.60	10.00	1000.00	53.20	---	1053.20	**1230.00**
	Inst	Set	Sm	2C	2.46	6.50	1170.00	81.80	---	1251.80	**1470.00**
6'-0" x 6'-8"	Inst	Set	Lg	2C	1.60	10.00	1210.00	53.20	---	1263.20	**1470.00**
	Inst	Set	Sm	2C	2.46	6.50	1410.00	81.80	---	1491.80	**1740.00**
7'-0" x 6'-8"	Inst	Set	Lg	2C	2.00	8.00	1400.00	66.50	---	1466.50	**1710.00**
	Inst	Set	Sm	2C	3.08	5.20	1640.00	102.00	---	1742.00	**2040.00**
8'-0" x 6'-8"	Inst	Set	Lg	2C	2.00	8.00	1540.00	66.50	---	1606.50	**1870.00**
	Inst	Set	Sm	2C	3.08	5.20	1800.00	102.00	---	1902.00	**2220.00**
9'-0" x 6'-8"	Inst	Set	Lg	2C	2.00	8.00	1740.00	66.50	---	1806.50	**2100.00**
	Inst	Set	Sm	2C	3.08	5.20	2030.00	102.00	---	2132.00	**2490.00**
10'-0" x 6'-8"	Inst	Set	Lg	2C	2.00	8.00	1930.00	66.50	---	1996.50	**2320.00**
	Inst	Set	Sm	2C	3.08	5.20	2260.00	102.00	---	2362.00	**2750.00**
For 8'-0"H, ADD	Inst	%	Lg	2C	---	---	20.0	---	---	---	**---**
	Inst	%	Sm	2C	---	---	20.0	---	---	---	**---**

Description	Oper	Unit	Vol	Crew Size	Man-hours per Unit	Crew Output per Day	Avg Mat'l Unit Cost	Avg Labor Unit Cost	Avg Equip Unit Cost	Avg Total Unit Cost	Avg Price Incl O&P

Track and hardware only

Description	Oper	Unit	Vol	Crew Size	Man-hours per Unit	Crew Output per Day	Avg Mat'l Unit Cost	Avg Labor Unit Cost	Avg Equip Unit Cost	Avg Total Unit Cost	Avg Price Incl O&P
4'-0" x 6'-8", 2 doors	Inst	Set	Lg	---	---	---	18.00	---	---	18.00	**20.70**
	Inst	Set	Sm	---	---	---	21.10	---	---	21.10	**24.20**
6'-0" x 6'-8", 2 doors	Inst	Set	Lg	---	---	---	24.00	---	---	24.00	**27.60**
	Inst	Set	Sm	---	---	---	28.10	---	---	28.10	**32.30**
8'-0" x 6'-8", 2 doors	Inst	Set	Lg	---	---	---	30.00	---	---	30.00	**34.50**
	Inst	Set	Sm	---	---	---	35.10	---	---	35.10	**40.40**
4'-0" x 8'-0", 2 doors	Inst	Set	Lg	---	---	---	18.00	---	---	18.00	**20.70**
	Inst	Set	Sm	---	---	---	21.10	---	---	21.10	**24.20**
6'-0" x 8'-0", 2 doors	Inst	Set	Lg	---	---	---	23.00	---	---	23.00	**26.50**
	Inst	Set	Sm	---	---	---	26.90	---	---	26.90	**31.00**
8'-0" x 8'-0", 2 doors	Inst	Set	Lg	---	---	---	30.00	---	---	30.00	**34.50**
	Inst	Set	Sm	---	---	---	35.10	---	---	35.10	**40.40**
10'-0" x 6'-8", 3 doors	Inst	Set	Lg	---	---	---	41.00	---	---	41.00	**47.20**
	Inst	Set	Sm	---	---	---	48.00	---	---	48.00	**55.20**
12'-0" x 6'-8", 3 doors	Inst	Set	Lg	---	---	---	51.00	---	---	51.00	**58.70**
	Inst	Set	Sm	---	---	---	59.70	---	---	59.70	**68.60**
10'-0" x 8'-0", 3 doors	Inst	Set	Lg	---	---	---	41.00	---	---	41.00	**47.20**
	Inst	Set	Sm	---	---	---	48.00	---	---	48.00	**55.20**
12'-0" x 8'-0", 3 doors	Inst	Set	Lg	---	---	---	51.00	---	---	51.00	**58.70**
	Inst	Set	Sm	---	---	---	59.70	---	---	59.70	**68.60**

Columns

See also Framing, page 178

Aluminum, extruded; self supporting

Designed as decorative, loadbearing elements for porches, entrances, colonnades, etc.; primed, knocked-down, and carton packed complete with cap and base

Column with standard cap and base

Description	Oper	Unit	Vol	Crew Size	Man-hours per Unit	Crew Output per Day	Avg Mat'l Unit Cost	Avg Labor Unit Cost	Avg Equip Unit Cost	Avg Total Unit Cost	Avg Price Incl O&P
8" dia. x 8' to 12' H	Inst	Ea	Lg	CS	4.00	6.00	144.00	125.00	53.60	322.60	**414.00**
	Inst	Ea	Sm	CS	6.15	3.90	169.00	193.00	82.50	444.50	**574.00**
10" dia. x 8' to 12' H	Inst	Ea	Lg	CS	4.00	6.00	192.00	125.00	53.60	370.60	**472.00**
	Inst	Ea	Sm	CS	6.15	3.90	225.00	193.00	82.50	500.50	**642.00**
10" dia. x 16' to 20' H	Inst	Ea	Lg	CS	5.33	4.50	217.00	167.00	71.50	455.50	**582.00**
	Inst	Ea	Sm	CS	8.19	2.93	255.00	257.00	110.00	622.00	**800.00**
12" dia. x 9' to 12' H	Inst	Ea	Lg	CS	4.00	6.00	348.00	125.00	53.60	526.60	**659.00**
	Inst	Ea	Sm	CS	6.15	3.90	408.00	193.00	82.50	683.50	**861.00**
12" dia. x 16' to 24' H	Inst	Ea	Lg	CS	5.33	4.50	401.00	167.00	71.50	639.50	**803.00**
	Inst	Ea	Sm	CS	8.19	2.93	470.00	257.00	110.00	837.00	**1060.00**

Column with Corinthian cap and decorative base

Description	Oper	Unit	Vol	Crew Size	Man-hours per Unit	Crew Output per Day	Avg Mat'l Unit Cost	Avg Labor Unit Cost	Avg Equip Unit Cost	Avg Total Unit Cost	Avg Price Incl O&P
8" dia. x 8' to 12' H	Inst	Ea	Lg	CS	4.00	6.00	250.00	125.00	53.60	428.60	**541.00**
	Inst	Ea	Sm	CS	6.15	3.90	293.00	193.00	82.50	568.50	**723.00**
10" dia. x 8' to 12' H	Inst	Ea	Lg	CS	4.00	6.00	314.00	125.00	53.60	492.60	**618.00**
	Inst	Ea	Sm	CS	6.15	3.90	368.00	193.00	82.50	643.50	**813.00**
10" dia. x 16' to 20' H	Inst	Ea	Lg	CS	5.33	4.50	343.00	167.00	71.50	581.50	**734.00**
	Inst	Ea	Sm	CS	8.19	2.93	403.00	257.00	110.00	770.00	**978.00**
12" dia. x 9' to 12' H	Inst	Ea	Lg	CS	4.00	6.00	514.00	125.00	53.60	692.60	**858.00**
	Inst	Ea	Sm	CS	6.15	3.90	603.00	193.00	82.50	878.50	**1090.00**
12" dia. x 16' to 24' H	Inst	Ea	Lg	CS	5.33	4.50	576.00	167.00	71.50	814.50	**1010.00**
	Inst	Ea	Sm	CS	8.19	2.93	676.00	257.00	110.00	1043.00	**1310.00**

Description	Oper	Unit	Vol	Crew Size	Man-hours per Unit	Crew Output per Day	Avg Mat'l Unit Cost	Avg Labor Unit Cost	Avg Equip Unit Cost	Avg Total Unit Cost	Avg Price Incl O&P

Brick columns. See Masonry, page 277

Wood, treated, No. 1 common and better white pine, T&G construction

Designed as decorative, loadbearing elements for porches, entrances, colonnades, etc.; primed, knocked-down, and carton packed complete with cap and base

Plain column with standard cap and base

Description	Oper	Unit	Vol	Crew Size	Man-hours per Unit	Crew Output per Day	Avg Mat'l Unit Cost	Avg Labor Unit Cost	Avg Equip Unit Cost	Avg Total Unit Cost	Avg Price Incl O&P
8" dia. x 8' to 12' H	Inst	Ea	Lg	CS	4.80	5.00	222.00	150.00	64.30	436.30	557.00
	Inst	Ea	Sm	CS	7.38	3.25	261.00	231.00	99.00	591.00	759.00
10" dia. x 8' to 12' H	Inst	Ea	Lg	CS	4.80	5.00	267.00	150.00	64.30	481.30	610.00
	Inst	Ea	Sm	CS	7.38	3.25	313.00	231.00	99.00	643.00	821.00
12" dia. x 8' to 12' H	Inst	Ea	Lg	CS	5.33	4.50	309.00	167.00	71.50	547.50	692.00
	Inst	Ea	Sm	CS	8.19	2.93	362.00	257.00	110.00	729.00	929.00
14" dia. x 12' to 16' H	Inst	Ea	Lg	CS	6.86	3.50	504.00	215.00	91.90	810.90	1020.00
	Inst	Ea	Sm	CS	10.5	2.28	591.00	329.00	141.00	1061.00	1340.00
16" dia. x 18' to 20' H	Inst	Ea	Lg	CS	7.50	3.20	740.00	235.00	101.00	1076.00	1340.00
	Inst	Ea	Sm	CS	11.5	2.08	868.00	360.00	155.00	1383.00	1740.00

Plain column with Corinthian cap and decorative base

Description	Oper	Unit	Vol	Crew Size	Man-hours per Unit	Crew Output per Day	Avg Mat'l Unit Cost	Avg Labor Unit Cost	Avg Equip Unit Cost	Avg Total Unit Cost	Avg Price Incl O&P
8" dia. x 8' to 12' H	Inst	Ea	Lg	CS	4.80	5.00	255.00	150.00	64.30	469.30	596.00
	Inst	Ea	Sm	CS	7.38	3.25	299.00	231.00	99.00	629.00	805.00
10" dia. x 8' to 12' H	Inst	Ea	Lg	CS	4.80	5.00	330.00	150.00	64.30	544.30	686.00
	Inst	Ea	Sm	CS	7.38	3.25	388.00	231.00	99.00	718.00	911.00
12" dia. x 8' to 12' H	Inst	Ea	Lg	CS	5.33	4.50	372.00	167.00	71.50	610.50	769.00
	Inst	Ea	Sm	CS	8.19	2.93	437.00	257.00	110.00	804.00	1020.00
14" dia. x 12' to 16' H	Inst	Ea	Lg	CS	6.86	3.50	561.00	215.00	91.90	867.90	1090.00
	Inst	Ea	Sm	CS	10.5	2.28	658.00	329.00	141.00	1128.00	1420.00
16" dia. x 18' to 20' H	Inst	Ea	Lg	CS	7.50	3.20	789.00	235.00	101.00	1125.00	1400.00
	Inst	Ea	Sm	CS	11.5	2.08	926.00	360.00	155.00	1441.00	1810.00

Concrete

Concrete Footings

1. **Dimensions.** 6" T, 8" T, 12" T x 12" W, 16" W, 20" W; 12" T x 24" W.

2. **Installation**

   a. Forms. 2" side forms equal in height to the thickness of the footing. 2" x 4" stakes 4'-0" oc, no less than the thickness of the footing. 2" x 4" bracing for stakes 8'-0" oc for 6" and 8" thick footings and 4'-0" oc for 12" thick footings. 1" x 2" or 1" x 3" spreaders 4'-0" oc

   b. Concrete. 1-2-4 mix is used in this section.

   c. Reinforcing steel. Various sizes, but usually only #3, #4, or #5 straight rods with end ties are used.

3. **Notes on Labor**

   a. Forming. Output based on a crew of two carpenters and one laborer.

   b. Grading, finish. Output based on what one laborer can do in one day.

   c. Reinforcing steel. Output based on what one laborer or one ironworker can do in one day.

   d. Concrete. Output based on a crew of two laborers and one carpenter.

   e. Forms, wrecking and cleaning. Output based on what one laborer can do in one day.

4. **Estimating Technique.** Determine the linear feet of footing.

Concrete Foundations

1. **Dimensions.** 8" T, 12" T x 4' H, 8' H, or 12' H.

2. **Installation**

   a. Forms. 4' x 8' panels made of 3/4" form grade plywood backed with 2" x 4" studs and sills (studs approximately 16" oc), three sets and six sets of 2" x 4" wales for 4', 8' and 12' high walls. 2" x 4" wales for 4', 8' and 12' high walls. 2" x 4" diagonal braces (with stakes) 12'-0" oc one side. Snap ties spaced 22" oc, 20" oc and 17" oc along each wale for 4', 8' and 12' high walls. Paraffin oil coating for forms. Twelve uses are estimated for panels; twenty uses for wales and braces; snap ties are used only once.

   b. Concrete. 1-2-4 mix.

   c. Reinforcing steel. Sizes #3 to #7. Bars are straight except dowels which may on occasion be bent rods.

3. **Notes on Labor**

   a. Concrete, placing. Output based on a crew of one carpenter and five laborers.

   b. Forming. Output based on a crew of four carpenters and one laborer.

   c. Reinforcing steel rods. Output based on what two laborers or two ironworkers can do in one day.

   d. Wrecking and cleaning forms. Output based on what two laborers can do in one day.

4. **Estimating Technique**

   a. Determine linear feet of wall if wall is 8" or 12" x 4', 8' or 12', or determine square feet of wall. Then calculate and add the linear feet of rods.

Concrete Interior Floor Finishes

1. **Dimensions.** $3^1/_2$", 4", 5", 6" thick x various areas.

2. **Installation**

   a. Forms. A wood form may or may not be required. A foundation wall may serve as a form for both basement and first floor slabs. In this section, only 2" x 4" and 2" x 6" side forms with stakes 4'-0" oc are considered.

   b. Finish grading. Dirt or gravel.

   c. Screeds (wood strips placed in area where concrete is to be placed). The concrete when placed will be finished even with top of the screeds. Screeds must be pulled before concrete sets up and the voids filled with concrete. 2" x 2" and 2" x 4" screeds with 2" x 2" stakes 6'-0" oc will be covered in this section.

   d. Steel reinforcing. Items to be covered are: #3, #4, #5 rods; 6 x 6/10-10 and 6 x 6/6-6 welded wire mesh.

   e. Concrete. 1-2-4 mix.

3. **Notes on Labor**

   a. Forms and screeds. Output based on a crew of two masons and one laborer.

   b. Finish grading. Output based on what one laborer can do in one day.

   c. Reinforcing. Output based on what two laborers or two ironworkers can do in one day.

   d. Concrete, place and finish. Output based on three cement masons and five laborers as a crew.

   e. Wrecking and cleaning forms. Output based on what one laborer can do in one day.

4. **Estimating Technique**

   a. Finish grading, mesh, and concrete. Determine the area and add waste.

   b. Forms, screeds and rods. Determine the linear feet.

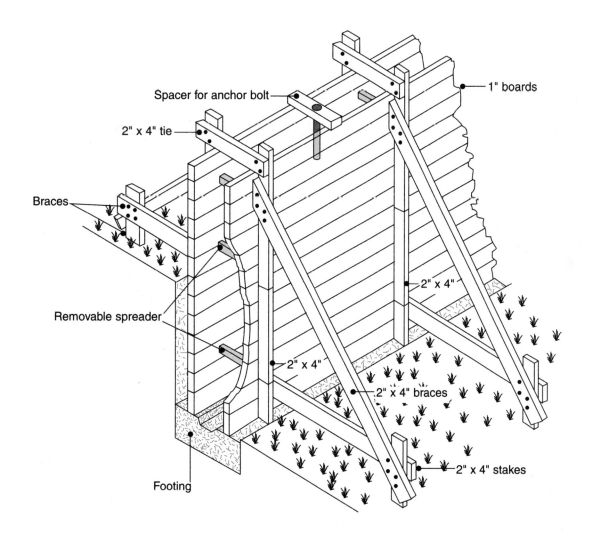

Description	Oper	Unit	Vol	Crew Size	Man-hours per Unit	Crew Output per Day	Avg Mat'l Unit Cost	Avg Labor Unit Cost	Avg Equip Unit Cost	Avg Total Unit Cost	Avg Price Incl O&P

Concrete, cast in place

Footings

Demo with air tools, reinforced

Description	Oper	Unit	Vol	Crew Size	Man-hours per Unit	Crew Output per Day	Avg Mat'l Unit Cost	Avg Labor Unit Cost	Avg Equip Unit Cost	Avg Total Unit Cost	Avg Price Incl O&P
8" T x 12" W	Demo	LF	Lg	AD	.171	140.0	---	5.16	.66	5.82	**8.30**
	Demo	LF	Sm	AD	.214	112.0	---	6.46	.83	7.29	**10.40**
8" T x 16" W	Demo	LF	Lg	AD	.182	132.0	---	5.49	.70	6.19	**8.83**
	Demo	LF	Sm	AD	.214	112.0	---	6.46	.83	7.29	**10.40**
8" T x 20" W	Demo	LF	Lg	AD	.194	124.0	---	5.85	.75	6.60	**9.41**
	Demo	LF	Sm	AD	.229	105.0	---	6.91	.88	7.79	**11.10**
12" T x 12" W	Demo	LF	Lg	AD	.190	126.0	---	5.73	.74	6.47	**9.23**
	Demo	LF	Sm	AD	.224	107.0	---	6.76	.87	7.63	**10.90**
12" T x 16" W	Demo	LF	Lg	AD	.200	120.0	---	6.04	.77	6.81	**9.70**
	Demo	LF	Sm	AD	.235	102.0	---	7.09	.91	8.00	**11.40**
12" T x 20" W	Demo	LF	Lg	AD	.211	114.0	---	6.37	.81	7.18	**10.20**
	Demo	LF	Sm	AD	.247	97.00	---	7.45	.96	8.41	**12.00**
12" T x 24" W	Demo	LF	Lg	AD	.222	108.0	---	6.70	.86	7.56	**10.80**
	Demo	LF	Sm	AD	.261	92.00	---	7.88	1.01	8.89	**12.70**

Place per LF poured footing

Forming, 4 uses

Description	Oper	Unit	Vol	Crew Size	Man-hours per Unit	Crew Output per Day	Avg Mat'l Unit Cost	Avg Labor Unit Cost	Avg Equip Unit Cost	Avg Total Unit Cost	Avg Price Incl O&P
2" x 6"	Inst	LF	Lg	CS	.107	225.0	.72	3.35	---	4.07	**5.75**
	Inst	LF	Sm	CS	.116	207.0	.98	3.63	---	4.61	**6.43**
2" x 8"	Inst	LF	Lg	CS	.120	200.0	.83	3.76	---	4.59	**6.47**
	Inst	LF	Sm	CS	.130	184.0	1.14	4.07	---	5.21	**7.25**
2" x 12"	Inst	LF	Lg	CS	.133	180.0	1.29	4.17	---	5.46	**7.54**
	Inst	LF	Sm	CS	.145	166.0	1.76	4.54	---	6.30	**8.57**

Grading, finish by hand

Description	Oper	Unit	Vol	Crew Size	Man-hours per Unit	Crew Output per Day	Avg Mat'l Unit Cost	Avg Labor Unit Cost	Avg Equip Unit Cost	Avg Total Unit Cost	Avg Price Incl O&P
6", 8", 12" T x 12" W	Inst	LF	Lg	1L	.023	345.0	---	.63	---	.63	**.94**
	Inst	LF	Sm	1L	.026	311.0	---	.71	---	.71	**1.06**
6", 8", 12" T x 16" W	Inst	LF	Lg	1L	.027	295.0	---	.74	---	.74	**1.10**
	Inst	LF	Sm	1L	.030	266.0	---	.82	---	.82	**1.23**
6", 8", 12" T x 20" W	Inst	LF	Lg	1L	.031	260.0	---	.85	---	.85	**1.27**
	Inst	LF	Sm	1L	.034	234.0	---	.93	---	.93	**1.39**
12" T x 24" W	Inst	LF	Lg	1L	.035	230.0	---	.96	---	.96	**1.43**
	Inst	LF	Sm	1L	.039	207.0	---	1.07	---	1.07	**1.59**

Description	Oper	Unit	Vol	Crew Size	Man-hours per Unit	Crew Output per Day	Avg Mat'l Unit Cost	Avg Labor Unit Cost	Avg Equip Unit Cost	Avg Total Unit Cost	Avg Price Incl O&P
Reinforcing steel in place, material costs include lap, waste, and tie wire											
Two (No. 3) 3/8" rods	Inst	LF	Lg	1L	.014	590.0	.46	.38	---	.84	**1.03**
	Inst	LF	Sm	1L	.015	531.0	.63	.41	---	1.04	**1.24**
Two (No. 4) 1/2" rods	Inst	LF	Lg	1L	.014	560.0	.53	.38	---	.91	**1.10**
	Inst	LF	Sm	1L	.016	504.0	.72	.44	---	1.16	**1.37**
Two (No. 5) 5/8" rods	Inst	LF	Lg	1L	.015	540.0	.82	.41	---	1.23	**1.43**
	Inst	LF	Sm	1L	.016	486.0	1.12	.44	---	1.56	**1.77**
Three (No. 3) 3/8" rods	Inst	LF	Lg	1L	.017	460.0	.69	.47	---	1.16	**1.38**
	Inst	LF	Sm	1L	.019	414.0	.94	.52	---	1.46	**1.72**
Three (No. 4) 1/2" rods	Inst	LF	Lg	1L	.018	440.0	.80	.49	---	1.29	**1.54**
	Inst	LF	Sm	1L	.020	396.0	1.08	.55	---	1.63	**1.90**
Three (No. 5) 5/8" rods	Inst	LF	Lg	1L	.019	420.0	1.24	.52	---	1.76	**2.02**
	Inst	LF	Sm	1L	.021	378.0	1.68	.58	---	2.26	**2.54**
Concrete, pour from truck into forms, using 3,000 PSI, 1-1/2" aggregate, 5.7 sack mix											
6" T x 12" W (54.0 LF/CY)	Inst	LF	Lg	LC	.039	615.0	1.85	1.15	---	3.00	**3.56**
	Inst	LF	Sm	LC	.043	554.0	2.53	1.26	---	3.79	**4.41**
6" T x 16" W (40.5 LF/CY)	Inst	LF	Lg	LC	.043	560.0	1.85	1.26	---	3.11	**3.73**
	Inst	LF	Sm	LC	.048	504.0	2.53	1.41	---	3.94	**4.63**
6" T x 20" W (32.3 LF/CY)	Inst	LF	Lg	LC	.044	545.0	2.77	1.29	---	4.06	**4.70**
	Inst	LF	Sm	LC	.049	491.0	3.80	1.44	---	5.24	**5.94**
8" T x 12" W (40.5 LF/CY)	Inst	LF	Lg	LC	.043	560.0	1.85	1.26	---	3.11	**3.73**
	Inst	LF	Sm	LC	.048	504.0	2.53	1.41	---	3.94	**4.63**
8" T x 16" W (30.4 LF/CY)	Inst	LF	Lg	LC	.047	510.0	2.77	1.38	---	4.15	**4.83**
	Inst	LF	Sm	LC	.052	459.0	3.80	1.53	---	5.33	**6.08**
8" T x 20" W (24.3 LF/CY)	Inst	LF	Lg	LC	.049	485.0	3.70	1.44	---	5.14	**5.84**
	Inst	LF	Sm	LC	.055	437.0	5.06	1.62	---	6.68	**7.47**
12" T x 12" W (27.0 LF/CY)	Inst	LF	Lg	LC	.045	535.0	3.70	1.32	---	5.02	**5.67**
	Inst	LF	Sm	LC	.050	482.0	5.06	1.47	---	6.53	**7.25**
12" T x 16" W (20.3 LF/CY)	Inst	LF	Lg	LC	.052	460.0	4.62	1.53	---	6.15	**6.90**
	Inst	LF	Sm	LC	.058	414.0	6.33	1.70	---	8.03	**8.87**
12" T x 20" W (16.2 LF/CY)	Inst	LF	Lg	LC	.057	420.0	5.54	1.67	---	7.21	**8.03**
	Inst	LF	Sm	LC	.063	378.0	7.60	1.85	---	9.45	**10.40**
12" T x 24" W (13.5 LF/CY)	Inst	LF	Lg	LC	.049	485.0	6.47	1.44	---	7.91	**8.61**
	Inst	LF	Sm	LC	.055	437.0	8.86	1.62	---	10.48	**11.30**
Forms, wreck, remove and clean											
2" x 6"	Inst	LF	Lg	1L	.046	175.0	---	1.26	---	1.26	**1.88**
	Inst	LF	Sm	1L	.051	158.0	---	1.40	---	1.40	**2.08**
2" x 8"	Inst	LF	Lg	1L	.050	160.0	---	1.37	---	1.37	**2.04**
	Inst	LF	Sm	1L	.056	144.0	---	1.54	---	1.54	**2.29**
2" x 12"	Inst	LF	Lg	1L	.055	145.0	---	1.51	---	1.51	**2.25**
	Inst	LF	Sm	1L	.061	131.0	---	1.67	---	1.67	**2.49**

Description	Oper	Unit	Vol	Crew Size	Man-hours per Unit	Crew Output per Day	Avg Mat'l Unit Cost	Avg Labor Unit Cost	Avg Equip Unit Cost	Avg Total Unit Cost	Avg Price Inc O&P

Foundations and retaining walls, demo with air tools, per LF wall

With reinforcing
4'-0" H

8" T	Demo	LF	Lg	AD	.300	80.00	---	9.05	1.16	10.21	**14.60**
	Demo	LF	Sm	AD	.353	68.00	---	10.70	1.36	12.06	**17.10**
12" T	Demo	LF	Lg	AD	.414	58.00	---	12.50	1.60	14.10	**20.10**
	Demo	LF	Sm	AD	.490	49.00	---	14.80	1.89	16.69	**23.80**

8'-0" H

8" T	Demo	LF	Lg	AD	.324	74.00	---	9.78	1.25	11.03	**15.70**
	Demo	LF	Sm	AD	.381	63.00	---	11.50	1.47	12.97	**18.50**
12" T	Demo	LF	Lg	AD	.462	52.00	---	13.90	1.78	15.68	**22.40**
	Demo	LF	Sm	AD	.545	44.00	---	16.50	2.11	18.61	**26.50**

12"-0" H

8" T	Demo	LF	Lg	AD	.353	68.00	---	10.70	1.36	12.06	**17.10**
	Demo	LF	Sm	AD	.414	58.00	---	12.50	1.60	14.10	**20.10**
12" T	Demo	LF	Lg	AD	.545	44.00	---	16.50	2.11	18.61	**26.50**
	Demo	LF	Sm	AD	.649	37.00	---	19.60	2.50	22.10	**31.50**

Without reinforcing
4'-0" H

8" T	Demo	LF	Lg	AD	.261	92.00	---	7.88	1.01	8.89	**12.70**
	Demo	LF	Sm	AD	.308	78.00	---	9.30	1.19	10.49	**15.00**
12" T	Demo	LF	Lg	AD	.353	68.00	---	10.70	1.36	12.06	**17.10**
	Demo	LF	Sm	AD	.414	58.00	---	12.50	1.60	14.10	**20.10**

8'-0" H

8" T	Demo	LF	Lg	AD	.286	84.00	---	8.63	1.10	9.73	**13.90**
	Demo	LF	Sm	AD	.338	71.00	---	10.20	1.31	11.51	**16.40**
12" T	Demo	LF	Lg	AD	.400	60.00	---	12.10	1.54	13.64	**19.40**
	Demo	LF	Sm	AD	.471	51.00	---	14.20	1.82	16.02	**22.90**

12"-0" H

8" T	Demo	LF	Lg	AD	.316	76.00	---	9.54	1.22	10.76	**15.30**
	Demo	LF	Sm	AD	.369	65.00	---	11.10	1.43	12.53	**17.90**
12" T	Demo	LF	Lg	AD	.462	52.00	---	13.90	1.78	15.68	**22.40**
	Demo	LF	Sm	AD	.545	44.00	---	16.50	2.11	18.61	**26.50**

Place foundation or wall

Forming only. Material price includes panel forms, wales, braces, snap ties, paraffin oil and nails
8" or 12" T x 4'-0" H

Make (@12 uses)	Inst	SF	Lg	CU	.058	685.0	1.13	1.86	---	2.99	**3.92**
	Inst	SF	Sm	CU	.063	630.0	1.55	2.02	---	3.57	**4.58**
Erect and coat	Inst	SF	Lg	CU	.098	410.0	.03	3.15	---	3.18	**4.75**
	Inst	SF	Sm	CU	.106	377.0	.03	3.40	---	3.43	**5.13**
Wreck and clean	Inst	SF	Lg	LB	.043	370.0	---	1.18	---	1.18	**1.76**
	Inst	SF	Sm	LB	.048	333.0	---	1.32	---	1.32	**1.96**

Description	Oper	Unit	Vol	Crew Size	Man-hours per Unit	Crew Output per Day	Avg Mat'l Unit Cost	Avg Labor Unit Cost	Avg Equip Unit Cost	Avg Total Unit Cost	Avg Price Incl O&P
8" or 12" T x 8'-0" H											
Make (@12 uses)	Inst	SF	Lg	CU	.061	655.0	1.04	1.96	---	3.00	**3.98**
	Inst	SF	Sm	CU	.066	603.0	1.43	2.12	---	3.55	**4.61**
Erect and coat	Inst	SF	Lg	CU	.118	340.0	.03	3.79	---	3.82	**5.71**
	Inst	SF	Sm	CU	.128	313.0	.03	4.11	---	4.14	**6.19**
Wreck and clean	Inst	SF	Lg	LB	.052	305.0	---	1.43	---	1.43	**2.13**
	Inst	SF	Sm	LB	.058	275.0	---	1.59	---	1.59	**2.37**
8" or 12" T x 12'-0" H											
Make (@12 uses)	Inst	SF	Lg	CU	.060	670.0	1.09	1.93	---	3.02	**3.98**
	Inst	SF	Sm	CU	.065	616.0	1.49	2.09	---	3.58	**4.62**
Erect and coat	Inst	SF	Lg	CU	.138	290.0	.03	4.43	---	4.46	**6.67**
	Inst	SF	Sm	CU	.150	267.0	.03	4.82	---	4.85	**7.25**
Wreck and clean	Inst	SF	Lg	LB	.062	260.0	---	1.70	---	1.70	**2.53**
	Inst	SF	Sm	LB	.068	234.0	---	1.87	---	1.87	**2.78**

Reinforcing steel rods. 5% waste included, pricing based on LF of rod

Description	Oper	Unit	Vol	Crew Size	Man-hours per Unit	Crew Output per Day	Avg Mat'l Unit Cost	Avg Labor Unit Cost	Avg Equip Unit Cost	Avg Total Unit Cost	Avg Price Incl O&P
No. 3 (3/8" rod)	Inst	LF	Lg	LB	.012	1390	.22	.33	---	.55	**.71**
	Inst	LF	Sm	LB	.013	1251	.30	.36	---	.66	**.83**
No. 4 (1/2" rod)	Inst	LF	Lg	LB	.012	1330	.25	.33	---	.58	**.74**
	Inst	LF	Sm	LB	.013	1197	.34	.36	---	.70	**.87**
No. 5 (5/8" rod)	Inst	LF	Lg	LB	.013	1260	.39	.36	---	.75	**.92**
	Inst	LF	Sm	LB	.014	1134	.54	.38	---	.92	**1.11**
No. 6 (3/4" rod)	Inst	LF	Lg	LB	.014	1130	.57	.38	---	.95	**1.14**
	Inst	LF	Sm	LB	.016	1017	.77	.44	---	1.21	**1.42**
No. 7 (7/8" rod)	Inst	LF	Lg	LB	.016	1000	.77	.44	---	1.21	**1.42**
	Inst	LF	Sm	LB	.018	900.0	1.05	.49	---	1.54	**1.79**

Concrete, placed from trucks into forms, material cost includes 5% waste and assumes use of 3,000 PSI, 1-1/2" aggregate, 5.7 sack mix

Description	Oper	Unit	Vol	Crew Size	Man-hours per Unit	Crew Output per Day	Avg Mat'l Unit Cost	Avg Labor Unit Cost	Avg Equip Unit Cost	Avg Total Unit Cost	Avg Price Incl O&P
8" T x 4'-0" H (10.12 LF/CY)	Inst	LF	Lg	LG	.070	690.0	9.24	1.99	---	11.23	**12.20**
	Inst	LF	Sm	LG	.077	621.0	12.70	2.19	---	14.89	**15.90**
8" T x 4'-0" H (40.5 SF/CY)	Inst	SF	Lg	LG	.017	2760	1.85	.48	---	2.33	**2.57**
	Inst	SF	Sm	LG	.019	2484	2.53	.54	---	3.07	**3.33**
8" T x 8'-0" H (5.06 LF/CY)	Inst	LF	Lg	LG	.139	345.0	18.50	3.95	---	22.45	**24.40**
	Inst	LF	Sm	LG	.154	311.0	25.30	4.37	---	29.67	**31.80**
8" T x 8'-0" H (40.5 SF/CY)	Inst	SF	Lg	LG	.017	2760	1.85	.48	---	2.33	**2.57**
	Inst	SF	Sm	LG	.019	2484	2.53	.54	---	3.07	**3.33**
8" T x 12'-0" H (3.37 LF/CY)	Inst	LF	Lg	LG	.209	230.0	27.70	5.94	---	33.64	**36.60**
	Inst	LF	Sm	LG	.232	207.0	38.00	6.59	---	44.59	**47.80**
8" T x 12'-0" H (40.5 SF/CY)	Inst	SF	Lg	LG	.017	2760	1.85	.48	---	2.33	**2.57**
	Inst	SF	Sm	LG	.019	2484	2.53	.54	---	3.07	**3.33**
12' T x 4'-0" H (6.75 LF/CY)	Inst	LF	Lg	LG	.104	463.0	13.90	2.95	---	16.85	**18.30**
	Inst	LF	Sm	LG	.115	417.0	19.00	3.27	---	22.27	**23.90**

Description	Oper	Unit	Vol	Crew Size	Man-hours per Unit	Crew Output per Day	Avg Mat'l Unit Cost	Avg Labor Unit Cost	Avg Equip Unit Cost	Avg Total Unit Cost	Avg Price Incl O&P
12" T x 4'-0" H (27.0 LF/CY)	Inst	SF	Lg	LG	.026	1850	3.70	.74	---	4.44	**4.80**
	Inst	SF	Sm	LG	.029	1665	5.06	.82	---	5.88	**6.29**
12" T x 8'-0" H (3.38 LF/CY)	Inst	LF	Lg	LG	.208	231.0	27.70	5.91	---	33.61	**36.50**
	Inst	LF	Sm	LG	.231	208.0	38.00	6.56	---	44.56	**47.80**
12" T x 8'-0" H (27.0 SF/CY)	Inst	SF	Lg	LG	.026	1850	3.70	.74	---	4.44	**4.80**
	Inst	SF	Sm	LG	.026	1850	5.06	.74	---	5.80	**6.16**
12" T x 12'-0" H (2.25 LF/CY)	Inst	LF	Lg	LG	.312	154.0	40.70	8.86	---	49.56	**53.90**
	Inst	LF	Sm	LG	.312	154.0	55.70	8.86	---	64.56	**68.90**
12" T x 12'-0" H (27.0 SF/CY)	Inst	SF	Lg	LG	.026	1850	3.70	.74	---	4.44	**4.80**
	Inst	SF	Sm	LG	.026	1850	5.06	.74	---	5.80	**6.16**

Slabs, sidewalks and driveways, demo with air tools

With reinforcing (6 x 6 / 10 x 10)

Description	Oper	Unit	Vol	Crew Size	Man-hours per Unit	Crew Output per Day	Avg Mat'l Unit Cost	Avg Labor Unit Cost	Avg Equip Unit Cost	Avg Total Unit Cost	Avg Price Incl O&P
4" T	Demo	SF	Lg	AD	.063	380.0	---	1.90	.17	2.07	**2.98**
	Demo	SF	Sm	AD	.074	323.0	---	2.23	.20	2.43	**3.50**
5" T	Demo	SF	Lg	AD	.073	330.0	---	2.20	.20	2.40	**3.46**
	Demo	SF	Sm	AD	.085	281.0	---	2.57	.23	2.80	**4.03**
6" T	Demo	SF	Lg	AD	.083	290.0	---	2.50	.22	2.72	**3.93**
	Demo	SF	Sm	AD	.097	247.0	---	2.93	.26	3.19	**4.59**

Without reinforcing

Description	Oper	Unit	Vol	Crew Size	Man-hours per Unit	Crew Output per Day	Avg Mat'l Unit Cost	Avg Labor Unit Cost	Avg Equip Unit Cost	Avg Total Unit Cost	Avg Price Incl O&P
4" T	Demo	SF	Lg	AD	.044	550.0	---	1.33	.12	1.45	**2.09**
	Demo	SF	Sm	AD	.051	468.0	---	1.54	.14	1.68	**2.42**
5" T	Demo	SF	Lg	AD	.051	475.0	---	1.54	.14	1.68	**2.42**
	Demo	SF	Sm	AD	.059	404.0	---	1.78	.16	1.94	**2.79**
6" T	Demo	SF	Lg	AD	.057	420.0	---	1.72	.15	1.87	**2.70**
	Demo	SF	Sm	AD	.067	357.0	---	2.02	.18	2.20	**3.17**

Place concrete

Forming, 4 uses

Description	Oper	Unit	Vol	Crew Size	Man-hours per Unit	Crew Output per Day	Avg Mat'l Unit Cost	Avg Labor Unit Cost	Avg Equip Unit Cost	Avg Total Unit Cost	Avg Price Incl O&P
2" x 4" (4" T slab)	Inst	LF	Lg	DD	.041	580.0	.25	1.24	---	1.49	**2.08**
	Inst	LF	Sm	DD	.045	534.0	.34	1.36	---	1.70	**2.35**
2" x 6" (5" T and 6" T slab)	Inst	LF	Lg	DD	.041	580.0	.31	1.24	---	1.55	**2.14**
	Inst	LF	Sm	DD	.045	534.0	.42	1.36	---	1.78	**2.43**

Grading

Description	Oper	Unit	Vol	Crew Size	Man-hours per Unit	Crew Output per Day	Avg Mat'l Unit Cost	Avg Labor Unit Cost	Avg Equip Unit Cost	Avg Total Unit Cost	Avg Price Incl O&P
Dirt, cut and fill, +/- 1/10 ft	Inst	SF	Lg	1L	.010	800.0	---	.27	---	.27	**.41**
	Inst	SF	Sm	1L	.011	720.0	---	.30	---	.30	**.45**
Gravel, 3/4" to 1-1/2" stone	Inst	SF	Lg	1L	.011	750.0	.36	.30	---	.66	**.81**
	Inst	SF	Sm	1L	.012	675.0	.49	.33	---	.82	**.98**

Screeds, 3 uses

Description	Oper	Unit	Vol	Crew Size	Man-hours per Unit	Crew Output per Day	Avg Mat'l Unit Cost	Avg Labor Unit Cost	Avg Equip Unit Cost	Avg Total Unit Cost	Avg Price Incl O&P
2" x 2"	Inst	LF	Lg	DD	.029	835.0	.10	.87	---	.97	**1.39**
	Inst	LF	Sm	DD	.031	768.0	.14	.94	---	1.08	**1.52**
2" x 4"	Inst	LF	Lg	DD	.032	760.0	.19	.97	---	1.16	**1.62**
	Inst	LF	Sm	DD	.034	699.0	.26	1.03	---	1.29	**1.78**

Description	Oper	Unit	Vol	Crew Size	Man-hours per Unit	Crew Output per Day	Avg Mat'l Unit Cost	Avg Labor Unit Cost	Avg Equip Unit Cost	Avg Total Unit Cost	Avg Price Incl O&P

Reinforcing steel rods, includes 5% waste, costs are per LF of rod or SF of mesh

Description	Oper	Unit	Vol	Crew Size	Man-hrs/Unit	Output/Day	Mat'l	Labor	Equip	Total	O&P
No. 3 (3/8" rod)	Inst	LF	Lg	LB	.011	1480	.22	.30	---	.52	.67
	Inst	LF	Sm	LB	.012	1332	.30	.33	---	.63	.79
No. 4 (1/2" rod)	Inst	LF	Lg	LB	.011	1410	.25	.30	---	.55	.70
	Inst	LF	Sm	LB	.013	1269	.34	.36	---	.70	.87
No. 5 (5/8" rod)	Inst	LF	Lg	LB	.012	1340	.39	.33	---	.72	.88
	Inst	LF	Sm	LB	.013	1206	.54	.36	---	.90	1.07
6 x 6 / 10-10 @ 21 lbs/CSF	Inst	SF	Lg	LB	.003	5250	.25	.08	---	.33	.37
	Inst	SF	Sm	LB	.003	4725	.34	.08	---	.42	.46
6 x 6 / 6-6 @ 42 lbs/CSF	Inst	SF	Lg	LB	.004	4500	.41	.11	---	.52	.57
	Inst	SF	Sm	LB	.004	4050	.55	.11	---	.66	.71

Concrete, pour and finish (steel trowel), material cost includes 5% waste and assumes use of 2500 PSI, 1" aggregate, 5.5 sack mix

Description	Oper	Unit	Vol	Crew Size	Man-hrs/Unit	Output/Day	Mat'l	Labor	Equip	Total	O&P
3-1/2" T (92.57 SF/CY)	Inst	SF	Lg	DF	.023	2805	.92	.67	---	1.59	1.91
	Inst	SF	Sm	DF	.025	2581	1.27	.72	---	1.99	2.34
4" T (81.00 SF/CY)	Inst	SF	Lg	DF	.024	2720	.92	.70	---	1.62	1.95
	Inst	SF	Sm	DF	.026	2502	1.27	.75	---	2.02	2.38
5" T (64.80 SF/CY)	Inst	SF	Lg	DF	.025	2590	1.85	.72	---	2.57	2.92
	Inst	SF	Sm	DF	.027	2383	2.53	.78	---	3.31	3.69
6" T (54.00 SF/CY)	Inst	SF	Lg	DF	.026	2460	1.85	.75	---	2.60	2.96
	Inst	SF	Sm	DF	.028	2263	2.53	.81	---	3.34	3.73

Forms, wreck and clean

Description	Oper	Unit	Vol	Crew Size	Man-hrs/Unit	Output/Day	Mat'l	Labor	Equip	Total	O&P
2" x 4"	Inst	LF	Lg	1L	.022	360.0	---	.60	---	.60	.90
	Inst	LF	Sm	1L	.025	324.0	---	.69	---	.69	1.02
2" x 6"	Inst	LF	Lg	1L	.024	340.0	---	.66	---	.66	.98
	Inst	LF	Sm	1L	.026	306.0	---	.71	---	.71	1.06

Material information

Ready mix delivered by truck

Material costs only, prices are typical for most cities and assumes delivery up to 20 miles for 10 CY or more, 3" to 4" slump

Footing and foundation, using 1-1/2" aggregate

Description	Oper	Unit	Vol	Crew Size	Man-hrs/Unit	Output/Day	Mat'l	Labor	Equip	Total	O&P
2000 PSI, 4.8 sack mix	Inst	CY	Lg	---	---	---	81.90	---	---	81.90	81.90
	Inst	CY	Sm	---	---	---	112.00	---	---	112.00	112.00
2500 PSI, 5.2 sack mix	Inst	CY	Lg	---	---	---	84.50	---	---	84.50	84.50
	Inst	CY	Sm	---	---	---	116.00	---	---	116.00	116.00
3000 PSI, 5.7 sack mix	Inst	CY	Lg	---	---	---	88.00	---	---	88.00	88.00
	Inst	CY	Sm	---	---	---	121.00	---	---	121.00	121.00
3500 PSI, 6.3 sack mix	Inst	CY	Lg	---	---	---	89.80	---	---	89.80	89.80
	Inst	CY	Sm	---	---	---	123.00	---	---	123.00	123.00
4000 PSI, 6.9 sack mix	Inst	CY	Lg	---	---	---	95.90	---	---	95.90	95.90
	Inst	CY	Sm	---	---	---	131.00	---	---	131.00	131.00

Description	Oper	Unit	Vol	Crew Size	Man-hours per Unit	Crew Output per Day	Avg Mat'l Unit Cost	Avg Labor Unit Cost	Avg Equip Unit Cost	Avg Total Unit Cost	Avg Price Incl O&P

Slab, sidewalk and driveway, using 1" aggregate

Description	Oper	Unit	Vol	Crew Size	Man-hours per Unit	Crew Output per Day	Avg Mat'l Unit Cost	Avg Labor Unit Cost	Avg Equip Unit Cost	Avg Total Unit Cost	Avg Price Incl O&P
2000 PSI, 5.0 sack mix	Inst	CY	Lg	---	---	---	83.60	---	---	83.60	**83.60**
	Inst	CY	Sm	---	---	---	115.00	---	---	115.00	**115.00**
2500 PSI, 5.5 sack mix	Inst	CY	Lg	---	---	---	87.10	---	---	87.10	**87.10**
	Inst	CY	Sm	---	---	---	119.00	---	---	119.00	**119.00**
3000 PSI, 6.0 sack mix	Inst	CY	Lg	---	---	---	89.80	---	---	89.80	**89.80**
	Inst	CY	Sm	---	---	---	123.00	---	---	123.00	**123.00**
3500 PSI, 6.6 sack mix	Inst	CY	Lg	---	---	---	94.20	---	---	94.20	**94.20**
	Inst	CY	Sm	---	---	---	129.00	---	---	129.00	**129.00**
4000 PSI, 7.1 sack mix	Inst	CY	Lg	---	---	---	96.80	---	---	96.80	**96.80**
	Inst	CY	Sm	---	---	---	133.00	---	---	133.00	**133.00**

Pump & grout mix, using pea gravel, 3/8" aggregate

Description	Oper	Unit	Vol	Crew Size	Man-hours per Unit	Crew Output per Day	Avg Mat'l Unit Cost	Avg Labor Unit Cost	Avg Equip Unit Cost	Avg Total Unit Cost	Avg Price Incl O&P
2000 PSI, 6.0 sack mix	Inst	CY	Lg	---	---	---	89.80	---	---	89.80	**89.80**
	Inst	CY	Sm	---	---	---	123.00	---	---	123.00	**123.00**
2500 PSI, 6.5 sack mix	Inst	CY	Lg	---	---	---	93.30	---	---	93.30	**93.30**
	Inst	CY	Sm	---	---	---	128.00	---	---	128.00	**128.00**
3000 PSI, 7.2 sack mix	Inst	CY	Lg	---	---	---	97.70	---	---	97.70	**97.70**
	Inst	CY	Sm	---	---	---	134.00	---	---	134.00	**134.00**
3500 PSI, 7.9 sack mix	Inst	CY	Lg	---	---	---	102.00	---	---	102.00	**102.00**
	Inst	CY	Sm	---	---	---	140.00	---	---	140.00	**140.00**
4000 PSI, 8.5 sack mix	Inst	CY	Lg	---	---	---	108.00	---	---	108.00	**108.00**
	Inst	CY	Sm	---	---	---	148.00	---	---	148.00	**148.00**

Adjustments

Extra delivery costs for ready-mix concrete

Delivery over 20 miles, ADD

Description	Oper	Unit	Vol	Crew Size	Man-hours per Unit	Crew Output per Day	Avg Mat'l Unit Cost	Avg Labor Unit Cost	Avg Equip Unit Cost	Avg Total Unit Cost	Avg Price Incl O&P
	Inst	Mi	Lg	---	---	---	2.24	---	---	2.24	**2.24**
	Inst	Mi	Sm	---	---	---	2.24	---	---	2.24	**2.24**

Standby charge in excess of 5 minutes per CY delivered per minute extra time, ADD

Description	Oper	Unit	Vol	Crew Size	Man-hours per Unit	Crew Output per Day	Avg Mat'l Unit Cost	Avg Labor Unit Cost	Avg Equip Unit Cost	Avg Total Unit Cost	Avg Price Incl O&P
	Inst	Min	Lg	---	---	---	1.12	---	---	1.12	**1.12**
	Inst	Min	Sm	---	---	---	1.12	---	---	1.12	**1.12**

Loads 7.25 CY or less, ADD

Description	Oper	Unit	Vol	Crew Size	Man-hours per Unit	Crew Output per Day	Avg Mat'l Unit Cost	Avg Labor Unit Cost	Avg Equip Unit Cost	Avg Total Unit Cost	Avg Price Incl O&P
7.25 CY	Inst	LS	Lg	---	---	---	6.90	---	---	6.90	**6.90**
	Inst	LS	Sm	---	---	---	6.90	---	---	6.90	**6.90**
6.0 CY	Inst	LS	Lg	---	---	---	24.60	---	---	24.60	**24.60**
	Inst	LS	Sm	---	---	---	24.60	---	---	24.60	**24.60**
5.0 CY	Inst	LS	Lg	---	---	---	39.40	---	---	39.40	**39.40**
	Inst	LS	Sm	---	---	---	39.40	---	---	39.40	**39.40**
4.0 CY	Inst	LS	Lg	---	---	---	54.20	---	---	54.20	**54.20**
	Inst	LS	Sm	---	---	---	54.20	---	---	54.20	**54.20**
3.0 CY	Inst	LS	Lg	---	---	---	69.00	---	---	69.00	**69.00**
	Inst	LS	Sm	---	---	---	69.00	---	---	69.00	**69.00**
2.0 CY	Inst	LS	Lg	---	---	---	83.80	---	---	83.80	**83.80**
	Inst	LS	Sm	---	---	---	83.80	---	---	83.80	**83.80**
1.0 CY	Inst	LS	Lg	---	---	---	98.60	---	---	98.60	**98.60**
	Inst	LS	Sm	---	---	---	98.60	---	---	98.60	**98.60**

Description	Oper	Unit	Vol	Crew Size	Man-hours per Unit	Crew Output per Day	Avg Mat'l Unit Cost	Avg Labor Unit Cost	Avg Equip Unit Cost	Avg Total Unit Cost	Avg Price Incl O&P

Extra costs for non-standard mix additives

High early strength

5.0 sack mix	Inst	CY	Lg	---	---	---	10.60	---	---	10.60	**10.60**
	Inst	CY	Sm	---	---	---	10.60	---	---	10.60	**10.60**
6.0 sack mix	Inst	CY	Lg	---	---	---	13.80	---	---	13.80	**13.80**
	Inst	CY	Sm	---	---	---	13.80	---	---	13.80	**13.80**

White cement (architectural)

	Inst	CY	Lg	---	---	---	47.90	---	---	47.90	**47.90**
	Inst	CY	Sm	---	---	---	47.90	---	---	47.90	**47.90**
For 1% calcium chloride	Inst	CY	Lg	---	---	---	1.60	---	---	1.60	**1.60**
	Inst	CY	Sm	---	---	---	1.60	---	---	1.60	**1.60**
For 2% calcium chloride	Inst	CY	Lg	---	---	---	2.24	---	---	2.24	**2.24**
	Inst	CY	Sm	---	---	---	2.24	---	---	2.24	**2.24**

Chemical compensated shrinkage (WRDA Admix)

	Inst	CY	Lg	---	---	---	14.40	---	---	14.40	**14.40**
	Inst	CY	Sm	---	---	---	14.40	---	---	14.40	**14.40**

Super plasticized mix, with 7"-8" slump

	Inst	%	Lg	---	---	---	8.0	---	---	8.0	**8.0**
	Inst	%	Sm	---	---	---	8.0	---	---	8.0	**8.0**

Coloring of concrete, ADD

Light or sand colors	Inst	CY	Lg	---	---	---	17.60	---	---	17.60	**17.60**
	Inst	CY	Sm	---	---	---	17.60	---	---	17.60	**17.60**
Medium or buff colors	Inst	CY	Lg	---	---	---	26.00	---	---	26.00	**26.00**
	Inst	CY	Sm	---	---	---	26.00	---	---	26.00	**26.00**
Dark colors	Inst	CY	Lg	---	---	---	39.20	---	---	39.20	**39.20**
	Inst	CY	Sm	---	---	---	39.20	---	---	39.20	**39.20**
Green	Inst	CY	Lg	---	---	---	43.70	---	---	43.70	**43.70**
	Inst	CY	Sm	---	---	---	43.70	---	---	43.70	**43.70**

Extra costs for non-standard aggregates

Lightweight aggregate, ADD

Mix from truck to forms	Inst	CY	Lg	---	---	---	38.10	---	---	38.10	**38.10**
	Inst	CY	Sm	---	---	---	38.10	---	---	38.10	**38.10**
Pump mix	Inst	CY	Lg	---	---	---	39.20	---	---	39.20	**39.20**
	Inst	CY	Sm	---	---	---	39.20	---	---	39.20	**39.20**

Granite aggregate, ADD

	Inst	CY	Lg	---	---	---	5.60	---	---	5.60	**5.60**
	Inst	CY	Sm	---	---	---	5.60	---	---	5.60	**5.60**

Concrete block. See Masonry, page 282

Description	Oper	Unit	Vol	Crew Size	Man-hours per Unit	Crew Output per Day	Avg Mat'l Unit Cost	Avg Labor Unit Cost	Avg Equip Unit Cost	Avg Total Unit Cost	Avg Price Incl O&P

Countertops

Formica

One-piece tops; straight, "L", or "U" shapes; surfaced with laminated plastic cemented to particleboard base

Post formed countertop with raised front drip edge

25" W, 1-1/2" H front edge, 4" H coved backsplash

Description	Oper	Unit	Vol	Crew Size	Man-hours per Unit	Crew Output per Day	Avg Mat'l Unit Cost	Avg Labor Unit Cost	Avg Equip Unit Cost	Avg Total Unit Cost	Avg Price Incl O&P
Satin/suede patterns	Inst	LF	Lg	2C	.485	33.00	24.40	16.10	---	40.50	**52.20**
	Inst	LF	Sm	2C	.808	19.80	26.70	26.90	---	53.60	**71.00**
Solid patterns	Inst	LF	Lg	2C	.485	33.00	27.10	16.10	---	43.20	**55.40**
	Inst	LF	Sm	2C	.808	19.80	29.70	26.90	---	56.60	**74.50**
Specialty finish patterns	Inst	LF	Lg	2C	.485	33.00	32.50	16.10	---	48.60	**61.60**
	Inst	LF	Sm	2C	.808	19.80	35.60	26.90	---	62.50	**81.30**
Wood tone patterns	Inst	LF	Lg	2C	.485	33.00	33.90	16.10	---	50.00	**63.20**
	Inst	LF	Sm	2C	.808	19.80	37.10	26.90	---	64.00	**83.00**

Double roll top, 1-1/2" H front and back edges, no backsplash

25" W

Description	Oper	Unit	Vol	Crew Size	Man-hours per Unit	Crew Output per Day	Avg Mat'l Unit Cost	Avg Labor Unit Cost	Avg Equip Unit Cost	Avg Total Unit Cost	Avg Price Incl O&P
Satin/suede patterns	Inst	LF	Lg	2C	.457	35.00	35.40	15.20	---	50.60	**63.50**
	Inst	LF	Sm	2C	.762	21.00	38.80	25.40	---	64.20	**82.60**
Solid patterns	Inst	LF	Lg	2C	.457	35.00	39.30	15.20	---	54.50	**68.00**
	Inst	LF	Sm	2C	.762	21.00	43.10	25.40	---	68.50	**87.60**
Specialty finish patterns	Inst	LF	Lg	2C	.457	35.00	47.20	15.20	---	62.40	**77.10**
	Inst	LF	Sm	2C	.762	21.00	51.70	25.40	---	77.10	**97.50**
Wood tone patterns	Inst	LF	Lg	2C	.457	35.00	49.10	15.20	---	64.30	**79.30**
	Inst	LF	Sm	2C	.762	21.00	53.80	25.40	---	79.20	**99.90**

36" W

Description	Oper	Unit	Vol	Crew Size	Man-hours per Unit	Crew Output per Day	Avg Mat'l Unit Cost	Avg Labor Unit Cost	Avg Equip Unit Cost	Avg Total Unit Cost	Avg Price Incl O&P
Satin/suede patterns	Inst	LF	Lg	2C	.500	32.00	43.60	16.60	---	60.20	**75.00**
	Inst	LF	Sm	2C	.833	19.20	47.70	27.70	---	75.40	**96.40**
Solid patterns	Inst	LF	Lg	2C	.500	32.00	48.40	16.60	---	65.00	**80.60**
	Inst	LF	Sm	2C	.833	19.20	53.00	27.70	---	80.70	**103.00**
Specialty finish patterns	Inst	LF	Lg	2C	.500	32.00	58.10	16.60	---	74.70	**91.70**
	Inst	LF	Sm	2C	.833	19.20	63.60	27.70	---	91.30	**115.00**
Wood tone patterns	Inst	LF	Lg	2C	.500	32.00	60.50	16.60	---	77.10	**94.50**
	Inst	LF	Sm	2C	.833	19.20	66.20	27.70	---	93.90	**118.00**

Post formed countertop with square edge veneer front

25" W, 1-1/2" H front edge, 4" H coved backsplash

Description	Oper	Unit	Vol	Crew Size	Man-hours per Unit	Crew Output per Day	Avg Mat'l Unit Cost	Avg Labor Unit Cost	Avg Equip Unit Cost	Avg Total Unit Cost	Avg Price Incl O&P
Satin/suede patterns	Inst	LF	Lg	2C	.485	33.00	30.50	16.10	---	46.60	**59.30**
	Inst	LF	Sm	2C	.808	19.80	33.40	26.90	---	60.30	**78.80**
Solid patterns	Inst	LF	Lg	2C	.485	33.00	33.90	16.10	---	50.00	**63.20**
	Inst	LF	Sm	2C	.808	19.80	37.10	26.90	---	64.00	**83.00**
Specialty finish patterns	Inst	LF	Lg	2C	.485	33.00	40.70	16.10	---	56.80	**71.00**
	Inst	LF	Sm	2C	.808	19.80	44.60	26.90	---	71.50	**91.60**
Wood tone patterns	Inst	LF	Lg	2C	.485	33.00	42.40	16.10	---	58.50	**72.90**
	Inst	LF	Sm	2C	.808	19.80	46.40	26.90	---	73.30	**93.70**

Description	Oper	Unit	Vol	Crew Size	Man-hours per Unit	Crew Output per Day	Avg Mat'l Unit Cost	Avg Labor Unit Cost	Avg Equip Unit Cost	Avg Total Unit Cost	Avg Price Incl O&P

Self-edge countertop with square edge veneer front
25" W, 1-1/2" H front edge, 4" H coved backsplash (@ 90-degree angle to deck)

Description	Oper	Unit	Vol	Crew	MH	Out	Mat'l	Labor	Equip	Total	Price
Satin/suede patterns	Inst	LF	Lg	2C	.485	33.00	73.30	16.10	---	89.40	**108.00**
	Inst	LF	Sm	2C	.808	19.80	80.20	26.90	---	107.10	**133.00**
Solid patterns	Inst	LF	Lg	2C	.485	33.00	81.40	16.10	---	97.50	**118.00**
	Inst	LF	Sm	2C	.808	19.80	89.20	26.90	---	116.10	**143.00**
Specialty finish patterns	Inst	LF	Lg	2C	.485	33.00	97.70	16.10	---	113.80	**137.00**
	Inst	LF	Sm	2C	.808	19.80	107.00	26.90	---	133.90	**163.00**
Wood tone patterns	Inst	LF	Lg	2C	.485	33.00	102.00	16.10	---	118.10	**141.00**
	Inst	LF	Sm	2C	.808	19.80	111.00	26.90	---	137.90	**168.00**

Material cost adjustments, ADD
Diagonal corner cut

Description	Oper	Unit	Vol				Mat'l			Total	Price
Standard	Inst	Ea	Lg	---	---	---	22.70	---	---	22.70	**22.70**
	Inst	Ea	Sm	---	---	---	24.80	---	---	24.80	**24.80**
With plateau shelf	Inst	Ea	Lg	---	---	---	107.00	---	---	107.00	**107.00**
	Inst	Ea	Sm	---	---	---	117.00	---	---	117.00	**117.00**

Radius corner cut

Description	Oper	Unit	Vol				Mat'l			Total	Price
3" or 6"	Inst	Ea	Lg	---	---	---	15.10	---	---	15.10	**15.10**
	Inst	Ea	Sm	---	---	---	16.60	---	---	16.60	**16.60**
12"	Inst	Ea	Lg	---	---	---	15.10	---	---	15.10	**15.10**
	Inst	Ea	Sm	---	---	---	16.60	---	---	16.60	**16.60**
Quarter radius end	Inst	Ea	Lg	---	---	---	15.10	---	---	15.10	**15.10**
	Inst	Ea	Sm	---	---	---	16.60	---	---	16.60	**16.60**
Half radius end	Inst	Ea	Lg	---	---	---	27.70	---	---	27.70	**27.70**
	Inst	Ea	Sm	---	---	---	30.40	---	---	30.40	**30.40**

Miter corner, shop assembled

	Inst	Ea	Lg	---	---	---	22.80	---	---	22.80	**22.80**
	Inst	Ea	Sm	---	---	---	25.00	---	---	25.00	**25.00**

Splicing any top or leg 12' or longer, shop assembled

	Inst	Ea	Lg	---	---	---	26.50	---	---	26.50	**26.50**
	Inst	Ea	Sm	---	---	---	29.00	---	---	29.00	**29.00**

Endsplash with finished sides and edges

	Inst	Ea	Lg	---	---	---	20.20	---	---	20.20	**20.20**
	Inst	Ea	Sm	---	---	---	22.10	---	---	22.10	**22.10**

Sink or range cutout	Inst	Ea	Lg	---	---	---	8.82	---	---	8.82	**8.82**
	Inst	Ea	Sm	---	---	---	9.66	---	---	9.66	**9.66**

Ceramic tile
Countertop/backsplash
Adhesive set with backmounted tile

Description	Oper	Unit	Vol	Crew	MH	Out	Mat'l	Labor	Equip	Total	Price
1" x 1"	Inst	SF	Lg	TB	.229	70.00	7.24	6.45	---	13.69	**19.80**
	Inst	SF	Sm	TB	.381	42.00	6.41	10.70	---	17.11	**23.10**
4-1/4" x 4-1/4"	Inst	SF	Lg	TB	.200	80.00	5.79	5.63	---	11.42	**16.60**
	Inst	SF	Sm	TB	.333	48.00	5.12	9.37	---	14.49	**19.70**

Description	Oper	Unit	Vol	Crew Size	Man-hours per Unit	Crew Output per Day	Avg Mat'l Unit Cost	Avg Labor Unit Cost	Avg Equip Unit Cost	Avg Total Unit Cost	Avg Price Incl O&P

Wood

Butcher block construction throughout top; custom, straight, "L", or "U" shapes

Self-edge top; 26" W, with 4" H backsplash, 1-1/2" H front and back edges

Description	Oper	Unit	Vol	Crew Size	Man-hours	Output	Mat'l	Labor	Equip	Total	Price O&P
	Inst	LF	Lg	2C	.485	33.00	66.80	16.10	---	82.90	101.00
	Inst	LF	Sm	2C	.746	21.45	74.80	24.80	---	99.60	123.00
Material cost adjustments, ADD											
Miter corner	Inst	Ea	Lg	---	---	---	16.70	---	---	16.70	16.70
	Inst	Ea	Sm	---	---	---	18.70	---	---	18.70	18.70
Sink or surface saver cutout	Inst	Ea	Lg	---	---	---	50.10	---	---	50.10	50.10
	Inst	Ea	Sm	---	---	---	56.10	---	---	56.10	56.10
45-degree plateau corner	Inst	Ea	Lg	---	---	---	251.00	---	---	251.00	251.00
	Inst	Ea	Sm	---	---	---	281.00	---	---	281.00	281.00
End splash	Inst	Ea	Lg	---	---	---	58.50	---	---	58.50	50.50
	Inst	Ea	Sm	---	---	---	65.50	---	---	65.50	65.50

Cupolas

Natural finish redwood, aluminum roof

Description	Oper	Unit	Vol	Crew Size	Man-hours	Output	Mat'l	Labor	Equip	Total	Price O&P
24" x 24" x 25" H	Inst	Ea	Lg	CA	1.14	7.00	262.00	37.90	---	299.90	385.00
	Inst	Ea	Sm	CA	1.63	4.90	305.00	54.20	---	359.20	463.00
30" x 30" x 30" H	Inst	Ea	Lg	CA	1.33	6.00	328.00	44.30	---	372.30	476.00
	Inst	Ea	Sm	CA	1.90	4.20	381.00	63.20	---	444.20	571.00
35" x 35" x 33" H	Inst	Ea	Lg	CA	1.60	5.00	448.00	53.20	---	501.20	640.00
	Inst	Ea	Sm	CA	2.29	3.50	522.00	76.20	---	598.20	766.00

Natural finish redwood, copper roof

Description	Oper	Unit	Vol	Crew Size	Man-hours	Output	Mat'l	Labor	Equip	Total	Price O&P
22" x 22" x 33" H	Inst	Ea	Lg	CA	1.14	7.00	307.00	37.90	---	344.90	440.00
	Inst	Ea	Sm	CA	1.63	4.90	357.00	54.20	---	411.20	528.00
25" x 25" x 39" H	Inst	Ea	Lg	CA	1.33	6.00	371.00	44.30	---	415.30	530.00
	Inst	Ea	Sm	CA	1.90	4.20	432.00	63.20	---	495.20	635.00
35" x 35" x 33" H	Inst	Ea	Lg	CA	1.60	5.00	525.00	53.20	---	578.20	736.00
	Inst	Ea	Sm	CA	2.29	3.50	611.00	76.20	---	687.20	878.00

Natural finish redwood, octagonal shape copper roof

Description	Oper	Unit	Vol	Crew Size	Man-hours	Output	Mat'l	Labor	Equip	Total	Price O&P
01" W, 07" H	Inst	Ea	Lg	CA	1.60	5.00	536.00	53.20	---	589.20	749.00
	Inst	Ea	Sm	CA	2.29	3.50	623.00	76.20	---	699.20	893.00
35" W, 43" H	Inst	Ea	Lg	CA	2.00	4.00	681.00	66.50	---	747.50	951.00
	Inst	Ea	Sm	CA	2.86	2.80	793.00	95.20	---	888.20	1130.00

Weathervanes for above, aluminum

Description	Oper	Unit	Vol	Crew Size	Man-hours	Output	Mat'l	Labor	Equip	Total	Price O&P
18" H, black finish	Inst	Ea	Lg	---	---	---	59.30	---	---	59.30	59.30
	Inst	Ea	Sm	---	---	---	69.00	---	---	69.00	69.00
24" H, black finish	Inst	Ea	Lg	---	---	---	78.00	---	---	78.00	78.00
	Inst	Ea	Sm	---	---	---	90.80	---	---	90.80	90.80
36" H, black and gold finish	Inst	Ea	Lg	---	---	---	175.00	---	---	175.00	175.00
	Inst	Ea	Sm	---	---	---	203.00	---	---	203.00	203.00

Demolition

Average Manhours per Unit is based on what the designated crew can do in one day. In this section, the crew might be a laborer, a carpenter, a floor layer, etc. Who does the wrecking depends on the quantity of wrecking to be done.

A contractor might use laborers exclusively on a large volume job, but use a carpenter or a carpenter and a laborer on a small volume job.

The choice of tools and equipment greatly affects the quantity of work that a worker can accomplish in a day. A person can remove more brick with a compressor and pneumatic tool than with a sledgehammer or pry bar. In this section, the phrase "by hand"

includes the use of hand tools, i.e., sledgehammers, wrecking bars, claw hammers, etc. When the Average Manhours per Unit is based on the use of equipment (not hand tools), a description of the equipment is provided.

The Average Manhours per Unit is not based on the use of "heavy" equipment such as bulldozers, cranes with wrecking balls, etc. The Average Manhours per Unit does include the labor involved in hauling wrecked material or debris to a dumpster located at the site. Average rental costs for dumpsters are: $450.00 for 40 CY (20' x 8' x 8' H); $365.00 for 30 CY (20' x 8' x 6' H). Rental period includes delivery and pickup when full.

Description	Oper	Unit	Vol	Crew Size	Man-hours per Unit	Crew Output per Day	Avg Mat'l Unit Cost	Avg Labor Unit Cost	Avg Equip Unit Cost	Avg Total Unit Cost	Avg Price Incl O&P

Demolition
Wreck and remove to dumpster

Concrete

Footings, with air tools; reinforced

Description	Oper	Unit	Vol	Crew Size	Man-hours per Unit	Crew Output per Day	Avg Mat'l Unit Cost	Avg Labor Unit Cost	Avg Equip Unit Cost	Avg Total Unit Cost	Avg Price Incl O&P
8" T x 12" W (.67 CF/LF)	Demo	LF	Lg	AB	.133	120.0	---	3.92	.77	4.69	**6.58**
	Demo	LF	Sm	AB	.205	78.00	---	6.05	1.19	7.24	**10.10**
8" T x 16" W (.89 CF/LF)	Demo	LF	Lg	AB	.160	100.0	---	4.72	.93	5.65	**7.91**
	Demo	LF	Sm	AB	.246	65.00	---	7.25	1.43	8.68	**12.20**
8" T x 20" W (1.11 CF/LF)	Demo	LF	Lg	AB	.178	90.00	---	5.25	1.03	6.28	**8.80**
	Demo	LF	Sm	AB	.274	58.50	---	8.08	1.58	9.66	**13.50**
12" T x 12" W (1.00 CF/LF)	Demo	LF	Lg	AB	.229	70.00	---	6.75	1.32	8.07	**11.30**
	Demo	LF	Sm	AB	.352	45.50	---	10.40	2.04	12.44	**17.40**
12" T x 16" W (1.33 CF/LF)	Demo	LF	Lg	AB	.267	60.00	---	7.87	1.54	9.41	**13.20**
	Demo	LF	Sm	AB	.410	39.00	---	12.10	2.38	14.48	**20.30**
12" T x 20" W (1.67 CF/LF)	Demo	LF	Lg	AB	.291	55.00	---	8.58	1.68	10.26	**14.40**
	Demo	LF	Sm	AB	.448	35.75	---	13.20	2.59	15.79	**22.20**
12" T x 24" W (2.00 CF/LF)	Demo	LF	Lg	AB	.320	50.00	---	9.44	1.85	11.29	**15.80**
	Demo	LF	Sm	AB	.492	32.50	---	14.50	2.85	17.35	**24.30**

Foundations and retaining walls, with air tools, per LF wall

With reinforcing
4'-0" H

Description	Oper	Unit	Vol	Crew Size	Man-hours per Unit	Crew Output per Day	Avg Mat'l Unit Cost	Avg Labor Unit Cost	Avg Equip Unit Cost	Avg Total Unit Cost	Avg Price Incl O&P
8" T (2.67 CF/LF)	Demo	SF	Lg	AB	.267	60.00	---	7.87	1.54	9.41	**13.20**
	Demo	SF	Sm	AB	.410	39.00	---	12.10	2.38	14.48	**20.30**
12" T (4.00 CF/LF)	Demo	SF	Lg	AB	.356	45.00	---	10.50	2.06	12.56	**17.60**
	Demo	SF	Sm	AB	.547	29.25	---	16.10	3.17	19.27	**27.10**

8'-0" H

Description	Oper	Unit	Vol	Crew Size	Man-hours per Unit	Crew Output per Day	Avg Mat'l Unit Cost	Avg Labor Unit Cost	Avg Equip Unit Cost	Avg Total Unit Cost	Avg Price Incl O&P
8" T (5.33 CF/LF)	Demo	SF	Lg	AB	.291	55.00	---	8.58	1.68	10.26	**14.40**
	Demo	SF	Sm	AB	.448	35.75	---	13.20	2.59	15.79	**22.20**
12" T (8.00 CF/LF)	Demo	SF	Lg	AB	.400	40.00	---	11.80	2.32	14.12	**19.80**
	Demo	SF	Sm	AB	.615	26.00	---	18.10	3.56	21.66	**30.40**

12'-0" H

Description	Oper	Unit	Vol	Crew Size	Man-hours per Unit	Crew Output per Day	Avg Mat'l Unit Cost	Avg Labor Unit Cost	Avg Equip Unit Cost	Avg Total Unit Cost	Avg Price Incl O&P
8" T (8.00 CF/LF)	Demo	SF	Lg	AB	.320	50.00	---	9.44	1.85	11.29	**15.80**
	Demo	SF	Sm	AB	.492	32.50	---	14.50	2.85	17.35	**24.30**
12" T (12.00 CF/LF)	Demo	SF	Lg	AB	.457	35.00	---	13.50	2.65	16.15	**22.60**
	Demo	SF	Sm	AB	.703	22.75	---	20.70	4.07	24.77	**34.80**

Without reinforcing
4'-0" H

Description	Oper	Unit	Vol	Crew Size	Man-hours per Unit	Crew Output per Day	Avg Mat'l Unit Cost	Avg Labor Unit Cost	Avg Equip Unit Cost	Avg Total Unit Cost	Avg Price Incl O&P
8" T (2.67 CF/LF)	Demo	SF	Lg	AB	.200	80.00	---	5.90	1.16	7.06	**9.89**
	Demo	SF	Sm	AB	.308	52.00	---	9.08	1.78	10.86	**15.20**
12" T (4.00 CF/LF)	Demo	SF	Lg	AB	.267	60.00	---	7.87	1.54	9.41	**13.20**
	Demo	SF	Sm	AB	.410	39.00	---	12.10	2.38	14.48	**20.30**

Description	Oper	Unit	Vol	Crew Size	Man-hours per Unit	Crew Output per Day	Avg Mat'l Unit Cost	Avg Labor Unit Cost	Avg Equip Unit Cost	Avg Total Unit Cost	Avg Price Incl O&P
8'-0" H											
8" T (5.33 CF/LF)	Demo	SF	Lg	AB	.229	70.00	---	6.75	1.32	8.07	**11.30**
	Demo	SF	Sm	AB	.352	45.50	---	10.40	2.04	12.44	**17.40**
12" T (8.00 CF/LF)	Demo	SF	Lg	AB	.320	50.00	---	9.44	1.85	11.29	**15.80**
	Demo	SF	Sm	AB	.492	32.50	---	14.50	2.85	17.35	**24.30**
12'-0" H											
8" T (8.00 CF/LF)	Demo	SF	Lg	AB	.267	60.00	---	7.87	1.54	9.41	**13.20**
	Demo	SF	Sm	AB	.410	39.00	---	12.10	2.38	14.48	**20.30**
12" T (12.00 CF/LF)	Demo	SF	Lg	AB	.400	40.00	---	11.80	2.32	14.12	**19.80**
	Demo	SF	Sm	AB	.615	26.00	---	18.10	3.56	21.66	**30.40**

Slabs with air tools
With reinforcing

Description	Oper	Unit	Vol	Crew Size	Man-hours per Unit	Crew Output per Day	Avg Mat'l Unit Cost	Avg Labor Unit Cost	Avg Equip Unit Cost	Avg Total Unit Cost	Avg Price Incl O&P
4" T	Demo	SF	Lg	AB	.038	425.0	---	1.12	.22	1.34	**1.88**
	Demo	SF	Sm	AB	.058	276.3	---	1.71	.34	2.05	**2.87**
5" T	Demo	SF	Lg	AB	.043	370.0	---	1.27	.25	1.52	**2.13**
	Demo	SF	Sm	AB	.067	240.5	---	1.98	.39	2.37	**3.31**
6" T	Demo	SF	Lg	AB	.046	350.0	---	1.36	.26	1.62	**2.27**
	Demo	SF	Sm	AB	.070	227.5	---	2.06	.41	2.47	**3.47**

Without reinforcing

Description	Oper	Unit	Vol	Crew Size	Man-hours per Unit	Crew Output per Day	Avg Mat'l Unit Cost	Avg Labor Unit Cost	Avg Equip Unit Cost	Avg Total Unit Cost	Avg Price Incl O&P
4" T	Demo	SF	Lg	AB	.032	500.0	---	.94	.19	1.13	**1.59**
	Demo	SF	Sm	AB	.049	325.0	---	1.45	.29	1.74	**2.43**
5" T	Demo	SF	Lg	AB	.038	425.0	---	1.12	.22	1.34	**1.88**
	Demo	SF	Sm	AB	.058	276.3	---	1.71	.34	2.05	**2.87**
6" T	Demo	SF	Lg	AB	.043	370.0	---	1.27	.25	1.52	**2.13**
	Demo	SF	Sm	AB	.067	240.5	---	1.98	.39	2.37	**3.31**

Masonry
Brick
Chimneys

Description	Oper	Unit	Vol	Crew Size	Man-hours per Unit	Crew Output per Day	Avg Mat'l Unit Cost	Avg Labor Unit Cost	Avg Equip Unit Cost	Avg Total Unit Cost	Avg Price Incl O&P
4" T wall	Demo	VLF	Lg	LB	.640	25.00	---	17.60	---	17.60	**26.20**
	Demo	VLF	Sm	LB	.985	16.25	---	27.00	---	27.00	**40.30**
8" T wall	Demo	VLF	Lg	LB	1.60	10.00	---	43.90	---	43.90	**65.40**
	Demo	VLF	Sm	LB	2.46	6.50	---	67.50	---	67.50	**101.00**
Columns, 12" x 12" o.d.											
	Demo	VLF	Lg	LB	.320	50.00	---	8.78	---	8.78	**13.10**
	Demo	VLF	Sm	LB	.492	32.50	---	13.50	---	13.50	**20.10**
Veneer, 4" T, with air tools	Demo	SF	Lg	AB	.059	270.0	---	1.74	.34	2.08	**2.92**
	Demo	SF	Sm	AB	.091	175.5	---	2.68	.53	3.21	**4.50**
Walls, with air tools											
8" T wall	Demo	SF	Lg	AB	.119	135.0	---	3.51	.69	4.20	**5.88**
	Demo	SF	Sm	AB	.182	87.75	---	5.37	1.06	6.43	**9.00**
12" T wall	Demo	SF	Lg	AB	.168	95.00	---	4.95	.98	5.93	**8.31**
	Demo	SF	Sm	AB	.259	61.75	---	7.64	1.50	9.14	**12.80**

Description	Oper	Unit	Vol	Crew Size	Man-hours per Unit	Crew Output per Day	Avg Mat'l Unit Cost	Avg Labor Unit Cost	Avg Equip Unit Cost	Avg Total Unit Cost	Avg Price Incl O&P

Concrete block; lightweight (haydite), standard or heavyweight

Foundations and retaining walls; no excavation included
Without reinforcing or with only lateral reinforcing
With air tools

Description	Oper	Unit	Vol	Crew Size	Man-hours per Unit	Crew Output per Day	Avg Mat'l Unit Cost	Avg Labor Unit Cost	Avg Equip Unit Cost	Avg Total Unit Cost	Avg Price Incl O&P
8" W x 8" H x 16" L	Demo	SF	Lg	AB	.057	280.0	---	1.68	.33	2.01	**2.82**
	Demo	SF	Sm	AB	.088	182.0	---	2.60	.51	3.11	**4.35**
12" W x 8" H x 16" L	Demo	SF	Lg	AB	.067	240.0	---	1.98	.39	2.37	**3.31**
	Demo	SF	Sm	AB	.103	156.0	---	3.04	.59	3.63	**5.09**

Without air tools

Description	Oper	Unit	Vol	Crew Size	Man-hours per Unit	Crew Output per Day	Avg Mat'l Unit Cost	Avg Labor Unit Cost	Avg Equip Unit Cost	Avg Total Unit Cost	Avg Price Incl O&P
8" W x 8" H x 16" L	Demo	SF	Lg	LB	.071	225.0	---	1.95	---	1.95	**2.90**
	Demo	SF	Sm	LB	.109	146.3	---	2.99	---	2.99	**4.45**
12" W x 8" H x 16" L	Demo	SF	Lg	LB	.084	190.0	---	2.30	---	2.30	**3.43**
	Demo	SF	Sm	LB	.130	123.5	---	3.57	---	3.57	**5.31**

With vertical reinforcing in every other core (2 cores per block) with cores filled
With air tools

Description	Oper	Unit	Vol	Crew Size	Man-hours per Unit	Crew Output per Day	Avg Mat'l Unit Cost	Avg Labor Unit Cost	Avg Equip Unit Cost	Avg Total Unit Cost	Avg Price Incl O&P
8" W x 8" H x 16" L	Demo	SF	Lg	AB	.094	170.0	---	2.77	.55	3.32	**4.65**
	Demo	SF	Sm	AB	.145	110.5	---	4.28	.84	5.12	**7.17**
12" W x 8" H x 16" L	Demo	SF	Lg	AB	.110	145.0	---	3.24	.64	3.88	**5.44**
	Demo	SF	Sm	AB	.170	94.25	---	5.01	.98	5.99	**8.40**

Exterior walls (above grade) and partitions, no shoring included

Without reinforcing or with only lateral reinforcing
With air tools

Description	Oper	Unit	Vol	Crew Size	Man-hours per Unit	Crew Output per Day	Avg Mat'l Unit Cost	Avg Labor Unit Cost	Avg Equip Unit Cost	Avg Total Unit Cost	Avg Price Incl O&P
8" W x 8" H x 16" L	Demo	SF	Lg	AB	.046	350.0	---	1.36	.26	1.62	**2.27**
	Demo	SF	Sm	AB	.070	227.5	---	2.06	.41	2.47	**3.47**
12" W x 8" H x 16" L	Demo	SF	Lg	AB	.053	300.0	---	1.56	.31	1.87	**2.62**
	Demo	SF	Sm	AB	.082	195.0	---	2.42	.48	2.90	**4.06**

Without air tools

Description	Oper	Unit	Vol	Crew Size	Man-hours per Unit	Crew Output per Day	Avg Mat'l Unit Cost	Avg Labor Unit Cost	Avg Equip Unit Cost	Avg Total Unit Cost	Avg Price Incl O&P
8" W x 8" H x 16" L	Demo	SF	Lg	LB	.057	280.0	---	1.56	---	1.56	**2.33**
	Demo	SF	Sm	LB	.088	182.0	---	2.41	---	2.41	**3.60**
12" W x 8" H x 16" L	Demo	SF	Lg	LB	.067	240.0	---	1.84	---	1.84	**2.74**
	Demo	SF	Sm	LB	.103	156.0	---	2.83	---	2.83	**4.21**

Fences

Without reinforcing or with only lateral reinforcing
With air tools

Description	Oper	Unit	Vol	Crew Size	Man-hours per Unit	Crew Output per Day	Avg Mat'l Unit Cost	Avg Labor Unit Cost	Avg Equip Unit Cost	Avg Total Unit Cost	Avg Price Incl O&P
6" W x 4" H x 16" L	Demo	SF	Lg	AB	.041	390.0	---	1.21	.24	1.45	**2.03**
	Demo	SF	Sm	AB	.063	253.5	---	1.86	.37	2.23	**3.12**
6" W x 6" H x 16" L	Demo	SF	Lg	AB	.043	370.0	---	1.27	.25	1.52	**2.13**
	Demo	SF	Sm	AB	.067	240.5	---	1.98	.39	2.37	**3.31**
8" W x 8" H x 16" L	Demo	SF	Lg	AB	.046	350.0	---	1.36	.26	1.62	**2.27**
	Demo	SF	Sm	AB	.070	227.5	---	2.06	.41	2.47	**3.47**
12" W x 8" H x 16" L	Demo	SF	Lg	AB	.053	300.0	---	1.56	.31	1.87	**2.62**
	Demo	SF	Sm	AB	.082	195.0	---	2.42	.48	2.90	**4.06**

Without air tools

Description	Oper	Unit	Vol	Crew Size	Man-hours per Unit	Crew Output per Day	Avg Mat'l Unit Cost	Avg Labor Unit Cost	Avg Equip Unit Cost	Avg Total Unit Cost	Avg Price Incl O&P
6" W x 4" H x 16" L	Demo	SF	Lg	LB	.052	310.0	---	1.43	---	1.43	**2.13**
	Demo	SF	Sm	LB	.079	201.5	---	2.17	---	2.17	**3.23**
6" W x 6" H x 16" L	Demo	SF	Lg	LB	.054	295.0	---	1.48	---	1.48	**2.21**
	Demo	SF	Sm	LB	.083	191.8	---	2.28	---	2.28	**3.39**
8" W x 8" H x 16" L	Demo	SF	Lg	LB	.057	280.0	---	1.56	---	1.56	**2.33**
	Demo	SF	Sm	LB	.088	182.0	---	2.41	---	2.41	**3.60**
12" W x 8" H x 16" L	Demo	SF	Lg	LB	.067	240.0	---	1.84	---	1.84	**2.74**
	Demo	SF	Sm	LB	.103	156.0	---	2.83	---	2.83	**4.21**

Description	Oper	Unit	Vol	Crew Size	Man-hours per Unit	Crew Output per Day	Avg Mat'l Unit Cost	Avg Labor Unit Cost	Avg Equip Unit Cost	Avg Total Unit Cost	Avg Price Incl O&P
Quarry tile, 6" or 9" squares											
Floors											
Conventional mortar set	Demo	SF	Lg	LB	.036	445.0	---	.99	---	.99	1.47
	Demo	SF	Sm	LB	.055	289.3	---	1.51	---	1.51	2.25
Dry-set mortar	Demo	SF	Lg	LB	.031	515.0	---	.85	---	.85	1.27
	Demo	SF	Sm	LB	.048	334.8	---	1.32	---	1.32	1.96

Rough carpentry (framing)

Dimension lumber

Beams, set on steel columns

Description	Oper	Unit	Vol	Crew Size	Man-hours per Unit	Crew Output per Day	Avg Mat'l Unit Cost	Avg Labor Unit Cost	Avg Equip Unit Cost	Avg Total Unit Cost	Avg Price Incl O&P
Built-up from 2" lumber											
4" T x 10" W - 10' L (2 pcs)	Demo	LF	Lg	LB	.019	855.0	---	.52	---	.52	.78
	Demo	LF	Sm	LB	.029	555.8	---	.80	---	.80	1.19
4" T x 12" W - 12' L (2 pcs)	Demo	LF	Lg	LB	.016	1025	---	.44	---	.44	.65
	Demo	LF	Sm	LB	.024	666.3	---	.66	---	.66	.98
6" T x 10" W - 10' L (3 pcs)	Demo	LF	Lg	LB	.019	855.0	---	.52	---	.52	.78
	Demo	LF	Sm	LB	.029	555.8	---	.80	---	.80	1.19
6" T x 12" W - 12' L (3 pcs)	Demo	LF	Lg	LB	.016	1025	---	.44	---	.44	.65
	Demo	LF	Sm	LB	.024	666.3	---	.66	---	.66	.98
Single member (solid lumber)											
3" T x 12" W - 12' L	Demo	LF	Lg	LB	.016	1025	---	.44	---	.44	.65
	Demo	LF	Sm	LB	.024	666.3	---	.66	---	.66	.98
4" T x 12" W - 12' L	Demo	LF	Lg	LB	.016	1025	---	.44	---	.44	.65
	Demo	LF	Sm	LB	.024	666.3	---	.66	---	.66	.98

Bracing, diagonal, notched-in, studs oc

Description	Oper	Unit	Vol	Crew Size	Man-hours per Unit	Crew Output per Day	Avg Mat'l Unit Cost	Avg Labor Unit Cost	Avg Equip Unit Cost	Avg Total Unit Cost	Avg Price Incl O&P
1" x 6" - 10' L	Demo	LF	Lg	LB	.026	605.0	---	.71	---	.71	1.06
	Demo	LF	Sm	LB	.041	393.3	---	1.12	---	1.12	1.68

Bridging, "X" type, 1" x 3", (8", 10", 12" T)

Description	Oper	Unit	Vol	Crew Size	Man-hours per Unit	Crew Output per Day	Avg Mat'l Unit Cost	Avg Labor Unit Cost	Avg Equip Unit Cost	Avg Total Unit Cost	Avg Price Incl O&P
16" oc	Demo	LF	Lg	LB	.046	350.0	---	1.26	---	1.26	1.88
	Demo	LF	Sm	LB	.070	227.5	---	1.92	---	1.92	2.86

Columns or posts

Description	Oper	Unit	Vol	Crew Size	Man-hours per Unit	Crew Output per Day	Avg Mat'l Unit Cost	Avg Labor Unit Cost	Avg Equip Unit Cost	Avg Total Unit Cost	Avg Price Incl O&P
4" x 4" -8' L	Demo	LF	Lg	LB	.021	770.0	---	.58	---	.58	.86
	Demo	LF	Sm	LB	.032	500.5	---	.88	---	.88	1.31
6" x 6" -8' L	Demo	LF	Lg	LB	.021	770.0	---	.58	---	.58	.86
	Demo	LF	Sm	LB	.032	500.5	---	.88	---	.88	1.31
6" x 8" -8' L	Demo	LF	Lg	LB	.025	650.0	---	.69	---	.69	1.02
	Demo	LF	Sm	LB	.038	422.5	---	1.04	---	1.04	1.55
8" x 8" -8' L	Demo	LF	Lg	LB	.026	615.0	---	.71	---	.71	1.06
	Demo	LF	Sm	LB	.040	399.8	---	1.10	---	1.10	1.63

Fascia

Description	Oper	Unit	Vol	Crew Size	Man-hours per Unit	Crew Output per Day	Avg Mat'l Unit Cost	Avg Labor Unit Cost	Avg Equip Unit Cost	Avg Total Unit Cost	Avg Price Incl O&P
1" x 4" - 12" L	Demo	LF	Lg	LB	.012	1315	---	.33	---	.33	.49
	Demo	LF	Sm	LB	.019	854.8	---	.52	---	.52	.78

Description	Oper	Unit	Vol	Crew Size	Man-hours per Unit	Crew Output per Day	Avg Mat'l Unit Cost	Avg Labor Unit Cost	Avg Equip Unit Cost	Avg Total Unit Cost	Avg Price Incl O&P
Firestops or stiffeners											
2" x 4" - 16"	Demo	LF	Lg	LB	.034	475.0	---	.93	---	.93	**1.39**
	Demo	LF	Sm	LB	.052	308.8	---	1.43	---	1.43	**2.13**
2" x 6" - 16"	Demo	LF	Lg	LB	.034	475.0	---	.93	---	.93	**1.39**
	Demo	LF	Sm	LB	.052	308.8	---	1.43	---	1.43	**2.13**
Furring strips, 1" x 4" - 8' L											
Walls; strips 12" oc											
Studs 16" oc	Demo	SF	Lg	LB	.013	1215	---	.36	---	.36	**.53**
	Demo	SF	Sm	LB	.020	789.8	---	.55	---	.55	**.82**
Studs 24" oc	Demo	SF	Lg	LB	.012	1365	---	.33	---	.33	**.49**
	Demo	SF	Sm	LB	.018	887.3	---	.49	---	.49	**.74**
Masonry (concrete blocks)	Demo	SF	Lg	LB	.015	1095	---	.41	---	.41	**.61**
	Demo	SF	Sm	LB	.022	711.8	---	.60	---	.60	**.90**
Concrete	Demo	SF	Lg	LB	.025	645.0	---	.69	---	.69	**1.02**
	Demo	SF	Sm	LB	.038	419.3	---	1.04	---	1.04	**1.55**
Ceiling; joists 16" oc											
Strips 12" oc	Demo	SF	Lg	LB	.019	840.0	---	.52	---	.52	**.78**
	Demo	SF	Sm	LB	.029	546.0	---	.80	---	.80	**1.19**
Strips 16" oc	Demo	SF	Lg	LB	.015	1075	---	.41	---	.41	**.61**
	Demo	SF	Sm	LB	.023	698.8	---	.63	---	.63	**.94**
Headers or lintels, over openings											
Built-up or single member											
4" T x 6" W - 4' L	Demo	LF	Lg	LB	.029	560.0	---	.80	---	.80	**1.19**
	Demo	LF	Sm	LB	.044	364.0	---	1.21	---	1.21	**1.80**
4" T x 8" W - 8' L	Demo	LF	Lg	LB	.022	720.0	---	.60	---	.60	**.90**
	Demo	LF	Sm	LB	.034	468.0	---	.93	---	.93	**1.39**
4" T x 10" W - 10' L	Demo	LF	Lg	LB	.019	855.0	---	.52	---	.52	**.78**
	Demo	LF	Sm	LB	.029	555.8	---	.80	---	.80	**1.19**
4" T x 12" W - 12' L	Demo	LF	Lg	LB	.016	1025	---	.44	---	.44	**.65**
	Demo	LF	Sm	LB	.024	666.3	---	.66	---	.66	**.98**
4" T x 14" W - 14' L	Demo	LF	Lg	LB	.016	1025	---	.44	---	.44	**.65**
	Demo	LF	Sm	LB	.024	666.3	---	.66	---	.66	**.98**
Joists											
Ceiling											
2" x 4" - 8' L	Demo	LF	Lg	LB	.013	1230	---	.36	---	.36	**.53**
	Demo	LF	Sm	LB	.020	799.5	---	.55	---	.55	**.82**
2" x 4" - 10' L	Demo	LF	Lg	LB	.011	1395	---	.30	---	.30	**.45**
	Demo	LF	Sm	LB	.018	906.8	---	.49	---	.49	**.74**
2" x 8" - 12' L	Demo	LF	Lg	LB	.011	1420	---	.30	---	.30	**.45**
	Demo	LF	Sm	LB	.017	923.0	---	.47	---	.47	**.69**
2" x 10" - 14' L	Demo	LF	Lg	LB	.010	1535	---	.27	---	.27	**.41**
	Demo	LF	Sm	LB	.016	997.8	---	.44	---	.44	**.65**
2" x 12" - 16' L	Demo	LF	Lg	LB	.010	1585	---	.27	---	.27	**.41**
	Demo	LF	Sm	LB	.016	1030	---	.44	---	.44	**.65**
Floor; seated on sill plate											
2" x 8" - 12' L	Demo	LF	Lg	LB	.010	1600	---	.27	---	.27	**.41**
	Demo	LF	Sm	LB	.015	1040	---	.41	---	.41	**.61**
2" x 10" - 14' L	Demo	LF	Lg	LB	.009	1725	---	.25	---	.25	**.37**
	Demo	LF	Sm	LB	.014	1121	---	.38	---	.38	**.57**
2" x 12" - 16' L	Demo	LF	Lg	LB	.009	1755	---	.25	---	.25	**.37**
	Demo	LF	Sm	LB	.014	1141	---	.38	---	.38	**.57**

Description	Oper	Unit	Vol	Crew Size	Man-hours per Unit	Crew Output per Day	Avg Mat'l Unit Cost	Avg Labor Unit Cost	Avg Equip Unit Cost	Avg Total Unit Cost	Avg Price Incl O&P
Ledgers											
Nailed, 2" x 6" - 12' L	Demo	LF	Lg	LB	.023	710.0	---	.63	---	.63	**.94**
	Demo	LF	Sm	LB	.035	461.5	---	.96	---	.96	**1.43**
Bolted, 3" x 8" - 12' L	Demo	LF	Lg	LB	.016	970.0	---	.44	---	.44	**.65**
	Demo	LF	Sm	LB	.025	630.5	---	.69	---	.69	**1.02**
Plates; 2" x 4" or 2" x 6"											
Double top nailed	Demo	LF	Lg	LB	.020	820.0	---	.55	---	.55	**.82**
	Demo	LF	Sm	LB	.030	533.0	---	.82	---	.82	**1.23**
Sill, nailed	Demo	LF	Lg	LB	.012	1360	---	.33	---	.33	**.49**
	Demo	LF	Sm	LB	.018	884.0	---	.49	---	.49	**.74**
Sill or bottom, bolted	Demo	LF	Lg	LB	.023	685.0	---	.63	---	.63	**.94**
	Demo	LF	Sm	LB	.036	445.3	---	.99	---	.99	**1.47**
Rafters											
Common											
2" x 4" - 14' L	Demo	LF	Lg	LB	.009	1795	---	.25	---	.25	**.37**
	Demo	LF	Sm	LB	.014	1167	---	.38	---	.38	**.57**
2" x 6" - 14' L	Demo	LF	Lg	LB	.010	1660	---	.27	---	.27	**.41**
	Demo	LF	Sm	LB	.015	1079	---	.41	---	.41	**.61**
2" x 8" - 14' L	Demo	LF	Lg	LB	.011	1435	---	.30	---	.30	**.45**
	Demo	LF	Sm	LB	.017	932.8	---	.47	---	.47	**.69**
2" x 10" - 14' L	Demo	LF	Lg	LB	.013	1195	---	.36	---	.36	**.53**
	Demo	LF	Sm	LB	.021	776.8	---	.58	---	.58	**.86**
Hip and/or valley											
2" x 4" - 16' L	Demo	LF	Lg	LB	.008	2050	---	.22	---	.22	**.33**
	Demo	LF	Sm	LB	.012	1333	---	.33	---	.33	**.49**
2" x 6" - 16' L	Demo	LF	Lg	LB	.008	1895	---	.22	---	.22	**.33**
	Demo	LF	Sm	LB	.013	1232	---	.36	---	.36	**.53**
2" x 8" - 16' L	Demo	LF	Lg	LB	.010	1640	---	.27	---	.27	**.41**
	Demo	LF	Sm	LB	.015	1066	---	.41	---	.41	**.61**
2" x 10" - 16' L	Demo	LF	Lg	LB	.012	1360	---	.33	---	.33	**.49**
	Demo	LF	Sm	LB	.018	884.0	---	.49	---	.49	**.74**
Jack											
2" x 4" - 6' L	Demo	LF	Lg	LB	.020	805.0	---	.55	---	.55	**.82**
	Demo	LF	Sm	LB	.031	523.3	---	.85	---	.85	**1.27**
2" x 6" - 6' L	Demo	LF	Lg	LB	.022	740.0	---	.60	---	.60	**.90**
	Demo	LF	Sm	LB	.033	481.0	---	.91	---	.91	**1.35**
2" x 8" - 6' L	Demo	LF	Lg	LB	.024	655.0	---	.66	---	.66	**.98**
	Demo	LF	Sm	LB	.038	425.8	---	1.04	---	1.04	**1.55**
2" x 10" - 6' L	Demo	LF	Lg	LB	.030	540.0	---	.82	---	.82	**1.23**
	Demo	LF	Sm	LB	.046	351.0	---	1.26	---	1.26	**1.88**

Description	Oper	Unit	Vol	Crew Size	Man-hours per Unit	Crew Output per Day	Avg Mat'l Unit Cost	Avg Labor Unit Cost	Avg Equip Unit Cost	Avg Total Unit Cost	Avg Price Incl O&P
Roof decking, solid T&G											
2" x 6" - 12' L	Demo	LF	Lg	LB	.025	645.0	---	.69	---	.69	**1.02**
	Demo	LF	Sm	LB	.038	419.3	---	1.04	---	1.04	**1.55**
2" x 8" - 12' L	Demo	LF	Lg	LB	.018	900.0	---	.49	---	.49	**.74**
	Demo	LF	Sm	LB	.027	585.0	---	.74	---	.74	**1.10**
Studs											
2" x 4" - 8' L	Demo	LF	Lg	LB	.012	1360	---	.33	---	.33	**.49**
	Demo	LF	Sm	LB	.018	884.0	---	.49	---	.49	**.74**
2" x 6" - 8' L	Demo	LF	Lg	LB	.012	1360	---	.33	---	.33	**.49**
	Demo	LF	Sm	LB	.018	884.0	---	.49	---	.49	**.74**
Stud partitions, studs 16" oc with bottom plate and double top plates and firestops; per LF partition											
2" x 4" - 8' L	Demo	LF	Lg	LB	.160	100.0	---	4.39	---	4.39	**6.54**
	Demo	LF	Sm	LB	.246	65.00	---	6.75	---	6.75	**10.10**
2" x 6" - 8' L	Demo	LF	Lg	LB	.160	100.0	---	4.39	---	4.39	**6.54**
	Demo	LF	Sm	LB	.246	65.00	---	6.75	---	6.75	**10.10**

Boards

Sheathing, regular or diagonal
1" x 8"

Description	Oper	Unit	Vol	Crew Size	Man-hours per Unit	Crew Output per Day	Avg Mat'l Unit Cost	Avg Labor Unit Cost	Avg Equip Unit Cost	Avg Total Unit Cost	Avg Price Incl O&P
Roof	Demo	SF	Lg	LB	.012	1340	---	.33	---	.33	**.49**
	Demo	SF	Sm	LB	.018	871.0	---	.49	---	.49	**.74**
Sidewall	Demo	SF	Lg	LB	.010	1615	---	.27	---	.27	**.41**
	Demo	SF	Sm	LB	.015	1050	---	.41	---	.41	**.61**

Subflooring, regular or diagonal

Description	Oper	Unit	Vol	Crew Size	Man-hours per Unit	Crew Output per Day	Avg Mat'l Unit Cost	Avg Labor Unit Cost	Avg Equip Unit Cost	Avg Total Unit Cost	Avg Price Incl O&P
1" x 8" - 16' L	Demo	SF	Lg	LB	.010	1525	---	.27	---	.27	**.41**
	Demo	SF	Sm	LB	.016	991.3	---	.44	---	.44	**.65**
1" x 10" - 16' L	Demo	SF	Lg	LB	.008	1930	---	.22	---	.22	**.33**
	Demo	SF	Sm	LB	.013	1255	---	.36	---	.36	**.53**

Plywood

Sheathing
Roof

Description	Oper	Unit	Vol	Crew Size	Man-hours per Unit	Crew Output per Day	Avg Mat'l Unit Cost	Avg Labor Unit Cost	Avg Equip Unit Cost	Avg Total Unit Cost	Avg Price Incl O&P
1/2" T, CDX	Demo	SF	Lg	LB	.008	1970	---	.22	---	.22	**.33**
	Demo	SF	Sm	LB	.012	1281	---	.33	---	.33	**.49**
5/8" T, CDX	Demo	SF	Lg	LB	.008	1935	---	.22	---	.22	**.33**
	Demo	SF	Sm	LB	.013	1258	---	.36	---	.36	**.53**
Wall											
3/8" or 1/2" T, CDX	Demo	SF	Lg	LB	.006	2520	---	.16	---	.16	**.25**
	Demo	SF	Sm	LB	.010	1638	---	.27	---	.27	**.41**
5/8" T, CDX	Demo	SF	Lg	LB	.007	2460	---	.19	---	.19	**.29**
	Demo	SF	Sm	LB	.010	1599	---	.27	---	.27	**.41**

Description	Oper	Unit	Vol	Crew Size	Man-hours per Unit	Crew Output per Day	Avg Mat'l Unit Cost	Avg Labor Unit Cost	Avg Equip Unit Cost	Avg Total Unit Cost	Avg Price Incl O&P
Subflooring											
5/8", 3/4" T, CDX	Demo	SF	Lg	LB	.007	2305	---	.19	---	.19	.29
	Demo	SF	Sm	LB	.011	1498	---	.30	---	.30	.45
1-1/8" T, 2-4-1, T&G long edges											
	Demo	SF	Lg	LB	.010	1615	---	.27	---	.27	.41
	Demo	SF	Sm	LB	.015	1050	---	.41	---	.41	.61
Trusses, "W" pattern with gin pole, 24' to 30' spans											
3 - in - 12 slope	Demo	Ea	Lg	LB	.615	26.00	---	16.90	---	16.90	25.10
	Demo	Ea	Sm	LB	.947	16.90	---	26.00	---	26.00	38.70
5 - in - 12 slope	Demo	Ea	Lg	LB	.615	26.00	---	16.90	---	16.90	25.10
	Demo	Ea	Sm	LB	.947	16.90	---	26.00	---	26.00	38.70

Finish carpentry

Description	Oper	Unit	Vol	Crew Size	Man-hours per Unit	Crew Output per Day	Avg Mat'l Unit Cost	Avg Labor Unit Cost	Avg Equip Unit Cost	Avg Total Unit Cost	Avg Price Incl O&P
Bath accessories, screwed											
	Demo	Ea	Lg	LB	.128	125.0	---	3.51	---	3.51	5.23
	Demo	Ea	Sm	LB	.197	81.25	---	5.40	---	5.40	8.05
Cabinets											
Kitchen, to 3' x 4', wood; base, wall, or peninsula											
	Demo	Ea	Lg	LB	.640	25.00	---	17.60	---	17.60	26.20
	Demo	Ea	Sm	LB	.985	16.25	---	27.00	---	27.00	40.30
Medicine, metal											
	Demo	Ea	Lg	LB	.533	30.00	---	14.60	---	14.60	21.80
	Demo	Ea	Sm	LB	.821	19.50	---	22.50	---	22.50	33.60
Vanity, cabinet and sink top											
Disconnect plumbing and remove to dumpster											
	Demo	Ea	Lg	LB	1.00	16.00	---	27.40	---	27.40	40.90
	Demo	Ea	Sm	LB	1.54	10.40	---	42.20	---	42.20	62.90
Remove old unit, replace with new unit, reconnect plumbing											
	Demo	Ea	Lg	SB	2.29	7.00	---	74.50	---	74.50	110.00
	Demo	Ea	Sm	SB	3.52	4.55	---	115.00	---	115.00	168.00
Hardwood flooring (over wood subfloors)											
Block, set in mastic	Demo	SF	Lg	LB	.013	1200	---	.36	---	.36	.53
	Demo	SF	Sm	LB	.021	780.0	---	.58	---	.58	.86
Strip, nailed	Demo	SF	Lg	LB	.018	900.0	---	.49	---	.49	.74
	Demo	SF	Sm	LB	.027	585.0	---	.74	---	.74	1.10
Marlite panels, 4' x 8',											
adhesive set	Demo	SF	Lg	LB	.019	850.0	---	.52	---	.52	.78
	Demo	SF	Sm	LB	.029	552.5	---	.80	---	.80	1.19
Molding and trim											
At base (floor)	Demo	LF	Lg	LB	.015	1060	---	.41	---	.41	.61
	Demo	LF	Sm	LB	.023	689.0	---	.63	---	.63	.94
At ceiling	Demo	LF	Lg	LB	.013	1200	---	.36	---	.36	.53
	Demo	LF	Sm	LB	.021	780.0	---	.58	---	.58	.86
On walls or cabinets	Demo	LF	Lg	LB	.010	1600	---	.27	---	.27	.41
	Demo	LF	Sm	LB	.015	1040	---	.41	---	.41	.61
Paneling											
Plywood, prefinished	Demo	SF	Lg	LB	.009	1850	---	.25	---	.25	.37
	Demo	SF	Sm	LB	.013	1203	---	.36	---	.36	.53
Wood	Demo	SF	Lg	LB	.010	1650	---	.27	---	.27	.41
	Demo	SF	Sm	LB	.015	1073	---	.41	---	.41	.61

Description	Oper	Unit	Vol	Crew Size	Man-hours per Unit	Crew Output per Day	Avg Mat'l Unit Cost	Avg Labor Unit Cost	Avg Equip Unit Cost	Avg Total Unit Cost	Avg Price Incl O&P

Weather protection
Insulation

Batt/roll, with wall or ceiling finish already removed
Joists, 16" or 24" oc

| | Demo | SF | Lg | LB | .005 | 2935 | --- | .14 | --- | .14 | .20 |
| | Demo | SF | Sm | LB | .008 | 1908 | --- | .22 | --- | .22 | .33 |

Rafters, 16" or 24" oc

| | Demo | SF | Lg | LB | .006 | 2560 | --- | .16 | --- | .16 | .25 |
| | Demo | SF | Sm | LB | .010 | 1664 | --- | .27 | --- | .27 | .41 |

Studs, 16" or 24" oc

| | Demo | SF | Lg | LB | .005 | 3285 | --- | .14 | --- | .14 | .20 |
| | Demo | SF | Sm | LB | .007 | 2135 | --- | .19 | --- | .19 | .29 |

Loose, with ceiling finish already removed
Joists, 16" or 24" oc

4" T	Demo	SF	Lg	LB	.004	3900	---	.11	---	.11	.16
	Demo	SF	Sm	LB	.006	2535	---	.16	---	.16	.25
6" T	Demo	SF	Lg	LB	.007	2340	---	.19	---	.19	.29
	Demo	SF	Sm	LB	.011	1521	---	.30	---	.30	.45

Rigid
Roofs

1/2" T	Demo	Sq	Lg	LB	.941	17.00	---	25.80	---	25.80	38.50
	Demo	Sq	Sm	LB	1.45	11.05	---	39.80	---	39.80	59.30
1" T	Demo	Sq	Lg	LB	1.07	15.00	---	29.40	---	29.40	43.70
	Demo	Sq	Sm	LB	1.64	9.75	---	45.00	---	45.00	67.00

Walls, 1/2" T

| | Demo | SF | Lg | LB | .007 | 2140 | --- | .19 | --- | .19 | .29 |
| | Demo | SF | Sm | LB | .012 | 1391 | --- | .33 | --- | .33 | .49 |

Sheet metal
Gutter and downspouts

Aluminum	Demo	LF	Lg	LB	.019	850.0	---	.52	---	.52	.78
	Demo	LF	Sm	LB	.029	552.5	---	.80	---	.80	1.19
Galvanized	Demo	LF	Lg	LB	.025	640.0	---	.69	---	.69	1.02
	Demo	LF	Sm	LB	.038	416.0	---	1.04	---	1.04	1.55

Roofing and siding
Aluminum

Roofing, nailed to wood
Corrugated (2-1/2"), 26" W with 3-3/4" side lap and 6" end lap

| | Demo | Sq | Lg | LB | 1.60 | 10.00 | --- | 43.90 | --- | 43.90 | 65.40 |
| | Demo | Sq | Sm | LB | 2.46 | 6.50 | --- | 67.50 | --- | 67.50 | 101.00 |

Siding, nailed to wood
Clapboard (i.e., lap drop)

8" exposure	Demo	SF	Lg	LB	.016	1025	---	.44	---	.44	.65
	Demo	SF	Sm	LB	.024	666.3	---	.66	---	.66	.98
10" exposure	Demo	SF	Lg	LB	.013	1280	---	.36	---	.36	.53
	Demo	SF	Sm	LB	.019	832.0	---	.52	---	.52	.78

Corrugated (2-1/2"), 26" W with 2-1/2" side lap and 4" end lap

	Demo	SF	Lg	LB	.013	1200	---	.36	---	.36	.53
	Demo	SF	Sm	LB	.021	780.0	---	.58	---	.58	.86
Panels, 4' x 8'	Demo	SF	Lg	LB	.007	2400	---	.19	---	.19	.29
	Demo	SF	Sm	LB	.010	1560	---	.27	---	.27	.41

Shingle, 24" L with 12" exposure

| | Demo | SF | Lg | LB | .011 | 1450 | --- | .30 | --- | .30 | .45 |
| | Demo | SF | Sm | LB | .017 | 942.5 | --- | .47 | --- | .47 | .69 |

Description	Oper	Unit	Vol	Crew Size	Man-hours per Unit	Crew Output per Day	Avg Mat'l Unit Cost	Avg Labor Unit Cost	Avg Equip Unit Cost	Avg Total Unit Cost	Avg Price Incl O&P
Asphalt shingle roofing											
240 lb/sq, strip, 3 tab 5" exposure											
	Demo	Sq	Lg	LB	1.00	16.00	---	27.40	---	27.40	**40.90**
	Demo	Sq	Sm	LB	1.54	10.40	---	42.20	---	42.20	**62.90**
Built-up/hot roofing (to wood deck)											
3 ply											
With gravel	Demo	Sq	Lg	LB	1.45	11.00	---	39.80	---	39.80	**59.30**
	Demo	Sq	Sm	LB	2.24	7.15	---	61.40	---	61.40	**91.60**
Without gravel	Demo	Sq	Lg	LB	1.14	14.00	---	31.30	---	31.30	**46.60**
	Demo	Sq	Sm	LB	1.76	9.10	---	48.30	---	48.30	**71.90**
5 ply											
With gravel	Demo	Sq	Lg	LB	1.60	10.00	---	43.90	---	43.90	**65.40**
	Demo	Sq	Sm	LB	2.46	6.50	---	67.50	---	67.50	**101.00**
Without gravel	Demo	Sq	Lg	LB	1.23	13.00	---	33.70	---	33.70	**50.30**
	Demo	Sq	Sm	LB	1.89	8.45	---	51.80	---	51.80	**77.20**
Clay tile roofing											
2 piece interlocking											
	Demo	Sq	Lg	LB	1.60	10.00	---	43.90	---	43.90	**65.40**
	Demo	Sq	Sm	LB	2.46	6.50	---	67.50	---	67.50	**101.00**
1 piece											
	Demo	Sq	Lg	LB	1.45	11.00	---	39.80	---	39.80	**59.30**
	Demo	Sq	Sm	LB	2.24	7.15	---	61.40	---	61.40	**91.60**
Hardboard siding											
Lap, 1/2" T x 12" W x 16' L, with 11" exposure											
	Demo	SF	Lg	LB	.012	1345	---	.33	---	.33	**.49**
	Demo	SF	Sm	LB	.018	874.3	---	.49	---	.49	**.74**
Panels, 7/16" T x 4' W x 8' H											
	Demo	SF	Lg	LB	.006	2520	---	.16	---	.16	**.25**
	Demo	SF	Sm	LB	.010	1638	---	.27	---	.27	**.41**
Mineral surfaced roll roofing											
Single coverage 90 lb/sq roll with 6" end lap and 2" headlap											
	Demo	Sq	Lg	LB	.500	32.00	---	13.70	---	13.70	**20.40**
	Demo	Sq	Sm	LB	.769	20.80	---	21.10	---	21.10	**31.40**
Double coverage selvage roll, with 6" end lap and 17" exposure											
	Demo	Sq	Lg	LB	.727	22.00	---	19.90	---	19.90	**29.70**
	Demo	Sq	Sm	LB	1.12	14.30	---	30.70	---	30.70	**45.80**
Wood											
Roofing											
Shakes											
24" L with 10" exposure											
1/2" to 3/4" T	Demo	Sq	Lg	LB	.640	25.00	---	17.60	---	17.60	**26.20**
	Demo	Sq	Sm	LB	.985	16.25	---	27.00	---	27.00	**40.30**
3/4" to 5/4" T	Demo	Sq	Lg	LB	.696	23.00	---	19.10	---	19.10	**28.50**
	Demo	Sq	Sm	LB	1.07	14.95	---	29.40	---	29.40	**43.70**
Shingles											
16" L with 5" exposure	Demo	Sq	Lg	LB	1.33	12.00	---	36.50	---	36.50	**54.40**
	Demo	Sq	Sm	LB	2.05	7.80	---	56.20	---	56.20	**83.80**
18" L with 5-1/2" exposure	Demo	Sq	Lg	LB	1.23	13.00	---	33.70	---	33.70	**50.30**
	Demo	Sq	Sm	LB	1.89	8.45	---	51.80	---	51.80	**77.20**
24" L with 7-1/2" exposure	Demo	Sq	Lg	LB	.889	18.00	---	24.40	---	24.40	**36.30**
	Demo	Sq	Sm	LB	1.37	11.70	---	37.60	---	37.60	**56.00**

Description	Oper	Unit	Vol	Crew Size	Man-hours per Unit	Crew Output per Day	Avg Mat'l Unit Cost	Avg Labor Unit Cost	Avg Equip Unit Cost	Avg Total Unit Cost	Avg Price Incl O&P
Siding											
Bevel											
1/2" x 8" with 6-3/4" exposure	Demo	SF	Lg	LB	.018	910.0	---	.49	---	.49	**.74**
	Demo	SF	Sm	LB	.027	591.5	---	.74	---	.74	**1.10**
5/8" x 10" with 8-3/4" exposure	Demo	SF	Lg	LB	.014	1180	---	.38	---	.38	**.57**
	Demo	SF	Sm	LB	.021	767.0	---	.58	---	.58	**.86**
3/4" x 12" with 10-3/4" exposure	Demo	SF	Lg	LB	.011	1450	---	.30	---	.30	**.45**
	Demo	SF	Sm	LB	.017	942.5	---	.47	---	.47	**.69**
Drop (horizontal), 1/4" T&G											
1" x 8" with 7" exposure	Demo	SF	Lg	LB	.017	945.0	---	.47	---	.47	**.69**
	Demo	SF	Sm	LB	.026	614.3	---	.71	---	.71	**1.06**
1" x 10" with 9" exposure	Demo	SF	Lg	LB	.013	1215	---	.36	---	.30	**.53**
	Demo	SF	Sm	LB	.020	789.8	---	.55	---	.55	**.82**
Board (1" x 12") and batten (1" x 2") @ 12" oc											
Horizontal											
	Demo	SF	Lg	LB	.013	1280	---	.36	---	.36	**.53**
	Demo	SF	Sm	LB	.019	832.0	---	.52	---	.52	**.78**
Vertical											
Standard	Demo	SF	Lg	LB	.016	1025	---	.44	---	.44	**.65**
	Demo	SF	Sm	LB	.024	666.3	---	.66	---	.66	**.98**
Reverse	Demo	SF	Lg	LB	.015	1065	---	.41	---	.41	**.61**
	Demo	SF	Sm	LB	.023	692.3	---	.63	---	.63	**.94**
Board on board (1" x 12" with 1-1/2" overlap), vertical											
	Demo	SF	Lg	LB	.017	950.0	---	.47	---	.47	**.69**
	Demo	SF	Sm	LB	.026	617.5	---	.71	---	.71	**1.06**
Plywood (1/2" T) with battens (1" x 2")											
16" oc battens	Demo	SF	Lg	LB	.007	2255	---	.19	---	.19	**.29**
	Demo	SF	Sm	LB	.011	1466	---	.30	---	.30	**.45**
24" oc battens	Demo	SF	Lg	LB	.007	2365	---	.19	---	.19	**.29**
	Demo	SF	Sm	LB	.010	1537	---	.27	---	.27	**.41**
Shakes											
24" L with 11-1/2" exposure											
1/2" to 3/4" T	Demo	SF	Lg	LB	.010	1600	---	.27	---	.27	**.41**
	Demo	SF	Sm	LB	.015	1040	---	41	---	.41	**.61**
3/4" to 5/4" T	Demo	SF	Lg	LB	.011	1440	---	.30	---	.30	**.45**
	Demo	SF	Sm	LB	.017	936.0	---	.47	---	.47	**.69**
Shingles											
16" L with 7-1/2" exposure	Demo	SF	Lg	LB	.016	1000	---	.44	---	.44	**.65**
	Demo	SF	Sm	LB	.025	650.0	---	.69	---	.69	**1.02**
18" L with 8-1/2" exposure	Demo	SF	Lg	LB	.014	1130	---	.38	---	.38	**.57**
	Demo	SF	Sm	LB	.022	734.5	---	.60	---	.60	**.90**
24" L with 11-1/2" exposure	Demo	SF	Lg	LB	.010	1530	---	.27	---	.27	**.41**
	Demo	SF	Sm	LB	.016	994.5	---	.44	---	.44	**.65**

Description	Oper	Unit	Vol	Crew Size	Man-hours per Unit	Crew Output per Day	Avg Mat'l Unit Cost	Avg Labor Unit Cost	Avg Equip Unit Cost	Avg Total Unit Cost	Avg Price Incl O&P

Doors, windows and glazing
Doors with related trim and frame

Closet, with track
Folding, 4 doors — Demo Set Lg LB 1.33 12.00 --- 36.50 --- 36.50 **54.40**
Demo Set Sm LB 2.05 7.80 --- 56.20 --- 56.20 **83.80**
Sliding, 2 or 3 doors — Demo Set Lg LB 1.33 12.00 --- 36.50 --- 36.50 **54.40**
Demo Set Sm LB 2.05 7.80 --- 56.20 --- 56.20 **83.80**

Entry, 3' x 7'
Demo Ea Lg LB 1.14 14.00 --- 31.30 --- 31.30 **46.60**
Demo Ea Sm LB 1.76 9.10 --- 48.30 --- 48.30 **71.90**

Fire, 3' x 7'
Demo Ea Lg LB 1.14 14.00 --- 31.30 --- 31.30 **46.60**
Demo Ea Sm LB 1.76 9.10 --- 48.30 --- 48.30 **71.90**

Garage
Wood, aluminum, or hardboard
Single — Demo Ea Lg LB 2.00 8.00 --- 54.90 --- 54.90 **81.70**
Demo Ea Sm LB 3.08 5.20 --- 84.50 --- 84.50 **126.00**
Double — Demo Ea Lg LB 2.67 6.00 --- 73.20 --- 73.20 **109.00**
Demo Ea Sm LB 4.10 3.90 --- 112.00 --- 112.00 **168.00**

Steel
Single — Demo Ea Lg LB 2.29 7.00 --- 62.80 --- 62.80 **93.60**
Demo Ea Sm LB 3.52 4.55 --- 96.60 --- 96.60 **144.00**
Double — Demo Ea Lg LB 3.20 5.00 --- 87.80 --- 87.80 **131.00**
Demo Ea Sm LB 4.92 3.25 --- 135.00 --- 135.00 **201.00**

Glass sliding, with track
2 lites wide — Demo Set Lg LB 2.00 8.00 --- 54.90 --- 54.90 **81.70**
Demo Set Sm LB 3.08 5.20 --- 84.50 --- 84.50 **126.00**
3 lites wide — Demo Set Lg LB 2.67 6.00 --- 73.20 --- 73.20 **109.00**
Demo Set Sm LB 4.10 3.90 --- 112.00 --- 112.00 **168.00**
4 lites wide — Demo Set Lg LB 4.00 4.00 --- 110.00 --- 110.00 **163.00**
Demo Set Sm LB 6.15 2.60 --- 169.00 --- 169.00 **251.00**

Interior, 3' x 7'
Demo Ea Lg LB 1.00 16.00 --- 27.40 --- 27.40 **40.90**
Demo Ea Sm LB 1.54 10.40 --- 42.20 --- 42.20 **62.90**

Screen, 3' x 7'
Demo Ea Lg LB .800 20.00 --- 21.90 --- 21.90 **32.70**
Demo Ea Sm LB 1.23 13.00 --- 33.70 --- 33.70 **50.30**

Storm combination, 3' x 7'
Demo Ea Lg LB .800 20.00 --- 21.90 --- 21.90 **32.70**
Demo Ea Sm LB 1.23 13.00 --- 33.70 --- 33.70 **50.30**

Windows, with related trim and frame

To 12 SF
Aluminum — Demo Ea Lg LB .762 21.00 --- 20.90 --- 20.90 **31.10**
Demo Ea Sm LB 1.17 13.65 --- 32.10 --- 32.10 **47.80**
Wood — Demo Ea Lg LB 1.00 16.00 --- 27.40 --- 27.40 **40.90**
Demo Ea Sm LB 1.54 10.40 --- 42.20 --- 42.20 **62.90**

13 SF to 50 SF
Aluminum — Demo Ea Lg LB 1.23 13.00 --- 33.70 --- 33.70 **50.30**
Demo Ea Sm LB 1.89 8.45 --- 51.80 --- 51.80 **77.20**
Wood — Demo Ea Lg LB 1.60 10.00 --- 43.90 --- 43.90 **65.40**
Demo Ea Sm LB 2.46 6.50 --- 67.50 --- 67.50 **101.00**

Description	Oper	Unit	Vol	Crew Size	Man-hours per Unit	Crew Output per Day	Avg Mat'l Unit Cost	Avg Labor Unit Cost	Avg Equip Unit Cost	Avg Total Unit Cost	Avg Price Incl O&P

Glazing, clean sash and remove old putty or rubber

3/32" T float, putty or rubber

Description	Oper	Unit	Vol	Crew Size	MH	Output	Mat'l	Labor	Equip	Total	Price
8" x 12" (0.667 SF)	Demo	SF	Lg	GA	.320	25.00	---	9.87	---	9.87	**14.60**
	Demo	SF	Sm	GA	.492	16.25	---	15.20	---	15.20	**22.50**
12" x 16" (1.333 SF)	Demo	SF	Lg	GA	.178	45.00	---	5.49	---	5.49	**8.12**
	Demo	SF	Sm	GA	.274	29.25	---	8.45	---	8.45	**12.50**
14" x 20" (1.944 SF)	Demo	SF	Lg	GA	.145	55.00	---	4.47	---	4.47	**6.62**
	Demo	SF	Sm	GA	.224	35.75	---	6.91	---	6.91	**10.20**
16" x 24" (2.667 SF)	Demo	SF	Lg	GA	.114	70.00	---	3.52	---	3.52	**5.20**
	Demo	SF	Sm	GA	.176	45.50	---	5.43	---	5.43	**8.03**
24" x 26" (4.333 SF)	Demo	SF	Lg	GA	.084	95.00	---	2.59	---	2.59	**3.83**
	Demo	SF	Sm	GA	.130	61.75	---	4.01	---	4.01	**5.93**
36" x 24" (6.000 SF)	Demo	SF	Lg	GA	.064	125.0	---	1.97	---	1.97	**2.92**
	Demo	SF	Sm	GA	.098	81.25	---	3.02	---	3.02	**4.47**

1/8" T float, putty, steel sash

Description	Oper	Unit	Vol	Crew Size	MH	Output	Mat'l	Labor	Equip	Total	Price
12" x 16" (1.333 SF)	Demo	SF	Lg	GA	.178	45.00	---	5.49	---	5.49	**8.12**
	Demo	SF	Sm	GA	.274	29.25	---	8.45	---	8.45	**12.50**
16" x 20" (2.222 SF)	Demo	SF	Lg	GA	.123	65.00	---	3.79	---	3.79	**5.61**
	Demo	SF	Sm	GA	.189	42.25	---	5.83	---	5.83	**8.63**
16" x 24" (2.667 SF)	Demo	SF	Lg	GA	.114	70.00	---	3.52	---	3.52	**5.20**
	Demo	SF	Sm	GA	.176	45.50	---	5.43	---	5.43	**8.03**
24" x 26" (4.333 SF)	Demo	SF	Lg	GA	.084	95.00	---	2.59	---	2.59	**3.83**
	Demo	SF	Sm	GA	.130	61.75	---	4.01	---	4.01	**5.93**
28" x 32" (6.222 SF)	Demo	SF	Lg	GA	.062	130.0	---	1.91	---	1.91	**2.83**
	Demo	SF	Sm	GA	.095	84.50	---	2.93	---	2.93	**4.34**
36" x 36" (9.000 SF)	Demo	SF	Lg	GA	.050	160.0	---	1.54	---	1.54	**2.28**
	Demo	SF	Sm	GA	.077	104.0	---	2.37	---	2.37	**3.51**
36" x 48" (12.000 SF)	Demo	SF	Lg	GA	.042	190.0	---	1.30	---	1.30	**1.92**
	Demo	SF	Sm	GA	.065	123.5	---	2.00	---	2.00	**2.97**

1/4" T float
Wood sash with putty

Description	Oper	Unit	Vol	Crew Size	MH	Output	Mat'l	Labor	Equip	Total	Price
72" x 48" (24.0 SF)	Demo	SF	Lg	GA	.043	185.0	---	1.33	---	1.33	**1.96**
	Demo	SF	Sm	GA	.067	120.3	---	2.07	---	2.07	**3.06**

Aluminum sash with aluminum channel and rigid neoprene rubber

Description	Oper	Unit	Vol	Crew Size	MH	Output	Mat'l	Labor	Equip	Total	Price
48" x 96" (32.0 SF)	Demo	SF	Lg	GA	.046	175.0	---	1.42	---	1.42	**2.10**
	Demo	SF	Sm	GA	.070	113.8	---	2.16	---	2.16	**3.19**
96" x 96" (64.0 SF)	Demo	SF	Lg	GA	.044	180.0	---	1.36	---	1.36	**2.01**
	Demo	SF	Sm	GA	.068	117.0	---	2.10	---	2.10	**3.10**

1" T insulating glass; with 2 pieces 1/4" float and 1/2" air space

Description	Oper	Unit	Vol	Crew Size	MH	Output	Mat'l	Labor	Equip	Total	Price
To 6.0 SF	Demo	SF	Lg	GA	.160	50.00	---	4.93	---	4.93	**7.30**
	Demo	SF	Sm	GA	.246	32.50	---	7.59	---	7.59	**11.20**
6.1 to 12.0 SF	Demo	SF	Lg	GA	.073	110.0	---	2.25	---	2.25	**3.33**
	Demo	SF	Sm	GA	.112	71.50	---	3.45	---	3.45	**5.11**
12.1 to 18.0 SF	Demo	SF	Lg	GA	.055	145.0	---	1.70	---	1.70	**2.51**
	Demo	SF	Sm	GA	.085	94.25	---	2.62	---	2.62	**3.88**
18.1 to 24.0 SF	Demo	SF	Lg	GA	.053	150.0	---	1.63	---	1.63	**2.42**
	Demo	SF	Sm	GA	.082	97.50	---	2.53	---	2.53	**3.74**

Description	Oper	Unit	Vol	Crew Size	Man-hours per Unit	Crew Output per Day	Avg Mat'l Unit Cost	Avg Labor Unit Cost	Avg Equip Unit Cost	Avg Total Unit Cost	Avg Price Incl O&P

Aluminum sliding door glass with aluminum channel and rigid neoprene rubber

34" x 76" (17.944 SF)

5/8" T insulating glass with 2 pieces 5/32" T (tempered with 1-1/4" air space)

	Demo	SF	Lg	GA	.047	170.0	---	1.45	---	1.45	**2.15**
	Demo	SF	Sm	GA	.072	110.5	---	2.22	---	2.22	**3.29**
5/32" T tempered	Demo	SF	Lg	GA	.041	195.0	---	1.26	---	1.26	**1.87**
	Demo	SF	Sm	GA	.063	126.8	---	1.94	---	1.94	**2.88**

46" x 76" (24.278 SF)

5/8" T insulating glass with 2 pieces 5/32" T (tempered with 1-1/4" air space)

	Demo	SF	Lg	GA	.037	215.0	---	1.14	---	1.14	**1.69**
	Demo	SF	Sm	GA	.057	139.8	---	1.76	---	1.76	**2.60**
5/32" T tempered	Demo	SF	Lg	GA	.033	245.0	---	1.02	---	1.02	**1.51**
	Demo	SF	Sm	GA	.050	159.3	---	1.54	---	1.54	**2.28**

Finishes

Exterior and interior, with hand tools; no insulation removal included

Plaster and stucco; remove to studs or sheathing

Lath (wood or metal) and plaster, walls and ceiling

2 coats	Demo	SY	Lg	LB	.123	130.0	---	3.37	---	3.37	**5.03**
	Demo	SY	Sm	LB	.189	84.50	---	5.18	---	5.18	**7.72**
3 coats	Demo	SY	Lg	LB	.133	120.0	---	3.65	---	3.65	**5.44**
	Demo	SY	Sm	LB	.205	78.00	---	5.62	---	5.62	**8.38**

Stucco and metal netting

2 coats	Demo	SY	Lg	LB	.168	95.00	---	4.61	---	4.61	**6.87**
	Demo	SY	Sm	LB	.259	61.75	---	7.10	---	7.10	**10.60**
3 coats	Demo	SY	Lg	LB	.200	80.00	---	5.49	---	5.49	**8.17**
	Demo	SY	Sm	LB	.308	52.00	---	8.45	---	8.45	**12.60**

Wallboard, gypsum (drywall)

| walls and ceilings | Demo | SF | Lg | LB | .010 | 1540 | --- | .27 | --- | .27 | **.41** |
| | Demo | SF | Sm | LB | .016 | 1001 | --- | .44 | --- | .44 | **.65** |

Ceramic, metal, plastic tile

Floors, 1" x 1"

Adhesive or dry-set base	Demo	SF	Lg	LB	.029	550.0	---	.80	---	.80	**1.19**
	Demo	SF	Sm	LB	.045	357.5	---	1.23	---	1.23	**1.84**
Conventional mortar base	Demo	SF	Lg	LB	.034	475.0	---	.93	---	.93	**1.39**
	Demo	SF	Sm	LB	.052	308.8	---	1.43	---	1.43	**2.13**

Walls, 1" x 1" or 4-1/4" x 4-1/4"

Adhesive or dry-set base	Demo	SF	Lg	LB	.033	480.0	---	.91	---	.91	**1.35**
	Demo	SF	Sm	LB	.051	312.0	---	1.40	---	1.40	**2.08**
Conventional mortar base	Demo	SF	Lg	LB	.040	400.0	---	1.10	---	1.10	**1.63**
	Demo	SF	Sm	LB	.062	260.0	---	1.70	---	1.70	**2.53**

Description	Oper	Unit	Vol	Crew Size	Man-hours per Unit	Crew Output per Day	Avg Mat'l Unit Cost	Avg Labor Unit Cost	Avg Equip Unit Cost	Avg Total Unit Cost	Avg Price Incl O&P

Acoustical or insulating ceiling tile

Description	Oper	Unit	Vol	Crew Size	Man-hrs/Unit	Crew Output/Day	Mat'l	Labor	Equip	Total	Price O&P
Adhesive set, tile only	Demo	SF	Lg	LB	.012	1300	---	.33	---	.33	.49
	Demo	SF	Sm	LB	.019	845.0	---	.52	---	.52	.78
Stapled, tile only	Demo	SF	Lg	LB	.014	1170	---	.38	---	.38	.57
	Demo	SF	Sm	LB	.021	760.5	---	.58	---	.58	.86
Stapled, tile and furring strips	Demo	SF	Lg	LB	.010	1540	---	.27	---	.27	.41
	Demo	SF	Sm	LB	.016	1001	---	.44	---	.44	.65

Suspended ceiling system;

Description	Oper	Unit	Vol	Crew Size	Man-hrs/Unit	Crew Output/Day	Mat'l	Labor	Equip	Total	Price O&P
panels and grid system	Demo	SF	Lg	LB	.009	1780	---	.25	---	.25	.37
	Demo	SF	Sm	LB	.014	1157	---	.38	---	.38	.57

Resilient flooring, adhesive set

Description	Oper	Unit	Vol	Crew Size	Man-hrs/Unit	Crew Output/Day	Mat'l	Labor	Equip	Total	Price O&P
Sheet products	Demo	SY	Lg	LB	.100	160.0	---	2.74	---	2.74	4.09
	Demo	SY	Sm	LB	.154	104.0	---	4.22	---	4.22	6.29
Tile products	Demo	SF	Lg	LB	.011	1500	---	.30	---	.30	.45
	Demo	SF	Sm	LB	.016	975.0	---	.44	---	.44	.65

Wallpaper

Average output is expressed in rolls (36.0 SF/single roll)

Description	Oper	Unit	Vol	Crew Size	Man-hrs/Unit	Crew Output/Day	Mat'l	Labor	Equip	Total	Price O&P
Single layer of paper from plaster with steaming equipment											
	Demo	Roll	Lg	1L	.400	20.00	---	11.00	---	11.00	16.40
	Demo	Roll	Sm	1L	.615	13.00	---	16.90	---	16.90	25.10
Several layers of paper from plaster with steaming equipment											
	Demo	Roll	Lg	1L	.667	12.00	---	18.30	---	18.30	27.30
	Demo	Roll	Sm	1L	1.03	7.80	---	28.30	---	28.30	42.10
Vinyls (with non-woven, woven, or synthetic fiber backings) from plaster with steaming equipment											
	Demo	Roll	Lg	1L	.267	30.00	---	7.32	---	7.32	10.90
	Demo	Roll	Sm	1L	.410	19.50	---	11.30	---	11.30	16.80
Single layer of paper from drywall with steaming equipment											
	Demo	Roll	Lg	1L	.400	20.00	---	11.00	---	11.00	16.40
	Demo	Roll	Sm	1L	.615	13.00	---	16.90	---	16.90	25.10
Several layers of paper from drywall with steaming equipment											
	Demo	Roll	Lg	1L	.667	12.00	---	18.30	---	18.30	27.30
	Demo	Roll	Sm	1L	1.03	7.80	---	28.30	---	28.30	42.10
Vinyls (with synthetic fiber backing) from drywall with steaming equipment											
	Demo	Roll	Lg	1L	.267	30.00	---	7.32	---	7.32	10.90
	Demo	Roll	Sm	1L	.410	19.50	---	11.30	---	11.30	16.80
Vinyls (with other backings) from drywall with steaming equipment											
	Demo	Roll	Lg	1L	.400	20.00	---	11.00	---	11.00	16.40
	Demo	Roll	Sm	1L	.615	13.00	---	16.90	---	16.90	25.10

Description	Oper	Unit	Vol	Crew Size	Man-hours per Unit	Crew Output per Day	Avg Mat'l Unit Cost	Avg Labor Unit Cost	Avg Equip Unit Cost	Avg Total Unit Cost	Avg Price Incl O&P

Dishwashers

High quality units. Labor cost includes rough-in

Frequently encountered applications

Description	Oper	Unit	Vol	Crew Size	Man-hours per Unit	Crew Output per Day	Avg Mat'l Unit Cost	Avg Labor Unit Cost	Avg Equip Unit Cost	Avg Total Unit Cost	Avg Price Incl O&P
Detach & reset unit	Reset	Ea	Lg	SA	1.33	6.00	---	50.10	---	50.10	**73.60**
	Reset	Ea	Sm	SA	1.78	4.50	---	67.10	---	67.10	**98.60**
Remove unit	Demo	Ea	Lg	SA	0.80	10.00	---	30.10	---	30.10	**44.30**
	Demo	Ea	Sm	SA	1.07	7.50	---	40.30	---	40.30	**59.30**
Install unit											
Standard											
	Inst	Ea	Lg	SC	6.67	2.40	482.00	244.00	---	726.00	**914.00**
	Inst	Ea	Sm	SC	8.9	1.80	586.00	326.00	---	912.00	**1150.00**
Average											
	Inst	Ea	Lg	SC	6.67	2.40	580.00	244.00	---	824.00	**1030.00**
	Inst	Ea	Sm	SC	8.9	1.80	705.00	326.00	---	1031.00	**1290.00**
High											
	Inst	Ea	Lg	SC	6.67	2.40	720.00	244.00	---	964.00	**1190.00**
	Inst	Ea	Sm	SC	8.9	1.80	875.00	326.00	---	1201.00	**1490.00**
Premium											
	Inst	Ea	Lg	SC	6.67	2.40	860.00	244.00	---	1104.00	**1350.00**
	Inst	Ea	Sm	SC	8.9	1.80	1040.00	326.00	---	1366.00	**1680.00**

Built-in front loading

Description	Oper	Unit	Vol	Crew Size	Man-hours per Unit	Crew Output per Day	Avg Mat'l Unit Cost	Avg Labor Unit Cost	Avg Equip Unit Cost	Avg Total Unit Cost	Avg Price Incl O&P
Six cycles	Inst	Ea	Lg	SC	6.67	2.40	860.00	244.00	---	1104.00	**1350.00**
	Inst	Ea	Sm	SC	8.9	1.80	1040.00	326.00	---	1366.00	**1680.00**
Five cycles	Inst	Ea	Lg	SC	6.67	2.40	720.00	244.00	---	964.00	**1190.00**
	Inst	Ea	Sm	SC	8.9	1.80	875.00	326.00	---	1201.00	**1490.00**
Four cycles	Inst	Ea	Lg	SC	6.67	2.40	580.00	244.00	---	824.00	**1030.00**
	Inst	Ea	Sm	SC	8.9	1.80	705.00	326.00	---	1031.00	**1290.00**
Three cycles	Inst	Ea	Lg	SC	6.67	2.40	482.00	244.00	---	726.00	**914.00**
	Inst	Ea	Sm	SC	8.9	1.80	586.00	326.00	---	912.00	**1150.00**
Two cycles	Inst	Ea	Lg	SC	6.67	2.40	279.00	244.00	---	523.00	**681.00**
	Inst	Ea	Sm	SC	8.9	1.80	339.00	326.00	---	665.00	**870.00**

Adjustments

Description	Oper	Unit	Vol	Crew Size	Man-hours per Unit	Crew Output per Day	Avg Mat'l Unit Cost	Avg Labor Unit Cost	Avg Equip Unit Cost	Avg Total Unit Cost	Avg Price Incl O&P
Remove and reset dishwasher only											
	Reset	Ea	Lg	SA	1.33	6.00	---	50.10	---	50.10	**73.60**
	Reset	Ea	Sm	SA	1.78	4.50	---	67.10	---	67.10	**98.60**
Remove dishwasher											
	Demo	Ea	Lg	SA	0.80	10.00	---	30.10	---	30.10	**44.30**
	Demo	Ea	Sm	SA	1.07	7.50	---	40.30	---	40.30	**59.30**
Front and side panel kits											
White or prime finish	Inst	Ea	Lg	---	---	---	36.40	---	---	36.40	**36.40**
	Inst	Ea	Sm	---	---	---	44.20	---	---	44.20	**44.20**
Regular color finish	Inst	Ea	Lg	---	---	---	36.40	---	---	36.40	**36.40**
	Inst	Ea	Sm	---	---	---	44.20	---	---	44.20	**44.20**
Brushed chrome finish	Inst	Ea	Lg	---	---	---	44.10	---	---	44.10	**44.10**
	Inst	Ea	Sm	---	---	---	53.60	---	---	53.60	**53.60**
Stainless steel finish	Inst	Ea	Lg	---	---	---	54.60	---	---	54.60	**54.60**
	Inst	Ea	Sm	---	---	---	66.30	---	---	66.30	**66.30**
Black-glass acrylic finish with trim kit											
	Inst	Ea	Lg	---	---	---	73.50	---	---	73.50	**73.50**
	Inst	Ea	Sm	---	---	---	89.30	---	---	89.30	**89.30**
Stainless steel trim kit	Inst	Ea	Lg	---	---	---	33.60	---	---	33.60	**33.60**
	Inst	Ea	Sm	---	---	---	40.80	---	---	40.80	**40.80**

Description	Oper	Unit	Vol	Crew Size	Man-hours per Unit	Crew Output per Day	Avg Mat'l Unit Cost	Avg Labor Unit Cost	Avg Equip Unit Cost	Avg Total Unit Cost	Avg Price Incl O&P

Portable front loading
Convertible, front loading, portable dishwashers; hardwood top, front & side panels

Six cycles

Description	Oper	Unit	Vol	Crew Size	Man-hours per Unit	Crew Output per Day	Avg Mat'l Unit Cost	Avg Labor Unit Cost	Avg Equip Unit Cost	Avg Total Unit Cost	Avg Price Incl O&P
White	Inst	Ea	Lg	SC	4.00	4.00	524.00	147.00	---	671.00	**818.00**
	Inst	Ea	Sm	SC	5.33	3.00	637.00	195.00	---	832.00	**1020.00**
Colors	Inst	Ea	Lg	SC	4.00	4.00	560.00	147.00	---	707.00	**860.00**
	Inst	Ea	Sm	SC	5.33	3.00	680.00	195.00	---	875.00	**1070.00**

Four cycles

Description	Oper	Unit	Vol	Crew Size	Man-hours per Unit	Crew Output per Day	Avg Mat'l Unit Cost	Avg Labor Unit Cost	Avg Equip Unit Cost	Avg Total Unit Cost	Avg Price Incl O&P
White	Inst	Ea	Lg	SC	4.00	4.00	461.00	147.00	---	608.00	**746.00**
	Inst	Ea	Sm	SC	5.33	3.00	560.00	195.00	---	755.00	**931.00**
Colors	Inst	Ea	Lg	SC	4.00	4.00	510.00	147.00	---	657.00	**802.00**
	Inst	Ea	Sm	SC	5.33	3.00	620.00	195.00	---	815.00	**1000.00**

Three cycles

Description	Oper	Unit	Vol	Crew Size	Man-hours per Unit	Crew Output per Day	Avg Mat'l Unit Cost	Avg Labor Unit Cost	Avg Equip Unit Cost	Avg Total Unit Cost	Avg Price Incl O&P
White	Inst	Ea	Lg	SC	4.00	4.00	391.00	147.00	---	538.00	**666.00**
	Inst	Ea	Sm	SC	5.33	3.00	475.00	195.00	---	670.00	**834.00**
Colors	Inst	Ea	Lg	SC	4.00	4.00	417.00	147.00	---	564.00	**695.00**
	Inst	Ea	Sm	SC	5.33	3.00	506.00	195.00	---	701.00	**869.00**

Front loading portables
Three cycles with hardwood top

Description	Oper	Unit	Vol	Crew Size	Man-hours per Unit	Crew Output per Day	Avg Mat'l Unit Cost	Avg Labor Unit Cost	Avg Equip Unit Cost	Avg Total Unit Cost	Avg Price Incl O&P
White	Inst	Ea	Lg	SC	4.00	4.00	391.00	147.00	---	538.00	**666.00**
	Inst	Ea	Sm	SC	5.33	3.00	475.00	195.00	---	670.00	**834.00**
Colors	Inst	Ea	Lg	SC	4.00	4.00	417.00	147.00	---	564.00	**695.00**
	Inst	Ea	Sm	SC	5.33	3.00	506.00	195.00	---	701.00	**869.00**

Two cycles with porcelain top

Description	Oper	Unit	Vol	Crew Size	Man-hours per Unit	Crew Output per Day	Avg Mat'l Unit Cost	Avg Labor Unit Cost	Avg Equip Unit Cost	Avg Total Unit Cost	Avg Price Incl O&P
White	Inst	Ea	Lg	SC	4.00	4.00	349.00	147.00	---	496.00	**617.00**
	Inst	Ea	Sm	SC	5.33	3.00	424.00	195.00	---	619.00	**775.00**
Colors	Inst	Ea	Lg	SC	4.00	4.00	398.00	147.00	---	545.00	**674.00**
	Inst	Ea	Sm	SC	5.33	3.00	484.00	195.00	---	679.00	**843.00**

Dishwasher - sink combination, includes good quality fittings, 36"-42" cabinet, faucets and water supply kit

Six cycles

Description	Oper	Unit	Vol	Crew Size	Man-hours per Unit	Crew Output per Day	Avg Mat'l Unit Cost	Avg Labor Unit Cost	Avg Equip Unit Cost	Avg Total Unit Cost	Avg Price Incl O&P
White	Inst	Ea	Lg	SC	10.7	1.50	1280.00	392.00	---	1672.00	**2050.00**
	Inst	Ea	Sm	SC	14.2	1.13	1550.00	520.00	---	2070.00	**2550.00**
Colors	Inst	Ea	Lg	SC	10.7	1.50	1350.00	392.00	---	1742.00	**2130.00**
	Inst	Ea	Sm	SC	14.2	1.13	1640.00	520.00	---	2160.00	**2650.00**

Three cycles

Description	Oper	Unit	Vol	Crew Size	Man-hours per Unit	Crew Output per Day	Avg Mat'l Unit Cost	Avg Labor Unit Cost	Avg Equip Unit Cost	Avg Total Unit Cost	Avg Price Incl O&P
White	Inst	Ea	Lg	SC	10.7	1.50	1060.00	392.00	---	1452.00	**1800.00**
	Inst	Ea	Sm	SC	14.2	1.13	1290.00	520.00	---	1810.00	**2250.00**

Remove and reset combination dishwasher - sink only

Description	Oper	Unit	Vol	Crew Size	Man-hours per Unit	Crew Output per Day	Avg Mat'l Unit Cost	Avg Labor Unit Cost	Avg Equip Unit Cost	Avg Total Unit Cost	Avg Price Incl O&P
	Reset	Ea	Lg	SA	2.22	3.60	---	83.60	---	83.60	**123.00**
	Reset	Ea	Sm	SA	2.96	2.70	---	112.00	---	112.00	**164.00**

Remove only, d/w - sink combo

Description	Oper	Unit	Vol	Crew Size	Man-hours per Unit	Crew Output per Day	Avg Mat'l Unit Cost	Avg Labor Unit Cost	Avg Equip Unit Cost	Avg Total Unit Cost	Avg Price Incl O&P
	Demo	Ea	Lg	SA	1.33	6.00	---	50.10	---	50.10	**73.60**
	Demo	Ea	Sm	SA	1.78	4.50	---	67.10	---	67.10	**98.60**

Material adjustments

Description	Oper	Unit	Vol	Crew Size	Man-hours per Unit	Crew Output per Day	Avg Mat'l Unit Cost	Avg Labor Unit Cost	Avg Equip Unit Cost	Avg Total Unit Cost	Avg Price Incl O&P
For standard quality, DEDUCT	Inst	%	Lg	---	---	---	-20.0	---	---	---	---
	Inst	%	Sm	---	---	---	-20.0	---	---	---	---
For economy quality, DEDUCT	Inst	%	Lg	---	---	---	-30.0	---	---	---	---
	Inst	%	Sm	---	---	---	-30.0	---	---	---	---

Door frames

Exterior wood door frames

Exterior frame with exterior trim

Description	Oper	Unit	Vol	Crew Size	Man-hours per Unit	Crew Output per Day	Avg Mat'l Unit Cost	Avg Labor Unit Cost	Avg Equip Unit Cost	Avg Total Unit Cost	Avg Price Incl O&P
5/4 x 4-9/16" deep											
Pine	Inst	LF	Lg	2C	.046	350.0	3.48	1.53	---	5.01	**6.30**
	Inst	LF	Sm	2C	.070	228.0	3.77	2.33	---	6.10	**7.83**
Oak	Inst	LF	Lg	2C	.049	325.0	5.95	1.63	---	7.58	**9.29**
	Inst	LF	Sm	2C	.076	211.0	6.45	2.53	---	8.98	**11.20**
Walnut	Inst	LF	Lg	2C	.053	300.0	7.44	1.76	---	9.20	**11.20**
	Inst	LF	Sm	2C	.082	195.0	8.06	2.73	---	10.79	**13.40**
5/4 x 5-3/16" deep											
Pine	Inst	LF	Lg	2C	.046	350.0	3.78	1.53	---	5.31	**6.64**
	Inst	LF	Sm	2C	.070	228.0	4.10	2.33	---	6.43	**8.21**
Oak	Inst	LF	Lg	2C	.049	325.0	6.42	1.63	---	8.05	**9.83**
	Inst	LF	Sm	2C	.076	211.0	6.96	2.53	---	9.49	**11.80**
Walnut	Inst	LF	Lg	2C	.053	300.0	9.66	1.76	---	11.42	**13.80**
	Inst	LF	Sm	2C	.082	195.0	10.50	2.73	---	13.23	**16.10**
5/4 x 6-9/16" deep											
Pine	Inst	LF	Lg	2C	.046	350.0	4.42	1.53	---	5.95	**7.38**
	Inst	LF	Sm	2C	.070	228.0	4.78	2.33	---	7.11	**8.99**
Oak	Inst	LF	Lg	2C	.049	325.0	7.56	1.63	---	9.19	**11.10**
	Inst	LF	Sm	2C	.076	211.0	8.19	2.53	---	10.72	**13.20**
Walnut	Inst	LF	Lg	2C	.053	300.0	11.30	1.76	---	13.06	**15.70**
	Inst	LF	Sm	2C	.082	195.0	12.30	2.73	---	15.03	**18.20**

Exterior sills

Description	Oper	Unit	Vol	Crew Size	Man-hours per Unit	Crew Output per Day	Avg Mat'l Unit Cost	Avg Labor Unit Cost	Avg Equip Unit Cost	Avg Total Unit Cost	Avg Price Incl O&P
8/4 x 8" deep											
No horns	Inst	LF	Lg	2C	.200	80.00	9.90	6.65	---	16.55	**21.40**
	Inst	LF	Sm	2C	.308	52.00	10.70	10.30	---	21.00	**27.70**
2" horns	Inst	LF	Lg	2C	.200	80.00	10.60	6.65	---	17.25	**22.20**
	Inst	LF	Sm	2C	.308	52.00	11.50	10.30	---	21.80	**28.60**
3" horns	Inst	LF	Lg	2C	.200	80.00	11.50	6.65	---	18.15	**23.20**
	Inst	LF	Sm	2C	.308	52.00	12.50	10.30	---	22.80	**29.70**
8/4 x 10" deep											
No horns	Inst	LF	Lg	2C	.267	60.00	12.80	8.88	---	21.68	**28.10**
	Inst	LF	Sm	2C	.410	39.00	13.90	13.60	---	27.50	**36.50**
2" horns	Inst	LF	Lg	2C	.267	60.00	14.10	8.88	---	22.98	**29.50**
	Inst	LF	Sm	2C	.410	39.00	15.30	13.60	---	28.90	**38.00**
3" horns	Inst	LF	Lg	2C	.267	60.00	15.40	8.88	---	24.28	**31.10**
	Inst	LF	Sm	2C	.410	39.00	16.70	13.60	---	30.30	**39.70**

Exterior, colonial frame and trim

Description	Oper	Unit	Vol	Crew Size	Man-hours per Unit	Crew Output per Day	Avg Mat'l Unit Cost	Avg Labor Unit Cost	Avg Equip Unit Cost	Avg Total Unit Cost	Avg Price Incl O&P
3' opening, in swing	Inst	Ea	Lg	2C	.800	20.00	498.00	26.60	---	524.60	**613.00**
	Inst	Ea	Sm	2C	1.23	13.00	540.00	40.90	---	580.90	**682.00**
5' 4" opening, in/out swing	Inst	Ea	Lg	2C	1.07	15.00	804.00	35.60	---	839.60	**978.00**
	Inst	Ea	Sm	2C	1.60	10.00	871.00	53.20	---	924.20	**1080.00**
6' opening, in/out swing	Inst	Ea	Lg	2C	1.60	10.00	996.00	53.20	---	1049.20	**1230.00**
	Inst	Ea	Sm	2C	2.29	7.00	1080.00	76.20	---	1156.20	**1360.00**

Description	Oper	Unit	Vol	Crew Size	Man-hours per Unit	Crew Output per Day	Avg Mat'l Unit Cost	Avg Labor Unit Cost	Avg Equip Unit Cost	Avg Total Unit Cost	Avg Price Incl O&P

Interior wood door frames
Interior frame
11/16" x 3-5/8" deep
Pine	Inst	LF	Lg	2C	.046	350.0	3.52	1.53	---	5.05	**6.34**
	Inst	LF	Sm	2C	.070	228.0	3.81	2.33	---	6.14	**7.88**
Oak	Inst	LF	Lg	2C	.049	325.0	4.32	1.63	---	5.95	**7.41**
	Inst	LF	Sm	2C	.076	211.0	4.68	2.53	---	7.21	**9.18**
Walnut	Inst	LF	Lg	2C	.053	300.0	6.48	1.76	---	8.24	**10.10**
	Inst	LF	Sm	2C	.082	195.0	7.02	2.73	---	9.75	**12.20**

11/16" x 4-9/16" deep
Pine	Inst	LF	Lg	2C	.046	350.0	3.62	1.53	---	5.15	**6.46**
	Inst	LF	Sm	2C	.070	228.0	3.93	2.33	---	6.26	**8.01**
Oak	Inst	LF	Lg	2C	.049	325.0	4.32	1.63	---	5.95	**7.41**
	Inst	LF	Sm	2C	.076	211.0	4.68	2.53	---	7.21	**9.18**
Walnut	Inst	LF	Lg	2C	.053	300.0	6.48	1.76	---	8.24	**10.10**
	Inst	LF	Sm	2C	.082	195.0	7.02	2.73	---	9.75	**12.20**

11/16" x 5-3/16" deep
Pine	Inst	LF	Lg	2C	.046	350.0	4.10	1.53	---	5.63	**7.01**
	Inst	LF	Sm	2C	.070	228.0	4.45	2.33	---	6.78	**8.61**
Oak	Inst	LF	Lg	2C	.049	325.0	4.44	1.63	---	6.07	**7.55**
	Inst	LF	Sm	2C	.076	211.0	4.81	2.53	---	7.34	**9.32**
Walnut	Inst	LF	Lg	2C	.053	300.0	6.72	1.76	---	8.48	**10.40**
	Inst	LF	Sm	2C	.082	195.0	7.28	2.73	---	10.01	**12.50**

Pocket door frame
2'-0" to 3'-0" x 6'-8"	Inst	Ea	Lg	2C	1.33	12.00	90.00	44.30	---	134.30	**170.00**
	Inst	Ea	Sm	2C	2.00	8.00	97.50	66.50	---	164.00	**212.00**
3'-6" x 6'-8"	Inst	Ea	Lg	2C	1.33	12.00	166.00	44.30	---	210.30	**257.00**
	Inst	Ea	Sm	2C	2.00	8.00	179.00	66.50	---	245.50	**306.00**
4'-0" x 6'-8"	Inst	Ea	Lg	2C	1.33	12.00	168.00	44.30	---	212.30	**260.00**
	Inst	Ea	Sm	2C	2.00	8.00	182.00	66.50	---	248.50	**309.00**

Threshold, oak
5/8" x 3-5/8" deep	Inst	LF	Lg	2C	.100	160.0	2.35	3.33	---	5.68	**7.69**
	Inst	LF	Sm	2C	.154	104.0	2.55	5.12	---	7.67	**10.60**
5/8" x 4-5/8" deep	Inst	LF	Lg	2C	.107	150.0	3.06	3.56	---	6.62	**8.86**
	Inst	LF	Sm	2C	.163	98.00	3.32	5.42	---	8.74	**12.00**
5/8" x 5-5/8" deep	Inst	LF	Lg	2C	.114	140.0	3.65	3.79	---	7.44	**9.89**
	Inst	LF	Sm	2C	.176	91.00	3.95	5.86	---	9.81	**13.30**

Description	Oper	Unit	Vol	Crew Size	Man-hours per Unit	Crew Output per Day	Avg Mat'l Unit Cost	Avg Labor Unit Cost	Avg Equip Unit Cost	Avg Total Unit Cost	Avg Price Incl O&P

Door hardware

Locksets

Outside locks

Knobs (pin tumbler)

Description	Oper	Unit	Vol	Crew Size	Man-hours per Unit	Crew Output per Day	Avg Mat'l Unit Cost	Avg Labor Unit Cost	Avg Equip Unit Cost	Avg Total Unit Cost	Avg Price Incl O&P
Excellent quality	Inst	Ea	Lg	CA	.667	12.00	139.00	22.20	---	161.20	**193.00**
	Inst	Ea	Sm	CA	1.00	8.00	162.00	33.30	---	195.30	**236.00**
Good quality	Inst	Ea	Lg	CA	.500	16.00	103.00	16.60	---	119.60	**143.00**
	Inst	Ea	Sm	CA	.800	10.00	120.00	26.60	---	146.60	**178.00**
Average quality	Inst	Ea	Lg	CA	.500	16.00	72.10	16.60	---	88.70	**108.00**
	Inst	Ea	Sm	CA	.800	10.00	84.00	26.60	---	110.60	**137.00**

Handlesets

Description	Oper	Unit	Vol	Crew Size	Man-hours per Unit	Crew Output per Day	Avg Mat'l Unit Cost	Avg Labor Unit Cost	Avg Equip Unit Cost	Avg Total Unit Cost	Avg Price Incl O&P
Excellent quality	Inst	Ea	Lg	CA	.667	12.00	180.00	22.20	---	202.20	**241.00**
	Inst	Ea	Sm	CA	1.00	8.00	210.00	33.30	---	243.30	**291.00**
Good quality	Inst	Ea	Lg	CA	.500	16.00	113.00	16.60	---	129.60	**155.00**
	Inst	Ea	Sm	CA	.800	10.00	132.00	26.60	---	158.60	**192.00**
Average quality	Inst	Ea	Lg	CA	.500	16.00	82.40	16.60	---	99.00	**120.00**
	Inst	Ea	Sm	CA	.800	10.00	96.00	26.60	---	122.60	**150.00**

Bath or bedroom locks

Description	Oper	Unit	Vol	Crew Size	Man-hours per Unit	Crew Output per Day	Avg Mat'l Unit Cost	Avg Labor Unit Cost	Avg Equip Unit Cost	Avg Total Unit Cost	Avg Price Incl O&P
Excellent quality	Inst	Ea	Lg	CA	.533	15.00	103.00	17.70	---	120.70	**145.00**
	Inst	Ea	Sm	CA	.800	10.00	120.00	26.60	---	146.60	**178.00**
Good quality	Inst	Ea	Lg	CA	.400	20.00	87.60	13.30	---	100.90	**121.00**
	Inst	Ea	Sm	CA	.615	13.00	102.00	20.50	---	122.50	**148.00**
Average quality	Inst	Ea	Lg	CA	.400	20.00	67.00	13.30	---	80.30	**97.00**
	Inst	Ea	Sm	CA	.615	13.00	78.00	20.50	---	98.50	**120.00**

Passage latches

Description	Oper	Unit	Vol	Crew Size	Man-hours per Unit	Crew Output per Day	Avg Mat'l Unit Cost	Avg Labor Unit Cost	Avg Equip Unit Cost	Avg Total Unit Cost	Avg Price Incl O&P
Excellent quality	Inst	Ea	Lg	CA	.400	20.00	118.00	13.30	---	131.30	**156.00**
	Inst	Ea	Sm	CA	.615	13.00	138.00	20.50	---	158.50	**189.00**
Good quality	Inst	Ea	Lg	CA	.308	26.00	88.60	10.30	---	98.90	**117.00**
	Inst	Ea	Sm	CA	.471	17.00	103.00	15.70	---	118.70	**142.00**
Average quality	Inst	Ea	Lg	CA	.308	26.00	61.80	10.30	---	72.10	**86.40**
	Inst	Ea	Sm	CA	.471	17.00	72.00	15.70	---	87.70	**106.00**

Dead locks (double key)

Description	Oper	Unit	Vol	Crew Size	Man-hours per Unit	Crew Output per Day	Avg Mat'l Unit Cost	Avg Labor Unit Cost	Avg Equip Unit Cost	Avg Total Unit Cost	Avg Price Incl O&P
Excellent quality	Inst	Ea	Lg	CA	.533	15.00	118.00	17.70	---	135.70	**163.00**
	Inst	Ea	Sm	CA	.800	10.00	138.00	26.60	---	164.60	**199.00**
Good quality	Inst	Ea	Lg	CA	.400	20.00	92.70	13.30	---	106.00	**127.00**
	Inst	Ea	Sm	CA	.615	13.00	108.00	20.50	---	128.50	**155.00**
Average quality	Inst	Ea	Lg	CA	.400	20.00	62.80	13.30	---	76.10	**92.20**
	Inst	Ea	Sm	CA	.615	13.00	73.20	20.50	---	93.70	**115.00**

Kickplates, 8" x 30"

Description	Oper	Unit	Vol	Crew Size	Man-hours per Unit	Crew Output per Day	Avg Mat'l Unit Cost	Avg Labor Unit Cost	Avg Equip Unit Cost	Avg Total Unit Cost	Avg Price Incl O&P
Aluminum	Inst	Ea	Lg	CA	.308	26.00	16.20	10.30	---	26.50	**34.00**
	Inst	Ea	Sm	CA	.471	17.00	18.80	15.70	---	34.50	**45.20**
Brass or bronze	Inst	Ea	Lg	CA	.308	26.00	23.10	10.30	---	33.40	**41.90**
	Inst	Ea	Sm	CA	.471	17.00	26.90	15.70	---	42.60	**54.40**

Description	Oper	Unit	Vol	Crew Size	Man-hours per Unit	Crew Output per Day	Avg Mat'l Unit Cost	Avg Labor Unit Cost	Avg Equip Unit Cost	Avg Total Unit Cost	Avg Price Incl O&P
Thresholds											
Aluminum, pre-notched, draft-proof, standard											
3/4" high											
32" long	Inst	Ea	Lg	CA	.250	32.00	10.60	8.32	---	18.92	**24.70**
	Inst	Ea	Sm	CA	.381	21.00	12.40	12.70	---	25.10	**33.20**
36" long	Inst	Ea	Lg	CA	.250	32.00	12.20	8.32	---	20.52	**26.50**
	Inst	Ea	Sm	CA	.381	21.00	14.20	12.70	---	26.90	**35.30**
42" long	Inst	Ea	Lg	CA	.250	32.00	14.50	8.32	---	22.82	**29.10**
	Inst	Ea	Sm	CA	.381	21.00	16.90	12.70	---	29.60	**38.40**
48" long	Inst	Ea	Lg	CA	.250	32.00	16.60	8.32	---	24.92	**31.50**
	Inst	Ea	Sm	CA	.381	21.00	19.30	12.70	---	32.00	**41.20**
60" long	Inst	Ea	Lg	CA	.250	32.00	20.60	8.32	---	28.92	**36.20**
	Inst	Ea	Sm	CA	.381	21.00	24.00	12.70	---	36.70	**46.60**
72" long	Inst	Ea	Lg	CA	.250	32.00	24.40	8.32	---	32.72	**40.60**
	Inst	Ea	Sm	CA	.381	21.00	28.40	12.70	---	41.10	**51.70**
For high rug-type, 1-1/8"											
ADD	Inst	%	Lg	CA	---	---	25.0	---	---	---	---
	Inst	%	Sm	CA	---	---	25.0	---	---	---	---
Wood, oak threshold with vinyl weather seal											
5/8" x 3-1/2"											
33" long	Inst	Ea	Lg	CA	.250	32.00	12.40	8.32	---	20.72	**26.70**
	Inst	Ea	Sm	CA	.381	21.00	14.40	12.70	---	27.10	**35.60**
37" long	Inst	Ea	Lg	CA	.250	32.00	13.90	8.32	---	22.22	**28.50**
	Inst	Ea	Sm	CA	.381	21.00	16.20	12.70	---	28.90	**37.70**
43" long	Inst	Ea	Lg	CA	.250	32.00	15.00	8.32	---	23.32	**29.70**
	Inst	Ea	Sm	CA	.381	21.00	17.50	12.70	---	30.20	**39.10**
49" long	Inst	Ea	Lg	CA	.250	32.00	17.10	8.32	---	25.42	**32.20**
	Inst	Ea	Sm	CA	.381	21.00	20.00	12.70	---	32.70	**42.00**
61" long	Inst	Ea	Lg	CA	.250	32.00	21.80	8.32	---	30.12	**37.50**
	Inst	Ea	Sm	CA	.381	21.00	25.40	12.70	---	38.10	**48.20**
73" long	Inst	Ea	Lg	CA	.250	32.00	26.10	8.32	---	34.42	**42.50**
	Inst	Ea	Sm	CA	.381	21.00	30.40	12.70	---	43.10	**53.90**
3/4" x 3-1/2"											
33" long	Inst	Ea	Lg	CA	.250	32.00	13.10	8.32	---	21.42	**27.60**
	Inst	Ea	Sm	CA	.381	21.00	15.30	12.70	---	28.00	**36.60**
37" long	Inst	Ea	Lg	CA	.250	32.00	14.70	8.32	---	23.02	**29.40**
	Inst	Ea	Sm	CA	.381	21.00	17.10	12.70	---	29.80	**38.70**

Doors

Interior Door Systems

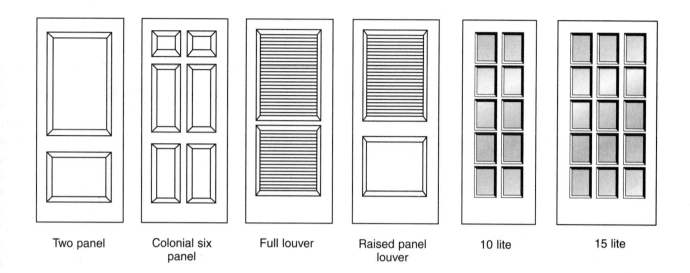

Two panel   Colonial six panel   Full louver   Raised panel louver   10 lite   15 lite

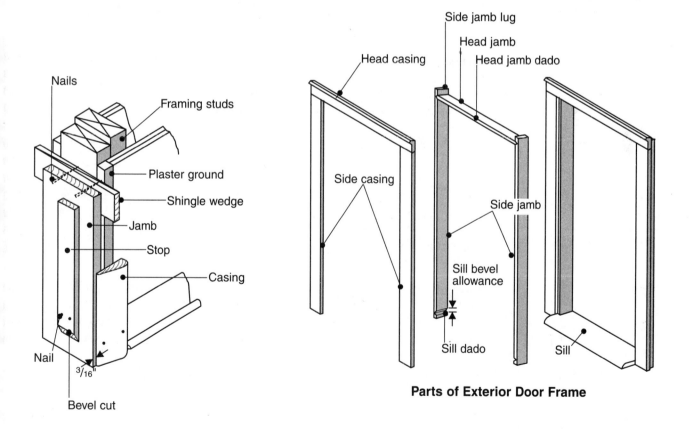

Parts of Exterior Door Frame

Assembled Package Door Units (Interior)

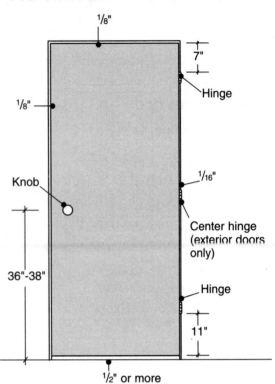

Door Clearances

Head jamb

3 nails (each side)

Door stops

Strike jamb

$\frac{1}{8}$"

7"

Hinge

$\frac{1}{8}$"

Knob

$\frac{1}{16}$"

Center hinge (exterior doors only)

36"-38"

Hinge

11"

$\frac{1}{2}$" or more

Unit Includes:

Jamb: $4\frac{9}{16}$" finger joint pine w/ stops applied.

Butts: 1 pair $3\frac{1}{2}$" x $3\frac{1}{2}$" applied to door and jamb.

Door: 3 degree bevel one side $\frac{3}{16}$" under std. width - net 80"
H center bored - $2\frac{1}{8}$" bore
$2\frac{3}{8}$" backset - 1" edge bore

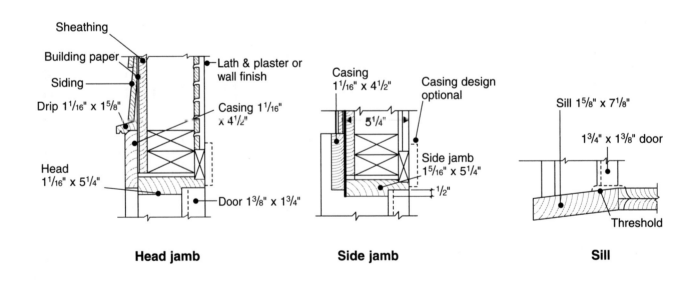

Sheathing

Building paper

Siding

Drip $1\frac{1}{16}$" x $1\frac{5}{8}$"

Lath & plaster or wall finish

Casing $1\frac{1}{16}$" x $4\frac{1}{2}$"

Head $1\frac{1}{16}$" x $5\frac{1}{4}$"

Door $1\frac{3}{8}$" x $1\frac{3}{4}$"

Casing $1\frac{1}{16}$" x $4\frac{1}{2}$"

Casing design optional

$5\frac{1}{4}$"

Side jamb $1\frac{5}{16}$" x $5\frac{1}{4}$"

$\frac{1}{2}$"

Sill $1\frac{5}{8}$" x $7\frac{1}{8}$"

$1\frac{3}{4}$" x $1\frac{3}{8}$" door

Threshold

Head jamb

Side jamb

Sill

Description	Oper	Unit	Vol	Crew Size	Man-hours per Unit	Crew Output per Day	Avg Mat'l Unit Cost	Avg Labor Unit Cost	Avg Equip Unit Cost	Avg Total Unit Cost	Avg Price Incl O&P

Doors

Entrance doors

Decorative glass doors and matching sidelights

Lafayette collection, 1-3/8" panels
Astoria style
2'-6", 2'-8", 3'-0" W x 6'-8" H doors

Description	Oper	Unit	Vol	Crew Size	Man-hours per Unit	Crew Output per Day	Avg Mat'l Unit Cost	Avg Labor Unit Cost	Avg Equip Unit Cost	Avg Total Unit Cost	Avg Price Incl O&P
Fir & pine	Inst	Ea	Lg	2C	1.33	12.00	1190.00	44.30	---	1234.30	**1440.00**
	Inst	Ea	Sm	2C	2.05	7.80	1390.00	68.20	---	1458.20	**1700.00**
Oak	Inst	Ea	Lg	2C	1.33	12.00	1560.00	44.30	---	1604.30	**1860.00**
	Inst	Ea	Sm	2C	2.05	7.80	1820.00	68.20	---	1888.20	**2200.00**

1'-0", 1'-2" x 6'-8-1/2" sidelights

Description	Oper	Unit	Vol	Crew Size	Man-hours per Unit	Crew Output per Day	Avg Mat'l Unit Cost	Avg Labor Unit Cost	Avg Equip Unit Cost	Avg Total Unit Cost	Avg Price Incl O&P
Fir & pine	Inst	Ea	Lg	2C	1.00	16.00	527.00	33.30	---	560.30	**656.00**
	Inst	Ea	Sm	2C	1.54	10.40	614.00	51.20	---	665.20	**783.00**
Oak	Inst	Ea	Lg	2C	1.00	16.00	671.00	33.30	---	704.30	**821.00**
	Inst	Ea	Sm	2C	1.54	10.40	781.00	51.20	---	832.20	**975.00**

Bourbon Royale style
2'-6", 2'-8", 3'-0" W x 6'-8" H doors

Description	Oper	Unit	Vol	Crew Size	Man-hours per Unit	Crew Output per Day	Avg Mat'l Unit Cost	Avg Labor Unit Cost	Avg Equip Unit Cost	Avg Total Unit Cost	Avg Price Incl O&P
Fir & pine	Inst	Ea	Lg	2C	1.33	12.00	1170.00	44.30	---	1214.30	**1410.00**
	Inst	Ea	Sm	2C	2.05	7.80	1360.00	68.20	---	1428.20	**1670.00**
Oak	Inst	Ea	Lg	2C	1.33	12.00	1550.00	44.30	---	1594.30	**1840.00**
	Inst	Ea	Sm	2C	2.05	7.80	1800.00	68.20	---	1868.20	**2170.00**

1'-0", 1'-2" x 6'-8-1/2" sidelights

Description	Oper	Unit	Vol	Crew Size	Man-hours per Unit	Crew Output per Day	Avg Mat'l Unit Cost	Avg Labor Unit Cost	Avg Equip Unit Cost	Avg Total Unit Cost	Avg Price Incl O&P
Fir & pine	Inst	Ea	Lg	2C	1.00	16.00	555.00	33.30	---	588.30	**688.00**
	Inst	Ea	Sm	2C	1.54	10.40	647.00	51.20	---	698.20	**821.00**
Oak	Inst	Ea	Lg	2C	1.00	16.00	688.00	33.30	---	721.30	**841.00**
	Inst	Ea	Sm	2C	1.54	10.40	802.00	51.20	---	853.20	**999.00**

Jubilee style
2'-6", 2'-8", 3'-0" W x 6'-8" H doors

Description	Oper	Unit	Vol	Crew Size	Man-hours per Unit	Crew Output per Day	Avg Mat'l Unit Cost	Avg Labor Unit Cost	Avg Equip Unit Cost	Avg Total Unit Cost	Avg Price Incl O&P
Fir & pine	Inst	Ea	Lg	2C	1.33	12.00	1230.00	44.30	---	1274.30	**1480.00**
	Inst	Ea	Sm	2C	2.05	7.80	1440.00	68.20	---	1508.20	**1750.00**
Oak	Inst	Ea	Lg	2C	1.33	12.00	1610.00	44.30	---	1654.30	**1910.00**
	Inst	Ea	Sm	2C	2.05	7.80	1870.00	68.20	---	1938.20	**2260.00**

1'-0", 1'-2" x 6'-8-1/2" sidelights

Description	Oper	Unit	Vol	Crew Size	Man-hours per Unit	Crew Output per Day	Avg Mat'l Unit Cost	Avg Labor Unit Cost	Avg Equip Unit Cost	Avg Total Unit Cost	Avg Price Incl O&P
Fir & pine	Inst	Ea	Lg	2C	1.00	16.00	517.00	33.30	---	550.30	**645.00**
	Inst	Ea	Sm	2C	1.54	10.40	602.00	51.20	---	653.20	**770.00**
Oak	Inst	Ea	Lg	2C	1.00	16.00	786.00	33.30	---	819.30	**954.00**
	Inst	Ea	Sm	2C	1.54	10.40	916.00	51.20	---	967.20	**1130.00**

Lexington style
2'-6", 2'-8", 3'-0" W x 6'-8" H doors

Description	Oper	Unit	Vol	Crew Size	Man-hours per Unit	Crew Output per Day	Avg Mat'l Unit Cost	Avg Labor Unit Cost	Avg Equip Unit Cost	Avg Total Unit Cost	Avg Price Incl O&P
Fir & pine	Inst	Ea	Lg	2C	1.33	12.00	1230.00	44.30	---	1274.30	**1480.00**
	Inst	Ea	Sm	2C	2.05	7.80	1430.00	68.20	---	1498.20	**1750.00**
Oak	Inst	Ea	Lg	2C	1.33	12.00	1600.00	44.30	---	1644.30	**1910.00**
	Inst	Ea	Sm	2C	2.05	7.80	1870.00	68.20	---	1938.20	**2250.00**

Description	Oper	Unit	Vol	Crew Size	Man- hours per Unit	Crew Output per Day	Avg Mat'l Unit Cost	Avg Labor Unit Cost	Avg Equip Unit Cost	Avg Total Unit Cost	Avg Price Incl O&P
1'-0", 1'-2" x 6'-8-1/2" sidelights											
Fir & pine	Inst	Ea	Lg	2C	1.00	16.00	551.00	33.30	---	584.30	**684.00**
	Inst	Ea	Sm	2C	1.54	10.40	642.00	51.20	---	693.20	**815.00**
Oak	Inst	Ea	Lg	2C	1.00	16.00	684.00	33.30	---	717.30	**836.00**
	Inst	Ea	Sm	2C	1.54	10.40	797.00	51.20	---	848.20	**993.00**
Marquis style											
2'-6", 2'-8", 3'-0" W x 6'-8" H doors											
Fir & pine	Inst	Ea	Lg	2C	1.33	12.00	1280.00	44.30	---	1324.30	**1540.00**
	Inst	Ea	Sm	2C	2.05	7.80	1500.00	68.20	---	1568.20	**1820.00**
Oak	Inst	Ea	Lg	2C	1.33	12.00	1660.00	44.30	---	1704.30	**1980.00**
	Inst	Ea	Sm	2C	2.05	7.80	1940.00	68.20	---	2008.20	**2330.00**
1'-0", 1'-2" x 6'-8-1/2" sidelights											
Fir & pine	Inst	Ea	Lg	2C	1.00	16.00	580.00	33.30	---	613.30	**717.00**
	Inst	Ea	Sm	2C	1.54	10.40	676.00	51.20	---	727.20	**854.00**
Oak	Inst	Ea	Lg	2C	1.00	16.00	646.00	33.30	---	679.30	**793.00**
	Inst	Ea	Sm	2C	1.54	10.40	752.00	51.20	---	803.20	**942.00**
Monaco style											
2'-6", 2'-8", 3'-0" W x 6'-8" H doors											
Fir & pine	Inst	Ea	Lg	2C	1.33	12.00	1510.00	44.30	---	1554.30	**1800.00**
	Inst	Ea	Sm	2C	2.05	7.80	1760.00	68.20	---	1828.20	**2120.00**
Oak	Inst	Ea	Lg	2C	1.33	12.00	1870.00	44.30	---	1914.30	**2220.00**
	Inst	Ea	Sm	2C	2.05	7.80	2180.00	68.20	---	2248.20	**2610.00**
1'-0", 1'-2" x 6'-8-1/2" sidelights											
Fir & pine	Inst	Ea	Lg	2C	1.00	16.00	691.00	33.30	---	724.30	**845.00**
	Inst	Ea	Sm	2C	1.54	10.40	805.00	51.20	---	856.20	**1000.00**
Oak	Inst	Ea	Lg	2C	1.00	16.00	824.00	33.30	---	857.30	**998.00**
	Inst	Ea	Sm	2C	1.54	10.40	960.00	51.20	---	1011.20	**1180.00**
Windsor style											
2'-6", 2'-8", 3'-0" W x 6'-8" H doors											
Fir & pine	Inst	Ea	Lg	2C	1.33	12.00	1270.00	44.30	---	1314.30	**1530.00**
	Inst	Ea	Sm	2C	2.05	7.80	1480.00	68.20	---	1548.20	**1800.00**
Oak	Inst	Ea	Lg	2C	1.33	12.00	1640.00	44.30	---	1684.30	**1950.00**
	Inst	Ea	Sm	2C	2.05	7.80	1910.00	68.20	---	1978.20	**2300.00**
1'-0", 1'-2" x 6'-8-1/2" sidelights											
Fir & pine	Inst	Ea	Lg	2C	1.00	16.00	545.00	33.30	---	578.30	**677.00**
	Inst	Ea	Sm	2C	1.54	10.40	635.00	51.20	---	686.20	**807.00**
Oak	Inst	Ea	Lg	2C	1.00	16.00	681.00	33.30	---	714.30	**833.00**
	Inst	Ea	Sm	2C	1.54	10.40	793.00	51.20	---	844.20	**989.00**

Description	Oper	Unit	Vol	Crew Size	Man-hours per Unit	Crew Output per Day	Avg Mat'l Unit Cost	Avg Labor Unit Cost	Avg Equip Unit Cost	Avg Total Unit Cost	Avg Price Incl O&P

Stile and rail raised panels

No lites

2 raised panels

Description	Oper	Unit	Vol	Crew Size	Man-hours per Unit	Crew Output per Day	Avg Mat'l Unit Cost	Avg Labor Unit Cost	Avg Equip Unit Cost	Avg Total Unit Cost	Avg Price Incl O&P
2'-6" to 3'-0" x 6'-8" x 1-3/4" T	Inst	Ea	Lg	2C	1.33	12.00	334.00	44.30	---	378.30	**450.00**
	Inst	Ea	Sm	2C	2.05	7.80	389.00	68.20	---	457.20	**549.00**
3'-6" x 6'-8" x 1-3/4" T	Inst	Ea	Lg	2C	1.45	11.00	334.00	48.20	---	382.20	**456.00**
	Inst	Ea	Sm	2C	2.22	7.20	389.00	73.90	---	462.90	**558.00**

4 raised panels

Description	Oper	Unit	Vol	Crew Size	Man-hours per Unit	Crew Output per Day	Avg Mat'l Unit Cost	Avg Labor Unit Cost	Avg Equip Unit Cost	Avg Total Unit Cost	Avg Price Incl O&P
2'-6" to 3'-0" x 6'-8" x 1-3/4" T	Inst	Ea	Lg	2C	1.33	12.00	304.00	44.30	---	348.30	**416.00**
	Inst	Ea	Sm	2C	2.05	7.80	354.00	68.20	---	422.20	**509.00**
3'-6" x 6'-8" x 1-3/4" T	Inst	Ea	Lg	2C	1.45	11.00	446.00	48.20	---	494.20	**585.00**
	Inst	Ea	Sm	2C	2.22	7.20	520.00	73.90	---	593.90	**708.00**

6 raised panels

Description	Oper	Unit	Vol	Crew Size	Man-hours per Unit	Crew Output per Day	Avg Mat'l Unit Cost	Avg Labor Unit Cost	Avg Equip Unit Cost	Avg Total Unit Cost	Avg Price Incl O&P
2'-0" to 2'-8" x 6'-8" x 1-3/4" T	Inst	Ea	Lg	2C	1.33	12.00	294.00	44.30	---	338.30	**404.00**
	Inst	Ea	Sm	2C	2.05	7.80	342.00	68.20	---	410.20	**496.00**
3'-0" x 6'-8" x 1-3/4" T	Inst	Ea	Lg	2C	1.45	11.00	302.00	48.20	---	350.20	**419.00**
	Inst	Ea	Sm	2C	2.22	7.20	352.00	73.90	---	425.90	**515.00**
3'-6" x 6'-8" x 1-3/4" T	Inst	Ea	Lg	2C	1.60	10.00	442.00	53.20	---	495.20	**588.00**
	Inst	Ea	Sm	2C	2.46	6.50	515.00	81.80	---	596.80	**715.00**

8 raised panels

Description	Oper	Unit	Vol	Crew Size	Man-hours per Unit	Crew Output per Day	Avg Mat'l Unit Cost	Avg Labor Unit Cost	Avg Equip Unit Cost	Avg Total Unit Cost	Avg Price Incl O&P
2'-6" to 2'-8" x 6'-8" x 1-3/4" T	Inst	Ea	Lg	2C	1.33	12.00	295.00	44.30	---	339.30	**405.00**
	Inst	Ea	Sm	2C	2.05	7.80	343.00	68.20	---	411.20	**497.00**
3'-0" x 6'-8" x 1-3/4" T	Inst	Ea	Lg	2C	1.45	11.00	302.00	48.20	---	350.20	**419.00**
	Inst	Ea	Sm	2C	2.22	7.20	352.00	73.90	---	425.90	**515.00**
3'-6" x 6'-8" x 1-3/4" T	Inst	Ea	Lg	2C	1.60	10.00	442.00	53.20	---	495.20	**588.00**
	Inst	Ea	Sm	2C	2.46	6.50	515.00	81.80	---	596.80	**715.00**
2'-6" to 2'-8" x 7'-0" x 1-3/4" T	Inst	Ea	Lg	2C	1.33	12.00	318.00	44.30	---	362.30	**432.00**
	Inst	Ea	Sm	2C	2.05	7.80	371.00	68.20	---	439.20	**529.00**
3'-0" x 7'-0" x 1-3/4" T	Inst	Ea	Lg	2C	1.45	11.00	343.00	48.20	---	391.20	**467.00**
	Inst	Ea	Sm	2C	2.22	7.20	400.00	73.90	---	473.90	**570.00**
3'-6" x 7'-0" x 1-3/4" T	Inst	Ea	Lg	2C	1.60	10.00	465.00	53.20	---	518.20	**614.00**
	Inst	Ea	Sm	2C	2.46	6.50	541.00	81.80	---	622.80	**745.00**

10 raised panels

Description	Oper	Unit	Vol	Crew Size	Man-hours per Unit	Crew Output per Day	Avg Mat'l Unit Cost	Avg Labor Unit Cost	Avg Equip Unit Cost	Avg Total Unit Cost	Avg Price Incl O&P
2'-6" to 2'-8" x 8'-0" x 1-3/4" T	Inst	Ea	Lg	2C	1.45	11.00	386.00	48.20	---	434.20	**517.00**
	Inst	Ea	Sm	2C	2.22	7.20	450.00	73.90	---	523.90	**628.00**
3'-0" x 8'-0" x 1-3/4" T	Inst	Ea	Lg	2C	1.60	10.00	405.00	53.20	---	458.20	**545.00**
	Inst	Ea	Sm	2C	2.46	6.50	472.00	81.80	---	553.80	**665.00**
3'-6" x 8'-0" x 1-3/4" T	Inst	Ea	Lg	2C	1.78	9.00	513.00	59.20	---	572.20	**679.00**
	Inst	Ea	Sm	2C	2.71	5.90	598.00	90.20	---	688.20	**823.00**

Description	Oper	Unit	Vol	Crew Size	Man-hours per Unit	Crew Output per Day	Avg Mat'l Unit Cost	Avg Labor Unit Cost	Avg Equip Unit Cost	Avg Total Unit Cost	Avg Price Incl O&P

Lites

1 raised panel, 1 lite
2'-6" to 3'-0" x 6'-8" x 1-3/4" T

Description	Oper	Unit	Vol	Crew Size	Man-hours per Unit	Crew Output per Day	Avg Mat'l Unit Cost	Avg Labor Unit Cost	Avg Equip Unit Cost	Avg Total Unit Cost	Avg Price Incl O&P
Open w/stops	Inst	Ea	Lg	2C	1.33	12.00	291.00	44.30	---	335.30	402.00
	Inst	Ea	Sm	2C	2.05	7.80	340.00	68.20	---	408.20	493.00
Tempered clear	Inst	Ea	Lg	2C	1.33	12.00	304.00	44.30	---	348.30	416.00
	Inst	Ea	Sm	2C	2.05	7.80	354.00	68.20	---	422.20	509.00
Acrylic or tempered amber	Inst	Ea	Lg	2C	1.33	12.00	365.00	44.30	---	409.30	486.00
	Inst	Ea	Sm	2C	2.05	7.80	425.00	68.20	---	493.20	591.00

1 raised panel, 6 lites
2'-6" to 3'-0" x 6'-8" x 1-3/4" T

Description	Oper	Unit	Vol	Crew Size	Man-hours per Unit	Crew Output per Day	Avg Mat'l Unit Cost	Avg Labor Unit Cost	Avg Equip Unit Cost	Avg Total Unit Cost	Avg Price Incl O&P
Open w/stops	Inst	Ea	Lg	2C	1.33	12.00	314.00	44.30		368.30	428.00
	Inst	Ea	Sm	2C	2.05	7.80	366.00	68.20	---	434.20	523.00
Tempered clear	Inst	Ea	Lg	2C	1.33	12.00	336.00	44.30	---	380.30	453.00
	Inst	Ea	Sm	2C	2.05	7.80	391.00	68.20	---	459.20	552.00
Acrylic or tempered amber	Inst	Ea	Lg	2C	1.33	12.00	381.00	44.30	---	425.30	505.00
	Inst	Ea	Sm	2C	2.05	7.80	444.00	68.20	---	512.20	613.00

1 raised panel, 9 lites
2'-6" to 3'-0" x 6'-8" x 1-3/4" T

Description	Oper	Unit	Vol	Crew Size	Man-hours per Unit	Crew Output per Day	Avg Mat'l Unit Cost	Avg Labor Unit Cost	Avg Equip Unit Cost	Avg Total Unit Cost	Avg Price Incl O&P
Open w/stops	Inst	Ea	Lg	2C	1.33	12.00	319.00	44.30	---	363.30	434.00
	Inst	Ea	Sm	2C	2.05	7.80	372.00	68.20	---	440.20	530.00
Tempered clear	Inst	Ea	Lg	2C	1.33	12.00	344.00	44.30	---	388.30	462.00
	Inst	Ea	Sm	2C	2.05	7.80	401.00	68.20	---	469.20	563.00
Acrylic or tempered amber	Inst	Ea	Lg	2C	1.33	12.00	417.00	44.30	---	461.30	546.00
	Inst	Ea	Sm	2C	2.05	7.80	486.00	68.20	---	554.20	661.00

1 raised panel, 12 diamond lites
2'-6" to 3'-0" x 6'-8" x 1-3/4" I

Description	Oper	Unit	Vol	Crew Size	Man-hours per Unit	Crew Output per Day	Avg Mat'l Unit Cost	Avg Labor Unit Cost	Avg Equip Unit Cost	Avg Total Unit Cost	Avg Price Incl O&P
Open w/stops	Inst	Ea	Lg	2C	1.33	12.00	328.00	44.30	---	372.30	443.00
	Inst	Ea	Sm	2C	2.05	7.80	382.00	68.20	---	450.20	541.00
Tempered clear	Inst	Ea	Lg	2C	1.33	12.00	417.00	44.30	---	461.30	546.00
	Inst	Ea	Sm	2C	2.05	7.80	486.00	68.20	---	554.20	661.00
Acrylic or tempered amber	Inst	Ea	Lg	2C	1.33	12.00	520.00	44.30	---	564.30	665.00
	Inst	Ea	Sm	2C	2.05	7.80	606.00	68.20	---	674.20	799.00

Dutch doors, country style, fir, stile and rail

Two raised panels in lower door
1 lite in upper door
2'-6" to 3'-0" x 6'-8" x 1-3/4" T

Description	Oper	Unit	Vol	Crew Size	Man-hours per Unit	Crew Output per Day	Avg Mat'l Unit Cost	Avg Labor Unit Cost	Avg Equip Unit Cost	Avg Total Unit Cost	Avg Price Incl O&P
Empty	Inst	Ea	Lg	2C	1.33	12.00	402.00	44.30	---	446.30	528.00
	Inst	Ea	Sm	2C	2.05	7.80	468.00	68.20	---	536.20	641.00
Tempered clear	Inst	Ea	Lg	2C	1.33	12.00	417.00	44.30	---	461.30	546.00
	Inst	Ea	Sm	2C	2.05	7.80	486.00	68.20	---	554.20	661.00
Tempered amber	Inst	Ea	Lg	2C	1.33	12.00	520.00	44.30	---	564.30	665.00
	Inst	Ea	Sm	2C	2.05	7.80	606.00	68.20	---	674.20	799.00

Description	Oper	Unit	Vol	Crew Size	Man-hours per Unit	Crew Output per Day	Avg Mat'l Unit Cost	Avg Labor Unit Cost	Avg Equip Unit Cost	Avg Total Unit Cost	Avg Price Incl O&P
4 lites in upper door											
2'-6" to 3'-0" x 6'-8" x 1-3/4" T											
Empty	Inst	Ea	Lg	2C	1.33	12.00	411.00	44.30	---	455.30	**539.00**
	Inst	Ea	Sm	2C	2.05	7.80	479.00	68.20	---	547.20	**653.00**
Tempered clear	Inst	Ea	Lg	2C	1.33	12.00	449.00	44.30	---	493.30	**583.00**
	Inst	Ea	Sm	2C	2.05	7.80	523.00	68.20	---	591.20	**704.00**
Tempered amber	Inst	Ea	Lg	2C	1.33	12.00	474.00	44.30	---	518.30	**611.00**
	Inst	Ea	Sm	2C	2.05	7.80	552.00	68.20	---	620.20	**737.00**
9 lites in upper door											
2'-6" to 3'-0" x 6'-8" x 1-3/4" T											
Empty	Inst	Ea	Lg	2C	1.33	12.00	407.00	44.30	---	451.30	**534.00**
	Inst	Ea	Sm	2C	2.05	7.80	474.00	68.20	---	542.20	**647.00**
Tempered clear	Inst	Ea	Lg	2C	1.33	12.00	422.00	44.30	---	466.30	**552.00**
	Inst	Ea	Sm	2C	2.05	7.80	492.00	68.20	---	560.20	**668.00**
Tempered amber	Inst	Ea	Lg	2C	1.33	12.00	524.00	44.30	---	568.30	**669.00**
	Inst	Ea	Sm	2C	2.05	7.80	611.00	68.20	---	679.20	**805.00**
12 diamond lites in upper door											
2'-6" to 3'-0" x 6'-8" x 1-3/4" T											
Empty	Inst	Ea	Lg	2C	1.33	12.00	447.00	44.30	---	491.30	**580.00**
	Inst	Ea	Sm	2C	2.05	7.80	521.00	68.20	---	589.20	**701.00**
Tempered clear	Inst	Ea	Lg	2C	1.33	12.00	480.00	44.30	---	524.30	**618.00**
	Inst	Ea	Sm	2C	2.05	7.80	559.00	68.20	---	627.20	**745.00**
Tempered amber	Inst	Ea	Lg	2C	1.33	12.00	593.00	44.30	---	637.30	**749.00**
	Inst	Ea	Sm	2C	2.05	7.80	691.00	68.20	---	759.20	**897.00**

Four diamond-shaped raised panels in lower door

Description	Oper	Unit	Vol	Crew Size	Man-hours per Unit	Crew Output per Day	Avg Mat'l Unit Cost	Avg Labor Unit Cost	Avg Equip Unit Cost	Avg Total Unit Cost	Avg Price Incl O&P
1 lite in upper door											
2'-6" to 3'-0" x 6'-8" x 1-3/4" T											
Empty	Inst	Ea	Lg	2C	1.33	12.00	405.00	44.30	---	449.30	**532.00**
	Inst	Ea	Sm	2C	2.05	7.80	472.00	68.20	---	540.20	**645.00**
Tempered clear	Inst	Ea	Lg	2C	1.33	12.00	420.00	44.30	---	464.30	**550.00**
	Inst	Ea	Sm	2C	2.05	7.80	490.00	68.20	---	558.20	**665.00**
Tempered amber	Inst	Ea	Lg	2C	1.33	12.00	523.00	44.30	---	567.30	**668.00**
	Inst	Ea	Sm	2C	2.05	7.80	610.00	68.20	---	678.20	**803.00**
6 lites in upper door											
2'-6" to 3'-0" x 6'-8" x 1-3/4" T											
Empty	Inst	Ea	Lg	2C	1.33	12.00	438.00	44.30	---	482.30	**570.00**
	Inst	Ea	Sm	2C	2.05	7.80	510.00	68.20	---	578.20	**689.00**
Tempered clear	Inst	Ea	Lg	2C	1.33	12.00	454.00	44.30	---	498.30	**589.00**
	Inst	Ea	Sm	2C	2.05	7.80	529.00	68.20	---	597.20	**711.00**
Tempered amber	Inst	Ea	Lg	2C	1.33	12.00	562.00	44.30	---	606.30	**713.00**
	Inst	Ea	Sm	2C	2.05	7.80	655.00	68.20	---	723.20	**856.00**

Description	Oper	Unit	Vol	Crew Size	Man-hours per Unit	Crew Output per Day	Avg Mat'l Unit Cost	Avg Labor Unit Cost	Avg Equip Unit Cost	Avg Total Unit Cost	Avg Price Incl O&P
9 lites in upper door											
2'-6" to 3'-0" x 6'-8" x 1-3/4" T											
Empty	Inst	Ea	Lg	2C	1.33	12.00	408.00	44.30	---	452.30	**535.00**
	Inst	Ea	Sm	2C	2.05	7.80	475.00	68.20	---	543.20	**649.00**
Tempered clear	Inst	Ea	Lg	2C	1.33	12.00	425.00	44.30	---	469.30	**556.00**
	Inst	Ea	Sm	2C	2.05	7.80	496.00	68.20	---	564.20	**672.00**
Tempered amber	Inst	Ea	Lg	2C	1.33	12.00	527.00	44.30	---	571.30	**673.00**
	Inst	Ea	Sm	2C	2.05	7.80	614.00	68.20	---	682.20	**809.00**
12 diamond lites in upper door											
2'-6" to 3'-0" x 6'-8" x 1-3/4" T											
Empty	Inst	Ea	Lg	2C	1.33	12.00	450.00	44.30	---	494.30	**584.00**
	Inst	Ea	Sm	2C	2.05	7.80	524.00	68.20	---	592.20	**705.00**
Tempered clear	Inst	Ea	Lg	2C	1.33	12.00	484.00	44.30	---	528.30	**823.00**
	Inst	Ea	Sm	2C	2.05	7.80	564.00	68.20	---	632.20	**751.00**
Tempered amber	Inst	Ea	Lg	2C	1.33	12.00	597.00	44.30	---	641.30	**753.00**
	Inst	Ea	Sm	2C	2.05	7.80	696.00	68.20	---	764.20	**903.00**

French doors

Douglas fir or hemlock 4-1/2" stiles and top rail, 9-1/2" bottom rail, tempered glass with stops

Description	Oper	Unit	Vol	Crew Size	Man-hours per Unit	Crew Output per Day	Avg Mat'l Unit Cost	Avg Labor Unit Cost	Avg Equip Unit Cost	Avg Total Unit Cost	Avg Price Incl O&P
1 lite											
1-3/8" T and 1-3/4" T											
2'-0" x 6'-8"	Inst	Ea	Lg	2C	1.14	14.00	219.00	37.90	---	256.90	**309.00**
	Inst	Ea	Sm	2C	1.76	9.10	256.00	58.60	---	314.60	**382.00**
2'-4", 2'-6", 2'-8" x 6'-8"	Inst	Ea	Lg	2C	1.33	12.00	225.00	44.30	---	269.30	**325.00**
	Inst	Ea	Sm	2C	2.05	7.80	262.00	68.20	---	330.20	**403.00**
3'-0" x 6'-8"	Inst	Ea	Lg	2C	1.60	10.00	228.00	53.20	---	281.20	**342.00**
	Inst	Ea	Sm	2C	2.46	6.50	265.00	81.80	---	346.80	**428.00**
5 lites, 5 high											
1-3/8" T and 1-3/4" T											
2'-0" x 6'-8"	Inst	Ea	Lg	2C	1.14	14.00	194.00	37.90	---	231.90	**280.00**
	Inst	Ea	Sm	2C	1.76	9.10	226.00	58.60	---	284.60	**347.00**
2'-4", 2'-6", 2'-8" x 6'-8"	Inst	Ea	Lg	2C	1.33	12.00	196.00	44.30	---	240.30	**291.00**
	Inst	Ea	Sm	2C	2.05	7.80	228.00	68.20	---	296.20	**365.00**
3'-0" x 6'-8"	Inst	Ea	Lg	2C	1.60	10.00	199.00	53.20	---	252.20	**308.00**
	Inst	Ea	Sm	2C	2.46	6.50	232.00	81.80	---	313.80	**389.00**
10 lites, 5 high											
1-3/8" T and 1-3/4" T											
2'-0" x 6'-8"	Inst	Ea	Lg	2C	1.14	14.00	215.00	37.90	---	252.90	**304.00**
	Inst	Ea	Sm	2C	1.76	9.10	251.00	58.60	---	309.60	**376.00**
2'-4", 2'-6", 2'-8" x 6'-8"	Inst	Ea	Lg	2C	1.33	12.00	221.00	44.30	---	265.30	**321.00**
	Inst	Ea	Sm	2C	2.05	7.80	258.00	68.20	---	326.20	**399.00**
3'-0" x 6'-8"	Inst	Ea	Lg	2C	1.60	10.00	227.00	53.20	---	280.20	**340.00**
	Inst	Ea	Sm	2C	2.46	6.50	264.00	81.80	---	345.80	**426.00**
15 lites, 5 high											
1-3/8" T and 1-3/4" T											
2'-6", 2'-8" x 6'-8"	Inst	Ea	Lg	2C	1.33	12.00	247.00	44.30	---	291.30	**351.00**
	Inst	Ea	Sm	2C	2.05	7.80	288.00	68.20	---	356.20	**434.00**
3'-0" x 6'-8"	Inst	Ea	Lg	2C	1.60	10.00	258.00	53.20	---	311.20	**376.00**
	Inst	Ea	Sm	2C	2.46	6.50	300.00	81.80	---	381.80	**468.00**

Description	Oper	Unit	Vol	Crew Size	Man-hours per Unit	Crew Output per Day	Avg Mat'l Unit Cost	Avg Labor Unit Cost	Avg Equip Unit Cost	Avg Total Unit Cost	Avg Price Incl O&P

French sidelites

Douglas fir or hemlock 4-1/2" stiles and top rail, 9-1/2" bottom rail, tempered glass with stops

1 lite
1-3/4" T

Description	Oper	Unit	Vol	Crew Size	Man-hrs	Crew Out	Mat'l	Labor	Equip	Total	Price O&P
1'-0", 1'-2" x 6'-8"	Inst	Ea	Lg	2C	1.00	16.00	149.00	33.30	---	182.30	**222.00**
	Inst	Ea	Sm	2C	1.54	10.40	174.00	51.20	---	225.20	**277.00**
1'-4", 1'-6" x 6'-8"	Inst	Ea	Lg	2C	1.00	16.00	160.00	33.30	---	193.30	**234.00**
	Inst	Ea	Sm	2C	1.54	10.40	186.00	51.20	---	237.20	**291.00**

5 lites, 5 high
1-3/4" T

Description	Oper	Unit	Vol	Crew Size	Man-hrs	Crew Out	Mat'l	Labor	Equip	Total	Price O&P
1'-0", 1'-2" x 6'-8"	Inst	Ea	Lg	2C	1.00	16.00	144.00	33.30	---	177.30	**216.00**
	Inst	Ea	Sm	2C	1.54	10.40	168.00	51.20	---	219.20	**270.00**
1'-4", 1'-6" x 6'-8"	Inst	Ea	Lg	2C	1.00	16.00	157.00	33.30	---	190.30	**230.00**
	Inst	Ea	Sm	2C	1.54	10.40	182.00	51.20	---	233.20	**287.00**

Garden doors

1 door, "X" unit

Description	Oper	Unit	Vol	Crew Size	Man-hrs	Crew Out	Mat'l	Labor	Equip	Total	Price O&P
2'-6" x 6'-8"	Inst	LS	Lg	2C	1.33	12.00	288.00	44.30	---	332.30	**398.00**
	Inst	LS	Sm	2C	2.05	7.80	336.00	68.20	---	404.20	**489.00**
3'-0" x 6'-8"	Inst	LS	Lg	2C	1.60	10.00	309.00	53.20	---	362.20	**435.00**
	Inst	LS	Sm	2C	2.46	6.50	360.00	81.80	---	441.80	**537.00**

2 doors, "XO/OX" unit

Description	Oper	Unit	Vol	Crew Size	Man-hrs	Crew Out	Mat'l	Labor	Equip	Total	Price O&P
5'-0" x 6'-8"	Inst	LS	Lg	2C	2.42	6.60	541.00	80.50	---	621.50	**743.00**
	Inst	LS	Sm	2C	3.72	4.30	630.00	124.00	---	754.00	**910.00**
6'-0" x 6'-8"	Inst	LS	Lg	2C	2.91	5.50	577.00	96.80	---	673.80	**809.00**
	Inst	LS	Sm	2C	4.44	3.60	672.00	148.00	---	820.00	**994.00**

3 doors, "XOX" unit

Description	Oper	Unit	Vol	Crew Size	Man-hrs	Crew Out	Mat'l	Labor	Equip	Total	Price O&P
7'-6" x 6'-8"	Inst	LS	Lg	2C	3.64	4.40	767.00	121.00	---	888.00	**1060.00**
	Inst	LS	Sm	2C	5.52	2.90	894.00	184.00	---	1078.00	**1300.00**
9'-0" x 6'-8"	Inst	LS	Lg	2C	4.32	3.70	824.00	144.00	---	968.00	**1160.00**
	Inst	LS	Sm	2C	6.67	2.40	960.00	222.00	---	1182.00	**1440.00**

Sliding doors

Glass sliding doors, with 5-1/2" anodized aluminum frame, trim, weatherstripping, tempered 3/16" T clear single-glazed glass, with screen

6'-8" H

Description	Oper	Unit	Vol	Crew Size	Man-hrs	Crew Out	Mat'l	Labor	Equip	Total	Price O&P
5' W, 2 lites, 1 sliding	Inst	Set	Lg	2C	4.00	4.00	422.00	133.00	---	555.00	**685.00**
	Inst	Set	Sm	2C	6.15	2.60	492.00	205.00	---	697.00	**873.00**
6' W, 2 lites, 1 sliding	Inst	Set	Lg	2C	4.00	4.00	471.00	133.00	---	604.00	**741.00**
	Inst	Set	Sm	2C	6.15	2.60	548.00	205.00	---	753.00	**938.00**
7' W, 2 lites, 1 sliding	Inst	Set	Lg	2C	4.00	4.00	506.00	133.00	---	639.00	**781.00**
	Inst	Set	Sm	2C	6.15	2.60	589.00	205.00	---	794.00	**985.00**
8' W, 2 lites, 1 sliding	Inst	Set	Lg	2C	4.00	4.00	543.00	133.00	---	676.00	**824.00**
	Inst	Set	Sm	2C	6.15	2.60	632.00	205.00	---	837.00	**1030.00**
10' W, 2 lites, 1 sliding	Inst	Set	Lg	2C	4.00	4.00	706.00	133.00	---	839.00	**1010.00**
	Inst	Set	Sm	2C	6.15	2.60	822.00	205.00	---	1027.00	**1250.00**

Description	Oper	Unit	Vol	Crew Size	Man-hours per Unit	Crew Output per Day	Avg Mat'l Unit Cost	Avg Labor Unit Cost	Avg Equip Unit Cost	Avg Total Unit Cost	Avg Price Incl O&P
9' W, 3 lites, 1 sliding	Inst	Set	Lg	2C	5.33	3.00	625.00	177.00	---	802.00	**985.00**
	Inst	Set	Sm	2C	8.00	2.00	728.00	266.00	---	994.00	**1240.00**
12' W, 3 lites, 1 sliding	Inst	Set	Lg	2C	5.33	3.00	733.00	177.00	---	910.00	**1110.00**
	Inst	Set	Sm	2C	8.00	2.00	854.00	266.00	---	1120.00	**1380.00**
15' W, 3 lites, 1 sliding	Inst	Set	Lg	2C	5.33	3.00	980.00	177.00	---	1157.00	**1390.00**
	Inst	Set	Sm	2C	8.00	2.00	1140.00	266.00	---	1406.00	**1710.00**

8'-0" H

Description	Oper	Unit	Vol	Crew Size	Man-hours per Unit	Crew Output per Day	Avg Mat'l Unit Cost	Avg Labor Unit Cost	Avg Equip Unit Cost	Avg Total Unit Cost	Avg Price Incl O&P
5' W, 2 lites, 1 sliding	Inst	Set	Lg	2C	4.00	4.00	497.00	133.00	---	630.00	**772.00**
	Inst	Set	Sm	2C	6.15	2.60	580.00	205.00	---	785.00	**973.00**
6' W, 2 lites, 1 sliding	Inst	Set	Lg	2C	4.00	4.00	535.00	133.00	---	668.00	**814.00**
	Inst	Set	Sm	2C	6.15	2.60	623.00	205.00	---	828.00	**1020.00**
7' W, 2 lites, 1 sliding	Inst	Set	Lg	2C	4.00	4.00	632.00	133.00	---	765.00	**927.00**
	Inst	Set	Sm	2C	6.15	2.60	737.00	205.00	---	942.00	**1150.00**
8' W, 2 lites, 1 sliding	Inst	Set	Lg	2C	4.00	4.00	731.00	133.00	---	864.00	**1040.00**
	Inst	Set	Sm	2C	6.15	2.60	852.00	205.00	---	1057.00	**1290.00**
10' W, 2 lites, 1 sliding	Inst	Set	Lg	2C	4.00	4.00	844.00	133.00	---	977.00	**1170.00**
	Inst	Set	Sm	2C	6.15	2.60	983.00	205.00	---	1188.00	**1440.00**
9' W, 3 lites, 1 sliding	Inst	Set	Lg	2C	5.33	3.00	735.00	177.00	---	912.00	**1110.00**
	Inst	Set	Sm	2C	8.00	2.00	857.00	266.00	---	1123.00	**1380.00**
12' W, 3 lites, 1 sliding	Inst	Set	Lg	2C	5.33	3.00	984.00	177.00	---	1161.00	**1400.00**
	Inst	Set	Sm	2C	8.00	2.00	1150.00	266.00	---	1416.00	**1720.00**
15' W, 3 lites, 1 sliding	Inst	Set	Lg	2C	5.33	3.00	1140.00	177.00	---	1317.00	**1580.00**
	Inst	Set	Sm	2C	8.00	2.00	1330.00	266.00	---	1596.00	**1930.00**

French sliding door units

Douglas fir doors, solid brass hardware, screen, frames, and exterior molding included

1 lite, 2 panels, single glazed
1-3/4" T

Description	Oper	Unit	Vol	Crew Size	Man-hours per Unit	Crew Output per Day	Avg Mat'l Unit Cost	Avg Labor Unit Cost	Avg Equip Unit Cost	Avg Total Unit Cost	Avg Price Incl O&P
5'-0" x 6'-8"	Inst	Ea	Lg	2C	5.33	3.00	2270.00	177.00	---	2447.00	**2880.00**
	Inst	Ea	Sm	2C	8.00	2.00	2640.00	266.00	---	2906.00	**3440.00**
6'-0" x 6'-8"	Inst	Ea	Lg	2C	5.33	3.00	2420.00	177.00	---	2597.00	**3050.00**
	Inst	Ea	Sm	2C	8.00	2.00	2820.00	266.00	---	3086.00	**3640.00**
7'-0" x 6'-8"	Inst	Ea	Lg	2C	5.33	3.00	2620.00	177.00	---	2797.00	**3280.00**
	Inst	Ea	Sm	2C	8.00	2.00	3050.00	266.00	---	3316.00	**3910.00**
8'-0" x 6'-8"	Inst	Ea	Lg	2C	5.33	3.00	2760.00	177.00	---	2937.00	**3440.00**
	Inst	Ea	Sm	2C	8.00	2.00	3210.00	266.00	---	3476.00	**4090.00**
5'-0" x 7'-0"	Inst	Ea	Lg	2C	5.33	3.00	2320.00	177.00	---	2497.00	**2940.00**
	Inst	Ea	Sm	2C	8.00	2.00	2710.00	266.00	---	2976.00	**3510.00**
6'-0" x 7'-0"	Inst	Ea	Lg	2C	5.33	3.00	2440.00	177.00	---	2617.00	**3080.00**
	Inst	Ea	Sm	2C	8.00	2.00	2850.00	266.00	---	3116.00	**3670.00**
7'-0" x 7'-0"	Inst	Ea	Lg	2C	5.33	3.00	2690.00	177.00	---	2867.00	**3360.00**
	Inst	Ea	Sm	2C	8.00	2.00	3140.00	266.00	---	3406.00	**4010.00**
8'-0" x 7'-0"	Inst	Ea	Lg	2C	5.33	3.00	2860.00	177.00	---	3037.00	**3550.00**
	Inst	Ea	Sm	2C	8.00	2.00	3330.00	266.00	---	3596.00	**4230.00**
5'-0" x 8'-0"	Inst	Ea	Lg	2C	5.33	3.00	2630.00	177.00	---	2807.00	**3290.00**
	Inst	Ea	Sm	2C	8.00	2.00	3060.00	266.00	---	3326.00	**3920.00**

Description	Oper	Unit	Vol	Crew Size	Man-hours per Unit	Crew Output per Day	Avg Mat'l Unit Cost	Avg Labor Unit Cost	Avg Equip Unit Cost	Avg Total Unit Cost	Avg Price Incl O&P
6'-0" x 8'-0"	Inst	Ea	Lg	2C	5.33	3.00	2820.00	177.00	---	2997.00	**3510.00**
	Inst	Ea	Sm	2C	8.00	2.00	3280.00	266.00	---	3546.00	**4170.00**
7'-0" x 8'-0"	Inst	Ea	Lg	2C	5.33	3.00	3030.00	177.00	---	3207.00	**3750.00**
	Inst	Ea	Sm	2C	8.00	2.00	3530.00	266.00	---	3796.00	**4460.00**
8'-0" x 8'-0"	Inst	Ea	Lg	2C	5.33	3.00	3220.00	177.00	---	3397.00	**3970.00**
	Inst	Ea	Sm	2C	8.00	2.00	3750.00	266.00	---	4016.00	**4710.00**

10 lites, 2 panels, single glazed
1-3/4" T

Description	Oper	Unit	Vol	Crew Size	Man-hours per Unit	Crew Output per Day	Avg Mat'l Unit Cost	Avg Labor Unit Cost	Avg Equip Unit Cost	Avg Total Unit Cost	Avg Price Incl O&P
5'-0" x 6'-8"	Inst	Ea	Lg	2C	5.33	3.00	2450.00	177.00	---	2627.00	**3090.00**
	Inst	Ea	Sm	2C	8.00	2.00	2860.00	266.00	---	3126.00	**3690.00**
6'-0" x 6'-8"	Inst	Ea	Lg	2C	5.33	3.00	2560.00	177.00	---	2737.00	**3210.00**
	Inst	Ea	Sm	2C	8.00	2.00	2980.00	266.00	---	3246.00	**3830.00**
5'-0" x 7'-0"	Inst	Ea	Lg	2C	5.33	3.00	2490.00	177.00	---	2667.00	**3130.00**
	Inst	Ea	Sm	2C	8.00	2.00	2910.00	266.00	---	3176.00	**3740.00**
6'-0" x 7'-0"	Inst	Ea	Lg	2C	5.33	3.00	2630.00	177.00	---	2807.00	**3290.00**
	Inst	Ea	Sm	2C	8.00	2.00	3060.00	266.00	---	3326.00	**3920.00**
5'-0" x 8'-0"	Inst	Ea	Lg	2C	5.33	3.00	2780.00	177.00	---	2957.00	**3460.00**
	Inst	Ea	Sm	2C	8.00	2.00	3240.00	266.00	---	3506.00	**4120.00**
6'-0" x 8'-0"	Inst	Ea	Lg	2C	5.33	3.00	2920.00	177.00	---	3097.00	**3630.00**
	Inst	Ea	Sm	2C	8.00	2.00	3410.00	266.00	---	3676.00	**4320.00**

15 lites, 2 panels, single glazed
1-3/4" T

Description	Oper	Unit	Vol	Crew Size	Man-hours per Unit	Crew Output per Day	Avg Mat'l Unit Cost	Avg Labor Unit Cost	Avg Equip Unit Cost	Avg Total Unit Cost	Avg Price Incl O&P
5'-0" x 6'-8"	Inst	Ea	Lg	2C	5.33	3.00	2580.00	177.00	---	2757.00	**3230.00**
	Inst	Ea	Sm	2C	8.00	2.00	3000.00	266.00	---	3266.00	**3850.00**
6'-0" x 6'-8"	Inst	Ea	Lg	2C	5.33	3.00	2690.00	177.00	---	2867.00	**3360.00**
	Inst	Ea	Sm	2C	8.00	2.00	3140.00	266.00	---	3406.00	**4010.00**
7'-0" x 6'-8"	Inst	Ea	Lg	2C	5.33	3.00	2870.00	177.00	---	3047.00	**3570.00**
	Inst	Ea	Sm	2C	8.00	2.00	3340.00	266.00	---	3606.00	**4240.00**
8'-0" x 6'-8"	Inst	Ea	Lg	2C	5.33	3.00	3000.00	177.00	---	3177.00	**3710.00**
	Inst	Ea	Sm	2C	8.00	2.00	3490.00	266.00	---	3756.00	**4420.00**
5'-0" x 7'-0"	Inst	Ea	Lg	2C	5.33	3.00	2630.00	177.00	---	2807.00	**3290.00**
	Inst	Ea	Sm	2C	8.00	2.00	3060.00	266.00	---	3326.00	**3920.00**
6'-0" x 7'-0"	Inst	Ea	Lg	2C	5.33	3.00	2760.00	177.00	---	2937.00	**3440.00**
	Inst	Ea	Sm	2C	8.00	2.00	3220.00	266.00	---	3486.00	**4100.00**
7'-0" x 7'-0"	Inst	Ea	Lg	2C	5.33	3.00	2950.00	177.00	---	3127.00	**3650.00**
	Inst	Ea	Sm	2C	8.00	2.00	3430.00	266.00	---	3696.00	**4350.00**
8'-0" x 7'-0"	Inst	Ea	Lg	2C	5.33	3.00	3090.00	177.00	---	3267.00	**3820.00**
	Inst	Ea	Sm	2C	8.00	2.00	3600.00	266.00	---	3866.00	**4540.00**
5'-0" x 8'-0"	Inst	Ea	Lg	2C	5.33	3.00	2910.00	177.00	---	3087.00	**3620.00**
	Inst	Ea	Sm	2C	8.00	2.00	3390.00	266.00	---	3656.00	**4300.00**
6'-0" x 8'-0"	Inst	Ea	Lg	2C	5.33	3.00	3060.00	177.00	---	3237.00	**3780.00**
	Inst	Ea	Sm	2C	8.00	2.00	3560.00	266.00	---	3826.00	**4500.00**
7'-0" x 8'-0"	Inst	Ea	Lg	2C	5.33	3.00	3280.00	177.00	---	3457.00	**4040.00**
	Inst	Ea	Sm	2C	8.00	2.00	3820.00	266.00	---	4086.00	**4790.00**
8'-0" x 8'-0"	Inst	Ea	Lg	2C	5.33	3.00	3410.00	177.00	---	3587.00	**4180.00**
	Inst	Ea	Sm	2C	8.00	2.00	3970.00	266.00	---	4236.00	**4960.00**

Description	Oper	Unit	Vol	Crew Size	Man-hours per Unit	Crew Output per Day	Avg Mat'l Unit Cost	Avg Labor Unit Cost	Avg Equip Unit Cost	Avg Total Unit Cost	Avg Price Incl O&P

Fire doors, natural birch

One hour rating

Description	Oper	Unit	Vol	Crew Size	Man-hours	Output	Mat'l	Labor	Equip	Total	Price
2'-6" x 6'-8" x 1-3/4" T	Inst	Ea	Lg	2C	1.23	13.00	220.00	40.90	---	260.90	**315.00**
	Inst	Ea	Sm	2C	1.88	8.50	257.00	62.60	---	319.60	**389.00**
3'-6" x 6'-8" x 1-3/4" T	Inst	Ea	Lg	2C	1.45	11.00	275.00	48.20	---	323.20	**389.00**
	Inst	Ea	Sm	2C	2.22	7.20	320.00	73.90	---	393.90	**479.00**
2'-6" x 7'-0" x 1-3/4" T	Inst	Ea	Lg	2C	1.23	13.00	231.00	40.90	---	271.90	**327.00**
	Inst	Ea	Sm	2C	1.88	8.50	269.00	62.60	---	331.60	**403.00**
3'-6" x 7'-0" x 1-3/4" T	Inst	Ea	Lg	2C	1.45	11.00	287.00	48.20	---	335.20	**403.00**
	Inst	Ea	Sm	2C	2.22	7.20	335.00	73.90	---	408.90	**496.00**

1.5 hour rating

Description	Oper	Unit	Vol	Crew Size	Man-hours	Output	Mat'l	Labor	Equip	Total	Price
2'-6" x 6'-8" x 1-3/4" T	Inst	Ea	Lg	2C	1.23	13.00	313.00	40.90	---	353.90	**421.00**
	Inst	Ea	Sm	2C	1.88	8.50	365.00	62.60	---	427.60	**513.00**
3'-6" x 6'-8" x 1-3/4" T	Inst	Ea	Lg	2C	1.45	11.00	349.00	48.20	---	397.20	**474.00**
	Inst	Ea	Sm	2C	2.22	7.20	407.00	73.90	---	480.90	**579.00**
2'-6" x 7'-0" x 1-3/4" T	Inst	Ea	Lg	2C	1.23	13.00	330.00	40.90	---	370.90	**440.00**
	Inst	Ea	Sm	2C	1.88	8.50	384.00	62.60	---	446.60	**535.00**
3'-6" x 7'-0" x 1-3/4" T	Inst	Ea	Lg	2C	1.45	11.00	412.00	48.20	---	460.20	**546.00**
	Inst	Ea	Sm	2C	2.22	7.20	480.00	73.90	---	553.90	**663.00**

Interior doors

Passage doors, flush face, includes hinge hardware

Hollow core prehung door units

Hardboard

Description	Oper	Unit	Vol	Crew Size	Man-hours	Output	Mat'l	Labor	Equip	Total	Price
2'-6" x 6'-8" x 1-3/8" T	Inst	Ea	Lg	2C	1.07	15.00	39.90	35.60	---	75.50	**99.30**
	Inst	Ea	Sm	2C	1.63	9.80	46.50	54.20	---	100.70	**135.00**
2'-8" x 6'-8" x 1-3/8" T	Inst	Ea	Lg	2C	1.14	14.00	42.40	37.90	---	80.30	**106.00**
	Inst	Ea	Sm	2C	1.76	9.10	49.40	58.60	---	108.00	**145.00**
3'-0" x 6'-8" x 1-3/8" T	Inst	Ea	Lg	2C	1.14	14.00	44.80	37.90	---	82.70	**108.00**
	Inst	Ea	Sm	2C	1.76	9.10	52.20	58.60	---	110.80	**148.00**
2'-6" x 7'-0" x 1-3/8" T	Inst	Ea	Lg	2C	1.07	15.00	50.80	35.60	---	86.40	**112.00**
	Inst	Ea	Sm	2C	1.63	9.80	59.20	54.20	---	113.40	**149.00**
2'-8" x 7'-0" x 1-3/8" T	Inst	Ea	Lg	2C	1.14	14.00	53.30	37.90	---	91.20	**118.00**
	Inst	Ea	Sm	2C	1.76	9.10	62.10	58.60	---	120.70	**159.00**
3'-0" x 7'-0" x 1-3/8" T	Inst	Ea	Lg	2C	1.14	14.00	55.70	37.90	---	93.60	**121.00**
	Inst	Ea	Sm	2C	1.76	9.10	64.90	58.60	---	123.50	**163.00**

Lauan

Description	Oper	Unit	Vol	Crew Size	Man-hours	Output	Mat'l	Labor	Equip	Total	Price
2'-6" x 6'-8" x 1-3/8" T	Inst	Ea	Lg	2C	1.07	15.00	52.90	35.60	---	88.50	**114.00**
	Inst	Ea	Sm	2C	1.63	9.80	61.60	54.20	---	115.80	**152.00**
2'-8" x 6'-8" x 1-3/8" T	Inst	Ea	Lg	2C	1.14	14.00	54.90	37.90	---	92.80	**120.00**
	Inst	Ea	Sm	2C	1.76	9.10	64.00	58.60	---	122.60	**161.00**
3'-0" x 6'-8" x 1-3/8" T	Inst	Ea	Lg	2C	1.14	14.00	58.90	37.90	---	96.80	**125.00**
	Inst	Ea	Sm	2C	1.76	9.10	68.60	58.60	---	127.20	**167.00**
2'-6" x 7'-0" x 1-3/8" T	Inst	Ea	Lg	2C	1.07	15.00	62.60	35.60	---	98.20	**125.00**
	Inst	Ea	Sm	2C	1.63	9.80	73.00	54.20	---	127.20	**165.00**
2'-8" x 7'-0" x 1-3/8" T	Inst	Ea	Lg	2C	1.14	14.00	64.90	37.90	---	102.80	**132.00**
	Inst	Ea	Sm	2C	1.76	9.10	75.60	58.60	---	134.20	**175.00**
3'-0" x 7'-0" x 1-3/8" T	Inst	Ea	Lg	2C	1.14	14.00	69.00	37.90	---	106.90	**136.00**
	Inst	Ea	Sm	2C	1.76	9.10	80.30	58.60	---	138.90	**180.00**

Description	Oper	Unit	Vol	Crew Size	Man-hours per Unit	Crew Output per Day	Avg Mat'l Unit Cost	Avg Labor Unit Cost	Avg Equip Unit Cost	Avg Total Unit Cost	Avg Price Incl O&P
Birch											
2'-6" x 6'-8" x 1-3/8" T	Inst	Ea	Lg	2C	1.07	15.00	63.90	35.60	---	99.50	**127.00**
	Inst	Ea	Sm	2C	1.63	9.80	74.50	54.20	---	128.70	**167.00**
2'-8" x 6'-8" x 1-3/8" T	Inst	Ea	Lg	2C	1.14	14.00	66.10	37.90	---	104.00	**133.00**
	Inst	Ea	Sm	2C	1.76	9.10	77.00	58.60	---	135.60	**176.00**
3'-0" x 6'-8" x 1-3/8" T	Inst	Ea	Lg	2C	1.14	14.00	70.00	37.90	---	107.90	**137.00**
	Inst	Ea	Sm	2C	1.76	9.10	81.50	58.60	---	140.10	**182.00**
2'-6" x 7'-0" x 1-3/8" T	Inst	Ea	Lg	2C	1.07	15.00	79.50	35.60	---	115.10	**145.00**
	Inst	Ea	Sm	2C	1.63	9.80	92.60	54.20	---	146.80	**188.00**
2'-8" x 7'-0" x 1-3/8" T	Inst	Ea	Lg	2C	1.14	14.00	82.50	37.90	---	120.40	**152.00**
	Inst	Ea	Sm	2C	1.76	9.10	96.10	58.60	---	154.70	**198.00**
3'-0" x 7'-0" x 1-3/8" T	Inst	Ea	Lg	2C	1.14	14.00	87.00	37.90	---	124.90	**157.00**
	Inst	Ea	Sm	2C	1.76	9.10	101.00	58.60	---	159.60	**204.00**

Solid core prehung door units

Description	Oper	Unit	Vol	Crew Size	Man-hours per Unit	Crew Output per Day	Avg Mat'l Unit Cost	Avg Labor Unit Cost	Avg Equip Unit Cost	Avg Total Unit Cost	Avg Price Incl O&P
Hardboard											
2'-6" x 6'-8" x 1-3/8" T	Inst	Ea	Lg	2C	1.14	14.00	81.40	37.90	---	119.30	**151.00**
	Inst	Ea	Sm	2C	1.76	9.10	94.90	58.60	---	153.50	**197.00**
2'-8" x 6'-8" x 1-3/8" T	Inst	Ea	Lg	2C	1.23	13.00	84.50	40.90	---	125.40	**159.00**
	Inst	Ea	Sm	2C	1.88	8.50	98.40	62.60	---	161.00	**207.00**
3'-0" x 6'-8" x 1-3/8" T	Inst	Ea	Lg	2C	1.23	13.00	86.40	40.90	---	127.30	**161.00**
	Inst	Ea	Sm	2C	1.88	8.50	101.00	62.60	---	163.60	**210.00**
2'-6" x 7'-0" x 1-3/8" T	Inst	Ea	Lg	2C	1.14	14.00	90.10	37.90	---	128.00	**161.00**
	Inst	Ea	Sm	2C	1.76	9.10	105.00	58.60	---	163.60	**209.00**
2'-8" x 7'-0" x 1-3/8" T	Inst	Ea	Lg	2C	1.23	13.00	92.80	40.90	---	133.70	**168.00**
	Inst	Ea	Sm	2C	1.88	8.50	108.00	62.60	---	170.60	**218.00**
3'-0" x 7'-0" x 1-3/8" T	Inst	Ea	Lg	2C	1.23	13.00	92.80	40.90	---	133.70	**168.00**
	Inst	Ea	Sm	2C	1.88	8.50	108.00	62.60	---	170.60	**218.00**
Lauan											
2'-6" x 6'-8" x 1-3/8" T	Inst	Ea	Lg	2C	1.14	14.00	93.00	37.90	---	130.90	**164.00**
	Inst	Ea	Sm	2C	1.76	9.10	108.00	58.60	---	166.60	**212.00**
2'-8" x 6'-8" x 1-3/8" T	Inst	Ea	Lg	2C	1.23	13.00	96.10	40.90	---	137.00	**172.00**
	Inst	Ea	Sm	2C	1.88	8.50	112.00	62.60	---	174.60	**223.00**
3'-0" x 6'-8" x 1-3/8" T	Inst	Ea	Lg	2C	1.23	13.00	99.80	40.90	---	140.70	**176.00**
	Inst	Ea	Sm	2C	1.88	8.50	116.00	62.60	---	178.60	**228.00**
2'-6" x 7'-0" x 1-3/8" T	Inst	Ea	Lg	2C	1.14	14.00	102.00	37.90	---	139.90	**174.00**
	Inst	Ea	Sm	2C	1.76	9.10	119.00	58.60	---	177.60	**224.00**
2'-8" x 7'-0" x 1-3/8" T	Inst	Ea	Lg	2C	1.23	13.00	104.00	40.90	---	144.90	**181.00**
	Inst	Ea	Sm	2C	1.88	8.50	122.00	62.60	---	184.60	**234.00**
3'-0" x 7'-0" x 1-3/8" T	Inst	Ea	Lg	2C	1.23	13.00	108.00	40.90	---	148.90	**186.00**
	Inst	Ea	Sm	2C	1.88	8.50	126.00	62.60	---	188.60	**239.00**
Birch											
2'-6" x 6'-8" x 1-3/8" T	Inst	Ea	Lg	2C	1.14	14.00	102.00	37.90	---	139.90	**174.00**
	Inst	Ea	Sm	2C	1.76	9.10	118.00	58.60	---	176.60	**224.00**
2'-8" x 6'-8" x 1-3/8" T	Inst	Ea	Lg	2C	1.23	13.00	104.00	40.90	---	144.90	**181.00**
	Inst	Ea	Sm	2C	1.88	8.50	121.00	62.60	---	183.60	**233.00**
3'-0" x 6'-8" x 1-3/8" T	Inst	Ea	Lg	2C	1.23	13.00	108.00	40.90	---	148.90	**185.00**
	Inst	Ea	Sm	2C	1.88	8.50	126.00	62.60	---	188.60	**238.00**
2'-6" x 7'-0" x 1-3/8" T	Inst	Ea	Lg	2C	1.14	14.00	110.00	37.90	---	147.90	**184.00**
	Inst	Ea	Sm	2C	1.76	9.10	129.00	58.60	---	187.60	**236.00**
2'-8" x 7'-0" x 1-3/8" T	Inst	Ea	Lg	2C	1.23	13.00	113.00	40.90	---	153.90	**191.00**
	Inst	Ea	Sm	2C	1.88	8.50	131.00	62.60	---	193.60	**245.00**
3'-0" x 7'-0" x 1-3/8" T	Inst	Ea	Lg	2C	1.23	13.00	117.00	40.90	---	157.90	**196.00**
	Inst	Ea	Sm	2C	1.88	8.50	137.00	62.60	---	199.60	**251.00**

Description	Oper	Unit	Vol	Crew Size	Man-hours per Unit	Crew Output per Day	Avg Mat'l Unit Cost	Avg Labor Unit Cost	Avg Equip Unit Cost	Avg Total Unit Cost	Avg Price Incl O&P

Prehung package door units

Hollow core, flush face door units

Hardboard

Description	Oper	Unit	Vol	Crew Size	Man-hours per Unit	Crew Output per Day	Avg Mat'l Unit Cost	Avg Labor Unit Cost	Avg Equip Unit Cost	Avg Total Unit Cost	Avg Price Incl O&P
2'-6" x 6'-8" x 1-3/8" T	Inst	Ea	Lg	2C	1.33	12.00	125.00	44.30	---	169.30	**210.00**
	Inst	Ea	Sm	2C	2.05	7.80	145.00	68.20	---	213.20	**270.00**
2'-8" x 6'-8" x 1-3/8" T	Inst	Ea	Lg	2C	1.33	12.00	127.00	44.30	---	171.30	**213.00**
	Inst	Ea	Sm	2C	2.05	7.80	148.00	68.20	---	216.20	**273.00**
3'-0" x 6'-8" x 1-3/8" T	Inst	Ea	Lg	2C	1.33	12.00	130.00	44.30	---	174.30	**216.00**
	Inst	Ea	Sm	2C	2.05	7.80	151.00	68.20	---	219.20	**276.00**

Lauan

Description	Oper	Unit	Vol	Crew Size	Man-hours per Unit	Crew Output per Day	Avg Mat'l Unit Cost	Avg Labor Unit Cost	Avg Equip Unit Cost	Avg Total Unit Cost	Avg Price Incl O&P
2'-6" x 6'-8" x 1-3/8" T	Inst	Ea	Lg	2C	1.33	12.00	138.00	44.30	---	182.30	**225.00**
	Inst	Ea	Sm	2C	2.05	7.80	161.00	68.20	---	229.20	**287.00**
2'-8" x 6'-8" x 1-3/8" T	Inst	Ea	Lg	2C	1.33	12.00	140.00	44.30	---	184.30	**227.00**
	Inst	Ea	Sm	2C	2.05	7.80	163.00	68.20	---	231.20	**290.00**
3'-0" x 6'-8" x 1-3/8" T	Inst	Ea	Lg	2C	1.33	12.00	144.00	44.30	---	188.30	**232.00**
	Inst	Ea	Sm	2C	2.05	7.80	167.00	68.20	---	235.20	**295.00**

Birch

Description	Oper	Unit	Vol	Crew Size	Man-hours per Unit	Crew Output per Day	Avg Mat'l Unit Cost	Avg Labor Unit Cost	Avg Equip Unit Cost	Avg Total Unit Cost	Avg Price Incl O&P
2'-6" x 6'-8" x 1-3/8" T	Inst	Ea	Lg	2C	1.33	12.00	149.00	44.30	---	193.30	**237.00**
	Inst	Ea	Sm	2C	2.05	7.80	173.00	68.20	---	241.20	**302.00**
2'-8" x 6'-8" x 1-3/8" T	Inst	Ea	Lg	2C	1.33	12.00	151.00	44.30	---	195.30	**240.00**
	Inst	Ea	Sm	2C	2.05	7.80	176.00	68.20	---	244.20	**305.00**
3'-0" x 6'-8" x 1-3/8" T	Inst	Ea	Lg	2C	1.33	12.00	155.00	44.30	---	199.30	**244.00**
	Inst	Ea	Sm	2C	2.05	7.80	180.00	68.20	---	248.20	**310.00**

Ash

Description	Oper	Unit	Vol	Crew Size	Man-hours per Unit	Crew Output per Day	Avg Mat'l Unit Cost	Avg Labor Unit Cost	Avg Equip Unit Cost	Avg Total Unit Cost	Avg Price Incl O&P
2'-6" x 6'-8" x 1-3/8" T	Inst	Ea	Lg	2C	1.33	12.00	167.00	44.30	---	211.30	**259.00**
	Inst	Ea	Sm	2C	2.05	7.80	195.00	68.20	---	263.20	**327.00**
2'-8" x 6'-8" x 1-3/8" T	Inst	Ea	Lg	2C	1.33	12.00	176.00	44.30	---	220.30	**268.00**
	Inst	Ea	Sm	2C	2.05	7.80	205.00	68.20	---	273.20	**338.00**
3'-0" x 6'-8" x 1-3/8" T	Inst	Ea	Lg	2C	1.33	12.00	180.00	44.30	---	224.30	**273.00**
	Inst	Ea	Sm	2C	2.05	7.80	210.00	68.20	---	278.20	**343.00**

Solid core, flush face door units

Hardboard

Description	Oper	Unit	Vol	Crew Size	Man-hours per Unit	Crew Output per Day	Avg Mat'l Unit Cost	Avg Labor Unit Cost	Avg Equip Unit Cost	Avg Total Unit Cost	Avg Price Incl O&P
2'-6" x 6'-8" x 1-3/8" T	Inst	Ea	Lg	2C	1.45	11.00	168.00	48.20	---	216.20	**266.00**
	Inst	Ea	Sm	2C	2.22	7.20	196.00	73.90	---	269.90	**336.00**
2'-8" x 6'-8" x 1-3/8" T	Inst	Ea	Lg	2C	1.45	11.00	171.00	48.20	---	219.20	**269.00**
	Inst	Ea	Sm	2C	2.22	7.20	199.00	73.90	---	272.90	**340.00**
3'-0" x 6'-8" x 1-3/8" T	Inst	Ea	Lg	2C	1.45	11.00	173.00	48.20	---	221.20	**271.00**
	Inst	Ea	Sm	2C	2.22	7.20	202.00	73.90	---	275.90	**343.00**

Lauan

Description	Oper	Unit	Vol	Crew Size	Man-hours per Unit	Crew Output per Day	Avg Mat'l Unit Cost	Avg Labor Unit Cost	Avg Equip Unit Cost	Avg Total Unit Cost	Avg Price Incl O&P
2'-6" x 6'-8" x 1-3/8" T	Inst	Ea	Lg	2C	1.45	11.00	180.00	48.20	---	228.20	**279.00**
	Inst	Ea	Sm	2C	2.22	7.20	209.00	73.90	---	282.90	**352.00**
2'-8" x 6'-8" x 1-3/8" T	Inst	Ea	Lg	2C	1.45	11.00	183.00	48.20	---	231.20	**283.00**
	Inst	Ea	Sm	2C	2.22	7.20	213.00	73.90	---	286.90	**356.00**
3'-0" x 6'-8" x 1-3/8" T	Inst	Ea	Lg	2C	1.45	11.00	186.00	48.20	---	234.20	**287.00**
	Inst	Ea	Sm	2C	2.22	7.20	217.00	73.90	---	290.90	**361.00**

Description	Oper	Unit	Vol	Crew Size	Man-hours per Unit	Crew Output per Day	Avg Mat'l Unit Cost	Avg Labor Unit Cost	Avg Equip Unit Cost	Avg Total Unit Cost	Avg Price Incl O&P
Birch											
2'-6" x 6'-8" x 1-3/8" T	Inst	Ea	Lg	2C	1.45	11.00	188.00	48.20	---	236.20	**289.00**
	Inst	Ea	Sm	2C	2.22	7.20	219.00	73.90	---	292.90	**363.00**
2'-8" x 6'-8" x 1-3/8" T	Inst	Ea	Lg	2C	1.45	11.00	190.00	48.20	---	238.20	**291.00**
	Inst	Ea	Sm	2C	2.22	7.20	222.00	73.90	---	295.90	**366.00**
3'-0" x 6'-8" x 1-3/8" T	Inst	Ea	Lg	2C	1.45	11.00	195.00	48.20	---	243.20	**296.00**
	Inst	Ea	Sm	2C	2.22	7.20	227.00	73.90	---	300.90	**371.00**
Ash											
2'-6" x 6'-8" x 1-3/8" T	Inst	Ea	Lg	2C	1.45	11.00	204.00	48.20	---	252.20	**307.00**
	Inst	Ea	Sm	2C	2.22	7.20	237.00	73.90	---	310.90	**384.00**
2'-8" x 6'-8" x 1-3/8" T	Inst	Ea	Lg	2C	1.45	11.00	208.00	48.20	---	256.20	**312.00**
	Inst	Ea	Sm	2C	2.22	7.20	242.00	73.90	---	315.90	**389.00**
3'-0" x 6'-8" x 1-3/8" T	Inst	Ea	Lg	2C	1.45	11.00	215.00	48.20	---	263.20	**320.00**
	Inst	Ea	Sm	2C	2.22	7.20	251.00	73.90	---	324.90	**399.00**

Adjustments

For mitered casing, both sides, supplied but not applied

Description	Oper	Unit	Vol	Crew Size	Man-hours per Unit	Crew Output per Day	Avg Mat'l Unit Cost	Avg Labor Unit Cost	Avg Equip Unit Cost	Avg Total Unit Cost	Avg Price Incl O&P
9/16" x 1-1/2" finger joint, ADD	Inst	Ea	Lg	---	---	---	25.00	---	---	25.00	**28.80**
	Inst	Ea	Sm	---	---	---	30.00	---	---	30.00	**34.50**
9/16" x 2-1/4" solid pine, ADD	Inst	Ea	Lg	---	---	---	36.10	---	---	36.10	**41.50**
	Inst	Ea	Sm	---	---	---	42.00	---	---	42.00	**48.30**
9/16" x 2-1/2" solid pine, ADD	Inst	Ea	Lg	---	---	---	34.00	---	---	34.00	**39.10**
	Inst	Ea	Sm	---	---	---	39.60	---	---	39.60	**45.50**
For solid fir jamb, 4-9/16", ADD											
	Inst	Ea	Lg	---	---	---	33.20	---	---	33.20	**38.20**
	Inst	Ea	Sm	---	---	---	38.60	---	---	38.60	**44.40**

Screen doors

Aluminum frame with fiberglass wire screen, and plain grille, includes hardware, hinges, closer and latch

Description	Oper	Unit	Vol	Crew Size	Man-hours per Unit	Crew Output per Day	Avg Mat'l Unit Cost	Avg Labor Unit Cost	Avg Equip Unit Cost	Avg Total Unit Cost	Avg Price Incl O&P
3'-0" x 7'-0"	Inst	Ea	Lg	2C	1.00	16.00	61.80	33.30	---	95.10	**121.00**
	Inst	Ea	Sm	2C	1.54	10.40	72.00	51.20	---	123.20	**160.00**

Wood screen doors, pine, aluminum wire screen, 1-1/8" T

Description	Oper	Unit	Vol	Crew Size	Man-hours per Unit	Crew Output per Day	Avg Mat'l Unit Cost	Avg Labor Unit Cost	Avg Equip Unit Cost	Avg Total Unit Cost	Avg Price Incl O&P
2'-6" x 6'-9"	Inst	Ea	Lg	2C	1.00	16.00	103.00	33.30	---	136.30	**168.00**
	Inst	Ea	Sm	2C	1.54	10.40	120.00	51.20	---	171.20	**215.00**
2'-8" x 6'-9"	Inst	Ea	Lg	2C	1.00	16.00	103.00	33.30	---	136.30	**168.00**
	Inst	Ea	Sm	2C	1.54	10.40	120.00	51.20	---	171.20	**215.00**
3'-0" x 6'-9"	Inst	Ea	Lg	2C	1.00	16.00	113.00	33.30	---	146.30	**180.00**
	Inst	Ea	Sm	2C	1.54	10.40	132.00	51.20	---	183.20	**229.00**
Half screen											
2'-6" x 6'-9"	Inst	Ea	Lg	2C	1.00	16.00	129.00	33.30	---	162.30	**198.00**
	Inst	Ea	Sm	2C	1.54	10.40	150.00	51.20	---	201.20	**249.00**
2'-8" x 6'-9"	Inst	Ea	Lg	2C	1.00	16.00	129.00	33.30	---	162.30	**198.00**
	Inst	Ea	Sm	2C	1.54	10.40	150.00	51.20	---	201.20	**249.00**
3'-0" x 6'-9"	Inst	Ea	Lg	2C	1.00	16.00	155.00	33.30	---	188.30	**228.00**
	Inst	Ea	Sm	2C	1.54	10.40	180.00	51.20	---	231.20	**284.00**

Description	Oper	Unit	Vol	Crew Size	Man-hours per Unit	Crew Output per Day	Avg Mat'l Unit Cost	Avg Labor Unit Cost	Avg Equip Unit Cost	Avg Total Unit Cost	Avg Price Incl O&P

Steel doors

24-gauge steel faces, fully primed, reversible, prices are for a standard 2-3/8" diameter bore

Decorative doors

Dual glazed, tempered clear

Description	Oper	Unit	Vol	Crew Size	Man-hours per Unit	Crew Output per Day	Avg Mat'l Unit Cost	Avg Labor Unit Cost	Avg Equip Unit Cost	Avg Total Unit Cost	Avg Price Incl O&P
2'-6", 2'-8", 3' x 6'-8" x 1-3/4" T	Inst	Ea	Lg	2C	1.45	11.00	244.00	48.20	---	292.20	**353.00**
	Inst	Ea	Sm	2C	2.22	7.20	284.00	73.90	---	357.90	**438.00**

Dual glazed, tempered clear, two bevel

Description	Oper	Unit	Vol	Crew Size	Man-hours per Unit	Crew Output per Day	Avg Mat'l Unit Cost	Avg Labor Unit Cost	Avg Equip Unit Cost	Avg Total Unit Cost	Avg Price Incl O&P
2'-6", 2'-8", 3' x 6'-8" x 1-3/4" T	Inst	Ea	Lg	2C	1.45	11.00	286.00	48.20	---	334.20	**402.00**
	Inst	Ea	Sm	2C	2.22	7.20	334.00	73.90	---	407.90	**494.00**

Embossed raised panel doors

6 raised panels

Description	Oper	Unit	Vol	Crew Size	Man-hours per Unit	Crew Output per Day	Avg Mat'l Unit Cost	Avg Labor Unit Cost	Avg Equip Unit Cost	Avg Total Unit Cost	Avg Price Incl O&P
2'-6", 2'-8", 3' x 6'-8" x 1-3/4" T	Inst	Ea	Lg	2C	1.45	11.00	138.00	48.20	---	186.20	**231.00**
	Inst	Ea	Sm	2C	2.22	7.20	160.00	73.90	---	233.90	**295.00**
2'-8" x 8' x 1-3/4" T	Inst	Ea	Lg	2C	1.60	10.00	232.00	53.20	---	285.20	**346.00**
	Inst	Ea	Sm	2C	2.46	6.50	270.00	81.80	---	351.80	**433.00**
3'-0" x 8' x 1-3/4" T	Inst	Ea	Lg	2C	1.60	10.00	253.00	53.20	---	306.20	**371.00**
	Inst	Ea	Sm	2C	2.46	6.50	295.00	81.80	---	376.80	**462.00**

6 raised panels, dual glazed, tempered clear

Description	Oper	Unit	Vol	Crew Size	Man-hours per Unit	Crew Output per Day	Avg Mat'l Unit Cost	Avg Labor Unit Cost	Avg Equip Unit Cost	Avg Total Unit Cost	Avg Price Incl O&P
2'-6", 2'-8", 3' x 6'-8" x 1-3/4" T	Inst	Ea	Lg	2C	1.45	11.00	214.00	48.20	---	262.20	**319.00**
	Inst	Ea	Sm	2C	2.22	7.20	250.00	73.90	---	323.90	**398.00**

8 raised panels

Description	Oper	Unit	Vol	Crew Size	Man-hours per Unit	Crew Output per Day	Avg Mat'l Unit Cost	Avg Labor Unit Cost	Avg Equip Unit Cost	Avg Total Unit Cost	Avg Price Incl O&P
2'-8", 3' x 6'-8" x 1-3/4" T	Inst	Ea	Lg	2C	1.45	11.00	140.00	48.20	---	188.20	**233.00**
	Inst	Ea	Sm	2C	2.22	7.20	163.00	73.90	---	236.90	**298.00**

Embossed raised panels and dual glazed tempered clear glass

2 raised panels with 1 lite

Description	Oper	Unit	Vol	Crew Size	Man-hours per Unit	Crew Output per Day	Avg Mat'l Unit Cost	Avg Labor Unit Cost	Avg Equip Unit Cost	Avg Total Unit Cost	Avg Price Incl O&P
2'-6", 2'-8", 3' x 6'-8" x 1-3/4" T	Inst	Ea	Lg	2C	1.45	11.00	231.00	48.20	---	279.20	**338.00**
	Inst	Ea	Sm	2C	2.22	7.20	269.00	73.90	---	342.90	**421.00**

2 raised panels with 9 lites

Description	Oper	Unit	Vol	Crew Size	Man-hours per Unit	Crew Output per Day	Avg Mat'l Unit Cost	Avg Labor Unit Cost	Avg Equip Unit Cost	Avg Total Unit Cost	Avg Price Incl O&P
2'-6", 2'-8", 3' x 6'-8" x 1-3/4" T	Inst	Ea	Lg	2C	1.45	11.00	235.00	48.20	---	283.20	**343.00**
	Inst	Ea	Sm	2C	2.22	7.20	274.00	73.90	---	347.90	**426.00**

2 raised panels with 12 diamond lites

Description	Oper	Unit	Vol	Crew Size	Man-hours per Unit	Crew Output per Day	Avg Mat'l Unit Cost	Avg Labor Unit Cost	Avg Equip Unit Cost	Avg Total Unit Cost	Avg Price Incl O&P
2'-6", 2'-8", 3' x 6'-8" x 1-3/4" T	Inst	Ea	Lg	2C	1.45	11.00	244.00	48.20	---	292.20	**353.00**
	Inst	Ea	Sm	2C	2.22	7.20	284.00	73.90	---	357.90	**438.00**

4 diamond raised panels with 1 lite

Description	Oper	Unit	Vol	Crew Size	Man-hours per Unit	Crew Output per Day	Avg Mat'l Unit Cost	Avg Labor Unit Cost	Avg Equip Unit Cost	Avg Total Unit Cost	Avg Price Incl O&P
2'-6", 2'-8", 3' x 6'-8" x 1-3/4" T	Inst	Ea	Lg	2C	1.45	11.00	231.00	48.20	---	279.20	**338.00**
	Inst	Ea	Sm	2C	2.22	7.20	269.00	73.90	---	342.90	**421.00**

4 diamond raised panels with 9 lites

Description	Oper	Unit	Vol	Crew Size	Man-hours per Unit	Crew Output per Day	Avg Mat'l Unit Cost	Avg Labor Unit Cost	Avg Equip Unit Cost	Avg Total Unit Cost	Avg Price Incl O&P
2'-6", 2'-8", 3' x 6'-8" x 1-3/4" T	Inst	Ea	Lg	2C	1.45	11.00	235.00	48.20	---	283.20	**343.00**
	Inst	Ea	Sm	2C	2.22	7.20	274.00	73.90	---	347.90	**426.00**

4 diamond raised panels with 12 diamond lites

Description	Oper	Unit	Vol	Crew Size	Man-hours per Unit	Crew Output per Day	Avg Mat'l Unit Cost	Avg Labor Unit Cost	Avg Equip Unit Cost	Avg Total Unit Cost	Avg Price Incl O&P
2'-6", 2'-8", 3' x 6'-8" x 1-3/4" T	Inst	Ea	Lg	2C	1.45	11.00	244.00	48.20	---	292.20	**353.00**
	Inst	Ea	Sm	2C	2.22	7.20	284.00	73.90	---	357.90	**438.00**

Description	Oper	Unit	Vol	Crew Size	Man-hours per Unit	Crew Output per Day	Avg Mat'l Unit Cost	Avg Labor Unit Cost	Avg Equip Unit Cost	Avg Total Unit Cost	Avg Price Incl O&P
Flush doors											
2' to 3' x 6'-8" x 1-3/4" T	Inst	Ea	Lg	2C	1.45	11.00	132.00	48.20	---	180.20	**224.00**
	Inst	Ea	Sm	2C	2.22	7.20	154.00	73.90	---	227.90	**287.00**
2' to 3' x 8'-0" x 1-3/4" T	Inst	Ea	Lg	2C	1.33	12.00	239.00	44.30	---	283.30	**341.00**
	Inst	Ea	Sm	2C	2.05	7.80	278.00	68.20	---	346.20	**422.00**
Dual glazed, tempered clear											
2'-6", 2'-8", 3' x 6'-8" x 1-3/4" T	Inst	Ea	Lg	2C	1.45	11.00	202.00	48.20	---	250.20	**305.00**
	Inst	Ea	Sm	2C	2.22	7.20	235.00	73.90	---	308.90	**381.00**
1-1/2 hour fire label											
2' to 3'-0" x 6'-8" x 1-3/4" T	Inst	Ea	Lg	2C	1.45	11.00	144.00	48.20	---	192.20	**238.00**
	Inst	Ea	Sm	2C	2.22	7.20	168.00	73.90	---	241.90	**304.00**
2' to 3' x 7' x 1-3/4" T	Inst	Ea	Lg	2C	1.33	12.00	175.00	44.30	---	219.30	**268.00**
	Inst	Ea	Sm	2C	2.05	7.80	204.00	68.20	---	272.20	**337.00**
French doors, flush face											
1 lite											
2'-6", 2'-8", 3' x 6'-8" x 1-3/4" T	Inst	Ea	Lg	2C	1.45	11.00	263.00	48.20	---	311.20	**374.00**
	Inst	Ea	Sm	2C	2.22	7.20	306.00	73.90	---	379.90	**463.00**
2'-6", 2'-8", 3' x 8, x 1-3/4" T	Inst	Ea	Lg	2C	1.60	10.00	433.00	53.20	---	486.20	**577.00**
	Inst	Ea	Sm	2C	2.46	6.50	504.00	81.80	---	585.80	**702.00**
10 lite											
2'-6", 2'-8", 3' x 6'-8" x 1-3/4" T	Inst	Ea	Lg	2C	1.45	11.00	278.00	48.20	---	326.20	**392.00**
	Inst	Ea	Sm	2C	2.22	7.20	324.00	73.90	---	397.90	**483.00**
2'-6", 2'-8", 3' x 8 x 1-3/4" T	Inst	Ea	Lg	2C	1.60	10.00	458.00	53.20	---	511.20	**607.00**
	Inst	Ea	Sm	2C	2.46	6.50	534.00	81.80	---	615.80	**737.00**
15 lite											
2'-6", 2'-8", 3' x 6'-8" x 1-3/4" T	Inst	Ea	Lg	2C	1.45	11.00	283.00	48.20	---	331.20	**398.00**
	Inst	Ea	Sm	2C	2.22	7.20	330.00	73.90	---	403.90	**490.00**
2'-6", 2'-8", 3' x 8 x 1-3/4" T	Inst	Ea	Lg	2C	1.60	10.00	469.00	53.20	---	522.20	**619.00**
	Inst	Ea	Sm	2C	2.46	6.50	546.00	81.80	---	627.80	**751.00**
French sidelites											
Embossed raised panel											
3 lite											
1', 1'-2" x 6'-8"	Inst	Ea	Lg	2C	1.00	16.00	144.00	33.30	---	177.30	**216.00**
	Inst	Ea	Sm	2C	1.54	10.40	168.00	51.20	---	219.20	**270.00**
Flush face											
1', 1'-2" x 6'-8"	Inst	Ea	Lg	2C	1.00	16.00	71.10	33.30	---	104.40	**132.00**
	Inst	Ea	Sm	2C	1.54	10.40	82.80	51.20	---	134.00	**172.00**
Dual glazed, tempered clear											
1', 1'-2" x 6'-8"	Inst	Ea	Lg	2C	1.00	16.00	149.00	33.30	---	182.30	**222.00**
	Inst	Ea	Sm	2C	1.54	10.40	174.00	51.20	---	225.20	**277.00**
Prehung package door units											
Flush steel door											
2'-6" x 6'-8" x 1-3/8" T	Inst	Ea	Lg	2C	1.33	12.00	245.00	44.30	---	289.30	**348.00**
	Inst	Ea	Sm	2C	2.05	7.80	286.00	68.20	---	354.20	**431.00**

Drywall (Sheetrock or wallboard)

The types of drywall are Standard, Fire Resistant, Water Resistant, and Fire and Water Resistant.

1. **Dimensions.** $1/4$", $3/8$" x 6' to 12'; $1/2$", $5/8$" x 6' to 14'.

2. **Installation**

   a. One ply. Sheets are nailed to studs and joists using $1 1/4$" or $1 3/8$" x .101 type 500 nails. Nails are usually spaced 8" oc on walls and 7" oc on ceilings. The joints between sheets are filled, taped and sanded. Tape and compound are available in kits containing 250 LF of tape and 18 lbs of compound or 75 LF of tape and 5 lbs of compound.

   b. Two plies. Initial ply is nailed to studs or joists. Second ply is laminated to first with taping compounds as the adhesive. Only the joints in the second ply are filled, taped and finished.

3. **Estimating Technique.** Determine area, deduct openings and add approximately 10% for waste.

Ceiling application

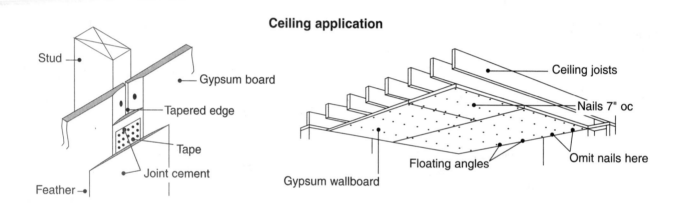

Stud — Gypsum board — Tapered edge — Tape — Joint cement — Feather

Ceiling joists — Nails 7" oc — Floating angles — Omit nails here — Gypsum wallboard

Wall application

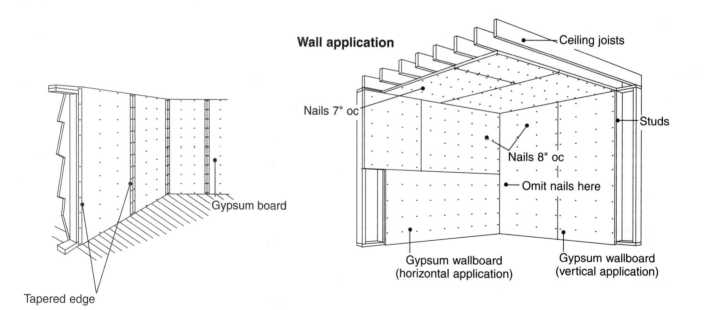

Tapered edge — Gypsum board

Ceiling joists — Nails 7" oc — Nails 8" oc — Omit nails here — Studs — Gypsum wallboard (horizontal application) — Gypsum wallboard (vertical application)

Description	Oper	Unit	Vol	Crew Size	Man-hours per Unit	Crew Output per Day	Avg Mat'l Unit Cost	Avg Labor Unit Cost	Avg Equip Unit Cost	Avg Total Unit Cost	Avg Price Incl O&P

Drywall

Demo Drywall or Sheetrock, walls and ceilings

| | Demo | SF | Lg | LB | .010 | 1540 | --- | .27 | --- | .27 | .41 |
| | Demo | SF | Sm | 1L | .025 | 647.0 | --- | .69 | --- | .69 | 1.02 |

Install Drywall or Sheetrock (gypsum plasterboard)

Includes tape, finish, nails or screws, and 5% waste
Applied to wood or metal frame

Walls only, 4' x 8', 10', 12'

3/8" T, standard	Inst	SF	Lg	2C	.022	736.0	.39	.73	---	1.12	1.55
	Inst	SF	Sm	CA	.026	309.0	.43	.87	---	1.30	1.79
1/2" T, standard	Inst	SF	Lg	2C	.022	736.0	.39	.73	---	1.12	1.55
	Inst	SF	Sm	CA	.026	309.0	.43	.87	---	1.30	1.79
5/8" T, standard	Inst	SF	Lg	2C	.023	704.0	.45	.77	---	1.22	1.67
	Inst	SF	Sm	CA	.027	296.0	.51	.90	---	1.41	1.93
Laminated 3/8" T sheets, std.	Inst	SF	Lg	2C	.029	544.0	.73	.96	---	1.69	2.29
	Inst	SF	Sm	CA	.035	228.0	.67	1.16	---	1.83	2.52

Ceilings only, 4' x 8', 10', 12'

3/8" T, standard	Inst	SF	Lg	2C	.026	608.0	.39	.87	---	1.26	1.75
	Inst	SF	Sm	CA	.031	255.0	.43	1.03	---	1.46	2.04
1/2" T, standard	Inst	SF	Lg	2C	.026	608.0	.39	.87	---	1.26	1.75
	Inst	SF	Sm	CA	.031	255.0	.43	1.03	---	1.46	2.04
5/8" T, standard	Inst	SF	Lg	2C	.028	576.0	.45	.93	---	1.38	1.91
	Inst	SF	Sm	CA	.033	242.0	.51	1.10	---	1.61	2.23
Laminated 3/8" T sheets, std.	Inst	SF	Lg	2C	.036	448.0	.73	1.20	---	1.93	2.64
	Inst	SF	Sm	CA	.043	188.0	.67	1.43	---	2.10	2.92

Average for ceilings and walls

3/8" T, standard	Inst	SF	Lg	2C	.023	710.0	.39	.77	---	1.16	1.60
	Inst	SF	Sm	CA	.027	298.0	.43	.90	---	1.33	1.84
1/2" T, standard	Inst	SF	Lg	2C	.023	710.0	.39	.77	---	1.16	1.60
	Inst	SF	Sm	CA	.027	298.0	.43	.90	---	1.33	1.84
5/8" T, standard	Inst	SF	Lg	2C	.024	678.0	.45	.80	---	1.25	1.72
	Inst	SF	Sm	CA	.028	285.0	.51	.93	---	1.44	1.98
Laminated 3/8" T sheets, std.	Inst	SF	Lg	2C	.030	525.0	.73	1.00	---	1.73	2.34
	Inst	SF	Sm	CA	.036	221.0	.67	1.20	---	1.87	2.57

Tape and bed joints on repaired sheetrock

Walls	Inst	SF	Lg	2C	.022	728.0	.12	.73	---	.85	1.24
	Inst	SF	Sm	CA	.026	306.0	.14	.87	---	1.01	1.46
Ceilings	Inst	SF	Lg	2C	.025	632.0	.12	.83	---	.95	1.39
	Inst	SF	Sm	CA	.030	265.0	.14	1.00	---	1.14	1.66

Thin coat plaster

| | Inst | SF | Lg | 2C | .014 | 1125 | .24 | .47 | --- | .71 | .97 |
| | Inst | SF | Sm | CA | .017 | 473.0 | .27 | .57 | --- | .84 | 1.16 |

Material and labor adjustments

| No tape and finish, DEDUCT | Inst | SF | Lg | 2C | -.011 | -1420 | -.12 | -.37 | --- | -.49 | --- |
| | Inst | SF | Sm | CA | -.013 | -596.00 | -.14 | -.43 | --- | -.57 | --- |

Description	Oper	Unit	Vol	Crew Size	Man-hours per Unit	Crew Output per Day	Avg Mat'l Unit Cost	Avg Labor Unit Cost	Avg Equip Unit Cost	Avg Total Unit Cost	Avg Price Incl O&P
For ceilings 9' H											
Ceiling only, ADD	Inst	SF	Lg	2C	.008	2000	---	.27	---	.27	**.40**
	Inst	SF	Sm	CA	.010	840.0	---	.33	---	.33	**.50**
Walls & ceilings (average), ADD											
	Inst	SF	Lg	2C	.003	5000	---	.10	---	.10	**.15**
	Inst	SF	Sm	CA	.004	2100	---	.13	---	.13	**.20**
Material adjustments											
For fire resistant (1/2" and 5/8" T) ADD											
	Inst	SF	Lg	2C	---	---	.11	---	---	.11	**.13**
	Inst	SF	Sm	CA	---	---	.13	---	---	.13	**.15**
For water resistant (1/2" and 5/8" T) ADD											
	Inst	SF	Lg	2C	---	---	.12	---	---	.12	**.14**
	Inst	SF	Sm	CA	---	---	.14	---	---	.14	**.16**
For fire and water resistant (5/8" T only) ADD											
	Inst	SF	Lg	2C	---	---	.23	---	---	.23	**.26**
	Inst	SF	Sm	CA	---	---	.27	---	---	.27	**.31**

Electrical

General work

Residential service, single phase system. Prices given on a cost per each basis for a unit price system which includes a weathercap, service entrance cable, meter socket, entrance disconnect switch, ground rod with clamp, ground cable, EMT, and panelboard

Weathercap

Description	Oper	Unit	Vol	Crew Size	Man-hours per Unit	Crew Output per Day	Avg Mat'l Unit Cost	Avg Labor Unit Cost	Avg Equip Unit Cost	Avg Total Unit Cost	Avg Price Incl O&P
100 AMP service	Inst	Ea	Lg	EA	1.00	8.00	6.60	35.60	---	42.20	**59.60**
	Inst	Fa	Sm	EA	1.43	5.60	7.17	51.00	---	58.17	**82.60**
200 AMP service	Inst	Ea	Lg	EA	1.60	5.00	19.50	57.00	---	76.50	**106.00**
	Inst	Ea	Sm	EA	2.29	3.50	21.10	81.60	---	102.70	**143.00**

Service entrance cable (typical allowance is 20 LF)

Description	Oper	Unit	Vol	Crew Size	Man-hours per Unit	Crew Output per Day	Avg Mat'l Unit Cost	Avg Labor Unit Cost	Avg Equip Unit Cost	Avg Total Unit Cost	Avg Price Incl O&P
100 AMP service	Inst	LF	Lg	EA	.123	65.00	.15	4.38	---	4.53	**6.57**
	Inst	LF	Sm	EA	.176	45.50	.17	6.27	---	6.44	**9.35**
200 AMP service	Inst	LF	Lg	EA	.178	45.00	.33	6.34	---	6.67	**9.64**
	Inst	LF	Sm	EA	.254	31.50	.36	9.05	---	9.41	**10.60**

Meter socket

Description	Oper	Unit	Vol	Crew Size	Man-hours per Unit	Crew Output per Day	Avg Mat'l Unit Cost	Avg Labor Unit Cost	Avg Equip Unit Cost	Avg Total Unit Cost	Avg Price Incl O&P
100 AMP service	Inst	Ea	Lg	EA	4.00	2.00	36.50	143.00	---	179.50	**250.00**
	Inst	Ea	Sm	EA	5.71	1.40	39.60	203.00	---	242.60	**343.00**
200 AMP service	Inst	Ea	Lg	EA	6.40	1.25	57.00	228.00	---	285.00	**398.00**
	Inst	Ea	Sm	EA	8.89	0.90	61.90	317.00	---	378.90	**534.00**

Entrance disconnect switch

Description	Oper	Unit	Vol	Crew Size	Man-hours per Unit	Crew Output per Day	Avg Mat'l Unit Cost	Avg Labor Unit Cost	Avg Equip Unit Cost	Avg Total Unit Cost	Avg Price Incl O&P
100 AMP service	Inst	Ea	Lg	EA	6.40	1.25	181.00	228.00	---	409.00	**541.00**
	Inst	Ea	Sm	EA	8.89	0.90	196.00	317.00	---	513.00	**688.00**
200 AMP service	Inst	Ea	Lg	EA	9.41	0.85	361.00	335.00	---	696.00	**905.00**
	Inst	Ea	Sm	EA	13.3	0.60	393.00	474.00	---	867.00	**1140.00**

Description	Oper	Unit	Vol	Crew Size	Man-hours per Unit	Crew Output per Day	Avg Mat'l Unit Cost	Avg Labor Unit Cost	Avg Equip Unit Cost	Avg Total Unit Cost	Avg Price Incl O&P
Ground rod, with clamp											
100 AMP service	Inst	Ea	Lg	EA	2.67	3.00	19.50	95.10	---	114.60	**161.00**
	Inst	Ea	Sm	EA	3.81	2.10	21.10	136.00	---	157.10	**223.00**
200 AMP service	Inst	Ea	Lg	EA	2.67	3.00	37.50	95.10	---	132.60	**182.00**
	Inst	Ea	Sm	EA	3.81	2.10	40.80	136.00	---	176.80	**245.00**
Ground cable (typical allowance is 10 LF)											
100 AMP service	Inst	LF	Lg	EA	.080	100.0	.18	2.85	---	3.03	**4.37**
	Inst	LF	Sm	EA	.114	70.00	.20	4.06	---	4.26	**6.16**
200 AMP service	Inst	LF	Lg	EA	.100	80.00	.29	3.56	---	3.85	**5.54**
	Inst	LF	Sm	EA	.143	56.00	.32	5.10	---	5.42	**7.81**
3/4" EMT (typical allowance is 10 LF)											
200 AMP service	Inst	LF	Lg	EA	.107	75.00	.72	3.81	---	4.53	**6.39**
	Inst	LF	Sm	EA	.152	52.50	.79	5.42	---	6.21	**8.82**
Panelboard											
100 AMP service - 12-circuit	Inst	Ea	Lg	EA	1.6	5.00	160.00	57.00	---	217.00	**267.00**
	Inst	Ea	Sm	EA	2.3	3.50	174.00	82.00	---	256.00	**319.00**
200 AMP service - 24-circuit	Inst	Ea	Lg	EA	2.3	3.50	382.00	82.00	---	464.00	**559.00**
	Inst	Ea	Sm	EA	3.2	2.50	415.00	114.00	---	529.00	**644.00**
Adjustments for other than normal working situations											
Cut and patch, ADD	Inst	%	Lg	EA	---	---	---	---	---	20.0	---
	Inst	%	Sm	EA	---	---	---	---	---	50.0	---
Dust protection, ADD	Inst	%	Lg	EA	---	---	---	---	---	10.0	---
	Inst	%	Sm	EA	---	---	---	---	---	40.0	---
Protect existing work, ADD	Inst	%	Lg	EA	---	---	---	---	---	20.0	---
	Inst	%	Sm	EA	---	---	---	---	---	50.0	---

Wiring per outlet or switch; wall or ceiling

Description	Oper	Unit	Vol	Crew Size	Man-hours per Unit	Crew Output per Day	Avg Mat'l Unit Cost	Avg Labor Unit Cost	Avg Equip Unit Cost	Avg Total Unit Cost	Avg Price Incl O&P
Romex, non-metallic sheathed cable, 600 volt, copper with ground wire											
	Inst	Ea	Lg	EA	.500	16.00	9.73	17.80	---	27.53	**37.20**
	Inst	Ea	Sm	EA	.714	11.20	10.60	25.40	---	36.00	**49.30**
BX, flexible armored cable, 600 volt, copper											
	Inst	Ea	Lg	EA	.667	12.00	16.70	23.80	---	40.50	**53.90**
	Inst	Ea	Sm	EA	.952	8.40	18.10	33.90	---	52.00	**70.40**
EMT with wire, electric metallic thinwall, 1/2"											
	Inst	Ea	Lg	EA	1.33	6.00	20.90	47.40	---	68.30	**93.20**
	Inst	Ea	Sm	EA	1.90	4.20	22.70	67.70	---	90.40	**125.00**
Rigid with wire, 1/2"											
	Inst	Ea	Lg	EA	2.00	4.00	27.80	71.30	---	99.10	**136.00**
	Inst	Ea	Sm	EA	2.86	2.80	30.20	102.00	---	132.20	**184.00**
Wiring, connection, and installation in closed wall structure, ADD											
	Inst	Ea	Lg	EA	1.00	8.00	---	35.60	---	35.60	**52.00**
	Inst	Ea	Sm	EA	1.43	5.60	---	51.00	---	51.00	**74.40**

Lighting. See Lighting fixtures, page 270

Description	Oper	Unit	Vol	Crew Size	Man-hours per Unit	Crew Output per Day	Avg Mat'l Unit Cost	Avg Labor Unit Cost	Avg Equip Unit Cost	Avg Total Unit Cost	Avg Price Incl O&P

Special systems
Burglary detection systems
Alarm bell

| | Inst | Ea | Lg | EA | 2.00 | 4.00 | 83.40 | 71.30 | --- | 154.70 | **200.00** |
| | Inst | Ea | Sm | EA | 2.86 | 2.80 | 90.60 | 102.00 | --- | 192.60 | **253.00** |

Burglar alarm, battery operated
Mechanical trigger

| | Inst | Ea | Lg | EA | 2.00 | 4.00 | 299.00 | 71.30 | --- | 370.30 | **448.00** |
| | Inst | Ea | Sm | EA | 2.86 | 2.80 | 325.00 | 102.00 | --- | 427.00 | **522.00** |

Electrical trigger

| | Inst | Ea | Lg | EA | 2.00 | 4.00 | 361.00 | 71.30 | --- | 432.30 | **520.00** |
| | Inst | Ea | Sm | EA | 2.86 | 2.80 | 393.00 | 102.00 | --- | 495.00 | **600.00** |

Adjustments, ADD
Outside key control

| | Inst | Ea | Lg | --- | --- | --- | 83.40 | --- | --- | 83.40 | **95.90** |
| | Inst | Ea | Sm | --- | --- | --- | 90.60 | --- | --- | 90.60 | **104.00** |

Remote signaling circuitry

| | Inst | Ea | Lg | --- | --- | --- | 135.00 | --- | --- | 135.00 | **155.00** |
| | Inst | Ea | Sm | --- | --- | --- | 146.00 | --- | --- | 146.00 | **168.00** |

Card reader
Standard

| | Inst | Ea | Lg | EA | 3.20 | 2.50 | 1010.00 | 114.00 | --- | 1124.00 | **1330.00** |
| | Inst | Ea | Sm | EA | 4.44 | 1.80 | 1090.00 | 158.00 | --- | 1248.00 | **1490.00** |

Multi-code

| | Inst | Ea | Lg | EA | 3.20 | 2.50 | 1300.00 | 114.00 | --- | 1414.00 | **1660.00** |
| | Inst | Ea | Sm | EA | 4.44 | 1.80 | 1410.00 | 158.00 | --- | 1568.00 | **1850.00** |

Detectors
Motion
Infrared photoelectric

| | Inst | Ea | Lg | EA | 4.00 | 2.00 | 209.00 | 143.00 | --- | 352.00 | **448.00** |
| | Inst | Ea | Sm | EA | 5.71 | 1.40 | 227.00 | 203.00 | --- | 430.00 | **558.00** |

Passive infrared

| | Inst | Ea | Lg | EA | 4.00 | 2.00 | 278.00 | 143.00 | --- | 421.00 | **528.00** |
| | Inst | Ea | Sm | EA | 5.71 | 1.40 | 302.00 | 203.00 | --- | 505.00 | **644.00** |

Ultrasonic, 12 volt

| | Inst | Ea | Lg | EA | 4.00 | 2.00 | 250.00 | 143.00 | --- | 393.00 | **496.00** |
| | Inst | Ea | Sm | EA | 5.71 | 1.40 | 272.00 | 203.00 | --- | 475.00 | **610.00** |

Microwave
10' to 200'

| | Inst | Ea | Lg | EA | 4.00 | 2.00 | 723.00 | 143.00 | --- | 866.00 | **1040.00** |
| | Inst | Ea | Sm | EA | 5.71 | 1.40 | 785.00 | 203.00 | --- | 988.00 | **1200.00** |

10' to 350'

| | Inst | Ea | Lg | EA | 4.00 | 2.00 | 2090.00 | 143.00 | --- | 2233.00 | **2610.00** |
| | Inst | Ea | Sm | EA | 5.71 | 1.40 | 2270.00 | 203.00 | --- | 2473.00 | **2900.00** |

Door switches
Hinge switch

| | Inst | Ea | Lg | EA | 1.60 | 5.00 | 63.90 | 57.00 | --- | 120.90 | **157.00** |
| | Inst | Ea | Sm | EA | 2.29 | 3.50 | 69.50 | 81.60 | --- | 151.10 | **199.00** |

Magnetic switch

| | Inst | Ea | Lg | EA | 1.60 | 5.00 | 76.50 | 57.00 | --- | 133.50 | **171.00** |
| | Inst | Ea | Sm | EA | 2.29 | 3.50 | 83.10 | 81.60 | --- | 164.70 | **215.00** |

Exit control locks
Horn alarm

| | Inst | Ea | Lg | EA | 2.00 | 4.00 | 375.00 | 71.30 | --- | 446.30 | **536.00** |
| | Inst | Ea | Sm | EA | 2.86 | 2.80 | 408.00 | 102.00 | --- | 510.00 | **618.00** |

Flashing light alarm

| | Inst | Ea | Lg | EA | 2.00 | 4.00 | 424.00 | 71.30 | --- | 495.30 | **592.00** |
| | Inst | Ea | Sm | EA | 2.86 | 2.80 | 461.00 | 102.00 | --- | 563.00 | **678.00** |

Glass break alarm switch

| | Inst | Ea | Lg | EA | 1.00 | 8.00 | 51.40 | 35.60 | --- | 87.00 | **111.00** |
| | Inst | Ea | Sm | EA | 1.43 | 5.60 | 55.90 | 51.00 | --- | 106.90 | **139.00** |

Indicating panels
1 channel

| | Inst | Ea | Lg | EA | 3.20 | 2.50 | 396.00 | 114.00 | --- | 510.00 | **622.00** |
| | Inst | Ea | Sm | EA | 4.44 | 1.80 | 430.00 | 158.00 | --- | 588.00 | **726.00** |

10 channel

| | Inst | Ea | Lg | EA | 5.33 | 1.50 | 1370.00 | 190.00 | --- | 1560.00 | **1850.00** |
| | Inst | Ea | Sm | EA | 7.27 | 1.10 | 1490.00 | 259.00 | --- | 1749.00 | **2090.00** |

20 channel

| | Inst | Ea | Lg | EA | 8.00 | 1.00 | 2640.00 | 285.00 | --- | 2925.00 | **3450.00** |
| | Inst | Ea | Sm | EA | 11.4 | 0.70 | 2870.00 | 406.00 | --- | 3276.00 | **3890.00** |

40 channel

| | Inst | Ea | Lg | EA | 16.0 | 0.50 | 4870.00 | 570.00 | --- | 5440.00 | **6430.00** |
| | Inst | Ea | Sm | EA | 20.0 | 0.40 | 5290.00 | 713.00 | --- | 6003.00 | **7120.00** |

Description	Oper	Unit	Vol	Crew Size	Man-hours per Unit	Crew Output per Day	Avg Mat'l Unit Cost	Avg Labor Unit Cost	Avg Equip Unit Cost	Avg Total Unit Cost	Avg Price Incl O&P
Police connect panel											
	Inst	Ea	Lg	EA	2.00	4.00	264.00	71.30	---	335.30	**408.00**
	Inst	Ea	Sm	EA	2.86	2.80	287.00	102.00	---	389.00	**479.00**
Siren											
	Inst	Ea	Lg	EA	2.00	4.00	158.00	71.30	---	229.30	**286.00**
	Inst	Ea	Sm	EA	2.86	2.80	172.00	102.00	---	274.00	**347.00**
Switchmats											
30" x 5'	Inst	Ea	Lg	EA	1.60	5.00	91.70	57.00	---	148.70	**189.00**
	Inst	Ea	Sm	EA	2.29	3.50	99.70	81.60	---	181.30	**234.00**
30" x 25'	Inst	Ea	Lg	EA	2.00	4.00	220.00	71.30	---	291.30	**357.00**
	Inst	Ea	Sm	EA	2.86	2.80	239.00	102.00	---	341.00	**423.00**
Telephone dialer											
	Inst	Ea	Lg	EA	1.60	5.00	417.00	57.00	---	474.00	**563.00**
	Inst	Ea	Sm	EA	2.29	3.50	453.00	81.60	---	534.60	**640.00**

Doorbell systems
Includes transformer, button, bell

Description	Oper	Unit	Vol	Crew Size	Man-hours per Unit	Crew Output per Day	Avg Mat'l Unit Cost	Avg Labor Unit Cost	Avg Equip Unit Cost	Avg Total Unit Cost	Avg Price Incl O&P
Door chimes, 2 notes											
	Inst	Ea	Lg	EA	.800	10.00	83.40	28.50	---	111.90	**138.00**
	Inst	Ea	Sm	EA	1.14	7.00	90.60	40.60	---	131.20	**163.00**
Tube-type chimes	Inst	Ea	Lg	EA	1.00	8.00	209.00	35.60	---	244.60	**292.00**
	Inst	Ea	Sm	EA	1.43	5.60	227.00	51.00	---	278.00	**335.00**
Transformer and button only											
	Inst	Ea	Lg	EA	.667	12.00	48.70	23.80	---	72.50	**90.60**
	Inst	Ea	Sm	EA	.952	8.40	52.90	33.90	---	86.80	**110.00**
Push button only											
	Inst	Ea	Lg	EA	.500	16.00	20.90	17.80	---	38.70	**50.00**
	Inst	Ea	Sm	EA	.714	11.20	22.70	25.40	---	48.10	**63.20**

Fire alarm systems

Description	Oper	Unit	Vol	Crew Size	Man-hours per Unit	Crew Output per Day	Avg Mat'l Unit Cost	Avg Labor Unit Cost	Avg Equip Unit Cost	Avg Total Unit Cost	Avg Price Incl O&P
Battery and rack											
	Inst	Ea	Lg	EA	2.67	3.00	834.00	95.10	---	929.10	**1100.00**
	Inst	Ea	Sm	EA	3.81	2.10	906.00	136.00	---	1042.00	**1240.00**
Automatic charger	Inst	Ea	Lg	EA	1.33	6.00	535.00	47.40	---	582.40	**685.00**
	Inst	Ea	Sm	EA	1.90	4.20	581.00	67.70	---	648.70	**767.00**
Detector											
Fixed temperature	Inst	Ea	Lg	EA	1.33	6.00	33.40	47.40	---	80.80	**108.00**
	Inst	Ea	Sm	EA	1.90	4.20	36.20	67.70	---	103.90	**141.00**
Rate of rise	Inst	Ea	Lg	EA	1.33	6.00	41.70	47.40	---	89.10	**117.00**
	Inst	Ea	Sm	EA	1.90	4.20	45.30	67.70	---	113.00	**151.00**
Door holder											
Electro-magnetic	Inst	Ea	Lg	EA	2.00	4.00	93.10	71.30	---	164.40	**211.00**
	Inst	Ea	Sm	EA	2.86	2.80	101.00	102.00	---	203.00	**265.00**
Combination holder/closer	Inst	Ea	Lg	EA	2.67	3.00	521.00	95.10	---	616.10	**738.00**
	Inst	Ea	Sm	EA	3.81	2.10	566.00	136.00	---	702.00	**849.00**
Fire drill switch	Inst	Ea	Lg	EA	1.33	6.00	104.00	47.40	---	151.40	**189.00**
	Inst	Ea	Sm	EA	1.90	4.20	113.00	67.70	---	180.70	**229.00**
Glass break alarm switch	Inst	Ea	Lg	EA	1.00	8.00	59.80	35.60	---	95.40	**121.00**
	Inst	Ea	Sm	EA	1.43	5.60	64.90	51.00	---	115.90	**149.00**
Signal bell	Inst	Ea	Lg	EA	1.33	6.00	59.80	47.40	---	107.20	**138.00**
	Inst	Ea	Sm	EA	1.90	4.20	64.90	67.70	---	132.60	**174.00**
Smoke detector											
Ceiling type	Inst	Ea	Lg	EA	1.33	6.00	77.80	47.40	---	125.20	**159.00**
	Inst	Ea	Sm	EA	1.90	4.20	84.60	67.70	---	152.30	**196.00**
Duct type	Inst	Ea	Lg	EA	2.67	3.00	306.00	95.10	---	401.10	**491.00**
	Inst	Ea	Sm	EA	3.81	2.10	332.00	136.00	---	468.00	**580.00**

Description	Oper	Unit	Vol	Crew Size	Man-hours per Unit	Crew Output per Day	Avg Mat'l Unit Cost	Avg Labor Unit Cost	Avg Equip Unit Cost	Avg Total Unit Cost	Avg Price Incl O&P

Intercom systems

Master stations, with digital AM/FM receiver, up to 20 remote stations, & telephone interface
Solid state with antenna, with 200' wire, 4 speakers

Description	Oper	Unit	Vol	Crew Size	Man-hours per Unit	Crew Output per Day	Avg Mat'l Unit Cost	Avg Labor Unit Cost	Avg Equip Unit Cost	Avg Total Unit Cost	Avg Price Incl O&P
	Inst	Set	Lg	EA	4.00	2.00	1670.00	143.00	---	1813.00	**2130.00**
	Inst	Set	Sm	EA	5.71	1.40	1810.00	203.00	---	2013.00	**2380.00**

Solid state with antenna, with 1000' wire, 8 speakers

	Inst	Set	Lg	EA	8.00	1.00	2220.00	285.00	---	2505.00	**2970.00**
	Inst	Set	Sm	EA	11.4	0.70	2420.00	406.00	---	2826.00	**3370.00**

Room to door intercoms
Master station, door station, with transformer, wire

	Inst	Set	Lg	EA	2.00	4.00	417.00	71.30	---	488.30	**584.00**
	Inst	Set	Sm	EA	2.86	2.80	453.00	102.00	---	555.00	**670.00**
Second door station, ADD	Inst	Ea	Lg	EA	1.00	8.00	55.60	35.60	---	91.20	**116.00**
	Inst	Ea	Sm	EA	1.43	5.60	60.40	51.00	---	111.40	**144.00**

Installation, wiring, and transformer, wire, ADD

	Inst	LS	Lg	EA	1.60	5.00	---	57.00	---	57.00	**83.20**
	Inst	LS	Sm	EA	2.29	3.50	---	81.60	---	81.60	**119.00**

Telephone, phone-jack wiring

Pre-wiring, per outlet or jack

	Inst	Ea	Lg	EA	.667	12.00	27.80	23.80	---	51.60	**66.70**
	Inst	Ea	Sm	EA	.952	8.40	30.20	33.90	---	64.10	**84.30**

Wiring, connection, and installation in closed wall structure, ADD

	Inst	Ea	Lg	EA	1.00	8.00	---	35.60	---	35.60	**52.00**
	Inst	Ea	Sm	EA	1.43	5.60	---	51.00	---	51.00	**74.40**

Television antenna

Television antenna outlet, with 300 OHM

	Inst	Ea	Lg	EA	.667	12.00	16.70	23.80	---	40.50	**53.90**
	Inst	Ea	Sm	EA	.952	8.40	18.10	33.90	---	52.00	**70.40**

Wiring, connection, and installation in closed wall structure, ADD

	Inst	Ea	Lg	EA	1.00	8.00	---	35.60	---	35.60	**52.00**
	Inst	Ea	Sm	EA	1.43	5.60	---	51.00	---	51.00	**74.40**

Thermostat wiring

For heating units located on the first floor
Thermostat on first floor

	Inst	Ea	Lg	EA	1.00	8.00	27.80	35.60	---	63.40	**84.00**
	Inst	Ea	Sm	EA	1.43	5.60	30.20	51.00	---	81.20	**109.00**

Wiring, connection, and installation in closed wall structure, ADD

	Inst	Ea	Lg	EA	1.00	8.00	---	35.60	---	35.60	**52.00**
	Inst	Ea	Sm	EA	1.43	5.60	---	51.00	---	51.00	**74.40**

Thermostat on second floor

	Inst	Ea	Lg	EA	1.60	5.00	34.80	57.00	---	91.80	**123.00**
	Inst	Ea	Sm	EA	2.29	3.50	37.80	81.60	---	119.40	**163.00**

Wiring, connection, and installation in closed wall structure, ADD

	Inst	Ea	Lg	EA	1.33	6.00	---	47.40	---	47.40	**69.20**
	Inst	Ea	Sm	EA	1.90	4.20	---	67.70	---	67.70	**98.80**

Description	Oper	Unit	Vol	Crew Size	Man-hours per Unit	Crew Output per Day	Avg Mat'l Unit Cost	Avg Labor Unit Cost	Avg Equip Unit Cost	Avg Total Unit Cost	Avg Price Incl O&P
Entrances											
Single & double door entrances											
	Demo	Ea	Lg	LB	.640	25.00	---	17.60	---	17.60	**26.20**
	Demo	Ea	Sm	LB	.889	18.00	---	24.40	---	24.40	**36.30**

Colonial design

White pine includes frames, pediments, and pilasters

Plain carved archway

Description	Oper	Unit	Vol	Crew Size	Man-hours per Unit	Crew Output per Day	Avg Mat'l Unit Cost	Avg Labor Unit Cost	Avg Equip Unit Cost	Avg Total Unit Cost	Avg Price Incl O&P
Single door units											
3'-0" W x 6'-8" H	Inst	Ea	Lg	2C	2.00	8.00	293.00	66.50	---	359.50	**393.00**
	Inst	Ea	Sm	2C	3.08	5.20	344.00	102.00	---	446.00	**498.00**
Double door units											
Two - 2'-6" W x 6'-8" H	Inst	Ea	Lg	2C	2.67	6.00	315.00	88.80	---	403.80	**448.00**
	Inst	Ea	Sm	2C	4.10	3.90	369.00	136.00	---	505.00	**574.00**
Two - 2'-8" W x 6'-8" H	Inst	Ea	Lg	2C	2.67	6.00	341.00	88.80	---	429.80	**474.00**
	Inst	Ea	Sm	2C	4.10	3.90	400.00	136.00	---	536.00	**605.00**
Two - 3'-0" W x 6'-8" H	Inst	Ea	Lg	2C	2.67	6.00	446.00	88.80	---	534.80	**579.00**
	Inst	Ea	Sm	2C	4.10	3.90	523.00	136.00	---	659.00	**728.00**

Decorative carved archway

Description	Oper	Unit	Vol	Crew Size	Man-hours per Unit	Crew Output per Day	Avg Mat'l Unit Cost	Avg Labor Unit Cost	Avg Equip Unit Cost	Avg Total Unit Cost	Avg Price Incl O&P
Single door units											
3'-0" W x 6'-8" H	Inst	Ea	Lg	2C	2.00	8.00	368.00	66.50	---	434.50	**467.00**
	Inst	Ea	Sm	2C	3.08	5.20	431.00	102.00	---	533.00	**585.00**
Double door units											
Two - 2'-6" W x 6'-8" H	Inst	Ea	Lg	2C	2.67	6.00	315.00	88.80	---	403.80	**448.00**
	Inst	Ea	Sm	2C	4.10	3.90	369.00	136.00	---	505.00	**574.00**
Two - 2'-8" W x 6'-8" H	Inst	Ea	Lg	2C	2.67	6.00	334.00	88.80	---	422.80	**467.00**
	Inst	Ea	Sm	2C	4.10	3.90	392.00	136.00	---	528.00	**597.00**
Two - 3 -0" W x 6'-8" H	Inst	Ea	Lg	2C	2.67	6.00	348.00	88.80	---	436.80	**481.00**
	Inst	Ea	Sm	2C	4.10	3.90	408.00	136.00	---	544.00	**613.00**

Excavation

Digging out or trenching

Description	Oper	Unit	Vol	Crew Size	Man-hours per Unit	Crew Output per Day	Avg Mat'l Unit Cost	Avg Labor Unit Cost	Avg Equip Unit Cost	Avg Total Unit Cost	Avg Price Incl O&P
	Demo	CY	Lg	LB	.034	475.0	---	.93	---	.93	**1.39**
	Demo	CY	Sm	LB	.056	285.0	---	1.54	---	1.54	**2.29**

Pits, medium earth, piled

With front end loader, track mounted, 1-1/2 CY capacity; 55 CY per hour

Description	Oper	Unit	Vol	Crew Size	Man-hours per Unit	Crew Output per Day	Avg Mat'l Unit Cost	Avg Labor Unit Cost	Avg Equip Unit Cost	Avg Total Unit Cost	Avg Price Incl O&P
	Demo	CY	Lg	VB	.036	440.0	---	1.16	.53	1.69	**2.25**
	Demo	CY	Sm	VB	.061	264.0	---	1.97	.88	2.85	**3.80**
By hand											
To 4'-0" D	Demo	CY	Lg	LB	1.07	15.00	---	29.40	---	29.40	**43.70**
	Demo	CY	Sm	LB	1.78	9.00	---	48.80	---	48.80	**72.80**
4'-0" to 6'-0" D	Demo	CY	Lg	LB	1.60	10.00	---	43.90	---	43.90	**65.40**
	Demo	CY	Sm	LB	2.67	6.00	---	73.20	---	73.20	**109.00**
6'-0" to 8'-0" D	Demo	CY	Lg	LB	2.67	6.00	---	73.20	---	73.20	**109.00**
	Demo	CY	Sm	LB	4.00	4.00	---	110.00	---	110.00	**163.00**

Description	Oper	Unit	Vol	Crew Size	Man-hours per Unit	Crew Output per Day	Avg Mat'l Unit Cost	Avg Labor Unit Cost	Avg Equip Unit Cost	Avg Total Unit Cost	Avg Price Incl O&P
Continuous footing or trench, medium earth, piled											
With tractor backhoe, 3/8 CY capacity (48 HP); 15 CY per hour											
	Demo	CY	Lg	VB	.133	120.0	---	4.30	1.93	6.23	**8.30**
	Demo	CY	Sm	VB	.222	72.00	---	7.18	3.21	10.39	**13.80**
By hand, to 4'-0" D											
	Demo	CY	Lg	LB	1.07	15.00	---	29.40	---	29.40	**43.70**
	Demo	CY	Sm	LB	1.78	9.00	---	48.80	---	48.80	**72.80**
Backfilling, by hand, medium soil											
Without compaction											
	Demo	CY	Lg	LB	.571	28.00	---	15.70	---	15.70	**23.30**
	Demo	CY	Sm	LB	.941	17.00	---	25.80	---	25.80	**38.50**
With hand compaction											
6" layers	Demo	CY	Lg	LB	.941	17.00	---	25.80	---	25.80	**38.50**
	Demo	CY	Sm	LB	1.60	10.00	---	43.90	---	43.90	**65.40**
12" layers	Demo	CY	Lg	LB	.727	22.00	---	19.90	---	19.90	**29.70**
	Demo	CY	Sm	LB	1.23	13.00	---	33.70	---	33.70	**50.30**
With vibrating plate compaction											
6" layers	Demo	CY	Lg	AB	.800	20.00	---	23.60	1.12	24.72	**36.00**
	Demo	CY	Sm	AB	1.33	12.00	---	39.20	1.86	41.06	**59.90**
12" layers	Demo	CY	Lg	AB	.667	24.00	---	19.70	.93	20.63	**30.00**
	Demo	CY	Sm	AB	1.14	14.00	---	33.60	1.60	35.20	**51.40**

Facebrick. See Masonry, page 276

Description	Oper	Unit	Vol	Crew Size	Man-hours per Unit	Crew Output per Day	Avg Mat'l Unit Cost	Avg Labor Unit Cost	Avg Equip Unit Cost	Avg Total Unit Cost	Avg Price Incl O&P

Fences

Basketweave

Redwood, preassembled, 8' L panels, includes 4" x 4" line posts, horizontal or vertical weave

Description	Oper	Unit	Vol	Crew Size	Man-hours per Unit	Crew Output per Day	Avg Mat'l Unit Cost	Avg Labor Unit Cost	Avg Equip Unit Cost	Avg Total Unit Cost	Avg Price Incl O&P
5' H	Inst	LF	Lg	CS	.100	240.0	12.10	3.13	---	15.23	**18.60**
	Inst	LF	Sm	CS	.133	180.0	13.60	4.17	---	17.77	**21.90**
6' H	Inst	LF	Lg	CS	.100	240.0	18.20	3.13	---	21.33	**25.60**
	Inst	LF	Sm	CS	.133	180.0	20.40	4.17	---	24.57	**29.70**

Adjustments

Corner or end posts, 8' H

Description	Oper	Unit	Vol	Crew Size	Man-hours per Unit	Crew Output per Day	Avg Mat'l Unit Cost	Avg Labor Unit Cost	Avg Equip Unit Cost	Avg Total Unit Cost	Avg Price Incl O&P
	Inst	Ea	Lg	CA	.615	13.00	24.20	20.50	---	44.70	**58.50**
	Inst	Ea	Sm	CA	.800	10.00	27.20	26.60	---	53.80	**71.20**

3-1/2' W x 5' H gate, with hardware

Description	Oper	Unit	Vol	Crew Size	Man-hours per Unit	Crew Output per Day	Avg Mat'l Unit Cost	Avg Labor Unit Cost	Avg Equip Unit Cost	Avg Total Unit Cost	Avg Price Incl O&P
	Inst	Ea	Lg	CA	.800	10.00	103.00	26.60	---	129.60	**158.00**
	Inst	Ea	Sm	CA	1.00	8.00	116.00	33.30	---	149.30	**183.00**

Board, per LF complete fence system

6' H boards nailed to wood frame, on 1 side only; 8' L redwood 4" x 4" (milled) posts, set 2' D in concrete filled holes @ 6' oc; frame members 2" x 4" (milled) as 2 rails between posts per 6' L fence section; costs are per LF of fence

Douglas fir frame members with redwood posts

Milled boards, "dog-eared" one end

Cedar

Description	Oper	Unit	Vol	Crew Size	Man-hours per Unit	Crew Output per Day	Avg Mat'l Unit Cost	Avg Labor Unit Cost	Avg Equip Unit Cost	Avg Total Unit Cost	Avg Price Incl O&P
1" x 6" - 6' H	Inst	LF	Lg	CS	.400	60.00	39.10	12.50	---	51.60	**63.70**
	Inst	LF	Sm	CS	.533	45.00	43.90	16.70	---	60.60	**75.50**
1" x 8" - 6' H	Inst	LF	Lg	CS	.343	70.00	39.80	10.70	---	50.50	**61.90**
	Inst	LF	Sm	CS	.453	53.00	44.70	14.20	---	58.90	**72.70**
1" x 10" - 6' H	Inst	LF	Lg	CS	.300	80.00	39.00	9.40	---	48.40	**58.90**
	Inst	LF	Sm	CS	.400	60.00	43.80	12.50	---	56.30	**69.20**
Douglas fir											
1" x 6" - 6' H	Inst	LF	Lg	CS	.400	60.00	37.60	12.50	---	50.10	**62.10**
	Inst	LF	Sm	CS	.533	45.00	42.30	16.70	---	59.00	**73.70**
1" x 8" - 6' H	Inst	LF	Lg	CS	.343	70.00	37.90	10.70	---	48.60	**59.70**
	Inst	LF	Sm	CS	.453	53.00	42.60	14.20	---	56.80	**70.30**
1" x 10" - 6' H	Inst	LF	Lg	CS	.300	80.00	37.60	9.40	---	47.00	**57.40**
	Inst	LF	Sm	CS	.400	60.00	42.30	12.50	---	54.80	**67.40**
Redwood											
1" x 6" - 6' H	Inst	LF	Lg	CS	.400	60.00	39.10	12.50	---	51.60	**63.70**
	Inst	LF	Sm	CS	.533	45.00	43.90	16.70	---	60.60	**75.50**
1" x 8" - 6' H	Inst	LF	Lg	CS	.343	70.00	39.80	10.70	---	50.50	**61.90**
	Inst	LF	Sm	CS	.453	53.00	44.70	14.20	---	58.90	**72.70**
1" x 10" - 6' H	Inst	LF	Lg	CS	.300	80.00	41.20	9.40	---	50.60	**61.50**
	Inst	LF	Sm	CS	.400	60.00	46.30	12.50	---	58.80	**72.10**

Description	Oper	Unit	Vol	Crew Size	Man-hours per Unit	Crew Output per Day	Avg Mat'l Unit Cost	Avg Labor Unit Cost	Avg Equip Unit Cost	Avg Total Unit Cost	Avg Price Incl O&P
Rough boards, both ends squared											
Cedar											
1" x 6" - 6' H	Inst	LF	Lg	CS	.369	65.00	38.30	11.60	---	49.90	**61.30**
	Inst	LF	Sm	CS	.490	49.00	43.00	15.40	---	58.40	**72.50**
1" x 8" - 6' H	Inst	LF	Lg	CS	.320	75.00	39.00	10.00	---	49.00	**59.90**
	Inst	LF	Sm	CS	.429	56.00	43.80	13.40	---	57.20	**70.60**
1" x 10" - 6' H	Inst	LF	Lg	CS	.282	85.00	38.20	8.83	---	47.03	**57.10**
	Inst	LF	Sm	CS	.375	64.00	42.90	11.80	---	54.70	**66.90**
Douglas fir											
1" x 6" - 6' H	Inst	LF	Lg	CS	.369	65.00	37.00	11.60	---	48.60	**59.90**
	Inst	LF	Sm	CS	.490	49.00	41.60	15.40	---	57.00	**70.90**
1" x 8" - 6' H	Inst	LF	Lg	CS	.320	75.00	37.40	10.00	---	47.40	**58.00**
	Inst	LF	Sm	CS	.429	56.00	42.00	13.40	---	55.40	**68.50**
1" x 10" - 6' H	Inst	LF	Lg	CS	.282	85.00	37.20	8.83	---	46.03	**56.00**
	Inst	LF	Sm	CS	.375	64.00	41.80	11.80	---	53.60	**65.70**
Redwood											
1" x 6" - 6' H	Inst	LF	Lg	CS	.369	65.00	38.30	11.60	---	49.90	**61.30**
	Inst	LF	Sm	CS	.490	49.00	43.00	15.40	---	58.40	**72.50**
1" x 8" - 6' H	Inst	LF	Lg	CS	.320	75.00	39.00	10.00	---	49.00	**59.90**
	Inst	LF	Sm	CS	.429	56.00	43.80	13.40	---	57.20	**70.60**
1" x 10" - 6' H	Inst	LF	Lg	CS	.282	85.00	40.40	8.83	---	49.23	**59.70**
	Inst	LF	Sm	CS	.375	64.00	45.40	11.80	---	57.20	**69.80**
Redwood frame members with redwood posts											
Milled boards, "dog-eared" one end											
Cedar											
1" x 6" - 6' H	Inst	LF	Lg	CS	.400	60.00	39.10	12.50	---	51.60	**63.70**
	Inst	LF	Sm	CS	.533	45.00	43.90	16.70	---	60.60	**75.60**
1" x 8" - 6' H	Inst	LF	Lg	CS	.343	70.00	39.80	10.70	---	50.50	**61.90**
	Inst	LF	Sm	CS	.453	53.00	44.80	14.20	---	59.00	**72.80**
1" x 10" - 6' H	Inst	LF	Lg	CS	.300	80.00	39.00	9.40	---	48.40	**59.00**
	Inst	LF	Sm	CS	.400	60.00	43.90	12.50	---	56.40	**69.20**
Douglas fir											
1" x 6" - 6' H	Inst	LF	Lg	CS	.400	60.00	37.70	12.50	---	50.20	**62.10**
	Inst	LF	Sm	CS	.533	45.00	42.30	16.70	---	59.00	**73.70**
1" x 8" - 6' H	Inst	LF	Lg	CS	.343	70.00	38.00	10.70	---	48.70	**59.80**
	Inst	LF	Sm	CS	.453	53.00	42.70	14.20	---	56.90	**70.40**
1" x 10" - 6' H	Inst	LF	Lg	CS	.300	80.00	37.70	9.40	---	47.10	**57.40**
	Inst	LF	Sm	CS	.400	60.00	42.30	12.50	---	54.80	**67.50**
Redwood											
1" x 6" - 6' H	Inst	LF	Lg	CS	.400	60.00	39.10	12.50	---	51.60	**63.70**
	Inst	LF	Sm	CS	.533	45.00	43.90	16.70	---	60.60	**75.60**
1" x 8" - 6' H	Inst	LF	Lg	CS	.343	70.00	39.80	10.70	---	50.50	**61.90**
	Inst	LF	Sm	CS	.453	53.00	44.80	14.20	---	59.00	**72.80**
1" x 10" - 6' H	Inst	LF	Lg	CS	.300	80.00	41.20	9.40	---	50.60	**61.50**
	Inst	LF	Sm	CS	.400	60.00	46.40	12.50	---	58.90	**72.10**

Description	Oper	Unit	Vol	Crew Size	Man-hours per Unit	Crew Output per Day	Avg Mat'l Unit Cost	Avg Labor Unit Cost	Avg Equip Unit Cost	Avg Total Unit Cost	Avg Price Incl O&P

Rough boards, both ends squared
Cedar
1" x 6" - 6' H	Inst	LF	Lg	CS	.369	65.00	38.30	11.60	---	49.90	**61.40**
	Inst	LF	Sm	CS	.490	49.00	43.00	15.40	---	58.40	**72.50**
1" x 8" - 6' H	Inst	LF	Lg	CS	.320	75.00	39.00	10.00	---	49.00	**59.90**
	Inst	LF	Sm	CS	.429	56.00	43.90	13.40	---	57.30	**70.60**
1" x 10" - 6' H	Inst	LF	Lg	CS	.282	85.00	38.20	8.83	---	47.03	**57.20**
	Inst	LF	Sm	CS	.375	64.00	42.90	11.80	---	54.70	**67.00**

Douglas fir
1" x 6" - 6' H	Inst	LF	Lg	CS	.369	65.00	37.00	11.60	---	48.60	**59.90**
	Inst	LF	Sm	CS	.490	49.00	41.60	15.40	---	57.00	**70.90**
1" x 8" - 6' H	Inst	LF	Lg	CS	.320	75.00	37.40	10.00	---	47.40	**58.10**
	Inst	LF	Sm	CS	.429	56.00	42.10	13.40	---	55.50	**68.50**
1" x 10" - 6' H	Inst	LF	Lg	CS	.282	85.00	37.20	8.83	---	46.03	**56.10**
	Inst	LF	Sm	CS	.375	64.00	41.80	11.80	---	53.60	**65.70**

Redwood
1" x 6" - 6' H	Inst	LF	Lg	CS	.369	65.00	38.30	11.60	---	49.90	**61.40**
	Inst	LF	Sm	CS	.490	49.00	43.00	15.40	---	58.40	**72.50**
1" x 8" - 6' H	Inst	LF	Lg	CS	.320	75.00	39.00	10.00	---	49.00	**59.90**
	Inst	LF	Sm	CS	.429	56.00	43.90	13.40	---	57.30	**70.60**
1" x 10" - 6' H	Inst	LF	Lg	CS	.282	85.00	40.40	8.83	---	49.23	**59.80**
	Inst	LF	Sm	CS	.375	64.00	45.50	11.80	---	57.30	**69.90**

Chain link
9 gauge galvanized steel, includes top rail (1-5/8" o.d.), line posts (2" o.d.) @ 10' oc and sleeves

36" H	Inst	LF	Lg	HB	.109	220.0	4.45	3.47	---	7.92	**10.30**
	Inst	LF	Sm	HB	.145	165.0	5.00	4.61	---	9.61	**12.60**
42" H	Inst	LF	Lg	HB	.114	210.0	4.77	3.63	---	8.40	**10.90**
	Inst	LF	Sm	HB	.152	158.0	5.36	4.84	---	10.20	**13.40**
48" H	Inst	LF	Lg	HB	.120	200.0	5.00	3.82	---	8.82	**11.40**
	Inst	LF	Sm	HB	.160	150.0	5.62	5.09	---	10.71	**14.10**
60" H	Inst	LF	Lg	HB	.133	180.0	5.76	4.23	---	9.99	**12.90**
	Inst	LF	Sm	HB	.178	135.0	6.47	5.66	---	12.13	**15.90**
72" H	Inst	LF	Lg	HB	.150	160.0	6.53	4.77	---	11.30	**14.60**
	Inst	LF	Sm	HB	.200	120.0	7.34	6.36	---	13.70	**17.90**

Description	Oper	Unit	Vol	Crew Size	Man-hours per Unit	Crew Output per Day	Avg Mat'l Unit Cost	Avg Labor Unit Cost	Avg Equip Unit Cost	Avg Total Unit Cost	Avg Price Incl O&P

Adjustments

11-1/2"-gauge galvanized steel fabric

DEDUCT	Inst	%	Lg	---	---	---	-27.0	---	---	---	---
	Inst	%	Sm	---	---	---	-27.0	---	---	---	---

12-gauge galvanized steel fabric

DEDUCT	Inst	%	Lg	---	---	---	-29.0	---	---	---	---
	Inst	%	Sm	---	---	---	-29.0	---	---	---	---

9-gauge green vinyl-coated fabric

DEDUCT	Inst	%	Lg	---	---	---	-4.0	---	---	---	---
	Inst	%	Sm	---	---	---	-4.0	---	---	---	---

11-gauge green vinyl-coated fabric

DEDUCT	Inst	%	Lg	---	---	---	-3.8	---	---	---	---
	Inst	%	Sm	---	---	---	-3.8	---	---	---	---

Filler strips, ADD

Aluminum, baked-on enamel finish
Diagonal, 1-7/8" W

48" H	Inst	LF	Lg	LB	.160	100.0	4.09	4.39	---	8.48	**11.20**
	Inst	LF	Sm	LB	.213	75.00	4.60	5.84	---	10.44	**14.00**
60" H	Inst	LF	Lg	LB	.160	100.0	4.96	4.39	---	9.35	**12.20**
	Inst	LF	Sm	LB	.213	75.00	5.58	5.84	---	11.42	**15.10**
72" H	Inst	LF	Lg	LB	.160	100.0	5.81	4.39	---	10.20	**13.20**
	Inst	LF	Sm	LB	.213	75.00	6.53	5.84	---	12.37	**16.20**

Vertical, 1-1/4" W

48" H	Inst	LF	Lg	LB	.160	100.0	4.09	4.39	---	8.48	**11.20**
	Inst	LF	Sm	LB	.213	75.00	4.60	5.84	---	10.44	**14.00**
60" H	Inst	LF	Lg	LB	.160	100.0	4.82	4.39	---	9.21	**12.10**
	Inst	LF	Sm	LB	.213	75.00	5.41	5.84	---	11.25	**14.90**
72" H	Inst	LF	Lg	LB	.160	100.0	5.66	4.39	---	10.05	**13.10**
	Inst	LF	Sm	LB	.213	75.00	6.36	5.84	---	12.20	**16.00**

Wood, redwood stain
Vertical, 1-1/4" W

48" H	Inst	LF	Lg	LB	.160	100.0	4.09	4.39	---	8.48	**11.20**
	Inst	LF	Sm	LB	.213	75.00	4.60	5.84	---	10.44	**14.00**
60" H	Inst	LF	Lg	LB	.160	100.0	4.96	4.39	---	9.35	**12.20**
	Inst	LF	Sm	LB	.213	75.00	5.58	5.84	---	11.42	**15.10**
72" H	Inst	LF	Lg	LB	.160	100.0	5.81	4.39	---	10.20	**13.20**
	Inst	LF	Sm	LB	.213	75.00	6.53	5.84	---	12.37	**16.20**

Description	Oper	Unit	Vol	Crew Size	Man-hours per Unit	Crew Output per Day	Avg Mat'l Unit Cost	Avg Labor Unit Cost	Avg Equip Unit Cost	Avg Total Unit Cost	Avg Price Incl O&P
Corner posts (2-1/2" o.d.), installed, heavyweight											
36" H	Inst	Ea	Lg	HA	.286	28.00	23.60	9.73	---	33.33	**41.60**
	Inst	Ea	Sm	HA	.381	21.00	26.50	13.00	---	39.50	**49.80**
42" H	Inst	Ea	Lg	HA	.296	27.00	26.70	10.10	---	36.80	**45.80**
	Inst	Ea	Sm	HA	.400	20.00	30.10	13.60	---	43.70	**54.80**
48" H	Inst	Ea	Lg	HA	.320	25.00	28.30	10.90	---	39.20	**48.80**
	Inst	Ea	Sm	HA	.421	19.00	31.80	14.30	---	46.10	**57.90**
60" H	Inst	Ea	Lg	HA	.364	22.00	33.80	12.40	---	46.20	**57.30**
	Inst	Ea	Sm	HA	.471	17.00	38.00	16.00	---	54.00	**67.60**
72" H	Inst	Ea	Lg	HA	.400	20.00	39.30	13.60	---	52.90	**65.50**
	Inst	Ea	Sm	HA	.533	15.00	44.20	18.10	---	62.30	**77.80**
End or gate posts (2-1/2" o.d.), installed, heavyweight											
36" H	Inst	Ea	Lg	HA	.286	28.00	18.10	9.73	---	27.83	**35.30**
	Inst	Ea	Sm	HA	.381	21.00	20.30	13.00	---	33.30	**42.70**
42" H	Inst	Ea	Lg	HA	.296	27.00	19.70	10.10	---	29.80	**37.60**
	Inst	Ea	Sm	HA	.400	20.00	22.10	13.60	---	35.70	**45.70**
48" H	Inst	Ea	Lg	HA	.320	25.00	22.00	10.90	---	32.90	**41.50**
	Inst	Ea	Sm	HA	.421	19.00	24.80	14.30	---	39.10	**49.80**
60" H	Inst	Ea	Lg	HA	.364	22.00	26.00	12.40	---	38.40	**48.30**
	Inst	Ea	Sm	HA	.471	17.00	29.20	16.00	---	45.20	**57.40**
72" H	Inst	Ea	Lg	HA	.400	20.00	29.90	13.60	---	43.50	**54.60**
	Inst	Ea	Sm	HA	.533	15.00	33.60	18.10	---	51.70	**65.60**
Gates, square corner frame, 9 gauge wire, installed											
3' W walkway gates											
36" H	Inst	Ea	Lg	HA	.615	13.00	66.10	20.90	---	87.00	**107.00**
	Inst	Ea	Sm	HA	.800	10.00	74.30	27.20	---	101.50	**126.00**
42" H	Inst	Ea	Lg	HA	.615	13.00	69.20	20.90	---	90.10	**111.00**
	Inst	Ea	Sm	HA	.800	10.00	77.80	27.20	---	105.00	**130.00**
48" H	Inst	Ea	Lg	HA	.667	12.00	70.80	22.70	---	93.50	**115.00**
	Inst	Ea	Sm	HA	.889	9.00	79.60	30.20	---	109.80	**137.00**
60" H	Inst	Ea	Lg	HA	.667	12.00	84.90	22.70	---	107.60	**131.00**
	Inst	Ea	Sm	HA	.889	9.00	95.50	30.20	---	125.70	**155.00**
72" H	Inst	Ea	Lg	HA	.727	11.00	99.10	24.70	---	123.80	**151.00**
	Inst	Ea	Sm	HA	1.00	8.00	111.00	34.00	---	145.00	**179.00**
12' W double driveway gates											
36" H	Inst	Ea	Lg	HA	1.33	6.00	175.00	45.20	---	220.20	**268.00**
	Inst	Ea	Sm	HA	1.60	5.00	196.00	54.40	---	250.40	**307.00**
42" H	Inst	Ea	Lg	HA	1.33	6.00	184.00	45.20	---	229.20	**279.00**
	Inst	Ea	Sm	HA	1.60	5.00	207.00	54.40	---	261.40	**319.00**
48" H	Inst	Ea	Lg	HA	1.60	5.00	190.00	54.40	---	244.40	**300.00**
	Inst	Ea	Sm	HA	2.00	4.00	214.00	68.00	---	282.00	**347.00**
60" H	Inst	Ea	Lg	HA	1.60	5.00	228.00	54.40	---	282.40	**343.00**
	Inst	Ea	Sm	HA	2.00	4.00	256.00	68.00	---	324.00	**396.00**
72" H	Inst	Ea	Lg	HA	2.00	4.00	252.00	68.00	---	320.00	**391.00**
	Inst	Ea	Sm	HA	2.67	3.00	283.00	90.80	---	373.80	**461.00**

Description	Oper	Unit	Vol	Crew Size	Man-hours per Unit	Crew Output per Day	Avg Mat'l Unit Cost	Avg Labor Unit Cost	Avg Equip Unit Cost	Avg Total Unit Cost	Avg Price Incl O&P

Split rail
Red cedar, 10' L sectional spans
Rails only

| | Inst | Ea | Lg | --- | --- | --- | 10.70 | --- | --- | 10.70 | **10.70** |
| | Inst | Ea | Sm | --- | --- | --- | 12.00 | --- | --- | 12.00 | **12.00** |

Bored 2 rail posts

5'-6" line or end posts	Inst	Ea	Lg	CA	.615	13.00	12.60	20.50	---	33.10	**45.20**
	Inst	Ea	Sm	CA	.800	10.00	14.10	26.60	---	40.70	**56.20**
5'-6" corner posts	Inst	Ea	Lg	CA	.615	13.00	14.20	20.50	---	34.70	**47.00**
	Inst	Ea	Sm	CA	.800	10.00	15.90	26.60	---	42.50	**58.20**

Bored 3 rail posts

6'-6" line or end posts	Inst	Ea	Lg	CA	.615	13.00	15.70	20.50	---	36.20	**48.80**
	Inst	Ea	Sm	CA	.800	10.00	17.70	26.60	---	44.30	**60.30**
6'-6" corner posts	Inst	Ea	Lg	CA	.615	13.00	17.30	20.50	---	37.80	**50.60**
	Inst	Ea	Sm	CA	.800	10.00	19.50	26.60	---	46.10	**62.30**

Complete fence estimate (does not include gates)
2 rail

| 36" H, 5'-6" post | Inst | LF | Lg | CJ | .042 | 380.0 | 3.56 | 1.27 | --- | 4.83 | **6.01** |
| | Inst | LF | Sm | CJ | .056 | 285.0 | 4.00 | 1.70 | --- | 5.70 | **7.15** |

3 rail

| 48" H, 6'-6" post | Inst | LF | Lg | CJ | .048 | 330.0 | 4.62 | 1.46 | --- | 6.08 | **7.50** |
| | Inst | LF | Sm | CJ | .065 | 248.0 | 5.20 | 1.97 | --- | 7.17 | **8.94** |

Gate
2 rails

3-1/2' W	Inst	Ea	Lg	CA	.615	13.00	62.90	20.50	---	83.40	**103.00**
	Inst	Ea	Sm	CA	.800	10.00	70.70	26.60	---	97.30	**121.00**
5' W	Inst	Ea	Lg	CA	.800	10.00	81.80	26.60	---	108.40	**134.00**
	Inst	Ea	Sm	CA	1.00	8.00	91.90	33.30	---	125.20	**156.00**

3 rails

3-1/2' W	Inst	Ea	Lg	CA	.615	13.00	78.70	20.50	---	99.20	**121.00**
	Inst	Ea	Sm	CA	.800	10.00	88.40	26.60	---	115.00	**142.00**
5' W	Inst	Ea	Lg	CA	.000	10.00	91.20	26.60	---	117.80	**145.00**
	Inst	Ea	Sm	CA	1.00	8.00	103.00	33.30	---	136.30	**168.00**

Description	Oper	Unit	Vol	Crew Size	Man-hours per Unit	Crew Output per Day	Avg Mat'l Unit Cost	Avg Labor Unit Cost	Avg Equip Unit Cost	Avg Total Unit Cost	Avg Price Incl O&P

Fiberglass panels

Corrugated; 8', 10', or 12' L panels; 2-1/2" W x 1/2" D corrugation

Nailed on wood frame

Description	Oper	Unit	Vol	Crew Size	Man-hours per Unit	Crew Output per Day	Avg Mat'l Unit Cost	Avg Labor Unit Cost	Avg Equip Unit Cost	Avg Total Unit Cost	Avg Price Incl O&P
4 oz., 0.03" T, 26" W	Inst	SF	Lg	CA	.040	200.0	1.51	1.33	---	2.84	**3.73**
	Inst	SF	Sm	CA	.053	150.0	1.64	1.76	---	3.40	**4.53**
Fire retardant	Inst	SF	Lg	CA	.040	200.0	2.09	1.33	---	3.42	**4.40**
	Inst	SF	Sm	CA	.053	150.0	2.26	1.76	---	4.02	**5.24**
5 oz., 0.037" T, 26" W	Inst	SF	Lg	CA	.040	200.0	1.86	1.33	---	3.19	**4.14**
	Inst	SF	Sm	CA	.053	150.0	2.02	1.76	---	3.78	**4.97**
Fire retardant	Inst	SF	Lg	CA	.040	200.0	2.57	1.33	---	3.90	**4.95**
	Inst	SF	Sm	CA	.053	150.0	2.78	1.76	---	4.54	**5.84**
6 oz., 0.045" T, 26" W	Inst	SF	Lg	CA	.040	200.0	2.21	1.33	---	3.54	**4.54**
	Inst	SF	Sm	CA	.053	150.0	2.39	1.76	---	4.15	**5.39**
Fire retardant	Inst	SF	Lg	CA	.040	200.0	3.05	1.33	---	4.38	**5.50**
	Inst	SF	Sm	CA	.053	150.0	3.30	1.76	---	5.06	**6.44**
8 oz., 0.06" T, 26" W	Inst	SF	Lg	CA	.040	200.0	2.56	1.33	---	3.89	**4.94**
	Inst	SF	Sm	CA	.053	150.0	2.77	1.76	---	4.53	**5.83**
Fire retardant	Inst	SF	Lg	CA	.040	200.0	3.53	1.33	---	4.86	**6.06**
	Inst	SF	Sm	CA	.053	150.0	3.82	1.76	---	5.58	**7.04**

Flat panels; 8', 10', or 12' L panels; clear, green, or white

Description	Oper	Unit	Vol	Crew Size	Man-hours per Unit	Crew Output per Day	Avg Mat'l Unit Cost	Avg Labor Unit Cost	Avg Equip Unit Cost	Avg Total Unit Cost	Avg Price Incl O&P
5 oz., 48" W	Inst	SF	Lg	CA	.040	200.0	1.01	1.33	---	2.34	**3.16**
	Inst	SF	Sm	CA	.053	150.0	1.09	1.76	---	2.85	**3.90**
Fire retardant	Inst	SF	Lg	CA	.040	200.0	1.01	1.33	---	2.34	**3.16**
	Inst	SF	Sm	CA	.053	150.0	1.09	1.76	---	2.85	**3.90**
6 oz., 48" W	Inst	SF	Lg	CA	.040	200.0	1.98	1.33	---	3.31	**4.27**
	Inst	SF	Sm	CA	.053	150.0	2.15	1.76	---	3.91	**5.12**
Fire retardant	Inst	SF	Lg	CA	.040	200.0	3.42	1.33	---	4.75	**5.93**
	Inst	SF	Sm	CA	.053	150.0	3.71	1.76	---	5.47	**6.91**
8 oz., 48" W	Inst	SF	Lg	CA	.040	200.0	2.56	1.33	---	3.89	**4.94**
	Inst	SF	Sm	CA	.053	150.0	2.77	1.76	---	4.53	**5.83**
Fire retardant	Inst	SF	Lg	CA	.040	200.0	4.42	1.33	---	5.75	**7.08**
	Inst	SF	Sm	CA	.053	150.0	4.78	1.76	---	6.54	**8.14**

Solar block; 8', 10', 12' L panels; 2-1/2" W x 1/2" D corrugation; nailed on wood frame

Description	Oper	Unit	Vol	Crew Size	Man-hours per Unit	Crew Output per Day	Avg Mat'l Unit Cost	Avg Labor Unit Cost	Avg Equip Unit Cost	Avg Total Unit Cost	Avg Price Incl O&P
5 oz., 26" W	Inst	SF	Lg	CA	.040	200.0	1.14	1.33	---	2.47	**3.31**
	Inst	SF	Sm	CA	.053	150.0	1.24	1.76	---	3.00	**4.07**

Description	Oper	Unit	Vol	Crew Size	Man- hours per Unit	Crew Output per Day	Avg Mat'l Unit Cost	Avg Labor Unit Cost	Avg Equip Unit Cost	Avg Total Unit Cost	Avg Price Incl O&P
Accessories											
Wood corrugated											
2-1/2" W x 1-1/2" D x 6' L	Inst	Ea	Lg	---	---	---	3.00	---	---	3.00	**3.00**
	Inst	Ea	Sm	---	---	---	3.25	---	---	3.25	**3.25**
2-1/2" W x 1-1/2" D x 8' L	Inst	Ea	Lg	---	---	---	3.65	---	---	3.65	**3.65**
	Inst	Ea	Sm	---	---	---	3.95	---	---	3.95	**3.95**
2-1/2" W x 3/4" D x 6' L	Inst	Ea	Lg	---	---	---	1.38	---	---	1.38	**1.38**
	Inst	Ea	Sm	---	---	---	1.50	---	---	1.50	**1.50**
2-1/2" W x 3/4" D x 8' L	Inst	Ea	Lg	---	---	---	1.78	---	---	1.78	**1.78**
	Inst	Ea	Sm	---	---	---	1.92	---	---	1.92	**1.92**
Rubber corrugated											
1" x 3"	Inst	Ea	Lg	---	---	---	1.38	---	---	1.38	**1.38**
	Inst	Ea	Sm	---	---	---	1.50	---	---	1.50	**1.50**
Polyfoam corrugated											
1" x 3"	Inst	Ea	Lg	---	---	---	.97	---	---	.97	**.97**
	Inst	Ea	Sm	---	---	---	1.05	---	---	1.05	**1.05**
Vertical crown molding											
Wood											
1-1/2" x 6' L	Inst	Ea	Lg	---	---	---	2.51	---	---	2.51	**2.51**
	Inst	Ea	Sm	---	---	---	2.72	---	---	2.72	**2.72**
1-1/2" x 8' L	Inst	Ea	Lg	---	---	---	3.31	---	---	3.31	**3.31**
	Inst	Ea	Sm	---	---	---	3.59	---	---	3.59	**3.59**
Polyfoam											
1" x 1" x 3' L	Inst	Ea	Lg	---	---	---	1.21	---	---	1.21	**1.21**
	Inst	Ea	Sm	---	---	---	1.31	---	---	1.31	**1.31**
Rubber											
1" x 1" x 3' L	Inst	Ea	Lg	---	---	---	2.29	---	---	2.29	**2.29**
	Inst	Ea	Sm	---	---	---	2.48	---	---	2.48	**2.48**

Description	Oper	Unit	Vol	Crew Size	Man-hours per Unit	Crew Output per Day	Avg Mat'l Unit Cost	Avg Labor Unit Cost	Avg Equip Unit Cost	Avg Total Unit Cost	Avg Price Incl O&P

Fireplaces

Woodburning, prefabricated. No masonry support required, installs directly on floor. Ceramic-backed firebox with black vitreous enamel side panels. No finish plastering or brick hearthwork included. Fire screen, 9" (i.d.) factory-built insulated chimneys with flue, lining, damper, and flashing with rain cap included. Chimney height from floor to where chimney exits through roof

36" W fireplace unit with:

Up to 9'-0" chimney height

Description	Oper	Unit	Vol	Crew Size	Man-hours per Unit	Crew Output per Day	Avg Mat'l Unit Cost	Avg Labor Unit Cost	Avg Equip Unit Cost	Avg Total Unit Cost	Avg Price Incl O&P
	Inst	LS	Lg	CJ	12.3	1.30	922.00	373.00	---	1295.00	**1620.00**
	Inst	LS	Sm	CJ	17.6	0.91	1000.00	534.00	---	1534.00	**1950.00**

9'-3" to 12'-2" chimney

| | Inst | LS | Lg | CJ | 13.3 | 1.20 | 970.00 | 404.00 | --- | 1374.00 | **1720.00** |
| | Inst | LS | Sm | CJ | 19.0 | 0.84 | 1050.00 | 577.00 | --- | 1627.00 | **2080.00** |

12'-3" to 15'-1" chimney

| | Inst | LS | Lg | CJ | 14.5 | 1.10 | 1020.00 | 440.00 | --- | 1460.00 | **1830.00** |
| | Inst | LS | Sm | CJ | 20.8 | 0.77 | 1100.00 | 631.00 | --- | 1731.00 | **2220.00** |

15'-2" to 18'-0" chimney

| | Inst | LS | Lg | CJ | 16.0 | 1.00 | 1070.00 | 486.00 | --- | 1556.00 | **1950.00** |
| | Inst | LS | Sm | CJ | 22.9 | 0.70 | 1160.00 | 695.00 | --- | 1855.00 | **2370.00** |

18'-1" to 20'-11" chimney

| | Inst | LS | Lg | CJ | 17.8 | 0.90 | 1110.00 | 540.00 | --- | 1650.00 | **2090.00** |
| | Inst | LS | Sm | CJ | 25.4 | 0.63 | 1210.00 | 771.00 | --- | 1981.00 | **2550.00** |

21'-3" to 23'-10" chimney

| | Inst | LS | Lg | CJ | 20.0 | 0.80 | 1160.00 | 607.00 | --- | 1767.00 | **2240.00** |
| | Inst | LS | Sm | CJ | 28.6 | 0.56 | 1260.00 | 868.00 | --- | 2128.00 | **2750.00** |

23'-11" to 24'-9" chimney

| | Inst | LS | Lg | CJ | 22.9 | 0.70 | 1210.00 | 695.00 | --- | 1905.00 | **2430.00** |
| | Inst | LS | Sm | CJ | 32.7 | | 1310.00 | 992.00 | --- | 2302.00 | **3000.00** |

42" W fireplace unit with:

Up to 9'-0" chimney height

| | Inst | LS | Lg | CJ | 12.3 | 1.30 | 1040.00 | 373.00 | --- | 1413.00 | **1760.00** |
| | Inst | LS | Sm | CJ | 17.6 | 0.91 | 1130.00 | 534.00 | --- | 1664.00 | **2100.00** |

9'-3" to 12'-2" chimney

| | Inst | LS | Lg | CJ | 13.3 | 1.20 | 1100.00 | 404.00 | --- | 1504.00 | **1870.00** |
| | Inst | LS | Sm | CJ | 19.0 | 0.84 | 1190.00 | 577.00 | --- | 1767.00 | **2240.00** |

12'-3" to 15'-1" chimney

| | Inst | LS | Lg | CJ | 14.5 | 1.10 | 1150.00 | 440.00 | --- | 1590.00 | **1980.00** |
| | Inst | LS | Sm | CJ | 20.8 | 0.77 | 1250.00 | 631.00 | --- | 1881.00 | **2380.00** |

15'-2" to 18'-0" chimney

| | Inst | LS | Lg | CJ | 16.0 | 1.00 | 1210.00 | 486.00 | --- | 1696.00 | **2120.00** |
| | Inst | LS | Sm | CJ | 22.9 | 0.70 | 1310.00 | 695.00 | --- | 2005.00 | **2550.00** |

18'-1" to 20'-11" chimney

| | Inst | LS | Lg | CJ | 17.8 | 0.90 | 1260.00 | 540.00 | --- | 1800.00 | **2260.00** |
| | Inst | LS | Sm | CJ | 25.4 | 0.63 | 1370.00 | 771.00 | --- | 2141.00 | **2730.00** |

21'-3" to 23'-10" chimney

| | Inst | LS | Lg | CJ | 20.0 | 0.80 | 1320.00 | 607.00 | --- | 1927.00 | **2420.00** |
| | Inst | LS | Sm | CJ | 28.6 | 0.56 | 1430.00 | 868.00 | --- | 2298.00 | **2940.00** |

23'-11" to 24'-9" chimney

| | Inst | LS | Lg | CJ | 22.9 | 0.70 | 1370.00 | 695.00 | --- | 2065.00 | **2620.00** |
| | Inst | LS | Sm | CJ | 32.7 | 0.49 | 1490.00 | 992.00 | --- | 2482.00 | **3200.00** |

Description	Oper	Unit	Vol	Crew Size	Man-hours per Unit	Crew Output per Day	Avg Mat'l Unit Cost	Avg Labor Unit Cost	Avg Equip Unit Cost	Avg Total Unit Cost	Avg Price Incl O&P
Accessories											
Log lighter with gas valve (straight or angle pattern)											
	Inst	Ea	Lg	SA	1.33	6.00	31.00	50.10	---	81.10	**109.00**
	Inst	Ea	Sm	SA	1.90	4.20	33.60	71.60	---	105.20	**144.00**
Log lighter, less gas valve (straight, angle, tee pattern)											
	Inst	Ea	Lg	SA	.667	12.00	11.60	25.10	---	36.70	**50.30**
	Inst	Ea	Sm	SA	0.95	8.40	12.60	35.80	---	48.40	**67.10**
Gas valve for log lighter											
	Inst	Ea	Lg	SA	.667	12.00	20.60	25.10	---	45.70	**60.70**
	Inst	Ea	Sm	SA	0.95	8.40	22.40	35.80	---	58.20	**78.40**
Spare parts											
Gas valve key											
	Inst	Ea	Lg	---	---	---	1.42	---	---	1.42	**1.42**
	Inst	Ea	Sm	---	---	---	1.54	---	---	1.54	**1.54**
Stem extender											
	Inst	Ea	Lg	---	---	---	2.58	---	---	2.58	**2.58**
	Inst	Ea	Sm	---	---	---	2.80	---	---	2.80	**2.80**
Extra long											
	Inst	Ea	Lg	---	---	---	3.55	---	---	3.55	**3.55**
	Inst	Ea	Sm	---	---	---	3.85	---	---	3.85	**3.85**
Valve floor plate											
	Inst	Ea	Lg	---	---	---	2.97	---	---	2.97	**2.97**
	Inst	Ea	Sm	---	---	---	3.22	---	---	3.22	**3.22**
Lighter burner tube (12" to 17")											
	Inst	Ea	Lg	---	---	---	5.93	---	---	5.93	**5.93**
	Inst	Ea	Sm	---	---	---	6.44	---	---	6.44	**6.44**

Fireplace mantels. See Mantels, fireplace, page 273

Flashing. See Sheet metal, page 364

Floor finishes. See individual items.

Floor joists. See Framing, page 180

Description	Oper	Unit	Vol	Crew Size	Man-hours per Unit	Crew Output per Day	Avg Mat'l Unit Cost	Avg Labor Unit Cost	Avg Equip Unit Cost	Avg Total Unit Cost	Avg Price Incl O&P

Food centers

Includes wiring, connection and installation in exposed drainboard only

Built-in models, 4-1/4" x 6-3/4" x 10" rough cut,

1/4 hp, 110 volts, 6 speed

Description	Oper	Unit	Vol	Crew Size	Man-hours per Unit	Crew Output per Day	Avg Mat'l Unit Cost	Avg Labor Unit Cost	Avg Equip Unit Cost	Avg Total Unit Cost	Avg Price Incl O&P
	Inst	Ea	Lg	EA	2.67	3.00	456.00	95.10	---	551.10	**664.00**
	Inst	Ea	Sm	EA	3.81	2.10	524.00	136.00	---	660.00	**801.00**
Options											
Blender	Inst	Ea	Lg	---	---	---	49.00	---	---	49.00	**49.00**
	Inst	Ea	Sm	---	---	---	56.30	---	---	56.30	**56.30**
Citrus fruit juicer	Inst	Ea	Lg	---	---	---	24.00	---	---	24.00	**24.00**
	Inst	Ea	Sm	---	---	---	27.60	---	---	27.60	**27.60**
Food processor	Inst	Ea	Lg	---	---	---	248.00	---	---	248.00	**248.00**
	Inst	Ea	Sm	---	---	---	285.00	---	---	285.00	**285.00**
Ice crusher	Inst	Ea	Lg	---	---	---	77.80	---	---	77.80	**77.80**
	Inst	Ea	Sm	---	---	---	89.30	---	---	89.30	**89.30**
Knife sharpener	Inst	Ea	Lg	---	---	---	51.10	---	---	51.10	**51.10**
	Inst	Ea	Sm	---	---	---	58.70	---	---	58.70	**58.70**
Meat grinder, shredder/slicer with power post											
	Inst	Ea	Lg	---	---	---	297.00	---	---	297.00	**297.00**
	Inst	Ea	Sm	---	---	---	341.00	---	---	341.00	**341.00**
Mixer	Inst	Ea	Lg	---	---	---	139.00	---	---	139.00	**139.00**
	Inst	Ea	Sm	---	---	---	160.00	---	---	160.00	**160.00**

Footings. See Concrete, page 98

Formica. See Countertops, page 106

Forming. See Concrete, page 98

Foundations. See Concrete, page 100

Framing

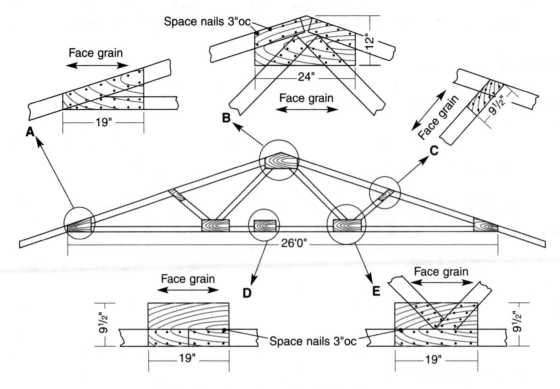

Space nails 3"oc

Face grain

12"

24"

Face grain

Face grain

9½"

Face grain

19"

A

B

C

26'0"

Face grain

D

E

Face grain

9½"

9½"

Space nails 3"oc

19"

19"

Construction of a 26 foot W truss:

A Bevel-heel gusset
B Peak gusset
C Upper chord intermediate gusset
D Splice of lower chord
E Lower chord intermediate gusset

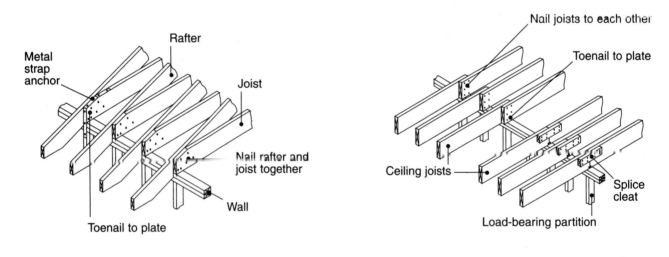

Metal strap anchor

Rafter

Joist

Nail rafter and joist together

Wall

Toenail to plate

Nail joists to each other

Toenail to plate

Ceiling joists

Splice cleat

Load-bearing partition

Wood hanger

Metal joist hanger

Flush ceiling framing

A

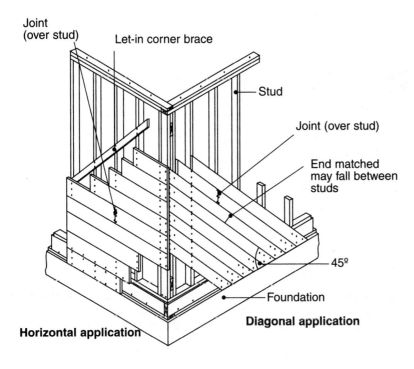

Joint (over stud)

Let-in corner brace

Stud

Joint (over stud)

End matched may fall between studs

45°

Foundation

Diagonal application

Horizontal application

Application of wood sheathing:
A Horizontal and diagonal
B Started at subfloor
C Started at foundation wall

B

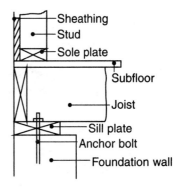

Sheathing

Stud

Sole plate

Subfloor

Joist

Sill plate

Anchor bolt

Foundation wall

C

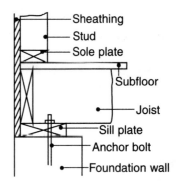

Sheathing

Stud

Sole plate

Subfloor

Joist

Sill plate

Anchor bolt

Foundation wall

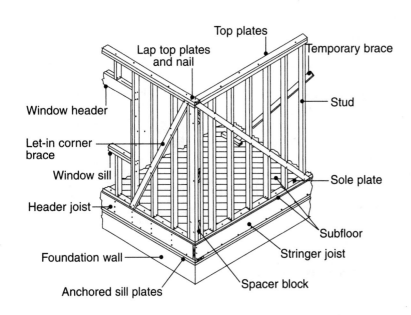

Top plates

Lap top plates and nail

Temporary brace

Stud

Window header

Let-in corner brace

Window sill

Sole plate

Header joist

Subfloor

Foundation wall

Stringer joist

Anchored sill plates

Spacer block

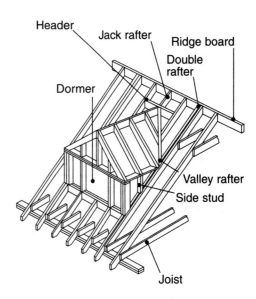

Header

Jack rafter

Ridge board

Double rafter

Dormer

Valley rafter

Side stud

Joist

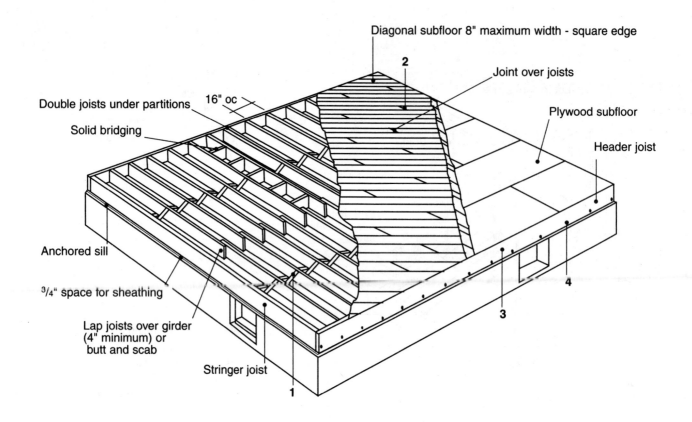

Diagonal subfloor 8" maximum width - square edge

Joint over joists

Plywood subfloor

Header joist

Double joists under partitions

16" oc

Solid bridging

Anchored sill

³/₄" space for sheathing

Lap joists over girder (4" minimum) or butt and scab

Stringer joist

Floor framing:
1. Nailing bridge to joists
2. Nailing board subfloor to joists
3. Nailing header to joists
4. Toenailing header to sill

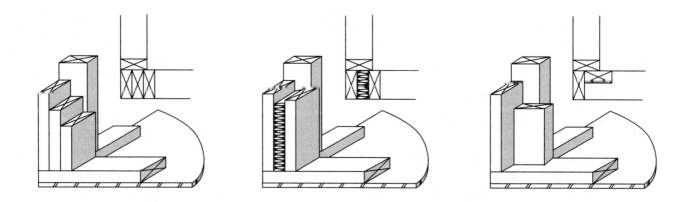

Stud arrangements at exterior corners

Description	Oper	Unit	Vol	Crew Size	Man-hours per Unit	Crew Output per Day	Avg Mat'l Unit Cost	Avg Labor Unit Cost	Avg Equip Unit Cost	Avg Total Unit Cost	Avg Price Incl O&P

Framing, rough carpentry

Dimension lumber

Beams; set on steel columns, not wood columns

Built-up (2 pieces)

Description	Oper	Unit	Vol	Crew Size	Man-hours per Unit	Crew Output per Day	Avg Mat'l Unit Cost	Avg Labor Unit Cost	Avg Equip Unit Cost	Avg Total Unit Cost	Avg Price Incl O&P
4" x 6" - 10'	Demo	LF	Lg	LB	.024	661.0	---	.66	---	.66	.98
	Demo	LF	Sm	LB	.028	562.0	---	.77	---	.77	1.14
4" x 6" - 10'	Inst	LF	Lg	2C	.047	339.0	1.11	1.56	.14	2.81	3.76
	Inst	LF	Sm	2C	.056	288.0	1.22	1.86	.16	3.24	4.36
4" x 8" - 10'	Demo	LF	Lg	LB	.027	595.0	---	.74	---	.74	1.10
	Demo	LF	Sm	LB	.032	506.0	---	.88	---	.88	1.31
4" x 8" - 10'	Inst	LF	Lg	2C	.046	350.0	1.52	1.53	.13	3.18	4.17
	Inst	LF	Sm	2C	.054	298.0	1.70	1.80	.16	3.66	4.81
4" x 10" - 10'	Demo	LF	Lg	LB	.030	540.0	---	.82	---	.82	1.23
	Demo	LF	Sm	LB	.035	459.0	---	.96	---	.96	1.43
4" x 10" - 10'	Inst	LF	Lg	2C	.049	325.0	2.18	1.63	.14	3.95	5.09
	Inst	LF	Sm	2C	.058	276.0	2.40	1.93	.17	4.50	5.82
4" x 12" - 12'	Demo	LF	Lg	LB	.032	495.0	---	.88	---	.88	1.31
	Demo	LF	Sm	LB	.038	421.0	---	1.04	---	1.04	1.55
4" x 12" - 12'	Inst	LF	Lg	2C	.053	304.0	2.65	1.76	.15	4.56	5.84
	Inst	LF	Sm	2C	.062	258.0	2.96	2.06	.18	5.20	6.68
6" x 8" - 10'	Demo	LF	Lg	LD	.043	744.0	---	1.31	---	1.31	1.94
	Demo	LF	Sm	LD	.051	632.0	---	1.55	---	1.55	2.31
6" x 8" - 10'	Inst	LF	Lg	CW	.074	435.0	3.88	2.25	.11	6.24	7.94
	Inst	LF	Sm	CW	.086	370.0	4.28	2.61	.13	7.02	8.97
6" x 10" - 10'	Demo	LF	Lg	LD	.047	676.0	---	1.43	---	1.43	2.13
	Demo	LF	Sm	LD	.056	575.0	---	1.70	---	1.70	2.53
6" x 10" - 10'	Inst	LF	Lg	CW	.079	405.0	4.84	2.40	.12	7.36	9.28
	Inst	LF	Sm	CW	.093	344.0	5.34	2.82	.14	8.30	10.50
6" x 12" - 12'	Demo	LF	Lg	LD	.052	619.0	---	1.58	---	1.58	2.35
	Demo	LF	Sm	LD	.061	526.0	---	1.85	---	1.85	2.76
6" x 12" - 12'	Inst	LF	Lg	CW	.084	379.0	5.80	2.55	.12	8.47	10.60
	Inst	LF	Sm	CW	.099	322.0	6.39	3.00	.14	9.53	12.00

Built-up (3 pieces)

Description	Oper	Unit	Vol	Crew Size	Man-hours per Unit	Crew Output per Day	Avg Mat'l Unit Cost	Avg Labor Unit Cost	Avg Equip Unit Cost	Avg Total Unit Cost	Avg Price Incl O&P
6" x 8" - 10'	Demo	LF	Lg	LD	.044	733.0	---	1.34	---	1.34	1.99
	Demo	LF	Sm	LD	.051	623.0	---	1.55	---	1.55	2.31
6" x 8" - 10'	Inst	LF	Lg	CW	.075	425.0	2.31	2.28	.11	4.70	6.18
	Inst	LF	Sm	CW	.089	361.0	2.58	2.70	.13	5.41	7.15
6" x 10" - 10'	Demo	LF	Lg	LD	.048	669.0	---	1.46	---	1.46	2.17
	Demo	LF	Sm	LD	.056	569.0	---	1.70	---	1.70	2.53
6" x 10" - 10'	Inst	LF	Lg	CW	.081	397.0	3.30	2.46	.12	5.88	7.60
	Inst	LF	Sm	CW	.095	337.0	3.64	2.88	.14	6.66	8.65

Description	Oper	Unit	Vol	Crew Size	Man-hours per Unit	Crew Output per Day	Avg Mat'l Unit Cost	Avg Labor Unit Cost	Avg Equip Unit Cost	Avg Total Unit Cost	Avg Price Incl O&P
6" x 12" - 12'	Demo	LF	Lg	LD	.053	609.0	---	1.61	---	1.61	2.40
	Demo	LF	Sm	LD	.062	518.0	---	1.88	---	1.88	2.80
6" x 12" - 12'	Inst	LF	Lg	CW	.087	368.0	4.00	2.64	.13	6.77	8.69
	Inst	LF	Sm	CW	.102	313.0	4.47	3.10	.15	7.72	9.93
9" x 10" - 12'	Demo	LF	Lg	LD	.059	541.0	---	1.79	---	1.79	2.67
	Demo	LF	Sm	LD	.070	460.0	---	2.12	---	2.12	3.17
9" x 10" - 12'	Inst	LF	Lg	CW	.095	338.0	7.28	2.88	.14	10.30	12.80
	Inst	LF	Sm	CW	.111	287.0	8.04	3.37	.16	11.57	14.50
9" x 12" - 12'	Demo	LF	Lg	LD	.065	493.0	---	1.97	---	1.97	2.94
	Demo	LF	Sm	LD	.076	419.0	---	2.31	---	2.31	3.44
9" x 12" - 12'	Inst	LF	Lg	CW	.102	315.0	8.72	3.10	.15	11.97	14.80
	Inst	LF	Sm	CW	.119	268.0	9.62	3.61	.17	13.40	16.70

Single member

Description	Oper	Unit	Vol	Crew Size	Man-hours per Unit	Crew Output per Day	Avg Mat'l Unit Cost	Avg Labor Unit Cost	Avg Equip Unit Cost	Avg Total Unit Cost	Avg Price Incl O&P
2" x 6"	Demo	LF	Lg	LB	.017	947.0	---	.47	---	.47	.69
	Demo	LF	Sm	LB	.020	805.0	---	.55	---	.55	.82
2" x 6"	Inst	LF	Lg	2C	.031	520.0	.63	1.03	.09	1.75	2.36
	Inst	LF	Sm	2C	.036	442.0	.70	1.20	.11	2.01	2.71
2" x 8"	Demo	LF	Lg	LB	.018	874.0	---	.49	---	.49	.74
	Demo	LF	Sm	LB	.022	743.0	---	.60	---	.60	.90
2" x 8"	Inst	LF	Lg	2C	.033	491.0	.84	1.10	.10	2.04	2.71
	Inst	LF	Sm	2C	.038	417.0	.94	1.26	.11	2.31	3.09
2" x 10"	Demo	LF	Lg	LB	.020	811.0	---	.55	---	.55	.82
	Demo	LF	Sm	LB	.023	689.0	---	.63	---	.63	.94
2" x 10"	Inst	LF	Lg	2C	.034	465.0	1.17	1.13	.10	2.40	3.14
	Inst	LF	Sm	2C	.041	395.0	1.29	1.36	.12	2.77	3.65
2" x 12"	Demo	LF	Lg	LB	.021	756.0	---	.58	---	.58	.86
	Demo	LF	Sm	LB	.025	643.0	---	.69	---	.69	1.02
2" x 12"	Inst	LF	Lg	2C	.036	440.0	1.40	1.20	.11	2.71	3.52
	Inst	LF	Sm	2C	.043	374.0	1.57	1.43	.12	3.12	4.07
3" x 6"	Demo	LF	Lg	LB	.019	832.0	---	.52	---	.52	.78
	Demo	LF	Sm	LB	.023	707.0	---	.63	---	.63	.94
3" x 6"	Inst	LF	Lg	2C	.034	472.0	1.51	1.13	.10	2.74	3.53
	Inst	LF	Sm	2C	.040	401.0	1.67	1.33	.12	3.12	4.04
3" x 8"	Demo	LF	Lg	LB	.021	756.0	---	.58	---	.58	.86
	Demo	LF	Sm	LB	.025	643.0	---	.69	---	.69	1.02
3" x 8"	Inst	LF	Lg	2C	.036	440.0	2.02	1.20	.11	3.33	4.23
	Inst	LF	Sm	2C	.043	374.0	2.23	1.43	.12	3.78	4.83
3" x 10"	Demo	LF	Lg	LB	.024	680.0	---	.66	---	.66	.98
	Demo	LF	Sm	LB	.028	578.0	---	.77	---	.77	1.14
3" x 10"	Inst	LF	Lg	2C	.039	408.0	2.50	1.30	.11	3.91	4.93
	Inst	LF	Sm	2C	.046	347.0	2.76	1.53	.13	4.42	5.60

Description	Oper	Unit	Vol	Crew Size	Man-hours per Unit	Crew Output per Day	Avg Mat'l Unit Cost	Avg Labor Unit Cost	Avg Equip Unit Cost	Avg Total Unit Cost	Avg Price Incl O&P
3" x 12"	Demo	LF	Lg	LB	.026	623.0	---	.71	---	.71	1.06
	Demo	LF	Sm	LB	.030	530.0	---	.82	---	.82	1.23
3" x 12"	Inst	LF	Lg	2C	.042	381.0	2.98	1.40	.12	4.50	5.64
	Inst	LF	Sm	2C	.049	324.0	3.29	1.63	.14	5.06	6.37
3" x 14"	Demo	LF	Lg	LB	.028	575.0	---	.77	---	.77	1.14
	Demo	LF	Sm	LB	.033	489.0	---	.91	---	.91	1.35
3" x 14"	Inst	LF	Lg	2C	.045	359.0	3.57	1.50	.13	5.20	6.48
	Inst	LF	Sm	2C	.052	305.0	3.97	1.73	.15	5.85	7.31
4" x 6"	Demo	LF	Lg	LB	.021	745.0	---	.58	---	.58	.86
	Demo	LF	Sm	LB	.025	633.0	---	.69	---	.69	1.02
4" x 6"	Inst	LF	Lg	2C	.037	433.0	1.82	1.23	.11	3.16	4.05
	Inst	LF	Sm	2C	.043	368.0	2.03	1.43	.13	3.59	4.61
4" x 8"	Demo	LF	Lg	LB	.025	653.0	---	.69	---	.69	1.02
	Demo	LF	Sm	LB	.029	555.0	---	.80	---	.80	1.19
4" x 8"	Inst	LF	Lg	2C	.041	394.0	2.43	1.36	.12	3.91	4.96
	Inst	LF	Sm	2C	.048	335.0	2.70	1.60	.14	4.44	5.64
4" x 10"	Demo	LF	Lg	LB	.027	586.0	---	.74	---	.74	1.10
	Demo	LF	Sm	LB	.032	498.0	---	.88	---	.88	1.31
4" x 10"	Inst	LF	Lg	2C	.044	363.0	3.00	1.46	.13	4.59	5.78
	Inst	LF	Sm	2C	.052	309.0	3.34	1.73	.15	5.22	6.59
4" x 12"	Demo	LF	Lg	LB	.031	522.0	---	.85	---	.85	1.27
	Demo	LF	Sm	LB	.036	444.0	---	.99	---	.99	1.47
4" x 12"	Inst	LF	Lg	2C	.048	333.0	3.58	1.60	.14	5.32	6.65
	Inst	LF	Sm	2C	.057	283.0	3.99	1.90	.16	6.05	7.59
4" x 14"	Demo	LF	Lg	LH	.037	642.0	---	1.01	---	1.01	1.51
	Demo	LF	Sm	LH	.044	546.0	---	1.21	---	1.21	1.80
4" x 14"	Inst	LF	Lg	CS	.059	404.0	4.37	1.85	.12	6.34	7.92
	Inst	LF	Sm	CS	.070	343.0	4.84	2.19	.14	7.17	9.00
4" x 16"	Demo	LF	Lg	LH	.040	594.0	---	1.10	---	1.10	1.63
	Demo	LF	Sm	LH	.048	505.0	---	1.32	---	1.32	1.96
4" x 16"	Inst	LF	Lg	CS	.063	379.0	5.21	1.97	.12	7.30	9.07
	Inst	LF	Sm	CS	.075	322.0	5.75	2.35	.14	8.24	10.30
6" x 8"	Demo	LF	Lg	LJ	.041	782.0	---	1.12	---	1.12	1.68
	Demo	LF	Sm	LJ	.048	665.0	---	1.32	---	1.32	1.96
6" x 8"	Inst	LF	Lg	CW	.069	466.0	4.68	2.09	.10	6.87	8.62
	Inst	LF	Sm	CW	.081	396.0	5.18	2.46	.12	7.76	9.76
6" x 10"	Demo	LF	Lg	LJ	.046	694.0	---	1.26	---	1.26	1.88
	Demo	LF	Sm	LJ	.054	590.0	---	1.48	---	1.48	2.21
6" x 10"	Inst	LF	Lg	CW	.075	427.0	5.94	2.28	.11	8.33	10.40
	Inst	LF	Sm	CW	.088	363.0	6.55	2.67	.13	9.35	11.70
6" x 12"	Demo	LF	Lg	LJ	.051	631.0	---	1.40	---	1.40	2.08
	Demo	LF	Sm	LJ	.060	536.0	---	1.65	---	1.65	2.45
6" x 12"	Inst	LF	Lg	CW	.081	396.0	7.37	2.46	.12	9.95	12.30
	Inst	LF	Sm	CW	.095	337.0	8.17	2.88	.14	11.19	13.90

Description	Oper	Unit	Vol	Crew Size	Man-hours per Unit	Crew Output per Day	Avg Mat'l Unit Cost	Avg Labor Unit Cost	Avg Equip Unit Cost	Avg Total Unit Cost	Avg Price Incl O&P
6" x 14"	Demo	LF	Lg	LJ	.056	567.0	---	1.54	---	1.54	2.29
	Demo	LF	Sm	LJ	.066	482.0	---	1.81	---	1.81	2.70
6" x 14"	Inst	LF	Lg	CW	.088	365.0	8.58	2.67	.13	11.38	14.00
	Inst	LF	Sm	CW	.103	310.0	9.51	3.13	.15	12.79	15.80
6" x 16"	Demo	LF	Lg	LJ	.061	523.0	---	1.67	---	1.67	2.49
	Demo	LF	Sm	LJ	.072	445.0	---	1.97	---	1.97	2.94
6" x 16"	Inst	LF	Lg	CW	.094	342.0	10.20	2.85	.14	13.19	16.20
	Inst	LF	Sm	CW	.110	291.0	11.40	3.34	.16	14.90	18.30
8" x 8"	Demo	LF	Lg	LJ	.048	668.0	---	1.32	---	1.32	1.96
	Demo	LF	Sm	LJ	.056	568.0	---	1.54	---	1.54	2.29
8" x 8"	Inst	LF	Lg	CW	.077	413.0	6.73	2.34	.11	9.18	11.40
	Inst	LF	Sm	CW	.091	351.0	7.50	2.76	.13	10.39	12.90
8" x 10"	Demo	LF	Lg	LJ	.058	556.0	---	1.59	---	1.59	2.37
	Demo	LF	Sm	LJ	.068	473.0	---	1.87	---	1.87	2.78
8" x 10"	Inst	LF	Lg	CW	.088	363.0	8.62	2.67	.13	11.42	14.10
	Inst	LF	Sm	CW	.104	309.0	9.58	3.16	.15	12.89	15.90
8" x 12"	Demo	LF	Lg	LJ	.062	515.0	---	1.70	---	1.70	2.53
	Demo	LF	Sm	LJ	.073	438.0	---	2.00	---	2.00	2.98
8" x 12"	Inst	LF	Lg	CW	.094	341.0	10.80	2.85	.14	13.79	16.80
	Inst	LF	Sm	CW	.110	290.0	11.90	3.34	.16	15.40	18.90
8" x 14"	Demo	LF	Lg	LJ	.068	468.0	---	1.87	---	1.87	2.78
	Demo	LF	Sm	LJ	.080	398.0	---	2.19	---	2.19	3.27
8" x 14"	Inst	LF	Lg	CW	.101	316.0	12.80	3.07	.15	16.02	19.50
	Inst	LF	Sm	CW	.119	269.0	14.30	3.61	.17	18.08	22.00
8" x 16"	Demo	LF	Lg	LJ	.076	423.0	---	2.08	---	2.08	3.11
	Demo	LF	Sm	LJ	.089	360.0	---	2.44	---	2.44	3.64
8" x 16"	Inst	LF	Lg	CW	.110	292.0	15.00	3.34	.16	18.50	22.40
	Inst	LF	Sm	CW	.129	248.0	16.70	3.92	.19	20.81	25.20

Blocking, horizontal, for studs

Description	Oper	Unit	Vol	Crew Size	Man-hours per Unit	Crew Output per Day	Avg Mat'l Unit Cost	Avg Labor Unit Cost	Avg Equip Unit Cost	Avg Total Unit Cost	Avg Price Incl O&P
2" x 4" - 12"	Demo	LF	Lg	1L	.011	710.0	---	.30	---	.30	.45
	Demo	LF	Sm	1L	.013	604.0	---	.36	---	.36	.53
2" x 4" - 12"	Inst	LF	Lg	CA	.036	224.0	.30	1.20	.21	1.00	2.40
	Inst	LF	Sm	CA	.042	190.0	.42	1.40	.25	2.07	2.83
2" x 4" - 16"	Demo	LF	Lg	1L	.011	761.0	---	.30	---	.30	.45
	Demo	LF	Sm	1L	.012	647.0	---	.33	---	.33	.49
2" x 4" - 16"	Inst	LF	Lg	CA	.029	274.0	.39	.96	.17	1.52	2.07
	Inst	LF	Sm	CA	.034	233.0	.42	1.13	.20	1.75	2.38
2" x 4" - 24"	Demo	LF	Lg	1L	.010	810.0	---	.27	---	.27	.41
	Demo	LF	Sm	1L	.012	689.0	---	.33	---	.33	.49
2" x 4" - 24"	Inst	LF	Lg	CA	.023	350.0	.39	.77	.13	1.29	1.73
	Inst	LF	Sm	CA	.027	298.0	.42	.90	.16	1.48	1.99

Description	Oper	Unit	Vol	Crew Size	Man-hours per Unit	Crew Output per Day	Avg Mat'l Unit Cost	Avg Labor Unit Cost	Avg Equip Unit Cost	Avg Total Unit Cost	Avg Price Incl O&P
2" x 6" - 12"	Demo	LF	Lg	1L	.013	608.0	---	.36	---	.36	**.53**
	Demo	LF	Sm	1L	.015	517.0	---	.41	---	.41	**.61**
2" x 6" - 12"	Inst	LF	Lg	CA	.039	207.0	.54	1.30	.23	2.07	**2.80**
	Inst	LF	Sm	CA	.045	176.0	.59	1.50	.27	2.36	**3.19**
2" x 6" - 16"	Demo	LF	Lg	1L	.012	647.0	---	.33	---	.33	**.49**
	Demo	LF	Sm	1L	.015	550.0	---	.41	---	.41	**.61**
2" x 6" - 16"	Inst	LF	Lg	CA	.032	252.0	.54	1.06	.19	1.79	**2.41**
	Inst	LF	Sm	CA	.037	214.0	.59	1.23	.22	2.04	**2.75**
2" x 6" - 24"	Demo	LF	Lg	1L	.012	683.0	---	.33	---	.33	**.49**
	Demo	LF	Sm	1L	.014	581.0	---	.38	---	.38	**.57**
2" x 6" - 24"	Inst	LF	Lg	CA	.025	323.0	.54	.83	.14	1.51	**2.01**
	Inst	LF	Sm	CA	.029	275.0	.59	.96	.17	1.72	**2.30**
2" x 8" - 12"	Demo	LF	Lg	1L	.015	532.0	---	.41	---	.41	**.61**
	Demo	LF	Sm	1L	.018	452.0	---	.49	---	.49	**.74**
2" x 8" - 12"	Inst	LF	Lg	CA	.041	193.0	.74	1.36	.24	2.34	**3.14**
	Inst	LF	Sm	CA	.049	164.0	.83	1.63	.28	2.74	**3.68**
2" x 8" - 16"	Demo	LF	Lg	1L	.014	564.0	---	.38	---	.38	**.57**
	Demo	LF	Sm	1L	.017	479.0	---	.47	---	.47	**.69**
2" x 8" - 16"	Inst	LF	Lg	CA	.034	235.0	.74	1.13	.20	2.07	**2.75**
	Inst	LF	Sm	CA	.040	200.0	.83	1.33	.23	2.39	**3.18**
2" x 8" - 24"	Demo	LF	Lg	1L	.014	592.0	---	.38	---	.38	**.57**
	Demo	LF	Sm	1L	.016	503.0	---	.44	---	.44	**.65**
2" x 8" - 24"	Inst	LF	Lg	CA	.027	299.0	.74	.90	.16	1.80	**2.36**
	Inst	LF	Sm	CA	.031	254.0	.83	1.03	.18	2.04	**2.68**

Bracing, diagonal let-ins

Studs, 12" oc

Description	Oper	Unit	Vol	Crew Size	Man-hours per Unit	Crew Output per Day	Avg Mat'l Unit Cost	Avg Labor Unit Cost	Avg Equip Unit Cost	Avg Total Unit Cost	Avg Price Incl O&P
1" x 6"	Demo	Set	Lg	1L	.011	714.0	---	.30	---	.30	**.45**
	Demo	Set	Sm	1L	.013	607.0	---	.36	---	.36	**.53**
1" x 6"	Inst	Set	Lg	CA	.056	143.0	.35	1.86	.33	2.54	**3.53**
	Inst	Set	Sm	CA	.066	122.0	.38	2.20	.38	2.96	**4.11**

Studs, 16" oc

Description	Oper	Unit	Vol	Crew Size	Man-hours per Unit	Crew Output per Day	Avg Mat'l Unit Cost	Avg Labor Unit Cost	Avg Equip Unit Cost	Avg Total Unit Cost	Avg Price Incl O&P
1" x 6"	Demo	Set	Lg	1L	.010	770.0	---	.27	---	.27	**.41**
	Demo	Set	Sm	1L	.012	655.0	---	.33	---	.33	**.49**
1" x 6"	Inst	Set	Lg	CA	.043	184.0	.35	1.43	.25	2.03	**2.80**
	Inst	Set	Sm	CA	.051	156.0	.38	1.70	.30	2.38	**3.28**

Studs, 24" oc

Description	Oper	Unit	Vol	Crew Size	Man-hours per Unit	Crew Output per Day	Avg Mat'l Unit Cost	Avg Labor Unit Cost	Avg Equip Unit Cost	Avg Total Unit Cost	Avg Price Incl O&P
1" x 6"	Demo	Set	Lg	1L	.010	825.0	---	.27	---	.27	**.41**
	Demo	Set	Sm	1L	.011	701.0	---	.30	---	.30	**.45**
1" x 6"	Inst	Set	Lg	CA	.033	242.0	.35	1.10	.19	1.64	**2.24**
	Inst	Set	Sm	CA	.039	206.0	.38	1.30	.23	1.91	**2.61**

Description	Oper	Unit	Vol	Crew Size	Man-hours per Unit	Crew Output per Day	Avg Mat'l Unit Cost	Avg Labor Unit Cost	Avg Equip Unit Cost	Avg Total Unit Cost	Avg Price Incl O&P
Bridging, "X" type											
For joists 12" oc											
1" x 3"	Demo	Set	Lg	1L	.015	550.0	---	.41	---	.41	**.61**
	Demo	Set	Sm	1L	.017	468.0	---	.47	---	.47	**.69**
1" x 3"	Inst	Set	Lg	CA	.057	140.0	.52	1.90	.33	2.75	**3.77**
	Inst	Set	Sm	CA	.067	119.0	.58	2.23	.39	3.20	**4.40**
2" x 2"	Demo	Set	Lg	1L	.015	536.0	---	.41	---	.41	**.61**
	Demo	Set	Sm	1L	.018	456.0	---	.49	---	.49	**.74**
2" x 2"	Inst	Set	Lg	CA	.058	139.0	.79	1.93	.34	3.06	**4.14**
	Inst	Set	Sm	CA	.068	118.0	.87	2.26	.40	3.53	**4.79**
For joists 16" oc											
1" x 3"	Demo	Set	Lg	1L	.015	530.0	---	.41	---	.41	**.61**
	Demo	Set	Sm	1L	.018	451.0	---	.49	---	.49	**.74**
1" x 3"	Inst	Set	Lg	CA	.058	137.0	.56	1.93	.34	2.83	**3.88**
	Inst	Set	Sm	CA	.069	116.0	.62	2.30	.40	3.32	**4.56**
2" x 2"	Demo	Set	Lg	1L	.016	515.0	---	.44	---	.44	**.65**
	Demo	Set	Sm	1L	.018	438.0	---	.49	---	.49	**.74**
2" x 2"	Inst	Set	Lg	CA	.059	136.0	.84	1.96	.34	3.14	**4.25**
	Inst	Set	Sm	CA	.069	116.0	.92	2.30	.40	3.62	**4.90**
For joists 24" oc											
1" x 3"	Demo	Set	Lg	1L	.016	507.0	---	.44	---	.44	**.65**
	Demo	Set	Sm	1L	.019	431.0	---	.52	---	.52	**.78**
1" x 3"	Inst	Set	Lg	CA	.016	492.0	.62	.53	.09	1.24	**1.60**
	Inst	Set	Sm	CA	.019	418.0	.69	.63	.11	1.43	**1.85**
2" x 2"	Demo	Set	Lg	1L	.060	134.0	---	1.65	---	1.65	**2.45**
	Demo	Set	Sm	1L	.070	114.0	---	1.92	---	1.92	**2.86**
2" x 2"	Inst	Set	Lg	CA	.000	133.0	.93	2.00	.35	3.28	**4.41**
	Inst	Set	Sm	CA	.071	113.0	1.03	2.36	.41	3.80	**5.14**
Bridging, solid, between joists											
2" x 6" - 12"	Demo	Set	Lg	1L	.014	590.0	---	.38	---	.38	**.57**
	Demo	Set	Sm	1L	.016	502.0	---	.44	---	.44	**.65**
2" x 6" - 12"	Inst	Set	Lg	CA	.040	199.0	.54	1.33	.23	2.10	**2.85**
	Inst	Set	Sm	CA	.047	169.0	.59	1.56	.28	2.43	**3.30**
2" x 8" - 12"	Demo	Set	Lg	1L	.015	517.0	---	.41	---	.41	**.61**
	Demo	Set	Sm	1L	.018	439.0	---	.49	---	.49	**.74**
2" x 8" - 12"	Inst	Set	Lg	CA	.043	185.0	.74	1.43	.25	2.42	**3.25**
	Inst	Set	Sm	CA	.051	157.0	.83	1.70	.30	2.83	**3.80**
2" x 10" - 12"	Demo	Set	Lg	1L	.018	455.0	---	.49	---	.49	**.74**
	Demo	Set	Sm	1L	.021	387.0	---	.58	---	.58	**.86**
2" x 10" - 12"	Inst	Set	Lg	CA	.046	174.0	1.08	1.53	.27	2.88	**3.81**
	Inst	Set	Sm	CA	.054	148.0	1.19	1.80	.32	3.31	**4.38**

Description	Oper	Unit	Vol	Crew Size	Man-hours per Unit	Crew Output per Day	Avg Mat'l Unit Cost	Avg Labor Unit Cost	Avg Equip Unit Cost	Avg Total Unit Cost	Avg Price Incl O&P
2" x 12" - 12"	Demo	Set	Lg	1L	.019	412.0	---	.52	---	.52	.78
	Demo	Set	Sm	1L	.023	350.0	---	.63	---	.63	.94
2" x 12" - 12"	Inst	Set	Lg	CA	.049	164.0	1.31	1.63	.28	3.22	4.23
	Inst	Set	Sm	CA	.058	139.0	1.46	1.93	.34	3.73	4.91
2" x 6" - 16"	Demo	Set	Lg	1L	.013	628.0	---	.36	---	.36	.53
	Demo	Set	Sm	1L	.015	534.0	---	.41	---	.41	.61
2" x 6" - 16"	Inst	Set	Lg	CA	.033	243.0	.54	1.10	.19	1.83	2.46
	Inst	Set	Sm	CA	.039	207.0	.59	1.30	.23	2.12	2.85
2" x 8" - 16"	Demo	Set	Lg	1L	.015	548.0	---	.41	---	.41	.61
	Demo	Set	Sm	1L	.017	466.0	---	.47	---	.47	.69
2" x 8" - 16"	Inst	Set	Lg	CA	.035	226.0	.74	1.16	.21	2.11	2.81
	Inst	Set	Sm	CA	.042	192.0	.83	1.40	.24	2.47	3.29
2" x 10" - 16"	Demo	Set	Lg	1L	.017	481.0	---	.47	---	.47	.69
	Demo	Set	Sm	1L	.020	409.0	---	.55	---	.55	.82
2" x 10" - 16"	Inst	Set	Lg	CA	.038	211.0	1.08	1.26	.22	2.56	3.36
	Inst	Set	Sm	CA	.045	179.0	1.19	1.50	.26	2.95	3.87
2" x 12" - 16"	Demo	Set	Lg	1L	.018	434.0	---	.49	---	.49	.74
	Demo	Set	Sm	1L	.022	369.0	---	.60	---	.60	.90
2" x 12" - 16"	Inst	Set	Lg	CA	.040	198.0	1.31	1.33	.24	2.88	3.74
	Inst	Set	Sm	CA	.048	168.0	1.46	1.60	.28	3.34	4.35
2" x 6" - 24"	Demo	Set	Lg	1L	.012	664.0	---	.33	---	.33	.49
	Demo	Set	Sm	1L	.014	564.0	---	.38	---	.38	.57
2" x 6" - 24"	Inst	Set	Lg	CA	.026	312.0	.54	.87	.15	1.56	2.07
	Inst	Set	Sm	CA	.030	265.0	.59	1.00	.18	1.77	2.36
2" x 8" - 24"	Demo	Set	Lg	1L	.014	577.0	---	.38	---	.38	.57
	Demo	Set	Sm	1L	.016	490.0	---	.44	---	.44	.65
2" x 8" - 24"	Inst	Set	Lg	CA	.028	289.0	.74	.93	.16	1.83	2.41
	Inst	Set	Sm	CA	.033	246.0	.83	1.10	.19	2.12	2.79
2" x 10" - 24"	Demo	Set	Lg	1L	.016	504.0	---	.44	---	.44	.65
	Demo	Set	Sm	1L	.019	428.0	---	.52	---	.52	.78
2" x 10" - 24"	Inst	Set	Lg	CA	.030	270.0	1.08	1.00	.17	2.25	2.91
	Inst	Set	Sm	CA	.035	230.0	1.19	1.16	.20	2.55	3.32
2" x 12" - 24"	Demo	Set	Lg	1L	.018	453.0	---	.49	---	.49	.74
	Demo	Set	Sm	1L	.021	385.0	---	.58	---	.58	.86
2" x 12" - 24"	Inst	Set	Lg	CA	.032	252.0	1.31	1.06	.19	2.56	3.29
	Inst	Set	Sm	CA	.037	214.0	1.46	1.23	.22	2.91	3.75

Description	Oper	Unit	Vol	Crew Size	Man-hours per Unit	Crew Output per Day	Avg Mat'l Unit Cost	Avg Labor Unit Cost	Avg Equip Unit Cost	Avg Total Unit Cost	Avg Price Incl O&P
Columns or posts, without base or cap, hardware, or chamfer corners											
4" x 4" - 8'	Demo	LF	Lg	LB	.027	584.0	---	.74	---	.74	**1.10**
	Demo	LF	Sm	LB	.032	496.0	---	.88	---	.88	**1.31**
4" x 4" - 8'	Inst	LF	Lg	2C	.051	315.0	1.16	1.70	.15	3.01	**4.03**
	Inst	LF	Sm	2C	.060	268.0	1.29	2.00	.17	3.46	**4.65**
4" x 6" - 8'	Demo	LF	Lg	LB	.030	529.0	---	.82	---	.82	**1.23**
	Demo	LF	Sm	LB	.036	450.0	---	.99	---	.99	**1.47**
4" x 6" - 8'	Inst	LF	Lg	2C	.054	294.0	1.75	1.80	.16	3.71	**4.87**
	Inst	LF	Sm	2C	.064	250.0	1.95	2.13	.19	4.27	**5.63**
4" x 8" - 8'	Demo	LF	Lg	LB	.033	479.0	---	.91	---	.91	**1.35**
	Demo	LF	Sm	LB	.039	407.0	---	1.07	---	1.07	**1.59**
4" x 0" - 0'	Inst	LF	Lg	2C	.058	274.0	2.35	1.93	.17	4.45	**5.77**
	Inst	LF	Sm	2C	.069	233.0	2.62	2.30	.20	5.12	**6.66**
6" x 6" - 8'	Demo	LF	Lg	LB	.037	432.0	---	1.01	---	1.01	**1.51**
	Demo	LF	Sm	LB	.044	367.0	---	1.21	---	1.21	**1.80**
6" x 6" - 8'	Inst	LF	Lg	2C	.065	248.0	3.46	2.16	.19	5.81	**7.41**
	Inst	LF	Sm	2C	.076	211.0	3.83	2.53	.22	6.58	**8.42**
6" x 8" - 8'	Demo	LF	Lg	LB	.042	385.0	---	1.15	---	1.15	**1.72**
	Demo	LF	Sm	LB	.049	327.0	---	1.34	---	1.34	**2.00**
6" x 8" - 8'	Inst	LF	Lg	2C	.070	229.0	4.61	2.33	.20	7.14	**9.00**
	Inst	LF	Sm	2C	.082	195.0	5.10	2.73	.24	8.07	**10.20**
6" x 10" - 8'	Demo	LF	Lg	LB	.046	345.0	---	1.26	---	1.26	**1.88**
	Demo	LF	Sm	LB	.055	293.0	---	1.51	---	1.51	**2.25**
6" x 10" - 8'	Inst	LF	Lg	2C	.075	212.0	5.86	2.50	.22	8.58	**10.70**
	Inst	LF	Sm	2C	.089	180.0	6.47	2.96	.26	9.69	**12.10**
8" x 8" - 8'	Demo	LF	Lg	LB	.050	319.0	---	1.37	---	1.37	**2.04**
	Demo	LF	Sm	LB	.059	271.0	---	1.62	---	1.62	**2.41**
8" x 8" - 8'	Inst	LF	Lg	2C	.082	196.0	6.66	2.73	.24	9.63	**12.00**
	Inst	LF	Sm	2C	.096	167.0	7.42	3.19	.28	10.89	**13.60**
8" x 10" - 8'	Demo	LF	Lg	LB	.059	269.0	---	1.62	---	1.62	**2.41**
	Demo	LF	Sm	LB	.070	229.0	---	1.92	---	1.92	**2.86**
8" x 10" - 8'	Inst	LF	Lg	2C	.092	174.0	8.54	3.06	.27	11.87	**14.70**
	Inst	LF	Sm	2C	.108	148.0	9.50	3.59	.32	13.41	**16.60**
Fascia											
1" x 4" - 12'	Demo	LF	Lg	LB	.012	1315	---	.33	---	.33	**.49**
	Demo	LF	Sm	LB	.014	1118	---	.38	---	.38	**.57**
1" x 4" - 12'	Inst	LF	Lg	2C	.041	386.0	.22	1.36	.12	1.70	**2.42**
	Inst	LF	Sm	2C	.049	328.0	.24	1.63	.14	2.01	**2.86**
Furring strips on ceilings											
1" x 3" on wood	Demo	LF	Lg	LB	.026	625.0	---	.71	---	.71	**1.06**
	Demo	LF	Sm	LB	.030	531.0	---	.82	---	.82	**1.23**
1" x 3" on wood	Inst	LF	Lg	2C	.028	574.0	.16	.93	.08	1.17	**1.66**
	Inst	LF	Sm	2C	.033	488.0	.18	1.10	.10	1.38	**1.95**

Description	Oper	Unit	Vol	Crew Size	Man-hours per Unit	Crew Output per Day	Avg Mat'l Unit Cost	Avg Labor Unit Cost	Avg Equip Unit Cost	Avg Total Unit Cost	Avg Price Incl O&P

Furring strips on walls

Description	Oper	Unit	Vol	Crew Size	Man-hours per Unit	Crew Output per Day	Avg Mat'l Unit Cost	Avg Labor Unit Cost	Avg Equip Unit Cost	Avg Total Unit Cost	Avg Price Incl O&P
1" x 3" on wood	Demo	LF	Lg	LB	.020	805.0	---	.55	---	.55	.82
	Demo	LF	Sm	LB	.023	684.0	---	.63	---	.63	.94
1" x 3" on wood	Inst	LF	Lg	2C	.022	741.0	.16	.73	.06	.95	1.34
	Inst	LF	Sm	2C	.025	630.0	.18	.83	.07	1.08	1.52
1" x 3" on masonry	Demo	LF	Lg	LB	.026	625.0	---	.71	---	.71	1.06
	Demo	LF	Sm	LB	.030	531.0	---	.82	---	.82	1.23
1" x 3" on masonry	Inst	LF	Lg	2C	.028	574.0	.16	.93	.08	1.17	1.66
	Inst	LF	Sm	2C	.033	488.0	.18	1.10	.10	1.38	1.95
1" x 3" on concrete	Demo	LF	Lg	LB	.035	456.0	---	.96	---	.96	1.43
	Demo	LF	Sm	LB	.041	388.0	---	1.12	---	1.12	1.68
1" x 3" on concrete	Inst	LF	Lg	2C	.038	418.0	.16	1.26	.11	1.53	2.19
	Inst	LF	Sm	2C	.045	355.0	.18	1.50	.13	1.81	2.58

Headers or lintels, over openings

4 feet wide

Description	Oper	Unit	Vol	Crew Size	Man-hours per Unit	Crew Output per Day	Avg Mat'l Unit Cost	Avg Labor Unit Cost	Avg Equip Unit Cost	Avg Total Unit Cost	Avg Price Incl O&P
4" x 6"	Demo	LF	Lg	1L	.024	338.0	---	.66	---	.66	.98
	Demo	LF	Sm	1L	.028	287.0	---	.77	---	.77	1.14
4" x 6"	Inst	LF	Lg	CA	.040	198.0	1.74	1.33	.24	3.31	4.24
	Inst	LF	Sm	CA	.048	168.0	1.94	1.60	.28	3.82	4.91
4" x 8"	Demo	LF	Lg	1L	.026	302.0	---	.71	---	.71	1.06
	Demo	LF	Sm	1L	.031	257.0	---	.85	---	.85	1.27
4" x 8"	Inst	LF	Lg	CA	.043	186.0	2.34	1.43	.25	4.02	5.09
	Inst	LF	Sm	CA	.051	158.0	2.61	1.70	.30	4.61	5.85
4" x 12"	Demo	LF	Lg	1L	.032	250.0	---	.88	---	.88	1.31
	Demo	LF	Sm	1L	.038	213.0	---	1.04	---	1.04	1.55
4" x 12"	Inst	LF	Lg	CA	.049	164.0	3.50	1.63	.28	5.41	6.75
	Inst	LF	Sm	CA	.058	139.0	3.90	1.93	.34	6.17	7.72
4" x 14"	Demo	LF	Lg	1L	.034	236.0	---	.93	---	.93	1.39
	Demo	LF	Sm	1L	.040	201.0	---	1.10	---	1.10	1.63
4" x 14"	Inst	LF	Lg	CA	.051	158.0	4.28	1.70	.30	6.28	7.77
	Inst	LF	Sm	CA	.060	134.0	4.75	2.00	.35	7.10	8.81

6 feet wide

Description	Oper	Unit	Vol	Crew Size	Man-hours per Unit	Crew Output per Day	Avg Mat'l Unit Cost	Avg Labor Unit Cost	Avg Equip Unit Cost	Avg Total Unit Cost	Avg Price Incl O&P
4" x 12"	Demo	LF	Lg	LB	.027	583.0	---	.74	---	.74	1.10
	Demo	LF	Sm	LB	.032	496.0	---	.88	---	.88	1.31
4" x 12"	Inst	LF	Lg	2C	.041	392.0	3.50	1.36	.12	4.98	6.19
	Inst	LF	Sm	2C	.048	333.0	3.90	1.60	.14	5.64	7.02

8 feet wide

Description	Oper	Unit	Vol	Crew Size	Man-hours per Unit	Crew Output per Day	Avg Mat'l Unit Cost	Avg Labor Unit Cost	Avg Equip Unit Cost	Avg Total Unit Cost	Avg Price Incl O&P
4" x 12"	Demo	LF	Lg	LB	.024	655.0	---	.66	---	.66	.98
	Demo	LF	Sm	LB	.029	557.0	---	.80	---	.80	1.19
4" x 12"	Inst	LF	Lg	2C	.035	453.0	3.50	1.16	.10	4.76	5.87
	Inst	LF	Sm	2C	.042	385.0	3.90	1.40	.12	5.42	6.70

Description	Oper	Unit	Vol	Crew Size	Man-hours per Unit	Crew Output per Day	Avg Mat'l Unit Cost	Avg Labor Unit Cost	Avg Equip Unit Cost	Avg Total Unit Cost	Avg Price Incl O&P
10 feet wide											
4" x 12"	Demo	LF	Lg	LB	.022	743.0	---	.60	---	.60	.90
	Demo	LF	Sm	LB	.025	632.0	---	.69	---	.69	1.02
4" x 12"	Inst	LF	Lg	2C	.030	535.0	3.50	1.00	.09	4.59	5.61
	Inst	LF	Sm	2C	.035	455.0	3.90	1.16	.10	5.16	6.33
4" x 14"	Demo	LF	Lg	LB	.026	627.0	---	.71	---	.71	1.06
	Demo	LF	Sm	LB	.030	533.0	---	.82	---	.82	1.23
4" x 14"	Inst	LF	Lg	2C	.035	455.0	4.28	1.16	.10	5.54	6.77
	Inst	LF	Sm	2C	.041	387.0	4.75	1.36	.12	6.23	7.63
12 feet wide											
4" x 14"	Demo	LF	Lg	LB	.024	667.0	---	.66	---	.66	.98
	Demo	LF	Sm	LB	.028	567.0	---	.77	---	.77	1.14
4" x 14"	Inst	LF	Lg	2C	.033	491.0	4.28	1.10	.10	5.48	6.67
	Inst	LF	Sm	2C	.038	417.0	4.75	1.26	.11	6.12	7.47
4" x 16"	Demo	LF	Lg	LB	.028	562.0	---	.77	---	.77	1.14
	Demo	LF	Sm	LB	.033	478.0	---	.91	---	.91	1.35
4" x 16"	Inst	LF	Lg	2C	.037	429.0	5.12	1.23	.11	6.46	7.84
	Inst	LF	Sm	2C	.044	365.0	5.65	1.46	.13	7.24	8.82
14 feet wide											
4" x 16"	Demo	LF	Lg	LB	.027	588.0	---	.74	---	.74	1.10
	Demo	LF	Sm	LB	.032	500.0	---	.88	---	.88	1.31
4" x 16"	Inst	LF	Lg	2C	.035	453.0	5.12	1.16	.10	6.38	7.73
	Inst	LF	Sm	2C	.042	385.0	5.65	1.40	.12	7.17	8.71
16 feet wide											
4" x 16"	Demo	LF	Lg	LB	.026	611.0	---	.71	---	.71	1.06
	Demo	LF	Sm	LB	.031	519.0	---	.85	---	.85	1.27
4" x 16"	Inst	LF	Lg	2C	.034	475.0	5.12	1.13	.10	6.35	7.68
	Inst	LF	Sm	2C	.040	404.0	5.65	1.33	.12	7.10	8.61
18 feet wide											
4" x 16"	Demo	LF	Lg	LB	.025	632.0	---	.69	---	.69	1.02
	Demo	LF	Sm	LB	.030	537.0	---	.82	---	.82	1.23
4" x 16"	Inst	LF	Lg	2C	.032	495.0	5.12	1.06	.09	6.27	7.58
	Inst	LF	Sm	2C	.038	421.0	5.65	1.26	.11	7.02	8.50

Joists, ceiling/floor, per LF of stick

Description	Oper	Unit	Vol	Crew Size	Man-hours per Unit	Crew Output per Day	Avg Mat'l Unit Cost	Avg Labor Unit Cost	Avg Equip Unit Cost	Avg Total Unit Cost	Avg Price Incl O&P
2" x 4" -6'	Demo	LF	Lg	LB	.016	1024	---	.44	---	.44	.65
	Demo	LF	Sm	LB	.018	870.0	---	.49	---	.49	.74
2" x 4" -6'	Inst	LF	Lg	2C	.019	844.0	.39	.63	.06	1.08	1.46
	Inst	LF	Sm	2C	.022	717.0	.42	.73	.07	1.22	1.65
2" x 4" -8'	Demo	LF	Lg	LB	.013	1213	---	.36	---	.36	.53
	Demo	LF	Sm	LB	.016	1031	---	.44	---	.44	.65
2" x 4" -8'	Inst	LF	Lg	2C	.016	1004	.39	.53	.05	.97	1.30
	Inst	LF	Sm	2C	.019	853.0	.42	.63	.05	1.10	1.48

Description	Oper	Unit	Vol	Crew Size	Man-hours per Unit	Crew Output per Day	Avg Mat'l Unit Cost	Avg Labor Unit Cost	Avg Equip Unit Cost	Avg Total Unit Cost	Avg Price Incl O&P
2" x 4" -10'	Demo	LF	Lg	LB	.012	1367	---	.33	---	.33	.49
	Demo	LF	Sm	LB	.014	1162	---	.38	---	.38	.57
2" x 4" -10'	Inst	LF	Lg	2C	.014	1134	.39	.47	.04	.90	1.19
	Inst	LF	Sm	2C	.017	964.0	.42	.57	.05	1.04	1.38
2" x 4" -12'	Demo	LF	Lg	LB	.011	1498	---	.30	---	.30	.45
	Demo	LF	Sm	LB	.013	1273	---	.36	---	.36	.53
2" x 4" -12'	Inst	LF	Lg	2C	.013	1245	.39	.43	.04	.86	1.14
	Inst	LF	Sm	2C	.015	1058	.42	.50	.04	.96	1.27
2" x 6" - 8'	Demo	LF	Lg	LB	.015	1064	---	.41	---	.41	.61
	Demo	LF	Sm	LB	.018	904.0	---	.49	---	.49	.74
2" x 6" - 8'	Inst	LF	Lg	2C	.018	891.0	.54	.60	.05	1.19	1.57
	Inst	LF	Sm	2C	.021	757.0	.59	.70	.06	1.35	1.79
2" x 6" - 10'	Demo	LF	Lg	LB	.013	1195	---	.36	---	.36	.53
	Demo	LF	Sm	LB	.016	1016	---	.44	---	.44	.65
2" x 6" - 10'	Inst	LF	Lg	2C	.016	1005	.54	.53	.05	1.12	1.47
	Inst	LF	Sm	2C	.019	854.0	.59	.63	.05	1.27	1.68
2" x 6" - 12'	Demo	LF	Lg	LB	.012	1307	---	.33	---	.33	.49
	Demo	LF	Sm	LB	.014	1111	---	.38	---	.38	.57
2" x 6" - 12'	Inst	LF	Lg	2C	.015	1102	.54	.50	.04	1.08	1.41
	Inst	LF	Sm	2C	.017	937.0	.59	.57	.05	1.21	1.58
2" x 6" - 14'	Demo	LF	Lg	LB	.011	1405	---	.30	---	.30	.45
	Demo	LF	Sm	LB	.013	1194	---	.36	---	.36	.53
2" x 6" - 14'	Inst	LF	Lg	2C	.013	1187	.54	.43	.04	1.01	1.31
	Inst	LF	Sm	2C	.016	1009	.59	.53	.05	1.17	1.53
2" x 8" - 10'	Demo	LF	Lg	LB	.015	1060	---	.41	---	.41	.61
	Demo	LF	Sm	LB	.018	901.0	---	.49	---	.49	.74
2" x 8" - 10'	Inst	LF	Lg	2C	.018	901.0	.74	.60	.05	1.39	1.80
	Inst	LF	Sm	2C	.021	766.0	.83	.70	.06	1.59	2.06
2" x 8" - 12'	Demo	LF	Lg	LB	.014	1156	---	.38	---	.38	.57
	Demo	LF	Sm	LB	.016	983.0	---	.44	---	.44	.65
2" x 8" - 12'	Inst	LF	Lg	2C	.016	986.0	.74	.53	.05	1.32	1.70
	Inst	LF	Sm	2C	.019	838.0	.83	.63	.06	1.52	1.96
2" x 8" - 14'	Demo	LF	Lg	LB	.013	1239	---	.36	---	.36	.53
	Demo	LF	Sm	LB	.015	1053	---	.41	---	.41	.61
2" x 8" - 14'	Inst	LF	Lg	2C	.015	1059	.74	.50	.04	1.28	1.64
	Inst	LF	Sm	2C	.018	900.0	.83	.60	.05	1.48	1.90
2" x 8" - 16'	Demo	LF	Lg	LB	.012	1313	---	.33	---	.33	.49
	Demo	LF	Sm	LB	.014	1116	---	.38	---	.38	.57
2" x 8" - 16'	Inst	LF	Lg	2C	.014	1124	.74	.47	.04	1.25	1.59
	Inst	LF	Sm	2C	.017	955.0	.83	.57	.05	1.45	1.85

Description	Oper	Unit	Vol	Crew Size	Man-hours per Unit	Crew Output per Day	Avg Mat'l Unit Cost	Avg Labor Unit Cost	Avg Equip Unit Cost	Avg Total Unit Cost	Avg Price Incl O&P
2" x 10" - 12'	Demo	LF	Lg	LB	.015	1034	---	.41	---	.41	.61
	Demo	LF	Sm	LB	.018	879.0	---	.49	---	.49	.74
2" x 10" - 12'	Inst	LF	Lg	2C	.018	890.0	1.08	.60	.05	1.73	2.19
	Inst	LF	Sm	2C	.021	757.0	1.19	.70	.06	1.95	2.48
2" x 10" - 14'	Demo	LF	Lg	LB	.014	1106	---	.38	---	.38	.57
	Demo	LF	Sm	LB	.017	940.0	---	.47	---	.47	.69
2" x 10" - 14'	Inst	LF	Lg	2C	.017	955.0	1.08	.57	.05	1.70	2.14
	Inst	LF	Sm	2C	.020	812.0	1.19	.67	.06	1.92	2.43
2" x 10" - 16'	Demo	LF	Lg	LB	.014	1171	---	.38	---	.38	.57
	Demo	LF	Sm	LB	.016	995.0	---	.44	---	.44	.65
2" x 10" - 16'	Inst	LF	Lg	2C	.016	1013	1.08	.53	.05	1.66	2.09
	Inst	LF	Sm	2C	.019	861.0	1.19	.63	.05	1.87	2.37
2" x 10" - 18'	Demo	LF	Lg	LB	.013	1229	---	.36	---	.36	.53
	Demo	LF	Sm	LB	.015	1045	---	.41	---	.41	.61
2" x 10" - 18'	Inst	LF	Lg	2C	.015	1065	1.08	.50	.04	1.62	2.03
	Inst	LF	Sm	2C	.018	905.0	1.19	.60	.05	1.84	2.32
2" x 12" - 14'	Demo	LF	Lg	LB	.016	998.0	---	.44	---	.44	.65
	Demo	LF	Sm	LB	.019	848.0	---	.52	---	.52	.78
2" x 12" - 14'	Inst	LF	Lg	2C	.018	869.0	1.31	.60	.05	1.96	2.45
	Inst	LF	Sm	2C	.022	739.0	1.46	.73	.06	2.25	2.84
2" x 12" - 16'	Demo	LF	Lg	LB	.015	1054	---	.41	---	.41	.61
	Demo	LF	Sm	LB	.018	896.0	---	.49	---	.49	.74
2" x 12" - 16'	Inst	LF	Lg	2C	.017	920.0	1.31	.57	.05	1.93	2.40
	Inst	LF	Sm	2C	.020	782.0	1.46	.67	.06	2.19	2.74
2" x 12" - 18'	Demo	LF	Lg	LB	.014	1106	---	.38	---	.38	.57
	Demo	LF	Sm	LB	.017	940.0	---	.47	---	.47	.69
2" x 12" - 18'	Inst	LF	Lg	2C	.017	967.0	1.31	.57	.05	1.93	2.40
	Inst	LF	Sm	2C	.019	822.0	1.46	.63	.06	2.15	2.69
2" x 12" - 20'	Demo	LF	Lg	LB	.014	1151	---	.38	---	.38	.57
	Demo	LF	Sm	LB	.016	978.0	---	.44	---	.44	.65
2" x 12" - 20'	Inst	LF	Lg	2C	.016	1009	1.31	.53	.05	1.89	2.36
	Inst	LF	Sm	2C	.019	858.0	1.46	.63	.05	2.14	2.68
3" x 8" - 12'	Demo	LF	Lg	LB	.017	942.0	---	47	---	47	.69
	Demo	LF	Sm	LB	.020	801.0	---	.55	---	.55	.82
3" x 8" - 12'	Inst	LF	Lg	2C	.020	816.0	1.92	.67	.06	2.65	3.27
	Inst	LF	Sm	2C	.023	694.0	2.12	.77	.07	2.96	3.66
3" x 8" - 14'	Demo	LF	Lg	LB	.016	1007	---	.44	---	.44	.65
	Demo	LF	Sm	LB	.019	856.0	---	.52	---	.52	.78
3" x 8" - 14'	Inst	LF	Lg	2C	.018	876.0	1.92	.60	.05	2.57	3.16
	Inst	LF	Sm	2C	.021	745.0	2.12	.70	.06	2.88	3.55
3" x 8" - 16'	Demo	LF	Lg	LB	.015	1065	---	.41	---	.41	.61
	Demo	LF	Sm	LB	.018	905.0	---	.49	---	.49	.74
3" x 8" - 16'	Inst	LF	Lg	2C	.017	929.0	1.92	.57	.05	2.54	3.11
	Inst	LF	Sm	2C	.020	790.0	2.12	.67	.06	2.85	3.50

Description	Oper	Unit	Vol	Crew Size	Man-hours per Unit	Crew Output per Day	Avg Mat'l Unit Cost	Avg Labor Unit Cost	Avg Equip Unit Cost	Avg Total Unit Cost	Avg Price Incl O&P
3" x 8" - 18'	Demo	LF	Lg	LB	.014	1117	---	.38	---	.38	**.57**
	Demo	LF	Sm	LB	.017	949.0	---	.47	---	.47	**.69**
3" x 8" - 18'	Inst	LF	Lg	2C	.016	976.0	1.92	.53	.05	2.50	**3.06**
	Inst	LF	Sm	2C	.019	830.0	2.12	.63	.06	2.81	**3.45**
3" x 10" - 16'	Demo	LF	Lg	LB	.017	918.0	---	.47	---	.47	**.69**
	Demo	LF	Sm	LB	.021	780.0	---	.58	---	.58	**.86**
3" x 10" - 16'	Inst	LF	Lg	2C	.020	811.0	2.40	.67	.06	3.13	**3.82**
	Inst	LF	Sm	2C	.023	689.0	2.65	.77	.07	3.49	**4.27**
3" x 10" - 18'	Demo	LF	Lg	LB	.017	961.0	---	.47	---	.47	**.69**
	Demo	LF	Sm	LB	.020	817.0	---	.55	---	.55	**.82**
3" x 10" - 18'	Inst	LF	Lg	2C	.019	851.0	2.40	.63	.05	3.08	**3.76**
	Inst	LF	Sm	2C	.022	723.0	2.65	.73	.06	3.44	**4.21**
3" x 10" - 20'	Demo	LF	Lg	LB	.016	1001	---	.44	---	.44	**.65**
	Demo	LF	Sm	LB	.019	851.0	---	.52	---	.52	**.78**
3" x 10" - 20'	Inst	LF	Lg	2C	.018	888.0	2.40	.60	.05	3.05	**3.71**
	Inst	LF	Sm	2C	.021	755.0	2.65	.70	.06	3.41	**4.16**
3" x 12" - 16'	Demo	LF	Lg	LB	.019	821.0	---	.52	---	.52	**.78**
	Demo	LF	Sm	LB	.023	698.0	---	.63	---	.63	**.94**
3" x 12" - 16'	Inst	LF	Lg	2C	.022	732.0	2.40	.73	.06	3.19	**3.92**
	Inst	LF	Sm	2C	.026	622.0	2.65	.87	.08	3.60	**4.43**
3" x 12" - 18'	Demo	LF	Lg	LB	.018	873.0	---	.49	---	.49	**.74**
	Demo	LF	Sm	LB	.022	742.0	---	.60	---	.60	**.90**
3" x 12" - 18'	Inst	LF	Lg	2C	.020	781.0	2.88	.67	.06	3.61	**4.37**
	Inst	LF	Sm	2C	.024	664.0	3.18	.80	.07	4.05	**4.92**
3" x 12" - 20'	Demo	LF	Lg	LB	.018	898.0	---	.49	---	.49	**.74**
	Demo	LF	Sm	LB	.021	763.0	---	.58	---	.58	**.86**
3" x 12" - 20'	Inst	LF	Lg	2C	.020	805.0	2.88	.67	.06	3.61	**4.37**
	Inst	LF	Sm	2C	.023	684.0	3.18	.77	.07	4.02	**4.87**
3" x 12" - 22'	Demo	LF	Lg	LB	.017	925.0	---	.47	---	.47	**.69**
	Demo	LF	Sm	LB	.020	786.0	---	.55	---	.55	**.82**
3" x 12" - 22'	Inst	LF	Lg	2C	.019	830.0	2.88	.63	.06	3.57	**4.32**
	Inst	LF	Sm	2C	.023	706.0	3.18	.77	.07	4.02	**4.87**

Joists, ceiling/floor, per SF of area

Description	Oper	Unit	Vol	Crew Size	Man-hours per Unit	Crew Output per Day	Avg Mat'l Unit Cost	Avg Labor Unit Cost	Avg Equip Unit Cost	Avg Total Unit Cost	Avg Price Incl O&P
2" x 4" - 6', 12" oc	Demo	SF	Lg	LB	.016	998.0	---	.44	---	.44	**.65**
	Demo	SF	Sm	LB	.019	848.0	---	.52	---	.52	**.78**
2" x 4" - 6', 12" oc	Inst	SF	Lg	2C	.019	823.0	.40	.63	.06	1.09	**1.47**
	Inst	SF	Sm	2C	.023	700.0	.44	.77	.07	1.28	**1.72**
2" x 4" - 6', 16" oc	Demo	SF	Lg	LB	.013	1188	---	.36	---	.36	**.53**
	Demo	SF	Sm	LB	.016	1010	---	.44	---	.44	**.65**
2" x 4" - 6', 16" oc	Inst	SF	Lg	2C	.016	979.0	.30	.53	.05	.88	**1.19**
	Inst	SF	Sm	2C	.019	832.0	.33	.63	.06	1.02	**1.39**

Description	Oper	Unit	Vol	Crew Size	Man-hours per Unit	Crew Output per Day	Avg Mat'l Unit Cost	Avg Labor Unit Cost	Avg Equip Unit Cost	Avg Total Unit Cost	Avg Price Incl O&P
2" x 4" - 6', 24" oc	Demo	SF	Lg	LB	.010	1559	---	.27	---	.27	.41
	Demo	SF	Sm	LB	.012	1325	---	.33	---	.33	.49
2" x 4" - 6', 24" oc	Inst	SF	Lg	2C	.012	1285	.21	.40	.04	.65	.88
	Inst	SF	Sm	2C	.015	1092	.23	.50	.04	.77	1.05
2" x 6" - 8', 12" oc	Demo	SF	Lg	LB	.015	1037	---	.41	---	.41	.61
	Demo	SF	Sm	LB	.018	881.0	---	.49	---	.49	.74
2" x 6" - 8', 12" oc	Inst	SF	Lg	2C	.018	869.0	.55	.60	.05	1.20	1.58
	Inst	SF	Sm	2C	.022	739.0	.61	.73	.06	1.40	1.86
2" x 6" - 8', 16" oc	Demo	SF	Lg	LB	.013	1235	---	.36	---	.36	.53
	Demo	SF	Sm	LB	.015	1050	---	.41	---	.41	.61
2" x 6" - 8', 16" oc	Inst	SF	Lg	2C	.015	1034	.42	.50	.05	.97	1.28
	Inst	SF	Sm	2C	.018	879.0	.47	.60	.05	1.12	1.49
2" x 6" - 8', 24" oc	Demo	SF	Lg	LB	.010	1622	---	.27	---	.27	.41
	Demo	SF	Sm	LB	.012	1379	---	.33	---	.33	.49
2" x 6" - 8', 24" oc	Inst	SF	Lg	2C	.012	1359	.29	.40	.03	.72	.96
	Inst	SF	Sm	2C	.014	1155	.32	.47	.04	.83	1.11
2" x 8" - 10', 12" oc	Demo	SF	Lg	LB	.015	1033	---	.41	---	.41	.61
	Demo	SF	Sm	LB	.018	878.0	---	.49	---	.49	.74
2" x 8" - 10', 12" oc	Inst	SF	Lg	2C	.018	878.0	.76	.60	.05	1.41	1.82
	Inst	SF	Sm	2C	.021	746.0	.86	.70	.06	1.62	2.10
2" x 8" - 10', 16" oc	Demo	SF	Lg	LB	.013	1231	---	.36	---	.36	.53
	Demo	SF	Sm	LB	.015	1046	---	.41	---	.41	.61
2" x 8" - 10', 16" oc	Inst	SF	Lg	2C	.015	1046	.58	.50	.04	1.12	1.46
	Inst	SF	Sm	2C	.018	889.0	.65	.60	.05	1.30	1.70
2" x 8" - 10', 24" oc	Demo	SF	Lg	LB	.010	1614	---	.27	---	.27	.41
	Demo	SF	Sm	LB	.012	1372	---	.33	---	.33	.49
2" x 8" - 10', 24" oc	Inst	SF	Lg	2C	.012	1372	.40	.40	.03	.83	1.09
	Inst	SF	Sm	2C	.014	1166	.44	.47	.04	.95	1.24
2" x 10" - 12', 12" oc	Demo	SF	Lg	LB	.016	1008	---	.44	---	.44	.65
	Demo	SF	Sm	LB	.019	857.0	---	.52	---	.52	.78
2" x 10" - 12', 12" oc	Inst	SF	Lg	2C	.018	868.0	1.10	.60	.05	1.75	2.21
	Inst	SF	Sm	2C	.022	738.0	1.22	.73	.06	2.01	2.56
2" x 10" - 12', 16" oc	Demo	SF	Lg	LB	.013	1200	---	.36	---	.36	.53
	Demo	SF	Sm	LB	.016	1020	---	.44	---	.44	.65
2" x 10" - 12', 16" oc	Inst	SF	Lg	2C	.015	1033	.83	.50	.05	1.38	1.75
	Inst	SF	Sm	2C	.018	878.0	.92	.60	.05	1.57	2.01

Description	Oper	Unit	Vol	Crew Size	Man-hours per Unit	Crew Output per Day	Avg Mat'l Unit Cost	Avg Labor Unit Cost	Avg Equip Unit Cost	Avg Total Unit Cost	Avg Price Incl O&P
2" x 10" - 12', 24" oc	Demo	SF	Lg	LB	.010	1575	---	.27	---	.27	**.41**
	Demo	SF	Sm	LB	.012	1339	---	.33	---	.33	**.49**
2" x 10" - 12', 24" oc	Inst	SF	Lg	2C	.012	1356	.57	.40	.03	1.00	**1.28**
	Inst	SF	Sm	2C	.014	1153	.63	.47	.04	1.14	**1.46**
2" x 12" - 14', 12" oc	Demo	SF	Lg	LB	.016	974.0	---	.44	---	.44	**.65**
	Demo	SF	Sm	LB	.019	828.0	---	.52	---	.52	**.78**
2" x 12" - 14', 12" oc	Inst	SF	Lg	2C	.019	848.0	1.34	.63	.06	2.03	**2.55**
	Inst	SF	Sm	2C	.022	721.0	1.50	.73	.06	2.29	**2.88**
2" x 12" - 14', 16" oc	Demo	SF	Lg	LB	.014	1159	---	.38	---	.38	**.57**
	Demo	SF	Sm	LB	.016	985.0	---	.44	---	.44	**.65**
2" x 12" - 14', 16" oc	Inst	SF	Lg	2C	.016	1009	1.02	.53	.05	1.60	**2.02**
	Inst	SF	Sm	2C	.019	858.0	1.14	.63	.05	1.82	**2.31**
2" x 12" - 14', 24" oc	Demo	SF	Lg	LB	.011	1521	---	.30	---	.30	**.45**
	Demo	SF	Sm	LB	.012	1293	---	.33	---	.33	**.49**
2" x 12" - 14', 24" oc	Inst	SF	Lg	2C	.012	1324	.69	.40	.04	1.13	**1.43**
	Inst	SF	Sm	2C	.014	1125	.77	.47	.04	1.28	**1.62**
3" x 8" - 14', 12" oc	Demo	SF	Lg	LB	.016	983.0	---	.44	---	.44	**.65**
	Demo	SF	Sm	LB	.019	836.0	---	.52	---	.52	**.78**
3" x 8" - 14', 12" oc	Inst	SF	Lg	2C	.019	855.0	1.97	.63	.05	2.65	**3.26**
	Inst	SF	Sm	2C	.022	727.0	2.18	.73	.06	2.97	**3.67**
3" x 8" - 14', 16" oc	Demo	SF	Lg	LB	.014	1170	---	.38	---	.38	**.57**
	Demo	SF	Sm	LB	.016	995.0	---	.44	---	.44	**.65**
3" x 8" - 14', 16" oc	Inst	SF	Lg	2C	.016	1017	1.49	.53	.05	2.07	**2.56**
	Inst	SF	Sm	2C	.019	864.0	1.65	.63	.05	2.33	**2.90**
3" x 8" - 14', 24" oc	Demo	SF	Lg	LB	.010	1535	---	.27	---	.27	**.41**
	Demo	SF	Sm	LB	.012	1305	---	.33	---	.33	**.49**
3" x 8" - 14', 24" oc	Inst	SF	Lg	2C	.012	1335	1.02	.40	.03	1.45	**1.80**
	Inst	SF	Sm	2C	.014	1135	1.12	.47	.04	1.63	**2.03**
3" x 10" - 16', 12" oc	Demo	SF	Lg	LB	.018	896.0	---	.49	---	.49	**.74**
	Demo	SF	Sm	LB	.021	762.0	---	.58	---	.58	**.86**
3" x 10" - 16', 12" oc	Inst	SF	Lg	2C	.020	792.0	2.46	.67	.06	3.19	**3.89**
	Inst	SF	Sm	2C	.024	673.0	2.72	.80	.07	3.59	**4.40**
3" x 10" - 16', 16" oc	Demo	SF	Lg	LB	.015	1066	---	.41	---	.41	**.61**
	Demo	SF	Sm	LB	.018	906.0	---	.49	---	.49	**.74**
3" x 10" - 16', 16" oc	Inst	SF	Lg	2C	.017	943.0	1.87	.57	.05	2.49	**3.05**
	Inst	SF	Sm	2C	.020	802.0	2.06	.67	.06	2.79	**3.43**
3" x 10" - 16', 24" oc	Demo	SF	Lg	LB	.011	1399	---	.30	---	.30	**.45**
	Demo	SF	Sm	LB	.013	1189	---	.36	---	.36	**.53**
3" x 10" - 16', 24" oc	Inst	SF	Lg	2C	.013	1236	1.26	.43	.04	1.73	**2.14**
	Inst	SF	Sm	2C	.015	1051	1.40	.50	.04	1.94	**2.40**

Description	Oper	Unit	Vol	Crew Size	Man-hours per Unit	Crew Output per Day	Avg Mat'l Unit Cost	Avg Labor Unit Cost	Avg Equip Unit Cost	Avg Total Unit Cost	Avg Price Incl O&P
3" x 12" - 18', 12" oc	Demo	SF	Lg	LB	.019	851.0	---	.52	---	.52	.78
	Demo	SF	Sm	LB	.022	723.0	---	.60	---	.60	.90
3" x 12" - 18', 12" oc	Inst	SF	Lg	2C	.021	762.0	2.96	.70	.06	3.72	4.51
	Inst	SF	Sm	2C	.025	648.0	3.26	.83	.07	4.16	5.07
3" x 12" - 18', 16" oc	Demo	SF	Lg	LB	.016	1013	---	.44	---	.44	.65
	Demo	SF	Sm	LB	.019	861.0	---	.52	---	.52	.78
3" x 12" - 18', 16" oc	Inst	SF	Lg	2C	.018	907.0	2.24	.60	.05	2.89	3.52
	Inst	SF	Sm	2C	.021	771.0	2.47	.70	.06	3.23	3.95
3" x 12" - 18', 24" oc	Demo	SF	Lg	LB	.012	1330	---	.33	---	.33	.49
	Demo	SF	Sm	LB	.014	1131	---	.38	---	.38	.57
3" x 12" - 18', 24" oc	Inst	SF	Lg	2C	.013	1190	1.52	.43	.04	1.99	2.44
	Inst	SF	Sm	2C	.016	1012	1.68	.53	.05	2.26	2.78

Ledgers

Nailed

Description	Oper	Unit	Vol	Crew Size	Man-hours per Unit	Crew Output per Day	Avg Mat'l Unit Cost	Avg Labor Unit Cost	Avg Equip Unit Cost	Avg Total Unit Cost	Avg Price Incl O&P
2" x 4" - 12'	Demo	LF	Lg	LB	.013	1220	---	.36	---	.36	.53
	Demo	LF	Sm	LB	.015	1037	---	.41	---	.41	.61
2" x 4" - 12'	Inst	LF	Lg	2C	.019	857.0	.44	.63	.05	1.12	1.50
	Inst	LF	Sm	2C	.022	728.0	.48	.73	.06	1.27	1.71
2" x 6" - 12'	Demo	LF	Lg	LB	.015	1082	---	.41	---	.41	.61
	Demo	LF	Sm	LB	.017	920.0	---	.47	---	.47	.69
2" x 6" - 12'	Inst	LF	Lg	2C	.021	774.0	.59	.70	.06	1.35	1.79
	Inst	LF	Sm	2C	.024	658.0	.65	.80	.07	1.52	2.02
2" x 8" - 12'	Demo	LF	Lg	LB	.016	973.0	---	.44	---	.44	.65
	Demo	LF	Sm	LB	.019	827.0	---	.52	---	.52	.78
2" x 8" - 12'	Inst	LF	Lg	2C	.023	706.0	.80	.77	.07	1.64	2.14
	Inst	LF	Sm	2C	.027	600.0	.89	.90	.08	1.87	2.45

Bolted

Labor/material costs are for ledgers with pre-embedded bolts. Labor costs include securing ledgers.

Description	Oper	Unit	Vol	Crew Size	Man-hours per Unit	Crew Output per Day	Avg Mat'l Unit Cost	Avg Labor Unit Cost	Avg Equip Unit Cost	Avg Total Unit Cost	Avg Price Incl O&P
3" x 6" - 12'	Demo	LF	Lg	LB	.022	718.0	---	.60	---	.60	.90
	Demo	LF	Sm	LB	.026	610.0	---	.71	---	.71	1.06
3" x 6" - 12'	Inst	LF	Lg	2C	.031	513.0	1.40	1.03	.09	2.52	3.25
	Inst	LF	Sm	2C	.037	436.0	1.55	1.23	.11	2.89	3.74
3" x 8" - 12'	Demo	LF	Lg	LB	.024	654.0	---	.66	---	.66	.98
	Demo	LF	Sm	LB	.029	556.0	---	.80	---	.80	1.19
3" x 8" - 12'	Inst	LF	Lg	2C	.034	475.0	1.91	1.13	.10	3.14	3.99
	Inst	LF	Sm	2C	.040	404.0	2.11	1.33	.12	3.56	4.54
3" x 10" - 12'	Demo	LF	Lg	LB	.027	593.0	---	.74	---	.74	1.10
	Demo	LF	Sm	LB	.032	504.0	---	.88	---	.88	1.31
3" x 10" - 12'	Inst	LF	Lg	2C	.037	438.0	2.39	1.23	.11	3.73	4.71
	Inst	LF	Sm	2C	.043	372.0	2.64	1.43	.13	4.20	5.31
3" x 12" - 12'	Demo	LF	Lg	LB	.029	547.0	---	.80	---	.80	1.19
	Demo	LF	Sm	LB	.034	465.0	---	.93	---	.93	1.39
3" x 12" - 12'	Inst	LF	Lg	2C	.039	408.0	2.87	1.30	.11	4.28	5.36
	Inst	LF	Sm	2C	.046	347.0	3.17	1.53	.13	4.83	6.07

Description	Oper	Unit	Vol	Crew Size	Man-hours per Unit	Crew Output per Day	Avg Mat'l Unit Cost	Avg Labor Unit Cost	Avg Equip Unit Cost	Avg Total Unit Cost	Avg Price Incl O&P

Patio framing

Wood deck: 4" x 4" rough sawn beams (4'-0" oc) leveled 1/16" to 1/8" and nailed to pre-set concrete piers with woodblock on top; 2" T x 4" W decking (S4S) with 1/2" spacing, double-nailed beam junctures

Description	Oper	Unit	Vol	Crew Size	Man-hours per Unit	Crew Output per Day	Avg Mat'l Unit Cost	Avg Labor Unit Cost	Avg Equip Unit Cost	Avg Total Unit Cost	Avg Price Incl O&P
Fir	Demo	SF	Lg	LB	.042	380.0	---	1.15	---	1.15	**1.72**
	Demo	SF	Sm	LB	.050	323.0	---	1.37	---	1.37	**2.04**
Fir	Inst	SF	Lg	CN	.061	330.0	4.02	1.96	.14	6.12	**7.70**
	Inst	SF	Sm	CN	.071	281.0	2.68	2.28	.17	5.13	**6.67**
Redwood	Demo	SF	Lg	LB	.042	380.0	---	1.15	---	1.15	**1.72**
	Demo	SF	Sm	LB	.050	323.0	---	1.37	---	1.37	**2.04**
Redwood	Inst	SF	Lg	CN	.061	330.0	4.30	1.96	.14	6.40	**8.02**
	Inst	SF	Sm	CN	.071	281.0	2.68	2.28	.17	5.13	**6.67**

Wood awning: 4" x 4" columns (10'-0" oc) nailed to wood; 2" x 6" beams nailed horizontally either side of columns; 2" x 6" ledger nailed to wall studs; 2" x 6" joists (4'-0" oc) nailed to ledger and toe-nailed on top of beams; 2" x 2" (4" oc) nailed to joists for sunscreen

Description	Oper	Unit	Vol	Crew Size	Man-hours per Unit	Crew Output per Day	Avg Mat'l Unit Cost	Avg Labor Unit Cost	Avg Equip Unit Cost	Avg Total Unit Cost	Avg Price Incl O&P
Redwood, rough sawn	Demo	SF	Lg	LB	.055	290.0	---	1.51	---	1.51	**2.25**
	Demo	SF	Sm	LB	.065	247.0	---	1.78	---	1.78	**2.66**
Redwood, rough sawn	Inst	SF	Lg	CS	.096	250.0	1.24	3.01	.19	4.44	**6.13**
	Inst	SF	Sm	CS	.113	213.0	.10	3.54	.22	3.86	**5.64**

Plates; joined with studs before setting

Double top, nailed

Description	Oper	Unit	Vol	Crew Size	Man-hours per Unit	Crew Output per Day	Avg Mat'l Unit Cost	Avg Labor Unit Cost	Avg Equip Unit Cost	Avg Total Unit Cost	Avg Price Incl O&P
2" x 4" - 8'	Demo	LF	Lg	LB	.020	820.0	---	.55	---	.55	**.82**
	Demo	LF	Sm	LB	.023	697.0	---	.63	---	.63	**.94**
2" x 4" - 8'	Inst	LF	Lg	2C	.039	410.0	.72	1.30	.11	2.13	**2.88**
	Inst	LF	Sm	2C	.046	349.0	.79	1.53	.13	2.45	**3.33**
2" x 6" - 8'	Demo	LF	Lg	LB	.020	820.0	---	.55	---	.55	**.82**
	Demo	LF	Sm	LB	.023	697.0	---	.63	---	.63	**.94**
2" x 6" - 8'	Inst	LF	Lg	2C	.039	410.0	1.10	1.30	.11	2.51	**3.32**
	Inst	LF	Sm	2C	.046	349.0	1.22	1.53	.13	2.88	**3.83**

Single bottom, nailed

Description	Oper	Unit	Vol	Crew Size	Man-hours per Unit	Crew Output per Day	Avg Mat'l Unit Cost	Avg Labor Unit Cost	Avg Equip Unit Cost	Avg Total Unit Cost	Avg Price Incl O&P
2" x 4" - 8'	Demo	LF	Lg	LB	.012	1360	---	.33	---	.33	**.49**
	Demo	LF	Sm	LB	.014	1156	---	.38	---	.38	**.57**
2" x 4" - 8'	Inst	LF	Lg	2C	.020	815.0	.40	.67	.06	1.13	**1.52**
	Inst	LF	Sm	2C	.023	693.0	.44	.77	.07	1.28	**1.72**
2" x 6" - 8'	Demo	LF	Lg	LB	.012	1360	---	.33	---	.33	**.49**
	Demo	LF	Sm	LB	.014	1156	---	.38	---	.38	**.57**
2" x 6" - 8'	Inst	LF	Lg	2C	.020	815.0	.55	.67	.06	1.28	**1.69**
	Inst	LF	Sm	2C	.023	693.0	.60	.77	.07	1.44	**1.91**

Sill or bottom, bolted. Labor/material costs are for plates with pre-embedded bolts.
Labor cost includes securing plates

Description	Oper	Unit	Vol	Crew Size	Man-hours per Unit	Crew Output per Day	Avg Mat'l Unit Cost	Avg Labor Unit Cost	Avg Equip Unit Cost	Avg Total Unit Cost	Avg Price Incl O&P
2" x 4" - 8'	Demo	LF	Lg	LB	.023	685.0	---	.63	---	.63	**.94**
	Demo	LF	Sm	LB	.027	582.0	---	.74	---	.74	**1.10**
2" x 4" - 8'	Inst	LF	Lg	2C	.029	560.0	.36	.96	.08	1.40	**1.94**
	Inst	LF	Sm	2C	.034	476.0	.39	1.13	.10	1.62	**2.25**
2" x 6" - 8'	Demo	LF	Lg	LB	.023	685.0	---	.63	---	.63	**.94**
	Demo	LF	Sm	LB	.027	582.0	---	.74	---	.74	**1.10**
2" x 6" - 8'	Inst	LF	Lg	2C	.029	560.0	.50	.96	.08	1.54	**2.10**
	Inst	LF	Sm	2C	.034	476.0	.56	1.13	.10	1.79	**2.44**

Description	Oper	Unit	Vol	Crew Size	Man-hours per Unit	Crew Output per Day	Avg Mat'l Unit Cost	Avg Labor Unit Cost	Avg Equip Unit Cost	Avg Total Unit Cost	Avg Price Incl O&P

Rafters, per LF of stick

Common, gable or hip, to 1/3 pitch

Description	Oper	Unit	Vol	Crew Size	Man-hours per Unit	Crew Output per Day	Avg Mat'l Unit Cost	Avg Labor Unit Cost	Avg Equip Unit Cost	Avg Total Unit Cost	Avg Price Incl O&P
2" x 4" - 8' Avg.	Demo	LF	Lg	LB	.022	744.0	---	.60	---	.60	.90
	Demo	LF	Sm	LB	.025	632.0	---	.69	---	.69	1.02
2" x 4" - 8' Avg.	Inst	LF	Lg	2C	.023	683.0	.39	.77	.07	1.23	1.67
	Inst	LF	Sm	2C	.028	581.0	.42	.93	.08	1.43	1.96
2" x 4" - 10' Avg.	Demo	LF	Lg	LB	.019	861.0	---	.52	---	.52	.78
	Demo	LF	Sm	LB	.022	732.0	---	.60	---	.60	.90
2" x 4" - 10' Avg.	Inst	LF	Lg	2C	.020	793.0	.39	.67	.06	1.12	1.51
	Inst	LF	Sm	2C	.024	674.0	.42	.80	.07	1.29	1.75
2" x 4" - 12' Avg.	Demo	LF	Lg	LB	.017	903.0	---	.47	---	.47	.69
	Demo	LF	Sm	LB	.020	819.0	---	.55	---	.55	.82
2" x 4" - 12' Avg.	Inst	LF	Lg	2C	.018	887.0	.39	.60	.05	1.04	1.40
	Inst	LF	Sm	2C	.021	754.0	.42	.70	.06	1.18	1.59
2" x 4" - 14' Avg.	Demo	LF	Lg	LB	.015	1053	---	.41	---	.41	.61
	Demo	LF	Sm	LB	.018	895.0	---	.49	---	.49	.74
2" x 4" - 14' Avg.	Inst	LF	Lg	2C	.016	971.0	.39	.53	.05	.97	1.30
	Inst	LF	Sm	2C	.019	825.0	.42	.63	.06	1.11	1.49
2" x 4" - 16' Avg.	Demo	LF	Lg	LB	.014	1134	---	.38	---	.38	.57
	Demo	LF	Sm	LB	.017	964.0	---	.47	---	.47	.69
2" x 4" - 16' Avg.	Inst	LF	Lg	2C	.015	1047	.39	.50	.04	.93	1.24
	Inst	LF	Sm	2C	.018	890.0	.42	.60	.05	1.07	1.43
2" x 6" - 10' Avg.	Demo	LF	Lg	LB	.021	748.0	---	.58	---	.58	.86
	Demo	LF	Sm	LB	.025	636.0	---	.69	---	.69	1.02
2" x 6" - 10' Avg.	Inst	LF	Lg	2C	.023	691.0	.54	.77	.07	1.38	1.84
	Inst	LF	Sm	2C	.027	587.0	.59	.90	.08	1.57	2.11
2" x 6" - 12' Avg.	Demo	LF	Lg	LB	.019	843.0	---	.52	---	.52	.78
	Demo	LF	Sm	LB	.022	717.0	---	.60	---	.60	.90
2" x 6" - 12' Avg.	Inst	LF	Lg	2C	.021	779.0	.54	.70	.06	1.30	1.73
	Inst	LF	Sm	2C	.024	662.0	.59	.80	.07	1.46	1.95
2" x 6" - 14' Avg.	Demo	LF	Lg	LB	.017	929.0	---	.47	---	.47	.69
	Demo	LF	Sm	LB	.020	790.0	---	.55	---	.55	.82
2" x 6" - 14' Avg.	Inst	LF	Lg	2C	.019	860.0	.54	.63	.05	1.22	1.62
	Inst	LF	Sm	2C	.022	731.0	.59	.73	.06	1.38	1.84
2" x 6" - 16' Avg.	Demo	LF	Lg	LB	.016	1008	---	.44	---	.44	.65
	Demo	LF	Sm	LB	.019	857.0	---	.52	---	.52	.78
2" x 6" - 16' Avg.	Inst	LF	Lg	2C	.017	933.0	.54	.57	.05	1.16	1.52
	Inst	LF	Sm	2C	.020	793.0	.59	.67	.06	1.32	1.74
2" x 6" - 18' Avg.	Demo	LF	Lg	LB	.015	1081	---	.41	---	.41	.61
	Demo	LF	Sm	LB	.017	919.0	---	.47	---	.47	.69
2" x 6" - 18' Avg.	Inst	LF	Lg	2C	.016	1001	.54	.53	.05	1.12	1.47
	Inst	LF	Sm	2C	.019	851.0	.59	.63	.05	1.27	1.68

Description	Oper	Unit	Vol	Crew Size	Man-hours per Unit	Crew Output per Day	Avg Mat'l Unit Cost	Avg Labor Unit Cost	Avg Equip Unit Cost	Avg Total Unit Cost	Avg Price Incl O&P
2" x 8" - 12' Avg.	Demo	LF	Lg	LB	.022	737.0	---	.60	---	.60	**.90**
	Demo	LF	Sm	LB	.026	626.0	---	.71	---	.71	**1.06**
2" x 8" - 12' Avg.	Inst	LF	Lg	2C	.023	684.0	.74	.77	.07	1.58	**2.07**
	Inst	LF	Sm	2C	.028	581.0	.83	.93	.08	1.84	**2.43**
2" x 8" - 14' Avg.	Demo	LF	Lg	LB	.020	804.0	---	.55	---	.55	**.82**
	Demo	LF	Sm	LB	.023	683.0	---	.63	---	.63	**.94**
2" x 8" - 14' Avg.	Inst	LF	Lg	2C	.021	748.0	.74	.70	.06	1.50	**1.96**
	Inst	LF	Sm	2C	.025	636.0	.83	.83	.07	1.73	**2.27**
2" x 8" - 16' Avg.	Demo	LF	Lg	LB	.019	864.0	---	.52	---	.52	**.78**
	Demo	LF	Sm	LB	.022	734.0	---	.60	---	.60	**.90**
2" x 8" - 16' Avg.	Inst	LF	Lg	2C	.020	805.0	.74	.67	.06	1.47	**1.91**
	Inst	LF	Sm	2C	.023	684.0	.83	.77	.07	1.67	**2.17**
2" x 8" - 18' Avg.	Demo	LF	Lg	LB	.017	918.0	---	.47	---	.47	**.69**
	Demo	LF	Sm	LB	.021	780.0	---	.58	---	.58	**.86**
2" x 8" - 18' Avg.	Inst	LF	Lg	2C	.019	856.0	.74	.63	.05	1.42	**1.85**
	Inst	LF	Sm	2C	.022	728.0	.83	.73	.06	1.62	**2.11**
2" x 8" - 20' Avg.	Demo	LF	Lg	LB	.017	967.0	---	.47	---	.47	**.69**
	Demo	LF	Sm	LB	.019	822.0	---	.52	---	.52	**.78**
2" x 8" - 20' Avg.	Inst	LF	Lg	2C	.018	902.0	.74	.60	.05	1.39	**1.80**
	Inst	LF	Sm	2C	.021	767.0	.83	.70	.06	1.59	**2.06**

Common, gable or hip, to 3/8-1/2 pitch

Description	Oper	Unit	Vol	Crew Size	Man-hours per Unit	Crew Output per Day	Avg Mat'l Unit Cost	Avg Labor Unit Cost	Avg Equip Unit Cost	Avg Total Unit Cost	Avg Price Incl O&P
2" x 4" - 8' Avg.	Demo	LF	Lg	LB	.025	632.0	---	.69	---	.69	**1.02**
	Demo	LF	Sm	LB	.030	537.0	---	.82	---	.82	**1.23**
2" x 4" - 8' Avg.	Inst	LF	Lg	2C	.028	580.0	.39	.93	.08	1.40	**1.93**
	Inst	LF	Sm	2C	.032	493.0	.42	1.06	.09	1.57	**2.17**
2" x 4" - 10' Avg.	Demo	LF	Lg	LB	.022	740.0	---	.60	---	.60	**.90**
	Demo	LF	Sm	LB	.025	629.0	---	.69	---	.69	**1.02**
2" x 4" - 10' Avg.	Inst	LF	Lg	2C	.024	680.0	.39	.80	.07	1.26	**1.72**
	Inst	LF	Sm	2C	.028	578.0	.42	.93	.08	1.43	**1.96**
2" x 4" - 12' Avg.	Demo	LF	Lg	LB	.019	836.0	---	.52	---	.52	**.78**
	Demo	LF	Sm	LB	.023	711.0	---	.63	---	.63	**.94**
2" x 4" - 12' Avg.	Inst	LF	Lg	2C	.021	769.0	.39	.70	.06	1.15	**1.56**
	Inst	LF	Sm	2C	.024	654.0	.42	.80	.07	1.29	**1.75**
2" x 4" - 14' Avg.	Demo	LF	Lg	LB	.017	922.0	---	.47	---	.47	**.69**
	Demo	LF	Sm	LB	.020	784.0	---	.55	---	.55	**.82**
2" x 4" - 14' Avg.	Inst	LF	Lg	2C	.019	849.0	.39	.63	.05	1.07	**1.45**
	Inst	LF	Sm	2C	.022	722.0	.42	.73	.06	1.21	**1.64**
2" x 4" - 16' Avg.	Demo	LF	Lg	LB	.016	999.0	---	.44	---	.44	**.65**
	Demo	LF	Sm	LB	.019	849.0	---	.52	---	.52	**.78**
2" x 4" - 16' Avg.	Inst	LF	Lg	2C	.017	921.0	.39	.57	.05	1.01	**1.35**
	Inst	LF	Sm	2C	.020	783.0	.42	.67	.06	1.15	**1.54**

Description	Oper	Unit	Vol	Crew Size	Man-hours per Unit	Crew Output per Day	Avg Mat'l Unit Cost	Avg Labor Unit Cost	Avg Equip Unit Cost	Avg Total Unit Cost	Avg Price Incl O&P
2" x 6" - 10' Avg.	Demo	LF	Lg	LB	.024	655.0	---	.66	---	.66	.98
	Demo	LF	Sm	LB	.029	557.0	---	.80	---	.80	1.19
2" x 6" - 10' Avg.	Inst	LF	Lg	2C	.026	604.0	.54	.87	.08	1.49	2.00
	Inst	LF	Sm	2C	.031	513.0	.59	1.03	.09	1.71	2.32
2" x 6" - 12' Avg.	Demo	LF	Lg	LB	.022	743.0	---	.60	---	.60	.90
	Demo	LF	Sm	LB	.025	632.0	---	.69	---	.69	1.02
2" x 6" - 12' Avg.	Inst	LF	Lg	2C	.023	686.0	.54	.77	.07	1.38	1.84
	Inst	LF	Sm	2C	.027	583.0	.59	.90	.08	1.57	2.11
2" x 6" - 14' Avg.	Demo	LF	Lg	LB	.019	825.0	---	.52	---	.52	.78
	Demo	LF	Sm	LB	.023	701.0	---	.63	---	.63	.94
2" x 6" - 14' Avg.	Inst	LF	Lg	2C	.021	762.0	.54	.70	.06	1.30	1.73
	Inst	LF	Sm	2C	.025	648.0	.59	.83	.07	1.49	2.00
2" x 6" - 16' Avg.	Demo	LF	Lg	LB	.018	901.0	---	.49	---	.49	.74
	Demo	LF	Sm	LB	.021	766.0	---	.58	---	.58	.86
2" x 6" - 16' Avg.	Inst	LF	Lg	2C	.019	833.0	.54	.63	.06	1.23	1.63
	Inst	LF	Sm	2C	.023	708.0	.59	.77	.07	1.43	1.90
2" x 6" - 18' Avg.	Demo	LF	Lg	LB	.016	971.0	---	.44	---	.44	.65
	Demo	LF	Sm	LB	.019	825.0	---	.52	---	.52	.78
2" x 6" - 18' Avg.	Inst	LF	Lg	2C	.018	898.0	.54	.60	.05	1.19	1.57
	Inst	LF	Sm	2C	.021	763.0	.59	.70	.06	1.35	1.79
2" x 8" - 12' Avg.	Demo	LF	Lg	LB	.024	660.0	---	.66	---	.66	.98
	Demo	LF	Sm	LB	.029	561.0	---	.80	---	.80	1.19
2" x 8" - 12' Avg.	Inst	LF	Lg	2C	.026	612.0	.74	.87	.08	1.69	2.23
	Inst	LF	Sm	2C	.031	520.0	.83	1.03	.09	1.95	2.59
2" x 8" - 14' Avg.	Demo	LF	Lg	LB	.022	725.0	---	.60	---	.60	.90
	Demo	LF	Sm	LB	.026	616.0	---	.71	---	.71	1.06
2" x 8" - 14' Avg.	Inst	LF	Lg	2C	.024	673.0	.74	.80	.07	1.61	2.12
	Inst	LF	Sm	2C	.028	572.0	.83	.93	.08	1.84	2.43
2" x 8" - 16' Avg.	Demo	LF	Lg	LB	.020	784.0	---	.55	---	.55	.82
	Demo	LF	Sm	LB	.024	666.0	---	.66	---	.66	.98
2" x 8" - 16' Avg.	Inst	LF	Lg	2C	.022	728.0	.74	.73	.06	1.53	2.01
	Inst	LF	Sm	2C	.026	619.0	.83	.87	.08	1.78	2.33
2" x 8" - 18' Avg.	Demo	LF	Lg	LB	.019	837.0	---	.52	---	.52	.78
	Demo	LF	Sm	LB	.023	711.0	---	.63	---	.63	.94
2" x 8" - 18' Avg.	Inst	LF	Lg	2C	.021	779.0	.74	.70	.06	1.50	1.96
	Inst	LF	Sm	2C	.024	662.0	.83	.80	.07	1.70	2.22
2" x 8" - 20' Avg.	Demo	LF	Lg	LB	.018	885.0	---	.49	---	.49	.74
	Demo	LF	Sm	LB	.021	752.0	---	.58	---	.58	.86
2" x 8" - 20' Avg.	Inst	LF	Lg	2C	.019	825.0	.74	.63	.06	1.43	1.86
	Inst	LF	Sm	2C	.023	701.0	.83	.77	.07	1.67	2.17

Description	Oper	Unit	Vol	Crew Size	Man-hours per Unit	Crew Output per Day	Avg Mat'l Unit Cost	Avg Labor Unit Cost	Avg Equip Unit Cost	Avg Total Unit Cost	Avg Price Incl O&P
Common, cut-up roofs, to 1/3 pitch											
2" x 4" - 8' Avg.	Demo	LF	Lg	LB	.027	588.0	---	.74	---	.74	**1.10**
	Demo	LF	Sm	LB	.032	500.0	---	.88	---	.88	**1.31**
2" x 4" - 8' Avg.	Inst	LF	Lg	2C	.030	539.0	.39	1.00	.09	1.48	**2.04**
	Inst	LF	Sm	2C	.035	458.0	.43	1.16	.10	1.69	**2.34**
2" x 4" - 10' Avg.	Demo	LF	Lg	LB	.023	692.0	---	.63	---	.63	**.94**
	Demo	LF	Sm	LB	.027	588.0	---	.74	---	.74	**1.10**
2" x 4" - 10' Avg.	Inst	LF	Lg	2C	.025	635.0	.39	.83	.07	1.29	**1.77**
	Inst	LF	Sm	2C	.030	540.0	.43	1.00	.09	1.52	**2.08**
2" x 4" - 12' Avg.	Demo	LF	Lg	LB	.020	783.0	---	.55	---	.55	**.82**
	Demo	LF	Sm	LB	.024	666.0	---	.66	---	.66	**.98**
2" x 4" - 12' Avg.	Inst	LF	Lg	2C	.022	720.0	.39	.73	.06	1.18	**1.61**
	Inst	LF	Sm	2C	.026	612.0	.43	.87	.08	1.38	**1.87**
2" x 4" - 14' Avg.	Demo	LF	Lg	LB	.018	868.0	---	.49	---	.49	**.74**
	Demo	LF	Sm	LB	.022	738.0	---	.60	---	.60	**.90**
2" x 4" - 14' Avg.	Inst	LF	Lg	2C	.020	799.0	.39	.67	.06	1.12	**1.51**
	Inst	LF	Sm	2C	.024	679.0	.43	.80	.07	1.30	**1.76**
2" x 4" - 16' Avg.	Demo	LF	Lg	LB	.017	945.0	---	.47	---	.47	**.69**
	Demo	LF	Sm	LB	.020	803.0	---	.55	---	.55	**.82**
2" x 4" - 16' Avg.	Inst	LF	Lg	2C	.018	870.0	.39	.60	.05	1.04	**1.40**
	Inst	LF	Sm	2C	.022	740.0	.43	.73	.06	1.22	**1.65**
2" x 6" - 10' Avg.	Demo	LF	Lg	LB	.026	617.0	---	.71	---	.71	**1.06**
	Demo	LF	Sm	LB	.031	524.0	---	.85	---	.85	**1.27**
2" x 6" - 10' Avg.	Inst	LF	Lg	2C	.028	569.0	.55	.93	.08	1.56	**2.11**
	Inst	LF	Sm	2C	.033	484.0	.61	1.10	.10	1.81	**2.45**
2" x 6" - 12' Avg.	Demo	LF	Lg	LB	.023	703.0	---	.63	---	.63	**.94**
	Demo	LF	Sm	LB	.027	598.0	---	.74	---	.74	**1.10**
2" x 6" - 12' Avg.	Inst	LF	Lg	2C	.025	648.0	---	.83	.07	.90	**1.32**
	Inst	LF	Sm	2C	.029	551.0	.61	.96	.08	1.65	**2.23**
2" x 6" - 14' Avg.	Demo	LF	Lg	LB	.020	781.0	---	.55	---	.55	**.82**
	Demo	LF	Sm	LB	.024	664.0	---	.66	---	.66	**.98**
2" x 6" - 14' Avg.	Inst	LF	Lg	2C	.022	721.0	---	.73	.06	.79	**1.16**
	Inst	LF	Sm	2C	.026	613.0	.61	.87	.08	1.56	**2.08**
2" x 6" - 16' Avg.	Demo	LF	Lg	LB	.019	855.0	---	.52	---	.52	**.78**
	Demo	LF	Sm	LB	.022	727.0	---	.60	---	.60	**.90**
2" x 6" - 16' Avg.	Inst	LF	Lg	2C	.020	789.0	.55	.67	.06	1.28	**1.69**
	Inst	LF	Sm	2C	.024	671.0	.61	.80	.07	1.48	**1.97**
2" x 6" - 18' Avg.	Demo	LF	Lg	LB	.017	923.0	---	.47	---	.47	**.69**
	Demo	LF	Sm	LB	.020	785.0	---	.55	---	.55	**.82**
2" x 6" - 18' Avg.	Inst	LF	Lg	2C	.019	853.0	.55	.63	.05	1.23	**1.63**
	Inst	LF	Sm	2C	.022	725.0	.61	.73	.06	1.40	**1.86**

Description	Oper	Unit	Vol	Crew Size	Man-hours per Unit	Crew Output per Day	Avg Mat'l Unit Cost	Avg Labor Unit Cost	Avg Equip Unit Cost	Avg Total Unit Cost	Avg Price Incl O&P
2" x 8" - 12' Avg.	Demo	LF	Lg	LB	.026	627.0	---	.71	---	.71	**1.06**
	Demo	LF	Sm	LB	.030	533.0	---	.82	---	.82	**1.23**
2" x 8" - 12' Avg.	Inst	LF	Lg	2C	.028	580.0	.76	.93	.08	1.77	**2.35**
	Inst	LF	Sm	2C	.032	493.0	.85	1.06	.09	2.00	**2.66**
2" x 8" - 14' Avg.	Demo	LF	Lg	LB	.023	691.0	---	.63	---	.63	**.94**
	Demo	LF	Sm	LB	.027	587.0	---	.74	---	.74	**1.10**
2" x 8" - 14' Avg.	Inst	LF	Lg	2C	.025	641.0	.76	.83	.07	1.66	**2.19**
	Inst	LF	Sm	2C	.029	545.0	.85	.96	.09	1.90	**2.51**
2" x 8" - 16' Avg.	Demo	LF	Lg	LB	.021	749.0	---	.58	---	.58	**.86**
	Demo	LF	Sm	LB	.025	637.0	---	.69	---	.69	**1.02**
2" x 8" - 16' Avg.	Inst	LF	Lg	2C	.023	695.0	.76	.77	.07	1.60	**2.09**
	Inst	LF	Sm	2C	.027	591.0	.85	.90	.08	1.83	**2.41**
2" x 8" - 18' Avg.	Demo	LF	Lg	LB	.020	802.0	---	.55	---	.55	**.82**
	Demo	LF	Sm	LB	.023	682.0	---	.63	---	.63	**.94**
2" x 8" - 18' Avg.	Inst	LF	Lg	2C	.021	745.0	.76	.70	.06	1.52	**1.98**
	Inst	LF	Sm	2C	.025	633.0	.85	.83	.07	1.75	**2.30**
2" x 8" - 20' Avg.	Demo	LF	Lg	LB	.019	850.0	---	.52	---	.52	**.78**
	Demo	LF	Sm	LB	.022	723.0	---	.60	---	.60	**.90**
2" x 8" - 20' Avg.	Inst	LF	Lg	2C	.020	791.0	.76	.67	.06	1.49	**1.93**
	Inst	LF	Sm	2C	.024	672.0	.85	.80	.07	1.72	**2.25**

Common, cut-up roofs, to 3/8-1/2 pitch

Description	Oper	Unit	Vol	Crew Size	Man-hours per Unit	Crew Output per Day	Avg Mat'l Unit Cost	Avg Labor Unit Cost	Avg Equip Unit Cost	Avg Total Unit Cost	Avg Price Incl O&P
2" x 4" - 8' Avg.	Demo	LF	Lg	LB	.031	516.0	---	.85	---	.85	**1.27**
	Demo	LF	Sm	LB	.036	439.0	---	.99	---	.99	**1.47**
2" x 4" - 8' Avg.	Inst	LF	Lg	2C	.034	473.0	.39	1.13	.10	1.62	**2.25**
	Inst	LF	Sm	2C	.040	402.0	.43	1.33	.12	1.88	**2.61**
2" x 4" - 10' Avg.	Demo	LF	Lg	LB	.026	612.0	---	.71	---	.71	**1.06**
	Demo	LF	Sm	LB	.031	520.0	---	.85	---	.85	**1.27**
2" x 4" - 10' Avg.	Inst	LF	Lg	2C	.028	562.0	.39	.93	.08	1.40	**1.93**
	Inst	LF	Sm	2C	.033	478.0	.43	1.10	.10	1.63	**2.24**
2" x 4" - 12' Avg.	Demo	LF	Lg	LB	.023	697.0	---	.63	---	.63	**.94**
	Demo	LF	Sm	LB	.027	592.0	---	.74	---	.74	**1.10**
2" x 4" - 12' Avg.	Inst	LF	Lg	2C	.025	640.0	.39	.83	.07	1.29	**1.77**
	Inst	LF	Sm	2C	.029	544.0	.43	.96	.09	1.48	**2.03**
2" x 4" - 14' Avg.	Demo	LF	Lg	LB	.021	777.0	---	.58	---	.58	**.86**
	Demo	LF	Sm	LB	.024	660.0	---	.66	---	.66	**.98**
2" x 4" - 14' Avg.	Inst	LF	Lg	2C	.022	714.0	.39	.73	.07	1.19	**1.62**
	Inst	LF	Sm	2C	.026	607.0	.43	.87	.08	1.38	**1.87**
2" x 4" - 16' Avg.	Demo	LF	Lg	LB	.019	849.0	---	.52	---	.52	**.78**
	Demo	LF	Sm	LB	.022	722.0	---	.60	---	.60	**.90**
2" x 4" - 16' Avg.	Inst	LF	Lg	2C	.021	780.0	.39	.70	.06	1.15	**1.56**
	Inst	LF	Sm	2C	.024	663.0	.43	.80	.07	1.30	**1.76**

Description	Oper	Unit	Vol	Crew Size	Man-hours per Unit	Crew Output per Day	Avg Mat'l Unit Cost	Avg Labor Unit Cost	Avg Equip Unit Cost	Avg Total Unit Cost	Avg Price Incl O&P
2" x 6" - 10' Avg.	Demo	LF	Lg	LB	.029	552.0	---	.80	---	.80	1.19
	Demo	LF	Sm	LB	.034	469.0	---	.93	---	.93	1.39
2" x 6" - 10' Avg.	Inst	LF	Lg	2C	.031	508.0	.55	1.03	.09	1.67	2.27
	Inst	LF	Sm	2C	.037	432.0	.61	1.23	.11	1.95	2.66
2" x 6" - 12' Avg.	Demo	LF	Lg	LB	.025	633.0	---	.69	---	.69	1.02
	Demo	LF	Sm	LB	.030	538.0	---	.82	---	.82	1.23
2" x 6" - 12' Avg.	Inst	LF	Lg	2C	.027	583.0	.55	.90	.08	1.53	2.06
	Inst	LF	Sm	2C	.032	496.0	.61	1.06	.09	1.76	2.39
2" x 6" - 14' Avg.	Demo	LF	Lg	LB	.023	706.0	---	.63	---	.63	.94
	Demo	LF	Sm	LB	.027	600.0	---	.74	---	.74	1.10
2" x 6" - 14' Avg.	Inst	LF	Lg	2C	.025	651.0	.55	.83	.07	1.45	1.95
	Inst	LF	Sm	2C	.029	553.0	.61	.96	.08	1.65	2.23
2" x 6" - 16' Avg.	Demo	LF	Lg	LB	.021	776.0	---	.58	---	.58	.86
	Demo	LF	Sm	LB	.024	660.0	---	.66	---	.66	.98
2" x 6" - 16' Avg.	Inst	LF	Lg	2C	.022	715.0	.55	.73	.07	1.35	1.80
	Inst	LF	Sm	2C	.026	608.0	.61	.87	.08	1.56	2.08
2" x 6" - 18' Avg.	Demo	LF	Lg	LB	.019	842.0	---	.52	---	.52	.78
	Demo	LF	Sm	LB	.022	716.0	---	.60	---	.60	.90
2" x 6" - 18' Avg.	Inst	LF	Lg	2C	.021	777.0	.55	.70	.06	1.31	1.74
	Inst	LF	Sm	2C	.024	660.0	.61	.80	.07	1.48	1.97
2" x 8" - 12' Avg.	Demo	LF	Lg	LB	.028	570.0	---	.77	---	.77	1.14
	Demo	LF	Sm	LB	.033	485.0	---	.91	---	.91	1.35
2" x 8" - 12' Avg.	Inst	LF	Lg	2C	.030	527.0	.76	1.00	.09	1.85	2.46
	Inst	LF	Sm	2C	.036	448.0	.85	1.20	.10	2.15	2.87
2" x 8" - 14' Avg.	Demo	LF	Lg	LB	.025	632.0	---	.69	---	.69	1.02
	Demo	LF	Sm	LB	.030	537.0	---	.82	---	.82	1.23
2" x 8" - 14' Avg.	Inst	LF	Lg	2C	.027	585.0	.76	.90	.08	1.74	2.30
	Inst	LF	Sm	2C	.032	497.0	.85	1.06	.09	2.00	2.66
2" x 8" - 16' Avg.	Demo	LF	Lg	LB	.023	688.0	---	.63	---	.63	.94
	Demo	LF	Sm	LB	.027	585.0	---	.74	---	.74	1.10
2" x 8" - 16' Avg.	Inst	LF	Lg	2C	.025	638.0	.76	.83	.07	1.66	2.19
	Inst	LF	Sm	2C	.030	542.0	.85	1.00	.09	1.94	2.56
2" x 8" - 18' Avg.	Demo	LF	Lg	LB	.022	739.0	---	.60	---	.60	.90
	Demo	LF	Sm	LB	.025	628.0	---	.69	---	.69	1.02
2" x 8" - 18' Avg.	Inst	LF	Lg	2C	.023	686.0	.76	.77	.07	1.60	2.09
	Inst	LF	Sm	2C	.027	583.0	.85	.90	.08	1.83	2.41
2" x 8" - 20' Avg.	Demo	LF	Lg	LB	.020	787.0	---	.55	---	.55	.82
	Demo	LF	Sm	LB	.024	669.0	---	.66	---	.66	.98
2" x 8" - 20' Avg.	Inst	LF	Lg	2C	.022	731.0	.76	.73	.06	1.55	2.03
	Inst	LF	Sm	2C	.026	621.0	.85	.87	.08	1.80	2.36

Description	Oper	Unit	Vol	Crew Size	Man-hours per Unit	Crew Output per Day	Avg Mat'l Unit Cost	Avg Labor Unit Cost	Avg Equip Unit Cost	Avg Total Unit Cost	Avg Price Incl O&P

Rafters, per SF of area

Gable or hip, to 1/3 pitch

Description	Oper	Unit	Vol	Crew Size	Man-hours per Unit	Crew Output per Day	Avg Mat'l Unit Cost	Avg Labor Unit Cost	Avg Equip Unit Cost	Avg Total Unit Cost	Avg Price Incl O&P
2" x 4" - 12" oc	Demo	SF	Lg	LB	.019	840.0	---	.52	---	.52	.78
	Demo	SF	Sm	LB	.022	714.0	---	.60	---	.60	.90
2" x 4" - 12" oc	Inst	SF	Lg	2C	.021	774.0	.44	.70	.06	1.20	1.61
	Inst	SF	Sm	2C	.024	658.0	.48	.80	.07	1.35	1.82
2" x 4" - 16" oc	Demo	SF	Lg	LB	.016	1000	---	.44	---	.44	.65
	Demo	SF	Sm	LB	.019	850.0	---	.52	---	.52	.78
2" x 4" - 16" oc	Inst	SF	Lg	2C	.017	921.0	.34	.57	.05	.96	1.29
	Inst	SF	Sm	2C	.020	783.0	.38	.67	.06	1.11	1.50
2" x 4" - 24" oc	Demo	SF	Lg	LB	.012	1312	---	.33	---	.33	.49
	Demo	SF	Sm	LB	.014	1115	---	.38	---	.38	.57
2" x 4" - 24" oc	Inst	SF	Lg	2C	.013	1208	.25	.43	.04	.72	.98
	Inst	SF	Sm	2C	.016	1027	.27	.53	.05	.85	1.16
2" x 6" - 12" oc	Demo	SF	Lg	LB	.019	821.0	---	.52	---	.52	.78
	Demo	SF	Sm	LB	.023	698.0	---	.63	---	.63	.94
2" x 6" - 12" oc	Inst	SF	Lg	2C	.021	760.0	.60	.70	.06	1.36	1.8
	Inst	SF	Sm	2C	.025	646.0	.66	.83	.07	1.56	2.08
2" x 6" - 16" oc	Demo	SF	Lg	LB	.016	978.0	---	.44	---	.44	.65
	Demo	SF	Sm	LB	.019	831.0	---	.52	---	.52	.78
2" x 6" - 16" oc	Inst	SF	Lg	2C	.018	905.0	.46	.60	.05	1.11	1.4
	Inst	SF	Sm	2C	.021	769.0	.51	.70	.06	1.27	1.6
2" x 6" - 24" oc	Demo	SF	Lg	LB	.012	1284	---	.33	---	.33	.49
	Demo	SF	Sm	LB	.015	1091	---	.41	---	.41	.6
2" x 6" - 24" oc	Inst	SF	Lg	2C	.013	1187	.33	.43	.04	.80	1.07
	Inst	SF	Sm	2C	.016	1009	.37	.53	.05	.95	1.2
2" x 8" - 12" oc	Demo	SF	Lg	LB	.020	785.0	---	.55	---	.55	.82
	Demo	SF	Sm	LB	.024	667.0	---	.66	---	.66	.9
2" x 8" - 12" oc	Inst	SF	Lg	2C	.022	730.0	.81	.73	.06	1.60	2.0
	Inst	SF	Sm	2C	.026	621.0	.90	.87	.08	1.85	2.4
2" x 8" - 16" oc	Demo	SF	Lg	LB	.017	933.0	---	.47	---	.47	.6
	Demo	SF	Sm	LB	.020	793.0	---	.55	---	.55	.8
2" x 8" - 16" oc	Inst	SF	Lg	2C	.018	868.0	.62	.60	.05	1.27	1.6
	Inst	SF	Sm	2C	.022	738.0	.69	.73	.06	1.48	1.9
2" x 8" - 24" oc	Demo	SF	Lg	LB	.013	1225	---	.36	---	.36	.5
	Demo	SF	Sm	LB	.015	1041	---	.41	---	.41	.6
2" x 8" - 24" oc	Inst	SF	Lg	2C	.014	1140	.44	.47	.04	.95	1.2
	Inst	SF	Sm	2C	.017	969.0	.49	.57	.05	1.11	1.4

Description	Oper	Unit	Vol	Crew Size	Man-hours per Unit	Crew Output per Day	Avg Mat'l Unit Cost	Avg Labor Unit Cost	Avg Equip Unit Cost	Avg Total Unit Cost	Avg Price Incl O&P
Gable or hip, to 3/8-1/2 pitch											
2" x 4" - 12" oc	Demo	SF	Lg	LB	.022	723.0	---	.60	---	.60	**.90**
	Demo	SF	Sm	LB	.026	615.0	---	.71	---	.71	**1.06**
2" x 4" - 12" oc	Inst	SF	Lg	2C	.024	664.0	.44	.80	.07	1.31	**1.77**
	Inst	SF	Sm	2C	.028	564.0	.48	.93	.08	1.49	**2.03**
2" x 4" - 16" oc	Demo	SF	Lg	LB	.019	859.0	---	.52	---	.52	**.78**
	Demo	SF	Sm	LB	.022	730.0	---	.60	---	.60	**.90**
2" x 4" - 16" oc	Inst	SF	Lg	2C	.020	789.0	.34	.67	.06	1.07	**1.45**
	Inst	SF	Sm	2C	.024	671.0	.38	.80	.07	1.25	**1.70**
2" x 4" - 24" oc	Demo	SF	Lg	LB	.014	1127	---	.38	---	.38	**.57**
	Demo	SF	Sm	LB	.017	958.0	---	.47	---	.47	**.69**
2" x 4" - 24" oc	Inst	SF	Lg	2C	.015	1036	.25	.50	.05	.80	**1.09**
	Inst	SF	Sm	2C	.018	881.0	.27	.60	.05	.92	**1.26**
2" x 6" - 12" oc	Demo	SF	Lg	LB	.022	725.0	---	.60	---	.60	**.90**
	Demo	SF	Sm	LB	.026	616.0	---	.71	---	.71	**1.06**
2" x 6" - 12" oc	Inst	SF	Lg	2C	.024	669.0	.60	.80	.07	1.47	**1.96**
	Inst	SF	Sm	2C	.028	569.0	.66	.93	.08	1.67	**2.24**
2" x 6" - 16" oc	Demo	SF	Lg	LB	.019	862.0	---	.52	---	.52	**.78**
	Demo	SF	Sm	LB	.022	733.0	---	.60	---	.60	**.90**
2" x 6" - 16" oc	Inst	SF	Lg	2C	.020	796.0	.46	.67	.06	1.19	**1.59**
	Inst	SF	Sm	2C	.024	677.0	.51	.80	.07	1.38	**1.85**
2" x 6" - 24" oc	Demo	SF	Lg	LB	.014	1132	---	.38	---	.38	**.57**
	Demo	SF	Sm	LB	.017	962.0	---	.47	---	.47	**.69**
2" x 6" - 24" oc	Inst	SF	Lg	2C	.015	1045	.33	.50	.04	.87	**1.17**
	Inst	SF	Sm	2C	.018	888.0	.37	.60	.05	1.02	**1.37**
2" x 8" - 12" oc	Demo	SF	Lg	LB	.023	708.0	---	.63	---	.63	**.94**
	Demo	SF	Sm	LB	.027	602.0	---	.74	---	.74	**1.10**
2" x 8" - 12" oc	Inst	SF	Lg	2C	.024	657.0	.81	.80	.07	1.68	**2.20**
	Inst	SF	Sm	2C	.029	558.0	.90	.96	.08	1.94	**2.56**
2" x 8" - 16" oc	Demo	SF	Lg	LB	.019	841.0	---	.52	---	.52	**.78**
	Demo	SF	Sm	LB	.022	715.0	---	.60	---	.60	**.90**
2" x 8" - 16" oc	Inst	SF	Lg	2C	.020	781.0	.62	.67	.06	1.35	**1.77**
	Inst	SF	Sm	2C	.024	664.0	.69	.80	.07	1.56	**2.06**
2" x 8" - 24" oc	Demo	SF	Lg	LB	.012	1351	---	.33	---	.33	**.49**
	Demo	SF	Sm	LB	.014	1148	---	.38	---	.38	**.57**
2" x 8" - 24" oc	Inst	SF	Lg	2C	.013	1260	.44	.43	.04	.91	**1.19**
	Inst	SF	Sm	2C	.015	1071	.49	.50	.04	1.03	**1.35**

Description	Oper	Unit	Vol	Crew Size	Man-hours per Unit	Crew Output per Day	Avg Mat'l Unit Cost	Avg Labor Unit Cost	Avg Equip Unit Cost	Avg Total Unit Cost	Avg Price Incl O&P
Cut-up roofs, to 1/3 pitch											
2" x 4" - 12" oc	Demo	SF	Lg	LB	.024	674.0	---	.66	---	.66	**.98**
	Demo	SF	Sm	LB	.028	573.0	---	.77	---	.77	**1.14**
2" x 4" - 12" oc	Inst	SF	Lg	2C	.026	619.0	.45	.87	.08	1.40	**1.90**
	Inst	SF	Sm	2C	.030	526.0	.49	1.00	.09	1.58	**2.15**
2" x 4" - 16" oc	Demo	SF	Lg	LB	.020	804.0	---	.55	---	.55	**.82**
	Demo	SF	Sm	LB	.023	683.0	---	.63	---	.63	**.94**
2" x 4" - 16" oc	Inst	SF	Lg	2C	.022	738.0	.35	.73	.06	1.14	**1.56**
	Inst	SF	Sm	2C	.026	627.0	.38	.87	.07	1.32	**1.80**
2" x 4" - 24" oc	Demo	SF	Lg	LB	.015	1055	---	.41	---	.41	**.61**
	Demo	SF	Sm	LB	.018	897.0	---	.49	---	.49	**.74**
2" x 4" - 24" oc	Inst	SF	Lg	2C	.017	968.0	.25	.57	.05	.87	**1.19**
	Inst	SF	Sm	2C	.019	823.0	.28	.63	.06	.97	**1.33**
2" x 6" - 12" oc	Demo	SF	Lg	LB	.023	686.0	---	.63	---	.63	**.94**
	Demo	SF	Sm	LB	.027	583.0	---	.74	---	.74	**1.10**
2" x 6" - 12" oc	Inst	SF	Lg	2C	.025	633.0	.61	.83	.07	1.51	**2.02**
	Inst	SF	Sm	2C	.030	538.0	.67	1.00	.09	1.76	**2.36**
2" x 6" - 16" oc	Demo	SF	Lg	LB	.020	816.0	---	.55	---	.55	**.82**
	Demo	SF	Sm	LB	.023	694.0	---	.63	---	.63	**.94**
2" x 6" - 16" oc	Inst	SF	Lg	2C	.021	752.0	.47	.70	.06	1.23	**1.65**
	Inst	SF	Sm	2C	.025	639.0	.52	.83	.07	1.42	**1.92**
2" x 6" - 24" oc	Demo	SF	Lg	LB	.015	1070	---	.41	---	.41	**.61**
	Demo	SF	Sm	LB	.018	910.0	---	.49	---	.49	**.74**
2" x 6" - 24" oc	Inst	SF	Lg	2C	.016	987.0	.34	.53	.05	.92	**1.24**
	Inst	SF	Sm	2C	.019	839.0	.37	.63	.06	1.06	**1.43**
2" x 8" - 12" oc	Demo	SF	Lg	LB	.024	674.0	---	.66	---	.66	**.98**
	Demo	SF	Sm	LB	.028	573.0	---	.77	---	.77	**1.14**
2" x 8" - 12" oc	Inst	SF	Lg	2C	.026	625.0	.82	.87	.07	1.76	**2.31**
	Inst	SF	Sm	2C	.030	531.0	.92	1.00	.09	2.01	**2.65**
2" x 8" - 16" oc	Demo	SF	Lg	LB	.020	802.0	---	.55	---	.55	**.82**
	Demo	SF	Sm	LB	.023	682.0	---	.63	---	.63	**.94**
2" x 8" - 16" oc	Inst	SF	Lg	2C	.022	744.0	.63	.73	.06	1.42	**1.88**
	Inst	SF	Sm	2C	.025	632.0	.70	.83	.07	1.60	**2.12**
2" x 8" - 24" oc	Demo	SF	Lg	LB	.015	1053	---	.41	---	.41	**.61**
	Demo	SF	Sm	LB	.018	895.0	---	.49	---	.49	**.74**
2" x 8" - 24" oc	Inst	SF	Lg	2C	.016	976.0	.45	.53	.05	1.03	**1.37**
	Inst	SF	Sm	2C	.019	830.0	.50	.63	.06	1.19	**1.58**

Description	Oper	Unit	Vol	Crew Size	Man-hours per Unit	Crew Output per Day	Avg Mat'l Unit Cost	Avg Labor Unit Cost	Avg Equip Unit Cost	Avg Total Unit Cost	Avg Price Incl O&P
Cut-up roofs, to 3/8-1/2 pitch											
2" x 4" - 12" oc	Demo	SF	Lg	LB	.027	596.0	---	.74	---	.74	**1.10**
	Demo	SF	Sm	LB	.032	507.0	---	.88	---	.88	**1.31**
2" x 4" - 12" oc	Inst	SF	Lg	2C	.029	547.0	.45	.96	.09	1.50	**2.05**
	Inst	SF	Sm	2C	.034	465.0	.49	1.13	.10	1.72	**2.36**
2" x 4" - 16" oc	Demo	SF	Lg	LB	.022	712.0	---	.60	---	.60	**.90**
	Demo	SF	Sm	LB	.026	605.0	---	.71	---	.71	**1.06**
2" x 4" - 16" oc	Inst	SF	Lg	2C	.025	653.0	.35	.83	.07	1.25	**1.72**
	Inst	SF	Sm	2C	.029	555.0	.38	.96	.08	1.42	**1.96**
2" x 4" - 24" oc	Demo	SF	Lg	LB	.017	932.0	---	.47	---	.47	**.69**
	Demo	SF	Sm	LB	.020	792.0	---	.55	---	.55	**.82**
2" x 4" - 24" oc	Inst	SF	Lg	2C	.019	855.0	.25	.63	.05	.93	**1.29**
	Inst	SF	Sm	2C	.022	727.0	.28	.73	.06	1.07	**1.48**
2" x 6" - 12" oc	Demo	SF	Lg	LB	.026	618.0	---	.71	---	.71	**1.06**
	Demo	SF	Sm	LB	.030	525.0	---	.82	---	.82	**1.23**
2" x 6" - 12" oc	Inst	SF	Lg	2C	.028	569.0	.61	.93	.08	1.62	**2.18**
	Inst	SF	Sm	2C	.033	484.0	.67	1.10	.10	1.87	**2.52**
2" x 6" - 16" oc	Demo	SF	Lg	LB	.022	735.0	---	.60	---	.60	**.90**
	Demo	SF	Sm	LB	.026	625.0	---	.71	---	.71	**1.06**
2" x 6" - 16" oc	Inst	SF	Lg	2C	.024	677.0	.47	.80	.07	1.34	**1.81**
	Inst	SF	Sm	2C	.028	575.0	.52	.93	.08	1.53	**2.08**
2" x 6" - 24" oc	Demo	SF	Lg	LB	.017	963.0	---	.47	---	.47	**.69**
	Demo	SF	Sm	LB	.020	819.0	---	.55	---	.55	**.82**
2" x 6" - 24" oc	Inst	SF	Lg	2C	.018	887.0	.34	.60	.05	.99	**1.34**
	Inst	SF	Sm	2C	.021	754.0	.37	.70	.06	1.13	**1.53**
2" x 8" - 12" oc	Demo	SF	Lg	LB	.026	616.0	---	.71	---	.71	**1.06**
	Demo	SF	Sm	LB	.031	524.0	---	.85	---	.85	**1.27**
2" x 8" - 12" oc	Inst	SF	Lg	2C	.028	570.0	.82	.93	.08	1.83	**2.42**
	Inst	SF	Sm	2C	.033	485.0	.92	1.10	.10	2.12	**2.81**
2" x 8" - 16" oc	Demo	SF	Lg	LB	.022	734.0	---	.60	---	.60	**.90**
	Demo	SF	Sm	LB	.026	624.0	---	.71	---	.71	**1.06**
2" x 8" - 16" oc	Inst	SF	Lg	2C	.024	679.0	.63	.80	.07	1.50	**1.99**
	Inst	SF	Sm	2C	.028	577.0	.70	.93	.08	1.71	**2.28**
2" x 8" - 24" oc	Demo	SF	Lg	LB	.017	962.0	---	.47	---	.47	**.69**
	Demo	SF	Sm	LB	.020	818.0	---	.55	---	.55	**.82**
2" x 8" - 24" oc	Inst	SF	Lg	2C	.018	891.0	.45	.60	.05	1.10	**1.47**
	Inst	SF	Sm	2C	.021	757.0	.50	.70	.06	1.26	**1.68**

Description	Oper	Unit	Vol	Crew Size	Man-hours per Unit	Crew Output per Day	Avg Mat'l Unit Cost	Avg Labor Unit Cost	Avg Equip Unit Cost	Avg Total Unit Cost	Avg Price Incl O&P

Roof decking, solid, T&G, dry for plank-and-beam construction

Description	Oper	Unit	Vol	Crew Size	Man-hours per Unit	Crew Output per Day	Avg Mat'l Unit Cost	Avg Labor Unit Cost	Avg Equip Unit Cost	Avg Total Unit Cost	Avg Price Incl O&P
2" x 6" - 12'	Demo	SF	Lg	LB	.025	645.0	---	.69	---	.69	1.02
	Demo	SF	Sm	LB	.029	548.0	---	.80	---	.80	1.19
2" x 6" - 12'	Inst	SF	Lg	2C	.033	480.0	.11	1.10	.10	1.31	1.87
	Inst	SF	Sm	2C	.039	408.0	.12	1.30	.11	1.53	2.19
2" x 8" - 12'	Demo	SF	Lg	LB	.018	900.0	---	.49	---	.49	.74
	Demo	SF	Sm	LB	.021	765.0	---	.58	---	.58	.86
2" x 8" - 12'	Inst	SF	Lg	2C	.024	670.0	.07	.80	.07	.94	1.35
	Inst	SF	Sm	2C	.028	570.0	.08	.93	.08	1.09	1.57

Studs/plates, per LF of stick

Walls or partitions

Description	Oper	Unit	Vol	Crew Size	Man-hours per Unit	Crew Output per Day	Avg Mat'l Unit Cost	Avg Labor Unit Cost	Avg Equip Unit Cost	Avg Total Unit Cost	Avg Price Incl O&P
2" x 4" - 8' Avg.	Demo	LF	Lg	LB	.016	1031	---	.44	---	.44	.65
	Demo	LF	Sm	LB	.018	876.0	---	.49	---	.49	.74
2" x 4" - 8' Avg.	Inst	LF	Lg	2C	.019	848.0	.37	.63	.06	1.06	1.43
	Inst	LF	Sm	2C	.022	721.0	.41	.73	.06	1.20	1.63
2" x 4" - 10' Avg.	Demo	LF	Lg	LB	.014	1172	---	.38	---	.38	.57
	Demo	LF	Sm	LB	.016	996.0	---	.44	---	.44	.65
2" x 4" - 10' Avg.	Inst	LF	Lg	2C	.017	969.0	.37	.57	.05	.99	1.32
	Inst	LF	Sm	2C	.019	824.0	.41	.63	.06	1.10	1.48
2" x 4" - 12' Avg.	Demo	LF	Lg	LB	.012	1308	---	.33	---	.33	.49
	Demo	LF	Sm	LB	.014	1112	---	.38	---	.38	.57
2" x 4" - 12' Avg.	Inst	LF	Lg	2C	.015	1082	.37	.50	.04	.91	1.21
	Inst	LF	Sm	2C	.017	920.0	.41	.57	.05	1.03	1.37
2" x 6" - 8' Avg.	Demo	LF	Lg	LB	.018	904.0	---	.49	---	.49	.74
	Demo	LF	Sm	LB	.021	768.0	---	.58	---	.58	.86
2" x 6" - 8' Avg.	Inst	LF	Lg	2C	.021	752.0	.51	.70	.06	1.27	1.69
	Inst	LF	Sm	2C	.025	639.0	.57	.83	.07	1.47	1.97
2" x 6" - 10' Avg.	Demo	LF	Lg	LB	.016	1032	---	.44	---	.44	.65
	Demo	LF	Sm	LB	.018	877.0	---	.49	---	.49	.74
2" x 6" - 10' Avg.	Inst	LF	Lg	2C	.019	861.0	.51	.63	.05	1.19	1.58
	Inst	LF	Sm	2C	.022	732.0	.57	.73	.06	1.36	1.81
2" x 6" - 12' Avg.	Demo	LF	Lg	LB	.014	1142	---	.38	---	.38	.57
	Demo	LF	Sm	LB	.016	971.0	---	.44	---	.44	.65
2" x 6" - 12' Avg.	Inst	LF	Lg	2C	.017	956.0	.51	.57	.05	1.13	1.48
	Inst	LF	Sm	2C	.020	813.0	.57	.67	.06	1.30	1.71

Description	Oper	Unit	Vol	Crew Size	Man-hours per Unit	Crew Output per Day	Avg Mat'l Unit Cost	Avg Labor Unit Cost	Avg Equip Unit Cost	Avg Total Unit Cost	Avg Price Incl O&P
2" x 8" - 8' Avg.	Demo	LF	Lg	LB	.020	802.0	---	.55	---	.55	.82
	Demo	LF	Sm	LB	.023	682.0	---	.63	---	.63	.94
2" x 8" - 8' Avg.	Inst	LF	Lg	2C	.024	673.0	.71	.80	.07	1.58	2.08
	Inst	LF	Sm	2C	.028	572.0	.79	.93	.08	1.80	2.39
2" x 8" - 10' Avg.	Demo	LF	Lg	LB	.017	915.0	---	.47	---	.47	.69
	Demo	LF	Sm	LB	.021	778.0	---	.58	---	.58	.86
2" x 8" - 10' Avg.	Inst	LF	Lg	2C	.021	771.0	.71	.70	.06	1.47	1.92
	Inst	LF	Sm	2C	.024	655.0	.79	.80	.07	1.66	2.18
2" x 8" - 12' Avg.	Demo	LF	Lg	LB	.016	1010	---	.44	---	.44	.65
	Demo	LF	Sm	LB	.019	859.0	---	.52	---	.52	.78
2" x 8" - 12' Avg.	Inst	LF	Lg	2C	.019	855.0	.71	.63	.05	1.39	1.81
	Inst	LF	Sm	2C	.022	727.0	.79	.73	.06	1.58	2.07

Gable ends

Description	Oper	Unit	Vol	Crew Size	Man-hours per Unit	Crew Output per Day	Avg Mat'l Unit Cost	Avg Labor Unit Cost	Avg Equip Unit Cost	Avg Total Unit Cost	Avg Price Incl O&P
2" x 4" - 3' Avg.	Demo	LF	Lg	LB	.024	660.0	---	.66	---	.66	.98
	Demo	LF	Sm	LB	.029	561.0	---	.80	---	.80	1.19
2" x 4" - 3' Avg.	Inst	LF	Lg	2C	.047	344.0	.39	1.56	.14	2.09	2.93
	Inst	LF	Sm	2C	.055	292.0	.42	1.83	.16	2.41	3.39
2" x 4" - 4' Avg.	Demo	LF	Lg	LB	.020	818.0	---	.55	---	.55	.82
	Demo	LF	Sm	LB	.023	695.0	---	.63	---	.63	.94
2" x 4" - 4' Avg.	Inst	LF	Lg	2C	.037	431.0	.39	1.23	.11	1.73	2.41
	Inst	LF	Sm	2C	.044	366.0	.42	1.46	.13	2.01	2.81
2" x 4" - 5' Avg.	Demo	LF	Lg	LB	.017	960.0	---	.47	---	.47	.69
	Demo	LF	Sm	LB	.020	816.0	---	.55	---	.55	.82
2" x 4" - 5' Avg.	Inst	LF	Lg	2C	.031	511.0	.39	1.03	.09	1.51	2.09
	Inst	LF	Sm	2C	.037	434.0	.42	1.23	.11	1.76	2.44
2" x 4" - 6' Avg.	Demo	LF	Lg	LB	.015	1082	---	.41	---	.41	.61
	Demo	LF	Sm	LB	.017	920.0	---	.47	---	.47	.69
2" x 4" - 6' Avg.	Inst	LF	Lg	2C	.028	580.0	.39	.93	.08	1.40	1.93
	Inst	LF	Sm	2C	.032	493.0	.42	1.06	.09	1.57	2.17
2" x 6" - 3' Avg.	Demo	LF	Lg	LB	.025	634.0	---	.69	---	.69	1.02
	Demo	LF	Sm	LB	.030	539.0	---	.82	---	.82	1.23
2" x 6" - 3' Avg.	Inst	LF	Lg	2C	.047	337.0	.54	1.56	.14	2.24	3.11
	Inst	LF	Sm	2C	.056	286.0	.59	1.86	.16	2.61	3.63
2" x 6" - 4' Avg.	Demo	LF	Lg	LB	.021	778.0	---	.58	---	.58	.86
	Demo	LF	Sm	LB	.024	661.0	---	.66	---	.66	.98
2" x 6" - 4' Avg.	Inst	LF	Lg	2C	.038	420.0	.54	1.26	.11	1.91	2.63
	Inst	LF	Sm	2C	.045	357.0	.59	1.50	.13	2.22	3.05

Description	Oper	Unit	Vol	Crew Size	Man-hours per Unit	Crew Output per Day	Avg Mat'l Unit Cost	Avg Labor Unit Cost	Avg Equip Unit Cost	Avg Total Unit Cost	Avg Price Incl O&P
2" x 6" - 5' Avg.	Demo	LF	Lg	LB	.018	906.0	---	.49	---	.49	.74
	Demo	LF	Sm	LB	.021	770.0	---	.58	---	.58	.86
2" x 6" - 5' Avg.	Inst	LF	Lg	2C	.032	495.0	.54	1.06	.09	1.69	2.31
	Inst	LF	Sm	2C	.038	421.0	.59	1.26	.11	1.96	2.69
2" x 6" - 6' Avg.	Demo	LF	Lg	LB	.016	1013	---	.44	---	.44	.65
	Demo	LF	Sm	LB	.019	861.0	---	.52	---	.52	.78
2" x 6" - 6' Avg.	Inst	LF	Lg	2C	.029	560.0	.54	.96	.08	1.58	2.15
	Inst	LF	Sm	2C	.034	476.0	.59	1.13	.10	1.82	2.48
2" x 8" - 3' Avg.	Demo	LF	Lg	LB	.026	610.0	---	.71	---	.71	1.06
	Demo	LF	Sm	LB	.031	519.0	---	.85	---	.85	1.27
2" x 8" - 3' Avg.	Inst	LF	Lg	2C	.048	330.0	.74	1.60	.14	2.48	3.39
	Inst	LF	Sm	2C	.057	281.0	.83	1.90	.17	2.90	3.97
2" x 8" - 4' Avg.	Demo	LF	Lg	LB	.022	742.0	---	.60	---	.60	.90
	Demo	LF	Sm	LB	.025	631.0	---	.69	---	.69	1.02
2" x 8" - 4' Avg.	Inst	LF	Lg	2C	.039	409.0	.74	1.30	.11	2.15	2.91
	Inst	LF	Sm	2C	.046	348.0	.83	1.53	.13	2.49	3.38
2" x 8" - 5' Avg.	Demo	LF	Lg	LB	.019	857.0	---	.52	---	.52	.78
	Demo	LF	Sm	LB	.022	728.0	---	.60	---	.60	.90
2" x 8" - 5' Avg.	Inst	LF	Lg	2C	.033	480.0	.74	1.10	.10	1.94	2.60
	Inst	LF	Sm	2C	.039	408.0	.83	1.30	.11	2.24	3.01
2" x 8" - 6' Avg.	Demo	LF	Lg	LB	.017	953.0	---	.47	---	.47	.69
	Demo	LF	Sm	LB	.020	810.0	---	.55	---	.55	.82
2" x 8" - 6' Avg.	Inst	LF	Lg	2C	.030	541.0	.74	1.00	.09	1.83	2.44
	Inst	LF	Sm	2C	.035	460.0	.83	1.16	.10	2.09	2.80

Studs/plates, per SF of area

Walls or partitions

Description	Oper	Unit	Vol	Crew Size	Man-hours per Unit	Crew Output per Day	Avg Mat'l Unit Cost	Avg Labor Unit Cost	Avg Equip Unit Cost	Avg Total Unit Cost	Avg Price Incl O&P
2" x 4" - 8', 12" oc	Demo	SF	Lg	LB	.027	601.0	---	.74	---	.74	1.10
	Demo	SF	Sm	LB	.031	511.0	---	.85	---	.85	1.27
2" x 4" - 8', 12" oc	Inst	SF	Lg	2C	.032	495.0	.64	1.06	.09	1.79	2.42
	Inst	SF	Sm	2C	.038	421.0	.71	1.26	.11	2.08	2.82
2" x 4" - 8', 16" oc	Demo	SF	Lg	LB	.024	674.0	---	.66	---	.66	.98
	Demo	SF	Sm	LB	.028	573.0	---	.77	---	.77	1.14
2" x 4" - 8', 16" oc	Inst	SF	Lg	2C	.029	555.0	.52	.96	.08	1.56	2.13
	Inst	SF	Sm	2C	.034	472.0	.58	1.13	.10	1.81	2.46
2" x 4" - 8', 24" oc	Demo	SF	Lg	LB	.020	793.0	---	.55	---	.55	.82
	Demo	SF	Sm	LB	.024	674.0	---	.66	---	.66	.98
2" x 4" - 8', 24" oc	Inst	SF	Lg	2C	.025	652.0	.40	.83	.07	1.30	1.78
	Inst	SF	Sm	2C	.029	554.0	.44	.96	.08	1.48	2.03

Description	Oper	Unit	Vol	Crew Size	Man-hours per Unit	Crew Output per Day	Avg Mat'l Unit Cost	Avg Labor Unit Cost	Avg Equip Unit Cost	Avg Total Unit Cost	Avg Price Incl O&P
2" x 4" - 10', 12" oc	Demo	SF	Lg	LB	.022	723.0	---	.60	---	.60	.90
	Demo	SF	Sm	LB	.026	615.0	---	.71	---	.71	1.06
2" x 4" - 10', 12" oc	Inst	SF	Lg	2C	.027	597.0	.62	.90	.08	1.60	2.14
	Inst	SF	Sm	2C	.032	507.0	.68	1.06	.09	1.83	2.47
2" x 4" - 10', 16" oc	Demo	SF	Lg	LB	.020	816.0	---	.55	---	.55	.82
	Demo	SF	Sm	LB	.023	694.0	---	.63	---	.63	.94
2" x 4" - 10', 16" oc	Inst	SF	Lg	2C	.024	674.0	.50	.80	.07	1.37	1.84
	Inst	SF	Sm	2C	.028	573.0	.55	.93	.08	1.56	2.11
2" x 4" - 10', 24" oc	Demo	SF	Lg	LB	.016	973.0	---	.44	---	.44	.65
	Demo	SF	Sm	LB	.019	827.0	---	.52	---	.52	.78
2" x 4" - 10', 24" oc	Inst	SF	Lg	2C	.020	803.0	.38	.67	.06	1.11	1.50
	Inst	SF	Sm	2C	.023	683.0	.42	.77	.07	1.26	1.70
2" x 4" - 12', 12" oc	Demo	SF	Lg	LB	.019	826.0	---	.52	---	.52	.78
	Demo	SF	Sm	LB	.023	702.0	---	.63	---	.63	.94
2" x 4" - 12', 12" oc	Inst	SF	Lg	2C	.023	683.0	.60	.77	.07	1.44	1.91
	Inst	SF	Sm	2C	.028	581.0	.66	.93	.08	1.67	2.24
2" x 4" - 12', 16" oc	Demo	SF	Lg	LB	.017	941.0	---	.47	---	.47	.69
	Demo	SF	Sm	LB	.020	800.0	---	.55	---	.55	.82
2" x 4" - 12', 16" oc	Inst	SF	Lg	2C	.021	778.0	.48	.70	.06	1.24	1.66
	Inst	SF	Sm	2C	.024	661.0	.52	.80	.07	1.39	1.87
2" x 4" - 12', 24" oc	Demo	SF	Lg	LB	.014	1142	---	.38	---	.38	.57
	Demo	SF	Sm	LB	.016	971.0	---	.44	---	.44	.65
2" x 4" - 12', 24" oc	Inst	SF	Lg	2C	.017	944.0	.36	.57	.05	.98	1.31
	Inst	SF	Sm	2C	.020	802.0	.39	.67	.06	1.12	1.51
2" x 6" - 8', 12" oc	Demo	SF	Lg	LB	.030	528.0	---	.82	---	.82	1.23
	Demo	SF	Sm	LB	.036	449.0	---	.99	---	.99	1.47
2" x 6" - 8', 12" oc	Inst	SF	Lg	2C	.036	439.0	.89	1.20	.11	2.20	2.93
	Inst	SF	Sm	2C	.043	373.0	.99	1.43	.13	2.55	3.41
2" x 6" - 8', 16" oc	Demo	SF	Lg	LB	.027	591.0	---	.74	---	.74	1.10
	Demo	SF	Sm	LB	.032	502.0	---	.88	---	.88	1.31
2" x 6" - 8', 16" oc	Inst	SF	Lg	2C	.033	491.0	.73	1.10	.10	1.93	2.59
	Inst	SF	Sm	2C	.038	417.0	.80	1.26	.11	2.17	2.93
2" x 6" - 8', 24" oc	Demo	SF	Lg	LB	.023	696.0	---	.63	---	.63	.94
	Demo	SF	Sm	LB	.027	592.0	---	.74	---	.74	1.10
2" x 6" - 8', 24" oc	Inst	SF	Lg	2C	.028	579.0	.56	.93	.08	1.57	2.12
	Inst	SF	Sm	2C	.033	492.0	.61	1.10	.09	1.80	2.44

Description	Oper	Unit	Vol	Crew Size	Man-hours per Unit	Crew Output per Day	Avg Mat'l Unit Cost	Avg Labor Unit Cost	Avg Equip Unit Cost	Avg Total Unit Cost	Avg Price Incl O&P
2" x 6" - 10', 12" oc	Demo	SF	Lg	LB	.025	632.0	---	.69	---	.69	**1.02**
	Demo	SF	Sm	LB	.030	537.0	---	.82	---	.82	**1.23**
2" x 6" - 10', 12" oc	Inst	SF	Lg	2C	.030	528.0	.85	1.00	.09	1.94	**2.56**
	Inst	SF	Sm	2C	.036	449.0	.94	1.20	.10	2.24	**2.98**
2" x 6" - 10', 16" oc	Demo	SF	Lg	LB	.022	714.0	---	.60	---	.60	**.90**
	Demo	SF	Sm	LB	.026	607.0	---	.71	---	.71	**1.06**
2" x 6" - 10', 16" oc	Inst	SF	Lg	2C	.027	596.0	.69	.90	.08	1.67	**2.22**
	Inst	SF	Sm	2C	.032	507.0	.76	1.06	.09	1.91	**2.56**
2" x 6" - 10', 24" oc	Demo	SF	Lg	LB	.019	852.0	---	.52	---	.52	**.78**
	Demo	SF	Sm	LB	.022	724.0	---	.60	---	.60	**.90**
2" x 6" - 10', 24" oc	Inst	SF	Lg	2C	.023	711.0	.52	.77	.07	1.36	**1.82**
	Inst	SF	Sm	2C	.026	604.0	.57	.87	.08	1.52	**2.03**
2" x 6" - 12', 12" oc	Demo	SF	Lg	LB	.022	721.0	---	.60	---	.60	**.90**
	Demo	SF	Sm	LB	.026	613.0	---	.71	---	.71	**1.06**
2" x 6" - 12', 12" oc	Inst	SF	Lg	2C	.027	603.0	.83	.90	.08	1.81	**2.38**
	Inst	SF	Sm	2C	.031	513.0	.91	1.03	.09	2.03	**2.68**
2" x 6" - 12', 16" oc	Demo	SF	Lg	LB	.019	822.0	---	.52	---	.52	**.78**
	Demo	SF	Sm	LB	.023	699.0	---	.63	---	.63	**.94**
2" x 6" - 12', 16" oc	Inst	SF	Lg	2C	.023	688.0	.66	.77	.07	1.50	**1.98**
	Inst	SF	Sm	2C	.027	585.0	.73	.90	.08	1.71	**2.27**
2" x 6" - 12', 24" oc	Demo	SF	Lg	LB	.016	996.0	---	.44	---	.44	**.65**
	Demo	SF	Sm	LB	.019	847.0	---	.52	---	.52	**.78**
2" x 6" - 12', 24" oc	Inst	SF	Lg	2C	.019	834.0	.50	.63	.06	1.19	**1.58**
	Inst	SF	Sm	2C	.023	709.0	.55	.77	.07	1.39	**1.85**
2" x 8" - 8', 12" oc	Demo	SF	Lg	LB	.034	468.0	---	.93	---	.93	**1.39**
	Demo	SF	Sm	LB	.040	398.0	---	1.10	---	1.10	**1.63**
2" x 8" - 8', 12" oc	Inst	SF	Lg	2C	.041	392.0	1.23	1.36	.12	2.71	**3.58**
	Inst	SF	Sm	2C	.048	333.0	1.38	1.60	.14	3.12	**4.12**
2" x 8" - 8', 16" oc	Demo	SF	Lg	LB	.030	525.0	---	.82	---	.82	**1.23**
	Demo	SF	Sm	LB	.036	446.0	---	.99	---	.99	**1.47**
2" x 8" - 8', 16" oc	Inst	SF	Lg	2C	.036	441.0	.99	1.20	.11	2.30	**3.05**
	Inst	SF	Sm	2C	.043	375.0	1.11	1.43	.12	2.66	**3.54**
2" x 8" - 8', 24" oc	Demo	SF	Lg	LB	.026	618.0	---	.71	---	.71	**1.06**
	Demo	SF	Sm	LB	.030	525.0	---	.82	---	.82	**1.23**
2" x 8" - 8', 24" oc	Inst	SF	Lg	2C	.031	518.0	.76	1.03	.09	1.88	**2.51**
	Inst	SF	Sm	2C	.036	440.0	.85	1.20	.11	2.16	**2.88**

Description	Oper	Unit	Vol	Crew Size	Man-hours per Unit	Crew Output per Day	Avg Mat'l Unit Cost	Avg Labor Unit Cost	Avg Equip Unit Cost	Avg Total Unit Cost	Avg Price Incl O&P
2" x 8" - 10', 12" oc	Demo	SF	Lg	LB	.029	560.0	---	.80	---	.80	**1.19**
	Demo	SF	Sm	LB	.034	476.0	---	.93	---	.93	**1.39**
2" x 8" - 10', 12" oc	Inst	SF	Lg	2C	.034	472.0	1.17	1.13	.10	2.40	**3.14**
	Inst	SF	Sm	2C	.040	401.0	1.31	1.33	.12	2.76	**3.62**
2" x 8" - 10', 16" oc	Demo	SF	Lg	LB	.025	633.0	---	.69	---	.69	**1.02**
	Demo	SF	Sm	LB	.030	538.0	---	.82	---	.82	**1.23**
2" x 8" - 10', 16" oc	Inst	SF	Lg	2C	.030	533.0	.94	1.00	.09	2.03	**2.67**
	Inst	SF	Sm	2C	.035	453.0	1.05	1.16	.10	2.31	**3.05**
2" x 8" - 10', 24" oc	Demo	SF	Lg	LB	.021	754.0	---	.58	---	.58	**.86**
	Demo	SF	Sm	LB	.025	641.0	---	.69	---	.69	**1.02**
2" x 8" - 10', 24" oc	Inst	SF	Lg	2C	.025	636.0	.71	.83	.07	1.61	**2.13**
	Inst	SF	Sm	2C	.030	541.0	.79	1.00	.09	1.88	**2.50**
2" x 8" - 12', 12" oc	Demo	SF	Lg	LB	.025	638.0	---	.69	---	.69	**1.02**
	Demo	SF	Sm	LB	.030	542.0	---	.82	---	.82	**1.23**
2" x 8" - 12', 12" oc	Inst	SF	Lg	2C	.030	540.0	1.14	1.00	.09	2.23	**2.90**
	Inst	SF	Sm	2C	.035	459.0	1.28	1.16	.10	2.54	**3.32**
2" x 8" - 12', 16" oc	Demo	SF	Lg	LB	.022	728.0	---	.60	---	.60	**.90**
	Demo	SF	Sm	LB	.026	619.0	---	.71	---	.71	**1.06**
2" x 8" - 12', 16" oc	Inst	SF	Lg	2C	.026	615.0	.91	.87	.08	1.86	**2.42**
	Inst	SF	Sm	2C	.031	523.0	1.02	1.03	.09	2.14	**2.81**
2" x 8" - 12', 24" oc	Demo	SF	Lg	LB	.018	881.0	---	.49	---	.49	**.74**
	Demo	SF	Sm	LB	.021	749.0	---	.58	---	.58	**.86**
2" x 8" - 12', 24" oc	Inst	SF	Lg	2C	.021	745.0	.67	.70	.06	1.43	**1.88**
	Inst	SF	Sm	2C	.025	633.0	.75	.83	.07	1.65	**2.18**

Gable ends

Description	Oper	Unit	Vol	Crew Size	Man-hours per Unit	Crew Output per Day	Avg Mat'l Unit Cost	Avg Labor Unit Cost	Avg Equip Unit Cost	Avg Total Unit Cost	Avg Price Incl O&P
2" x 4" - 3' Avg., 12" oc	Demo	SF	Lg	LB	.074	216.0	---	2.03	---	2.03	**3.02**
	Demo	SF	Sm	LB	.087	184.0	---	2.39	---	2.39	**3.56**
2" x 4" - 3' Avg., 12" oc	Inst	SF	Lg	2C	.142	113.0	1.18	4.72	.41	6.31	**8.85**
	Inst	SF	Sm	2C	.167	96.00	1.29	5.56	.49	7.34	**10.30**
2" x 4" - 3' Avg., 16" oc	Demo	SF	Lg	LB	.069	231.0	---	1.89	---	1.89	**2.82**
	Demo	SF	Sm	LB	.082	196.0	---	2.25	---	2.25	**3.35**
2" x 4" - 3' Avg., 16" oc	Inst	SF	Lg	2C	.132	121.0	1.10	4.39	.39	5.88	**8.24**
	Inst	SF	Sm	2C	.155	103.0	1.21	5.16	.45	6.82	**9.58**
2" x 4" - 3' Avg., 24" oc	Demo	SF	Lg	LB	.063	253.0	---	1.73	---	1.73	**2.57**
	Demo	SF	Sm	LB	.074	215.0	---	2.03	---	2.03	**3.02**
2" x 4" - 3' Avg., 24" oc	Inst	SF	Lg	2C	.121	132.0	1.01	4.03	.35	5.39	**7.55**
	Inst	SF	Sm	2C	.143	112.0	1.11	4.76	.42	6.29	**8.83**

Description	Oper	Unit	Vol	Crew Size	Man-hours per Unit	Crew Output per Day	Avg Mat'l Unit Cost	Avg Labor Unit Cost	Avg Equip Unit Cost	Avg Total Unit Cost	Avg Price Incl O&P
2" x 4" - 4' Avg., 12" oc	Demo	SF	Lg	LB	.050	323.0	---	1.37	---	1.37	2.04
	Demo	SF	Sm	LB	.058	275.0	---	1.59	---	1.59	2.37
2" x 4" - 4' Avg., 12" oc	Inst	SF	Lg	2C	.094	170.0	.98	3.13	.27	4.38	6.09
	Inst	SF	Sm	2C	.110	145.0	1.08	3.66	.32	5.06	7.05
2" x 4" - 4' Avg., 16" oc	Demo	SF	Lg	LB	.046	348.0	---	1.26	---	1.26	1.88
	Demo	SF	Sm	LB	.054	296.0	---	1.48	---	1.48	2.21
2" x 4" - 4' Avg., 16" oc	Inst	SF	Lg	2C	.087	183.0	.91	2.89	.25	4.05	5.64
	Inst	SF	Sm	2C	.103	156.0	1.00	3.43	.30	4.73	6.59
2" x 4" - 4' Avg., 24" oc	Demo	SF	Lg	LB	.042	384.0	---	1.15	---	1.15	1.72
	Demo	SF	Sm	LB	.049	326.0	---	1.34	---	1.34	2.00
2" x 4" - 4' Avg., 24" oc	Inst	SF	Lg	2C	.079	202.0	.82	2.63	.23	3.68	5.12
	Inst	SF	Sm	2C	.093	172.0	.90	3.09	.27	4.26	5.95
2" x 4" - 5' Avg., 12" oc	Demo	SF	Lg	LB	.039	412.0	---	1.07	---	1.07	1.59
	Demo	SF	Sm	LB	.046	350.0	---	1.26	---	1.26	1.88
2" x 4" - 5' Avg., 12" oc	Inst	SF	Lg	2C	.073	219.0	.90	2.43	.21	3.54	4.89
	Inst	SF	Sm	2C	.086	186.0	.99	2.86	.25	4.10	5.68
2" x 4" - 5' Avg., 16" oc	Demo	SF	Lg	LB	.035	454.0	---	.96	---	.96	1.43
	Demo	SF	Sm	LB	.041	386.0	---	1.12	---	1.12	1.68
2" x 4" - 5' Avg., 16" oc	Inst	SF	Lg	2C	.066	241.0	.82	2.20	.19	3.21	4.43
	Inst	SF	Sm	2C	.078	205.0	.90	2.60	.23	3.73	5.16
2" x 4" - 5' Avg., 24" oc	Demo	SF	Lg	LB	.031	514.0	---	.85	---	.85	1.27
	Demo	SF	Sm	LB	.037	437.0	---	1.01	---	1.01	1.51
2" x 4" - 5' Avg., 24" oc	Inst	SF	Lg	2C	.059	273.0	.73	1.96	.17	2.86	3.95
	Inst	SF	Sm	2C	.069	232.0	.81	2.30	.20	3.31	4.58
2" x 4" - 6' Avg., 12" oc	Demo	SF	Lg	LB	.032	501.0	---	.88	---	.88	1.31
	Demo	SF	Sm	LB	.038	426.0	---	1.04	---	1.04	1.55
2" x 4" - 6' Avg., 12" oc	Inst	SF	Lg	2C	.059	269.0	.84	1.96	.17	2.97	4.08
	Inst	SF	Sm	2C	.070	229.0	.92	2.33	.20	3.45	4.75
2" x 4" - 6' Avg., 16" oc	Demo	SF	Lg	LB	.028	564.0	---	.77	---	.77	1.14
	Demo	SF	Sm	LB	.033	479.0	---	.91	---	.91	1.35
2" x 4" - 6' Avg., 16" oc	Inst	SF	Lg	2C	.053	302.0	.74	1.76	.15	2.65	3.65
	Inst	SF	Sm	2C	.062	257.0	.82	2.06	.18	3.06	4.22
2" x 4" - 6' Avg., 24" oc	Demo	SF	Lg	LB	.025	650.0	---	.69	---	.69	1.02
	Demo	SF	Sm	LB	.029	553.0	---	.80	---	.80	1.19
2" x 4" - 6' Avg., 24" oc	Inst	SF	Lg	2C	.046	349.0	.65	1.53	.13	2.31	3.17
	Inst	SF	Sm	2C	.054	297.0	.72	1.80	.16	2.68	3.68

Description	Oper	Unit	Vol	Crew Size	Man-hours per Unit	Crew Output per Day	Avg Mat'l Unit Cost	Avg Labor Unit Cost	Avg Equip Unit Cost	Avg Total Unit Cost	Avg Price Incl O&P
2" x 6" - 3' Avg., 12" oc	Demo	SF	Lg	LB	.077	208.0	---	2.11	---	2.11	**3.15**
	Demo	SF	Sm	LB	.090	177.0	---	2.47	---	2.47	**3.68**
2" x 6" - 3' Avg., 12" oc	Inst	SF	Lg	2C	.145	110.0	1.65	4.82	.42	6.89	**9.55**
	Inst	SF	Sm	2C	.170	94.00	1.82	5.66	.50	7.98	**11.10**
2" x 6" - 3' Avg., 16" oc	Demo	SF	Lg	LB	.072	222.0	---	1.97	---	1.97	**2.94**
	Demo	SF	Sm	LB	.085	189.0	---	2.33	---	2.33	**3.47**
2" x 6" - 3' Avg., 16" oc	Inst	SF	Lg	2C	.136	118.0	1.55	4.52	.40	6.47	**8.97**
	Inst	SF	Sm	2C	.160	100.0	1.71	5.32	.47	7.50	**10.40**
2" x 6" - 3' Avg., 24" oc	Demo	SF	Lg	LB	.066	243.0	---	1.81	---	1.81	**2.70**
	Demo	SF	Sm	LB	.077	207.0	---	2.11	---	2.11	**3.15**
2" x 6" - 3' Avg., 24" oc	Inst	SF	Lg	2C	.124	129.0	1.42	4.13	.36	5.91	**8.18**
	Inst	SF	Sm	2C	.145	110.0	1.57	4.82	.42	6.81	**9.46**
2" x 6" - 4' Avg., 12" oc	Demo	SF	Lg	LB	.052	307.0	---	1.43	---	1.43	**2.13**
	Demo	SF	Sm	LB	.061	261.0	---	1.67	---	1.67	**2.49**
2" x 6" - 4' Avg., 12" oc	Inst	SF	Lg	2C	.096	166.0	1.37	3.19	.28	4.84	**6.65**
	Inst	SF	Sm	2C	.113	141.0	1.52	3.76	.33	5.61	**7.72**
2" x 6" - 4' Avg., 16" oc	Demo	SF	Lg	LB	.048	331.0	---	1.32	---	1.32	**1.96**
	Demo	SF	Sm	LB	.057	281.0	---	1.56	---	1.56	**2.33**
2" x 6" - 4' Avg., 16" oc	Inst	SF	Lg	2C	.089	179.0	1.27	2.96	.26	4.49	**6.16**
	Inst	SF	Sm	2C	.105	152.0	1.41	3.49	.31	5.21	**7.17**
2" x 6" - 4' Avg., 24" oc	Demo	SF	Lg	LB	.044	366.0	---	1.21	---	1.21	**1.80**
	Demo	SF	Sm	LB	.051	311.0	---	1.40	---	1.40	**2.08**
2" x 6" - 4' Avg., 24" oc	Inst	SF	Lg	2C	.081	197.0	1.15	2.69	.24	4.08	**5.61**
	Inst	SF	Sm	2C	.096	167.0	1.27	3.19	.28	4.74	**6.53**
2" x 6" - 5' Avg., 12" oc	Demo	SF	Lg	LB	.041	389.0	---	1.12	---	1.12	**1.68**
	Demo	SF	Sm	LB	.048	331.0	---	1.32	---	1.32	**1.96**
2" x 6" - 5' Avg., 12" oc	Inst	SF	Lg	2C	.075	212.0	1.26	2.50	.22	3.98	**5.41**
	Inst	SF	Sm	2C	.089	180.0	1.39	2.96	.26	4.61	**6.30**
2" x 6" - 5' Avg., 16" oc	Demo	SF	Lg	LB	.037	428.0	---	1.01	---	1.01	**1.51**
	Demo	SF	Sm	LB	.044	364.0	---	1.21	---	1.21	**1.80**
2" x 6" - 5' Avg., 16" oc	Inst	SF	Lg	2C	.068	234.0	1.15	2.26	.20	3.61	**4.92**
	Inst	SF	Sm	2C	.080	199.0	1.27	2.66	.23	4.16	**5.68**
2" x 6" - 5' Avg., 24" oc	Demo	SF	Lg	LB	.033	484.0	---	.91	---	.91	**1.35**
	Demo	SF	Sm	LB	.039	411.0	---	1.07	---	1.07	**1.59**
2" x 6" - 5' Avg., 24" oc	Inst	SF	Lg	2C	.060	265.0	1.02	2.00	.18	3.20	**4.35**
	Inst	SF	Sm	2C	.071	225.0	1.13	2.36	.21	3.70	**5.05**

Description	Oper	Unit	Vol	Crew Size	Man-hours per Unit	Crew Output per Day	Avg Mat'l Unit Cost	Avg Labor Unit Cost	Avg Equip Unit Cost	Avg Total Unit Cost	Avg Price Incl O&P
2" x 6" - 6' Avg., 12" oc	Demo	SF	Lg	LB	.034	469.0	---	.93	---	.93	**1.39**
	Demo	SF	Sm	LB	.040	399.0	---	1.10	---	1.10	**1.63**
2" x 6" - 6' Avg., 12" oc	Inst	SF	Lg	2C	.062	259.0	1.17	2.06	.18	3.41	**4.62**
	Inst	SF	Sm	2C	.073	220.0	1.29	2.43	.21	3.93	**5.34**
2" x 6" - 6' Avg., 16" oc	Demo	SF	Lg	LB	.030	528.0	---	.82	---	.82	**1.23**
	Demo	SF	Sm	LB	.036	449.0	---	.99	---	.99	**1.47**
2" x 6" - 6' Avg., 16" oc	Inst	SF	Lg	2C	.055	292.0	1.04	1.83	.16	3.03	**4.10**
	Inst	SF	Sm	2C	.065	248.0	1.15	2.16	.19	3.50	**4.76**
2" x 6" - 6' Avg., 24" oc	Demo	SF	Lg	LB	.026	609.0	---	.71	---	.71	**1.06**
	Demo	SF	Sm	LB	.031	518.0	---	.85	---	.85	**1.27**
2" x 6" - 6' Avg., 24" oc	Inst	SF	Lg	2C	.048	336.0	.91	1.60	.14	2.65	**3.58**
	Inst	SF	Sm	2C	.056	286.0	1.00	1.86	.16	3.02	**4.10**
2" x 8" - 3' Avg., 12" oc	Demo	SF	Lg	LB	.080	200.0	---	2.19	---	2.19	**3.27**
	Demo	SF	Sm	LB	.094	170.0	---	2.58	---	2.58	**3.84**
2" x 8" - 3' Avg., 12" oc	Inst	SF	Lg	2C	.148	108.0	2.27	4.92	.43	7.62	**10.40**
	Inst	SF	Sm	2C	.174	92.00	2.54	5.79	.51	8.84	**12.10**
2" x 8" - 3' Avg., 16" oc	Demo	SF	Lg	LB	.075	213.0	---	2.06	---	2.06	**3.07**
	Demo	SF	Sm	LB	.088	181.0	---	2.41	---	2.41	**3.60**
2" x 8" - 3' Avg., 16" oc	Inst	SF	Lg	2C	.138	116.0	2.13	4.59	.40	7.12	**9.74**
	Inst	SF	Sm	2C	.162	99.00	2.39	5.39	.47	8.25	**11.30**
2" x 8" - 3' Avg., 24" oc	Demo	SF	Lg	LB	.068	234.0	---	1.87	---	1.87	**2.78**
	Demo	SF	Sm	LB	.080	199.0	---	2.19	---	2.19	**3.27**
2" x 8" - 3' Avg., 24" oc	Inst	SF	Lg	2C	.126	127.0	1.95	4.19	.37	6.51	**8.90**
	Inst	SF	Sm	2C	.148	108.0	2.18	4.92	.43	7.53	**10.30**
2" x 8" - 4' Avg., 12" oc	Demo	SF	Lg	LB	.055	293.0	---	1.51	---	1.51	**2.25**
	Demo	SF	Sm	LB	.064	249.0	---	1.76	---	1.76	**2.62**
2" x 8" - 4' Avg., 12" oc	Inst	SF	Lg	2C	.099	161.0	1.89	3.29	.29	5.47	**7.40**
	Inst	SF	Sm	2C	.117	137.0	2.12	3.89	.34	6.35	**8.62**
2" x 8" - 4' Avg., 16" oc	Demo	SF	Lg	LB	.051	316.0	---	1.40	---	1.40	**2.08**
	Demo	SF	Sm	LB	.059	269.0	---	1.62	---	1.62	**2.41**
2" x 8" - 4' Avg., 16" oc	Inst	SF	Lg	2C	.092	174.0	1.75	3.06	.27	5.08	**6.87**
	Inst	SF	Sm	2C	.108	148.0	1.96	3.59	.32	5.87	**7.96**
2" x 8" - 4' Avg., 24" oc	Demo	SF	Lg	LB	.046	349.0	---	1.26	---	1.26	**1.88**
	Demo	SF	Sm	LB	.054	297.0	---	1.48	---	1.48	**2.21**
2" x 8" - 4' Avg., 24" oc	Inst	SF	Lg	2C	.083	192.0	1.58	2.76	.24	4.58	**6.20**
	Inst	SF	Sm	2C	.098	163.0	1.77	3.26	.29	5.32	**7.22**

Description	Oper	Unit	Vol	Crew Size	Man-hours per Unit	Crew Output per Day	Avg Mat'l Unit Cost	Avg Labor Unit Cost	Avg Equip Unit Cost	Avg Total Unit Cost	Avg Price Incl O&P
2" x 8" - 5' Avg., 12" oc	Demo	SF	Lg	LB	.043	368.0	---	1.18	---	1.18	**1.76**
	Demo	SF	Sm	LB	.051	313.0	---	1.40	---	1.40	**2.08**
2" x 8" - 5' Avg., 12" oc	Inst	SF	Lg	2C	.078	206.0	1.73	2.60	.23	4.56	**6.11**
	Inst	SF	Sm	2C	.091	175.0	1.94	3.03	.27	5.24	**7.04**
2" x 8" - 5' Avg., 16" oc	Demo	SF	Lg	LB	.040	405.0	---	1.10	---	1.10	**1.63**
	Demo	SF	Sm	LB	.047	344.0	---	1.29	---	1.29	**1.92**
2" x 8" - 5' Avg., 16" oc	Inst	SF	Lg	2C	.070	227.0	1.55	2.33	.21	4.09	**5.49**
	Inst	SF	Sm	2C	.083	193.0	1.74	2.76	.24	4.74	**6.38**
2" x 8" - 5' Avg., 24" oc	Demo	SF	Lg	LB	.035	458.0	---	.96	---	.96	**1.43**
	Demo	SF	Sm	LB	.041	389.0	---	1.12	---	1.12	**1.68**
2" x 8" - 5' Avg., 24" oc	Inst	SF	Lg	2C	.062	257.0	1.37	2.06	.18	3.61	**4.85**
	Inst	SF	Sm	2C	.073	218.0	1.53	2.43	.21	4.17	**5.61**
2" x 8" - 6' Avg., 12" oc	Demo	SF	Lg	LB	.036	441.0	---	.99	---	.99	**1.47**
	Demo	SF	Sm	LB	.043	375.0	---	1.18	---	1.18	**1.76**
2" x 8" - 6' Avg., 12" oc	Inst	SF	Lg	2C	.064	251.0	1.61	2.13	.19	3.93	**5.24**
	Inst	SF	Sm	2C	.075	213.0	1.80	2.50	.22	4.52	**6.03**
2" x 8" - 6' Avg., 16" oc	Demo	SF	Lg	LB	.032	497.0	---	.88	---	.88	**1.31**
	Demo	SF	Sm	LB	.038	422.0	---	1.04	---	1.04	**1.55**
2" x 8" - 6' Avg., 16" oc	Inst	SF	Lg	2C	.057	282.0	1.43	1.90	.17	3.50	**4.66**
	Inst	SF	Sm	2C	.067	240.0	1.60	2.23	.19	4.02	**5.37**
2" x 8" - 6' Avg., 24" oc	Demo	SF	Lg	LB	.028	573.0	---	.77	---	.77	**1.14**
	Demo	SF	Sm	LB	.033	487.0	---	.91	---	.91	**1.35**
2" x 8" - 6' Avg., 24" oc	Inst	SF	Lg	2C	.049	325.0	1.25	1.63	.14	3.02	**4.02**
	Inst	SF	Sm	2C	.058	276.0	1.40	1.93	.17	3.50	**4.67**

Studs/plates, per LF of wall or partition

Walls or partitions

Description	Oper	Unit	Vol	Crew Size	Man-hours per Unit	Crew Output per Day	Avg Mat'l Unit Cost	Avg Labor Unit Cost	Avg Equip Unit Cost	Avg Total Unit Cost	Avg Price Incl O&P
2" x 4" - 8', 12" oc	Demo	LF	Lg	LB	.213	75.00	---	5.84	---	5.84	**8.71**
	Demo	LF	Sm	LB	.250	64.00	---	6.86	---	6.86	**10.20**
2" x 4" - 8', 12" oc	Inst	LF	Lg	2C	.258	62.00	4.92	8.58	.75	14.25	**19.30**
	Inst	LF	Sm	2C	.302	53.00	5.40	10.10	.88	16.38	**22.20**
2" x 4" - 10', 12" oc	Demo	LF	Lg	LB	.222	72.00	---	6.09	---	6.09	**9.07**
	Demo	LF	Sm	LB	.262	61.00	---	7.19	---	7.19	**10.70**
2" x 4" - 10', 12" oc	Inst	LF	Lg	2C	.271	59.00	5.87	9.02	.79	15.68	**21.10**
	Inst	LF	Sm	2C	.320	50.00	6.44	10.70	.93	18.07	**24.30**
2" x 4" - 12', 12" oc	Demo	LF	Lg	LB	.232	69.00	---	6.36	---	6.36	**9.48**
	Demo	LF	Sm	LB	.271	59.00	---	7.43	---	7.43	**11.10**
2" x 4" - 12', 12" oc	Inst	LF	Lg	2C	.281	57.00	6.84	9.35	.82	17.01	**22.70**
	Inst	LF	Sm	2C	.333	48.00	7.51	11.10	.97	19.58	**26.20**

Description	Oper	Unit	Vol	Crew Size	Man-hours per Unit	Crew Output per Day	Avg Mat'l Unit Cost	Avg Labor Unit Cost	Avg Equip Unit Cost	Avg Total Unit Cost	Avg Price Incl O&P
2" x 4" - 8', 16" oc	Demo	LF	Lg	LB	.188	85.00	---	5.16	---	5.16	7.68
	Demo	LF	Sm	LB	.222	72.00	---	6.09	---	6.09	9.07
2" x 4" - 8', 16" oc	Inst	LF	Lg	2C	.229	70.00	3.97	7.62	.67	12.26	16.70
	Inst	LF	Sm	2C	.267	60.00	4.36	8.88	.78	14.02	19.10
2" x 4" - 10', 16" oc	Demo	LF	Lg	LB	.198	81.00	---	5.43	---	5.43	8.09
	Demo	LF	Sm	LB	.232	69.00	---	6.36	---	6.36	9.48
2" x 4" - 10', 16" oc	Inst	LF	Lg	2C	.239	67.00	4.69	7.95	.70	13.34	18.00
	Inst	LF	Sm	2C	.281	57.00	5.15	9.35	.82	15.32	20.80
2" x 4" - 12', 16" oc	Demo	LF	Lg	LB	.205	78.00	---	5.62	---	5.62	8.38
	Demo	LF	Sm	LB	.242	66.00	---	6.64	---	6.64	9.89
2" x 4" - 12', 16" oc	Inst	LF	Lg	2C	.246	65.00	5.37	8.18	.72	14.27	19.20
	Inst	LF	Sm	2C	.291	55.00	5.89	9.68	.85	16.42	22.20
2" x 4" - 8', 24" oc	Demo	LF	Lg	LB	.162	99.00	---	4.44	---	4.44	6.62
	Demo	LF	Sm	LB	.190	84.00	---	5.21	---	5.21	7.77
2" x 4" - 8', 24" oc	Inst	LF	Lg	2C	.198	81.00	2.99	6.59	.58	10.16	13.90
	Inst	LF	Sm	2C	.232	69.00	3.28	7.72	.68	11.68	16.00
2" x 4" - 10', 24" oc	Demo	LF	Lg	LB	.165	97.00	---	4.53	---	4.53	6.74
	Demo	LF	Sm	LB	.195	82.00	---	5.35	---	5.35	7.97
2" x 4" - 10', 24" oc	Inst	LF	Lg	2C	.200	80.00	3.51	6.65	.58	10.74	14.60
	Inst	LF	Sm	2C	.235	68.00	3.86	7.82	.69	12.37	16.90
2" x 4" - 12', 24" oc	Demo	LF	Lg	LB	.168	95.00	---	4.61	---	4.61	6.87
	Demo	LF	Sm	LB	.198	81.00	---	5.43	---	5.43	8.09
2" x 4" - 12', 24" oc	Inst	LF	Lg	2C	.203	79.00	3.95	6.75	.59	11.29	15.30
	Inst	LF	Sm	2C	.239	67.00	4.34	7.95	.70	12.99	17.60
2" x 6" - 8', 12" oc	Demo	LF	Lg	LB	.242	66.00	---	6.64	---	6.64	9.89
	Demo	LF	Sm	LB	.286	56.00	---	7.84	---	7.84	11.70
2" x 6" - 8', 12" oc	Inst	LF	Lg	2C	.291	55.00	6.93	9.68	.85	17.46	23.30
	Inst	LF	Sm	2C	.340	47.00	7.65	11.30	.99	19.94	26.80
2" x 6" - 10', 12" oc	Demo	LF	Lg	LB	.254	63.00	---	6.97	---	6.97	10.40
	Demo	LF	Sm	LB	.296	54.00	---	8.12	---	8.12	12.10
2" x 6" - 10', 12" oc	Inst	LF	Lg	2C	.302	53.00	8.25	10.10	.88	19.23	25.40
	Inst	LF	Sm	2C	.366	45.00	9.11	11.80	1.04	21.05	28.30
2" x 6" - 12', 12" oc	Demo	LF	Lg	LB	.267	60.00	---	7.32	---	7.32	10.90
	Demo	LF	Sm	LB	.314	51.00	---	8.61	---	8.61	12.80
2" x 6" - 12', 12" oc	Inst	LF	Lg	2C	.320	50.00	9.59	10.70	.93	21.22	27.90
	Inst	LF	Sm	2C	.372	43.00	10.60	12.40	1.09	24.09	31.80
2" x 6" - 8', 16" oc	Demo	LF	Lg	LB	.216	74.00	---	5.92	---	5.92	8.83
	Demo	LF	Sm	LB	.254	63.00	---	6.97	---	6.97	10.40
2" x 6" - 8', 16" oc	Inst	LF	Lg	2C	.262	61.00	5.60	8.72	.76	15.08	20.30
	Inst	LF	Sm	2C	.308	52.00	6.18	10.30	.90	17.38	23.40

Description	Oper	Unit	Vol	Crew Size	Man-hours per Unit	Crew Output per Day	Avg Mat'l Unit Cost	Avg Labor Unit Cost	Avg Equip Unit Cost	Avg Total Unit Cost	Avg Price Incl O&P
2" x 6" - 10', 16" oc	Demo	LF	Lg	LB	.225	71.00	---	6.17	---	6.17	**9.20**
	Demo	LF	Sm	LB	.267	60.00	---	7.32	---	7.32	**10.90**
2" x 6" - 10', 16" oc	Inst	LF	Lg	2C	.271	59.00	6.58	9.02	.79	16.39	**21.90**
	Inst	LF	Sm	2C	.320	50.00	7.27	10.70	.93	18.90	**25.30**
2" x 6" - 12', 16" oc	Demo	LF	Lg	LB	.232	69.00	---	6.36	---	6.36	**9.48**
	Demo	LF	Sm	LB	.271	59.00	---	7.43	---	7.43	**11.10**
2" x 6" - 12', 16" oc	Inst	LF	Lg	2C	.276	58.00	7.59	9.18	.80	17.57	**23.30**
	Inst	LF	Sm	2C	.327	49.00	8.38	10.90	.95	20.23	**26.90**
2" x 6" - 8', 24" oc	Demo	LF	Lg	LB	.184	87.00	---	5.05	---	5.05	**7.52**
	Demo	LF	Sm	LB	.216	74.00	---	5.92	---	5.92	**8.83**
2" x 6" - 8', 24" oc	Inst	LF	Lg	2C	.222	72.00	4.23	7.39	.65	12.27	**16.60**
	Inst	LF	Sm	2C	.262	61.00	4.67	8.72	.76	14.15	**19.20**
2" x 6" - 10', 24" oc	Demo	LF	Lg	LB	.188	85.00	---	5.16	---	5.16	**7.68**
	Demo	LF	Sm	LB	.222	72.00	---	6.09	---	6.09	**9.07**
2" x 6" - 10', 24" oc	Inst	LF	Lg	2C	.225	71.00	4.92	7.49	.66	13.07	**17.60**
	Inst	LF	Sm	2C	.267	60.00	5.43	8.88	.78	15.09	**20.40**
2" x 6" - 12', 24" oc	Demo	LF	Lg	LB	.190	84.00	---	5.21	---	5.21	**7.77**
	Demo	LF	Sm	LB	.225	71.00	---	6.17	---	6.17	**9.20**
2" x 6" - 12', 24" oc	Inst	LF	Lg	2C	.229	70.00	5.60	7.62	.67	13.89	**18.50**
	Inst	LF	Sm	2C	.267	60.00	6.18	8.88	.78	15.84	**21.20**
2" x 8" - 8', 12" oc	Demo	LF	Lg	LB	.271	59.00	---	7.43	---	7.43	**11.10**
	Demo	LF	Sm	LB	.320	50.00	---	8.78	---	8.78	**13.10**
2" x 8" - 8', 12" oc	Inst	LF	Lg	2C	.327	49.00	9.61	10.90	.95	21.46	**28.30**
	Inst	LF	Sm	2C	.381	42.00	10.80	12.70	1.11	24.61	**32.50**
2" x 8" - 10', 12" oc	Demo	LF	Lg	LB	.286	56.00	---	7.84	---	7.84	**11.70**
	Demo	LF	Sm	LB	.333	48.00	---	9.13	---	9.13	**13.60**
2" x 8" - 10', 12" oc	Inst	LF	Lg	2C	.340	47.00	11.40	11.30	.99	23.69	**31.10**
	Inst	LF	Sm	2C	.400	40.00	12.80	13.30	1.17	27.27	**35.80**
2" x 8" - 12', 12" oc	Demo	LF	Lg	LB	.302	53.00	---	8.28	---	8.28	**12.30**
	Demo	LF	Sm	LB	.356	45.00	---	9.77	---	9.77	**14.60**
2" x 8" - 12', 12" oc	Inst	LF	Lg	2C	.356	45.00	13.30	11.80	1.04	26.14	**34.10**
	Inst	LF	Sm	2C	.421	38.00	14.90	14.00	1.23	30.13	**39.40**
2" x 8" - 8', 16" oc	Demo	LF	Lg	LB	.242	66.00	---	6.64	---	6.64	**9.89**
	Demo	LF	Sm	LB	.286	56.00	---	7.84	---	7.84	**11.70**
2" x 8" - 8', 16" oc	Inst	LF	Lg	2C	.291	55.00	7.72	9.68	.85	18.25	**24.30**
	Inst	LF	Sm	2C	.340	47.00	8.64	11.30	.99	20.93	**27.90**
2" x 8" - 10', 16" oc	Demo	LF	Lg	LB	.254	63.00	---	6.97	---	6.97	**10.40**
	Demo	LF	Sm	LB	.296	54.00	---	8.12	---	8.12	**12.10**
2" x 8" - 10', 16" oc	Inst	LF	Lg	2C	.302	53.00	9.11	10.10	.88	20.09	**26.40**
	Inst	LF	Sm	2C	.356	45.00	10.20	11.80	1.04	23.04	**30.60**
2" x 8" - 12', 16" oc	Demo	LF	Lg	LB	.262	61.00	---	7.19	---	7.19	**10.70**
	Demo	LF	Sm	LB	.308	52.00	---	8.45	---	8.45	**12.60**
2" x 8" - 12', 16" oc	Inst	LF	Lg	2C	.308	52.00	10.60	10.30	.90	21.80	**28.40**
	Inst	LF	Sm	2C	.364	44.00	11.80	12.10	1.06	24.96	**32.80**

Description	Oper	Unit	Vol	Crew Size	Man-hours per Unit	Crew Output per Day	Avg Mat'l Unit Cost	Avg Labor Unit Cost	Avg Equip Unit Cost	Avg Total Unit Cost	Avg Price Incl O&P
2" x 8" - 8', 24" oc	Demo	LF	Lg	LB	.205	78.00	---	5.62	---	5.62	**8.38**
	Demo	LF	Sm	LB	.242	66.00	---	6.64	---	6.64	**9.89**
2" x 8" - 8', 24" oc	Inst	LF	Lg	2C	.246	65.00	5.87	8.18	.72	14.77	**19.80**
	Inst	LF	Sm	2C	.291	55.00	6.57	9.68	.85	17.10	**22.90**
2" x 8" - 10', 24" oc	Demo	LF	Lg	LB	.211	76.00	---	5.79	---	5.79	**8.62**
	Demo	LF	Sm	LB	.246	65.00	---	6.75	---	6.75	**10.10**
2" x 8" - 10', 24" oc	Inst	LF	Lg	2C	.250	64.00	6.80	8.32	.73	15.85	**21.00**
	Inst	LF	Sm	2C	.296	54.00	7.62	9.85	.86	18.33	**24.40**
2" x 8" - 12', 24" oc	Demo	LF	Lg	LB	.216	74.00	---	5.92	---	5.92	**8.83**
	Demo	LF	Sm	LB	.254	63.00	---	6.97	---	6.97	**10.40**
2" x 8" - 12', 24" oc	Inst	LF	Lg	2C	.258	62.00	7.72	8.58	.75	17.05	**22.50**
	Inst	LF	Sm	2C	.302	53.00	8.64	10.10	.88	19.62	**25.90**

Sheathing, walls

Boards, 1" x 8"

Description	Oper	Unit	Vol	Crew Size	Man-hours per Unit	Crew Output per Day	Avg Mat'l Unit Cost	Avg Labor Unit Cost	Avg Equip Unit Cost	Avg Total Unit Cost	Avg Price Incl O&P
Horizontal	Demo	SF	Lg	LB	.021	755.0	---	.58	---	.58	**.86**
	Demo	SF	Sm	LB	.025	642.0	---	.69	---	.69	**1.02**
Horizontal	Inst	SF	Lg	2C	.023	695.0	.73	.77	.07	1.57	**2.06**
	Inst	SF	Sm	2C	.027	591.0	.80	.90	.08	1.78	**2.35**
Diagonal	Demo	SF	Lg	LB	.024	672.0	---	.66	---	.66	**.98**
	Demo	SF	Sm	LB	.028	571.0	---	.77	---	.77	**1.14**
Diagonal	Inst	SF	Lg	2C	.026	618.0	.79	.87	.08	1.74	**2.29**
	Inst	SF	Sm	2C	.030	525.0	.87	1.00	.09	1.96	**2.59**

Plywood

Description	Oper	Unit	Vol	Crew Size	Man-hours per Unit	Crew Output per Day	Avg Mat'l Unit Cost	Avg Labor Unit Cost	Avg Equip Unit Cost	Avg Total Unit Cost	Avg Price Incl O&P
3/8"	Demo	SF	Lg	LB	.014	1145	---	.38	---	.38	**.57**
	Demo	SF	Sm	LB	.016	973.0	---	.44	---	.44	**.65**
3/8"	Inst	SF	Lg	2C	.015	1059	.41	.50	.04	.95	**1.26**
	Inst	SF	Sm	2C	.018	900.0	.45	.60	.05	1.10	**1.47**
1/2"	Demo	SF	Lg	LB	.014	1145	---	.38	---	.38	**.57**
	Demo	SF	Sm	LB	.016	973.0	---	.44	---	.44	**.65**
1/2"	Inst	SF	Lg	2C	.015	1059	.55	.50	.04	1.09	**1.42**
	Inst	SF	Sm	2C	.018	900.0	.61	.60	.05	1.26	**1.65**
5/8"	Demo	SF	Lg	LB	.014	1145	---	.38	---	.38	**.57**
	Demo	SF	Sm	LB	.016	973.0	---	.44	---	.44	**.65**
5/8"	Inst	SF	Lg	2C	.015	1059	.67	.50	.04	1.21	**1.56**
	Inst	SF	Sm	2C	.018	900.0	.74	.60	.05	1.39	**1.80**

Particleboard

Description	Oper	Unit	Vol	Crew Size	Man-hours per Unit	Crew Output per Day	Avg Mat'l Unit Cost	Avg Labor Unit Cost	Avg Equip Unit Cost	Avg Total Unit Cost	Avg Price Incl O&P
1/2"	Demo	SF	Lg	LB	.014	1145	---	.38	---	.38	**.57**
	Demo	SF	Sm	LB	.016	973.0	---	.44	---	.44	**.65**
1/2"	Inst	SF	Lg	2C	.015	1059	.46	.50	.04	1.00	**1.32**
	Inst	SF	Sm	2C	.018	900.0	.52	.60	.05	1.17	**1.55**

Description	Oper	Unit	Vol	Crew Size	Man-hours per Unit	Crew Output per Day	Avg Mat'l Unit Cost	Avg Labor Unit Cost	Avg Equip Unit Cost	Avg Total Unit Cost	Avg Price Incl O&P
OSB strand board											
3/8"	Demo	SF	Lg	LB	.014	1145	---	.38	---	.38	.57
	Demo	SF	Sm	LB	.016	973.0	---	.44	---	.44	.65
3/8"	Inst	SF	Lg	2C	.015	1059	.36	.50	.04	.90	1.20
	Inst	SF	Sm	2C	.018	900.0	.41	.60	.05	1.06	1.42
1/2"	Demo	SF	Lg	LB	.014	1145	---	.38	---	.38	.57
	Demo	SF	Sm	LB	.016	973.0	---	.44	---	.44	.65
1/2"	Inst	SF	Lg	2C	.015	1059	.42	.50	.04	.96	1.27
	Inst	SF	Sm	2C	.018	900.0	.46	.60	.05	1.11	1.48
5/8"	Demo	SF	Lg	LB	.014	1145	---	.38	---	.38	.57
	Demo	SF	Sm	LB	.016	973.0	---	.44	---	.44	.65
5/8"	Inst	SF	Lg	2C	.015	1059	.67	.50	.04	1.21	1.56
	Inst	SF	Sm	2C	.018	900.0	.74	.60	.05	1.39	1.80
Sheathing, roof											
Boards, 1" x 8"											
Horizontal	Demo	SF	Lg	LB	.018	884.0	---	.49	---	.49	.74
	Demo	SF	Sm	LB	.021	751.0	---	.58	---	.58	.86
Horizontal	Inst	SF	Lg	2C	.020	819.0	.73	.67	.06	1.46	1.90
	Inst	SF	Sm	2C	.023	696.0	.80	.77	.07	1.64	2.14
Diagonal	Demo	SF	Lg	LB	.021	775.0	---	.58	---	.58	.86
	Demo	SF	Sm	LB	.024	659.0	---	.66	---	.66	.98
Diagonal	Inst	SF	Lg	2C	.022	715.0	.79	.73	.07	1.59	2.08
	Inst	SF	Sm	2C	.026	608.0	.87	.87	.08	1.82	2.38
Plywood											
1/2"	Demo	SF	Lg	LB	.013	1277	---	.36	---	.36	.53
	Demo	SF	Sm	LB	.015	1085	---	.41	---	.41	.61
1/2"	Inst	SF	Lg	2C	.013	1187	.53	.43	.04	1.00	1.30
	Inst	SF	Sm	2C	.016	1009	.58	.53	.05	1.16	1.52
5/8"	Demo	SF	Lg	LB	.013	1277	---	.36	---	.36	.53
	Demo	SF	Sm	LB	.015	1085	---	.41	---	.41	.61
5/8"	Inst	SF	Lg	2C	.013	1187	.64	.43	.04	1.11	1.42
	Inst	SF	Sm	2C	.016	1009	.70	.53	.05	1.28	1.65
3/4"	Demo	SF	Lg	LB	.013	1187	---	.36	---	.36	.53
	Demo	SF	Sm	LB	.016	1009	---	.44	---	.44	.65
3/4"	Inst	SF	Lg	2C	.015	1102	.75	.50	.04	1.29	1.65
	Inst	SF	Sm	2C	.017	937.0	.83	.57	.05	1.45	1.85

Description	Oper	Unit	Vol	Crew Size	Man-hours per Unit	Crew Output per Day	Avg Mat'l Unit Cost	Avg Labor Unit Cost	Avg Equip Unit Cost	Avg Total Unit Cost	Avg Price Incl O&P

Sheathing, subfloor

Boards, 1" x 8"

Description	Oper	Unit	Vol	Crew Size	Man-hrs	Crew Output	Mat'l	Labor	Equip	Total	Price O&P
Horizontal	Demo	SF	Lg	LB	.019	828.0	---	.52	---	.52	.78
	Demo	SF	Sm	LB	.023	704.0	---	.63	---	.63	.94
Horizontal	Inst	SF	Lg	2C	.021	764.0	.74	.70	.06	1.50	1.96
	Inst	SF	Sm	2C	.025	649.0	.82	.83	.07	1.72	2.26
Diagonal	Demo	SF	Lg	LB	.022	729.0	---	.60	---	.60	.90
	Demo	SF	Sm	LB	.026	620.0	---	.71	---	.71	1.06
Diagonal	Inst	SF	Lg	2C	.024	672.0	.81	.80	.07	1.68	2.20
	Inst	SF	Sm	2C	.028	571.0	.89	.93	.08	1.90	2.50

Plywood

Description	Oper	Unit	Vol	Crew Size	Man-hrs	Crew Output	Mat'l	Labor	Equip	Total	Price O&P
5/8"	Demo	SF	Lg	LB	.013	1216	---	.36	---	.36	.53
	Demo	SF	Sm	LB	.015	1034	---	.41	---	.41	.61
5/8"	Inst	SF	Lg	2C	.014	1127	.67	.47	.04	1.18	1.51
	Inst	SF	Sm	2C	.017	958.0	.74	.57	.05	1.36	1.75
3/4"	Demo	SF	Lg	LB	.014	1134	---	.38	---	.38	.57
	Demo	SF	Sm	LB	.017	964.0	---	.47	---	.47	.69
3/4"	Inst	SF	Lg	2C	.015	1050	.78	.50	.04	1.32	1.69
	Inst	SF	Sm	2C	.018	893.0	.87	.60	.05	1.52	1.95
1-1/8"	Demo	SF	Lg	LB	.019	854.0	---	.52	---	.52	.78
	Demo	SF	Sm	LB	.022	726.0	---	.60	---	.60	.90
1-1/8"	Inst	SF	Lg	2C	.020	798.0	1.14	.67	.06	1.87	2.37
	Inst	SF	Sm	2C	.024	678.0	1.27	.80	.07	2.14	2.73

Particleboard

Description	Oper	Unit	Vol	Crew Size	Man-hrs	Crew Output	Mat'l	Labor	Equip	Total	Price O&P
5/8"	Demo	SF	Lg	LB	.013	1216	---	.36	---	.36	.53
	Demo	SF	Sm	LB	.015	1034	---	.41	---	.41	.61
5/8"	Inst	SF	Lg	2C	.014	1127	.53	.47	.04	1.04	1.35
	Inst	SF	Sm	2C	.017	958.0	.58	.57	.05	1.20	1.57
3/4"	Demo	SF	Lg	LB	.014	1134	---	.38	---	.38	.57
	Demo	SF	Sm	LB	.017	964.0	---	.47	---	.47	.69
3/4"	Inst	SF	Lg	2C	.015	1050	.67	.50	.04	1.21	1.56
	Inst	SF	Sm	2C	.018	893.0	.75	.60	.05	1.40	1.81

Underlayment

Description	Oper	Unit	Vol	Crew Size	Man-hrs	Crew Output	Mat'l	Labor	Equip	Total	Price O&P
Plywood, 3/8"	Demo	SF	Lg	LB	.013	1216	---	.36	---	.36	.53
	Demo	SF	Sm	LB	.015	1034	---	.41	---	.41	.61
Plywood, 3/8"	Inst	SF	Lg	2C	.014	1127	.45	.47	.04	.96	1.26
	Inst	SF	Sm	2C	.017	958.0	.50	.57	.05	1.12	1.47
Hardboard, 0.215"	Demo	SF	Lg	LB	.013	1216	---	.36	---	.36	.53
	Demo	SF	Sm	LB	.015	1034	---	.41	---	.41	.61
Hardboard, 0.215"	Inst	SF	Lg	2C	.014	1127	.32	.47	.04	.83	1.11
	Inst	SF	Sm	2C	.017	958.0	.35	.57	.05	.97	1.30

Description	Oper	Unit	Vol	Crew Size	Man-hours per Unit	Crew Output per Day	Avg Mat'l Unit Cost	Avg Labor Unit Cost	Avg Equip Unit Cost	Avg Total Unit Cost	Avg Price Incl O&P
Trusses, shop fabricated, wood "W" type											
1/8 pitch											
3" rise in 12" run											
20' span	Demo	Ea	Lg	LJ	.865	37.00	---	23.70	---	23.70	**35.40**
	Demo	Ea	Sm	LJ	1.032	31.00	---	28.30	---	28.30	**42.20**
20' span	Inst	Ea	Lg	CX	.941	34.00	34.20	31.30	1.37	66.87	**87.70**
	Inst	Ea	Sm	CX	1.10	29.00	37.60	36.60	1.61	75.81	**99.70**
22' span	Demo	Ea	Lg	LJ	.865	37.00	---	23.70	---	23.70	**35.40**
	Demo	Ea	Sm	LJ	1.032	31.00	---	28.30	---	28.30	**42.20**
22' span	Inst	Ea	Lg	CX	.941	34.00	38.10	31.30	1.37	70.77	**92.20**
	Inst	Ea	Sm	CX	1.10	29.00	41.90	36.60	1.61	80.11	**105.00**
24' span	Demo	Ea	Lg	LJ	.865	37.00	---	23.70	---	23.70	**35.40**
	Demo	Ea	Sm	LJ	1.032	31.00	---	28.30	---	28.30	**42.20**
24' span	Inst	Ea	Lg	CX	.941	34.00	39.90	31.30	1.37	72.57	**94.20**
	Inst	Ea	Sm	CX	1.10	29.00	43.80	36.60	1.61	82.01	**107.00**
26' span	Demo	Ea	Lg	LJ	.865	37.00	---	23.70	---	23.70	**35.40**
	Demo	Ea	Sm	LJ	1.032	31.00	---	28.30	---	28.30	**42.20**
26' span	Inst	Ea	Lg	CX	.941	34.00	43.80	31.30	1.37	76.47	**98.70**
	Inst	Ea	Sm	CX	1.10	29.00	48.10	36.60	1.61	86.31	**112.00**
28' span	Demo	Ea	Lg	LJ	.865	37.00	---	23.70	---	23.70	**35.40**
	Demo	Ea	Sm	LJ	1.032	31.00	---	28.30	---	28.30	**42.20**
28' span	Inst	Ea	Lg	CX	.941	34.00	45.60	31.30	1.37	78.27	**101.00**
	Inst	Ea	Sm	CX	1.10	29.00	50.10	36.60	1.61	88.31	**114.00**
30' span	Demo	Ea	Lg	LJ	.914	35.00	---	25.10	---	25.10	**37.40**
	Demo	Ea	Sm	LJ	1.07	30.00	---	29.40	---	29.40	**43.70**
30' span	Inst	Ea	Lg	CX	1.00	32.00	49.50	33.30	1.46	84.26	**108.00**
	Inst	Ea	Sm	CX	1.19	27.00	54.40	39.60	1.73	95.73	**124.00**
32' span	Demo	Ea	Lg	LJ	.914	35.00	---	25.10	---	25.10	**37.40**
	Demo	Ea	Sm	LJ	1.07	30.00	---	29.40	---	29.40	**43.70**
32' span	Inst	Ea	Lg	CX	1.00	32.00	50.80	33.30	1.46	85.56	**110.00**
	Inst	Ea	Sm	CX	1.19	27.00	55.80	39.60	1.73	97.13	**125.00**
34' span	Demo	Ea	Lg	LJ	.914	35.00	---	25.10	---	25.10	**37.40**
	Demo	Ea	Sm	LJ	1.07	30.00	---	29.40	---	29.40	**43.70**
34' span	Inst	Ea	Lg	CX	1.00	32.00	53.10	33.30	1.46	87.86	**112.00**
	Inst	Ea	Sm	CX	1.19	27.00	58.30	39.60	1.73	99.63	**128.00**
36' span	Demo	Ea	Lg	LJ	.970	33.00	---	26.60	---	26.60	**39.60**
	Demo	Ea	Sm	LJ	1.14	28.00	---	31.30	---	31.30	**46.60**
36' span	Inst	Ea	Lg	CX	1.03	31.00	57.00	34.30	1.51	92.81	**118.00**
	Inst	Ea	Sm	CX	1.23	26.00	62.60	40.90	1.79	105.29	**135.00**

Description	Oper	Unit	Vol	Crew Size	Man-hours per Unit	Crew Output per Day	Avg Mat'l Unit Cost	Avg Labor Unit Cost	Avg Equip Unit Cost	Avg Total Unit Cost	Avg Price Incl O&P
5/24 pitch											
5" rise in 12" run											
20' span	Demo	Ea	Lg	LJ	.865	37.00	---	23.70	---	23.70	**35.40**
	Demo	Ea	Sm	LJ	1.032	31.00	---	28.30	---	28.30	**42.20**
20' span	Inst	Ea	Lg	CX	.941	34.00	27.30	31.30	1.37	59.97	**79.70**
	Inst	Ea	Sm	CX	1.10	29.00	29.90	36.60	1.61	68.11	**90.90**
22' span	Demo	Ea	Lg	LJ	.865	37.00	---	23.70	---	23.70	**35.40**
	Demo	Ea	Sm	LJ	1.032	31.00	---	28.30	---	28.30	**42.20**
22' span	Inst	Ea	Lg	CX	.941	34.00	30.60	31.30	1.37	63.27	**83.50**
	Inst	Ea	Sm	CX	1.10	29.00	33.60	36.60	1.61	71.81	**95.10**
24' span	Demo	Ea	Lg	LJ	.865	37.00	---	23.70	---	23.70	**35.40**
	Demo	Ea	Sm	LJ	1.032	31.00	---	28.30	---	28.30	**42.20**
24' span	Inst	Ea	Lg	CX	.941	34.00	32.30	31.30	1.37	64.97	**85.50**
	Inst	Ea	Sm	CX	1.10	29.00	35.50	36.60	1.61	73.71	**97.30**
26' span	Demo	Ea	Lg	LJ	.865	37.00	---	23.70	---	23.70	**35.40**
	Demo	Ea	Sm	LJ	1.032	31.00	---	28.30	---	28.30	**42.20**
26' span	Inst	Ea	Lg	CX	.941	34.00	36.10	31.30	1.37	68.77	**89.90**
	Inst	Ea	Sm	CX	1.10	29.00	39.70	36.60	1.61	77.91	**102.00**
28' span	Demo	Ea	Lg	LJ	.865	37.00	---	23.70	---	23.70	**35.40**
	Demo	Ea	Sm	LJ	1.032	31.00	---	28.30	---	28.30	**42.20**
28' span	Inst	Ea	Lg	CX	.941	34.00	43.10	31.30	1.37	75.77	**97.80**
	Inst	Ea	Sm	CX	1.10	29.00	47.30	36.60	1.61	85.51	**111.00**
30' span	Demo	Ea	Lg	LJ	.914	35.00	---	25.10	---	25.10	**37.40**
	Demo	Ea	Sm	LJ	1.07	30.00	---	29.40	---	29.40	**43.70**
30' span	Inst	Ea	Lg	CX	1.00	32.00	46.40	33.30	1.46	81.16	**105.00**
	Inst	Ea	Sm	CX	1.19	27.00	50.90	39.60	1.73	92.23	**120.00**
32' span	Demo	Ea	Lg	LJ	.914	35.00	---	25.10	---	25.10	**37.40**
	Demo	Ea	Sm	LJ	1.07	30.00	---	29.40	---	29.40	**43.70**
32' span	Inst	Ea	Lg	CX	1.00	32.00	47.60	33.30	1.46	82.36	**106.00**
	Inst	Ea	Sm	CX	1.19	27.00	52.30	39.60	1.73	93.63	**121.00**
34' span	Demo	Ea	Lg	LJ	.914	35.00	---	25.10	---	25.10	**37.40**
	Demo	Ea	Sm	LJ	1.07	30.00	---	29.40	---	29.40	**43.70**
34' span	Inst	Ea	Lg	CX	1.00	32.00	52.40	33.30	1.46	87.16	**112.00**
	Inst	Ea	Sm	CX	1.19	27.00	57.60	39.60	1.73	98.93	**127.00**
36' span	Demo	Ea	Lg	LJ	.970	33.00	---	26.60	---	26.60	**39.60**
	Demo	Ea	Sm	LJ	1.14	28.00	---	31.30	---	31.30	**46.60**
36' span	Inst	Ea	Lg	CX	1.03	31.00	54.20	34.30	1.51	90.01	**115.00**
	Inst	Ea	Sm	CX	1.23	26.00	59.50	40.90	1.79	102.19	**132.00**

Description	Oper	Unit	Vol	Crew Size	Man-hours per Unit	Crew Output per Day	Avg Mat'l Unit Cost	Avg Labor Unit Cost	Avg Equip Unit Cost	Avg Total Unit Cost	Avg Price Incl O&P
1/4 pitch											
6" rise in 12" run											
20' span	Demo	Ea	Lg	LJ	.865	37.00	---	23.70	---	23.70	**35.40**
	Demo	Ea	Sm	LJ	1.032	31.00	---	28.30	---	28.30	**42.20**
20' span	Inst	Ea	Lg	CX	.941	34.00	27.80	31.30	1.37	60.47	**80.30**
	Inst	Ea	Sm	CX	1.10	29.00	30.50	36.60	1.61	68.71	**91.60**
22' span	Demo	Ea	Lg	LJ	.865	37.00	---	23.70	---	23.70	**35.40**
	Demo	Ea	Sm	LJ	1.032	31.00	---	28.30	---	28.30	**42.20**
22' span	Inst	Ea	Lg	CX	.941	34.00	31.10	31.30	1.37	63.77	**84.10**
	Inst	Ea	Sm	CX	1.10	29.00	34.10	36.60	1.61	72.31	**95.80**
24' span	Demo	Ea	Lg	LJ	.865	37.00	---	23.70	---	23.70	**35.40**
	Demo	Ea	Sm	LJ	1.032	31.00	---	28.30	---	28.30	**42.20**
24' span	Inst	Ea	Lg	CX	.941	34.00	34.90	31.30	1.37	67.57	**88.50**
	Inst	Ea	Sm	CX	1.10	29.00	38.30	36.60	1.61	76.51	**101.00**
26' span	Demo	Ea	Lg	LJ	.865	37.00	---	23.70	---	23.70	**35.40**
	Demo	Ea	Sm	LJ	1.032	31.00	---	28.30	---	28.30	**42.20**
26' span	Inst	Ea	Lg	CX	.941	34.00	36.10	31.30	1.37	68.77	**89.90**
	Inst	Ea	Sm	CX	1.10	29.00	39.70	36.60	1.61	77.91	**102.00**
28' span	Demo	Ea	Lg	LJ	.865	37.00	---	23.70	---	23.70	**35.40**
	Demo	Ea	Sm	LJ	1.032	31.00	---	28.30	---	28.30	**42.20**
28' span	Inst	Ea	Lg	CX	.941	34.00	43.60	31.30	1.37	76.27	**98.40**
	Inst	Ea	Sm	CX	1.10	29.00	47.90	36.60	1.61	86.11	**112.00**
30' span	Demo	Ea	Lg	LJ	.914	35.00	---	25.10	---	25.10	**37.40**
	Demo	Ea	Sm	LJ	1.07	30.00	---	29.40	---	29.40	**43.70**
30' span	Inst	Ea	Lg	CX	1.00	32.00	45.30	33.30	1.46	80.06	**103.00**
	Inst	Ea	Sm	CX	1.19	27.00	49.80	39.60	1.73	91.13	**118.00**
32' span	Demo	Ea	Lg	LJ	.914	35.00	---	25.10	---	25.10	**37.40**
	Demo	Ea	Sm	LJ	1.07	30.00	---	29.40	---	29.40	**43.70**
32' span	Inst	Ea	Lg	CX	1.00	32.00	49.10	33.30	1.46	83.86	**108.00**
	Inst	Ea	Sm	CX	1.19	27.00	54.00	39.60	1.73	95.33	**123.00**
34' span	Demo	Ea	Lg	LJ	.914	35.00	---	25.10	---	25.10	**37.40**
	Demo	Ea	Sm	LJ	1.07	30.00	---	29.40	---	29.40	**43.70**
34' span	Inst	Ea	Lg	CX	1.00	32.00	52.90	33.30	1.46	87.66	**112.00**
	Inst	Ea	Sm	CX	1.19	27.00	58.20	39.60	1.73	99.53	**128.00**
36' span	Demo	Ea	Lg	LJ	.970	33.00	---	26.60	---	26.60	**39.60**
	Demo	Ea	Sm	LJ	1.14	28.00	---	31.30	---	31.30	**46.60**
36' span	Inst	Ea	Lg	CX	1.03	31.00	55.70	34.30	1.51	91.51	**117.00**
	Inst	Ea	Sm	CX	1.23	26.00	61.20	40.90	1.79	103.89	**134.00**

Description	Oper	Unit	Vol	Crew Size	Man-hours per Unit	Crew Output per Day	Avg Mat'l Unit Cost	Avg Labor Unit Cost	Avg Equip Unit Cost	Avg Total Unit Cost	Avg Price Incl O&P

Garage doors

Detach & reset operations

Wood, aluminum, or hardboard

Description	Oper	Unit	Vol	Crew Size	Man-hours per Unit	Crew Output per Day	Avg Mat'l Unit Cost	Avg Labor Unit Cost	Avg Equip Unit Cost	Avg Total Unit Cost	Avg Price Incl O&P
Single	Reset	Ea	Lg	LB	2.67	6.00	---	73.20	---	73.20	**109.00**
	Reset	Ea	Sm	LB	3.56	4.50	---	97.70	---	97.70	**146.00**
Double	Reset	Ea	Lg	LB	3.56	4.50	---	97.70	---	97.70	**146.00**
	Reset	Ea	Sm	LB	4.73	3.38	---	130.00	---	130.00	**193.00**

Steel

Description	Oper	Unit	Vol	Crew Size	Man-hours per Unit	Crew Output per Day	Avg Mat'l Unit Cost	Avg Labor Unit Cost	Avg Equip Unit Cost	Avg Total Unit Cost	Avg Price Incl O&P
Single	Reset	Ea	Lg	LB	3.20	5.00	---	87.80	---	87.80	**131.00**
	Reset	Ea	Sm	LB	4.27	3.75	---	117.00	---	117.00	**175.00**
Double	Reset	Ea	Lg	LB	4.00	4.00	---	110.00	---	110.00	**163.00**
	Reset	Ea	Sm	LB	4.92	3.25	---	135.00	---	135.00	**201.00**

Remove operations

Wood, aluminum, or hardboard

Description	Oper	Unit	Vol	Crew Size	Man-hours per Unit	Crew Output per Day	Avg Mat'l Unit Cost	Avg Labor Unit Cost	Avg Equip Unit Cost	Avg Total Unit Cost	Avg Price Incl O&P
Single	Demo	Ea	Lg	LB	2.00	8.00	---	54.90	---	54.90	**81.70**
	Demo	Ea	Sm	LB	3.08	5.20	---	84.50	---	84.50	**126.00**
Double	Demo	Ea	Lg	LB	2.67	6.00	---	73.20	---	73.20	**109.00**
	Demo	Ea	Sm	LB	4.10	3.90	---	112.00	---	112.00	**168.00**

Steel

Description	Oper	Unit	Vol	Crew Size	Man-hours per Unit	Crew Output per Day	Avg Mat'l Unit Cost	Avg Labor Unit Cost	Avg Equip Unit Cost	Avg Total Unit Cost	Avg Price Incl O&P
Single	Demo	Ea	Lg	LB	2.29	7.00	---	62.80	---	62.80	**93.60**
	Demo	Ea	Sm	LB	3.52	4.55	---	96.60	---	96.60	**144.00**
Double	Demo	Ea	Lg	LB	3.20	5.00	---	87.80	---	87.80	**131.00**
	Demo	Ea	Sm	LB	4.92	3.25	---	135.00	---	135.00	**201.00**

Replace operations

Aluminum frame with plastic skin bonded to polystyrene foam core

Jamb type with hardware and deluxe lock

Description	Oper	Unit	Vol	Crew Size	Man-hours per Unit	Crew Output per Day	Avg Mat'l Unit Cost	Avg Labor Unit Cost	Avg Equip Unit Cost	Avg Total Unit Cost	Avg Price Incl O&P
8' x 7', single	Inst	Ea	Lg	2C	4.00	4.00	294.00	133.00	---	427.00	**537.00**
	Inst	Ea	Sm	2C	5.71	2.80	337.00	190.00	---	527.00	**673.00**
8' x 8', single	Inst	Ea	Lg	2C	4.00	4.00	382.00	133.00	---	515.00	**639.00**
	Inst	Ea	Sm	2C	5.71	2.80	439.00	190.00	---	629.00	**790.00**
9' x 7', single	Inst	Ea	Lg	2C	4.00	4.00	321.00	133.00	---	454.00	**569.00**
	Inst	Ea	Sm	2C	5.71	2.80	368.00	190.00	---	558.00	**709.00**
9' x 8', single	Inst	Ea	Lg	2C	4.00	4.00	413.00	133.00	---	546.00	**674.00**
	Inst	Ea	Sm	2C	5.71	2.80	474.00	190.00	---	664.00	**830.00**
16' x 7', double	Inst	Ea	Lg	2C	5.33	3.00	529.00	177.00	---	706.00	**875.00**
	Inst	Ea	Sm	2C	7.62	2.10	608.00	254.00	---	862.00	**1080.00**
16' x 8', double	Inst	Ea	Lg	2C	5.33	3.00	686.00	177.00	---	863.00	**1050.00**
	Inst	Ea	Sm	2C	7.62	2.10	787.00	254.00	---	1041.00	**1290.00**

Description	Oper	Unit	Vol	Crew Size	Man-hours per Unit	Crew Output per Day	Avg Mat'l Unit Cost	Avg Labor Unit Cost	Avg Equip Unit Cost	Avg Total Unit Cost	Avg Price Incl O&P
Track type with hardware and deluxe lock											
8' x 7', single	Inst	Ea	Lg	2C	4.00	4.00	347.00	133.00	---	480.00	**598.00**
	Inst	Ea	Sm	2C	5.71	2.80	398.00	190.00	---	588.00	**743.00**
9' x 7', single	Inst	Ea	Lg	2C	4.00	4.00	381.00	133.00	---	514.00	**638.00**
	Inst	Ea	Sm	2C	5.71	2.80	438.00	190.00	---	628.00	**788.00**
16' x 7', double	Inst	Ea	Lg	2C	5.33	3.00	620.00	177.00	---	797.00	**979.00**
	Inst	Ea	Sm	2C	7.62	2.10	712.00	254.00	---	966.00	**1200.00**
Sectional type with hardware and key lock											
8' x 7', single	Inst	Ea	Lg	2C	4.00	4.00	496.00	133.00	---	629.00	**770.00**
	Inst	Ea	Sm	2C	5.71	2.80	569.00	190.00	---	759.00	**940.00**
9' x 7', single	Inst	Ea	Lg	2C	4.00	4.00	542.00	133.00	---	675.00	**823.00**
	Inst	Ea	Sm	2C	5.71	2.80	622.00	190.00	---	812.00	**1000.00**
16' x 7', double	Inst	Ea	Lg	2C	5.33	3.00	903.00	177.00	---	1080.00	**1300.00**
	Inst	Ea	Sm	2C	7.62	2.10	1040.00	254.00	---	1294.00	**1570.00**

Fiberglass

Description	Oper	Unit	Vol	Crew Size	Man-hours per Unit	Crew Output per Day	Avg Mat'l Unit Cost	Avg Labor Unit Cost	Avg Equip Unit Cost	Avg Total Unit Cost	Avg Price Incl O&P
Jamb type with hardware and deluxe lock											
8' x 7', single	Inst	Ea	Lg	2C	4.00	4.00	259.00	133.00	---	392.00	**498.00**
	Inst	Ea	Sm	2C	5.71	2.80	298.00	190.00	---	488.00	**627.00**
8' x 8', single	Inst	Ea	Lg	2C	4.00	4.00	364.00	133.00	---	497.00	**618.00**
	Inst	Ea	Sm	2C	5.71	2.80	418.00	190.00	---	608.00	**766.00**
9' x 7', single	Inst	Ea	Lg	2C	4.00	4.00	310.00	133.00	---	443.00	**556.00**
	Inst	Ea	Sm	2C	5.71	2.80	356.00	190.00	---	546.00	**694.00**
9' x 8', single	Inst	Ea	Lg	2C	4.00	4.00	404.00	133.00	---	537.00	**664.00**
	Inst	Ea	Sm	2C	5.71	2.80	464.00	190.00	---	654.00	**818.00**
16' x 7', double	Inst	Ea	Lg	2C	5.33	3.00	550.00	177.00	---	727.00	**898.00**
	Inst	Ea	Sm	2C	7.62	2.10	631.00	254.00	---	885.00	**1110.00**
16' x 8', double	Inst	Ea	Lg	2C	5.33	3.00	731.00	177.00	---	908.00	**1110.00**
	Inst	Ea	Sm	2C	7.62	2.10	839.00	254.00	---	1093.00	**1350.00**
Track type with hardware and deluxe lock											
8' x 7', single	Inst	Ea	Lg	2C	4.00	4.00	347.00	133.00	---	480.00	**598.00**
	Inst	Ea	Sm	2C	5.71	2.80	398.00	190.00	---	588.00	**743.00**
9' x 7', single	Inst	Ea	Lg	2C	4.00	4.00	381.00	133.00	---	514.00	**638.00**
	Inst	Ea	Sm	2C	5.71	2.80	438.00	190.00	---	628.00	**788.00**
16' x 7', double	Inst	Ea	Lg	2C	5.33	3.00	590.00	177.00	---	767.00	**944.00**
	Inst	Ea	Sm	2C	7.62	2.10	677.00	254.00	---	931.00	**1160.00**
Sectional type with hardware and key lock											
8' x 7', single	Inst	Ea	Lg	2C	4.00	4.00	308.00	133.00	---	441.00	**554.00**
	Inst	Ea	Sm	2C	5.71	2.80	353.00	190.00	---	543.00	**691.00**
9' x 7', single	Inst	Ea	Lg	2C	4.00	4.00	335.00	133.00	---	468.00	**585.00**
	Inst	Ea	Sm	2C	5.71	2.80	384.00	190.00	---	574.00	**727.00**
16' x 7', double	Inst	Ea	Lg	2C	5.33	3.00	526.00	177.00	---	703.00	**871.00**
	Inst	Ea	Sm	2C	7.62	2.10	604.00	254.00	---	858.00	**1070.00**

Description	Oper	Unit	Vol	Crew Size	Man-hours per Unit	Crew Output per Day	Avg Mat'l Unit Cost	Avg Labor Unit Cost	Avg Equip Unit Cost	Avg Total Unit Cost	Avg Price Incl O&P
Steel											
Jamb type with hardware and deluxe lock											
8' x 7', single	Inst	Ea	Lg	2C	4.00	4.00	225.00	133.00	---	358.00	**458.00**
	Inst	Ea	Sm	2C	5.71	2.80	258.00	190.00	---	448.00	**582.00**
8' x 8', single	Inst	Ea	Lg	2C	4.00	4.00	300.00	133.00	---	433.00	**545.00**
	Inst	Ea	Sm	2C	5.71	2.80	345.00	190.00	---	535.00	**681.00**
9' x 7', single	Inst	Ea	Lg	2C	4.00	4.00	246.00	133.00	---	379.00	**483.00**
	Inst	Ea	Sm	2C	5.71	2.80	283.00	190.00	---	473.00	**610.00**
9' x 8', single	Inst	Ea	Lg	2C	4.00	4.00	335.00	133.00	---	468.00	**585.00**
	Inst	Ea	Sm	2C	5.71	2.80	384.00	190.00	---	574.00	**727.00**
16' x 7', double	Inst	Ea	Lg	2C	5.33	3.00	418.00	177.00	---	595.00	**747.00**
	Inst	Ea	Sm	2C	7.62	2.10	480.00	254.00	---	734.00	**932.00**
16' x 8', double	Inst	Ea	Lg	2C	5.33	3.00	569.00	177.00	---	746.00	**921.00**
	Inst	Ea	Sm	2C	7.62	2.10	653.00	254.00	---	907.00	**1130.00**
Track type with hardware and deluxe lock											
8' x 7', single	Inst	Ea	Lg	2C	4.00	4.00	254.00	133.00	---	387.00	**492.00**
	Inst	Ea	Sm	2C	5.71	2.80	291.00	190.00	---	481.00	**620.00**
9' x 7', single	Inst	Ea	Lg	2C	4.00	4.00	283.00	133.00	---	416.00	**525.00**
	Inst	Ea	Sm	2C	5.71	2.80	325.00	190.00	---	515.00	**659.00**
16' x 7', double	Inst	Ea	Lg	2C	5.33	3.00	454.00	177.00	---	631.00	**788.00**
	Inst	Ea	Sm	2C	7.62	2.10	521.00	254.00	---	775.00	**979.00**
Sectional type with hardware and key lock											
8' x 7', single	Inst	Ea	Lg	2C	4.00	4.00	247.00	133.00	---	380.00	**484.00**
	Inst	Ea	Sm	2C	5.71	2.80	284.00	190.00	---	474.00	**612.00**
9' x 7', single	Inst	Ea	Lg	2C	4.00	4.00	258.00	133.00	---	391.00	**496.00**
	Inst	Ea	Sm	2C	5.71	2.80	296.00	190.00	---	486.00	**626.00**
16' x 7', double	Inst	Ea	Lg	2C	5.33	3.00	497.00	177.00	---	674.00	**837.00**
	Inst	Ea	Sm	2C	7.62	2.10	570.00	254.00	---	824.00	**1040.00**
Wood											
Jamb type with hardware and deluxe lock											
8' x 7', single	Inst	Ea	Lg	2C	4.00	4.00	309.00	133.00	---	442.00	**555.00**
	Inst	Ea	Sm	2C	5.71	2.80	355.00	190.00	---	545.00	**693.00**
8' x 8', single	Inst	Ea	Lg	2C	4.00	4.00	321.00	133.00	---	454.00	**509.00**
	Inst	Ea	Sm	2C	5.71	2.80	368.00	190.00	---	558.00	**709.00**
9' x 7', single	Inst	Ea	Lg	2C	4.00	4.00	337.00	133.00	---	470.00	**587.00**
	Inst	Ea	Sm	2C	5.71	2.80	387.00	190.00	---	577.00	**730.00**
9' x 8', single	Inst	Ea	Lg	2C	4.00	4.00	417.00	133.00	---	550.00	**679.00**
	Inst	Ea	Sm	2C	5.71	2.80	479.00	190.00	---	669.00	**835.00**
16' x 7', double	Inst	Ea	Lg	2C	5.33	3.00	546.00	177.00	---	723.00	**894.00**
	Inst	Ea	Sm	2C	7.62	2.10	627.00	254.00	---	881.00	**1100.00**
16' x 8', double	Inst	Ea	Lg	2C	5.33	3.00	705.00	177.00	---	882.00	**1080.00**
	Inst	Ea	Sm	2C	7.62	2.10	810.00	254.00	---	1064.00	**1310.00**

Description	Oper	Unit	Vol	Crew Size	Man-hours per Unit	Crew Output per Day	Avg Mat'l Unit Cost	Avg Labor Unit Cost	Avg Equip Unit Cost	Avg Total Unit Cost	Avg Price Incl O&P
Track type with hardware and deluxe lock											
8' x 7', single	Inst	Ea	Lg	2C	4.00	4.00	352.00	133.00	---	485.00	**605.00**
	Inst	Ea	Sm	2C	5.71	2.80	404.00	190.00	---	594.00	**750.00**
9' x 7', single	Inst	Ea	Lg	2C	4.00	4.00	381.00	133.00	---	514.00	**638.00**
	Inst	Ea	Sm	2C	5.71	2.80	438.00	190.00	---	628.00	**788.00**
16' x 7', double	Inst	Ea	Lg	2C	5.33	3.00	609.00	177.00	---	786.00	**967.00**
	Inst	Ea	Sm	2C	7.62	2.10	699.00	254.00	---	953.00	**1180.00**
Sectional type with hardware and key lock											
8' x 7', single	Inst	Ea	Lg	2C	4.00	4.00	342.00	133.00	---	475.00	**593.00**
	Inst	Ea	Sm	2C	5.71	2.80	393.00	190.00	---	583.00	**737.00**
9' x 7', single	Inst	Ea	Lg	2C	4.00	4.00	367.00	133.00	---	500.00	**622.00**
	Inst	Ea	Sm	2C	5.71	2.80	422.00	190.00	---	612.00	**770.00**
16' x 7', double	Inst	Ea	Lg	2C	5.33	3.00	683.00	177.00	---	860.00	**1050.00**
	Inst	Ea	Sm	2C	7.62	2.10	784.00	254.00	---	1038.00	**1280.00**

Garage door operators

Radio controlled for single or double doors.
Labor includes wiring, connection and installation.

Chain drive, 1/4 hp, with receiver and one transmitter

Description	Oper	Unit	Vol	Crew Size	Man-hours per Unit	Crew Output per Day	Avg Mat'l Unit Cost	Avg Labor Unit Cost	Avg Equip Unit Cost	Avg Total Unit Cost	Avg Price Incl O&P
	Inst	LS	Lg	ED	5.33	3.00	176.00	184.00	---	360.00	**471.00**
	Inst	LS	Sm	ED	7.11	2.25	206.00	245.00	---	451.00	**595.00**

Screw-worm drive, 1/3 hp, with receiver and one transmitter

| | Inst | LS | Lg | ED | 5.33 | 3.00 | 207.00 | 184.00 | --- | 391.00 | **506.00** |
| | Inst | LS | Sm | ED | 7.11 | 2.25 | 242.00 | 245.00 | --- | 487.00 | **636.00** |

Deluxe models, 1/2 hp, with receiver, transmitter, and time delay light

Chain drive, not for vault-type garages

| | Inst | LS | Lg | ED | 5.33 | 3.00 | 233.00 | 184.00 | --- | 417.00 | **536.00** |
| | Inst | LS | Sm | ED | 7.11 | 2.25 | 273.00 | 245.00 | --- | 518.00 | **672.00** |

Screw drive with threaded worm screw

| | Inst | LS | Lg | ED | 5.33 | 3.00 | 264.00 | 184.00 | --- | 448.00 | **572.00** |
| | Inst | LS | Sm | ED | 7.11 | 2.25 | 309.00 | 245.00 | --- | 554.00 | **713.00** |

Additional transmitters, ADD

	Inst	Ea	Lg	EA	---	---	37.00	---	---	37.00	**42.50**
	Inst	Ea	Sm	EA	---	---	43.30	---	---	43.30	**49.80**
Exterior key switch	Inst	Ea	Lg	EA	.667	12.00	22.00	23.80	---	45.80	**60.00**
	Inst	Ea	Sm	EA	.889	9.00	25.80	31.70	---	57.50	**75.90**
To remove and replace unit and receiver											
	Inst	LS	Lg	2C	2.67	6.00	---	88.80	---	88.80	**133.00**
	Inst	LS	Sm	2C	3.56	4.50	---	118.00	---	118.00	**178.00**

Description	Oper	Unit	Vol	Crew Size	Man-hours per Unit	Crew Output per Day	Avg Mat'l Unit Cost	Avg Labor Unit Cost	Avg Equip Unit Cost	Avg Total Unit Cost	Avg Price Incl O&P

Garbage disposers

Includes wall switch and labor includes rough-in.
See also Trash compactors, page 434

Frequently encountered applications

Detach & reset operations

Description	Oper	Unit	Vol	Crew Size	Man-hours per Unit	Crew Output per Day	Avg Mat'l Unit Cost	Avg Labor Unit Cost	Avg Equip Unit Cost	Avg Total Unit Cost	Avg Price Incl O&P
Garbage disposer	Reset	Ea	Lg	SA	2.67	3.00	---	101.00	---	101.00	**148.00**
	Reset	Ea	Sm	SA	3.56	2.25	---	134.00	---	134.00	**197.00**

Remove operations

Garbage disposer	Demo	Ea	Lg	SA	2.67	3.00	---	101.00	---	101.00	**148.00**
	Demo	Ea	Sm	SA	3.56	2.25	---	134.00	---	134.00	**197.00**

Replace operations

Standard, 1/3 HP	Inst	Ea	Lg	SA	2.67	3.00	96.30	101.00	---	197.30	**259.00**
	Inst	Ea	Sm	SA	3.56	2.25	117.00	134.00	---	251.00	**332.00**
Average, 1/2 HP	Inst	Ea	Lg	SA	2.67	3.00	111.00	101.00	---	212.00	**275.00**
	Inst	Ea	Sm	SA	3.56	2.25	134.00	134.00	---	268.00	**352.00**
High, 3/4 HP	Inst	Ea	Lg	SA	2.67	3.00	165.00	101.00	---	266.00	**337.00**
	Inst	Ea	Sm	SA	3.56	2.25	200.00	134.00	---	334.00	**427.00**
Premium, 1 HP	Inst	Ea	Lg	SA	2.67	3.00	268.00	101.00	---	369.00	**456.00**
	Inst	Ea	Sm	SA	3.56	2.25	326.00	134.00	---	460.00	**572.00**

In-Sink-Erator Products

Model "Badger 1," 1/3 HP, continuous feed, 1 year parts protection

	Inst	Ea	Lg	SA	2.67	3.00	96.30	101.00	---	197.30	**259.00**
	Inst	Ea	Sm	SA	3.56	2.25	117.00	134.00	---	251.00	**332.00**

Model "Badger V," 1/2 HP, continuous feed, 1 year parts protection

	Inst	Ea	Lg	SA	2.67	3.00	111.00	101.00	---	212.00	**275.00**
	Inst	Ea	Sm	SA	3.56	2.25	134.00	134.00	---	268.00	**352.00**

Model 333, 3/4 HP, continuous feed, 4 year parts protection

	Inst	Ea	Lg	SA	2.67	3.00	165.00	101.00	---	266.00	**337.00**
	Inst	Ea	Sm	SA	3.56	2.25	200.00	134.00	---	334.00	**427.00**

Model 77, 1 HP, automatic reversing feed, 7 year parts protection, stainless steel construction

	Inst	Ea	Lg	SA	2.67	3.00	268.00	101.00	---	369.00	**456.00**
	Inst	Ea	Sm	SA	3.56	2.25	326.00	134.00	---	460.00	**572.00**

Model 777SS 1 HP, continuous feed, 7 year parts protection, stainless steel construction

	Inst	Ea	Lg	SA	2.67	3.00	265.00	101.00	---	366.00	**453.00**
	Inst	Ea	Sm	SA	3.56	2.25	322.00	134.00	---	456.00	**567.00**

Model 17, 3/4 HP, batch feed, auto reversing, 5 year parts protection, stainless steel construction

	Inst	Ea	Lg	SA	2.67	3.00	345.00	101.00	---	446.00	**545.00**
	Inst	Ea	Sm	SA	3.56	2.25	419.00	134.00	---	553.00	**679.00**

Septic, 3 year parts protection, stainless steel grind elements

	Inst	Ea	Lg	SA	2.67	3.00	218.00	101.00	---	319.00	**398.00**
	Inst	Ea	Sm	SA	3.56	2.25	265.00	134.00	---	399.00	**501.00**

Description	Oper	Unit	Vol	Crew Size	Man-hours per Unit	Crew Output per Day	Avg Mat'l Unit Cost	Avg Labor Unit Cost	Avg Equip Unit Cost	Avg Total Unit Cost	Avg Price Incl O&P
Adjustments											
To only remove and reset garbage disposer											
	Reset	Ea	Lg	SA	2.67	3.00	---	101.00	---	101.00	**148.00**
	Reset	Ea	Sm	SA	3.56	2.25	---	134.00	---	134.00	**197.00**
To only remove garbage disposer											
	Demo	Ea	Lg	SA	2.67	3.00	---	101.00	---	101.00	**148.00**
	Demo	Ea	Sm	SA	3.56	2.25	---	134.00	---	134.00	**197.00**
Parts and accessories											
Stainless steel stopper	Inst	Ea	Lg	---	---	---	8.40	---	---	8.40	**8.40**
	Inst	Ea	Sm	---	---	---	10.20	---	---	10.20	**10.20**
Dishwasher connector kit	Inst	Ea	Lg	---	---	---	7.70	---	---	7.70	**7.70**
	Inst	Ea	Sm	---	---	---	9.35	---	---	9.35	**9.35**
Flexible tail pipe	Inst	Ea	Lg	---	---	---	9.80	---	---	9.80	**9.80**
	Inst	Ea	Sm	---	---	---	11.90	---	---	11.90	**11.90**
Power cord accessory kit	Inst	Ea	Lg	---	---	---	8.40	---	---	8.40	**8.40**
	Inst	Ea	Sm	---	---	---	10.20	---	---	10.20	**10.20**
Service wrench	Inst	Ea	Lg	---	---	---	4.20	---	---	4.20	**4.20**
	Inst	Ea	Sm	---	---	---	5.10	---	---	5.10	**5.10**
Plastic stopper	Inst	Ea	Lg	---	---	---	4.20	---	---	4.20	**4.20**
	Inst	Ea	Sm	---	---	---	5.10	---	---	5.10	**5.10**
Deluxe mounting gasket	Inst	Ea	Lg	---	---	---	6.13	---	---	6.13	**6.13**
	Inst	Ea	Sm	---	---	---	7.44	---	---	7.44	**7.44**

Girders. See Framing, page 168

Description	Oper	Unit	Vol	Crew Size	Man-hours per Unit	Crew Output per Day	Avg Mat'l Unit Cost	Avg Labor Unit Cost	Avg Equip Unit Cost	Avg Total Unit Cost	Avg Price Incl O&P

Glass and glazing
3/16" T float with putty in wood sash

Description	Oper	Unit	Vol	Crew Size	MH	Output	Mat'l	Labor	Equip	Total	O&P
8" x 12"	Demo	SF	Lg	GA	.320	25.00	---	9.87	---	9.87	14.60
	Demo	SF	Sm	GA	.457	17.50	---	14.10	---	14.10	20.90
8" x 12"	Inst	SF	Lg	GA	.267	30.00	5.96	8.23	---	14.19	19.00
	Inst	SF	Sm	GA	.381	21.00	6.60	11.80	---	18.40	25.00
12" x 16"	Demo	SF	Lg	GA	.178	45.00	---	5.49	---	5.49	8.12
	Demo	SF	Sm	GA	.254	31.50	---	7.83	---	7.83	11.60
12" x 16"	Inst	SF	Lg	GA	.145	55.00	4.89	4.47	---	9.36	12.20
	Inst	SF	Sm	GA	.208	38.50	5.41	6.41	---	11.82	15.70
14" x 20"	Demo	SF	Lg	GA	.145	55.00	---	4.47	---	4.47	6.62
	Demo	SF	Sm	GA	.208	38.50	---	6.41	---	6.41	9.49
14" x 20"	Inst	SF	Lg	GA	.114	70.00	4.45	3.52	---	7.97	10.30
	Inst	SF	Sm	GA	.163	49.00	4.93	5.03	---	9.96	13.10
16" x 24"	Demo	SF	Lg	GA	.114	70.00	---	3.52	---	3.52	5.20
	Demo	SF	Sm	GA	.163	49.00	---	5.03	---	5.03	7.44
16" x 24"	Inst	SF	Lg	GA	.089	90.00	4.16	2.74	---	6.90	8.85
	Inst	SF	Sm	GA	.127	63.00	4.60	3.92	---	8.52	11.10
24" x 26"	Demo	SF	Lg	GA	.084	95.00	---	2.59	---	2.59	3.83
	Demo	SF	Sm	GA	.120	66.50	---	3.70	---	3.70	5.48
24" x 26"	Inst	SF	Lg	GA	.067	120.0	3.74	2.07	---	5.81	7.36
	Inst	SF	Sm	GA	.095	84.00	4.14	2.93	---	7.07	9.10
36" x 24"	Demo	SF	Lg	GA	.064	125.0	---	1.97	---	1.97	2.92
	Demo	SF	Sm	GA	.091	87.50	---	2.81	---	2.81	4.15
36" x 24"	Inst	SF	Lg	GA	.052	155.0	3.56	1.60	---	5.16	6.47
	Inst	SF	Sm	GA	.074	108.5	3.94	2.28	---	6.22	7.91

1/8" T float with putty in steel sash

Description	Oper	Unit	Vol	Crew Size	MH	Output	Mat'l	Labor	Equip	Total	O&P
12" x 16"	Demo	SF	Lg	GA	.178	45.00	---	5.49	---	5.49	8.12
	Demo	SF	Sm	GA	.254	31.50	---	7.83	---	7.83	11.60
12" x 16"	Inst	SF	Lg	GA	.145	55.00	6.87	4.47	---	11.34	14.50
	Inst	SF	Sm	GA	.208	38.50	7.61	6.41	---	14.02	18.20
16" x 20"	Demo	SF	Lg	GA	.123	65.00	---	3.79	---	3.79	5.61
	Demo	SF	Sm	GA	.176	45.50	---	5.43	---	5.43	8.03
16" x 20"	Inst	SF	Lg	GA	.100	80.00	5.94	3.08	---	9.02	11.40
	Inst	SF	Sm	GA	.143	56.00	6.57	4.41	---	10.98	14.10
16" x 24"	Demo	SF	Lg	GA	.114	70.00	---	3.52	---	3.52	5.20
	Demo	SF	Sm	GA	.163	49.00	---	5.03	---	5.03	7.44
16" x 24"	Inst	SF	Lg	GA	.089	90.00	5.71	2.74	---	8.45	10.60
	Inst	SF	Sm	GA	.127	63.00	6.32	3.92	---	10.24	13.10
24" x 24"	Demo	SF	Lg	GA	.084	95.00	---	2.59	---	2.59	3.83
	Demo	SF	Sm	GA	.120	66.50	---	3.70	---	3.70	5.48
24" x 24"	Inst	SF	Lg	GA	.067	120.0	5.04	2.07	---	7.11	8.85
	Inst	SF	Sm	GA	.095	84.00	5.58	2.93	---	8.51	10.80
28" x 32"	Demo	SF	Lg	GA	.062	130.0	---	1.91	---	1.91	2.83
	Demo	SF	Sm	GA	.088	91.00	---	2.71	---	2.71	4.02

Description	Oper	Unit	Vol	Crew Size	Man-hours per Unit	Crew Output per Day	Avg Mat'l Unit Cost	Avg Labor Unit Cost	Avg Equip Unit Cost	Avg Total Unit Cost	Avg Price Incl O&P
28" x 32"	Inst	SF	Lg	GA	.050	160.0	4.68	1.54	---	6.22	**7.66**
	Inst	SF	Sm	GA	.071	112.0	5.18	2.19	---	7.37	**9.20**
36" x 36"	Demo	SF	Lg	GA	.050	160.0	---	1.54	---	1.54	**2.28**
	Demo	SF	Sm	GA	.071	112.0	---	2.19	---	2.19	**3.24**
36" x 36"	Inst	SF	Lg	GA	.039	205.0	4.35	1.20	---	5.55	**6.78**
	Inst	SF	Sm	GA	.056	143.5	4.82	1.73	---	6.55	**8.10**
36" x 48"	Demo	SF	Lg	GA	.042	190.0	---	1.30	---	1.30	**1.92**
	Demo	SF	Sm	GA	.060	133.0	---	1.85	---	1.85	**2.74**
36" x 48"	Inst	SF	Lg	GA	.032	250.0	4.14	.99	---	5.13	**6.22**
	Inst	SF	Sm	GA	.046	175.0	4.59	1.42	---	6.01	**7.38**

1/4" T float

With putty and points in wood sash

Description	Oper	Unit	Vol	Crew Size	Man-hours per Unit	Crew Output per Day	Avg Mat'l Unit Cost	Avg Labor Unit Cost	Avg Equip Unit Cost	Avg Total Unit Cost	Avg Price Incl O&P
72" x 48"	Demo	SF	Lg	GA	.043	185.0	---	1.33	---	1.33	**1.96**
	Demo	SF	Sm	GA	.062	129.5	---	1.91	---	1.91	**2.83**
72" x 48"	Inst	SF	Lg	GB	.052	305.0	4.12	1.60	---	5.72	**7.11**
	Inst	SF	Sm	GB	.075	213.5	4.56	2.31	---	6.87	**8.67**

With aluminum channel and rigid neoprene rubber in aluminum sash

Description	Oper	Unit	Vol	Crew Size	Man-hours per Unit	Crew Output per Day	Avg Mat'l Unit Cost	Avg Labor Unit Cost	Avg Equip Unit Cost	Avg Total Unit Cost	Avg Price Incl O&P
48" x 96"	Demo	SF	Lg	GA	.046	175.0	---	1.42	---	1.42	**2.10**
	Demo	SF	Sm	GA	.065	122.5	---	2.00	---	2.00	**2.97**
48" x 96"	Inst	SF	Lg	GB	.046	350.0	3.65	1.42	---	5.07	**6.30**
	Inst	SF	Sm	GB	.065	245.0	4.05	2.00	---	6.05	**7.62**
96" x 96"	Demo	SF	Lg	GA	.044	180.0	---	1.36	---	1.36	**2.01**
	Demo	SF	Sm	GA	.063	126.0	---	1.94	---	1.94	**2.88**
96" x 96"	Inst	SF	Lg	GC	.037	645.0	3.61	1.14	---	4.75	**5.84**
	Inst	SF	Sm	GC	.053	451.5	4.00	1.63	---	5.63	**7.02**

1" T insulating glass (2 pieces 1/4" T float with 1/2" air space) with putty and points in wood sash

Description	Oper	Unit	Vol	Crew Size	Man-hours per Unit	Crew Output per Day	Avg Mat'l Unit Cost	Avg Labor Unit Cost	Avg Equip Unit Cost	Avg Total Unit Cost	Avg Price Incl O&P
To 6.0 SF	Demo	SF	Lg	GA	.160	50.00	---	4.93	---	4.93	**7.30**
	Demo	SF	Sm	GA	.229	35.00	---	7.06	---	7.06	**10.50**
To 6.0 SF	Inst	SF	Lg	GA	.133	60.00	9.91	4.10	---	14.01	**17.50**
	Inst	SF	Sm	GA	.190	42.00	11.00	5.86	---	16.86	**21.30**
6.1 SF to 12.0 SF	Demo	SF	Lg	GA	.073	110.0	---	2.25	---	2.25	**3.33**
	Demo	SF	Sm	GA	.104	77.00	---	3.21	---	3.21	**4.75**
6.1 SF to 12.0 SF	Inst	SF	Lg	GA	.059	135.0	8.93	1.82	---	10.75	**13.00**
	Inst	SF	Sm	GA	.085	94.50	9.89	2.62	---	12.51	**15.30**
12.1 SF to 18.0 SF	Demo	SF	Lg	GA	.053	150.0	---	1.63	---	1.63	**2.42**
	Demo	SF	Sm	GA	.076	105.0	---	2.34	---	2.34	**3.47**
12.1 SF to 18.0 SF	Inst	SF	Lg	GA	.044	180.0	8.53	1.36	---	9.89	**11.80**
	Inst	SF	Sm	GA	.063	126.0	9.44	1.94	---	11.38	**13.70**
18.1 SF to 24.0 SF	Demo	SF	Lg	GA	.055	145.0	---	1.70	---	1.70	**2.51**
	Demo	SF	Sm	GA	.079	101.5	---	2.44	---	2.44	**3.61**
18.1 SF to 24.0 SF	Inst	SF	Lg	GB	.064	250.0	8.25	1.97	---	10.22	**12.40**
	Inst	SF	Sm	GB	.091	175.0	9.13	2.81	---	11.94	**14.70**

Description	Oper	Unit	Vol	Crew Size	Man-hours per Unit	Crew Output per Day	Avg Mat'l Unit Cost	Avg Labor Unit Cost	Avg Equip Unit Cost	Avg Total Unit Cost	Avg Price Incl O&P

Aluminum sliding door glass with aluminum channel and rigid neoprene rubber

3/16" T tempered glass

Description	Oper	Unit	Vol	Crew Size	MH	Output	Mat'l	Labor	Equip	Total	Price
34" W x 76" H	Demo	SF	Lg	GA	.041	195.0	---	1.26	---	1.26	1.87
	Demo	SF	Sm	GA	.059	136.5	---	1.82	---	1.82	2.69
34" W x 76" H	Inst	SF	Lg	GA	.033	245.0	3.01	1.02	---	4.03	4.97
	Inst	SF	Sm	GA	.047	171.5	3.33	1.45	---	4.78	5.97
46" W x 76" H	Demo	SF	Lg	GA	.033	245.0	---	1.02	---	1.02	1.51
	Demo	SF	Sm	GA	.047	171.5	---	1.45	---	1.45	2.15
46" W x 76" H	Inst	SF	Lg	GB	.048	330.0	2.98	1.48	---	4.46	5.62
	Inst	SF	Sm	GB	.069	231.0	3.30	2.13	---	5.43	6.94

5/8" T insulating glass (2 pieces 3/16" T tempered with 1/4" T air space)

Description	Oper	Unit	Vol	Crew Size	MH	Output	Mat'l	Labor	Equip	Total	Price
34" W x 76" H	Demo	SF	Lg	GA	.047	170.0	---	1.45	---	1.45	2.15
	Demo	SF	Sm	GA	.067	119.0	---	2.07	---	2.07	3.06
34" W x 76" H	Inst	SF	Lg	GA	.035	230.0	6.03	1.08	---	7.11	8.53
	Inst	SF	Sm	GA	.050	161.0	6.68	1.54	---	8.22	9.96
46" W x 76" H	Demo	SF	Lg	GA	.037	215.0	---	1.14	---	1.14	1.69
	Demo	SF	Sm	GA	.053	150.5	---	1.63	---	1.63	2.42
46" W x 76" H	Inst	SF	Lg	GB	.052	310.0	5.95	1.60	---	7.55	9.22
	Inst	SF	Sm	GB	.074	217.0	6.59	2.28	---	8.87	11.00

Grading. See Concrete, page 98

Description	Oper	Unit	Vol	Crew Size	Man-hours per Unit	Crew Output per Day	Avg Mat'l Unit Cost	Avg Labor Unit Cost	Avg Equip Unit Cost	Avg Total Unit Cost	Avg Price Incl O&P
Glu-lam products											
Beams											
3-1/8" thick, SF pricing based on 16' oc											
9" deep, 20' long	Demo	LF	Lg	LK	.133	360.0	---	4.13	.42	4.55	**6.57**
	Demo	LF	Sm	LK	.190	252.0	---	5.89	.60	6.49	**9.38**
9" deep, 20' long	Inst	LF	Lg	CY	.156	360.0	9.38	4.89	.42	14.69	**18.50**
	Inst	LF	Sm	CY	.222	252.0	14.30	6.96	.60	21.86	**27.40**
9" deep, 20' long	Demo	BF	Lg	LK	.044	1080	---	1.36	.14	1.50	**2.17**
	Demo	BF	Sm	LK	.063	756.0	---	1.95	.20	2.15	**3.11**
9" deep, 20' long	Inst	BF	Lg	CY	.052	1080	3.12	1.63	.14	4.89	**6.17**
	Inst	BF	Sm	CY	.074	756.0	4.74	2.32	.20	7.26	**9.13**
9" deep, 20' long	Demo	SF	Lg	LK	.007	6545	---	.22	.02	.24	**.34**
	Demo	SF	Sm	LK	.010	4582	---	.31	.03	.34	**.49**
9" deep, 20' long	Inst	SF	Lg	CY	.009	6545	.51	.28	.02	.81	**1.03**
	Inst	SF	Sm	CY	.012	4582	.78	.38	.03	1.19	**1.49**
10-1/2" deep, 20' long	Demo	LF	Lg	LK	.133	360.0	---	4.13	.42	4.55	**6.57**
	Demo	LF	Sm	LK	.190	252.0	---	5.89	.60	6.49	**9.38**
10-1/2" deep, 20' long	Inst	LF	Lg	CY	.156	360.0	10.90	4.89	.42	16.21	**20.30**
	Inst	LF	Sm	CY	.222	252.0	16.50	6.96	.60	24.06	**30.10**
10-1/2" deep, 20' long	Demo	BF	Lg	LK	.038	1260	---	1.18	.12	1.30	**1.88**
	Demo	BF	Sm	LK	.054	882.0	---	1.68	.17	1.85	**2.67**
10-1/2" deep, 20' long	Inst	BF	Lg	CY	.044	1260	3.11	1.38	.12	4.61	**5.76**
	Inst	BF	Sm	CY	.063	882.0	4.73	1.97	.17	6.87	**8.57**
10-1/2" deep, 20' long	Demo	SF	Lg	LK	.007	6545	---	.22	.02	.24	**.34**
	Demo	SF	Sm	LK	.010	4582	---	.31	.03	.34	**.49**
10-1/2" deep, 20' long	Inst	SF	Lg	CY	.009	6545	.59	.28	.02	.89	**1.12**
	Inst	SF	Sm	CY	.012	4582	.90	.38	.03	1.31	**1.63**
12" deep, 20' long	Demo	LF	Lg	LK	.133	360.0	---	4.13	.42	4.55	**6.57**
	Demo	LF	Sm	LK	.190	252.0	---	5.89	.60	6.49	**9.38**
12" deep, 20' long	Inst	LF	Lg	CY	.156	360.0	12.10	4.89	.42	17.41	**21.70**
	Inst	LF	Sm	CY	.222	252.0	18.40	6.96	.60	25.96	**32.20**
12" deep, 20' long	Demo	BF	Lg	LK	.033	1440	---	1.02	.10	1.12	**1.62**
	Demo	BF	Sm	LK	.048	1008	---	1.49	.15	1.64	**2.37**
12" deep, 20' long	Inst	BF	Lg	CY	.039	1440	3.01	1.22	.10	4.33	**5.39**
	Inst	BF	Sm	CY	.056	1008	4.58	1.76	.15	6.49	**8.05**
12" deep, 20' long	Demo	SF	Lg	LK	.007	6545	---	.22	.02	.24	**.34**
	Demo	SF	Sm	LK	.010	4582	---	.31	.03	.34	**.49**
12" deep, 20' long	Inst	SF	Lg	CY	.009	6545	.66	.28	.02	.96	**1.20**
	Inst	SF	Sm	CY	.012	4582	1.00	.38	.03	1.41	**1.74**

Description	Oper	Unit	Vol	Crew Size	Man-hours per Unit	Crew Output per Day	Avg Mat'l Unit Cost	Avg Labor Unit Cost	Avg Equip Unit Cost	Avg Total Unit Cost	Avg Price Incl O&P

3-1/8" thick beams, SF pricing based on 16' oc (continued)

Description	Oper	Unit	Vol	Crew Size	Man-hours per Unit	Crew Output per Day	Avg Mat'l Unit Cost	Avg Labor Unit Cost	Avg Equip Unit Cost	Avg Total Unit Cost	Avg Price Incl O&P
13-1/2" deep, 20' long	Demo	LF	Lg	LK	.133	360.0	---	4.13	.42	4.55	6.57
	Demo	LF	Sm	LK	.190	252.0	---	5.89	.60	6.49	9.38
13-1/2" deep, 20' long	Inst	LF	Lg	CY	.156	360.0	13.50	4.89	.42	18.81	23.20
	Inst	LF	Sm	CY	.222	252.0	20.40	6.96	.60	27.96	34.50
13-1/2" deep, 20' long	Demo	BF	Lg	LK	.030	1620	---	.93	.09	1.02	1.48
	Demo	BF	Sm	LK	.042	1134	---	1.30	.13	1.43	2.07
13-1/2" deep, 20' long	Inst	BF	Lg	CY	.035	1620	2.99	1.10	.09	4.18	5.17
	Inst	BF	Sm	CY	.049	1134	4.54	1.54	.13	6.21	7.65
13-1/2" deep, 20' long	Demo	SF	Lg	LK	.007	6545	---	.22	.02	.24	.34
	Demo	SF	Sm	LK	.010	4582	---	.31	.03	.34	.49
13-1/2" deep, 20' long	Inst	SF	Lg	CY	.009	6545	.74	.28	.02	1.04	1.29
	Inst	SF	Sm	CY	.012	4582	1.12	.38	.03	1.53	1.88
15" deep, 20' long	Demo	LF	Lg	LK	.133	360.0	---	4.13	.42	4.55	6.57
	Demo	LF	Sm	LK	.190	252.0	---	5.89	.60	6.49	9.38
15" deep, 20' long	Inst	LF	Lg	CY	.156	360.0	14.80	4.89	.42	20.11	24.80
	Inst	LF	Sm	CY	.222	252.0	22.50	6.96	.60	30.06	36.90
15" deep, 20' long	Demo	BF	Lg	LK	.027	1800	---	.84	.08	.92	1.33
	Demo	BF	Sm	LK	.038	1260	---	1.18	.12	1.30	1.88
15" deep, 20' long	Inst	BF	Lg	s	.031	1800	2.96	.97	.08	4.01	4.94
	Inst	BF	Sm	CY	.044	1260	4.50	1.38	.12	6.00	7.36
15" deep, 20' long	Demo	SF	Lg	LK	.007	6545	---	.22	.02	.24	.34
	Demo	SF	Sm	LK	.010	4582	---	.31	.03	.34	.49
15" deep, 20' long	Inst	SF	Lg	CY	.009	6545	.81	.28	.02	1.11	1.37
	Inst	SF	Sm	CY	.012	4582	1.23	.38	.03	1.64	2.01
16-1/2" deep, 20' long	Demo	LF	Lg	LK	.133	360.0	---	4.13	.42	4.55	6.57
	Demo	LF	Sm	LK	.190	252.0	---	5.89	.60	6.49	9.38
16-1/2" deep, 20' long	Inst	LF	Lg	CY	.156	360.0	16.10	4.89	.42	21.41	26.36
	Inst	LF	Sm	CY	.222	252.0	24.50	6.96	.60	32.06	39.20
16-1/2" deep, 20' long	Demo	BF	Lg	LK	.024	1980	---	.74	.08	.82	1.19
	Demo	BF	Sm	LK	.035	1386	---	1.09	.11	1.20	1.73
16-1/2" deep, 20' long	Inst	BF	Lg	CY	.028	1980	2.93	.88	.08	3.89	4.77
	Inst	BF	Sm	CY	.040	1386	4.45	1.25	.11	5.81	7.11
16-1/2" deep, 20' long	Demo	SF	Lg	LK	.007	6545	---	.22	.02	.24	.3
	Demo	SF	Sm	LK	.010	4582	---	.31	.03	.34	.4
16-1/2" deep, 20' long	Inst	SF	Lg	CY	.009	6545	.88	.28	.02	1.18	1.4
	Inst	SF	Sm	CY	.012	4582	1.34	.38	.03	1.75	2.1
18" deep, 20' long	Demo	LF	Lg	LK	.133	360.0	---	4.13	.42	4.55	6.5
	Demo	LF	Sm	LK	.190	252.0	---	5.89	.60	6.49	9.3
18" deep, 20' long	Inst	LF	Lg	CY	.156	360.0	17.50	4.89	.42	22.81	27.9
	Inst	LF	Sm	CY	.222	252.0	26.60	6.96	.60	34.16	41.6
18" deep, 20' long	Demo	BF	Lg	LK	.022	2160	---	.68	.07	.75	1.0
	Demo	BF	Sm	LK	.032	1512	---	.99	.10	1.09	1.5
18" deep, 20' long	Inst	BF	Lg	CY	.026	2160	2.92	.81	.07	3.80	4.6
	Inst	BF	Sm	CY	.037	1512	4.44	1.16	.10	5.70	6.9
18" deep, 20' long	Demo	SF	Lg	LK	.007	6545	---	.22	.02	.24	.3
	Demo	SF	Sm	LK	.010	4582	---	.31	.03	.34	.4
18" deep, 20' long	Inst	SF	Lg	CY	.009	6545	.96	.28	.02	1.26	1.5
	Inst	SF	Sm	CY	.012	4582	1.46	.38	.03	1.87	2.2

Description	Oper	Unit	Vol	Crew Size	Man-hours per Unit	Crew Output per Day	Avg Mat'l Unit Cost	Avg Labor Unit Cost	Avg Equip Unit Cost	Avg Total Unit Cost	Avg Price Incl O&P
19-1/2" deep, 30' long	Demo	LF	Lg	LK	.133	360.0	---	4.13	.42	4.55	**6.57**
	Demo	LF	Sm	LK	.190	252.0	---	5.89	.60	6.49	**9.38**
19-1/2" deep, 30' long	Inst	LF	Lg	CY	.156	360.0	18.00	4.89	.42	23.31	**28.50**
	Inst	LF	Sm	CY	.222	252.0	27.40	6.96	.60	34.96	**42.50**
19-1/2" deep, 30' long	Demo	BF	Lg	LK	.021	2340	---	.65	.06	.71	**1.03**
	Demo	BF	Sm	LK	.029	1638	---	.90	.09	.99	**1.43**
19-1/2" deep, 30' long	Inst	BF	Lg	CY	.024	2340	2.77	.75	.06	3.58	**4.37**
	Inst	BF	Sm	CY	.034	1638	4.21	1.07	.09	5.37	**6.53**
19-1/2" deep, 30' long	Demo	SF	Lg	LK	.007	6545	---	.22	.02	.24	**.34**
	Demo	SF	Sm	LK	.010	4582	---	.31	.03	.34	**.49**
19-1/2" deep, 30' long	Inst	SF	Lg	CY	.009	6545	.99	.28	.02	1.29	**1.58**
	Inst	SF	Sm	CY	.012	4582	1.50	.38	.03	1.91	**2.32**
21" deep, 30' long	Demo	LF	Lg	LK	.133	360.0	---	4.13	.42	4.55	**6.57**
	Demo	LF	Sm	LK	.190	252.0	---	5.89	.60	6.49	**9.38**
21" deep, 30' long	Inst	LF	Lg	CY	.156	360.0	19.30	4.89	.42	24.61	**30.00**
	Inst	LF	Sm	CY	.222	252.0	29.40	6.96	.60	36.96	**44.80**
21" deep, 30' long	Demo	BF	Lg	LK	.019	2520	---	.59	.06	.65	**.94**
	Demo	BF	Sm	LK	.027	1764	---	.84	.09	.93	**1.34**
21" deep, 30' long	Inst	BF	Lg	CY	.022	2520	2.77	.69	.06	3.52	**4.28**
	Inst	BF	Sm	CY	.032	1764	4.21	1.00	.09	5.30	**6.44**
21" deep, 30' long	Demo	SF	Lg	LK	.007	6545	---	.22	.02	.24	**.34**
	Demo	SF	Sm	LK	.010	4582	---	.31	.03	.34	**.49**
21" deep, 30' long	Inst	SF	Lg	CY	.009	6545	1.05	.28	.02	1.35	**1.65**
	Inst	SF	Sm	CY	.012	4582	1.60	.38	.03	2.01	**2.43**
22-1/2" deep, 30' long	Demo	LF	Lg	LK	.133	360.0	---	4.13	.42	4.55	**6.57**
	Demo	LF	Sm	LK	.190	252.0	---	5.89	.60	6.49	**9.38**
22-1/2" deep, 30' long	Inst	LF	Lg	CY	.156	360.0	20.60	4.89	.42	25.91	**31.50**
	Inst	LF	Sm	CY	.222	252.0	31.40	6.96	.60	38.96	**47.10**
22-1/2" deep, 30' long	Demo	BF	Lg	LK	.018	2700	---	.56	.06	.62	**.89**
	Demo	BF	Sm	LK	.025	1890	---	.78	.08	.86	**1.24**
22-1/2" deep, 30' long	Inst	BF	Lg	CY	.021	2700	2.75	.66	.06	3.47	**4.21**
	Inst	BF	Sm	CY	.030	1890	4.18	.94	.08	5.20	**6.30**
22-1/2" deep, 30' long	Demo	SF	Lg	LK	.007	6545	---	.22	.02	.24	**.34**
	Demo	SF	Sm	LK	.010	4582	---	.31	.03	.34	**.49**
22-1/2" deep, 30' long	Inst	SF	Lg	CY	.009	6545	1.12	.28	.02	1.42	**1.73**
	Inst	SF	Sm	CY	.012	4582	1.70	.38	.03	2.11	**2.55**
24" deep, 30' long	Demo	LF	Lg	LK	.133	360.0	---	4.13	.42	4.55	**6.57**
	Demo	LF	Sm	LK	.190	252.0	---	5.89	.60	6.49	**9.38**
24" deep, 30' long	Inst	LF	Lg	CY	.156	360.0	22.00	4.89	.42	27.31	**33.00**
	Inst	LF	Sm	CY	.222	252.0	33.40	6.96	.60	40.96	**49.40**
24" deep, 30' long	Demo	BF	Lg	LK	.017	2880	---	.53	.05	.58	**.84**
	Demo	BF	Sm	LK	.024	2016	---	.74	.07	.81	**1.18**

Description	Oper	Unit	Vol	Crew Size	Man-hours per Unit	Crew Output per Day	Avg Mat'l Unit Cost	Avg Labor Unit Cost	Avg Equip Unit Cost	Avg Total Unit Cost	Avg Price Incl O&P
24" deep, 30' long	Inst	BF	Lg	CY	.019	2880	2.74	.60	.05	3.39	**4.09**
	Inst	BF	Sm	CY	.028	2016	4.16	.88	.07	5.11	**6.17**
24" deep, 30' long	Demo	SF	Lg	LK	.007	6545	---	.22	.02	.24	**.34**
	Demo	SF	Sm	LK	.010	4582	---	.31	.03	.34	**.49**
24" deep, 30' long	Inst	SF	Lg	CY	.009	6545	1.19	.28	.02	1.49	**1.81**
	Inst	SF	Sm	CY	.012	4582	1.81	.38	.03	2.22	**2.68**
25-1/2" deep, 30' long	Demo	LF	Lg	LK	.133	360.0	---	4.13	.42	4.55	**6.57**
	Demo	LF	Sm	LK	.190	252.0	---	5.89	.60	6.49	**9.38**
25-1/2" deep, 30' long	Inst	LF	Lg	CY	.156	360.0	23.30	4.89	.42	28.61	**34.60**
	Inst	LF	Sm	CY	.222	252.0	35.50	6.96	.60	43.06	**51.80**
25-1/2" deep, 30' long	Demo	BF	Lg	LK	.016	3060	---	.50	.05	.55	**.79**
	Demo	BF	Sm	LK	.022	2142	---	.68	.07	.75	**1.09**
25-1/2" deep, 30' long	Inst	BF	Lg	CY	.018	3060	2.74	.56	.05	3.35	**4.05**
	Inst	BF	Sm	CY	.026	2142	4.16	.81	.07	5.04	**6.08**
25-1/2" deep, 30' long	Demo	SF	Lg	LK	.007	6545	---	.22	.02	.24	**.34**
	Demo	SF	Sm	LK	.010	4582	---	.31	.03	.34	**.49**
25-1/2" deep, 30' long	Inst	SF	Lg	CY	.009	6545	1.27	.28	.02	1.57	**1.90**
	Inst	SF	Sm	CY	.012	4582	1.93	.38	.03	2.34	**2.81**
27" deep, 30' long	Demo	LF	Lg	LK	.133	360.0	---	4.13	.42	4.55	**6.57**
	Demo	LF	Sm	LK	.190	252.0	---	5.89	.60	6.49	**9.38**
27" deep, 30' long	Inst	LF	Lg	CY	.156	360.0	24.70	4.89	.42	30.01	**36.10**
	Inst	LF	Sm	CY	.222	252.0	37.50	6.96	.60	45.06	**54.10**
27" deep, 30' long	Demo	BF	Lg	LK	.015	3240	---	.47	.05	.52	**.74**
	Demo	BF	Sm	LK	.021	2268	---	.65	.07	.72	**1.04**
27" deep, 30' long	Inst	BF	Lg	CY	.017	3240	2.74	.53	.05	3.32	**4.00**
	Inst	BF	Sm	CY	.025	2268	4.16	.78	.07	5.01	**6.03**
27" deep, 30' long	Demo	SF	Lg	LK	.007	6545	---	.22	.02	.24	**.34**
	Demo	SF	Sm	LK	.010	4582	---	.31	.03	.34	**.49**
27" deep, 30' long	Inst	SF	Lg	CY	.009	6545	1.34	.28	.02	1.64	**1.98**
	Inst	SF	Sm	CY	.012	4582	2.04	.38	.03	2.45	**2.94**

5-1/8" thick, SF pricing based on 16' oc

Description	Oper	Unit	Vol	Crew Size	Man-hours per Unit	Crew Output per Day	Avg Mat'l Unit Cost	Avg Labor Unit Cost	Avg Equip Unit Cost	Avg Total Unit Cost	Avg Price Incl O&P
12" deep, 20' long	Demo	LF	Lg	LK	.133	360.0	---	4.13	.42	4.55	**6.57**
	Demo	LF	Sm	LK	.190	252.0	---	5.89	.60	6.49	**9.38**
12" deep, 20' long	Inst	LF	Lg	CY	.156	360.0	14.80	4.89	.42	20.11	**24.70**
	Inst	LF	Sm	CY	.222	252.0	22.50	6.96	.60	30.06	**36.90**
12" deep, 20' long	Demo	BF	Lg	LK	.022	2160	---	.68	.07	.75	**1.09**
	Demo	BF	Sm	LK	.032	1512	---	.99	.10	1.09	**1.58**
12" deep, 20' long	Inst	BF	Lg	CY	.026	2160	2.47	.81	.07	3.35	**4.13**
	Inst	BF	Sm	CY	.037	1512	3.75	1.16	.10	5.01	**6.15**
12" deep, 20' long	Demo	SF	Lg	LK	.007	6545	---	.22	.02	.24	**.34**
	Demo	SF	Sm	LK	.010	4582	---	.31	.03	.34	**.49**
12" deep, 20' long	Inst	SF	Lg	CY	.009	6545	.81	.28	.02	1.11	**1.37**
	Inst	SF	Sm	CY	.012	4582	1.23	.38	.03	1.64	**2.01**

Description	Oper	Unit	Vol	Crew Size	Man-hours per Unit	Crew Output per Day	Avg Mat'l Unit Cost	Avg Labor Unit Cost	Avg Equip Unit Cost	Avg Total Unit Cost	Avg Price Incl O&P
13-1/2" deep, 20' long	Demo	LF	Lg	LK	.133	360.0	---	4.13	.42	4.55	6.57
	Demo	LF	Sm	LK	.190	252.0	---	5.89	.60	6.49	9.38
13-1/2" deep, 20' long	Inst	LF	Lg	CY	.156	360.0	16.40	4.89	.42	21.71	26.60
	Inst	LF	Sm	CY	.222	252.0	24.90	6.96	.60	32.46	39.70
13-1/2" deep, 20' long	Demo	BF	Lg	LK	.020	2430	---	.62	.06	.68	.98
	Demo	BF	Sm	LK	.028	1701	---	.87	.09	.96	1.38
13-1/2" deep, 20' long	Inst	BF	Lg	CY	.023	2430	2.42	.72	.06	3.20	3.92
	Inst	BF	Sm	CY	.033	1701	3.68	1.03	.09	4.80	5.87
13-1/2" deep, 20' long	Demo	SF	Lg	LK	.007	6545	---	.22	.02	.24	.34
	Demo	SF	Sm	LK	.010	4582	---	.31	.03	.34	.49
13-1/2" deep, 20' long	Inst	SF	Lg	CY	.009	6545	.89	.28	.02	1.19	1.47
	Inst	SF	Sm	CY	.012	4582	1.35	.38	.03	1.76	2.15
15" deep, 20' long	Demo	LF	Lg	LK	.133	360.0	---	4.13	.42	4.55	6.57
	Demo	LF	Sm	I K	.190	252.0	---	5.89	.60	6.49	9.38
15" deep, 20' long	Inst	LF	Lg	CY	.156	360.0	18.00	4.89	.42	23.31	28.50
	Inst	LF	Sm	CY	.222	252.0	27.40	6.96	.60	34.96	42.60
15" deep, 20' long	Demo	BF	Lg	LK	.018	2700	---	.56	.06	.62	.89
	Demo	BF	Sm	LK	.025	1890	---	.78	.08	.86	1.24
15" deep, 20' long	Inst	BF	Lg	CY	.021	2700	2.41	.66	.06	3.13	3.82
	Inst	BF	Sm	CY	.030	1890	3.66	.94	.08	4.68	5.70
15" deep, 20' long	Demo	SF	Lg	LK	.007	6545	---	.22	.02	.24	.34
	Demo	SF	Sm	LK	.010	4582	---	.31	.03	.34	.49
15" deep, 20' long	Inst	SF	Lg	CY	.009	6545	.99	.28	.02	1.29	1.58
	Inst	SF	Sm	CY	.012	4582	1.50	.38	.03	1.91	2.32
16-1/2" deep, 20' long	Demo	LF	Lg	LK	.133	360.0	---	4.13	.42	4.55	6.57
	Demo	LF	Sm	LK	.190	252.0	---	5.89	.60	6.49	9.38
16-1/2" deep, 20' long	Inst	LF	Lg	CY	.156	360.0	19.70	4.89	.42	25.01	30.40
	Inst	LF	Sm	CY	.222	252.0	29.90	6.96	.60	37.46	45.40
16-1/2" deep, 20' long	Demo	BF	Lg	LK	.016	2970	---	.50	.05	.55	.79
	Demo	BF	Sm	LK	.023	2079	---	.71	.07	.78	1.13
16-1/2" deep, 20' long	Inst	BF	Lg	CY	.019	2970	2.38	.60	.05	3.03	3.68
	Inst	BF	Sm	CY	.027	2079	3.62	.85	.07	4.54	5.50
16-1/2" deep, 20' long	Demo	SF	Lg	LK	.007	6545	---	.22	.02	.24	.34
	Demo	SF	Sm	LK	.010	4582	---	.31	.03	.34	.49
16-1/2" deep, 20' long	Inst	SF	Lg	CY	.009	6545	1.07	.28	.02	1.37	1.67
	Inst	SF	Sm	CY	.012	4582	1.63	.38	.03	2.04	2.47
18" deep, 20' long	Demo	LF	Lg	LK	.133	360.0	---	4.13	.42	4.55	6.57
	Demo	LF	Sm	LK	.190	252.0	---	5.89	.60	6.49	9.38
18" deep, 20' long	Inst	LF	Lg	CY	.156	360.0	21.30	4.89	.42	26.61	32.30
	Inst	LF	Sm	CY	.222	252.0	32.40	6.96	.60	39.96	48.30
18" deep, 20' long	Demo	BF	Lg	LK	.015	3240	---	.47	.05	.52	.74
	Demo	BF	Sm	LK	.021	2268	---	.65	.07	.72	1.04

Description	Oper	Unit	Vol	Crew Size	Man-hours per Unit	Crew Output per Day	Avg Mat'l Unit Cost	Avg Labor Unit Cost	Avg Equip Unit Cost	Avg Total Unit Cost	Avg Price Incl O&P
5-1/8" thick beams, SF pricing based on 16' oc (continued)											
18" deep, 20' long	Inst	BF	Lg	CY	.017	3240	2.37	.53	.05	2.95	**3.57**
	Inst	BF	Sm	CY	.025	2268	3.60	.78	.07	4.45	**5.39**
18" deep, 20' long	Demo	SF	Lg	LK	.007	6545	---	.22	.02	.24	**.34**
	Demo	SF	Sm	LK	.010	4582	---	.31	.03	.34	**.49**
18" deep, 20' long	Inst	SF	Lg	CY	.009	6545	1.16	.28	.02	1.46	**1.78**
	Inst	SF	Sm	CY	.012	4582	1.76	.38	.03	2.17	**2.62**
19-1/2" deep, 20' long	Demo	LF	Lg	LK	.133	360.0	---	4.13	.42	4.55	**6.57**
	Demo	LF	Sm	LK	.190	252.0	---	5.89	.60	6.49	**9.38**
19-1/2" deep, 20' long	Inst	LF	Lg	CY	.156	360.0	22.90	4.89	.42	28.21	**34.10**
	Inst	LF	Sm	CY	.222	252.0	34.80	6.96	.60	42.36	**51.00**
19-1/2" deep, 20' long	Demo	BF	Lg	LK	.014	3510	---	.43	.04	.47	**.69**
	Demo	BF	Sm	LK	.020	2457	---	.62	.06	.68	**.98**
19-1/2" deep, 20' long	Inst	BF	Lg	CY	.016	3510	2.34	.50	.04	2.88	**3.48**
	Inst	BF	Sm	CY	.023	2457	3.56	.72	.06	4.34	**5.24**
19-1/2" deep, 20' long	Demo	SF	Lg	LK	.007	6545	---	.22	.02	.24	**.34**
	Demo	SF	Sm	LK	.010	4582	---	.31	.03	.34	**.49**
19-1/2" deep, 20' long	Inst	SF	Lg	CY	.009	6545	1.25	.28	.02	1.55	**1.88**
	Inst	SF	Sm	CY	.012	4582	1.90	.38	.03	2.31	**2.78**
21" deep, 30' long	Demo	LF	Lg	LK	.133	360.0	---	4.13	.42	4.55	**6.57**
	Demo	LF	Sm	LK	.190	252.0	---	5.89	.60	6.49	**9.38**
21" deep, 30' long	Inst	LF	Lg	CY	.156	360.0	23.70	4.89	.42	29.01	**35.00**
	Inst	LF	Sm	CY	.222	252.0	36.00	6.96	.60	43.56	**52.50**
21" deep, 30' long	Demo	BF	Lg	LK	.013	3780	---	.40	.04	.44	**.64**
	Demo	BF	Sm	LK	.018	2646	---	.56	.06	.62	**.89**
21" deep, 30' long	Inst	BF	Lg	CY	.015	3780	2.26	.47	.04	2.77	**3.34**
	Inst	BF	Sm	CY	.021	2646	3.44	.66	.06	4.16	**5.00**
21" deep, 30' long	Demo	SF	Lg	LK	.007	6545	---	.22	.02	.24	**.34**
	Demo	SF	Sm	LK	.010	4582	---	.31	.03	.34	**.49**
21" deep, 30' long	Inst	SF	Lg	CY	.009	6545	1.30	.28	.02	1.60	**1.94**
	Inst	SF	Sm	CY	.012	4582	1.98	.38	.03	2.39	**2.87**
22-1/2" deep, 30' long	Demo	LF	Lg	LK	.133	360.0	---	4.13	.42	4.55	**6.57**
	Demo	LF	Sm	LK	.190	252.0	---	5.89	.60	6.49	**9.38**
22-1/2" deep, 30' long	Inst	LF	Lg	CY	.156	360.0	25.30	4.89	.42	30.61	**36.90**
	Inst	LF	Sm	CY	.222	252.0	38.50	6.96	.60	46.06	**55.30**
22-1/2" deep, 30' long	Demo	BF	Lg	LK	.012	4050	---	.37	.04	.41	**.59**
	Demo	BF	Sm	LK	.017	2835	---	.53	.05	.58	**.84**
22-1/2" deep, 30' long	Inst	BF	Lg	CY	.014	4050	2.25	.44	.04	2.73	**3.29**
	Inst	BF	Sm	CY	.020	2835	3.42	.63	.05	4.10	**4.92**
22-1/2" deep, 30' long	Demo	SF	Lg	LK	.007	6545	---	.22	.02	.24	**.34**
	Demo	SF	Sm	LK	.010	4582	---	.31	.03	.34	**.49**
22-1/2" deep, 30' long	Inst	SF	Lg	CY	.009	6545	1.38	.28	.02	1.68	**2.03**
	Inst	SF	Sm	CY	.012	4582	2.10	.38	.03	2.51	**3.01**

Description	Oper	Unit	Vol	Crew Size	Man-hours per Unit	Crew Output per Day	Avg Mat'l Unit Cost	Avg Labor Unit Cost	Avg Equip Unit Cost	Avg Total Unit Cost	Avg Price Incl O&P
24" deep, 30' long	Demo	LF	Lg	LK	.133	360.0	---	4.13	.42	4.55	**6.57**
	Demo	LF	Sm	LK	.190	252.0	---	5.89	.60	6.49	**9.38**
24" deep, 30' long	Inst	LF	Lg	CY	.156	360.0	26.90	4.89	.42	32.21	**38.70**
	Inst	LF	Sm	CY	.222	252.0	40.90	6.96	.60	48.46	**58.10**
24" deep, 30' long	Demo	BF	Lg	LK	.011	4320	---	.34	.03	.37	**.54**
	Demo	BF	Sm	LK	.016	3024	---	.50	.05	.55	**.79**
24" deep, 30' long	Inst	BF	Lg	CY	.013	4320	2.25	.41	.03	2.69	**3.23**
	Inst	BF	Sm	CY	.019	3024	3.42	.60	.05	4.07	**4.88**
24" deep, 30' long	Demo	SF	Lg	LK	.007	6545	---	.22	.02	.24	**.34**
	Demo	SF	Sm	LK	.010	4582	---	.31	.03	.34	**.49**
24" deep, 30' long	Inst	SF	Lg	CY	.009	6545	1.47	.28	.02	1.77	**2.13**
	Inst	SF	Sm	CY	.012	4582	2.23	.38	.03	2.64	**3.16**
25-1/2" deep, 30' long	Demo	LF	Lg	LK	.133	360.0	---	4.13	.42	4.55	**6.57**
	Demo	LF	Sm	LK	.190	252.0	---	5.89	.60	6.49	**9.38**
25-1/2" deep, 30' long	Inst	LF	Lg	CY	.156	360.0	28.50	4.89	.42	33.81	**40.50**
	Inst	LF	Sm	CY	.222	252.0	43.30	6.96	.60	50.86	**60.90**
25-1/2" deep, 30' long	Demo	BF	Lg	LK	.010	4590	---	.31	.03	.34	**.49**
	Demo	BF	Sm	LK	.015	3213	---	.47	.05	.52	**.74**
25-1/2" deep, 30' long	Inst	BF	Lg	CY	.012	4590	2.23	.38	.03	2.64	**3.16**
	Inst	BF	Sm	CY	.017	3213	3.39	.53	.05	3.97	**4.75**
25-1/2" deep, 30' long	Demo	SF	Lg	LK	.007	6545	---	.22	.02	.24	**.34**
	Demo	SF	Sm	LK	.010	4582	---	.31	.03	.34	**.49**
25-1/2" deep, 30' long	Inst	SF	Lg	CY	.009	6545	1.56	.28	.02	1.86	**2.24**
	Inst	SF	Sm	CY	.012	4582	2.37	.38	.03	2.78	**3.32**
27" deep, 40' long	Demo	LF	Lg	LK	.133	360.0	---	4.13	.42	4.55	**6.57**
	Demo	LF	Sm	LK	.190	252.0	---	5.89	.60	6.49	**9.38**
27" deep, 40' long	Inst	LF	Lg	CY	.156	360.0	29.50	4.89	.42	34.81	**41.70**
	Inst	LF	Sm	CY	.222	252.0	44.90	6.96	.60	52.46	**62.60**
27" deep, 40' long	Demo	BF	Lg	LK	.010	4860	---	.31	.03	.34	**.49**
	Demo	BF	Sm	LK	.014	3402	---	.43	.04	.47	**.69**
27" deep, 40' long	Inst	BF	Lg	CY	.012	4860	2.19	.38	.03	2.60	**3.11**
	Inst	BF	Sm	CY	.016	3402	3.33	.50	.04	3.87	**4.62**
27" deep, 40' long	Demo	SF	Lg	LK	.007	6545	---	.22	.02	.24	**.34**
	Demo	SF	Sm	LK	.010	4582	---	.31	.03	.34	**.49**
27" deep, 40' long	Inst	SF	Lg	CY	.009	6545	1.62	.28	.02	1.92	**2.31**
	Inst	SF	Sm	CY	.012	4582	2.46	.38	.03	2.87	**3.42**
28-1/2" deep, 40' long	Demo	LF	Lg	LK	.133	360.0	---	4.13	.42	4.55	**6.57**
	Demo	LF	Sm	LK	.190	252.0	---	5.89	.60	6.49	**9.38**
28-1/2" deep, 40' long	Inst	LF	Lg	CY	.156	360.0	31.10	4.89	.42	36.41	**43.50**
	Inst	LF	Sm	CY	.222	252.0	47.30	6.96	.60	54.86	**65.40**
28-1/2" deep, 40' long	Demo	BF	Lg	LK	.009	5130	---	.28	.03	.31	**.45**
	Demo	BF	Sm	LK	.013	3591	---	.40	.04	.44	**.64**
28-1/2" deep, 40' long	Inst	BF	Lg	CY	.011	5130	2.18	.34	.03	2.55	**3.05**
	Inst	BF	Sm	CY	.016	3591	3.31	.50	.04	3.85	**4.60**
28-1/2" deep, 40' long	Demo	SF	Lg	LK	.007	6545	---	.22	.02	.24	**.34**
	Demo	SF	Sm	LK	.010	4582	---	.31	.03	.34	**.49**
28-1/2" deep, 40' long	Inst	SF	Lg	CY	.009	6545	1.70	.28	.02	2.00	**2.40**
	Inst	SF	Sm	CY	.012	4582	2.58	.38	.03	2.99	**3.56**

Description	Oper	Unit	Vol	Crew Size	Man-hours per Unit	Crew Output per Day	Avg Mat'l Unit Cost	Avg Labor Unit Cost	Avg Equip Unit Cost	Avg Total Unit Cost	Avg Price Incl O&P

5-1/8" thick beams, SF pricing based on 16' oc (continued)

Description	Oper	Unit	Vol	Crew Size	Man-hours per Unit	Crew Output per Day	Avg Mat'l Unit Cost	Avg Labor Unit Cost	Avg Equip Unit Cost	Avg Total Unit Cost	Avg Price Incl O&P
30" deep, 40' long	Demo	LF	Lg	LK	.133	360.0	---	4.13	.42	4.55	6.57
	Demo	LF	Sm	LK	.190	252.0	---	5.89	.60	6.49	9.38
30" deep, 40' long	Inst	LF	Lg	CY	.156	360.0	32.70	4.89	.42	38.01	45.40
	Inst	LF	Sm	CY	.222	252.0	49.70	6.96	.60	57.26	68.20
30" deep, 40' long	Demo	BF	Lg	LK	.009	5400	---	.28	.03	.31	.45
	Demo	BF	Sm	LK	.013	3780	---	.40	.04	.44	.64
30" deep, 40' long	Inst	BF	Lg	CY	.010	5400	s	.31	.03	2.52	3.01
	Inst	BF	Sm	CY	.015	3780	3.31	.47	.04	3.82	4.55
30" deep, 40' long	Demo	SF	Lg	LK	.007	6545	---	.22	.02	.24	.34
	Demo	SF	Sm	LK	.010	4582	---	.31	.03	.34	.49
30" deep, 40' long	Inst	SF	Lg	CY	.009	6545	1.78	.28	.02	2.08	2.49
	Inst	SF	Sm	CY	.012	4582	2.71	.38	.03	3.12	3.71
31-1/2" deep, 40' long	Demo	LF	Lg	LK	.133	360.0	---	4.13	.42	4.55	6.57
	Demo	LF	Sm	LK	.190	252.0	---	5.89	.60	6.49	9.38
31-1/2" deep, 40' long	Inst	LF	Lg	CY	.156	360.0	34.30	4.89	.42	39.61	47.10
	Inst	LF	Sm	CY	.222	252.0	52.10	6.96	.60	59.66	70.90
31-1/2" deep, 40' long	Demo	BF	Lg	LK	.008	5670	---	.25	.03	.28	.40
	Demo	BF	Sm	LK	.012	3969	---	.37	.04	.41	.59
31-1/2" deep, 40' long	Inst	BF	Lg	CY	.010	5670	2.18	.31	.03	2.52	3.01
	Inst	BF	Sm	CY	.014	3969	3.31	.44	.04	3.79	4.50
31-1/2" deep, 40' long	Demo	SF	Lg	LK	.007	6545	---	.22	.02	.24	.34
	Demo	SF	Sm	LK	.010	4582	---	.31	.03	.34	.49
31-1/2" deep, 40' long	Inst	SF	Lg	CY	.009	6545	1.88	.28	.02	2.18	2.61
	Inst	SF	Sm	CY	.012	4582	2.86	.38	.03	3.27	3.88
33" deep, 40' long	Demo	LF	Lg	LK	.133	360.0	---	4.13	.42	4.55	6.57
	Demo	LF	Sm	LK	.190	252.0	---	5.89	.60	6.49	9.38
33" deep, 40' long	Inst	LF	Lg	CY	.156	360.0	35.80	4.89	.42	41.11	49.00
	Inst	LF	Sm	CY	.222	252.0	54.50	6.96	.60	62.06	73.70
33" deep, 40' long	Demo	BF	Lg	LK	.008	5940	---	.25	.03	.28	.40
	Demo	BF	Sm	LK	.012	4158	---	.37	.04	.41	.59
33" deep, 40' long	Inst	BF	Lg	CY	.009	5940	2.16	.28	.03	2.47	2.94
	Inst	BF	Sm	CY	.013	4158	3.28	.41	.04	3.73	4.42
33" deep, 40' long	Demo	SF	Lg	LK	.007	6545	---	.22	.02	.24	.34
	Demo	SF	Sm	LK	.010	4582	---	.31	.03	.34	.49
33" deep, 40' long	Inst	SF	Lg	CY	.009	6545	1.96	.28	.02	2.26	2.70
	Inst	SF	Sm	CY	.012	4582	2.98	.38	.03	3.39	4.02
34-1/2" deep, 40' long	Demo	LF	Lg	LK	.133	360.0	---	4.13	.42	4.55	6.57
	Demo	LF	Sm	LK	.190	252.0	---	5.89	.60	6.49	9.38
34-1/2" deep, 40' long	Inst	LF	Lg	CY	.156	360.0	37.40	4.89	.42	42.71	50.80
	Inst	LF	Sm	CY	.222	252.0	56.90	6.96	.60	64.46	76.40
34-1/2" deep, 40' long	Demo	BF	Lg	LK	.008	6210	---	.25	.02	.27	.39
	Demo	BF	Sm	LK	.011	4347	---	.34	.03	.37	.54

Description	Oper	Unit	Vol	Crew Size	Man-hours per Unit	Crew Output per Day	Avg Mat'l Unit Cost	Avg Labor Unit Cost	Avg Equip Unit Cost	Avg Total Unit Cost	Avg Price Incl O&P
34-1/2" deep, 40' long	Inst	BF	Lg	CY	.009	6210	2.16	.28	.02	2.46	**2.93**
	Inst	BF	Sm	CY	.013	4347	3.28	.41	.03	3.72	**4.41**
34-1/2" deep, 40' long	Demo	SF	Lg	LK	.007	6545	---	.22	.02	.24	**.34**
	Demo	SF	Sm	LK	.010	4582	---	.31	.03	.34	**.49**
34-1/2" deep, 40' long	Inst	SF	Lg	CY	.009	6545	2.04	.28	.02	2.34	**2.79**
	Inst	SF	Sm	CY	.012	4582	3.10	.38	.03	3.51	**4.16**
36" deep, 50' long	Demo	LF	Lg	LK	.133	360.0	---	4.13	.42	4.55	**6.57**
	Demo	LF	Sm	LK	.190	252.0	---	5.89	.60	6.49	**9.38**
36" deep, 50' long	Inst	LF	Lg	CY	.156	360.0	38.50	4.89	.42	43.81	**52.10**
	Inst	LF	Sm	CY	.222	252.0	58.60	6.96	.60	66.16	**78.40**
36" deep, 50' long	Demo	BF	Lg	LK	.007	6480	---	.22	.02	.24	**.34**
	Demo	BF	Sm	LK	.011	4536	---	.34	.03	.37	**.54**
36" deep, 50' long	Inst	BF	Lg	CY	.009	6480	2.14	.28	.02	2.44	**2.90**
	Inst	BF	Sm	CY	.012	4536	3.25	.38	.03	3.66	**4.33**
36" deep, 50' long	Demo	SF	Lg	LK	.007	6545	---	.22	.02	.24	**.34**
	Demo	SF	Sm	LK	.010	4582	---	.31	.03	.34	**.49**
36" deep, 50' long	Inst	SF	Lg	CY	.009	6545	2.11	.28	.02	2.41	**2.87**
	Inst	SF	Sm	CY	.012	4582	3.21	.38	.03	3.62	**4.29**
37-1/2" deep, 50' long	Demo	LF	Lg	LK	.133	360.0	---	4.13	.42	4.55	**6.57**
	Demo	LF	Sm	LK	.190	252.0	---	5.89	.60	6.49	**9.38**
37-1/2" deep, 50' long	Inst	LF	Lg	CY	.156	360.0	40.10	4.89	.42	45.41	**53.90**
	Inst	LF	Sm	CY	.222	252.0	61.00	6.96	.60	68.56	**81.20**
37-1/2" deep, 50' long	Demo	BF	Lg	LK	.007	6750	---	.22	.02	.24	**.34**
	Demo	BF	Sm	LK	.010	4725	---	.31	.03	.34	**.49**
37-1/2" deep, 50' long	Inst	BF	Lg	CY	.008	6750	2.14	.25	.02	2.41	**2.86**
	Inst	BF	Sm	CY	.012	4725	3.25	.38	.03	3.66	**4.33**
37-1/2" deep, 50' long	Demo	SF	Lg	LK	.007	6545	---	.22	.02	.24	**.34**
	Demo	SF	Sm	LK	.010	4582	---	.31	.03	.34	**.49**
37-1/2" deep, 50' long	Inst	SF	Lg	CY	.009	6545	2.19	.28	.02	2.49	**2.96**
	Inst	SF	Sm	CY	.012	4582	3.33	.38	.03	3.74	**4.42**
39" deep, 50' long	Demo	LF	Lg	LK	.133	360.0	---	4.13	.42	4.55	**6.57**
	Demo	LF	Sm	LK	.190	252.0	---	5.89	.60	6.49	**9.38**
39" deep, 50' long	Inst	LF	Lg	CY	.156	360.0	41.70	4.89	.42	47.01	**55.80**
	Inst	LF	Sm	CY	.222	252.0	63.40	6.96	.60	70.96	**84.00**
39" deep, 50' long	Demo	BF	Lg	LK	.007	7020	---	.22	.02	.24	**.34**
	Demo	BF	Sm	LK	.010	4914	---	.31	.03	.34	**.49**
39" deep, 50' long	Inst	BF	Lg	CY	.008	7020	2.14	.25	.02	2.41	**2.86**
	Inst	BF	Sm	CY	.011	4914	3.25	.34	.03	3.62	**4.28**
39" deep, 50' long	Demo	SF	Lg	LK	.007	6545	---	.22	.02	.24	**.34**
	Demo	SF	Sm	LK	.010	4582	---	.31	.03	.34	**.49**
39" deep, 50' long	Inst	SF	Lg	CY	.009	6545	2.27	.28	.02	2.57	**3.05**
	Inst	SF	Sm	CY	.012	4582	3.45	.38	.03	3.86	**4.56**

Description	Oper	Unit	Vol	Crew Size	Man-hours per Unit	Crew Output per Day	Avg Mat'l Unit Cost	Avg Labor Unit Cost	Avg Equip Unit Cost	Avg Total Unit Cost	Avg Price Incl O&P

6-3/4" thick, SF pricing based on 16' oc

Description	Oper	Unit	Vol	Crew Size	Man-hours per Unit	Crew Output per Day	Avg Mat'l Unit Cost	Avg Labor Unit Cost	Avg Equip Unit Cost	Avg Total Unit Cost	Avg Price Incl O&P
30" deep, 30' long	Demo	LF	Lg	LK	.133	360.0	---	4.13	.42	4.55	6.57
	Demo	LF	Sm	LK	.190	252.0	---	5.89	.60	6.49	9.38
30" deep, 30' long	Inst	LF	Lg	CY	.156	360.0	39.30	4.89	.42	44.61	52.90
	Inst	LF	Sm	CY	.222	252.0	59.70	6.96	.60	67.26	79.70
30" deep, 30' long	Demo	BF	Lg	LK	.007	7200	---	.22	.02	.24	.34
	Demo	BF	Sm	LK	.010	5040	---	.31	.03	.34	.49
30" deep, 30' long	Inst	BF	Lg	CY	.008	7200	1.96	.25	.02	2.23	2.65
	Inst	BF	Sm	CY	.011	5040	2.98	.34	.03	3.35	3.97
30" deep, 30' long	Demo	SF	Lg	LK	.007	6545	---	.22	.02	.24	.34
	Demo	SF	Sm	LK	.010	4582	---	.31	.03	.34	.49
30" deep, 30' long	Inst	SF	Lg	CY	.009	6545	2.15	.28	.02	2.45	2.92
	Inst	SF	Sm	CY	.012	4582	3.27	.38	.03	3.68	4.35
31-1/2" deep, 30' long	Demo	LF	Lg	LK	.133	360.0	---	4.13	.42	4.55	6.57
	Demo	LF	Sm	LK	.190	252.0	---	5.89	.60	6.49	9.38
31-1/2" deep, 30' long	Inst	LF	Lg	CY	.156	360.0	41.20	4.89	.42	46.51	55.10
	Inst	LF	Sm	CY	.222	252.0	62.60	6.96	.60	70.16	83.00
31-1/2" deep, 30' long	Demo	BF	Lg	LK	.006	7560	---	.19	.02	.21	.30
	Demo	BF	Sm	LK	.009	5292	---	.28	.03	.31	.45
31-1/2" deep, 30' long	Inst	BF	Lg	CY	.007	7560	1.96	.22	.02	2.20	2.60
	Inst	BF	Sm	CY	.011	5292	2.98	.34	.03	3.35	3.97
31-1/2" deep, 30' long	Demo	SF	Lg	LK	.007	6545	---	.22	.02	.24	.34
	Demo	SF	Sm	LK	.010	4582	---	.31	.03	.34	.49
31-1/2" deep, 30' long	Inst	SF	Lg	CY	.009	6545	2.25	.28	.02	2.55	3.03
	Inst	SF	Sm	CY	.012	4582	3.42	.38	.03	3.83	4.53
33" deep, 30' long	Demo	LF	Lg	LK	.133	360.0	---	4.13	.42	4.55	6.57
	Demo	LF	Sm	LK	.190	252.0	---	5.89	.60	6.49	9.38
33" deep, 30' long	Inst	LF	Lg	CY	.156	360.0	43.10	4.89	.42	48.41	57.30
	Inst	LF	Sm	CY	.222	252.0	65.40	6.96	.60	72.96	86.30
33" deep, 30' long	Demo	BF	Lg	LK	.006	7920	---	.19	.02	.21	.30
	Demo	BF	Sm	LK	.009	5544	---	.28	.03	.31	.45
33" deep, 30' long	Inst	BF	Lg	CY	.007	7920	1.96	.22	.02	2.20	2.60
	Inst	BF	Sm	CY	.010	5544	2.08	.31	.03	3.32	3.93
33" deep, 30' long	Demo	SF	Lg	LK	.007	6545	---	.22	.02	.24	.34
	Demo	SF	Sm	LK	.010	4582	---	.31	.03	.34	.49
33" deep, 30' long	Inst	SF	Lg	CY	.009	6545	2.36	.28	.02	2.66	3.16
	Inst	SF	Sm	CY	.012	4582	3.59	.38	.03	4.00	4.72
34-1/2" deep, 40' long	Demo	LF	Lg	LK	.133	360.0	---	4.13	.42	4.55	6.57
	Demo	LF	Sm	LK	.190	252.0	---	5.89	.60	6.49	9.38
34-1/2" deep, 40' long	Inst	LF	Lg	CY	.156	360.0	44.20	4.89	.42	49.51	58.60
	Inst	LF	Sm	CY	.222	252.0	67.20	6.96	.60	74.76	88.30
34-1/2" deep, 40' long	Demo	BF	Lg	LK	.006	8280	---	.19	.02	.21	.30
	Demo	BF	Sm	LK	.008	5796	---	.25	.03	.28	.40

Description	Oper	Unit	Vol	Crew Size	Man-hours per Unit	Crew Output per Day	Avg Mat'l Unit Cost	Avg Labor Unit Cost	Avg Equip Unit Cost	Avg Total Unit Cost	Avg Price Incl O&P
34-1/2" deep, 40' long	Inst	BF	Lg	CY	.007	8280	1.92	.22	.02	2.16	2.56
	Inst	BF	Sm	CY	.010	5796	2.92	.31	.03	3.26	3.86
34-1/2" deep, 40' long	Demo	SF	Lg	LK	.007	6545	---	.22	.02	.24	.34
	Demo	SF	Sm	LK	.010	4582	---	.31	.03	.34	.49
34-1/2" deep, 40' long	Inst	SF	Lg	CY	.009	6545	2.41	.28	.02	2.71	3.21
	Inst	SF	Sm	CY	.012	4582	3.66	.38	.03	4.07	4.80
36" deep, 40' long	Demo	LF	Lg	LK	.133	360.0	---	4.13	.42	4.55	6.57
	Demo	LF	Sm	LK	.190	252.0	---	5.89	.60	6.49	9.38
36" deep, 40' long	Inst	LF	Lg	CY	.156	360.0	46.10	4.89	.42	51.41	60.70
	Inst	LF	Sm	CY	.222	252.0	70.00	6.96	.60	77.56	91.60
36" deep, 40' long	Demo	BF	Lg	LK	.006	8640	---	.19	.02	.21	.30
	Demo	BF	Sm	LK	.008	6048	---	.25	.02	.27	.39
36" deep, 40' long	Inst	BF	Lg	CY	.006	8640	1.92	.19	.02	2.13	2.51
	Inst	BF	Sm	CY	.009	6048	2.92	.28	.02	3.22	3.80
36" deep, 40' long	Demo	SF	Lg	LK	.007	6545	---	.22	.02	.24	.34
	Demo	SF	Sm	LK	.010	4582	---	.31	.03	.34	.49
36" deep, 40' long	Inst	SF	Lg	CY	.009	6545	2.52	.28	.02	2.82	3.34
	Inst	SF	Sm	CY	.012	4582	3.83	.38	.03	4.24	5.00
37-1/2" deep, 40' long	Demo	LF	Lg	LK	.133	360.0	---	4.13	.42	4.55	6.57
	Demo	LF	Sm	LK	.190	252.0	---	5.89	.60	6.49	9.38
37-1/2" deep, 40' long	Inst	LF	Lg	CY	.156	360.0	47.90	4.89	.42	53.21	62.90
	Inst	LF	Sm	CY	.222	252.0	72.80	6.96	.60	80.36	94.80
37-1/2" deep, 40' long	Demo	BF	Lg	LK	.005	9000	---	.16	.02	.18	.25
	Demo	BF	Sm	LK	.008	6300	---	.25	.02	.27	.39
37-1/2" deep, 40' long	Inst	BF	Lg	CY	.006	9000	1.92	.19	.02	2.13	2.51
	Inst	BF	Sm	CY	.009	6300	2.92	.28	.02	3.22	3.80
37-1/2" deep, 40' long	Demo	SF	Lg	LK	.007	6545	---	.22	.02	.24	.34
	Demo	SF	Sm	LK	.010	4582	---	.31	.03	.34	.49
37-1/2" deep, 40' long	Inst	SF	Lg	CY	.009	6545	2.62	.28	.02	2.92	3.46
	Inst	SF	Sm	CY	.012	4582	3.98	.38	.03	4.39	5.17
39" deep, 50' long	Demo	LF	Lg	LK	.133	360.0	---	4.13	.42	4.55	6.57
	Demo	LF	Sm	LK	.190	252.0	---	5.89	.60	6.49	9.38
39" deep, 50' long	Inst	LF	Lg	CY	.156	360.0	49.30	4.89	.42	54.61	64.50
	Inst	LF	Sm	CY	.222	252.0	75.00	6.96	.60	82.56	97.20
39" deep, 50' long	Demo	BF	Lg	LK	.005	9360	---	.16	.02	.18	.25
	Demo	BF	Sm	LK	.007	6552	---	.22	.02	.24	.34
39" deep, 50' long	Inst	BF	Lg	CY	.006	9360	1.89	.19	.02	2.10	2.48
	Inst	BF	Sm	CY	.009	6552	2.87	.28	.02	3.17	3.74
39" deep, 50' long	Demo	SF	Lg	LK	.007	6545	---	.22	.02	.24	.34
	Demo	SF	Sm	LK	.010	4582	---	.31	.03	.34	.49
39" deep, 50' long	Inst	SF	Lg	CY	.009	6545	2.70	.28	.02	3.00	3.55
	Inst	SF	Sm	CY	.012	4582	4.10	.38	.03	4.51	5.31

Description	Oper	Unit	Vol	Crew Size	Man-hours per Unit	Crew Output per Day	Avg Mat'l Unit Cost	Avg Labor Unit Cost	Avg Equip Unit Cost	Avg Total Unit Cost	Avg Price Inc O&P
40-1/2" deep, 50' long	Demo	LF	Lg	LK	.133	360.0	---	4.13	.42	4.55	6.57
	Demo	LF	Sm	LK	.190	252.0	---	5.89	.60	6.49	9.38
40-1/2" deep, 50' long	Inst	LF	Lg	CY	.156	360.0	51.20	4.89	.42	56.51	66.60
	Inst	LF	Sm	CY	.222	252.0	77.80	6.96	.60	85.36	100.00
40-1/2" deep, 50' long	Demo	BF	Lg	LK	.005	9720	---	.16	.02	.18	.25
	Demo	BF	Sm	LK	.007	6804	---	.22	.02	.24	.34
40-1/2" deep, 50' long	Inst	BF	Lg	CY	.006	9720	1.89	.19	.02	2.10	2.48
	Inst	BF	Sm	CY	.008	6804	2.87	.25	.02	3.14	3.70
40-1/2" deep, 50' long	Demo	SF	Lg	LK	.007	6545	---	.22	.02	.24	.34
	Demo	SF	Sm	LK	.010	4582	---	.31	.03	.34	.49
40-1/2" deep, 50' long	Inst	SF	Lg	CY	.009	6545	2.79	.28	.02	3.09	3.65
	Inst	SF	Sm	CY	.012	4582	4.24	.38	.03	4.65	5.47
42" deep, 50' long	Demo	LF	Lg	LK	.133	360.0	---	4.13	.42	4.55	6.57
	Demo	LF	Sm	LK	.190	252.0	---	5.89	.60	6.49	9.38
42" deep, 50' long	Inst	LF	Lg	CY	.156	360.0	53.00	4.89	.42	58.31	68.70
	Inst	LF	Sm	CY	.222	252.0	80.60	6.96	.60	88.16	104.00
42" deep, 50' long	Demo	BF	Lg	LK	.005	10080	---	.16	.01	.17	.24
	Demo	BF	Sm	LK	.007	7056	---	.22	.02	.24	.34
42" deep, 50' long	Inst	BF	Lg	CY	.006	10080	1.89	.19	.01	2.09	2.47
	Inst	BF	Sm	CY	.008	7056	2.87	.25	.02	3.14	3.70
42" deep, 50' long	Demo	SF	Lg	LK	.007	6545	---	.22	.02	.24	.34
	Demo	SF	Sm	LK	.010	4582	---	.31	.03	.34	.49
42" deep, 50' long	Inst	SF	Lg	CY	.009	6545	2.89	.28	.02	3.19	3.77
	Inst	SF	Sm	CY	.012	4582	4.39	.38	.03	4.80	5.64
43-1/2" deep, 50' long	Demo	LF	Lg	LK	.133	360.0	---	4.13	.42	4.55	6.57
	Demo	LF	Sm	LK	.190	252.0	---	5.89	.60	6.49	9.38
43-1/2" deep, 50' long	Inst	LF	Lg	CY	.156	360.0	54.80	4.89	.42	60.11	70.80
	Inst	LF	Sm	CY	.222	252.0	83.40	6.96	.60	90.96	107.00
43-1/2" deep, 50' long	Demo	BF	Lg	LK	.005	10440	---	.16	.01	.17	.24
	Demo	BF	Sm	LK	.007	7308	---	.22	.02	.24	.34
43-1/2" deep, 50' long	Inst	BF	Lg	CY	.005	10440	1.89	.16	.01	2.06	2.42
	Inst	BF	Sm	CY	.008	7308	2.87	.25	.02	3.14	3.70
43-1/2" deep, 50' long	Demo	SF	Lg	LK	.007	6545	---	.22	.02	.24	.34
	Demo	SF	Sm	LK	.010	4582	---	.31	.03	.34	.49
43-1/2" deep, 50' long	Inst	SF	Lg	CY	.009	6545	3.00	.28	.02	3.30	3.89
	Inst	SF	Sm	CY	.012	4582	4.56	.38	.03	4.97	5.84
45" deep, 50' long	Demo	LF	Lg	LK	.133	360.0	---	4.13	.42	4.55	6.57
	Demo	LF	Sm	LK	.190	252.0	---	5.89	.60	6.49	9.38
45" deep, 50' long	Inst	LF	Lg	CY	.156	360.0	56.70	4.89	.42	62.01	73.00
	Inst	LF	Sm	CY	.222	252.0	86.20	6.96	.60	93.76	110.00
45" deep, 50' long	Demo	BF	Lg	LK	.004	10800	---	.12	.01	.13	.19
	Demo	BF	Sm	LK	.006	7560	---	.19	.02	.21	.30
45" deep, 50' long	Inst	BF	Lg	CY	.005	10800	1.89	.16	.01	2.06	2.42
	Inst	BF	Sm	CY	.007	7560	2.87	.22	.02	3.11	3.65
45" deep, 50' long	Demo	SF	Lg	LK	.007	6545	---	.22	.02	.24	.34
	Demo	SF	Sm	LK	.010	4582	---	.31	.03	.34	.49
45" deep, 50' long	Inst	SF	Lg	CY	.009	6545	3.10	.28	.02	3.40	4.01
	Inst	SF	Sm	CY	.012	4582	4.71	.38	.03	5.12	6.01

Description	Oper	Unit	Vol	Crew Size	Man-hours per Unit	Crew Output per Day	Avg Mat'l Unit Cost	Avg Labor Unit Cost	Avg Equip Unit Cost	Avg Total Unit Cost	Avg Price Incl O&P

8-3/4" thick, SF pricing based on 16' oc

Description	Oper	Unit	Vol	Crew Size	Man-hours per Unit	Crew Output per Day	Avg Mat'l Unit Cost	Avg Labor Unit Cost	Avg Equip Unit Cost	Avg Total Unit Cost	Avg Price Incl O&P
36" deep, 30' long	Demo	LF	Lg	LK	.133	360.0	---	4.13	.42	4.55	**6.57**
	Demo	LF	Sm	LK	.190	252.0	---	5.89	.60	6.49	**9.38**
36" deep, 30' long	Inst	LF	Lg	CY	.156	360.0	53.70	4.89	.42	59.01	**69.50**
	Inst	LF	Sm	CY	.222	252.0	81.70	6.96	.60	89.26	**105.00**
36" deep, 30' long	Demo	BF	Lg	LK	.004	10800	---	.12	.01	.13	**.19**
	Demo	BF	Sm	LK	.006	7560	---	.19	.02	.21	**.30**
36" deep, 30' long	Inst	BF	Lg	CY	.005	10800	1.79	.16	.01	1.96	**2.30**
	Inst	BF	Sm	CY	.007	7560	2.72	.22	.02	2.96	**3.48**
36" deep, 30' long	Demo	SF	Lg	LK	.007	6545	---	.22	.02	.24	**.34**
	Demo	SF	Sm	LK	.010	4582	---	.31	.03	.34	**.49**
36" deep, 30' long	Inst	SF	Lg	CY	.009	6545	2.93	.28	.02	3.23	**3.81**
	Inst	SF	Sm	CY	.012	4582	4.45	.38	.03	4.86	**5.71**
37-1/2" deep, 30' long	Demo	LF	Lg	LK	.133	360.0	---	4.13	.42	4.55	**6.57**
	Demo	LF	Sm	LK	.190	252.0	---	5.89	.60	6.49	**9.38**
37-1/2" deep, 30' long	Inst	LF	Lg	CY	.156	360.0	55.80	4.89	.42	61.11	**71.90**
	Inst	LF	Sm	CY	.222	252.0	84.80	6.96	.60	92.36	**109.00**
37-1/2" deep, 30' long	Demo	BF	Lg	LK	.004	11250	---	.12	.01	.13	**.19**
	Demo	BF	Sm	LK	.006	7875	---	.19	.02	.21	**.30**
37-1/2" deep, 30' long	Inst	BF	Lg	CY	.005	11250	1.78	.16	.01	1.95	**2.29**
	Inst	BF	Sm	CY	.007	7875	2.71	.22	.02	2.95	**3.47**
37-1/2" deep, 30' long	Demo	SF	Lg	LK	.007	6545	---	.22	.02	.24	**.34**
	Demo	SF	Sm	LK	.010	4582	---	.31	.03	.34	**.49**
37-1/2" deep, 30' long	Inst	SF	Lg	CY	.009	6545	3.04	.28	.02	3.34	**3.94**
	Inst	SF	Sm	CY	.012	4582	4.62	.38	.03	5.03	**5.91**
39" deep, 30' long	Demo	LF	Lg	LK	.133	360.0	---	4.13	.42	4.55	**6.57**
	Demo	LF	Sm	LK	.190	252.0	---	5.89	.60	6.49	**9.38**
39" deep, 30' long	Inst	LF	Lg	CY	.156	360.0	58.10	4.89	.42	63.41	**74.50**
	Inst	LF	Sm	CY	.222	252.0	88.30	6.96	.60	95.86	**113.00**
39" deep, 30' long	Demo	BF	Lg	LK	.004	11700	---	.12	.01	.13	**.19**
	Demo	BF	Sm	LK	.006	8190	---	.19	.02	.21	**.30**
39" deep, 30' long	Inst	BF	Lg	CY	.005	11700	1.78	.16	.01	1.95	**2.29**
	Inst	BF	Sm	CY	.007	8190	2.71	.22	.02	2.95	**3.47**
39" deep, 30' long	Demo	SF	Lg	LK	.007	6545	---	.22	.02	.24	**.34**
	Demo	SF	Sm	LK	.010	4582	---	.31	.03	.34	**.49**
39" deep, 30' long	Inst	SF	Lg	CY	.009	6545	3.16	.28	.02	3.46	**4.08**
	Inst	SF	Sm	CY	.012	4582	4.80	.38	.03	5.21	**6.11**
40-1/2" deep, 30' long	Demo	LF	Lg	LK	.133	360.0	---	4.13	.42	4.55	**6.57**
	Demo	LF	Sm	LK	.190	252.0	---	5.89	.60	6.49	**9.38**
40-1/2" deep, 30' long	Inst	LF	Lg	CY	.156	360.0	60.20	4.89	.42	65.51	**77.00**
	Inst	LF	Sm	CY	.222	252.0	91.60	6.96	.60	99.16	**116.00**
40-1/2" deep, 30' long	Demo	BF	Lg	LK	.004	12150	---	.12	.01	.13	**.19**
	Demo	BF	Sm	LK	.006	8505	---	.19	.02	.21	**.30**
40-1/2" deep, 30' long	Inst	BF	Lg	CY	.005	12150	1.78	.16	.01	1.95	**2.29**
	Inst	BF	Sm	CY	.007	8505	2.71	.22	.02	2.95	**3.47**
40-1/2" deep, 30' long	Demo	SF	Lg	LK	.007	6545	---	.22	.02	.24	**.34**
	Demo	SF	Sm	LK	.010	4582	---	.31	.03	.34	**.49**
40-1/2" deep, 30' long	Inst	SF	Lg	CY	.009	6545	3.29	.28	.02	3.59	**4.23**
	Inst	SF	Sm	CY	.012	4582	5.00	.38	.03	5.41	**6.34**

8-3/4" thick beams, SF pricing based on 16' oc (continued)

Description	Oper	Unit	Vol	Crew Size	Man-hours per Unit	Crew Output per Day	Avg Mat'l Unit Cost	Avg Labor Unit Cost	Avg Equip Unit Cost	Avg Total Unit Cost	Avg Price Incl O&P
42" deep, 30' long	Demo	LF	Lg	LK	.133	360.0	---	4.13	.42	4.55	6.57
	Demo	LF	Sm	LK	.190	252.0	---	5.89	.60	6.49	9.38
42" deep, 30' long .	Inst	LF	Lg	CY	.156	360.0	62.40	4.89	.42	67.71	79.50
	Inst	LF	Sm	CY	.222	252.0	94.80	6.96	.60	102.36	120.00
42" deep, 30' long	Demo	BF	Lg	LK	.004	12600	---	.12	.01	.13	.19
	Demo	BF	Sm	LK	.005	8820	---	.16	.02	.18	.25
42" deep, 30' long	Inst	BF	Lg	CY	.004	12600	1.78	.13	.01	1.92	2.25
	Inst	BF	Sm	CY	.006	8820	2.71	.19	.02	2.92	3.42
42" deep, 30' long	Demo	SF	Lg	LK	.007	6545	---	.22	.02	.24	.34
	Demo	SF	Sm	LK	.010	4582	---	.31	.03	.34	.49
42" deep, 30' long	Inst	SF	Lg	CY	.009	6545	3.40	.28	.02	3.70	4.35
	Inst	SF	Sm	CY	.012	4582	5.17	.38	.03	5.58	6.54
43-1/2" deep, 40' long	Demo	LF	Lg	LK	.133	360.0	---	4.13	.42	4.55	6.57
	Demo	LF	Sm	LK	.190	252.0	---	5.89	.60	6.49	9.38
43-1/2" deep, 40' long	Inst	LF	Lg	CY	.156	360.0	63.80	4.89	.42	69.11	81.10
	Inst	LF	Sm	CY	.222	252.0	97.00	6.96	.60	104.56	123.00
43-1/2" deep, 40' long	Demo	BF	Lg	LK	.004	13050	---	.12	.01	.13	.19
	Demo	BF	Sm	LK	.005	9135	---	.16	.02	.18	.25
43-1/2" deep, 40' long	Inst	BF	Lg	CY	.004	13050	1.75	.13	.01	1.89	2.21
	Inst	BF	Sm	CY	.006	9135	2.66	.19	.02	2.87	3.36
43-1/2" deep, 40' long	Demo	SF	Lg	LK	.007	6545	---	.22	.02	.24	.34
	Demo	SF	Sm	LK	.010	4582	---	.31	.03	.34	.49
43-1/2" deep, 40' long	Inst	SF	Lg	CY	.009	6545	3.48	.28	.02	3.78	4.45
	Inst	SF	Sm	CY	.012	4582	5.29	.38	.03	5.70	6.68
45" deep, 40' long	Demo	LF	Lg	LK	.133	360.0	---	4.13	.42	4.55	6.57
	Demo	LF	Sm	LK	.190	252.0	---	5.89	.60	6.49	9.38
45" deep, 40' long	Inst	LF	Lg	CY	.156	360.0	65.90	4.89	.42	71.21	83.60
	Inst	LF	Sm	CY	.222	252.0	100.00	6.96	.60	107.56	126.00
45" deep, 40' long	Demo	BF	Lg	LK	.004	13500	---	.12	.01	.13	.19
	Demo	BF	Sm	LK	.005	9450	---	.16	.02	.18	.25
45" deep, 40' long	Inst	BF	Lg	CY	.004	13500	1.75	.13	.01	1.89	2.21
	Inst	BF	Sm	CY	.006	9450	2.66	.19	.02	2.87	3.36
45" deep, 40' long	Demo	SF	Lg	LK	.007	6545	---	.22	.02	.24	.34
	Demo	SF	Sm	LK	.010	4582	---	.31	.03	.34	.49
45" deep, 40' long	Inst	SF	Lg	CY	.009	6545	3.60	.28	.02	3.90	4.58
	Inst	SF	Sm	CY	.012	4582	5.47	.38	.03	5.88	6.88
46-1/2" deep, 40' long	Demo	LF	Lg	LK	.133	360.0	---	4.13	.42	4.55	6.57
	Demo	LF	Sm	LK	.190	252.0	---	5.89	.60	6.49	9.38
46-1/2" deep, 40' long	Inst	LF	Lg	CY	.156	360.0	68.10	4.89	.42	73.41	86.10
	Inst	LF	Sm	CY	.222	252.0	104.00	6.96	.60	111.56	130.00
46-1/2" deep, 40' long	Demo	BF	Lg	LK	.003	13950	---	.09	.01	.10	.15
	Demo	BF	Sm	LK	.005	9765	---	.16	.02	.18	.25

Description	Oper	Unit	Vol	Crew Size	Man-hours per Unit	Crew Output per Day	Avg Mat'l Unit Cost	Avg Labor Unit Cost	Avg Equip Unit Cost	Avg Total Unit Cost	Avg Price Incl O&P
46-1/2" deep, 40' long	Inst	BF	Lg	CY	.004	13950	1.75	.13	.01	1.89	2.21
	Inst	BF	Sm	CY	.006	9765	2.66	.19	.02	2.87	3.36
46-1/2" deep, 40' long	Demo	SF	Lg	LK	.007	6545	---	.22	.02	.24	.34
	Demo	SF	Sm	LK	.010	4582	---	.31	.03	.34	.49
46-1/2" deep, 40' long	Inst	SF	Lg	CY	.009	6545	3.71	.28	.02	4.01	4.71
	Inst	SF	Sm	CY	.012	4582	5.64	.38	.03	6.05	7.08
48" deep, 50' long	Demo	LF	Lg	LK	.133	360.0	---	4.13	.42	4.55	6.57
	Demo	LF	Sm	LK	.190	252.0	---	5.89	.60	6.49	9.38
48" deep, 50' long	Inst	LF	Lg	CY	.156	360.0	69.70	4.89	.42	75.01	87.90
	Inst	LF	Sm	CY	.222	252.0	106.00	6.96	.60	113.56	133.00
48" deep, 50' long	Demo	BF	Lg	LK	.003	14400	---	.09	.01	.10	.15
	Demo	BF	Sm	LK	.005	10080	---	.16	.01	.17	.24
48" deep, 50' long	Inst	BF	Lg	CY	.004	14400	1.74	.13	.01	1.88	2.20
	Inst	BF	Sm	CY	.006	10080	2.64	.19	.01	2.84	3.33
48" deep, 50' long	Demo	SF	Lg	LK	.007	6545	---	.22	.02	.24	.34
	Demo	SF	Sm	LK	.010	4582	---	.31	.03	.34	.49
48" deep, 50' long	Inst	SF	Lg	CY	.009	6545	3.81	.28	.02	4.11	4.82
	Inst	SF	Sm	CY	.012	4582	5.79	.38	.03	6.20	7.25
49-1/2" deep, 50' long	Demo	LF	Lg	LK	.133	360.0	---	4.13	.42	4.55	6.57
	Demo	LF	Sm	LK	.190	252.0	---	5.89	.60	6.49	9.38
49-1/2" deep, 50' long	Inst	LF	Lg	CY	.156	360.0	71.90	4.89	.42	77.21	90.40
	Inst	LF	Sm	CY	.222	252.0	109.00	6.96	.60	116.56	137.00
49-1/2" deep, 50' long	Demo	BF	Lg	LK	.003	14850	---	.09	.01	.10	.15
	Demo	BF	Sm	LK	.005	10395	---	.16	.01	.17	.24
49-1/2" deep, 50' long	Inst	BF	Lg	CY	.004	14850	1.74	.13	.01	1.88	2.20
	Inst	BF	Sm	CY	.005	10395	2.64	.16	.01	2.81	3.28
49-1/2" deep, 50' long	Demo	SF	Lg	LK	.007	6545	---	.22	.02	.24	.34
	Demo	SF	Sm	LK	.010	4582	---	.31	.03	.34	.49
49-1/2" deep, 50' long	Inst	SF	Lg	CY	.009	6545	3.92	.28	.02	4.22	4.95
	Inst	SF	Sm	CY	.012	4582	5.96	.38	.03	6.37	7.45
51" deep, 50' long	Demo	LF	Lg	LK	.133	360.0	---	4.13	.42	4.55	6.57
	Demo	LF	Sm	LK	.190	252.0	---	5.89	.60	6.49	9.38
51" deep, 50' long	Inst	LF	Lg	CY	.156	360.0	74.00	4.89	.42	79.31	92.80
	Inst	LF	Sm	CY	.222	252.0	112.00	6.96	.60	119.56	140.00
51" deep, 50' long	Demo	BF	Lg	LK	.003	15300	---	.09	.01	.10	.15
	Demo	BF	Sm	LK	.004	10710	---	.12	.01	.13	.19
51" deep, 50' long	Inst	BF	Lg	CY	.004	15300	1.74	.13	.01	1.88	2.20
	Inst	BF	Sm	CY	.005	10710	2.64	.16	.01	2.81	3.28
51" deep, 50' long	Demo	SF	Lg	LK	.007	6545	---	.22	.02	.24	.34
	Demo	SF	Sm	LK	.010	4582	---	.31	.03	.34	.49
51" deep, 50' long	Inst	SF	Lg	CY	.009	6545	4.04	.28	.02	4.34	5.09
	Inst	SF	Sm	CY	.012	4582	6.14	.38	.03	6.55	7.66

Description	Oper	Unit	Vol	Crew Size	Man-hours per Unit	Crew Output per Day	Avg Mat'l Unit Cost	Avg Labor Unit Cost	Avg Equip Unit Cost	Avg Total Unit Cost	Avg Price Incl O&P

8-3/4" thick beams, SF pricing based on 16' oc (continued)

Description	Oper	Unit	Vol	Crew Size	Man-hours per Unit	Crew Output per Day	Avg Mat'l Unit Cost	Avg Labor Unit Cost	Avg Equip Unit Cost	Avg Total Unit Cost	Avg Price Incl O&P
52-1/2" deep, 50' long	Demo	LF	Lg	LK	.133	360.0	---	4.13	.42	4.55	6.57
	Demo	LF	Sm	LK	.190	252.0	---	5.89	.60	6.49	9.38
52-1/2" deep, 50' long	Inst	LF	Lg	CY	.156	360.0	76.10	4.89	.42	81.41	95.30
	Inst	LF	Sm	CY	.222	252.0	116.00	6.96	.60	123.56	144.00
52-1/2" deep, 50' long	Demo	BF	Lg	LK	.003	15750	---	.09	.01	.10	.15
	Demo	BF	Sm	LK	.004	11025	---	.12	.01	.13	.19
52-1/2" deep, 50' long	Inst	BF	Lg	CY	.004	15750	1.74	.13	.01	1.88	2.20
	Inst	BF	Sm	CY	.005	11025	2.64	.16	.01	2.81	3.28
52-1/2" deep, 50' long	Demo	SF	Lg	LK	.007	6545	---	.22	.02	.24	.34
	Demo	SF	Sm	LK	.010	4582	---	.31	.03	.34	.49
52-1/2" deep, 50' long	Inst	SF	Lg	CY	.009	6545	4.15	.28	.02	4.45	5.22
	Inst	SF	Sm	CY	.012	4582	6.31	.38	.03	6.72	7.85
54" deep, 50' long	Demo	LF	Lg	LK	.133	360.0	---	4.13	.42	4.55	6.57
	Demo	LF	Sm	LK	.190	252.0	---	5.89	.60	6.49	9.38
54" deep, 50' long	Inst	LF	Lg	CY	.156	360.0	78.30	4.89	.42	83.61	97.70
	Inst	LF	Sm	CY	.222	252.0	119.00	6.96	.60	126.56	148.00
54" deep, 50' long	Demo	BF	Lg	LK	.003	16200	---	.09	.01	.10	.15
	Demo	BF	Sm	LK	.004	11340	---	.12	.01	.13	.19
54" deep, 50' long	Inst	BF	Lg	CY	.003	16200	1.74	.09	.01	1.84	2.15
	Inst	BF	Sm	CY	.005	11340	2.64	.16	.01	2.81	3.28
54" deep, 50' long	Demo	SF	Lg	LK	.007	6545	---	.22	.02	.24	.34
	Demo	SF	Sm	LK	.010	4582	---	.31	.03	.34	.49
54" deep, 50' long	Inst	SF	Lg	CY	.009	6545	4.26	.28	.02	4.56	5.34
	Inst	SF	Sm	CY	.012	4582	6.48	.38	.03	6.89	8.05
55-1/2" deep, 50' long	Demo	LF	Lg	LK	.133	360.0	---	4.13	.42	4.55	6.57
	Demo	LF	Sm	LK	.190	252.0	---	5.89	.60	6.49	9.38
55-1/2" deep, 50' long	Inst	LF	Lg	CY	.156	360.0	80.40	4.89	.42	85.71	100.00
	Inst	LF	Sm	CY	.222	252.0	122.00	6.96	.60	129.56	152.00
55-1/2" deep, 50' long	Demo	BF	Lg	LK	.003	16650	---	.09	.01	.10	.15
	Demo	BF	Sm	LK	.004	11655	---	.12	.01	.13	.19
55-1/2" deep, 50' long	Inst	BF	Lg	CY	.003	16650	1.74	.09	.01	1.84	2.15
	Inst	BF	Sm	CY	.005	11655	2.64	.16	.01	2.81	3.28
55-1/2" deep, 50' long	Demo	SF	Lg	LK	.007	6545	---	.22	.02	.24	.34
	Demo	SF	Sm	LK	.010	4582	---	.31	.03	.34	.49
55-1/2" deep, 50' long	Inst	SF	Lg	CY	.009	6545	4.38	.28	.02	4.68	5.48
	Inst	SF	Sm	CY	.012	4582	6.66	.38	.03	7.07	8.25
57" deep, 50' long	Demo	LF	Lg	LK	.133	360.0	---	4.13	.42	4.55	6.57
	Demo	LF	Sm	LK	.190	252.0	---	5.89	.60	6.49	9.38
57" deep, 50' long	Inst	LF	Lg	CY	.156	360.0	82.50	4.89	.42	87.81	103.00
	Inst	LF	Sm	CY	.222	252.0	125.00	6.96	.60	132.56	155.00
57" deep, 50' long	Demo	BF	Lg	LK	.003	17100	---	.09	.01	.10	.15
	Demo	BF	Sm	LK	.004	11970	---	.12	.01	.13	.19
57" deep, 50' long	Inst	BF	Lg	CY	.003	17100	1.74	.09	.01	1.84	2.15
	Inst	BF	Sm	CY	.005	11970	2.64	.16	.01	2.81	3.28
57" deep, 50' long	Demo	SF	Lg	LK	.007	6545	---	.22	.02	.24	.34
	Demo	SF	Sm	LK	.010	4582	---	.31	.03	.34	.49
57" deep, 50' long	Inst	SF	Lg	CY	.009	6545	4.51	.28	.02	4.81	5.63
	Inst	SF	Sm	CY	.012	4582	6.86	.38	.03	7.27	8.48

Description	Oper	Unit	Vol	Crew Size	Man-hours per Unit	Crew Output per Day	Avg Mat'l Unit Cost	Avg Labor Unit Cost	Avg Equip Unit Cost	Avg Total Unit Cost	Avg Price Incl O&P
10-3/4" thick, SF pricing based on 16' oc											
42" deep, 50' long	Demo	LF	Lg	LK	.133	360.0	---	4.13	.42	4.55	**6.57**
	Demo	LF	Sm	LK	.190	252.0	---	5.89	.60	6.49	**9.38**
42" deep, 50' long	Inst	LF	Lg	CY	.156	360.0	69.60	4.89	.42	74.91	**87.80**
	Inst	LF	Sm	CY	.222	252.0	106.00	6.96	.60	113.56	**133.00**
42" deep, 50' long	Demo	BF	Lg	LK	.003	15120	---	.09	.01	.10	**.15**
	Demo	BF	Sm	LK	.005	10584	---	.16	.01	.17	**.24**
42" deep, 50' long	Inst	BF	Lg	CY	.004	15120	1.66	.13	.01	1.80	**2.11**
	Inst	BF	Sm	CY	.005	10584	2.52	.16	.01	2.69	**3.14**
42" deep, 50' long	Demo	SF	Lg	LK	.007	6545	---	.22	.02	.24	**.34**
	Demo	SF	Sm	LK	.010	4582	---	.31	.03	.34	**.49**
42" deep, 50' long	Inst	SF	Lg	CY	.009	6545	3.79	.28	.02	4.09	**4.80**
	Inst	SF	Sm	CY	.012	4582	5.76	.38	.03	6.17	**7.22**
43-1/2" deep, 50' long	Demo	LF	Lg	LK	.133	360.0	---	4.13	.42	4.55	**6.57**
	Demo	LF	Sm	LK	.190	252.0	---	5.89	.60	6.49	**9.38**
43-1/2" deep, 50' long	Inst	LF	Lg	CY	.156	360.0	72.00	4.89	.42	77.31	**90.60**
	Inst	LF	Sm	CY	.222	252.0	109.00	6.96	.60	116.56	**137.00**
43-1/2" deep, 50' long	Demo	BF	Lg	LK	.003	15660	---	.09	.01	.10	**.15**
	Demo	BF	Sm	LK	.004	10962	---	.12	.01	.13	**.19**
43-1/2" deep, 50' long	Inst	BF	Lg	CY	.004	15660	1.66	.13	.01	1.80	**2.11**
	Inst	BF	Sm	CY	.005	10962	2.52	.16	.01	2.69	**3.14**
43-1/2" deep, 50' long	Demo	SF	Lg	LK	.007	6545	---	.22	.02	.24	**.34**
	Demo	SF	Sm	LK	.010	4582	---	.31	.03	.34	**.49**
43-1/2" deep, 50' long	Inst	SF	Lg	CY	.009	6545	3.93	.28	.02	4.23	**4.96**
	Inst	SF	Sm	CY	.012	4582	5.97	.38	.03	6.38	**7.46**
45" deep, 50' long	Demo	LF	Lg	LK	.133	360.0	---	4.13	.42	4.55	**6.57**
	Demo	LF	Sm	LK	.190	252.0	---	5.89	.60	6.49	**9.38**
45" deep, 50' long	Inst	LF	Lg	CY	.156	360.0	74.50	4.89	.42	79.81	**93.40**
	Inst	LF	Sm	CY	.222	252.0	113.00	6.96	.60	120.56	**141.00**
45" deep, 50' long	Demo	BF	Lg	LK	.003	16200	---	.09	.01	.10	**.15**
	Demo	BF	Sm	LK	.004	11340	---	.12	.01	.13	**.19**
45" deep, 50' long	Inst	BF	Lg	CY	.003	16200	1.66	.09	.01	1.76	**2.06**
	Inst	BF	Sm	CY	.005	11340	2.52	.16	.01	2.69	**3.14**
45" deep, 50' long	Demo	SF	Lg	LK	.007	6545	---	.22	.02	.24	**.34**
	Demo	SF	Sm	LK	.010	4582	---	.31	.03	.34	**.49**
45" deep, 50' long	Inst	SF	Lg	CY	.009	6545	4.07	.28	.02	4.37	**5.12**
	Inst	SF	Sm	CY	.012	4582	6.19	.38	.03	6.60	**7.71**
46-1/2" deep, 50' long	Demo	LF	Lg	LK	.133	360.0	---	4.13	.42	4.55	**6.57**
	Demo	LF	Sm	LK	.190	252.0	---	5.89	.60	6.49	**9.38**
46-1/2" deep, 50' long	Inst	LF	Lg	CY	.156	360.0	76.90	4.89	.42	82.21	**96.20**
	Inst	LF	Sm	CY	.222	252.0	117.00	6.96	.60	124.56	**145.00**
46-1/2" deep, 50' long	Demo	BF	Lg	LK	.003	16740	---	.09	.01	.10	**.15**
	Demo	BF	Sm	LK	.004	11718	---	.12	.01	.13	**.19**
46-1/2" deep, 50' long	Inst	BF	Lg	CY	.003	16740	1.66	.09	.01	1.76	**2.06**
	Inst	BF	Sm	CY	.005	11718	2.52	.16	.01	2.69	**3.14**
46-1/2" deep, 50' long	Demo	SF	Lg	LK	.007	6545	---	.22	.02	.24	**.34**
	Demo	SF	Sm	LK	.010	4582	---	.31	.03	.34	**.49**
46-1/2" deep, 50' long	Inst	SF	Lg	CY	.009	6545	4.19	.28	.02	4.49	**5.26**
	Inst	SF	Sm	CY	.012	4582	6.37	.38	.03	6.78	**7.92**

Description	Oper	Unit	Vol	Crew Size	Man-hours per Unit	Crew Output per Day	Avg Mat'l Unit Cost	Avg Labor Unit Cost	Avg Equip Unit Cost	Avg Total Unit Cost	Avg Price Incl O&P

10-3/4" thick beams, SF pricing based on 16' oc (continued)

Description	Oper	Unit	Vol	Crew Size	Man-hours	Crew Output	Mat'l	Labor	Equip	Total	Price
48" deep, 50' long	Demo	LF	Lg	LK	.133	360.0	---	4.13	.42	4.55	6.57
	Demo	LF	Sm	LK	.190	252.0	---	5.89	.60	6.49	9.38
48" deep, 50' long	Inst	LF	Lg	CY	.156	360.0	79.30	4.89	.42	84.61	99.00
	Inst	LF	Sm	CY	.222	252.0	121.00	6.96	.60	128.56	150.00
48" deep, 50' long	Demo	BF	Lg	LK	.003	17280	---	.09	.01	.10	.15
	Demo	BF	Sm	LK	.004	12096	---	.12	.01	.13	.19
48" deep, 50' long	Inst	BF	Lg	CY	.003	17280	1.66	.09	.01	1.76	2.06
	Inst	BF	Sm	CY	.005	12096	2.52	.16	.01	2.69	3.14
48" deep, 50' long	Demo	SF	Lg	LK	.007	6545	---	.22	.02	.24	.34
	Demo	SF	Sm	LK	.010	4582	---	.31	.03	.34	.49
48" deep, 50' long	Inst	SF	Lg	CY	.009	6545	4.33	.28	.02	4.63	5.42
	Inst	SF	Sm	CY	.012	4582	6.58	.38	.03	6.99	8.16
49-1/2" deep, 50' long	Demo	LF	Lg	LK	.133	360.0	---	4.13	.42	4.55	6.57
	Demo	LF	Sm	LK	.190	252.0	---	5.89	.60	6.49	9.38
49-1/2" deep, 50' long	Inst	LF	Lg	CY	.156	360.0	81.70	4.89	.42	87.01	102.00
	Inst	LF	Sm	CY	.222	252.0	124.00	6.96	.60	131.56	154.00
49-1/2" deep, 50' long	Demo	BF	Lg	LK	.003	17820	---	.09	.01	.10	.15
	Demo	BF	Sm	LK	.004	12474	---	.12	.01	.13	.19
49-1/2" deep, 50' long	Inst	BF	Lg	CY	.003	17820	1.64	.09	.01	1.74	2.04
	Inst	BF	Sm	CY	.004	12474	2.49	.13	.01	2.63	3.06
49-1/2" deep, 50' long	Demo	SF	Lg	LK	.007	6545	---	.22	.02	.24	.34
	Demo	SF	Sm	LK	.010	4582	---	.31	.03	.34	.49
49-1/2" deep, 50' long	Inst	SF	Lg	CY	.009	6545	4.47	.28	.02	4.77	5.58
	Inst	SF	Sm	CY	.012	4582	6.79	.38	.03	7.20	8.40
51" deep, 50' long	Demo	LF	Lg	LK	.133	360.0	---	4.13	.42	4.55	6.57
	Demo	LF	Sm	LK	.190	252.0	---	5.89	.60	6.49	9.38
51" deep, 50' long	Inst	LF	Lg	CY	.156	360.0	84.10	4.89	.42	89.41	104.00
	Inst	LF	Sm	CY	.222	252.0	128.00	6.96	.60	135.56	158.00
51" deep, 50' long	Demo	BF	Lg	LK	.003	18360	---	.09	.01	.10	.15
	Demo	BF	Sm	LK	.004	12852	---	.12	.01	.13	.19
51" deep, 50' long	Inst	BF	Lg	CY	.003	18360	1.64	.09	.01	1.74	2.04
	Inst	BF	Sm	CY	.004	12852	2.49	.13	.01	2.63	3.06
51" deep, 50' longs	Demo	SF	Lg	LK	.007	6545	---	.22	.02	.24	.34
	Demo	SF	Sm	LK	.010	4582	---	.31	.03	.34	.49
51" deep, 50' long	Inst	SF	Lg	CY	.009	6545	4.59	.28	.02	4.89	5.72
	Inst	SF	Sm	CY	.012	4582	6.98	.38	.03	7.39	8.62
52-1/2" deep, 60' long	Demo	LF	Lg	LK	.133	360.0	---	4.13	.42	4.55	6.57
	Demo	LF	Sm	LK	.190	252.0	---	5.89	.60	6.49	9.38
52-1/2" deep, 60' long	Inst	LF	Lg	CY	.156	360.0	86.40	4.89	.42	91.71	107.00
	Inst	LF	Sm	CY	.222	252.0	131.00	6.96	.60	138.56	162.00
52-1/2" deep, 60' long	Demo	BF	Lg	LK	.003	18900	---	.09	.01	.10	.15
	Demo	BF	Sm	LK	.004	13230	---	.12	.01	.13	.19
52-1/2" deep, 60' long	Inst	BF	Lg	CY	.003	18900	1.64	.09	.01	1.74	2.04
	Inst	BF	Sm	CY	.004	13230	2.49	.13	.01	2.63	3.06
52-1/2" deep, 60' long	Demo	SF	Lg	LK	.007	6545	---	.22	.02	.24	.34
	Demo	SF	Sm	LK	.010	4582	---	.31	.03	.34	.49
52-1/2" deep, 60' long	Inst	SF	Lg	CY	.009	6545	4.71	.28	.02	5.01	5.86
	Inst	SF	Sm	CY	.012	4582	7.16	.38	.03	7.57	8.83

Description	Oper	Unit	Vol	Crew Size	Man-hours per Unit	Crew Output per Day	Avg Mat'l Unit Cost	Avg Labor Unit Cost	Avg Equip Unit Cost	Avg Total Unit Cost	Avg Price Incl O&P
54" deep, 60' long	Demo	LF	Lg	LK	.133	360.0	---	4.13	.42	4.55	**6.57**
	Demo	LF	Sm	LK	.190	252.0	---	5.89	.60	6.49	**9.38**
54" deep, 60' long	Inst	LF	Lg	CY	.156	360.0	88.80	4.89	.42	94.11	**110.00**
	Inst	LF	Sm	CY	.222	252.0	135.00	6.96	.60	142.56	**166.00**
54" deep, 60' long	Demo	BF	Lg	LK	.002	19440	---	.06	.01	.07	**.10**
	Demo	BF	Sm	LK	.004	13608	---	.12	.01	.13	**.19**
54" deep, 60' long	Inst	BF	Lg	CY	.003	19440	1.64	.09	.01	1.74	**2.04**
	Inst	BF	Sm	CY	.004	13608	2.49	.13	.01	2.63	**3.06**
54" deep, 60' long	Demo	SF	Lg	LK	.007	6545	---	.22	.02	.24	**.34**
	Demo	SF	Sm	LK	.010	4582	---	.31	.03	.34	**.49**
54" deep, 60' long	Inst	SF	Lg	CY	.009	6545	4.85	.28	.02	5.15	**6.02**
	Inst	SF	Sm	CY	.012	4582	7.37	.38	.03	7.78	**9.07**
55-1/2" deep, 60' long	Demo	LF	Lg	LK	.133	360.0	---	4.13	.42	4.55	**6.57**
	Demo	LF	Sm	LK	.190	252.0	---	5.89	.60	6.49	**9.38**
55-1/2" deep, 60' long	Inst	LF	Lg	CY	.156	360.0	91.20	4.89	.42	96.51	**113.00**
	Inst	LF	Sm	CY	.222	252.0	139.00	6.96	.60	146.56	**170.00**
55-1/2" deep, 60' long	Demo	BF	Lg	LK	.002	19980	---	.06	.01	.07	**.10**
	Demo	BF	Sm	LK	.003	13986	---	.09	.01	.10	**.15**
55-1/2" deep, 60' long	Inst	BF	Lg	CY	.003	19980	1.64	.09	.01	1.74	**2.04**
	Inst	BF	Sm	CY	.004	13986	2.49	.13	.01	2.63	**3.06**
55-1/2" deep, 60' long	Demo	SF	Lg	LK	.007	6545	---	.22	.02	.24	**.34**
	Demo	SF	Sm	LK	.010	4582	---	.31	.03	.34	**.49**
55-1/2" deep, 60' long	Inst	SF	Lg	CY	.009	6545	4.97	.28	.02	5.27	**6.16**
	Inst	SF	Sm	CY	.012	4582	7.55	.38	.03	7.96	**9.28**
57" deep, 60' long	Demo	LF	Lg	LK	.133	360.0	---	4.13	.42	4.55	**6.57**
	Demo	LF	Sm	LK	.190	252.0	---	5.89	.60	6.49	**9.38**
57" deep, 60' long	Inst	LF	Lg	CY	.156	360.0	93.60	4.89	.42	98.91	**115.00**
	Inst	LF	Sm	CY	.222	252.0	142.00	6.96	.60	149.56	**175.00**
57" deep, 60' long	Demo	BF	Lg	LK	.002	20520	---	.06	.01	.07	**.10**
	Demo	BF	Sm	LK	.003	14364	---	.09	.01	.10	**.15**
57" deep, 60' long	Inst	BF	Lg	CY	.003	20520	1.64	.09	.01	1.74	**2.04**
	Inst	BF	Sm	CY	.004	14364	2.49	.13	.01	2.63	**3.06**
57" deep, 60' long	Demo	SF	Lg	LK	.007	6545	---	.22	.02	.24	**.34**
	Demo	SF	Sm	LK	.010	4582	---	.31	.03	.34	**.49**
57" deep, 60' long	Inst	SF	Lg	CY	.009	6545	5.11	.28	.02	5.41	**6.32**
	Inst	SF	Sm	CY	.012	4582	7.77	.38	.03	8.18	**9.53**
58-1/2" deep, 60' long	Demo	LF	Lg	LK	.133	360.0	---	4.13	.42	4.55	**6.57**
	Demo	LF	Sm	LK	.190	252.0	---	5.89	.60	6.49	**9.38**
58-1/2" deep, 60' long	Inst	LF	Lg	CY	.156	360.0	96.10	4.89	.42	101.41	**118.00**
	Inst	LF	Sm	CY	.222	252.0	146.00	6.96	.60	153.56	**179.00**
58-1/2" deep, 60' long	Demo	BF	Lg	LK	.002	21060	---	.06	.01	.07	**.10**
	Demo	BF	Sm	LK	.003	14742	---	.09	.01	.10	**.15**
58-1/2" deep, 60' long	Inst	BF	Lg	CY	.003	21060	1.64	.09	.01	1.74	**2.04**
	Inst	BF	Sm	CY	.004	14742	2.49	.13	.01	2.63	**3.06**
58-1/2" deep, 60' long	Demo	SF	Lg	LK	.007	6545	---	.22	.02	.24	**.34**
	Demo	SF	Sm	LK	.010	4582	---	.31	.03	.34	**.49**
58-1/2" deep, 60' long	Inst	SF	Lg	CY	.009	6545	5.25	.28	.02	5.55	**6.48**
	Inst	SF	Sm	CY	.012	4582	7.98	.38	.03	8.39	**9.77**

Description	Oper	Unit	Vol	Crew Size	Man-hours per Unit	Crew Output per Day	Avg Mat'l Unit Cost	Avg Labor Unit Cost	Avg Equip Unit Cost	Avg Total Unit Cost	Avg Price Incl O&P
60" deep, 60' long	Demo	LF	Lg	LK	.133	360.0	---	4.13	.42	4.55	6.57
	Demo	LF	Sm	LK	.190	252.0	---	5.89	.60	6.49	9.38
60" deep, 60' long	Inst	LF	Lg	CY	.156	360.0	98.50	4.89	.42	103.81	121.00
	Inst	LF	Sm	CY	.222	252.0	150.00	6.96	.60	157.56	183.00
60" deep, 60' long	Demo	BF	Lg	LK	.002	21600	---	.06	.01	.07	.10
	Demo	BF	Sm	LK	.003	15120	---	.09	.01	.10	.15
60" deep, 60' long	Inst	BF	Lg	CY	.003	21600	1.64	.09	.01	1.74	2.04
	Inst	BF	Sm	CY	.004	15120	2.49	.13	.01	2.63	3.06
60" deep, 60' long	Demo	SF	Lg	LK	.007	6545	---	.22	.02	.24	.34
	Demo	SF	Sm	LK	.010	4582	---	.31	.03	.34	.49
60" deep, 60' long	Inst	SF	Lg	CY	.009	6545	5.37	.28	.02	5.67	6.62
	Inst	SF	Sm	CY	.012	4582	8.16	.38	.03	8.57	9.98

Purlins
16' long, STR #1, SF pricing based on 8' oc

Description	Oper	Unit	Vol	Crew Size	Man-hours per Unit	Crew Output per Day	Avg Mat'l Unit Cost	Avg Labor Unit Cost	Avg Equip Unit Cost	Avg Total Unit Cost	Avg Price Incl O&P
2" x 8"	Demo	LF	Lg	LL	.030	1620	---	.90	.09	.99	1.43
	Demo	LF	Sm	LL	.042	1134	---	1.26	.13	1.39	2.01
2" x 8"	Inst	LF	Lg	CZ	.067	1080	2.62	2.09	.14	4.85	6.28
	Inst	LF	Sm	CZ	.095	756.0	3.98	2.96	.20	7.14	9.21
2" x 8"	Demo	BF	Lg	LL	.022	2155	---	.66	.07	.73	1.05
	Demo	BF	Sm	LL	.032	1509	---	.96	.10	1.06	1.53
2" x 8"	Inst	BF	Lg	CZ	.050	1436	1.97	1.56	.10	3.63	4.70
	Inst	BF	Sm	CZ	.072	1005	2.99	2.24	.15	5.38	6.95
2" x 8"	Demo	SF	Lg	LL	.003	14727	---	.09	.01	.10	.14
	Demo	SF	Sm	LL	.005	10309	---	.15	.01	.16	.23
2" x 8"	Inst	SF	Lg	CZ	.007	9818	.29	.22	.02	.53	.68
	Inst	SF	Sm	CZ	.010	6873	.44	.31	.02	.77	.99
2" x 10"	Demo	LF	Lg	LL	.030	1620	---	.90	.09	.99	1.43
	Demo	LF	Sm	LL	.042	1134	---	1.26	.13	1.39	2.01
2" x 10"	Inst	LF	Lg	CZ	.067	1080	3.10	2.09	.14	5.33	6.83
	Inst	LF	Sm	CZ	.095	756.0	4.71	2.96	.20	7.87	10.10
2" x 10"	Demo	BF	Lg	LL	.018	2705	---	.54	.06	.60	.87
	Demo	BF	Sm	LL	.025	1894	---	.75	.08	.83	1.20
2" x 10"	Inst	BF	Lg	CZ	.040	1804	1.30	1.24	.08	2.62	3.44
	Inst	BF	Sm	CZ	.057	1263	1.98	1.77	.12	3.87	5.06
2" x 10"	Demo	SF	Lg	LL	.003	14727	---	.09	.01	.10	.14
	Demo	SF	Sm	LL	.005	10309	---	.15	.01	.16	.23
2" x 10"	Inst	SF	Lg	CZ	.007	9818	.34	.22	.02	.58	.74
	Inst	SF	Sm	CZ	.010	6873	.52	.31	.02	.85	1.08

Description	Oper	Unit	Vol	Crew Size	Man-hours per Unit	Crew Output per Day	Avg Mat'l Unit Cost	Avg Labor Unit Cost	Avg Equip Unit Cost	Avg Total Unit Cost	Avg Price Incl O&P
2" x 12"	Demo	LF	Lg	LL	.030	1620	---	.90	.09	.99	1.43
	Demo	LF	Sm	LL	.042	1134	---	1.26	.13	1.39	2.01
2" x 12"	Inst	LF	Lg	CZ	.067	1080	3.47	2.09	.14	5.70	7.26
	Inst	LF	Sm	CZ	.095	756.0	5.27	2.96	.20	8.43	10.70
2" x 12"	Demo	BF	Lg	LL	.015	3240	---	.45	.05	.50	.72
	Demo	BF	Sm	LL	.021	2268	---	.63	.07	.70	1.01
2" x 12"	Inst	BF	Lg	CZ	.033	2160	1.12	1.03	.07	2.22	2.90
	Inst	BF	Sm	CZ	.048	1512	1.70	1.49	.10	3.29	4.30
2" x 12"	Demo	SF	Lg	LL	.003	14727	---	.09	.01	.10	.14
	Demo	SF	Sm	LL	.005	10309	---	.15	.01	.16	.23
2" x 12"	Inst	SF	Lg	CZ	.007	9818	.38	.22	.02	.62	.78
	Inst	SF	Sm	CZ	.010	6873	.58	.31	.02	.91	1.15
3" x 8"	Demo	LF	Lg	LL	.030	1620	---	.90	.09	.99	1.43
	Demo	LF	Sm	LL	.042	1134	---	1.26	.13	1.39	2.01
3" x 8"	Inst	LF	Lg	CZ	.067	1080	3.26	2.09	.14	5.49	7.02
	Inst	LF	Sm	CZ	.095	756.0	4.96	2.96	.20	8.12	10.30
3" x 8"	Demo	BF	Lg	LL	.015	3240	---	.45	.05	.50	.72
	Demo	BF	Sm	LL	.021	2268	---	.63	.07	.70	1.01
3" x 8"	Inst	BF	Lg	CZ	.033	2160	1.01	1.03	.07	2.11	2.77
	Inst	BF	Sm	CZ	.048	1512	1.54	1.49	.10	3.13	4.11
3" x 8"	Demo	SF	Lg	LL	.003	14727	---	.09	.01	.10	.14
	Demo	SF	Sm	LL	.005	10309	---	.15	.01	.16	.23
3" x 8"	Inst	SF	Lg	CZ	.007	9818	.36	.22	.02	.60	.76
	Inst	SF	Sm	CZ	.010	6873	.55	.31	.02	.88	1.12
3" x 10"	Demo	LF	Lg	LL	.030	1620	---	.90	.09	.99	1.43
	Demo	LF	Sm	LL	.042	1134	---	1.26	.13	1.39	2.01
3" x 10"	Inst	LF	Lg	CZ	.067	1080	3.75	2.09	.14	5.98	7.58
	Inst	LF	Sm	CZ	.095	756.0	5.70	2.96	.20	8.86	11.20
3" x 10"	Demo	BF	Lg	LL	.012	4050	---	.36	.04	.40	.58
	Demo	BF	Sm	LL	.017	2835	---	.51	.05	.56	.81
3" x 10"	Inst	BF	Lg	CZ	.027	2700	.89	.84	.06	1.79	2.34
	Inst	BF	Sm	CZ	.038	1890	1.35	1.18	.08	2.61	3.41
3" x 10"	Demo	SF	Lg	LL	.003	14727	---	.09	.01	.10	.14
	Demo	SF	Sm	LL	.005	10309	---	.15	.01	.16	.23
3" x 10"	Inst	SF	Lg	CZ	.007	9818	.41	.22	.02	.65	.82
	Inst	SF	Sm	CZ	.010	6873	.62	.31	.02	.95	1.20
3" x 12"	Demo	LF	Lg	LL	.030	1620	---	.90	.09	.99	1.43
	Demo	LF	Sm	LL	.042	1134	---	1.26	.13	1.39	2.01
3" x 12"	Inst	LF	Lg	CZ	.067	1080	4.30	2.09	.14	6.53	8.21
	Inst	LF	Sm	CZ	.095	756.0	6.54	2.96	.20	9.70	12.20
3" x 12"	Demo	BF	Lg	LL	.010	4860	---	.30	.03	.33	.48
	Demo	BF	Sm	LL	.014	3402	---	.42	.04	.46	.67
3" x 12"	Inst	BF	Lg	CZ	.022	3240	.82	.68	.05	1.55	2.02
	Inst	BF	Sm	CZ	.032	2268	1.25	1.00	.07	2.32	3.00
3" x 12"	Demo	SF	Lg	LL	.003	14727	---	.09	.01	.10	.14
	Demo	SF	Sm	LL	.005	10309	---	.15	.01	.16	.23
3" x 12"	Inst	SF	Lg	CZ	.007	9818	.47	.22	.02	.71	.89
	Inst	SF	Sm	CZ	.010	6873	.71	.31	.02	1.04	1.30

Description	Oper	Unit	Vol	Crew Size	Man-hours per Unit	Crew Output per Day	Avg Mat'l Unit Cost	Avg Labor Unit Cost	Avg Equip Unit Cost	Avg Total Unit Cost	Avg Price Incl O&P
3" x 14"	Demo	LF	Lg	LL	.030	1620	---	.90	.09	.99	1.43
	Demo	LF	Sm	LL	.042	1134	---	1.26	.13	1.39	2.01
3" x 14"	Inst	LF	Lg	CZ	.067	1080	5.12	2.09	.14	7.35	9.16
	Inst	LF	Sm	CZ	.095	756.0	7.78	2.96	.20	10.94	13.60
3" x 14"	Demo	BF	Lg	LL	.008	5670	---	.24	.03	.27	.39
	Demo	BF	Sm	LL	.012	3969	---	.36	.04	.40	.58
3" x 14"	Inst	BF	Lg	CZ	.019	3780	1.48	.59	.04	2.11	2.63
	Inst	BF	Sm	CZ	.027	2646	2.25	.84	.06	3.15	3.91
3" x 14"	Demo	SF	Lg	LL	.003	14727	---	.09	.01	.10	.14
	Demo	SF	Sm	LL	.005	10309	---	.15	.01	.16	.23
3" x 14"	Inst	SF	Lg	CZ	.007	9818	.56	.22	.02	.80	.99
	Inst	SF	Sm	CZ	.010	6873	.85	.31	.02	1.18	1.46

Sub-purlins
8' long, STR #1, SF pricing based on 2' oc

Description	Oper	Unit	Vol	Crew Size	Man-hours per Unit	Crew Output per Day	Avg Mat'l Unit Cost	Avg Labor Unit Cost	Avg Equip Unit Cost	Avg Total Unit Cost	Avg Price Incl O&P
2" x 4"	Demo	LF	Lg	LL	.013	3630	---	.39	.04	.43	.62
	Demo	LF	Sm	LL	.019	2541	---	.57	.06	.63	.91
2" x 4"	Inst	LF	Lg	CZ	.030	2420	.37	.93	.06	1.36	1.89
	Inst	LF	Sm	CZ	.043	1694	.56	1.34	.09	1.99	2.74
2" x 4"	Demo	BF	Lg	LL	.020	2432	---	.60	.06	.66	.96
	Demo	BF	Sm	LL	.028	1702	---	.84	.09	.93	1.34
2" x 4"	Inst	BF	Lg	CZ	.044	1621	.55	1.37	.09	2.01	2.78
	Inst	BF	Sm	CZ	.063	1135	.84	1.96	.13	2.93	4.04
2" x 4"	Demo	SF	Lg	LL	.006	8067	---	.18	.02	.20	.29
	Demo	SF	Sm	LL	.009	5647	---	.27	.03	.30	.43
2" x 4"	Inst	SF	Lg	CZ	.013	5378	.16	.40	.03	.59	.82
	Inst	SF	Sm	CZ	.019	3765	.24	.59	.04	.87	1.20
2" x 6"	Demo	LF	Lg	LL	.013	3630	---	.39	.04	.43	.62
	Demo	LF	Sm	LL	.019	2541	---	.57	.06	.63	.91
2" x 6"	Inst	LF	Lg	CZ	.030	2420	.55	.93	.06	1.54	2.09
	Inst	LF	Sm	CZ	.043	1694	.84	1.34	.09	2.27	3.06
2" x 6"	Demo	BF	Lg	LL	.013	3630	---	.39	.04	.43	.62
	Demo	BF	Sm	LL	.019	2541	---	.57	.06	.63	.91
2" x 6"	Inst	BF	Lg	CZ	.030	2420	.55	.93	.06	1.54	2.09
	Inst	BF	Sm	CZ	.043	1694	.84	1.34	.09	2.27	3.06
2" x 6"	Demo	SF	Lg	LL	.006	8067	---	.18	.02	.20	.29
	Demo	SF	Sm	LL	.009	5647	---	.27	.03	.30	.43
2" x 6"	Inst	SF	Lg	CZ	.013	5378	.25	.40	.03	.68	.92
	Inst	SF	Sm	CZ	.019	3765	.38	.59	.04	1.01	1.36

Description	Oper	Unit	Vol	Crew Size	Man-hours per Unit	Crew Output per Day	Avg Mat'l Unit Cost	Avg Labor Unit Cost	Avg Equip Unit Cost	Avg Total Unit Cost	Avg Price Incl O&P
2" x 8"	Demo	LF	Lg	LL	.013	3630	---	.39	.04	.43	.62
	Demo	LF	Sm	LL	.019	2541	---	.57	.06	.63	.91
2" x 8"	Inst	LF	Lg	CZ	.030	2420	.82	.93	.06	1.81	2.40
	Inst	LF	Sm	CZ	.043	1694	1.25	1.34	.09	2.68	3.53
2" x 8"	Demo	BF	Lg	LL	.010	5445	---	.30	.03	.33	.48
	Demo	BF	Sm	LL	.014	3812	---	.42	.04	.46	.67
2" x 8"	Inst	BF	Lg	CZ	.022	3630	.62	.68	.05	1.35	1.79
	Inst	BF	Sm	CZ	.032	2541	.94	1.00	.07	2.01	2.64
2" x 8"	Demo	SF	Lg	LL	.006	8067	---	.18	.02	.20	.29
	Demo	SF	Sm	LL	.009	5647	---	.27	.03	.30	.43
2" x 8"	Inst	SF	Lg	CZ	.013	5378	.36	.40	.03	.79	1.05
	Inst	SF	Sm	CZ	.019	3765	.55	.59	.04	1.18	1.56
3" x 4"	Demo	LF	Lg	LL	.013	3630	---	.39	.04	.43	.62
	Demo	LF	Sm	LL	.019	2541	---	.57	.06	.63	.91
3" x 4"	Inst	LF	Lg	CZ	.030	2420	.62	.93	.06	1.61	2.17
	Inst	LF	Sm	CZ	.043	1694	.94	1.34	.09	2.37	3.18
3" x 4"	Demo	BF	Lg	LL	.013	7260	---	.39	.04	.43	.62
	Demo	BF	Sm	LL	.019	5082	---	.57	.06	.63	.91
3" x 4"	Inst	BF	Lg	CZ	.030	4840	.62	.93	.06	1.61	2.17
	Inst	BF	Sm	CZ	.043	3388	.94	1.34	.09	2.37	3.18
3" x 4"	Demo	SF	Lg	LL	.006	8067	---	.18	.02	.20	.29
	Demo	SF	Sm	LL	.009	5647	---	.27	.03	.30	.43
3" x 4"	Inst	SF	Lg	CZ	.013	5378	.27	.40	.03	.70	.95
	Inst	SF	Sm	CZ	.019	3765	.41	.59	.04	1.04	1.40
3" x 6"	Demo	LF	Lg	LL	.013	3630	---	.39	.04	.43	.62
	Demo	LF	Sm	LL	.019	2541	---	.57	.06	.63	.91
3" x 6"	Inst	LF	Lg	CZ	.030	2420	.93	.93	.06	1.92	2.53
	Inst	LF	Sm	CZ	.043	1694	1.41	1.34	.09	2.84	3.72
3" x 6"	Demo	BF	Lg	LL	.009	4828	---	.27	.03	.30	.43
	Demo	BF	Sm	LL	.013	3380	---	.39	.04	.43	.62
3" x 6"	Inst	BF	Lg	CZ	.020	3219	.62	.62	.04	1.28	1.69
	Inst	BF	Sm	CZ	.028	2253	.94	.87	.06	1.87	2.45
3" x 6"	Demo	SF	Lg	LL	.006	8067	---	.18	.02	.20	.29
	Demo	SF	Sm	LL	.009	5647	---	.27	.03	.30	.43
3" x 6"	Inst	SF	Lg	CZ	.013	5378	.41	.40	.03	.84	1.11
	Inst	SF	Sm	CZ	.019	3765	.62	.59	.04	1.25	1.64
3" x 8"	Demo	LF	Lg	LL	.013	3630	---	.39	.04	.43	.62
	Demo	LF	Sm	LL	.019	2541	---	.57	.06	.63	.91
3" x 8"	Inst	LF	Lg	CZ	.030	2420	1.23	.93	.06	2.22	2.87
	Inst	LF	Sm	CZ	.043	1694	1.87	1.34	.09	3.30	4.25
3" x 8"	Demo	BF	Lg	LL	.007	3630	---	.21	.02	.23	.33
	Demo	BF	Sm	LL	.009	2541	---	.27	.03	.30	.43
3" x 8"	Inst	BF	Lg	CZ	.015	2420	.62	.47	.03	1.12	1.44
	Inst	BF	Sm	CZ	.021	1694	.94	.65	.04	1.63	2.10
3" x 8"	Demo	SF	Lg	LL	.006	8067	---	.18	.02	.20	.29
	Demo	SF	Sm	LL	.009	5647	---	.27	.03	.30	.43
3" x 8"	Inst	SF	Lg	CZ	.013	5378	.53	.40	.03	.96	1.25
	Inst	SF	Sm	CZ	.019	3765	.81	.59	.04	1.44	1.86

Description	Oper	Unit	Vol	Crew Size	Man-hours per Unit	Crew Output per Day	Avg Mat'l Unit Cost	Avg Labor Unit Cost	Avg Equip Unit Cost	Avg Total Unit Cost	Avg Price Incl O&P

Ledgers
Bolts 2' oc

Description	Oper	Unit	Vol	Crew Size	Man-hours per Unit	Crew Output per Day	Avg Mat'l Unit Cost	Avg Labor Unit Cost	Avg Equip Unit Cost	Avg Total Unit Cost	Avg Price Incl O&P
3" x 8"	Demo	LF	Lg	LC	.037	648.0	---	1.09	.23	1.32	**1.85**
	Demo	LF	Sm	LC	.053	454.0	---	1.56	.33	1.89	**2.65**
3" x 8"	Inst	LF	Lg	CS	.056	432.0	2.40	1.75	.35	4.50	**5.74**
	Inst	LF	Sm	CS	.079	302.0	3.65	2.47	.50	6.62	**8.41**
3" x 8"	Demo	BF	Lg	LC	.019	1296	---	.56	.12	.68	**.95**
	Demo	BF	Sm	LC	.026	907.0	---	.76	.17	.93	**1.31**
3" x 8"	Inst	BF	Lg	CS	.028	864.0	1.21	.88	.17	2.26	**2.88**
	Inst	BF	Sm	CS	.040	605.0	1.84	1.25	.25	3.34	**4.25**
3" x 8"	Demo	SF	Lg	LC	.002	12597	---	.06	.01	.07	**.10**
	Demo	SF	Sm	LC	.003	8818	---	.09	.02	.11	**.15**
3" x 8"	Inst	SF	Lg	CS	.003	8398	.12	.09	.02	.23	**.30**
	Inst	SF	Sm	CS	.004	5879	.18	.13	.03	.34	**.42**
3" x 10"	Demo	LF	Lg	LC	.037	648.0	---	1.09	.23	1.32	**1.85**
	Demo	LF	Sm	LC	.053	454.0	---	1.56	.33	1.89	**2.65**
3" x 10"	Inst	LF	Lg	CS	.056	432.0	2.74	1.75	.35	4.84	**6.13**
	Inst	LF	Sm	CS	.079	302.0	4.16	2.47	.50	7.13	**9.00**
3" x 10"	Demo	BF	Lg	LC	.015	1620	---	.44	.09	.53	**.75**
	Demo	BF	Sm	LC	.021	1134	---	.62	.13	.75	**1.05**
3" x 10"	Inst	BF	Lg	CS	.022	1080	1.10	.69	.14	1.93	**2.44**
	Inst	BF	Sm	CS	.032	756.0	1.67	1.00	.20	2.87	**3.62**
3" x 10"	Demo	SF	Lg	LC	.002	12597	---	.06	.01	.07	**.10**
	Demo	SF	Sm	LC	.003	8818	---	.09	.02	.11	**.15**
3" x 10"	Inst	SF	Lg	CS	.003	8398	.14	.09	.02	.25	**.32**
	Inst	SF	Sm	CS	.004	5879	.21	.13	.03	.37	**.46**
3" x 12"	Demo	LF	Lg	LC	.037	648.0	---	1.09	.23	1.32	**1.85**
	Demo	LF	Sm	LC	.053	454.0	---	1.56	.33	1.89	**2.65**
3" x 12"	Inst	LF	Lg	CS	.056	432.0	3.08	1.75	.35	5.18	**6.52**
	Inst	LF	Sm	CS	.079	302.0	4.68	2.47	.50	7.65	**9.59**
3" x 12"	Demo	BF	Lg	LC	.012	1944	---	.35	.08	.43	**.61**
	Demo	BF	Sm	LC	.018	1361	---	.53	.11	.64	**.90**
3" x 12"	Inst	BF	Lg	CS	.019	1296	1.03	.60	.12	1.75	**2.20**
	Inst	BF	Sm	CS	.026	907.0	1.57	.81	.17	2.55	**3.20**
3" x 12"	Demo	SF	Lg	LC	.002	12597	---	.06	.01	.07	**.10**
	Demo	SF	Sm	LC	.003	8818	---	.09	.02	.11	**.15**
3" x 12"	Inst	SF	Lg	CS	.003	8398	.15	.09	.02	.26	**.33**
	Inst	SF	Sm	CS	.004	5879	.23	.13	.03	.39	**.48**

Description	Oper	Unit	Vol	Crew Size	Man-hours per Unit	Crew Output per Day	Avg Mat'l Unit Cost	Avg Labor Unit Cost	Avg Equip Unit Cost	Avg Total Unit Cost	Avg Price Incl O&P
4" x 8"	Demo	LF	Lg	LC	.037	648.0	---	1.09	.23	1.32	**1.85**
	Demo	LF	Sm	LC	.053	454.0	---	1.56	.33	1.89	**2.65**
4" x 8"	Inst	LF	Lg	CS	.056	432.0	3.04	1.75	.35	5.14	**6.48**
	Inst	LF	Sm	CS	.079	302.0	4.62	2.47	.50	7.59	**9.53**
4" x 8"	Demo	BF	Lg	LC	.014	1730	---	.41	.09	.50	**.70**
	Demo	BF	Sm	LC	.020	1211	---	.59	.12	.71	**1.00**
4" x 8"	Inst	BF	Lg	CS	.021	1153	1.14	.66	.13	1.93	**2.43**
	Inst	BF	Sm	CS	.030	807.0	1.73	.94	.19	2.86	**3.59**
4" x 8"	Demo	SF	Lg	LC	.002	12597	---	.06	.01	.07	**.10**
	Demo	SF	Sm	LC	.003	8818	---	.09	.02	.11	**.15**
4" x 8"	Inst	SF	Lg	CS	.003	8398	.15	.09	.02	.26	**.33**
	Inst	SF	Sm	CS	.004	5879	.23	.13	.03	.39	**.48**
4" x 10"	Demo	LF	Lg	LC	.037	648.0	---	1.09	.23	1.32	**1.85**
	Demo	LF	Sm	LC	.053	454.0	---	1.56	.33	1.89	**2.65**
4" x 10"	Inst	LF	Lg	CS	.056	432.0	3.53	1.75	.35	5.63	**7.04**
	Inst	LF	Sm	CS	.079	302.0	5.37	2.47	.50	8.34	**10.40**
4" x 10"	Demo	BF	Lg	LC	.011	2158	---	.32	.07	.39	**.55**
	Demo	BF	Sm	LC	.016	1511	---	.47	.10	.57	**.80**
4" x 10"	Inst	BF	Lg	CS	.017	1439	1.07	.53	.10	1.70	**2.13**
	Inst	BF	Sm	CS	.024	1007	1.63	.75	.15	2.53	**3.15**
4" x 10"	Demo	SF	Lg	LC	.002	12597	---	.06	.01	.07	**.10**
	Demo	SF	Sm	LC	.003	8818	---	.09	.02	.11	**.15**
4" x 10"	Inst	SF	Lg	CS	.003	8398	.18	.09	.02	.29	**.37**
	Inst	SF	Sm	CS	.004	5879	.27	.13	.03	.43	**.53**
4" x 12"	Demo	LF	Lg	LC	.037	648.0	---	1.09	.23	1.32	**1.85**
	Demo	LF	Sm	LC	.053	454.0	---	1.56	.33	1.89	**2.65**
4" x 12"	Inst	LF	Lg	CS	.056	432.0	4.04	1.75	.35	6.14	**7.63**
	Inst	LF	Sm	CS	.079	302.0	6.14	2.47	.50	9.11	**11.30**
4" x 12"	Demo	BF	Lg	LC	.009	2592	---	.26	.06	.32	**.45**
	Demo	BF	Sm	LC	.013	1814	---	.38	.08	.46	**.65**
4" x 12"	Inst	BF	Lg	CS	.014	1728	1.01	.44	.09	1.54	**1.91**
	Inst	BF	Sm	CS	.020	1210	1.54	.63	.12	2.29	**2.83**
4" x 12"	Demo	SF	Lg	LC	.002	12597	---	.06	.01	.07	**.10**
	Demo	SF	Sm	LC	.003	8818	---	.09	.02	.11	**.15**
4" x 12"	Inst	SF	Lg	CS	.003	8398	.19	.09	.02	.30	**.38**
	Inst	SF	Sm	CS	.004	5879	.29	.13	.03	.45	**.55**

Description	Oper	Unit	Vol	Crew Size	Man-hours per Unit	Crew Output per Day	Avg Mat'l Unit Cost	Avg Labor Unit Cost	Avg Equip Unit Cost	Avg Total Unit Cost	Avg Price Incl O&P

Gutters and downspouts

Aluminum, baked on painted finish (white or brown)
Gutter, 5" box type

Description	Oper	Unit	Vol	Crew Size	Man-hours per Unit	Crew Output per Day	Avg Mat'l Unit Cost	Avg Labor Unit Cost	Avg Equip Unit Cost	Avg Total Unit Cost	Avg Price Incl O&P
Heavyweight gauge	Inst	LF	Lg	UB	.074	215.0	1.12	2.73	---	3.85	**5.34**
	Inst	LF	Sm	UB	.106	150.5	1.24	3.92	---	5.16	**7.22**
Standard weight gauge	Inst	LF	Lg	UB	.074	215.0	.90	2.73	---	3.63	**5.08**
	Inst	LF	Sm	UB	.106	150.5	.99	3.92	---	4.91	**6.94**
End caps	Inst	Ea	Lg	UB	---	---	.65	---	---	.65	**.65**
	Inst	Ea	Sm	UB	---	---	.72	---	---	.72	**.72**
Drop outlet for downspouts	Inst	Ea	Lg	UB	---	---	3.68	---	---	3.68	**3.68**
	Inst	Ea	Sm	UB	---	---	4.07	---	---	4.07	**4.07**
Inside/outside corner	Inst	Ea	Lg	UB	---	---	4.62	---	---	4.62	**4.62**
	Inst	Ea	Sm	UB	---	---	5.11	---	---	5.11	**5.11**
Joint connector (4 each package)											
	Inst	Pkg	Lg	UB	---	---	5.00	---	---	5.00	**5.00**
	Inst	Pkg	Sm	UB	---	---	5.53	---	---	5.53	**5.53**
Strap hanger (10 each package)											
	Inst	Pkg	Lg	UB	---	---	6.60	---	---	6.60	**6.60**
	Inst	Pkg	Sm	UB	---	---	7.30	---	---	7.30	**7.30**
Fascia bracket (4 each package)											
	Inst	Pkg	Lg	UB	---	---	6.60	---	---	6.60	**6.60**
	Inst	Pkg	Sm	UB	---	---	7.30	---	---	7.30	**7.30**
Spike and ferrule (10 each package)											
	Inst	Pkg	Lg	UB	---	---	5.93	---	---	5.93	**5.93**
	Inst	Pkg	Sm	UB	---	---	6.56	---	---	6.56	**6.56**

Downspout, corrugated square

Description	Oper	Unit	Vol	Crew Size	Man-hours per Unit	Crew Output per Day	Avg Mat'l Unit Cost	Avg Labor Unit Cost	Avg Equip Unit Cost	Avg Total Unit Cost	Avg Price Incl O&P
Standard weight gauge	Inst	LF	Lg	UB	.064	250.0	.92	2.36	---	3.28	**4.56**
	Inst	LF	Sm	UB	.091	175.0	1.02	3.36	---	4.38	**6.15**
Regular/side elbow	Inst	Ea	Lg	UB	---	---	1.57	---	---	1.57	**1.57**
	Inst	Ea	Sm	UB	---	---	1.74	---	---	1.74	**1.74**
Downspout holder (4 each package)											
	Inst	Pkg	Lg	UB	---	---	4.61	---	---	4.61	**4.61**
	Inst	Pkg	Sm	UB	---	---	5.10	---	---	5.10	**5.10**

Steel, natural finish
Gutter, 4" box type

Description	Oper	Unit	Vol	Crew Size	Man-hours per Unit	Crew Output per Day	Avg Mat'l Unit Cost	Avg Labor Unit Cost	Avg Equip Unit Cost	Avg Total Unit Cost	Avg Price Incl O&P
Heavyweight gauge	Inst	LF	Lg	UB	.074	215.0	.73	2.73	---	3.46	**4.89**
	Inst	LF	Sm	UB	.106	150.5	.80	3.92	---	4.72	**6.72**
End caps	Inst	Ea	Lg	UB	---	---	.65	---	---	.65	**.65**
	Inst	Ea	Sm	UB	---	---	.72	---	---	.72	**.72**
Drop outlet for downspouts	Inst	Ea	Lg	UB	---	---	2.36	---	---	2.36	**2.36**
	Inst	Ea	Sm	UB	---	---	2.61	---	---	2.61	**2.61**
Inside/outside corner	Inst	Ea	Lg	UB	---	---	3.55	---	---	3.55	**3.55**
	Inst	Ea	Sm	UB	---	---	3.93	---	---	3.93	**3.93**
Joint connector (4 each package)											
	Inst	Pkg	Lg	UB	---	---	3.68	---	---	3.68	**3.68**
	Inst	Pkg	Sm	UB	---	---	4.07	---	---	4.07	**4.07**
Strap hanger (10 each package)											
	Inst	Pkg	Lg	UB	---	---	5.00	---	---	5.00	**5.00**
	Inst	Pkg	Sm	UB	---	---	5.53	---	---	5.53	**5.53**

Description	Oper	Unit	Vol	Crew Size	Man-hours per Unit	Crew Output per Day	Avg Mat'l Unit Cost	Avg Labor Unit Cost	Avg Equip Unit Cost	Avg Total Unit Cost	Avg Price Incl O&P
Fascia bracket (4 each package)											
	Inst	Pkg	Lg	UB	---	---	4.95	---	---	4.95	**4.95**
	Inst	Pkg	Sm	UB	---	---	5.48	---	---	5.48	**5.48**
Spike and ferrule (10 each package)											
	Inst	Pkg	Lg	UB	---	---	4.34	---	---	4.34	**4.34**
	Inst	Pkg	Sm	UB	---	---	4.80	---	---	4.80	**4.80**

Downspout, 30 gauge galvanized steel

Description	Oper	Unit	Vol	Crew Size	Man-hours per Unit	Crew Output per Day	Avg Mat'l Unit Cost	Avg Labor Unit Cost	Avg Equip Unit Cost	Avg Total Unit Cost	Avg Price Incl O&P
Standard weight gauge	Inst	LF	Lg	UB	.074	215.0	.99	2.73	---	3.72	**5.19**
	Inst	LF	Sm	UB	.106	150.5	1.10	3.92	---	5.02	**7.06**
Regular/side elbow	Inst	Ea	Lg	UB	---	---	1.57	---	---	1.57	**1.57**
	Inst	Ea	Sm	UB	---	---	1.74	---	---	1.74	**1.74**
Downspout holder (4 each package)											
	Inst	Pkg	Lg	UB	---	---	4.21	---	---	4.21	**4.21**
	Inst	Pkg	Sm	UB	---	---	4.66	---	---	4.66	**4.66**

Vinyl, extruded 5" PVC
White

Description	Oper	Unit	Vol	Crew Size	Man-hours per Unit	Crew Output per Day	Avg Mat'l Unit Cost	Avg Labor Unit Cost	Avg Equip Unit Cost	Avg Total Unit Cost	Avg Price Incl O&P
Gutter	Inst	LF	Lg	UB	.074	215.0	1.12	2.73	---	3.85	**5.34**
	Inst	LF	Sm	UB	.106	150.5	1.24	3.92	---	5.16	**7.22**
End caps	Inst	Ea	Lg	UB	---	---	1.57	---	---	1.57	**1.57**
	Inst	Ea	Sm	UB	---	---	1.74	---	---	1.74	**1.74**
Drop outlet for downspouts	Inst	Ea	Lg	UB	---	---	5.94	---	---	5.94	**5.94**
	Inst	Ea	Sm	UB	---	---	6.57	---	---	6.57	**6.57**
Inside/outside corner	Inst	Ea	Lg	UB	---	---	5.94	---	---	5.94	**5.94**
	Inst	Ea	Sm	UB	---	---	6.57	---	---	6.57	**6.57**
Joint connector	Inst	Ea	Lg	UB	---	---	1.97	---	---	1.97	**1.97**
	Inst	Ea	Sm	UB	---	---	2.18	---	---	2.18	**2.18**
Fascia bracket (4 each package)											
	Inst	Pkg	Lg	UB	---	---	7.92	---	---	7.92	**7.92**
	Inst	Pkg	Sm	UB	---	---	8.76	---	---	8.76	**8.76**
Downspout	Inst	LF	Lg	UB	.064	250.0	1.25	2.36	---	3.61	**4.94**
	Inst	LF	Sm	UB	.091	175.0	1.39	3.36	---	4.75	**6.58**
Downspout driplet	Inst	Ea	Lg	UB	---	---	7.92	---	---	7.92	**7.92**
	Inst	Ea	Sm	UB	---	---	8.76	---	---	8.76	**8.76**
Downspout joiner	Inst	Ea	Lg	UB	---	---	2.63	---	---	2.63	**2.63**
	Inst	Ea	Sm	UB	---	---	2.91	---	---	2.91	**2.91**
Downspout holder (2 each package)											
	Inst	Pkg	Lg	UB	---	---	7.92	---	---	7.92	**7.92**
	Inst	Pkg	Sm	UB	---	---	8.76	---	---	8.76	**8.76**
Regular elbow	Inst	Ea	Lg	UB	---	---	2.89	---	---	2.89	**2.89**
	Inst	Ea	Sm	UB	---	---	3.20	---	---	3.20	**3.20**
Well cap	Inst	Ea	Lg	UB	---	---	3.68	---	---	3.68	**3.68**
	Inst	Ea	Sm	UB	---	---	4.07	---	---	4.07	**4.07**
Well outlet	Inst	Ea	Lg	UB	---	---	5.00	---	---	5.00	**5.00**
	Inst	Ea	Sm	UB	---	---	5.53	---	---	5.53	**5.53**
Expansion joint connector	Inst	Ea	Lg	UB	---	---	7.26	---	---	7.26	**7.26**
	Inst	Ea	Sm	UB	---	---	8.03	---	---	8.03	**8.03**
Rafter adapter (4 each package)											
	Inst	Pkg	Lg	UB	---	---	5.27	---	---	5.27	**5.27**
	Inst	Pkg	Sm	UB	---	---	5.83	---	---	5.83	**5.83**

Description	Oper	Unit	Vol	Crew Size	Man-hours per Unit	Crew Output per Day	Avg Mat'l Unit Cost	Avg Labor Unit Cost	Avg Equip Unit Cost	Avg Total Unit Cost	Avg Price Incl O&P
Brown											
Gutter	Inst	LF	Lg	UB	.074	215.0	1.06	2.73	---	3.79	**5.27**
	Inst	LF	Sm	UB	.106	150.5	1.17	3.92	---	5.09	**7.14**
End caps	Inst	Ea	Lg	UB	---	---	1.97	---	---	1.97	**1.97**
	Inst	Ea	Sm	UB	---	---	2.18	---	---	2.18	**2.18**
Drop outlet for downspouts	Inst	Ea	Lg	UB	---	---	7.26	---	---	7.26	**7.26**
	Inst	Ea	Sm	UB	---	---	8.03	---	---	8.03	**8.03**
Inside/outside corner	Inst	Ea	Lg	UB	---	---	7.92	---	---	7.92	**7.92**
	Inst	Ea	Sm	UB	---	---	8.76	---	---	8.76	**8.76**
Joint connector	Inst	Ea	Lg	UB	---	---	2.63	---	---	2.63	**2.63**
	Inst	Ea	Sm	UB	---	---	2.91	---	---	2.91	**2.91**
Fascia bracket (4 each package)											
	Inst	Pkg	Lg	UB	---	---	9.89	---	---	9.89	**9.89**
	Inst	Pkg	Sm	UB	---	---	10.90	---	---	10.90	**10.90**
Downspout	Inst	LF	Lg	UB	.064	250.0	1.39	2.36	---	3.75	**5.10**
	Inst	LF	Sm	UB	.091	175.0	1.53	3.36	---	4.89	**6.74**
Downspout driplet	Inst	Ea	Lg	UB	---	---	9.24	---	---	9.24	**9.24**
	Inst	Ea	Sm	UB	---	---	10.20	---	---	10.20	**10.20**
Downspout joiner	Inst	Ea	Lg	UB	---	---	3.15	---	---	3.15	**3.15**
	Inst	Ea	Sm	UB	---	---	3.49	---	---	3.49	**3.49**
Downspout holder (2 each package)											
	Inst	Pkg	Lg	UB	---	---	9.90	---	---	9.90	**9.90**
	Inst	Pkg	Sm	UB	---	---	11.00	---	---	11.00	**11.00**
Regular elbow	Inst	Ea	Lg	UB	---	---	3.68	---	---	3.68	**3.68**
	Inst	Ea	Sm	UB	---	---	4.07	---	---	4.07	**4.07**
Well cap	Inst	Ea	Lg	UB	---	---	4.34	---	---	4.34	**4.34**
	Inst	Ea	Sm	UB	---	---	4.80	---	---	4.80	**4.80**
Well outlet	Inst	Ea	Lg	UB	---	---	6.32	---	---	6.32	**6.32**
	Inst	Ea	Sm	UB	---	---	6.99	---	---	6.99	**6.99**
Expansion joint connector	Inst	Ea	Lg	UB	---	---	9.24	---	---	9.24	**9.24**
	Inst	Ea	Sm	UB	---	---	10.20	---	---	10.20	**10.20**
Rafter adapter (4 each package)											
	Inst	Pkg	Lg	UB	---	---	5.27	---	---	5.27	**5.27**
	Inst	Pkg	Sm	UB	---	---	5.83	---	---	5.83	**5.83**

Hardboard. See Paneling, page 332

Hardwood flooring

1. Strip flooring is nailed into place over wood sub-flooring or over wood sleeper strips. Using 3¼" W strips leaves 25% cutting and fitting waste; 2¼" W strips leave 33% waste. Nails and the respective cutting and fitting waste have been included in the unit costs.

2. Block flooring is laid in mastic applied to felt-covered wood subfloor. Mastic, 5% block waste and felt are included in material unit costs for block or parquet flooring.

Strip flooring:
A Side and end matched
B Side matched
C Square edged

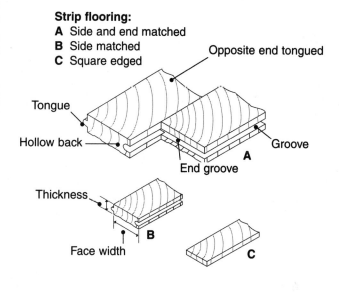

Two types of wood block flooring

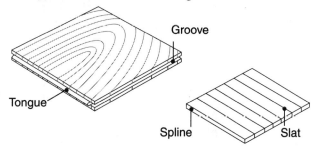

Installation of first strip of flooring

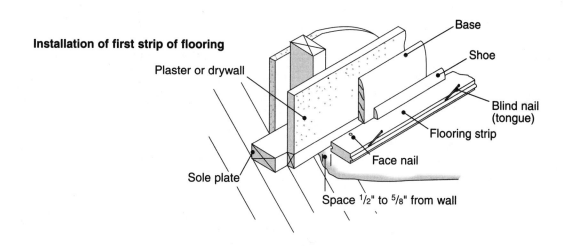

Nailing of flooring:
A Angle of nailing
B Setting the nail without damage to the flooring

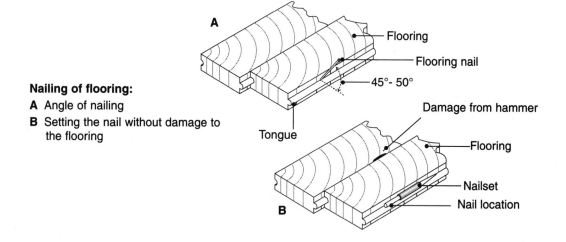

Description	Oper	Unit	Vol	Crew Size	Man-hours per Unit	Crew Output per Day	Avg Mat'l Unit Cost	Avg Labor Unit Cost	Avg Equip Unit Cost	Avg Total Unit Cost	Avg Price Incl O&P

Hardwood flooring
Includes waste and nails

Strip; installed over wood subfloor

Prefinished oak, prime

25/32" x 3-1/4"

Description	Oper	Unit	Vol	Crew Size	Man-hours per Unit	Crew Output per Day	Avg Mat'l Unit Cost	Avg Labor Unit Cost	Avg Equip Unit Cost	Avg Total Unit Cost	Avg Price Incl O&P
Lay floor	Inst	SF	Lg	FD	.045	400.0	9.77	1.42	---	11.19	**13.40**
	Inst	SF	Sm	FD	.069	260.0	10.90	2.18	---	13.08	**15.70**
Wax, polish, machine buff	Inst	SF	Lg	FC	.009	860.0	.04	.29	.03	.36	**.50**
	Inst	SF	Sm	FC	.014	559.0	.04	.45	.04	.53	**.75**

25/32" x 2-1/4"

Description	Oper	Unit	Vol	Crew Size	Man-hours per Unit	Crew Output per Day	Avg Mat'l Unit Cost	Avg Labor Unit Cost	Avg Equip Unit Cost	Avg Total Unit Cost	Avg Price Incl O&P
Lay floor	Inst	SF	Lg	FD	.060	300.0	8.79	1.90	---	10.69	**12.90**
	Inst	SF	Sm	FD	.092	195.0	9.80	2.91	---	12.71	**15.50**
Wax, polish, machine buff	Inst	SF	Lg	FC	.009	860.0	.04	.29	.03	.36	**.50**
	Inst	SF	Sm	FC	.014	559.0	.04	.45	.04	.53	**.75**

Unfinished

25/32" x 3-1/4", lay floor only
Fir

Description	Oper	Unit	Vol	Crew Size	Man-hours per Unit	Crew Output per Day	Avg Mat'l Unit Cost	Avg Labor Unit Cost	Avg Equip Unit Cost	Avg Total Unit Cost	Avg Price Incl O&P
Vertical grain	Inst	SF	Lg	FD	.038	480.0	3.02	1.20	---	4.22	**5.24**
	Inst	SF	Sm	FD	.058	312.0	3.37	1.83	---	5.20	**6.57**
Flat grain	Inst	SF	Lg	FD	.038	480.0	3.43	1.20	---	4.63	**5.71**
	Inst	SF	Sm	FD	.058	312.0	3.82	1.83	---	5.65	**7.09**

Yellow pine

Description	Oper	Unit	Vol	Crew Size	Man-hours per Unit	Crew Output per Day	Avg Mat'l Unit Cost	Avg Labor Unit Cost	Avg Equip Unit Cost	Avg Total Unit Cost	Avg Price Incl O&P
	Inst	SF	Lg	FD	.038	480.0	2.94	1.20	---	4.14	**5.15**
	Inst	SF	Sm	FD	.058	312.0	3.28	1.83	---	5.11	**6.47**

25/32" x 2-1/4", lay floor only
Maple

Description	Oper	Unit	Vol	Crew Size	Man-hours per Unit	Crew Output per Day	Avg Mat'l Unit Cost	Avg Labor Unit Cost	Avg Equip Unit Cost	Avg Total Unit Cost	Avg Price Incl O&P
	Inst	SF	Lg	FD	.056	320.0	4.41	1.77	---	6.18	**7.67**
	Inst	SF	Sm	FD	.087	208.0	4.91	2.75	---	7.66	**9.69**

Oak

Description	Oper	Unit	Vol	Crew Size	Man-hours per Unit	Crew Output per Day	Avg Mat'l Unit Cost	Avg Labor Unit Cost	Avg Equip Unit Cost	Avg Total Unit Cost	Avg Price Incl O&P
	Inst	SF	Lg	FD	.056	320.0	4.93	1.77	---	6.70	**8.27**
	Inst	SF	Sm	FD	.087	208.0	5.49	2.75	---	8.24	**10.40**

Yellow pine

Description	Oper	Unit	Vol	Crew Size	Man-hours per Unit	Crew Output per Day	Avg Mat'l Unit Cost	Avg Labor Unit Cost	Avg Equip Unit Cost	Avg Total Unit Cost	Avg Price Incl O&P
	Inst	SF	Lg	FD	.047	385.0	2.99	1.49	---	4.48	**5.62**
	Inst	SF	Sm	FD	.072	250.0	3.33	2.28	---	5.61	**7.17**

Machine sand, fill and finish

Description	Oper	Unit	Vol	Crew Size	Man-hours per Unit	Crew Output per Day	Avg Mat'l Unit Cost	Avg Labor Unit Cost	Avg Equip Unit Cost	Avg Total Unit Cost	Avg Price Incl O&P
New floors	Inst	SF	Lg	FC	.019	430.0	.11	.61	.05	.77	**1.07**
	Inst	SF	Sm	FC	.029	280.0	.12	.93	.08	1.13	**1.59**
Damaged floors	Inst	SF	Lg	FC	.028	285.0	.15	.90	.08	1.13	**1.57**
	Inst	SF	Sm	FC	.043	185.0	.16	1.38	.12	1.66	**2.33**
Wax, polish and machine buff	Inst	SF	Lg	FC	.009	860.0	.04	.29	.03	.36	**.50**
	Inst	SF	Sm	FC	.014	559.0	.04	.45	.04	.53	**.75**

Description	Oper	Unit	Vol	Crew Size	Man-hours per Unit	Crew Output per Day	Avg Mat'l Unit Cost	Avg Labor Unit Cost	Avg Equip Unit Cost	Avg Total Unit Cost	Avg Price Incl O&P
Block											
Laid in mastic over wood subfloor covered with felt											
Oak, 5/16" x 12" x 12"											
Lay floor only											
Prefinished	Inst	SF	Lg	FD	.035	520.0	10.80	1.11	---	11.91	**14.10**
	Inst	SF	Sm	FD	.053	338.0	12.10	1.67	---	13.77	**16.30**
Unfinished	Inst	SF	Lg	FD	.033	550.0	5.78	1.04	---	6.82	**8.18**
	Inst	SF	Sm	FD	.050	358.0	6.45	1.58	---	8.03	**9.74**
Teak, 5/16" x 12" x 12"											
Lay floor only											
Prefinished	Inst	SF	Lg	FD	.035	520.0	13.00	1.11	---	14.11	**16.50**
	Inst	SF	Sm	FD	.053	338.0	14.40	1.67	---	16.07	**19.10**
Unfinished	Inst	SF	Lg	FD	.033	550.0	7.92	1.04	---	8.96	**10.60**
	Inst	SF	Sm	FD	.050	358.0	8.84	1.58	---	10.42	**12.50**
Oak, 13/16" x 12" x 12"											
Lay floor only											
Prefinished	Inst	SF	Lg	FD	.035	520.0	19.40	1.11	---	20.51	**23.90**
	Inst	SF	Sm	FD	.053	338.0	21.60	1.67	---	23.27	**27.40**
Unfinished	Inst	SF	Lg	FD	.036	500.0	14.40	1.14	---	15.54	**18.20**
	Inst	SF	Sm	FD	.055	325.0	16.00	1.74	---	17.74	**21.00**
Machine sand, fill and finish											
New floors											
	Inst	SF	Lg	FC	.019	430.0	.11	.61	.05	.77	**1.07**
	Inst	SF	Sm	FC	.029	280.0	.12	.93	.08	1.13	**1.59**
Damaged floors											
	Inst	SF	Lg	FC	.028	285.0	.15	.90	.08	1.13	**1.57**
	Inst	SF	Sm	FC	.043	185.0	.16	1.38	.12	1.66	**2.33**
Wax, polish and machine buff											
	Inst	SF	Lg	FC	.009	860.0	.04	.29	.03	.36	**.50**
	Inst	SF	Sm	FC	.014	559.0	.04	.45	.04	.53	**.75**
Parquetry, 5/16" x 9" x 9"											
Lay floor only											
Oak	Inst	SF	Lg	FD	.036	500.0	5.78	1.14	---	6.92	**8.32**
	Inst	SF	Sm	FD	.055	325.0	6.45	1.74	---	8.19	**9.97**
Walnut	Inst	SF	Lg	FD	.036	500.0	14.90	1.14	---	16.04	**18.80**
	Inst	SF	Sm	FD	.055	325.0	16.60	1.74	---	18.34	**21.60**
Teak	Inst	SF	Lg	FD	.036	500.0	7.92	1.14	---	9.06	**10.80**
	Inst	SF	Sm	FD	.055	325.0	8.84	1.74	---	10.58	**12.70**

Description	Oper	Unit	Vol	Crew Size	Man-hours per Unit	Crew Output per Day	Avg Mat'l Unit Cost	Avg Labor Unit Cost	Avg Equip Unit Cost	Avg Total Unit Cost	Avg Price Incl O&P
Machine sand, fill and finish											
New floors											
	Inst	SF	Lg	FC	.019	430.0	.11	.61	.05	.77	**1.07**
	Inst	SF	Sm	FC	.029	280.0	.12	.93	.08	1.13	**1.59**
Damaged floors											
	Inst	SF	Lg	FC	.028	285.0	.15	.90	.08	1.13	**1.57**
	Inst	SF	Sm	FC	.043	185.0	.16	1.38	.12	1.66	**2.33**
Wax, polish and machine buff											
	Inst	SF	Lg	FC	.009	860.0	.04	.29	.03	.36	**.50**
	Inst	SF	Sm	FC	.014	559.0	.04	.45	.04	.53	**.75**
Acrylic wood parquet blocks											
5/16" x 12" x 12" set in epoxy											
	Inst	SF	Lg	FD	.036	500.0	7.01	1.14	---	8.15	**9.73**
	Inst	SF	Sm	FD	.055	325.0	7.82	1.74	---	9.56	**11.60**
Wax, polish and machine buff											
	Inst	SF	Lg	FC	.009	860.0	.04	.29	.03	.36	**.50**
	Inst	SF	Sm	FC	.014	559.0	.04	.45	.04	.53	**.75**

Heating

Boilers

Electric-fired heaters, includes standard controls and trim, ASME

Description	Oper	Unit	Vol	Crew Size	Man-hours per Unit	Crew Output per Day	Avg Mat'l Unit Cost	Avg Labor Unit Cost	Avg Equip Unit Cost	Avg Total Unit Cost	Avg Price Incl O&P
Hot water											
12 KW, 40 MBHP	Inst	Ea	Lg	SD	20.0	1.20	4120.00	672.00	---	4792.00	**5730.00**
	Inst	Ea	Sm	SD	26.7	0.90	4560.00	897.00	---	5457.00	**6560.00**
24 KW, 82 MBHP	Inst	Ea	Lg	SD	20.9	1.15	4400.00	702.00	---	5102.00	**6090.00**
	Inst	Ea	Sm	SD	27.9	0.86	4860.00	937.00	---	5797.00	**6970.00**
45 KW, 154 MBHP	Inst	Ea	Lg	SD	21.4	1.12	4840.00	719.00	---	5559.00	**6620.00**
	Inst	Ea	Sm	SD	28.6	0.84	5340.00	960.00	---	6300.00	**7560.00**
60 KW, 205 MBHP	Inst	Ea	Lg	SD	21.8	1.10	5740.00	732.00	---	6472.00	**7670.00**
	Inst	Ea	Sm	SD	28.9	0.83	6340.00	970.00	---	7310.00	**8710.00**
240 KW, 820 MBHP	Inst	Ea	Lg	SD	24.0	1.00	12400.00	806.00	---	13206.00	**15400.00**
	Inst	Ea	Sm	SD	32.0	0.75	13700.00	1070.00	---	14770.00	**17300.00**
480 KW, 1,635 MBHP	Inst	Ea	Lg	SD	25.3	0.95	21300.00	850.00	---	22150.00	**25800.00**
	Inst	Ea	Sm	SD	33.8	0.71	23500.00	1140.00	---	24640.00	**28700.00**
1,200 KW, 4,095 MBHP	Inst	Ea	Lg	SD	26.7	0.90	38900.00	897.00	---	39797.00	**46000.00**
	Inst	Ea	Sm	SD	35.3	0.68	42900.00	1190.00	---	44090.00	**51100.00**
2,400 KW, 8,190 MBHP	Inst	Ea	Lg	SD	28.2	0.85	66300.00	947.00		67247.00	**77600.00**
	Inst	Ea	Sm	SD	37.5	0.64	73200.00	1260.00	---	74460.00	**86100.00**
Steam											
6 KW, 20 MBHP	Inst	Ea	Lg	SD	21.8	1.10	9800.00	732.00	---	10532.00	**12300.00**
	Inst	Ea	Sm	SD	28.9	0.83	10800.00	970.00	---	11770.00	**13900.00**
60 KW, 205 MBHP	Inst	Ea	Lg	SD	26.7	0.90	11900.00	897.00	---	12797.00	**15000.00**
	Inst	Ea	Sm	SD	35.3	0.68	13100.00	1190.00	---	14290.00	**16800.00**
240 KW, 815 MBHP	Inst	Ea	Lg	SD	60.0	0.40	18700.00	2010.00	---	20710.00	**24500.00**
	Inst	Ea	Sm	SD	80.0	0.30	20700.00	2690.00	---	23390.00	**27700.00**
600 KW, 2,047 MBHP	Inst	Ea	Lg	SE	160	0.20	28900.00	5540.00	---	34440.00	**41400.00**
	Inst	Ea	Lg	SE	213	0.15	31900.00	7370.00	---	39270.00	**47500.00**
Minimum Job Charge											
	Inst	Job	Lg	SD	12.0	2.00	---	403.00	---	403.00	**592.00**
	Inst	Job	Sm	SD	17.1	1.40	---	574.00	---	574.00	**844.00**

Description	Oper	Unit	Vol	Crew Size	Man-hours per Unit	Crew Output per Day	Avg Mat'l Unit Cost	Avg Labor Unit Cost	Avg Equip Unit Cost	Avg Total Unit Cost	Avg Price Incl O&P

Gas-fired, natural or propane, heaters; includes standard controls; MBHP gross output

Cast iron with insulated jacket
Hot water

Description	Oper	Unit	Vol	Crew Size	Man-hours per Unit	Crew Output per Day	Avg Mat'l Unit Cost	Avg Labor Unit Cost	Avg Equip Unit Cost	Avg Total Unit Cost	Avg Price Incl O&P
80 MBHP	Inst	Ea	Lg	SF	17.1	1.40	1300.00	586.00	---	1886.00	2360.00
	Inst	Ea	Sm	SF	22.9	1.05	1440.00	785.00	---	2225.00	2810.00
100 MBHP	Inst	Ea	Lg	SF	17.5	1.37	1490.00	600.00	---	2090.00	2590.00
	Inst	Ea	Sm	SF	23.3	1.03	1640.00	798.00	---	2438.00	3060.00
122 MBHP	Inst	Ea	Lg	SF	17.8	1.35	1640.00	610.00	---	2250.00	2790.00
	Inst	Ea	Sm	SF	23.8	1.01	1820.00	815.00	---	2635.00	3290.00
163 MBHP	Inst	Ea	Lg	SF	18.2	1.32	2020.00	624.00	---	2644.00	3230.00
	Inst	Ea	Sm	SF	24.2	0.99	2230.00	829.00	---	3059.00	3780.00
440 MBHP	Inst	Ea	Lg	SF	18.5	1.30	3940.00	634.00	---	4574.00	5460.00
	Inst	Ea	Sm	SF	24.5	0.98	4350.00	839.00	---	5189.00	6240.00
2,000 MBHP	Inst	Ea	Lg	SF	24.0	1.00	12800.00	822.00	---	13622.00	15900.00
	Inst	Ea	Sm	SF	32.0	0.75	14100.00	1100.00	---	15200.00	17800.00
6,970 MBHP	Inst	Ea	Lg	SF	60.0	0.40	86800.00	2060.00	---	88860.00	103000.00
	Inst	Ea	Sm	SF	80.0	0.30	95900.00	2740.00	---	98640.00	114000.00

Steam

Description	Oper	Unit	Vol	Crew Size	Man-hours per Unit	Crew Output per Day	Avg Mat'l Unit Cost	Avg Labor Unit Cost	Avg Equip Unit Cost	Avg Total Unit Cost	Avg Price Incl O&P
80 MBHP	Inst	Ea	Lg	SF	17.1	1.40	1300.00	586.00	---	1886.00	2360.00
	Inst	Ea	Sm	SF	22.9	1.05	1440.00	785.00	---	2225.00	2810.00
163 MBHP	Inst	Ea	Lg	SF	18.2	1.32	2020.00	624.00	---	2644.00	3230.00
	Inst	Ea	Sm	SF	24.2	0.99	2230.00	829.00	---	3059.00	3780.00
440 MBHP	Inst	Ea	Lg	SF	18.5	1.30	3940.00	634.00	---	4574.00	5460.00
	Inst	Ea	Sm	SF	24.5	0.98	4350.00	839.00	---	5189.00	6240.00
3,570 MBHP	Inst	Ea	Lg	SF	26.7	0.90	20100.00	915.00	---	21015.00	24400.00
	Inst	Ea	Sm	SF	35.3	0.68	22200.00	1210.00	---	23410.00	27300.00
6,970 MBHP	Inst	Ea	Lg	SF	60.0	0.40	65100.00	2060.00	---	67160.00	77900.00
	Inst	Ea	Sm	SF	80.0	0.30	71900.00	2740.00	---	74640.00	86700.00

Steel with insulated jacket; includes burner and one zone valve
Hot water

Description	Oper	Unit	Vol	Crew Size	Man-hours per Unit	Crew Output per Day	Avg Mat'l Unit Cost	Avg Labor Unit Cost	Avg Equip Unit Cost	Avg Total Unit Cost	Avg Price Incl O&P
50 MBHP	Inst	Ea	Lg	SF	10.9	2.20	1950.00	373.00	---	2323.00	2790.00
	Inst	Ea	Sm	SF	14.5	1.65	2160.00	497.00	---	2657.00	3210.00
70 MBHP	Inst	Ea	Lg	SF	10.9	2.20	2170.00	373.00	---	2543.00	3040.00
	Inst	Ea	Sm	SF	14.5	1.65	2400.00	497.00	---	2897.00	3490.00
90 MBHP	Inst	Ea	Lg	SF	11.4	2.10	2200.00	391.00	---	2591.00	3110.00
	Inst	Ea	Sm	SF	15.2	1.58	2430.00	521.00	---	2951.00	3560.00
105 MBHP	Inst	Ea	Lg	SF	12.0	2.00	2480.00	411.00	---	2891.00	3460.00
	Inst	Ea	Sm	SF	16.0	1.50	2740.00	548.00	---	3288.00	3960.00
130 MBHP	Inst	Ea	Lg	SF	12.6	1.90	2820.00	432.00	---	3252.00	3880.00
	Inst	Ea	Sm	SF	16.8	1.43	3120.00	576.00	---	3696.00	4430.00

Description	Oper	Unit	Vol	Crew Size	Man-hours per Unit	Crew Output per Day	Avg Mat'l Unit Cost	Avg Labor Unit Cost	Avg Equip Unit Cost	Avg Total Unit Cost	Avg Price Incl O&P
150 MBHP	Inst	Ea	Lg	SF	13.3	1.80	3290.00	456.00	---	3746.00	**4450.00**
	Inst	Ea	Sm	SF	17.8	1.35	3630.00	610.00	---	4240.00	**5070.00**
185 MBHP	Inst	Ea	Lg	SF	14.5	1.65	3970.00	497.00	---	4467.00	**5290.00**
	Inst	Ea	Sm	SF	19.4	1.24	4380.00	665.00	---	5045.00	**6020.00**
235 MBHP	Inst	Ea	Lg	SF	16.0	1.50	5020.00	548.00	---	5568.00	**6580.00**
	Inst	Ea	Sm	SF	21.2	1.13	5550.00	726.00	---	6276.00	**7450.00**
290 MBHP	Inst	Ea	Lg	SF	17.1	1.40	5670.00	586.00	---	6256.00	**7390.00**
	Inst	Ea	Sm	SF	22.9	1.05	6270.00	785.00	---	7055.00	**8360.00**
480 MBHP	Inst	Ea	Lg	SF	26.7	0.90	8180.00	915.00	---	9095.00	**10800.00**
	Inst	Ea	Sm	SF	35.3	0.68	9040.00	1210.00	---	10250.00	**12200.00**
640 MBHP	Inst	Ea	Lg	SF	34.3	0.70	9730.00	1180.00	---	10910.00	**12900.00**
	Inst	Ea	Sm	SF	45.3	0.53	10800.00	1550.00	---	12350.00	**14600.00**
800 MBHP	Inst	Ea	Lg	SF	41.4	0.58	11300.00	1420.00	---	12720.00	**15100.00**
	Inst	Ea	Sm	SF	54.5	0.44	12500.00	1870.00	---	14370.00	**17200.00**
960 MBHP	Inst	Ea	Lg	SF	43.6	0.55	14100.00	1490.00	---	15590.00	**18500.00**
	Inst	Ea	Sm	SF	58.5	0.41	15600.00	2000.00	---	17600.00	**20900.00**
Minimum Job Charge											
	Inst	Job	Lg	SF	12.0	2.00	---	411.00	---	411.00	**604.00**
	Inst	Job	Sm	SF	17.1	1.40	---	586.00	---	586.00	**861.00**

Oil-fired heaters; includes standard controls; flame retention burner; MBHP gross output

Cast iron with insulated jacket

Hot water

Description	Oper	Unit	Vol	Crew Size	Man-hours per Unit	Crew Output per Day	Avg Mat'l Unit Cost	Avg Labor Unit Cost	Avg Equip Unit Cost	Avg Total Unit Cost	Avg Price Incl O&P
110 MBHP	Inst	Ea	Lg	SF	24.0	1.00	1460.00	822.00	---	2282.00	**2880.00**
	Inst	Ea	Sm	SF	32.0	0.75	1610.00	1100.00	---	2710.00	**3460.00**
200 MBHP	Inst	Ea	Lg	SF	34.3	0.70	2020.00	1180.00	---	3200.00	**4040.00**
	Inst	Ea	Sm	SF	45.3	0.53	2230.00	1550.00	---	3780.00	**4840.00**
1,080 MBHP	Inst	Ea	Lg	SF	80.0	0.30	10300.00	2740.00	---	13040.00	**15900.00**
	Inst	Ea	Sm	SF	104	0.23	11400.00	3560.00	---	14960.00	**18300.00**
1,320 MBHP	Inst	Ea	Lg	SF	96.0	0.25	9320.00	3290.00	---	12610.00	**15600.00**
	Inst	Ea	Sm	SF	126	0.19	10300.00	4320.00	---	14620.00	**18200.00**
2,100 MBHP	Inst	Ea	Lg	SF	120	0.20	12400.00	4110.00	---	16510.00	**20300.00**
	Inst	Ea	Sm	SF	160	0.15	13700.00	5480.00	---	19180.00	**23800.00**
4,360 MBHP	Inst	Ea	Lg	SF	160	0.15	22200.00	5480.00	---	27680.00	**33600.00**
	Inst	Ea	Sm	SF	218	0.11	24500.00	7470.00	---	31970.00	**39200.00**
6,970 MBHP	Inst	Ea	Lg	SF	300	0.08	64500.00	10300.00	---	74800.00	**89300.00**
	Inst	Ea	Sm	SF	400	0.06	71200.00	13700.00	---	84900.00	**102000.00**

Steam

Description	Oper	Unit	Vol	Crew Size	Man-hours per Unit	Crew Output per Day	Avg Mat'l Unit Cost	Avg Labor Unit Cost	Avg Equip Unit Cost	Avg Total Unit Cost	Avg Price Incl O&P
110 MBHP	Inst	Ea	Lg	SF	24.0	1.00	1460.00	822.00	---	2282.00	**2880.00**
	Inst	Ea	Sm	SF	32.0	0.75	1610.00	1100.00	---	2710.00	**3460.00**
205 MBHP	Inst	Ea	Lg	SF	34.3	0.70	2020.00	1180.00	---	3200.00	**4040.00**
	Inst	Ea	Sm	SF	45.3	0.53	2230.00	1550.00	---	3780.00	**4840.00**
1,085 MBHP	Inst	Ea	Lg	SF	80.0	0.30	10300.00	2740.00	---	13040.00	**15900.00**
	Inst	Ea	Sm	SF	104.3	0.23	11400.00	3570.00	---	14970.00	**18300.00**

Description	Oper	Unit	Vol	Crew Size	Man-hours per Unit	Crew Output per Day	Avg Mat'l Unit Cost	Avg Labor Unit Cost	Avg Equip Unit Cost	Avg Total Unit Cost	Avg Price Incl O&P
1,360 MBHP	Inst	Ea	Lg	SF	96.0	0.25	9320.00	3290.00	---	12610.00	**15600.00**
	Inst	Ea	Sm	SF	126	0.19	10300.00	4320.00	---	14620.00	**18200.00**
2,175 MBHP	Inst	Ea	Lg	SF	120	0.20	12400.00	4110.00	---	16510.00	**20300.00**
	Inst	Ea	Sm	SF	160	0.15	13700.00	5480.00	---	19180.00	**23800.00**
4,360 MBHP	Inst	Ea	Lg	SF	160	0.15	22200.00	5480.00	---	27680.00	**33600.00**
	Inst	Ea	Sm	SF	218	0.11	24500.00	7470.00	---	31970.00	**39200.00**
6,970 MBHP	Inst	Ea	Lg	SF	300	0.08	64500.00	10300.00	---	74800.00	**89300.00**
	Inst	Ea	Sm	SF	400	0.06	71200.00	13700.00	---	84900.00	**102000.00**

Steel insulated jacket burner
Hot water

Description	Oper	Unit	Vol	Crew Size	Man-hours per Unit	Crew Output per Day	Avg Mat'l Unit Cost	Avg Labor Unit Cost	Avg Equip Unit Cost	Avg Total Unit Cost	Avg Price Incl O&P
105 MBHP	Inst	Ea	Lg	SF	13.3	1.80	2570.00	456.00	---	3026.00	**3630.00**
	Inst	Ea	Sm	SF	17.8	1.35	2840.00	610.00	---	3450.00	**4170.00**
120 MBHP	Inst	Ea	Lg	SF	15.0	1.60	2600.00	514.00	---	3114.00	**3750.00**
	Inst	Ea	Sm	SF	20.0	1.20	2880.00	685.00	---	3565.00	**4320.00**
140 MBHP	Inst	Ea	Lg	SF	17.1	1.40	2730.00	586.00	---	3316.00	**4000.00**
	Inst	Ea	Sm	SF	22.9	1.05	3010.00	785.00	---	3795.00	**4620.00**
170 MBHP	Inst	Ea	Lg	SF	20.0	1.20	3260.00	685.00	---	3945.00	**4750.00**
	Inst	Ea	Sm	SF	26.7	0.90	3600.00	915.00	---	4515.00	**5480.00**
225 MBHP	Inst	Ea	Lg	SF	26.7	0.90	3880.00	915.00	---	4795.00	**5800.00**
	Inst	Ea	Sm	SF	35.3	0.68	4280.00	1210.00	---	5490.00	**6700.00**
315 MBHP	Inst	Ea	Lg	SF	30.0	0.80	5700.00	1030.00	---	6730.00	**8070.00**
	Inst	Ea	Sm	SF	40.0	0.60	6300.00	1370.00	---	7670.00	**9260.00**
420 MBHP	Inst	Ea	Lg	SF	34.3	0.70	6420.00	1180.00	---	7600.00	**9110.00**
	Inst	Ea	Sm	SF	45.3	0.53	7090.00	1550.00	---	8640.00	**10400.00**
Minimum Job Charge											
	Inst	Job	Lg	SF	12.0	2.00	---	411.00	---	411.00	**604.00**
	Inst	Job	Sm	SF	17.1	1.40	---	586.00	---	586.00	**861.00**

Boiler accessories
Burners

Conversion, gas-fired, LP or natural
Residential, gun type, atmospheric input

Description	Oper	Unit	Vol	Crew Size	Man-hours per Unit	Crew Output per Day	Avg Mat'l Unit Cost	Avg Labor Unit Cost	Avg Equip Unit Cost	Avg Total Unit Cost	Avg Price Incl O&P
72 to 200 MBHP	Inst	Ea	Lg	SB	8.00	2.00	812.00	260.00	---	1072.00	**1320.00**
	Inst	Ea	Sm	SB	10.7	1.50	897.00	348.00	---	1245.00	**1540.00**
120 to 360 MBHP	Inst	Ea	Lg	SB	9.14	1.75	899.00	298.00	---	1197.00	**1470.00**
	Inst	Ea	Sm	SB	12.2	1.31	993.00	397.00	---	1390.00	**1730.00**
280 to 800 MBHP	Inst	Ea	Lg	SB	10.7	1.50	1740.00	348.00	---	2088.00	**2510.00**
	Inst	Ea	Sm	SB	14.2	1.13	1920.00	462.00	---	2382.00	**2890.00**

Flame retention, oil fired assembly

Description	Oper	Unit	Vol	Crew Size	Man-hours per Unit	Crew Output per Day	Avg Mat'l Unit Cost	Avg Labor Unit Cost	Avg Equip Unit Cost	Avg Total Unit Cost	Avg Price Incl O&P
2.0 to 5.0 GPH	Inst	Ea	Lg	SB	9.14	1.75	620.00	298.00	---	918.00	**1150.00**
	Inst	Ea	Sm	SB	12.2	1.31	685.00	397.00	---	1082.00	**1370.00**

Description	Oper	Unit	Vol	Crew Size	Man-hours per Unit	Crew Output per Day	Avg Mat'l Unit Cost	Avg Labor Unit Cost	Avg Equip Unit Cost	Avg Total Unit Cost	Avg Price Incl O&P

Forced warm air systems
Duct furnaces
Furnace includes burner, controls, stainless steel heat exchanger
Gas-fired with an electric ignition
Outdoor installation, includes vent cap

Description	Oper	Unit	Vol	Crew Size	Man-hours per Unit	Crew Output per Day	Avg Mat'l Unit Cost	Avg Labor Unit Cost	Avg Equip Unit Cost	Avg Total Unit Cost	Avg Price Incl O&P
225 MBHP output	Inst	Ea	Lg	SB	8.00	2.00	3470.00	260.00	---	3730.00	**4380.00**
	Inst	Ea	Sm	SB	10.7	1.50	3840.00	348.00	---	4188.00	**4920.00**
375 MBHP output	Inst	Ea	Lg	SB	10.7	1.50	5020.00	348.00	---	5368.00	**6290.00**
	Inst	Ea	Sm	SB	14.2	1.13	5550.00	462.00	---	6012.00	**7060.00**
450 MBHP output	Inst	Ea	Lg	SB	12.3	1.30	5270.00	400.00	---	5670.00	**6650.00**
	Inst	Ea	Sm	SB	16.3	0.98	5820.00	531.00	---	6351.00	**7480.00**

Furnaces, hot air heating with blowers and standard controls

Gas or oil lines and couplings not included (see below and next page), flue piping not included (see page 435)

Electric-fired, UL listed, heat staging, 240 volt run and connection

Description	Oper	Unit	Vol	Crew Size	Man-hours per Unit	Crew Output per Day	Avg Mat'l Unit Cost	Avg Labor Unit Cost	Avg Equip Unit Cost	Avg Total Unit Cost	Avg Price Incl O&P
30 MBHP	Inst	Ea	Lg	UE	5.56	3.60	502.00	183.00	---	685.00	**848.00**
	Inst	Ea	Sm	UE	7.41	2.70	555.00	244.00	---	799.00	**999.00**
75 MBHP	Inst	Ea	Lg	UE	5.71	3.50	639.00	188.00	---	827.00	**1010.00**
	Inst	Ea	Sm	UE	7.60	2.63	706.00	250.00	---	956.00	**1180.00**
85 MBHP	Inst	Ea	Lg	UE	6.25	3.20	713.00	206.00	---	919.00	**1120.00**
	Inst	Ea	Sm	UE	8.33	2.40	788.00	274.00	---	1062.00	**1310.00**
90 MBHP	Inst	Ea	Lg	UE	6.67	3.00	893.00	219.00	---	1112.00	**1350.00**
	Inst	Ea	Sm	UE	8.89	2.25	986.00	292.00	---	1278.00	**1570.00**
Minimum Job Charge											
	Inst	Job	Lg	UD	8.00	2.00	---	258.00	---	258.00	**381.00**
	Inst	Job	Sm	UD	11.4	1.40	---	367.00	---	367.00	**543.00**

Gas-fired, AGA certified, direct drive models

Description	Oper	Unit	Vol	Crew Size	Man-hours per Unit	Crew Output per Day	Avg Mat'l Unit Cost	Avg Labor Unit Cost	Avg Equip Unit Cost	Avg Total Unit Cost	Avg Price Incl O&P
40 MBHP	Inst	Ea	Lg	UD	4.44	3.60	539.00	143.00	---	682.00	**832.00**
	Inst	Ea	Sm	UD	5.93	2.70	596.00	191.00	---	787.00	**968.00**
65 MBHP	Inst	Ea	Lg	UD	4.57	3.50	719.00	147.00	---	866.00	**1040.00**
	Inst	Ea	Sm	UD	6.08	2.63	795.00	196.00	---	991.00	**1200.00**
80 MBHP	Inst	Ea	Lg	UD	4.71	3.40	763.00	152.00	---	915.00	**1100.00**
	Inst	Ea	Sm	UD	6.27	2.55	843.00	202.00	---	1045.00	**1270.00**
85 MBHP	Inst	Ea	Lg	UD	5.00	3.20	787.00	161.00	---	948.00	**1140.00**
	Inst	Ea	Sm	UD	6.67	2.40	870.00	215.00	---	1085.00	**1320.00**
105 MBHP	Inst	Ea	Lg	UD	5.33	3.00	812.00	172.00	---	984.00	**1190.00**
	Inst	Ea	Sm	UD	7.11	2.25	897.00	229.00	---	1126.00	**1370.00**
125 MBHP	Inst	Ea	Lg	UD	5.71	2.80	936.00	184.00	---	1120.00	**1350.00**
	Inst	Ea	Sm	UD	7.62	2.10	1030.00	245.00	---	1275.00	**1550.00**
160 MBHP	Inst	Ea	Lg	UD	6.15	2.60	961.00	198.00	---	1159.00	**1400.00**
	Inst	Ea	Sm	UD	8.21	1.95	1060.00	264.00	---	1324.00	**1610.00**
200 MBHP	Inst	Ea	Lg	UD	6.67	2.40	2290.00	215.00	---	2505.00	**2960.00**
	Inst	Ea	Sm	UD	8.89	1.80	2530.00	286.00	---	2816.00	**3340.00**
Minimum Job Charge											
	Inst	Job	Lg	UD	8.00	2.00	---	258.00	---	258.00	**381.00**
	Inst	Job	Sm	UD	11.4	1.40	---	367.00	---	367.00	**543.00**
Gas line with couplings											
	Inst	LF	Lg	SB	.356	45.00	---	11.60	---	11.60	**17.00**
	Inst	LF	Sm	SB	.474	33.75	---	15.40	---	15.40	**22.70**

Description	Oper	Unit	Vol	Crew Size	Man-hours per Unit	Crew Output per Day	Avg Mat'l Unit Cost	Avg Labor Unit Cost	Avg Equip Unit Cost	Avg Total Unit Cost	Avg Price Incl O&P
Oil-fired, UL listed, gun-type burner											
55 MBHP	Inst	Ea	Lg	UD	4.57	3.50	924.00	147.00	---	1071.00	**1280.00**
	Inst	Ea	Sm	UD	6.08	2.63	1020.00	196.00	---	1216.00	**1460.00**
100 MBHP	Inst	Ea	Lg	UD	5.33	3.00	980.00	172.00	---	1152.00	**1380.00**
	Inst	Ea	Sm	UD	7.11	2.25	1080.00	229.00	---	1309.00	**1580.00**
125 MBHP	Inst	Ea	Lg	UD	5.71	2.80	1360.00	184.00	---	1544.00	**1840.00**
	Inst	Ea	Sm	UD	7.62	2.10	1510.00	245.00	---	1755.00	**2100.00**
150 MBHP	Inst	Ea	Lg	UD	6.15	2.60	1520.00	198.00	---	1718.00	**2040.00**
	Inst	Ea	Sm	UD	8.21	1.95	1680.00	264.00	---	1944.00	**2320.00**
200 MBHP	Inst	Ea	Lg	UD	6.67	2.40	2540.00	215.00	---	2755.00	**3240.00**
	Inst	Ea	Sm	UD	8.89	1.80	2810.00	286.00	---	3096.00	**3650.00**
Minimum Job Charge											
	Inst	Job	Lg	UD	8.00	2.00	---	258.00	---	258.00	**381.00**
	Inst	Job	Sm	UD	11.4	1.40	---	367.00	---	367.00	**543.00**
Oil line with couplings											
	Inst	LF	Lg	SB	.200	80.00	---	6.51	---	6.51	**9.57**
	Inst	LF	Sm	SB	.267	60.00	---	8.69	---	8.69	**12.80**
Combo fired (wood, coal, oil combination) complete with burner											
115 MBHP (based on oil)	Inst	Ea	Lg	UD	5.52	2.90	2760.00	178.00	---	2938.00	**3440.00**
	Inst	Ea	Sm	UD	7.34	2.18	3050.00	236.00	---	3286.00	**3860.00**
140 MBHP (based on oil)	Inst	Ea	Lg	UD	5.93	2.70	4430.00	191.00	---	4621.00	**5380.00**
	Inst	Ea	Sm	UD	7.88	2.03	4900.00	254.00	---	5154.00	**6010.00**
150 MBHP (based on oil)	Inst	Ea	Lg	UD	6.15	2.60	4430.00	198.00	---	4628.00	**5390.00**
	Inst	Ea	Sm	UD	8.21	1.95	4900.00	264.00	---	5164.00	**6020.00**
170 MBHP (based on oil)	Inst	Ea	Lg	UD	6.40	2.50	4680.00	206.00	---	4886.00	**5690.00**
	Inst	Ea	Sm	UD	8.51	1.88	5170.00	274.00	---	5444.00	**6350.00**
Minimum Job Charge											
	Inst	Job	Lg	UD	8.00	2.00	---	258.00	---	258.00	**381.00**
	Inst	Job	Sm	UD	11.4	1.40	---	367.00	---	367.00	**543.00**
Oil line with couplings											
	Inst	LF	Lg	SB	.200	80.00	---	6.51	---	6.51	**9.57**
	Inst	LF	Sm	SB	.267	60.00	---	8.69	---	8.69	**12.80**

Space heaters, gas-fired

Unit includes cabinet, grilles, fan, controls, burner and thermostat; no flue piping included (see page 435)

Floor mounted

Description	Oper	Unit	Vol	Crew Size	Man-hours per Unit	Crew Output per Day	Avg Mat'l Unit Cost	Avg Labor Unit Cost	Avg Equip Unit Cost	Avg Total Unit Cost	Avg Price Incl O&P
60 MBHP	Inst	Ea	Lg	SB	2.00	8.00	663.00	65.10	---	728.10	**859.00**
	Inst	Ea	Sm	SB	2.67	6.00	733.00	86.90	---	819.90	**971.00**
180 MBHP	Inst	Ea	Lg	SB	4.00	4.00	986.00	130.00	---	1116.00	**1330.00**
	Inst	Ea	Sm	SB	5.33	3.00	1090.00	173.00	---	1263.00	**1510.00**

Suspension mounted, propeller fan

Description	Oper	Unit	Vol	Crew Size	Man-hours per Unit	Crew Output per Day	Avg Mat'l Unit Cost	Avg Labor Unit Cost	Avg Equip Unit Cost	Avg Total Unit Cost	Avg Price Incl O&P
20 MBHP	Inst	Ea	Lg	SB	2.67	6.00	496.00	86.90	---	582.90	**698.00**
	Inst	Ea	Sm	SB	3.56	4.50	548.00	116.00	---	664.00	**801.00**
60 MBHP	Inst	Ea	Lg	SB	3.20	5.00	614.00	104.00	---	718.00	**859.00**
	Inst	Ea	Sm	SB	4.27	3.75	678.00	139.00	---	817.00	**984.00**
130 MBHP	Inst	Ea	Lg	SB	4.00	4.00	905.00	130.00	---	1035.00	**1230.00**
	Inst	Ea	Sm	SB	5.33	3.00	1000.00	173.00	---	1173.00	**1410.00**
320 MBHP	Inst	Ea	Lg	SB	8.00	2.00	1860.00	260.00	---	2120.00	**2520.00**
	Inst	Ea	Sm	SB	10.7	1.50	2060.00	348.00	---	2408.00	**2880.00**
Powered venter, adapter	ADD	Ea	Lg	SB	2.00	8.00	315.00	65.10	---	380.10	**458.00**
	ADD	Ea	Sm	SB	2.67	6.00	348.00	86.90	---	434.90	**528.00**

Description	Oper	Unit	Vol	Crew Size	Man-hours per Unit	Crew Output per Day	Avg Mat'l Unit Cost	Avg Labor Unit Cost	Avg Equip Unit Cost	Avg Total Unit Cost	Avg Price Incl O&P

Wall furnace, self-contained thermostat

Single capacity, recessed or surface mounted, 1-speed fan

15 MBHP	Inst	Ea	Lg	SB	3.20	5.00	546.00	104.00	---	650.00	**781.00**
	Inst	Ea	Sm	SB	4.27	3.75	603.00	139.00	---	742.00	**898.00**
25 MBHP	Inst	Ea	Lg	SB	4.00	4.00	564.00	130.00	---	694.00	**840.00**
	Inst	Ea	Sm	SB	5.33	3.00	623.00	173.00	---	796.00	**972.00**
35 MBHP	Inst	Ea	Lg	SB	5.33	3.00	756.00	173.00	---	929.00	**1120.00**
	Inst	Ea	Sm	SB	7.11	2.25	836.00	231.00	---	1067.00	**1300.00**

Dual capacity, recessed or surface mounted, 2-speed blowers

50 MBHP (direct vent)	Inst	Ea	Lg	SB	8.00	2.00	756.00	260.00	---	1016.00	**1250.00**
	Inst	Ea	Sm	SB	10.7	1.50	836.00	348.00	---	1184.00	**1470.00**
60 MBHP (up vent)	Inst	Ea	Lg	SB	8.00	2.00	670.00	260.00	---	930.00	**1150.00**
	Inst	Ea	Sm	SB	10.7	1.50	740.00	348.00	---	1088.00	**1360.00**

Register kit for circulating heat to second room

	Inst	Ea	Lg	SB	2.00	8.00	68.20	65.10	---	133.30	**174.00**
	Inst	Ea	Sm	SB	2.67	6.00	75.40	86.90	---	162.30	**214.00**

Minimum Job Charge

	Inst	Job	Lg	SB	8.00	2.00	---	260.00	---	260.00	**383.00**
	Inst	Job	Sm	SB	11.4	1.40	---	371.00	---	371.00	**545.00**

Bathroom heaters, electric-fired

Ceiling Heat-A-Ventlite, includes grille, blower, 4" round duct 13" x 7" dia., 3-way switch

1,500 watt	Inst	Ea	Lg	EA	4.00	2.00	294.00	143.00	---	437.00	**546.00**
	Inst	Ea	Sm	EA	5.33	1.50	325.00	190.00	---	515.00	**651.00**
1,800 watt	Inst	Ea	Lg	EA	4.00	2.00	326.00	143.00	---	469.00	**583.00**
	Inst	Ea	Sm	EA	5.33	1.50	360.00	190.00	---	550.00	**692.00**

Ceiling Heat-A-Lite, includes grille, airotor wheel, 13" x 7" dia., 2-way switch

1,500 watt	Inst	Ea	Lg	EA	4.00	2.00	217.00	143.00	---	360.00	**458.00**
	Inst	Ea	Sm	EA	5.33	1.50	240.00	190.00	---	430.00	**553.00**

Ceiling radiant heating using infrared lamps

Recessed, Heat-A-Lamp

One bulb, 250 watt lamp	Inst	Ea	Lg	EA	3.33	2.40	55.80	119.00	---	174.80	**237.00**
	Inst	Ea	Sm	EA	4.44	1.80	61.70	158.00	---	219.70	**302.00**
Two bulb, 500 watt lamp	Inst	Ea	Lg	EA	3.33	2.40	95.50	119.00	---	214.50	**283.00**
	Inst	Ea	Sm	EA	4.44	1.80	105.00	158.00	---	263.00	**352.00**
Three bulb, 750 watt lamp	Inst	Ea	Lg	EA	3.33	2.40	166.00	119.00	---	285.00	**364.00**
	Inst	Ea	Sm	EA	4.44	1.80	184.00	158.00	---	342.00	**442.00**

Recessed, Heat-A-Vent

One bulb, 250 watt lamp	Inst	Ea	Lg	EA	3.33	2.40	115.00	119.00	---	234.00	**306.00**
	Inst	Ea	Sm	EA	4.44	1.80	127.00	158.00	---	285.00	**377.00**
Two bulb, 500 watt lamp	Inst	Ea	Lg	EA	3.33	2.40	134.00	119.00	---	253.00	**327.00**
	Inst	Ea	Sm	EA	4.44	1.80	148.00	158.00	---	306.00	**401.00**

Description	Oper	Unit	Vol	Crew Size	Man-hours per Unit	Crew Output per Day	Avg Mat'l Unit Cost	Avg Labor Unit Cost	Avg Equip Unit Cost	Avg Total Unit Cost	Avg Price Incl O&P
Wall heaters, recessed											
Fan forced											
1250 watt heating element	Inst	Ea	Lg	EA	2.50	3.20	154.00	89.10	---	243.10	**307.00**
	Inst	Ea	Sm	EA	3.33	2.40	170.00	119.00	---	289.00	**369.00**
Radiant heating											
1200 watt heating element	Inst	Ea	Lg	EA	2.50	3.20	122.00	89.10	---	211.10	**270.00**
	Inst	Ea	Sm	EA	3.33	2.40	134.00	119.00	---	253.00	**328.00**
1500 watt heating element	Inst	Ea	Lg	EA	2.50	3.20	128.00	89.10	---	217.10	**277.00**
	Inst	Ea	Sm	EA	3.33	2.40	141.00	119.00	---	260.00	**336.00**
Wiring, connection, and installation in closed wall or ceiling structure ADD											
	Inst	Ea	Lg	EA	2.42	3.30	---	86.20	---	86.20	**126.00**
	Inst	Ea	Sm	EA	3.46	2.31	---	123.00	---	123.00	**180.00**

Insulation
Batt or roll
With wall or ceiling finish already removed

Description	Oper	Unit	Vol	Crew Size	Man-hours per Unit	Crew Output per Day	Avg Mat'l Unit Cost	Avg Labor Unit Cost	Avg Equip Unit Cost	Avg Total Unit Cost	Avg Price Incl O&P
Joists, 16" or 24" oc	Demo	SF	Lg	LB	.005	2935	---	.14	---	.14	**.20**
	Demo	SF	Sm	LB	.007	2201	---	.19	---	.19	**.29**
Rafters, 16" or 24" oc	Demo	SF	Lg	LB	.006	2560	---	.16	---	.16	**.25**
	Demo	SF	Sm	LB	.008	1920	---	.22	---	.22	**.33**
Studs, 16" or 24" oc	Demo	SF	Lg	LB	.005	3285	---	.14	---	.14	**.20**
	Demo	SF	Sm	LB	.006	2464	---	.16	---	.16	**.25**

Place and/or staple, Johns-Manville fiberglass; allowance made for joists, rafters, studs

Joists
Unfaced
3-1/2" T (R-13)

Description	Oper	Unit	Vol	Crew Size	Man-hours per Unit	Crew Output per Day	Avg Mat'l Unit Cost	Avg Labor Unit Cost	Avg Equip Unit Cost	Avg Total Unit Cost	Avg Price Incl O&P
16" oc	Inst	SF	Lg	CA	.008	975.0	.44	.27	---	.71	**.91**
	Inst	SF	Sm	CA	.011	731.0	.49	.37	---	.86	**1.11**
6-1/2" T (R-19)											
16" oc	Inst	SF	Lg	CA	.008	975.0	.51	.27	---	.78	**.99**
	Inst	SF	Sm	CA	.011	731.0	.56	.37	---	.93	**1.19**
24" oc	Inst	SF	Lg	CA	.005	1460	.51	.17	---	.68	**.84**
	Inst	SF	Sm	CA	.007	1095	.56	.23	---	.79	**.99**
7" T (R-22)											
16" oc	Inst	SF	Lg	CA	.008	975.0	.65	.27	---	.92	**1.15**
	Inst	SF	Sm	CA	.011	731.0	.72	.37	---	1.09	**1.38**
24" oc	Inst	SF	Lg	CA	.005	1460	.65	.17	---	.82	**1.00**
	Inst	SF	Sm	CA	.007	1095	.72	.23	---	.95	**1.18**
9-1/4" T (R-30)											
16" oc	Inst	SF	Lg	CA	.008	975.0	.75	.27	---	1.02	**1.26**
	Inst	SF	Sm	CA	.011	731.0	.83	.37	---	1.20	**1.50**
24" oc	Inst	SF	Lg	CA	.005	1460	.75	.17	---	.92	**1.11**
	Inst	SF	Sm	CA	.007	1095	.83	.23	---	1.06	**1.30**

Description	Oper	Unit	Vol	Crew Size	Man-hours per Unit	Crew Output per Day	Avg Mat'l Unit Cost	Avg Labor Unit Cost	Avg Equip Unit Cost	Avg Total Unit Cost	Avg Price Incl O&P
Kraft-faced											
3-1/2" T (R-11)											
16" oc	Inst	SF	Lg	CA	.008	975.0	.38	.27	---	.65	.84
	Inst	SF	Sm	CA	.011	731.0	.42	.37	---	.79	1.03
24" oc	Inst	SF	Lg	CA	.005	1460	.38	.17	---	.55	.69
	Inst	SF	Sm	CA	.007	1095	.42	.23	---	.65	.83
3-1/2" T (R-13)											
16" oc	Inst	SF	Lg	CA	.008	975.0	.44	.27	---	.71	.91
	Inst	SF	Sm	CA	.011	731.0	.49	.37	---	.86	1.11
6-1/2" T (R-19)											
16" oc	Inst	SF	Lg	CA	.008	975.0	.51	.27	---	.78	.99
	Inst	SF	Sm	CA	.011	731.0	.56	.37	---	.93	1.19
24" oc	Inst	SF	Lg	CA	.005	1460	.51	.17	---	.68	.84
	Inst	SF	Sm	CA	.007	1095	.56	.23	---	.79	.99
7" T (R-22)											
16" oc	Inst	SF	Lg	CA	.008	975.0	.65	.27	---	.92	1.15
	Inst	SF	Sm	CA	.011	731.0	.72	.37	---	1.09	1.38
24" oc	Inst	SF	Lg	CA	.005	1460	.65	.17	---	.82	1.00
	Inst	SF	Sm	CA	.007	1095	.72	.23	---	.95	1.18
9-1/4" T (R-30)											
16" oc	Inst	SF	Lg	CA	.008	975.0	.75	.27	---	1.02	1.26
	Inst	SF	Sm	CA	.011	731.0	.83	.37	---	1.20	1.50
24" oc	Inst	SF	Lg	CA	.005	1460	.75	.17	---	.92	1.11
	Inst	SF	Sm	CA	.007	1095	.83	.23	---	1.06	1.30
Foil-faced											
4" T (R-11)											
16" oc	Inst	SF	Lg	CA	.008	975.0	.38	.27	---	.65	.84
	Inst	SF	Sm	CA	.011	731.0	.42	.37	---	.79	1.03
24" oc	Inst	SF	Lg	CA	.005	1460	.38	.17	---	.55	.69
	Inst	SF	Sm	CA	.007	1095	.42	.23	---	.65	.83
6-1/2" T (R-19)											
16" oc	Inst	SF	Lg	CA	.008	975.0	.51	.27	---	.78	.99
	Inst	SF	Sm	CA	.011	731.0	.56	.37	---	.93	1.1
24" oc	Inst	SF	Lg	CA	.005	1460	.51	.17	---	.68	.84
	Inst	SF	Sm	CA	.007	1095	.56	.23	---	.79	.99

Description	Oper	Unit	Vol	Crew Size	Man-hours per Unit	Crew Output per Day	Avg Mat'l Unit Cost	Avg Labor Unit Cost	Avg Equip Unit Cost	Avg Total Unit Cost	Avg Price Incl O&P
Rafters											
Unfaced											
3-1/2" T (R-13)											
16" oc	Inst	SF	Lg	CA	.012	650.0	.44	.40	---	.84	**1.10**
	Inst	SF	Sm	CA	.016	488.0	.49	.53	---	1.02	**1.36**
6-1/2" T (R-19)											
16" oc	Inst	SF	Lg	CA	.012	650.0	.51	.40	---	.91	**1.19**
	Inst	SF	Sm	CA	.016	488.0	.56	.53	---	1.09	**1.44**
24" oc	Inst	SF	Lg	CA	.008	975.0	.51	.27	---	.78	**.99**
	Inst	SF	Sm	CA	.011	731.0	.56	.37	---	.93	**1.19**
7" T (R-22)											
16" oc	Inst	SF	Lg	CA	.012	650.0	.65	.40	---	1.05	**1.35**
	Inst	SF	Sm	CA	.016	488.0	.72	.53	---	1.25	**1.63**
24" oc	Inst	SF	Lg	CA	.008	975.0	.65	.27	---	.92	**1.15**
	Inst	SF	Sm	CA	.011	731.0	.72	.37	---	1.09	**1.38**
9-1/4" T (R-30)											
16" oc	Inst	SF	Lg	CA	.012	650.0	.75	.40	---	1.15	**1.46**
	Inst	SF	Sm	CA	.016	488.0	.83	.53	---	1.36	**1.75**
24" oc	Inst	SF	Lg	CA	.008	975.0	.75	.27	---	1.02	**1.26**
	Inst	SF	Sm	CA	.011	731.0	.83	.37	---	1.20	**1.50**
Kraft-faced											
3-1/2" T (R-11)											
16" oc	Inst	SF	Lg	CA	.012	650.0	.38	.40	---	.78	**1.04**
	Inst	SF	Sm	CA	.016	488.0	.42	.53	---	.95	**1.28**
24" oc	Inst	SF	Lg	CA	.008	975.0	.38	.27	---	.65	**.84**
	Inst	SF	Sm	CA	.011	731.0	.42	.37	---	.79	**1.03**
3-1/2" T (R-13)											
16" oc	Inst	SF	Lg	CA	.012	650.0	.44	.40	---	.84	**1.10**
	Inst	SF	Sm	CA	.016	488.0	.49	.53	---	1.02	**1.36**
6-1/2" T (R-19)											
16" oc	Inst	SF	Lg	CA	.012	650.0	.51	.40	---	.91	**1.19**
	Inst	SF	Sm	CA	.016	488.0	.56	.53	---	1.09	**1.44**
24" oc	Inst	SF	Lg	CA	.008	975.0	.51	.27	---	.78	**.99**
	Inst	SF	Sm	CA	.011	731.0	.56	.37	---	.93	**1.19**
7" T (R-22)											
16" oc	Inst	SF	Lg	CA	.012	650.0	.65	.40	---	1.05	**1.35**
	Inst	SF	Sm	CA	.016	488.0	.72	.53	---	1.25	**1.63**
24" oc	Inst	SF	Lg	CA	.008	975.0	.65	.27	---	.92	**1.15**
	Inst	SF	Sm	CA	.011	731.0	.72	.37	---	1.09	**1.38**
9-1/4" T (R-30)											
16" oc	Inst	SF	Lg	CA	.012	650.0	.75	.40	---	1.15	**1.46**
	Inst	SF	Sm	CA	.016	488.0	.83	.53	---	1.36	**1.75**
24" oc	Inst	SF	Lg	CA	.008	975.0	.75	.27	---	1.02	**1.26**
	Inst	SF	Sm	CA	.011	731.0	.83	.37	---	1.20	**1.50**

Description	Oper	Unit	Vol	Crew Size	Man-hours per Unit	Crew Output per Day	Avg Mat'l Unit Cost	Avg Labor Unit Cost	Avg Equip Unit Cost	Avg Total Unit Cost	Avg Price Incl O&P
Foil-faced											
4" T (R-11)											
16" oc	Inst	SF	Lg	CA	.012	650.0	.38	.40	---	.78	1.04
	Inst	SF	Sm	CA	.016	488.0	.42	.53	---	.95	1.28
24" oc	Inst	SF	Lg	CA	.008	975.0	.38	.27	---	.65	.84
	Inst	SF	Sm	CA	.011	731.0	.42	.37	---	.79	1.03
6-1/2" T (R-19)											
16" oc	Inst	SF	Lg	CA	.012	650.0	.51	.40	---	.91	1.19
	Inst	SF	Sm	CA	.016	488.0	.56	.53	---	1.09	1.44
24" oc	Inst	SF	Lg	CA	.008	975.0	.51	.27	---	.78	.99
	Inst	SF	Sm	CA	.011	731.0	.56	.37	---	.93	1.19
Studs											
Unfaced											
3-1/2" T (R-13)											
16" oc	Inst	SF	Lg	CA	.010	815.0	.44	.33	---	.77	1.01
	Inst	SF	Sm	CA	.013	611.0	.49	.43	---	.92	1.21
6-1/2" T (R-19)											
16" oc	Inst	SF	Lg	CA	.010	815.0	.51	.33	---	.84	1.09
	Inst	SF	Sm	CA	.013	611.0	.56	.43	---	.99	1.29
24" oc	Inst	SF	Lg	CA	.007	1220	.51	.23	---	.74	.94
	Inst	SF	Sm	CA	.009	915.0	.56	.30	---	.86	1.09
7" T (R-22)											
16" oc	Inst	SF	Lg	CA	.010	815.0	.65	.33	---	.98	1.25
	Inst	SF	Sm	CA	.013	611.0	.72	.43	---	1.15	1.48
24" oc	Inst	SF	Lg	CA	.007	1220	.65	.23	---	.88	1.10
	Inst	SF	Sm	CA	.009	915.0	.72	.30	---	1.02	1.28
9-1/4" T (R-30)											
16" oc	Inst	SF	Lg	CA	.010	815.0	.75	.33	---	1.08	1.36
	Inst	SF	Sm	CA	.013	611.0	.83	.43	---	1.26	1.60
24" oc	Inst	SF	Lg	CA	.007	1220	.75	.23	---	.98	1.21
	Inst	SF	Sm	CA	.009	915.0	.83	.30	---	1.13	1.40
Kraft-faced											
3-1/2" T (R-11)											
16" oc	Inst	SF	Lg	CA	.010	815.0	.38	.33	---	.71	.94
	Inst	SF	Sm	CA	.013	611.0	.42	.43	---	.85	1.13
24" oc	Inst	SF	Lg	CA	.007	1220	.38	.23	---	.61	.79
	Inst	SF	Sm	CA	.009	915.0	.42	.30	---	.72	.93
3-1/2" T (R-13)											
16" oc	Inst	SF	Lg	CA	.010	815.0	.44	.33	---	.77	1.01
	Inst	SF	Sm	CA	.013	611.0	.49	.43	---	.92	1.21
6-1/2" T (R-19)											
16" oc	Inst	SF	Lg	CA	.010	815.0	.51	.33	---	.84	1.09
	Inst	SF	Sm	CA	.013	611.0	.56	.43	---	.99	1.29
24" oc	Inst	SF	Lg	CA	.007	1220	.51	.23	---	.74	.94
	Inst	SF	Sm	CA	.009	915.0	.56	.30	---	.86	1.09

Description	Oper	Unit	Vol	Crew Size	Man-hours per Unit	Crew Output per Day	Avg Mat'l Unit Cost	Avg Labor Unit Cost	Avg Equip Unit Cost	Avg Total Unit Cost	Avg Price Incl O&P
7" T (R-22)											
16" oc	Inst	SF	Lg	CA	.010	815.0	.65	.33	---	.98	**1.25**
	Inst	SF	Sm	CA	.013	611.0	.72	.43	---	1.15	**1.48**
24" oc	Inst	SF	Lg	CA	.007	1220	.65	.23	---	.88	**1.10**
	Inst	SF	Sm	CA	.009	915.0	.72	.30	---	1.02	**1.28**
9-1/4" T (R-30)											
16" oc	Inst	SF	Lg	CA	.010	815.0	.75	.33	---	1.08	**1.36**
	Inst	SF	Sm	CA	.013	611.0	.83	.43	---	1.26	**1.60**
24" oc	Inst	SF	Lg	CA	.007	1220	.75	.23	---	.98	**1.21**
	Inst	SF	Sm	CA	.009	915.0	.83	.30	---	1.13	**1.40**
Foil-faced											
4" T (R-11)											
16" oc	Inst	SF	Lg	CA	.010	815.0	.38	.33	---	.71	**.94**
	Inst	SF	Sm	CA	.013	611.0	.42	.43	---	.85	**1.13**
24" oc	Inst	SF	Lg	CA	.007	1220	.38	.23	---	.61	**.79**
	Inst	SF	Sm	CA	.009	915.0	.42	.30	---	.72	**.93**
6-1/2" T (R-19)											
16" oc	Inst	SF	Lg	CA	.010	815.0	.51	.33	---	.84	**1.09**
	Inst	SF	Sm	CA	.013	611.0	.56	.43	---	.99	**1.29**
24" oc	Inst	SF	Lg	CA	.007	1220	.51	.23	---	.74	**.94**
	Inst	SF	Sm	CA	.009	915.0	.56	.30	---	.86	**1.09**

Loose fill

With ceiling finish already removed

Joists, 16" or 24" oc

Description	Oper	Unit	Vol	Crew Size	Man-hours per Unit	Crew Output per Day	Avg Mat'l Unit Cost	Avg Labor Unit Cost	Avg Equip Unit Cost	Avg Total Unit Cost	Avg Price Incl O&P
4" T	Demo	SF	Lg	LB	.004	3900	---	.11	---	.11	**.16**
	Demo	SF	Sm	LB	.005	2925	---	.14	---	.14	**.20**
6" T	Demo	SF	Lg	LB	.007	2340	---	.19	---	.19	**.29**
	Demo	SF	Sm	LB	.009	1755	---	.25	---	.25	**.37**

Allowance made for joists, cavities, and cores

Insulating wood, granule or pellet (40 lbs/bag, 4 CF/bag)

Joists, @ 7 lbs/CF density

Description	Oper	Unit	Vol	Crew Size	Man-hours per Unit	Crew Output per Day	Avg Mat'l Unit Cost	Avg Labor Unit Cost	Avg Equip Unit Cost	Avg Total Unit Cost	Avg Price Incl O&P
16" oc											
4" T	Inst	SF	Lg	CH	.013	960.0	.66	.41	---	1.07	**1.37**
	Inst	SF	Sm	CH	.017	720.0	.73	.53	---	1.26	**1.64**
6" T	Inst	SF	Lg	CH	.017	720.0	.99	.53	---	1.52	**1.94**
	Inst	SF	Sm	CH	.022	540.0	1.09	.69	---	1.78	**2.29**
24" oc											
4" T	Inst	SF	Lg	CH	.013	930.0	.68	.41	---	1.09	**1.39**
	Inst	SF	Sm	CH	.017	698.0	.75	.53	---	1.28	**1.66**
6" T	Inst	SF	Lg	CH	.017	690.0	1.02	.53	---	1.55	**1.97**
	Inst	SF	Sm	CH	.023	518.0	1.13	.72	---	1.85	**2.38**

Vermiculite/Perlite (approximately 10 lbs/bag, 4 CF/bag)

Joists

Description	Oper	Unit	Vol	Crew Size	Man-hours per Unit	Crew Output per Day	Avg Mat'l Unit Cost	Avg Labor Unit Cost	Avg Equip Unit Cost	Avg Total Unit Cost	Avg Price Incl O&P
16" oc											
4" T	Inst	SF	Lg	CH	.009	1340	.52	.28	---	.80	**1.02**
	Inst	SF	Sm	CH	.012	1005	.57	.38	---	.95	**1.22**
6" T	Inst	SF	Lg	CH	.012	1000	.78	.38	---	1.16	**1.46**
	Inst	SF	Sm	CH	.016	750.0	.86	.50	---	1.36	**1.74**

Description	Oper	Unit	Vol	Crew Size	Man-hours per Unit	Crew Output per Day	Avg Mat'l Unit Cost	Avg Labor Unit Cost	Avg Equip Unit Cost	Avg Total Unit Cost	Avg Price Incl O&P
24" oc											
4" T	Inst	SF	Lg	CH	.009	1300	.53	.28	---	.81	1.03
	Inst	SF	Sm	CH	.012	975.0	.59	.38	---	.97	1.24
6" T	Inst	SF	Lg	CH	.013	960.0	.80	.41	---	1.21	1.53
	Inst	SF	Sm	CH	.017	720.0	.88	.53	---	1.41	1.81
Cavity walls											
1" T	Inst	SF	Lg	CH	.004	3070	.12	.13	---	.25	.33
	Inst	SF	Sm	CH	.005	2303	.14	.16	---	.30	.40
2" T	Inst	SF	Lg	CH	.007	1690	.25	.22	---	.47	.62
	Inst	SF	Sm	CH	.009	1268	.27	.28	---	.55	.73
Block walls (2 cores/block)											
8" T block	Inst	SF	Lg	CH	.015	815.0	.43	.47	---	.90	1.20
	Inst	SF	Sm	CH	.020	611.0	.48	.63	---	1.11	1.49
12" T block	Inst	SF	Lg	CH	.020	610.0	.81	.63	---	1.44	1.87
	Inst	SF	Sm	CH	.026	458.0	.89	.81	---	1.70	2.25

Rigid

Description	Oper	Unit	Vol	Crew Size	Man-hours per Unit	Crew Output per Day	Avg Mat'l Unit Cost	Avg Labor Unit Cost	Avg Equip Unit Cost	Avg Total Unit Cost	Avg Price Incl O&P
Roofs											
1/2" T	Demo	Sq	Lg	LB	.941	17.00	---	25.80	---	25.80	38.50
	Demo	Sq	Sm	LB	1.23	13.00	---	33.70	---	33.70	50.30
1" T	Demo	Sq	Lg	LB	1.07	15.00	---	29.40	---	29.40	43.70
	Demo	Sq	Sm	LB	1.45	11.00	---	39.80	---	39.80	59.30
Walls											
1/2" T	Demo	SF	Lg	LB	.007	2140	---	.19	---	.19	.29
	Demo	SF	Sm	LB	.010	1605	---	.27	---	.27	.41

Rigid insulating board

Roofs, over wood decks, 5% waste included

Normal (dry) moisture conditions within building

Nail one ply 15 lb felt, set and nail:

Description	Oper	Unit	Vol	Crew Size	Man-hours per Unit	Crew Output per Day	Avg Mat'l Unit Cost	Avg Labor Unit Cost	Avg Equip Unit Cost	Avg Total Unit Cost	Avg Price Incl O&P
2' x 8' x 1/2" T&G asphalt sheathing											
	Inst	SF	Lg	CN	.020	1020	.60	.64	---	1.24	1.65
	Inst	SF	Sm	CN	.026	765.0	.66	.83	---	1.49	2.01
4' x 8' x 1/2" asphalt sheathing											
	Inst	SF	Lg	2C	.014	1120	.60	.47	---	1.07	1.39
	Inst	SF	Sm	2C	.019	840.0	.66	.63	---	1.29	1.71
4' x 8' x 1/2" building block	Inst	SF	Lg	2C	.014	1120	.60	.47	---	1.07	1.39
	Inst	SF	Sm	2C	.019	840.0	.66	.63	---	1.29	1.71

Excessive (humid) moisture conditions within building

Nail and overlap three plies 15 lb felt, mop laps and surface one coat and embed:

Description	Oper	Unit	Vol	Crew Size	Man-hours per Unit	Crew Output per Day	Avg Mat'l Unit Cost	Avg Labor Unit Cost	Avg Equip Unit Cost	Avg Total Unit Cost	Avg Price Incl O&P
2' x 8' x 1/2" T&G asphalt sheathing											
	Inst	SF	Lg	RT	.031	780.0	.69	1.07	---	1.76	2.46
	Inst	SF	Sm	RT	.041	585.0	.77	1.42	---	2.19	3.08
4' x 8' x 1/2" asphalt sheathing											
	Inst	SF	Lg	RT	.030	810.0	.69	1.04	---	1.73	2.40
	Inst	SF	Sm	RT	.039	608.0	.77	1.35	---	2.12	2.98
4' x 8' x 1/2" building block	Inst	SF	Lg	RT	.030	810.0	.69	1.04	---	1.73	2.40
	Inst	SF	Sm	RT	.039	608.0	.77	1.35	---	2.12	2.98

Description	Oper	Unit	Vol	Crew Size	Man-hours per Unit	Crew Output per Day	Avg Mat'l Unit Cost	Avg Labor Unit Cost	Avg Equip Unit Cost	Avg Total Unit Cost	Avg Price Incl O&P

Over noncombustible decks, 5% waste included

Normal (dry) moisture conditions within building
Mop one coat and embed:

Description	Oper	Unit	Vol	Crew Size	Man-hours per Unit	Crew Output per Day	Avg Mat'l Unit Cost	Avg Labor Unit Cost	Avg Equip Unit Cost	Avg Total Unit Cost	Avg Price Incl O&P
2' x 8' x 1/2" T&G asphalt sheathing											
	Inst	SF	Lg	RT	.022	1100	.72	.76	---	1.48	**2.01**
	Inst	SF	Sm	RT	.029	825.0	.80	1.00	---	1.80	**2.47**
4' x 8' x 1/2" asphalt sheathing											
	Inst	SF	Lg	RT	.021	1160	.72	.73	---	1.45	**1.95**
	Inst	SF	Sm	RT	.028	870.0	.80	.97	---	1.77	**2.42**
4' x 8' x 1/2" building block	Inst	SF	Lg	RT	.021	1160	.72	.73	---	1.45	**1.95**
	Inst	SF	Sm	RT	.028	870.0	.80	.97	---	1.77	**2.42**

Excessive (humid) moisture conditions within building
Mop one coat, embed two plies 15 lb felt, mop and embed:

Description	Oper	Unit	Vol	Crew Size	Man-hours per Unit	Crew Output per Day	Avg Mat'l Unit Cost	Avg Labor Unit Cost	Avg Equip Unit Cost	Avg Total Unit Cost	Avg Price Incl O&P
2' x 8' x 1/2" T&G asphalt sheathing											
	Inst	SF	Lg	RT	.038	630.0	.92	1.31	---	2.23	**3.10**
	Inst	SF	Sm	RT	.051	473.0	1.02	1.76	---	2.78	**3.91**
4' x 8' x 1/2" asphalt sheathing											
	Inst	SF	Lg	RT	.037	650.0	.92	1.28	---	2.20	**3.04**
	Inst	SF	Sm	RT	.049	488.0	1.02	1.69	---	2.71	**3.80**
4' x 8' x 1/2" building block	Inst	SF	Lg	RT	.037	650.0	.92	1.28	---	2.20	**3.04**
	Inst	SF	Sm	RT	.049	488.0	1.02	1.69	---	2.71	**3.80**

Walls, nailed, 5% waste included

Description	Oper	Unit	Vol	Crew Size	Man-hours per Unit	Crew Output per Day	Avg Mat'l Unit Cost	Avg Labor Unit Cost	Avg Equip Unit Cost	Avg Total Unit Cost	Avg Price Incl O&P
4' x 8' x 1/2" asphalt sheathing											
Straight wall	Inst	SF	Lg	2C	.011	1440	.55	.37	---	.92	**1.18**
	Inst	SF	Sm	2C	.015	1080	.61	.50	---	1.11	**1.45**
Cut-up wall	Inst	SF	Lg	2C	.013	1200	.55	.43	---	.98	**1.28**
	Inst	SF	Sm	2C	.018	900.0	.61	.60	---	1.21	**1.60**
4' x 8' x 1/2" building board											
Straight wall	Inst	SF	Lg	2C	.011	1440	.55	.37	---	.92	**1.18**
	Inst	SF	Sm	2C	.015	1080	.61	.50	---	1.11	**1.45**
Cut-up wall	Inst	SF	Lg	2C	.013	1200	.55	.43	---	.98	**1.28**
	Inst	SF	Sm	2C	.018	900.0	.61	.60	---	1.21	**1.60**

Intercom systems. See Electrical, page 154
Jacuzzi whirlpools. See Spas, page 417
Lath & plaster. See Plaster, page 337

Description	Oper	Unit	Vol	Crew Size	Man-hours per Unit	Crew Output per Day	Avg Mat'l Unit Cost	Avg Labor Unit Cost	Avg Equip Unit Cost	Avg Total Unit Cost	Avg Price Incl O&P

Lighting fixtures

Labor includes hanging and connecting fixtures

Indoor lighting

Fluorescent

Pendant mounted worklights

Description	Oper	Unit	Vol	Crew Size	Man-hours per Unit	Crew Output per Day	Avg Mat'l Unit Cost	Avg Labor Unit Cost	Avg Equip Unit Cost	Avg Total Unit Cost	Avg Price Incl O&P
4' L, two 40 watt RS	Inst	Ea	Lg	EA	.800	10.00	52.00	28.50	---	80.50	**101.00**
	Inst	Ea	Sm	EA	1.14	7.00	58.50	40.60	---	99.10	**127.00**
4' L, two 60 watt RS	Inst	Ea	Lg	EA	.800	10.00	84.70	28.50	---	113.20	**139.00**
	Inst	Ea	Sm	EA	1.14	7.00	95.20	40.60	---	135.80	**169.00**
8' L, two 75 watt RS	Inst	Ea	Lg	EA	1.00	8.00	98.00	35.60	---	133.60	**165.00**
	Inst	Ea	Sm	EA	1.43	5.60	110.00	51.00	---	161.00	**201.00**

Recessed mounted light fixture with acrylic diffuser

Description	Oper	Unit	Vol	Crew Size	Man-hours per Unit	Crew Output per Day	Avg Mat'l Unit Cost	Avg Labor Unit Cost	Avg Equip Unit Cost	Avg Total Unit Cost	Avg Price Incl O&P
1' W x 4' L, two 40 watt RS	Inst	Ea	Lg	EA	.800	10.00	55.70	28.50	---	84.20	**106.00**
	Inst	Ea	Sm	EA	1.14	7.00	62.60	40.60	---	103.20	**131.00**
2' W x 2' L, two U 40 watt RS	Inst	Ea	Lg	EA	.800	10.00	60.50	28.50	---	89.00	**111.00**
	Inst	Ea	Sm	EA	1.14	7.00	68.00	40.60	---	108.60	**138.00**
2' W x 4' L, four 40 watt RS	Inst	Ea	Lg	EA	1.00	8.00	67.80	35.60	---	103.40	**130.00**
	Inst	Ea	Sm	EA	1.43	5.60	76.20	51.00	---	127.20	**162.00**

Strip lighting fixtures

Description	Oper	Unit	Vol	Crew Size	Man-hours per Unit	Crew Output per Day	Avg Mat'l Unit Cost	Avg Labor Unit Cost	Avg Equip Unit Cost	Avg Total Unit Cost	Avg Price Incl O&P
4' L, one 40 watt RS	Inst	Ea	Lg	EA	.800	10.00	31.50	28.50	---	60.00	**77.80**
	Inst	Ea	Sm	EA	1.14	7.00	35.40	40.60	---	76.00	**100.00**
4' L, two 40 watt RS	Inst	Ea	Lg	EA	.800	10.00	33.90	28.50	---	62.40	**80.60**
	Inst	Ea	Sm	EA	1.14	7.00	38.10	40.60	---	78.70	**103.00**
8' L, one 75 watt SL	Inst	Ea	Lg	EA	1.00	8.00	47.20	35.60	---	82.80	**106.00**
	Inst	Ea	Sm	EA	1.43	5.60	53.00	51.00	---	104.00	**135.00**
8' L, two 75 watt SL	Inst	Ea	Lg	EA	1.00	8.00	56.90	35.60	---	92.50	**117.00**
	Inst	Ea	Sm	EA	1.43	5.60	63.90	51.00	---	114.90	**148.00**

Surface mounted, acrylic diffuser

Description	Oper	Unit	Vol	Crew Size	Man-hours per Unit	Crew Output per Day	Avg Mat'l Unit Cost	Avg Labor Unit Cost	Avg Equip Unit Cost	Avg Total Unit Cost	Avg Price Incl O&P
1' W x 4' L, two 40 watt RS	Inst	Ea	Lg	EA	.800	10.00	82.30	28.50	---	110.80	**136.00**
	Inst	Ea	Sm	EA	1.14	7.00	92.50	40.60	---	133.10	**166.00**
2' W x 2' L, two U 40 watt RS	Inst	Ea	Lg	EA	.800	10.00	103.00	28.50	---	131.50	**160.00**
	Inst	Ea	Sm	EA	1.14	7.00	116.00	40.60	---	156.60	**192.00**
2' W x 4' L, four 40 watt RS	Inst	Ea	Lg	EA	1.00	8.00	105.00	35.60	---	140.60	**173.00**
	Inst	Ea	Sm	EA	1.43	5.60	118.00	51.00	---	160.00	**210.00**

White enameled circline steel ceiling fixtures

Description	Oper	Unit	Vol	Crew Size	Man-hours per Unit	Crew Output per Day	Avg Mat'l Unit Cost	Avg Labor Unit Cost	Avg Equip Unit Cost	Avg Total Unit Cost	Avg Price Incl O&P
8" W, 22 watt RS	Inst	Ea	Lg	EA	.667	12.00	29.50	23.80	---	53.30	**68.60**
	Inst	Ea	Sm	EA	0.95	8.40	33.10	33.90	---	67.00	**87.50**
12" W, 22 to 32 watt RS	Inst	Ea	Lg	EA	.667	12.00	44.20	23.80	---	68.00	**85.50**
	Inst	Ea	Sm	EA	0.95	8.40	49.70	33.90	---	83.60	**107.00**
16" W, 22 to 40 watt RS	Inst	Ea	Lg	EA	.667	12.00	59.00	23.80	---	82.80	**102.00**
	Inst	Ea	Sm	EA	0.95	8.40	66.30	33.90	---	100.20	**126.00**

Decorative circline fixtures

Description	Oper	Unit	Vol	Crew Size	Man-hours per Unit	Crew Output per Day	Avg Mat'l Unit Cost	Avg Labor Unit Cost	Avg Equip Unit Cost	Avg Total Unit Cost	Avg Price Incl O&P
12" W, 22 to 32 watt RS	Inst	Ea	Lg	EA	.667	12.00	115.00	23.80	---	138.80	**167.00**
	Inst	Ea	Sm	EA	0.95	8.40	129.00	33.90	---	162.90	**198.00**
16" W, 22 to 40 watt RS	Inst	Ea	Lg	EA	.667	12.00	199.00	23.80	---	222.80	**264.00**
	Inst	Ea	Sm	EA	0.95	8.40	224.00	33.90	---	257.90	**307.00**

Incandescent

Description	Oper	Unit	Vol	Crew Size	Man-hours per Unit	Crew Output per Day	Avg Mat'l Unit Cost	Avg Labor Unit Cost	Avg Equip Unit Cost	Avg Total Unit Cost	Avg Price Incl O&P
Ceiling fixture, surface mounted											
15" x 5", white bent glass	Inst	Ea	Lg	EA	.667	12.00	33.90	23.80	---	57.70	73.70
	Inst	Ea	Sm	EA	0.95	8.40	38.10	33.90	---	72.00	93.20
10" x 7", two light, circular	Inst	Ea	Lg	EA	.667	12.00	42.70	23.80	---	66.50	83.90
	Inst	Ea	Sm	EA	0.95	8.40	48.00	33.90	---	81.90	105.00
8" x 8", one light, screw-in	Inst	Ea	Lg	EA	.667	12.00	28.00	23.80	---	51.80	66.90
	Inst	Ea	Sm	EA	0.95	8.40	31.50	33.90	---	65.40	85.60
Ceiling fixture, recessed											
Square fixture, drop or flat lens											
8" frame	Inst	Ea	Lg	EA	1.00	8.00	42.70	35.60	---	78.30	101.00
	Inst	Ea	Sm	EA	1.43	5.60	48.00	51.00	---	99.00	130.00
10" frame	Inst	Ea	Lg	EA	1.00	8.00	48.60	35.60	---	84.20	108.00
	Inst	Ea	Sm	EA	1.43	5.60	54.70	51.00	---	105.70	137.00
12" frame	Inst	Ea	Lg	EA	1.00	8.00	51.60	35.60	---	87.20	111.00
	Inst	Ea	Sm	EA	1.43	5.60	58.00	51.00	---	109.00	141.00
Round fixture for concentrated light over small areas											
7" shower fixture frame	Inst	Ea	Lg	EA	1.00	8.00	42.70	35.60	---	78.30	101.00
	Inst	Ea	Sm	EA	1.43	5.60	48.00	51.00	---	99.00	130.00
8" spotlight fixture	Inst	Ea	Lg	EA	1.00	8.00	41.30	35.60	---	76.90	99.50
	Inst	Ea	Sm	EA	1.43	5.60	46.40	51.00	---	97.40	128.00
8" flat lens or stepped baffle frame											
	Inst	Ea	Lg	EA	1.00	8.00	42.70	35.60	---	78.30	101.00
	Inst	Ea	Sm	EA	1.43	5.60	48.00	51.00	---	99.00	130.00
Track lighting for highlighting effects from a ceiling or wall											
Swivel track heads with the ability to slide head along track to a new position											
Track heads											
Large cylinder	Inst	Ea	Lg	EA	---	---	50.10	---	---	50.10	50.10
	Inst	Ea	Sm	EA	---	---	56.30	---	---	56.30	56.30
Small cylinder	Inst	Ea	Lg	EA	---	---	41.30	---	---	41.30	41.30
	Inst	Ea	Sm	EA	---	---	46.40	---	---	46.40	46.40
Sphere cylinder	Inst	Ea	Lg	EA	---	---	50.10	---	---	50.10	50.10
	Inst	Ea	Sm	EA	---	---	56.30	---	---	56.30	56.30
Track, 1-7/16" W x 3/4" D											
2' track	Inst	Ea	Lg	EA	.333	24.00	20.60	11.90	---	32.50	41.10
	Inst	Ea	Sm	EA	.476	16.80	23.20	17.00	---	40.20	51.40
4' track	Inst	Ea	Lg	EA	.500	16.00	51.60	17.80	---	69.40	85.30
	Inst	Ea	Sm	EA	.714	11.20	58.00	25.40	---	83.40	104.00
8' track	Inst	Ea	Lg	EA	.667	12.00	76.60	23.80	---	100.40	123.00
	Inst	Ea	Sm	EA	0.95	8.40	86.10	33.90	---	120.00	148.00
Straight connector; joins two track sections end to end											
7-5/8" L	Inst	Ea	Lg	EA	---	---	14.70	---	---	14.70	14.70
	Inst	Ea	Sm	EA	---	---	16.60	---	---	16.60	16.60
L - connector, joins two track sections for 90-degree angle turns											
7-5/8" L	Inst	Ea	Lg	EA	---	---	14.70	---	---	14.70	14.70
	Inst	Ea	Sm	EA	---	---	16.60	---	---	16.60	16.60
Feed in unit, attaches to ceiling or wall outlet box, supplies electrical current to all heads on track(s)											
	Inst	Ea	Lg	EA	.667	12.00	16.20	23.80	---	40.00	53.30
	Inst	Ea	Sm	EA	0.95	8.40	18.20	33.90	---	52.10	70.40

Description	Oper	Unit	Vol	Crew Size	Man-hours per Unit	Crew Output per Day	Avg Mat'l Unit Cost	Avg Labor Unit Cost	Avg Equip Unit Cost	Avg Total Unit Cost	Avg Price Incl O&P
Wall fixtures											
White glass with on/off switch											
1 light, 5" W x 5" H	Inst	Ea	Lg	EA	.667	12.00	16.20	23.80	---	40.00	**53.30**
	Inst	Ea	Sm	EA	0.95	8.40	18.20	33.90	---	52.10	**70.40**
2 light, 14" W x 5" H	Inst	Ea	Lg	EA	.667	12.00	25.10	23.80	---	48.90	**63.50**
	Inst	Ea	Sm	EA	0.95	8.40	28.20	33.90	---	62.10	**81.80**
4 light, 24" W x 4" H	Inst	Ea	Lg	EA	.667	12.00	36.80	23.80	---	60.60	**77.10**
	Inst	Ea	Sm	EA	0.95	8.40	41.40	33.90	---	75.30	**97.00**
Swivel wall fixture											
1 light, 4" W x 8" H	Inst	Ea	Lg	EA	.667	12.00	25.40	23.80	---	49.20	**63.90**
	Inst	Ea	Sm	EA	0.95	8.40	28.60	33.90	---	62.50	**82.30**
2 light, 9" W x 9" H	Inst	Ea	Lg	EA	.667	12.00	44.20	23.80	---	68.00	**85.50**
	Inst	Ea	Sm	EA	0.95	8.40	49.70	33.90	---	83.60	**107.00**

Outdoor lighting

Description	Oper	Unit	Vol	Crew Size	Man-hours per Unit	Crew Output per Day	Avg Mat'l Unit Cost	Avg Labor Unit Cost	Avg Equip Unit Cost	Avg Total Unit Cost	Avg Price Incl O&P
Ceiling fixture for porch											
7" W x 13" L, one 75 watt RS	Inst	Ea	Lg	EA	.667	12.00	53.10	23.80	---	76.90	**95.70**
	Inst	Ea	Sm	EA	0.95	8.40	59.60	33.90	---	93.50	**118.00**
11" W x 4" L, two 60 watt RS	Inst	Ea	Lg	EA	.667	12.00	64.80	23.80	---	88.60	**109.00**
	Inst	Ea	Sm	EA	0.95	8.40	72.90	33.90	---	106.80	**133.00**
15" W x 4" L, three 60 watt RS											
	Inst	Ea	Lg	EA	.667	12.00	66.30	23.80	---	90.10	**111.00**
	Inst	Ea	Sm	EA	0.95	8.40	74.50	33.90	---	108.40	**135.00**
Wall fixture for porch											
6" W x 10" L, one 75 watt RS	Inst	Ea	Lg	EA	.667	12.00	57.50	23.80	---	81.30	**101.00**
	Inst	Ea	Sm	EA	0.95	8.40	64.60	33.90	---	98.50	**124.00**
6" W x 16" L, one 75 watt RS	Inst	Ea	Lg	EA	.667	12.00	72.20	23.80	---	96.00	**118.00**
	Inst	Ea	Sm	EA	0.95	8.40	81.20	33.90	---	115.10	**143.00**
Post lantern fixture											
Aluminum cast posts											
84" H post with lantern, set in concrete, includes 50' of conduit, circuit, and cement											
	Inst	Ea	Lg	EA	8.00	1.00	339.00	285.00	---	624.00	**000.00**
	Inst	Ea	Sm	EA	11.4	0.70	381.00	406.00	---	787.00	**1030.00**
Urethane (in form of simulated redwood) over steel post											
Post with matching lantern, set in concrete, includes 50' of conduit, circuit, and cement											
	Inst	Ea	Lg	EA	8.00	1.00	405.00	285.00	---	690.00	**882.00**
	Inst	Ea	Sm	EA	11.4	0.70	456.00	406.00	---	862.00	**1120.00**

Linoleum. See Resilient flooring, page 342

Lumber. See Framing, page 171

Description	Oper	Unit	Vol	Crew Size	Man-hours per Unit	Crew Output per Day	Avg Mat'l Unit Cost	Avg Labor Unit Cost	Avg Equip Unit Cost	Avg Total Unit Cost	Avg Price Incl O&P

Mantels, fireplace

Ponderosa pine, kiln-dried, unfinished, assembled

Versailles, ornate, French design

Description	Oper	Unit	Vol	Crew Size	Man-hours per Unit	Crew Output per Day	Avg Mat'l Unit Cost	Avg Labor Unit Cost	Avg Equip Unit Cost	Avg Total Unit Cost	Avg Price Incl O&P
59" W x 46" H	Inst	Ea	Lg	CJ	4.00	4.00	1750.00	121.00	---	1871.00	**2190.00**
	Inst	Ea	Sm	CJ	5.71	2.80	2090.00	173.00	---	2263.00	**2660.00**

Victorian, ornate, English design

Description	Oper	Unit	Vol	Crew Size	Man-hours per Unit	Crew Output per Day	Avg Mat'l Unit Cost	Avg Labor Unit Cost	Avg Equip Unit Cost	Avg Total Unit Cost	Avg Price Incl O&P
63" W x 52" H	Inst	Ea	Lg	CJ	4.00	4.00	1180.00	121.00	---	1301.00	**1540.00**
	Inst	Ea	Sm	CJ	5.71	2.80	1410.00	173.00	---	1583.00	**1880.00**

Chelsea, plain, English design

Description	Oper	Unit	Vol	Crew Size	Man-hours per Unit	Crew Output per Day	Avg Mat'l Unit Cost	Avg Labor Unit Cost	Avg Equip Unit Cost	Avg Total Unit Cost	Avg Price Incl O&P
70" W x 52" H	Inst	Ea	Lg	CJ	4.00	4.00	938.00	121.00	---	1059.00	**1260.00**
	Inst	Ea	Sm	CJ	5.71	2.80	1120.00	173.00	---	1293.00	**1550.00**

Jamestown, plain, Early American

Description	Oper	Unit	Vol	Crew Size	Man-hours per Unit	Crew Output per Day	Avg Mat'l Unit Cost	Avg Labor Unit Cost	Avg Equip Unit Cost	Avg Total Unit Cost	Avg Price Incl O&P
68" W x 53" H	Inst	Ea	Lg	CJ	4.00	4.00	585.00	121.00	---	706.00	**855.00**
	Inst	Ea	Sm	CJ	5.71	2.80	699.00	173.00	---	872.00	**1060.00**

Marlite panels

Panels, 4' x 8', adhesive set

Description	Oper	Unit	Vol	Crew Size	Man-hours per Unit	Crew Output per Day	Avg Mat'l Unit Cost	Avg Labor Unit Cost	Avg Equip Unit Cost	Avg Total Unit Cost	Avg Price Incl O&P
	Demo	SF	Lg	LB	.009	1850	---	.25	---	.25	**.37**
	Demo	SF	Sm	LB	.012	1295	---	.33	---	.33	**.49**

Plastic-coated masonite panels; 4' x 8', 5' x 5'; screw applied; channel molding around perimeter; 1/8" T

Description	Oper	Unit	Vol	Crew Size	Man-hours per Unit	Crew Output per Day	Avg Mat'l Unit Cost	Avg Labor Unit Cost	Avg Equip Unit Cost	Avg Total Unit Cost	Avg Price Incl O&P
Solid colors	Inst	SF	Lg	CJ	.047	340.0	2.48	1.43	---	3.91	**4.99**
	Inst	SF	Sm	CJ	.067	238.0	2.84	2.03	---	4.87	**6.32**
Patterned panels	Inst	SF	Lg	CJ	.047	340.0	2.71	1.43	---	4.14	**5.26**
	Inst	SF	Sm	CJ	.067	238.0	3.11	2.03	---	5.14	**6.63**

Molding, 1/8" panels; corners, divisions, or edging; nailed to framing or sheathing

Description	Oper	Unit	Vol	Crew Size	Man-hours per Unit	Crew Output per Day	Avg Mat'l Unit Cost	Avg Labor Unit Cost	Avg Equip Unit Cost	Avg Total Unit Cost	Avg Price Incl O&P
Bright anodized	Inst	LF	Lg	CJ	.063	255.0	.15	1.91	---	2.06	**3.04**
	Inst	LF	Sm	CJ	.090	178.5	.18	2.73	---	2.91	**4.30**
Gold anodized	Inst	LF	Lg	CJ	.063	255.0	.17	1.91	---	2.08	**3.06**
	Inst	LF	Sm	CJ	.090	178.5	.20	2.73	---	2.93	**4.33**
Colors	Inst	LF	Lg	CJ	.063	255.0	.22	1.91	---	2.13	**3.12**
	Inst	LF	Sm	CJ	.090	178.5	.25	2.73	---	2.98	**4.39**

Masonry

Brick. All material costs include mortar and waste, 5% on brick and 30% on mortar.

1. **Dimensions**

    a. Standard or regular: 8" L x $2^1/4$" H x $3^3/4$" W.

    b. Modular: $7^5/8$" L x $2^1/4$" H x $3^5/8$" W.

    c. Norman: $11^5/8$" L x $2^2/3$" H x $3^5/8$" W.

    d. Roman: $11^5/8$" L x $1^5/8$" H x $3^5/8$" W.

2. **Installation**

    a. Mortar joints are $3/8$" thick both horizontally and vertically.

    b. Mortar mix is 1:3, 1 part masonry cement and 3 parts sand.

    c. Galvanized, corrugated wall ties are used on veneers at 1 tie per SF wall area. The tie is $7/8$" x 7" x 16 ga.

    d. Running bond used on veneers and walls.

3. **Notes on Labor.** Output is based on a crew composed of bricklayers and bricktenders at a 1:1 ratio.

4. **Estimating Technique.** Chimneys and columns figured per vertical linear foot with allowances already made for brick waste and mortar waste. Veneers and walls are computed per square foot of wall area.

Concrete (Masonry) Block. All material costs include 3% block waste and 30% mortar waste.

1. **Dimension.** All blocks are two core.

    a. Heavyweight: 8" T blocks weigh approximately 46 lbs/block; 12" T blocks weigh approximately 65 lbs/block.

    b. Lightweight: Also known as haydite blocks. 8" T blocks weigh approximately 30 lbs/block; 12" T blocks weigh approximately 41 lbs/block.

2. **Installation**

    a. Mortar joints are $3/8$" T both horizontally and vertically.

    b. Mortar mix is 1:3, 1 part masonry cement and 3 parts sand.

    c. Reinforcing: Lateral metal is regular truss with 9 gauge sides and ties. Vertical steel is #4 ($1/2$" dia.) rods, at 0.668 lbs/LF.

3. **Notes on Labor.** Output is based on a crew composed of bricklayers and bricktenders.

4. **Estimating Technique.** Figure chimneys and columns per vertical linear foot, with allowances already made for block waste and mortar waste. Veneers and walls are computed per square foot of wall area.

Quarry Tile (on Floor). Includes 5% tile waste.

1. **Dimensions:** 6" square tile is $1/2$" T and 9" square tile is $3/4$" T.

2. **Installation**

    a. Conventional mortar set utilizes portland cement, mortar mix, sand, and water. The mortar dry-cures and bonds to tile.

    b. Dry-set mortar utilizes dry-set portland cement, mortar mix, sand, and water. The mortar dry-cures and bonds to tile.

3. **Notes on Labor.** Output is based on a crew composed of bricklayers and bricktenders.

4. **Estimating Technique.** Compute square feet of floor area.

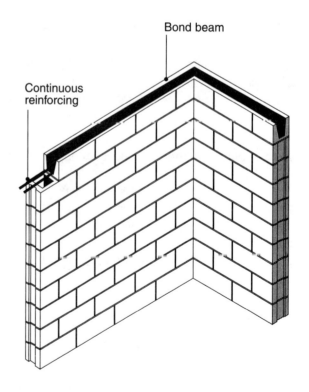

Typical concrete block wall

4" wall thickness

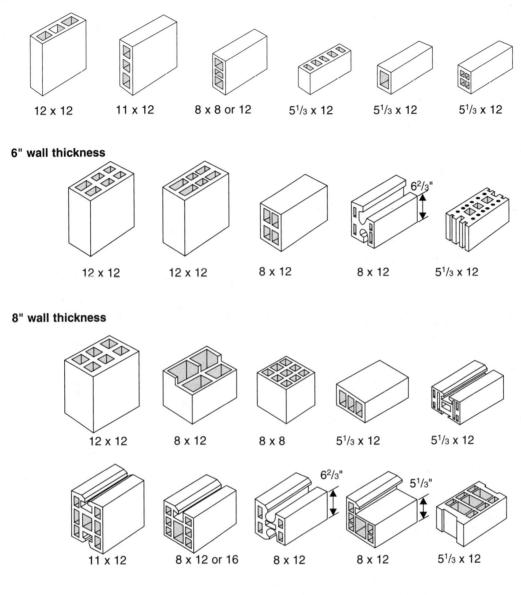

12 x 12   11 x 12   8 x 8 or 12   5$\frac{1}{3}$ x 12   5$\frac{1}{3}$ x 12   5$\frac{1}{3}$ x 12

6" wall thickness

12 x 12   12 x 12   8 x 12   8 x 12   5$\frac{1}{3}$ x 12

6$\frac{2}{3}$"

8" wall thickness

12 x 12   8 x 12   8 x 8   5$\frac{1}{3}$ x 12   5$\frac{1}{3}$ x 12

11 x 12   8 x 12 or 16   8 x 12   8 x 12   5$\frac{1}{3}$ x 12

6$\frac{2}{3}$"   5$\frac{1}{3}$"

10" wall thickness

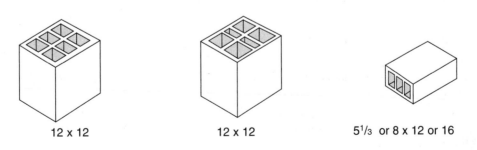

12 x 12   12 x 12   5$\frac{1}{3}$ or 8 x 12 or 16

Common sizes and shapes for clay tile

Description	Oper	Unit	Vol	Crew Size	Man-hours per Unit	Crew Output per Day	Avg Mat'l Unit Cost	Avg Labor Unit Cost	Avg Equip Unit Cost	Avg Total Unit Cost	Avg Price Incl O&P

Masonry

Brick, standard

Running bond, 3/8" mortar joints

Chimneys

Flue lining included; no scaffolding included

4" T wall with standard brick

Description	Oper	Unit	Vol	Crew Size	Man-hours per Unit	Crew Output per Day	Avg Mat'l Unit Cost	Avg Labor Unit Cost	Avg Equip Unit Cost	Avg Total Unit Cost	Avg Price Incl O&P
16" x 16" with one 8" x 8" flue											
	Demo	VLF	Lg	AB	1.04	15.40	---	30.70	3.27	33.97	**48.70**
	Demo	VLF	Sm	LB	1.48	10.78	---	40.60	---	40.60	**60.50**
20" x 16" with one 12" x 8" flue											
	Demo	VLF	Lg	AB	1.20	13.30	---	35.40	3.78	39.18	**56.20**
	Demo	VLF	Sm	LB	1.72	9.31	---	47.20	---	47.20	**70.30**
20" x 20" with one 12" x 12" flue											
	Demo	VLF	Lg	AB	1.31	12.20	---	38.60	4.13	42.73	**61.30**
	Demo	VLF	Sm	LB	1.87	8.54	---	51.30	---	51.30	**76.40**
28" x 16" with two 8" x 8" flues											
	Demo	VLF	Lg	AB	1.60	10.00	---	47.20	5.03	52.23	**74.90**
	Demo	VLF	Sm	LB	2.29	7.00	---	62.80	---	62.80	**93.60**
32" x 16" with one 8" x 8" and one 12" x 8" flue											
	Demo	VLF	Lg	AB	1.70	9.40	---	50.10	5.36	55.46	**79.60**
	Demo	VLF	Sm	LB	2.43	6.58	---	66.70	---	66.70	**99.30**
36" x 16" with two 12" x 8" flues											
	Demo	VLF	Lg	AB	1.82	8.80	---	53.70	5.72	59.42	**85.20**
	Demo	VLF	Sm	LB	2.60	6.16	---	71.30	---	71.30	**106.00**
36" x 20" with two 12" x 12" flues											
	Demo	VLF	Lg	AB	2.03	7.90	---	59.90	6.37	66.27	**95.00**
	Demo	VLF	Sm	LB	2.89	5.53	---	79.30	---	79.30	**118.00**
16" x 16" with one 8" x 8" flue											
	Inst	VLF	Lg	BK	1.12	14.30	24.30	33.30	---	57.60	**77.50**
	Inst	VLF	Sm	BK	1.60	10.01	28.10	47.50	---	75.60	**103.00**
20" x 16" with one 12" x 8" flue											
	Inst	VLF	Lg	BK	1.27	12.60	32.10	37.70	---	69.80	**93.10**
	Inst	VLF	Sm	BK	1.81	8.82	37.00	53.80	---	90.80	**123.00**
20" x 20" with one 12" x 12" flue											
	Inst	VLF	Lg	BK	1.42	11.30	36.40	42.20	---	78.60	**105.00**
	Inst	VLF	Sm	BK	2.02	7.91	42.00	60.00	---	102.00	**138.00**
28" x 16" with two 8" x 8" flues											
	Inst	VLF	Lg	BK	1.70	9.40	44.30	50.50	---	94.80	**126.00**
	Inst	VLF	Sm	BK	2.43	6.58	51.10	72.20	---	123.30	**166.00**
32" x 16" with one 8" x 8" and one 12" x 8" flue											
	Inst	VLF	Lg	BK	1.82	8.80	51.60	54.10	---	105.70	**140.00**
	Inst	VLF	Sm	BK	2.60	6.16	59.60	77.30	---	136.90	**184.00**
36" x 16" with two 12" x 8" flues											
	Inst	VLF	Lg	BK	1.95	8.20	59.40	57.90	---	117.30	**155.00**
	Inst	VLF	Sm	BK	2.79	5.74	68.50	82.90	---	151.40	**202.00**
36" x 20" with two 12" x 12" flues											
	Inst	VLF	Lg	BK	2.16	7.40	68.60	64.20	---	132.80	**174.00**
	Inst	VLF	Sm	BK	3.09	5.18	79.10	91.80	---	170.90	**228.00**

Description	Oper	Unit	Vol	Crew Size	Man-hours per Unit	Crew Output per Day	Avg Mat'l Unit Cost	Avg Labor Unit Cost	Avg Equip Unit Cost	Avg Total Unit Cost	Avg Price Incl O&P
8" T wall with standard brick											
24" x 24" with one 8" x 8" flue	Demo	VLF	Lg	AB	2.22	7.20	---	65.50	6.99	72.49	**104.00**
	Demo	VLF	Sm	LB	3.17	5.04	---	87.00	---	87.00	**130.00**
28" x 24" with one 12" x 8" flue	Demo	VLF	Lg	AB	2.39	6.70	---	70.50	7.51	78.01	**112.00**
	Demo	VLF	Sm	LB	3.41	4.69	---	93.50	---	93.50	**139.00**
28" x 28" with one 12" x 12" flue	Demo	VLF	Lg	AB	2.62	6.10	---	77.30	8.25	85.55	**123.00**
	Demo	VLF	Sm	LB	3.75	4.27	---	103.00	---	103.00	**153.00**
36" x 24" with two 8" x 8" flues	Demo	VLF	Lg	AB	2.91	5.50	---	85.80	9.15	94.95	**136.00**
	Demo	VLF	Sm	LB	4.16	3.85	---	114.00	---	114.00	**170.00**
40" x 24" with one 8" x 8" and one 12" x 8" flue											
	Demo	VLF	Lg	AB	3.20	5.00	---	94.40	10.10	104.50	**150.00**
	Demo	VLF	Sm	LB	4.57	3.50	---	125.00	---	125.00	**187.00**
44" x 24" with two 12" x 8" flues	Demo	VLF	Lg	AB	3.33	4.80	---	98.20	10.50	108.70	**156.00**
	Demo	VLF	Sm	LB	4.76	3.36	---	131.00	---	131.00	**195.00**
44" x 28" with two 12" x 12" flues	Demo	VLF	Lg	AB	3.56	4.50	---	105.00	11.20	116.20	**167.00**
	Demo	VLF	Sm	LB	5.08	3.15	---	139.00	---	139.00	**208.00**
24" x 24" with one 8" x 8" flue	Inst	VLF	Lg	BK	2.50	6.40	46.30	74.30	---	120.60	**164.00**
	Inst	VLF	Sm	BK	3.57	4.48	53.50	106.00	---	159.50	**220.00**
28" x 24" with one 12" x 8" flue	Inst	VLF	Lg	BK	2.76	5.80	55.90	82.00	---	137.90	**186.00**
	Inst	VLF	Sm	BK	3.94	4.06	64.70	117.00	---	181.70	**249.00**
28" x 28" with one 12" x 12" flue	Inst	VLF	Lg	BK	2.96	5.40	62.60	87.90	---	150.50	**203.00**
	Inst	VLF	Sm	BK	4.23	3.78	72.40	126.00	---	198.40	**270.00**
36" x 24" with two 8" x 8" flues	Inst	VLF	Lg	BK	3.33	4.80	72.40	98.90	---	171.30	**231.00**
	Inst	VLF	Sm	BK	4.76	3.36	83.70	141.00	---	224.70	**307.00**
40" x 24" with one 8" x 8" and one 12" x 8" flue											
	Inst	VLF	Lg	BK	3.72	4.30	84.40	111.00	---	195.40	**262.00**
	Inst	VLF	Sm	BK	5.32	3.01	97.60	158.00	---	255.60	**348.00**
44" x 24" with two 12" x 8" flues	Inst	VLF	Lg	BK	3.81	4.20	92.20	113.00	---	205.20	**275.00**
	Inst	VLF	Sm	BK	5.44	2.94	106.00	162.00	---	268.00	**363.00**
44" x 28" with two 12" x 12" flues	Inst	VLF	Lg	BK	4.10	3.90	104.00	122.00	---	226.00	**301.00**
	Inst	VLF	Sm	BK	5.86	2.73	120.00	174.00	---	294.00	**397.00**

Columns

Outside dimension; no shoring; solid centers; no scaffolding included

Description	Oper	Unit	Vol	Crew Size	Man-hours per Unit	Crew Output per Day	Avg Mat'l Unit Cost	Avg Labor Unit Cost	Avg Equip Unit Cost	Avg Total Unit Cost	Avg Price Incl O&P
8" x 8"	Demo	VLF	Lg	AB	.331	48.30	---	9.76	1.04	10.80	**15.50**
	Demo	VLF	Sm	LB	.473	33.81	---	13.00	---	13.00	**19.30**
12" x 8"	Demo	VLF	Lg	AB	.546	29.30	---	16.10	1.72	17.82	**25.60**
	Demo	VLF	Sm	LB	.780	20.51	---	21.40	---	21.40	**31.90**
16" x 8"	Demo	VLF	Lg	AB	.711	22.50	---	21.00	2.24	23.24	**33.30**
	Demo	VLF	Sm	LB	1.02	15.75	---	28.00	---	28.00	**41.70**
20" x 8"	Demo	VLF	Lg	AB	.870	18.40	---	25.70	2.74	28.44	**40.70**
	Demo	VLF	Sm	LB	1.24	12.88	---	34.00	---	34.00	**50.70**
24" x 8"	Demo	VLF	Lg	AB	1.02	15.70	---	30.10	3.21	33.31	**47.70**
	Demo	VLF	Sm	LB	1.46	10.99	---	40.10	---	40.10	**59.70**

Description	Oper	Unit	Vol	Crew Size	Man-hours per Unit	Crew Output per Day	Avg Mat'l Unit Cost	Avg Labor Unit Cost	Avg Equip Unit Cost	Avg Total Unit Cost	Avg Price Incl O&P
Columns (continued)											
12" x 12"	Demo	VLF	Lg	AB	.792	20.20	---	23.40	2.49	25.89	**37.10**
	Demo	VLF	Sm	LB	1.13	14.14	---	31.00	---	31.00	**46.20**
16" x 12"	Demo	VLF	Lg	AB	1.02	15.70	---	30.10	3.21	33.31	**47.70**
	Demo	VLF	Sm	LB	1.46	10.99	---	40.10	---	40.10	**59.70**
20" x 12"	Demo	VLF	Lg	AB	1.23	13.00	---	36.30	3.87	40.17	**57.60**
	Demo	VLF	Sm	LB	1.76	9.10	---	48.30	---	48.30	**71.90**
24" x 12"	Demo	VLF	Lg	AB	1.43	11.20	---	42.20	4.49	46.69	**66.90**
	Demo	VLF	Sm	LB	2.04	7.84	---	56.00	---	56.00	**83.40**
28" x 12"	Demo	VLF	Lg	AB	1.62	9.90	---	47.80	5.08	52.88	**75.80**
	Demo	VLF	Sm	LB	2.31	6.93	---	63.40	---	63.40	**94.40**
32" x 12"	Demo	VLF	Lg	AB	1.80	8.90	---	53.10	5.66	58.76	**84.20**
	Demo	VLF	Sm	LB	2.57	6.23	---	70.50	---	70.50	**105.00**
16" x 16"	Demo	VLF	Lg	AB	1.30	12.30	---	38.30	4.09	42.39	**60.80**
	Demo	VLF	Sm	AB	1.86	8.61	---	54.90	5.85	60.75	**87.00**
20" x 16"	Demo	VLF	Lg	AB	1.55	10.30	---	45.70	4.89	50.59	**72.60**
	Demo	VLF	Sm	AB	2.22	7.21	---	65.50	6.98	72.48	**104.00**
24" x 16"	Demo	VLF	Lg	AB	1.80	8.90	---	53.10	5.66	58.76	**84.20**
	Demo	VLF	Sm	AB	2.57	6.23	---	75.80	8.08	83.88	**120.00**
28" x 16"	Demo	VLF	Lg	AB	2.00	8.00	---	59.00	6.29	65.29	**93.60**
	Demo	VLF	Sm	AB	2.86	5.60	---	84.30	8.99	93.29	**134.00**
32" x 16"	Demo	VLF	Lg	AB	2.22	7.20	---	65.50	6.99	72.49	**104.00**
	Demo	VLF	Sm	AB	3.17	5.04	---	93.50	9.99	103.49	**148.00**
36" x 16"	Demo	VLF	Lg	AB	2.39	6.70	---	70.50	7.51	78.01	**112.00**
	Demo	VLF	Sm	AB	3.41	4.69	---	101.00	10.70	111.70	**160.00**
20" x 20"	Demo	VLF	Lg	AB	1.84	8.70	---	54.30	5.79	60.09	**86.10**
	Demo	VLF	Sm	AB	2.63	6.09	---	77.60	8.27	85.87	**123.00**
24" x 20"	Demo	VLF	Lg	AB	2.16	7.40	---	63.70	6.80	70.50	**101.00**
	Demo	VLF	Sm	AB	3.09	5.18	---	91.10	9.72	100.82	**145.00**
28" x 20"	Demo	VLF	Lg	AB	2.39	6.70	---	70.50	7.51	78.01	**112.00**
	Demo	VLF	Sm	AB	3.41	4.69	---	101.00	10.70	111.70	**160.00**
32" x 20"	Demo	VLF	Lg	AB	2.62	6.10	---	77.30	8.25	85.55	**123.00**
	Demo	VLF	Sm	AB	3.75	4.27	---	111.00	11.80	122.80	**175.00**
36" x 20"	Demo	VLF	Lg	AB	2.81	5.70	---	82.90	8.83	91.73	**131.00**
	Demo	VLF	Sm	AB	4.01	3.99	---	118.00	12.60	130.60	**188.00**
24" x 24"	Demo	VLF	Lg	AB	2.39	6.70	---	70.50	7.51	78.01	**112.00**
	Demo	VLF	Sm	AB	3.41	4.69	---	101.00	10.70	111.70	**160.00**
28" x 24"	Demo	VLF	Lg	AB	2.67	6.00	---	78.70	8.39	87.09	**125.00**
	Demo	VLF	Sm	AB	3.81	4.20	---	112.00	12.00	124.00	**178.00**
32" x 24"	Demo	VLF	Lg	AB	2.91	5.50	---	85.80	9.15	94.95	**136.00**
	Demo	VLF	Sm	AB	4.16	3.85	---	123.00	13.10	136.10	**195.00**
36" x 24"	Demo	VLF	Lg	AB	3.14	5.10	---	92.60	9.87	102.47	**147.00**
	Demo	VLF	Sm	AB	4.48	3.57	---	132.00	14.10	146.10	**210.00**

Description	Oper	Unit	Vol	Crew Size	Man-hours per Unit	Crew Output per Day	Avg Mat'l Unit Cost	Avg Labor Unit Cost	Avg Equip Unit Cost	Avg Total Unit Cost	Avg Price Incl O&P
28" x 28"	Demo	VLF	Lg	AB	2.96	5.40	---	87.30	9.32	96.62	**139.00**
	Demo	VLF	Sm	AB	4.23	3.78	---	125.00	13.30	138.30	**198.00**
32" x 28"	Demo	VLF	Lg	AB	3.27	4.90	---	96.40	10.30	106.70	**153.00**
	Demo	VLF	Sm	AB	4.66	3.43	---	137.00	14.70	151.70	**218.00**
36" x 28"	Demo	VLF	Lg	AB	3.56	4.50	---	105.00	11.20	116.20	**167.00**
	Demo	VLF	Sm	AB	5.08	3.15	---	150.00	16.00	166.00	**238.00**
32" x 32"	Demo	VLF	Lg	AB	3.56	4.50	---	105.00	11.20	116.20	**167.00**
	Demo	VLF	Sm	AB	5.08	3.15	---	150.00	16.00	166.00	**238.00**
36" x 32"	Demo	VLF	Lg	AB	3.90	4.10	---	115.00	12.30	127.30	**183.00**
	Demo	VLF	Sm	AB	5.57	2.87	---	164.00	17.50	181.50	**261.00**
36" x 36"	Demo	VLF	Lg	AB	4.21	3.80	---	124.00	13.30	137.30	**197.00**
	Demo	VLF	Sm	AB	6.02	2.66	---	178.00	18.90	196.90	**282.00**
8" x 8" (9.33 Brick/VLF)	Inst	VLF	Lg	BK	.415	38.60	4.33	12.30	---	16.63	**23.40**
	Inst	VLF	Sm	BK	.592	27.02	5.02	17.60	---	22.62	**32.00**
12" x 8" (14.00 Brick/VLF)	Inst	VLF	Lg	BK	.606	26.40	6.59	18.00	---	24.59	**34.40**
	Inst	VLF	Sm	BK	.866	18.48	7.63	25.70	---	33.33	**47.10**
16" x 8" (18.67 Brick/VLF)	Inst	VLF	Lg	BK	.784	20.40	8.76	23.30	---	32.06	**44.80**
	Inst	VLF	Sm	BK	1.12	14.28	10.20	33.30	---	43.50	**61.20**
20" x 8" (23.33 Brick/VLF)	Inst	VLF	Lg	BK	.958	16.70	10.90	28.50	---	39.40	**55.00**
	Inst	VLF	Sm	BK	1.37	11.69	12.70	40.70	---	53.40	**75.20**
24" x 8" (28.00 Brick/VLF)	Inst	VLF	Lg	BK	1.12	14.30	13.10	33.30	---	46.40	**64.60**
	Inst	VLF	Sm	BK	1.60	10.01	15.20	47.50	---	62.70	**88.30**
12" x 12" (21.00 Brick/VLF)	Inst	VLF	Lg	BK	.874	18.30	9.80	26.00	---	35.80	**50.00**
	Inst	VLF	Sm	BK	1.25	12.81	11.40	37.10	---	48.50	**68.40**
16" x 12" (28.00 Brick/VLF)	Inst	VLF	Lg	BK	1.12	14.30	13.10	33.30	---	46.40	**64.60**
	Inst	VLF	Sm	BK	1.60	10.01	15.20	47.50	---	62.70	**88.30**
20" x 12" (35.00 Brick/VLF)	Inst	VLF	Lg	BK	1.34	11.90	16.40	39.80	---	56.20	**78.20**
	Inst	VLF	Sm	BK	1.92	8.33	19.00	57.00	---	76.00	**107.00**
24" x 12" (42.00 Brick/VLF)	Inst	VLF	Lg	BK	1.57	10.20	19.60	46.60	---	66.20	**92.00**
	Inst	VLF	Sm	BK	2.24	7.14	22.70	66.60	---	89.30	**125.00**
28" x 12" (49.00 Brick/VLF)	Inst	VLF	Lg	BK	1.76	9.10	22.90	52.30	---	75.20	**104.00**
	Inst	VLF	Sm	BK	2.51	6.37	26.50	74.60	---	101.10	**142.00**
32" x 12" (56.00 Brick/VLF)	Inst	VLF	Lg	BK	1.95	8.20	26.20	57.90	---	84.10	**116.00**
	Inst	VLF	Sm	BK	2.79	5.74	30.30	82.90	---	113.20	**158.00**
16" x 16" (37.33 Brick/VLF)	Inst	VLF	Lg	BK	1.42	11.30	17.40	42.20	---	59.60	**82.90**
	Inst	VLF	Sm	BK	2.02	7.91	20.20	60.00	---	80.20	**113.00**
20" x 16" (46.67 Brick/VLF)	Inst	VLF	Lg	BK	1.70	9.40	21.90	50.50	---	72.40	**100.00**
	Inst	VLF	Sm	BK	2.43	6.58	25.30	72.20	---	97.50	**137.00**
24" x 16" (56.00 Brick/VLF)	Inst	VLF	Lg	BK	1.95	8.20	26.20	57.90	---	84.10	**116.00**
	Inst	VLF	Sm	BK	2.79	5.74	30.30	82.90	---	113.20	**158.00**
28" x 16" (65.33 Brick/VLF)	Inst	VLF	Lg	BK	2.16	7.40	30.50	64.20	---	94.70	**131.00**
	Inst	VLF	Sm	BK	3.09	5.18	35.40	91.80	---	127.20	**177.00**
32" x 16" (74.67 Brick/VLF)	Inst	VLF	Lg	BK	2.39	6.70	34.90	71.00	---	105.90	**146.00**
	Inst	VLF	Sm	BK	3.41	4.69	40.50	101.00	---	141.50	**197.00**
36" x 16" (84.00 Brick/VLF)	Inst	VLF	Lg	BK	2.58	6.20	39.30	76.70	---	116.00	**159.00**
	Inst	VLF	Sm	BK	3.69	4.34	45.50	110.00	---	155.50	**216.00**

Description	Oper	Unit	Vol	Crew Size	Man-hours per Unit	Crew Output per Day	Avg Mat'l Unit Cost	Avg Labor Unit Cost	Avg Equip Unit Cost	Avg Total Unit Cost	Avg Price Incl O&P
20" x 20" (58.33 Brick/VLF)	Inst	VLF	Lg	BK	2.00	8.00	27.30	59.40	---	86.70	**120.00**
	Inst	VLF	Sm	BK	2.86	5.60	31.60	85.00	---	116.60	**163.00**
24" x 20" (70.00 Brick/VLF)	Inst	VLF	Lg	BK	2.32	6.90	32.70	68.90	---	101.60	**140.00**
	Inst	VLF	Sm	BK	3.31	4.83	37.90	98.30	---	136.20	**190.00**
28" x 20" (81.67 Brick/VLF)	Inst	VLF	Lg	BK	2.58	6.20	38.10	76.70	---	114.80	**158.00**
	Inst	VLF	Sm	BK	3.69	4.34	44.20	110.00	---	154.20	**214.00**
32" x 20" (93.33 Brick/VLF)	Inst	VLF	Lg	BK	2.81	5.70	43.60	83.50	---	127.10	**175.00**
	Inst	VLF	Sm	BK	4.01	3.99	50.50	119.00	---	169.50	**236.00**
36" x 20" (105.00 Brick/VLF)	Inst	VLF	Lg	BK	3.02	5.30	49.10	89.70	---	138.80	**190.00**
	Inst	VLF	Sm	BK	4.31	3.71	56.90	128.00	---	184.90	**256.00**
24" x 24" (84.00 Brick/VLF)	Inst	VLF	Lg	BK	2.58	6.20	39.30	76.70	---	116.00	**159.00**
	Inst	VLF	Sm	BK	3.69	4.34	45.50	110.00	---	155.50	**216.00**
28" x 24" (98.00 Brick/VLF)	Inst	VLF	Lg	BK	2.86	5.60	45.80	85.00	---	130.80	**179.00**
	Inst	VLF	Sm	BK	4.08	3.92	53.00	121.00	---	174.00	**242.00**
32" x 24" (112.00 Brick/VLF)	Inst	VLF	Lg	BK	3.08	5.20	52.40	91.50	---	143.90	**197.00**
	Inst	VLF	Sm	BK	4.40	3.64	60.70	131.00	---	191.70	**265.00**
36" x 24" (126.00 Brick/VLF)	Inst	VLF	Lg	BK	3.33	4.80	58.90	98.90	---	157.80	**215.00**
	Inst	VLF	Sm	BK	4.76	3.36	68.20	141.00	---	209.20	**289.00**
28" x 28" (114.33 Brick/VLF)	Inst	VLF	Lg	BK	3.20	5.00	53.50	95.10	---	148.60	**203.00**
	Inst	VLF	Sm	BK	4.57	3.50	62.00	136.00	---	198.00	**274.00**
32" x 28" (130.67 Brick/VLF)	Inst	VLF	Lg	BK	3.48	4.60	61.10	103.00	---	164.10	**224.00**
	Inst	VLF	Sm	BK	4.97	3.22	70.80	148.00	---	218.80	**301.00**
36" x 28" (147.00 Brick/VLF)	Inst	VLF	Lg	BK	3.90	4.10	68.70	116.00	---	184.70	**252.00**
	Inst	VLF	Sm	BK	5.57	2.87	79.60	165.00	---	244.60	**338.00**
32" x 32" (149.33 Brick/VLF)	Inst	VLF	Lg	BK	4.00	4.00	69.80	119.00	---	188.80	**257.00**
	Inst	VLF	Sm	BK	5.71	2.80	80.90	170.00	---	250.90	**346.00**
36" x 32" (168.00 Brick/VLF)	Inst	VLF	Lg	BK	4.44	3.60	78.50	132.00	---	210.50	**287.00**
	Inst	VLF	Sm	BK	6.35	2.52	91.00	189.00	---	280.00	**386.00**
36" x 36" (189.00 Brick/VLF)	Inst	VLF	Lg	BK	5.00	3.20	88.30	149.00	---	237.30	**323.00**
	Inst	VLF	Sm	BK	7.14	2.24	102.00	212.00	---	314.00	**434.00**

Veneers

4" T, with air tools

Description	Oper	Unit	Vol	Crew Size	Man-hours per Unit	Crew Output per Day	Avg Mat'l Unit Cost	Avg Labor Unit Cost	Avg Equip Unit Cost	Avg Total Unit Cost	Avg Price Incl O&P
	Demo	SF	Lg	AB	.049	325.0	---	1.45	.15	1.60	**2.29**
	Demo	SF	Sm	AB	.070	227.5	---	2.06	.22	2.28	**3.28**

4" T, with wall ties; no scaffolding included

Description	Oper	Unit	Vol	Crew Size	Man-hours per Unit	Crew Output per Day	Avg Mat'l Unit Cost	Avg Labor Unit Cost	Avg Equip Unit Cost	Avg Total Unit Cost	Avg Price Incl O&P
Common, 8" x 2-2/3" x 4"	Inst	SF	Lg	BD	.170	235.0	2.37	5.19	.15	7.71	**10.60**
	Inst	SF	Sm	BO	.195	164.5	2.74	5.79	.22	8.75	**12.00**
Standard face, 8" x 2-2/3" x 4"	Inst	SF	Lg	BD	.182	220.0	3.18	5.55	.16	8.89	**12.10**
	Inst	SF	Sm	BO	.208	154.0	3.69	6.18	.23	10.10	**13.70**
Glazed, 8" x 2-2/3" x 4"	Inst	SF	Lg	BD	.190	210.0	7.23	5.80	.17	13.20	**17.10**
	Inst	SF	Sm	BO	.218	147.0	8.35	6.48	.24	15.07	**19.50**

Description	Oper	Unit	Vol	Crew Size	Man-hours per Unit	Crew Output per Day	Avg Mat'l Unit Cost	Avg Labor Unit Cost	Avg Equip Unit Cost	Avg Total Unit Cost	Avg Price Incl O&P
Other brick types/sizes											
8" x 4" x 4"	Inst	SF	Lg	BD	.127	315.0	2.49	3.87	.11	6.47	**8.75**
	Inst	SF	Sm	BO	.145	220.5	2.87	4.31	.16	7.34	**9.88**
8" x 3-1/5" x 4"	Inst	SF	Lg	BD	.151	265.0	3.66	4.61	.13	8.40	**11.20**
	Inst	SF	Sm	BO	.173	185.5	4.19	5.14	.19	9.52	**12.70**
12" x 4" x 6"	Inst	SF	Lg	BD	.091	440.0	3.88	2.78	.08	6.74	**8.68**
	Inst	SF	Sm	BO	.104	308.0	4.47	3.09	.12	7.68	**9.86**
12" x 2-2/3" x 4"	Inst	SF	Lg	BD	.123	325.0	3.17	3.75	.11	7.03	**9.35**
	Inst	SF	Sm	BO	.141	227.5	3.64	4.19	.16	7.99	**10.60**
12" x 3-1/5" x 4"	Inst	SF	Lg	BD	.107	375.0	2.47	3.26	.10	5.83	**7.80**
	Inst	SF	Sm	BO	.122	262.5	2.83	3.62	.14	6.59	**8.79**
12" x 2" x 4"	Inst	SF	Lg	BD	.160	250.0	4.29	4.88	.14	9.31	**12.40**
	Inst	SF	Sm	BO	.183	175.0	4.93	5.44	.20	10.57	**14.00**
12" x 2-2/3" x 6"	Inst	SF	Lg	BD	.127	315.0	3.66	3.87	.11	7.64	**10.10**
	Inst	SF	Sm	BO	.145	220.5	4.22	4.31	.16	8.69	**11.40**
12" x 4" x 4"	Inst	SF	Lg	BD	.089	450.0	2.89	2.72	.08	5.69	**7.45**
	Inst	SF	Sm	BO	.102	315.0	3.33	3.03	.11	6.47	**8.45**

Walls

With air tools

Description	Oper	Unit	Vol	Crew Size	Man-hours per Unit	Crew Output per Day	Avg Mat'l Unit Cost	Avg Labor Unit Cost	Avg Equip Unit Cost	Avg Total Unit Cost	Avg Price Incl O&P
8" T	Demo	SF	Lg	AB	.086	185.0	---	2.54	.27	2.81	**4.02**
	Demo	SF	Sm	AB	.124	129.5	---	3.66	.39	4.05	**5.80**
12" T	Demo	SF	Lg	AB	.123	130.0	---	3.63	.39	4.02	**5.76**
	Demo	SF	Sm	AB	.176	91.00	---	5.19	.55	5.74	**8.23**
16" T	Demo	SF	Lg	AB	.160	100.0	---	4.72	.50	5.22	**7.48**
	Demo	SF	Sm	AB	.229	70.00	---	6.75	.72	7.47	**10.70**
24" T	Demo	SF	Lg	AB	.188	85.00	---	5.54	.59	6.13	**8.80**
	Demo	SF	Sm	AB	.269	59.50	---	7.93	.85	8.78	**12.60**

With common brick; no scaffolding included

Description	Oper	Unit	Vol	Crew Size	Man-hours per Unit	Crew Output per Day	Avg Mat'l Unit Cost	Avg Labor Unit Cost	Avg Equip Unit Cost	Avg Total Unit Cost	Avg Price Incl O&P
8" T (13.50 Brick/SF)	Inst	SF	Lg	BO	.337	95.00	4.66	10.00	.38	15.04	**20.70**
	Inst	SF	Sm	ML	.361	66.50	5.38	11.20	.54	17.12	**23.40**
12" T (20.25 Brick/SF)	Inst	SF	Lg	BO	.457	70.00	7.04	13.60	.51	21.15	**28.80**
	Inst	SF	Sm	ML	.490	49.00	8.13	15.20	.73	24.06	**32.70**
16" T (27.00 Brick/SF)	Inst	SF	Lg	BO	.582	55.00	9.41	17.30	.65	27.36	**37.20**
	Inst	SF	Sm	ML	.623	38.50	10.90	19.30	.93	31.13	**42.30**
24" T (40.50 Brick/SF)	Inst	SF	Lg	BO	.800	40.00	14.10	23.80	.89	38.79	**52.50**
	Inst	SF	Sm	ML	.857	28.00	16.30	26.60	1.27	44.17	**59.60**

Description	Oper	Unit	Vol	Crew Size	Man-hours per Unit	Crew Output per Day	Avg Mat'l Unit Cost	Avg Labor Unit Cost	Avg Equip Unit Cost	Avg Total Unit Cost	Avg Price Incl O&P

Brick, adobe
Running bond, 3/8" mortar joints

Walls, with air tools

Description	Oper	Unit	Vol	Crew Size	MH	Output	Mat'l	Labor	Equip	Total	O&P
4" T	Demo	SF	Lg	AB	.041	390.0	---	1.21	.13	1.34	1.92
	Demo	SF	Sm	AB	.059	273.0	---	1.74	.18	1.92	2.76
6" T	Demo	SF	Lg	AB	.042	380.0	---	1.24	.13	1.37	1.96
	Demo	SF	Sm	AB	.060	266.0	---	1.77	.19	1.96	2.81
8" T	Demo	SF	Lg	AB	.043	370.0	---	1.27	.14	1.41	2.02
	Demo	SF	Sm	AB	.062	259.0	---	1.83	.19	2.02	2.90
12" T	Demo	SF	Lg	AB	.046	350.0	---	1.36	.14	1.50	2.15
	Demo	SF	Sm	AB	.065	245.0	---	1.92	.21	2.13	3.05

Walls, no scaffolding included

Description	Oper	Unit	Vol	Crew Size	MH	Output	Mat'l	Labor	Equip	Total	O&P
4" x 4" x 16"	Inst	SF	Lg	BO	.139	230.0	1.76	4.13	.16	6.05	8.34
	Inst	SF	Sm	ML	.149	161.0	2.01	4.62	.22	6.85	9.42
6" x 4" x 16"	Inst	SF	Lg	BO	.145	220.0	2.65	4.31	.16	7.12	9.63
	Inst	SF	Sm	ML	.156	154.0	2.01	4.84	.23	7.08	9.76
8" x 4" x 16"	Inst	SF	Lg	BO	.152	210.0	3.40	4.52	.17	8.09	10.80
	Inst	SF	Sm	ML	.163	147.0	3.91	5.06	.24	9.21	12.30
12" x 4" x 16"	Inst	SF	Lg	BO	.168	190.0	5.36	4.99	.19	10.54	13.80
	Inst	SF	Sm	ML	.180	133.0	6.16	5.59	.27	12.02	15.70

Concrete block
Lightweight (haydite) or heavyweight blocks; 2 cores/block, solid face; includes allowances for lintels, bond beams

Foundations and retaining walls

No excavation included

Without reinforcing or with lateral reinforcing only

With air tools

Description	Oper	Unit	Vol	Crew Size	MH	Output	Mat'l	Labor	Equip	Total	O&P
8" W x 8" H x 16" L	Demo	SF	Lg	AB	.048	335.0	---	1.42	.15	1.57	2.25
	Demo	SF	Sm	AB	.068	234.5	---	2.01	.21	2.22	3.18
10" W x 8" H x 16" L	Demo	SF	Lg	AB	.052	310.0	---	1.53	.16	1.69	2.43
	Demo	SF	Sm	AB	.074	217.0	---	2.18	.23	2.41	3.46
12" W x 8" H x 16" L	Demo	SF	Lg	AB	.055	290.0	---	1.62	.17	1.79	2.57
	Demo	SF	Sm	AB	.079	203.0	---	2.33	.25	2.58	3.70

Without air tools

Description	Oper	Unit	Vol	Crew Size	MH	Output	Mat'l	Labor	Equip	Total	O&P
8" W x 8" H x 16" L	Demo	SF	Lg	LB	.059	270.0	---	1.62	---	1.62	2.41
	Demo	SF	Sm	LB	.085	189.0	---	2.33	---	2.33	3.47
10" W x 8" H x 16" L	Demo	SF	Lg	LB	.064	250.0	---	1.76	---	1.76	2.62
	Demo	SF	Sm	LB	.091	175.0	---	2.50	---	2.50	3.72
12" W x 8" H x 16" L	Demo	SF	Lg	LB	.070	230.0	---	1.92	---	1.92	2.86
	Demo	SF	Sm	LB	.099	161.0	---	2.72	---	2.72	4.05

With vertical reinforcing in every other core (2 core blocks) with core concrete filled

With air tools

Description	Oper	Unit	Vol	Crew Size	MH	Output	Mat'l	Labor	Equip	Total	O&P
8" W x 8" H x 16" L	Demo	SF	Lg	AB	.078	205.0	---	2.30	.25	2.55	3.65
	Demo	SF	Sm	AB	.111	143.5	---	3.27	.35	3.62	5.20
10" W x 8" H x 16" L	Demo	SF	Lg	AB	.084	190.0	---	2.48	.26	2.74	3.93
	Demo	SF	Sm	AB	.120	133.0	---	3.54	.38	3.92	5.62
12" W x 8" H x 16" L	Demo	SF	Lg	AB	.091	175.0	---	2.68	.29	2.97	4.26
	Demo	SF	Sm	AB	.131	122.5	---	3.86	.41	4.27	6.13

Description	Oper	Unit	Vol	Crew Size	Man-hours per Unit	Crew Output per Day	Avg Mat'l Unit Cost	Avg Labor Unit Cost	Avg Equip Unit Cost	Avg Total Unit Cost	Avg Price Incl O&P
No reinforcing											
Heavyweight blocks											
8" x 8" x 16"	Inst	SF	Lg	BD	.127	315.0	1.35	3.87	.11	5.33	7.44
	Inst	SF	Sm	BO	.145	220.5	1.55	4.31	.16	6.02	8.36
10" x 8" x 16"	Inst	SF	Lg	BD	.138	290.0	1.67	4.21	.12	6.00	8.31
	Inst	SF	Sm	BO	.158	203.0	1.92	4.69	.18	6.79	9.38
12" x 8" x 16"	Inst	SF	Lg	BD	.151	265.0	2.03	4.61	.13	6.77	9.33
	Inst	SF	Sm	BO	.173	185.5	2.35	5.14	.19	7.68	10.60
Lightweight blocks											
8" x 8" x 16"	Inst	SF	Lg	BD	.114	350.0	1.54	3.48	.10	5.12	7.05
	Inst	SF	Sm	BO	.131	245.0	1.77	3.89	.15	5.81	7.98
10" x 8" x 16"	Inst	SF	Lg	BD	.125	320.0	1.86	3.81	.11	5.78	7.93
	Inst	SF	Sm	BO	.143	224.0	2.14	4.25	.16	6.55	8.95
12" x 8" x 16"	Inst	SF	Lg	BD	.136	295.0	2.22	4.15	.12	6.49	8.86
	Inst	SF	Sm	BO	.155	206.5	2.55	4.61	.17	7.33	9.96
Lateral reinforcing every second course											
Heavyweight blocks											
8" x 8" x 16"	Inst	SF	Lg	BD	.133	300.0	1.35	4.06	.12	5.53	7.72
	Inst	SF	Sm	BO	.152	210.0	1.55	4.52	.17	6.24	8.68
10" x 8" x 16"	Inst	SF	Lg	BD	.145	275.0	1.67	4.42	.13	6.22	8.64
	Inst	SF	Sm	BO	.166	192.5	1.92	4.93	.19	7.04	9.75
12" x 8" x 16"	Inst	SF	Lg	BD	.160	250.0	2.03	4.88	.14	7.05	9.75
	Inst	SF	Sm	BO	.183	175.0	2.35	5.44	.20	7.99	11.00
Lightweight blocks											
8" x 8" x 16"	Inst	SF	Lg	BD	.119	335.0	1.54	3.63	.11	5.28	7.29
	Inst	SF	Sm	BO	.136	234.5	1.77	4.04	.15	5.96	8.20
10" x 8" x 16"	Inst	SF	Lg	BD	.131	305.0	1.86	4.00	.12	5.98	8.21
	Inst	SF	Sm	BO	.150	213.5	2.14	4.46	.17	6.77	9.27
12" x 8" x 16"	Inst	SF	Lg	BD	.143	280.0	2.22	4.36	.13	6.71	9.18
	Inst	SF	Sm	BO	.163	196.0	2.55	4.84	.18	7.57	10.30
Vertical reinforcing (No. 4 rod) every second core with core concrete filled											
Heavyweight blocks											
8" x 8" x 16"	Inst	SF	Lg	BD	.182	220.0	2.14	5.55	.16	7.85	10.90
	Inst	SF	Sm	BD	.260	154.0	2.47	7.93	.23	10.63	14.90
10" x 8" x 16"	Inst	SF	Lg	BD	.200	200.0	2.56	6.10	.18	8.84	12.20
	Inst	SF	Sm	BD	.286	140.0	2.94	8.73	.25	11.92	16.60
12" x 8" x 16"	Inst	SF	Lg	BD	.216	185.0	3.06	6.59	.19	9.84	13.50
	Inst	SF	Sm	BD	.309	129.5	3.52	9.43	.28	13.23	18.40
Lightweight blocks											
8" x 8" x 16"	Inst	SF	Lg	BD	.167	240.0	2.34	5.10	.15	7.59	10.40
	Inst	SF	Sm	BD	.238	168.0	2.69	7.26	.21	10.16	14.10
10" x 8" x 16"	Inst	SF	Lg	BD	.182	220.0	2.74	5.55	.16	8.45	11.60
	Inst	SF	Sm	BD	.260	154.0	3.16	7.93	.23	11.32	15.70
12" x 8" x 16"	Inst	SF	Lg	BD	.200	200.0	3.24	6.10	.18	9.52	13.00
	Inst	SF	Sm	BD	.286	140.0	3.73	8.73	.25	12.71	17.50

Description	Oper	Unit	Vol	Crew Size	Man-hours per Unit	Crew Output per Day	Avg Mat'l Unit Cost	Avg Labor Unit Cost	Avg Equip Unit Cost	Avg Total Unit Cost	Avg Price Incl O&P

Exterior walls (above grade)

No bracing or shoring included

Without reinforcing or with lateral reinforcing only

With air tools

Description	Oper	Unit	Vol	Crew Size	Man-hours per Unit	Crew Output per Day	Avg Mat'l Unit Cost	Avg Labor Unit Cost	Avg Equip Unit Cost	Avg Total Unit Cost	Avg Price Incl O&P
8" W x 8" H x 16" L	Demo	SF	Lg	AB	.048	335.0	---	1.42	.15	1.57	**2.25**
	Demo	SF	Sm	AB	.068	234.5	---	2.01	.21	2.22	**3.18**
10" W x 8" H x 16" L	Demo	SF	Lg	AB	.052	305.0	---	1.53	.17	1.70	**2.44**
	Demo	SF	Sm	AB	.075	213.5	---	2.21	.24	2.45	**3.51**
12" W x 8" H x 16" L	Demo	SF	Lg	AB	.058	275.0	---	1.71	.18	1.89	**2.71**
	Demo	SF	Sm	AB	.083	192.5	---	2.45	.26	2.71	**3.88**

Without air tools

Description	Oper	Unit	Vol	Crew Size	Man-hours per Unit	Crew Output per Day	Avg Mat'l Unit Cost	Avg Labor Unit Cost	Avg Equip Unit Cost	Avg Total Unit Cost	Avg Price Incl O&P
8" W x 8" H x 16" L	Demo	SF	Lg	LB	.060	265.0	---	1.65	---	1.65	**2.45**
	Demo	SF	Sm	LB	.086	185.5	---	2.36	---	2.36	**3.51**
10" W x 8" H x 16" L	Demo	SF	Lg	LB	.067	240.0	---	1.84	---	1.84	**2.74**
	Demo	SF	Sm	LB	.095	168.0	---	2.61	---	2.61	**3.88**
12" W x 8" H x 16" L	Demo	SF	Lg	LB	.073	220.0	---	2.00	---	2.00	**2.98**
	Demo	SF	Sm	LB	.104	154.0	---	2.85	---	2.85	**4.25**

No reinforcing

Heavyweight blocks

Description	Oper	Unit	Vol	Crew Size	Man-hours per Unit	Crew Output per Day	Avg Mat'l Unit Cost	Avg Labor Unit Cost	Avg Equip Unit Cost	Avg Total Unit Cost	Avg Price Incl O&P
8" x 8" x 16"	Inst	SF	Lg	BD	.136	295.0	1.35	4.15	.12	5.62	**7.86**
	Inst	SF	Sm	BO	.155	206.5	1.55	4.61	.17	6.33	**8.81**
10" x 8" x 16"	Inst	SF	Lg	BD	.148	270.0	1.67	4.52	.13	6.32	**8.78**
	Inst	SF	Sm	BO	.169	189.0	1.92	5.02	.19	7.13	**9.88**
12" x 8" x 16"	Inst	SF	Lg	BD	.160	250.0	2.03	4.88	.14	7.05	**9.75**
	Inst	SF	Sm	BO	.183	175.0	2.35	5.44	.20	7.99	**11.00**

Lightweight blocks

Description	Oper	Unit	Vol	Crew Size	Man-hours per Unit	Crew Output per Day	Avg Mat'l Unit Cost	Avg Labor Unit Cost	Avg Equip Unit Cost	Avg Total Unit Cost	Avg Price Incl O&P
8" x 8" x 16"	Inst	SF	Lg	BD	.121	330.0	1.54	3.69	.11	5.34	**7.38**
	Inst	SF	Sm	BO	.139	231.0	1.77	4.13	.15	6.05	**8.34**
10" x 8" x 16"	Inst	SF	Lg	BD	.131	305.0	1.86	4.00	.12	5.98	**8.21**
	Inst	SF	Sm	BO	.150	213.5	1.87	4.46	.17	6.50	**8.96**
12" x 8" x 16"	Inst	SF	Lg	BD	.143	280.0	2.22	4.36	.13	6.71	**9.18**
	Inst	SF	Sm	BO	.163	196.0	2.55	4.84	.18	7.57	**10.30**

Lateral reinforcing every second course

Heavyweight blocks

Description	Oper	Unit	Vol	Crew Size	Man-hours per Unit	Crew Output per Day	Avg Mat'l Unit Cost	Avg Labor Unit Cost	Avg Equip Unit Cost	Avg Total Unit Cost	Avg Price Incl O&P
8" x 8" x 16"	Inst	SF	Lg	BD	.143	280.0	1.35	4.36	.13	5.84	**8.18**
	Inst	SF	Sm	BO	.163	196.0	1.55	4.84	.18	6.57	**9.10**
10" x 8" x 16"	Inst	SF	Lg	BD	.154	260.0	1.67	4.70	.14	6.51	**9.06**
	Inst	SF	Sm	BO	.176	182.0	1.92	5.23	.20	7.35	**10.20**
12" x 8" x 16"	Inst	SF	Lg	BD	.167	240.0	2.03	5.10	.15	7.28	**10.10**
	Inst	SF	Sm	BO	.190	168.0	2.35	5.64	.21	8.20	**11.30**

Lightweight blocks

Description	Oper	Unit	Vol	Crew Size	Man-hours per Unit	Crew Output per Day	Avg Mat'l Unit Cost	Avg Labor Unit Cost	Avg Equip Unit Cost	Avg Total Unit Cost	Avg Price Incl O&P
8" x 8" x 16"	Inst	SF	Lg	BD	.127	315.0	1.54	3.87	.11	5.52	**7.65**
	Inst	SF	Sm	BO	.145	220.5	1.77	4.31	.16	6.24	**8.61**
10" x 8" x 16"	Inst	SF	Lg	BD	.138	290.0	2.12	4.21	.12	6.45	**8.83**
	Inst	SF	Sm	BO	.158	203.0	1.87	4.69	.18	6.74	**9.32**
12" x 8" x 16"	Inst	SF	Lg	BD	.151	265.0	2.22	4.61	.13	6.96	**9.55**
	Inst	SF	Sm	BO	.173	185.5	2.55	5.14	.19	7.88	**10.80**

Description	Oper	Unit	Vol	Crew Size	Man-hours per Unit	Crew Output per Day	Avg Mat'l Unit Cost	Avg Labor Unit Cost	Avg Equip Unit Cost	Avg Total Unit Cost	Avg Price Incl O&P
Partitions (above grade)											
No bracing or shoring included											
Without reinforcing or with lateral reinforcing only											
With air tools											
4" W x 8" H x 16" L	Demo	SF	Lg	AB	.048	330.0	---	1.42	.15	1.57	**2.25**
	Demo	SF	Sm	AB	.069	231.0	---	2.03	.22	2.25	**3.23**
6" W x 8" H x 16" L	Demo	SF	Lg	AB	.052	310.0	---	1.53	.16	1.69	**2.43**
	Demo	SF	Sm	AB	.074	217.0	---	2.18	.23	2.41	**3.46**
8" W x 8" H x 16" L	Demo	SF	Lg	AB	.054	295.0	---	1.59	.17	1.76	**2.53**
	Demo	SF	Sm	AB	.077	206.5	---	2.27	.24	2.51	**3.60**
10" W x 8" H x 16" L	Demo	SF	Lg	AB	.060	265.0	---	1.77	.19	1.96	**2.81**
	Demo	SF	Sm	AB	.086	185.5	---	2.54	.27	2.81	**4.02**
12" W x 8" H x 16" L	Demo	SF	Lg	AB	.068	235.0	---	2.01	.21	2.22	**3.18**
	Demo	SF	Sm	AB	.097	164.5	---	2.86	.31	3.17	**4.54**
No reinforcing											
Heavyweight blocks											
4" W x 8" H x 16" L	Inst	SF	Lg	BD	.129	310.0	.99	3.94	.12	5.05	**7.12**
	Inst	SF	Sm	BO	.147	217.0	1.14	4.37	.16	5.67	**7.98**
6" W x 8" H x 16" L	Inst	SF	Lg	BD	.138	290.0	1.24	4.21	.12	5.57	**7.82**
	Inst	SF	Sm	BO	.158	203.0	1.43	4.69	.18	6.30	**8.82**
8" W x 8" H x 16" L	Inst	SF	Lg	BD	.151	265.0	1.35	4.61	.13	6.09	**8.55**
	Inst	SF	Sm	BO	.173	185.5	1.55	5.14	.19	6.88	**9.63**
10" W x 8" H x 16" L	Inst	SF	Lg	BD	.163	245.0	1.67	4.97	.15	6.79	**9.48**
	Inst	SF	Sm	BO	.187	171.5	1.92	5.56	.21	7.69	**10.70**
12" W x 8" H x 16" L	Inst	SF	Lg	BD	.178	225.0	2.03	5.43	.16	7.62	**10.60**
	Inst	SF	Sm	BO	.203	157.5	2.35	6.03	.23	8.61	**11.90**
Lightweight blocks											
4" W x 8" H x 16" L	Inst	SF	Lg	BD	.119	335.0	1.19	3.63	.11	4.93	**6.89**
	Inst	SF	Sm	BO	.136	234.5	1.37	4.04	.15	5.56	**7.74**
6" W x 8" H x 16" L	Inst	SF	Lg	BD	.127	315.0	1.43	3.87	.11	5.41	**7.53**
	Inst	SF	Sm	BO	.145	220.5	1.64	4.31	.16	6.11	**8.46**
8" W x 8" H x 16" L	Inst	SF	Lg	BD	.136	295.0	1.54	4.15	.12	5.81	**8.07**
	Inst	SF	Sm	BO	.155	206.5	1.77	4.61	.17	6.55	**9.07**
10" W x 8" H x 16" L	Inst	SF	Lg	BD	.148	270.0	1.86	4.52	.13	6.51	**9.00**
	Inst	SF	Sm	BO	.169	189.0	1.87	5.02	.19	7.08	**9.82**
12" W x 8" H x 16" L	Inst	SF	Lg	BD	.160	250.0	2.22	4.88	.14	7.24	**9.97**
	Inst	SF	Sm	BO	.183	175.0	2.55	5.44	.20	8.19	**11.20**
Lateral reinforcing every second course											
Heavyweight blocks											
4" W x 8" H x 16" L	Inst	SF	Lg	BD	.136	295.0	.99	4.15	.12	5.26	**7.44**
	Inst	SF	Sm	BO	.155	206.5	1.14	4.61	.17	5.92	**8.34**
6" W x 8" H x 16" L	Inst	SF	Lg	BD	.145	275.0	1.24	4.42	.13	5.79	**8.15**
	Inst	SF	Sm	BO	.166	192.5	1.43	4.93	.19	6.55	**9.18**

Description	Oper	Unit	Vol	Crew Size	Man-hours per Unit	Crew Output per Day	Avg Mat'l Unit Cost	Avg Labor Unit Cost	Avg Equip Unit Cost	Avg Total Unit Cost	Avg Price Incl O&P
8" W x 8" H x 16" L	Inst	SF	Lg	BD	.160	250.0	1.35	4.88	.14	6.37	**8.97**
	Inst	SF	Sm	BO	.183	175.0	1.55	5.44	.20	7.19	**10.10**
10" W x 8" H x 16" L	Inst	SF	Lg	BD	.174	230.0	1.67	5.31	.16	7.14	**9.99**
	Inst	SF	Sm	BO	.199	161.0	1.92	5.91	.22	8.05	**11.20**
12" W x 8" H x 16" L	Inst	SF	Lg	BD	.190	210.0	2.03	5.80	.17	8.00	**11.10**
	Inst	SF	Sm	BO	.218	147.0	2.35	6.48	.24	9.07	**12.60**

Lightweight blocks

Description	Oper	Unit	Vol	Crew Size	Man-hours per Unit	Crew Output per Day	Avg Mat'l Unit Cost	Avg Labor Unit Cost	Avg Equip Unit Cost	Avg Total Unit Cost	Avg Price Incl O&P
4" W x 8" H x 16" L	Inst	SF	Lg	BD	.125	320.0	1.19	3.81	.11	5.11	**7.16**
	Inst	SF	Sm	BO	.143	224.0	1.37	4.25	.16	5.78	**8.06**
6" W x 8" H x 16" L	Inst	SF	Lg	BD	.133	300.0	1.43	4.06	.12	5.61	**7.81**
	Inst	SF	Sm	BO	.152	210.0	1.64	4.52	.17	6.33	**8.78**
8" W x 8" H x 16" L	Inst	SF	Lg	BD	.143	280.0	1.54	4.36	.13	6.03	**8.40**
	Inst	SF	Sm	BO	.163	196.0	1.77	4.84	.18	6.79	**9.43**
10" W x 8" H x 16" L	Inst	SF	Lg	BD	.154	260.0	1.86	4.70	.14	6.70	**9.28**
	Inst	SF	Sm	BO	.176	182.0	1.87	5.23	.20	7.30	**10.10**
12" W x 8" H x 16" L	Inst	SF	Lg	BD	.167	240.0	2.22	5.10	.15	7.47	**10.30**
	Inst	SF	Sm	BO	.190	168.0	2.55	5.64	.21	8.40	**11.60**

Fences

Without reinforcing or with lateral reinforcing only

With air tools

Description	Oper	Unit	Vol	Crew Size	Man-hours per Unit	Crew Output per Day	Avg Mat'l Unit Cost	Avg Labor Unit Cost	Avg Equip Unit Cost	Avg Total Unit Cost	Avg Price Incl O&P
6" W x 4" H x 16" L	Demo	SF	Lg	AB	.041	390.0	---	1.21	.13	1.34	**1.92**
	Demo	SF	Sm	AB	.059	273.0	---	1.74	.18	1.92	**2.76**
6" W x 6" H x 16" L	Demo	SF	Lg	AB	.043	370.0	---	1.27	.14	1.41	**2.02**
	Demo	SF	Sm	AB	.062	259.0	---	1.83	.19	2.02	**2.90**
8" W x 8" H x 16" L	Demo	SF	Lg	AB	.046	350.0	---	1.36	.14	1.50	**2.15**
	Demo	SF	Sm	AB	.065	245.0	---	1.92	.21	2.13	**3.05**
10" W x 8" H x 16" L	Demo	SF	Lg	AB	.049	325.0	---	1.45	.15	1.60	**2.29**
	Demo	SF	Sm	AB	.070	227.5	---	2.06	.22	2.28	**3.28**
12" W x 8" H x 16" L	Demo	SF	Lg	AB	.053	300.0	---	1.56	.17	1.73	**2.48**
	Demo	SF	Sm	AB	.076	210.0	---	2.24	.24	2.48	**3.56**

Without air tools

Description	Oper	Unit	Vol	Crew Size	Man-hours per Unit	Crew Output per Day	Avg Mat'l Unit Cost	Avg Labor Unit Cost	Avg Equip Unit Cost	Avg Total Unit Cost	Avg Price Incl O&P
6" W x 4" H x 16" L	Demo	SF	Lg	LB	.052	310.0	---	1.43	---	1.43	**2.13**
	Demo	SF	Sm	LB	.074	217.0	---	2.03	---	2.03	**3.02**
6" W x 6" H x 16" L	Demo	SF	Lg	LB	.054	295.0	---	1.48	---	1.48	**2.21**
	Demo	SF	Sm	LB	.077	206.5	---	2.11	---	2.11	**3.15**
8" W x 8" H x 16" L	Demo	SF	Lg	LB	.057	280.0	---	1.56	---	1.56	**2.33**
	Demo	SF	Sm	LB	.082	196.0	---	2.25	---	2.25	**3.35**
10" W x 8" H x 16" L	Demo	SF	Lg	LB	.073	220.0	---	2.00	---	2.00	**2.98**
	Demo	SF	Sm	LB	.104	154.0	---	2.85	---	2.85	**4.25**
12" W x 8" H x 16" L	Demo	SF	Lg	LB	.067	240.0	---	1.84	---	1.84	**2.74**
	Demo	SF	Sm	LB	.095	168.0	---	2.61	---	2.61	**3.88**

Description	Oper	Unit	Vol	Crew Size	Man-hours per Unit	Crew Output per Day	Avg Mat'l Unit Cost	Avg Labor Unit Cost	Avg Equip Unit Cost	Avg Total Unit Cost	Avg Price Incl O&P
Fences, lightweight blocks											
No reinforcing											
4" W x 8" H x 16" L	Inst	SF	Lg	BD	.108	370.0	1.19	3.30	.10	4.59	**6.38**
	Inst	SF	Sm	BO	.124	259.0	1.37	3.68	.14	5.19	**7.20**
6" W x 4" H x 16" L	Inst	SF	Lg	BD	.222	180.0	2.44	6.77	.20	9.41	**13.10**
	Inst	SF	Sm	BO	.254	126.0	2.81	7.55	.28	10.64	**14.80**
6" W x 6" H x 16" L	Inst	SF	Lg	BD	.154	260.0	1.95	4.70	.14	6.79	**9.38**
	Inst	SF	Sm	BO	.176	182.0	2.24	5.23	.20	7.67	**10.60**
6" W x 8" H x 16" L	Inst	SF	Lg	BD	.114	350.0	1.43	3.48	.10	5.01	**6.93**
	Inst	SF	Sm	BO	.131	245.0	1.64	3.89	.15	5.68	**7.83**
8" W x 8" H x 16" L	Inst	SF	Lg	BD	.121	330.0	1.54	3.69	.11	5.34	**7.38**
	Inst	SF	Sm	BO	.139	231.0	1.77	4.13	.15	6.05	**8.34**
10" W x 8" H x 16" L	Inst	SF	Lg	BD	.131	305.0	1.86	4.00	.12	5.98	**8.21**
	Inst	SF	Sm	BO	.150	213.5	1.87	4.46	.17	6.50	**8.96**
12" W x 8" H x 16" L	Inst	SF	Lg	BD	.143	280.0	2.22	4.36	.13	6.71	**9.18**
	Inst	SF	Sm	BO	.163	196.0	2.55	4.84	.18	7.57	**10.30**
Lateral reinforcing every third course											
4" W x 8" H x 16" L	Inst	SF	Lg	BD	.113	355.0	1.19	3.45	.10	4.74	**6.61**
	Inst	SF	Sm	BO	.129	248.5	1.37	3.83	.14	5.34	**7.43**
6" W x 4" H x 16" L	Inst	SF	Lg	BD	.229	175.0	2.44	6.99	.20	9.63	**13.40**
	Inst	SF	Sm	BO	.261	122.5	2.81	7.75	.29	10.85	**15.10**
6" W x 6" H x 16" L	Inst	SF	Lg	BD	.163	245.0	1.95	4.97	.15	7.07	**9.80**
	Inst	SF	Sm	BO	.187	171.5	2.24	5.56	.21	8.01	**11.10**
6" W x 8" H x 16" L	Inst	SF	Lg	BD	.119	335.0	1.43	3.63	.11	5.17	**7.16**
	Inst	SF	Sm	BO	.136	234.5	1.64	4.04	.15	5.83	**8.06**
8" W x 8" H x 16" L	Inst	SF	Lg	BD	.127	315.0	1.54	3.87	.11	5.52	**7.65**
	Inst	SF	Sm	BO	.145	220.5	1.77	4.31	.16	6.24	**8.61**
10" W x 8" H x 16" L	Inst	SF	Lg	BD	.138	290.0	1.86	4.21	.12	6.19	**8.53**
	Inst	SF	Sm	BO	.158	203.0	1.87	4.69	.18	6.74	**9.32**
12" W x 8" H x 16" L	Inst	SF	Lg	BD	.151	265.0	2.22	4.61	.13	6.96	**9.55**
	Inst	SF	Sm	BO	.173	185.5	2.55	5.14	.19	7.88	**10.80**

Concrete slump block

Running bond, 3/8 " mortar joints

Walls, with air tools

Description	Oper	Unit	Vol	Crew Size	Man-hours per Unit	Crew Output per Day	Avg Mat'l Unit Cost	Avg Labor Unit Cost	Avg Equip Unit Cost	Avg Total Unit Cost	Avg Price Incl O&P
4" T	Demo	SF	Lg	AB	.043	370.0	---	1.27	.14	1.41	**2.02**
	Demo	SF	Sm	AB	.062	259.0	---	1.83	.19	2.02	**2.90**
6" T	Demo	SF	Lg	AB	.046	350.0	---	1.36	.14	1.50	**2.15**
	Demo	SF	Sm	AB	.065	245.0	---	1.92	.21	2.13	**3.05**
8" T	Demo	SF	Lg	AB	.048	335.0	---	1.42	.15	1.57	**2.25**
	Demo	SF	Sm	AB	.068	234.5	---	2.01	.21	2.22	**3.18**
12" T	Demo	SF	Lg	AB	.058	275.0	---	1.71	.18	1.89	**2.71**
	Demo	SF	Sm	AB	.083	192.5	---	2.45	.26	2.71	**3.88**

Description	Oper	Unit	Vol	Crew Size	Man-hours per Unit	Crew Output per Day	Avg Mat'l Unit Cost	Avg Labor Unit Cost	Avg Equip Unit Cost	Avg Total Unit Cost	Avg Price Incl O&P
Walls, no scaffolding included											
4" x 4" x 16"	Inst	SF	Lg	BO	.128	250.0	1.74	3.80	.14	5.68	**7.81**
	Inst	SF	Sm	ML	.137	175.0	1.99	4.25	.20	6.44	**8.82**
6" x 4" x 16"	Inst	SF	Lg	BO	.133	240.0	2.12	3.95	.15	6.22	**8.47**
	Inst	SF	Sm	ML	.143	168.0	2.44	4.44	.21	7.09	**9.63**
6" x 6" x 16"	Inst	SF	Lg	BO	.119	270.0	1.58	3.54	.13	5.25	**7.21**
	Inst	SF	Sm	ML	.127	189.0	1.83	3.94	.19	5.96	**8.17**
8" x 4" x 16"	Inst	SF	Lg	BO	.139	230.0	2.64	4.13	.16	6.93	**9.35**
	Inst	SF	Sm	ML	.149	161.0	3.03	4.62	.22	7.87	**10.60**
8" x 6" x 16"	Inst	SF	Lg	BO	.123	260.0	2.02	3.65	.14	5.81	**7.91**
	Inst	SF	Sm	ML	.132	182.0	2.32	4.10	.20	6.62	**8.97**
12" x 4" x 16"	Inst	SF	Lg	BO	.152	210.0	4.35	4.52	.17	9.04	**11.90**
	Inst	SF	Sm	ML	.163	147.0	5.01	5.06	.24	10.31	**13.50**
12" x 6" x 16"	Inst	SF	Lg	BO	.133	240.0	3.14	3.95	.15	7.24	**9.65**
	Inst	SF	Sm	ML	.143	168.0	3.61	4.44	.21	8.26	**11.00**

Concrete slump brick

Running bond, 3/8 " mortar joints

Walls, with air tools

Description	Oper	Unit	Vol	Crew Size	Man-hours per Unit	Crew Output per Day	Avg Mat'l Unit Cost	Avg Labor Unit Cost	Avg Equip Unit Cost	Avg Total Unit Cost	Avg Price Incl O&P
4" T	Demo	SF	Lg	AB	.043	370.0	---	1.27	.14	1.41	**2.02**
	Demo	SF	Sm	AB	.062	259.0	---	1.83	.19	2.02	**2.90**

Walls, no scaffolding included

Description	Oper	Unit	Vol	Crew Size	Man-hours per Unit	Crew Output per Day	Avg Mat'l Unit Cost	Avg Labor Unit Cost	Avg Equip Unit Cost	Avg Total Unit Cost	Avg Price Incl O&P
4" x 4" x 8"	Inst	SF	Lg	BO	.145	220.0	11.80	4.31	.16	16.27	**20.20**
	Inst	SF	Sm	ML	.156	154.0	13.60	4.84	.23	18.67	**23.00**
4" x 4" x 12"	Inst	SF	Lg	BO	.133	240.0	6.11	3.95	.15	10.21	**13.10**
	Inst	SF	Sm	ML	.143	168.0	7.03	4.44	.21	11.68	**14.90**

Concrete textured screen block

Running bond, 3/8 " mortar joints

Walls, with air tools

Description	Oper	Unit	Vol	Crew Size	Man-hours per Unit	Crew Output per Day	Avg Mat'l Unit Cost	Avg Labor Unit Cost	Avg Equip Unit Cost	Avg Total Unit Cost	Avg Price Incl O&P
4" T	Demo	SF	Lg	AB	.033	480.0	---	.97	.10	1.07	**1.54**
	Demo	SF	Sm	AB	.048	336.0	---	1.42	.15	1.57	**2.25**

Walls, no scaffolding included

Description	Oper	Unit	Vol	Crew Size	Man-hours per Unit	Crew Output per Day	Avg Mat'l Unit Cost	Avg Labor Unit Cost	Avg Equip Unit Cost	Avg Total Unit Cost	Avg Price Incl O&P
4" x 6" x 6"	Inst	SF	Lg	BO	.200	160.0	2.64	5.94	.22	8.80	**12.10**
	Inst	SF	Sm	ML	.214	112.0	3.05	6.64	.32	10.01	**13.70**
4" x 8" x 8"	Inst	SF	Lg	BO	.133	240.0	3.56	3.95	.15	7.66	**10.10**
	Inst	SF	Sm	ML	.143	168.0	4.11	4.44	.21	8.76	**11.60**
4" x 12" x 12"	Inst	SF	Lg	BO	.089	360.0	2.12	2.64	.10	4.86	**6.48**
	Inst	SF	Sm	ML	.095	252.0	2.44	2.95	.14	5.53	**7.34**

Description	Oper	Unit	Vol	Crew Size	Man-hours per Unit	Crew Output per Day	Avg Mat'l Unit Cost	Avg Labor Unit Cost	Avg Equip Unit Cost	Avg Total Unit Cost	Avg Price Incl O&P

Glass block
Plain, 4" thick

Description	Oper	Unit	Vol	Crew Size	Man-hours per Unit	Crew Output per Day	Avg Mat'l Unit Cost	Avg Labor Unit Cost	Avg Equip Unit Cost	Avg Total Unit Cost	Avg Price Incl O&P
6" x 6"	Demo	SF	Lg	LB	.064	250.0	---	1.76	---	1.76	2.62
	Demo	SF	Sm	LB	.091	175.0	---	2.50	---	2.50	3.72
8" x 8"	Demo	SF	Lg	LB	.057	280.0	---	1.56	---	1.56	2.33
	Demo	SF	Sm	LB	.082	196.0	---	2.25	---	2.25	3.35
12" x 12"	Demo	SF	Lg	LB	.052	310.0	---	1.43	---	1.43	2.13
	Demo	SF	Sm	LB	.074	217.0	---	2.03	---	2.03	3.02
6" x 6"	Inst	SF	Lg	ML	.320	75.00	18.50	9.93	---	28.43	36.00
	Inst	SF	Sm	BK	.305	52.50	21.20	9.06	---	30.26	37.90
8" x 8"	Inst	SF	Lg	ML	.218	110.0	11.50	6.77	---	18.27	23.30
	Inst	SF	Sm	BK	.208	77.00	13.20	6.18	---	19.38	24.40
12" x 12"	Inst	SF	Lg	ML	.145	165.0	14.60	4.50	---	19.10	23.50
	Inst	SF	Sm	BK	.139	115.5	16.80	4.13	---	20.93	25.50

Glazed tile
6-T Series. Includes normal allowance for special shapes

Description	Oper	Unit	Vol	Crew Size	Man-hours per Unit	Crew Output per Day	Avg Mat'l Unit Cost	Avg Labor Unit Cost	Avg Equip Unit Cost	Avg Total Unit Cost	Avg Price Incl O&P
All sizes	Demo	SF	Lg	LB	.052	310.0	---	1.43	---	1.43	2.13
Glazed 1 side	Demo	SF	Sm	LB	.074	217.0	---	2.03	---	2.03	3.02
2" x 5-1/3" x 12"	Inst	SF	Lg	ML	.286	84.00	3.18	8.88	---	12.06	16.90
	Inst	SF	Sm	BK	.379	42.22	3.64	11.30	---	14.94	21.00
4" x 5-1/3" x 12"	Inst	SF	Lg	ML	.300	80.00	4.80	9.31	---	14.11	19.40
	Inst	SF	Sm	BK	.398	40.20	5.49	11.80	---	17.29	23.90
6" x 5-1/3" x 12"	Inst	SF	Lg	ML	.316	76.00	6.29	9.81	---	16.10	21.90
	Inst	SF	Sm	BK	.419	38.19	7.19	12.50	---	19.69	26.80
8" x 5-1/3" x 12"	Inst	SF	Lg	ML	.353	68.00	7.47	11.00	---	18.47	24.90
	Inst	SF	Sm	BK	.468	34.19	8.52	13.90	---	22.42	30.50
Glazed 2 sides											
4" x 5-1/3" x 12"	Inst	SF	Lg	ML	.375	64.00	7.23	11.60	---	18.83	25.70
	Inst	SF	Sm	BK	.498	32.13	8.28	14.80	---	23.08	31.60
6" x 5-1/3" x 12"	Inst	SF	Lg	ML	.400	60.00	8.86	12.40	---	21.26	28.70
	Inst	SF	Sm	BK	.531	30.13	10.10	15.80	---	25.90	35.20
8" x 5-1/3" x 12"	Inst	SF	Lg	ML	.462	52.00	10.70	14.30	---	25.00	33.60
	Inst	SF	Sm	BK	.612	26.14	12.20	18.20	---	30.40	41.10

Quarry tile
Floors

Description	Oper	Unit	Vol	Crew Size	Man-hours per Unit	Crew Output per Day	Avg Mat'l Unit Cost	Avg Labor Unit Cost	Avg Equip Unit Cost	Avg Total Unit Cost	Avg Price Incl O&P
Conventional mortar set	Demo	SF	Lg	LB	.030	535.0	---	.82	---	.82	1.23
	Demo	SF	Sm	LB	.043	372.1	---	1.18	---	1.18	1.76
Dry-set mortar	Demo	SF	Lg	LB	.026	620.0	---	.71	---	.71	1.06
	Demo	SF	Sm	LB	.037	432.4	---	1.01	---	1.01	1.51
Conventional mortar set with unmounted tile											
6" x 6" x 1/2" T	Inst	SF	Lg	BO	.123	260.0	1.18	3.65	.14	4.97	6.94
	Inst	SF	Sm	ML	.164	146.3	1.36	5.09	.24	6.69	9.39
9" x 9" x 3/4" T	Inst	SF	Lg	BO	.103	310.0	1.65	3.06	.12	4.83	6.58
	Inst	SF	Sm	ML	.138	173.9	1.89	4.28	.21	6.38	8.77
Dry-set mortar with unmounted tile											
6" x 6" x 1/2" T	Inst	SF	Lg	BO	.102	315.0	4.13	3.03	.11	7.27	9.37
	Inst	SF	Sm	ML	.135	177.8	4.76	4.19	.20	9.15	11.90
9" x 9" x 3/4" T	Inst	SF	Lg	BO	.082	390.0	4.61	2.44	.09	7.14	9.02
	Inst	SF	Sm	ML	.109	220.2	5.29	3.38	.16	8.83	11.30

Molding

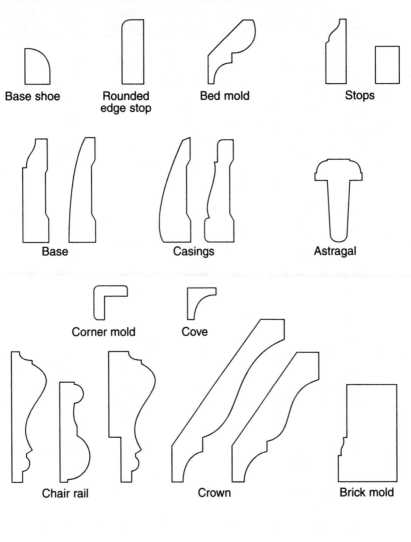

Base shoe

Rounded edge stop

Bed mold

Stops

Base

Casings

Astragal

Corner mold

Cove

Chair rail

Crown

Brick mold

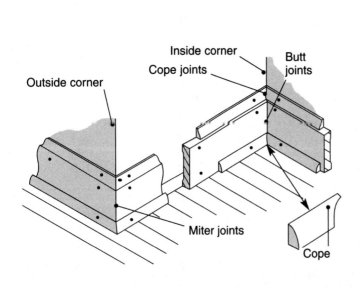

Outside corner

Inside corner

Cope joints

Butt joints

Miter joints

Cope

Base molding installation

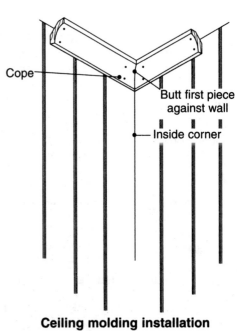

Cope

Butt first piece against wall

Inside corner

Ceiling molding installation

Description	Oper	Unit	Vol	Crew Size	Man-hours per Unit	Crew Output per Day	Avg Mat'l Unit Cost	Avg Labor Unit Cost	Avg Equip Unit Cost	Avg Total Unit Cost	Avg Price Incl O&P

Molding and trim
Removal

Description	Oper	Unit	Vol	Crew Size	Man-hours per Unit	Crew Output per Day	Avg Mat'l Unit Cost	Avg Labor Unit Cost	Avg Equip Unit Cost	Avg Total Unit Cost	Avg Price Incl O&P
At base (floor)	Demo	LF	Lg	LB	.015	1060	---	.41	---	.41	**.61**
	Demo	LF	Sm	LB	.022	742.0	---	.60	---	.60	**.90**
At ceiling	Demo	LF	Lg	LB	.013	1200	---	.36	---	.36	**.53**
	Demo	LF	Sm	LB	.019	840.0	---	.52	---	.52	**.78**
On wall or cabinets	Demo	LF	Lg	LB	.010	1600	---	.27	---	.27	**.41**
	Demo	LF	Sm	LB	.014	1120	---	.38	---	.38	**.57**

Solid pine
Unfinished

Casing

Flat streamline

Description	Oper	Unit	Vol	Crew Size	Man-hours per Unit	Crew Output per Day	Avg Mat'l Unit Cost	Avg Labor Unit Cost	Avg Equip Unit Cost	Avg Total Unit Cost	Avg Price Incl O&P
1/2" x 1-5/8"	Inst	LF	Lg	CA	.030	266.0	.88	1.00	---	1.88	**2.51**
	Inst	LF	Sm	CA	.043	186.0	1.03	1.43	---	2.46	**3.33**
5/8" x 1-5/8"	Inst	LF	Lg	CA	.030	266.0	.96	1.00	---	1.96	**2.60**
	Inst	LF	Sm	CA	.043	186.0	1.11	1.43	---	2.54	**3.42**
5/8" x 2-1/2"	Inst	LF	Lg	CA	.030	266.0	1.23	1.00	---	2.23	**2.91**
	Inst	LF	Sm	CA	.043	186.0	1.43	1.43	---	2.86	**3.79**

Oval streamline

Description	Oper	Unit	Vol	Crew Size	Man-hours per Unit	Crew Output per Day	Avg Mat'l Unit Cost	Avg Labor Unit Cost	Avg Equip Unit Cost	Avg Total Unit Cost	Avg Price Incl O&P
1/2" x 1-5/8"	Inst	LF	Lg	CA	.030	266.0	.90	1.00	---	1.90	**2.53**
	Inst	LF	Sm	CA	.043	186.0	1.05	1.43	---	2.48	**3.35**
5/8" x 1-5/8"	Inst	LF	Lg	CA	.030	266.0	.96	1.00	---	1.96	**2.60**
	Inst	LF	Sm	CA	.043	186.0	1.11	1.43	---	2.54	**3.42**
5/8" x 2-1/2"	Inst	LF	Lg	CA	.030	266.0	1.41	1.00	---	2.41	**3.12**
	Inst	LF	Sm	CA	.043	186.0	1.64	1.43	---	3.07	**4.03**

711

Description	Oper	Unit	Vol	Crew Size	Man-hours per Unit	Crew Output per Day	Avg Mat'l Unit Cost	Avg Labor Unit Cost	Avg Equip Unit Cost	Avg Total Unit Cost	Avg Price Incl O&P
5/8" x 1-5/8"	Inst	LF	Lg	CA	.030	266.0	.96	1.00	---	1.96	**2.60**
	Inst	LF	Sm	CA	.043	186.0	1.11	1.43	---	2.54	**3.42**
5/8" x 2-1/2"	Inst	LF	Lg	CA	.030	266.0	1.41	1.00	---	2.41	**3.12**
	Inst	LF	Sm	CA	.043	186.0	1.64	1.43	---	3.07	**4.03**

3 step

Description	Oper	Unit	Vol	Crew Size	Man-hours per Unit	Crew Output per Day	Avg Mat'l Unit Cost	Avg Labor Unit Cost	Avg Equip Unit Cost	Avg Total Unit Cost	Avg Price Incl O&P
5/8" x 1-5/8"	Inst	LF	Lg	CA	.030	266.0	.86	1.00	---	1.86	**2.49**
	Inst	LF	Sm	CA	.043	186.0	1.00	1.43	---	2.43	**3.30**
5/8" x 2-1/2"	Inst	LF	Lg	CA	.030	266.0	1.26	1.00	---	2.26	**2.95**
	Inst	LF	Sm	CA	.043	186.0	1.47	1.43	---	2.90	**3.84**

2 round edge

Description	Oper	Unit	Vol	Crew Size	Man-hours per Unit	Crew Output per Day	Avg Mat'l Unit Cost	Avg Labor Unit Cost	Avg Equip Unit Cost	Avg Total Unit Cost	Avg Price Incl O&P
5/8" x 2-1/2"	Inst	LF	Lg	CA	.030	266.0	.93	1.00	---	1.93	**2.57**
	Inst	LF	Sm	CA	.043	186.0	1.08	1.43	---	2.51	**3.39**

B & B

Description	Oper	Unit	Vol	Crew Size	Man-hours per Unit	Crew Output per Day	Avg Mat'l Unit Cost	Avg Labor Unit Cost	Avg Equip Unit Cost	Avg Total Unit Cost	Avg Price Incl O&P
1/2" x 1-5/8"	Inst	LF	Lg	CA	.030	266.0	.86	1.00	---	1.86	**2.49**
	Inst	LF	Sm	CA	.043	186.0	1.00	1.43	---	2.43	**3.30**

4 bead

Description	Oper	Unit	Vol	Crew Size	Man-hours per Unit	Crew Output per Day	Avg Mat'l Unit Cost	Avg Labor Unit Cost	Avg Equip Unit Cost	Avg Total Unit Cost	Avg Price Incl O&P
1/2" x 1-5/8"	Inst	LF	Lg	CA	.030	266.0	.81	1.00	---	1.81	**2.43**
	Inst	LF	Sm	CA	.043	186.0	.94	1.43	---	2.37	**3.23**

Description	Oper	Unit	Vol	Crew Size	Man-hours per Unit	Crew Output per Day	Avg Mat'l Unit Cost	Avg Labor Unit Cost	Avg Equip Unit Cost	Avg Total Unit Cost	Avg Price Incl O&P
1550											
1/2" x 1-5/8"	Inst	LF	Lg	CA	.030	266.0	.71	1.00	---	1.71	**2.31**
	Inst	LF	Sm	CA	.043	186.0	.82	1.43	---	2.25	**3.09**
Mullion											
1/4" x 3-1/2"	Inst	LF	Lg	CA	.030	266.0	.90	1.00	---	1.90	**2.53**
	Inst	LF	Sm	CA	.043	186.0	1.05	1.43	---	2.48	**3.35**
Universal											
5/8" x 2-1/2"	Inst	LF	Lg	CA	.030	266.0	1.00	1.00	---	2.00	**2.65**
	Inst	LF	Sm	CA	.043	186.0	1.17	1.43	---	2.60	**3.49**
Colonial											
11/16" x 2-1/2"	Inst	LF	Lg	CA	.030	266.0	1.14	1.00	---	2.14	**2.81**
	Inst	LF	Sm	CA	.043	186.0	1.33	1.43	---	2.76	**3.68**
11/16" x 3-1/2"	Inst	LF	Lg	CA	.030	266.0	1.44	1.00	---	2.44	**3.15**
	Inst	LF	Sm	CA	.043	186.0	1.67	1.43	---	3.10	**4.07**
11/16" x 4-1/2"	Inst	LF	Lg	CA	.030	266.0	2.64	1.00	---	3.64	**4.53**
	Inst	LF	Sm	CA	.043	186.0	3.07	1.43	---	4.50	**5.68**
11/16" x 2-7/8"	Inst	LF	Lg	CA	.030	266.0	1.42	1.00	---	2.42	**3.13**
	Inst	LF	Sm	CA	.043	186.0	1.65	1.43	---	3.08	**4.04**
356											
11/16" x 2-1/4"	Inst	LF	Lg	CA	.030	266.0	1.16	1.00	---	2.16	**2.83**
	Inst	LF	Sm	CA	.043	186.0	1.35	1.43	---	2.78	**3.70**
Back band											
11/16" x 1-1/8"	Inst	LF	Lg	CA	.030	266.0	.65	1.00	---	1.65	**2.24**
	Inst	LF	Sm	CA	.043	186.0	.76	1.43	---	2.19	**3.02**
Cape Cod											
5/8" x 2-1/2"	Inst	LF	Lg	CA	.030	266.0	1.26	1.00	---	2.26	**2.95**
	Inst	LF	Sm	CA	.043	186.0	1.47	1.43	---	2.90	**3.84**
11/16" x 3-1/2"	Inst	LF	Lg	CA	.030	266.0	1.95	1.00	---	2.95	**3.74**
	Inst	LF	Sm	CA	.043	186.0	2.27	1.43	---	3.70	**4.76**
Rosette plinth block w/ edge detail											
3/4" x 2-3/4"	Inst	Ea	Lg	CA	.250	32.00	3.42	8.32	---	11.74	**16.40**
	Inst	Ea	Sm	CA	.364	22.00	3.98	12.10	---	16.08	**22.70**
3/4" x 3-1/2"	Inst	Ea	Lg	CA	.250	32.00	3.78	8.32	---	12.10	**16.80**
	Inst	Ea	Sm	CA	.364	22.00	4.40	12.10	---	16.50	**23.20**
3/4" x 3-3/4"	Inst	Ea	Lg	CA	.250	32.00	4.38	8.32	---	12.70	**17.50**
	Inst	Ea	Sm	CA	.364	22.00	5.09	12.10	---	17.19	**24.00**

Description	Oper	Unit	Vol	Crew Size	Man-hours per Unit	Crew Output per Day	Avg Mat'l Unit Cost	Avg Labor Unit Cost	Avg Equip Unit Cost	Avg Total Unit Cost	Avg Price Incl O&P
Base & base shoe											
Shoe											
3/8" x 3/4"	Inst	LF	Lg	CA	.030	266.0	.50	1.00	---	1.50	**2.07**
	Inst	LF	Sm	CA	.043	186.0	.58	1.43	---	2.01	**2.81**
Flat streamline											
7/16" x 1-1/2"	Inst	LF	Lg	CA	.030	266.0	.58	1.00	---	1.58	**2.16**
	Inst	LF	Sm	CA	.043	186.0	.67	1.43	---	2.10	**2.92**
Streamline											
7/16" x 1-5/8"	Inst	LF	Lg	CA	.030	266.0	.67	1.00	---	1.67	**2.27**
	Inst	LF	Sm	CA	.043	186.0	.78	1.43	---	2.21	**3.04**
1/2" x 1-5/8"	Inst	LF	Lg	CA	.030	266.0	.73	1.00	---	1.73	**2.34**
	Inst	LF	Sm	CA	.043	186.0	.85	1.43	---	2.28	**3.12**
3/8" x 2-1/4"	Inst	LF	Lg	CA	.030	266.0	.88	1.00	---	1.88	**2.51**
	Inst	LF	Sm	CA	.043	186.0	1.03	1.43	---	2.46	**3.33**
7/16" x 2-1/2"	Inst	LF	Lg	CA	.030	266.0	1.23	1.00	---	2.23	**2.91**
	Inst	LF	Sm	CA	.043	186.0	1.43	1.43	---	2.86	**3.79**
7/16" x 3-1/2"	Inst	LF	Lg	CA	.030	266.0	1.49	1.00	---	2.49	**3.21**
	Inst	LF	Sm	CA	.043	186.0	1.73	1.43	---	3.16	**4.14**
1/2" x 2-1/2"	Inst	LF	Lg	CA	.030	266.0	1.21	1.00	---	2.21	**2.89**
	Inst	LF	Sm	CA	.043	186.0	1.40	1.43	---	2.83	**3.76**
Combination											
1/2" x 1-5/8"	Inst	LF	Lg	CA	.030	266.0	.67	1.00	---	1.67	**2.27**
	Inst	LF	Sm	CA	.043	186.0	.78	1.43	---	2.21	**3.04**
1/2" x 2-1/2"	Inst	LF	Lg	CA	.030	266.0	.99	1.00	---	1.99	**2.64**
	Inst	LF	Sm	CA	.043	186.0	1.16	1.43	---	2.59	**3.48**
711											
1/2" x 2-1/2"	Inst	LF	Lg	CA	.030	266.0	.97	1.00	---	1.97	**2.61**
	Inst	LF	Sm	CA	.043	186.0	1.12	1.43	---	2.55	**3.43**
7/16" x 3-1/2"	Inst	LF	Lg	CA	.030	266.0	1.06	1.00	---	2.06	**2.72**
	Inst	LF	Sm	CA	.043	186.0	1.23	1.43	---	2.66	**3.56**
3 step											
1/2" x 2-1/2"	Inst	LF	Lg	CA	.030	266.0	.78	1.00	---	1.78	**2.39**
	Inst	LF	Sm	CA	.043	186.0	.91	1.43	---	2.34	**3.19**
1/2" x 3-1/2"	Inst	LF	Lg	CA	.030	266.0	1.13	1.00	---	2.13	**2.80**
	Inst	LF	Sm	CA	.043	186.0	1.32	1.43	---	2.75	**3.66**
1550											
7/16" x 2-1/2"	Inst	LF	Lg	CA	.030	266.0	1.23	1.00	---	2.23	**2.91**
	Inst	LF	Sm	CA	.043	186.0	1.43	1.43	---	2.86	**3.79**
Cape Cod											
9/16" x 4"	Inst	LF	Lg	CA	.030	266.0	1.29	1.00	---	2.29	**2.98**
	Inst	LF	Sm	CA	.043	186.0	1.50	1.43	---	2.93	**3.87**

Description	Oper	Unit	Vol	Crew Size	Man-hours per Unit	Crew Output per Day	Avg Mat'l Unit Cost	Avg Labor Unit Cost	Avg Equip Unit Cost	Avg Total Unit Cost	Avg Price Incl O&P
Colonial											
11/16" x 3-1/2"	Inst	LF	Lg	CA	.030	266.0	1.58	1.00	---	2.58	**3.31**
	Inst	LF	Sm	CA	.043	186.0	1.84	1.43	---	3.27	**4.26**
11/16" x 4-1/2"	Inst	LF	Lg	CA	.030	266.0	1.79	1.00	---	2.79	**3.56**
	Inst	LF	Sm	CA	.043	186.0	2.09	1.43	---	3.52	**4.55**
Base cap											
11/16" x 1-3/8"	Inst	LF	Lg	CA	.030	266.0	.75	1.00	---	1.75	**2.36**
	Inst	LF	Sm	CA	.043	186.0	.87	1.43	---	2.30	**3.15**
Rosette plinth block w/ edge detail											
3/4" x 2-3/4" x 6"	Inst	Ea	Lg	CA	.250	32.00	2.81	8.32	---	11.13	**15.70**
	Inst	Ea	Sm	CA	.364	22.00	3.26	12.10	---	15.36	**21.90**
3/4" x 3-1/2" x 6"	Inst	Ea	Lg	CA	.250	32.00	3.10	8.32	---	11.42	**16.00**
	Inst	Ea	Sm	CA	.364	22.00	3.61	12.10	---	15.71	**22.30**
3/4" x 3-3/4" x 6"	Inst	Ea	Lg	CA	.250	32.00	3.59	8.32	---	11.91	**16.60**
	Inst	Ea	Sm	CA	.364	22.00	4.17	12.10	---	16.27	**23.00**
Colonial plinth block w/ edge detail											
1" x 2-3/4" x 6"	Inst	Ea	Lg	CA	.250	32.00	2.81	8.32	---	11.13	**15.70**
	Inst	Ea	Sm	CA	.364	22.00	3.26	12.10	---	15.36	**21.90**
1" x 3-3/4" x 6"	Inst	Ea	Lg	CA	.250	32.00	3.59	8.32	---	11.91	**16.60**
	Inst	Ea	Sm	CA	.364	22.00	4.17	12.10	---	16.27	**23.00**

Stops

Description	Oper	Unit	Vol	Crew Size	Man-hours per Unit	Crew Output per Day	Avg Mat'l Unit Cost	Avg Labor Unit Cost	Avg Equip Unit Cost	Avg Total Unit Cost	Avg Price Incl O&P
Round edge											
3/8" x 1/2"	Inst	LF	Lg	CA	.030	266.0	.52	1.00	---	1.52	**2.10**
	Inst	LF	Sm	CA	.043	186.0	.60	1.43	---	2.03	**2.84**
3/8" x 3/4"	Inst	LF	Lg	CA	.030	266.0	.54	1.00	---	1.54	**2.12**
	Inst	LF	Sm	CA	.043	186.0	.63	1.43	---	2.06	**2.87**
3/8" x 1"	Inst	LF	Lg	CA	.030	266.0	.67	1.00	---	1.67	**2.27**
	Inst	LF	Sm	CA	.043	186.0	.78	1.43	---	2.21	**3.04**
3/8" x 1-1/4"	Inst	LF	Lg	CA	.030	266.0	.80	1.00	---	1.80	**2.42**
	Inst	LF	Sm	CA	.043	186.0	.93	1.43	---	2.36	**3.22**
3/8" x 1-5/8"	Inst	LF	Lg	CA	.030	266.0	.82	1.00	---	1.82	**2.44**
	Inst	LF	Sm	CA	.043	186.0	.95	1.43	---	2.38	**3.24**
1/2" x 3/4"	Inst	LF	Lg	CA	.030	266.0	.63	1.00	---	1.63	**2.22**
	Inst	LF	Sm	CA	.043	186.0	.74	1.43	---	2.17	**3.00**
1/2" x 1"	Inst	LF	Lg	CA	.030	266.0	.65	1.00	---	1.65	**2.24**
	Inst	LF	Sm	CA	.043	186.0	.76	1.43	---	2.19	**3.02**
1/2" x 1-1/4"	Inst	LF	Lg	CA	.030	266.0	.79	1.00	---	1.79	**2.41**
	Inst	LF	Sm	CA	.043	186.0	.92	1.43	---	2.35	**3.20**
1/2" x 1-5/8"	Inst	LF	Lg	CA	.030	266.0	.91	1.00	---	1.91	**2.54**
	Inst	LF	Sm	CA	.043	186.0	1.06	1.43	---	2.49	**3.37**
Ogee stop											
7/16" x 1-5/8"	Inst	LF	Lg	CA	.030	266.0	.57	1.00	---	1.57	**2.15**
	Inst	LF	Sm	CA	.043	186.0	.66	1.43	---	2.09	**2.91**

Description	Oper	Unit	Vol	Crew Size	Man-hours per Unit	Crew Output per Day	Avg Mat'l Unit Cost	Avg Labor Unit Cost	Avg Equip Unit Cost	Avg Total Unit Cost	Avg Price Incl O&P
Crown & cornice											
Crown											
9/16" x 1-5/8"	Inst	LF	Lg	CA	.040	200.0	1.24	1.33	---	2.57	**3.42**
	Inst	LF	Sm	CA	.057	140.0	1.44	1.90	---	3.34	**4.50**
9/16" x 2-1/4"	Inst	LF	Lg	CA	.040	200.0	1.38	1.33	---	2.71	**3.58**
	Inst	LF	Sm	CA	.057	140.0	1.61	1.90	---	3.51	**4.70**
9/16" x 3-1/2"	Inst	LF	Lg	CA	.040	200.0	1.52	1.33	---	2.85	**3.74**
	Inst	LF	Sm	CA	.057	140.0	1.77	1.90	---	3.67	**4.88**
9/16" x 4-1/4"	Inst	LF	Lg	CA	.040	200.0	1.70	1.33	---	3.03	**3.95**
	Inst	LF	Sm	CA	.057	140.0	1.98	1.90	---	3.88	**5.12**
9/16" x 5-1/2"	Inst	LF	Lg	CA	.040	200.0	1.92	1.33	---	3.25	**4.20**
	Inst	LF	Sm	CA	.057	140.0	2.24	1.90	---	4.14	**5.42**
Colonial crown											
9/16" x 3-3/8"	Inst	LF	Lg	CA	.040	200.0	1.32	1.33	---	2.65	**3.51**
	Inst	LF	Sm	CA	.057	140.0	1.54	1.90	---	3.44	**4.62**
9/16" x 4-1/2"	Inst	LF	Lg	CA	.040	200.0	1.70	1.33	---	3.03	**3.95**
	Inst	LF	Sm	CA	.057	140.0	1.98	1.90	---	3.88	**5.12**
Cornice											
1-1/8" x 4-1/2"	Inst	LF	Lg	CA	.040	200.0	7.06	1.33	---	8.39	**10.10**
	Inst	LF	Sm	CA	.057	140.0	8.21	1.90	---	10.11	**12.30**
1-1/4" x 6"	Inst	LF	Lg	CA	.040	200.0	7.70	1.33	---	9.03	**10.90**
	Inst	LF	Sm	CA	.057	140.0	8.96	1.90	---	10.86	**13.20**
Chair rail											
Chair rail											
1/2" x 1-1/2"	Inst	LF	Lg	CA	.030	266.0	.65	1.00	---	1.65	**2.24**
	Inst	LF	Sm	CA	.043	186.0	.76	1.43	---	2.19	**3.02**
9/16" x 2-1/2"	Inst	LF	Lg	CA	.030	266.0	1.44	1.00	---	2.44	**3.15**
	Inst	LF	Sm	CA	.043	186.0	1.67	1.43	---	3.10	**4.07**
Colonial chair rail											
11/16" x 2-1/2"	Inst	LF	Lg	CA	.030	266.0	1.10	1.00	---	2.10	**2.76**
	Inst	LF	Sm	CA	.043	186.0	1.28	1.43	---	2.71	**3.62**
Outside corner guard											
3/4" x 3/4"	Inst	LF	Lg	CA	.030	266.0	.66	1.00	---	1.66	**2.26**
	Inst	LF	Sm	CA	.043	186.0	.77	1.43	---	2.20	**3.03**
1" x 1"	Inst	LF	Lg	CA	.030	266.0	1.23	1.00	---	2.23	**2.91**
	Inst	LF	Sm	CA	.043	186.0	1.43	1.43	---	2.86	**3.79**
1-1/8" x 1-1/8"	Inst	LF	Lg	CA	.030	266.0	1.69	1.00	---	2.69	**3.44**
	Inst	LF	Sm	CA	.043	186.0	1.97	1.43	---	3.40	**4.41**
1-3/8" x 1-3/8"	Inst	LF	Lg	CA	.030	266.0	1.85	1.00	---	2.85	**3.62**
	Inst	LF	Sm	CA	.043	186.0	2.15	1.43	---	3.58	**4.62**

Description	Oper	Unit	Vol	Crew Size	Man-hours per Unit	Crew Output per Day	Avg Mat'l Unit Cost	Avg Labor Unit Cost	Avg Equip Unit Cost	Avg Total Unit Cost	Avg Price Incl O&P
Astragal											
Flat astragal											
3/4" x 1-5/8"	Inst	LF	Lg	CA	.030	266.0	1.64	1.00	---	2.64	**3.38**
	Inst	LF	Sm	CA	.043	186.0	1.90	1.43	---	3.33	**4.33**
T astragal											
1-1/4" x 2-1/4"	Inst	LF	Lg	CA	.030	266.0	2.66	1.00	---	3.66	**4.56**
	Inst	LF	Sm	CA	.043	186.0	3.09	1.43	---	4.52	**5.70**
Cap											
Panel cap											
1/2" x 1-1/2"	Inst	LF	Lg	CA	.030	266.0	1.00	1.00	---	2.00	**2.65**
	Inst	LF	Sm	CA	.043	186.0	1.17	1.43	---	2.60	**3.49**
Wainscot cap											
5/8" x 1-1/4"	Inst	LF	Lg	CA	.030	266.0	1.00	1.00	---	2.00	**2.65**
	Inst	LF	Sm	CA	.043	186.0	1.17	1.43	---	2.60	**3.49**
Glass bead & screen											
Insert											
3/16" x 1/2"	Inst	LF	Lg	CA	.030	266.0	.48	1.00	---	1.48	**2.05**
	Inst	LF	Sm	CA	.043	186.0	.56	1.43	---	1.99	**2.79**
Flat											
1/4" x 3/4"	Inst	LF	Lg	CA	.030	266.0	.48	1.00	---	1.48	**2.05**
	Inst	LF	Sm	CA	.043	186.0	.56	1.43	---	1.99	**2.79**
Beaded											
1/4" x 3/4"	Inst	LF	Lg	CA	.030	266.0	.48	1.00	---	1.48	**2.05**
	Inst	LF	Sm	CA	.043	186.0	.56	1.43	---	1.99	**2.79**
Scribe											
3/16" x 3/4"	Inst	LF	Lg	CA	.030	266.0	.29	1.00	---	1.29	**1.83**
	Inst	LF	Sm	CA	.043	186.0	.34	1.43	---	1.77	**2.54**
Cloverlcaf											
3/8" x 3/4"	Inst	LF	Lg	CA	.030	266.0	.57	1.00	---	1.57	**2.15**
	Inst	LF	Sm	CA	.043	186.0	.66	1.43	---	2.09	**2.91**
Glass bead											
1/4" x 7/16"	Inst	LF	Lg	CA	.030	266.0	.44	1.00	---	1.44	**2.00**
	Inst	LF	Sm	CA	.043	186.0	.51	1.43	---	1.94	**2.73**
3/8" x 3/8"	Inst	LF	Lg	CA	.030	266.0	.44	1.00	---	1.44	**2.00**
	Inst	LF	Sm	CA	.043	186.0	.51	1.43	---	1.94	**2.73**
Quarter round											
1/4" x 1/4"	Inst	LF	Lg	CA	.030	266.0	.32	1.00	---	1.32	**1.87**
	Inst	LF	Sm	CA	.043	186.0	.37	1.43	---	1.80	**2.57**
3/8" x 3/8"	Inst	LF	Lg	CA	.030	266.0	.40	1.00	---	1.40	**1.96**
	Inst	LF	Sm	CA	.043	186.0	.46	1.43	---	1.89	**2.68**
1/2" x 1/2"	Inst	LF	Lg	CA	.030	266.0	.48	1.00	---	1.48	**2.05**
	Inst	LF	Sm	CA	.043	186.0	.56	1.43	---	1.99	**2.79**
5/8" x 5/8"	Inst	LF	Lg	CA	.030	266.0	.52	1.00	---	1.52	**2.10**
	Inst	LF	Sm	CA	.043	186.0	.60	1.43	---	2.03	**2.84**
3/4" x 3/4"	Inst	LF	Lg	CA	.030	266.0	.66	1.00	---	1.66	**2.26**
	Inst	LF	Sm	CA	.043	186.0	.77	1.43	---	2.20	**3.03**
1" x 1"	Inst	LF	Lg	CA	.030	266.0	1.32	1.00	---	2.32	**3.02**
	Inst	LF	Sm	CA	.043	186.0	1.54	1.43	---	2.97	**3.92**

Description	Oper	Unit	Vol	Crew Size	Man-hours per Unit	Crew Output per Day	Avg Mat'l Unit Cost	Avg Labor Unit Cost	Avg Equip Unit Cost	Avg Total Unit Cost	Avg Price Incl O&P

Cove

Cove

3/8" x 3/8"	Inst	LF	Lg	CA	.040	200.0	.37	1.33	---	1.70	**2.42**
	Inst	LF	Sm	CA	.057	140.0	.43	1.90	---	2.33	**3.34**
1/2" x 1/2"	Inst	LF	Lg	CA	.040	200.0	.44	1.33	---	1.77	**2.50**
	Inst	LF	Sm	CA	.057	140.0	.51	1.90	---	2.41	**3.43**
5/8" x 5/8"	Inst	LF	Lg	CA	.040	200.0	.66	1.33	---	1.99	**2.76**
	Inst	LF	Sm	CA	.057	140.0	.77	1.90	---	2.67	**3.73**
3/4" x 3/4"	Inst	LF	Lg	CA	.040	200.0	.66	1.33	---	1.99	**2.76**
	Inst	LF	Sm	CA	.057	140.0	.77	1.90	---	2.67	**3.73**
1" x 1"	Inst	LF	Lg	CA	.040	200.0	1.32	1.33	---	2.65	**3.51**
	Inst	LF	Sm	CA	.057	140.0	1.54	1.90	---	3.44	**4.62**

Sprung cove

5/8" x 1-5/8"	Inst	LF	Lg	CA	.040	200.0	1.28	1.33	---	2.61	**3.47**
	Inst	LF	Sm	CA	.057	140.0	1.49	1.90	---	3.39	**4.56**
5/8" x 2-1/4"	Inst	LF	Lg	CA	.040	200.0	1.56	1.33	---	2.89	**3.79**
	Inst	LF	Sm	CA	.057	140.0	1.82	1.90	---	3.72	**4.94**

Linoleum cove

7/16" x 1-1/4"	Inst	LF	Lg	CA	.040	200.0	.75	1.33	---	2.08	**2.86**
	Inst	LF	Sm	CA	.057	140.0	.88	1.90	---	2.78	**3.86**

Half round

1/4" x 1/2"	Inst	LF	Lg	CA	.030	266.0	.37	1.00	---	1.37	**1.92**
	Inst	LF	Sm	CA	.043	186.0	.43	1.43	---	1.86	**2.64**
1/4" x 5/8"	Inst	LF	Lg	CA	.030	266.0	.37	1.00	---	1.37	**1.92**
	Inst	LF	Sm	CA	.043	186.0	.43	1.43	---	1.86	**2.64**
3/8" x 3/4"	Inst	LF	Lg	CA	.030	266.0	.44	1.00	---	1.44	**2.00**
	Inst	LF	Sm	CA	.043	186.0	.51	1.43	---	1.94	**2.73**
1/2" x 1"	Inst	LF	Lg	CA	.030	266.0	.81	1.00	---	1.81	**2.43**
	Inst	LF	Sm	CA	.043	186.0	.94	1.43	---	2.37	**3.23**

Lattice - batts - S4S

Lattice

5/16" x 1-1/4"	Inst	LF	Lg	CA	.030	266.0	.52	1.00	---	1.52	**2.10**
	Inst	LF	Sm	CA	.043	186.0	.61	1.43	---	2.04	**2.85**
5/16" x 1-5/8"	Inst	LF	Lg	CA	.030	266.0	.70	1.00	---	1.70	**2.30**
	Inst	LF	Sm	CA	.043	186.0	.81	1.43	---	2.24	**3.08**
5/16" x 2-1/2"	Inst	LF	Lg	CA	.030	266.0	1.00	1.00	---	2.00	**2.65**
	Inst	LF	Sm	CA	.043	186.0	1.17	1.43	---	2.60	**3.49**

Batts

5/16" x 2-1/2"	Inst	LF	Lg	CA	.030	266.0	.73	1.00	---	1.73	**2.34**
	Inst	LF	Sm	CA	.043	186.0	.85	1.43	---	2.28	**3.12**
5/16" x 3-1/2"	Inst	LF	Lg	CA	.030	266.0	1.07	1.00	---	2.07	**2.73**
	Inst	LF	Sm	CA	.043	186.0	1.24	1.43	---	2.67	**3.57**

Description	Oper	Unit	Vol	Crew Size	Man-hours per Unit	Crew Output per Day	Avg Mat'l Unit Cost	Avg Labor Unit Cost	Avg Equip Unit Cost	Avg Total Unit Cost	Avg Price Incl O&P
S4S											
1/2" x 1/2"	Inst	LF	Lg	CA	.030	266.0	.34	1.00	---	1.34	**1.89**
	Inst	LF	Sm	CA	.043	186.0	.40	1.43	---	1.83	**2.61**
1/2" x 3/4"	Inst	LF	Lg	CA	.030	266.0	.40	1.00	---	1.40	**1.96**
	Inst	LF	Sm	CA	.043	186.0	.46	1.43	---	1.89	**2.68**
1/2" x 1-1/2"	Inst	LF	Lg	CA	.030	266.0	.81	1.00	---	1.81	**2.43**
	Inst	LF	Sm	CA	.043	186.0	.94	1.43	---	2.37	**3.23**
1/2" x 2-1/2"	Inst	LF	Lg	CA	.030	266.0	1.14	1.00	---	2.14	**2.81**
	Inst	LF	Sm	CA	.043	186.0	1.33	1.43	---	2.76	**3.68**
1/2" x 3-1/2"	Inst	LF	Lg	CA	.030	266.0	1.55	1.00	---	2.55	**3.28**
	Inst	LF	Sm	CA	.043	186.0	1.81	1.43	---	3.24	**4.23**
1/2" x 5-1/2"	Inst	LF	Lg	CA	.030	266.0	2.77	1.00	---	3.77	**4.68**
	Inst	LF	Sm	CA	.043	186.0	3.22	1.43	---	4.65	**5.85**
11/16" x 1-1/4"	Inst	LF	Lg	CA	.030	266.0	1.16	1.00	---	2.16	**2.83**
	Inst	LF	Sm	CA	.043	186.0	1.35	1.43	---	2.78	**3.70**
11/16" x 1-5/8"	Inst	LF	Lg	CA	.030	266.0	1.44	1.00	---	2.44	**3.15**
	Inst	LF	Sm	CA	.043	186.0	1.67	1.43	---	3.10	**4.07**
11/16" x 2-1/2"	Inst	LF	Lg	CA	.030	266.0	1.85	1.00	---	2.85	**3.62**
	Inst	LF	Sm	CA	.043	186.0	2.15	1.43	---	3.58	**4.62**
11/16" x 3-1/2"	Inst	LF	Lg	CA	.030	266.0	2.53	1.00	---	3.53	**4.41**
	Inst	LF	Sm	CA	.043	186.0	2.94	1.43	---	4.37	**5.53**
3/4" x 3/4"	Inst	LF	Lg	CA	.030	266.0	.73	1.00	---	1.73	**2.34**
	Inst	LF	Sm	CA	.043	186.0	.85	1.43	---	2.28	**3.12**
1" x 1"	Inst	LF	Lg	CA	.030	266.0	1.31	1.00	---	2.31	**3.00**
	Inst	LF	Sm	CA	.043	186.0	1.52	1.43	---	2.95	**3.89**
1-9/16" x 1-9/16"	Inst	LF	Lg	CA	.030	266.0	2.48	1.00	---	3.48	**4.35**
	Inst	LF	Sm	CA	.043	186.0	2.89	1.43	---	4.32	**5.47**
S4S 4EE											
1-1/2" x 1-1/2"	Inst	LF	Lg	CA	.030	266.0	2.41	1.00	---	3.41	**4.27**
	Inst	LF	Sm	CA	.043	186.0	2.80	1.43	---	4.23	**5.37**

Window stool

Flat stool

Description	Oper	Unit	Vol	Crew Size	Man-hours per Unit	Crew Output per Day	Avg Mat'l Unit Cost	Avg Labor Unit Cost	Avg Equip Unit Cost	Avg Total Unit Cost	Avg Price Incl O&P
11/16" x 4-3/4"	Inst	LF	Lg	CA	.030	266.0	4.14	1.00	---	5.14	**6.26**
	Inst	LF	Sm	CA	.043	186.0	4.82	1.43	---	6.25	**7.69**
11/16" x 5-1/2"	Inst	LF	Lg	CA	.030	266.0	4.51	1.00	---	5.51	**6.68**
	Inst	LF	Sm	CA	.043	186.0	5.24	1.43	---	6.67	**8.17**
11/16" x 7-1/4"	Inst	LF	Lg	CA	.030	266.0	6.72	1.00	---	7.72	**9.23**
	Inst	LF	Sm	CA	.043	186.0	7.81	1.43	---	9.24	**11.10**

Rabbeted stool

Description	Oper	Unit	Vol	Crew Size	Man-hours per Unit	Crew Output per Day	Avg Mat'l Unit Cost	Avg Labor Unit Cost	Avg Equip Unit Cost	Avg Total Unit Cost	Avg Price Incl O&P
7/8" x 2-1/2"	Inst	LF	Lg	CA	.030	266.0	3.23	1.00	---	4.23	**5.21**
	Inst	LF	Sm	CA	.043	186.0	3.76	1.43	---	5.19	**6.47**
7/8" x 3-1/2"	Inst	LF	Lg	CA	.030	266.0	4.51	1.00	---	5.51	**6.68**
	Inst	LF	Sm	CA	.043	186.0	5.24	1.43	---	6.67	**8.17**
7/8" x 5-1/2"	Inst	LF	Lg	CA	.030	266.0	5.46	1.00	---	6.46	**7.78**
	Inst	LF	Sm	CA	.043	186.0	6.35	1.43	---	7.78	**9.45**

Description	Oper	Unit	Vol	Crew Size	Man-hours per Unit	Crew Output per Day	Avg Mat'l Unit Cost	Avg Labor Unit Cost	Avg Equip Unit Cost	Avg Total Unit Cost	Avg Price Incl O&P
Panel & decorative molds											
Panel											
3/8" x 5/8"	Inst	LF	Lg	CA	.030	266.0	.63	1.00	---	1.63	**2.22**
	Inst	LF	Sm	CA	.043	186.0	.73	1.43	---	2.16	**2.99**
5/8" x 3/4"	Inst	LF	Lg	CA	.030	266.0	.93	1.00	---	1.93	**2.57**
	Inst	LF	Sm	CA	.043	186.0	1.08	1.43	---	2.51	**3.39**
3/4" x 1"	Inst	LF	Lg	CA	.030	266.0	1.34	1.00	---	2.34	**3.04**
	Inst	LF	Sm	CA	.043	186.0	1.56	1.43	---	2.99	**3.94**
Panel or brick											
1-1/16" x 1-1/2"	Inst	LF	Lg	CA	.030	266.0	1.74	1.00	---	2.74	**3.50**
	Inst	LF	Sm	CA	.043	186.0	2.02	1.43	---	3.45	**4.47**
Panel											
3/8" x 1-1/8"	Inst	LF	Lg	CA	.030	266.0	.93	1.00	---	1.93	**2.57**
	Inst	IF	Sm	CA	.043	186.0	1.08	1.43	---	2.51	**3.39**
9/16" x 1-3/8"	Inst	LF	Lg	CA	.030	266.0	.88	1.00	---	1.88	**2.51**
	Inst	LF	Sm	CA	.043	186.0	1.03	1.43	---	2.46	**3.33**
3/4" x 1-5/8"	Inst	LF	Lg	CA	.030	266.0	1.76	1.00	---	2.76	**3.52**
	Inst	LF	Sm	CA	.043	186.0	2.04	1.43	---	3.47	**4.49**
Beauty M-340											
3/4" x 1-5/8"	Inst	LF	Lg	CA	.030	266.0	1.32	1.00	---	2.32	**3.02**
	Inst	LF	Sm	CA	.043	186.0	1.54	1.43	---	2.97	**3.92**
Beauty M-400											
3/4" x 1-7/8"	Inst	LF	Lg	CA	.030	266.0	2.09	1.00	---	3.09	**3.90**
	Inst	LF	Sm	CA	.043	186.0	2.43	1.43	---	3.86	**4.94**
Decorative											
1-3/8" x 3-1/4"	Inst	LF	Lg	CA	.030	266.0	5.23	1.00	---	6.23	**7.51**
	Inst	LF	Sm	CA	.043	186.0	6.09	1.43	---	7.52	**9.15**
Base cap											
11/16" x 1-3/8"	Inst	LF	Lg	CA	.030	266.0	1.52	1.00	---	2.52	**3.25**
	Inst	LF	Sm	CA	.043	186.0	1.77	1.43	---	3.20	**4.18**
M340 corner arcs											
3/4" x 1-5/8"	Inst	LF	Lg	CA	.250	32.00	3.31	8.32	---	11.63	**16.30**
	Inst	LF	Sm	CA	.364	22.00	3.85	12.10	---	15.95	**22.60**
Miscellaneous											
Band mold											
3/4" x 1-5/8"	Inst	LF	Lg	CA	.030	266.0	1.18	1.00	---	2.18	**2.85**
	Inst	LF	Sm	CA	.043	186.0	1.37	1.43	---	2.80	**3.72**
Parting bead											
3/8" x 3/4"	Inst	LF	Lg	CA	.030	266.0	1.62	1.00	---	2.62	**3.36**
	Inst	LF	Sm	CA	.043	186.0	1.88	1.43	---	3.31	**4.31**

Description	Oper	Unit	Vol	Crew Size	Man-hours per Unit	Crew Output per Day	Avg Mat'l Unit Cost	Avg Labor Unit Cost	Avg Equip Unit Cost	Avg Total Unit Cost	Avg Price Incl O&P
Apron											
3/8" x 1-1/2"	Inst	LF	Lg	CA	.030	266.0	.83	1.00	---	1.83	**2.45**
	Inst	LF	Sm	CA	.043	186.0	.96	1.43	---	2.39	**3.25**
Flat Hoffco											
3/8" x 1"	Inst	LF	Lg	CA	.030	266.0	.58	1.00	---	1.58	**2.16**
	Inst	LF	Sm	CA	.043	186.0	.67	1.43	---	2.10	**2.92**
Drip											
3/4" x 1-1/4"	Inst	LF	Lg	CA	.030	266.0	1.21	1.00	---	2.21	**2.89**
	Inst	LF	Sm	CA	.043	186.0	1.40	1.43	---	2.83	**3.76**
1139 picture											
3/4" x 1-5/8"	Inst	LF	Lg	CA	.030	266.0	1.29	1.00	---	2.29	**2.98**
	Inst	LF	Sm	CA	.043	186.0	1.50	1.43	---	2.93	**3.87**
Chamfer											
1/2" x 1/2"	Inst	LF	Lg	CA	.030	266.0	.32	1.00	---	1.32	**1.87**
	Inst	LF	Sm	CA	.043	186.0	.37	1.43	---	1.80	**2.57**
3/4" x 3/4"	Inst	LF	Lg	CA	.030	266.0	.41	1.00	---	1.41	**1.97**
	Inst	LF	Sm	CA	.043	186.0	.48	1.43	---	1.91	**2.70**
1" x 1"	Inst	LF	Lg	CA	.030	266.0	.91	1.00	---	1.91	**2.54**
	Inst	LF	Sm	CA	.043	186.0	1.06	1.43	---	2.49	**3.37**
Nose & cove											
1" x 1-5/8"	Inst	LF	Lg	CA	.030	266.0	2.15	1.00	---	3.15	**3.97**
	Inst	LF	Sm	CA	.043	186.0	2.50	1.43	---	3.93	**5.02**
Cabinet crown											
5/16" x 1-1/4"	Inst	LF	Lg	CA	.030	266.0	.46	1.00	---	1.46	**2.03**
	Inst	LF	Sm	CA	.043	186.0	.54	1.43	---	1.97	**2.77**

Finger joint pine

Paint grade

Casing

Prefit

Description	Oper	Unit	Vol	Crew Size	Man-hours per Unit	Crew Output per Day	Avg Mat'l Unit Cost	Avg Labor Unit Cost	Avg Equip Unit Cost	Avg Total Unit Cost	Avg Price Incl O&P
9/16" x 1-1/2"	Inst	LF	Lg	CA	.030	266.0	.46	1.00	---	1.46	**2.03**
	Inst	LF	Sm	CA	.043	186.0	.54	1.43	---	1.97	**2.77**
Flat streamline											
1/2" x 1-5/8"	Inst	LF	Lg	CA	.030	266.0	.61	1.00	---	1.61	**2.20**
	Inst	LF	Sm	CA	.043	186.0	.71	1.43	---	2.14	**2.96**
5/8" x 1-5/8"	Inst	LF	Lg	CA	.030	266.0	.63	1.00	---	1.63	**2.22**
	Inst	LF	Sm	CA	.043	186.0	.73	1.43	---	2.16	**2.99**
11/16" x 2-1/4"	Inst	LF	Lg	CA	.030	266.0	.88	1.00	---	1.88	**2.51**
	Inst	LF	Sm	CA	.043	186.0	1.03	1.43	---	2.46	**3.33**
711											
5/8" x 1-5/8"	Inst	LF	Lg	CA	.030	266.0	.63	1.00	---	1.63	**2.22**
	Inst	LF	Sm	CA	.043	186.0	.74	1.43	---	2.17	**3.00**
5/8" x 2-1/2"	Inst	LF	Lg	CA	.030	266.0	.96	1.00	---	1.96	**2.60**
	Inst	LF	Sm	CA	.043	186.0	1.11	1.43	---	2.54	**3.42**

Description	Oper	Unit	Vol	Crew Size	Man-hours per Unit	Crew Output per Day	Avg Mat'l Unit Cost	Avg Labor Unit Cost	Avg Equip Unit Cost	Avg Total Unit Cost	Avg Price Incl O&P
Monterey											
5/8" x 2-1/2"	Inst	LF	Lg	CA	.030	266.0	1.02	1.00	---	2.02	**2.67**
	Inst	LF	Sm	CA	.043	186.0	1.19	1.43	---	2.62	**3.51**
3 step											
5/8" x 2-1/2"	Inst	LF	Lg	CA	.030	266.0	.96	1.00	---	1.96	**2.60**
	Inst	LF	Sm	CA	.043	186.0	1.11	1.43	---	2.54	**3.42**
Universal											
5/8" x 2-1/2"	Inst	LF	Lg	CA	.030	266.0	.97	1.00	---	1.97	**2.61**
	Inst	LF	Sm	CA	.043	186.0	1.12	1.43	---	2.55	**3.43**
Colonial											
9/16" x 4-1/4"	Inst	LF	Lg	CA	.030	266.0	1.60	1.00	---	2.60	**3.34**
	Inst	LF	Sm	CA	.043	186.0	1.86	1.43	---	3.29	**4.29**
11/16" x 2-7/8"	Inst	LF	Lg	CA	.030	266.0	1.15	1.00	---	2.15	**2.82**
	Inst	LF	Sm	CA	.043	186.0	1.34	1.43	---	2.77	**3.69**
Clam shell											
11/16" x 2"	Inst	LF	Lg	CA	.030	266.0	.77	1.00	---	1.77	**2.38**
	Inst	LF	Sm	CA	.043	186.0	.90	1.43	---	2.33	**3.18**
356											
9/16" x 2"	Inst	LF	Lg	CA	.030	266.0	.77	1.00	---	1.77	**2.38**
	Inst	LF	Sm	CA	.043	186.0	.90	1.43	---	2.33	**3.18**
11/16" x 2-1/4"	Inst	LF	Lg	CA	.030	266.0	.85	1.00	---	1.85	**2.47**
	Inst	LF	Sm	CA	.043	186.0	.98	1.43	---	2.41	**3.27**
WM 366											
11/16" x 2-1/4"	Inst	LF	Lg	CA	.030	266.0	.85	1.00	---	1.85	**2.47**
	Inst	LF	Sm	CA	.043	186.0	.98	1.43	---	2.41	**3.27**
3 fluted											
1/2" x 3-1/4"	Inst	LF	Lg	CA	.030	266.0	1.21	1.00	---	2.21	**2.89**
	Inst	LF	Sm	CA	.043	186.0	1.41	1.43	---	2.84	**3.77**
5 fluted											
1/2" x 3-1/2"	Inst	LF	Lg	CA	.030	266.0	1.30	1.00	---	2.30	**2.99**
	Inst	LF	Sm	CA	.043	186.0	1.51	1.43	---	2.94	**3.88**
Cape Cod											
5/8" x 2-1/2"	Inst	LF	Lg	CA	.030	266.0	1.01	1.00	---	2.01	**2.66**
	Inst	LF	Sm	CA	.043	186.0	1.18	1.43	---	2.61	**3.50**
11/16" x 3-1/2"	Inst	LF	Lg	CA	.030	266.0	1.68	1.00	---	2.68	**3.43**
	Inst	LF	Sm	CA	.043	186.0	1.96	1.43	---	3.39	**4.40**
Chesapeake											
3/4" x 3-1/2"	Inst	LF	Lg	CA	.030	266.0	1.70	1.00	---	2.70	**3.45**
	Inst	LF	Sm	CA	.043	186.0	1.98	1.43	---	3.41	**4.42**
Coronado											
1-1/4" x 2-1/2"	Inst	LF	Lg	CA	.030	266.0	1.99	1.00	---	2.99	**3.79**
	Inst	LF	Sm	CA	.043	186.0	2.31	1.43	---	3.74	**4.80**

Description	Oper	Unit	Vol	Crew Size	Man-hours per Unit	Crew Output per Day	Avg Mat'l Unit Cost	Avg Labor Unit Cost	Avg Equip Unit Cost	Avg Total Unit Cost	Avg Price Incl O&P
Bell court											
1-1/4" x 2-1/2"	Inst	LF	Lg	CA	.030	266.0	1.99	1.00	---	2.99	**3.79**
	Inst	LF	Sm	CA	.043	186.0	2.31	1.43	---	3.74	**4.80**
Hermosa											
1-1/4" x 3-1/4"	Inst	LF	Lg	CA	.030	266.0	2.70	1.00	---	3.70	**4.60**
	Inst	LF	Sm	CA	.043	186.0	3.14	1.43	---	4.57	**5.76**
Cambridge											
11/16" x 3-1/4"	Inst	LF	Lg	CA	.030	266.0	1.66	1.00	---	2.66	**3.41**
	Inst	LF	Sm	CA	.043	186.0	1.93	1.43	---	3.36	**4.37**
275											
5/8" x 3"	Inst	LF	Lg	CA	.030	266.0	1.24	1.00	---	2.24	**2.92**
	Inst	LF	Sm	CA	.043	186.0	1.44	1.43	---	2.87	**3.80**

Stops

Description	Oper	Unit	Vol	Crew Size	Man-hours per Unit	Crew Output per Day	Avg Mat'l Unit Cost	Avg Labor Unit Cost	Avg Equip Unit Cost	Avg Total Unit Cost	Avg Price Incl O&P
2 round edge											
3/8" x 1-1/4"	Inst	LF	Lg	CA	.030	266.0	.36	1.00	---	1.36	**1.91**
	Inst	LF	Sm	CA	.043	186.0	.42	1.43	---	1.85	**2.63**
Round edge											
1/2" x 1-1/4"	Inst	LF	Lg	CA	.030	266.0	.52	1.00	---	1.52	**2.10**
	Inst	LF	Sm	CA	.043	186.0	.60	1.43	---	2.03	**2.84**
1/2" x 1-1/2"	Inst	LF	Lg	CA	.030	266.0	.66	1.00	---	1.66	**2.26**
	Inst	LF	Sm	CA	.043	186.0	.77	1.43	---	2.20	**3.03**
Ogee											
1/2" x 1-3/4"	Inst	LF	Lg	CA	.030	266.0	.71	1.00	---	1.71	**2.31**
	Inst	LF	Sm	CA	.043	186.0	.82	1.43	---	2.25	**3.09**

Base

Description	Oper	Unit	Vol	Crew Size	Man-hours per Unit	Crew Output per Day	Avg Mat'l Unit Cost	Avg Labor Unit Cost	Avg Equip Unit Cost	Avg Total Unit Cost	Avg Price Incl O&P
Base shoe											
3/8" x 3/4"	Inst	LF	Lg	CA	.030	266.0	.33	1.00	---	1.33	**1.88**
	Inst	LF	Sm	CA	.043	186.0	.39	1.43	---	1.82	**2.59**
Flat											
7/16" x 1-1/2"	Inst	LF	Lg	CA	.030	266.0	.41	1.00	---	1.41	**1.97**
	Inst	LF	Sm	CA	.043	186.0	.48	1.43	---	1.91	**2.70**
Streamline or reversible											
3/8" x 2-1/4"	Inst	LF	Lg	CA	.030	266.0	.55	1.00	---	1.55	**2.13**
	Inst	LF	Sm	CA	.043	186.0	.64	1.43	---	2.07	**2.88**
Streamline											
7/16" x 2-1/2"	Inst	LF	Lg	CA	.030	266.0	.68	1.00	---	1.68	**2.28**
	Inst	LF	Sm	CA	.043	186.0	.79	1.43	---	2.22	**3.05**
7/16" x 3-1/2"	Inst	LF	Lg	CA	.030	266.0	1.04	1.00	---	2.04	**2.69**
	Inst	LF	Sm	CA	.043	186.0	1.21	1.43	---	2.64	**3.54**

Description	Oper	Unit	Vol	Crew Size	Man-hours per Unit	Crew Output per Day	Avg Mat'l Unit Cost	Avg Labor Unit Cost	Avg Equip Unit Cost	Avg Total Unit Cost	Avg Price Incl O&P
711											
7/16" x 2-1/2"	Inst	LF	Lg	CA	.040	200.0	.72	1.33	---	2.05	**2.82**
	Inst	LF	Sm	CA	.057	140.0	.83	1.90	---	2.73	**3.80**
7/16" x 3-1/2"	Inst	LF	Lg	CA	.040	200.0	.97	1.33	---	2.30	**3.11**
	Inst	LF	Sm	CA	.057	140.0	1.12	1.90	---	3.02	**4.13**
3 step											
7/16" x 2-1/2"	Inst	LF	Lg	CA	.040	200.0	.79	1.33	---	2.12	**2.90**
	Inst	LF	Sm	CA	.057	140.0	.92	1.90	---	2.82	**3.90**
3 step base											
1/2" x 4-1/2"	Inst	LF	Lg	CA	.040	200.0	1.78	1.33	---	3.11	**4.04**
	Inst	LF	Sm	CA	.057	140.0	2.08	1.90	---	3.98	**5.24**
WM 623											
9/16" x 3-1/4"	Inst	LF	Lg	CA	.040	200.0	1.26	1.33	---	2.59	**3.45**
	Inst	LF	Sm	CA	.057	140.0	1.47	1.90	---	3.37	**4.54**
WM 618											
9/16" x 5-1/4"	Inst	LF	Lg	CA	.040	200.0	1.95	1.33	---	3.28	**4.24**
	Inst	LF	Sm	CA	.057	140.0	2.27	1.90	---	4.17	**5.46**
Monterey											
7/16" x 4-1/4"	Inst	LF	Lg	CA	.040	200.0	1.29	1.33	---	2.62	**3.48**
	Inst	LF	Sm	CA	.057	140.0	1.50	1.90	---	3.40	**4.57**
Cape Cod											
9/16" x 4"	Inst	LF	Lg	CA	.040	200.0	1.62	1.33	---	2.95	**3.86**
	Inst	LF	Sm	CA	.057	140.0	1.88	1.90	---	3.78	**5.01**
9/16" x 5-1/4"	Inst	LF	Lg	CA	.040	200.0	2.28	1.33	---	3.61	**4.62**
	Inst	LF	Sm	CA	.057	140.0	2.65	1.90	---	4.55	**5.89**
"B" Cape Cod base											
9/16" x 5-1/4"	Inst	LF	Lg	CA	.040	200.0	2.28	1.33	---	3.61	**4.62**
	Inst	LF	Sm	CA	.057	140.0	2.65	1.90	---	4.55	**5.89**
356 base											
1/2" x 2-1/2"	Inst	LF	Lg	CA	.030	266.0	.91	1.00	---	1.91	**2.54**
	Inst	LF	Sm	CA	.043	186.0	1.06	1.43	---	2.49	**3.37**
1/2" x 3-1/4"	Inst	LF	Lg	CA	.030	266.0	1.19	1.00	---	2.19	**2.87**
	Inst	LF	Sm	CA	.043	186.0	1.38	1.43	---	2.81	**3.73**
175 base											
9/16" x 4-1/4"	Inst	LF	Lg	CA	.040	200.0	1.67	1.33	---	3.00	**3.92**
	Inst	LF	Sm	CA	.057	140.0	1.95	1.90	---	3.85	**5.09**
Colonial											
9/16" x 4-1/4"	Inst	LF	Lg	CA	.040	200.0	2.45	1.33	---	3.78	**4.81**
	Inst	LF	Sm	CA	.057	140.0	2.85	1.90	---	4.75	**6.12**

Description	Oper	Unit	Vol	Crew Size	Man-hours per Unit	Crew Output per Day	Avg Mat'l Unit Cost	Avg Labor Unit Cost	Avg Equip Unit Cost	Avg Total Unit Cost	Avg Price Incl O&P
Crown & cornice											
Colonial crown											
9/16" x 2-1/4"	Inst	LF	Lg	CA	.040	200.0	.88	1.33	---	2.21	**3.01**
	Inst	LF	Sm	CA	.057	140.0	1.03	1.90	---	2.93	**4.03**
9/16" x 3-3/8"	Inst	LF	Lg	CA	.040	200.0	1.43	1.33	---	2.76	**3.64**
	Inst	LF	Sm	CA	.057	140.0	1.66	1.90	---	3.56	**4.75**
9/16" x 4-1/2"	Inst	LF	Lg	CA	.040	200.0	1.95	1.33	---	3.28	**4.24**
	Inst	LF	Sm	CA	.057	140.0	2.27	1.90	---	4.17	**5.46**
Cornice											
13/16" x 4-5/8"	Inst	LF	Lg	CA	.040	200.0	3.56	1.33	---	4.89	**6.09**
	Inst	LF	Sm	CA	.057	140.0	4.14	1.90	---	6.04	**7.61**
1-1/4" x 5-7/8"	Inst	LF	Lg	CA	.040	200.0	5.48	1.33	---	6.81	**8.30**
	Inst	LF	Sm	CA	.057	140.0	6.38	1.90	---	8.28	**10.20**
Crown											
3/4" x 6"	Inst	LF	Lg	CA	.040	200.0	3.73	1.33	---	5.06	**6.29**
	Inst	LF	Sm	CA	.057	140.0	4.33	1.90	---	6.23	**7.82**
Cornice											
1-1/8" x 3-1/2"	Inst	LF	Lg	CA	.040	200.0	2.70	1.33	---	4.03	**5.10**
	Inst	LF	Sm	CA	.057	140.0	3.14	1.90	---	5.04	**6.46**
Georgian crown											
11/16" x 4-1/4"	Inst	LF	Lg	CA	.040	200.0	2.17	1.33	---	3.50	**4.49**
	Inst	LF	Sm	CA	.057	140.0	2.53	1.90	---	4.43	**5.75**
Cove											
11/16" x 11/16"	Inst	LF	Lg	CA	.030	266.0	.43	1.00	---	1.43	**1.99**
	Inst	LF	Sm	CA	.043	186.0	.50	1.43	---	1.93	**2.72**
Chair rail											
Chair rail											
9/16" x 2-1/2"	Inst	LF	Lg	CA	.030	266.0	.99	1.00	---	1.99	**2.64**
	Inst	LF	Sm	CA	.043	186.0	1.16	1.43	---	2.59	**3.48**
Colonial chair rail											
3/4" x 3"	Inst	LF	Lg	CA	.030	266.0	1.51	1.00	---	2.51	**3.23**
	Inst	LF	Sm	CA	.043	186.0	1.75	1.43	---	3.18	**4.16**
Exterior mold											
Stucco mold											
7/8" x 1-1/4"	Inst	LF	Lg	CA	.030	266.0	.79	1.00	---	1.79	**2.41**
	Inst	LF	Sm	CA	.043	186.0	.92	1.43	---	2.35	**3.20**
WM 180 eastern brick											
1-1/4" x 2"	Inst	LF	Lg	CA	.030	266.0	1.70	1.00	---	2.70	**3.45**
	Inst	LF	Sm	CA	.043	186.0	1.98	1.43	---	3.41	**4.42**

Description	Oper	Unit	Vol	Crew Size	Man-hours per Unit	Crew Output per Day	Avg Mat'l Unit Cost	Avg Labor Unit Cost	Avg Equip Unit Cost	Avg Total Unit Cost	Avg Price Incl O&P

Stool & apron

Flat stool

11/16" x 4-3/4"	Inst	LF	Lg	CA	.030	266.0	2.48	1.00	---	3.48	**4.35**
	Inst	LF	Sm	CA	.043	186.0	2.89	1.43	---	4.32	**5.47**
11/16" x 5-1/2"	Inst	LF	Lg	CA	.030	266.0	3.31	1.00	---	4.31	**5.30**
	Inst	LF	Sm	CA	.043	186.0	3.85	1.43	---	5.28	**6.57**
11/16" x 7-1/4"	Inst	LF	Lg	CA	.030	266.0	4.14	1.00	---	5.14	**6.26**
	Inst	LF	Sm	CA	.043	186.0	4.82	1.43	---	6.25	**7.69**

Apron

3/8" x 1-1/2"	Inst	LF	Lg	CA	.030	266.0	.52	1.00	---	1.52	**2.10**
	Inst	LF	Sm	CA	.043	186.0	.60	1.43	---	2.03	**2.84**

Oak

Hardwood

Casing

Streamline

5/8" x 1-5/8"	Inst	LF	Lg	CA	.030	266.0	1.21	1.00	---	2.21	**2.89**
	Inst	LF	Sm	CA	.043	186.0	1.41	1.43	---	2.84	**3.77**

711

1/2" x 1-5/8"	Inst	LF	Lg	CA	.030	266.0	1.15	1.00	---	2.15	**2.82**
	Inst	LF	Sm	CA	.043	186.0	1.34	1.43	---	2.77	**3.69**
5/8" x 2-3/8"	Inst	LF	Lg	CA	.030	266.0	1.44	1.00	---	2.44	**3.15**
	Inst	LF	Sm	CA	.043	186.0	1.67	1.43	---	3.10	**4.07**

Universal

5/8" x 2-1/2"	Inst	LF	Lg	CA	.030	266.0	1.73	1.00	---	2.73	**3.49**
	Inst	LF	Sm	CA	.043	186.0	2.01	1.43	---	3.44	**4.46**

Colonial

11/16" x 2-7/8"	Inst	LF	Lg	CA	.030	266.0	1.33	1.00	---	2.33	**3.03**
	Inst	LF	Sm	CA	.043	186.0	1.55	1.43	---	2.98	**3.93**
11/16" x 3-1/2"	Inst	LF	Lg	CA	.030	266.0	1.87	1.00	---	2.87	**3.65**
	Inst	LF	Sm	CA	.043	186.0	2.17	1.43	---	3.60	**4.64**
11/16" x 4-1/4"	Inst	LF	Lg	CA	.030	266.0	2.06	1.00	---	3.06	**3.87**
	Inst	LF	Sm	CA	.043	186.0	2.40	1.43	---	3.83	**4.91**

Mull

5/16" x 3-1/4"	Inst	LF	Lg	CA	.030	266.0	1.58	1.00	---	2.58	**3.31**
	Inst	LF	Sm	CA	.043	186.0	1.84	1.43	---	3.27	**4.26**

356

9/16" x 2-1/4"	Inst	LF	Lg	CA	.030	266.0	.97	1.00	---	1.97	**2.61**
	Inst	LF	Sm	CA	.043	186.0	1.12	1.43	---	2.55	**3.43**

Cape Cod

5/8" x 2-1/2"	Inst	LF	Lg	CA	.030	266.0	1.02	1.00	---	2.02	**2.67**
	Inst	LF	Sm	CA	.043	186.0	1.19	1.43	---	2.62	**3.51**

Description	Oper	Unit	Vol	Crew Size	Man-hours per Unit	Crew Output per Day	Avg Mat'l Unit Cost	Avg Labor Unit Cost	Avg Equip Unit Cost	Avg Total Unit Cost	Avg Price Incl O&P
Fluted											
1/2" x 3-1/2"	Inst	LF	Lg	CA	.030	266.0	1.45	1.00	---	2.45	**3.16**
	Inst	LF	Sm	CA	.043	186.0	1.69	1.43	---	3.12	**4.09**
Victorian											
11/16" x 4-1/4"	Inst	LF	Lg	CA	.030	266.0	4.00	1.00	---	5.00	**6.10**
	Inst	LF	Sm	CA	.043	186.0	4.65	1.43	---	6.08	**7.49**
Cambridge											
11/16" x 3-1/4"	Inst	LF	Lg	CA	.030	266.0	2.35	1.00	---	3.35	**4.20**
	Inst	LF	Sm	CA	.043	186.0	2.73	1.43	---	4.16	**5.29**
Rosette plinth block w/ edge detail											
3/4" x 2-3/4"	Inst	Ea	Lg	CA	.250	32.00	3.45	8.32	---	11.77	**16.50**
	Inst	Ea	Sm	CA	.364	22.00	4.01	12.10	---	16.11	**22.80**
3/4" x 3-3/4"	Inst	Ea	Lg	CA	.250	32.00	4.14	8.32	---	12.46	**17.20**
	Inst	Ea	Sm	CA	.364	22.00	4.82	12.10	---	16.92	**23.70**

Base

Description	Oper	Unit	Vol	Crew Size	Man-hours per Unit	Crew Output per Day	Avg Mat'l Unit Cost	Avg Labor Unit Cost	Avg Equip Unit Cost	Avg Total Unit Cost	Avg Price Incl O&P
Base shoe											
1/2" x 3/4"	Inst	LF	Lg	CA	.030	266.0	.63	1.00	---	1.63	**2.22**
	Inst	LF	Sm	CA	.043	186.0	.74	1.43	---	2.17	**3.00**
Colonial											
11/16" x 4-1/4"	Inst	LF	Lg	CA	.030	266.0	2.44	1.00	---	3.44	**4.30**
	Inst	LF	Sm	CA	.043	186.0	2.84	1.43	---	4.27	**5.41**
Streamline											
1/2" x 2-1/2"	Inst	LF	Lg	CA	.030	266.0	1.28	1.00	---	2.28	**2.97**
	Inst	LF	Sm	CA	.043	186.0	1.49	1.43	---	2.92	**3.86**
1/2" x 3-1/2"	Inst	LF	Lg	CA	.030	266.0	1.45	1.00	---	2.45	**3.16**
	Inst	LF	Sm	CA	.043	186.0	1.69	1.43	---	3.12	**4.09**
711											
1/2" x 2-1/2"	Inst	LF	Lg	CA	.030	266.0		1.00	---	2.34	**3.04**
	Inst	LF	Sm	CA	.043	186.0	1.56	1.43	---	2.99	**3.94**
1/2" x 3-1/2"	Inst	LF	Lg	CA	.030	266.0	1.78	1.00	---	2.78	**3.54**
	Inst	LF	Sm	CA	.043	186.0	2.08	1.43	---	3.51	**4.54**
Cape Cod											
7/16" x 4-1/4"	Inst	LF	Lg	CA	.030	266.0	2.26	1.00	---	3.26	**4.10**
	Inst	LF	Sm	CA	.043	186.0	2.63	1.43	---	4.06	**5.17**
Montoroy											
7/16" x 4-1/4"	Inst	LF	Lg	CA	.030	266.0	2.26	1.00	---	3.26	**4.10**
	Inst	LF	Sm	CA	.043	186.0	2.63	1.43	---	4.06	**5.17**
Rosette plinth block w/ edge detail											
3/4" x 2-3/4"	Inst	Ea	Lg	CA	.250	32.00	5.31	8.32	---	13.63	**18.60**
	Inst	Ea	Sm	CA	.364	22.00	6.17	12.10	---	18.27	**25.30**
3/4" x 3-1/2"	Inst	Ea	Lg	CA	.250	32.00	6.35	8.32	---	14.67	**19.80**
	Inst	Ea	Sm	CA	.364	22.00	7.38	12.10	---	19.48	**26.70**
Colonial plinth block w/ edge detail											
1" x 2-3/4" x 6"	Inst	Ea	Lg	CA	.250	32.00	4.00	8.32	---	12.32	**17.10**
	Inst	Ea	Sm	CA	.364	22.00	4.65	12.10	---	16.75	**23.50**
1" x 3-3/4" x 6"	Inst	Ea	Lg	CA	.250	32.00	4.89	8.32	---	13.21	**18.10**
	Inst	Ea	Sm	CA	.364	22.00	5.69	12.10	---	17.79	**24.70**

Description	Oper	Unit	Vol	Crew Size	Man-hours per Unit	Crew Output per Day	Avg Mat'l Unit Cost	Avg Labor Unit Cost	Avg Equip Unit Cost	Avg Total Unit Cost	Avg Price Incl O&P
Stops											
Round edge											
3/8" x 1-1/4"	Inst	LF	Lg	CA	.030	266.0	.89	1.00	---	1.89	**2.52**
	Inst	LF	Sm	CA	.043	186.0	1.04	1.43	---	2.47	**3.34**
Ogee											
1/2" x 1-3/4"	Inst	LF	Lg	CA	.030	266.0	1.66	1.00	---	2.66	**3.41**
	Inst	LF	Sm	CA	.043	186.0	1.93	1.43	---	3.36	**4.37**
Crown & cornice											
Crown											
1/2" x 1-5/8"	Inst	LF	Lg	CA	.040	200.0	1.13	1.33	---	2.46	**3.30**
	Inst	LF	Sm	CA	.057	140.0	1.32	1.90	---	3.22	**4.36**
1/2" x 2-1/4"	Inst	LF	Lg	CA	.040	200.0	1.58	1.33	---	2.91	**3.81**
	Inst	LF	Sm	CA	.057	140.0	1.84	1.90	---	3.74	**4.96**
1/2" x 3-1/2"	Inst	LF	Lg	CA	.040	200.0	2.27	1.33	---	3.60	**4.61**
	Inst	LF	Sm	CA	.057	140.0	2.64	1.90	---	4.54	**5.88**
Colonial crown											
5/8" x 3-1/2"	Inst	LF	Lg	CA	.040	200.0	2.35	1.33	---	3.68	**4.70**
	Inst	LF	Sm	CA	.057	140.0	2.73	1.90	---	4.63	**5.98**
5/8" x 4-1/2"	Inst	LF	Lg	CA	.040	200.0	3.59	1.33	---	4.92	**6.12**
	Inst	LF	Sm	CA	.057	140.0	4.17	1.90	---	6.07	**7.64**
Cornice											
13/16" x 4-5/8"	Inst	LF	Lg	CA	.040	200.0	3.65	1.33	---	4.98	**6.19**
	Inst	LF	Sm	CA	.057	140.0	4.25	1.90	---	6.15	**7.73**
1-1/4" x 6"	Inst	LF	Lg	CA	.040	200.0	7.31	1.33	---	8.64	**10.40**
	Inst	LF	Sm	CA	.057	140.0	8.51	1.90	---	10.41	**12.60**
Quarter round											
1/2" x 1/2"	Inst	LF	Lg	CA	.030	266.0	.58	1.00	---	1.58	**2.16**
	Inst	LF	Sm	CA	.043	186.0	.67	1.43	---	2.10	**2.92**
3/4 x 3/4"	Inst	LF	Lg	CA	.030	266.0	1.02	1.00	---	2.02	**2.67**
	Inst	LF	Sm	CA	.043	186.0	1.19	1.43	---	2.62	**3.51**
Cove											
1/2" x 1/2"	Inst	LF	Lg	CA	.040	200.0	.58	1.33	---	1.91	**2.66**
	Inst	LF	Sm	CA	.057	140.0	.67	1.90	---	2.57	**3.62**
3/4 x 3/4"	Inst	LF	Lg	CA	.040	200.0	1.02	1.33	---	2.35	**3.17**
	Inst	LF	Sm	CA	.057	140.0	1.19	1.90	---	3.09	**4.21**
Half round											
3/8" x 3/4"	Inst	LF	Lg	CA	.030	266.0	.62	1.00	---	1.62	**2.21**
	Inst	LF	Sm	CA	.043	186.0	.72	1.43	---	2.15	**2.97**

25

Description	Oper	Unit	Vol	Crew Size	Man-hours per Unit	Crew Output per Day	Avg Mat'l Unit Cost	Avg Labor Unit Cost	Avg Equip Unit Cost	Avg Total Unit Cost	Avg Price Incl O&P
Screen mold											
Flat screen											
3/8" x 3/4"	Inst	LF	Lg	CA	.030	266.0	.64	1.00	---	1.64	**2.23**
	Inst	LF	Sm	CA	.043	186.0	.75	1.43	---	2.18	**3.01**
Scribe											
3/16" x 3/4"	Inst	LF	Lg	CA	.030	266.0	.56	1.00	---	1.56	**2.14**
	Inst	LF	Sm	CA	.043	186.0	.65	1.43	---	2.08	**2.89**
Cloverleaf											
3/8" x 3/4"	Inst	LF	Lg	CA	.030	266.0	.77	1.00	---	1.77	**2.38**
	Inst	LF	Sm	CA	.043	186.0	.90	1.43	---	2.33	**3.18**
Outside corner guard											
3/4" x 3/4"	Inst	LF	Lg	CA	.030	266.0	1.03	1.00	---	2.03	**2.68**
	Inst	LF	Sm	CA	.043	186.0	1.20	1.43	---	2.63	**3.53**
1" x 1"	Inst	LF	Lg	CA	.030	266.0	1.19	1.00	---	2.19	**2.87**
	Inst	LF	Sm	CA	.043	186.0	1.38	1.43	---	2.81	**3.73**
1-1/8" x 1-1/8"	Inst	LF	Lg	CA	.030	266.0	1.77	1.00	---	2.77	**3.53**
	Inst	LF	Sm	CA	.043	186.0	2.05	1.43	---	3.48	**4.50**
Chair rail											
Chair rail											
5/8" x 2-1/2"	Inst	LF	Lg	CA	.030	266.0	1.86	1.00	---	2.86	**3.64**
	Inst	LF	Sm	CA	.043	186.0	2.16	1.43	---	3.59	**4.63**
Colonial chair rail											
5/8" x 2-1/2"	Inst	LF	Lg	CA	.030	266.0	2.21	1.00	---	3.21	**4.04**
	Inst	LF	Sm	CA	.043	186.0	2.57	1.43	---	4.00	**5.10**
Victorian											
11/16" x 4-1/4"	Inst	LF	Lg	CA	.030	266.0	4.94	1.00	---	5.94	**7.18**
	Inst	LF	Sm	CA	.043	186.0	5.75	1.43	---	7.18	**8.76**
Panel cap											
1/2" x 1-1/2"	Inst	LF	Lg	CA	.030	266.0	1.27	1.00	---	2.27	**2.96**
	Inst	LF	Sm	CA	.043	186.0	1.40	1.43	---	2.91	**3.85**
Astragal											
Flat astragal											
3/4" x 1-5/8"	Inst	LF	Lg	CA	.030	266.0	1.85	1.00	---	2.85	**3.62**
	Inst	LF	Sm	CA	.043	186.0	2.15	1.43	---	3.58	**4.62**
1-3/4 T astragal											
1-1/4" x 2-1/4"	Inst	LF	Lg	CA	.030	266.0	6.76	1.00	---	7.76	**9.27**
	Inst	LF	Sm	CA	.043	186.0	7.86	1.43	---	9.29	**11.20**
T astragal mahogany											
1-1/4" x 2-1/4" (1-3/8" door)	Inst	LF	Lg	CA	.030	266.0	3.44	1.00	---	4.44	**5.45**
	Inst	LF	Sm	CA	.043	186.0	4.00	1.43	---	5.43	**6.75**

Description	Oper	Unit	Vol	Crew Size	Man-hours per Unit	Crew Output per Day	Avg Mat'l Unit Cost	Avg Labor Unit Cost	Avg Equip Unit Cost	Avg Total Unit Cost	Avg Price Incl O&P
Nose & cove											
1" x 1-5/8"	Inst	LF	Lg	CA	.040	200.0	2.27	1.33	---	3.60	4.61
	Inst	LF	Sm	CA	.057	140.0	2.64	1.90	---	4.54	5.88
Hand rail											
Hand rail											
1-1/2" x 2-1/2"	Inst	LF	Lg	CA	.030	266.0	4.00	1.00	---	5.00	6.10
	Inst	LF	Sm	CA	.043	186.0	4.65	1.43	---	6.08	7.49
Plowed											
1-1/2" x 2-1/2"	Inst	LF	Lg	CA	.030	266.0	4.42	1.00	---	5.42	6.58
	Inst	LF	Sm	CA	.043	186.0	5.14	1.43	---	6.57	8.06
Panel mold											
Panel mold											
3/8" x 5/8"	Inst	LF	Lg	CA	.030	266.0	.66	1.00	---	1.66	2.26
	Inst	LF	Sm	CA	.043	186.0	.77	1.43	---	2.20	3.03
5/8" x 3/4"	Inst	LF	Lg	CA	.030	266.0	.95	1.00	---	1.95	2.59
	Inst	LF	Sm	CA	.043	186.0	1.10	1.43	---	2.53	3.41
3/4" x 1"	Inst	LF	Lg	CA	.030	266.0	1.31	1.00	---	2.31	3.00
	Inst	LF	Sm	CA	.043	186.0	1.52	1.43	---	2.95	3.89
3/4" x 1-5/8"	Inst	LF	Lg	CA	.030	266.0	1.90	1.00	---	2.90	3.68
	Inst	LF	Sm	CA	.043	186.0	2.21	1.43	---	3.64	4.69
Beauty mold											
9/16" x 1-3/8"	Inst	LF	Lg	CA	.030	266.0	1.58	1.00	---	2.58	3.31
	Inst	LF	Sm	CA	.043	186.0	1.84	1.43	---	3.27	4.26
3/4" x 1-7/8"	Inst	LF	Lg	CA	.030	266.0	2.43	1.00	---	3.43	4.29
	Inst	LF	Sm	CA	.043	186.0	2.82	1.43	---	4.25	5.39
Raised panel											
1-1/8" x 1-1/4"	Inst	LF	Lg	CA	.030	266.0	1.90	1.00	---	2.90	3.68
	Inst	LF	Sm	CA	.043	186.0	2.21	1.43	---	3.64	4.69
Full round											
1-3/8"	Inst	LF	Lg	CA	.030	266.0	2.69	1.00	---	3.69	4.59
	Inst	LF	Sm	CA	.043	186.0	3.12	1.43	---	4.55	5.73
S4S											
1" x 4"	Inst	LF	Lg	CA	.040	200.0	3.65	1.33	---	4.98	6.19
	Inst	LF	Sm	CA	.057	140.0	4.25	1.90	---	6.15	7.73
1" x 6"	Inst	LF	Lg	CA	.040	200.0	5.52	1.33	---	6.85	8.34
	Inst	LF	Sm	CA	.057	140.0	6.42	1.90	---	8.32	10.20
1" x 8"	Inst	LF	Lg	CA	.040	200.0	7.31	1.33	---	8.64	10.40
	Inst	LF	Sm	CA	.057	140.0	8.51	1.90	---	10.41	12.60
1" x 10"	Inst	LF	Lg	CA	.040	200.0	9.45	1.33	---	10.78	12.90
	Inst	LF	Sm	CA	.057	140.0	11.00	1.90	---	12.90	15.50
1" x 12"	Inst	LF	Lg	CA	.040	200.0	11.70	1.33	---	13.03	15.50
	Inst	LF	Sm	CA	.057	140.0	13.60	1.90	---	15.50	18.50

Description	Oper	Unit	Vol	Crew Size	Man-hours per Unit	Crew Output per Day	Avg Mat'l Unit Cost	Avg Labor Unit Cost	Avg Equip Unit Cost	Avg Total Unit Cost	Avg Price Incl O&P
Oak bar nosing											
1-1/4" x 3-1/2"	Inst	LF	Lg	CA	.030	266.0	6.62	1.00	---	7.62	**9.11**
	Inst	LF	Sm	CA	.043	186.0	7.70	1.43	---	9.13	**11.00**
1-5/8" x 5"	Inst	LF	Lg	CA	.030	266.0	9.45	1.00	---	10.45	**12.40**
	Inst	LF	Sm	CA	.043	186.0	11.00	1.43	---	12.43	**14.80**

Redwood
Lattice
Lattice

Description	Oper	Unit	Vol	Crew Size	Man-hours per Unit	Crew Output per Day	Avg Mat'l Unit Cost	Avg Labor Unit Cost	Avg Equip Unit Cost	Avg Total Unit Cost	Avg Price Incl O&P
5/16" x 1-1/4"	Inst	LF	Lg	CA	.030	266.0	.30	1.00	---	1.30	**1.84**
	Inst	LF	Sm	CA	.043	186.0	.35	1.43	---	1.78	**2.55**
5/16" x 1-5/8"	Inst	LF	Lg	CA	.030	266.0	.36	1.00	---	1.36	**1.91**
	Inst	LF	Sm	CA	.043	186.0	.42	1.43	---	1.85	**2.63**

Batts

Description	Oper	Unit	Vol	Crew Size	Man-hours per Unit	Crew Output per Day	Avg Mat'l Unit Cost	Avg Labor Unit Cost	Avg Equip Unit Cost	Avg Total Unit Cost	Avg Price Incl O&P
5/16" x 2-1/2"	Inst	LF	Lg	CA	.030	266.0	.56	1.00	---	1.56	**2.14**
	Inst	LF	Sm	CA	.043	186.0	.65	1.43	---	2.08	**2.89**
5/16" x 3-1/2"	Inst	LF	Lg	CA	.030	266.0	.78	1.00	---	1.78	**2.39**
	Inst	LF	Sm	CA	.043	186.0	.91	1.43	---	2.34	**3.19**
5/16" x 5-1/2"	Inst	LF	Lg	CA	.030	266.0	1.33	1.00	---	2.33	**3.03**
	Inst	LF	Sm	CA	.043	186.0	1.55	1.43	---	2.98	**3.93**

Miscellaneous exterior molds
Bricks

Description	Oper	Unit	Vol	Crew Size	Man-hours per Unit	Crew Output per Day	Avg Mat'l Unit Cost	Avg Labor Unit Cost	Avg Equip Unit Cost	Avg Total Unit Cost	Avg Price Incl O&P
1-1/2" x 1-1/2"	Inst	LF	Lg	CA	.040	200.0	1.36	1.33	---	2.69	**3.56**
	Inst	LF	Sm	CA	.057	140.0	1.58	1.90	---	3.48	**4.66**

Siding

Description	Oper	Unit	Vol	Crew Size	Man-hours per Unit	Crew Output per Day	Avg Mat'l Unit Cost	Avg Labor Unit Cost	Avg Equip Unit Cost	Avg Total Unit Cost	Avg Price Incl O&P
7/8" x 1-5/8"	Inst	LF	Lg	CA	.040	200.0	1.00	1.33	---	2.33	**3.15**
	Inst	LF	Sm	CA	.057	140.0	1.17	1.90	---	3.07	**4.19**

Stucco

Description	Oper	Unit	Vol	Crew Size	Man-hours per Unit	Crew Output per Day	Avg Mat'l Unit Cost	Avg Labor Unit Cost	Avg Equip Unit Cost	Avg Total Unit Cost	Avg Price Incl O&P
7/8" x 1-1/2"	Inst	LF	Lg	CA	.040	200.0	.59	1.33	---	1.92	**2.67**
	Inst	LF	Sm	CA	.057	140.0	.68	1.90	---	2.58	**3.63**

Watertable without lip

Description	Oper	Unit	Vol	Crew Size	Man-hours per Unit	Crew Output per Day	Avg Mat'l Unit Cost	Avg Labor Unit Cost	Avg Equip Unit Cost	Avg Total Unit Cost	Avg Price Incl O&P
1-1/2" x 2-3/8"	Inst	LF	Lg	CA	.040	200.0	2.18	1.33	---	3.51	**4.50**
	Inst	LF	Sm	CA	.057	140.0	2.54	1.90	---	4.44	**5.77**

Watertable with lip

Description	Oper	Unit	Vol	Crew Size	Man-hours per Unit	Crew Output per Day	Avg Mat'l Unit Cost	Avg Labor Unit Cost	Avg Equip Unit Cost	Avg Total Unit Cost	Avg Price Incl O&P
1-1/2" x 2-3/8"	Inst	LF	Lg	CA	.040	200.0	2.18	1.33	---	3.51	**4.50**
	Inst	LF	Sm	CA	.057	140.0	2.54	1.90	---	4.44	**5.77**

Corrugated

Description	Oper	Unit	Vol	Crew Size	Man-hours per Unit	Crew Output per Day	Avg Mat'l Unit Cost	Avg Labor Unit Cost	Avg Equip Unit Cost	Avg Total Unit Cost	Avg Price Incl O&P
3/4" x 1-5/8"	Inst	LF	Lg	CA	.040	200.0	.41	1.33	---	1.74	**2.47**
	Inst	LF	Sm	CA	.057	140.0	.48	1.90	---	2.38	**3.40**

Crest

Description	Oper	Unit	Vol	Crew Size	Man-hours per Unit	Crew Output per Day	Avg Mat'l Unit Cost	Avg Labor Unit Cost	Avg Equip Unit Cost	Avg Total Unit Cost	Avg Price Incl O&P
3/4" x 1-5/8"	Inst	LF	Lg	CA	.040	200.0	.41	1.33	---	1.74	**2.47**
	Inst	LF	Sm	CA	.057	140.0	.48	1.90	---	2.38	**3.40**

Description	Oper	Unit	Vol	Crew Size	Man-hours per Unit	Crew Output per Day	Avg Mat'l Unit Cost	Avg Labor Unit Cost	Avg Equip Unit Cost	Avg Total Unit Cost	Avg Price Incl O&P

Resin flexible molding
Primed paint grade

Diameter casing

711

5/8" x 2-1/2"	Inst	LF	Lg	CA	.030	266.0	9.79	1.00	---	10.79	**12.80**
	Inst	LF	Sm	CA	.043	186.0	11.40	1.43	---	12.83	**15.20**

Universal

5/8" x 2-1/2"	Inst	LF	Lg	CA	.030	266.0	9.79	1.00	---	10.79	**12.80**
	Inst	LF	Sm	CA	.043	186.0	11.40	1.43	---	12.83	**15.20**

356

5/8" x 2-1/2"	Inst	LF	Lg	CA	.030	266.0	9.79	1.00	---	10.79	**12.80**
	Inst	LF	Sm	CA	.043	186.0	11.40	1.43	---	12.83	**15.20**

Cape Cod

5/8" x 2-1/2"	Inst	LF	Lg	CA	.030	266.0	9.79	1.00	---	10.79	**12.80**
	Inst	LF	Sm	CA	.043	186.0	11.40	1.43	---	12.83	**15.20**

Bases & round corner blocks

Base shoe

3/8" x 3/4"	Inst	LF	Lg	CA	.030	266.0	1.85	1.00	---	2.85	**3.62**
	Inst	LF	Sm	CA	.043	186.0	2.15	1.43	---	3.58	**4.62**

Streamline

3/8" x 2-1/4"	Inst	LF	Lg	CA	.030	266.0	3.02	1.00	---	4.02	**4.97**
	Inst	LF	Sm	CA	.043	186.0	3.51	1.43	---	4.94	**6.18**
7/16" x 3-1/2"	Inst	LF	Lg	CA	.030	266.0	4.87	1.00	---	5.87	**7.10**
	Inst	LF	Sm	CA	.043	186.0	5.66	1.43	---	7.09	**8.66**

711

7/16" x 2-1/2"	Inst	LF	Lg	CA	.030	266.0	2.91	1.00	---	3.91	**4.84**
	Inst	LF	Sm	CA	.043	186.0	3.38	1.43	---	4.81	**6.03**
7/16" x 3-1/2"	Inst	LF	Lg	CA	.030	266.0	4.39	1.00	---	5.39	**6.55**
	Inst	LF	Sm	CA	.043	186.0	5.10	1.43	---	6.53	**8.01**

WM 623

9/16" x 3-1/4"	Inst	LF	Lg	CA	.030	266.0	4.02	1.00	---	5.02	**6.12**
	Inst	LF	Sm	CA	.043	186.0	4.68	1.43	---	6.11	**7.53**

WM 618

9/16" x 5-1/4"	Inst	LF	Lg	CA	.030	266.0	9.31	1.00	---	10.31	**12.20**
	Inst	LF	Sm	CA	.043	186.0	10.80	1.43	---	12.23	**14.60**

Cape Cod

9/16" x 5-1/4"	Inst	LF	Lg	CA	.030	266.0	7.41	1.00	---	8.41	**10.00**
	Inst	LF	Sm	CA	.043	186.0	8.61	1.43	---	10.04	**12.10**
9/16" x 4"	Inst	LF	Lg	CA	.030	266.0	5.29	1.00	---	6.29	**7.58**
	Inst	LF	Sm	CA	.043	186.0	6.15	1.43	---	7.58	**9.22**

Description	Oper	Unit	Vol	Crew Size	Man-hours per Unit	Crew Output per Day	Avg Mat'l Unit Cost	Avg Labor Unit Cost	Avg Equip Unit Cost	Avg Total Unit Cost	Avg Price Incl O&P
"B" Cape Cod											
9/16" x 5-1/4"	Inst	LF	Lg	CA	.030	266.0	8.46	1.00	---	9.46	**11.20**
	Inst	LF	Sm	CA	.043	186.0	9.84	1.43	---	11.27	**13.50**
Colonial											
11/16" x 4-1/4"	Inst	LF	Lg	CA	.030	266.0	6.35	1.00	---	7.35	**8.80**
	Inst	LF	Sm	CA	.043	186.0	7.38	1.43	---	8.81	**10.60**
Monterey											
7/16" x 4-1/4"	Inst	LF	Lg	CA	.030	266.0	6.35	1.00	---	7.35	**8.80**
	Inst	LF	Sm	CA	.043	186.0	7.38	1.43	---	8.81	**10.60**
356											
1-1/2" x 2-1/2"	Inst	LF	Lg	CA	.030	266.0	3.17	1.00	---	4.17	**5.14**
	Inst	LF	Sm	CA	.043	186.0	3.69	1.43	---	5.12	**6.39**
1-1/2" x 3-1/4"	Inst	LF	Lg	CA	.030	266.0	4.87	1.00	---	5.87	**7.10**
	Inst	LF	Sm	CA	.043	186.0	5.66	1.43	---	7.09	**8.66**
Step											
7/16" x 2-1/2"	Inst	LF	Lg	CA	.030	266.0	3.39	1.00	---	4.39	**5.40**
	Inst	LF	Sm	CA	.043	186.0	3.94	1.43	---	5.37	**6.68**
1/2" x 4-1/2"	Inst	LF	Lg	CA	.030	266.0	6.67	1.00	---	7.67	**9.17**
	Inst	LF	Sm	CA	.043	186.0	7.76	1.43	---	9.19	**11.10**

Oak grain

Diameter casing

711

Description	Oper	Unit	Vol	Crew Size	Man-hours per Unit	Crew Output per Day	Avg Mat'l Unit Cost	Avg Labor Unit Cost	Avg Equip Unit Cost	Avg Total Unit Cost	Avg Price Incl O&P
5/8" x 2-1/2"	Inst	LF	Lg	CA	.030	266.0	10.60	1.00	---	11.60	**13.70**
	Inst	LF	Sm	CA	.043	186.0	12.30	1.43	---	13.73	**16.30**
356											
11/16" x 2-1/2"	Inst	LF	Lg	CA	.030	266.0	10.60	1.00	---	11.60	**13.70**
	Inst	LF	Sm	CA	.043	186.0	12.30	1.43	---	13.73	**16.30**

Bases & base shoe

711 base

Description	Oper	Unit	Vol	Crew Size	Man-hours per Unit	Crew Output per Day	Avg Mat'l Unit Cost	Avg Labor Unit Cost	Avg Equip Unit Cost	Avg Total Unit Cost	Avg Price Incl O&P
3/8" x 2-1/2"	Inst	LF	Lg	CA	.030	266.0	4.23	1.00	---	5.23	**6.36**
	Inst	LF	Sm	CA	.043	186.0	4.92	1.43	---	6.35	**7.80**
3/8" x 3-1/2"	Inst	LF	Lg	CA	.030	266.0	6.35	1.00	---	7.35	**8.80**
	Inst	LF	Sm	CA	.043	186.0	7.38	1.43	---	8.81	**10.60**
Base shoe											
3/8" x 3/4"	Inst	LF	Lg	CA	.030	266.0	2.12	1.00	---	3.12	**3.94**
	Inst	LF	Sm	CA	.043	186.0	2.46	1.43	---	3.89	**4.98**

Description	Oper	Unit	Vol	Crew Size	Man-hours per Unit	Crew Output per Day	Avg Mat'l Unit Cost	Avg Labor Unit Cost	Avg Equip Unit Cost	Avg Total Unit Cost	Avg Price Incl O&P
Spindles											

Western hemlock, clear, kiln dried, turned for decorative applications

Planter design

1-11/16" x 1-11/16"

Description	Oper	Unit	Vol	Crew Size	Man-hours per Unit	Crew Output per Day	Avg Mat'l Unit Cost	Avg Labor Unit Cost	Avg Equip Unit Cost	Avg Total Unit Cost	Avg Price Incl O&P
3'-0" H	Inst	Ea	Lg	CA	.333	24.00	3.92	11.10	---	15.02	**21.10**
	Inst	Ea	Sm	CA	.471	17.00	4.56	15.70	---	20.26	**28.80**
4'-0" H	Inst	Ea	Lg	CA	.333	24.00	5.02	11.10	---	16.12	**22.40**
	Inst	Ea	Sm	CA	.471	17.00	5.84	15.70	---	21.54	**30.20**

2-3/8" x 2-3/8"

Description	Oper	Unit	Vol	Crew Size	Man-hours per Unit	Crew Output per Day	Avg Mat'l Unit Cost	Avg Labor Unit Cost	Avg Equip Unit Cost	Avg Total Unit Cost	Avg Price Incl O&P
3'-0" H	Inst	Ea	Lg	CA	.333	24.00	8.15	11.10	---	19.25	**26.00**
	Inst	Ea	Sm	CA	.471	17.00	9.48	15.70	---	25.18	**34.40**
4'-0" H	Inst	Ea	Lg	CA	.333	24.00	11.20	11.10	---	22.30	**29.50**
	Inst	Ea	Sm	CA	.471	17.00	13.00	15.70	---	28.70	**38.40**
5'-0" H	Inst	Ea	Lg	CA	.333	24.00	13.50	11.10	---	24.60	**32.20**
	Inst	Ea	Sm	CA	.471	17.00	15.80	15.70	---	31.50	**41.60**
6'-0" H	Inst	Ea	Lg	CA	.400	20.00	17.30	13.30	---	30.60	**39.80**
	Inst	Ea	Sm	CA	.571	14.00	20.10	19.00	---	39.10	**51.60**
8'-0" H	Inst	Ea	Lg	CA	.400	20.00	33.50	13.30	---	46.80	**58.50**
	Inst	Ea	Sm	CA	.571	14.00	39.00	19.00	---	58.00	**73.40**

3-1/4" x 3-1/4"

Description	Oper	Unit	Vol	Crew Size	Man-hours per Unit	Crew Output per Day	Avg Mat'l Unit Cost	Avg Labor Unit Cost	Avg Equip Unit Cost	Avg Total Unit Cost	Avg Price Incl O&P
3'-0" H	Inst	Ea	Lg	CA	.333	24.00	15.50	11.10	---	26.60	**34.40**
	Inst	Ea	Sm	CA	.471	17.00	18.00	15.70	---	33.70	**44.20**
4'-0" H	Inst	Ea	Lg	CA	.333	24.00	20.60	11.10	---	31.70	**40.40**
	Inst	Ea	Sm	CA	.471	17.00	24.00	15.70	---	39.70	**51.10**
5'-0" H	Inst	Ea	Lg	CA	.333	24.00	26.10	11.10	---	37.20	**46.70**
	Inst	Ea	Sm	CA	.471	17.00	30.40	15.70	---	46.10	**58.50**
6'-0" H	Inst	Ea	Lg	CA	.400	20.00	31.30	13.30	---	44.60	**55.90**
	Inst	Ea	Sm	CA	.571	14.00	36.40	19.00	---	55.40	**70.30**
8'-0" H	Inst	Ea	Lg	CA	.400	20.00	45.10	13.30	---	58.40	**71.90**
	Inst	Ea	Sm	CA	.571	14.00	52.50	19.00	---	71.50	**88.90**

Colonial design

1-11/16" x 1-11/16"

Description	Oper	Unit	Vol	Crew Size	Man-hours per Unit	Crew Output per Day	Avg Mat'l Unit Cost	Avg Labor Unit Cost	Avg Equip Unit Cost	Avg Total Unit Cost	Avg Price Incl O&P
1'-0" H	Inst	Ea	Lg	CA	.286	28.00	1.27	9.52	---	10.79	**15.70**
	Inst	Ea	Sm	CA	.400	20.00	1.48	13.30	---	14.78	**21.70**
1'-6" H	Inst	Ea	Lg	CA	.286	28.00	2.27	9.52	---	11.79	**16.90**
	Inst	Ea	Sm	CA	.400	20.00	2.64	13.30	---	15.94	**23.00**
2'-0" H	Inst	Ea	Lg	CA	.286	28.00	2.91	9.52	---	12.43	**17.60**
	Inst	Ea	Sm	CA	.400	20.00	3.38	13.30	---	16.68	**23.90**
2'-4" H	Inst	Ea	Lg	CA	.333	24.00	3.02	11.10	---	14.12	**20.10**
	Inst	Ea	Sm	CA	.471	17.00	3.51	15.70	---	19.21	**27.50**
2'-8" H	Inst	Ea	Lg	CA	.333	24.00	3.39	11.10	---	14.49	**20.50**
	Inst	Ea	Sm	CA	.471	17.00	3.94	15.70	---	19.64	**28.00**
3'-0" H	Inst	Ea	Lg	CA	.333	24.00	4.02	11.10	---	15.12	**21.20**
	Inst	Ea	Sm	CA	.471	17.00	4.68	15.70	---	20.38	**28.90**

Description	Oper	Unit	Vol	Crew Size	Man-hours per Unit	Crew Output per Day	Avg Mat'l Unit Cost	Avg Labor Unit Cost	Avg Equip Unit Cost	Avg Total Unit Cost	Avg Price Incl O&P
2-3/8" x 2-3/8"											
1'-0" H	Inst	Ea	Lg	CA	.286	28.00	2.59	9.52	---	12.11	**17.30**
	Inst	Ea	Sm	CA	.400	20.00	3.02	13.30	---	16.32	**23.40**
1'-6" H	Inst	Ea	Lg	CA	.286	28.00	3.60	9.52	---	13.12	**18.40**
	Inst	Ea	Sm	CA	.400	20.00	4.18	13.30	---	17.48	**24.80**
2'-0" H	Inst	Ea	Lg	CA	.286	28.00	5.45	9.52	---	14.97	**20.50**
	Inst	Ea	Sm	CA	.400	20.00	6.33	13.30	---	19.63	**27.20**
2'-4" H	Inst	Ea	Lg	CA	.333	24.00	5.82	11.10	---	16.92	**23.30**
	Inst	Ea	Sm	CA	.471	17.00	6.77	15.70	---	22.47	**31.30**
2'-8" H	Inst	Ea	Lg	CA	.333	24.00	7.41	11.10	---	18.51	**25.10**
	Inst	Ea	Sm	CA	.471	17.00	8.61	15.70	---	24.31	**33.40**
3'-0" H	Inst	Ea	Lg	CA	.333	24.00	8.15	11.10	---	19.25	**26.00**
	Inst	Ea	Sm	CA	.471	17.00	9.48	15.70	---	25.18	**34.40**
3-1/4" x 3-1/4"											
1'-6" H	Inst	Ea	Lg	CA	.286	28.00	6.08	9.52	---	15.60	**21.30**
	Inst	Ea	Sm	CA	.400	20.00	7.07	13.30	---	20.37	**28.10**
2'-0" H	Inst	Ea	Lg	CA	.286	28.00	8.25	9.52	---	17.77	**23.80**
	Inst	Ea	Sm	CA	.400	20.00	9.60	13.30	---	22.90	**31.00**
2'-4" H	Inst	Ea	Lg	CA	.333	24.00	9.68	11.10	---	20.78	**27.80**
	Inst	Ea	Sm	CA	.471	17.00	11.30	15.70	---	27.00	**36.50**
2'-8" H	Inst	Ea	Lg	CA	.333	24.00	11.00	11.10	---	22.10	**29.20**
	Inst	Ea	Sm	CA	.471	17.00	12.70	15.70	---	28.40	**38.20**
3'-0" H	Inst	Ea	Lg	CA	.333	24.00	15.50	11.10	---	26.60	**34.40**
	Inst	Ea	Sm	CA	.471	17.00	18.00	15.70	---	33.70	**44.20**
8'-0" H	Inst	Ea	Lg	CA	.400	20.00	45.10	13.30	---	58.40	**71.90**
	Inst	Ea	Sm	CA	.571	14.00	52.50	19.00	---	71.50	**88.90**

Mediterranean design

1-11/16" x 1-11/16"											
1'-0" H	Inst	Ea	Lg	CA	.286	28.00	1.32	9.52	---	10.84	**15.80**
	Inst	Ea	Sm	CA	.400	20.00	1.54	13.30	---	14.84	**21.70**
1'-6" H	Inst	Ea	Lg	CA	.286	28.00	2.17	9.52	---	11.69	**16.80**
	Inst	Ea	Sm	CA	.400	20.00	2.53	13.30	---	15.83	**22.90**
2'-0" H	Inst	Ea	Lg	CA	.286	28.00	2.30	9.52	---	11.90	**17.00**
	Inst	Ea	Sm	CA	.400	20.00	2.77	13.30	---	16.07	**23.20**
2'-4" H	Inst	Ea	Lg	CA	.333	24.00	2.81	11.10	---	13.91	**19.90**
	Inst	Ea	Sm	CA	.471	17.00	3.26	15.70	---	18.96	**27.30**
2'-8" H	Inst	Ea	Lg	CA	.333	24.00	3.12	11.10	---	14.22	**20.20**
	Inst	Ea	Sm	CA	.471	17.00	3.63	15.70	---	19.33	**27.70**
3'-0" H	Inst	Ea	Lg	CA	.333	24.00	3.92	11.10	---	15.02	**21.10**
	Inst	Ea	Sm	CA	.471	17.00	4.56	15.70	---	20.26	**28.80**
4'-0" H	Inst	Ea	Lg	CA	.333	24.00	5.02	11.10	---	16.12	**22.40**
	Inst	Ea	Sm	CA	.471	17.00	5.84	15.70	---	21.54	**30.20**
5'-0" H	Inst	Ea	Lg	CA	.333	24.00	9.04	11.10	---	20.14	**27.00**
	Inst	Ea	Sm	CA	.471	17.00	10.50	15.70	---	26.20	**35.60**

Description	Oper	Unit	Vol	Crew Size	Man-hours per Unit	Crew Output per Day	Avg Mat'l Unit Cost	Avg Labor Unit Cost	Avg Equip Unit Cost	Avg Total Unit Cost	Avg Price Incl O&P
2-3/8" x 2-3/8"											
1'-0" H	Inst	Ea	Lg	CA	.286	28.00	2.38	9.52	---	11.90	**17.00**
	Inst	Ea	Sm	CA	.400	20.00	2.77	13.30	---	16.07	**23.20**
1'-6" H	Inst	Ea	Lg	CA	.286	28.00	3.44	9.52	---	12.96	**18.20**
	Inst	Ea	Sm	CA	.400	20.00	4.00	13.30	---	17.30	**24.60**
2'-0" H	Inst	Ea	Lg	CA	.286	28.00	4.34	9.52	---	13.86	**19.30**
	Inst	Ea	Sm	CA	.400	20.00	5.05	13.30	---	18.35	**25.80**
2'-4" H	Inst	Ea	Lg	CA	.333	24.00	5.56	11.10	---	16.66	**23.00**
	Inst	Ea	Sm	CA	.471	17.00	6.46	15.70	---	22.16	**30.90**
2'-8" H	Inst	Ea	Lg	CA	.333	24.00	6.93	11.10	---	18.03	**24.60**
	Inst	Ea	Sm	CA	.471	17.00	8.06	15.70	---	23.76	**32.80**
3'-0" H	Inst	Ea	Lg	CA	.333	24.00	7.94	11.10	---	19.04	**25.80**
	Inst	Ea	Sm	CA	.471	17.00	9.23	15.70	---	24.93	**34.10**
4'-0" H	Inst	Ea	Lg	CA	.333	24.00	9.42	11.10	---	20.52	**27.50**
	Inst	Ea	Sm	CA	.471	17.00	11.00	15.70	---	26.70	**36.10**
5'-0" H	Inst	Ea	Lg	CA	.333	24.00	13.00	11.10	---	24.10	**31.50**
	Inst	Ea	Sm	CA	.471	17.00	15.10	15.70	---	30.80	**40.90**
6'-0" H	Inst	Ea	Lg	CA	.400	20.00	18.20	13.30	---	31.50	**40.90**
	Inst	Ea	Sm	CA	.571	14.00	21.20	19.00	---	40.20	**52.80**
8'-0" H	Inst	Ea	Lg	CA	.400	20.00	33.50	13.30	---	46.80	**58.50**
	Inst	Ea	Sm	CA	.571	14.00	39.00	19.00	---	58.00	**73.40**
3-1/4" x 3-1/4"											
3'-0" H	Inst	Ea	Lg	CA	.333	24.00	15.00	11.10	---	26.10	**33.80**
	Inst	Ea	Sm	CA	.471	17.00	17.40	15.70	---	33.10	**43.50**
4'-0" H	Inst	Ea	Lg	CA	.333	24.00	17.60	11.10	---	28.70	**36.90**
	Inst	Ea	Sm	CA	.471	17.00	20.50	15.70	---	36.20	**47.10**
5'-0" H	Inst	Ea	Lg	CA	.333	24.00	23.70	11.10	---	34.80	**43.90**
	Inst	Ea	Sm	CA	.471	17.00	27.60	15.70	---	43.30	**55.20**
6'-0" H	Inst	Ea	Lg	CA	.400	20.00	33.30	13.30	---	46.60	**58.30**
	Inst	Ea	Sm	CA	.571	14.00	38.80	19.00	---	57.80	**73.10**
8'-0" H	Inst	Ea	Lg	CA	.400	20.00	52.60	13.30	---	65.90	**80.40**
	Inst	Ea	Sm	CA	.571	14.00	61.20	19.00	---	80.20	**98.80**

Spindle rails, 8'-0" H pieces

Description	Oper	Unit	Vol	Crew Size	Man-hours per Unit	Crew Output per Day	Avg Mat'l Unit Cost	Avg Labor Unit Cost	Avg Equip Unit Cost	Avg Total Unit Cost	Avg Price Incl O&P
For 1-11/16" spindles											
	Inst	Ea	Lg	CA	---	---	13.30	---	---	13.30	**15.30**
	Inst	Ea	Sm	CA	---	---	15.50	---	---	15.50	**17.80**
For 2-3/8" spindles											
	Inst	Ea	Lg	CA	---	---	19.30	---	---	19.30	**22.20**
	Inst	Ea	Sm	CA	---	---	22.40	---	---	22.40	**25.80**
For 3-1/4" spindles											
	Inst	Ea	Lg	CA	---	---	22.40	---	---	22.40	**25.70**
	Inst	Ea	Sm	CA	---	---	26.00	---	---	26.00	**29.90**

Description	Oper	Unit	Vol	Crew Size	Man-hours per Unit	Crew Output per Day	Avg Mat'l Unit Cost	Avg Labor Unit Cost	Avg Equip Unit Cost	Avg Total Unit Cost	Avg Price Incl O&P

Wood bullnose round corners

Pine base R/C

Colonial

Description	Oper	Unit	Vol	Crew Size	Man-hours per Unit	Crew Output per Day	Avg Mat'l Unit Cost	Avg Labor Unit Cost	Avg Equip Unit Cost	Avg Total Unit Cost	Avg Price Incl O&P
11/16" x 4-1/4"	Inst	LF	Lg	CA	.030	266.0	1.08	1.00	---	2.08	**2.74**
	Inst	LF	Sm	CA	.043	186.0	1.25	1.43	---	2.68	**3.58**

Streamline

Description	Oper	Unit	Vol	Crew Size	Man-hours per Unit	Crew Output per Day	Avg Mat'l Unit Cost	Avg Labor Unit Cost	Avg Equip Unit Cost	Avg Total Unit Cost	Avg Price Incl O&P
3/8" x 2-1/4"	Inst	LF	Lg	CA	.030	266.0	.56	1.00	---	1.56	**2.14**
	Inst	LF	Sm	CA	.043	186.0	.65	1.43	---	2.08	**2.89**
7/16" x 3-1/2"	Inst	LF	Lg	CA	.030	266.0	.80	1.00	---	1.80	**2.42**
	Inst	LF	Sm	CA	.043	186.0	.93	1.43	---	2.36	**3.22**

711

Description	Oper	Unit	Vol	Crew Size	Man-hours per Unit	Crew Output per Day	Avg Mat'l Unit Cost	Avg Labor Unit Cost	Avg Equip Unit Cost	Avg Total Unit Cost	Avg Price Incl O&P
7/16" x 2-1/2"	Inst	LF	Lg	CA	.030	266.0	.66	1.00	---	1.66	**2.26**
	Inst	LF	Sm	CA	.043	186.0	.77	1.43	---	2.20	**3.03**
7/16" x 3-1/2"	Inst	LF	Lg	CA	.030	266.0	.94	1.00	---	1.94	**2.58**
	Inst	LF	Sm	CA	.043	186.0	1.09	1.43	---	2.52	**3.40**

3 step

Description	Oper	Unit	Vol	Crew Size	Man-hours per Unit	Crew Output per Day	Avg Mat'l Unit Cost	Avg Labor Unit Cost	Avg Equip Unit Cost	Avg Total Unit Cost	Avg Price Incl O&P
7/16" x 2-1/2"	Inst	LF	Lg	CA	.030	266.0	.59	1.00	---	1.59	**2.18**
	Inst	LF	Sm	CA	.043	186.0	.68	1.43	---	2.11	**2.93**
1/2" x 4-1/2"	Inst	LF	Lg	CA	.030	266.0	1.74	1.00	---	2.74	**3.50**
	Inst	LF	Sm	CA	.043	186.0	2.02	1.43	---	3.45	**4.47**

WM 623

Description	Oper	Unit	Vol	Crew Size	Man-hours per Unit	Crew Output per Day	Avg Mat'l Unit Cost	Avg Labor Unit Cost	Avg Equip Unit Cost	Avg Total Unit Cost	Avg Price Incl O&P
9/16" x 3-1/4"	Inst	LF	Lg	CA	.030	266.0	.82	1.00	---	1.82	**2.44**
	Inst	LF	Sm	CA	.043	186.0	.95	1.43	---	2.38	**3.24**

WM 618

Description	Oper	Unit	Vol	Crew Size	Man-hours per Unit	Crew Output per Day	Avg Mat'l Unit Cost	Avg Labor Unit Cost	Avg Equip Unit Cost	Avg Total Unit Cost	Avg Price Incl O&P
9/16" x 5-1/4"	Inst	LF	Lg	CA	.030	266.0	1.26	1.00	---	2.26	**2.95**
	Inst	LF	Sm	CA	.043	186.0	1.47	1.43	---	2.90	**3.84**

Monterey

Description	Oper	Unit	Vol	Crew Size	Man-hours per Unit	Crew Output per Day	Avg Mat'l Unit Cost	Avg Labor Unit Cost	Avg Equip Unit Cost	Avg Total Unit Cost	Avg Price Incl O&P
7/16" x 4-1/4"	Inst	LF	Lg	CA	.030	266.0	.87	1.00	---	1.87	**2.50**
	Inst	LF	Sm	CA	.043	186.0	1.02	1.43	---	2.45	**3.32**

Cape Cod

Description	Oper	Unit	Vol	Crew Size	Man-hours per Unit	Crew Output per Day	Avg Mat'l Unit Cost	Avg Labor Unit Cost	Avg Equip Unit Cost	Avg Total Unit Cost	Avg Price Incl O&P
9/16" x 4"	Inst	LF	Lg	CA	.030	266.0	1.06	1.00	---	2.06	**2.72**
	Inst	LF	Sm	CA	.043	186.0	1.23	1.43	---	2.66	**3.56**
9/16" x 5-1/4"	Inst	LF	Lg	CA	.030	266.0	1.27	1.00	---	2.27	**2.96**
	Inst	LF	Sm	CA	.043	186.0	1.48	1.43	---	2.91	**3.85**

"B" Cape Cod

Description	Oper	Unit	Vol	Crew Size	Man-hours per Unit	Crew Output per Day	Avg Mat'l Unit Cost	Avg Labor Unit Cost	Avg Equip Unit Cost	Avg Total Unit Cost	Avg Price Incl O&P
9/16" x 5-1/4"	Inst	LF	Lg	CA	.030	266.0	1.27	1.00	---	2.27	**2.96**
	Inst	LF	Sm	CA	.043	186.0	1.48	1.43	---	2.91	**3.85**

356

Description	Oper	Unit	Vol	Crew Size	Man-hours per Unit	Crew Output per Day	Avg Mat'l Unit Cost	Avg Labor Unit Cost	Avg Equip Unit Cost	Avg Total Unit Cost	Avg Price Incl O&P
1/2" x 2-1/4"	Inst	LF	Lg	CA	.030	266.0	.77	1.00	---	1.77	**2.38**
	Inst	LF	Sm	CA	.043	186.0	.90	1.43	---	2.33	**3.18**
1/2" x 3-1/4"	Inst	LF	Lg	CA	.030	266.0	.98	1.00	---	1.98	**2.62**
	Inst	LF	Sm	CA	.043	186.0	1.13	1.43	---	2.56	**3.45**

Base shoe

Description	Oper	Unit	Vol	Crew Size	Man-hours per Unit	Crew Output per Day	Avg Mat'l Unit Cost	Avg Labor Unit Cost	Avg Equip Unit Cost	Avg Total Unit Cost	Avg Price Incl O&P
3/8" x 3/4"	Inst	LF	Lg	CA	.030	266.0	.20	1.00	---	1.20	**1.73**
	Inst	LF	Sm	CA	.043	186.0	.24	1.43	---	1.67	**2.42**

Description	Oper	Unit	Vol	Crew Size	Man-hours per Unit	Crew Output per Day	Avg Mat'l Unit Cost	Avg Labor Unit Cost	Avg Equip Unit Cost	Avg Total Unit Cost	Avg Price Incl O&P

Oak base R/C

Base shoe
| 3/8" x 3/4" | Inst | LF | Lg | CA | .030 | 266.0 | .33 | 1.00 | --- | 1.33 | **1.88** |
| | Inst | LF | Sm | CA | .043 | 186.0 | .39 | 1.43 | --- | 1.82 | **2.59** |

711
3/8" x 2-1/2"	Inst	LF	Lg	CA	.030	266.0	.97	1.00	---	1.97	**2.61**
	Inst	LF	Sm	CA	.043	186.0	1.12	1.43	---	2.55	**3.43**
3/8" x 3-1/2"	Inst	LF	Lg	CA	.030	266.0	1.21	1.00	---	2.21	**2.89**
	Inst	LF	Sm	CA	.043	186.0	1.41	1.43	---	2.84	**3.77**

Pine chair rail R/C

Chair rail
| 9/16" x 2-1/2" | Inst | LF | Lg | CA | .030 | 266.0 | .86 | 1.00 | --- | 1.86 | **2.49** |
| | Inst | LF | Sm | CA | .043 | 186.0 | 1.00 | 1.43 | --- | 2.43 | **3.30** |

Colonial chair rail
11/16" x 2-1/2"	Inst	LF	Lg	CA	.030	266.0	.86	1.00	---	1.86	**2.49**
	Inst	LF	Sm	CA	.043	186.0	1.01	1.43	---	2.44	**3.31**
3/4" x 3"	Inst	LF	Lg	CA	.030	266.0	.99	1.00	---	1.99	**2.64**
	Inst	LF	Sm	CA	.043	186.0	1.16	1.43	---	2.59	**3.48**

Pine crown & cornice R/C

Colonial crown
9/16" x 2-1/4"	Inst	LF	Lg	CA	.030	266.0	.77	1.00	---	1.77	**2.38**
	Inst	LF	Sm	CA	.043	186.0	.90	1.43	---	2.33	**3.18**
11/16" x 3-3/8"	Inst	LF	Lg	CA	.030	266.0	1.04	1.00	---	2.04	**2.69**
	Inst	LF	Sm	CA	.043	186.0	1.21	1.43	---	2.64	**3.54**
11/16" x 4-1/2"	Inst	LF	Lg	CA	.030	266.0	1.81	1.00	---	2.81	**3.58**
	Inst	LF	Sm	CA	.043	186.0	2.11	1.43	---	3.54	**4.57**

Cornice
| 13/16" x 4-5/8" | Inst | LF | Lg | CA | .030 | 266.0 | 2.05 | 1.00 | --- | 3.05 | **3.85** |
| | Inst | LF | Sm | CA | .043 | 186.0 | 2.39 | 1.43 | --- | 3.82 | **4.89** |

Georgian crown
| 11/16" x 4-1/4" | Inst | LF | Lg | CA | .030 | 266.0 | 2.24 | 1.00 | --- | 3.24 | **4.07** |
| | Inst | LF | Sm | CA | .043 | 186.0 | 2.61 | 1.43 | --- | 4.04 | **5.15** |

Painting

Interior and Exterior. There is a paint for almost every type of surface and surface condition. The large variety makes it impractical to consider each paint individually. For this reason, average output and average material cost/unit are based on the paints and prices listed below.

1. **Installation.** Paint can be applied by brush, roller or spray gun. Only application by brush or roller is considered in this section.

2. **Notes on Labor.** Average Manhours per Unit, for both roller and brush, is based on what one painter can do in one day. The output for cleaning is also based on what one painter can do in one day.

3. **Estimating Technique.** Use these techniques to determine quantities for interior and exterior painting before you apply unit costs.

Interior

a. Floors, walls and ceilings. Figure actual area. No deductions for openings.

b. Doors and windows. **Only openings to be painted:** Figure 36 SF or 4 SY for each side of each door and 27 SF for each side of each window. Based on doors 3'-0" x 7'-0" or smaller and windows 3'-0" x 4'-0" or smaller. **Openings to be painted with walls:** Figure wall area plus 27 SF or 3 SY for each side of each door and 18 SF or 2 SY for each side of each window. Based on doors 3'-0" x 7'-0" or smaller and windows 3'-0" x 4'-0" or smaller.

For larger doors and windows, add 1'-0" to height and width and figure area.

c. Base or picture moldings and chair rails. Less than 1'-0" wide, figure one SF/LF. On 1'-0" or larger, figure actual area.

d. Stairs (including treads, risers, cove and stringers). Add 2'-0" width (for treads, risers, etc.) times length plus 2'-0".

e. Balustrades. Add 1'-0" to height, figure two times area to paint two sides.

Exterior

a. Walls. (No deductions for openings.)

Siding. Figure actual area plus 10%.

Shingles. Figure actual area plus 40%.

Characteristics - Interior			Characteristics - Exterior		
Type	Coverage SF/Gal.	Surface	Type	Coverage SF/Gal.	Surface
Latex, flat	450	Plaster/drywall	Oil base*	300	Plain siding & stucco
Latex, enamel	450	Doors, windows, trim	Oil base*	450	Door, windows, trim
Shellac	500	Woodwork	Oil base*	300	Shingle siding
Varnish	500	Woodwork	Stain	200	Shingle siding
Stain	500	Woodwork	Latex, masonry	400	Stucco & masonry

*Certain latex paints may also be used on exterior work.

Brick, stucco, concrete and smooth wood surfaces. Figure actual area.

b. Doors and windows. See Interior, doors and windows.

c. Eaves (including soffit or exposed rafter ends and fascia). **Enclosed.** If sidewalls are to be painted the same color, figure 1.5 times actual area. If sidewalls are to be painted a different color, figure 2.0 times actual area. If sidewalls are not to be painted, figure 2.5 times actual area. **Rafter ends exposed:** If sidewalls are to be painted same color, figure 2.5 times actual area. If sidewalls are to be painted a different color, figure 3.0 times actual area. If sidewalls are not to be painted, figure 4.0 times actual area.

d. Porch rails. See Interior, balustrades (previous page).

e. Gutters and downspouts. Figure 2.0 SF/LF or $^2/_9$ SY/LF.

f. Latticework. Figure 2.0 times actual area for each side.

g. Fences. **Solid fence:** Figure actual area of each side to be painted. **Basketweave:** Figure 1.5 times actual area for each side to be painted.

Calculating Square Foot Coverage

Triangle

To find the number of square feet in any shape triangle or 3 sided surface, multiply the height by the width and divide the total by 2.

Square

Multiply the base measurement in feet times the height in feet.

Rectangle

Multiply the base measurement in feet times the height in feet.

Arch Roof

Multiply length (B) by width (A) and add one-half the total.

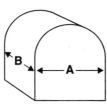

Circle

To find the number of square feet in a circle multiply the diameter (distance across) by itself and them multiply this total by .7854.

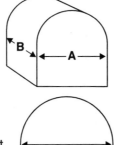

Cylinder

When the circumference (distance around the cylinder) is known, multiply height by circumference. When the diameter (distance across) is known, multiply diameter by 3.1416. This gives circumference. Then multiply by height.

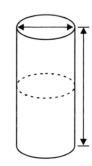

Gambrel Roof

Multiply length (B) by width (A) and add one-third of the total.

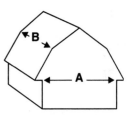

Cone

Determine area of base by multiplying 3.1416 times radius (A) in feet.

Determine the surface area of a cone by multiplying circumference of base (in feet) times one-half of the slant height (B) in feet.

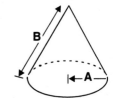

Add the square foot area of the base to the square foot area of the cone for total square foot area.

Description	Oper	Unit	Vol	Crew Size	Man-hours per Unit	Crew Output per Day	Avg Mat'l Unit Cost	Avg Labor Unit Cost	Avg Equip Unit Cost	Avg Total Unit Cost	Avg Price Incl O&P

Painting and cleaning

Frequently encountered applications

Interior

Description	Oper	Unit	Vol	Crew Size	Man-hours per Unit	Crew Output per Day	Avg Mat'l Unit Cost	Avg Labor Unit Cost	Avg Equip Unit Cost	Avg Total Unit Cost	Avg Price Incl O&P
Wet clean, walls	Inst	SF	Lg	NA	.005	1680	.02	.17	---	.19	.27
	Inst	SF	Sm	NA	.006	1260	.02	.20	---	.22	.32
Wet clean, floors	Inst	SF	Lg	NA	.004	2240	.01	.13	---	.14	.21
	Inst	SF	Sm	NA	.005	1680	.02	.17	---	.19	.27
Wet clean, millwork and trim	Inst	SF	Lg	NA	.005	1600	.02	.17	---	.19	.27
	Inst	SF	Sm	NA	.007	1200	.02	.23	---	.25	.37
Prime or seal, one coat	Inst	SF	Lg	NA	.004	1840	.06	.13	---	.19	.26
	Inst	SF	Sm	NA	.006	1380	.07	.20	---	.27	.37
Paint, one coat	Inst	SF	Lg	NA	.005	1680	.09	.17	---	.26	.35
	Inst	SF	Sm	NA	.006	1260	.10	.20	---	.30	.41
Paint, two coats	Inst	SF	Lg	NA	.008	1040	.17	.27	---	.44	.58
	Inst	SF	Sm	NA	.010	780.0	.19	.34	---	.53	.71

Exterior

Description	Oper	Unit	Vol	Crew Size	Man-hours per Unit	Crew Output per Day	Avg Mat'l Unit Cost	Avg Labor Unit Cost	Avg Equip Unit Cost	Avg Total Unit Cost	Avg Price Incl O&P
Wet clean, walls	Inst	SF	Lg	NA	.005	1480	.02	.17	---	.19	.27
	Inst	SF	Sm	NA	.007	1110	.02	.23	---	.25	.37
Paint, one coat	Inst	SF	Lg	NA	.007	1200	.21	.23	---	.44	.58
	Inst	SF	Sm	NA	.009	900.0	.24	.30	---	.54	.71
Paint, two coats	Inst	SF	Lg	NA	.012	680.0	.31	.40	---	.71	.94
	Inst	SF	Sm	NA	.016	510.0	.35	.54	---	.89	1.18

Interior

Time calculations include normal materials handling and protection of furniture and other property not to be painted

Preparation

Excluding openings, unless otherwise indicated

Cleaning, wet

Smooth finishes

Description	Oper	Unit	Vol	Crew Size	Man-hours per Unit	Crew Output per Day	Avg Mat'l Unit Cost	Avg Labor Unit Cost	Avg Equip Unit Cost	Avg Total Unit Cost	Avg Price Incl O&P
Plaster and drywall	Inst	SF	Lg	NA	.005	1680	.02	.17	---	.19	.27
	Inst	SF	Sm	NA	.006	1260	.02	.20	---	.22	.32
Paneling	Inst	SF	Lg	NA	.004	1800	.02	.13	---	.15	.22
	Inst	SF	Sm	NA	.006	1350	.02	.20	---	.22	.32
Millwork and trim	Inst	SF	Lg	NA	.005	1600	.02	.17	---	.19	.27
	Inst	SF	Sm	NA	.007	1200	.02	.23	---	.25	.37
Floors	Inst	SF	Lg	NA	.004	2240	.01	.13	---	.14	.21
	Inst	SF	Sm	NA	.005	1680	.02	.17	---	.19	.27

Sand finishes

Description	Oper	Unit	Vol	Crew Size	Man-hours per Unit	Crew Output per Day	Avg Mat'l Unit Cost	Avg Labor Unit Cost	Avg Equip Unit Cost	Avg Total Unit Cost	Avg Price Incl O&P
	Inst	SF	Lg	NA	.007	1160	.01	.23	---	.24	.36
	Inst	SF	Sm	NA	.009	870.0	.02	.30	---	.32	.47

Description	Oper	Unit	Vol	Crew Size	Man-hours per Unit	Crew Output per Day	Avg Mat'l Unit Cost	Avg Labor Unit Cost	Avg Equip Unit Cost	Avg Total Unit Cost	Avg Price Incl O&P

Sheetrock or drywall, repair joints and cracks or pops
| Tape, fill, and finish | Inst | SF | Lg | NA | .009 | 880.0 | .03 | .30 | --- | .33 | .48 |
| | Inst | SF | Sm | NA | .012 | 660.0 | .04 | .40 | --- | .44 | .64 |

Sheetrock or drywall, repair
Thin coat plaster in lieu of taping
| | Inst | SF | Lg | NA | .012 | 680.0 | .10 | .40 | --- | .50 | .71 |
| | Inst | SF | Sm | NA | .016 | 510.0 | .11 | .54 | --- | .65 | .92 |

Light sanding
Before first coat	Inst	SF	Lg	NA	.005	1600	.01	.17	---	.18	.26
	Inst	SF	Sm	NA	.007	1200	.01	.23	---	.24	.36
Before second coat	Inst	SF	Lg	NA	.005	1760	.01	.17	---	.18	.26
	Inst	SF	Sm	NA	.006	1320	.01	.20	---	.21	.31
Before third coat	Inst	SF	Lg	NA	.004	1920	.01	.13	---	.14	.21
	Inst	SF	Sm	NA	.006	1440	.01	.20	---	.21	.31

Liquid removal of paint or varnish
Paneling (170 SF/gal)	Inst	SF	Lg	NA	.025	320.0	.16	.84	---	1.00	1.42
	Inst	SF	Sm	NA	.033	240.0	.17	1.11	---	1.28	1.83
Millwork & trim (170 SF/gal)	Inst	SF	Lg	NA	.029	280.0	.16	.97	---	1.13	1.62
	Inst	SF	Sm	NA	.038	210.0	.17	1.27	---	1.44	2.07
Floors (170 SF/gal)	Inst	SF	Lg	NA	.017	480.0	.16	.57	---	.73	1.02
	Inst	SF	Sm	NA	.022	360.0	.17	.74	---	.91	1.28

Burning off paint
| | Inst | SF | Lg | NA | .040 | 200.0 | .02 | 1.34 | --- | 1.36 | 2.01 |
| | Inst | SF | Sm | NA | .053 | 150.0 | .03 | 1.78 | --- | 1.81 | 2.66 |

One coat application
Excluding openings unless otherwise indicated

Sizing, on sheetrock or plaster
Smooth finish
Brush (650 SF/gal)	Inst	SF	Lg	NA	.003	2400	.02	.10	---	.12	.17
	Inst	SF	Sm	NA	.004	1800	.03	.13	---	.16	.23
Roller (625 SF/gal)	Inst	SF	Lg	NA	.003	3200	.03	.10	---	.13	.18
	Inst	SF	Sm	NA	.003	2400	.03	.10	---	.13	.18

Sand finish
Brush (550 SF/gal)	Inst	SF	Lg	NA	.005	1720	.03	.17	---	.20	.28
	Inst	SF	Sm	NA	.006	1290	.03	.20	---	.23	.33
Roller (525 SF/gal)	Inst	SF	Lg	NA	.003	2320	.03	.10	---	.13	.18
	Inst	SF	Sm	NA	.005	1740	.03	.17	---	.20	.28

Description	Oper	Unit	Vol	Crew Size	Man-hours per Unit	Crew Output per Day	Avg Mat'l Unit Cost	Avg Labor Unit Cost	Avg Equip Unit Cost	Avg Total Unit Cost	Avg Price Incl O&P
Sealer											
Sheetrock or plaster											
Smooth finish											
Brush (300 SF/gal)	Inst	SF	Lg	NA	.006	1440	.06	.20	---	.26	**.36**
	Inst	SF	Sm	NA	.007	1080	.07	.23	---	.30	**.42**
Roller (285 SF/gal)	Inst	SF	Lg	NA	.004	1840	.06	.13	---	.19	**.26**
	Inst	SF	Sm	NA	.006	1380	.07	.20	---	.27	**.37**
Spray (250 SF/gal)	Inst	SF	Lg	NC	.003	2600	.07	.10	---	.17	**.23**
	Inst	SF	Sm	NC	.004	1950	.08	.14	---	.22	**.29**
Sand finish											
Brush (250 SF/gal)	Inst	SF	Lg	NA	.008	1040	.07	.27	---	.34	**.47**
	Inst	SF	Sm	NA	.010	780.0	.08	.34	---	.42	**.58**
Roller (235 SF/gal)	Inst	SF	Lg	NA	.006	1360	.08	.20	---	.28	**.39**
	Inst	SF	Sm	NA	.008	1020	.09	.27	---	.36	**.50**
Spray (210 SF/gal)	Inst	SF	Lg	NC	.003	2600	.09	.10	---	.19	**.25**
	Inst	SF	Sm	NC	.004	1950	.10	.14	---	.24	**.31**
Acoustical tile or panels											
Brush (225 SF/gal)	Inst	SF	Lg	NA	.006	1240	.08	.20	---	.28	**.39**
	Inst	SF	Sm	NA	.009	930.0	.09	.30	---	.39	**.55**
Roller (200 SF/gal)	Inst	SF	Lg	NA	.005	1600	.09	.17	---	.26	**.35**
	Inst	SF	Sm	NA	.007	1200	.10	.23	---	.33	**.46**
Spray (160 SF/gal)	Inst	SF	Lg	NC	.004	2200	.12	.14	---	.26	**.34**
	Inst	SF	Sm	NC	.005	1650	.13	.17	---	.30	**.40**
Latex											
Drywall or plaster, latex flat											
Smooth finish											
Brush (300 SF/gal)	Inst	SF	Lg	NA	.006	1320	.09	.20	---	.29	**.40**
	Inst	SF	Sm	NA	.008	990.0	.10	.27	---	.37	**.51**
Roller (285 SF/gal)	Inst	SF	Lg	NA	.005	1680	.09	.17	---	.26	**.35**
	Inst	SF	Sm	NA	.006	1260	.10	.20	---	.30	**.41**
Spray (260 SF/gal)	Inst	SF	Lg	NC	.003	2400	.10	.10	---	.20	**.26**
	Inst	SF	Sm	NC	.004	1800	.11	.14	---	.25	**.33**
Sand finish											
Brush (250 SF/gal)	Inst	SF	Lg	NA	.008	960.0	.11	.27	---	.38	**.52**
	Inst	SF	Sm	NA	.011	720.0	.12	.37	---	.49	**.68**
Roller (235 SF/gal)	Inst	SF	Lg	NA	.007	1200	.11	.23	---	.34	**.47**
	Inst	SF	Sm	NA	.009	900.0	.13	.30	---	.43	**.59**
Spray (210 SF/gal)	Inst	SF	Lg	NC	.003	2400	.13	.10	---	.23	**.30**
	Inst	SF	Sm	NC	.004	1800	.14	.14	---	.28	**.36**
Texture or stipple applied to drywall, one coat											
Brush (125 SF/gal)	Inst	SF	Lg	NA	.007	1200	.21	.23	---	.44	**.58**
	Inst	SF	Sm	NA	.009	900.0	.24	.30	---	.54	**.71**
Roller (120 SF/gal)	Inst	SF	Lg	NA	.005	1600	.22	.17	---	.39	**.49**
	Inst	SF	Sm	NA	.007	1200	.25	.23	---	.48	**.62**

Description	Oper	Unit	Vol	Crew Size	Man-hours per Unit	Crew Output per Day	Avg Mat'l Unit Cost	Avg Labor Unit Cost	Avg Equip Unit Cost	Avg Total Unit Cost	Avg Price Incl O&P
Paneling, latex enamel											
Brush (300 SF/gal)	Inst	SF	Lg	NA	.006	1320	.09	.20	---	.29	**.40**
	Inst	SF	Sm	NA	.008	990.0	.10	.27	---	.37	**.51**
Roller (285 SF/gal)	Inst	SF	Lg	NA	.005	1680	.09	.17	---	.26	**.35**
	Inst	SF	Sm	NA	.006	1260	.10	.20	---	.30	**.41**
Spray (260 SF/gal)	Inst	SF	Lg	NC	.003	2400	.10	.10	---	.20	**.26**
	Inst	SF	Sm	NC	.004	1800	.11	.14	---	.25	**.33**
Acoustical tile or panels, latex flat											
Brush (225 SF/gal)	Inst	SF	Lg	NA	.007	1120	.12	.23	---	.35	**.48**
	Inst	SF	Sm	NA	.010	840.0	.13	.34	---	.47	**.64**
Roller (210 SF/gal)	Inst	SF	Lg	NA	.006	1440	.13	.20	---	.33	**.44**
	Inst	SF	Sm	NA	.007	1080	.14	.23	---	.37	**.50**
Spray (185 SF/gal)	Inst	SF	Lg	NC	.004	2000	.14	.14	---	.28	**.36**
	Inst	SF	Sm	NC	.005	1500	.16	.17	---	.33	**.43**

Millwork and trim, latex enamel

Doors and windows

Description	Oper	Unit	Vol	Crew Size	Man-hours per Unit	Crew Output per Day	Avg Mat'l Unit Cost	Avg Labor Unit Cost	Avg Equip Unit Cost	Avg Total Unit Cost	Avg Price Incl O&P
Roller and/or brush (360 SF/gal)	Inst	SF	Lg	NA	.013	640.0	.07	.44	---	.51	**.72**
	Inst	SF	Sm	NA	.017	480.0	.08	.57	---	.65	**.93**
Spray, doors only (325 SF / gal)	Inst	SF	Lg	NC	.006	1360	.08	.21	---	.29	**.39**
	Inst	SF	Sm	NC	.008	1020	.09	.28	---	.37	**.51**
Cabinets											
Roller and/or brush (360 SF/gal)	Inst	SF	Lg	NA	.013	600.0	.07	.44	---	.51	**.72**
	Inst	SF	Sm	NA	.018	450.0	.08	.60	---	.68	**.98**
Spray, doors only (325 SF / gal)	Inst	SF	Lg	NC	.006	1300	.08	.21	---	.29	**.39**
	Inst	SF	Sm	NC	.008	975.0	.09	.28	---	.37	**.51**
Louvers, spray (300 SF/gal)											
	Inst	SF	Lg	NC	.018	440.0	.09	.62	---	.71	**1.02**
	Inst	SF	Sm	NC	.024	330.0	.10	.83	---	.93	**1.34**

Picture molding, chair rail, base, ceiling mold etc., less than 6" high

Note: SF equals LF on trim less than 6" high

Description	Oper	Unit	Vol	Crew Size	Man-hours per Unit	Crew Output per Day	Avg Mat'l Unit Cost	Avg Labor Unit Cost	Avg Equip Unit Cost	Avg Total Unit Cost	Avg Price Incl O&P
Brush (900 SF/gal)	Inst	SF	Lg	NA	.008	960.0	.03	.27	---	.30	**.43**
	Inst	SF	Sm	NA	.011	720.0	.03	.37	---	.40	**.58**
Floors, wood											
Brush (405 SF/gal)	Inst	SF	Lg	NA	.004	2000	.07	.13	---	.20	**.28**
	Inst	SF	Sm	NA	.005	1500	.07	.17	---	.24	**.33**
Roller (385 SF/gal)	Inst	SF	Lg	NA	.003	2400	.07	.10	---	.17	**.23**
	Inst	SF	Sm	NA	.004	1800	.08	.13	---	.21	**.29**
For custom colors, ADD	Inst	%	Lg	---	---	---	10.0	---	---	---	---
	Inst	%	Sm	---	---	---	10.0	---	---	---	---
Floor seal											
Brush (450 SF/gal)	Inst	SF	Lg	NA	.003	2800	.06	.10	---	.16	**.21**
	Inst	SF	Sm	NA	.004	2100	.07	.13	---	.20	**.28**
Roller (430 SF/gal)	Inst	SF	Lg	NA	.003	3200	.07	.10	---	.17	**.23**
	Inst	SF	Sm	NA	.003	2400	.08	.10	---	.18	**.24**
Penetrating stainwax (hardwood floors)											
Brush (450 SF/gal)	Inst	SF	Lg	NA	.004	2000	.06	.13	---	.19	**.26**
	Inst	SF	Sm	NA	.005	1500	.07	.17	---	.24	**.33**
Roller (425 SF/gal)	Inst	SF	Lg	NA	.003	2400	.07	.10	---	.17	**.23**
	Inst	SF	Sm	NA	.004	1800	.08	.13	---	.21	**.29**

Description	Oper	Unit	Vol	Crew Size	Man-hours per Unit	Crew Output per Day	Avg Mat'l Unit Cost	Avg Labor Unit Cost	Avg Equip Unit Cost	Avg Total Unit Cost	Avg Price Incl O&P

Natural finishes

Paneling, brush work unless otherwise indicated
Stain, brush on - wipe off (360 SF/gal)

| | Inst | SF | Lg | NA | .014 | 560.0 | .08 | .47 | --- | .55 | .78 |
| | Inst | SF | Sm | NA | .019 | 420.0 | .09 | .64 | --- | .73 | 1.04 |

Varnish (380 SF/gal)

| | Inst | SF | Lg | NA | .006 | 1360 | .08 | .20 | --- | .28 | .39 |
| | Inst | SF | Sm | NA | .008 | 1020 | .09 | .27 | --- | .36 | .50 |

Shellac (630 SF/gal)

| | Inst | SF | Lg | NA | .005 | 1600 | .06 | .17 | --- | .23 | .31 |
| | Inst | SF | Sm | NA | .007 | 1200 | .06 | .23 | --- | .29 | .41 |

Lacquer
  Brush (450 SF/gal)

| | Inst | SF | Lg | NA | .004 | 1920 | .06 | .13 | --- | .19 | .26 |
| | Inst | SF | Sm | NA | .006 | 1440 | .07 | .20 | --- | .27 | .37 |

  Spray (300 SF/gal)

| | Inst | SF | Lg | NC | .003 | 2400 | .10 | .10 | --- | .20 | .26 |
| | Inst | SF | Sm | NC | .004 | 1800 | .11 | .14 | --- | .25 | .33 |

Doors and windows, brush work unless otherwise indicated
Stain, brush on - wipe off (450 SF/gal)

| | Inst | SF | Lg | NA | .032 | 250.0 | .06 | 1.07 | --- | 1.13 | 1.65 |
| | Inst | SF | Sm | NA | .043 | 188.0 | .07 | 1.44 | --- | 1.51 | 2.21 |

Varnish (550 SF/gal)

| | Inst | SF | Lg | NA | .024 | 340.0 | .06 | .80 | --- | .86 | 1.26 |
| | Inst | SF | Sm | NA | .031 | 255.0 | .07 | 1.04 | --- | 1.11 | 1.62 |

Shellac (550 SF/gal)

| | Inst | SF | Lg | NA | .022 | 365.0 | .06 | .74 | --- | .80 | 1.16 |
| | Inst | SF | Sm | NA | .029 | 274.0 | .07 | .97 | --- | 1.04 | 1.52 |

Lacquer
  Brush (550 SF/gal)

| | Inst | SF | Lg | NA | .020 | 405.0 | .05 | .67 | --- | .72 | 1.05 |
| | Inst | SF | Sm | NA | .026 | 304.0 | .06 | .87 | --- | .93 | 1.36 |

  Spray doors (300 SF/gal)

| | Inst | SF | Lg | NC | .010 | 800.0 | .10 | .35 | --- | .45 | .62 |
| | Inst | SF | Sm | NC | .013 | 600.0 | .11 | .45 | --- | .56 | .79 |

Cabinets, brush work unless otherwise indicated
Stain, brush on - wipe off (450 SF/gal)

| | Inst | SF | Lg | NA | .034 | 235.0 | .06 | 1.14 | --- | 1.20 | 1.75 |
| | Inst | SF | Sm | NA | .045 | 176.0 | .07 | 1.51 | --- | 1.58 | 2.31 |

Varnish (550 SF/gal)

| | Inst | SF | Lg | NA | .025 | 320.0 | .06 | .84 | --- | .90 | 1.31 |
| | Inst | SF | Sm | NA | .033 | 240.0 | .07 | 1.11 | --- | 1.18 | 1.72 |

Shellac (550 SF/gal)

| | Inst | SF | Lg | NA | .023 | 345.0 | .06 | .77 | --- | .83 | 1.21 |
| | Inst | SF | Sm | NA | .031 | 259.0 | .07 | 1.04 | --- | 1.11 | 1.62 |

Lacquer
  Brush (550 SF/gal)

| | Inst | SF | Lg | NA | .021 | 385.0 | .05 | .70 | --- | .75 | 1.10 |
| | Inst | SF | Sm | NA | .028 | 289.0 | .06 | .94 | --- | 1.00 | 1.46 |

  Spray (300 SF/gal)

| | Inst | SF | Lg | NC | .011 | 750.0 | .10 | .38 | --- | .48 | .67 |
| | Inst | SF | Sm | NC | .014 | 563.0 | .11 | .48 | --- | .59 | .84 |

Louvers, lacquer, spray (300 SF/gal)

| | Inst | SF | Lg | NC | .017 | 480.0 | .10 | .59 | --- | .69 | .98 |
| | Inst | SF | Sm | NC | .022 | 360.0 | .11 | .76 | --- | .87 | 1.25 |

Description	Oper	Unit	Vol	Crew Size	Man-hours per Unit	Crew Output per Day	Avg Mat'l Unit Cost	Avg Labor Unit Cost	Avg Equip Unit Cost	Avg Total Unit Cost	Avg Price Incl O&P

Picture molding, chair rail, base, ceiling mold etc., less than 6" high
Note: SF equals LF on trim less than 6" high
Varnish, brush (900 SF/gal)

| | Inst | SF | Lg | NA | .008 | 1040 | .04 | .27 | --- | .31 | .44 |
| | Inst | SF | Sm | NA | .010 | 780.0 | .04 | .34 | --- | .38 | .54 |

Shellac, brush (900 SF/gal)

| | Inst | SF | Lg | NA | .008 | 1040 | .04 | .27 | --- | .31 | .44 |
| | Inst | SF | Sm | NA | .010 | 780.0 | .04 | .34 | --- | .38 | .54 |

Lacquer, spray (700 SF/gal)

| | Inst | SF | Lg | NA | .006 | 1280 | .04 | .20 | --- | .24 | .34 |
| | Inst | SF | Sm | NA | .008 | 960.0 | .05 | .27 | --- | .32 | .45 |

Floors, wood, brush work unless otherwise indicated
Shellac (450 SF/gal)

| | Inst | SF | Lg | NA | .004 | 2000 | .08 | .13 | --- | .21 | .29 |
| | Inst | SF | Sm | NA | .005 | 1500 | .09 | .17 | --- | .26 | .35 |

Varnish (500 SF/gal)

| | Inst | SF | Lg | NA | .004 | 2000 | .06 | .13 | --- | .19 | .26 |
| | Inst | SF | Sm | NA | .005 | 1500 | .07 | .17 | --- | .24 | .33 |

Buffing, by machine

| | Inst | SF | Lg | NA | .003 | 2800 | .02 | .10 | --- | .12 | .17 |
| | Inst | SF | Sm | NA | .004 | 2100 | .02 | .13 | --- | .15 | .22 |

Waxing and polishing, by hand (1,000 SF/gal)

| | Inst | SF | Lg | NA | .005 | 1520 | .03 | .17 | --- | .20 | .28 |
| | Inst | SF | Sm | NA | .007 | 1140 | .04 | .23 | --- | .27 | .39 |

Two coat application
Excluding openings, unless otherwise indicated
For sizing or sealer, see One coat application, page 321

Latex
No spray work included, see One coat application, page 322
Drywall or plaster, latex flat
Smooth finish
Brush (170 SF/gal)

| | Inst | SF | Lg | NA | .009 | 880.0 | .16 | .30 | --- | .46 | .62 |
| | Inst | SF | Sm | NA | .012 | 660.0 | .17 | .40 | --- | .57 | .78 |

Roller (160 SF/gal)

| | Inst | SF | Lg | NA | .008 | 1040 | .17 | .27 | --- | .44 | .58 |
| | Inst | SF | Sm | NA | .010 | 780.0 | .19 | .34 | --- | .53 | .71 |

Sand finish
Brush (170 SF/gal)

| | Inst | SF | Lg | NA | .013 | 600.0 | .16 | .44 | --- | .60 | .82 |
| | Inst | SF | Sm | NA | .018 | 450.0 | .17 | .60 | --- | .77 | 1.08 |

Roller (160 SF/gal)

| | Inst | SF | Lg | NA | .011 | 760.0 | .17 | .37 | --- | .54 | .73 |
| | Inst | SF | Sm | NA | .014 | 570.0 | .19 | .47 | --- | .66 | .90 |

Paneling, latex enamel
Brush (170 SF/gal)

| | Inst | SF | Lg | NA | .009 | 880.0 | .16 | .30 | --- | .46 | .62 |
| | Inst | SF | Sm | NA | .012 | 660.0 | .17 | .40 | --- | .57 | .78 |

Roller (160 SF/gal)

| | Inst | SF | Lg | NA | .008 | 1040 | .17 | .27 | --- | .44 | .58 |
| | Inst | SF | Sm | NA | .010 | 780.0 | .19 | .34 | --- | .53 | .71 |

Acoustical tile or panels, latex flat
Brush (130 SF/gal)

| | Inst | SF | Lg | NA | .012 | 680.0 | .20 | .40 | --- | .60 | .82 |
| | Inst | SF | Sm | NA | .016 | 510.0 | .23 | .54 | --- | .77 | 1.05 |

Roller (120 SF/gal)

| | Inst | SF | Lg | NA | .009 | 880.0 | .22 | .30 | --- | .52 | .69 |
| | Inst | SF | Sm | NA | .012 | 660.0 | .25 | .40 | --- | .65 | .87 |

Description	Oper	Unit	Vol	Crew Size	Man-hours per Unit	Crew Output per Day	Avg Mat'l Unit Cost	Avg Labor Unit Cost	Avg Equip Unit Cost	Avg Total Unit Cost	Avg Price Incl O&P

Millwork and trim, enamel

Doors and windows

Roller and/or brush (200 SF/gal)

| | Inst | SF | Lg | NA | .024 | 335.0 | .13 | .80 | --- | .93 | **1.33** |
| | Inst | SF | Sm | NA | .032 | 251.0 | .15 | 1.07 | --- | 1.22 | **1.75** |

Cabinets

Roller and/or brush (200 SF/gal)

| | Inst | SF | Lg | NA | .025 | 315.0 | .13 | .84 | --- | .97 | **1.38** |
| | Inst | SF | Sm | NA | .034 | 236.0 | .15 | 1.14 | --- | 1.29 | **1.85** |

Louvers, see One coat application, page 323

Picture molding, chair rail, base, ceiling mold etc., less than 6" high

Note: SF equals LF on trim less than 6" high

| Brush (510 SF/gal) | Inst | SF | Lg | NA | .016 | 500.0 | .05 | .54 | --- | .59 | **.85** |
| | Inst | SF | Sm | NA | .021 | 375.0 | .06 | .70 | --- | .76 | **1.11** |

Floors, wood

Brush (230 SF/gal)	Inst	SF	Lg	NA	.007	1120	.11	.23	---	.34	**.47**
	Inst	SF	Sm	NA	.010	840.0	.13	.34	---	.47	**.64**
Roller (220 SF/gal)	Inst	SF	Lg	NA	.006	1280	.12	.20	---	.32	**.43**
	Inst	SF	Sm	NA	.008	960.0	.14	.27	---	.41	**.55**

For custom colors, ADD

| | Inst | % | Lg | --- | --- | --- | 10.0 | --- | --- | --- | **---** |
| | Inst | % | Sm | --- | --- | --- | 10.0 | --- | --- | --- | **---** |

For wood floor seal, penetrating stainwax, or natural finish, see One coat application, page 323

Exterior

Time calculations include normal materials handling and protection of property not to be painted

Preparation

Excluding openings, unless otherwise indicated

Cleaning, wet

| Plain siding | Inst | SF | Lg | NA | .005 | 1480 | .02 | .17 | --- | .19 | **.27** |
| | Inst | SF | Sm | NA | .007 | 1110 | .02 | .23 | --- | .25 | **.37** |

Exterior doors and trim

Note: SF equals LF on trim less than 6" high

| | Inst | SF | Lg | NA | .006 | 1400 | .02 | .20 | --- | .22 | **.32** |
| | Inst | SF | Sm | NA | .008 | 1050 | .02 | .27 | --- | .29 | **.42** |

Windows, wash and clean glass

	Inst	SF	Lg	NA	.006	1280	.01	.20	---	.21	**.31**
	Inst	SF	Sm	NA	.008	960.0	.02	.27	---	.29	**.42**
Porch floors and steps	Inst	SF	Lg	NA	.004	2240	.02	.13	---	.15	**.22**
	Inst	SF	Sm	NA	.005	1680	.02	.17	---	.19	**.27**

Acid wash

| Gutters and downspouts | Inst | SF | Lg | NA | .008 | 1040 | .01 | .27 | --- | .28 | **.41** |
| | Inst | SF | Sm | NA | .010 | 780.0 | .02 | .34 | --- | .36 | **.52** |

Sanding, light

| Porch floors and steps | Inst | SF | Lg | NA | .005 | 1760 | .01 | .17 | --- | .18 | **.26** |
| | Inst | SF | Sm | NA | .006 | 1320 | .02 | .20 | --- | .22 | **.32** |

Description	Oper	Unit	Vol	Crew Size	Man-hours per Unit	Crew Output per Day	Avg Mat'l Unit Cost	Avg Labor Unit Cost	Avg Equip Unit Cost	Avg Total Unit Cost	Avg Price Incl O&P
Sanding and puttying											
Plain siding	Inst	SF	Lg	NA	.006	1440	.01	.20	---	.21	.31
	Inst	SF	Sm	NA	.007	1080	.02	.23	---	.25	.37
Exterior doors and trim											
Note: SF equals LF on trim less than 6" high											
	Inst	SF	Lg	NA	.011	720.0	.01	.37	---	.38	.56
	Inst	SF	Sm	NA	.015	540.0	.02	.50	---	.52	.77
Puttying sash or reglazing windows (30 SF glass/lb glazing compound)											
	Inst	SF	Lg	NA	.033	240.0	.48	1.11	---	1.59	2.17
	Inst	SF	Sm	NA	.044	180.0	.54	1.48	---	2.02	2.78

One coat application

Excluding openings, unless otherwise indicated

Latex, flat (unless otherwise indicated)

Description	Oper	Unit	Vol	Crew Size	Man-hours per Unit	Crew Output per Day	Avg Mat'l Unit Cost	Avg Labor Unit Cost	Avg Equip Unit Cost	Avg Total Unit Cost	Avg Price Incl O&P
Plain siding											
Brush (300 SF/gal)	Inst	SF	Lg	NA	.010	800.0	.09	.34	---	.43	.60
	Inst	SF	Sm	NA	.013	600.0	.10	.44	---	.54	.76
Roller (275 SF/gal)	Inst	SF	Lg	NA	.008	1040	.10	.27	---	.37	.51
	Inst	SF	Sm	NA	.010	780.0	.11	.34	---	.45	.62
Spray (325 SF/gal)	Inst	SF	Lg	NC	.005	1750	.08	.17	---	.25	.34
	Inst	SF	Sm	NC	.006	1313	.09	.21	---	.30	.41
Shingle siding											
Brush (270 SF/gal)	Inst	SF	Lg	NA	.009	880.0	.10	.30	---	.40	.56
	Inst	SF	Sm	NA	.012	660.0	.11	.40	---	.51	.72
Roller (260 SF/gal)	Inst	SF	Lg	NA	.007	1200	.10	.23	---	.33	.46
	Inst	SF	Sm	NA	.009	900.0	.11	.30	---	.41	.57
Spray (300 SF/gal)	Inst	SF	Lg	NC	.005	1600	.09	.17	---	.26	.35
	Inst	SF	Sm	NC	.007	1200	.10	.24	---	.34	.47
Stucco											
Brush (135 SF/gal)	Inst	SF	Lg	NA	.010	800.0	.20	.34	---	.54	.72
	Inst	SF	Sm	NA	.013	600.0	.22	.44	---	.66	.89
Roller (125 SF/gal)	Inst	SF	Lg	NA	.007	1200	.21	.23	---	.44	.58
	Inst	SF	Sm	NA	.009	900.0	.24	.30	---	.54	.71
Spray (150 SF/gal)	Inst	SF	Lg	NC	.005	1600	.18	.17	---	.35	.45
	Inst	SF	Sm	NC	.007	1200	.20	.24	---	.44	.58

Cement walls, see Cement base paint, page 329

Description	Oper	Unit	Vol	Crew Size	Man-hours per Unit	Crew Output per Day	Avg Mat'l Unit Cost	Avg Labor Unit Cost	Avg Equip Unit Cost	Avg Total Unit Cost	Avg Price Incl O&P
Masonry block, brick, tile; masonry latex											
Brush (180 SF/gal)	Inst	SF	Lg	NA	.009	880.0	.15	.30	---	.45	.61
	Inst	SF	Sm	NA	.012	660.0	.17	.40	---	.57	.78
Roller (125 SF/gal)	Inst	SF	Lg	NA	.006	1440	.21	.20	---	.41	.53
	Inst	SF	Sm	NA	.007	1080	.24	.23	---	.47	.61
Spray (160 SF/gal)	Inst	SF	Lg	NC	.004	1840	.17	.14	---	.31	.39
	Inst	SF	Sm	NC	.006	1380	.19	.21	---	.40	.52

Description	Oper	Unit	Vol	Crew Size	Man-hours per Unit	Crew Output per Day	Avg Mat'l Unit Cost	Avg Labor Unit Cost	Avg Equip Unit Cost	Avg Total Unit Cost	Avg Price Incl O&P
Doors, exterior side only											
Brush (375 SF/gal)	Inst	SF	Lg	NA	.013	640.0	.07	.44	---	.51	.72
	Inst	SF	Sm	NA	.017	480.0	.08	.57	---	.65	.93
Roller (375 SF/gal)	Inst	SF	Lg	NA	.009	865.0	.07	.30	---	.37	.52
	Inst	SF	Sm	NA	.012	649.0	.08	.40	---	.48	.68
Windows, exterior side only, brush work											
(450 SF/gal)	Inst	SF	Lg	NA	.013	640.0	.04	.44	---	.48	.69
	Inst	SF	Sm	NA	.017	480.0	.02	.57	---	.59	.87
Trim, less than 6" high, brush											
Note: SF equals LF on trim less than 6" high											
High gloss (300 SF/gal)	Inst	SF	Lg	NA	.009	900.0	.09	.30	---	.39	.55
	Inst	SF	Sm	NA	.012	675.0	.10	.40	---	.50	.71
Screens, full; high gloss											
Paint applied to wood only, brush work											
(700 SF/gal)	Inst	SF	Lg	NA	.015	540.0	.04	.50	---	.54	.79
	Inst	SF	Sm	NA	.020	405.0	.04	.67	---	.71	1.04
Paint applied to wood (brush) and wire (spray)											
(475 SF/gal)	Inst	SF	Lg	NC	.018	450.0	.06	.62	---	.68	.99
	Inst	SF	Sm	NC	.024	338.0	.06	.83	---	.89	1.29
Storm windows and doors, 2 lites, brush work											
(340 SF/gal)	Inst	SF	Lg	NA	.024	340.0	.08	.80	---	.88	1.28
	Inst	SF	Sm	NA	.031	255.0	.09	1.04	---	1.13	1.64
Blinds or shutters											
Brush (120 SF/gal)	Inst	SF	Lg	NA	.062	130.0	.22	2.08	---	2.30	3.32
	Inst	SF	Sm	NA	.082	98.00	.25	2.75	---	3.00	4.36
Spray (300 SF/gal)	Inst	SF	Lg	NC	.020	400.0	.09	.69	---	.78	1.12
	Inst	SF	Sm	NC	.027	300.0	.10	.93	---	1.03	1.49
Gutters and downspouts, brush work											
(225 LF/gal), galvanized	Inst	SF	Lg	NA	.013	600.0	.12	.44	---	.56	.78
	Inst	SF	Sm	NA	.018	450.0	.13	.60	---	.73	1.04
Porch floors and steps, wood											
Brush (340 SF/gal)	Inst	SF	Lg	NA	.005	1520	.08	.17	---	.25	.34
	Inst	SF	Sm	NA	.007	1140	.09	.23	---	.32	.45
Roller (325 SF/gal)	Inst	SF	Lg	NA	.004	1920	.08	.13	---	.21	.29
	Inst	SF	Sm	NA	.006	1440	.09	.20	---	.29	.40
Shingle roofs											
Brush (135 SF/gal)	Inst	SF	Lg	NA	.008	1040	.20	.27	---	.47	.62
	Inst	SF	Sm	NA	.010	780.0	.22	.34	---	.56	.74
Roller (125 SF/gal)	Inst	SF	Lg	NA	.006	1360	.21	.20	---	.41	.53
	Inst	SF	Sm	NA	.008	1020	.24	.27	---	.51	.66
Spray (150 SF/gal)	Inst	SF	Lg	NC	.004	2080	.18	.14	---	.32	.40
	Inst	SF	Sm	NC	.005	1560	.20	.17	---	.37	.48
For custom colors, ADD											
	Inst	%	Lg	---	---	---	10.0	---	---	---	---
	Inst	%	Sm	---	---	---	10.0	---	---	---	---

Description	Oper	Unit	Vol	Crew Size	Man-hours per Unit	Crew Output per Day	Avg Mat'l Unit Cost	Avg Labor Unit Cost	Avg Equip Unit Cost	Avg Total Unit Cost	Avg Price Incl O&P
Cement base paint (epoxy concrete enamel)											
Cement walls, smooth finish											
Brush (120 SF/gal)	Inst	SF	Lg	NA	.006	1320	.30	.20	---	.50	**.63**
	Inst	SF	Sm	NA	.008	990.0	.34	.27	---	.61	**.77**
Roller (110 SF/gal)	Inst	SF	Lg	NA	.004	1920	.33	.13	---	.46	**.56**
	Inst	SF	Sm	NA	.006	1440	.37	.20	---	.57	**.70**
Concrete porch floors and steps											
Brush (400 SF/gal)	Inst	SF	Lg	NA	.004	2000	.09	.13	---	.22	**.30**
	Inst	SF	Sm	NA	.005	1500	.10	.17	---	.27	**.36**
Roller (375 SF/gal)	Inst	SF	Lg	NA	.003	2400	.10	.10	---	.20	**.26**
	Inst	SF	Sm	NA	.004	1800	.11	.13	---	.24	**.32**
Stain											
Shingle siding											
Brush (180 SF/gal)	Inst	SF	Lg	NA	.010	800.0	.16	.34	---	.50	**.67**
	Inst	SF	Sm	NA	.013	600.0	.18	.44	---	.62	**.84**
Roller (170 SF/gal)	Inst	SF	Lg	NA	.007	1120	.17	.23	---	.40	**.53**
	Inst	SF	Sm	NA	.010	840.0	.19	.34	---	.53	**.71**
Spray (200 SF/gal)	Inst	SF	Lg	NC	.005	1520	.14	.17	---	.31	**.41**
	Inst	SF	Sm	NC	.007	1140	.16	.24	---	.40	**.53**
Shingle roofs											
Brush (180 SF/gal)	Inst	SF	Lg	NA	.008	960.0	.16	.27	---	.43	**.57**
	Inst	SF	Sm	NA	.011	720.0	.18	.37	---	.55	**.74**
Roller (170 SF/gal)	Inst	SF	Lg	NA	.006	1280	.17	.20	---	.37	**.48**
	Inst	SF	Sm	NA	.008	960.0	.19	.27	---	.46	**.61**
Spray (200 SF/gal)	Inst	SF	Lg	NC	.004	1920	.14	.14	---	.28	**.36**
	Inst	SF	Sm	NC	.006	1440	.16	.21	---	.37	**.48**

Two coat application

Excluding openings, unless otherwise indicated

Latex, flat (unless otherwise indicated)

Description	Oper	Unit	Vol	Crew Size	Man-hours per Unit	Crew Output per Day	Avg Mat'l Unit Cost	Avg Labor Unit Cost	Avg Equip Unit Cost	Avg Total Unit Cost	Avg Price Incl O&P
Plain siding											
Brush (170 SF/gal)	Inst	SF	Lg	NA	.018	440.0	.16	.60	---	.76	**1.07**
	Inst	SF	Sm	NA	.024	330.0	.17	.80	---	.97	**1.38**
Roller (155 SF/gal)	Inst	SF	Lg	NA	.014	560.0	.17	.47	---	.64	**.88**
	Inst	SF	Sm	NA	.019	420.0	.19	.64	---	.83	**1.15**
Spray (185 SF/gal)	Inst	SF	Lg	NC	.008	960.0	.14	.28	---	.42	**.56**
	Inst	SF	Sm	NC	.011	720.0	.16	.38	---	.54	**.74**
Shingle siding											
Brush (150 SF/gal)	Inst	SF	Lg	NA	.017	480.0	.18	.57	---	.75	**1.04**
	Inst	SF	Sm	NA	.022	360.0	.20	.74	---	.94	**1.31**
Roller (150 SF/gal)	Inst	SF	Lg	NA	.013	640.0	.18	.44	---	.62	**.84**
	Inst	SF	Sm	NA	.017	480.0	.20	.57	---	.77	**1.06**
Spray (170 SF/gal)	Inst	SF	Lg	NC	.009	880.0	.16	.31	---	.47	**.64**
	Inst	SF	Sm	NC	.012	660.0	.17	.41	---	.58	**.80**

Description	Oper	Unit	Vol	Crew Size	Man-hours per Unit	Crew Output per Day	Avg Mat'l Unit Cost	Avg Labor Unit Cost	Avg Equip Unit Cost	Avg Total Unit Cost	Avg Price Incl O&P
Stucco											
Brush (90 SF/gal)	Inst	SF	Lg	NA	.017	480.0	.29	.57	---	.86	**1.16**
	Inst	SF	Sm	NA	.022	360.0	.33	.74	---	1.07	**1.46**
Roller (85 SF/gal)	Inst	SF	Lg	NA	.012	680.0	.31	.40	---	.71	**.94**
	Inst	SF	Sm	NA	.016	510.0	.35	.54	---	.89	**1.18**
Spray (100 SF/gal)	Inst	SF	Lg	NC	.009	880.0	.26	.31	---	.57	**.75**
	Inst	SF	Sm	NC	.012	660.0	.30	.41	---	.71	**.94**
Cement wall, see Cement base paint, page 331											
Masonry block, brick, tile; masonry latex											
Brush (120 SF/gal)	Inst	SF	Lg	NA	.015	520.0	.22	.50	---	.72	**.99**
	Inst	SF	Sm	NA	.021	390.0	.25	.70	---	.95	**1.32**
Roller (85 SF/gal)	Inst	SF	Lg	NA	.011	760.0	.31	.37	---	.68	**.89**
	Inst	SF	Sm	NA	.014	570.0	.35	.47	---	.82	**1.08**
Spray (105 SF/gal)	Inst	SF	Lg	NC	.008	1000	.25	.28	---	.53	**.68**
	Inst	SF	Sm	NC	.011	750.0	.28	.38	---	.66	**.87**
Doors, exterior side only											
Brush (215 SF/gal)	Inst	SF	Lg	NA	.024	335.0	.12	.80	---	.92	**1.32**
	Inst	SF	Sm	NA	.032	251.0	.14	1.07	---	1.21	**1.74**
Roller (215 SF/gal)	Inst	SF	Lg	NA	.018	455.0	.12	.60	---	.72	**1.03**
	Inst	SF	Sm	NA	.023	341.0	.14	.77	---	.91	**1.30**
Windows, exterior side only, brush work (255 SF/gal)											
	Inst	SF	Lg	NA	.024	335.0	.10	.80	---	.90	**1.30**
	Inst	SF	Sm	NA	.032	251.0	.12	1.07	---	1.19	**1.72**
Trim, less than 6" high, brush. High gloss (230 SF/gal)											
Note: SF equals LF on trim less than 6" high											
	Inst	SF	Lg	NA	.017	480.0	.11	.57	---	.68	**.96**
	Inst	SF	Sm	NA	.022	360.0	.13	.74	---	.87	**1.24**
Screens, full; high gloss											
Paint applied to wood only, brush work (400 SF/gal)											
	Inst	SF	Lg	NA	.028	285.0	.07	.94	---	1.01	**1.47**
	Inst	SF	Sm	NA	.037	214.0	.07	1.24	---	1.31	**1.01**
Paint applied to wood (brush) and wire (spray) (270 SF/gal)											
White	Inst	SF	Lg	NC	.033	240.0	.10	1.14	---	1.24	**1.80**
	Inst	SF	Sm	NC	.044	180.0	.11	1.52	---	1.63	**2.37**
Blinds or shutters											
Brush (65 SF/gal)	Inst	SF	Lg	NA	.114	70.00	.41	3.82	---	4.23	**6.11**
	Inst	SF	Sm	NA	.151	53.00	.46	5.06	---	5.52	**8.00**
Spray (170 SF/gal)	Inst	SF	Lg	NC	.038	210.0	.16	1.31	---	1.47	**2.12**
	Inst	SF	Sm	NC	.051	158.0	.17	1.76	---	1.93	**2.80**
Gutters and downspouts, brush work (130 LF/gal), galvanized											
	Inst	SF	Lg	NA	.025	315.0	.20	.84	---	1.04	**1.46**
	Inst	SF	Sm	NA	.034	236.0	.23	1.14	---	1.37	**1.94**

Description	Oper	Unit	Vol	Crew Size	Man-hours per Unit	Crew Output per Day	Avg Mat'l Unit Cost	Avg Labor Unit Cost	Avg Equip Unit Cost	Avg Total Unit Cost	Avg Price Incl O&P
Porch floors and steps, wood											
Brush (195 SF/gal)	Inst	SF	Lg	NA	.010	800.0	.14	.34	---	.48	.65
	Inst	SF	Sm	NA	.013	600.0	.15	.44	---	.59	.81
Roller (185 SF/gal)	Inst	SF	Lg	NA	.008	1000	.14	.27	---	.41	.55
	Inst	SF	Sm	NA	.011	750.0	.16	.37	---	.53	.72
Shingle roofs											
Brush (75 SF/gal)	Inst	SF	Lg	NA	.013	640.0	.35	.44	---	.79	1.03
	Inst	SF	Sm	NA	.017	480.0	.40	.57	---	.97	1.28
Roller (70 SF/gal)	Inst	SF	Lg	NA	.010	800.0	.38	.34	---	.72	.91
	Inst	SF	Sm	NA	.013	600.0	.42	.44	---	.86	1.11
Spray (85 SF/gal)	Inst	SF	Lg	NC	.008	1000	.31	.28	---	.59	.75
	Inst	SF	Sm	NC	.011	750.0	.35	.38	---	.73	.95
For custom colors, ADD	Inst	%	Lg	---	---	---	10.0	---	---	---	---
	Inst	%	Sm	---	---	---	10.0	---	---	---	---

Cement base paint (epoxy concrete enamel)

Cement walls, smooth finish

Description	Oper	Unit	Vol	Crew Size	Man-hours per Unit	Crew Output per Day	Avg Mat'l Unit Cost	Avg Labor Unit Cost	Avg Equip Unit Cost	Avg Total Unit Cost	Avg Price Incl O&P
Brush (80 SF/gal)	Inst	SF	Lg	NA	.011	720.0	.45	.37	---	.82	1.04
	Inst	SF	Sm	NA	.015	540.0	.51	.50	---	1.01	1.31
Roller (75 SF/gal)	Inst	SF	Lg	NA	.008	1040	.48	.27	---	.75	.93
	Inst	SF	Sm	NA	.010	780.0	.54	.34	---	.88	1.09
Concrete porch floors and steps											
Brush (225 SF/gal)	Inst	SF	Lg	NA	.008	1040	.16	.27	---	.43	.57
	Inst	SF	Sm	NA	.010	780.0	.18	.34	---	.52	.69
Roller (210 SF/gal)	Inst	SF	Lg	NA	.008	1000	.17	.27	---	.44	.58
	Inst	SF	Sm	NA	.011	750.0	.19	.37	---	.56	.76

Stain

Shingle siding

Description	Oper	Unit	Vol	Crew Size	Man-hours per Unit	Crew Output per Day	Avg Mat'l Unit Cost	Avg Labor Unit Cost	Avg Equip Unit Cost	Avg Total Unit Cost	Avg Price Incl O&P
Brush (105 SF/gal)	Inst	SF	Lg	NA	.017	480.0	.27	.57	---	.84	1.14
	Inst	SF	Sm	NA	.022	360.0	.31	.74	---	1.05	1.43
Roller (100 SF/gal)	Inst	SF	Lg	NA	.013	600.0	.29	.44	---	.73	.96
	Inst	SF	Sm	NA	.018	450.0	.32	.60	---	.92	1.25
Spray (115 SF/gal)	Inst	SF	Lg	NC	.009	880.0	.25	.31	---	.56	.74
	Inst	SF	Sm	NC	.012	660.0	.28	.41	---	.69	.92
Shingle roofs											
Brush (105 SF/gal)	Inst	SF	Lg	NA	.014	560.0	.27	.47	---	.74	.99
	Inst	SF	Sm	NA	.019	420.0	.31	.64	---	.95	1.28
Roller (100 SF/gal)	Inst	SF	Lg	NA	.011	760.0	.29	.37	---	.66	.87
	Inst	SF	Sm	NA	.014	570.0	.32	.47	---	.79	1.05
Spray (115 SF/gal)	Inst	SF	Lg	NC	.007	1120	.25	.24	---	.49	.63
	Inst	SF	Sm	NC	.010	840.0	.28	.35	---	.63	.82

Description	Oper	Unit	Vol	Crew Size	Man-hours per Unit	Crew Output per Day	Avg Mat'l Unit Cost	Avg Labor Unit Cost	Avg Equip Unit Cost	Avg Total Unit Cost	Avg Price Incl O&P

Paneling
Hardboard and plywood

Demolition

Sheets, plywood or hardboard	Demo	SF	Lg	LB	.009	1850	---	.25	---	.25	.37
	Demo	SF	Sm	LB	.012	1295	---	.33	---	.33	.49
Boards, wood	Demo	SF	Lg	LB	.010	1650	---	.27	---	.27	.41
	Demo	SF	Sm	LB	.014	1155	---	.38	---	.38	.57

Installation

Waste, nails, and adhesives not included

Economy hardboard

Presdwood, 4' x 8' sheets
Standard

1/8" T	Inst	SF	Lg	2C	.016	1025	.23	.53	---	.76	1.06
	Inst	SF	Sm	2C	.022	718.0	.27	.73	---	1.00	1.41
1/4" T	Inst	SF	Lg	2C	.016	1025	.36	.53	---	.89	1.21
	Inst	SF	Sm	2C	.022	718.0	.41	.73	---	1.14	1.57

Tempered

1/8" T	Inst	SF	Lg	2C	.016	1025	.39	.53	---	.92	1.25
	Inst	SF	Sm	2C	.022	718.0	.46	.73	---	1.19	1.63
1/4" T	Inst	SF	Lg	2C	.016	1025	.63	.53	---	1.16	1.52
	Inst	SF	Sm	2C	.022	718.0	.73	.73	---	1.46	1.94

Duolux, 4' x 8' sheets
Standard

1/8" T	Inst	SF	Lg	2C	.016	1025	.24	.53	---	.77	1.07
	Inst	SF	Sm	2C	.022	718.0	.28	.73	---	1.01	1.42
1/4" T	Inst	SF	Lg	2C	.016	1025	.37	.53	---	.90	1.22
	Inst	SF	Sm	2C	.022	718.0	.42	.73	---	1.15	1.58

Tempered

1/8" T	Inst	SF	Lg	2C	.016	1025	.43	.53	---	.96	1.29
	Inst	SF	Sm	2C	.022	718.0	.50	.73	---	1.23	1.67
1/4" T	Inst	SF	Lg	2C	.016	1025	.49	.53	---	1.02	1.36
	Inst	SF	Sm	2C	.022	718.0	.55	.73	---	1.28	1.73

Particleboard, 40 lb interior underlayment

Nailed to floors

3/8" T	Inst	SF	Lg	2C	.010	1535	.43	.33	---	.76	.99
	Inst	SF	Sm	2C	.015	1075	.50	.50	---	1.00	1.32
1/2" T	Inst	SF	Lg	2C	.010	1535	.51	.33	---	.84	1.09
	Inst	SF	Sm	2C	.015	1075	.58	.50	---	1.08	1.42
5/8" T	Inst	SF	Lg	2C	.010	1535	.57	.33	---	.90	1.15
	Inst	SF	Sm	2C	.015	1075	.66	.50	---	1.16	1.51
3/4" T	Inst	SF	Lg	2C	.010	1535	.65	.33	---	.98	1.25
	Inst	SF	Sm	2C	.015	1075	.75	.50	---	1.25	1.61

Nailed to walls

3/8" T	Inst	SF	Lg	2C	.011	1440	.43	.37	---	.80	1.04
	Inst	SF	Sm	2C	.016	1008	.50	.53	---	1.03	1.37
1/2" T	Inst	SF	Lg	2C	.011	1440	.51	.37	---	.88	1.14
	Inst	SF	Sm	2C	.016	1008	.58	.53	---	1.11	1.47
5/8" T	Inst	SF	Lg	2C	.011	1440	.57	.37	---	.94	1.20
	Inst	SF	Sm	2C	.016	1008	.66	.53	---	1.19	1.56
3/4" T	Inst	SF	Lg	2C	.011	1440	.65	.37	---	1.02	1.30
	Inst	SF	Sm	2C	.016	1008	.75	.53	---	1.28	1.66

Description	Oper	Unit	Vol	Crew Size	Man-hours per Unit	Crew Output per Day	Avg Mat'l Unit Cost	Avg Labor Unit Cost	Avg Equip Unit Cost	Avg Total Unit Cost	Avg Price Incl O&P

Masonite prefinished 4' x 8' panels

Description	Oper	Unit	Vol	Crew Size	Man-hours	Crew Output	Avg Mat'l	Avg Labor	Avg Equip	Avg Total	Avg Price O&P
1/4" T, oak and maple designs											
	Inst	SF	Lg	2C	.020	800.0	.58	.67	---	1.25	**1.67**
	Inst	SF	Sm	2C	.029	560.0	.66	.96	---	1.62	**2.21**
1/4" T, nutwood designs	Inst	SF	Lg	2C	.020	800.0	.67	.67	---	1.34	**1.77**
	Inst	SF	Sm	2C	.029	560.0	.78	.96	---	1.74	**2.34**
1/4" T, weathered white	Inst	SF	Lg	2C	.020	800.0	.67	.67	---	1.34	**1.77**
	Inst	SF	Sm	2C	.029	560.0	.78	.96	---	1.74	**2.34**
1/4" T, brick or stone designs	Inst	SF	Lg	2C	.020	800.0	.94	.67	---	1.61	**2.08**
	Inst	SF	Sm	2C	.029	560.0	1.09	.96	---	2.05	**2.70**

Pegboard, 4' x 8' sheets

Presdwood, tempered

Description	Oper	Unit	Vol	Crew Size	Man-hours	Crew Output	Avg Mat'l	Avg Labor	Avg Equip	Avg Total	Avg Price O&P
1/8" T	Inst	SF	Lg	2C	.016	1025	.51	.53	---	1.04	**1.39**
	Inst	SF	Sm	2C	.022	718.0	.58	.73	---	1.31	**1.77**
1/4" T	Inst	SF	Lg	2C	.016	1025	.67	.53	---	1.20	**1.57**
	Inst	SF	Sm	2C	.022	718.0	.77	.73	---	1.50	**1.98**

Duolux, tempered

Description	Oper	Unit	Vol	Crew Size	Man-hours	Crew Output	Avg Mat'l	Avg Labor	Avg Equip	Avg Total	Avg Price O&P
1/8" T	Inst	SF	Lg	2C	.016	1025	.74	.53	---	1.27	**1.65**
	Inst	SF	Sm	2C	.022	718.0	.84	.73	---	1.57	**2.06**
1/4" T	Inst	SF	Lg	2C	.016	1025	1.06	.53	---	1.59	**2.02**
	Inst	SF	Sm	2C	.022	718.0	1.22	.73	---	1.95	**2.50**

Unfinished hardwood plywood, applied with nails (nail heads filled)

Ash, flush face

Description	Oper	Unit	Vol	Crew Size	Man-hours	Crew Output	Avg Mat'l	Avg Labor	Avg Equip	Avg Total	Avg Price O&P
1/8" x 4' x 7', 8'	Inst	SF	Lg	CS	.027	900.0	1.12	.85	---	1.97	**2.56**
	Inst	SF	Sm	CS	.038	630.0	1.29	1.19	---	2.48	**3.27**
3/16" x 4' x 8'	Inst	SF	Lg	CS	.027	900.0	1.35	.85	---	2.20	**2.82**
	Inst	SF	Sm	CS	.038	630.0	1.55	1.19	---	2.74	**3.57**
1/4" x 4' x 8'	Inst	SF	Lg	CS	.027	900.0	1.58	.85	---	2.43	**3.09**
	Inst	SF	Sm	CS	.038	630.0	1.81	1.19	---	3.00	**3.87**
1/4" x 4' x 10'	Inst	SF	Lg	CS	.027	900.0	1.78	.85	---	2.63	**3.32**
	Inst	SF	Sm	CS	.038	630.0	2.04	1.19	---	3.23	**4.13**
1/2" x 4' x 10'	Inst	SF	Lg	CS	.027	900.0	2.02	.85	---	2.87	**3.59**
	Inst	SF	Sm	CS	.038	630.0	2.33	1.19	---	3.52	**4.47**
3/4" x 4' x 8' vertical core	Inst	SF	Lg	CS	.027	900.0	2.21	.85	---	3.06	**3.81**
	Inst	SF	Sm	CS	.038	630.0	2.54	1.19	---	3.73	**4.71**
3/4" x 4' x 8' lumber core	Inst	SF	Lg	CS	.027	900.0	3.14	.85	---	3.99	**4.88**
	Inst	SF	Sm	CS	.038	630.0	3.61	1.19	---	4.80	**5.94**

Ash, V-grooved

Description	Oper	Unit	Vol	Crew Size	Man-hours	Crew Output	Avg Mat'l	Avg Labor	Avg Equip	Avg Total	Avg Price O&P
3/16" x 4' x 8'	Inst	SF	Lg	CS	.027	900.0	1.22	.85	---	2.07	**2.67**
	Inst	SF	Sm	CS	.038	630.0	1.40	1.19	---	2.59	**3.40**
1/4" x 4' x 8'	Inst	SF	Lg	CS	.027	900.0	1.50	.85	---	2.35	**2.99**
	Inst	SF	Sm	CS	.038	630.0	1.73	1.19	---	2.92	**3.78**
1/4" x 4' x 10'	Inst	SF	Lg	CS	.027	900.0	1.72	.85	---	2.57	**3.25**
	Inst	SF	Sm	CS	.038	630.0	1.99	1.19	---	3.18	**4.07**

Description	Oper	Unit	Vol	Crew Size	Man-hours per Unit	Crew Output per Day	Avg Mat'l Unit Cost	Avg Labor Unit Cost	Avg Equip Unit Cost	Avg Total Unit Cost	Avg Price Incl O&P
Birch, natural, "A" grade face											
Flush face											
1/8" x 4' x 8'	Inst	SF	Lg	CS	.027	900.0	.86	.85	---	1.71	**2.26**
	Inst	SF	Sm	CS	.038	630.0	.99	1.19	---	2.18	**2.92**
3/16" x 4' x 8'	Inst	SF	Lg	CS	.027	900.0	1.35	.85	---	2.20	**2.82**
	Inst	SF	Sm	CS	.038	630.0	1.55	1.19	---	2.74	**3.57**
1/4" x 4' x 8'	Inst	SF	Lg	CS	.027	900.0	1.36	.85	---	2.21	**2.83**
	Inst	SF	Sm	CS	.038	630.0	1.56	1.19	---	2.75	**3.58**
1/4" x 4' x 10'	Inst	SF	Lg	CS	.027	900.0	1.78	.85	---	2.63	**3.32**
	Inst	SF	Sm	CS	.038	630.0	2.04	1.19	---	3.23	**4.13**
3/8" x 4' x 8'	Inst	SF	Lg	CS	.027	900.0	1.96	.85	---	2.81	**3.52**
	Inst	SF	Sm	CS	.038	630.0	2.25	1.19	---	3.44	**4.37**
1/2" x 4' x 8'	Inst	SF	Lg	CS	.027	900.0	2.26	.85	---	3.11	**3.87**
	Inst	SF	Sm	CS	.038	630.0	2.60	1.19	---	3.79	**4.78**
3/4" x 4' x 8' lumber core	Inst	SF	Lg	CS	.027	900.0	3.09	.85	---	3.94	**4.82**
	Inst	SF	Sm	CS	.038	630.0	3.54	1.19	---	4.73	**5.86**
V-grooved											
1/4" x 4' x 8', mismatched	Inst	SF	Lg	CS	.027	900.0	1.69	.85	---	2.54	**3.21**
	Inst	SF	Sm	CS	.038	630.0	1.94	1.19	---	3.13	**4.02**
1/4" x 4' x 8'	Inst	SF	Lg	CS	.027	900.0	1.70	.85	---	2.55	**3.22**
	Inst	SF	Sm	CS	.038	630.0	1.96	1.19	---	3.15	**4.04**
1/4" x 4' x 10'	Inst	SF	Lg	CS	.027	900.0	1.93	.85	---	2.78	**3.49**
	Inst	SF	Sm	CS	.038	630.0	2.22	1.19	---	3.41	**4.34**
Birch, select red											
1/4" x 4' x 8'	Inst	SF	Lg	CS	.027	900.0	1.70	.85	---	2.55	**3.22**
	Inst	SF	Sm	CS	.038	630.0	1.96	1.19	---	3.15	**4.04**
Birch, select white											
1/4" x 4' x 8'	Inst	SF	Lg	CS	.027	900.0	1.70	.85	---	2.55	**3.22**
	Inst	SF	Sm	CS	.038	630.0	1.96	1.19	---	3.15	**4.04**
Oak, flush face											
1/8" x 4' x 8'	Inst	SF	Lg	CS	.027	900.0	.85	.85	---	1.70	**2.25**
	Inst	SF	Sm	CS	.038	630.0	.98	1.19	---	2.17	**2.91**
1/4" x 4' x 8'	Inst	SF	Lg	CS	.027	900.0	1.08	.85	---	1.93	**2.51**
	Inst	SF	Sm	CS	.038	630.0	1.24	1.19	---	2.43	**3.21**
1/2" x 4' x 8'	Inst	SF	Lg	CS	.027	900.0	2.09	.85	---	2.94	**3.67**
	Inst	SF	Sm	CS	.038	630.0	2.41	1.19	---	3.60	**4.56**

Description	Oper	Unit	Vol	Crew Size	Man-hours per Unit	Crew Output per Day	Avg Mat'l Unit Cost	Avg Labor Unit Cost	Avg Equip Unit Cost	Avg Total Unit Cost	Avg Price Incl O&P
Philippine mahogany											
Rotary cut											
1/8" x 4' x 8'	Inst	SF	Lg	CS	.027	900.0	.41	.85	---	1.26	**1.74**
	Inst	SF	Sm	CS	.038	630.0	.47	1.19	---	1.66	**2.33**
3/16" x 4' x 8'	Inst	SF	Lg	CS	.027	900.0	.60	.85	---	1.45	**1.96**
	Inst	SF	Sm	CS	.038	630.0	.69	1.19	---	1.88	**2.58**
1/4" x 4' x 8'	Inst	SF	Lg	CS	.027	900.0	.60	.85	---	1.45	**1.96**
	Inst	SF	Sm	CS	.038	630.0	.69	1.19	---	1.88	**2.58**
1/4" x 4' x 10'	Inst	SF	Lg	CS	.027	900.0	1.04	.85	---	1.89	**2.46**
	Inst	SF	Sm	CS	.038	630.0	1.20	1.19	---	2.39	**3.17**
1/2" x 4' x 8'	Inst	SF	Lg	CS	.027	900.0	1.21	.85	---	2.06	**2.66**
	Inst	SF	Sm	CS	.038	630.0	1.39	1.19	---	2.58	**3.38**
3/4" x 4' x 8' vert. core	Inst	SF	Lg	CS	.027	900.0	1.84	.85	---	2.69	**3.38**
	Inst	SF	Sm	CS	.038	630.0	2.12	1.19	---	3.31	**4.22**
V-grooved											
3/16" x 4' x 8'	Inst	SF	Lg	CS	.027	900.0	.63	.85	---	1.48	**1.99**
	Inst	SF	Sm	CS	.038	630.0	.73	1.19	---	1.92	**2.63**
1/4" x 4' x 8'	Inst	SF	Lg	CS	.027	900.0	.72	.85	---	1.57	**2.10**
	Inst	SF	Sm	CS	.038	630.0	.81	1.19	---	2.00	**2.72**
1/4" x 4' x 10'	Inst	SF	Lg	CS	.027	900.0	1.11	.85	---	1.96	**2.55**
	Inst	SF	Sm	CS	.038	630.0	1.28	1.19	---	2.47	**3.26**

Plaster & Stucco

1. Dimensions (Lath)

a. Gypsum. Plain, perforated and insulating. Normally, each lath is 16" x 48" in $3/8$" or $1/2$" thicknesses; a five-piece bundle covers 3 SY.

b. Wire. Only diamond and riblash are discussed here. Diamond lath is furnished in 27" x 96"-wide sheets covering 16 SY and 20 SY respectively.

c. Wood. Can be fir, pine, redwood, spruce, etc. Bundles may consist of 50 or 100 pieces of $3/8$" x $1^1/2$" x 48" lath covering 3.4 SY and 6.8 SY respectively. There is usually a $3/8$" gap between lath.

2. Dimensions (Plaster)

Only two and three coat gypsum cement plaster are discussed here.

3. Installation (Lath)

Laths are nailed. The types and size of nails vary with the type and thickness of lath. Quantity will vary with oc spacing of studs or joists. Only lath applied to wood will be considered here.

a. Nails for gypsum lath. Common type is 13 gauge blued $19/64$" flathead, $1^1/8$" long, spaced approximately 4" oc, and $1^1/4$" long spaced approximately 5" oc for $3/8$" and $1/2$" lath respectively.

b. Nails for wire lath. For ceiling, common is $1^1/2$" long 11 gauge barbed galvanized with a $7/16$" head diameter.

c. Nails for wood lath. 3d fine common.

d. Gypsum lath may be attached by the use of air-driven staples. Wire lath may be tied to support, usually with 18 gauge wire.

4. Installation (Plaster)

Quantities of materials used to plaster vary with the type of lath and thickness of plaster.

a. Two coat. Brown and finish coat.

b. Three coat. Scratch, brown and finish coat.

For types and quantities of material used, see cost tables under **Plaster**.

5. Notes on Labor

a. Lath. Average Manhour per Unit is based on what one lather can do in one day.

b. Plaster. Average Manhour per Unit is based on what two plasterers and one laborer can do in one day.

Stucco

1. Dimensions

a. 18 gauge wire

b. 15 lb. felt paper

c. 1" x 18 gauge galvanized netting

d. Mortar of 1:3 mix

2. Installation (Lathing)

a. 18 gauge wire stretched taut horizontally across studs at approximately 8" oc.

b. 15 lb. felt paper placed over wire.

c. 1" x 18 gauge galvanized netting placed over felt.

3. Installation (Mortar)

a. The mortar mix used in this section is a 1:3 mix.

b. One CY of mortar is comprised of 1 CY sand, 9 CF portland cement and 100 lbs. hydrated lime.

c. For mortar requirements for 100 SY of stucco, see table below

4. Estimating Technique.
Determine area and deduct area of window and door openings. No waste has been included in the following figures unless otherwise noted. For waste, add 10% to total area.

	Cubic Yards Per CSY	
Stucco Thickness	**On Masonry**	**On Netting**
1/2"	1.50	1.75
5/8"	1.90	2.20
3/4"	2.20	2.60
1"	2.90	3.40

Description	Oper	Unit	Vol	Crew Size	Man-hours per Unit	Crew Output per Day	Avg Mat'l Unit Cost	Avg Labor Unit Cost	Avg Equip Unit Cost	Avg Total Unit Cost	Avg Price Incl O&P

Plaster and stucco

Remove both plaster or stucco and lath or netting to studs or sheathing

Lath (wood or metal) and plaster, walls and ceilings

Description	Oper	Unit	Vol	Crew Size	Man-hours per Unit	Crew Output per Day	Avg Mat'l Unit Cost	Avg Labor Unit Cost	Avg Equip Unit Cost	Avg Total Unit Cost	Avg Price Incl O&P
2 coats	Demo	SY	Lg	LB	.123	130.0	---	3.37	---	3.37	**5.03**
	Demo	SY	Sm	LB	.176	91.00	---	4.83	---	4.83	**7.19**
3 coats	Demo	SY	Lg	LB	.133	120.0	---	3.65	---	3.65	**5.44**
	Demo	SY	Sm	LB	.190	84.00	---	5.21	---	5.21	**7.77**

Lath (only), nails included

Gypsum lath, 16" x 48", applied with nails to ceilings or walls, 5% waste included, nails included (.067 lbs per CY)

Description	Oper	Unit	Vol	Crew Size	Man-hours per Unit	Crew Output per Day	Avg Mat'l Unit Cost	Avg Labor Unit Cost	Avg Equip Unit Cost	Avg Total Unit Cost	Avg Price Incl O&P
3/8" T, perforated or plain	Inst	SY	Lg	LR	.100	80.00	2.52	3.23	---	5.75	**7.71**
	Inst	SY	Sm	LR	.143	56.00	2.77	4.62	---	7.39	**10.10**
1/2" T, perforated or plain	Inst	SY	Lg	LR	.100	80.00	2.75	3.23	---	5.98	**7.98**
	Inst	SY	Sm	LR	.143	56.00	3.03	4.62	---	7.65	**10.40**
3/8" T, insulating, aluminum foil back											
	Inst	SY	Lg	LR	.100	80.00	3.11	3.23	---	6.34	**8.39**
	Inst	SY	Sm	LR	.143	56.00	3.42	4.62	---	8.04	**10.80**
1/2" T, insulating, aluminum foil back											
	Inst	SY	Lg	LR	.100	80.00	3.34	3.23	---	6.57	**8.66**
	Inst	SY	Sm	LR	.143	56.00	3.68	4.62	---	8.30	**11.10**
For installation with staples,											
DEDUCT	Inst	SY	Lg	LR	-.025	-325.00	---	.81	---	.81	---
	Inst	SY	Sm	LR	-.035	-228.00	---	1.13	---	1.13	---

Metal lath, nailed to ceilings or walls, 5% waste and nails (0.067 lbs per SY) included

Diamond lath (junior mesh), 27" x 96" sheets, nailed to wood members @ 16" oc

Description	Oper	Unit	Vol	Crew Size	Man-hours per Unit	Crew Output per Day	Avg Mat'l Unit Cost	Avg Labor Unit Cost	Avg Equip Unit Cost	Avg Total Unit Cost	Avg Price Incl O&P
3.4 lb black painted	Inst	SY	Lg	LR	.067	120.0	3.56	2.16	---	5.72	**7.32**
	Inst	SY	Sm	LR	.095	84.00	3.92	3.07	---	6.99	**9.08**
3.4 lb galvanized	Inst	SY	Lg	LR	.067	120.0	3.87	2.16	---	6.03	**7.68**
	Inst	SY	Sm	LR	.095	84.00	4.25	3.07	---	7.32	**9.46**

Riblath, 3/8" high rib, 27" x 96" sheets, nailed to wood members @ 24" oc

Description	Oper	Unit	Vol	Crew Size	Man-hours per Unit	Crew Output per Day	Avg Mat'l Unit Cost	Avg Labor Unit Cost	Avg Equip Unit Cost	Avg Total Unit Cost	Avg Price Incl O&P
3.4 lb painted	Inst	SY	Lg	LR	.050	160.0	3.97	1.62	---	5.59	**6.97**
	Inst	SY	Sm	LR	.071	112.0	4.37	2.29	---	6.66	**8.44**
3.4 lb galvanized	Inst	SY	Lg	LR	.050	160.0	4.67	1.62	---	6.29	**7.78**
	Inst	SY	Sm	LR	.071	112.0	5.15	2.29	---	7.44	**9.34**

Wood lath, nailed to ceilings or walls, 5% waste and nails included, redwood, "A" grade and better

3/8" x 1-1/2" x 48" @ 3/8" spacing

Description	Oper	Unit	Vol	Crew Size	Man-hours per Unit	Crew Output per Day	Avg Mat'l Unit Cost	Avg Labor Unit Cost	Avg Equip Unit Cost	Avg Total Unit Cost	Avg Price Incl O&P
	Inst	SY	Lg	LR	.100	80.00	5.95	3.23	---	9.18	**11.70**
	Inst	SY	Sm	LR	.143	56.00	6.54	4.62	---	11.16	**14.40**

Description	Oper	Unit	Vol	Crew Size	Man-hours per Unit	Crew Output per Day	Avg Mat'l Unit Cost	Avg Labor Unit Cost	Avg Equip Unit Cost	Avg Total Unit Cost	Avg Price Incl O&P

Labor adjustments, lath

For gypsum or wood lath above second floor

| ADD | Inst | SY | Lg | LR | .007 | 1140 | --- | .23 | --- | .23 | .34 |
| | Inst | SY | Sm | LR | .010 | 798.0 | --- | .32 | --- | .32 | .48 |

For metal lath above second floor

| ADD | Inst | SY | Lg | LR | .017 | 475.0 | --- | .55 | --- | .55 | .82 |
| | Inst | SY | Sm | LR | .024 | 333.0 | --- | .78 | --- | .78 | 1.16 |

Plaster (only), applied to ceilings and walls, 10% waste included

Material price includes gypsum plaster, sand, hydrated lime and gauging plaster

Two coats gypsum plaster

On gypsum lath	Inst	SY	Lg	P3	.209	115.0	2.67	6.61	.31	9.59	13.20
	Inst	SY	Sm	P3	.296	81.00	2.98	9.36	.44	12.78	17.80
On unit masonry, no lath	Inst	SY	Lg	P3	.214	112.0	2.90	6.77	.32	9.99	13.70
	Inst	SY	Sm	P3	.308	78.00	3.23	9.74	.46	13.43	18.70

Three coats gypsum plaster

On gypsum lath	Inst	SY	Lg	P3	.289	83.00	2.87	9.14	.43	12.44	17.40
	Inst	SY	Sm	P3	.414	58.00	3.21	13.10	.61	16.92	23.80
On unit masonry, no lath	Inst	SY	Lg	P3	.289	83.00	3.12	9.14	.43	12.69	17.60
	Inst	SY	Sm	P3	.414	58.00	3.49	13.10	.61	17.20	24.10
On wire lath	Inst	SY	Lg	P3	.300	80.00	4.60	9.49	.45	14.54	19.90
	Inst	SY	Sm	P3	.429	56.00	5.15	13.60	.64	19.39	26.80
On wood lath	Inst	SY	Lg	P3	.300	80.00	3.06	9.49	.45	13.00	18.10
	Inst	SY	Sm	P3	.429	56.00	3.42	13.60	.64	17.66	24.80

Labor adjustments, plaster

For plaster above second floor

| ADD | Inst | SY | Lg | P3 | .063 | 384.0 | --- | 1.99 | --- | 1.99 | 2.97 |
| | Inst | SY | Sm | P3 | .089 | 269.0 | --- | 2.81 | --- | 2.81 | 4.19 |

Thin coat plaster over sheetrock (in lieu of taping)

| | Inst | SY | Lg | P3 | .192 | 125.0 | .75 | 6.07 | --- | 6.82 | 9.91 |
| | Inst | SY | Sm | P3 | .273 | 88.00 | .82 | 8.63 | --- | 9.45 | 13.80 |

Description	Oper	Unit	Vol	Crew Size	Man-hours per Unit	Crew Output per Day	Avg Mat'l Unit Cost	Avg Labor Unit Cost	Avg Equip Unit Cost	Avg Total Unit Cost	Avg Price Incl O&P
Stucco, exterior walls											
Netting, galvanized, 1" x 18 ga x 48", with 18 ga wire and 15 lb felt											
	Inst	SY	Lg	LR	.133	60.00	3.25	4.30	---	7.55	**10.10**
	Inst	SY	Sm	LR	.190	42.00	3.57	6.14	---	9.71	**13.30**
Steel-Tex, 49" W x 11-1/2' L rolls, with felt backing											
	Inst	SY	Lg	LR	.100	80.00	5.19	3.23	---	8.42	**10.80**
	Inst	SY	Sm	LR	.143	56.00	5.70	4.62	---	10.32	**13.40**
1 coat work with float finish											
Over masonry	Inst	SY	Lg	P3	.209	115.0	.63	6.61	.31	7.55	**10.90**
	Inst	SY	Sm	P3	.296	81.00	.69	9.36	.44	10.49	**15.20**
2 coat work with float finish											
Over masonry	Inst	SY	Lg	P3	.400	60.00	1.84	12.70	.59	15.13	**21.60**
	Inst	SY	Sm	P3	.571	42.00	2.02	18.10	.85	20.97	**30.10**
Over metal netting	Inst	SY	Lg	PE	.471	85.00	2.59	14.70	.42	17.71	**25.30**
	Inst	SY	Sm	PE	.667	60.00	2.85	20.80	.59	24.24	**34.90**
3 coat work with float finish											
Over metal netting	Inst	SY	Lg	PE	.615	65.00	3.50	19.20	.55	23.25	**33.20**
	Inst	SY	Sm	PE	.870	46.00	3.85	27.10	.78	31.73	**45.70**

Plumbing. See individual items

Description	Oper	Unit	Vol	Crew Size	Man-hours per Unit	Crew Output per Day	Avg Mat'l Unit Cost	Avg Labor Unit Cost	Avg Equip Unit Cost	Avg Total Unit Cost	Avg Price Incl O&P

Range hoods
Metal finishes

Labor includes wiring and connection by electrician and installation by carpenter in stud-exposed structure only

Economy model, UL approved; mitered, welded construction; completely assembled and wired; includes fan, motor, washable aluminum filter, and light

24" wide

Description	Oper	Unit	Vol	Crew Size	MH	Output	Mat'l	Labor	Equip	Total	O&P
3-1/4" x 10" duct, 160 CFM	Inst	Ea	Lg	ED	4.00	4.00	74.90	138.00	---	212.90	**287.00**
	Inst	Ea	Sm	ED	5.71	2.80	84.50	197.00	---	281.50	**384.00**
Non-ducted, 160 CFM	Inst	Ea	Lg	ED	4.00	4.00	81.30	138.00	---	219.30	**295.00**
	Inst	Ea	Sm	ED	5.71	2.80	91.70	197.00	---	288.70	**393.00**

30" wide

Description	Oper	Unit	Vol	Crew Size	MH	Output	Mat'l	Labor	Equip	Total	O&P
7" round duct, 240 CFM	Inst	Ea	Lg	ED	4.00	4.00	74.90	138.00	---	212.90	**287.00**
	Inst	Ea	Sm	ED	5.71	2.80	84.50	197.00	---	281.50	**384.00**
3-1/4" x 10" duct, 160 CFM	Inst	Ea	Lg	ED	4.00	4.00	81.40	138.00	---	219.40	**295.00**
	Inst	Ea	Sm	ED	5.71	2.80	91.90	197.00	---	288.90	**393.00**
Non-ducted, 160 CFM	Inst	Ea	Lg	ED	4.00	4.00	91.70	138.00	---	229.70	**307.00**
	Inst	Ea	Sm	ED	5.71	2.80	103.00	197.00	---	300.00	**406.00**

36" wide

Description	Oper	Unit	Vol	Crew Size	MH	Output	Mat'l	Labor	Equip	Total	O&P
7" round duct, 240 CFM	Inst	Ea	Lg	ED	4.00	4.00	86.10	138.00	---	224.10	**300.00**
	Inst	Ea	Sm	ED	5.71	2.80	97.20	197.00	---	294.20	**399.00**
3-1/4" x 10" duct, 160 CFM	Inst	Ea	Lg	ED	4.00	4.00	91.70	138.00	---	229.70	**307.00**
	Inst	Ea	Sm	ED	5.71	2.80	103.00	197.00	---	300.00	**406.00**
Non-ducted, 160 CFM	Inst	Ea	Lg	ED	4.00	4.00	109.00	138.00	---	247.00	**326.00**
	Inst	Ea	Sm	ED	5.71	2.80	123.00	197.00	---	320.00	**428.00**

Standard model, UL approved; deluxe mitered wrap-around styling; solid state fan control, infinite speed settings; removable filter; built-in damper and dual light assembly

30" wide x 9" deep

Description	Oper	Unit	Vol	Crew Size	MH	Output	Mat'l	Labor	Equip	Total	O&P
3-1/4" x 10" duct	Inst	Ea	Lg	ED	4.00	4.00	146.00	138.00	---	284.00	**369.00**
	Inst	Ea	Sm	ED	5.71	2.80	165.00	197.00	---	362.00	**477.00**
Non-ducted	Inst	Ea	Lg	ED	4.00	4.00	250.00	138.00	---	388.00	**489.00**
	Inst	Ea	Sm	ED	5.71	2.80	282.00	197.00	---	479.00	**612.00**

36" wide x 9" deep

Description	Oper	Unit	Vol	Crew Size	MH	Output	Mat'l	Labor	Equip	Total	O&P
3-1/4" x 10" duct	Inst	Ea	Lg	ED	4.00	4.00	166.00	138.00	---	304.00	**392.00**
	Inst	Ea	Sm	ED	5.71	2.80	187.00	197.00	---	384.00	**503.00**
Non-ducted	Inst	Ea	Lg	ED	4.00	4.00	178.00	138.00	---	316.00	**406.00**
	Inst	Ea	Sm	ED	5.71	2.80	201.00	197.00	---	398.00	**518.00**

42" wide x 9" deep

Description	Oper	Unit	Vol	Crew Size	MH	Output	Mat'l	Labor	Equip	Total	O&P
3-1/4" x 10" duct	Inst	Ea	Lg	ED	5.33	3.00	143.00	184.00	---	327.00	**433.00**
	Inst	Ea	Sm	ED	7.62	2.10	162.00	263.00	---	425.00	**569.00**
Non-ducted	Inst	Ea	Lg	ED	5.33	3.00	157.00	184.00	---	341.00	**449.00**
	Inst	Ea	Sm	ED	7.62	2.10	177.00	263.00	---	440.00	**587.00**

Description	Oper	Unit	Vol	Crew Size	Man-hours per Unit	Crew Output per Day	Avg Mat'l Unit Cost	Avg Labor Unit Cost	Avg Equip Unit Cost	Avg Total Unit Cost	Avg Price Incl O&P

Decorator/designer model; solid state fan control, infinite speed settings; easy clean aluminum mesh grease filters; enclosed light assembly and switches

Single faced hood - fan exhausts horizontally or vertically, 24" D canopy x 21" front to back

Description	Oper	Unit	Vol	Crew Size	Man-hours per Unit	Crew Output per Day	Avg Mat'l Unit Cost	Avg Labor Unit Cost	Avg Equip Unit Cost	Avg Total Unit Cost	Avg Price Incl O&P
30" wide	Inst	Ea	Lg	ED	5.33	3.00	287.00	184.00	---	471.00	**599.00**
	Inst	Ea	Sm	ED	7.62	2.10	324.00	263.00	---	587.00	**756.00**
36" wide	Inst	Ea	Lg	ED	5.33	3.00	309.00	184.00	---	493.00	**623.00**
	Inst	Ea	Sm	ED	7.62	2.10	348.00	263.00	---	611.00	**784.00**

Single faced, contemporary style, duct horizontally or vertically, 9" D canopy

30" wide

Description	Oper	Unit	Vol	Crew Size	Man-hours per Unit	Crew Output per Day	Avg Mat'l Unit Cost	Avg Labor Unit Cost	Avg Equip Unit Cost	Avg Total Unit Cost	Avg Price Incl O&P
With 330 cfm power unit	Inst	Ea	Lg	ED	5.33	3.00	271.00	184.00	---	455.00	**580.00**
	Inst	Ea	Sm	ED	7.62	2.10	306.00	263.00	---	569.00	**735.00**
With 410 cfm power unit	Inst	Ea	Lg	ED	5.33	3.00	309.00	184.00	---	493.00	**623.00**
	Inst	Ea	Sm	ED	7.62	2.10	348.00	263.00	---	611.00	**784.00**

36" wide

Description	Oper	Unit	Vol	Crew Size	Man-hours per Unit	Crew Output per Day	Avg Mat'l Unit Cost	Avg Labor Unit Cost	Avg Equip Unit Cost	Avg Total Unit Cost	Avg Price Incl O&P
With 330 cfm power unit	Inst	Ea	Lg	ED	5.33	3.00	287.00	184.00	---	471.00	**598.00**
	Inst	Ea	Sm	ED	7.62	2.10	323.00	263.00	---	586.00	**755.00**
With 410 cfm power unit	Inst	Ea	Lg	ED	5.33	3.00	319.00	184.00	---	503.00	**635.00**
	Inst	Ea	Sm	ED	7.62	2.10	360.00	263.00	---	623.00	**798.00**

Material adjustments

Description	Oper	Unit	Vol	Crew Size	Man-hours per Unit	Crew Output per Day	Avg Mat'l Unit Cost	Avg Labor Unit Cost	Avg Equip Unit Cost	Avg Total Unit Cost	Avg Price Incl O&P
Stainless steel, ADD	Inst	%	Lg	---	---	---	33.0	---	---	---	---
	Inst	%	Sm	---	---	---	33.0	---	---	---	---

Reinforcing steel. See Concrete, page 99

Resilient flooring

Sheet Products

Linoleum or vinyl sheet are normally installed either over a wood subfloor or a smooth concrete subfloor. When laid over wood, a layer of felt must first be laid in paste. This keeps irregularities in the wood from showing through. The amount of paste required to bond both felt and sheet is approximately 16 gallons (5% waste included) per 100 SY. When laid over smooth concrete subfloor, the sheet products can be bonded directly to the floor. However, the concrete subfloor should not be in direct contact with the ground because of excessive moisture. For bonding to concrete, 8 gallons of paste is required (5% waste included) per 100 SY. After laying the flooring over concrete or wood, wax is applied using 0.5 gallon per 100 SY. Paste, wax, felt (as needed), and 10% sheet waste are included in material costs.

Tile Products

All resilient tile can be bonded the same way as sheet products, either to smooth concrete or to felt over wood subfloor. When bonded to smooth concrete, a concrete primer (0.5 gal/100 SF) is first applied to seal the concrete. The tiles are then bonded to the sealed floor with resilient tile cement (0.6 gal/100 SF). On wood subfloors, felt is first laid in paste (0.9 gal/100 SF) and then the tiles are bonded to the felt with resilient tile emulsion (0.7 gal/100 SF). Bonding materials, felt (as needed), and 10% tile waste are included in material costs.

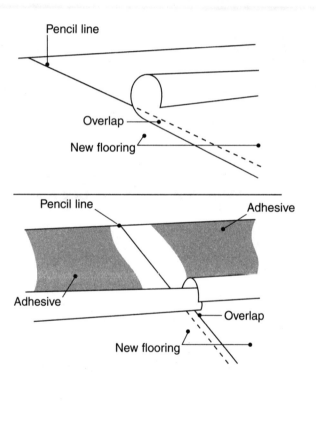

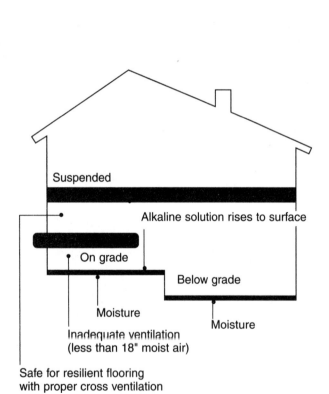

Description	Oper	Unit	Vol	Crew Size	Man-hours per Unit	Crew Output per Day	Avg Mat'l Unit Cost	Avg Labor Unit Cost	Avg Equip Unit Cost	Avg Total Unit Cost	Avg Price Incl O&P
Resilient flooring											
Adhesive set sheet products											
	Demo	SY	Lg	LB	.100	160.0	---	2.74	---	2.74	**4.09**
	Demo	SY	Sm	LB	.143	112.0	---	3.92	---	3.92	**5.84**
Adhesive set tile products											
	Demo	SF	Lg	LB	.011	1500	---	.30	---	.30	**.45**
	Demo	SF	Sm	LB	.015	1050	---	.41	---	.41	**.61**

Install over smooth concrete subfloor
Tile

Asphalt, 9" x 9", 1/8" thick, marbleized

Description	Oper	Unit	Vol	Crew Size	Man-hours per Unit	Crew Output per Day	Avg Mat'l Unit Cost	Avg Labor Unit Cost	Avg Equip Unit Cost	Avg Total Unit Cost	Avg Price Incl O&P
Group B colors, dark	Inst	SF	Lg	FB	.028	640.0	1.43	.85	---	2.28	**2.89**
	Inst	SF	Sm	FB	.040	448.0	1.58	1.21	---	2.79	**3.60**
Group C colors, medium	Inst	SF	Lg	FB	.028	640.0	1.62	.85	---	2.47	**3.11**
	Inst	SF	Sm	FB	.040	448.0	1.80	1.21	---	3.01	**3.85**
Group D colors, light	Inst	SF	Lg	FB	.028	640.0	1.62	.85	---	2.47	**3.11**
	Inst	SF	Sm	FB	.040	448.0	1.80	1.21	---	3.01	**3.85**

Vinyl tile, 12" x 12", including adhesive
1/16" thick

Description	Oper	Unit	Vol	Crew Size	Man-hours per Unit	Crew Output per Day	Avg Mat'l Unit Cost	Avg Labor Unit Cost	Avg Equip Unit Cost	Avg Total Unit Cost	Avg Price Incl O&P
Vega II	Inst	SF	Lg	FB	.020	900.0	1.30	.61	---	1.91	**2.38**
	Inst	SF	Sm	FB	.029	630.0	1.44	.88	---	2.32	**2.95**

.080" thick

Description	Oper	Unit	Vol	Crew Size	Man-hours per Unit	Crew Output per Day	Avg Mat'l Unit Cost	Avg Labor Unit Cost	Avg Equip Unit Cost	Avg Total Unit Cost	Avg Price Incl O&P
Designer slate	Inst	SF	Lg	FB	.020	900.0	2.78	.61	---	3.39	**4.09**
	Inst	SF	Sm	FB	.029	630.0	3.09	.88	---	3.97	**4.84**

1/8" thick

Description	Oper	Unit	Vol	Crew Size	Man-hours per Unit	Crew Output per Day	Avg Mat'l Unit Cost	Avg Labor Unit Cost	Avg Equip Unit Cost	Avg Total Unit Cost	Avg Price Incl O&P
Embassy oak	Inst	SF	Lg	FB	.020	900.0	3.43	.61	---	4.04	**4.83**
	Inst	SF	Sm	FB	.029	630.0	3.80	.88	---	4.68	**5.66**
Majestic slate	Inst	SF	Lg	FB	.020	900.0	2.78	.61	---	3.39	**4.09**
	Inst	SF	Sm	FB	.029	630.0	3.09	.88	---	3.97	**4.84**
Mediterranean marble	Inst	SF	Lg	FB	.020	900.0	3.05	.61	---	3.66	**4.40**
	Inst	SF	Sm	FB	.029	630.0	3.38	.88	---	4.26	**5.18**
Pecan	Inst	SF	Lg	FB	.020	900.0	3.56	.61	---	4.17	**4.98**
	Inst	SF	Sm	FB	.029	630.0	3.95	.88	---	4.83	**5.83**
Plaza brick	Inst	SF	Lg	FB	.020	900.0	5.38	.61	---	5.99	**7.08**
	Inst	SF	Sm	FB	.029	630.0	5.96	.88	---	6.84	**8.14**
Plymouth plank	Inst	SF	Lg	FB	.020	900.0	3.89	.61	---	4.50	**5.36**
	Inst	SF	Sm	FB	.029	630.0	4.31	.88	---	5.19	**6.25**
Teak	Inst	SF	Lg	FB	.020	900.0	3.30	.61	---	3.91	**4.68**
	Inst	SF	Sm	FB	.029	630.0	3.66	.88	---	4.54	**5.50**
Terrazzo	Inst	SF	Lg	FB	.020	900.0	2.65	.61	---	3.26	**3.94**
	Inst	SF	Sm	FB	.029	630.0	2.94	.88	---	3.82	**4.67**

Vinyl tile, 12" x 12", self-stick
.080" thick

Description	Oper	Unit	Vol	Crew Size	Man-hours per Unit	Crew Output per Day	Avg Mat'l Unit Cost	Avg Labor Unit Cost	Avg Equip Unit Cost	Avg Total Unit Cost	Avg Price Incl O&P
Elite	Inst	SF	Lg	FB	.025	725.0	2.08	.76	---	2.84	**3.50**
	Inst	SF	Sm	FB	.035	508.0	2.30	1.06	---	3.36	**4.20**
Stylglo	Inst	SF	Lg	FB	.025	725.0	1.56	.76	---	2.32	**2.91**
	Inst	SF	Sm	FB	.035	508.0	1.73	1.06	---	2.79	**3.55**

1/16" thick

Description	Oper	Unit	Vol	Crew Size	Man-hours per Unit	Crew Output per Day	Avg Mat'l Unit Cost	Avg Labor Unit Cost	Avg Equip Unit Cost	Avg Total Unit Cost	Avg Price Incl O&P
Decorator	Inst	SF	Lg	FB	.025	725.0	1.03	.76	---	1.79	**2.30**
	Inst	SF	Sm	FB	.035	508.0	1.14	1.06	---	2.20	**2.87**
Proclaim	Inst	SF	Lg	FB	.025	725.0	1.36	.76	---	2.12	**2.68**
	Inst	SF	Sm	FB	.035	508.0	1.50	1.06	---	2.56	**3.28**

Description	Oper	Unit	Vol	Crew Size	Man-hours per Unit	Crew Output per Day	Avg Mat'l Unit Cost	Avg Labor Unit Cost	Avg Equip Unit Cost	Avg Total Unit Cost	Avg Price Incl O&P
Composition vinyl tile, 12" x 12"											
3/32" thick											
Designer colors	Inst	SF	Lg	FB	.025	725.0	1.16	.76	---	1.92	**2.45**
	Inst	SF	Sm	FB	.035	508.0	1.29	1.06	---	2.35	**3.04**
Standard colors (marbleized)	Inst	SF	Lg	FB	.025	725.0	1.49	.76	---	2.25	**2.83**
	Inst	SF	Sm	FB	.035	508.0	1.65	1.06	---	2.71	**3.45**
1/8" thick											
Designer colors, pebbled	Inst	SF	Lg	FB	.025	725.0	2.21	.76	---	2.97	**3.65**
	Inst	SF	Sm	FB	.035	508.0	2.45	1.06	---	3.51	**4.37**
Standard colors (marbleized)	Inst	SF	Lg	FB	.025	725.0	1.30	.76	---	2.06	**2.61**
	Inst	SF	Sm	FB	.035	508.0	1.44	1.06	---	2.50	**3.21**
Solid black or white	Inst	SF	Lg	FB	.025	725.0	2.08	.76	---	2.84	**3.50**
	Inst	SF	Sm	FB	.035	508.0	2.30	1.06	---	3.36	**4.20**
Solid colors	Inst	SF	Lg	FB	.025	725.0	2.27	.76	---	3.03	**3.72**
	Inst	SF	Sm	FB	.035	508.0	2.51	1.06	---	3.57	**4.44**

Sheet vinyl

Description	Oper	Unit	Vol	Crew Size	Man-hours per Unit	Crew Output per Day	Avg Mat'l Unit Cost	Avg Labor Unit Cost	Avg Equip Unit Cost	Avg Total Unit Cost	Avg Price Incl O&P
Armstrong, no-wax, 6' wide											
Designer Solarian (.070")	Inst	SY	Lg	FB	.257	70.00	36.30	7.77	---	44.07	**53.20**
	Inst	SY	Sm	FB	.367	49.00	40.20	11.10	---	51.30	**62.60**
Designer Solarian II (.090")	Inst	SY	Lg	FB	.257	70.00	46.70	7.77	---	54.47	**65.10**
	Inst	SY	Sm	FB	.367	49.00	51.70	11.10	---	62.80	**75.80**
Imperial Accotone (.065")	Inst	SY	Lg	FB	.257	70.00	11.70	7.77	---	19.47	**24.80**
	Inst	SY	Sm	FB	.367	49.00	12.90	11.10	---	24.00	**31.20**
Solarian Supreme (.090")	Inst	SY	Lg	FB	.257	70.00	49.30	7.77	---	57.07	**68.10**
	Inst	SY	Sm	FB	.367	49.00	54.60	11.10	---	65.70	**79.10**
Sundial Solarian (.077")	Inst	SY	Lg	FB	.257	70.00	18.10	7.77	---	25.87	**32.30**
	Inst	SY	Sm	FB	.367	49.00	20.10	11.10	---	31.20	**39.40**
Mannington, 6' wide											
Vega (.080")	Inst	SY	Lg	FB	.257	70.00	10.40	7.77	---	18.17	**23.40**
	Inst	SY	Sm	FB	.367	49.00	11.50	11.10	---	22.60	**29.50**
Vinyl Ease (.073")	Inst	SY	Lg	FB	.257	70.00	9.07	7.77	---	16.84	**21.90**
	Inst	SY	Sm	FB	.367	49.00	10.10	11.10	---	21.20	**27.90**
Tarkett, 6' wide											
Preference (.065")	Inst	SY	Lg	FB	.257	70.00	8.42	7.77	---	16.19	**21.10**
	Inst	SY	Sm	FB	.367	49.00	9.34	11.10	---	20.44	**27.10**
Softred (.062")	Inst	SY	Lg	FB	.257	70.00	10.40	7.77	---	18.17	**23.40**
	Inst	SY	Sm	FB	.367	49.00	11.50	11.10	---	22.60	**29.50**

Description	Oper	Unit	Vol	Crew Size	Man- hours per Unit	Crew Output per Day	Avg Mat'l Unit Cost	Avg Labor Unit Cost	Avg Equip Unit Cost	Avg Total Unit Cost	Avg Price Incl O&P

Install over wood subfloor

Tile

Asphalt, 9" x 9", 1/8" thick, marbleized

Description	Oper	Unit	Vol	Crew Size	Man- hours per Unit	Crew Output per Day	Avg Mat'l Unit Cost	Avg Labor Unit Cost	Avg Equip Unit Cost	Avg Total Unit Cost	Avg Price Incl O&P
Group B colors, dark	Inst	SF	Lg	FB	.032	555.0	1.43	.97	---	2.40	3.07
	Inst	SF	Sm	FB	.046	389.0	1.58	1.39	---	2.97	3.86
Group C colors, medium	Inst	SF	Lg	FB	.032	555.0	1.62	.97	---	2.59	3.29
	Inst	SF	Sm	FB	.046	389.0	1.80	1.39	---	3.19	4.12
Group D colors, light	Inst	SF	Lg	FB	.032	555.0	1.62	.97	---	2.59	3.29
	Inst	SF	Sm	FB	.046	389.0	1.80	1.39	---	3.19	4.12

Vinyl tile, 12" x 12", including adhesive
1/16" thick

Description	Oper	Unit	Vol	Crew Size	Man- hours per Unit	Crew Output per Day	Avg Mat'l Unit Cost	Avg Labor Unit Cost	Avg Equip Unit Cost	Avg Total Unit Cost	Avg Price Incl O&P
Vega II	Inst	SF	Lg	FB	.024	750.0	1.30	.73	---	2.03	2.56
	Inst	SF	Sm	FB	.034	525.0	1.44	1.03	---	2.47	3.17

.080" thick

Description	Oper	Unit	Vol	Crew Size	Man- hours per Unit	Crew Output per Day	Avg Mat'l Unit Cost	Avg Labor Unit Cost	Avg Equip Unit Cost	Avg Total Unit Cost	Avg Price Incl O&P
Designer slate	Inst	SF	Lg	FB	.024	750.0	2.78	.73	---	3.51	4.26
	Inst	SF	Sm	FB	.034	525.0	3.09	1.03	---	4.12	5.07

1/8" thick

Description	Oper	Unit	Vol	Crew Size	Man- hours per Unit	Crew Output per Day	Avg Mat'l Unit Cost	Avg Labor Unit Cost	Avg Equip Unit Cost	Avg Total Unit Cost	Avg Price Incl O&P
Embassy oak	Inst	SF	Lg	FB	.024	750.0	3.43	.73	---	4.16	5.01
	Inst	SF	Sm	FB	.034	525.0	3.80	1.03	---	4.83	5.88
Majestic slate	Inst	SF	Lg	FB	.024	750.0	2.78	.73	---	3.51	4.26
	Inst	SF	Sm	FB	.034	525.0	3.09	1.03	---	4.12	5.07
Mediterranean marble	Inst	SF	Lg	FB	.024	750.0	3.05	.73	---	3.78	4.57
	Inst	SF	Sm	FB	.034	525.0	3.38	1.03	---	4.41	5.40
Pecan	Inst	SF	Lg	FB	.024	750.0	3.56	.73	---	4.29	5.16
	Inst	SF	Sm	FB	.034	525.0	3.95	1.03	---	4.98	6.05
Plaza brick	Inst	SF	Lg	FB	.024	750.0	5.38	.73	---	6.11	7.25
	Inst	SF	Sm	FB	.034	525.0	5.96	1.03	---	6.99	8.37
Plymouth plank	Inst	SF	Lg	FB	.024	750.0	3.89	.73	---	4.62	5.54
	Inst	SF	Sm	FB	.034	525.0	4.31	1.03	---	5.34	6.47
Teak	Inst	SF	Lg	FB	.024	750.0	3.30	.73	---	4.03	4.86
	Inst	SF	Sm	FB	.034	525.0	3.66	1.03	---	4.69	5.72
Terrazzo	Inst	SF	Lg	FB	.024	750.0	2.65	.73	---	3.38	4.11
	Inst	SF	Sm	FB	.034	525.0	2.94	1.03	---	3.97	4.89

Vinyl tile, 12" x 12", self-stick
.080" thick

Description	Oper	Unit	Vol	Crew Size	Man- hours per Unit	Crew Output per Day	Avg Mat'l Unit Cost	Avg Labor Unit Cost	Avg Equip Unit Cost	Avg Total Unit Cost	Avg Price Incl O&P
Elite	Inst	SF	Lg	FB	.024	750.0	2.08	.73	---	2.81	3.46
	Inst	SF	Sm	FB	.034	525.0	2.30	1.03	---	3.33	4.16
Stylglo	Inst	SF	Lg	FB	.024	750.0	1.56	.73	---	2.29	2.86
	Inst	SF	Sm	FB	.034	525.0	1.73	1.03	---	2.76	3.50

1/16" thick

Description	Oper	Unit	Vol	Crew Size	Man- hours per Unit	Crew Output per Day	Avg Mat'l Unit Cost	Avg Labor Unit Cost	Avg Equip Unit Cost	Avg Total Unit Cost	Avg Price Incl O&P
Decorator	Inst	SF	Lg	FB	.024	750.0	1.03	.73	---	1.76	2.25
	Inst	SF	Sm	FB	.034	525.0	1.14	1.03	---	2.17	2.82
Proclaim	Inst	SF	Lg	FB	.024	750.0	1.36	.73	---	2.09	2.63
	Inst	SF	Sm	FB	.034	525.0	1.50	1.03	---	2.53	3.24

Description	Oper	Unit	Vol	Crew Size	Man-hours per Unit	Crew Output per Day	Avg Mat'l Unit Cost	Avg Labor Unit Cost	Avg Equip Unit Cost	Avg Total Unit Cost	Avg Price Incl O&P
Composition vinyl tile, 12" x 12"											
3/32" thick											
Designer colors	Inst	SF	Lg	FB	.024	750.0	1.16	.73	---	1.89	**2.40**
	Inst	SF	Sm	FB	.034	525.0	1.29	1.03	---	2.32	**3.00**
Standard colors (marbleized)	Inst	SF	Lg	FB	.024	750.0	1.49	.73	---	2.22	**2.78**
	Inst	SF	Sm	FB	.034	525.0	1.65	1.03	---	2.68	**3.41**
1/8" thick											
Designer colors, pebbled	Inst	SF	Lg	FB	.024	750.0	2.21	.73	---	2.94	**3.61**
	Inst	SF	Sm	FB	.034	525.0	2.45	1.03	---	3.48	**4.33**
Standard colors (marbleized)	Inst	SF	Lg	FB	.024	750.0	1.30	.73	---	2.03	**2.56**
	Inst	SF	Sm	FB	.034	525.0	1.44	1.03	---	2.47	**3.17**
Solid black or white	Inst	SF	Lg	FB	.024	750.0	2.08	.73	---	2.81	**3.46**
	Inst	SF	Sm	FB	.034	525.0	2.30	1.03	---	3.33	**4.16**
Solid colors	Inst	SF	Lg	FB	.024	750.0	2.27	.73	---	3.00	**3.68**
	Inst	SF	Sm	FB	.034	525.0	2.51	1.03	---	3.54	**4.40**

Sheet vinyl

Description	Oper	Unit	Vol	Crew Size	Man-hours per Unit	Crew Output per Day	Avg Mat'l Unit Cost	Avg Labor Unit Cost	Avg Equip Unit Cost	Avg Total Unit Cost	Avg Price Incl O&P
Armstrong, no-wax, 6' wide											
Designer Solarian (.070")	Inst	SY	Lg	FB	.300	60.00	36.30	9.08	---	45.38	**55.10**
	Inst	SY	Sm	FB	.429	42.00	40.20	13.00	---	53.20	**65.30**
Designer Solarian II (.090")	Inst	SY	Lg	FB	.300	60.00	46.70	9.08	---	55.78	**67.00**
	Inst	SY	Sm	FB	.429	42.00	51.70	13.00	---	64.70	**78.50**
Imperial Accotone (.065")	Inst	SY	Lg	FB	.300	60.00	11.70	9.08	---	20.78	**26.80**
	Inst	SY	Sm	FB	.429	42.00	12.90	13.00	---	25.90	**33.90**
Solarian Supreme (.090")	Inst	SY	Lg	FB	.300	60.00	49.30	9.08	---	58.38	**70.00**
	Inst	SY	Sm	FB	.429	42.00	54.60	13.00	---	67.60	**81.80**
Sundial Solarian (.077")	Inst	SY	Lg	FB	.300	60.00	18.10	9.08	---	27.18	**34.20**
	Inst	SY	Sm	FB	.429	42.00	20.10	13.00	---	33.10	**42.20**
Mannington, 6' wide											
Vega (.080")	Inst	SY	Lg	FB	.300	60.00	10.40	9.08	---	19.48	**25.30**
	Inst	SY	Sm	FB	.429	42.00	11.50	13.00	---	24.50	**32.30**
Vinyl Ease (.073")	Inst	SY	Lg	FB	.300	60.00	9.07	9.08	---	18.15	**23.80**
	Inst	SY	Sm	FB	.429	42.00	10.10	13.00	---	23.10	**30.60**
Tarkett, 6' wide											
Preference (.065")	Inst	SY	Lg	FB	.300	60.00	8.42	9.08	---	17.50	**23.00**
	Inst	SY	Sm	FB	.429	42.00	9.34	13.00	---	22.34	**29.80**
Softred (.062")	Inst	SY	Lg	FB	.300	60.00	10.40	9.08	---	19.48	**25.30**
	Inst	SY	Sm	FB	.429	42.00	11.50	13.00	---	24.50	**32.30**

Description	Oper	Unit	Vol	Crew Size	Man-hours per Unit	Crew Output per Day	Avg Mat'l Unit Cost	Avg Labor Unit Cost	Avg Equip Unit Cost	Avg Total Unit Cost	Avg Price Incl O&P

Related materials and operations

Top set base
Vinyl
   2-1/2" H

Description	Oper	Unit	Vol	Crew Size	Man-hours per Unit	Crew Output per Day	Avg Mat'l Unit Cost	Avg Labor Unit Cost	Avg Equip Unit Cost	Avg Total Unit Cost	Avg Price Incl O&P
Colors	Inst	LF	Lg	FB	.036	500.0	.49	1.09	---	1.58	**2.16**
	Inst	LF	Sm	FB	.051	350.0	.55	1.54	---	2.09	**2.90**
Wood grain	Inst	LF	Lg	FB	.036	500.0	.66	1.09	---	1.75	**2.36**
	Inst	LF	Sm	FB	.051	350.0	.73	1.54	---	2.27	**3.11**

   4" H

Description	Oper	Unit	Vol	Crew Size	Man-hours per Unit	Crew Output per Day	Avg Mat'l Unit Cost	Avg Labor Unit Cost	Avg Equip Unit Cost	Avg Total Unit Cost	Avg Price Incl O&P
Colors	Inst	LF	Lg	FB	.036	500.0	.58	1.09	---	1.67	**2.27**
	Inst	LF	Sm	FB	.051	350.0	.64	1.54	---	2.18	**3.00**
Wood grain	Inst	LF	Lg	FB	.036	500.0	.78	1.09	---	1.87	**2.50**
	Inst	LF	Sm	FB	.051	350.0	.86	1.54	---	2.40	**3.26**

Linoleum cove, 7/16" x 1-1/4"

Description	Oper	Unit	Vol	Crew Size	Man-hours per Unit	Crew Output per Day	Avg Mat'l Unit Cost	Avg Labor Unit Cost	Avg Equip Unit Cost	Avg Total Unit Cost	Avg Price Incl O&P
Softwood	Inst	LF	Lg	FB	.051	350.0	.58	1.54	---	2.12	**2.93**
	Inst	LF	Sm	FB	.073	245.0	.64	2.21	---	2.85	**3.98**

Retaining walls, see Concrete, page 100, or Masonry, page 282

Roofing

Fiberglass Shingles, 225 lb., three tab strip. Three bundle/square; 20 year

1. **Dimensions.** Each shingle is 12" x 36". With a 5" exposure to the weather, 80 shingles are required to cover one square (100 SF).

2. **Installation**

   a. Over wood. After scatter-nailing one ply of 15 lb. felt, the shingles are installed. Four nails per shingle is customary.

   b. Over existing roofing. 15 lb. felt is not required. Shingles are installed the same as over wood, except longer nails are used, i.e., $1^1/4$" in lieu of 1".

3. **Estimating Technique.** Determine roof area and add a percentage for starters, ridge, and valleys. The percent to be added varies, but generally:

   a. For plain gable and hip – add 10%.

   b. For gable or hip with dormers or intersecting roof(s) – add 15%.

   c. For gable or hip with dormers and intersecting roof(s) – add 20%.

When ridge or hip shingles are special ordered, reduce the above percentages by approximately 50%. Then apply unit costs to the area calculated (including the allowance above).

Mineral Surfaced Roll Roofing, 90 lb.

1. **Dimensions.** A roll is 36" wide x 36'-0" long, or 108 SF. It covers one square (100 SF), after including 8 SF for head and end laps.

2. **Installation.** Usually, lap cement and $7/8$" galvanized nails are furnished with each roll.

   a. Over wood. Roll roofing is usually applied directly to sheathing with $7/8$" galvanized nails. End and head laps are normally 6" and 2" or 3" respectively.

   b. Over existing roofing. Applied the same as over wood, except nails are not less than $1^1/4$" long.

   c. Roll roofing may be installed by the exposed nail or the concealed nail method. In the exposed nail method, nails are visible at laps, and head laps are usually 2". In the concealed nail method, no nails are visible, the head lap is a minimum of 3", and there is a 9"-wide strip of roll installed along rakes and eaves.

3. **Estimating Technique.** Determine the area and add a percentage for hip and/or ridge. The percentage to be added varies, but generally:

   a. For plain gable or hip – add 5%.

   b. For cut-up roof – add 10%.

If metal ridge and/or hip are used, reduce the above percentages by approximately 50%. Then apply unit costs to the area calculated (including the allowance above).

Wood Shingles

1. **Dimensions.** Shingles are available in 16", 18", and 24" lengths and in uniform widths of 4", 5", or 6". But they are commonly furnished in random widths averaging 4" wide.

2. **Installation.** The normal exposure to weather for 16", 18" and 24" shingles on roofs with $1/4$ or steeper pitch is 5", $5^1/2$", and $7^1/2$" respectively. Where the slope is less than $1/4$, the exposure is usually reduced. Generally, 3d commons are used when shingles are applied to strip sheathing. 5d commons are used when shingles are applied over existing shingles. Two nails per shingle is the general rule, but on some roofs, one nail per shingle is used.

3. **Estimating Technique.** Determine the roof area and add a percentage for waste, starters, ridge, and hip shingles.

   a. Wood shingles. The amount of exposure determines the percentage of 1 square of shingles required to cover 100 SF of roof. Multiply the material cost per square of shingles by the appropriate percent (from the table on the next page) to determine the cost to cover 100 SF of roof with wood shingles. The table does not include cutting waste. NOTE: Nails and the exposure factors in this table have already been calculated into the Average Unit Material Costs.

   b. Nails. The weight of nails required per square varies with the size of the nail and with the shingle exposure. Multiply cost per pound of nails by the appropriate pounds per square. In the table on the next page, pounds of nails per square are based on two nails per shingle. For one nail per shingle, deduct 50%.

   c. The percentage to be added for starters, hip, and ridge shingles varies, but generally:

      1) Plain gable and hip – add 10%.

      2) Gable or hip with dormers or intersecting roof(s) – add 15%.

      3) Gable or hip with dormers and intersecting roof(s) – add 20%.

When ridge and/or hip shingles are special ordered, reduce these percentages by 50%. Then

Shingle Length	Exposure	% of Sq. Required to Cover 100 SF	Lbs of Nails per Square (2 Nails/Shingle)	
			3d	5d
24"	7½"	100	2.3	2.7
24"	7"	107	2.4	2.9
24"	6½"	115	2.6	3.2
24"	6"	125	2.8	3.4
24"	5¾"	130	2.9	3.6
18"	5½"	100	3.1	3.7
18"	5"	110	3.4	4.1
18"	4½"	122	3.8	4.6
16"	5"	100	3.4	4.1
16"	4½"	111	3.8	4.6
16"	4"	125	4.2	5.1
16"	3¾"	133	4.5	5.5

apply unit costs to the area calculated (including the allowance above).

Built-Up Roofing

1. **Dimensions**

   a. 15 lb. felt. Rolls are 36" wide x 144'-0" long or 432 SF – 4 squares per roll. 32 SF (or 8 SF/sq) is for laps.

   b. 30 lb. felt. Rolls are 36" wide x 72'-0" long or 216 SF – 2 squares per roll. 16 SF (or 8 SF/sq) is for laps.

   c. Asphalt. Available in 100 lb. cartons. Average asphalt usage is:

      1) Top coat without gravel – 35 lbs. per square.

      2) Top coat with gravel – 60 lbs. per square.

      3) Each coat (except top coat) – 25 lbs. per square.

   d. Tar. Available in 550 lb. kegs. Average tar usage is:

      1) Top coat without gravel – 40 lbs. per square.

      2) Top coat without gravel – 75 lbs. per square.

      3) Each coat (except top coat) – 30 lbs. per square.

   e. Gravel or slag. Normally, 400 lbs. of gravel or 275 lbs. of slag are used to cover one square.

2. **Installation**

   a. Over wood. Normally, one or two plies of felt are applied by scatter nailing before hot mopping is commenced. Subsequent plies of felt and gravel or slag are imbedded in hot asphalt or tar.

   b. Over concrete. Every ply of felt and gravel or slag is imbedded in hot asphalt or tar.

3. **Estimating Technique**

   a. For buildings with parapet walls, use the outside dimensions of the building to determine the area. This area is usually sufficient to include the flashing.

   b. For buildings without parapet walls, determine the area.

   c. Don't deduct any opening less than 10'-0" x 10'-0" (100 SF). Deduct 50% of the opening for openings larger than 100 SF but smaller than 300 SF, and 100% of the openings exceeding 300 SF.

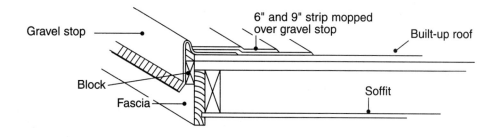

Gravel stop — Block — Fascia — 6" and 9" strip mopped over gravel stop — Built-up roof — Soffit

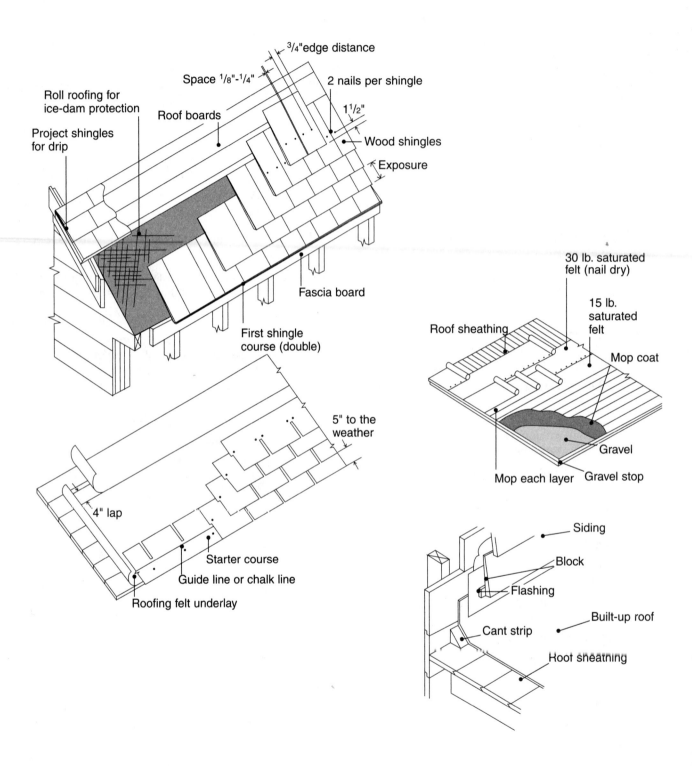

³/₄" edge distance

Space ¹/₈"-¹/₄"

2 nails per shingle

1¹/₂"

Roll roofing for ice-dam protection

Roof boards

Wood shingles

Project shingles for drip

Exposure

Fascia board

First shingle course (double)

5" to the weather

4" lap

Starter course

Guide line or chalk line

Roofing felt underlay

30 lb. saturated felt (nail dry)

15 lb. saturated felt

Roof sheathing

Mop coat

Gravel

Mop each layer

Gravel stop

Siding

Block

Flashing

Built-up roof

Cant strip

Roof sheathing

Description	Oper	Unit	Vol	Crew Size	Man-hours per Unit	Crew Output per Day	Avg Mat'l Unit Cost	Avg Labor Unit Cost	Avg Equip Unit Cost	Avg Total Unit Cost	Avg Price Incl O&P

Roofing

Aluminum, nailed to wood

Corrugated (2-1/2"), 26" W, with 3-3/4" side lap and

Description	Oper	Unit	Vol	Crew Size	Man-hours per Unit	Crew Output per Day	Avg Mat'l Unit Cost	Avg Labor Unit Cost	Avg Equip Unit Cost	Avg Total Unit Cost	Avg Price Incl O&P
6" end lap	Demo	Sq	Lg	LB	1.60	10.00	---	43.90	---	43.90	**65.40**
	Demo	Sq	Sm	LB	2.00	8.00	---	54.90	---	54.90	**81.70**

Corrugated (2-1/2"), 26" W, with 3-3/4" side lap and 6" end lap
0.0175" thick

Description	Oper	Unit	Vol	Crew Size	Man-hours per Unit	Crew Output per Day	Avg Mat'l Unit Cost	Avg Labor Unit Cost	Avg Equip Unit Cost	Avg Total Unit Cost	Avg Price Incl O&P
Natural	Inst	Sq	Lg	UC	2.13	15.00	153.00	68.60	---	221.60	**277.00**
	Inst	Sq	Sm	UC	2.67	12.00	167.00	86.00	---	253.00	**319.00**
Painted	Inst	Sq	Lg	UC	2.13	15.00	197.00	68.60	---	265.60	**328.00**
	Inst	Sq	Sm	UC	2.67	12.00	214.00	86.00	---	300.00	**374.00**

0.019" thick

Description	Oper	Unit	Vol	Crew Size	Man-hours per Unit	Crew Output per Day	Avg Mat'l Unit Cost	Avg Labor Unit Cost	Avg Equip Unit Cost	Avg Total Unit Cost	Avg Price Incl O&P
Natural	Inst	Sq	Lg	UC	2.13	15.00	153.00	68.60	---	221.60	**277.00**
	Inst	Sq	Sm	UC	2.67	12.00	167.00	86.00	---	253.00	**319.00**
Painted	Inst	Sq	Lg	UC	2.13	15.00	197.00	68.60	---	265.60	**328.00**
	Inst	Sq	Sm	UC	2.67	12.00	214.00	86.00	---	300.00	**374.00**

Composition shingle roofing, 3 tab strip, 12" x 36", nailed, seal down

Demo Shingle, to deck

Description	Oper	Unit	Vol	Crew Size	Man-hours per Unit	Crew Output per Day	Avg Mat'l Unit Cost	Avg Labor Unit Cost	Avg Equip Unit Cost	Avg Total Unit Cost	Avg Price Incl O&P
5" exposure	Demo	Sq	Lg	LB	1.00	16.00	---	27.40	---	27.40	**40.90**
	Demo	Sq	Sm	LB	1.25	12.80	---	34.30	---	34.30	**51.10**

Install 25-year shingles, 240 lb/Sq

Over existing roofing

Description	Oper	Unit	Vol	Crew Size	Man-hours per Unit	Crew Output per Day	Avg Mat'l Unit Cost	Avg Labor Unit Cost	Avg Equip Unit Cost	Avg Total Unit Cost	Avg Price Incl O&P
Gable, plain	Inst	Sq	Lg	RL	1.33	15.00	38.60	46.70	---	85.30	**117.00**
	Inst	Sq	Sm	RL	1.67	12.00	42.10	58.70	---	100.80	**139.00**
Gable with dormers	Inst	Sq	Lg	RL	1.48	13.50	39.00	52.00	---	91.00	**125.00**
	Inst	Sq	Sm	RL	1.85	10.80	42.50	65.00	---	107.50	**150.00**
Gable with intersecting roofs	Inst	Sq	Lg	RL	1.48	13.50	39.30	52.00	---	91.30	**126.00**
	Inst	Sq	Sm	RL	1.85	10.80	42.90	65.00	---	107.90	**150.00**
Gable w/dormers & intersections											
	Inst	Sq	Lg	RL	1.67	12.00	39.70	58.70	---	98.40	**137.00**
	Inst	Sq	Sm	RL	2.08	9.60	43.20	73.10	---	116.30	**163.00**
Hip, plain	Inst	Sq	Lg	RL	1.40	14.33	39.00	49.20	---	88.20	**121.00**
	Inst	Sq	Sm	RL	1.75	11.46	42.50	61.50	---	104.00	**144.00**
Hip with dormers	Inst	Sq	Lg	RL	1.56	12.80	39.30	54.80	---	94.10	**130.00**
	Inst	Sq	Sm	RL	1.95	10.24	42.90	68.50	---	111.40	**155.00**
Hip with intersecting roofs	Inst	Sq	Lg	RL	1.56	12.80	39.70	54.80	---	94.50	**131.00**
	Inst	Sq	Sm	RL	1.95	10.24	43.20	68.50	---	111.70	**156.00**
Hip with dormers & intersections											
	Inst	Sq	Lg	RL	1.77	11.30	40.00	62.20	---	102.20	**142.00**
	Inst	Sq	Sm	RL	2.21	9.04	43.60	77.60	---	121.20	**171.00**

Description	Oper	Unit	Vol	Crew Size	Man-hours per Unit	Crew Output per Day	Avg Mat'l Unit Cost	Avg Labor Unit Cost	Avg Equip Unit Cost	Avg Total Unit Cost	Avg Price Incl O&P
Over wood decks, includes felt											
Gable, plain	Inst	Sq	Lg	RL	1.43	14.00	43.00	50.20	---	93.20	**127.00**
	Inst	Sq	Sm	RL	1.79	11.20	46.80	62.90	---	109.70	**151.00**
Gable with dormers	Inst	Sq	Lg	RL	1.60	12.50	43.30	56.20	---	99.50	**137.00**
	Inst	Sq	Sm	RL	2.00	10.00	47.20	70.30	---	117.50	**163.00**
Gable with intersecting roofs	Inst	Sq	Lg	RL	1.60	12.50	43.70	56.20	---	99.90	**137.00**
	Inst	Sq	Sm	RL	2.00	10.00	47.60	70.30	---	117.90	**164.00**
Gable w/dormers & intersections											
	Inst	Sq	Lg	RL	1.82	11.00	44.00	63.90	---	107.90	**150.00**
	Inst	Sq	Sm	RL	2.27	8.80	48.00	79.80	---	127.80	**179.00**
Hip, plain	Inst	Sq	Lg	RL	1.50	13.33	43.30	52.70	---	96.00	**131.00**
	Inst	Sq	Sm	RL	1.88	10.66	47.20	66.00	---	113.20	**157.00**
Hip with dormers	Inst	Sq	Lg	RL	1.69	11.80	43.70	59.40	---	103.10	**142.00**
	Inst	Sq	Sm	RL	2.12	9.44	47.60	74.50	---	122.10	**170.00**
Hip with intersecting roofs	Inst	Sq	Lg	RL	1.69	11.80	44.00	59.40	---	103.40	**143.00**
	Inst	Sq	Sm	RL	2.12	9.44	48.00	74.50	---	122.50	**171.00**
Hip with dormers & intersections											
	Inst	Sq	Lg	RL	1.94	10.30	44.30	68.20	---	112.50	**157.00**
	Inst	Sq	Sm	RL	2.43	8.24	48.30	85.40	---	133.70	**188.00**

Install 30-year laminated shingles

Over existing roofing

Description	Oper	Unit	Vol	Crew Size	Man-hours per Unit	Crew Output per Day	Avg Mat'l Unit Cost	Avg Labor Unit Cost	Avg Equip Unit Cost	Avg Total Unit Cost	Avg Price Incl O&P
Gable, plain	Inst	Sq	Lg	RL	1.33	15.00	54.50	46.70	---	101.20	**135.00**
	Inst	Sq	Sm	RL	1.67	12.00	59.40	58.70	---	118.10	**159.00**
Gable with dormers	Inst	Sq	Lg	RL	1.48	13.50	55.00	52.00	---	107.00	**144.00**
	Inst	Sq	Sm	RL	1.85	10.80	60.00	65.00	---	125.00	**170.00**
Gable with intersecting roofs	Inst	Sq	Lg	RL	1.48	13.50	55.50	52.00	---	107.50	**144.00**
	Inst	Sq	Sm	RL	1.85	10.80	60.50	65.00	---	125.50	**170.00**
Gable w/dormers & intersections											
	Inst	Sq	Lg	RL	1.67	12.00	56.00	58.70	---	114.70	**155.00**
	Inst	Sq	Sm	RL	2.08	9.60	61.10	73.10	---	134.20	**183.00**
Hip, plain	Inst	Sq	Lg	RL	1.40	14.33	55.00	49.20	---	104.20	**140.00**
	Inst	Sq	Sm	RL	1.75	11.46	60.00	61.50	---	121.50	**164.00**
Hip with dormers	Inst	Sq	Lg	RL	1.56	12.80	55.50	54.80	---	110.30	**149.00**
	Inst	Sq	Sm	RL	1.95	10.24	60.50	68.50	---	129.00	**176.00**
Hip with intersecting roofs	Inst	Sq	Lg	RL	1.56	12.80	56.00	54.80	---	110.80	**149.00**
	Inst	Sq	Sm	RL	1.95	10.24	61.10	68.50	---	129.60	**176.00**
Hip with dormers & intersections											
	Inst	Sq	Lg	RL	1.77	11.30	56.50	62.20	---	118.70	**161.00**
	Inst	Sq	Sm	RL	2.21	9.04	61.60	77.60	---	139.20	**191.00**

Over wood decks, includes felt

Description	Oper	Unit	Vol	Crew Size	Man-hours per Unit	Crew Output per Day	Avg Mat'l Unit Cost	Avg Labor Unit Cost	Avg Equip Unit Cost	Avg Total Unit Cost	Avg Price Incl O&P
Gable, plain	Inst	Sq	Lg	RL	1.43	14.00	58.90	50.20	---	109.10	**146.00**
	Inst	Sq	Sm	RL	1.79	11.20	64.20	62.90	---	127.10	**171.00**
Gable with dormers	Inst	Sq	Lg	RL	1.60	12.50	59.40	56.20	---	115.60	**155.00**
	Inst	Sq	Sm	RL	2.00	10.00	64.70	70.30	---	135.00	**183.00**
Gable with intersecting roofs	Inst	Sq	Lg	RL	1.60	12.50	59.90	56.20	---	116.10	**156.00**
	Inst	Sq	Sm	RL	2.00	10.00	65.30	70.30	---	135.60	**184.00**
Gable w/dormers & intersections											
	Inst	Sq	Lg	RL	1.82	11.00	60.40	63.90	---	124.30	**169.00**
	Inst	Sq	Sm	RL	2.27	8.80	65.80	79.80	---	145.60	**199.00**
Hip, plain	Inst	Sq	Lg	RL	1.50	13.33	59.40	52.70	---	112.10	**150.00**
	Inst	Sq	Sm	RL	1.88	10.66	64.70	66.00	---	130.70	**177.00**

Description	Oper	Unit	Vol	Crew Size	Man-hours per Unit	Crew Output per Day	Avg Mat'l Unit Cost	Avg Labor Unit Cost	Avg Equip Unit Cost	Avg Total Unit Cost	Avg Price Incl O&P
Hip with dormers	Inst	Sq	Lg	RL	1.69	11.80	59.90	59.40	---	119.30	**161.00**
	Inst	Sq	Sm	RL	2.12	9.44	65.30	74.50	---	139.80	**190.00**
Hip with intersecting roofs	Inst	Sq	Lg	RL	1.69	11.80	60.40	59.40	---	119.80	**161.00**
	Inst	Sq	Sm	RL	2.12	9.44	65.80	74.50	---	140.30	**191.00**
Hip with dormers & intersections											
	Inst	Sq	Lg	RL	1.94	10.30	60.90	68.20	---	129.10	**176.00**
	Inst	Sq	Sm	RL	2.43	8.24	66.30	85.40	---	151.70	**209.00**

Install 40-year laminated shingles

Over existing roofing

Description	Oper	Unit	Vol	Crew Size	Man-hours per Unit	Crew Output per Day	Avg Mat'l Unit Cost	Avg Labor Unit Cost	Avg Equip Unit Cost	Avg Total Unit Cost	Avg Price Incl O&P
Gable, plain	Inst	Sq	Lg	RL	1.33	15.00	68.50	46.70	---	115.20	**151.00**
	Inst	Sq	Sm	RL	1.67	12.00	74.60	58.70	---	133.30	**177.00**
Gable with dormers	Inst	Sq	Lg	RL	1.48	13.50	69.10	52.00	---	121.10	**160.00**
	Inst	Sq	Sm	RL	1.85	10.80	75.30	65.00	---	140.30	**187.00**
Gable with intersecting roofs	Inst	Sq	Lg	RL	1.48	13.50	69.70	52.00	---	121.70	**161.00**
	Inst	Sq	Sm	RL	1.85	10.80	76.00	65.00	---	141.00	**188.00**
Gable w/dormers & intersections											
	Inst	Sq	Lg	RL	1.67	12.00	70.40	58.70	---	129.10	**172.00**
	Inst	Sq	Sm	RL	2.08	9.60	76.70	73.10	---	149.80	**201.00**
Hip, plain	Inst	Sq	Lg	RL	1.40	14.33	69.10	49.20	---	118.30	**156.00**
	Inst	Sq	Sm	RL	1.75	11.46	75.30	61.50	---	136.80	**182.00**
Hip with dormers	Inst	Sq	Lg	RL	1.56	12.80	69.70	54.80	---	124.50	**165.00**
	Inst	Sq	Sm	RL	1.95	10.24	76.00	68.50	---	144.50	**194.00**
Hip with intersecting roofs	Inst	Sq	Lg	RL	1.56	12.80	70.40	54.80	---	125.20	**166.00**
	Inst	Sq	Sm	RL	1.95	10.24	76.70	68.50	---	145.20	**194.00**
Hip with dormers & intersections											
	Inst	Sq	Lg	RL	1.77	11.30	71.00	62.20	---	133.20	**178.00**
	Inst	Sq	Sm	RL	2.21	9.04	77.40	77.60	---	155.00	**209.00**

Over wood decks, includes felt

Description	Oper	Unit	Vol	Crew Size	Man-hours per Unit	Crew Output per Day	Avg Mat'l Unit Cost	Avg Labor Unit Cost	Avg Equip Unit Cost	Avg Total Unit Cost	Avg Price Incl O&P
Gable, plain	Inst	Sq	Lg	RL	1.43	14.00	72.80	50.20	---	123.00	**162.00**
	Inst	Sq	Sm	RL	1.79	11.20	79.40	62.90	---	142.30	**189.00**
Gable with dormers	Inst	Sq	Lg	RL	1.60	12.50	73.40	56.20	---	129.60	**172.00**
	Inst	Sq	Sm	RL	2.00	10.00	80.10	70.30	---	150.40	**201.00**
Gable with intersecting roofs	Inst	Sq	Lg	RL	1.60	12.50	74.10	56.20	---	130.30	**172.00**
	Inst	Sq	Sm	RL	2.00	10.00	80.70	70.30	---	151.00	**202.00**
Gable w/dormers & intersections											
	Inst	Sq	Lg	RL	1.82	11.00	74.70	63.90	---	138.60	**185.00**
	Inst	Sq	Sm	RL	2.27	8.80	81.40	79.80	---	161.20	**217.00**
Hip, plain	Inst	Sq	Lg	RL	1.50	13.33	73.40	52.70	---	126.10	**166.00**
	Inst	Sq	Sm	RL	1.88	10.66	80.10	66.00	---	146.10	**194.00**
Hip with dormers	Inst	Sq	Lg	RL	1.69	11.80	74.10	59.40	---	133.50	**177.00**
	Inst	Sq	Sm	RL	2.12	9.44	80.70	74.50	---	155.20	**208.00**
Hip with intersecting roofs	Inst	Sq	Lg	RL	1.69	11.80	74.70	59.40	---	134.10	**178.00**
	Inst	Sq	Sm	RL	2.12	9.44	81.40	74.50	---	155.90	**209.00**
Hip with dormers & intersections											
	Inst	Sq	Lg	RL	1.94	10.30	75.30	68.20	---	143.50	**192.00**
	Inst	Sq	Sm	RL	2.43	8.24	82.10	85.40	---	167.50	**227.00**

Description	Oper	Unit	Vol	Crew Size	Man-hours per Unit	Crew Output per Day	Avg Mat'l Unit Cost	Avg Labor Unit Cost	Avg Equip Unit Cost	Avg Total Unit Cost	Avg Price Incl O&P

Install 50-year lifetime laminated shingles

Over existing roofing

Description	Oper	Unit	Vol	Crew Size	Man-hours per Unit	Crew Output per Day	Avg Mat'l Unit Cost	Avg Labor Unit Cost	Avg Equip Unit Cost	Avg Total Unit Cost	Avg Price Incl O&P
Gable, plain	Inst	Sq	Lg	RL	1.33	15.00	79.40	46.70	---	126.10	**164.00**
	Inst	Sq	Sm	RL	1.67	12.00	86.60	58.70	---	145.30	**191.00**
Gable with dormers	Inst	Sq	Lg	RL	1.48	13.50	80.20	52.00	---	132.20	**173.00**
	Inst	Sq	Sm	RL	1.85	10.80	87.40	65.00	---	152.40	**201.00**
Gable with intersecting roofs	Inst	Sq	Lg	RL	1.48	13.50	80.90	52.00	---	132.90	**174.00**
	Inst	Sq	Sm	RL	1.85	10.80	88.20	65.00	---	153.20	**202.00**
Gable w/dormers & intersection	Inst	Sq	Lg	RL	1.67	12.00	81.60	58.70	---	140.30	**185.00**
	Inst	Sq	Sm	RL	2.08	9.60	89.00	73.10	---	162.10	**216.00**
Hip, plain	Inst	Sq	Lg	RL	1.40	14.33	80.20	49.20	---	129.40	**168.00**
	Inst	Sq	Sm	RL	1.75	11.46	87.40	61.50	---	148.90	**196.00**
Hip with dormers	Inst	Sq	Lg	RL	1.56	12.80	80.90	54.80	---	135.70	**178.00**
	Inst	Sq	Sm	RL	1.95	10.24	88.20	68.50	---	156.70	**208.00**
Hip with intersecting roofs	Inst	Sq	Lg	RL	1.56	12.80	81.60	54.80	---	136.40	**179.00**
	Inst	Sq	Sm	RL	1.95	10.24	89.00	68.50	---	157.50	**209.00**
Hip with dormers & intersections											
	Inst	Sq	Lg	RL	1.77	11.30	82.40	62.20	---	144.60	**191.00**
	Inst	Sq	Sm	RL	2.21	9.04	89.80	77.60	---	167.40	**224.00**

Over wood decks, includes felt

Description	Oper	Unit	Vol	Crew Size	Man-hours per Unit	Crew Output per Day	Avg Mat'l Unit Cost	Avg Labor Unit Cost	Avg Equip Unit Cost	Avg Total Unit Cost	Avg Price Incl O&P
Gable, plain	Inst	Sq	Lg	RL	1.43	14.00	83.80	50.20	---	134.00	**174.00**
	Inst	Sq	Sm	RL	1.79	11.20	91.30	62.90	---	154.20	**202.00**
Gable with dormers	Inst	Sq	Lg	RL	1.60	12.50	84.50	56.20	---	140.70	**184.00**
	Inst	Sq	Sm	RL	2.00	10.00	92.10	70.30	---	162.40	**215.00**
Gable with intersecting roofs	Inst	Sq	Lg	RL	1.60	12.50	85.20	56.20	---	141.40	**185.00**
	Inst	Sq	Sm	RL	2.00	10.00	92.90	70.30	---	163.20	**216.00**
Gable w/dormers & intersection	Inst	Sq	Lg	RL	1.82	11.00	86.00	63.90	---	149.90	**198.00**
	Inst	Sq	Sm	RL	2.27	8.80	93.70	79.80	---	173.50	**231.00**
Hip, plain	Inst	Sq	Lg	RL	1.50	13.33	84.50	52.70	---	137.20	**179.00**
	Inst	Sq	Sm	RL	1.88	10.66	92.10	66.00	---	158.10	**208.00**
Hip with dormers	Inst	Sq	Lg	RL	1.69	11.80	85.20	59.40	---	144.60	**190.00**
	Inst	Sq	Sm	RL	2.12	9.44	92.90	74.50	---	167.40	**222.00**
Hip with intersecting roofs	Inst	Sq	Lg	RL	1.69	11.80	86.00	59.40	---	145.40	**191.00**
	Inst	Sq	Sm	RL	2.12	9.44	93.70	74.50	---	168.20	**223.00**
Hip with dormers & intersections											
	Inst	Sq	Lg	RL	1.94	10.30	86.70	68.20	---	154.90	**205.00**
	Inst	Sq	Sm	RL	2.43	8.24	94.50	85.40	---	179.90	**241.00**

Related materials and operations

Ridge or hip roll

Description	Oper	Unit	Vol	Crew Size	Man-hours per Unit	Crew Output per Day	Avg Mat'l Unit Cost	Avg Labor Unit Cost	Avg Equip Unit Cost	Avg Total Unit Cost	Avg Price Incl O&P
90 lb mineral surfaced	Inst	LF	Lg	2R	.020	800.0	---	.74	---	.74	**1.15**
	Inst	LF	Sm	2R	.025	640.0	---	.93	---	.93	**1.44**

Valley roll

Description	Oper	Unit	Vol	Crew Size	Man-hours per Unit	Crew Output per Day	Avg Mat'l Unit Cost	Avg Labor Unit Cost	Avg Equip Unit Cost	Avg Total Unit Cost	Avg Price Incl O&P
90 lb mineral surfaced	Inst	LF	Lg	2R	.020	800.0	.24	.74	---	.98	**1.42**
	Inst	LF	Sm	2R	.025	640.0	.26	.93	---	1.19	**1.73**

Description	Oper	Unit	Vol	Crew Size	Man-hours per Unit	Crew Output per Day	Avg Mat'l Unit Cost	Avg Labor Unit Cost	Avg Equip Unit Cost	Avg Total Unit Cost	Avg Price Incl O&P

Ridge / hip units, 9" x 12" with 5" exp.; 80 pcs / bdle @ 33.3 LF/bdle

Standard

Over existing roofing	Inst	LF	Lg	2R	.023	700.0	1.06	.85	---	1.91	**2.54**
	Inst	LF	Sm	2R	.029	560.0	1.16	1.07	---	2.23	**3.00**
Over wood decks	Inst	LF	Lg	2R	.024	660.0	1.06	.89	---	1.95	**2.60**
	Inst	LF	Sm	2R	.030	528.0	1.16	1.11	---	2.27	**3.06**

Architectural

Over existing roofing	Inst	LF	Lg	2R	.023	700.0	1.18	.85	---	2.03	**2.68**
	Inst	LF	Sm	2R	.029	560.0	1.29	1.07	---	2.36	**3.15**
Over wood decks	Inst	LF	Lg	2R	.024	660.0	1.18	.89	---	2.07	**2.74**
	Inst	LF	Sm	2R	.030	528.0	1.29	1.11	---	2.40	**3.21**

Built-up or membrane roofing

Install over existing roofing

One mop coat over smooth surface

With asphalt	Inst	Sq	Lg	RS	.356	135.0	10.80	12.50	---	23.30	**31.70**
	Inst	Sq	Sm	RS	.444	108.0	11.70	15.60	---	27.30	**37.60**
With tar	Inst	Sq	Lg	RS	.356	135.0	18.80	12.50	---	31.30	**41.00**
	Inst	Sq	Sm	RS	.444	108.0	20.50	15.60	---	36.10	**47.70**

Remove gravel, mop one coat, redistribute old gravel

With asphalt	Inst	Sq	Lg	RS	.906	53.00	16.80	31.70	---	48.50	**68.50**
	Inst	Sq	Sm	RS	1.13	42.40	18.20	39.60	---	57.80	**82.30**
With tar	Inst	Sq	Lg	RS	.906	53.00	29.00	31.70	---	60.70	**82.60**
	Inst	Sq	Sm	RS	1.13	42.40	31.70	39.60	---	71.30	**97.80**

Mop in one 30 lb cap sheet, plus smooth top mop coat

With asphalt	Inst	Sq	Lg	RS	.842	57.00	26.20	29.50	---	55.70	**75.80**
	Inst	Sq	Sm	RS	1.05	45.60	28.40	36.80	---	65.20	**89.70**
With tar	Inst	Sq	Lg	RS	.842	57.00	38.80	29.50	---	68.30	**90.30**
	Inst	Sq	Sm	RS	1.05	45.60	42.30	36.80	---	79.10	**106.00**

Remove gravel, mop in one 30 lb cap sheet, mop in and redistribute old gravel

With asphalt	Inst	Sq	Lg	RS	1.45	33.00	32.20	50.80	---	83.00	**116.00**
	Inst	Sq	Sm	RS	1.82	26.40	34.90	63.70	---	98.60	**139.00**
With tar	Inst	Sq	Lg	RS	1.45	33.00	49.00	50.80	---	99.80	**135.00**
	Inst	Sq	Sm	RS	1.82	26.40	53.40	63.70	---	117.10	**160.00**

Mop in two 15 lb felt plies, plus smooth top mop coat

With asphalt	Inst	Sq	Lg	RS	1.33	36.00	32.20	46.60	---	78.80	**109.00**
	Inst	Sq	Sm	RS	1.67	28.80	34.90	58.50	---	93.40	**131.00**
With tar	Inst	Sq	Lg	RS	1.33	36.00	49.30	46.60	---	95.90	**129.00**
	Inst	Sq	Sm	RS	1.67	28.80	53.80	58.50	---	112.30	**152.00**

Remove gravel, mop in two 15 lb felt plies, mop in and redistribute old gravel

With asphalt	Inst	Sq	Lg	RS	1.92	25.00	38.20	67.20	---	105.40	**148.00**
	Inst	Sq	Sm	RS	2.40	20.00	41.40	84.10	---	125.50	**178.00**
With tar	Inst	Sq	Lg	RS	1.92	25.00	59.50	67.20	---	126.70	**173.00**
	Inst	Sq	Sm	RS	2.40	20.00	64.90	84.10	---	149.00	**205.00**

Description	Oper	Unit	Vol	Crew Size	Man-hours per Unit	Crew Output per Day	Avg Mat'l Unit Cost	Avg Labor Unit Cost	Avg Equip Unit Cost	Avg Total Unit Cost	Avg Price Incl O&P

Demo over smooth wood or concrete deck
3 ply

Description	Oper	Unit	Vol	Crew Size	MH	CO	Mat'l	Labor	Equip	Total	Price
With gravel	Demo	Sq	Lg	LB	1.45	11.00	---	39.80	---	39.80	**59.30**
	Demo	Sq	Sm	LB	1.82	8.80	---	49.90	---	49.90	**74.40**
Without gravel	Demo	Sq	Lg	LB	1.14	14.00	---	31.30	---	31.30	**46.60**
	Demo	Sq	Sm	LB	1.43	11.20	---	39.20	---	39.20	**58.40**

5 ply

Description	Oper	Unit	Vol	Crew Size	MH	CO	Mat'l	Labor	Equip	Total	Price
With gravel	Demo	Sq	Lg	LB	1.60	10.00	---	43.90	---	43.90	**65.40**
	Demo	Sq	Sm	LB	2.00	8.00	---	54.90	---	54.90	**81.70**
Without gravel	Demo	Sq	Lg	LB	1.23	13.00	---	33.70	---	33.70	**50.30**
	Demo	Sq	Sm	LB	1.54	10.40	---	42.20	---	42.20	**62.90**

Install over smooth wood decks
Nail one 15 lb felt ply, plus mop in one 90 lb mineral surfaced ply

Description	Oper	Unit	Vol	Crew Size	MH	CO	Mat'l	Labor	Equip	Total	Price
With asphalt	Inst	Sq	Lg	RS	.600	80.00	34.40	21.00	---	55.40	**72.10**
	Inst	Sq	Sm	RS	.750	64.00	37.50	26.30	---	63.80	**83.80**
With tar	Inst	Sq	Lg	RS	.600	80.00	39.70	21.00	---	60.70	**78.30**
	Inst	Sq	Sm	RS	.750	64.00	43.30	26.30	---	69.60	**90.50**

Nail one and mop in two 15 lb felt plies, plus smooth top mop coat

Description	Oper	Unit	Vol	Crew Size	MH	CO	Mat'l	Labor	Equip	Total	Price
With asphalt	Inst	Sq	Lg	RS	1.45	33.00	38.30	50.80	---	89.10	**123.00**
	Inst	Sq	Sm	RS	1.82	26.40	41.60	63.70	---	105.30	**147.00**
With tar	Inst	Sq	Lg	RS	1.45	33.00	55.40	50.80	---	106.20	**142.00**
	Inst	Sq	Sm	RS	1.82	26.40	60.40	63.70	---	124.10	**168.00**

Nail one and mop in two 15 lb felt plies, plus mop in and distribute gravel

Description	Oper	Unit	Vol	Crew Size	MH	CO	Mat'l	Labor	Equip	Total	Price
With asphalt	Inst	Sq	Lg	RS	1.92	25.00	64.30	67.20	---	131.50	**178.00**
	Inst	Sq	Sm	RS	2.40	20.00	68.10	84.10	---	152.20	**209.00**
With tar	Inst	Sq	Lg	RS	1.92	25.00	85.60	67.20	---	152.80	**203.00**
	Inst	Sq	Sm	RS	2.40	20.00	91.60	84.10	---	175.70	**236.00**

Nail one and mop in three 15 lb felt plies, plus mop in and distribute gravel

Description	Oper	Unit	Vol	Crew Size	MH	CO	Mat'l	Labor	Equip	Total	Price
With asphalt	Inst	Sq	Lg	RS	2.40	20.00	75.00	84.10	---	159.10	**216.00**
	Inst	Sq	Sm	RS	3.00	16.00	79.70	105.00	---	184.70	**254.00**
With tar	Inst	Sq	Lg	RS	2.40	20.00	101.00	84.10	---	185.10	**246.00**
	Inst	Sq	Sm	RS	3.00	16.00	108.00	105.00	---	213.00	**287.00**

Nail one and mop in four 15 lb plies, plus mop in and distribute gravel

Description	Oper	Unit	Vol	Crew Size	MH	CO	Mat'l	Labor	Equip	Total	Price
With asphalt	Inst	Sq	Lg	RS	3.00	16.00	85.70	105.00	---	190.70	**261.00**
	Inst	Sq	Sm	RS	3.75	12.80	91.30	131.00	---	222.30	**309.00**
With tar	Inst	Sq	Lg	RS	3.00	16.00	116.00	105.00	---	221.00	**296.00**
	Inst	Sq	Sm	RS	3.75	12.80	125.00	131.00	---	256.00	**347.00**

Description	Oper	Unit	Vol	Crew Size	Man-hours per Unit	Crew Output per Day	Avg Mat'l Unit Cost	Avg Labor Unit Cost	Avg Equip Unit Cost	Avg Total Unit Cost	Avg Price Incl O&P
Install over smooth concrete decks											
Nail one 15 lb felt ply, plus mop in one 90 lb mineral surfaced ply											
With asphalt	Inst	Sq	Lg	RS	.774	62.00	40.40	27.10	---	67.50	**88.50**
	Inst	Sq	Sm	RS	.968	49.60	44.00	33.90	---	77.90	**103.00**
With tar	Inst	Sq	Lg	RS	.774	62.00	50.00	27.10	---	77.10	**99.50**
	Inst	Sq	Sm	RS	.968	49.60	54.50	33.90	---	88.40	**115.00**
Nail one and mop in two 15 lb felt plies, plus smooth top mop coat											
With asphalt	Inst	Sq	Lg	RS	1.78	27.00	42.90	62.30	---	105.20	**146.00**
	Inst	Sq	Sm	RS	2.22	21.60	46.50	77.70	---	124.20	**174.00**
With tar	Inst	Sq	Lg	RS	1.78	27.00	64.60	62.30	---	126.90	**171.00**
	Inst	Sq	Sm	RS	2.22	21.60	70.40	77.70	---	148.10	**201.00**
Nail one and mop in two 15 lb felt plies, plus mop in and distribute gravel											
With asphalt	Inst	Sq	Lg	RS	2.29	21.00	68.90	80.20	---	149.10	**203.00**
	Inst	Sq	Sm	RS	2.86	16.80	73.00	100.00	---	173.00	**239.00**
With tar	Inst	Sq	Lg	RS	2.29	21.00	94.80	80.20	---	175.00	**233.00**
	Inst	Sq	Sm	RS	2.86	16.80	102.00	100.00	---	202.00	**272.00**
Nail one and mop in three 15 lb felt plies, plus mop in and distribute gravel											
With asphalt	Inst	Sq	Lg	RS	2.82	17.00	79.60	98.80	---	178.40	**245.00**
	Inst	Sq	Sm	RS	3.53	13.60	84.60	124.00	---	208.60	**289.00**
With tar	Inst	Sq	Lg	RS	2.82	17.00	110.00	98.80	---	208.80	**280.00**
	Inst	Sq	Sm	RS	3.53	13.60	118.00	124.00	---	242.00	**328.00**
Nail one and mop in four 15 lb plies, plus mop in and distribute gravel											
With asphalt	Inst	Sq	Lg	RS	3.20	15.00	90.20	112.00	---	202.20	**277.00**
	Inst	Sq	Sm	RS	4.00	12.00	96.30	140.00	---	236.30	**328.00**
With tar	Inst	Sq	Lg	RS	3.20	15.00	125.00	112.00	---	237.00	**318.00**
	Inst	Sq	Sm	RS	4.00	12.00	135.00	140.00	---	275.00	**372.00**

Tile, clay or concrete
Demo tile over wood

Description	Oper	Unit	Vol	Crew Size	Man-hours per Unit	Crew Output per Day	Avg Mat'l Unit Cost	Avg Labor Unit Cost	Avg Equip Unit Cost	Avg Total Unit Cost	Avg Price Incl O&P
2 piece (interlocking)	Demo	Sq	Lg	LB	1.60	10.00	---	43.90	---	43.90	**65.40**
	Demo	Sq	Sm	LB	2.00	8.00	---	54.90	---	54.90	**81.70**
1 piece	Demo	Sq	Lg	LB	1.45	11.00	---	39.80	---	39.80	**59.30**
	Demo	Sq	Sm	LB	1.82	8.80	---	49.90	---	49.90	**74.40**

Install tile over wood; includes felt

Description	Oper	Unit	Vol	Crew Size	Man-hours per Unit	Crew Output per Day	Avg Mat'l Unit Cost	Avg Labor Unit Cost	Avg Equip Unit Cost	Avg Total Unit Cost	Avg Price Incl O&P
2 piece (interlocking)											
Red	Inst	Sq	Lg	RG	4.44	5.40	200.00	150.00	---	350.00	**463.00**
	Inst	Sq	Sm	RG	5.56	4.32	218.00	188.00	---	406.00	**543.00**
Other colors	Inst	Sq	Lg	RG	4.44	5.40	247.00	150.00	---	397.00	**517.00**
	Inst	Sq	Sm	RG	5.56	4.32	269.00	188.00	---	457.00	**601.00**
Tile shingles											
Mission	Inst	Sq	Lg	RG	8.00	3.00	109.00	271.00	---	380.00	**545.00**
	Inst	Sq	Sm	RG	10.0	2.40	119.00	338.00	---	457.00	**661.00**
Spanish	Inst	Sq	Lg	RG	8.00	3.00	133.00	271.00	---	404.00	**573.00**
	Inst	Sq	Sm	RG	10.0	2.40	145.00	338.00	---	483.00	**691.00**

Description	Oper	Unit	Vol	Crew Size	Man-hours per Unit	Crew Output per Day	Avg Mat'l Unit Cost	Avg Labor Unit Cost	Avg Equip Unit Cost	Avg Total Unit Cost	Avg Price Incl O&P

Mineral surfaced roll

Single coverage 90 lb/sq roll, with 6" end lap and 2" head lap

| | Demo | Sq | Lg | LB | .500 | 32.00 | --- | 13.70 | --- | 13.70 | **20.40** |
| | Demo | Sq | Sm | LB | .625 | 25.60 | --- | 17.10 | --- | 17.10 | **25.50** |

Single coverage roll on

Plain gable over

Existing roofing with

Nails concealed	Inst	Sq	Lg	2R	.696	23.00	26.00	25.80	---	51.80	**69.90**
	Inst	Sq	Sm	2R	.870	18.40	28.30	32.20	---	60.50	**82.60**
Nails exposed	Inst	Sq	Lg	2R	.640	25.00	23.60	23.70	---	47.30	**63.90**
	Inst	Sq	Sm	2R	.800	20.00	25.80	29.60	---	55.40	**75.60**

Wood deck with

Nails concealed	Inst	Sq	Lg	2R	.727	22.00	26.70	26.90	---	53.60	**72.50**
	Inst	Sq	Sm	2R	.909	17.60	29.20	33.70	---	62.90	**85.70**
Nails exposed	Inst	Sq	Lg	2R	.667	24.00	24.30	24.70	---	49.00	**66.30**
	Inst	Sq	Sm	2R	.833	19.20	26.50	30.90	---	57.40	**78.30**

Plain hip over

Existing roofing with

Nails concealed	Inst	Sq	Lg	2R	.727	22.00	26.70	26.90	---	53.60	**72.50**
	Inst	Sq	Sm	2R	.909	17.60	29.20	33.70	---	62.90	**85.70**
Nails exposed	Inst	Sq	Lg	2R	.667	24.00	24.30	24.70	---	49.00	**66.30**
	Inst	Sq	Sm	2R	.833	19.20	26.50	30.90	---	57.40	**78.30**

Wood deck with

Nails concealed	Inst	Sq	Lg	2R	.800	20.00	27.20	29.60	---	56.80	**77.30**
	Inst	Sq	Sm	2R	1.00	16.00	29.70	37.10	---	66.80	**91.60**
Nails exposed	Inst	Sq	Lg	2R	.696	23.00	24.80	25.80	---	50.60	**68.50**
	Inst	Sq	Sm	2R	.870	18.40	27.00	32.20	---	59.20	**81.00**

Double coverage selvage roll, with 19" lap and 17" exposure

| | Demo | Sq | Lg | LB | .727 | 22.00 | --- | 19.90 | --- | 19.90 | **29.70** |
| | Demo | Sq | Sm | LB | .909 | 17.60 | --- | 24.90 | --- | 24.90 | **37.20** |

Double coverage selvage roll, with 19" lap and 17" exposure (2 rolls/Sq)

Plain gable over

Wood deck with nails exposed

| | Inst | Sq | Lg | 2R | 1.14 | 14.00 | 54.40 | 42.20 | --- | 96.60 | **128.00** |
| | Inst | Sq | Sm | 2R | 1.43 | 11.20 | 59.30 | 53.00 | --- | 112.30 | **150.00** |

Related materials and operations

Starter strips (36' L rolls) along eaves and up rakes and gable ends on new roofing over wood decks

9" W	Inst	LF	Lg	---	---	---	.45	---	---	.45	**.52**
	Inst	LF	Sm	---	---	---	.49	---	---	.49	**.56**
12" W	Inst	LF	Lg	---	---	---	.56	---	---	.56	**.64**
	Inst	LF	Sm	---	---	---	.61	---	---	.61	**.70**
18" W	Inst	LF	Lg	---	---	---	.73	---	---	.73	**.84**
	Inst	LF	Sm	---	---	---	.80	---	---	.80	**.92**
24" W	Inst	LF	Lg	---	---	---	.90	---	---	.90	**1.04**
	Inst	LF	Sm	---	---	---	.98	---	---	.98	**1.13**

Description	Oper	Unit	Vol	Crew Size	Man-hours per Unit	Crew Output per Day	Avg Mat'l Unit Cost	Avg Labor Unit Cost	Avg Equip Unit Cost	Avg Total Unit Cost	Avg Price Incl O&P

Wood
Shakes
24" L with 10" exposure

Description	Oper	Unit	Vol	Crew Size	Man-hours per Unit	Crew Output per Day	Avg Mat'l Unit Cost	Avg Labor Unit Cost	Avg Equip Unit Cost	Avg Total Unit Cost	Avg Price Incl O&P
1/2" to 3/4" T	Demo	Sq	Lg	LB	.640	25.00	---	17.60	---	17.60	**26.20**
	Demo	Sq	Sm	LB	.800	20.00	---	21.90	---	21.90	**32.70**
3/4" to 5/4" T	Demo	Sq	Lg	LB	.711	22.50	---	19.50	---	19.50	**29.10**
	Demo	Sq	Sm	LB	.889	18.00	---	24.40	---	24.40	**36.30**

Shakes, over wood deck; 2 nails/shake, 24" L with 10" exp., 6" W (avg.), red cedar, sawn one side
Gable

Description	Oper	Unit	Vol	Crew Size	Man-hours per Unit	Crew Output per Day	Avg Mat'l Unit Cost	Avg Labor Unit Cost	Avg Equip Unit Cost	Avg Total Unit Cost	Avg Price Incl O&P
1/2" to 3/4" T	Inst	Sq	Lg	RQ	1.83	12.60	183.00	64.80	---	247.80	**310.00**
	Inst	Sq	Sm	RQ	2.28	10.08	199.00	80.70	---	279.70	**354.00**
3/4" to 5/4" T	Inst	Sq	Lg	RQ	2.04	11.30	264.00	72.20	---	336.20	**415.00**
	Inst	Sq	Sm	RQ	2.54	9.04	288.00	89.90	---	377.90	**470.00**

Gable with dormers

Description	Oper	Unit	Vol	Crew Size	Man-hours per Unit	Crew Output per Day	Avg Mat'l Unit Cost	Avg Labor Unit Cost	Avg Equip Unit Cost	Avg Total Unit Cost	Avg Price Incl O&P
1/2" to 3/4" T	Inst	Sq	Lg	RQ	1.92	12.00	186.00	68.00	---	254.00	**319.00**
	Inst	Sq	Sm	RQ	2.40	9.60	203.00	85.00	---	288.00	**365.00**
3/4" to 5/4" T	Inst	Sq	Lg	RQ	2.15	10.70	269.00	76.10	---	345.10	**427.00**
	Inst	Sq	Sm	RQ	2.69	8.56	293.00	95.30	---	388.30	**485.00**

Gable with valleys

Description	Oper	Unit	Vol	Crew Size	Man-hours per Unit	Crew Output per Day	Avg Mat'l Unit Cost	Avg Labor Unit Cost	Avg Equip Unit Cost	Avg Total Unit Cost	Avg Price Incl O&P
1/2" to 3/4" T	Inst	Sq	Lg	RQ	1.92	12.00	186.00	68.00	---	254.00	**319.00**
	Inst	Sq	Sm	RQ	2.40	9.60	203.00	85.00	---	288.00	**365.00**
3/4" to 5/4" T	Inst	Sq	Lg	RQ	2.15	10.70	269.00	76.10	---	345.10	**427.00**
	Inst	Sq	Sm	RQ	2.69	8.56	293.00	95.30	---	388.30	**485.00**

Hip

Description	Oper	Unit	Vol	Crew Size	Man-hours per Unit	Crew Output per Day	Avg Mat'l Unit Cost	Avg Labor Unit Cost	Avg Equip Unit Cost	Avg Total Unit Cost	Avg Price Incl O&P
1/2" to 3/4" T	Inst	Sq	Lg	RQ	1.92	12.00	186.00	68.00	---	254.00	**319.00**
	Inst	Sq	Sm	RQ	2.40	9.60	203.00	85.00	---	288.00	**365.00**
3/4" to 5/4" T	Inst	Sq	Lg	RQ	2.15	10.70	269.00	76.10	---	345.10	**427.00**
	Inst	Sq	Sm	RQ	2.69	8.56	293.00	95.30	---	388.30	**485.00**

Hip with valleys

Description	Oper	Unit	Vol	Crew Size	Man-hours per Unit	Crew Output per Day	Avg Mat'l Unit Cost	Avg Labor Unit Cost	Avg Equip Unit Cost	Avg Total Unit Cost	Avg Price Incl O&P
1/2" to 3/4" T	Inst	Sq	Lg	RQ	2.04	11.30	190.00	72.20	---	262.20	**330.00**
	Inst	Sq	Sm	RQ	2.54	9.04	207.00	89.90	---	296.90	**377.00**
3/4" to 5/4" T	Inst	Sq	Lg	RQ	2.25	10.20	274.00	79.70	---	353.70	**438.00**
	Inst	Sq	Sm	RQ	2.82	8.16	298.00	99.90	---	397.90	**498.00**

Related materials and operations
Ridge/hip units, 10" W with 10" exp., 20 pieces / bundle

Description	Oper	Unit	Vol	Crew Size	Man-hours per Unit	Crew Output per Day	Avg Mat'l Unit Cost	Avg Labor Unit Cost	Avg Equip Unit Cost	Avg Total Unit Cost	Avg Price Incl O&P
	Inst	LF	Lg	RJ	.018	900.0	3.41	.70	---	4.11	**5.01**
	Inst	LF	Sm	RJ	.022	720.0	3.72	.86	---	4.58	**5.60**

Roll valley, galvanized, unpainted, 28 gauge, 50' L rolls

Description	Oper	Unit	Vol	Crew Size	Man-hours per Unit	Crew Output per Day	Avg Mat'l Unit Cost	Avg Labor Unit Cost	Avg Equip Unit Cost	Avg Total Unit Cost	Avg Price Incl O&P
14" W	Inst	LF	Lg	UA	.067	120.0	1.83	2.48	---	4.31	**5.77**
	Inst	LF	Sm	UA	.083	96.00	1.99	3.07	---	5.06	**6.83**
20" W	Inst	LF	Lg	UA	.067	120.0	1.97	2.48	---	4.45	**5.93**
	Inst	LF	Sm	UA	.083	96.00	2.15	3.07	---	5.22	**7.01**

Rosin sized sheathing paper 36" W, 500 SF/roll; nailed

Description	Oper	Unit	Vol	Crew Size	Man-hours per Unit	Crew Output per Day	Avg Mat'l Unit Cost	Avg Labor Unit Cost	Avg Equip Unit Cost	Avg Total Unit Cost	Avg Price Incl O&P
Over open sheathing	Inst	Sq	Lg	RJ	.168	95.00	2.16	6.54	---	8.70	**12.60**
	Inst	Sq	Sm	RJ	.211	76.00	2.36	8.21	---	10.57	**15.40**
Over solid sheathing	Inst	Sq	Lg	RJ	.114	140.0	2.16	4.43	---	6.59	**9.36**
	Inst	Sq	Sm	RJ	.143	112.0	2.36	5.56	---	7.92	**11.30**

Description	Oper	Unit	Vol	Crew Size	Man-hours per Unit	Crew Output per Day	Avg Mat'l Unit Cost	Avg Labor Unit Cost	Avg Equip Unit Cost	Avg Total Unit Cost	Avg Price Incl O&P

Shingles

Description	Oper	Unit	Vol	Crew Size	Man-hours	Output	Mat'l	Labor	Equip	Total	Price
16" L with 5" exposure	Demo	Sq	Lg	LB	1.33	12.00	---	36.50	---	36.50	**54.40**
	Demo	Sq	Sm	LB	1.67	9.60	---	45.80	---	45.80	**68.30**
18" L with 5-1/2" exposure	Demo	Sq	Lg	LB	1.27	12.60	---	34.80	---	34.80	**51.90**
	Demo	Sq	Sm	LB	1.59	10.08	---	43.60	---	43.60	**65.00**
24" L with 7-1/2" exposure	Demo	Sq	Lg	LB	.889	18.00	---	24.40	---	24.40	**36.30**
	Demo	Sq	Sm	LB	1.11	14.40	---	30.50	---	30.50	**45.40**

Shingles, red cedar, No. 1 perfect, 4" W (avg.), 2 nails per shingle

Over existing roofing materials on
Gable, plain

Description	Oper	Unit	Vol	Crew Size	Man-hours	Output	Mat'l	Labor	Equip	Total	Price
16" L with 5" exposure	Inst	Sq	Lg	RM	3.57	5.60	270.00	131.00	---	401.00	**513.00**
	Inst	Sq	Sm	RM	4.46	4.48	294.00	163.00	---	457.00	**591.00**
18" L with 5-1/2" exposure	Inst	Sq	Lg	RM	3.23	6.20	300.00	118.00	---	418.00	**529.00**
	Inst	Sq	Sm	RM	4.03	4.96	327.00	148.00	---	475.00	**605.00**
24" L with 7-1/2" exposure	Inst	Sq	Lg	RM	2.33	8.60	281.00	85.30	---	366.30	**455.00**
	Inst	Sq	Sm	RM	2.91	6.88	306.00	107.00	---	413.00	**517.00**

Gable with dormers

Description	Oper	Unit	Vol	Crew Size	Man-hours	Output	Mat'l	Labor	Equip	Total	Price
16" L with 5" exposure	Inst	Sq	Lg	RM	3.70	5.40	272.00	135.00	---	407.00	**523.00**
	Inst	Sq	Sm	RM	4.63	4.32	297.00	169.00	---	466.00	**604.00**
18" L with 5-1/2" exposure	Inst	Sq	Lg	RM	3.33	6.00	303.00	122.00	---	425.00	**537.00**
	Inst	Sq	Sm	RM	4.17	4.80	330.00	153.00	---	483.00	**616.00**
24" L with 7-1/2" exposure	Inst	Sq	Lg	RM	3.17	6.30	283.00	116.00	---	399.00	**505.00**
	Inst	Sq	Sm	RM	3.97	5.04	309.00	145.00	---	454.00	**580.00**

Gable with intersecting roofs

Description	Oper	Unit	Vol	Crew Size	Man-hours	Output	Mat'l	Labor	Equip	Total	Price
16" L with 5" exposure	Inst	Sq	Lg	RM	3.70	5.40	272.00	135.00	---	407.00	**523.00**
	Inst	Sq	Sm	RM	4.63	4.32	297.00	169.00	---	466.00	**604.00**
18" L with 5-1/2" exposure	Inst	Sq	Lg	RM	3.33	6.00	303.00	122.00	---	425.00	**537.00**
	Inst	Sq	Sm	RM	4.17	4.80	330.00	153.00	---	483.00	**616.00**
24" L with 7-1/2" exposure	Inst	Sq	Lg	RM	3.17	6.30	283.00	116.00	---	399.00	**505.00**
	Inst	Sq	Sm	RM	3.97	5.04	309.00	145.00	---	454.00	**580.00**

Gable with dormers & intersecting roofs

Description	Oper	Unit	Vol	Crew Size	Man-hours	Output	Mat'l	Labor	Equip	Total	Price
16" L with 5" exposure	Inst	Sq	Lg	RM	3.85	5.20	277.00	141.00	---	418.00	**537.00**
	Inst	Sq	Sm	RM	4.81	4.16	302.00	176.00	---	478.00	**620.00**
18" L with 5-1/2" exposure	Inst	Sq	Lg	RM	3.45	5.80	300.00	126.00	---	434.00	**550.00**
	Inst	Sq	Sm	RM	4.31	4.64	336.00	158.00	---	494.00	**631.00**
24" L with 7-1/2" exposure	Inst	Sq	Lg	RM	2.47	8.10	288.00	90.40	---	378.40	**471.00**
	Inst	Sq	Sm	RM	3.09	6.48	314.00	113.00	---	427.00	**536.00**

Hip, plain

Description	Oper	Unit	Vol	Crew Size	Man-hours	Output	Mat'l	Labor	Equip	Total	Price
16" L with 5" exposure	Inst	Sq	Lg	RM	3.70	5.40	272.00	135.00	---	407.00	**523.00**
	Inst	Sq	Sm	RM	4.63	4.32	297.00	169.00	---	466.00	**604.00**
18" L with 5-1/2" exposure	Inst	Sq	Lg	RM	3.33	6.00	303.00	122.00	---	425.00	**537.00**
	Inst	Sq	Sm	RM	4.17	4.80	330.00	153.00	---	483.00	**616.00**
24" L with 7-1/2" exposure	Inst	Sq	Lg	RM	2.41	8.30	283.00	88.20	---	371.20	**462.00**
	Inst	Sq	Sm	RM	3.01	6.64	309.00	110.00	---	419.00	**526.00**

Description	Oper	Unit	Vol	Crew Size	Man-hours per Unit	Crew Output per Day	Avg Mat'l Unit Cost	Avg Labor Unit Cost	Avg Equip Unit Cost	Avg Total Unit Cost	Avg Price Incl O&P
Hip with dormers											
16" L with 5" exposure	Inst	Sq	Lg	RM	3.85	5.20	275.00	141.00	---	416.00	**534.00**
	Inst	Sq	Sm	RM	4.81	4.16	299.00	176.00	---	475.00	**617.00**
18" L with 5-1/2" exposure	Inst	Sq	Lg	RM	3.51	5.70	306.00	128.00	---	434.00	**551.00**
	Inst	Sq	Sm	RM	4.39	4.56	333.00	161.00	---	494.00	**632.00**
24" L with 7-1/2" exposure	Inst	Sq	Lg	RM	2.53	7.90	286.00	92.60	---	378.60	**472.00**
	Inst	Sq	Sm	RM	3.16	6.32	311.00	116.00	---	427.00	**537.00**
Hip with intersecting roofs											
16" L with 5" exposure	Inst	Sq	Lg	RM	3.85	5.20	275.00	141.00	---	416.00	**534.00**
	Inst	Sq	Sm	RM	4.81	4.16	299.00	176.00	---	475.00	**617.00**
18" L with 5-1/2" exposure	Inst	Sq	Lg	RM	3.51	5.70	306.00	128.00	---	434.00	**551.00**
	Inst	Sq	Sm	RM	4.39	4.56	333.00	161.00	---	494.00	**632.00**
24" L with 7-1/2" exposure	Inst	Sq	Lg	RM	2.53	7.90	286.00	92.60	---	378.60	**472.00**
	Inst	Sq	Sm	RM	3.16	6.32	311.00	116.00	---	427.00	**537.00**
Hip with dormers & intersecting roofs											
16" L with 5" exposure	Inst	Sq	Lg	RM	4.00	5.00	279.00	146.00	---	425.00	**548.00**
	Inst	Sq	Sm	RM	5.00	4.00	305.00	183.00	---	488.00	**634.00**
18" L with 5-1/2" exposure	Inst	Sq	Lg	RM	3.64	5.50	311.00	133.00	---	444.00	**564.00**
	Inst	Sq	Sm	RM	4.55	4.40	339.00	167.00	---	506.00	**648.00**
24" L with 7-1/2" exposure	Inst	Sq	Lg	RM	2.60	7.70	291.00	95.20	---	386.20	**482.00**
	Inst	Sq	Sm	RM	3.25	6.16	317.00	119.00	---	436.00	**549.00**
Over wood decks on											
Gable, plain											
16" L with 5" exposure	Inst	Sq	Lg	RM	3.33	6.00	266.00	122.00	---	388.00	**494.00**
	Inst	Sq	Sm	RM	4.17	4.80	289.00	153.00	---	442.00	**569.00**
18" L with 5-1/2" exposure	Inst	Sq	Lg	RM	3.03	6.60	296.00	111.00	---	407.00	**513.00**
	Inst	Sq	Sm	RM	3.79	5.28	323.00	139.00	---	462.00	**587.00**
24" L with 7-1/2" exposure	Inst	Sq	Lg	RM	2.22	9.00	277.00	81.30	---	358.30	**445.00**
	Inst	Sq	Sm	RM	2.78	7.20	302.00	102.00	---	404.00	**505.00**
Gable with dormers											
16" L with 5" exposure	Inst	Sq	Lg	RM	3.45	5.80	268.00	126.00	---	394.00	**504.00**
	Inst	Sq	Sm	RM	4.31	4.64	292.00	158.00	---	450.00	**580.00**
18" L with 5-1/2" exposure	Inst	Sq	Lg	RM	3.13	6.40	299.00	115.00	---	414.00	**521.00**
	Inst	Sq	Sm	RM	3.91	5.12	326.00	143.00	---	469.00	**597.00**
24" L with 7-1/2" exposure	Inst	Sq	Lg	RM	2.30	8.70	280.00	84.20	---	364.20	**452.00**
	Inst	Sq	Sm	RM	2.87	6.96	305.00	105.00	---	410.00	**513.00**
Gable with intersecting roofs											
16" L with 5" exposure	Inst	Sq	Lg	RM	3.45	5.80	268.00	126.00	---	394.00	**504.00**
	Inst	Sq	Sm	RM	4.31	4.64	292.00	158.00	---	450.00	**580.00**
18" L with 5-1/2" exposure	Inst	Sq	Lg	RM	3.13	6.40	299.00	115.00	---	414.00	**521.00**
	Inst	Sq	Sm	RM	3.91	5.12	326.00	143.00	---	469.00	**597.00**
24" L with 7-1/2" exposure	Inst	Sq	Lg	RM	2.30	8.70	280.00	84.20	---	364.20	**452.00**
	Inst	Sq	Sm	RM	2.87	6.96	305.00	105.00	---	410.00	**513.00**

Description	Oper	Unit	Vol	Crew Size	Man-hours per Unit	Crew Output per Day	Avg Mat'l Unit Cost	Avg Labor Unit Cost	Avg Equip Unit Cost	Avg Total Unit Cost	Avg Price Incl O&P
Gable with dormers & intersecting roofs											
16" L with 5" exposure	Inst	Sq	Lg	RM	3.57	5.60	273.00	131.00	---	404.00	**516.00**
	Inst	Sq	Sm	RM	4.46	4.48	297.00	163.00	---	460.00	**595.00**
18" L with 5-1/2" exposure	Inst	Sq	Lg	RM	3.23	6.20	304.00	118.00	---	422.00	**533.00**
	Inst	Sq	Sm	RM	4.03	4.96	332.00	148.00	---	480.00	**610.00**
24" L with 7-1/2" exposure	Inst	Sq	Lg	RM	2.35	8.50	285.00	86.00	---	371.00	**461.00**
	Inst	Sq	Sm	RM	2.94	6.80	310.00	108.00	---	418.00	**523.00**
Hip, plain											
16" L with 5" exposure	Inst	Sq	Lg	RM	3.45	5.80	268.00	126.00	---	394.00	**504.00**
	Inst	Sq	Sm	RM	4.31	4.64	292.00	158.00	---	450.00	**580.00**
18" L with 5-1/2" exposure	Inst	Sq	Lg	RM	3.13	6.40	299.00	115.00	---	414.00	**521.00**
	Inst	Sq	Sm	RM	3.91	5.12	326.00	143.00	---	469.00	**597.00**
24" L with 7-1/2" exposure	Inst	Sq	Lg	RM	2.30	8.70	280.00	84.20	---	364.20	**452.00**
	Inst	Sq	Sm	RM	2.87	6.96	305.00	105.00	---	410.00	**513.00**
Hip with dormers											
16" L with 5" exposure	Inst	Sq	Lg	RM	3.57	5.60	270.00	131.00	---	401.00	**513.00**
	Inst	Sq	Sm	RM	4.46	4.48	295.00	163.00	---	458.00	**592.00**
18" L with 5-1/2" exposure	Inst	Sq	Lg	RM	3.28	6.10	302.00	120.00	---	422.00	**533.00**
	Inst	Sq	Sm	RM	4.10	4.88	329.00	150.00	---	479.00	**611.00**
24" L with 7-1/2" exposure	Inst	Sq	Lg	RM	2.41	8.30	282.00	88.20	---	370.20	**461.00**
	Inst	Sq	Sm	RM	3.01	6.64	307.00	110.00	---	417.00	**524.00**
Hip with intersecting roofs											
16" L with 5" exposure	Inst	Sq	Lg	RM	3.57	5.60	270.00	131.00	---	401.00	**513.00**
	Inst	Sq	Sm	RM	4.46	4.48	295.00	163.00	---	458.00	**592.00**
18" L with 5-1/2" exposure	Inst	Sq	Lg	RM	3.28	6.10	302.00	120.00	---	422.00	**533.00**
	Inst	Sq	Sm	RM	4.10	4.88	329.00	150.00	---	479.00	**611.00**
24" L with 7-1/2" exposure	Inst	Sq	Lg	RM	2.41	8.30	282.00	88.20	---	370.20	**461.00**
	Inst	Sq	Sm	RM	3.01	6.64	307.00	110.00	---	417.00	**524.00**
Hip with dormers & intersecting roofs											
16" L with 5" exposure	Inst	Sq	Lg	RM	3.70	5.40	275.00	135.00	---	410.00	**526.00**
	Inst	Sq	Sm	RM	4.63	4.32	300.00	169.00	---	469.00	**607.00**
18" L with 5-1/2" exposure	Inst	Sq	Lg	RM	3.39	5.90	307.00	124.00	---	431.00	**545.00**
	Inst	Sq	Sm	RM	4.24	4.72	335.00	155.00	---	490.00	**625.00**
24" L with 7-1/2" exposure	Inst	Sq	Lg	RM	2.47	8.10	287.00	90.40	---	377.40	**470.00**
	Inst	Sq	Sm	RM	3.09	6.48	313.00	113.00	---	426.00	**535.00**

Description	Oper	Unit	Vol	Crew Size	Man-hours per Unit	Crew Output per Day	Avg Mat'l Unit Cost	Avg Labor Unit Cost	Avg Equip Unit Cost	Avg Total Unit Cost	Avg Price Incl O&P
Related materials and operations											
Ridge/hip units, 40 pieces / bundle											
Over existing roofing											
5" exposure	Inst	LF	Lg	RJ	.029	560.0	2.88	1.13	---	4.01	**5.06**
	Inst	LF	Sm	RJ	.036	448.0	3.14	1.40	---	4.54	**5.78**
5-1/2" exposure	Inst	LF	Lg	RJ	.027	590.0	2.88	1.05	---	3.93	**4.94**
	Inst	LF	Sm	RJ	.034	472.0	3.14	1.32	---	4.46	**5.66**
7-1/2" exposure	Inst	LF	Lg	RJ	.023	700.0	1.93	.89	---	2.82	**3.61**
	Inst	LF	Sm	RJ	.029	560.0	2.10	1.13	---	3.23	**4.16**
Over wood decks											
5" exposure	Inst	LF	Lg	RJ	.027	600.0	2.88	1.05	---	3.93	**4.94**
	Inst	LF	Sm	RJ	.033	480.0	3.14	1.28	---	4.42	**5.60**
5-1/2" exposure	Inst	LF	Lg	RJ	.025	630.0	2.88	.97	---	3.85	**4.82**
	Inst	LF	Sm	RJ	.032	504.0	3.14	1.24	---	4.38	**5.54**
7-1/2" exposure	Inst	LF	Lg	RJ	.021	750.0	1.93	.82	---	2.75	**3.49**
	Inst	LF	Sm	RJ	.027	600.0	2.10	1.05	---	3.15	**4.04**
Roll valley, galvanized, 28 gauge 50' L rolls, unpainted											
18" W	Inst	LF	Lg	UA	.067	120.0	1.83	2.48	---	4.31	**5.77**
	Inst	LF	Sm	UA	.083	96.00	1.99	3.07	---	5.06	**6.83**
24" W	Inst	LF	Lg	UA	.067	120.0	1.97	2.48	---	4.45	**5.93**
	Inst	LF	Sm	UA	.083	96.00	2.15	3.07	---	5.22	**7.01**
Rosin sized sheathing paper, 36" W, 500 SF/roll; nailed											
Over open sheathing	Inst	Sq	Lg	RJ	.168	95.00	2.16	6.54	---	8.70	**12.60**
	Inst	Sq	Sm	RJ	.211	76.00	2.36	8.21	---	10.57	**15.40**
Over solid sheathing	Inst	Sq	Lg	RJ	.114	140.0	2.16	4.43	---	6.59	**9.36**
	Inst	Sq	Sm	RJ	.143	112.0	2.36	5.56	---	7.92	**11.30**
For No. 2 (red label) grade											
DEDUCT	Inst	%	Lg	---	---	---	-10.0	---	---	---	---
	Inst	%	Sm	---	---	---	-10.0	---	---	---	---

Sheathing. See Framing, page 210

Description	Oper	Unit	Vol	Crew Size	Man-hours per Unit	Crew Output per Day	Avg Mat'l Unit Cost	Avg Labor Unit Cost	Avg Equip Unit Cost	Avg Total Unit Cost	Avg Price Incl O&P

Sheet metal
Flashing, general; galvanized

Vertical chimney flashing
	Inst	LF	Lg	UA	.059	135.0	.41	2.18	---	2.59	**3.70**
	Inst	LF	Sm	UA	.079	101.0	.47	2.92	---	3.39	**4.86**

"Z" bar flashing
Standard	Inst	LF	Lg	UA	.019	420.0	.41	.70	---	1.11	**1.51**
	Inst	LF	Sm	UA	.025	315.0	.47	.92	---	1.39	**1.91**
Old style	Inst	LF	Lg	UA	.019	420.0	.41	.70	---	1.11	**1.51**
	Inst	LF	Sm	UA	.025	315.0	.47	.92	---	1.39	**1.91**
For plywood siding	Inst	LF	Lg	UA	.019	420.0	.31	.70	---	1.01	**1.40**
	Inst	LF	Sm	UA	.025	315.0	.35	.92	---	1.27	**1.77**

Rain diverter, 1" x 3"
4' long	Inst	Ea	Lg	UA	.160	50.00	2.34	5.91	---	8.25	**11.40**
	Inst	Ea	Sm	UA	.211	38.00	2.64	7.80	---	10.44	**14.60**
5' long	Inst	Ea	Lg	UA	.160	50.00	2.63	5.91	---	8.54	**11.80**
	Inst	Ea	Sm	UA	.211	38.00	2.97	7.80	---	10.77	**15.00**
10' long	Inst	Ea	Lg	UA	.229	35.00	4.43	8.46	---	12.89	**17.60**
	Inst	Ea	Sm	UA	.308	26.00	5.00	11.40	---	16.40	**22.60**

Roof flashing
Nosing, roof edging at 90-degree angle or open at 105-degree; galvanized

3/4" x 3/4"	Inst	LF	Lg	UA	.016	500.0	.19	.59	---	.78	**1.09**
	Inst	LF	Sm	UA	.021	375.0	.21	.78	---	.99	**1.39**
1" x 1"	Inst	LF	Lg	UA	.016	500.0	.19	.59	---	.78	**1.09**
	Inst	LF	Sm	UA	.021	375.0	.21	.78	---	.99	**1.39**
1" x 2"	Inst	LF	Lg	UA	.016	500.0	.23	.59	---	.82	**1.14**
	Inst	LF	Sm	UA	.021	375.0	.26	.78	---	1.04	**1.45**
1-1/2" x 1-1/2"	Inst	LF	Lg	UA	.016	500.0	.23	.59	---	.82	**1.14**
	Inst	LF	Sm	UA	.021	375.0	.26	.78	---	1.04	**1.45**
2" x 2"	Inst	LF	Lg	UA	.016	500.0	.28	.59	---	.87	**1.20**
	Inst	LF	Sm	UA	.021	375.0	.32	.78	---	1.10	**1.52**
2" x 3"	Inst	LF	Lg	UA	.016	500.0	.35	.59	---	.94	**1.28**
	Inst	LF	Sm	UA	.021	375.0	.39	.78	---	1.17	**1.60**
2" x 4"	Inst	LF	Lg	UA	.016	500.0	.48	.59	---	1.07	**1.43**
	Inst	LF	Sm	UA	.021	375.0	.54	.78	---	1.32	**1.77**
3" x 3"	Inst	LF	Lg	UA	.020	400.0	.48	.74	---	1.22	**1.65**
	Inst	LF	Sm	UA	.027	300.0	.54	1.00	---	1.54	**2.10**
3" x 4"	Inst	LF	Lg	UA	.020	400.0	.56	.74	---	1.30	**1.74**
	Inst	LF	Sm	UA	.027	300.0	.63	1.00	---	1.63	**2.20**
3" x 5"	Inst	LF	Lg	UA	.027	300.0	.67	1.00	---	1.67	**2.25**
	Inst	LF	Sm	UA	.036	225.0	.75	1.33	---	2.08	**2.83**
4" x 4"	Inst	LF	Lg	UA	.027	300.0	.67	1.00	---	1.67	**2.25**
	Inst	LF	Sm	UA	.036	225.0	.75	1.33	---	2.08	**2.83**
4" x 6"	Inst	LF	Lg	UA	.027	300.0	.84	1.00	---	1.84	**2.44**
	Inst	LF	Sm	UA	.036	225.0	.95	1.33	---	2.28	**3.06**
5" x 5"	Inst	LF	Lg	UA	.027	300.0	.84	1.00	---	1.84	**2.44**
	Inst	LF	Sm	UA	.036	225.0	.95	1.33	---	2.28	**3.06**
6" x 6"	Inst	LF	Lg	UA	.027	300.0	.98	1.00	---	1.98	**2.60**
	Inst	LF	Sm	UA	.036	225.0	1.11	1.33	---	2.44	**3.25**

Description	Oper	Unit	Vol	Crew Size	Man-hours per Unit	Crew Output per Day	Avg Mat'l Unit Cost	Avg Labor Unit Cost	Avg Equip Unit Cost	Avg Total Unit Cost	Avg Price Incl O&P
Roll valley, 50' rolls											
Galvanized, 28 gauge, unpainted											
8" wide	Inst	LF	Lg	UA	.027	300.0	.55	1.00	---	1.55	**2.11**
	Inst	LF	Sm	UA	.036	225.0	.62	1.33	---	1.95	**2.68**
10" wide	Inst	LF	Lg	UA	.027	300.0	.89	1.00	---	1.89	**2.50**
	Inst	LF	Sm	UA	.036	225.0	1.01	1.33	---	2.34	**3.13**
20" wide	Inst	LF	Lg	UA	.027	300.0	1.25	1.00	---	2.25	**2.91**
	Inst	LF	Sm	UA	.036	225.0	1.41	1.33	---	2.74	**3.59**
Tin seamless, painted											
8" wide	Inst	LF	Lg	UA	.027	300.0	.43	1.00	---	1.43	**1.97**
	Inst	LF	Sm	UA	.036	225.0	.48	1.33	---	1.81	**2.52**
10" wide	Inst	LF	Lg	UA	.027	300.0	.67	1.00	---	1.67	**2.25**
	Inst	LF	Sm	UA	.036	225.0	.75	1.33	---	2.08	**2.83**
20" wide	Inst	LF	Lg	UA	.027	300.0	.93	1.00	---	1.93	**2.55**
	Inst	LF	Sm	UA	.036	225.0	1.05	1.33	---	2.38	**3.18**
Aluminum seamless, .016 gauge											
8" wide	Inst	LF	Lg	UA	.027	300.0	.51	1.00	---	1.51	**2.06**
	Inst	LF	Sm	UA	.036	225.0	.57	1.33	---	1.90	**2.62**
10" wide	Inst	LF	Lg	UA	.027	300.0	.85	1.00	---	1.85	**2.45**
	Inst	LF	Sm	UA	.036	225.0	.96	1.33	---	2.29	**3.07**
20" wide	Inst	LF	Lg	UA	.027	300.0	1.21	1.00	---	2.21	**2.87**
	Inst	LF	Sm	UA	.036	225.0	1.37	1.33	---	2.70	**3.54**
"W" valley											
Galvanized hemmed tin											
18" girth	Inst	LF	Lg	UA	.027	300.0	1.32	1.00	---	2.32	**2.99**
	Inst	LF	Sm	UA	.036	225.0	1.49	1.33	---	2.82	**3.68**
24" girth	Inst	LF	Lg	UA	.027	300.0	1.74	1.00	---	2.74	**3.48**
	Inst	LF	Sm	UA	.036	225.0	1.97	1.33	---	3.30	**4.23**

Gravel stop

Description	Oper	Unit	Vol	Crew Size	Man-hours per Unit	Crew Output per Day	Avg Mat'l Unit Cost	Avg Labor Unit Cost	Avg Equip Unit Cost	Avg Total Unit Cost	Avg Price Incl O&P
4-1/2" galvanized	Inst	LF	Lg	UA	.020	400.0	.31	.74	---	1.05	**1.45**
	Inst	LF	Sm	UA	.027	300.0	.35	1.00	---	1.35	**1.88**
6" galvanized	Inst	LF	Lg	UA	.023	350.0	.40	.85	---	1.25	**1.72**
	Inst	LF	Sm	UA	.030	263.0	.45	1.11	---	1.56	**2.16**
7-1/4" galvanized	Inst	LF	Lg	UA	.027	300.0	.44	1.00	---	1.44	**1.98**
	Inst	LF	Sm	UA	.036	225.0	.50	1.33	---	1.83	**2.54**
4-1/2" bonderized	Inst	LF	Lg	UA	.020	400.0	.39	.74	---	1.13	**1.54**
	Inst	LF	Sm	UA	.027	300.0	.44	1.00	---	1.44	**1.98**
6" bonderized	Inst	LF	Lg	UA	.023	350.0	.52	.85	---	1.37	**1.86**
	Inst	LF	Sm	UA	.030	263.0	.59	1.11	---	1.70	**2.32**
7-1/4" bonderized	Inst	LF	Lg	UA	.027	300.0	.57	1.00	---	1.57	**2.13**
	Inst	LF	Sm	UA	.036	225.0	.65	1.33	---	1.98	**2.72**

Gutters and downspouts. See page 250

Description	Oper	Unit	Vol	Crew Size	Man-hours per Unit	Crew Output per Day	Avg Mat'l Unit Cost	Avg Labor Unit Cost	Avg Equip Unit Cost	Avg Total Unit Cost	Avg Price Incl O&P

Roof edging (drip edge) with 1/4" kickout, galvanized

Description	Oper	Unit	Vol	Crew Size	Man-hours per Unit	Crew Output per Day	Avg Mat'l Unit Cost	Avg Labor Unit Cost	Avg Equip Unit Cost	Avg Total Unit Cost	Avg Price Incl O&P
1" x 1-1/2"	Inst	LF	Lg	UA	.016	500.0	.24	.59	---	.83	1.15
	Inst	LF	Sm	UA	.021	375.0	.27	.78	---	1.05	1.46
1-1/2" x 1-1/2"	Inst	LF	Lg	UA	.016	500.0	.24	.59	---	.83	1.15
	Inst	LF	Sm	UA	.021	375.0	.27	.78	---	1.05	1.46
2" x 2"	Inst	LF	Lg	UA	.016	500.0	.29	.59	---	.88	1.21
	Inst	LF	Sm	UA	.021	375.0	.33	.78	---	1.11	1.53

Vents

Clothes dryer vent set, aluminum, hood, duct, inside plate

Description	Oper	Unit	Vol	Crew Size	Man-hours per Unit	Crew Output per Day	Avg Mat'l Unit Cost	Avg Labor Unit Cost	Avg Equip Unit Cost	Avg Total Unit Cost	Avg Price Incl O&P
3" diameter	Inst	Set	Lg	UA	.500	16.00	3.26	18.50	---	21.76	31.10
	Inst	Set	Sm	UA	.667	12.00	3.68	24.70	---	28.38	40.70
4" diameter	Inst	Set	Lg	UA	.500	16.00	3.59	18.50	---	22.09	31.50
	Inst	Set	Sm	UA	.667	12.00	4.05	24.70	---	28.75	41.10

Dormer louvers, half round, 1/8" or 1/4" mesh

18" x 9"

Description	Oper	Unit	Vol	Crew Size	Man-hours per Unit	Crew Output per Day	Avg Mat'l Unit Cost	Avg Labor Unit Cost	Avg Equip Unit Cost	Avg Total Unit Cost	Avg Price Incl O&P
Galvanized, 5-12 pitch	Inst	Ea	Lg	UA	.667	12.00	20.30	24.70	---	45.00	59.80
	Inst	Ea	Sm	UA	0.89	9.00	22.90	32.90	---	55.80	75.00
Painted, 3-12 pitch	Inst	Ea	Lg	UA	.667	12.00	23.90	24.70	---	48.60	64.00
	Inst	Ea	Sm	UA	0.89	9.00	27.00	32.90	---	59.90	79.70
24" x 12"											
Galvanized, 5-12 pitch	Inst	Ea	Lg	UA	.667	12.00	25.40	24.70	---	50.10	65.70
	Inst	Ea	Sm	UA	0.89	9.00	28.70	32.90	---	61.60	81.70
Painted, 3-12 pitch	Inst	Ea	Lg	UA	.667	12.00	28.90	24.70	---	53.60	69.70
	Inst	Ea	Sm	UA	0.89	9.00	32.50	32.90	---	65.40	86.10

Foundation vents, galvanized, 1/4" mesh

Description	Oper	Unit	Vol	Crew Size	Man-hours per Unit	Crew Output per Day	Avg Mat'l Unit Cost	Avg Labor Unit Cost	Avg Equip Unit Cost	Avg Total Unit Cost	Avg Price Incl O&P
6" x 14", stucco	Inst	Ea	Lg	UA	.333	24.00	1.56	12.30	---	13.86	20.00
	Inst	Ea	Sm	UA	.444	18.00	1.76	16.40	---	18.16	26.30
8" x 14", stucco	Inst	Ea	Lg	UA	.333	24.00	1.86	12.30	---	14.16	20.40
	Inst	Ea	Sm	UA	.444	18.00	2.10	16.40	---	18.50	26.70
6" x 14", two-way	Inst	Ea	Lg	UA	.333	24.00	1.86	12.30	---	14.16	20.40
	Inst	Ea	Sm	UA	.444	18.00	2.10	16.40	---	18.50	26.70
6" x 14", for siding (flat type)	Inst	Ea	Lg	UA	.333	24.00	1.56	12.30	---	13.86	20.00
	Inst	Ea	Sm	UA	.444	18.00	1.76	16.40	---	18.16	26.30
6" x 14", louver type	Inst	Ea	Lg	UA	.333	24.00	1.80	12.30	---	14.10	20.30
	Inst	Ea	Sm	UA	.444	18.00	2.03	16.40	---	18.43	26.60
6" x 14", foundation insert	Inst	Ea	Lg	UA	.333	24.00	1.56	12.30	---	13.86	20.00
	Inst	Ea	Sm	UA	.444	18.00	1.76	16.40	---	18.16	26.30

Description	Oper	Unit	Vol	Crew Size	Man-hours per Unit	Crew Output per Day	Avg Mat'l Unit Cost	Avg Labor Unit Cost	Avg Equip Unit Cost	Avg Total Unit Cost	Avg Price Incl O&P
Louver vents, round, 1/8" or 1/4" mesh, galvanized											
12" diameter	Inst	Ea	Lg	UA	.500	16.00	20.80	18.50	---	39.30	**51.20**
	Inst	Ea	Sm	UA	.667	12.00	23.40	24.70	---	48.10	**63.40**
14" diameter	Inst	Ea	Lg	UA	.500	16.00	22.00	18.50	---	40.50	**52.70**
	Inst	Ea	Sm	UA	.667	12.00	24.80	24.70	---	49.50	**65.00**
16" diameter	Inst	Ea	Lg	UA	.500	16.00	23.90	18.50	---	42.40	**54.80**
	Inst	Ea	Sm	UA	.667	12.00	26.90	24.70	---	51.60	**67.50**
18" diameter	Inst	Ea	Lg	UA	.500	16.00	27.40	18.50	---	45.90	**58.90**
	Inst	Ea	Sm	UA	.667	12.00	30.90	24.70	---	55.60	**72.00**
24" diameter	Inst	Ea	Lg	UA	.500	16.00	46.60	18.50	---	65.10	**80.90**
	Inst	Ea	Sm	UA	.667	12.00	52.50	24.70	---	77.20	**96.90**
Access doors											
Multi-purpose, galvanized											
24" x 18", nail-on type	Inst	Ea	Lg	UA	.500	16.00	13.20	18.50	---	31.70	**42.50**
	Inst	Ea	Sm	UA	.667	12.00	14.90	24.70	---	39.60	**53.60**
24" x 24", nail-on type	Inst	Ea	Lg	UA	.500	16.00	14.80	18.50	---	33.30	**44.40**
	Inst	Ea	Sm	UA	.667	12.00	16.70	24.70	---	41.40	**55.70**
24" x 18" x 6" deep box type	Inst	Ea	Lg	UA	.500	16.00	18.00	18.50	---	36.50	**48.00**
	Inst	Ea	Sm	UA	.667	12.00	20.30	24.70	---	45.00	**59.80**
Attic access doors, galvanized											
22" x 22"	Inst	Ea	Lg	UA	.667	12.00	18.60	24.70	---	43.30	**57.90**
	Inst	Ea	Sm	UA	0.89	9.00	21.00	32.90	---	53.90	**72.80**
22" x 30"	Inst	Ea	Lg	UA	.667	12.00	21.60	24.70	---	46.30	**61.30**
	Inst	Ea	Sm	UA	0.89	9.00	24.30	32.90	---	57.20	**76.60**
30" x 30"	Inst	Ea	Lg	UA	.667	12.00	22.70	24.70	---	47.40	**62.60**
	Inst	Ea	Sm	UA	0.89	9.00	25.70	32.90	---	58.60	**78.20**
Tub access doors, galvanized											
14" x 12"	Inst	Ea	Lg	UA	.500	16.00	9.24	18.50	---	27.74	**38.00**
	Inst	Ea	Sm	UA	.667	12.00	10.40	24.70	---	35.10	**48.50**
14" x 15"	Inst	Ea	Lg	UA	.500	16.00	9.58	18.50	---	28.08	**38.40**
	Inst	Ea	Sm	UA	.667	12.00	10.80	24.70	---	35.50	**48.90**
Utility louvers, aluminum or galvanized											
4" x 6"	Inst	Ea	Lg	UA	.200	40.00	1.10	7.39	---	8.49	**12.20**
	Inst	Ea	Sm	UA	.267	30.00	1.25	9.87	---	11.12	**16.00**
8" x 6"	Inst	Ea	Lg	UA	.200	40.00	1.56	7.39	---	8.95	**12.70**
	Inst	Ea	Sm	UA	.267	30.00	1.76	9.87	---	11.63	**16.60**
8" x 8"	Inst	Ea	Lg	UA	.200	40.00	1.80	7.39	---	9.19	**13.00**
	Inst	Ea	Sm	UA	.267	30.00	2.03	9.87	---	11.90	**16.90**
8" x 12"	Inst	Ea	Lg	UA	.200	40.00	2.10	7.39	---	9.49	**13.40**
	Inst	Ea	Sm	UA	.267	30.00	2.37	9.87	---	12.24	**17.30**
8" x 18"	Inst	Ea	Lg	UA	.200	40.00	2.51	7.39	---	9.90	**13.80**
	Inst	Ea	Sm	UA	.267	30.00	2.84	9.87	---	12.71	**17.90**
12" x 12"	Inst	Ea	Lg	UA	.200	40.00	2.51	7.39	---	9.90	**13.80**
	Inst	Ea	Sm	UA	.267	30.00	2.84	9.87	---	12.71	**17.90**
14" x 14"	Inst	Ea	Lg	UA	.200	40.00	3.18	7.39	---	10.57	**14.60**
	Inst	Ea	Sm	UA	.267	30.00	3.59	9.87	---	13.46	**18.70**

Description	Oper	Unit	Vol	Crew Size	Man-hours per Unit	Crew Output per Day	Avg Mat'l Unit Cost	Avg Labor Unit Cost	Avg Equip Unit Cost	Avg Total Unit Cost	Avg Price Incl O&P

Shower and tub doors
Shower doors
Hinged shower doors, anodized aluminum frame, tempered hammered glass, hardware included

67" H door, adjustable
29-3/4" to 31-1/2"

Obscure glass

Description	Oper	Unit	Vol	Crew Size	Man-hours per Unit	Crew Output per Day	Avg Mat'l Unit Cost	Avg Labor Unit Cost	Avg Equip Unit Cost	Avg Total Unit Cost	Avg Price Incl O&P
Chrome	Inst	Ea	Lg	CA	1.33	6.00	256.00	44.30	---	300.30	**361.00**
	Inst	Ea	Sm	CA	1.78	4.50	311.00	59.20	---	370.20	**447.00**
Polished brass	Inst	Ea	Lg	CA	1.33	6.00	287.00	44.30	---	331.30	**396.00**
	Inst	Ea	Sm	CA	1.78	4.50	349.00	59.20	---	408.20	**490.00**
White	Inst	Ea	Lg	CA	1.33	6.00	366.00	44.30	---	410.30	**487.00**
	Inst	Ea	Sm	CA	1.78	4.50	445.00	59.20	---	504.20	**600.00**
Bone	Inst	Ea	Lg	CA	1.33	6.00	366.00	44.30	---	410.30	**487.00**
	Inst	Ea	Sm	CA	1.78	4.50	445.00	59.20	---	504.20	**600.00**

Clear glass

Description	Oper	Unit	Vol	Crew Size	Man-hours per Unit	Crew Output per Day	Avg Mat'l Unit Cost	Avg Labor Unit Cost	Avg Equip Unit Cost	Avg Total Unit Cost	Avg Price Incl O&P
Chrome	Inst	Ea	Lg	CA	1.33	6.00	293.00	44.30	---	337.30	**404.00**
	Inst	Ea	Sm	CA	1.78	4.50	356.00	59.20	---	415.20	**498.00**
Polished brass	Inst	Ea	Lg	CA	1.33	6.00	321.00	44.30	---	365.30	**435.00**
	Inst	Ea	Sm	CA	1.78	4.50	389.00	59.20	---	448.20	**537.00**
White	Inst	Ea	Lg	CA	1.33	6.00	399.00	44.30	---	443.30	**525.00**
	Inst	Ea	Sm	CA	1.78	4.50	485.00	59.20	---	544.20	**646.00**
Bone	Inst	Ea	Lg	CA	1.33	6.00	399.00	44.30	---	443.30	**525.00**
	Inst	Ea	Sm	CA	1.78	4.50	485.00	59.20	---	544.20	**646.00**

Etched glass

Description	Oper	Unit	Vol	Crew Size	Man-hours per Unit	Crew Output per Day	Avg Mat'l Unit Cost	Avg Labor Unit Cost	Avg Equip Unit Cost	Avg Total Unit Cost	Avg Price Incl O&P
Chrome	Inst	Ea	Lg	CA	1.33	6.00	314.00	44.30	---	358.30	**427.00**
	Inst	Ea	Sm	CA	1.78	4.50	381.00	59.20	---	440.20	**527.00**
Polished brass	Inst	Ea	Lg	CA	1.33	6.00	343.00	44.30	---	387.30	**461.00**
	Inst	Ea	Sm	CA	1.78	4.50	417.00	59.20	---	476.20	**568.00**
White	Inst	Ea	Lg	CA	1.33	6.00	423.00	44.30	---	467.30	**553.00**
	Inst	Ea	Sm	CA	1.78	4.50	513.00	59.20	---	572.20	**679.00**
Bone	Inst	Ea	Lg	CA	1.33	6.00	423.00	44.30	---	467.30	**553.00**
	Inst	Ea	Sm	CA	1.78	4.50	513.00	59.20	---	572.20	**679.00**

Fluted

Description	Oper	Unit	Vol	Crew Size	Man-hours per Unit	Crew Output per Day	Avg Mat'l Unit Cost	Avg Labor Unit Cost	Avg Equip Unit Cost	Avg Total Unit Cost	Avg Price Incl O&P
Chrome	Inst	Ea	Lg	CA	1.33	6.00	401.00	44.30	---	445.30	**528.00**
	Inst	Ea	Sm	CA	1.78	4.50	487.00	59.20	---	546.20	**649.00**
Polished brass	Inst	Ea	Lg	CA	1.33	6.00	433.00	44.30	---	477.30	**564.00**
	Inst	Ea	Sm	CA	1.78	4.50	525.00	59.20	---	584.20	**693.00**
White	Inst	Ea	Lg	CA	1.33	6.00	510.00	44.30	---	554.30	**653.00**
	Inst	Ea	Sm	CA	1.78	4.50	620.00	59.20	---	679.20	**801.00**
Bone	Inst	Ea	Lg	CA	1.33	6.00	510.00	44.30	---	554.30	**653.00**
	Inst	Ea	Sm	CA	1.78	4.50	620.00	59.20	---	679.20	**801.00**

Description	Oper	Unit	Vol	Crew Size	Man-hours per Unit	Crew Output per Day	Avg Mat'l Unit Cost	Avg Labor Unit Cost	Avg Equip Unit Cost	Avg Total Unit Cost	Avg Price Incl O&P
33-1/8" to 34-7/8"											
Obscure glass											
Chrome	Inst	Ea	Lg	CA	1.33	6.00	256.00	44.30	---	300.30	**361.00**
	Inst	Ea	Sm	CA	1.78	4.50	311.00	59.20	---	370.20	**447.00**
Polished brass	Inst	Ea	Lg	CA	1.33	6.00	287.00	44.30	---	331.30	**396.00**
	Inst	Ea	Sm	CA	1.78	4.50	349.00	59.20	---	408.20	**490.00**
White	Inst	Ea	Lg	CA	1.33	6.00	366.00	44.30	---	410.30	**487.00**
	Inst	Ea	Sm	CA	1.78	4.50	445.00	59.20	---	504.20	**600.00**
Bone	Inst	Ea	Lg	CA	1.33	6.00	399.00	44.30	---	443.30	**525.00**
	Inst	Ea	Sm	CA	1.78	4.50	485.00	59.20	---	544.20	**646.00**
Clear glass											
Chrome	Inst	Ea	Lg	CA	1.33	6.00	293.00	44.30	---	337.30	**404.00**
	Inst	Ea	Sm	CA	1.78	4.50	356.00	59.20	---	415.20	**498.00**
Polished brass	Inst	Ea	Lg	CA	1.33	6.00	327.00	44.30	---	371.30	**442.00**
	Inst	Ea	Sm	CA	1.78	4.50	397.00	59.20	---	456.20	**545.00**
White	Inst	Ea	Lg	CA	1.33	6.00	399.00	44.30	---	443.30	**525.00**
	Inst	Ea	Sm	CA	1.78	4.50	485.00	59.20	---	544.20	**646.00**
Bone	Inst	Ea	Lg	CA	1.33	6.00	399.00	44.30	---	443.30	**525.00**
	Inst	Ea	Sm	CA	1.78	4.50	485.00	59.20	---	544.20	**646.00**
Etched glass											
Chrome	Inst	Ea	Lg	CA	1.33	6.00	314.00	44.30	---	358.30	**427.00**
	Inst	Ea	Sm	CA	1.78	4.50	381.00	59.20	---	440.20	**527.00**
Polished brass	Inst	Ea	Lg	CA	1.33	6.00	343.00	44.30	---	387.30	**461.00**
	Inst	Ea	Sm	CA	1.78	4.50	417.00	59.20	---	476.20	**568.00**
White	Inst	Ea	Lg	CA	1.33	6.00	423.00	44.30	---	467.30	**553.00**
	Inst	Ea	Sm	CA	1.78	4.50	513.00	59.20	---	572.20	**679.00**
Bone	Inst	Ea	Lg	CA	1.33	6.00	423.00	44.30	---	467.30	**553.00**
	Inst	Ea	Sm	CA	1.78	4.50	513.00	59.20	---	572.20	**679.00**
Fluted											
Chrome	Inst	Ea	Lg	CA	1.33	6.00	401.00	44.30	---	445.30	**528.00**
	Inst	Ea	Sm	CA	1.78	4.50	487.00	59.20	---	546.20	**649.00**
Polished brass	Inst	Ea	Lg	CA	1.33	6.00	433.00	44.30	---	477.30	**564.00**
	Inst	Ea	Sm	CA	1.78	4.50	525.00	59.20	---	584.20	**693.00**
White	Inst	Ea	Lg	CA	1.33	6.00	510.00	44.30	---	554.30	**653.00**
	Inst	Ea	Sm	CA	1.78	4.50	620.00	59.20	---	679.20	**801.00**
Bone	Inst	Ea	Lg	CA	1.33	6.00	510.00	44.30	---	554.30	**653.00**
	Inst	Ea	Sm	CA	1.78	4.50	620.00	59.20	---	679.20	**801.00**
23" x 24-1/2" x 23 x 72" neo angle											
Obscure glass											
Chrome	Inst	Ea	Lg	CA	1.33	6.00	531.00	44.30	---	575.30	**677.00**
	Inst	Ea	Sm	CA	1.78	4.50	645.00	59.20	---	704.20	**831.00**
Polished brass	Inst	Ea	Lg	CA	1.33	6.00	651.00	44.30	---	695.30	**815.00**
	Inst	Ea	Sm	CA	1.78	4.50	791.00	59.20	---	850.20	**998.00**
White	Inst	Ea	Lg	CA	1.33	6.00	805.00	44.30	---	849.30	**992.00**
	Inst	Ea	Sm	CA	1.78	4.50	978.00	59.20	---	1037.20	**1210.00**
Bone	Inst	Ea	Lg	CA	1.33	6.00	805.00	44.30	---	849.30	**992.00**
	Inst	Ea	Sm	CA	1.78	4.50	978.00	59.20	---	1037.20	**1210.00**

Description	Oper	Unit	Vol	Crew Size	Man-hours per Unit	Crew Output per Day	Avg Mat'l Unit Cost	Avg Labor Unit Cost	Avg Equip Unit Cost	Avg Total Unit Cost	Avg Price Incl O&P
Clear glass											
Chrome	Inst	Ea	Lg	CA	1.33	6.00	589.00	44.30	---	633.30	**743.00**
	Inst	Ea	Sm	CA	1.78	4.50	715.00	59.20	---	774.20	**911.00**
Polished brass	Inst	Ea	Lg	CA	1.33	6.00	709.00	44.30	---	753.30	**882.00**
	Inst	Ea	Sm	CA	1.78	4.50	861.00	59.20	---	920.20	**1080.00**
White	Inst	Ea	Lg	CA	1.33	6.00	850.00	44.30	---	894.30	**1040.00**
	Inst	Ea	Sm	CA	1.78	4.50	1030.00	59.20	---	1089.20	**1280.00**
Bone	Inst	Ea	Lg	CA	1.33	6.00	850.00	44.30	---	894.30	**1040.00**
	Inst	Ea	Sm	CA	1.78	4.50	1030.00	59.20	---	1089.20	**1280.00**
Etched glass											
Chrome	Inst	Ea	Lg	CA	1.33	6.00	600.00	44.30	---	644.30	**756.00**
	Inst	Ea	Sm	CA	1.78	4.50	728.00	59.20	---	787.20	**927.00**
Polished brass	Inst	Ea	Lg	CA	1.33	6.00	719.00	44.30	---	763.30	**893.00**
	Inst	Ea	Sm	CA	1.78	4.50	873.00	59.20	---	932.20	**1090.00**
White	Inst	Ea	Lg	CA	1.33	6.00	873.00	44.30	---	917.30	**1070.00**
	Inst	Ea	Sm	CA	1.78	4.50	1060.00	59.20	---	1119.20	**1310.00**
Bone	Inst	Ea	Lg	CA	1.33	6.00	873.00	44.30	---	917.30	**1070.00**
	Inst	Ea	Sm	CA	1.78	4.50	1060.00	59.20	---	1119.20	**1310.00**
Fluted											
Chrome	Inst	Ea	Lg	CA	1.33	6.00	881.00	44.30	---	925.30	**1080.00**
	Inst	Ea	Sm	CA	1.78	4.50	1070.00	59.20	---	1129.20	**1320.00**
Polished brass	Inst	Ea	Lg	CA	1.33	6.00	1000.00	44.30	---	1044.30	**1220.00**
	Inst	Ea	Sm	CA	1.78	4.50	1210.00	59.20	---	1269.20	**1490.00**
White	Inst	Ea	Lg	CA	1.33	6.00	1160.00	44.30	---	1204.30	**1390.00**
	Inst	Ea	Sm	CA	1.78	4.50	1400.00	59.20	---	1459.20	**1700.00**
Bone	Inst	Ea	Lg	CA	1.33	6.00	1160.00	44.30	---	1204.30	**1390.00**
	Inst	Ea	Sm	CA	1.78	4.50	1400.00	59.20	---	1459.20	**1700.00**

Sliding shower doors, 2 bypassing panels, anodized aluminum frame, tempered hammered glass, hardware included

71-1/2" H doors

40" to 42" wide opening

Description	Oper	Unit	Vol	Crew Size	Man-hours per Unit	Crew Output per Day	Avg Mat'l Unit Cost	Avg Labor Unit Cost	Avg Equip Unit Cost	Avg Total Unit Cost	Avg Price Incl O&P
Obscure glass											
Chrome	Inst	Ea	Lg	CA	1.45	5.50	300.00	48.20	---	348.20	**418.00**
	Inst	Ea	Sm	CA	1.94	4.13	365.00	64.50	---	429.50	**516.00**
Polished brass	Inst	Ea	Lg	CA	1.45	5.50	338.00	48.20	---	386.20	**461.00**
	Inst	Ea	Sm	CA	1.94	4.13	411.00	64.50	---	475.50	**569.00**
White	Inst	Ea	Lg	CA	1.45	5.50	454.00	48.20	---	502.20	**594.00**
	Inst	Ea	Sm	CA	1.94	4.13	551.00	64.50	---	615.50	**730.00**
Bone	Inst	Ea	Lg	CA	1.45	5.50	454.00	48.20	---	502.20	**594.00**
	Inst	Ea	Sm	CA	1.94	4.13	551.00	64.50	---	615.50	**730.00**
Clear glass											
Chrome	Inst	Ea	Lg	CA	1.45	5.50	347.00	48.20	---	395.20	**472.00**
	Inst	Ea	Sm	CA	1.94	4.13	422.00	64.50	---	486.50	**582.00**
Polished brass	Inst	Ea	Lg	CA	1.45	5.50	387.00	48.20	---	435.20	**518.00**
	Inst	Ea	Sm	CA	1.94	4.13	470.00	64.50	---	534.50	**637.00**
White	Inst	Ea	Lg	CA	1.45	5.50	501.00	48.20	---	549.20	**649.00**
	Inst	Ea	Sm	CA	1.94	4.13	609.00	64.50	---	673.50	**797.00**
Bone	Inst	Ea	Lg	CA	1.45	5.50	501.00	48.20	---	549.20	**649.00**
	Inst	Ea	Sm	CA	1.94	4.13	609.00	64.50	---	673.50	**797.00**

Description	Oper	Unit	Vol	Crew Size	Man-hours per Unit	Crew Output per Day	Avg Mat'l Unit Cost	Avg Labor Unit Cost	Avg Equip Unit Cost	Avg Total Unit Cost	Avg Price Incl O&P
Etched glass											
Chrome	Inst	Ea	Lg	CA	1.45	5.50	363.00	48.20	---	411.20	**490.00**
	Inst	Ea	Sm	CA	1.94	4.13	441.00	64.50	---	505.50	**604.00**
Polished brass	Inst	Ea	Lg	CA	1.45	5.50	400.00	48.20	---	448.20	**532.00**
	Inst	Ea	Sm	CA	1.94	4.13	485.00	64.50	---	549.50	**655.00**
White	Inst	Ea	Lg	CA	1.45	5.50	520.00	48.20	---	568.20	**670.00**
	Inst	Ea	Sm	CA	1.94	4.13	632.00	64.50	---	696.50	**823.00**
Bone	Inst	Ea	Lg	CA	1.45	5.50	520.00	48.20	---	568.20	**670.00**
	Inst	Ea	Sm	CA	1.94	4.13	632.00	64.50	---	696.50	**823.00**
Fluted											
Chrome	Inst	Ea	Lg	CA	1.45	5.50	533.00	48.20	---	581.20	**686.00**
	Inst	Ea	Sm	CA	1.94	4.13	648.00	64.50	---	712.50	**842.00**
Polished brass	Inst	Ea	Lg	CA	1.45	5.50	571.00	48.20	---	619.20	**728.00**
	Inst	Ea	Sm	CA	1.94	4.13	693.00	64.50	---	757.50	**893.00**
White	Inst	Ea	Lg	CA	1.45	5.50	691.00	48.20	---	739.20	**867.00**
	Inst	Ea	Sm	CA	1.94	4.13	839.00	64.50	---	903.50	**1060.00**
Bone	Inst	Ea	Lg	CA	1.45	5.50	691.00	48.20	---	739.20	**867.00**
	Inst	Ea	Sm	CA	1.94	4.13	839.00	64.50	---	903.50	**1060.00**
44" to 48" wide opening											
Obscure glass											
Chrome	Inst	Ea	Lg	CA	1.45	5.50	300.00	48.20	---	348.20	**418.00**
	Inst	Ea	Sm	CA	1.94	4.13	365.00	64.50	---	429.50	**516.00**
Polished brass	Inst	Ea	Lg	CA	1.45	5.50	338.00	48.20	---	386.20	**461.00**
	Inst	Ea	Sm	CA	1.94	4.13	411.00	64.50	---	475.50	**569.00**
White	Inst	Ea	Lg	CA	1.45	5.50	454.00	48.20	---	502.20	**594.00**
	Inst	Ea	Sm	CA	1.94	4.13	551.00	64.50	---	615.50	**730.00**
Bone	Inst	Ea	Lg	CA	1.45	5.50	454.00	48.20	---	502.20	**594.00**
	Inst	Ea	Sm	CA	1.94	4.13	551.00	64.50	---	615.50	**730.00**
Clear glass											
Chrome	Inst	Ea	Lg	CA	1.45	5.50	347.00	48.20	---	395.20	**472.00**
	Inst	Ea	Sm	CA	1.94	4.13	422.00	64.50	---	486.50	**582.00**
Polished brass	Inst	Ea	Lg	CA	1.45	5.50	387.00	48.20	---	435.20	**518.00**
	Inst	Ea	Sm	CA	1.94	4.13	470.00	64.50	---	534.50	**637.00**
White	Inst	Ea	Lg	CA	1.45	5.50	501.00	48.20	---	549.20	**649.00**
	Inst	Ea	Sm	CA	1.94	4.13	609.00	64.50	---	673.50	**797.00**
Bone	Inst	Ea	Lg	CA	1.45	5.50	501.00	48.20	---	549.20	**649.00**
	Inst	Ea	Sm	CA	1.94	4.13	609.00	64.50	---	673.50	**797.00**
Etched glass											
Chrome	Inst	Ea	Lg	CA	1.45	5.50	363.00	48.20	---	411.20	**490.00**
	Inst	Ea	Sm	CA	1.94	4.13	441.00	64.50	---	505.50	**604.00**
Polished brass	Inst	Ea	Lg	CA	1.45	5.50	400.00	48.20	---	448.20	**532.00**
	Inst	Ea	Sm	CA	1.94	4.13	485.00	64.50	---	549.50	**655.00**
White	Inst	Ea	Lg	CA	1.45	5.50	520.00	48.20	---	568.20	**670.00**
	Inst	Ea	Sm	CA	1.94	4.13	632.00	64.50	---	696.50	**823.00**
Bone	Inst	Ea	Lg	CA	1.45	5.50	520.00	48.20	---	568.20	**670.00**
	Inst	Ea	Sm	CA	1.94	4.13	632.00	64.50	---	696.50	**823.00**

Description	Oper	Unit	Vol	Crew Size	Man-hours per Unit	Crew Output per Day	Avg Mat'l Unit Cost	Avg Labor Unit Cost	Avg Equip Unit Cost	Avg Total Unit Cost	Avg Price Incl O&P
Fluted											
Chrome	Inst	Ea	Lg	CA	1.45	5.50	533.00	48.20	---	581.20	**686.00**
	Inst	Ea	Sm	CA	1.94	4.13	648.00	64.50	---	712.50	**842.00**
Polished brass	Inst	Ea	Lg	CA	1.45	5.50	571.00	48.20	---	619.20	**728.00**
	Inst	Ea	Sm	CA	1.94	4.13	693.00	64.50	---	757.50	**893.00**
White	Inst	Ea	Lg	CA	1.45	5.50	691.00	48.20	---	739.20	**867.00**
	Inst	Ea	Sm	CA	1.94	4.13	839.00	64.50	---	903.50	**1060.00**
Bone	Inst	Ea	Lg	CA	1.45	5.50	691.00	48.20	---	739.20	**867.00**
	Inst	Ea	Sm	CA	1.94	4.13	839.00	64.50	---	903.50	**1060.00**
56" to 60" wide opening											
Obscure glass											
Chrome	Inst	Ea	Lg	CA	1.45	5.50	351.00	48.20	---	399.20	**476.00**
	Inst	Ea	Sm	CA	1.94	4.13	427.00	64.50	---	491.50	**588.00**
Polished brass	Inst	Ea	Lg	CA	1.45	5.50	388.00	48.20	---	436.20	**518.00**
	Inst	Ea	Sm	CA	1.94	4.13	471.00	64.50	---	535.50	**638.00**
White	Inst	Ea	Lg	CA	1.45	5.50	472.00	48.20	---	520.20	**615.00**
	Inst	Ea	Sm	CA	1.94	4.13	573.00	64.50	---	637.50	**756.00**
Bone	Inst	Ea	Lg	CA	1.45	5.50	472.00	48.20	---	520.20	**615.00**
	Inst	Ea	Sm	CA	1.94	4.13	573.00	64.50	---	637.50	**756.00**
Clear glass											
Chrome	Inst	Ea	Lg	CA	1.45	5.50	400.00	48.20	---	448.20	**532.00**
	Inst	Ea	Sm	CA	1.94	4.13	485.00	64.50	---	549.50	**655.00**
Polished brass	Inst	Ea	Lg	CA	1.45	5.50	437.00	48.20	---	485.20	**575.00**
	Inst	Ea	Sm	CA	1.94	4.13	530.00	64.50	---	594.50	**707.00**
White	Inst	Ea	Lg	CA	1.45	5.50	550.00	48.20	---	598.20	**704.00**
	Inst	Ea	Sm	CA	1.94	4.13	667.00	64.50	---	731.50	**864.00**
Bone	Inst	Ea	Lg	CA	1.45	5.50	550.00	48.20	---	598.20	**704.00**
	Inst	Ea	Sm	CA	1.94	4.13	667.00	64.50	---	731.50	**864.00**
Etched glass											
Chrome	Inst	Ea	Lg	CA	1.45	5.50	413.00	48.20	---	461.20	**547.00**
	Inst	Ea	Sm	CA	1.94	4.13	502.00	64.50	---	566.50	**674.00**
Polished brass	Inst	Ea	Lg	CA	1.45	5.50	449.00	48.20	---	497.20	**589.00**
	Inst	Ea	Sm	CA	1.94	4.13	546.00	64.50	---	610.50	**724.00**
White	Inst	Ea	Lg	CA	1.45	5.50	571.00	48.20	---	619.20	**729.00**
	Inst	Ea	Sm	CA	1.94	4.13	694.00	64.50	---	758.50	**894.00**
Bone	Inst	Ea	Lg	CA	1.45	5.50	571.00	48.20	---	619.20	**729.00**
	Inst	Ea	Sm	CA	1.94	4.13	694.00	64.50	---	758.50	**894.00**
Fluted											
Chrome	Inst	Ea	Lg	CA	1.45	5.50	569.00	48.20	---	617.20	**727.00**
	Inst	Ea	Sm	CA	1.94	4.13	691.00	64.50	---	755.50	**892.00**
Polished brass	Inst	Ea	Lg	CA	1.45	5.50	639.00	48.20	---	687.20	**807.00**
	Inst	Ea	Sm	CA	1.94	4.13	776.00	64.50	---	840.50	**989.00**
White	Inst	Ea	Lg	CA	1.45	5.50	740.00	48.20	---	788.20	**923.00**
	Inst	Ea	Sm	CA	1.94	4.13	898.00	64.50	---	962.50	**1130.00**
Bone	Inst	Ea	Lg	CA	1.45	5.50	740.00	48.20	---	788.20	**923.00**
	Inst	Ea	Sm	CA	1.94	4.13	898.00	64.50	---	962.50	**1130.00**

Description	Oper	Unit	Vol	Crew Size	Man-hours per Unit	Crew Output per Day	Avg Mat'l Unit Cost	Avg Labor Unit Cost	Avg Equip Unit Cost	Avg Total Unit Cost	Avg Price Incl O&P

Tub doors

Sliding tub doors, 2 bypassing panels

60" W x 58-1/4" H

Obscure glass

Description	Oper	Unit	Vol	Crew Size	Man-hours per Unit	Crew Output per Day	Avg Mat'l Unit Cost	Avg Labor Unit Cost	Avg Equip Unit Cost	Avg Total Unit Cost	Avg Price Incl O&P
Chrome	Inst	Ea	Lg	CA	1.45	5.50	284.00	48.20	---	332.20	**398.00**
	Inst	Ea	Sm	CA	1.94	4.13	344.00	64.50	---	408.50	**493.00**
Polished brass	Inst	Ea	Lg	CA	1.45	5.50	314.00	48.20	---	362.20	**433.00**
	Inst	Ea	Sm	CA	1.94	4.13	381.00	64.50	---	445.50	**535.00**

Clear glass

Description	Oper	Unit	Vol	Crew Size	Man-hours per Unit	Crew Output per Day	Avg Mat'l Unit Cost	Avg Labor Unit Cost	Avg Equip Unit Cost	Avg Total Unit Cost	Avg Price Incl O&P
Chrome	Inst	Ea	Lg	CA	1.45	5.50	323.00	48.20	---	371.20	**444.00**
	Inst	Ea	Sm	CA	1.94	4.13	393.00	64.50	---	457.50	**548.00**
Polished brass	Inst	Ea	Lg	CA	1.45	5.50	354.00	48.20	---	402.20	**479.00**
	Inst	Ea	Sm	CA	1.94	4.13	429.00	64.50	---	493.50	**590.00**

Shower bases or receptors

American Standard Products, Americast with drain strainer

Detach & reset operations

Description	Oper	Unit	Vol	Crew Size	Man-hours per Unit	Crew Output per Day	Avg Mat'l Unit Cost	Avg Labor Unit Cost	Avg Equip Unit Cost	Avg Total Unit Cost	Avg Price Incl O&P
Any size	Reset	Ea	Lg	SB	2.67	6.00	29.30	86.90	---	116.20	**161.00**
	Reset	Ea	Sm	SB	3.81	4.20	33.20	124.00	---	157.20	**220.00**

Remove operations

Description	Oper	Unit	Vol	Crew Size	Man-hours per Unit	Crew Output per Day	Avg Mat'l Unit Cost	Avg Labor Unit Cost	Avg Equip Unit Cost	Avg Total Unit Cost	Avg Price Incl O&P
Any size	Demo	Ea	Lg	SB	1.60	10.00	---	52.10	---	52.10	**76.60**
	Demo	Ea	Sm	SB	2.29	7.00	---	74.50	---	74.50	**110.00**

Install rough-in

Shower base

Description	Oper	Unit	Vol	Crew Size	Man-hours per Unit	Crew Output per Day	Avg Mat'l Unit Cost	Avg Labor Unit Cost	Avg Equip Unit Cost	Avg Total Unit Cost	Avg Price Incl O&P
Shower base, any size	Inst	Ea	Lg	SB	13.3	1.20	63.80	433.00	---	496.80	**710.00**
	Inst	Ea	Sm	SB	19.0	0.84	72.30	618.00	---	690.30	**992.00**

Replace operations

32" x 32" x 6-5/8" H

Description	Oper	Unit	Vol	Crew Size	Man-hours per Unit	Crew Output per Day	Avg Mat'l Unit Cost	Avg Labor Unit Cost	Avg Equip Unit Cost	Avg Total Unit Cost	Avg Price Incl O&P
White	Reset	Ea	Lg	SB	4.00	4.00	349.00	130.00	---	479.00	**592.00**
	Reset	Ea	Sm	SB	5.71	2.80	395.00	186.00	---	581.00	**728.00**
Colors	Reset	Ea	Lg	SB	4.00	4.00	378.00	130.00	---	508.00	**626.00**
	Reset	Ea	Sm	SB	5.71	2.80	428.00	186.00	---	614.00	**766.00**
Premium colors	Reset	Ea	Lg	SB	4.00	4.00	401.00	130.00	---	531.00	**652.00**
	Reset	Ea	Sm	SB	5.71	2.80	454.00	186.00	---	640.00	**795.00**

36" x 36-3/16" x 6-5/8" H

Description	Oper	Unit	Vol	Crew Size	Man-hours per Unit	Crew Output per Day	Avg Mat'l Unit Cost	Avg Labor Unit Cost	Avg Equip Unit Cost	Avg Total Unit Cost	Avg Price Incl O&P
White	Reset	Ea	Lg	SB	4.00	4.00	359.00	130.00	---	489.00	**604.00**
	Reset	Ea	Sm	SB	5.71	2.80	406.00	186.00	---	592.00	**740.00**
Colors	Reset	Ea	Lg	SB	4.00	4.00	389.00	130.00	---	519.00	**638.00**
	Reset	Ea	Sm	SB	5.71	2.80	440.00	186.00	---	626.00	**780.00**
Premium colors	Reset	Ea	Lg	SB	4.00	4.00	412.00	130.00	---	542.00	**665.00**
	Reset	Ea	Sm	SB	5.71	2.80	467.00	186.00	---	653.00	**810.00**

Description	Oper	Unit	Vol	Crew Size	Man-hours per Unit	Crew Output per Day	Avg Mat'l Unit Cost	Avg Labor Unit Cost	Avg Equip Unit Cost	Avg Total Unit Cost	Avg Price Incl O&P
42-1/8" x 42-1/8" x 6-5/8" H											
White	Reset	Ea	Lg	SB	4.00	4.00	569.00	130.00	---	699.00	**845.00**
	Reset	Ea	Sm	SB	5.71	2.80	644.00	186.00	---	830.00	**1010.00**
Colors	Reset	Ea	Lg	SB	4.00	4.00	605.00	130.00	---	735.00	**887.00**
	Reset	Ea	Sm	SB	5.71	2.80	685.00	186.00	---	871.00	**1060.00**
Premium colors	Reset	Ea	Lg	SB	4.00	4.00	626.00	130.00	---	756.00	**912.00**
	Reset	Ea	Sm	SB	5.71	2.80	710.00	186.00	---	896.00	**1090.00**
48-1/8" x 34-1/4" x 6-5/8" H											
White	Reset	Ea	Lg	SB	4.00	4.00	482.00	130.00	---	612.00	**746.00**
	Reset	Ea	Sm	SB	5.71	2.80	547.00	186.00	---	733.00	**902.00**
Colors	Reset	Ea	Lg	SB	4.00	4.00	517.00	130.00	---	647.00	**786.00**
	Reset	Ea	Sm	SB	5.71	2.80	586.00	186.00	---	772.00	**947.00**
Premium colors	Reset	Ea	Lg	SB	4.00	4.00	538.00	130.00	---	668.00	**810.00**
	Reset	Ea	Sm	SB	5.71	2.80	609.00	186.00	---	795.00	**974.00**
60-1/8" x 34-1/8" x 6-5/8" H											
White	Reset	Ea	Lg	SB	4.00	4.00	596.00	130.00	---	726.00	**877.00**
	Reset	Ea	Sm	SB	5.71	2.80	676.00	186.00	---	862.00	**1050.00**
Colors	Reset	Ea	Lg	SB	4.00	4.00	632.00	130.00	---	762.00	**918.00**
	Reset	Ea	Sm	SB	5.71	2.80	717.00	186.00	---	903.00	**1100.00**
Premium colors	Reset	Ea	Lg	SB	4.00	4.00	656.00	130.00	---	786.00	**945.00**
	Reset	Ea	Sm	SB	5.71	2.80	743.00	186.00	---	929.00	**1130.00**
36-1/4" x 36-1/8" x 6-5/8" H, neo angle											
White	Reset	Ea	Lg	SB	4.00	4.00	372.00	130.00	---	502.00	**619.00**
	Reset	Ea	Sm	SB	5.71	2.80	422.00	186.00	---	608.00	**758.00**
Colors	Reset	Ea	Lg	SB	4.00	4.00	404.00	130.00	---	534.00	**655.00**
	Reset	Ea	Sm	SB	5.71	2.80	457.00	186.00	---	643.00	**799.00**
Premium colors	Reset	Ea	Lg	SB	4.00	4.00	426.00	130.00	---	556.00	**681.00**
	Reset	Ea	Sm	SB	5.71	2.80	483.00	186.00	---	669.00	**828.00**
38-1/8" x 38-1/16" x 6-5/8" H, neo angle											
White	Reset	Ea	Lg	SB	4.00	4.00	390.00	130.00	---	520.00	**640.00**
	Reset	Ea	Sm	SB	5.71	2.80	442.00	186.00	---	628.00	**782.00**
Colors	Reset	Ea	Lg	SB	4.00	4.00	421.00	130.00	---	551.00	**675.00**
	Reset	Ea	Sm	SB	5.71	2.80	477.00	186.00	---	663.00	**822.00**
Premium colors	Reset	Ea	Lg	SB	4.00	4.00	443.00	130.00	---	573.00	**701.00**
	Reset	Ea	Sm	SB	5.71	2.80	502.00	186.00	---	688.00	**851.00**
42-1/4" x 42-1/8" x 6-5/8" H, neo angle											
White	Reset	Ea	Lg	SB	4.00	4.00	596.00	130.00	---	726.00	**877.00**
	Reset	Ea	Sm	SB	5.71	2.80	676.00	186.00	---	862.00	**1050.00**
Colors	Reset	Ea	Lg	SB	4.00	4.00	632.00	130.00	---	762.00	**918.00**
	Reset	Ea	Sm	SB	5.71	2.80	717.00	186.00	---	903.00	**1100.00**
Premium colors	Reset	Ea	Lg	SB	4.00	4.00	656.00	130.00	---	786.00	**945.00**
	Reset	Ea	Sm	SB	5.71	2.80	743.00	186.00	---	929.00	**1130.00**

Description	Oper	Unit	Vol	Crew Size	Man-hours per Unit	Crew Output per Day	Avg Mat'l Unit Cost	Avg Labor Unit Cost	Avg Equip Unit Cost	Avg Total Unit Cost	Avg Price Incl O&P

Shower stalls

Shower stall units with slip-resistant fiberglass floors, and reinforced plastic integral wall surrounds (Aqua Glass), available in white or color, with good quality fittings, single control faucets, and shower head sprayer

Detach & reset operations

Description	Oper	Unit	Vol	Crew Size	Man-hours per Unit	Crew Output per Day	Avg Mat'l Unit Cost	Avg Labor Unit Cost	Avg Equip Unit Cost	Avg Total Unit Cost	Avg Price Incl O&P
Single integral piece	Reset	Ea	Lg	SB	5.33	3.00	38.50	173.00	---	211.50	**299.00**
	Reset	Ea	Sm	SB	7.11	2.25	46.80	231.00	---	277.80	**394.00**
Multi-piece, non-integral	Reset	Ea	Lg	SB	4.00	4.00	38.50	130.00	---	168.50	**236.00**
	Reset	Ea	Sm	SB	5.33	3.00	46.80	173.00	---	219.80	**309.00**

Remove operations

Description	Oper	Unit	Vol	Crew Size	Man-hours per Unit	Crew Output per Day	Avg Mat'l Unit Cost	Avg Labor Unit Cost	Avg Equip Unit Cost	Avg Total Unit Cost	Avg Price Incl O&P
Single integral piece	Demo	Ea	Lg	SB	2.67	6.00	21.00	86.90	---	107.90	**152.00**
	Demo	Ea	Sm	SB	3.56	4.50	25.50	116.00	---	141.50	**200.00**
Multi-piece, non-integral	Demo	Ea	Lg	SB	2.00	8.00	21.00	65.10	---	86.10	**120.00**
	Demo	Ea	Sm	SB	2.67	6.00	25.50	86.90	---	112.40	**157.00**

Install rough-in

Description	Oper	Unit	Vol	Crew Size	Man-hours per Unit	Crew Output per Day	Avg Mat'l Unit Cost	Avg Labor Unit Cost	Avg Equip Unit Cost	Avg Total Unit Cost	Avg Price Incl O&P
Shower stall, any size	Inst	Ea	Lg	SB	10.0	1.60	59.50	326.00	---	385.50	**547.00**
	Inst	Ea	Sm	SB	13.3	1.20	72.30	433.00	---	505.30	**719.00**

Replace operations

Three-wall stall, one integral piece, no door, with plastic drain

Lasco Compact, Paloma-I
32" x 32" x 72" H

Description	Oper	Unit	Vol	Crew Size	Man-hours per Unit	Crew Output per Day	Avg Mat'l Unit Cost	Avg Labor Unit Cost	Avg Equip Unit Cost	Avg Total Unit Cost	Avg Price Incl O&P
Stock color	Inst	Ea	Lg	SB	8.00	2.00	463.00	260.00	---	723.00	**916.00**
	Inst	Ea	Sm	SB	10.7	1.50	563.00	348.00	---	911.00	**1160.00**

Lasco Compact, Paloma-II
34" x 34" x 72" H

Description	Oper	Unit	Vol	Crew Size	Man-hours per Unit	Crew Output per Day	Avg Mat'l Unit Cost	Avg Labor Unit Cost	Avg Equip Unit Cost	Avg Total Unit Cost	Avg Price Incl O&P
Stock color	Inst	Ea	Lg	SB	8.00	2.00	515.00	260.00	---	775.00	**975.00**
	Inst	Ea	Sm	SB	10.7	1.50	626.00	348.00	---	974.00	**1230.00**

Lasco Compact, Paloma-III
36" x 36" x 72" H

Description	Oper	Unit	Vol	Crew Size	Man-hours per Unit	Crew Output per Day	Avg Mat'l Unit Cost	Avg Labor Unit Cost	Avg Equip Unit Cost	Avg Total Unit Cost	Avg Price Incl O&P
Stock color	Inst	Ea	Lg	SB	8.00	2.00	497.00	260.00	---	757.00	**954.00**
	Inst	Ea	Sm	SB	10.7	1.50	604.00	348.00	---	952.00	**1210.00**

Lasco Compact, Paloma-IV
42" x 34" x 72" H

Description	Oper	Unit	Vol	Crew Size	Man-hours per Unit	Crew Output per Day	Avg Mat'l Unit Cost	Avg Labor Unit Cost	Avg Equip Unit Cost	Avg Total Unit Cost	Avg Price Incl O&P
Stock color	Inst	Ea	Lg	SB	8.00	2.00	524.00	260.00	---	784.00	**986.00**
	Inst	Ea	Sm	SB	10.7	1.50	637.00	348.00	---	985.00	**1240.00**

Lasco Deluxe, Larissa, 1 seat
47-3/4" x 33-1/2" x 72" H

Description	Oper	Unit	Vol	Crew Size	Man-hours per Unit	Crew Output per Day	Avg Mat'l Unit Cost	Avg Labor Unit Cost	Avg Equip Unit Cost	Avg Total Unit Cost	Avg Price Incl O&P
Stock color	Inst	Ea	Lg	SB	8.00	2.00	568.00	260.00	---	828.00	**1040.00**
	Inst	Ea	Sm	SB	10.7	1.50	689.00	348.00	---	1037.00	**1300.00**

Description	Oper	Unit	Vol	Crew Size	Man-hours per Unit	Crew Output per Day	Avg Mat'l Unit Cost	Avg Labor Unit Cost	Avg Equip Unit Cost	Avg Total Unit Cost	Avg Price Incl O&P
Lasco Deluxe, Camila-I, 2 seat											
48" x 35" x 72" H											
Stock color	Inst	Ea	Lg	SB	8.00	2.00	547.00	260.00	---	807.00	**1010.00**
	Inst	Ea	Sm	SB	10.7	1.50	664.00	348.00	---	1012.00	**1280.00**
Lasco Deluxe, Camila-II, 2 seat											
54" x 35" x 72" H											
Stock color	Inst	Ea	Lg	SB	8.89	1.80	561.00	289.00	---	850.00	**1070.00**
	Inst	Ea	Sm	SB	11.9	1.35	682.00	387.00	---	1069.00	**1350.00**
Lasco Deluxe, Camila-III, 2 seat											
60" x 35" x 72" H											
Stock color	Inst	Ea	Lg	SB	8.89	1.80	568.00	289.00	---	857.00	**1080.00**
	Inst	Ea	Sm	SB	11.9	1.35	609.00	307.00	---	1076.00	**1360.00**

Two-wall stall, one integral piece, no door, with plastic drain

Description	Oper	Unit	Vol	Crew Size	Man-hours per Unit	Crew Output per Day	Avg Mat'l Unit Cost	Avg Labor Unit Cost	Avg Equip Unit Cost	Avg Total Unit Cost	Avg Price Incl O&P
Lasco Varina, space saver											
36" x 36" x 72" H											
Stock color	Inst	Ea	Lg	SB	8.00	2.00	521.00	260.00	---	781.00	**982.00**
	Inst	Ea	Sm	SB	10.7	1.50	632.00	348.00	---	980.00	**1240.00**
Lasco Galina-I, space saver, neo angle											
38" x 38" x 72" H											
Stock color	Inst	Ea	Lg	SB	8.00	2.00	486.00	260.00	---	746.00	**941.00**
	Inst	Ea	Sm	SB	10.7	1.50	590.00	348.00	---	938.00	**1190.00**
Lasco Galina-II, space saver, neo angle											
38" x 38" x 80" H											
Stock color	Inst	Ea	Lg	SB	8.00	2.00	527.00	260.00	---	787.00	**989.00**
	Inst	Ea	Sm	SB	10.7	1.50	640.00	348.00	---	988.00	**1250.00**
Lasco Galina-III, space saver, neo angle											
36" x 36" x 77" H											
Stock color	Inst	Ea	Lg	SB	8.00	2.00	700.00	260.00	---	960.00	**1190.00**
	Inst	Ea	Sm	SB	10.7	1.50	850.00	348.00	---	1198.00	**1490.00**
Lasco Neopolitan-I, space saver, neo angle											
38" x 38" x 72" H											
Stock color	Inst	Ea	Lg	SB	8.00	2.00	558.00	260.00	---	818.00	**1020.00**
	Inst	Ea	Sm	SB	10.7	1.50	677.00	348.00	---	1025.00	**1290.00**

Description	Oper	Unit	Vol	Crew Size	Man-hours per Unit	Crew Output per Day	Avg Mat'l Unit Cost	Avg Labor Unit Cost	Avg Equip Unit Cost	Avg Total Unit Cost	Avg Price Incl O&P

Multi-piece, nonintegral, nonassembled units with drain

Lasco Caledon-I, 2-piece
32" x 32" x 72-3/4" H

Description	Oper	Unit	Vol	Crew Size	Man-hours per Unit	Crew Output per Day	Avg Mat'l Unit Cost	Avg Labor Unit Cost	Avg Equip Unit Cost	Avg Total Unit Cost	Avg Price Incl O&P
Stock color	Inst	Ea	Lg	SB	8.00	2.00	503.00	260.00	---	763.00	**961.00**
	Inst	Ea	Sm	SB	10.7	1.50	610.00	348.00	---	958.00	**1210.00**

Lasco Caledon-II, 3-piece
36" x 36" x 72-3/4" H

Description	Oper	Unit	Vol	Crew Size	Man-hours per Unit	Crew Output per Day	Avg Mat'l Unit Cost	Avg Labor Unit Cost	Avg Equip Unit Cost	Avg Total Unit Cost	Avg Price Incl O&P
Stock color	Inst	Ea	Lg	SB	8.00	2.00	568.00	260.00	---	828.00	**1040.00**
	Inst	Ea	Sm	SB	10.7	1.50	689.00	348.00	---	1037.00	**1300.00**

Lasco Caledon-III, 2-piece
48" x 34" x 72" H

Description	Oper	Unit	Vol	Crew Size	Man-hours per Unit	Crew Output per Day	Avg Mat'l Unit Cost	Avg Labor Unit Cost	Avg Equip Unit Cost	Avg Total Unit Cost	Avg Price Incl O&P
Stock color	Inst	Ea	Lg	SB	8.89	1.80	566.00	289.00	---	855.00	**1080.00**
	Inst	Ea	Sm	SB	11.9	1.35	687.00	387.00	---	1074.00	**1360.00**

Lasco Caledon-III, 3-piece
48" x 34" x 72-3/4" H

Description	Oper	Unit	Vol	Crew Size	Man-hours per Unit	Crew Output per Day	Avg Mat'l Unit Cost	Avg Labor Unit Cost	Avg Equip Unit Cost	Avg Total Unit Cost	Avg Price Incl O&P
Stock color	Inst	Ea	Lg	SB	8.89	1.80	611.00	289.00	---	900.00	**1130.00**
	Inst	Ea	Sm	SB	11.9	1.35	742.00	387.00	---	1129.00	**1420.00**

Lasco Talia-II, 2-piece
36" x 36" x 72" H

Description	Oper	Unit	Vol	Crew Size	Man-hours per Unit	Crew Output per Day	Avg Mat'l Unit Cost	Avg Labor Unit Cost	Avg Equip Unit Cost	Avg Total Unit Cost	Avg Price Incl O&P
Stock color	Inst	Ea	Lg	SB	8.00	2.00	534.00	260.00	---	794.00	**997.00**
	Inst	Ea	Sm	SB	10.7	1.50	649.00	348.00	---	997.00	**1260.00**

Description	Oper	Unit	Vol	Crew Size	Man-hours per Unit	Crew Output per Day	Avg Mat'l Unit Cost	Avg Labor Unit Cost	Avg Equip Unit Cost	Avg Total Unit Cost	Avg Price Incl O&P

Shower tub units

Shower tub three-wall stall combinations are fiberglass reinforced plastic with integral bath/shower and wall surrounds; includes good quality fittings, single control faucets, and shower head sprayer

Detach & reset operations

Description	Oper	Unit	Vol	Crew Size	Man-hours per Unit	Crew Output per Day	Avg Mat'l Unit Cost	Avg Labor Unit Cost	Avg Equip Unit Cost	Avg Total Unit Cost	Avg Price Incl O&P
Single integral piece	Reset	Ea	Lg	SB	5.33	3.00	38.50	173.00	---	211.50	**299.00**
	Reset	Ea	Sm	SB	6.96	2.30	46.80	227.00	---	273.80	**387.00**
Multi-piece, non-integral	Reset	Ea	Lg	SB	4.00	4.00	38.50	130.00	---	168.50	**236.00**
	Reset	Ea	Sm	SB	5.33	3.00	46.80	173.00	---	219.80	**309.00**

Remove operations

Description	Oper	Unit	Vol	Crew Size	Man-hours per Unit	Crew Output per Day	Avg Mat'l Unit Cost	Avg Labor Unit Cost	Avg Equip Unit Cost	Avg Total Unit Cost	Avg Price Incl O&P
Single integral piece	Demo	Ea	Lg	SB	2.67	6.00	21.00	86.90	---	107.90	**152.00**
	Demo	Ea	Sm	SB	3.56	4.50	25.50	116.00	---	141.50	**200.00**
Multi-piece, non-integral	Demo	Ea	Lg	SB	2.00	8.00	21.00	65.10	---	86.10	**120.00**
	Demo	Ea	Sm	SB	2.67	6.00	25.50	86.90	---	112.40	**157.00**

Install rough-in

Description	Oper	Unit	Vol	Crew Size	Man-hours per Unit	Crew Output per Day	Avg Mat'l Unit Cost	Avg Labor Unit Cost	Avg Equip Unit Cost	Avg Total Unit Cost	Avg Price Incl O&P
Shower tub	Inst	Ea	Lg	SB	11.4	1.40	59.50	371.00	---	430.50	**614.00**
	Inst	Ea	Sm	SB	14.5	1.10	72.30	472.00	---	544.30	**777.00**

Replace operations
One-piece combination, (unitized), no door

Lasco Kassia, with grab bar
54-1/2" x 29" x 72" H

Description	Oper	Unit	Vol	Crew Size	Man-hours per Unit	Crew Output per Day	Avg Mat'l Unit Cost	Avg Labor Unit Cost	Avg Equip Unit Cost	Avg Total Unit Cost	Avg Price Incl O&P
Stock color	Inst	Ea	Lg	SB	8.00	2.00	553.00	260.00	---	813.00	**1020.00**
	Inst	Ea	Sm	SB	10.7	1.50	672.00	348.00	---	1020.00	**1280.00**

Lasco Capriel-I, with grab bar
60" x 30" x 72" H

Description	Oper	Unit	Vol	Crew Size	Man-hours per Unit	Crew Output per Day	Avg Mat'l Unit Cost	Avg Labor Unit Cost	Avg Equip Unit Cost	Avg Total Unit Cost	Avg Price Incl O&P
Stock color	Inst	Ea	Lg	SB	8.00	2.00	555.00	260.00	---	815.00	**1020.00**
	Inst	Ea	Sm	SB	10.7	1.50	674.00	348.00	---	1022.00	**1290.00**

Lasco Capriel-II, with grab bar
60" x 32" x 74" H

Description	Oper	Unit	Vol	Crew Size	Man-hours per Unit	Crew Output per Day	Avg Mat'l Unit Cost	Avg Labor Unit Cost	Avg Equip Unit Cost	Avg Total Unit Cost	Avg Price Incl O&P
Stock color	Inst	Ea	Lg	SB	8.00	2.00	585.00	260.00	---	845.00	**1060.00**
	Inst	Ea	Sm	SB	10.7	1.50	711.00	348.00	---	1059.00	**1330.00**

Lasco Capriel-III, with grab bar
60" x 32" x 78" H

Description	Oper	Unit	Vol	Crew Size	Man-hours per Unit	Crew Output per Day	Avg Mat'l Unit Cost	Avg Labor Unit Cost	Avg Equip Unit Cost	Avg Total Unit Cost	Avg Price Incl O&P
Stock color	Inst	Ea	Lg	SB	8.00	2.00	615.00	260.00	---	875.00	**1090.00**
	Inst	Ea	Sm	SB	10.7	1.50	747.00	348.00	---	1095.00	**1370.00**

Multi-piece combination, no door

Lasco Mali, 3-piece
60" x 30" x 72" H

Description	Oper	Unit	Vol	Crew Size	Man-hours per Unit	Crew Output per Day	Avg Mat'l Unit Cost	Avg Labor Unit Cost	Avg Equip Unit Cost	Avg Total Unit Cost	Avg Price Incl O&P
Stock color	Inst	Ea	Lg	SB	8.00	2.00	568.00	260.00	---	828.00	**1040.00**
	Inst	Ea	Sm	SB	10.7	1.50	689.00	348.00	---	1037.00	**1300.00**

Lasco Oriana, 2-piece
60" x 30" x 72" H

Description	Oper	Unit	Vol	Crew Size	Man-hours per Unit	Crew Output per Day	Avg Mat'l Unit Cost	Avg Labor Unit Cost	Avg Equip Unit Cost	Avg Total Unit Cost	Avg Price Incl O&P
Stock color	Inst	Ea	Lg	SB	8.00	2.00	552.00	260.00	---	812.00	**1020.00**
	Inst	Ea	Sm	SB	10.7	1.50	670.00	348.00	---	1018.00	**1280.00**

Description	Oper	Unit	Vol	Crew Size	Man-hours per Unit	Crew Output per Day	Avg Mat'l Unit Cost	Avg Labor Unit Cost	Avg Equip Unit Cost	Avg Total Unit Cost	Avg Price Incl O&P

Shutters
Aluminum, louvered, 14" W

Description	Oper	Unit	Vol	Crew Size	Man-hours per Unit	Crew Output per Day	Avg Mat'l Unit Cost	Avg Labor Unit Cost	Avg Equip Unit Cost	Avg Total Unit Cost	Avg Price Incl O&P
3'-0" long	Inst	Pair	Lg	CA	1.00	8.00	48.20	33.30	---	81.50	**105.00**
	Inst	Pair	Sm	CA	1.33	6.00	54.90	44.30	---	99.20	**130.00**
6'-8" long	Inst	Pair	Lg	CA	1.00	8.00	80.30	33.30	---	113.60	**142.00**
	Inst	Pair	Sm	CA	1.33	6.00	91.50	44.30	---	135.80	**172.00**

Pine or fir, provincial raised panel, primed, in stock, 1-1/16" T
12" W

Description	Oper	Unit	Vol	Crew Size	Man-hours per Unit	Crew Output per Day	Avg Mat'l Unit Cost	Avg Labor Unit Cost	Avg Equip Unit Cost	Avg Total Unit Cost	Avg Price Incl O&P
2'-1" long	Inst	Pair	Lg	CA	1.00	8.00	29.90	33.30	---	63.20	**84.20**
	Inst	Pair	Sm	CA	1.33	6.00	34.00	44.30	---	78.30	**106.00**
3'-1" long	Inst	Pair	Lg	CA	1.00	8.00	31.60	33.30	---	64.90	**86.20**
	Inst	Pair	Sm	CA	1.33	6.00	36.00	44.30	---	80.30	**108.00**
4'-1" long	Inst	Pair	Lg	CA	1.00	8.00	36.20	33.30	---	69.50	**91.60**
	Inst	Pair	Sm	CA	1.33	6.00	41.30	44.30	---	85.60	**114.00**
5'-1" long	Inst	Pair	Lg	CA	1.00	8.00	43.90	33.30	---	77.20	**100.00**
	Inst	Pair	Sm	CA	1.33	6.00	50.00	44.30	---	94.30	**124.00**
6'-1" long	Inst	Pair	Lg	CA	1.00	8.00	51.80	33.30	---	85.10	**110.00**
	Inst	Pair	Sm	CA	1.33	6.00	59.10	44.30	---	103.40	**134.00**

18" W

Description	Oper	Unit	Vol	Crew Size	Man-hours per Unit	Crew Output per Day	Avg Mat'l Unit Cost	Avg Labor Unit Cost	Avg Equip Unit Cost	Avg Total Unit Cost	Avg Price Incl O&P
2'-1" long	Inst	Pair	Lg	CA	1.00	8.00	35.50	33.30	---	68.80	**90.70**
	Inst	Pair	Sm	CA	1.33	6.00	40.40	44.30	---	84.70	**113.00**
3'-1" long	Inst	Pair	Lg	CA	1.00	8.00	37.30	33.30	---	70.60	**92.90**
	Inst	Pair	Sm	CA	1.33	6.00	42.60	44.30	---	86.90	**115.00**
4'-1" long	Inst	Pair	Lg	CA	1.00	8.00	44.40	33.30	---	77.70	**101.00**
	Inst	Pair	Sm	CA	1.33	6.00	50.60	44.30	---	94.90	**125.00**
5'-1" long	Inst	Pair	Lg	CA	1.00	8.00	52.10	33.30	---	85.40	**110.00**
	Inst	Pair	Sm	CA	1.33	6.00	59.40	44.30	---	103.70	**135.00**
6'-1" long	Inst	Pair	Lg	CA	1.00	8.00	62.50	33.30	---	95.80	**122.00**
	Inst	Pair	Sm	CA	1.33	6.00	71.30	44.30	---	115.60	**148.00**

Door blinds, 6'-9" long

Description	Oper	Unit	Vol	Crew Size	Man-hours per Unit	Crew Output per Day	Avg Mat'l Unit Cost	Avg Labor Unit Cost	Avg Equip Unit Cost	Avg Total Unit Cost	Avg Price Incl O&P
1'-3" wide	Inst	Pair	Lg	CA	1.00	8.00	64.20	33.30	---	97.50	**124.00**
	Inst	Pair	Sm	CA	1.33	6.00	73.20	44.30	---	117.50	**151.00**
1'-6" wide	Inst	Pair	Lg	CA	1.00	8.00	80.30	33.30	---	113.60	**142.00**
	Inst	Pair	Sm	CA	1.33	6.00	91.50	44.30	---	135.80	**172.00**

Description	Oper	Unit	Vol	Crew Size	Man-hours per Unit	Crew Output per Day	Avg Mat'l Unit Cost	Avg Labor Unit Cost	Avg Equip Unit Cost	Avg Total Unit Cost	Avg Price Incl O&P

Birch, special order stationary slat blinds 1-1/16" T

12" W

Description	Oper	Unit	Vol	Crew Size	Man-hours per Unit	Crew Output per Day	Avg Mat'l Unit Cost	Avg Labor Unit Cost	Avg Equip Unit Cost	Avg Total Unit Cost	Avg Price Incl O&P
2'-1" long	Inst	Pair	Lg	CA	1.00	8.00	58.20	33.30	---	91.50	**117.00**
	Inst	Pair	Sm	CA	1.33	6.00	66.30	44.30	---	110.60	**143.00**
3'-1" long	Inst	Pair	Lg	CA	1.00	8.00	69.20	33.30	---	102.50	**129.00**
	Inst	Pair	Sm	CA	1.33	6.00	78.90	44.30	---	123.20	**157.00**
4'-1" long	Inst	Pair	Lg	CA	1.00	8.00	81.10	33.30	---	114.40	**143.00**
	Inst	Pair	Sm	CA	1.33	6.00	92.40	44.30	---	136.70	**173.00**
5'-1" long	Inst	Pair	Lg	CA	1.00	8.00	93.40	33.30	---	126.70	**157.00**
	Inst	Pair	Sm	CA	1.33	6.00	106.00	44.30	---	150.30	**189.00**
6'-1" long	Inst	Pair	Lg	CA	1.00	8.00	107.00	33.30	---	140.30	**173.00**
	Inst	Pair	Sm	CA	1.33	6.00	122.00	44.30	---	166.30	**207.00**

18" W

Description	Oper	Unit	Vol	Crew Size	Man-hours per Unit	Crew Output per Day	Avg Mat'l Unit Cost	Avg Labor Unit Cost	Avg Equip Unit Cost	Avg Total Unit Cost	Avg Price Incl O&P
2'-1" long	Inst	Pair	Lg	CA	1.00	8.00	63.20	33.30	---	96.50	**123.00**
	Inst	Pair	Sm	CA	1.33	6.00	72.00	44.30	---	116.30	**149.00**
3'-1" long	Inst	Pair	Lg	CA	1.00	8.00	77.80	33.30	---	111.10	**139.00**
	Inst	Pair	Sm	CA	1.33	6.00	88.80	44.30	---	133.10	**168.00**
4'-1" long	Inst	Pair	Lg	CA	1.00	8.00	92.30	33.30	---	125.60	**156.00**
	Inst	Pair	Sm	CA	1.33	6.00	105.00	44.30	---	149.30	**187.00**
5'-1" long	Inst	Pair	Lg	CA	1.00	8.00	109.00	33.30	---	142.30	**175.00**
	Inst	Pair	Sm	CA	1.33	6.00	124.00	44.30	---	168.30	**209.00**
6'-1" long	Inst	Pair	Lg	CA	1.00	8.00	123.00	33.30	---	156.30	**191.00**
	Inst	Pair	Sm	CA	1.33	6.00	140.00	44.30	---	184.30	**227.00**

Western hemlock, colonial style, primed, 1-1/8" T

14" wide

Description	Oper	Unit	Vol	Crew Size	Man-hours per Unit	Crew Output per Day	Avg Mat'l Unit Cost	Avg Labor Unit Cost	Avg Equip Unit Cost	Avg Total Unit Cost	Avg Price Incl O&P
5'-7" long	Inst	Pair	Lg	CA	1.00	8.00	56.70	33.30	---	90.00	**115.00**
	Inst	Pair	Sm	CA	1.33	6.00	64.70	44.30	---	109.00	**141.00**

16" wide

Description	Oper	Unit	Vol	Crew Size	Man-hours per Unit	Crew Output per Day	Avg Mat'l Unit Cost	Avg Labor Unit Cost	Avg Equip Unit Cost	Avg Total Unit Cost	Avg Price Incl O&P
3'-0" long	Inst	Pair	Lg	CA	1.00	8.00	35.30	33.30	---	68.60	**90.50**
	Inst	Pair	Sm	CA	1.33	6.00	40.30	44.30	---	84.60	**113.00**
4'-3" long	Inst	Pair	Lg	CA	1.00	8.00	42.80	33.30	---	76.10	**99.10**
	Inst	Pair	Sm	CA	1.33	6.00	48.80	44.30	---	93.10	**123.00**
5'-0" long	Inst	Pair	Lg	CA	1.00	8.00	49.20	33.30	---	82.50	**107.00**
	Inst	Pair	Sm	CA	1.33	6.00	56.10	44.30	---	100.40	**131.00**
6'-0" long	Inst	Pair	Lg	CA	1.00	8.00	54.60	33.30	---	87.90	**113.00**
	Inst	Pair	Sm	CA	1.33	6.00	62.20	44.30	---	106.50	**138.00**

Door blinds, 6'-9" long

Description	Oper	Unit	Vol	Crew Size	Man-hours per Unit	Crew Output per Day	Avg Mat'l Unit Cost	Avg Labor Unit Cost	Avg Equip Unit Cost	Avg Total Unit Cost	Avg Price Incl O&P
1'-3" wide	Inst	Pair	Lg	CA	1.00	8.00	70.60	33.30	---	103.90	**131.00**
	Inst	Pair	Sm	CA	1.33	6.00	80.50	44.30	---	124.80	**159.00**
1'-6" wide	Inst	Pair	Lg	CA	1.00	8.00	72.80	33.30	---	106.10	**134.00**
	Inst	Pair	Sm	CA	1.33	6.00	83.00	44.30	---	127.30	**162.00**

Description	Oper	Unit	Vol	Crew Size	Man-hours per Unit	Crew Output per Day	Avg Mat'l Unit Cost	Avg Labor Unit Cost	Avg Equip Unit Cost	Avg Total Unit Cost	Avg Price Incl O&P

Cellwood shutters, molded structural foam polystyrene

Prefinished louver, 16" wide

3'-3" long	Inst	Pair	Lg	CA	1.00	8.00	34.60	33.30	---	67.90	**89.70**
	Inst	Pair	Sm	CA	1.33	6.00	39.50	44.30	---	83.80	**112.00**
4'-7" long	Inst	Pair	Lg	CA	1.00	8.00	43.70	33.30	---	77.00	**100.00**
	Inst	Pair	Sm	CA	1.33	6.00	49.80	44.30	---	94.10	**124.00**
5'-3" long	Inst	Pair	Lg	CA	1.00	8.00	49.30	33.30	---	82.60	**107.00**
	Inst	Pair	Sm	CA	1.33	6.00	56.20	44.30	---	100.50	**131.00**
6'-3" long	Inst	Pair	Lg	CA	1.00	8.00	66.80	33.30	---	100.10	**127.00**
	Inst	Pair	Sm	CA	1.33	6.00	76.10	44.30	---	120.40	**154.00**

Door blinds

16" wide x 6'-9" long	Inst	Pair	Lg	CA	1.00	8.00	69.00	33.30	---	102.30	**129.00**
	Inst	Pair	Sm	CA	1.33	6.00	78.60	44.30	---	122.90	**157.00**

Prefinished panel, 16" wide

3'-3" long	Inst	Pair	Lg	CA	1.00	8.00	36.00	33.30	---	69.30	**91.30**
	Inst	Pair	Sm	CA	1.33	6.00	41.00	44.30	---	85.30	**114.00**
4'-7" long	Inst	Pair	Lg	CA	1.00	8.00	45.90	33.30	---	79.20	**103.00**
	Inst	Pair	Sm	CA	1.33	6.00	52.30	44.30	---	96.60	**127.00**
5'-3" long	Inst	Pair	Lg	CA	1.00	8.00	51.20	33.30	---	84.50	**109.00**
	Inst	Pair	Sm	CA	1.33	6.00	58.30	44.30	---	102.60	**133.00**
6'-3" long	Inst	Pair	Lg	CA	1.00	8.00	66.80	33.30	---	100.10	**127.00**
	Inst	Pair	Sm	CA	1.33	6.00	76.10	44.30	---	120.40	**154.00**

Door blinds

16" wide x 6'-9" long	Inst	Pair	Lg	CA	1.00	8.00	72.70	33.30	---	106.00	**134.00**
	Inst	Pair	Sm	CA	1.33	6.00	82.90	44.30	---	127.20	**162.00**

Siding

Aluminum Siding

1. **Dimensions and Descriptions.** The types of aluminum siding discussed are: (1) painted (2) available in various colors, (3) either 0.024" gauge or 0.019" gauge. The types of aluminum siding discussed in this section are:

   a. Clapboard: 12'-6" long; 8" exposure; $5/8$" butt; 0.024" gauge.

   b. Board and batten: 10'-0" long; 10" exposure; 1/2" butt; 0.024" gauge.

   c. "Double-five-inch" clapboard: 12'-0" long; 10" exposure; $1/2$" butt; 0.024" gauge.

   d. Sculpture clapboard: 12'-6" long; 8" exposure; $1/2$" butt; 0.024" gauge.

   e. Shingle siding: 24" long; 12" exposure; $1^1/8$" butt; 0.019" gauge.

Most types of aluminum siding may be available in a smooth or rough texture.

2. **Installation.** Applied horizontally with nails. See chart below.

3. **Estimating Technique.** Determine area and deduct area of window and door openings then add for cutting and fitting waste. In the figures on aluminum siding, no waste has been included.

Wood Siding

The types of wood exterior finishes covered in this section are:

   a. Bevel siding

   b. Drop siding

   c. Vertical siding

   d. Batten siding

   e. Plywood siding with battens

   f. Wood shingle siding

Bevel Siding

1. **Dimension.** Boards are 4", 6", 8", or 12" wide with a thickness of the top edge of $3/16$". The thickness of the bottom edge is $9/16$" for 4" and 6" widths and $11/16$" for 8", 10", 12" widths.

2. **Installation.** Applied horizontally with 6d or 8d common nails. See exposure and waste table below.

3. **Estimating Technique.** Determine area and deduct area of window and door openings, then add for lap, exposure, and cutting and fitting waste. In the figures on bevel siding, waste has been included, unless otherwise noted.

Type	Exposure	Butt	Normal Range of Cut & Fit Waste*
Clapboard	8"	$5/8$"	6% - 10%
Board & Batten	10"	$1/2$"	4% - 8%
Double-Five-Inch	10"	$1/2$"	6% - 10%
Sculptured	8"	$1/2$"	6% - 10%
Shingle Siding	12"	$1^1/8$"	4% - 6%

*The amount of waste will vary with the methods of installing siding and can be affected by dimension of the area to be covered.

Board Size	Actual Width	Lap	Exposure	Milling & Lap Waste	Normal Range of Cut & Fit Waste*
$1/2$" x 4"	$3^1/4$"	$1/2$"	$2^3/4$"	31.25%	19% - 23%
$1/2$" x 6"	$5^1/4$"	$1/2$"	$4^3/4$"	21.00%	10% - 13%
$1/2$" x 8"	$7^1/4$"	$1/2$"	$6^3/4$"	15.75%	7% - 10%
$5/8$" x 4"	$3^1/4$"	$1/2$"	$2^3/4$"	31.25%	19% - 23%
$5/8$" x 6"	$5^1/4$"	$1/2$"	$4^3/4$"	21.00%	10% - 13%
$5/8$" x 8"	$7^1/4$"	$1/2$"	$6^3/4$"	15.75%	7% - 10%
$5/8$" x 10"	$9^1/4$"	$1/2$"	$8^3/4$"	12.50%	6% - 10%
$3/4$" x 6"	$5^1/4$"	$1/2$"	$4^3/4$"	21.00%	10% - 13%
$3/4$" x 8"	$7^1/4$"	$1/2$"	$6^3/4$"	15.75%	7% - 10%
$3/4$" x 10"	$9^1/4$"	$1/2$"	$8^3/4$"	12.50%	6% - 10%
$3/4$" x 12"	$11^1/4$"	$1/2$"	$10^3/4$"	10.50%	6% - 10%

*The amount of waste will vary with the methods of installing siding and can be affected by dimension of the area to be covered.

Drop Siding, Tongue and Grooved

1. **Dimension.** The boards are 4", 6", 8", 10", or 12" wide, and all boards are 3/4" thick at both edges.

2. **Installation.** Applied horizontally, usually with 8d common nails. See exposure and waste table below:

Board Size	Actual Width	Lap +1/16"	Exposure	Milling & Lap Waste	Normal Range of Cut & Waste*
1" x 4"	3 1/4"	3/16"	3 1/16"	23.50%	7% - 10%
1" x 6"	5 1/4"	3/16"	5 1/16"	15.75%	6% - 8%
1" x 8"	7 1/4"	3/16"	7 1/16"	11.75%	5% - 6%
1" x 10"	9 1/4"	3/16"	9 1/16"	9.50%	4% - 5%
1" x 12"	11 1/4"	3/16"	11 1/16"	8.00%	4% - 5%

*The amount of waste will vary with the methods of installing siding and can be affected by dimensions of the area to be covered.

3. **Estimating Technique.** Determine area and deduct area of window and door openings; then add for lap, exposure, butting and fitting waste. In the figures on drop siding, waste has been included, unless otherwise noted.

Vertical Siding, Tongue and Grooved

1. **Dimensions.** Boards are 8", 10", or 12" wide, and all boards are 3/4" thick at both top and bottom edges.

2. **Installation.** Applied vertically over horizontal furring strips usually with 8d common nails. See exposure and waste table below:

Board Size	Actual Width	Lap 1/16"	Exposure	Milling & Lap Waste	Normal Range of Cut & Fit Waste
1" x 8"	7 1/4"	3/16"	7 1/16"	11.75%	5% - 7%
1" x 10"	9 1/4"	3/16"	9 1/16"	9.50%	4% - 5%
1" x 12"	11 1/4"	3/16"	11 1/16"	8.00%	4% - 5%

3. **Estimating Technique.** Determine area and deduct area of window and door openings; then add for lap, exposure, cutting and fitting waste. In the figures on vertical siding, waste has been included, unless otherwise noted.

Batten Siding

1. **Dimension.** The boards are 8", 10", or 12" wide. If the boards are rough, they are 1" thick; if the boards are dressed, then they are 25/32" thick. The battens are 1" thick and 2" wide.

2. **Installation.** The 8", 10", or 12"-wide boards are installed vertically over 1" x 3" or 1" x 4" horizontal furring strips. The battens are then installed over the seams of the vertical boards. See exposure and waste table below:

Board Size & Type		Actual Width & Exposure	Milling Waste	Normal Range of Cut & Fit Waste*
1" x 8"	rough sawn	8"	—	3% - 5%
	dressed	7 1/2"	6.25%	3% - 5%
1" x 10"	rough sawn	10"	—	3% - 5%
	dressed	9 1/2"	5.00%	3% - 5%
1" x 12"	rough sawn	12"	—	3% - 5%
	dressed	11 1/2"	4.25%	3% - 5%

*The amount of waste will vary with the methods of installing siding and can be affected by dimension of the area to be covered.

3. **Estimating Technique.** Determine area and deduct area of window and door openings; then add for exposure, cutting and fitting waste. In the figures on batten siding, waste has been included, unless otherwise noted.

Quantity of Battens - No Cut and Fit Waste Included		
Board Size and Type		LF per SF or BF
1" x 6"	rough sawn	2.00
	dressed	2.25
1" x 8"	rough sawn	1.55
	dressed	1.60
1" x 10"	rough sawn	1.20
	dressed	1.30
1" x 12"	rough sawn	1.00
	dressed	1.05

Plywood Siding with Battens

1. **Dimensions.** The plywood panels are 4' wide and 8', 9', or 10' long. The panels are $3/8$", $1/2$" or $5/8$" thick. The battens are 1" thick and 2" wide.

2. **Installation.** The panels are applied directly to the studs with the 4'-0" width parallel to the plate. Nails used are 6d commons. The waste will normally range from 3% to 5%.

Quantity of Battens - No Cut and Fit Waste Included	
Spacing	LF per SF or BF
16" oc	.75
24" oc	.50

3. **Estimating Technique.** Determine area and deduct area of window and door openings; then add for cutting and fitting waste. In the figures on plywood siding with battens, waste has been included, unless otherwise noted.

Wood Shingle Siding

1. **Dimension.** The wood shingles have an average width of 4" and length of 16", 18" or 24". The thickness may vary in all cases.

2. **Installation.** On roofs with wood shingles, there should be three thicknesses of wood shingle at every point on the roof. But on exterior walls with wood shingles, only two thicknesses of wood siding shingles are needed. Nail to 1" x 3" or 1" x 4" furring strips using 3d rust-resistant nails. Use 5d rust-resistant nails when applying new shingles over old shingles. See table below:

Shingle Length	Exposure	Shingles per Square, No Waste Included	Lbs. of Nails at Two Nails per Shingle	
			3d	5d
16"	$7^1/_2$"	480	2.3	2.7
18"	$8^1/_2$"	424	2.0	2.4
24"	$11^1/_2$"	313	1.5	1.8

3. **Estimating Technique.** Determine the area and deduct the area of window and door openings. Then add approximately 10% for waste. In the figures on wood shingle siding, waste has been included, unless otherwise noted. In the above table, pounds of nails per square are based on two nails per shingle. For one nail per shingle, deduct 50%.

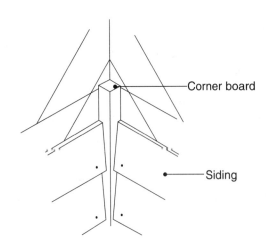

Corner board for application of horizontal siding at interior corner

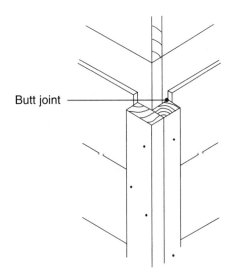

Corner boards for application of horizontal siding at exterior corner

Description	Oper	Unit	Vol	Crew Size	Man-hours per Unit	Crew Output per Day	Avg Mat'l Unit Cost	Avg Labor Unit Cost	Avg Equip Unit Cost	Avg Total Unit Cost	Avg Price Incl O&P

Siding

Aluminum, nailed to wood

Clapboard (i.e. lap, drop)

Description	Oper	Unit	Vol	Crew Size	Man-hours per Unit	Crew Output per Day	Avg Mat'l Unit Cost	Avg Labor Unit Cost	Avg Equip Unit Cost	Avg Total Unit Cost	Avg Price Incl O&P
8" exposure	Demo	SF	Lg	LB	.016	1025	---	.44	---	.44	.65
	Demo	SF	Sm	LB	.022	718.0	---	.60	---	.60	.90
10" exposure	Demo	SF	Lg	LB	.013	1280	---	.36	---	.36	.53
	Demo	SF	Sm	LB	.018	896.0	---	.49	---	.49	.74

Clapboard, horizontal, 1/2" butt

Straight wall

Description	Oper	Unit	Vol	Crew Size	Man-hours per Unit	Crew Output per Day	Avg Mat'l Unit Cost	Avg Labor Unit Cost	Avg Equip Unit Cost	Avg Total Unit Cost	Avg Price Incl O&P
8" exposure	Inst	SF	Lg	2C	.026	620.0	2.17	.87	---	3.04	3.79
	Inst	SF	Sm	2C	.037	434.0	2.54	1.23	---	3.77	4.77
10" exposure	Inst	SF	Lg	2C	.021	770.0	1.80	.70	---	2.50	3.12
	Inst	SF	Sm	2C	.030	539.0	2.11	1.00	---	3.11	3.92

Cut-up wall

Description	Oper	Unit	Vol	Crew Size	Man-hours per Unit	Crew Output per Day	Avg Mat'l Unit Cost	Avg Labor Unit Cost	Avg Equip Unit Cost	Avg Total Unit Cost	Avg Price Incl O&P
8" exposure	Inst	SF	Lg	2C	.031	520.0	2.85	1.03	---	3.88	4.82
	Inst	SF	Sm	2C	.044	364.0	3.34	1.46	---	4.80	6.04
10" exposure	Inst	SF	Lg	2C	.025	640.0	2.47	.83	---	3.30	4.09
	Inst	SF	Sm	2C	.036	448.0	2.90	1.20	---	4.10	5.13

Corrugated (2-1/2"), 26" W, 2-1/2" side lap, 4" end lap

Description	Oper	Unit	Vol	Crew Size	Man-hours per Unit	Crew Output per Day	Avg Mat'l Unit Cost	Avg Labor Unit Cost	Avg Equip Unit Cost	Avg Total Unit Cost	Avg Price Incl O&P
	Demo	SF	Lg	LB	.013	1200	---	.36	---	.36	.53
	Demo	SF	Sm	LB	.019	840.0	---	.52	---	.52	.78

0.0175" thick

Description	Oper	Unit	Vol	Crew Size	Man-hours per Unit	Crew Output per Day	Avg Mat'l Unit Cost	Avg Labor Unit Cost	Avg Equip Unit Cost	Avg Total Unit Cost	Avg Price Incl O&P
Natural	Inst	SF	Lg	UC	.019	1660	2.49	.61	---	3.10	3.77
	Inst	SF	Sm	UC	.028	1162	2.92	.90	---	3.82	4.69
Painted	Inst	SF	Lg	UC	.019	1660	2.70	.61	---	3.31	4.01
	Inst	SF	Sm	UC	.028	1162	3.17	.90	---	4.07	4.98

0.019" thick

Description	Oper	Unit	Vol	Crew Size	Man-hours per Unit	Crew Output per Day	Avg Mat'l Unit Cost	Avg Labor Unit Cost	Avg Equip Unit Cost	Avg Total Unit Cost	Avg Price Incl O&P
Natural	Inst	SF	Lg	UC	.019	1660	2.53	.61	---	3.14	3.81
	Inst	SF	Sm	UC	.028	1162	2.98	.90	---	3.88	4.76
Painted	Inst	SF	Lg	UC	.019	1660	2.75	.61	---	3.36	4.07
	Inst	SF	Sm	UC	.028	1162	3.23	.90	---	4.13	5.05

Panels, 4' x 8' board and batten (12" oc)

Description	Oper	Unit	Vol	Crew Size	Man-hours per Unit	Crew Output per Day	Avg Mat'l Unit Cost	Avg Labor Unit Cost	Avg Equip Unit Cost	Avg Total Unit Cost	Avg Price Incl O&P
	Demo	SF	Lg	LB	.007	2400	---	.19	---	.19	.29
	Demo	SF	Sm	LB	.010	1680	---	.27	---	.27	.41

Straight wall

Description	Oper	Unit	Vol	Crew Size	Man-hours per Unit	Crew Output per Day	Avg Mat'l Unit Cost	Avg Labor Unit Cost	Avg Equip Unit Cost	Avg Total Unit Cost	Avg Price Incl O&P
Smooth	Inst	SF	Lg	2C	.013	1220	2.23	.43	---	2.66	3.21
	Inst	SF	Sm	2C	.019	854.0	2.62	.63	---	3.25	3.96
Rough sawn	Inst	SF	Lg	2C	.013	1220	2.44	.43	---	2.87	3.45
	Inst	SF	Sm	2C	.019	854.0	2.87	.63	---	3.50	4.25

Description	Oper	Unit	Vol	Crew Size	Man-hours per Unit	Crew Output per Day	Avg Mat'l Unit Cost	Avg Labor Unit Cost	Avg Equip Unit Cost	Avg Total Unit Cost	Avg Price Incl O&P
Cut-up wall											
Smooth	Inst	SF	Lg	2C	.016	1025	2.91	.53	---	3.44	**4.15**
	Inst	SF	Sm	2C	.022	718.0	3.42	.73	---	4.15	**5.03**
Rough sawn	Inst	SF	Lg	2C	.016	1025	3.13	.53	---	3.66	**4.40**
	Inst	SF	Sm	2C	.022	718.0	3.68	.73	---	4.41	**5.33**

Shingle, 24" L with 12" exposure

Description	Oper	Unit	Vol	Crew Size	Man-hours per Unit	Crew Output per Day	Avg Mat'l Unit Cost	Avg Labor Unit Cost	Avg Equip Unit Cost	Avg Total Unit Cost	Avg Price Incl O&P
	Demo	SF	Lg	LB	.011	1450	---	.30	---	.30	**.45**
	Demo	SF	Sm	LB	.016	1015	---	.44	---	.44	**.65**
Straight wall	Inst	SF	Lg	2C	.018	890.0	.05	.60	---	.65	**.96**
	Inst	SF	Sm	2C	.026	623.0	.06	.87	---	.93	**1.37**
Cut-up wall	Inst	SF	Lg	2C	.023	710.0	.05	.77	---	.82	**1.21**
	Inst	SF	Sm	2C	.032	497.0	.06	1.06	---	1.12	**1.67**

Hardboard siding, Masonite
Lap, 1/2" T x 12" W x 16' L, with 11" exposure

Description	Oper	Unit	Vol	Crew Size	Man-hours per Unit	Crew Output per Day	Avg Mat'l Unit Cost	Avg Labor Unit Cost	Avg Equip Unit Cost	Avg Total Unit Cost	Avg Price Incl O&P
	Demo	SF	Lg	LB	.012	1345	---	.33	---	.33	**.49**
	Demo	SF	Sm	LB	.017	942.0	---	.47	---	.47	**.69**
Straight wall											
Smooth finish	Inst	SF	Lg	2C	.023	700.0	1.41	.77	---	2.18	**2.77**
	Inst	SF	Sm	2C	.033	490.0	1.66	1.10	---	2.76	**3.56**
Textured finish	Inst	SF	Lg	2C	.023	700.0	1.67	.77	---	2.44	**3.07**
	Inst	SF	Sm	2C	.033	490.0	1.96	1.10	---	3.06	**3.90**
Wood-like finish	Inst	SF	Lg	2C	.023	700.0	1.78	.77	---	2.55	**3.19**
	Inst	SF	Sm	2C	.033	490.0	2.09	1.10	---	3.19	**4.05**
Cut-up wall											
Smooth finish	Inst	SF	Lg	2C	.027	585.0	1.45	.90	---	2.35	**3.02**
	Inst	SF	Sm	2C	.039	410.0	1.70	1.30	---	3.00	**3.90**
Textured finish	Inst	SF	Lg	2C	.027	585.0	1.71	.90	---	2.61	**3.31**
	Inst	SF	Sm	2C	.039	410.0	2.01	1.30	---	3.31	**4.26**
Wood-like finish	Inst	SF	Lg	2C	.027	585.0	1.83	.90	---	2.73	**3.45**
	Inst	SF	Sm	2C	.039	410.0	2.14	1.30	---	3.44	**4.41**

Description	Oper	Unit	Vol	Crew Size	Man-hours per Unit	Crew Output per Day	Avg Mat'l Unit Cost	Avg Labor Unit Cost	Avg Equip Unit Cost	Avg Total Unit Cost	Avg Price Incl O&P
Panels, 7/16" T x 4' W x 8' H											
	Demo	SF	Lg	LB	.006	2520	---	.16	---	.16	.25
	Demo	SF	Sm	LB	.009	1764	---	.25	---	.25	.37
Grooved/planked panels, textured or wood-like finish											
Primecote	Inst	SF	Lg	2C	.013	1215	1.33	.43	---	1.76	2.18
	Inst	SF	Sm	2C	.019	851.0	1.56	.63	---	2.19	2.74
Unprimed	Inst	SF	Lg	2C	.013	1215	1.44	.43	---	1.87	2.30
	Inst	SF	Sm	2C	.019	851.0	1.70	.63	---	2.33	2.90
Panel with batten preattached											
Unprimed	Inst	SF	Lg	2C	.013	1215	1.52	.43	---	1.95	2.40
	Inst	SF	Sm	2C	.019	851.0	1.78	.63	---	2.41	3.00
Stucco textured panel											
Primecote	Inst	SF	Lg	2C	.013	1215	1.55	.43	---	1.98	2.43
	Inst	SF	Sm	2C	.019	851.0	1.83	.63	---	2.46	3.05

Vinyl

Solid vinyl panels .024 gauge, woodgrained or colors

Description	Oper	Unit	Vol	Crew Size	Man-hours per Unit	Crew Output per Day	Avg Mat'l Unit Cost	Avg Labor Unit Cost	Avg Equip Unit Cost	Avg Total Unit Cost	Avg Price Incl O&P
10" or double 5" panels	Demo	SF	Lg	LB	.012	1345	---	.33	---	.33	.49
	Demo	SF	Sm	LB	.017	942.0	---	.47	---	.47	.69
10" or double 5" panels, standard											
	Inst	SF	Lg	2C	.032	500.0	.75	1.06	---	1.81	2.46
	Inst	SF	Sm	2C	.046	350.0	.88	1.53	---	2.41	3.31
10" or double 5" panels, insulated											
	Inst	SF	Lg	2C	.032	500.0	1.05	1.06	---	2.11	2.80
	Inst	SF	Sm	2C	.046	350.0	1.24	1.53	---	2.77	3.72
Starter strip	Inst	LF	Lg	2C	.040	400.0	.27	1.33	---	1.60	2.31
	Inst	LF	Sm	2C	.057	280.0	.32	1.90	---	2.22	3.21
J channel for corners	Inst	LF	Lg	2C	.036	450.0	.63	1.20	---	1.83	2.52
	Inst	LF	Sm	2C	.051	315.0	.74	1.70	---	2.44	3.40
Outside corner and corner post											
	Inst	LF	Lg	2C	.040	400.0	.93	1.33	---	2.26	3.07
	Inst	LF	Sm	2C	.057	280.0	1.10	1.90	---	3.00	4.11
Casing and trim	Inst	LF	Lg	2C	.046	350.0	.49	1.53	---	2.02	2.86
	Inst	LF	Sm	2C	.065	245.0	.58	2.16	---	2.74	3.91
Soffit and fascia	Inst	LF	Lg	2C	.040	400.0	.69	1.33	---	2.02	2.79
	Inst	LF	Sm	2C	.057	280.0	.81	1.90	---	2.71	3.78

Wood

Bevel

Description	Oper	Unit	Vol	Crew Size	Man-hours per Unit	Crew Output per Day	Avg Mat'l Unit Cost	Avg Labor Unit Cost	Avg Equip Unit Cost	Avg Total Unit Cost	Avg Price Incl O&P
1/2" x 8" with 6-3/4" exp.	Demo	SF	Lg	LB	.018	910.0	---	.49	---	.49	.74
	Demo	SF	Sm	LB	.025	637.0	---	.69	---	.69	1.02
5/8" x 10" with 8-3/4" exp.	Demo	SF	Lg	LB	.014	1180	---	.38	---	.38	.57
	Demo	SF	Sm	LB	.019	826.0	---	.52	---	.52	.78
3/4" x 12" with 10-3/4" exp.	Demo	SF	Lg	LB	.011	1450	---	.30	---	.30	.45
	Demo	SF	Sm	LB	.016	1015	---	.44	---	.44	.65

Description	Oper	Unit	Vol	Crew Size	Man-hours per Unit	Crew Output per Day	Avg Mat'l Unit Cost	Avg Labor Unit Cost	Avg Equip Unit Cost	Avg Total Unit Cost	Avg Price Incl O&P

Bevel (horizontal), 1/2" lap, 10' L

Redwood, kiln-dried, clear VG, pattern

1/2" x 8", with 6-3/4" exp.

Description	Oper	Unit	Vol	Crew Size	Man-hours per Unit	Crew Output per Day	Avg Mat'l Unit Cost	Avg Labor Unit Cost	Avg Equip Unit Cost	Avg Total Unit Cost	Avg Price Incl O&P
Rough ends	Inst	SF	Lg	2C	.037	435.0	2.63	1.23	---	3.86	**4.87**
	Inst	SF	Sm	2C	.052	305.0	3.09	1.73	---	4.82	**6.15**
Fitted ends	Inst	SF	Lg	2C	.048	335.0	2.70	1.60	---	4.30	**5.50**
	Inst	SF	Sm	2C	.068	235.0	3.17	2.26	---	5.43	**7.04**
Mitered ends	Inst	SF	Lg	2C	.056	285.0	2.76	1.86	---	4.62	**5.97**
	Inst	SF	Sm	2C	.080	200.0	3.24	2.66	---	5.90	**7.72**

5/8" x 10", with 8-3/4" exp.

Description	Oper	Unit	Vol	Crew Size	Man-hours per Unit	Crew Output per Day	Avg Mat'l Unit Cost	Avg Labor Unit Cost	Avg Equip Unit Cost	Avg Total Unit Cost	Avg Price Incl O&P
Rough ends	Inst	SF	Lg	2C	.029	560.0	2.57	.96	---	3.53	**4.40**
	Inst	SF	Sm	2C	.041	392.0	3.02	1.36	---	4.38	**5.52**
Fitted ends	Inst	SF	Lg	2C	.037	430.0	2.63	1.23	---	3.86	**4.87**
	Inst	SF	Sm	2C	.053	301.0	3.10	1.76	---	4.86	**6.21**
Mitered ends	Inst	SF	Lg	2C	.043	370.0	2.69	1.43	---	4.12	**5.24**
	Inst	SF	Sm	2C	.062	259.0	3.17	2.06	---	5.23	**6.74**

3/4" x 12", with 10-3/4" exp.

Description	Oper	Unit	Vol	Crew Size	Man-hours per Unit	Crew Output per Day	Avg Mat'l Unit Cost	Avg Labor Unit Cost	Avg Equip Unit Cost	Avg Total Unit Cost	Avg Price Incl O&P
Rough ends	Inst	SF	Lg	2C	.023	690.0	4.45	.77	---	5.22	**6.27**
	Inst	SF	Sm	2C	.033	483.0	5.24	1.10	---	6.34	**7.67**
Fitted ends	Inst	SF	Lg	2C	.030	530.0	4.57	1.00	---	5.57	**6.75**
	Inst	SF	Sm	2C	.043	371.0	5.37	1.43	---	6.80	**8.32**
Mitered ends	Inst	SF	Lg	2C	.035	455.0	4.68	1.16	---	5.84	**7.13**
	Inst	SF	Sm	2C	.050	319.0	5.50	1.66	---	7.16	**8.82**

Red cedar, clear, kiln-dried

1/2" x 8", with 6-3/4" exp.

Description	Oper	Unit	Vol	Crew Size	Man-hours per Unit	Crew Output per Day	Avg Mat'l Unit Cost	Avg Labor Unit Cost	Avg Equip Unit Cost	Avg Total Unit Cost	Avg Price Incl O&P
Rough ends	Inst	SF	Lg	2C	.037	435.0	2.33	1.23	---	3.56	**4.53**
	Inst	SF	Sm	2C	.052	305.0	2.74	1.73	---	4.47	**5.75**
Fitted ends	Inst	SF	Lg	2C	.048	335.0	2.38	1.60	---	3.98	**5.13**
	Inst	SF	Sm	2C	.068	235.0	2.80	2.26	---	5.06	**6.61**
Mitered ends	Inst	SF	Lg	2C	.056	285.0	2.44	1.86	---	4.30	**5.60**
	Inst	SF	Sm	2C	.080	200.0	2.87	2.66	---	5.53	**7.29**

5/8" x 10", with 8-3/4" exp.

Description	Oper	Unit	Vol	Crew Size	Man-hours per Unit	Crew Output per Day	Avg Mat'l Unit Cost	Avg Labor Unit Cost	Avg Equip Unit Cost	Avg Total Unit Cost	Avg Price Incl O&P
Rough ends	Inst	SF	Lg	2C	.029	560.0	2.44	.96	---	3.40	**4.25**
	Inst	SF	Sm	2C	.041	392.0	2.87	1.36	---	4.23	**5.35**
Fitted ends	Inst	SF	Lg	2C	.037	430.0	2.50	1.23	---	3.73	**4.72**
	Inst	SF	Sm	2C	.053	301.0	2.95	1.76	---	4.71	**6.04**
Mitered ends	Inst	SF	Lg	2C	.043	370.0	2.56	1.43	---	3.99	**5.09**
	Inst	SF	Sm	2C	.062	259.0	3.02	2.06	---	5.08	**6.57**

3/4" x 12", with 10-3/4" exp.

Description	Oper	Unit	Vol	Crew Size	Man-hours per Unit	Crew Output per Day	Avg Mat'l Unit Cost	Avg Labor Unit Cost	Avg Equip Unit Cost	Avg Total Unit Cost	Avg Price Incl O&P
Rough ends	Inst	SF	Lg	2C	.023	690.0	4.03	.77	---	4.80	**5.78**
	Inst	SF	Sm	2C	.033	483.0	4.74	1.10	---	5.84	**7.10**
Fitted ends	Inst	SF	Lg	2C	.030	530.0	4.13	1.00	---	5.13	**6.25**
	Inst	SF	Sm	2C	.043	371.0	4.86	1.43	---	6.29	**7.74**
Mitered ends	Inst	SF	Lg	2C	.035	455.0	4.23	1.16	---	5.39	**6.61**
	Inst	SF	Sm	2C	.050	319.0	4.98	1.66	---	6.64	**8.22**

Description	Oper	Unit	Vol	Crew Size	Man-hours per Unit	Crew Output per Day	Avg Mat'l Unit Cost	Avg Labor Unit Cost	Avg Equip Unit Cost	Avg Total Unit Cost	Avg Price Incl O&P
Drop (horizontal) T&G, 1/4" lap											
1" x 8" with 7" exp.	Demo	SF	Lg	LB	.017	945.0	---	.47	---	.47	.69
	Demo	SF	Sm	LB	.024	662.0	---	.66	---	.66	.98
1" x 10" with 9" exp.	Demo	SF	Lg	LB	.013	1215	---	.36	---	.36	.53
	Demo	SF	Sm	LB	.019	851.0	---	.52	---	.52	.78
Drop (horizontal) T&G, 1/4" lap, 10' L (avg.)											
Redwood, kiln-dried, clear VG, patterns											
1" x 8", with 7" exp.											
Rough ends	Inst	SF	Lg	2C	.034	465.0	4.31	1.13	---	5.44	6.65
	Inst	SF	Sm	2C	.049	326.0	5.07	1.63	---	6.70	8.28
Fitted ends	Inst	SF	Lg	2C	.044	360.0	4.42	1.46	---	5.88	7.28
	Inst	SF	Sm	2C	.063	252.0	5.20	2.10	---	7.30	9.12
Mitered ends	Inst	SF	Lg	2C	.051	315.0	4.52	1.70	---	6.22	7.74
	Inst	SF	Sm	2C	.072	221.0	5.32	2.40	---	7.72	9.71
1" x 10", with 9" exp.											
Rough ends	Inst	SF	Lg	2C	.027	600.0	4.53	.90	---	5.43	6.56
	Inst	SF	Sm	2C	.038	420.0	5.32	1.26	---	6.58	8.01
Fitted ends	Inst	SF	Lg	2C	.034	465.0	4.64	1.13	---	5.77	7.03
	Inst	SF	Sm	2C	.049	326.0	5.45	1.63	---	7.08	8.71
Mitered ends	Inst	SF	Lg	2C	.040	405.0	4.75	1.33	---	6.08	7.46
	Inst	SF	Sm	2C	.056	284.0	5.59	1.86	---	7.45	9.22
1" x 8", with 7-3/4" exp.											
Rough ends	Inst	SF	Lg	2C	.031	515.0	3.91	1.03	---	4.94	6.04
	Inst	SF	Sm	2C	.044	361.0	4.61	1.46	---	6.07	7.50
Fitted ends	Inst	SF	Lg	2C	.040	400.0	4.02	1.33	---	5.35	6.62
	Inst	SF	Sm	2C	.057	280.0	4.73	1.90	---	6.63	8.28
Mitered ends	Inst	SF	Lg	2C	.046	350.0	4.12	1.53	---	5.65	7.03
	Inst	SF	Sm	2C	.065	245.0	4.85	2.16	---	7.01	8.82
1" x 10", with 9-3/4" exp.											
Rough ends	Inst	SF	Lg	2C	.025	650.0	4.20	.83	---	5.03	6.08
	Inst	SF	Sm	2C	.035	455.0	4.94	1.16	---	6.10	7.43
Fitted ends	Inst	SF	Lg	2C	.032	505.0	4.32	1.06	---	5.38	6.57
	Inst	SF	Sm	2C	.045	354.0	5.07	1.50	---	6.57	8.08
Mitered ends	Inst	SF	Lg	2C	.036	440.0	4.43	1.20	---	5.63	6.89
	Inst	SF	Sm	2C	.052	308.0	5.21	1.73	---	6.94	8.59

Description	Oper	Unit	Vol	Crew Size	Man-hours per Unit	Crew Output per Day	Avg Mat'l Unit Cost	Avg Labor Unit Cost	Avg Equip Unit Cost	Avg Total Unit Cost	Avg Price Incl O&P

Red cedar, clear, kiln-dried
1" x 8", with 7" exp.

Rough ends	Inst	SF	Lg	2C	.034	465.0	3.94	1.13	---	5.07	**6.23**
	Inst	SF	Sm	2C	.049	326.0	4.63	1.63	---	6.26	**7.77**
Fitted ends	Inst	SF	Lg	2C	.044	360.0	4.04	1.46	---	5.50	**6.84**
	Inst	SF	Sm	2C	.063	252.0	4.74	2.10	---	6.84	**8.60**
Mitered ends	Inst	SF	Lg	2C	.051	315.0	4.13	1.70	---	5.83	**7.29**
	Inst	SF	Sm	2C	.072	221.0	4.85	2.40	---	7.25	**9.17**

1" x 10", with 9" exp.

Rough ends	Inst	SF	Lg	2C	.027	600.0	4.07	.90	---	4.97	**6.03**
	Inst	SF	Sm	2C	.038	420.0	4.80	1.26	---	6.06	**7.42**
Fitted ends	Inst	SF	Lg	2C	.034	465.0	4.17	1.13	---	5.30	**6.49**
	Inst	SF	Sm	2C	.049	326.0	4.92	1.63	---	6.55	**8.10**
Mitered ends	Inst	SF	Lg	2C	.040	405.0	4.28	1.33	---	5.61	**6.92**
	Inst	SF	Sm	2C	.056	284.0	5.04	1.86	---	6.90	**8.59**

Pine, kiln-dried, select
Rough sawn
1" x 8", with 7-3/4" exp.

Rough ends	Inst	SF	Lg	2C	.031	515.0	2.75	1.03	---	3.78	**4.71**
	Inst	SF	Sm	2C	.044	361.0	3.23	1.46	---	4.69	**5.91**
Fitted ends	Inst	SF	Lg	2C	.040	400.0	2.82	1.33	---	4.15	**5.24**
	Inst	SF	Sm	2C	.057	280.0	3.31	1.90	---	5.21	**6.65**
Mitered ends	Inst	SF	Lg	2C	.046	350.0	2.89	1.53	---	4.42	**5.62**
	Inst	SF	Sm	2C	.065	245.0	3.40	2.16	---	5.56	**7.15**

1" x 10", with 9-3/4" exp.

Rough ends	Inst	SF	Lg	2C	.025	650.0	3.47	.83	---	4.30	**5.24**
	Inst	SF	Sm	2C	.035	455.0	4.08	1.16	---	5.24	**6.44**
Fitted ends	Inst	SF	Lg	2C	.032	505.0	3.56	1.06	---	4.62	**5.69**
	Inst	SF	Sm	2C	.045	354.0	4.19	1.50	---	5.69	**7.06**
Mitered ends	Inst	SF	Lg	2C	.036	440.0	3.65	1.20	---	4.85	**5.99**
	Inst	SF	Sm	2C	.052	308.0	4.30	1.73	---	6.03	**7.54**

Dressed/milled
1" x 8", with 7" exp.

Rough ends	Inst	SF	Lg	2C	.034	465.0	3.29	1.13	---	4.42	**5.48**
	Inst	SF	Sm	2C	.049	326.0	3.87	1.63	---	5.50	**6.90**
Fitted ends	Inst	SF	Lg	2C	.044	360.0	3.37	1.46	---	4.83	**6.07**
	Inst	SF	Sm	2C	.063	252.0	3.96	2.10		6.06	**7.70**
Mitered ends	Inst	SF	Lg	2C	.051	315.0	3.45	1.70	---	5.15	**6.51**
	Inst	SF	Sm	2C	.072	221.0	4.05	2.40	---	6.45	**8.25**

1" x 10", with 9" exp.

Rough ends	Inst	SF	Lg	2C	.027	600.0	4.10	.90	---	5.00	**6.06**
	Inst	SF	Sm	2C	.038	420.0	4.82	1.26	---	6.08	**7.44**
Fitted ends	Inst	SF	Lg	2C	.034	465.0	4.20	1.13	---	5.33	**6.53**
	Inst	SF	Sm	2C	.049	326.0	4.94	1.63	---	6.57	**8.13**
Mitered ends	Inst	SF	Lg	2C	.040	405.0	4.30	1.33	---	5.63	**6.94**
	Inst	SF	Sm	2C	.056	284.0	5.06	1.86	---	6.92	**8.61**

Description	Oper	Unit	Vol	Crew Size	Man-hours per Unit	Crew Output per Day	Avg Mat'l Unit Cost	Avg Labor Unit Cost	Avg Equip Unit Cost	Avg Total Unit Cost	Avg Price Incl O&P

Board (1" x 12") and batten (1" x 2", 12" oc)

Horizontal
| | Demo | SF | Lg | LB | .013 | 1280 | --- | .36 | --- | .36 | .53 |
| | Demo | SF | Sm | LB | .018 | 896.0 | --- | .49 | --- | .49 | .74 |

Vertical
Standard	Demo	SF	Lg	LB	.016	1025	---	.44	---	.44	.65
	Demo	SF	Sm	LB	.022	718.0	---	.60	---	.60	.90
Reverse	Demo	SF	Lg	LB	.015	1065	---	.41	---	.41	.61
	Demo	SF	Sm	LB	.021	746.0	---	.58	---	.58	.86

Board (1" x 12") and batten (1" x 2", 12" oc); rough sawn, common, standard and better, green
Horizontal application, @ 10' L
Redwood	Inst	SF	Lg	2C	.029	560.0	2.02	.96	---	2.98	3.77
	Inst	SF	Sm	2C	.041	392.0	2.38	1.36	---	3.74	4.78
Red cedar	Inst	SF	Lg	2C	.029	560.0	1.43	.96	---	2.39	3.09
	Inst	SF	Sm	2C	.041	392.0	1.75	1.36	---	3.11	4.06
Pine	Inst	SF	Lg	2C	.029	560.0	2.01	.96	---	2.97	3.76
	Inst	SF	Sm	2C	.041	392.0	2.37	1.36	---	3.73	4.77

Vertical, std. (batten on board), @ 8' L
Redwood	Inst	SF	Lg	2C	.036	450.0	2.02	1.20	---	3.22	4.12
	Inst	SF	Sm	2C	.051	315.0	2.38	1.70	---	4.08	5.28
Red cedar	Inst	SF	Lg	2C	.036	450.0	1.43	1.20	---	2.63	3.44
	Inst	SF	Sm	2C	.051	315.0	1.75	1.70	---	3.45	4.56
Pine	Inst	SF	Lg	2C	.036	450.0	2.01	1.20	---	3.21	4.11
	Inst	SF	Sm	2C	.051	315.0	2.37	1.70	---	4.07	5.27

Vertical, reverse (board on batten), @ 8' L
Redwood	Inst	SF	Lg	2C	.036	440.0	2.02	1.20	---	3.22	4.12
	Inst	SF	Sm	2C	.052	308.0	2.38	1.73	---	4.11	5.33
Red cedar	Inst	SF	Lg	2C	.036	440.0	1.43	1.20	---	2.63	3.44
	Inst	SF	Sm	2C	.052	308.0	1.75	1.73	---	3.48	4.61
Pine	Inst	SF	Lg	2C	.036	440.0	2.01	1.20	---	3.21	4.11
	Inst	SF	Sm	2C	.052	308.0	2.37	1.73	---	4.10	5.32

Board on board (1" x 12" with 1-1/2" overlap), vertical
| | Demo | SF | Lg | LB | .017 | 950.0 | --- | .47 | --- | .47 | .69 |
| | Demo | SF | Sm | LB | .024 | 665.0 | --- | .66 | --- | .66 | .98 |

Board on board (1" x 12" with 1-1/2" overlap) rough sawn common, standard & better green, 8' L; vertical
Redwood	Inst	SF	Lg	2C	.027	590.0	1.94	.90	---	2.84	3.58
	Inst	SF	Sm	2C	.039	413.0	2.29	1.30	---	3.59	4.58
Red cedar	Inst	SF	Lg	2C	.027	590.0	1.39	.90	---	2.29	2.95
	Inst	SF	Sm	2C	.039	413.0	1.70	1.30	---	3.00	3.90
Pine	Inst	SF	Lg	2C	.027	590.0	2.05	.90	---	2.95	3.71
	Inst	SF	Sm	2C	.039	413.0	2.41	1.30	---	3.71	4.72

Description	Oper	Unit	Vol	Crew Size	Man-hours per Unit	Crew Output per Day	Avg Mat'l Unit Cost	Avg Labor Unit Cost	Avg Equip Unit Cost	Avg Total Unit Cost	Avg Price Incl O&P
Plywood (1/2" T) with battens (1" x 2")											
16" oc battens	Demo	SF	Lg	LB	.007	2255	---	.19	---	.19	.29
	Demo	SF	Sm	LB	.010	1579	---	.27	---	.27	.41
24" oc battens	Demo	SF	Lg	LB	.007	2365	---	.19	---	.19	.29
	Demo	SF	Sm	LB	.010	1656	---	.27	---	.27	.41

Plywood (1/2" T) sanded exterior, AB grade with battens (1" x 2") S4S, common, standard & better green

Description	Oper	Unit	Vol	Crew Size	Man-hours per Unit	Crew Output per Day	Avg Mat'l Unit Cost	Avg Labor Unit Cost	Avg Equip Unit Cost	Avg Total Unit Cost	Avg Price Incl O&P
16" oc											
Fir	Inst	SF	Lg	2C	.022	735.0	1.45	.73	---	2.18	2.77
	Inst	SF	Sm	2C	.031	515.0	1.71	1.03	---	2.74	3.51
Pine	Inst	SF	Lg	2C	.022	735.0	1.49	.73	---	2.22	2.81
	Inst	SF	Sm	2C	.031	515.0	1.75	1.03	---	2.78	3.56
24" oc											
Fir	Inst	SF	Lg	2C	.018	865.0	1.40	.60	---	2.00	2.51
	Inst	SF	Sm	2C	.026	606.0	1.65	.87	---	2.52	3.20
Pine	Inst	SF	Lg	2C	.018	865.0	1.43	.60	---	2.03	2.54
	Inst	SF	Sm	2C	.026	606.0	1.69	.87	---	2.56	3.24

Shakes only, no sheathing included

24" L with 11-1/2" exposure

Description	Oper	Unit	Vol	Crew Size	Man-hours per Unit	Crew Output per Day	Avg Mat'l Unit Cost	Avg Labor Unit Cost	Avg Equip Unit Cost	Avg Total Unit Cost	Avg Price Incl O&P
1/2" to 3/4" T	Demo	SF	Lg	LB	.010	1600	---	.27	---	.27	.41
	Demo	SF	Sm	LB	.014	1120	---	.38	---	.38	.57
3/4" to 5/4" T	Demo	SF	Lg	LB	.011	1440	---	.30	---	.30	.45
	Demo	SF	Sm	LB	.016	1008	---	.44	---	.44	.65

Shakes only, over sheathing, 2 nails/shake

24" L with 11-1/2" exp., @ 6" W, red cedar, sawn one side

Description	Oper	Unit	Vol	Crew Size	Man-hours per Unit	Crew Output per Day	Avg Mat'l Unit Cost	Avg Labor Unit Cost	Avg Equip Unit Cost	Avg Total Unit Cost	Avg Price Incl O&P
Straight wall											
1/2" to 3/4" T	Inst	SF	Lg	CN	.021	940.0	1.53	.67	---	2.20	2.77
	Inst	SF	Sm	CN	.030	658.0	1.80	.96	---	2.76	3.51
3/4" to 5/4" T	Inst	SF	Lg	CN	.024	850.0	2.03	.77	---	2.80	3.49
	Inst	SF	Sm	CN	.034	595.0	2.39	1.09	---	3.48	4.39
Cut-up wall											
1/2" to 3/4" T	Inst	SF	Lg	CN	.027	750.0	1.56	.87	---	2.43	3.09
	Inst	SF	Sm	CN	.038	525.0	1.84	1.22	---	3.06	3.95
3/4" to 5/4" T	Inst	SF	Lg	CN	.029	680.0	2.07	.93	---	3.00	3.78
	Inst	SF	Sm	CN	.042	476.0	2.43	1.35	---	3.78	4.82

Related materials and operations

Rosin sized sheathing paper, 36" W 500 SF/roll; nailed

Description	Oper	Unit	Vol	Crew Size	Man-hours per Unit	Crew Output per Day	Avg Mat'l Unit Cost	Avg Labor Unit Cost	Avg Equip Unit Cost	Avg Total Unit Cost	Avg Price Incl O&P
Over solid sheathing	Inst	SF	Lg	2C	.001	12500	---	.03	---	.03	.05
	Inst	SF	Sm	2C	.002	8750	---	.07	---	.07	.10

Description	Oper	Unit	Vol	Crew Size	Man-hours per Unit	Crew Output per Day	Avg Mat'l Unit Cost	Avg Labor Unit Cost	Avg Equip Unit Cost	Avg Total Unit Cost	Avg Price Incl O&P
Shingles only, no sheathing included											
16" L with 7-1/2" exposure	Demo	SF	Lg	LB	.016	1000	---	.44	---	.44	**.65**
	Demo	SF	Sm	LB	.023	700.0	---	.63	---	.63	**.94**
18" L with 8-1/2" exposure	Demo	SF	Lg	LB	.014	1130	---	.38	---	.38	**.57**
	Demo	SF	Sm	LB	.020	791.0	---	.55	---	.55	**.82**
24" L with 11-1/2" exposure	Demo	SF	Lg	LB	.010	1530	---	.27	---	.27	**.41**
	Demo	SF	Sm	LB	.015	1071	---	.41	---	.41	**.61**
Shingles only, red cedar, No 1. perfect, @ 4" W, 2 nails/shake, over sheathing											
Straight wall											
16" L with 7-1/2" exposure	Inst	SF	Lg	2C	.027	585.0	1.79	.90	---	2.69	**3.41**
	Inst	SF	Sm	2C	.039	410.0	2.10	1.30	---	3.40	**4.36**
18" L with 8-1/2" exposure	Inst	SF	Lg	2C	.024	660.0	1.37	.80	---	2.17	**2.77**
	Inst	SF	Sm	2C	.035	462.0	1.61	1.16	---	2.77	**3.60**
24" L with 11-1/2" exposure	Inst	SF	Lg	2C	.018	895.0	1.75	.60	---	2.35	**2.91**
	Inst	SF	Sm	2C	.026	627.0	2.06	.87	---	2.93	**3.67**
Cut-up wall											
16" L with 7-1/2" exposure	Inst	SF	Lg	2C	.034	470.0	1.82	1.13	---	2.95	**3.79**
	Inst	SF	Sm	2C	.049	329.0	2.14	1.63	---	3.77	**4.91**
18" L with 8-1/2" exposure	Inst	SF	Lg	2C	.030	530.0	1.40	1.00	---	2.40	**3.11**
	Inst	SF	Sm	2C	.043	371.0	1.64	1.43	---	3.07	**4.03**
24" L with 11-1/2" exposure	Inst	SF	Lg	2C	.022	715.0	1.79	.73	---	2.52	**3.16**
	Inst	SF	Sm	2C	.032	501.0	2.10	1.06	---	3.16	**4.01**
Related materials and operations											
Rosin sized sheathing paper, 36" W 500 SF/roll; nailed											
Over solid sheathing	Inst	SF	Lg	2C	.001	12500	---	.03	---	.03	**.05**
	Inst	SF	Sm	2C	.002	8750	---	.07	---	.07	**.10**

Description	Oper	Unit	Vol	Crew Size	Man-hours per Unit	Crew Output per Day	Avg Mat'l Unit Cost	Avg Labor Unit Cost	Avg Equip Unit Cost	Avg Total Unit Cost	Avg Price Incl O&P

Sinks

Bathroom

Lavatories (bathroom sinks), with good quality fittings and faucets, concealed overflow

Vitreous china

Countertop units, with self rim (American Standard Products)

Antiquity
24-1/2" x 18" with 8" or 4" cc

Description	Oper	Unit	Vol	Crew Size	MH	Output	Mat'l	Labor	Equip	Total	Price
Color	Inst	Ea	Lg	SA	2.67	3.00	318.00	101.00	---	419.00	513.00
	Inst	Ea	Sm	SA	3.81	2.10	383.00	144.00	---	527.00	651.00
White	Inst	Ea	Lg	SA	2.67	3.00	284.00	101.00	---	385.00	474.00
	Inst	Ea	Sm	SA	3.81	2.10	342.00	144.00	---	486.00	604.00

Aqualyn
20" x 17" with 8" or 4" cc

Color	Inst	Ea	Lg	SA	2.67	3.00	262.00	101.00	---	363.00	449.00
	Inst	Ea	Sm	SA	3.81	2.10	315.00	144.00	---	459.00	574.00
White	Inst	Ea	Lg	SA	2.67	3.00	236.00	101.00	---	337.00	420.00
	Inst	Ea	Sm	SA	3.81	2.10	285.00	144.00	---	429.00	538.00

Ellisse Petite
24" x 19" with 8" or 4" cc

Color	Inst	Ea	Lg	SA	2.67	3.00	390.00	101.00	---	491.00	596.00
	Inst	Ea	Sm	SA	3.81	2.10	470.00	144.00	---	614.00	751.00
White	Inst	Ea	Lg	SA	2.67	3.00	340.00	101.00	---	441.00	539.00
	Inst	Ea	Sm	SA	3.81	2.10	410.00	144.00	---	554.00	683.00

Heritage
26" x 20" with 8" or 4" cc

Color	Inst	Ea	Lg	SA	2.67	3.00	694.00	101.00	---	795.00	945.00
	Inst	Ea	Sm	SA	3.81	2.10	836.00	144.00	---	980.00	1170.00
White	Inst	Ea	Lg	SA	2.67	3.00	585.00	101.00	---	686.00	820.00
	Inst	Ea	Sm	SA	3.81	2.10	705.00	144.00	---	849.00	1020.00

Hexalyn
22" x 19" with 8" or 4" cc

Color	Inst	Ea	Lg	SA	2.67	3.00	350.00	101.00	---	451.00	550.00
	Inst	Ea	Sm	SA	3.81	2.10	422.00	144.00	---	566.00	696.00
White	Inst	Ea	Lg	SA	2.67	3.00	309.00	101.00	---	410.00	503.00
	Inst	Ea	Sm	SA	3.81	2.10	372.00	144.00	---	516.00	639.00

Lexington
22" x 19" with 8" or 4" cc

Color	Inst	Ea	Lg	SA	2.67	3.00	440.00	101.00	---	549.00	663.00
	Inst	Ea	Sm	SA	3.81	2.10	540.00	144.00	---	684.00	832.00
White	Inst	Ea	Lg	SA	2.67	3.00	388.00	101.00	---	489.00	594.00
	Inst	Ea	Sm	SA	3.81	2.10	467.00	144.00	---	611.00	749.00

Pallas Petite
24" x 18" with 8" or 4" cc

Color	Inst	Ea	Lg	SA	2.67	3.00	478.00	101.00	---	579.00	698.00
	Inst	Ea	Sm	SA	3.81	2.10	576.00	144.00	---	720.00	874.00
White	Inst	Ea	Lg	SA	2.67	3.00	412.00	101.00	---	513.00	622.00
	Inst	Ea	Sm	SA	3.81	2.10	497.00	144.00	---	641.00	783.00

Description	Oper	Unit	Vol	Crew Size	Man-hours per Unit	Crew Output per Day	Avg Mat'l Unit Cost	Avg Labor Unit Cost	Avg Equip Unit Cost	Avg Total Unit Cost	Avg Price Incl O&P
Renaissance											
19" diameter with 8" or 4" cc											
Color	Inst	Ea	Lg	SA	2.67	3.00	258.00	101.00	---	359.00	**445.00**
	Inst	Ea	Sm	SA	3.81	2.10	311.00	144.00	---	455.00	**569.00**
White	Inst	Ea	Lg	SA	2.67	3.00	234.00	101.00	---	335.00	**416.00**
	Inst	Ea	Sm	SA	3.81	2.10	282.00	144.00	---	426.00	**535.00**
Roma											
26" x 20" with 8" or 4" cc w/ sealant											
Color	Inst	Ea	Lg	SA	2.67	3.00	658.00	101.00	---	759.00	**904.00**
	Inst	Ea	Sm	SA	3.81	2.10	793.00	144.00	---	937.00	**1120.00**
White	Inst	Ea	Lg	SA	2.67	3.00	556.00	101.00	---	657.00	**787.00**
	Inst	Ea	Sm	SA	3.81	2.10	670.00	144.00	---	814.00	**982.00**
Rondalyn											
19" diameter with 8" or 4" cc											
Color	Inst	Ea	Lg	SA	2.67	3.00	249.00	101.00	---	350.00	**435.00**
	Inst	Ea	Sm	SA	3.81	2.10	301.00	144.00	---	445.00	**557.00**
White	Inst	Ea	Lg	SA	2.67	3.00	227.00	101.00	---	328.00	**409.00**
	Inst	Ea	Sm	SA	3.81	2.10	274.00	144.00	---	418.00	**526.00**

Counter and undercounter units (Kohler Products, Artists Edition)

Description	Oper	Unit	Vol	Crew Size	Man-hours per Unit	Crew Output per Day	Avg Mat'l Unit Cost	Avg Labor Unit Cost	Avg Equip Unit Cost	Avg Total Unit Cost	Avg Price Incl O&P
Ankara, 17" 14"											
Self-Rimming	Inst	Ea	Lg	SA	2.67	3.00	1100.00	101.00	---	1201.00	**1410.00**
	Inst	Ea	Sm	SA	3.81	2.10	1330.00	144.00	---	1474.00	**1740.00**
Undercounter	Inst	Ea	Lg	SA	2.67	3.00	954.00	101.00	---	1055.00	**1240.00**
	Inst	Ea	Sm	SA	3.81	2.10	1150.00	144.00	---	1294.00	**1530.00**
Briar Rose, 17" x 14"											
Self-Rimming	Inst	Ea	Lg	SA	2.67	3.00	1050.00	101.00	---	1151.00	**1350.00**
	Inst	Ea	Sm	SA	3.81	2.10	1260.00	144.00	---	1404.00	**1660.00**
Undercounter	Inst	Ea	Lg	SA	2.67	3.00	951.00	101.00	---	1052.00	**1240.00**
	Inst	Ea	Sm	SA	3.81	2.10	1150.00	144.00	---	1294.00	**1530.00**
Crimson Topaz, 17" x 14"											
Self-Rimming	Inst	Ea	Lg	SA	2.67	3.00	926.00	101.00	---	1027.00	**1210.00**
	Inst	Ea	Sm	SA	3.81	2.10	1120.00	144.00	---	1264.00	**1500.00**
Undercounter	Inst	Ea	Lg	SA	2.67	3.00	926.00	101.00	---	1027.00	**1210.00**
	Inst	Ea	Sm	SA	3.81	2.10	1120.00	144.00	---	1264.00	**1500.00**
English Trellis, 17" x 14"											
Self-Rimming	Inst	Ea	Lg	SA	2.67	3.00	1050.00	101.00	---	1151.00	**1350.00**
	Inst	Ea	Sm	SA	3.81	2.10	1260.00	144.00	---	1404.00	**1660.00**
Undercounter	Inst	Ea	Lg	SA	2.67	3.00	951.00	101.00	---	1052.00	**1240.00**
	Inst	Ea	Sm	SA	3.81	2.10	1150.00	144.00	---	1294.00	**1530.00**
Peonies and Ivy, 17" x 14"											
Self-Rimming	Inst	Ea	Lg	SA	2.67	3.00	1100.00	101.00	---	1201.00	**1410.00**
	Inst	Ea	Sm	SA	3.81	2.10	1330.00	144.00	---	1474.00	**1740.00**
Undercounter	Inst	Ea	Lg	SA	2.67	3.00	954.00	101.00	---	1055.00	**1240.00**
	Inst	Ea	Sm	SA	3.81	2.10	1150.00	144.00	---	1294.00	**1530.00**
Provencial, 17" x 14"											
Self-Rimming	Inst	Ea	Lg	SA	2.67	3.00	1100.00	101.00	---	1201.00	**1410.00**
	Inst	Ea	Sm	SA	3.81	2.10	1330.00	144.00	---	1474.00	**1740.00**
Undercounter	Inst	Ea	Lg	SA	2.67	3.00	954.00	101.00	---	1055.00	**1240.00**
	Inst	Ea	Sm	SA	3.81	2.10	1150.00	144.00	---	1294.00	**1530.00**

Description	Oper	Unit	Vol	Crew Size	Man-hours per Unit	Crew Output per Day	Avg Mat'l Unit Cost	Avg Labor Unit Cost	Avg Equip Unit Cost	Avg Total Unit Cost	Avg Price Incl O&P

Pedestal sinks (American Standard Products)

Antiquity w/ pedestal
24-1/2" x 19" with 8" or 4" cc

Color	Inst	Ea	Lg	SA	2.67	3.00	306.00	101.00	---	407.00	**499.00**
	Inst	Ea	Sm	SA	3.81	2.10	369.00	144.00	---	513.00	**635.00**
White	Inst	Ea	Lg	SA	2.67	3.00	390.00	101.00	---	491.00	**596.00**
	Inst	Ea	Sm	SA	3.81	2.10	470.00	144.00	---	614.00	**751.00**

Lavatory slab only

Color	Inst	Ea	Lg	SA	1.60	5.00	185.00	60.30	---	245.30	**301.00**
	Inst	Ea	Sm	SA	2.00	4.00	223.00	75.30	---	298.30	**367.00**
White	Inst	Ea	Lg	SA	1.60	5.00	146.00	60.30	---	206.30	**257.00**
	Inst	Ea	Sm	SA	2.00	4.00	176.00	75.30	---	251.30	**314.00**

Pedestal only

Color	Inst	Ea	Lg	SA	1.60	5.00	121.00	60.30	---	181.30	**228.00**
	Inst	Ea	Sm	SA	2.00	4.00	146.00	75.30	---	221.30	**278.00**
White	Inst	Ea	Lg	SA	1.60	5.00	97.20	60.30	---	157.50	**200.00**
	Inst	Ea	Sm	SA	2.00	4.00	117.00	75.30	---	192.30	**246.00**

Cadet w/ pedestal
24" x 20" with 8" or 4" cc

Color	Inst	Ea	Lg	SA	2.67	3.00	406.00	101.00	---	507.00	**615.00**
	Inst	Ea	Sm	SA	3.81	2.10	490.00	144.00	---	634.00	**774.00**
White	Inst	Ea	Lg	SA	2.67	3.00	357.00	101.00	---	458.00	**559.00**
	Inst	Ea	Sm	SA	3.81	2.10	431.00	144.00	---	575.00	**706.00**

Lavatory slab only

Color	Inst	Ea	Lg	SA	1.60	5.00	190.00	60.30	---	250.30	**307.00**
	Inst	Ea	Sm	SA	2.00	4.00	229.00	75.30	---	304.30	**374.00**
White	Inst	Ea	Lg	SA	1.60	5.00	154.00	60.30	---	214.30	**266.00**
	Inst	Ea	Sm	SA	2.00	4.00	186.00	75.30	---	261.30	**324.00**

Pedestal only

Color	Inst	Ea	Lg	SA	1.60	5.00	75.30	60.30	---	135.60	**175.00**
	Inst	Ea	Sm	SA	2.00	4.00	90.80	75.30	---	166.10	**215.00**
White	Inst	Ea	Lg	SA	1.60	5.00	61.30	60.30	---	121.60	**159.00**
	Inst	Ea	Sm	SA	2.00	4.00	73.90	75.30	---	149.20	**196.00**

Ellisse Petite w/ pedestal
25" x 21" with 8" or 4" cc

Color	Inst	Ea	Lg	SA	2.67	3.00	723.00	101.00	---	824.00	**980.00**
	Inst	Ea	Sm	SA	3.81	2.10	872.00	144.00	---	1016.00	**1210.00**
White	Inst	Ea	Lg	SA	2.67	3.00	608.00	101.00	---	709.00	**847.00**
	Inst	Ea	Sm	SA	3.81	2.10	733.00	144.00	---	877.00	**1050.00**

Lavatory slab only

Color	Inst	Ea	Lg	SA	1.60	5.00	344.00	60.30	---	404.30	**485.00**
	Inst	Ea	Sm	SA	2.00	4.00	415.00	75.30	---	490.30	**588.00**
White	Inst	Ea	Lg	SA	1.60	5.00	276.00	60.30	---	336.30	**406.00**
	Inst	Ea	Sm	SA	2.00	4.00	333.00	75.30	---	408.30	**493.00**

Description	Oper	Unit	Vol	Crew Size	Man-hours per Unit	Crew Output per Day	Avg Mat'l Unit Cost	Avg Labor Unit Cost	Avg Equip Unit Cost	Avg Total Unit Cost	Avg Price Incl O&P
Pedestal only											
Color	Inst	Ea	Lg	SA	1.60	5.00	233.00	60.30	---	293.30	**357.00**
	Inst	Ea	Sm	SA	2.00	4.00	281.00	75.30	---	356.30	**434.00**
White	Inst	Ea	Lg	SA	1.60	5.00	189.00	60.30	---	249.30	**306.00**
	Inst	Ea	Sm	SA	2.00	4.00	228.00	75.30	---	303.30	**373.00**
Lexington w/ pedestal											
24" x 18" with 8" or 4" cc											
Color	Inst	Ea	Lg	SA	2.67	3.00	606.00	101.00	---	707.00	**845.00**
	Inst	Ea	Sm	SA	3.81	2.10	730.00	144.00	---	874.00	**1050.00**
White	Inst	Ea	Lg	SA	2.67	3.00	516.00	101.00	---	617.00	**741.00**
	Inst	Ea	Sm	SA	3.81	2.10	622.00	144.00	---	766.00	**926.00**
Lavatory slab only											
Color	Inst	Ea	Lg	SA	1.60	5.00	314.00	60.30	---	374.30	**450.00**
	Inst	Ea	Sm	SA	2.00	4.00	379.00	75.30	---	454.30	**547.00**
White	Inst	Ea	Lg	SA	1.60	5.00	257.00	60.30	---	317.30	**384.00**
	Inst	Ea	Sm	SA	2.00	4.00	309.00	75.30	---	384.30	**467.00**
Pedestal only											
Color	Inst	Ea	Lg	SA	1.60	5.00	145.00	60.30	---	205.30	**256.00**
	Inst	Ea	Sm	SA	2.00	4.00	175.00	75.30	---	250.30	**312.00**
White	Inst	Ea	Lg	SA	1.60	5.00	113.00	60.30	---	173.30	**219.00**
	Inst	Ea	Sm	SA	2.00	4.00	136.00	75.30	---	211.30	**267.00**
Repertoire w/ pedestal											
23" x 18-1/4" with 8" or 4" cc											
Color	Inst	Ea	Lg	SA	2.67	3.00	481.00	101.00	---	582.00	**701.00**
	Inst	Ea	Sm	SA	3.81	2.10	579.00	144.00	---	723.00	**877.00**
White	Inst	Ea	Lg	SA	2.67	3.00	424.00	101.00	---	525.00	**635.00**
	Inst	Ea	Sm	SA	3.81	2.10	511.00	144.00	---	655.00	**798.00**
Lavatory slab only											
Color	Inst	Ea	Lg	SA	1.60	5.00	244.00	60.30	---	304.30	**369.00**
	Inst	Ea	Sm	SA	2.00	4.00	294.00	75.30	---	369.30	**448.00**
White	Inst	Ea	Lg	SA	1.60	5.00	201.00	60.30	---	261.30	**320.00**
	Inst	Ea	Sm	SA	2.00	4.00	243.00	75.30	---	318.30	**390.00**
Pedestal only											
Color	Inst	Ea	Lg	SA	1.60	5.00	91.10	60.30	---	151.40	**193.00**
	Inst	Ea	Sm	SA	2.00	4.00	110.00	75.30	---	185.30	**237.00**
White	Inst	Ea	Lg	SA	1.60	5.00	73.60	60.30	---	133.90	**173.00**
	Inst	Ea	Sm	SA	2.00	4.00	88.70	75.30	---	164.00	**213.00**

Pedestal sinks (Kohler Products)

Anatole

21" x 18" with 8" or 4" cc

Description	Oper	Unit	Vol	Crew Size	Man-hours per Unit	Crew Output per Day	Avg Mat'l Unit Cost	Avg Labor Unit Cost	Avg Equip Unit Cost	Avg Total Unit Cost	Avg Price Incl O&P
Color	Inst	Ea	Lg	SA	2.67	3.00	821.00	101.00	---	922.00	**1090.00**
	Inst	Ea	Sm	SA	3.81	2.10	989.00	144.00	---	1133.00	**1350.00**
Premium color	Inst	Ea	Lg	SA	2.67	3.00	922.00	101.00	---	1023.00	**1210.00**
	Inst	Ea	Sm	SA	3.81	2.10	1110.00	144.00	---	1254.00	**1490.00**
White	Inst	Ea	Lg	SA	2.67	3.00	665.00	101.00	---	766.00	**913.00**
	Inst	Ea	Sm	SA	3.81	2.10	802.00	144.00	---	946.00	**1130.00**

Description	Oper	Unit	Vol	Crew Size	Man-hours per Unit	Crew Output per Day	Avg Mat'l Unit Cost	Avg Labor Unit Cost	Avg Equip Unit Cost	Avg Total Unit Cost	Avg Price Incl O&P
Cabernet											
22" x 18" x 33" with 8" or 4" cc											
Color	Inst	Ea	Lg	SA	2.67	3.00	591.00	101.00	---	692.00	**828.00**
	Inst	Ea	Sm	SA	3.81	2.10	713.00	144.00	---	857.00	**1030.00**
Premium color	Inst	Ea	Lg	SA	2.67	3.00	658.00	101.00	---	759.00	**905.00**
	Inst	Ea	Sm	SA	3.81	2.10	794.00	144.00	---	938.00	**1120.00**
White	Inst	Ea	Lg	SA	2.67	3.00	489.00	101.00	---	590.00	**710.00**
	Inst	Ea	Sm	SA	3.81	2.10	589.00	144.00	---	733.00	**889.00**
Chablis											
24" x 19" x 32" with 8" or 4" cc											
Color	Inst	Ea	Lg	SA	2.67	3.00	670.00	101.00	---	771.00	**918.00**
	Inst	Ea	Sm	SA	3.81	2.10	807.00	144.00	---	951.00	**1140.00**
Premium color	Inst	Ea	Lg	SA	2.67	3.00	748.00	101.00	---	849.00	**1010.00**
	Inst	Ea	Sm	SA	3.81	2.10	902.00	144.00	---	1046.00	**1250.00**
White	Inst	Ea	Lg	SA	2.67	3.00	549.00	101.00	---	650.00	**779.00**
	Inst	Ea	Sm	SA	3.81	2.10	662.00	144.00	---	806.00	**972.00**
Chardonnay											
28" x 22" x 33" with 8" or 4" cc											
Color	Inst	Ea	Lg	SA	2.67	3.00	1170.00	101.00	---	1271.00	**1490.00**
	Inst	Ea	Sm	SA	3.81	2.10	1410.00	144.00	---	1554.00	**1830.00**
Premium color	Inst	Ea	Lg	SA	2.67	3.00	1320.00	101.00	---	1421.00	**1670.00**
	Inst	Ea	Sm	SA	3.81	2.10	1590.00	144.00	---	1734.00	**2040.00**
White	Inst	Ea	Lg	SA	2.67	3.00	931.00	101.00	---	1032.00	**1220.00**
	Inst	Ea	Sm	SA	3.81	2.10	1120.00	144.00	---	1264.00	**1500.00**
Fleur											
30" x 23" x 35" with 8" or 4" cc											
Color	Inst	Ea	Lg	SA	2.67	3.00	1060.00	101.00	---	1161.00	**1370.00**
	Inst	Ea	Sm	SA	3.81	2.10	1280.00	144.00	---	1424.00	**1680.00**
White	Inst	Ea	Lg	SA	2.67	3.00	848.00	101.00	---	949.00	**1120.00**
	Inst	Ea	Sm	SA	3.81	2.10	1020.00	144.00	---	1164.00	**1390.00**
Pillow Talk											
28" x 21" x 33" with 8" or 4" cc											
Color	Inst	Ea	Lg	SA	2.67	3.00	1400.00	101.00	---	1501.00	**1750.00**
	Inst	Ea	Sm	SA	3.81	2.10	1680.00	144.00	---	1824.00	**2150.00**
Premium color	Inst	Ea	Lg	SA	2.67	3.00	1580.00	101.00	---	1681.00	**1970.00**
	Inst	Ea	Sm	SA	3.81	2.10	1910.00	144.00	---	2054.00	**2410.00**
White	Inst	Ea	Lg	SA	2.67	3.00	1110.00	101.00	---	1211.00	**1420.00**
	Inst	Ea	Sm	SA	3.81	2.10	1340.00	144.00	---	1484.00	**1750.00**
Pinoir											
22" x 18" x 33" with 8" or 4" cc											
Color	Inst	Ea	Lg	SA	2.67	3.00	436.00	101.00	---	537.00	**650.00**
	Inst	Ea	Sm	SA	3.81	2.10	526.00	144.00	---	670.00	**816.00**
Premium color	Inst	Ea	Lg	SA	2.67	3.00	480.00	101.00	---	581.00	**700.00**
	Inst	Ea	Sm	SA	3.81	2.10	579.00	144.00	---	723.00	**876.00**
White	Inst	Ea	Lg	SA	2.67	3.00	369.00	101.00	---	470.00	**573.00**
	Inst	Ea	Sm	SA	3.81	2.10	445.00	144.00	---	589.00	**723.00**

Description	Oper	Unit	Vol	Crew Size	Man-hours per Unit	Crew Output per Day	Avg Mat'l Unit Cost	Avg Labor Unit Cost	Avg Equip Unit Cost	Avg Total Unit Cost	Avg Price Incl O&P
Revival											
24" x 18" x 34" with 8" or 4" cc											
Color	Inst	Ea	Lg	SA	2.67	3.00	789.00	101.00	---	890.00	**1050.00**
	Inst	Ea	Sm	SA	3.81	2.10	951.00	144.00	---	1095.00	**1300.00**
Premium color	Inst	Ea	Lg	SA	2.67	3.00	885.00	101.00	---	986.00	**1170.00**
	Inst	Ea	Sm	SA	3.81	2.10	1070.00	144.00	---	1214.00	**1440.00**
White	Inst	Ea	Lg	SA	2.67	3.00	641.00	101.00	---	742.00	**884.00**
	Inst	Ea	Sm	SA	3.81	2.10	772.00	144.00	---	916.00	**1100.00**

Wall hung units (American Standard Products)

Corner Minette

Description	Oper	Unit	Vol	Crew Size	Man-hours per Unit	Crew Output per Day	Avg Mat'l Unit Cost	Avg Labor Unit Cost	Avg Equip Unit Cost	Avg Total Unit Cost	Avg Price Incl O&P
11" x 16" with 8" or 4" cc											
Color	Inst	Ea	Lg	SA	2.00	4.00	434.00	75.30	---	509.30	**610.00**
	Inst	Ea	Sm	SA	2.86	2.80	523.00	108.00	---	631.00	**760.00**
White	Inst	Ea	Lg	SA	2.00	4.00	362.00	75.30	---	437.30	**526.00**
	Inst	Ea	Sm	SA	2.86	2.80	436.00	108.00	---	544.00	**660.00**
Declyn											
19" x 17" with 8" or 4" cc											
Color	Inst	Ea	Lg	SA	2.67	3.00	250.00	101.00	---	351.00	**436.00**
	Inst	Ea	Sm	SA	3.81	2.10	302.00	144.00	---	446.00	**558.00**
White	Inst	Ea	Lg	SA	2.67	3.00	231.00	101.00	---	332.00	**413.00**
	Inst	Ea	Sm	SA	3.81	2.10	278.00	144.00	---	422.00	**531.00**
Penlyn											
18" x 16" with 8" or 4" cc											
White	Inst	Ea	Lg	SA	2.67	3.00	323.00	101.00	---	424.00	**519.00**
	Inst	Ea	Sm	SA	3.81	2.10	389.00	144.00	---	533.00	**659.00**
Roxalyn											
20" x 18" with 8" or 4" cc											
Color	Inst	Ea	Lg	SA	2.67	3.00	391.00	101.00	---	492.00	**598.00**
	Inst	Ea	Sm	SA	3.81	2.10	472.00	144.00	---	616.00	**753.00**
White	Inst	Ea	Lg	SA	2.67	3.00	326.00	101.00	---	427.00	**522.00**
	Inst	Ea	Sm	SA	3.81	2.10	392.00	144.00	---	536.00	**662.00**
Wheelchair, meets ADA requirements											
27" x 20" with 8" or 4" cc											
Color	Inst	Ea	Lg	SA	2.67	3.00	670.00	101.00	---	771.00	**918.00**
	Inst	Ea	Sm	SA	3.81	2.10	807.00	144.00	---	951.00	**1140.00**
White	Inst	Ea	Lg	SA	2.67	3.00	524.00	101.00	---	625.00	**751.00**
	Inst	Ea	Sm	SA	3.81	2.10	632.00	144.00	---	776.00	**938.00**

Description	Oper	Unit	Vol	Crew Size	Man-hours per Unit	Crew Output per Day	Avg Mat'l Unit Cost	Avg Labor Unit Cost	Avg Equip Unit Cost	Avg Total Unit Cost	Avg Price Incl O&P
Wall hung units (Kohler Products)											
Cabernet											
27" x 20-1/2" with 8" or 4" cc											
Color	Inst	Ea	Lg	SA	2.67	3.00	858.00	101.00	---	959.00	**1130.00**
	Inst	Ea	Sm	SA	3.81	2.10	1030.00	144.00	---	1174.00	**1400.00**
White	Inst	Ea	Lg	SA	2.67	3.00	694.00	101.00	---	795.00	**945.00**
	Inst	Ea	Sm	SA	3.81	2.10	836.00	144.00	---	980.00	**1170.00**
Chablis											
24" x 19" with 8" or 4" cc											
Color	Inst	Ea	Lg	SA	2.67	3.00	655.00	101.00	---	756.00	**901.00**
	Inst	Ea	Sm	SA	3.81	2.10	789.00	144.00	---	933.00	**1120.00**
White	Inst	Ea	Lg	SA	2.67	3.00	537.00	101.00	---	638.00	**766.00**
	Inst	Ea	Sm	SA	3.81	2.10	648.00	144.00	---	792.00	**956.00**
Chesapeake											
19" x 17" with 4" cc											
Color	Inst	Ea	Lg	SA	2.67	3.00	263.00	101.00	---	364.00	**450.00**
	Inst	Ea	Sm	SA	3.81	2.10	317.00	144.00	---	461.00	**575.00**
White	Inst	Ea	Lg	SA	2.67	3.00	236.00	101.00	---	337.00	**419.00**
	Inst	Ea	Sm	SA	3.81	2.10	284.00	144.00	---	428.00	**538.00**
20" x 18" with 4" cc											
Color	Inst	Ea	Lg	SA	2.67	3.00	299.00	101.00	---	400.00	**491.00**
	Inst	Ea	Sm	SA	3.81	2.10	360.00	144.00	---	504.00	**625.00**
White	Inst	Ea	Lg	SA	2.67	3.00	263.00	101.00	---	364.00	**451.00**
	Inst	Ea	Sm	SA	3.81	2.10	318.00	144.00	---	462.00	**576.00**
Fleur											
19-7/8" x 14-3/8" with single hole											
Color	Inst	Ea	Lg	SA	2.67	3.00	796.00	101.00	---	897.00	**1060.00**
	Inst	Ea	Sm	SA	3.81	2.10	959.00	144.00	---	1103.00	**1310.00**
White	Inst	Ea	Lg	SA	2.67	3.00	646.00	101.00	---	747.00	**891.00**
	Inst	Ea	Sm	SA	3.81	2.10	779.00	144.00	---	923.00	**1110.00**
Greenwich											
18" x 16" with 8" or 4" cc											
White	Inst	Ea	Lg	SA	2.67	3.00	270.00	101.00	---	371.00	**459.00**
	Inst	Ea	Sm	SA	3.81	2.10	326.00	144.00	---	470.00	**586.00**
20" x 18" with 8" or 4" cc											
Color	Inst	Ea	Lg	SA	2.67	3.00	278.00	101.00	---	379.00	**468.00**
	Inst	Ea	Sm	SA	3.81	2.10	335.00	144.00	---	479.00	**596.00**
White	Inst	Ea	Lg	SA	2.67	3.00	248.00	101.00	---	349.00	**433.00**
	Inst	Ea	Sm	SA	3.81	2.10	298.00	144.00	---	442.00	**554.00**
Pinoir											
22" x 18" with 8" or 4" cc											
Color	Inst	Ea	Lg	SA	2.67	3.00	430.00	101.00	---	531.00	**642.00**
	Inst	Ea	Sm	SA	3.81	2.10	518.00	144.00	---	662.00	**807.00**
White	Inst	Ea	Lg	SA	2.67	3.00	364.00	101.00	---	465.00	**567.00**
	Inst	Ea	Sm	SA	3.81	2.10	439.00	144.00	---	583.00	**716.00**

Description	Oper	Unit	Vol	Crew Size	Man-hours per Unit	Crew Output per Day	Avg Mat'l Unit Cost	Avg Labor Unit Cost	Avg Equip Unit Cost	Avg Total Unit Cost	Avg Price Incl O&P
Portrait											
20" x 15"											
White	Inst	Ea	Lg	SA	2.67	3.00	469.00	101.00	---	570.00	**687.00**
	Inst	Ea	Sm	SA	3.81	2.10	565.00	144.00	---	709.00	**861.00**
27" x 19-3/8" with 8" cc											
Color	Inst	Ea	Lg	SA	2.67	3.00	773.00	101.00	---	874.00	**1040.00**
	Inst	Ea	Sm	SA	3.81	2.10	932.00	144.00	---	1076.00	**1280.00**
White	Inst	Ea	Lg	SA	2.67	3.00	628.00	101.00	---	729.00	**871.00**
	Inst	Ea	Sm	SA	3.81	2.10	758.00	144.00	---	902.00	**1080.00**
Adjustments											
To only remove and reset lavatory											
	Reset	Ea	Lg	SA	2.00	4.00	---	75.30	---	75.30	**111.00**
	Reset	Ea	Sm	SA	2.86	2.80	---	108.00	---	108.00	**158.00**
To install rough-in											
	Inst	Ea	Lg	SA	6.67	1.20	---	251.00	---	251.00	**369.00**
	Inst	Ea	Sm	SA	10.0	0.80	---	377.00	---	377.00	**554.00**
For colors, ADD	Inst	%	Lg	---	---	---	30.0	---	---	---	**---**
	Inst	%	Sm	---	---	---	30.0	---	---	---	**---**
For two 15" L towel bars, ADD											
	Inst	Pr	Lg	---	---	---	39.40	---	---	39.40	**45.30**
	Inst	Pr	Sm	---	---	---	47.50	---	---	47.50	**54.70**

Enameled cast iron

Countertop units (Kohler Products)

Description	Oper	Unit	Vol	Crew Size	Man-hours per Unit	Crew Output per Day	Avg Mat'l Unit Cost	Avg Labor Unit Cost	Avg Equip Unit Cost	Avg Total Unit Cost	Avg Price Incl O&P
Farmington											
19" x 16" with 8" or 4" cc											
Color	Inst	Ea	Lg	SA	2.67	3.00	261.00	101.00	---	362.00	**448.00**
	Inst	Ea	Sm	SA	3.81	2.10	315.00	144.00	---	459.00	**573.00**
White	Inst	Ea	Lg	SA	2.67	3.00	235.00	101.00	---	336.00	**418.00**
	Inst	Ea	Sm	SA	3.81	2.10	283.00	144.00	---	427.00	**536.00**
Radiant											
19" diameter with 4" cc											
Color	Inst	Ea	Lg	SA	2.67	3.00	257.00	101.00	---	358.00	**443.00**
	Inst	Ea	Sm	SA	3.81	2.10	310.00	144.00	---	454.00	**567.00**
White	Inst	Ea	Lg	SA	2.67	3.00	231.00	101.00	---	332.00	**414.00**
	Inst	Ea	Sm	SA	3.81	2.10	279.00	144.00	---	423.00	**532.00**
Tahoe											
20" x 19" with 8" or 4" cc											
Color	Inst	Ea	Lg	SA	2.67	3.00	345.00	101.00	---	446.00	**545.00**
	Inst	Ea	Sm	SA	3.81	2.10	416.00	144.00	---	560.00	**690.00**
White	Inst	Ea	Lg	SA	2.67	3.00	299.00	101.00	---	400.00	**492.00**
	Inst	Ea	Sm	SA	3.81	2.10	361.00	144.00	---	505.00	**626.00**

Description	Oper	Unit	Vol	Crew Size	Man-hours per Unit	Crew Output per Day	Avg Mat'l Unit Cost	Avg Labor Unit Cost	Avg Equip Unit Cost	Avg Total Unit Cost	Avg Price Incl O&P

Countertop units, with self rim (Kohler Products)

Ellington
19-1/4" x 16-1/4" with 8" or 4" cc

Description	Oper	Unit	Vol	Crew Size	Man-hours per Unit	Crew Output per Day	Avg Mat'l Unit Cost	Avg Labor Unit Cost	Avg Equip Unit Cost	Avg Total Unit Cost	Avg Price Incl O&P
Color	Inst	Ea	Lg	SA	2.67	3.00	269.00	101.00	---	370.00	**457.00**
	Inst	Ea	Sm	SA	3.81	2.10	324.00	144.00	---	468.00	**584.00**
White	Inst	Ea	Lg	SA	2.67	3.00	241.00	101.00	---	342.00	**424.00**
	Inst	Ea	Sm	SA	3.81	2.10	290.00	144.00	---	434.00	**544.00**

Ellipse
33" x 19" with 8" cc

Description	Oper	Unit	Vol	Crew Size	Man-hours per Unit	Crew Output per Day	Avg Mat'l Unit Cost	Avg Labor Unit Cost	Avg Equip Unit Cost	Avg Total Unit Cost	Avg Price Incl O&P
Color	Inst	Ea	Lg	SA	2.67	3.00	670.00	101.00	---	771.00	**918.00**
	Inst	Ea	Sm	SA	3.81	2.10	808.00	144.00	---	952.00	**1140.00**
White	Inst	Ea	Lg	SA	2.67	3.00	549.00	101.00	---	650.00	**779.00**
	Inst	Ea	Sm	SA	3.81	2.10	662.00	144.00	---	806.00	**972.00**

Farmington
19-1/4" x 16-1/4" with 8" or 4" cc

Description	Oper	Unit	Vol	Crew Size	Man-hours per Unit	Crew Output per Day	Avg Mat'l Unit Cost	Avg Labor Unit Cost	Avg Equip Unit Cost	Avg Total Unit Cost	Avg Price Incl O&P
Color	Inst	Ea	Lg	SA	2.67	3.00	261.00	101.00	---	362.00	**448.00**
	Inst	Ea	Sm	SA	3.81	2.10	315.00	144.00	---	459.00	**573.00**
White	Inst	Ea	Lg	SA	2.67	3.00	235.00	101.00	---	336.00	**418.00**
	Inst	Ea	Sm	SA	3.81	2.10	283.00	144.00	---	427.00	**536.00**

Hexsign
22-1/2" x 19" with 8" or 4" cc

Description	Oper	Unit	Vol	Crew Size	Man-hours per Unit	Crew Output per Day	Avg Mat'l Unit Cost	Avg Labor Unit Cost	Avg Equip Unit Cost	Avg Total Unit Cost	Avg Price Incl O&P
Color	Inst	Ea	Lg	SA	2.67	3.00	363.00	101.00	---	464.00	**565.00**
	Inst	Ea	Sm	SA	3.81	2.10	438.00	144.00	---	582.00	**714.00**
White	Inst	Ea	Lg	SA	2.67	3.00	313.00	101.00	---	414.00	**508.00**
	Inst	Ea	Sm	SA	3.81	2.10	377.00	144.00	---	521.00	**645.00**

Man's
28" x 19" with 8" cc

Description	Oper	Unit	Vol	Crew Size	Man-hours per Unit	Crew Output per Day	Avg Mat'l Unit Cost	Avg Labor Unit Cost	Avg Equip Unit Cost	Avg Total Unit Cost	Avg Price Incl O&P
Color	Inst	Ea	Lg	SA	2.67	3.00	458.00	101.00	---	559.00	**674.00**
	Inst	Ea	Sm	SA	3.81	2.10	552.00	144.00	---	696.00	**846.00**
White	Inst	Ea	Lg	SA	2.67	3.00	386.00	101.00	---	487.00	**592.00**
	Inst	Ea	Sm	SA	3.81	2.10	465.00	144.00	---	609.00	**746.00**

Terragon
19-3/4" x 15-3/4"

Description	Oper	Unit	Vol	Crew Size	Man-hours per Unit	Crew Output per Day	Avg Mat'l Unit Cost	Avg Labor Unit Cost	Avg Equip Unit Cost	Avg Total Unit Cost	Avg Price Incl O&P
Color	Inst	Ea	Lg	SA	2.67	3.00	445.00	101.00	---	546.00	**659.00**
	Inst	Ea	Sm	SA	3.81	2.10	536.00	144.00	---	680.00	**827.00**
White	Inst	Ea	Lg	SA	2.67	3.00	376.00	101.00	---	477.00	**580.00**
	Inst	Ea	Sm	SA	3.81	2.10	453.00	144.00	---	597.00	**732.00**

Description	Oper	Unit	Vol	Crew Size	Man-hours per Unit	Crew Output per Day	Avg Mat'l Unit Cost	Avg Labor Unit Cost	Avg Equip Unit Cost	Avg Total Unit Cost	Avg Price Incl O&P

Wall hung units, with shelf back (Kohler Products)

Hampton

19" x 17"

Description	Oper	Unit	Vol	Crew Size	Man-hours per Unit	Crew Output per Day	Avg Mat'l Unit Cost	Avg Labor Unit Cost	Avg Equip Unit Cost	Avg Total Unit Cost	Avg Price Incl O&P
White	Inst	Ea	Lg	SA	2.67	3.00	373.00	101.00	---	474.00	**577.00**
	Inst	Ea	Sm	SA	3.81	2.10	450.00	144.00	---	594.00	**728.00**

22" x 19"

White	Inst	Ea	Lg	SA	2.67	3.00	414.00	101.00	---	515.00	**624.00**
	Inst	Ea	Sm	SA	3.81	2.10	499.00	144.00	---	643.00	**785.00**

Marston, corner unit

16" x 16"

Color	Inst	Ea	Lg	SA	2.67	3.00	559.00	101.00	---	660.00	**791.00**
	Inst	Ea	Sm	SA	3.81	2.10	674.00	144.00	---	818.00	**986.00**
White	Inst	Ea	Lg	SA	2.67	3.00	464.00	101.00	---	565.00	**681.00**
	Inst	Ea	Sm	SA	3.81	2.10	559.00	144.00	---	703.00	**854.00**

Taunton

16" x 14"

White	Inst	Ea	Lg	SA	2.67	3.00	437.00	101.00	---	538.00	**650.00**
	Inst	Ea	Sm	SA	3.81	2.10	527.00	144.00	---	671.00	**817.00**

Trailer

13" x 13"

White	Inst	Ea	Lg	SA	2.00	4.00	441.00	75.30	---	516.30	**618.00**
	Inst	Ea	Sm	SA	4.00	2.00	532.00	151.00	---	683.00	**833.00**

Adjustments

To only remove and reset lavatory

	Reset	Ea	Lg	SA	2.00	4.00	---	75.30	---	75.30	**111.00**
	Reset	Ea	Sm	SA	2.86	2.80	---	108.00	---	108.00	**158.00**

To install rough-in

	Inst	Ea	Lg	SA	6.67	1.20	---	251.00	---	251.00	**369.00**
	Inst	Ea	Sm	SA	10.0	0.80	---	377.00	---	377.00	**554.00**
For colors, ADD	Inst	%	Lg	---	---	---	25.0	---	---	---	**---**
	Inst	%	Sm	---	---	---	25.0	---	---	---	**---**
For two 15" L towel bars, ADD	Inst	Pr	Lg	---	---	---	39.40	---	---	39.40	**45.30**
	Inst	Pr	Sm	---	---	---	47.50	---	---	47.50	**54.70**

Description	Oper	Unit	Vol	Crew Size	Man-hours per Unit	Crew Output per Day	Avg Mat'l Unit Cost	Avg Labor Unit Cost	Avg Equip Unit Cost	Avg Total Unit Cost	Avg Price Incl O&P

Americast - Porcelain enameled finish over steel composite material
Countertop units, with self rim (American Standard Products)
Acclivity.
  19" diameter with 8" or 4" cc

Description	Oper	Unit	Vol	Crew Size	Man-hours per Unit	Crew Output per Day	Avg Mat'l Unit Cost	Avg Labor Unit Cost	Avg Equip Unit Cost	Avg Total Unit Cost	Avg Price Incl O&P
Color	Inst	Ea	Lg	SA	2.67	3.00	107.00	101.00	---	208.00	**271.00**
	Inst	Ea	Sm	SA	3.81	2.10	129.00	144.00	---	273.00	**359.00**
White	Inst	Ea	Lg	SA	2.67	3.00	227.00	101.00	---	328.00	**409.00**
	Inst	Ea	Sm	SA	3.81	2.10	274.00	144.00	---	418.00	**526.00**

Affinity
  20" x 17" with 8" or 4" cc

Description	Oper	Unit	Vol	Crew Size	Man-hours per Unit	Crew Output per Day	Avg Mat'l Unit Cost	Avg Labor Unit Cost	Avg Equip Unit Cost	Avg Total Unit Cost	Avg Price Incl O&P
Color	Inst	Ea	Lg	SA	2.67	3.00	277.00	101.00	---	378.00	**466.00**
	Inst	Ea	Sm	SA	3.81	2.10	334.00	144.00	---	478.00	**595.00**
White	Inst	Ea	Lg	SA	2.67	3.00	243.00	101.00	---	344.00	**427.00**
	Inst	Ea	Sm	SA	3.81	2.10	293.00	144.00	---	437.00	**548.00**

Adjustments
  To only remove and reset lavatory

Description	Oper	Unit	Vol	Crew Size	Man-hours per Unit	Crew Output per Day	Avg Mat'l Unit Cost	Avg Labor Unit Cost	Avg Equip Unit Cost	Avg Total Unit Cost	Avg Price Incl O&P
	Reset	Ea	Lg	SA	2.00	4.00	---	75.30	---	75.30	**111.00**
	Reset	Ea	Sm	SA	2.86	2.80	---	108.00	---	108.00	**158.00**
To install rough-in											
	Inst	Ea	Lg	SA	6.67	1.20	---	251.00	---	251.00	**369.00**
	Inst	Ea	Sm	SA	10.0	0.80	---	377.00	---	377.00	**554.00**
For colors, ADD	Inst	%	Lg	---	s	---	15.0	---	---	---	**---**
	Inst	%	Sm	---	---	---	15.0	---	---	---	**---**

Kitchen

(Kitchen and utility), with good quality fittings, faucets, and sprayer

Americast - Porcelain enameled finish over steel composite material (American Standard)
Silhouette, double bowl with self rim
  33" x 22"

Description	Oper	Unit	Vol	Crew Size	Man-hours per Unit	Crew Output per Day	Avg Mat'l Unit Cost	Avg Labor Unit Cost	Avg Equip Unit Cost	Avg Total Unit Cost	Avg Price Incl O&P
Color	Inst	Ea	Lg	SA	5.00	1.60	470.00	188.00	---	658.00	**818.00**
	Inst	Ea	Sm	SA	7.27	1.10	567.00	274.00	---	841.00	**1050.00**
Premium color	Inst	Ea	Lg	SA	5.00	1.60	518.00	188.00	---	706.00	**872.00**
	Inst	Ea	Sm	SA	7.27	1.10	624.00	274.00	---	898.00	**1120.00**
White	Inst	Ea	Lg	SA	5.00	1.60	404.00	188.00	---	592.00	**741.00**
	Inst	Ea	Sm	SA	7.27	1.10	487.00	274.00	---	761.00	**962.00**

Silhouette, double bowl with tile edge
  33" x 22"

Description	Oper	Unit	Vol	Crew Size	Man-hours per Unit	Crew Output per Day	Avg Mat'l Unit Cost	Avg Labor Unit Cost	Avg Equip Unit Cost	Avg Total Unit Cost	Avg Price Incl O&P
Color	Inst	Ea	Lg	SA	5.00	1.60	394.00	188.00	---	582.00	**730.00**
	Inst	Ea	Sm	SA	7.27	1.10	475.00	274.00	---	749.00	**949.00**
Premium color	Inst	Ea	Lg	SA	5.00	1.60	430.00	188.00	---	618.00	**771.00**
	Inst	Ea	Sm	SA	7.27	1.10	518.00	274.00	---	792.00	**998.00**
White	Inst	Ea	Lg	SA	5.00	1.60	340.00	188.00	---	528.00	**668.00**
	Inst	Ea	Sm	SA	7.27	1.10	410.00	274.00	---	684.00	**874.00**

Description	Oper	Unit	Vol	Crew Size	Man-hours per Unit	Crew Output per Day	Avg Mat'l Unit Cost	Avg Labor Unit Cost	Avg Equip Unit Cost	Avg Total Unit Cost	Avg Price Incl O&P
Silhouette, dual level double bowl with self rim											
38" x 22"											
Color	Inst	Ea	Lg	SA	5.00	1.60	612.00	188.00	---	800.00	**981.00**
	Inst	Ea	Sm	SA	7.27	1.10	738.00	274.00	---	1012.00	**1250.00**
Premium color	Inst	Ea	Lg	SA	5.00	1.60	675.00	188.00	---	863.00	**1050.00**
	Inst	Ea	Sm	SA	7.27	1.10	814.00	274.00	---	1088.00	**1340.00**
White	Inst	Ea	Lg	SA	5.00	1.60	518.00	188.00	---	706.00	**873.00**
	Inst	Ea	Sm	SA	7.27	1.10	625.00	274.00	---	899.00	**1120.00**
Silhouette, dual level double bowl with tile edge or color matched rim											
38" x 22"											
Color	Inst	Ea	Lg	SA	5.00	1.60	631.00	188.00	---	819.00	**1000.00**
	Inst	Ea	Sm	SA	7.27	1.10	760.00	274.00	---	1034.00	**1280.00**
Premium color	Inst	Ea	Lg	SA	5.00	1.60	697.00	188.00	---	885.00	**1080.00**
	Inst	Ea	Sm	SA	7.27	1.10	840.00	274.00	---	1114.00	**1370.00**
White	Inst	Ea	Lg	SA	5.00	1.60	532.00	188.00	---	720.00	**889.00**
	Inst	Ea	Sm	SA	7.27	1.10	642.00	274.00	---	916.00	**1140.00**
Silhouette, dual level double bowl with self rim											
33" x 22"											
Color	Inst	Ea	Lg	SA	5.00	1.60	494.00	188.00	---	682.00	**845.00**
	Inst	Ea	Sm	SA	7.27	1.10	595.00	274.00	---	869.00	**1090.00**
Premium color	Inst	Ea	Lg	SA	5.00	1.60	541.00	188.00	---	729.00	**899.00**
	Inst	Ea	Sm	SA	7.27	1.10	652.00	274.00	---	926.00	**1150.00**
White	Inst	Ea	Lg	SA	5.00	1.60	420.00	188.00	---	608.00	**760.00**
	Inst	Ea	Sm	SA	7.27	1.10	507.00	274.00	---	781.00	**985.00**
Silhouette, dual level double bowl with tile edge or color matched rim											
33" x 22"											
Color	Inst	Ea	Lg	SA	5.00	1.60	513.00	188.00	---	701.00	**867.00**
	Inst	Ea	Sm	SA	7.27	1.10	619.00	274.00	---	893.00	**1110.00**
Premium color	Inst	Ea	Lg	SA	5.00	1.60	565.00	188.00	---	753.00	**926.00**
	Inst	Ea	Sm	SA	7.27	1.10	681.00	274.00	---	955.00	**1190.00**
White	Inst	Ea	Lg	SA	5.00	1.60	437.00	188.00	---	625.00	**779.00**
	Inst	Ea	Sm	SA	7.27	1.10	527.00	274.00	---	801.00	**1010.00**
Silhouette, single bowl with self rim											
25" x 22"											
Color	Inst	Ea	Lg	SA	4.00	2.00	361.00	151.00	---	512.00	**636.00**
	Inst	Ea	Sm	SA	5.71	1.40	435.00	215.00	---	650.00	**816.00**
Premium color	Inst	Ea	Lg	SA	4.00	2.00	392.00	151.00	---	543.00	**673.00**
	Inst	Ea	Sm	SA	5.71	1.40	473.00	215.00	---	688.00	**860.00**
White	Inst	Ea	Lg	SA	4.00	2.00	313.00	151.00	---	464.00	**582.00**
	Inst	Ea	Sm	SA	5.71	1.40	378.00	215.00	---	593.00	**751.00**
Silhouette, single bowl with tile edge											
25" x 22"											
Color	Inst	Ea	Lg	SA	4.00	2.00	382.00	151.00	---	533.00	**661.00**
	Inst	Ea	Sm	SA	5.71	1.40	460.00	215.00	---	675.00	**845.00**
Premium color	Inst	Ea	Lg	SA	4.00	2.00	415.00	151.00	---	566.00	**699.00**
	Inst	Ea	Sm	SA	5.71	1.40	500.00	215.00	---	715.00	**892.00**
White	Inst	Ea	Lg	SA	4.00	2.00	330.00	151.00	---	481.00	**601.00**
	Inst	Ea	Sm	SA	5.71	1.40	398.00	215.00	---	613.00	**774.00**

Description	Oper	Unit	Vol	Crew Size	Man-hours per Unit	Crew Output per Day	Avg Mat'l Unit Cost	Avg Labor Unit Cost	Avg Equip Unit Cost	Avg Total Unit Cost	Avg Price Incl O&P
Silhouette, single bowl with self rim											
18" x 18"											
Color	Inst	Ea	Lg	SA	2.67	3.00	324.00	101.00	---	425.00	**520.00**
	Inst	Ea	Sm	SA	3.81	2.10	391.00	144.00	---	535.00	**660.00**
Premium color	Inst	Ea	Lg	SA	2.67	3.00	350.00	101.00	---	451.00	**551.00**
	Inst	Ea	Sm	SA	3.81	2.10	422.00	144.00	---	566.00	**697.00**
White	Inst	Ea	Lg	SA	2.67	3.00	285.00	101.00	---	386.00	**475.00**
	Inst	Ea	Sm	SA	3.81	2.10	343.00	144.00	---	487.00	**605.00**
Silhouette, single bowl with tile edge											
18" x 18"											
Color	Inst	Ea	Lg	SA	2.67	3.00	344.00	101.00	---	445.00	**544.00**
	Inst	Ea	Sm	SA	3.81	2.10	415.00	144.00	---	559.00	**688.00**
Premium color	Inst	Ea	Lg	SA	2.67	3.00	374.00	101.00	---	475.00	**578.00**
	Inst	Ea	Sm	SA	3.81	2.10	451.00	144.00	---	595.00	**729.00**
White	Inst	Ea	Lg	SA	2.67	3.00	299.00	101.00	---	400.00	**491.00**
	Inst	Ea	Sm	SA	3.81	2.10	360.00	144.00	---	504.00	**625.00**

Enameled cast iron (Kohler Products)

Description	Oper	Unit	Vol	Crew Size	Man-hours per Unit	Crew Output per Day	Avg Mat'l Unit Cost	Avg Labor Unit Cost	Avg Equip Unit Cost	Avg Total Unit Cost	Avg Price Incl O&P
Bakersfield, single bowl											
31" x 22"											
Color	Inst	Ea	Lg	SA	5.00	1.60	445.00	188.00	---	633.00	**788.00**
	Inst	Ea	Sm	SA	7.27	1.10	536.00	274.00	---	810.00	**1020.00**
Premium color	Inst	Ea	Lg	SA	5.00	1.60	495.00	188.00	---	683.00	**847.00**
	Inst	Ea	Sm	SA	7.27	1.10	597.00	274.00	---	871.00	**1090.00**
White	Inst	Ea	Lg	SA	5.00	1.60	377.00	188.00	---	565.00	**710.00**
	Inst	Ea	Sm	SA	7.27	1.10	454.00	274.00	---	728.00	**925.00**
Bon Vivant, triple bowl, 2 large, 1 small with tile or self rim											
48" x 22"											
Color	Inst	Ea	Lg	SA	5.00	1.60	916.00	188.00	---	1104.00	**1330.00**
	Inst	Ea	Sm	SA	7.27	1.10	1100.00	274.00	---	1374.00	**1670.00**
Premium color	Inst	Ea	Lg	SA	5.00	1.60	1040.00	188.00	---	1228.00	**1470.00**
	Inst	Ea	Sm	SA	7.27	1.10	1250.00	274.00	---	1524.00	**1840.00**
White	Inst	Ea	Lg	SA	5.00	1.60	754.00	188.00	---	942.00	**1140.00**
	Inst	Ea	Sm	SA	7.27	1.10	908.00	274.00	---	1182.00	**1450.00**
Cantina, double with tile or self rim, 2 large											
43" x 22"											
Color	Inst	Ea	Lg	SA	5.00	1.60	670.00	188.00	---	858.00	**1050.00**
	Inst	Ea	Sm	SA	7.27	1.10	807.00	274.00	---	1081.00	**1330.00**
Premium color	Inst	Ea	Lg	SA	5.00	1.60	754.00	188.00	---	942.00	**1140.00**
	Inst	Ea	Sm	SA	7.27	1.10	909.00	274.00	---	1183.00	**1450.00**
White	Inst	Ea	Lg	SA	5.00	1.60	557.00	188.00	---	745.00	**917.00**
	Inst	Ea	Sm	SA	7.27	1.10	671.00	274.00	---	945.00	**1170.00**

Description	Oper	Unit	Vol	Crew Size	Man-hours per Unit	Crew Output per Day	Avg Mat'l Unit Cost	Avg Labor Unit Cost	Avg Equip Unit Cost	Avg Total Unit Cost	Avg Price Incl O&P
Ecocycle, double bowl with regular rim											
43" x 22"											
Color	Inst	Ea	Lg	SA	5.00	1.60	1600.00	188.00	---	1788.00	**2110.00**
	Inst	Ea	Sm	SA	7.27	1.10	1920.00	274.00	---	2194.00	**2610.00**
Premium color	Inst	Ea	Lg	SA	5.00	1.60	1820.00	188.00	---	2008.00	**2370.00**
	Inst	Ea	Sm	SA	7.27	1.10	2190.00	274.00	---	2464.00	**2920.00**
White	Inst	Ea	Lg	SA	5.00	1.60	1300.00	188.00	---	1488.00	**1770.00**
	Inst	Ea	Sm	SA	7.27	1.10	1560.00	274.00	---	1834.00	**2200.00**
Epicurean, double bowl with cutting board and drainboard, 1 large, 1 small											
43" x 22"											
Color	Inst	Ea	Lg	SA	5.00	1.60	800.00	188.00	---	988.00	**1200.00**
	Inst	Ea	Sm	SA	7.27	1.10	965.00	274.00	---	1239.00	**1510.00**
White	Inst	Ea	Lg	SA	5.00	1.60	669.00	188.00	---	857.00	**1050.00**
	Inst	Ea	Sm	SA	7.27	1.10	807.00	274.00	---	1081.00	**1330.00**
Executive Chef, double bowl with tile or self rim, 1 large, 1 medium											
33" x 22"											
Color	Inst	Ea	Lg	SA	5.00	1.60	601.00	188.00	---	789.00	**968.00**
	Inst	Ea	Sm	SA	7.27	1.10	724.00	274.00	---	998.00	**1240.00**
Premium color	Inst	Ea	Lg	SA	5.00	1.60	675.00	188.00	---	863.00	**1050.00**
	Inst	Ea	Sm	SA	7.27	1.10	814.00	274.00	---	1088.00	**1340.00**
White	Inst	Ea	Lg	SA	5.00	1.60	502.00	188.00	---	690.00	**854.00**
	Inst	Ea	Sm	SA	7.27	1.10	605.00	274.00	---	879.00	**1100.00**
Lakefield double bowl with tile or self rim, 1 large, 1 small											
33" x 22"											
Color	Inst	Ea	Lg	SA	5.00	1.60	498.00	188.00	---	686.00	**849.00**
	Inst	Ea	Sm	SA	7.27	1.10	600.00	274.00	---	874.00	**1090.00**
Premium color	Inst	Ea	Lg	SA	5.00	1.60	557.00	188.00	---	745.00	**917.00**
	Inst	Ea	Sm	SA	7.27	1.10	671.00	274.00	---	945.00	**1170.00**
White	Inst	Ea	Lg	SA	5.00	1.60	419.00	188.00	---	607.00	**759.00**
	Inst	Ea	Sm	SA	7.27	1.10	506.00	274.00	---	780.00	**984.00**
Marsala Hi-Low with self or tile rim											
33" x 22"											
Color	Inst	Ea	Lg	SA	5.00	1.60	601.00	188.00	---	789.00	**968.00**
	Inst	Ea	Sm	SA	7.27	1.10	724.00	274.00	---	998.00	**1240.00**
Premium color	Inst	Ea	Lg	SA	5.00	1.60	675.00	188.00	---	863.00	**1050.00**
	Inst	Ea	Sm	SA	7.27	1.10	814.00	274.00	---	1088.00	**1340.00**
White	Inst	Ea	Lg	SA	5.00	1.60	502.00	188.00	---	690.00	**854.00**
	Inst	Ea	Sm	SA	7.27	1.10	605.00	274.00	---	879.00	**1100.00**
Mayfield, single for frame mount with self or tile rim											
24" x 21"											
Color	Inst	Ea	Lg	SA	4.00	2.00	321.00	151.00	---	472.00	**591.00**
	Inst	Ea	Sm	SA	5.71	1.40	387.00	215.00	---	602.00	**762.00**
White	Inst	Ea	Lg	SA	4.00	2.00	283.00	151.00	---	434.00	**546.00**
	Inst	Ea	Sm	SA	5.71	1.40	341.00	215.00	---	556.00	**708.00**

Description	Oper	Unit	Vol	Crew Size	Man-hours per Unit	Crew Output per Day	Avg Mat'l Unit Cost	Avg Labor Unit Cost	Avg Equip Unit Cost	Avg Total Unit Cost	Avg Price Incl O&P

Porcelain on steel

Double bowl, with self rim
33" x 22"

Description	Oper	Unit	Vol	Crew Size	Man-hours per Unit	Crew Output per Day	Avg Mat'l Unit Cost	Avg Labor Unit Cost	Avg Equip Unit Cost	Avg Total Unit Cost	Avg Price Incl O&P
Color	Inst	Ea	Lg	SA	5.00	1.60	206.00	188.00	---	394.00	**513.00**
	Inst	Ea	Sm	SA	7.27	1.10	248.00	274.00	---	522.00	**688.00**
White	Inst	Ea	Lg	SA	5.00	1.60	194.00	188.00	---	382.00	**500.00**
	Inst	Ea	Sm	SA	7.27	1.10	234.00	274.00	---	508.00	**672.00**

Double bowl, dual level with self rim
33" x 22"

Description	Oper	Unit	Vol	Crew Size	Man-hours per Unit	Crew Output per Day	Avg Mat'l Unit Cost	Avg Labor Unit Cost	Avg Equip Unit Cost	Avg Total Unit Cost	Avg Price Incl O&P
Color	Inst	Ea	Lg	SA	5.00	1.60	234.00	188.00	---	422.00	**546.00**
	Inst	Ea	Sm	SA	7.27	1.10	282.00	274.00	---	556.00	**727.00**
White	Inst	Ea	Lg	SA	5.00	1.60	222.00	188.00	---	410.00	**533.00**
	Inst	Ea	Sm	SA	7.27	1.10	268.00	274.00	---	542.00	**711.00**

Single bowl
25" x 22"

Description	Oper	Unit	Vol	Crew Size	Man-hours per Unit	Crew Output per Day	Avg Mat'l Unit Cost	Avg Labor Unit Cost	Avg Equip Unit Cost	Avg Total Unit Cost	Avg Price Incl O&P
Color	Inst	Ea	Lg	SA	4.00	2.00	196.00	151.00	---	347.00	**447.00**
	Inst	Ea	Sm	SA	5.71	1.40	236.00	215.00	---	451.00	**588.00**
White	Inst	Ea	Lg	SA	4.00	2.00	186.00	151.00	---	337.00	**435.00**
	Inst	Ea	Sm	SA	5.71	1.40	224.00	215.00	---	439.00	**573.00**

Stainless steel (Kohler)

Triple bowl with self rim
Ravinia

Description	Oper	Unit	Vol	Crew Size	Man-hours per Unit	Crew Output per Day	Avg Mat'l Unit Cost	Avg Labor Unit Cost	Avg Equip Unit Cost	Avg Total Unit Cost	Avg Price Incl O&P
43" x 22"	Inst	Ea	Lg	SA	5.00	1.60	1210.00	188.00	---	1398.00	**1670.00**
	Inst	Ea	Sm	SA	7.27	1.10	1460.00	274.00	---	1734.00	**2080.00**

Double bowl with self rim
Ravinia

Description	Oper	Unit	Vol	Crew Size	Man-hours per Unit	Crew Output per Day	Avg Mat'l Unit Cost	Avg Labor Unit Cost	Avg Equip Unit Cost	Avg Total Unit Cost	Avg Price Incl O&P
33" x 22"	Inst	Ea	Lg	SA	5.00	1.60	724.00	188.00	---	912.00	**1110.00**
	Inst	Ea	Sm	SA	7.27	1.10	873.00	274.00	---	1147.00	**1410.00**
42" x 22"	Inst	Ea	Lg	SA	5.00	1.60	899.00	188.00	---	1087.00	**1310.00**
	Inst	Ea	Sm	SA	7.27	1.10	1080.00	274.00	---	1354.00	**1650.00**

Single bowl with self rim
Ravinia

Description	Oper	Unit	Vol	Crew Size	Man-hours per Unit	Crew Output per Day	Avg Mat'l Unit Cost	Avg Labor Unit Cost	Avg Equip Unit Cost	Avg Total Unit Cost	Avg Price Incl O&P
25" x 22"	Inst	Ea	Lg	SA	4.00	2.00	506.00	151.00	---	657.00	**804.00**
	Inst	Ea	Sm	SA	5.71	1.40	610.00	215.00	---	825.00	**1020.00**

Ballad

Description	Oper	Unit	Vol	Crew Size	Man-hours per Unit	Crew Output per Day	Avg Mat'l Unit Cost	Avg Labor Unit Cost	Avg Equip Unit Cost	Avg Total Unit Cost	Avg Price Incl O&P
43" x 22"	Inst	Ea	Lg	SA	4.00	2.00	1040.00	151.00	---	1191.00	**1420.00**
	Inst	Ea	Sm	SA	5.71	1.40	1260.00	215.00	---	1475.00	**1760.00**
33" x 22"	Inst	Ea	Lg	SA	4.00	2.00	659.00	151.00	---	010.00	**979.00**
	Inst	Ea	Sm	SA	5.71	1.40	794.00	215.00	---	1009.00	**1230.00**
25" x 22"	Inst	Ea	Lg	SA	4.00	2.00	410.00	151.00	---	561.00	**693.00**
	Inst	Ea	Sm	SA	5.71	1.40	495.00	215.00	---	710.00	**885.00**

Description	Oper	Unit	Vol	Crew Size	Man-hours per Unit	Crew Output per Day	Avg Mat'l Unit Cost	Avg Labor Unit Cost	Avg Equip Unit Cost	Avg Total Unit Cost	Avg Price Incl O&P

Bar
Enameled cast iron (Kohler)
Addison
13" x 13"

Description	Oper	Unit	Vol	Crew Size	Man-hours per Unit	Crew Output per Day	Avg Mat'l Unit Cost	Avg Labor Unit Cost	Avg Equip Unit Cost	Avg Total Unit Cost	Avg Price Incl O&P
Color	Inst	Ea	Lg	SA	2.67	3.00	288.00	101.00	---	389.00	**479.00**
	Inst	Ea	Sm	SA	3.81	2.10	347.00	144.00	---	491.00	**610.00**
Premium color	Inst	Ea	Lg	SA	2.67	3.00	315.00	101.00	---	416.00	**510.00**
	Inst	Ea	Sm	SA	3.81	2.10	380.00	144.00	---	524.00	**648.00**
White	Inst	Ea	Lg	SA	2.67	3.00	251.00	101.00	---	352.00	**437.00**
	Inst	Ea	Sm	SA	3.81	2.10	303.00	144.00	---	447.00	**559.00**

Apertif with tile or self rim
16" x 19"

Description	Oper	Unit	Vol	Crew Size	Man-hours per Unit	Crew Output per Day	Avg Mat'l Unit Cost	Avg Labor Unit Cost	Avg Equip Unit Cost	Avg Total Unit Cost	Avg Price Incl O&P
Color	Inst	Ea	Lg	SA	2.67	3.00	346.00	101.00	---	447.00	**546.00**
	Inst	Ea	Sm	SA	3.81	2.10	417.00	144.00	---	561.00	**690.00**
Premium color	Inst	Ea	Lg	SA	2.67	3.00	382.00	101.00	---	483.00	**587.00**
	Inst	Ea	Sm	SA	3.81	2.10	460.00	144.00	---	604.00	**740.00**
White	Inst	Ea	Lg	SA	2.67	3.00	298.00	101.00	---	399.00	**490.00**
	Inst	Ea	Sm	SA	3.81	2.10	359.00	144.00	---	503.00	**624.00**

Entertainer with semicircular self rim or undercounter self rim
22" x 18"

Description	Oper	Unit	Vol	Crew Size	Man-hours per Unit	Crew Output per Day	Avg Mat'l Unit Cost	Avg Labor Unit Cost	Avg Equip Unit Cost	Avg Total Unit Cost	Avg Price Incl O&P
Color	Inst	Ea	Lg	SA	4.00	2.00	305.00	151.00	---	456.00	**573.00**
	Inst	Ea	Sm	SA	5.71	1.40	368.00	215.00	---	583.00	**739.00**
Premium color	Inst	Ea	Lg	SA	4.00	2.00	335.00	151.00	---	486.00	**607.00**
	Inst	Ea	Sm	SA	5.71	1.40	404.00	215.00	---	619.00	**781.00**
White	Inst	Ea	Lg	SA	4.00	2.00	265.00	151.00	---	416.00	**527.00**
	Inst	Ea	Sm	SA	5.71	1.40	320.00	215.00	---	535.00	**684.00**

Sorbet with tile or undercounter rim or self rim
15" x 15"

Description	Oper	Unit	Vol	Crew Size	Man-hours per Unit	Crew Output per Day	Avg Mat'l Unit Cost	Avg Labor Unit Cost	Avg Equip Unit Cost	Avg Total Unit Cost	Avg Price Incl O&P
Color	Inst	Ea	Lg	SA	2.67	3.00	315.00	101.00	---	416.00	**510.00**
	Inst	Ea	Sm	SA	3.81	2.10	380.00	144.00	---	524.00	**648.00**
Premium color	Inst	Ea	Lg	SA	2.67	3.00	347.00	101.00	---	448.00	**547.00**
	Inst	Ea	Sm	SA	3.81	2.10	418.00	144.00	---	562.00	**692.00**
White	Inst	Ea	Lg	SA	2.67	3.00	273.00	101.00	---	374.00	**462.00**
	Inst	Ea	Sm	SA	3.81	2.10	330.00	144.00	---	474.00	**590.00**

Stainless steel (Kohler)
Ravinia
22" x 15"

Description	Oper	Unit	Vol	Crew Size	Man-hours per Unit	Crew Output per Day	Avg Mat'l Unit Cost	Avg Labor Unit Cost	Avg Equip Unit Cost	Avg Total Unit Cost	Avg Price Incl O&P
	Inst	Ea	Lg	SA	2.67	3.00	562.00	101.00	---	663.00	**795.00**
	Inst	Ea	Sm	SA	3.81	2.10	678.00	144.00	---	822.00	**991.00**

Lyric
15" x 15"

Description	Oper	Unit	Vol	Crew Size	Man-hours per Unit	Crew Output per Day	Avg Mat'l Unit Cost	Avg Labor Unit Cost	Avg Equip Unit Cost	Avg Total Unit Cost	Avg Price Incl O&P
	Inst	Ea	Lg	SA	2.67	3.00	213.00	101.00	---	314.00	**393.00**
	Inst	Ea	Sm	SA	3.81	2.10	257.00	144.00	---	401.00	**507.00**

Description	Oper	Unit	Vol	Crew Size	Man-hours per Unit	Crew Output per Day	Avg Mat'l Unit Cost	Avg Labor Unit Cost	Avg Equip Unit Cost	Avg Total Unit Cost	Avg Price Incl O&P

Utility/service
Enameled cast iron (Kohler)
Sutton high back

Description	Oper	Unit	Vol	Crew Size	Man-hours per Unit	Crew Output per Day	Avg Mat'l Unit Cost	Avg Labor Unit Cost	Avg Equip Unit Cost	Avg Total Unit Cost	Avg Price Incl O&P
24" x 18", 8" high back	Inst	Ea	Lg	SA	4.00	2.00	641.00	151.00	---	792.00	**958.00**
	Inst	Ea	Sm	SA	5.71	1.40	772.00	215.00	---	987.00	**1200.00**
Tech											
24" x 18", bubbler mound	Inst	Ea	Lg	SA	4.00	2.00	352.00	151.00	---	503.00	**626.00**
	Inst	Ea	Sm	SA	5.71	1.40	425.00	215.00	---	640.00	**804.00**
Whitby											
28" x 28" corner	Inst	Ea	Lg	SA	4.00	2.00	715.00	151.00	---	866.00	**1040.00**
	Inst	Ea	Sm	SA	5.71	1.40	862.00	215.00	---	1077.00	**1310.00**
Glen Falls, laundry sink 25" x 22"											
Color	Inst	Ea	Lg	SA	4.00	2.00	516.00	151.00	---	667.00	**815.00**
	Inst	Ea	Sm	SA	5.71	1.40	623.00	215.00	---	838.00	**1030.00**
White	Inst	Ea	Lg	SA	4.00	2.00	431.00	151.00	---	582.00	**718.00**
	Inst	Ea	Sm	SA	5.71	1.40	520.00	215.00	---	735.00	**914.00**
River Falls, laundry sink 25" x 22"											
Color	Inst	Ea	Lg	SA	4.00	2.00	568.00	151.00	---	719.00	**874.00**
	Inst	Ea	Sm	SA	5.71	1.40	684.00	215.00	---	899.00	**1100.00**
White	Inst	Ea	Lg	SA	4.00	2.00	472.00	151.00	---	623.00	**765.00**
	Inst	Ea	Sm	SA	5.71	1.40	570.00	215.00	---	785.00	**971.00**
Westover, double deep, laundry sink 42" x 21"											
Color	Inst	Ea	Lg	SA	5.00	1.60	832.00	188.00	---	1020.00	**1230.00**
	Inst	Ea	Sm	SA	7.27	1.10	1000.00	274.00	---	1274.00	**1560.00**
White	Inst	Ea	Lg	SA	5.00	1.60	684.00	188.00	---	872.00	**1060.00**
	Inst	Ea	Sm	SA	7.27	1.10	825.00	274.00	---	1099.00	**1350.00**

Vitreous china (Kohler)
Hollister

Description	Oper	Unit	Vol	Crew Size	Man-hours per Unit	Crew Output per Day	Avg Mat'l Unit Cost	Avg Labor Unit Cost	Avg Equip Unit Cost	Avg Total Unit Cost	Avg Price Incl O&P
28" x 22"	Inst	Ea	Lg	SA	4.00	2.00	851.00	151.00	---	1002.00	**1200.00**
	Inst	Ea	Sm	SA	5.71	1.40	1030.00	215.00	---	1245.00	**1500.00**
Sudbury											
22" x 20"	Inst	Ea	Lg	SA	4.00	2.00	511.00	151.00	---	662.00	**809.00**
	Inst	Ea	Sm	SA	5.71	1.40	616.00	215.00	---	831.00	**1020.00**
Tyrrell											
20" x 20" w/ siphon jet	Inst	Ea	Lg	SA	5.00	1.60	908.00	188.00	---	1096.00	**1320.00**
	Inst	Ea	Sm	SA	7.27	1.10	1090.00	274.00	---	1364.00	**1660.00**
Adjustments											
To only remove and reset sink											
Single bowl	Reset	Ea	Lg	SA	2.00	4.00	---	75.30	---	75.30	**111.00**
	Reset	Ea	Sm	SA	2.86	2.80	---	108.00	---	108.00	**158.00**
Double bowl	Reset	Ea	Lg	SA	2.11	3.80	---	79.50	---	79.50	**117.00**
	Reset	Ea	Sm	SA	2.96	2.70	---	112.00	---	112.00	**164.00**
Triple bowl	Reset	Ea	Lg	SA	2.29	3.50	---	86.30	---	86.30	**127.00**
	Reset	Ea	Sm	SA	3.20	2.50	---	121.00	---	121.00	**177.00**
To install rough-in											
	Inst	Ea	Lg	SA	6.67	1.20	---	251.00	---	251.00	**369.00**
	Inst	Ea	Sm	SA	10.0	0.80	---	377.00	---	377.00	**554.00**
For 18" gauge steel, ADD	Inst	%	Lg	---	---	---	40.0	---	---	---	---
	Inst	%	Sm	---	---	---	40.0	---	---	---	---

Description	Oper	Unit	Vol	Crew Size	Man-hours per Unit	Crew Output per Day	Avg Mat'l Unit Cost	Avg Labor Unit Cost	Avg Equip Unit Cost	Avg Total Unit Cost	Avg Price Incl O&P

Skylights, skywindows, roof windows

Labor costs are for installation of skylight, flashing and roller shade only; no carpentry or roofing work included

Polycarbonate dome, clear transparent or tinted, roof opening sizes, curb or flush mount, self-flashing

Description	Oper	Unit	Vol	Crew Size	Man-hours per Unit	Crew Output per Day	Avg Mat'l Unit Cost	Avg Labor Unit Cost	Avg Equip Unit Cost	Avg Total Unit Cost	Avg Price Incl O&P
Single dome											
22" x 22"	Inst	Ea	Lg	CA	1.57	5.10	43.40	52.20	---	95.60	**128.00**
	Inst	Ea	Sm	CA	2.41	3.32	48.10	80.20	---	128.30	**176.00**
22" x 46"	Inst	Ea	Lg	CA	2.58	3.10	86.80	85.80	---	172.60	**229.00**
	Inst	Ea	Sm	CA	3.96	2.02	96.20	132.00	---	228.20	**308.00**
30" x 30"	Inst	Ea	Lg	CA	2.58	3.10	72.30	85.80	---	158.10	**212.00**
	Inst	Ea	Sm	CA	3.96	2.02	80.20	132.00	---	212.20	**290.00**
30" x 46"	Inst	Ea	Lg	CA	2.96	2.70	126.00	98.50	---	224.50	**292.00**
	Inst	Ea	Sm	CA	4.55	1.76	139.00	151.00	---	290.00	**387.00**
46" x 46"	Inst	Ea	Lg	CA	3.48	2.30	217.00	116.00	---	333.00	**424.00**
	Inst	Ea	Sm	CA	5.33	1.50	241.00	177.00	---	418.00	**543.00**
Double dome											
22" x 22"	Inst	Ea	Lg	CA	1.57	5.10	57.80	52.20	---	110.00	**145.00**
	Inst	Ea	Sm	CA	2.41	3.32	64.10	80.20	---	144.30	**194.00**
22" x 46"	Inst	Ea	Lg	CA	2.58	3.10	126.00	85.80	---	211.80	**273.00**
	Inst	Ea	Sm	CA	3.96	2.02	139.00	132.00	---	271.00	**358.00**
30" x 30"	Inst	Ea	Lg	CA	2.58	3.10	110.00	85.80	---	195.80	**255.00**
	Inst	Ea	Sm	CA	3.96	2.02	122.00	132.00	---	254.00	**338.00**
30" x 46"	Inst	Ea	Lg	CA	2.96	2.70	164.00	98.50	---	262.50	**337.00**
	Inst	Ea	Sm	CA	4.55	1.76	182.00	151.00	---	333.00	**436.00**
46" x 46"	Inst	Ea	Lg	CA	3.48	2.30	265.00	116.00	---	381.00	**478.00**
	Inst	Ea	Sm	CA	5.33	1.50	293.00	177.00	---	470.00	**603.00**
Triple dome											
22" x 22"	Inst	Ea	Lg	CA	1.57	5.10	77.10	52.20	---	129.30	**167.00**
	Inst	Ea	Sm	CA	2.41	3.32	85.50	80.20	---	165.70	**219.00**
22" x 46"	Inst	Ea	Lg	CA	2.58	3.10	174.00	85.80	---	259.80	**329.00**
	Inst	Ea	Sm	CA	3.96	2.02	193.00	132.00	---	325.00	**419.00**
30" x 30"	Inst	Ea	Lg	CA	2.58	3.10	145.00	85.80	---	230.80	**295.00**
	Inst	Ea	Sm	CA	3.96	2.02	161.00	132.00	---	293.00	**382.00**
30" x 46"	Inst	Ea	Lg	CA	2.96	2.70	217.00	98.50	---	315.50	**398.00**
	Inst	Ea	Sm	CA	4.55	1.76	241.00	151.00	---	392.00	**504.00**
46" x 46"	Inst	Ea	Lg	CA	3.48	2.30	332.00	116.00	---	448.00	**556.00**
	Inst	Ea	Sm	CA	5.33	1.50	368.00	177.00	---	545.00	**690.00**

Description	Oper	Unit	Vol	Crew Size	Man-hours per Unit	Crew Output per Day	Avg Mat'l Unit Cost	Avg Labor Unit Cost	Avg Equip Unit Cost	Avg Total Unit Cost	Avg Price Incl O&P
Operable skylight, Velux model VS											
21-9/16" x 27-1/2"	Inst	Ea	Lg	CA	3.42	2.34	291.00	114.00	---	405.00	**505.00**
	Inst	Ea	Sm	CA	5.26	1.52	322.00	175.00	---	497.00	**633.00**
21-9/16" x 38-1/2"	Inst	Ea	Lg	CA	3.42	2.34	323.00	114.00	---	437.00	**542.00**
	Inst	Ea	Sm	CA	5.26	1.52	358.00	175.00	---	533.00	**674.00**
21-9/16" x 46-3/8"	Inst	Ea	Lg	CA	3.42	2.34	351.00	114.00	---	465.00	**574.00**
	Inst	Ea	Sm	CA	5.26	1.52	389.00	175.00	---	564.00	**709.00**
21-9/16" x 55"	Inst	Ea	Lg	CA	3.42	2.34	372.00	114.00	---	486.00	**598.00**
	Inst	Ea	Sm	CA	5.26	1.52	412.00	175.00	---	587.00	**736.00**
30-5/8" x 38-1/2"	Inst	Ea	Lg	CA	3.42	2.34	369.00	114.00	---	483.00	**595.00**
	Inst	Ea	Sm	CA	5.26	1.52	409.00	175.00	---	584.00	**733.00**
30-5/8" x 55"	Inst	Ea	Lg	CA	3.56	2.25	443.00	118.00	---	561.00	**688.00**
	Inst	Ea	Sm	CA	5.48	1.46	492.00	182.00	---	674.00	**839.00**
44-3/4" x 27-1/2"	Inst	Ea	Lg	CA	3.56	2.25	389.00	118.00	---	507.00	**625.00**
	Inst	Ea	Sm	CA	5.48	1.46	431.00	182.00	---	613.00	**770.00**
44-3/4" x 46-1/2"	Inst	Ea	Lg	CA	3.56	2.25	492.00	118.00	---	610.00	**744.00**
	Inst	Ea	Sm	CA	5.48	1.46	546.00	182.00	---	728.00	**901.00**
Fixed skylight, Velux model FS											
21-9/16" x 27-1/2"	Inst	Ea	Lg	CA	3.42	2.34	211.00	114.00	---	325.00	**413.00**
	Inst	Ea	Sm	CA	5.26	1.52	234.00	175.00	---	409.00	**531.00**
21-9/16" x 38-1/2"	Inst	Ea	Lg	CA	3.42	2.34	238.00	114.00	---	352.00	**445.00**
	Inst	Ea	Sm	CA	5.26	1.52	264.00	175.00	---	439.00	**566.00**
21-9/16" x 46-3/8"	Inst	Ea	Lg	CA	3.42	2.34	264.00	114.00	---	378.00	**474.00**
	Inst	Ea	Sm	CA	5.26	1.52	293.00	175.00	---	468.00	**599.00**
21-9/16" x 55"	Inst	Ea	Lg	CA	3.42	2.34	297.00	114.00	---	411.00	**512.00**
	Inst	Ea	Sm	CA	5.26	1.52	329.00	175.00	---	504.00	**641.00**
21-9/16" x 70-7/8"	Inst	Ea	Lg	CA	3.42	2.34	348.00	114.00	---	462.00	**571.00**
	Inst	Ea	Sm	CA	5.26	1.52	386.00	175.00	---	561.00	**706.00**
30-5/8" x 38-1/2"	Inst	Ea	Lg	CA	3.42	2.34	266.00	114.00	---	380.00	**476.00**
	Inst	Ea	Sm	CA	5.26	1.52	295.00	175.00	---	470.00	**602.00**
30-5/8" x 55"	Inst	Ea	Lg	CA	3.56	2.25	330.00	118.00	---	448.00	**558.00**
	Inst	Ea	Sm	CA	5.48	1.46	366.00	182.00	---	548.00	**695.00**
44-3/4" x 27-1/2"	Inst	Ea	Lg	CA	3.56	2.25	311.00	118.00	---	429.00	**535.00**
	Inst	Ea	Sm	CA	5.48	1.46	345.00	182.00	---	527.00	**670.00**
44-3/4" x 46-1/2"	Inst	Ea	Lg	CA	3.56	2.25	369.00	118.00	---	487.00	**602.00**
	Inst	Ea	Sm	CA	5.48	1.46	409.00	182.00	---	591.00	**744.00**
Fixed ventilation with removable filter on ventilation flap, Velux model FSF											
21-9/16" x 27-1/2"	Inst	Ea	Lg	CA	3.14	2.55	220.00	104.00	---	324.00	**409.00**
	Inst	Ea	Sm	CA	4.82	1.66	243.00	160.00	---	403.00	**520.00**
21-9/16" x 38-1/2"	Inst	Ea	Lg	CA	3.14	2.55	252.00	104.00	---	356.00	**446.00**
	Inst	Ea	Sm	CA	4.82	1.66	279.00	160.00	---	439.00	**562.00**
21-9/16" x 46-3/8"	Inst	Ea	Lg	CA	3.14	2.55	266.00	104.00	---	370.00	**462.00**
	Inst	Ea	Sm	CA	4.82	1.66	294.00	160.00	---	454.00	**579.00**
21-9/16" x 55"	Inst	Ea	Lg	CA	3.42	2.34	364.00	114.00	---	478.00	**589.00**
	Inst	Ea	Sm	CA	5.26	1.52	403.00	175.00	---	578.00	**726.00**

Description	Oper	Unit	Vol	Crew Size	Man-hours per Unit	Crew Output per Day	Avg Mat'l Unit Cost	Avg Labor Unit Cost	Avg Equip Unit Cost	Avg Total Unit Cost	Avg Price Incl O&P
21-9/16" x 70-7/8"	Inst	Ea	Lg	CA	3.39	2.36	294.00	113.00	---	407.00	**507.00**
	Inst	Ea	Sm	CA	5.23	1.53	326.00	174.00	---	500.00	**636.00**
30-5/8" x 38-1/2"	Inst	Ea	Lg	CA	3.14	2.55	293.00	104.00	---	397.00	**494.00**
	Inst	Ea	Sm	CA	4.82	1.66	325.00	160.00	---	485.00	**614.00**
30-5/8" x 55"	Inst	Ea	Lg	CA	3.39	2.36	340.00	113.00	---	453.00	**560.00**
	Inst	Ea	Sm	CA	5.23	1.53	377.00	174.00	---	551.00	**694.00**
44-3/4" x 27-1/2"	Inst	Ea	Lg	CA	3.36	2.38	312.00	112.00	---	424.00	**526.00**
	Inst	Ea	Sm	CA	5.16	1.55	346.00	172.00	---	518.00	**655.00**
44-3/4" x 46-1/2"	Inst	Ea	Lg	CA	3.39	2.36	377.00	113.00	---	490.00	**602.00**
	Inst	Ea	Sm	CA	5.23	1.53	418.00	174.00	---	592.00	**741.00**

Low profile, acrylic double domes with molded edge. Preassembled units include plywood curb-liner, 16 oz. copper flashing for pitched or flat roof, screen, and all hardware

Ventarama

Ventilating type

Description	Oper	Unit	Vol	Crew Size	Man-hours per Unit	Crew Output per Day	Avg Mat'l Unit Cost	Avg Labor Unit Cost	Avg Equip Unit Cost	Avg Total Unit Cost	Avg Price Incl O&P
22" x 30"	Inst	Ea	Lg	CA	2.07	3.87	225.00	68.90	---	293.90	**363.00**
	Inst	Ea	Sm	CA	3.17	2.52	250.00	105.00	---	355.00	**446.00**
22" x 45-1/2"	Inst	Ea	Lg	CA	2.07	3.87	271.00	68.90	---	339.90	**415.00**
	Inst	Ea	Sm	CA	3.17	2.52	301.00	105.00	---	406.00	**504.00**
30" x 22"	Inst	Ea	Lg	CA	2.07	3.87	225.00	68.90	---	293.90	**363.00**
	Inst	Ea	Sm	CA	3.17	2.52	250.00	105.00	---	355.00	**446.00**
30" x 30"	Inst	Ea	Lg	CA	2.07	3.87	248.00	68.90	---	316.90	**389.00**
	Inst	Ea	Sm	CA	3.17	2.52	275.00	105.00	---	380.00	**475.00**
30" x 45-1/2"	Inst	Ea	Lg	CA	2.58	3.10	304.00	85.80	---	389.80	**478.00**
	Inst	Ea	Sm	CA	3.96	2.02	337.00	132.00	---	469.00	**585.00**
30" x 72"	Inst	Ea	Lg	CA	2.58	3.10	409.00	85.80	---	494.80	**600.00**
	Inst	Ea	Sm	CA	3.96	2.02	454.00	132.00	---	586.00	**720.00**
45-1/2" x 30"	Inst	Ea	Lg	CA	2.58	3.10	304.00	85.80	---	389.80	**478.00**
	Inst	Ea	Sm	CA	3.96	2.02	337.00	132.00	---	469.00	**585.00**
45-1/2" x 45-1/2"	Inst	Ea	Lg	CA	2.58	3.10	368.00	85.80	---	453.80	**552.00**
	Inst	Ea	Sm	CA	3.96	2.02	408.00	132.00	---	540.00	**667.00**
Add for insulated glass											
	Inst	%	Lg	---	---	---	30.0	---	---	---	---
	Inst	%	Sm	---	---	---	30.0	---	---	---	---
Add for bronze tinted outer dome											
	Inst	Ea	Lg	---	---	---	32.20	---	---	---	**37.00**
	Inst	Ea	Sm	---	---	---	35.70	---	---	---	**41.10**
Add for white inner dome											
	Inst	Ea	Lg	---	---	---	15.60	---	---	---	**18.00**
	Inst	Ea	Sm	---	---	---	17.30	---	---	---	**19.90**
Add for roll shade											
22" width	Inst	Ea	Lg	---	---	---	92.00	---	---	---	**106.00**
	Inst	Ea	Sm	---	---	---	102.00	---	---	---	**117.00**
30" width	Inst	Ea	Lg	---	---	---	99.40	---	---	---	**114.00**
	Inst	Ea	Sm	---	---	---	110.00	---	---	---	**127.00**
45-1/2" width	Inst	Ea	Lg	---	---	---	106.00	---	---	---	**122.00**
	Inst	Ea	Sm	---	---	---	117.00	---	---	---	**135.00**

Description	Oper	Unit	Vol	Crew Size	Man-hours per Unit	Crew Output per Day	Avg Mat'l Unit Cost	Avg Labor Unit Cost	Avg Equip Unit Cost	Avg Total Unit Cost	Avg Price Incl O&P
Add for storm panel with weatherstripping											
30" x 30"	Inst	Ea	Lg	---	---	---	46.00	---	---	---	**52.90**
	Inst	Ea	Sm	---	---	---	51.00	---	---	---	**58.70**
45-1/2" x 45-1/2"	Inst	Ea	Lg	---	---	---	63.50	---	---	---	**73.00**
	Inst	Ea	Sm	---	---	---	70.40	---	---	---	**80.90**
Add for motorization											
	Inst	Ea	Lg	CA	3.23	2.48	184.00	107.00	---	291.00	**373.00**
	Inst	Ea	Sm	CA	4.97	1.61	204.00	165.00	---	369.00	**483.00**
Fixed type											
22" x 30"	Inst	Ea	Lg	CA	2.07	3.87	175.00	68.90	---	243.90	**304.00**
	Inst	Ea	Sm	CA	3.17	2.52	194.00	105.00	---	299.00	**381.00**
22" x 45-1/2"	Inst	Ea	Lg	CA	2.07	3.87	221.00	68.90	---	289.90	**357.00**
	Inst	Ea	Sm	CA	3.17	2.52	245.00	105.00	---	350.00	**440.00**
30" x 22"	Inst	Ea	Lg	CA	2.07	3.87	175.00	68.90	---	243.90	**304.00**
	Inst	Ea	Sm	CA	3.17	2.52	194.00	105.00	---	299.00	**381.00**
30" x 30"	Inst	Ea	Lg	CA	2.07	3.87	198.00	68.90	---	266.90	**331.00**
	Inst	Ea	Sm	CA	3.17	2.52	219.00	105.00	---	324.00	**410.00**
30" x 45-1/2"	Inst	Ea	Lg	CA	2.58	3.10	244.00	85.80	---	329.80	**409.00**
	Inst	Ea	Sm	CA	3.96	2.02	270.00	132.00	---	402.00	**508.00**
30" x 72"	Inst	Ea	Lg	CA	2.58	3.10	322.00	85.80	---	407.80	**499.00**
	Inst	Ea	Sm	CA	3.96	2.02	357.00	132.00	---	489.00	**608.00**
45-1/2" x 30"	Inst	Ea	Lg	CA	2.58	3.10	244.00	85.80	---	329.80	**409.00**
	Inst	Ea	Sm	CA	3.96	2.02	270.00	132.00	---	402.00	**508.00**
45-1/2" x 45-1/2"	Inst	Ea	Lg	CA	2.58	3.10	276.00	85.80	---	361.80	**446.00**
	Inst	Ea	Sm	CA	3.96	2.02	306.00	132.00	---	438.00	**550.00**
Add for insulated laminated glass											
	Inst	%	Lg	---	---	---	30.0	---	---	---	---
	Inst	%	Sm	---	---	---	30.0	---	---	---	---

Description	Oper	Unit	Vol	Crew Size	Man-hours per Unit	Crew Output per Day	Avg Mat'l Unit Cost	Avg Labor Unit Cost	Avg Equip Unit Cost	Avg Total Unit Cost	Avg Price Incl O&P

Skywindows

Labor costs are for installation of skywindows only, no carpentry or roofing work included

Ultraseal self-flashing units. Clear insulated glass, flexible flange, no mastic or step flashing required. Other glazings available

Skywindow E-Class

Description	Oper	Unit	Vol	Crew Size	Man-hrs/Unit	Output/Day	Mat'l	Labor	Equip	Total	Price O&P
22-1/2" x 22-1/2" fixed	Inst	Ea	Lg	CA	.800	10.00	199.00	26.60	---	225.60	268.00
	Inst	Ea	Sm	CA	1.23	6.50	220.00	40.90	---	260.90	315.00
22-1/2" x 38-1/2" fixed	Inst	Ea	Lg	CA	.800	10.00	228.00	26.60	---	254.60	302.00
	Inst	Ea	Sm	CA	1.23	6.50	253.00	40.90	---	293.90	352.00
22-1/2" x 46-1/2" fixed	Inst	Ea	Lg	CA	.800	10.00	269.00	26.60	---	295.60	349.00
	Inst	Ea	Sm	CA	1.23	6.50	298.00	40.90	---	338.90	404.00
22-1/2" x 70-1/2" fixed	Inst	Ea	Lg	CA	.800	10.00	294.00	26.60	---	320.60	378.00
	Inst	Ea	Sm	CA	1.23	6.50	326.00	40.90	---	366.90	437.00
30-1/2" x 30-1/2" fixed	Inst	Ea	Lg	CA	.800	10.00	269.00	26.60	---	295.60	349.00
	Inst	Ea	Sm	CA	1.23	6.50	298.00	40.90	---	338.90	404.00
30-1/2" x 46-1/2" fixed	Inst	Ea	Lg	CA	.800	10.00	328.00	26.60	---	354.60	417.00
	Inst	Ea	Sm	CA	1.23	6.50	363.00	40.90	---	403.90	479.00
46-1/2" x 46-1/2" fixed	Inst	Ea	Lg	CA	.800	10.00	368.00	26.60	---	394.60	463.00
	Inst	Ea	Sm	CA	1.23	6.50	408.00	40.90	---	448.90	531.00
22-1/2" x 22-1/2" venting	Inst	Ea	Lg	CA	.800	10.00	376.00	26.60	---	402.60	473.00
	Inst	Ea	Sm	CA	1.23	6.50	417.00	40.90	---	457.90	541.00
22-1/2" x 30-1/2" venting	Inst	Ea	Lg	CA	.800	10.00	408.00	26.60	---	434.60	509.00
	Inst	Ea	Sm	CA	1.23	6.50	452.00	40.90	---	492.90	581.00
22-1/2" x 46-1/2" venting	Inst	Ea	Lg	CA	.800	10.00	461.00	26.60	---	487.60	570.00
	Inst	Ea	Sm	CA	1.23	6.50	511.00	40.90	---	551.90	649.00
30-1/2" x 30-1/2" venting	Inst	Ea	Lg	CA	.800	10.00	436.00	26.60	---	462.60	541.00
	Inst	Ea	Sm	CA	1.23	6.50	483.00	40.90	---	523.90	617.00
30-1/2" x 46-1/2" venting	Inst	Ea	Lg	CA	.800	10.00	500.00	26.60	---	526.60	615.00
	Inst	Ea	Sm	CA	1.23	6.50	555.00	40.90	---	595.90	700.00

Step flash-pan flashed units. Insulated clear tempered glass. Deck-mounted step flash unit. Other glazings available

Skywindow Excel-10

Description	Oper	Unit	Vol	Crew Size	Man-hrs/Unit	Output/Day	Mat'l	Labor	Equip	Total	Price O&P
22-1/2" x 22-1/2" fixed	Inst	Ea	Lg	CA	2.58	3.10	184.00	85.80	---	269.80	340.00
	Inst	Ea	Sm	CA	3.96	2.02	204.00	132.00	---	336.00	432.00
22-1/2" x 38-1/2" fixed	Inst	Ea	Lg	CA	2.58	3.10	193.00	85.80	---	278.80	351.00
	Inst	Ea	Sm	CA	3.96	2.02	214.00	132.00	---	346.00	444.00
22-1/2" x 46-1/2" fixed	Inst	Ea	Lg	CA	2.58	3.10	203.00	85.80	---	288.80	363.00
	Inst	Ea	Sm	CA	3.96	2.02	225.00	132.00	---	357.00	457.00
22-1/2" x 70-1/2" fixed	Inst	Ea	Lg	CA	2.58	3.10	236.00	85.80	---	321.80	400.00
	Inst	Ea	Sm	CA	3.96	2.02	261.00	132.00	---	393.00	498.00
30-1/2" x 30-1/2" fixed	Inst	Ea	Lg	CA	2.58	3.10	218.00	85.80	---	303.80	380.00
	Inst	Ea	Sm	CA	3.96	2.02	242.00	132.00	---	374.00	476.00

Description	Oper	Unit	Vol	Crew Size	Man-hours per Unit	Crew Output per Day	Avg Mat'l Unit Cost	Avg Labor Unit Cost	Avg Equip Unit Cost	Avg Total Unit Cost	Avg Price Incl O&P
30-1/2" x 46-1/2" fixed	Inst	Ea	Lg	CA	2.58	3.10	263.00	85.80	---	348.80	**431.00**
	Inst	Ea	Sm	CA	3.96	2.02	292.00	132.00	---	424.00	**533.00**
22-1/2" x 22-1/2" vented	Inst	Ea	Lg	CA	2.58	3.10	308.00	85.80	---	393.80	**483.00**
	Inst	Ea	Sm	CA	3.96	2.02	342.00	132.00	---	474.00	**591.00**
22-1/2" x 30-1/2" vented	Inst	Ea	Lg	CA	2.58	3.10	332.00	85.80	---	417.80	**511.00**
	Inst	Ea	Sm	CA	3.96	2.02	368.00	132.00	---	500.00	**621.00**
22-1/2" x 46-1/2" vented	Inst	Ea	Lg	CA	2.58	3.10	397.00	85.80	---	482.80	**585.00**
	Inst	Ea	Sm	CA	3.96	2.02	440.00	132.00	---	572.00	**703.00**
30-1/2" x 30-1/2" vented	Inst	Ea	Lg	CA	2.58	3.10	382.00	85.80	---	467.80	**568.00**
	Inst	Ea	Sm	CA	3.96	2.02	423.00	132.00	---	555.00	**684.00**
30-1/2" x 46-1/2" vented	Inst	Ea	Lg	CA	2.58	3.10	426.00	85.80	---	511.80	**619.00**
	Inst	Ea	Sm	CA	3.96	2.02	472.00	132.00	---	604.00	**741.00**
Add for flashing kit	Inst	Ea	Lg	---	---	---	60.00	---	---	---	**60.00**
	Inst	Ea	Sm	---	---	---	60.00	---	---	---	**60.00**

Low-profile insulated glass skywindow. Deck mount, clear tempered glass, self-flashing models. Other glazings available

Skywindow Genra-1

Description	Oper	Unit	Vol	Crew Size	Man-hours per Unit	Crew Output per Day	Avg Mat'l Unit Cost	Avg Labor Unit Cost	Avg Equip Unit Cost	Avg Total Unit Cost	Avg Price Incl O&P
22-1/2" x 22-1/2" fixed	Inst	Ea	Lg	CA	2.58	3.10	164.00	85.80	---	249.80	**317.00**
	Inst	Ea	Sm	CA	3.96	2.02	182.00	132.00	---	314.00	**406.00**
22-1/2" x 30-1/2" fixed	Inst	Ea	Lg	CA	2.58	3.10	193.00	85.80	---	278.80	**351.00**
	Inst	Ea	Sm	CA	3.96	2.02	214.00	132.00	---	346.00	**444.00**
22-1/2" x 46-1/2" fixed	Inst	Ea	Lg	CA	2.58	3.10	237.00	85.80	---	322.80	**402.00**
	Inst	Ea	Sm	CA	3.96	2.02	263.00	132.00	---	395.00	**500.00**
22-1/2" x 70-1/2" fixed	Inst	Ea	Lg	CA	2.58	3.10	313.00	85.80	---	398.80	**488.00**
	Inst	Ea	Sm	CA	3.96	2.02	347.00	132.00	---	479.00	**596.00**
30-1/2" x 30-1/2" fixed	Inst	Ea	Lg	CA	2.58	3.10	224.00	85.80	---	309.80	**386.00**
	Inst	Ea	Sm	CA	3.96	2.02	248.00	132.00	---	300.00	**403.00**
30-1/2" x 46-1/2" fixed	Inst	Ea	Lg	CA	2.58	3.10	272.00	85.80	---	357.80	**442.00**
	Inst	Ea	Sm	CA	3.96	2.02	302.00	132.00	---	434.00	**545.00**
46-1/2" x 46-1/2" fixed	Inst	Ea	Lg	CA	2.58	3.10	348.00	85.80	---	433.80	**529.00**
	Inst	Ea	Sm	CA	3.96	2.02	386.00	132.00	---	518.00	**641.00**
22-1/2" x 22-1/2" vented	Inst	Ea	Lg	CA	2.58	3.10	321.00	85.80	---	406.80	**498.00**
	Inst	Ea	Sm	CA	3.96	2.02	356.00	132.00	---	488.00	**607.00**
22-1/2" x 30-1/2" vented	Inst	Ea	Lg	CA	2.58	3.10	340.00	85.80	---	425.80	**520.00**
	Inst	Ea	Sm	CA	3.96	2.02	377.00	132.00	---	509.00	**632.00**
22-1/2" x 46-1/2" vented	Inst	Ea	Lg	CA	2.58	3.10	405.00	85.80	---	490.80	**594.00**
	Inst	Ea	Sm	CA	3.96	2.02	449.00	132.00	---	581.00	**714.00**
30-1/2" x 30-1/2" vented	Inst	Ea	Lg	CA	2.58	3.10	377.00	85.80	---	462.80	**563.00**
	Inst	Ea	Sm	CA	3.96	2.02	418.00	132.00	---	550.00	**679.00**
30-1/2" x 46-1/2" vented	Inst	Ea	Lg	CA	2.58	3.10	442.00	85.80	---	527.80	**637.00**
	Inst	Ea	Sm	CA	3.96	2.02	490.00	132.00	---	622.00	**761.00**
46-1/2" x 46-1/2" vented	Inst	Ea	Lg	CA	2.58	3.10	543.00	85.80	---	628.80	**753.00**
	Inst	Ea	Sm	CA	3.96	2.02	602.00	132.00	---	734.00	**890.00**

Description	Oper	Unit	Vol	Crew Size	Man-hours per Unit	Crew Output per Day	Avg Mat'l Unit Cost	Avg Labor Unit Cost	Avg Equip Unit Cost	Avg Total Unit Cost	Avg Price Incl O&P

Roof windows

Includes prefabricated flashing, exterior awning, interior rollerblind and insect screen. Sash rotates 180 degrees. Labor costs are for installation on roofs with 10- to 85-degree slope and include installation of roof window, flashing and sun screen accessories. Labor costs do not include carpentry or roofing work other than curb. Add for interior trim. Listed by actual dimensions, top hung

Aluminum-clad wood frame, double-insulated tempered glass

Description	Oper	Unit	Vol	Crew Size	Man-hours per Unit	Crew Output per Day	Avg Mat'l Unit Cost	Avg Labor Unit Cost	Avg Equip Unit Cost	Avg Total Unit Cost	Avg Price Incl O&P
30-1/2" x 46-1/2" fixed	Inst	Ea	Lg	CA	3.56	2.25	543.00	118.00	---	661.00	802.00
	Inst	Ea	Sm	CA	5.48	1.46	602.00	182.00	---	784.00	966.00
46-1/2" x 46-1/2" fixed	Inst	Ea	Lg	CA	3.56	2.25	607.00	118.00	---	725.00	876.00
	Inst	Ea	Sm	CA	5.48	1.46	673.00	182.00	---	855.00	1050.00

Spas

Spas, with good quality fittings and faucets
Also see Bathtubs

Detach & reset operations

Description	Oper	Unit	Vol	Crew Size	Man-hours per Unit	Crew Output per Day	Avg Mat'l Unit Cost	Avg Labor Unit Cost	Avg Equip Unit Cost	Avg Total Unit Cost	Avg Price Incl O&P
Whirlpool spa	Reset	Ea	Lg	SB	5.33	3.00	38.50	173.00	---	211.50	299.00
	Reset	Ea	Sm	SB	7.0	2.30	46.80	228.00	---	274.80	389.00

Remove operations

Description	Oper	Unit	Vol	Crew Size	Man-hours per Unit	Crew Output per Day	Avg Mat'l Unit Cost	Avg Labor Unit Cost	Avg Equip Unit Cost	Avg Total Unit Cost	Avg Price Incl O&P
Whirlpool spa	Demo	Ea	Lg	SB	2.67	6.00	21.00	86.90	---	107.90	152.00
	Demo	Ea	Sm	SB	3.6	4.50	25.50	117.00	---	142.50	202.00

Install rough-in

Description	Oper	Unit	Vol	Crew Size	Man-hours per Unit	Crew Output per Day	Avg Mat'l Unit Cost	Avg Labor Unit Cost	Avg Equip Unit Cost	Avg Total Unit Cost	Avg Price Incl O&P
Whirlpool spa	Inst	Ea	Lg	SB	10.0	1.60	59.50	326.00	---	385.50	547.00
	Inst	Ea	Sm	SB	13.3	1.20	72.30	433.00	---	505.30	719.00

Replace operations
Jacuzzi Products, whirlpools and baths with color matched trim on jets and suction
Builder Series

Nova model, oval tub, with ledge
60" x 42" x 18-1/2" H

Description	Oper	Unit	Vol	Crew Size	Man-hours per Unit	Crew Output per Day	Avg Mat'l Unit Cost	Avg Labor Unit Cost	Avg Equip Unit Cost	Avg Total Unit Cost	Avg Price Incl O&P
White / stock color	Inst	Ea	Lg	SB	8.0	2.00	1420.00	260.00	---	1680.00	2020.00
	Inst	Ea	Sm	SB	10.7	1.50	1730.00	348.00	---	2078.00	2500.00
Premium color	Inst	Ea	Lg	SB	8.0	2.00	1490.00	260.00	---	1750.00	2090.00
	Inst	Ea	Sm	SB	10.7	1.50	1810.00	348.00	---	2158.00	2590.00

72" x 42" x 20-1/2" H

Description	Oper	Unit	Vol	Crew Size	Man-hours per Unit	Crew Output per Day	Avg Mat'l Unit Cost	Avg Labor Unit Cost	Avg Equip Unit Cost	Avg Total Unit Cost	Avg Price Incl O&P
White / stock color	Inst	Ea	Lg	SB	8.0	2.00	1730.00	260.00	---	1990.00	2370.00
	Inst	Ea	Sm	SB	10.7	1.50	2100.00	348.00	---	2448.00	2920.00
Premium color	Inst	Ea	Lg	SB	8.0	2.00	1770.00	260.00	---	2030.00	2410.00
	Inst	Ea	Sm	SB	10.7	1.50	2150.00	348.00	---	2498.00	2980.00

Description	Oper	Unit	Vol	Crew Size	Man-hours per Unit	Crew Output per Day	Avg Mat'l Unit Cost	Avg Labor Unit Cost	Avg Equip Unit Cost	Avg Total Unit Cost	Avg Price Incl O&P
Riva model, oval tub											
62" x 43" x 18-1/2" H											
White / stock color	Inst	Ea	Lg	SB	8.0	2.00	1460.00	260.00	---	1720.00	**2060.00**
	Inst	Ea	Sm	SB	10.7	1.50	1770.00	348.00	---	2118.00	**2550.00**
Premium color	Inst	Ea	Lg	SB	8.0	2.00	1570.00	260.00	---	1830.00	**2190.00**
	Inst	Ea	Sm	SB	10.7	1.50	1910.00	348.00	---	2258.00	**2710.00**
72" x 42" x 20-1/2" H											
White / stock color	Inst	Ea	Lg	SB	8.0	2.00	1700.00	260.00	---	1960.00	**2330.00**
	Inst	Ea	Sm	SB	10.7	1.50	2060.00	348.00	---	2408.00	**2880.00**
Premium color	Inst	Ea	Lg	SB	8.0	2.00	1730.00	260.00	---	1990.00	**2380.00**
	Inst	Ea	Sm	SB	10.7	1.50	2100.00	348.00	---	2448.00	**2930.00**
Tara model, corner unit, angled tub											
60" x 60" x 20-3/4" H											
White / stock color	Inst	Ea	Lg	SB	8.0	2.00	1460.00	260.00	---	1720.00	**2060.00**
	Inst	Ea	Sm	SB	10.7	1.50	1770.00	348.00	---	2118.00	**2550.00**
Premium color	Inst	Ea	Lg	SB	8.0	2.00	1570.00	260.00	---	1830.00	**2190.00**
	Inst	Ea	Sm	SB	10.7	1.50	1910.00	348.00	---	2258.00	**2710.00**
Torino model, rectangle tub											
66" x 42" x 20-1/2" H											
White / stock color	Inst	Ea	Lg	SB	8.0	2.00	1460.00	260.00	---	1720.00	**2060.00**
	Inst	Ea	Sm	SB	10.7	1.50	1770.00	348.00	---	2118.00	**2550.00**
Premium color	Inst	Ea	Lg	SB	8.0	2.00	1570.00	260.00	---	1830.00	**2190.00**
	Inst	Ea	Sm	SB	10.7	1.50	1910.00	348.00	---	2258.00	**2710.00**

Designer Collection

Description	Oper	Unit	Vol	Crew Size	Man-hours per Unit	Crew Output per Day	Avg Mat'l Unit Cost	Avg Labor Unit Cost	Avg Equip Unit Cost	Avg Total Unit Cost	Avg Price Incl O&P
Allusion model, rectangle tub											
66" x 36" x 26" H											
White / stock color	Inst	Ea	Lg	SB	8.0	2.00	2800.00	260.00	---	3060.00	**3600.00**
	Inst	Ea	Sm	SB	10.7	1.50	3400.00	348.00		3748.00	**4420.00**
Premium color	Inst	Ea	Lg	SB	8.0	2.00	2910.00	260.00	---	3170.00	**3730.00**
	Inst	Ea	Sm	SB	10.7	1.50	3530.00	348.00	---	3878.00	**4570.00**
72" x 36" x 26" H											
White / stock color	Inst	Ea	Lg	SB	8.0	2.00	3060.00	260.00	---	3320.00	**3900.00**
	Inst	Ea	Sm	SB	10.7	1.50	3710.00	348.00	---	4058.00	**4780.00**
Premium color	Inst	Ea	Lg	SB	8.0	2.00	3090.00	260.00	---	3350.00	**3940.00**
	Inst	Ea	Sm	SB	10.7	1.50	3750.00	348.00	---	4098.00	**4830.00**
72" x 42" x 26" H											
White / stock color	Inst	Ea	Lg	SB	8.0	2.00	3210.00	260.00	---	3470.00	**4070.00**
	Inst	Ea	Sm	SB	10.7	1.50	3890.00	348.00	---	4238.00	**4990.00**
Premium color	Inst	Ea	Lg	SB	8.0	2.00	3240.00	260.00	---	3500.00	**4110.00**
	Inst	Ea	Sm	SB	10.7	1.50	3940.00	348.00	---	4288.00	**5040.00**
Aura model											
72" x 60" x 20-1/2"											
White / stock color	Inst	Ea	Lg	SB	8.0	2.00	5290.00	260.00	---	5550.00	**6460.00**
	Inst	Ea	Sm	SB	10.7	1.50	6420.00	348.00	---	6768.00	**7900.00**
Premium color	Inst	Ea	Lg	SB	8.0	2.00	5320.00	260.00	---	5580.00	**6500.00**
	Inst	Ea	Sm	SB	10.7	1.50	6460.00	348.00	---	6808.00	**7950.00**

Description	Oper	Unit	Vol	Crew Size	Man-hours per Unit	Crew Output per Day	Avg Mat'l Unit Cost	Avg Labor Unit Cost	Avg Equip Unit Cost	Avg Total Unit Cost	Avg Price Incl O&P
Ciprea model, with removable skirt											
72" x 48" x 20-1/2"											
White / stock color	Inst	Ea	Lg	SB	8.0	2.00	4960.00	260.00	---	5220.00	**6090.00**
	Inst	Ea	Sm	SB	10.7	1.50	6020.00	348.00	---	6368.00	**7440.00**
Premium color	Inst	Ea	Lg	SB	8.0	2.00	5110.00	260.00	---	5370.00	**6260.00**
	Inst	Ea	Sm	SB	10.7	1.50	6210.00	348.00	---	6558.00	**7650.00**
Fiore model, with removable skirt											
66" x 66" x 27-1/4"											
White / stock color	Inst	Ea	Lg	SB	8.0	2.00	6440.00	260.00	---	6700.00	**7790.00**
	Inst	Ea	Sm	SB	10.7	1.50	7820.00	348.00	---	8168.00	**9510.00**
Premium color	Inst	Ea	Lg	SB	8.0	2.00	6580.00	260.00	---	6840.00	**7950.00**
	Inst	Ea	Sm	SB	10.7	1.50	7990.00	348.00	---	8338.00	**9700.00**
Fontana model											
72" x 54" x 28" H, angled back, with RapidHeat											
White / stock color	Inst	Ea	Lg	SB	8.0	2.00	5570.00	260.00	---	5830.00	**6790.00**
	Inst	Ea	Sm	SB	10.7	1.50	6760.00	348.00	---	7108.00	**8290.00**
Premium color	Inst	Ea	Lg	SB	8.0	2.00	5740.00	260.00	---	6000.00	**6980.00**
	Inst	Ea	Sm	SB	10.7	1.50	6970.00	348.00	---	7318.00	**8520.00**

Options for whirlpool baths and spas

Description	Oper	Unit	Vol	Crew Size	Man-hours per Unit	Crew Output per Day	Avg Mat'l Unit Cost	Avg Labor Unit Cost	Avg Equip Unit Cost	Avg Total Unit Cost	Avg Price Incl O&P
Timer kit, 30 min, wall mount											
Polished chrome	Inst	Ea	Lg	---	---	---	60.90	---	---	60.90	**60.90**
	Inst	Ea	Sm	---	---	---	74.00	---	---	74.00	**74.00**
Gold	Inst	Ea	Lg	---	---	---	71.40	---	---	71.40	**71.40**
	Inst	Ea	Sm	---	---	---	86.70	---	---	86.70	**86.70**
Trim kit											
Polished chrome	Inst	Ea	Lg	---	---	---	107.00	---	---	107.00	**107.00**
	Inst	Ea	Sm	---	---	---	130.00	---	---	130.00	**130.00**
Bright brass	Inst	Ea	Lg	---	---	---	142.00	---	---	142.00	**142.00**
	Inst	Ea	Sm	---	---	---	173.00	---	---	173.00	**173.00**
Trip lever drain with vent											
Polished chrome	Inst	Ea	Lg	---	---	---	94.50	---	---	94.50	**94.50**
	Inst	Ea	Sm	---	---	---	115.00	---	---	115.00	**115.00**
Bright brass	Inst	Ea	Lg	---	---	---	127.00	---	---	127.00	**127.00**
	Inst	Ea	Sm	---	---	---	155.00	---	---	155.00	**155.00**
Gold	Inst	Ea	Lg	---	---	---	140.00	---	---	140.00	**140.00**
	Inst	Ea	Sm	---	---	---	170.00	---	---	170.00	**170.00**
Turn drain with vent											
Polished chrome	Inst	Ea	Lg	---	---	---	125.00	---	---	125.00	**125.00**
	Inst	Ea	Sm	---	---	---	152.00	---	---	152.00	**152.00**
Bright brass	Inst	Ea	Lg	---	---	---	160.00	---	---	160.00	**160.00**
	Inst	Ea	Sm	---	---	---	195.00	---	---	195.00	**195.00**
Gold	Inst	Ea	Lg	---	---	---	151.00	---	---	151.00	**151.00**
	Inst	Ea	Sm	---	---	---	184.00	---	---	184.00	**184.00**

Stairs

Stair parts

Balusters, stock pine

Description	Oper	Unit	Vol	Crew Size	Man-hours per Unit	Crew Output per Day	Avg Mat'l Unit Cost	Avg Labor Unit Cost	Avg Equip Unit Cost	Avg Total Unit Cost	Avg Price Incl O&P
1 -1/16" x 1-1/16"	Inst	LF	Lg	CA	.040	200.0	2.48	1.33	---	3.81	**4.85**
	Inst	LF	Sm	CA	.062	130.0	2.72	2.06	---	4.78	**6.22**
1-5/8" x 1-5/8"	Inst	LF	Lg	CA	.050	160.0	2.29	1.66	---	3.95	**5.13**
	Inst	LF	Sm	CA	.077	104.0	2.51	2.56	---	5.07	**6.73**

Balusters, turned

30" high

Description	Oper	Unit	Vol	Crew Size	Man-hours per Unit	Crew Output per Day	Avg Mat'l Unit Cost	Avg Labor Unit Cost	Avg Equip Unit Cost	Avg Total Unit Cost	Avg Price Incl O&P
Pine	Inst	Ea	Lg	CA	.333	24.00	9.32	11.10	---	20.42	**27.30**
	Inst	Ea	Sm	CA	.513	15.60	10.20	17.10	---	27.30	**37.40**
Birch	Inst	Ea	Lg	CA	.333	24.00	10.90	11.10	---	22.00	**29.20**
	Inst	Ea	Sm	CA	.513	15.60	11.90	17.10	---	29.00	**39.30**

42" high

Description	Oper	Unit	Vol	Crew Size	Man-hours per Unit	Crew Output per Day	Avg Mat'l Unit Cost	Avg Labor Unit Cost	Avg Equip Unit Cost	Avg Total Unit Cost	Avg Price Incl O&P
Pine	Inst	Ea	Lg	CA	.400	20.00	10.90	13.30	---	24.20	**32.50**
	Inst	Ea	Sm	CA	.615	13.00	11.90	20.50	---	32.40	**44.40**
Birch	Inst	Ea	Lg	CA	.400	20.00	14.10	13.30	---	27.40	**36.10**
	Inst	Ea	Sm	CA	.615	13.00	15.40	20.50	---	35.90	**48.40**

Newels, 3-1/4" wide

Description	Oper	Unit	Vol	Crew Size	Man-hours per Unit	Crew Output per Day	Avg Mat'l Unit Cost	Avg Labor Unit Cost	Avg Equip Unit Cost	Avg Total Unit Cost	Avg Price Incl O&P
Starting	Inst	Ea	Lg	CA	1.33	6.00	151.00	44.30	---	195.30	**240.00**
	Inst	Ea	Sm	CA	2.05	3.90	166.00	68.20	---	234.20	**293.00**
Landing	Inst	Ea	Lg	CA	2.00	4.00	214.00	66.50	---	280.50	**346.00**
	Inst	Ea	Sm	CA	3.08	2.60	235.00	102.00	---	337.00	**424.00**

Railings, built-up

Description	Oper	Unit	Vol	Crew Size	Man-hours per Unit	Crew Output per Day	Avg Mat'l Unit Cost	Avg Labor Unit Cost	Avg Equip Unit Cost	Avg Total Unit Cost	Avg Price Incl O&P
Oak	Inst	LF	Lg	CA	.160	50.00	15.10	5.32	---	20.42	**25.40**
	Inst	LF	Sm	CA	.246	32.50	16.60	8.18	---	24.78	**31.30**

Railings, subrail

Description	Oper	Unit	Vol	Crew Size	Man-hours per Unit	Crew Output per Day	Avg Mat'l Unit Cost	Avg Labor Unit Cost	Avg Equip Unit Cost	Avg Total Unit Cost	Avg Price Incl O&P
Oak	Inst	LF	Lg	CA	.080	100.0	20.20	2.66	---	22.86	**27.20**
	Inst	LF	Sm	CA	.123	65.00	22.10	4.09	---	26.19	**31.50**

Risers, 3/4" x 7-1/2" high

Description	Oper	Unit	Vol	Crew Size	Man-hours per Unit	Crew Output per Day	Avg Mat'l Unit Cost	Avg Labor Unit Cost	Avg Equip Unit Cost	Avg Total Unit Cost	Avg Price Incl O&P
Beech	Inst	LF	Lg	CA	.133	60.00	6.55	4.42	---	10.97	**14.20**
	Inst	LF	Sm	CA	.205	39.00	7.18	6.82	---	14.00	**18.50**
Fir	Inst	LF	Lg	CA	.133	60.00	1.89	4.42	---	6.31	**8.81**
	Inst	LF	Sm	CA	.205	39.00	2.07	6.82	---	8.89	**12.60**
Oak	Inst	LF	Lg	CA	.133	60.00	5.99	4.42	---	10.41	**13.50**
	Inst	LF	Sm	CA	.205	39.00	6.56	6.82	---	13.38	**17.80**
Pine	Inst	LF	Lg	CA	.133	60.00	1.89	4.42	---	6.31	**8.81**
	Inst	LF	Sm	CA	.205	39.00	2.07	6.82	---	8.89	**12.60**

Description	Oper	Unit	Vol	Crew Size	Man-hours per Unit	Crew Output per Day	Avg Mat'l Unit Cost	Avg Labor Unit Cost	Avg Equip Unit Cost	Avg Total Unit Cost	Avg Price Incl O&P
Skirt board, pine											
1" x 10"	Inst	LF	Lg	CA	.160	50.00	2.08	5.32	---	7.40	**10.40**
	Inst	LF	Sm	CA	.246	32.50	2.28	8.18	---	10.46	**14.90**
1" x 12"	Inst	LF	Lg	CA	.178	45.00	2.52	5.92	---	8.44	**11.80**
	Inst	LF	Sm	CA	.274	29.25	2.76	9.12	---	11.88	**16.90**
Treads, oak											
1-16" x 9-1/2" wide											
3' long	Inst	Ea	Lg	CA	.500	16.00	27.70	16.60	---	44.30	**56.80**
	Inst	Ea	Sm	CA	.769	10.40	30.40	25.60	---	56.00	**73.30**
4' long	Inst	Ea	Lg	CA	.533	15.00	36.50	17.70	---	54.20	**68.60**
	Inst	Ea	Sm	CA	.821	9.75	40.00	27.30	---	67.30	**87.00**
1-1/16" x 11-1/2" wide											
3' long	Inst	Ea	Lg	CA	.500	16.00	29.60	16.60	---	46.20	**59.00**
	Inst	Ea	Sm	CA	.769	10.40	32.40	25.60	---	58.00	**75.70**
6' long	Inst	Ea	Lg	CA	.667	12.00	70.60	22.20	---	92.80	**114.00**
	Inst	Ea	Sm	CA	1.03	7.80	77.30	34.30	---	111.60	**140.00**
For beech treads, ADD	Inst	%	Lg	---	---	---	40.0	---	---	---	---
	Inst	%	Sm	---	---	---	40.0	---	---	---	---
For mitered return nosings, ADD	Inst	LF	Lg	CA	.133	60.00	10.60	4.42	---	15.02	**18.80**
	Inst	LF	Sm	CA	.205	39.00	11.60	6.82	---	18.42	**23.60**

Stairs, shop fabricated, per flight
Basement stairs, soft wood, open risers

Description	Oper	Unit	Vol	Crew Size	Man-hours per Unit	Crew Output per Day	Avg Mat'l Unit Cost	Avg Labor Unit Cost	Avg Equip Unit Cost	Avg Total Unit Cost	Avg Price Incl O&P
3' wide, 8' high	Inst	Flt	Lg	2C	4.00	4.00	473.00	133.00	---	606.00	**743.00**
	Inst	Flt	Sm	2C	6.15	2.60	518.00	205.00	---	723.00	**902.00**

Box stairs, no handrails, 3' wide
Oak treads

Description	Oper	Unit	Vol	Crew Size	Man-hours per Unit	Crew Output per Day	Avg Mat'l Unit Cost	Avg Labor Unit Cost	Avg Equip Unit Cost	Avg Total Unit Cost	Avg Price Incl O&P
2' high	Inst	Flt	Lg	2C	3.20	5.00	188.00	106.00	---	294.00	**376.00**
	Inst	Flt	Sm	2C	4.92	3.25	206.00	164.00	---	370.00	**482.00**
4' high	Inst	Flt	Lg	2C	4.00	4.00	403.00	133.00	---	536.00	**663.00**
	Inst	Flt	Sm	2C	6.15	2.60	442.00	205.00	---	647.00	**815.00**
6' high	Inst	Flt	Lg	2C	4.57	3.50	655.00	152.00	---	807.00	**982.00**
	Inst	Flt	Sm	2C	7.03	2.28	718.00	234.00	---	952.00	**1180.00**
8' high	Inst	Flt	Lg	2C	5.33	3.00	819.00	177.00	---	996.00	**1210.00**
	Inst	Flt	Sm	2C	8.21	1.95	897.00	273.00	---	1170.00	**1440.00**

Description	Oper	Unit	Vol	Crew Size	Man-hours per Unit	Crew Output per Day	Avg Mat'l Unit Cost	Avg Labor Unit Cost	Avg Equip Unit Cost	Avg Total Unit Cost	Avg Price Incl O&P
Pine treads for carpet											
2' high	Inst	Flt	Lg	2C	3.20	5.00	149.00	106.00	---	255.00	**331.00**
	Inst	Flt	Sm	2C	4.92	3.25	163.00	164.00	---	327.00	**433.00**
4' high	Inst	Flt	Lg	2C	4.00	4.00	253.00	133.00	---	386.00	**491.00**
	Inst	Flt	Sm	2C	6.15	2.60	277.00	205.00	---	482.00	**626.00**
6' high	Inst	Flt	Lg	2C	4.57	3.50	384.00	152.00	---	536.00	**670.00**
	Inst	Flt	Sm	2C	7.03	2.28	421.00	234.00	---	655.00	**835.00**
8' high	Inst	Flt	Lg	2C	5.33	3.00	479.00	177.00	---	656.00	**817.00**
	Inst	Flt	Sm	2C	8.21	1.95	524.00	273.00	---	797.00	**1010.00**
Stair rail with balusters, 5 risers											
	Inst	Ea	Lg	2C	1.07	15.00	221.00	35.60	---	256.60	**307.00**
	Inst	Ea	Sm	2C	1.64	9.75	242.00	54.60	---	296.60	**360.00**
For 4' wide stairs, ADD											
	Inst	%	Lg	2C	---	---	25.0	5.0	---	---	**---**
	Inst	%	Sm	2C	---	---	25.0	5.0	---	---	**---**

Open stairs, prefinished, metal stringers, 3'-6" wide treads, no railings

Description	Oper	Unit	Vol	Crew Size	Man-hours per Unit	Crew Output per Day	Avg Mat'l Unit Cost	Avg Labor Unit Cost	Avg Equip Unit Cost	Avg Total Unit Cost	Avg Price Incl O&P
3' high	Inst	Flt	Lg	2C	3.20	5.00	548.00	106.00	---	654.00	**790.00**
	Inst	Flt	Sm	2C	4.92	3.25	600.00	164.00	---	764.00	**936.00**
4' high	Inst	Flt	Lg	2C	4.00	4.00	687.00	133.00	---	820.00	**989.00**
	Inst	Flt	Sm	2C	6.15	2.60	752.00	205.00	---	957.00	**1170.00**
6' high	Inst	Flt	Lg	2C	4.57	3.50	1200.00	152.00	---	1352.00	**1600.00**
	Inst	Flt	Sm	2C	7.03	2.28	1310.00	234.00	---	1544.00	**1860.00**
8' high	Inst	Flt	Lg	2C	5.33	3.00	1920.00	177.00	---	2097.00	**2480.00**
	Inst	Flt	Sm	2C	8.21	1.95	2100.00	273.00	---	2373.00	**2830.00**

Adjustments

For 3-piece wood railings and balusters, ADD

Description	Oper	Unit	Vol	Crew Size	Man-hours per Unit	Crew Output per Day	Avg Mat'l Unit Cost	Avg Labor Unit Cost	Avg Equip Unit Cost	Avg Total Unit Cost	Avg Price Incl O&P
3' high	Inst	Ea	Lg	2C	1.07	15.00	188.00	35.60	---	223.60	**269.00**
	Inst	Ea	Sm	2C	1.64	9.75	206.00	54.60	---	260.60	**318.00**
4' high	Inst	Ea	Lg	2C	1.14	14.00	241.00	37.90	---	278.90	**334.00**
	Inst	Ea	Sm	2C	1.76	9.10	264.00	58.60	---	322.60	**391.00**
6' high	Inst	Ea	Lg	2C	1.23	13.00	370.00	40.90	---	410.90	**487.00**
	Inst	Ea	Sm	2C	1.89	8.45	406.00	62.90	---	468.90	**561.00**
8' high	Inst	Ea	Lg	2C	1.33	12.00	460.00	44.30	---	504.30	**595.00**
	Inst	Ea	Sm	2C	2.05	7.80	504.00	68.20	---	572.20	**682.00**
For 3'-6" x 3'-6" platform, ADD											
	Inst	Ea	Lg	2C	4.00	4.00	214.00	133.00	---	347.00	**446.00**
	Inst	Ea	Sm	2C	6.15	2.60	235.00	205.00	---	440.00	**577.00**

Description	Oper	Unit	Vol	Crew Size	Man- hours per Unit	Crew Output per Day	Avg Mat'l Unit Cost	Avg Labor Unit Cost	Avg Equip Unit Cost	Avg Total Unit Cost	Avg Price Incl O&P
Curved stairways, oak, unfinished, curved balustrade											
Open one side											
9' high	Inst	Flt	Lg	2C	22.9	0.70	8030.00	762.00	---	8792.00	**10400.00**
	Inst	Flt	Sm	2C	35.2	0.46	8800.00	1170.00	---	9970.00	**11900.00**
10' high	Inst	Flt	Lg	2C	22.9	0.70	9070.00	762.00	---	9832.00	**11600.00**
	Inst	Flt	Sm	2C	35.2	0.46	9940.00	1170.00	---	11110.00	**13200.00**
Open both sides											
9' high	Inst	Flt	Lg	2C	32.0	0.50	12600.00	1060.00	---	13660.00	**16100.00**
	Inst	Flt	Sm	2C	49.2	0.33	13800.00	1640.00	---	15440.00	**18300.00**
10' high	Inst	Flt	Lg	2C	32.0	0.50	13600.00	1060.00	---	14660.00	**17200.00**
	Inst	Flt	Sm	2C	49.2	0.33	14900.00	1640.00	---	16540.00	**19600.00**

Steel, reinforcing. See Concrete, page 99

Stucco. See Plaster, page 337

Studs. See Framing, page 198

Subflooring. See Framing, page 212

Suspended Ceilings

Suspended ceilings consist of a grid of small metal hangers which are hung from the ceiling framing with wire or strap, and drop-in panels sized to fit the grid system. The main advantage to this type of ceiling finish is that it can be adjusted to any height desired. It can cover a lot of flaws and obstructions (uneven plaster, pipes, wiring, and ductwork). With a high ceiling, the suspended ceiling reduces the sound transfer from the floor above and increases the ceiling's insulating value. One great advantage of suspended ceilings is that the area directly above the tiles is readily accessible for repair and maintenance work; the tiles merely have to be lifted out from the grid. This system also eliminates the need for any other ceiling finish.

1. Installation Procedure

a. Decide on ceiling height. It must clear the lowest obstacles but remain above windows and doors. Snap a chalk line around walls at the new height. Check it with a carpenter's level.

b. Install wall molding/angle at the marked line; this supports the tiles around the room's perimeter. Check with a level. Use nails (6d) or screws to fasten molding to studs (16" or 24" oc); on masonry, use concrete nails 24" oc. The molding should be flush along the chalk line with the bottom leg of the L-angle facing into the room.

c. Fit wall molding at inside corners by straight cutting (90 degrees) and overlapping; for outside corners, miter cut corner (45 degrees) and butt together. Cut the molding to required lengths with tin snips or a hacksaw.

d. Mark center line of room above the wall molding. Decide where the first main runner should be to get the desired border width. Position main runners by running taut string lines across room. Repeat for the first row of cross tees closest to one wall. Then attach hanger wires with screw eyes to the joists at 4' intervals.

e. Trim main runners at wall to line up the slot with cross tee string. Rest trimmed end on wall molding; insert wire through holes; recheck with level; twist wire several times.

f. Insert cross tee and tabs into main runner's slots; push down to lock. Check with level. Repeat.

g. Drop 2' x 4' panels into place.

2. Estimating Materials

a. Measure the ceiling and plot it on graph paper. Mark the direction of ceiling joists. On the ceiling itself, snap chalk lines along the joist lines.

b. Draw ceiling layout for grid system onto graph paper. Plan the ceiling layout, figuring full panels across the main ceiling and evenly trimmed partial panels at the edges. To calculate the width of border panels in each direction, determine the width of the gap left after full panels are placed all across the dimension; then divide by 2.

c. For purposes of ordering material, determine the room size in even number of feet. If the room length or width is not divisible by 2 feet, increase the dimension to the next larger unit divisible by 2. For example, a room that measured 11'-6" x 14'-4" would be considered a 12' x 16' room. This allows for waste and/or breakage. In this example, you would order 192 SF of material.

d. Main runners and tees must run perpendicular to ceiling joists. For a 2' x 2' grid, cross and main tees are 2' apart. For a 2' x 4' grid, main tees are 4' apart and 4' cross tees connect the main tees; then 2' cross tees can be used to connect the 4' cross tees. The long panels of the grid are set parallel to the ceiling joists, so the T-shaped main runner is attached perpendicular to joists at 4' oc.

e. Wall molding/angle is manufactured in 10' lengths. Main runners and tees are manufactured in 12' lengths. Cross tees are either 2' long or 4' long. Hanger wire is 12 gauge, and attached to joists with either screw eyes or hook-and-nail. Drop-in panels are either 8 SF (2' x 4') or 4 SF (2' x 2').

f. To find the number of drop-in panels required, divide the nominal room size in square feet (e.g., 192 SF in above example) by square feet of panel to be used.

g. The quantity of 12 gauge wire depends on the "drop" distance of the ceiling. Figure one suspension wire and screw eye for each 4' of main runner; if any main run is longer than 12' then splice plates are needed, with a wire and screw eye on each side of the splice. The length of each wire should be at least 2" longer than the "drop" distance, to allow for a twist after passing through runner or tee.

Description	Oper	Unit	Vol	Crew Size	Man-hours per Unit	Crew Output per Day	Avg Mat'l Unit Cost	Avg Labor Unit Cost	Avg Equip Unit Cost	Avg Total Unit Cost	Avg Price Incl O&P

Suspended ceiling systems

Suspended ceiling panels and grid system

| | Demo | SF | Lg | LB | .009 | 1780 | --- | .25 | --- | .25 | .37 |
| | Demo | SF | Sm | LB | .013 | 1246 | --- | .36 | --- | .36 | .53 |

Complete 2' x 4' slide lock grid system (baked enamel)

Labor includes installing grid and laying in panels

Luminous panels

Prismatic panels

Acrylic	Inst	SF	Lg	CA	.021	380.0	1.98	.70	---	2.68	3.33
	Inst	SF	Sm	CA	.030	266.0	2.17	1.00	---	3.17	3.99
Polystyrene	Inst	SF	Lg	CA	.021	380.0	1.22	.70	---	1.92	2.45
	Inst	SF	Sm	CA	.030	266.0	1.33	1.00	---	2.33	3.03

Ribbed panels

Acrylic	Inst	SF	Lg	CA	.021	380.0	2.11	.70	---	2.81	3.47
	Inst	SF	Sm	CA	.030	266.0	2.31	1.00	---	3.31	4.15
Polystyrene	Inst	SF	Lg	CA	.021	380.0	1.35	.70	---	2.05	2.60
	Inst	SF	Sm	CA	.030	266.0	1.47	1.00	---	2.47	3.19

Fiberglass panels

1/2" T

Plain white	Inst	SF	Lg	CA	.020	405.0	1.09	.67	---	1.76	2.25
	Inst	SF	Sm	CA	.028	284.0	1.20	.93	---	2.13	2.78
Sculptured white	Inst	SF	Lg	CA	.020	405.0	1.22	.67	---	1.89	2.40
	Inst	SF	Sm	CA	.028	284.0	1.33	.93	---	2.26	2.93

5/8" T

Fissured	Inst	SF	Lg	CA	.020	405.0	1.28	.67	---	1.95	2.47
	Inst	SF	Sm	CA	.028	284.0	1.40	.93	---	2.33	3.01
Fissured, fire rated	Inst	SF	Lg	CA	.020	405.0	1.41	.67	---	2.08	2.62
	Inst	SF	Sm	CA	.028	284.0	1.54	.93	---	2.47	3.17

Mineral fiber panels

1/2" T

| | Inst | SF | Lg | CA | .020 | 405.0 | .90 | .67 | --- | 1.57 | 2.03 |
| | Inst | SF | Sm | CA | .028 | 284.0 | .99 | .93 | --- | 1.92 | 2.54 |

9/16" T

| | Inst | SF | Lg | CA | .020 | 405.0 | 1.45 | .67 | --- | 2.12 | 2.67 |
| | Inst | SF | Sm | CA | .028 | 284.0 | 1.58 | .93 | --- | 2.51 | 3.21 |

For wood grained grid components

| ADD | Inst | % | Lg | --- | --- | --- | 7.0 | --- | --- | --- | --- |
| | Inst | % | Sm | --- | --- | --- | 7.0 | --- | --- | --- | --- |

Description	Oper	Unit	Vol	Crew Size	Man-hours per Unit	Crew Output per Day	Avg Mat'l Unit Cost	Avg Labor Unit Cost	Avg Equip Unit Cost	Avg Total Unit Cost	Avg Price Incl O&P

Panels only (labor to lay panels in grid system)

Luminous panels

Prismatic panels

Acrylic	Inst	SF	Lg	CA	.008	1020	1.65	.27	---	1.92	**2.30**
	Inst	SF	Sm	CA	.011	714.0	1.81	.37	---	2.18	**2.63**
Polystyrene	Inst	SF	Lg	CA	.008	1020	.89	.27	---	1.16	**1.42**
	Inst	SF	Sm	CA	.011	714.0	.97	.37	---	1.34	**1.66**

Ribbed panels

Acrylic	Inst	SF	Lg	CA	.008	1020	1.78	.27	---	2.05	**2.45**
	Inst	SF	Sm	CA	.011	714.0	1.95	.37	---	2.32	**2.79**
Polystyrene	Inst	SF	Lg	CA	.008	1020	1.02	.27	---	1.29	**1.57**
	Inst	SF	Sm	CA	.011	714.0	1.11	.37	---	1.48	**1.83**

Fiberglass panels

1/2" T

Plain white	Inst	SF	Lg	CA	.007	1220	.76	.23	---	.99	**1.22**
	Inst	SF	Sm	CA	.009	854.0	.83	.30	---	1.13	**1.40**
Sculptured white	Inst	SF	Lg	CA	.007	1220	.89	.23	---	1.12	**1.37**
	Inst	SF	Sm	CA	.009	854.0	.97	.30	---	1.27	**1.56**

5/8" T

Fissured	Inst	SF	Lg	CA	.007	1220	.95	.23	---	1.18	**1.44**
	Inst	SF	Sm	CA	.009	854.0	1.04	.30	---	1.34	**1.65**
Fissured, fire rated	Inst	SF	Lg	CA	.007	1220	1.08	.23	---	1.31	**1.59**
	Inst	SF	Sm	CA	.009	854.0	1.18	.30	---	1.48	**1.81**

Mineral fiber panels

5/8" T

	Inst	SF	Lg	CA	.007	1220	.57	.23	---	.80	**1.00**
	Inst	SF	Sm	CA	.009	854.0	.63	.30	---	.93	**1.17**

3/4" T

	Inst	SF	Lg	CA	.007	1220	1.12	.23	---	1.35	**1.64**
	Inst	SF	Sm	CA	.009	854.0	1.22	.30	---	1.52	**1.85**

Description	Oper	Unit	Vol	Crew Size	Man-hours per Unit	Crew Output per Day	Avg Mat'l Unit Cost	Avg Labor Unit Cost	Avg Equip Unit Cost	Avg Total Unit Cost	Avg Price Incl O&P

Grid system components
Main runner, 12' L pieces

Description	Oper	Unit	Vol	Crew Size	Man-hours per Unit	Crew Output per Day	Avg Mat'l Unit Cost	Avg Labor Unit Cost	Avg Equip Unit Cost	Avg Total Unit Cost	Avg Price Incl O&P
Enamel	Inst	Ea	Lg	---	---	---	5.65	---	---	5.65	**5.65**
	Inst	Ea	Sm	---	---	---	6.19	---	---	6.19	**6.19**
Wood grain	Inst	Ea	Lg	---	---	---	6.48	---	---	6.48	**6.48**
	Inst	Ea	Sm	---	---	---	7.09	---	---	7.09	**7.09**

Cross tees
2' L pieces

Description	Oper	Unit	Vol	Crew Size	Man-hours per Unit	Crew Output per Day	Avg Mat'l Unit Cost	Avg Labor Unit Cost	Avg Equip Unit Cost	Avg Total Unit Cost	Avg Price Incl O&P
Enamel	Inst	Ea	Lg	---	---	---	1.02	---	---	1.02	**1.02**
	Inst	Ea	Sm	---	---	---	1.11	---	---	1.11	**1.11**
Wood grain	Inst	Ea	Lg	---	---	---	1.14	---	---	1.14	**1.14**
	Inst	Ea	Sm	---	---	---	1.25	---	---	1.25	**1.25**

4' L pieces

Description	Oper	Unit	Vol	Crew Size	Man-hours per Unit	Crew Output per Day	Avg Mat'l Unit Cost	Avg Labor Unit Cost	Avg Equip Unit Cost	Avg Total Unit Cost	Avg Price Incl O&P
Enamel	Inst	Ea	Lg	---	---	---	1.91	---	---	1.91	**1.91**
	Inst	Ea	Sm	---	---	---	2.09	---	---	2.09	**2.09**
Wood grain	Inst	Ea	Lg	---	---	---	2.16	---	---	2.16	**2.16**
	Inst	Ea	Sm	---	---	---	2.36	---	---	2.36	**2.36**

Wall molding, 10' L pieces

Description	Oper	Unit	Vol	Crew Size	Man-hours per Unit	Crew Output per Day	Avg Mat'l Unit Cost	Avg Labor Unit Cost	Avg Equip Unit Cost	Avg Total Unit Cost	Avg Price Incl O&P
Enamel	Inst	Ea	Lg	---	---	---	3.62	---	---	3.62	**3.62**
	Inst	Ea	Sm	---	---	---	3.96	---	---	3.96	**3.96**
Wood grain	Inst	Ea	Lg	---	---	---	4.76	---	---	4.76	**4.76**
	Inst	Ea	Sm	---	---	---	5.21	---	---	5.21	**5.21**

Main runner hold-down clips
1/2" T panels (1000/carton)

Description	Oper	Unit	Vol	Crew Size	Man-hours per Unit	Crew Output per Day	Avg Mat'l Unit Cost	Avg Labor Unit Cost	Avg Equip Unit Cost	Avg Total Unit Cost	Avg Price Incl O&P
	Inst	Ctn	Lg	---	---	---	10.20	---	---	10.20	**10.20**
	Inst	Ctn	Sm	---	---	---	11.10	---	---	11.10	**11.10**

5/8" T panels (1000/carton)

Description	Oper	Unit	Vol	Crew Size	Man-hours per Unit	Crew Output per Day	Avg Mat'l Unit Cost	Avg Labor Unit Cost	Avg Equip Unit Cost	Avg Total Unit Cost	Avg Price Incl O&P
	Inst	Ctn	Lg	---	---	---	14.00	---	---	14.00	**14.00**
	Inst	Ctn	Sm	---	---	---	15.30	---	---	15.30	**15.30**

Telephone prewiring. See Electrical, page 154

Television antenna. See Electrical, page 154

Thermostat. See Electrical, page 154

Tile, ceiling. See Acoustical treatment, page 22

Tile, ceramic. See Ceramic tile, page 83

Tile, quarry. See Masonry, page 289

Tile, vinyl or asphalt. See Resilient flooring, page 343

Description	Oper	Unit	Vol	Crew Size	Man-hours per Unit	Crew Output per Day	Avg Mat'l Unit Cost	Avg Labor Unit Cost	Avg Equip Unit Cost	Avg Total Unit Cost	Avg Price Incl O&P

Toilets, bidets, urinals

Toilets (water closets), vitreous china, 1.6 GPF, includes seat, shut-off valve, connectors, flanges, water supply valve

Frequently encountered applications

Detach & reset operations

Description	Oper	Unit	Vol	Crew Size	Man-hours per Unit	Crew Output per Day	Avg Mat'l Unit Cost	Avg Labor Unit Cost	Avg Equip Unit Cost	Avg Total Unit Cost	Avg Price Incl O&P
Toilet	Reset	Ea	Lg	SA	1.33	6.00	35.00	50.10	---	85.10	**114.00**
	Reset	Ea	Sm	SA	1.78	4.50	42.50	67.10	---	109.60	**147.00**
Bidet	Reset	Ea	Lg	SA	1.33	6.00	35.00	50.10	---	85.10	**114.00**
	Reset	Ea	Sm	SA	1.78	4.50	42.50	67.10	---	109.60	**147.00**
Urinal	Reset	Ea	Lg	SA	1.33	6.00	35.00	50.10	---	85.10	**114.00**
	Reset	Ea	Sm	SA	1.78	4.50	42.50	67.10	---	109.60	**147.00**

Remove operations

Description	Oper	Unit	Vol	Crew Size	Man-hours per Unit	Crew Output per Day	Avg Mat'l Unit Cost	Avg Labor Unit Cost	Avg Equip Unit Cost	Avg Total Unit Cost	Avg Price Incl O&P
Toilet	Demo	Ea	Lg	SA	0.67	12.00	10.50	25.20	---	35.70	**49.20**
	Demo	Ea	Sm	SA	0.89	9.00	12.80	33.50	---	46.30	**63.90**
Bidet	Demo	Ea	Lg	SA	0.67	12.00	10.50	25.20	---	35.70	**49.20**
	Demo	Ea	Sm	SA	0.89	9.00	12.80	33.50	---	46.30	**63.90**
Urinal	Demo	Ea	Lg	SA	0.67	12.00	10.50	25.20	---	35.70	**49.20**
	Demo	Ea	Sm	SA	0.89	9.00	12.80	33.50	---	46.30	**63.90**

Install rough-in

Description	Oper	Unit	Vol	Crew Size	Man-hours per Unit	Crew Output per Day	Avg Mat'l Unit Cost	Avg Labor Unit Cost	Avg Equip Unit Cost	Avg Total Unit Cost	Avg Price Incl O&P
Toilet	Inst	Ea	Lg	SA	8.00	1.00	140.00	301.00	---	441.00	**604.00**
	Inst	Ea	Sm	SA	10.0	0.80	170.00	377.00	---	547.00	**749.00**
Bidet	Inst	Ea	Lg	SA	8.00	1.00	123.00	301.00	---	424.00	**584.00**
	Inst	Ea	Sm	SA	10.0	0.80	149.00	377.00	---	526.00	**725.00**

Replace operations
Toilet / water closet

Description	Oper	Unit	Vol	Crew Size	Man-hours per Unit	Crew Output per Day	Avg Mat'l Unit Cost	Avg Labor Unit Cost	Avg Equip Unit Cost	Avg Total Unit Cost	Avg Price Incl O&P
One piece, floor mounted	Inst	Ea	Lg	SA	2.00	4.00	340.00	75.30	---	415.30	**501.00**
	Inst	Ea	Sm	SA	2.67	3.00	412.00	101.00	---	513.00	**622.00**
Two piece, floor mounted	Inst	Ea	Lg	SA	2.00	4.00	298.00	75.30	---	373.30	**454.00**
	Inst	Ea	Sm	SA	2.67	3.00	362.00	101.00	---	463.00	**564.00**
Two piece, wall mounted	Inst	Ea	Lg	SA	2.00	4.00	609.00	75.30	---	684.30	**811.00**
	Inst	Ea	Sm	SA	2.67	3.00	740.00	101.00	---	841.00	**998.00**

Bidet

Description	Oper	Unit	Vol	Crew Size	Man-hours per Unit	Crew Output per Day	Avg Mat'l Unit Cost	Avg Labor Unit Cost	Avg Equip Unit Cost	Avg Total Unit Cost	Avg Price Incl O&P
Deck-mounted fittings	Inst	Ea	Lg	SA	1.60	5.00	484.00	60.30	---	544.30	**645.00**
	Inst	Ea	Sm	SA	2.11	3.80	587.00	79.50	---	666.50	**792.00**

Urinal

Description	Oper	Unit	Vol	Crew Size	Man-hours per Unit	Crew Output per Day	Avg Mat'l Unit Cost	Avg Labor Unit Cost	Avg Equip Unit Cost	Avg Total Unit Cost	Avg Price Incl O&P
Wall mounted	Inst	Ea	Lg	SA	2.67	3.00	727.00	101.00	---	828.00	**984.00**
	Inst	Ea	Sm	SA	3.48	2.30	883.00	131.00	---	1014.00	**1210.00**

Description	Oper	Unit	Vol	Crew Size	Man-hours per Unit	Crew Output per Day	Avg Mat'l Unit Cost	Avg Labor Unit Cost	Avg Equip Unit Cost	Avg Total Unit Cost	Avg Price Incl O&P

One piece, floor mounted
American Standard Products

Cadet, round

Description	Oper	Unit	Vol	Crew Size	Man-hours per Unit	Crew Output per Day	Avg Mat'l Unit Cost	Avg Labor Unit Cost	Avg Equip Unit Cost	Avg Total Unit Cost	Avg Price Incl O&P
White	Inst	Ea	Lg	SA	2.00	4.00	263.00	75.30	---	338.30	**413.00**
	Inst	Ea	Sm	SA	2.67	3.00	319.00	101.00	---	420.00	**514.00**
Color	Inst	Ea	Lg	SA	2.00	4.00	310.00	75.30	---	385.30	**467.00**
	Inst	Ea	Sm	SA	2.67	3.00	377.00	101.00	---	478.00	**581.00**
Premium color	Inst	Ea	Lg	SA	2.00	4.00	368.00	75.30	---	443.30	**533.00**
	Inst	Ea	Sm	SA	2.67	3.00	446.00	101.00	---	547.00	**661.00**

Cadet, elongated

Description	Oper	Unit	Vol	Crew Size	Man-hours per Unit	Crew Output per Day	Avg Mat'l Unit Cost	Avg Labor Unit Cost	Avg Equip Unit Cost	Avg Total Unit Cost	Avg Price Incl O&P
White	Inst	Ea	Lg	SA	2.00	4.00	310.00	75.30	---	385.30	**467.00**
	Inst	Ea	Sm	SA	2.67	3.00	377.00	101.00	---	478.00	**581.00**
Color	Inst	Ea	Lg	SA	2.00	4.00	371.00	75.30	---	446.30	**537.00**
	Inst	Ea	Sm	SA	2.67	3.00	451.00	101.00	---	552.00	**666.00**
Premium color	Inst	Ea	Lg	SA	2.00	4.00	431.00	75.30	---	506.30	**607.00**
	Inst	Ea	Sm	SA	2.67	3.00	524.00	101.00	---	625.00	**750.00**

Hamilton, elongated, space saving

Description	Oper	Unit	Vol	Crew Size	Man-hours per Unit	Crew Output per Day	Avg Mat'l Unit Cost	Avg Labor Unit Cost	Avg Equip Unit Cost	Avg Total Unit Cost	Avg Price Incl O&P
White	Inst	Ea	Lg	SA	2.00	4.00	260.00	75.30	---	335.30	**410.00**
	Inst	Ea	Sm	SA	2.67	3.00	316.00	101.00	---	417.00	**511.00**
Color	Inst	Ea	Lg	SA	2.00	4.00	340.00	75.30	---	415.30	**501.00**
	Inst	Ea	Sm	SA	2.67	3.00	412.00	101.00	---	513.00	**622.00**
Premium color	Inst	Ea	Lg	SA	2.00	4.00	396.00	75.30	---	471.30	**566.00**
	Inst	Ea	Sm	SA	2.67	3.00	481.00	101.00	---	582.00	**701.00**

Savona, elongated

Description	Oper	Unit	Vol	Crew Size	Man-hours per Unit	Crew Output per Day	Avg Mat'l Unit Cost	Avg Labor Unit Cost	Avg Equip Unit Cost	Avg Total Unit Cost	Avg Price Incl O&P
White	Inst	Ea	Lg	SA	2.00	4.00	474.00	75.30	---	549.30	**656.00**
	Inst	Ea	Sm	SA	2.67	3.00	575.00	101.00	---	676.00	**810.00**
Color	Inst	Ea	Lg	SA	2.00	4.00	594.00	75.30	---	669.30	**794.00**
	Inst	Ea	Sm	SA	2.67	3.00	722.00	101.00	---	823.00	**978.00**
Premium color	Inst	Ea	Lg	SA	2.00	4.00	665.00	75.30	---	740.30	**875.00**
	Inst	Ea	Sm	SA	2.67	3.00	808.00	101.00	---	909.00	**1080.00**

Lexington, elongated, low profile model

Description	Oper	Unit	Vol	Crew Size	Man-hours per Unit	Crew Output per Day	Avg Mat'l Unit Cost	Avg Labor Unit Cost	Avg Equip Unit Cost	Avg Total Unit Cost	Avg Price Incl O&P
White	Inst	Ea	Lg	SA	2.00	4.00	529.00	75.30	---	604.30	**719.00**
	Inst	Ea	Sm	SA	2.67	3.00	643.00	101.00	---	744.00	**887.00**
Color	Inst	Ea	Lg	SA	2.00	4.00	661.00	75.30	---	736.30	**871.00**
	Inst	Ea	Sm	SA	2.67	3.00	802.00	101.00	---	903.00	**1070.00**
Premium color	Inst	Ea	Lg	SA	2.00	4.00	744.00	75.30	---	819.30	**966.00**
	Inst	Ea	Sm	SA	2.67	3.00	904.00	101.00	---	1005.00	**1190.00**

Fontaine, elongated, pressure-assisted model

Description	Oper	Unit	Vol	Crew Size	Man-hours per Unit	Crew Output per Day	Avg Mat'l Unit Cost	Avg Labor Unit Cost	Avg Equip Unit Cost	Avg Total Unit Cost	Avg Price Incl O&P
White	Inst	Ea	Lg	SA	2.00	4.00	681.00	75.30	---	756.30	**894.00**
	Inst	Ea	Sm	SA	2.67	3.00	827.00	101.00	---	928.00	**1100.00**
Color	Inst	Ea	Lg	SA	2.00	4.00	853.00	75.30	---	928.30	**1090.00**
	Inst	Ea	Sm	SA	2.67	3.00	1040.00	101.00	---	1141.00	**1340.00**
Premium color	Inst	Ea	Lg	SA	2.00	4.00	956.00	75.30	---	1031.30	**1210.00**
	Inst	Ea	Sm	SA	2.67	3.00	1160.00	101.00	---	1261.00	**1480.00**

Description	Oper	Unit	Vol	Crew Size	Man-hours per Unit	Crew Output per Day	Avg Mat'l Unit Cost	Avg Labor Unit Cost	Avg Equip Unit Cost	Avg Total Unit Cost	Avg Price Incl O&P
Kohler Products											
Rialto, round											
White	Inst	Ea	Lg	SA	2.00	4.00	274.00	75.30	---	349.30	**426.00**
	Inst	Ea	Sm	SA	2.67	3.00	332.00	101.00	---	433.00	**530.00**
Color	Inst	Ea	Lg	SA	2.00	4.00	356.00	75.30	---	431.30	**520.00**
	Inst	Ea	Sm	SA	2.67	3.00	432.00	101.00	---	533.00	**645.00**
Premium color	Inst	Ea	Lg	SA	2.00	4.00	409.00	75.30	---	484.30	**581.00**
	Inst	Ea	Sm	SA	2.67	3.00	497.00	101.00	---	598.00	**719.00**
Gabrielle, elongated											
White	Inst	Ea	Lg	SA	2.00	4.00	402.00	75.30	---	477.30	**574.00**
	Inst	Ea	Sm	SA	2.67	3.00	489.00	101.00	---	590.00	**710.00**
Color	Inst	Ea	Lg	SA	2.00	4.00	523.00	75.30	---	598.30	**712.00**
	Inst	Ea	Sm	SA	2.67	3.00	635.00	101.00	---	736.00	**879.00**
Premium color	Inst	Ea	Lg	SA	2.00	4.00	602.00	75.30	---	677.30	**803.00**
	Inst	Ea	Sm	SA	2.67	3.00	731.00	101.00	---	832.00	**988.00**
San Martine, elongated											
White	Inst	Ea	Lg	SA	2.00	4.00	583.00	75.30	---	658.30	**781.00**
	Inst	Ea	Sm	SA	2.67	3.00	707.00	101.00	---	808.00	**961.00**
Color	Inst	Ea	Lg	SA	2.00	4.00	757.00	75.30	---	832.30	**982.00**
	Inst	Ea	Sm	SA	2.67	3.00	920.00	101.00	---	1021.00	**1210.00**
Premium color	Inst	Ea	Lg	SA	2.00	4.00	871.00	75.30	---	946.30	**1110.00**
	Inst	Ea	Sm	SA	2.67	3.00	1060.00	101.00	---	1161.00	**1360.00**
Revival, elongated, with lift knob											
White	Inst	Ea	Lg	SA	2.00	4.00	797.00	75.30	---	872.30	**1030.00**
	Inst	Ea	Sm	SA	2.67	3.00	968.00	101.00	---	1069.00	**1260.00**
Color	Inst	Ea	Lg	SA	2.00	4.00	1050.00	75.30	---	1125.30	**1320.00**
	Inst	Ea	Sm	SA	2.67	3.00	1270.00	101.00	---	1371.00	**1610.00**
Premium color	Inst	Ea	Lg	SA	2.00	4.00	1180.00	75.30	---	1255.30	**1470.00**
	Inst	Ea	Sm	SA	2.67	3.00	1430.00	101.00	---	1531.00	**1790.00**

Two piece, floor mounted

American Standard Products

Description	Oper	Unit	Vol	Crew Size	Man-hours per Unit	Crew Output per Day	Avg Mat'l Unit Cost	Avg Labor Unit Cost	Avg Equip Unit Cost	Avg Total Unit Cost	Avg Price Incl O&P
Cadet, elongated											
White	Inst	Ea	Lg	SA	2.00	4.00	238.00	75.30	---	313.30	**384.00**
	Inst	Ea	Sm	SA	2.67	3.00	289.00	101.00	---	390.00	**480.00**
Color	Inst	Ea	Lg	SA	2.00	4.00	298.00	75.30	---	373.30	**454.00**
	Inst	Ea	Sm	SA	2.67	3.00	362.00	101.00	---	463.00	**564.00**
Premium color	Inst	Ea	Lg	SA	2.00	4.00	330.00	75.30	---	405.30	**491.00**
	Inst	Ea	Sm	SA	2.67	3.00	401.00	101.00	---	502.00	**609.00**
Cadet, elongated, pressure assisted											
White	Inst	Ea	Lg	SA	2.00	4.00	379.00	75.30	---	454.30	**546.00**
	Inst	Ea	Sm	SA	2.67	3.00	460.00	101.00	---	561.00	**677.00**
Color	Inst	Ea	Lg	SA	2.00	4.00	465.00	75.30	---	540.30	**645.00**
	Inst	Ea	Sm	SA	2.67	3.00	564.00	101.00	---	665.00	**797.00**
Premium color	Inst	Ea	Lg	SA	2.00	4.00	520.00	75.30	---	595.30	**709.00**
	Inst	Ea	Sm	SA	2.67	3.00	632.00	101.00	---	733.00	**874.00**

Description	Oper	Unit	Vol	Crew Size	Man-hours per Unit	Crew Output per Day	Avg Mat'l Unit Cost	Avg Labor Unit Cost	Avg Equip Unit Cost	Avg Total Unit Cost	Avg Price Incl O&P
Cadet, elongated, hi-boy, 16-1/2" H											
White	Inst	Ea	Lg	SA	2.00	4.00	238.00	75.30	---	313.30	**384.00**
	Inst	Ea	Sm	SA	2.67	3.00	289.00	101.00	---	390.00	**480.00**
Color	Inst	Ea	Lg	SA	2.00	4.00	298.00	75.30	---	373.30	**454.00**
	Inst	Ea	Sm	SA	2.67	3.00	362.00	101.00	---	463.00	**564.00**
Premium color	Inst	Ea	Lg	SA	2.00	4.00	330.00	75.30	---	405.30	**491.00**
	Inst	Ea	Sm	SA	2.67	3.00	401.00	101.00	---	502.00	**609.00**
Ravenna											
Round											
White	Inst	Ea	Lg	SA	2.00	4.00	167.00	75.30	---	242.30	**303.00**
	Inst	Ea	Sm	SA	2.67	3.00	203.00	101.00	---	304.00	**381.00**
Color	Inst	Ea	Lg	SA	2.00	4.00	207.00	75.30	---	282.30	**348.00**
	Inst	Ea	Sm	SA	2.67	3.00	251.00	101.00	---	352.00	**436.00**
Premium color	Inst	Ea	Lg	SA	2.00	4.00	235.00	75.30	---	310.30	**381.00**
	Inst	Ea	Sm	SA	2.67	3.00	286.00	101.00	---	387.00	**476.00**
Elongated											
White	Inst	Ea	Lg	SA	2.00	4.00	203.00	75.30	---	278.30	**344.00**
	Inst	Ea	Sm	SA	2.67	3.00	247.00	101.00	---	348.00	**431.00**
Color	Inst	Ea	Lg	SA	2.00	4.00	251.00	75.30	---	326.30	**399.00**
	Inst	Ea	Sm	SA	2.67	3.00	304.00	101.00	---	405.00	**498.00**
Premium color	Inst	Ea	Lg	SA	2.00	4.00	282.00	75.30	---	357.30	**435.00**
	Inst	Ea	Sm	SA	2.67	3.00	343.00	101.00	---	444.00	**542.00**
Town Square											
Round											
White	Inst	Ea	Lg	SA	2.00	4.00	270.00	75.30	---	345.30	**421.00**
	Inst	Ea	Sm	SA	2.67	3.00	328.00	101.00	---	429.00	**525.00**
Color	Inst	Ea	Lg	SA	2.00	4.00	351.00	75.30	---	426.30	**514.00**
	Inst	Ea	Sm	SA	2.67	3.00	426.00	101.00	---	527.00	**638.00**
Premium color	Inst	Ea	Lg	SA	2.00	4.00	405.00	75.30	---	480.30	**577.00**
	Inst	Ea	Sm	SA	2.67	3.00	492.00	101.00	---	593.00	**714.00**
Elongated											
White	Inst	Ea	Lg	SA	2.00	4.00	312.00	75.30	---	387.30	**470.00**
	Inst	Ea	Sm	SA	2.67	3.00	379.00	101.00	---	480.00	**584.00**
Color	Inst	Ea	Lg	SA	2.00	4.00	405.00	75.30	---	480.30	**577.00**
	Inst	Ea	Sm	SA	2.67	3.00	492.00	101.00	---	593.00	**714.00**
Premium color	Inst	Ea	Lg	SA	2.00	4.00	468.00	75.30	---	543.30	**648.00**
	Inst	Ea	Sm	SA	2.67	3.00	568.00	101.00	---	669.00	**801.00**
Yorkville, elongated, pressure assisted											
White	Inst	Ea	Lg	SA	2.00	4.00	629.00	75.30	---	704.30	**834.00**
	Inst	Ea	Sm	SA	2.67	3.00	764.00	101.00	---	865.00	**1030.00**
Color	Inst	Ea	Lg	SA	2.00	4.00	739.00	75.30	---	814.30	**960.00**
	Inst	Ea	Sm	SA	2.67	3.00	897.00	101.00	---	998.00	**1180.00**
Premium color	Inst	Ea	Lg	SA	2.00	4.00	815.00	75.30	---	890.30	**1050.00**
	Inst	Ea	Sm	SA	2.67	3.00	989.00	101.00	---	1090.00	**1290.00**

Description	Oper	Unit	Vol	Crew Size	Man-hours per Unit	Crew Output per Day	Avg Mat'l Unit Cost	Avg Labor Unit Cost	Avg Equip Unit Cost	Avg Total Unit Cost	Avg Price Incl O&P

Kohler Products

Memoirs, round
White	Inst	Ea	Lg	SA	2.00	4.00	258.00	75.30	---	333.30	**408.00**
	Inst	Ea	Sm	SA	2.67	3.00	314.00	101.00	---	415.00	**509.00**
Color	Inst	Ea	Lg	SA	2.00	4.00	336.00	75.30	---	411.30	**497.00**
	Inst	Ea	Sm	SA	2.67	3.00	408.00	101.00	---	509.00	**617.00**
Premium color	Inst	Ea	Lg	SA	2.00	4.00	386.00	75.30	---	461.30	**555.00**
	Inst	Ea	Sm	SA	2.67	3.00	469.00	101.00	---	570.00	**687.00**

Memoirs, elongated
White	Inst	Ea	Lg	SA	2.00	4.00	301.00	75.30	---	376.30	**457.00**
	Inst	Ea	Sm	SA	2.67	3.00	365.00	101.00	---	466.00	**568.00**
Color	Inst	Ea	Lg	SA	2.00	4.00	391.00	75.30	---	466.30	**560.00**
	Inst	Ea	Sm	SA	2.67	3.00	474.00	101.00	---	575.00	**693.00**
Premium color	Inst	Ea	Lg	SA	2.00	4.00	449.00	75.30	---	524.30	**627.00**
	Inst	Ea	Sm	SA	2.67	3.00	546.00	101.00	---	647.00	**775.00**

Highline, elongated
White	Inst	Ea	Lg	SA	2.00	4.00	175.00	75.30	---	250.30	**312.00**
	Inst	Ea	Sm	SA	2.67	3.00	212.00	101.00	---	313.00	**392.00**
Color	Inst	Ea	Lg	SA	2.00	4.00	224.00	75.30	---	299.30	**368.00**
	Inst	Ea	Sm	SA	2.67	3.00	272.00	101.00	---	373.00	**461.00**
Premium color	Inst	Ea	Lg	SA	2.00	4.00	258.00	75.30	---	333.30	**407.00**
	Inst	Ea	Sm	SA	2.67	3.00	313.00	101.00	---	414.00	**507.00**

Revival, elongated, with lift knob
White	Inst	Ea	Lg	SA	2.00	4.00	455.00	75.30	---	530.30	**634.00**
	Inst	Ea	Sm	SA	2.67	3.00	553.00	101.00	---	654.00	**783.00**
Color	Inst	Ea	Lg	SA	2.00	4.00	599.00	75.30	---	674.30	**800.00**
	Inst	Ea	Sm	SA	2.67	3.00	728.00	101.00	---	829.00	**985.00**
Premium color	Inst	Ea	Lg	SA	2.00	4.00	695.00	75.30	---	770.30	**910.00**
	Inst	Ea	Sm	SA	2.67	3.00	844.00	101.00	---	945.00	**1120.00**

Two piece, wall mounted
American Standard Products

Glenwall, elongated
White	Inst	Ea	Lg	SA	2.00	4.00	497.00	75.30	---	572.30	**682.00**
	Inst	Ea	Sm	SA	2.67	3.00	604.00	101.00	---	705.00	**842.00**
Color	Inst	Ea	Lg	SA	2.00	4.00	609.00	75.30	---	684.30	**811.00**
	Inst	Ea	Sm	SA	2.67	3.00	740.00	101.00	---	841.00	**998.00**
Premium color	Inst	Ea	Lg	SA	2.00	4.00	682.00	75.30	---	757.30	**895.00**
	Inst	Ea	Sm	SA	2.67	3.00	828.00	101.00	---	929.00	**1100.00**

Description	Oper	Unit	Vol	Crew Size	Man-hours per Unit	Crew Output per Day	Avg Mat'l Unit Cost	Avg Labor Unit Cost	Avg Equip Unit Cost	Avg Total Unit Cost	Avg Price Incl O&P

Bidet, deck-mounted fitting
American Standard Products
Cadet

Description	Oper	Unit	Vol	Crew Size	Man-hours per Unit	Crew Output per Day	Avg Mat'l Unit Cost	Avg Labor Unit Cost	Avg Equip Unit Cost	Avg Total Unit Cost	Avg Price Incl O&P
White	Inst	Ea	Lg	SA	1.60	5.00	335.00	60.30	---	395.30	**473.00**
	Inst	Ea	Sm	SA	2.11	3.80	406.00	79.50	---	485.50	**584.00**
Color	Inst	Ea	Lg	SA	1.60	5.00	418.00	60.30	---	478.30	**569.00**
	Inst	Ea	Sm	SA	2.11	3.80	507.00	79.50	---	586.50	**700.00**
Premium Color	Inst	Ea	Lg	SA	1.60	5.00	468.00	60.30	---	528.30	**626.00**
	Inst	Ea	Sm	SA	2.11	3.80	568.00	79.50	---	647.50	**770.00**

Heritage

Description	Oper	Unit	Vol	Crew Size	Man-hours per Unit	Crew Output per Day	Avg Mat'l Unit Cost	Avg Labor Unit Cost	Avg Equip Unit Cost	Avg Total Unit Cost	Avg Price Incl O&P
White	Inst	Ea	Lg	SA	1.60	5.00	495.00	60.30	---	555.30	**658.00**
	Inst	Ea	Sm	SA	2.11	3.80	601.00	79.50	---	680.50	**808.00**
Color	Inst	Ea	Lg	SA	1.60	5.00	619.00	60.30	---	679.30	**800.00**
	Inst	Ea	Sm	SA	2.11	3.80	751.00	79.50	---	830.50	**981.00**

Savona

Description	Oper	Unit	Vol	Crew Size	Man-hours per Unit	Crew Output per Day	Avg Mat'l Unit Cost	Avg Labor Unit Cost	Avg Equip Unit Cost	Avg Total Unit Cost	Avg Price Incl O&P
White	Inst	Ea	Lg	SA	1.60	5.00	339.00	60.30	---	399.30	**478.00**
	Inst	Ea	Sm	SA	2.11	3.80	411.00	79.50	---	490.50	**590.00**
Color	Inst	Ea	Lg	SA	1.60	5.00	425.00	60.30	---	485.30	**577.00**
	Inst	Ea	Sm	SA	2.11	3.80	516.00	79.50	---	595.50	**710.00**
Premium color	Inst	Ea	Lg	SA	1.60	5.00	472.00	60.30	---	532.30	**631.00**
	Inst	Ea	Sm	SA	2.11	3.80	573.00	79.50	---	652.50	**776.00**

Kohler Products
Portrait

Description	Oper	Unit	Vol	Crew Size	Man-hours per Unit	Crew Output per Day	Avg Mat'l Unit Cost	Avg Labor Unit Cost	Avg Equip Unit Cost	Avg Total Unit Cost	Avg Price Incl O&P
White	Inst	Ea	Lg	SA	1.60	5.00	372.00	60.30	---	432.30	**517.00**
	Inst	Ea	Sm	SA	2.11	3.80	452.00	79.50	---	531.50	**637.00**
Color	Inst	Ea	Lg	SA	1.60	5.00	484.00	60.30	---	544.30	**645.00**
	Inst	Ea	Sm	SA	2.11	3.80	587.00	79.50	---	666.50	**792.00**
Premium Color	Inst	Ea	Lg	SA	1.60	5.00	556.00	60.30	---	616.30	**728.00**
	Inst	Ea	Sm	SA	2.11	3.80	676.00	79.50	---	755.50	**894.00**

Revival

Description	Oper	Unit	Vol	Crew Size	Man-hours per Unit	Crew Output per Day	Avg Mat'l Unit Cost	Avg Labor Unit Cost	Avg Equip Unit Cost	Avg Total Unit Cost	Avg Price Incl O&P
White	Inst	Ea	Lg	SA	1.60	5.00	433.00	60.30	---	493.30	**586.00**
	Inst	Ea	Sm	SA	2.11	3.80	526.00	79.50	---	605.50	**721.00**
Color	Inst	Ea	Lg	SA	1.60	5.00	563.00	60.30	---	623.30	**736.00**
	Inst	Ea	Sm	SA	2.11	3.80	683.00	79.50	---	762.50	**903.00**
Premium color	Inst	Ea	Lg	SA	1.60	5.00	647.00	60.30	---	707.30	**832.00**
	Inst	Ea	Sm	SA	2.11	3.80	785.00	79.50	---	864.50	**1020.00**

Description	Oper	Unit	Vol	Crew Size	Man-hours per Unit	Crew Output per Day	Avg Mat'l Unit Cost	Avg Labor Unit Cost	Avg Equip Unit Cost	Avg Total Unit Cost	Avg Price Incl O&P

Urinal, wall mounted, 1.0 GPF
American Standard Products

Lynnbrook

Description	Oper	Unit	Vol	Crew Size	Man-hours per Unit	Crew Output per Day	Avg Mat'l Unit Cost	Avg Labor Unit Cost	Avg Equip Unit Cost	Avg Total Unit Cost	Avg Price Incl O&P
White	Inst	Ea	Lg	SA	2.67	3.00	416.00	101.00	---	517.00	**626.00**
	Inst	Ea	Sm	SA	3.48	2.30	505.00	131.00	---	636.00	**773.00**
Color	Inst	Ea	Lg	SA	2.67	3.00	512.00	101.00	---	613.00	**736.00**
	Inst	Ea	Sm	SA	3.48	2.30	621.00	131.00	---	752.00	**907.00**

Innsbrook

Description	Oper	Unit	Vol	Crew Size	Man-hours per Unit	Crew Output per Day	Avg Mat'l Unit Cost	Avg Labor Unit Cost	Avg Equip Unit Cost	Avg Total Unit Cost	Avg Price Incl O&P
White	Inst	Ea	Lg	SA	2.67	3.00	590.00	101.00	---	691.00	**826.00**
	Inst	Ea	Sm	SA	3.48	2.30	717.00	131.00	---	848.00	**1020.00**
Color	Inst	Ea	Lg	SA	2.67	3.00	727.00	101.00	---	828.00	**984.00**
	Inst	Ea	Sm	SA	3.48	2.30	883.00	131.00	---	1014.00	**1210.00**

Trash compactors

Includes wiring, connection, and installation in a pre-cutout area

Description	Oper	Unit	Vol	Crew Size	Man-hours per Unit	Crew Output per Day	Avg Mat'l Unit Cost	Avg Labor Unit Cost	Avg Equip Unit Cost	Avg Total Unit Cost	Avg Price Incl O&P
White	Inst	Ea	Lg	EA	2.00	4.00	450.00	71.30	---	521.30	**622.00**
	Inst	Ea	Sm	EA	2.67	3.00	544.00	95.10	---	639.10	**764.00**
Colors	Inst	Ea	Lg	EA	2.00	4.00	486.00	71.30	---	557.30	**663.00**
	Inst	Ea	Sm	EA	2.67	3.00	587.00	95.10	---	682.10	**814.00**
To remove and reset											
	Reset	Ea	Lg	EA	.727	11.00	---	25.90	---	25.90	**37.80**
	Reset	Ea	Sm	EA	.964	8.30	---	34.40	---	34.40	**50.20**
Material adjustments											
Stainless steel trim	R&R	LS	Lg	---	---	---	25.20	---	---	25.20	**29.00**
	R&R	LS	Sm	---	---	---	30.50	---	---	30.50	**35.00**
Stainless steel panel / trim	R&R	LS	Lg	---	---	---	54.00	---	---	54.00	**62.10**
	R&R	LS	Sm	---	---	---	65.30	---	---	65.30	**75.00**
"Black Glas" acrylic front panel / trim											
	R&R	LS	Lg	---	---	---	54.00	---	---	54.00	**62.10**
	R&R	LS	Sm	---	---	---	65.30	---	---	65.30	**75.00**
Hardwood top	R&R	LS	Lg	---	---	---	36.00	---	---	36.00	**41.40**
	R&R	LS	Sm	---	---	---	43.50	---	---	43.50	**50.00**

Trim. See Molding, page 291

Trusses. See Framing, page 213

Vanity units. See Cabinetry, page 73

Description	Oper	Unit	Vol	Crew Size	Man-hours per Unit	Crew Output per Day	Avg Mat'l Unit Cost	Avg Labor Unit Cost	Avg Equip Unit Cost	Avg Total Unit Cost	Avg Price Incl O&P

Ventilation

Flue piping or vent chimney

Prefabricated metal, UL listed

Gas, double wall, galv. steel

Description	Oper	Unit	Vol	Crew Size	Man-hours per Unit	Crew Output per Day	Avg Mat'l Unit Cost	Avg Labor Unit Cost	Avg Equip Unit Cost	Avg Total Unit Cost	Avg Price Incl O&P
3" diameter	Inst	VLF	Lg	UB	.229	70.00	3.89	8.46	---	12.35	**17.00**
	Inst	VLF	Sm	UB	.286	56.00	4.39	10.60	---	14.99	**20.70**
4" diameter	Inst	VLF	Lg	UB	.242	66.00	4.76	8.94	---	13.70	**18.70**
	Inst	VLF	Sm	UB	.302	53.00	5.38	11.20	---	16.58	**22.70**
5" diameter	Inst	VLF	Lg	UB	.258	62.00	5.65	9.53	---	15.18	**20.60**
	Inst	VLF	Sm	UB	.320	50.00	6.38	11.80	---	18.18	**24.80**
6" diameter	Inst	VLF	Lg	UB	.276	58.00	6.56	10.20	---	16.76	**22.60**
	Inst	VLF	Sm	UB	.348	46.00	7.41	12.90	---	20.31	**27.60**
7" diameter	Inst	VLF	Lg	UB	.296	54.00	9.66	10.90	---	20.56	**27.30**
	Inst	VLF	Sm	UB	.372	43.00	10.90	13.80	---	24.70	**32.90**
8" diameter	Inst	VLF	Lg	UB	.320	50.00	10.80	11.80	---	22.60	**29.90**
	Inst	VLF	Sm	UB	.400	40.00	12.20	14.80	---	27.00	**35.90**
10" diameter	Inst	VLF	Lg	UB	.348	46.00	22.70	12.90	---	35.60	**45.20**
	Inst	VLF	Sm	UB	.432	37.00	25.70	16.00	---	41.70	**53.20**
12" diameter	Inst	VLF	Lg	UB	.381	42.00	30.50	14.10	---	44.60	**55.90**
	Inst	VLF	Sm	UB	.471	34.00	34.50	17.40	---	51.90	**65.40**
16" diameter	Inst	VLF	Lg	UB	.421	38.00	69.00	15.60	---	84.60	**102.00**
	Inst	VLF	Sm	UB	.533	30.00	78.00	19.70	---	97.70	**119.00**
20" diameter	Inst	VLF	Lg	UB	.471	34.00	106.00	17.40	---	123.40	**147.00**
	Inst	VLF	Sm	UB	.593	27.00	120.00	21.90	---	141.90	**170.00**
24" diameter	Inst	VLF	Lg	UB	.533	30.00	164.00	19.70	---	183.70	**218.00**
	Inst	VLF	Sm	UB	.667	24.00	186.00	24.70	---	210.70	**250.00**

Vent damper, bi-metal, 6" flue

Description	Oper	Unit	Vol	Crew Size	Man-hours per Unit	Crew Output per Day	Avg Mat'l Unit Cost	Avg Labor Unit Cost	Avg Equip Unit Cost	Avg Total Unit Cost	Avg Price Incl O&P
Gas, auto., electric	Inst	Ea	Lg	UB	2.67	6.00	150.00	98.70	---	248.70	**318.00**
	Inst	Ea	Sm	UB	3.20	5.00	169.00	118.00	---	287.00	**369.00**
Oil, auto., electric	Inst	Ea	Lg	UB	2.67	6.00	179.00	98.70	---	277.70	**352.00**
	Inst	Ea	Sm	UB	3.20	5.00	203.00	118.00	---	321.00	**408.00**

All fuel, double wall, stainless steel

Description	Oper	Unit	Vol	Crew Size	Man-hours per Unit	Crew Output per Day	Avg Mat'l Unit Cost	Avg Labor Unit Cost	Avg Equip Unit Cost	Avg Total Unit Cost	Avg Price Incl O&P
6" diameter	Inst	VLF	Lg	UB	.267	60.00	32.20	9.87	---	42.07	**51.60**
	Inst	VLF	Sm	UB	.333	48.00	36.40	12.30	---	48.70	**60.10**
7" diameter	Inst	VLF	Lg	UB	.286	56.00	41.40	10.60	---	52.00	**63.30**
	Inst	VLF	Sm	UB	.356	45.00	46.80	13.20	---	60.00	**73.30**
8" diameter	Inst	VLF	Lg	UB	.308	52.00	48.30	11.40	---	59.70	**72.40**
	Inst	VLF	Sm	UB	.381	42.00	54.60	14.10	---	68.70	**83.60**
10" diameter	Inst	VLF	Lg	UB	.333	48.00	70.20	12.30	---	82.50	**98.90**
	Inst	VLF	Sm	UB	.421	38.00	79.30	15.60	---	94.90	**114.00**
12" diameter	Inst	VLF	Lg	UB	.364	44.00	94.30	13.50	---	107.80	**128.00**
	Inst	VLF	Sm	UB	.457	35.00	107.00	16.90	---	123.90	**148.00**
14" diameter	Inst	VLF	Lg	UB	.400	40.00	124.00	14.80	---	138.80	**165.00**
	Inst	VLF	Sm	UB	.500	32.00	140.00	18.50	---	158.50	**189.00**

Description	Oper	Unit	Vol	Crew Size	Man-hours per Unit	Crew Output per Day	Avg Mat'l Unit Cost	Avg Labor Unit Cost	Avg Equip Unit Cost	Avg Total Unit Cost	Avg Price Incl O&P
All fuel, double wall, stainless steel fittings											
Roof support											
6" diameter	Inst	Ea	Lg	UB	.533	30.00	82.20	19.70	---	101.90	**124.00**
	Inst	Ea	Sm	UB	.667	24.00	93.00	24.70	---	117.70	**143.00**
7" diameter	Inst	Ea	Lg	UB	.571	28.00	92.60	21.10	---	113.70	**138.00**
	Inst	Ea	Sm	UB	.727	22.00	105.00	26.90	---	131.90	**160.00**
8" diameter	Inst	Ea	Lg	UB	.615	26.00	101.00	22.70	---	123.70	**149.00**
	Inst	Ea	Sm	UB	.762	21.00	114.00	28.20	---	142.20	**172.00**
10" diameter	Inst	Ea	Lg	UB	.667	24.00	132.00	24.70	---	156.70	**189.00**
	Inst	Ea	Sm	UB	.842	19.00	150.00	31.10	---	181.10	**218.00**
12" diameter	Inst	Ea	Lg	UB	.727	22.00	160.00	26.90	---	186.90	**224.00**
	Inst	Ea	Sm	UB	0.89	18.00	181.00	32.90	---	213.90	**256.00**
14" diameter	Inst	Ea	Lg	UB	.800	20.00	202.00	29.60	---	231.60	**277.00**
	Inst	Ea	Sm	UB	1.00	16.00	229.00	37.00	---	266.00	**318.00**
Elbow, 15 degree											
6" diameter	Inst	Ea	Lg	UB	.533	30.00	71.90	19.70	---	91.60	**112.00**
	Inst	Ea	Sm	UB	.667	24.00	81.30	24.70	---	106.00	**130.00**
7" diameter	Inst	Ea	Lg	UB	.571	28.00	80.50	21.10	---	101.60	**124.00**
	Inst	Ea	Sm	UB	.727	22.00	91.00	26.90	---	117.90	**144.00**
8" diameter	Inst	Ea	Lg	UB	.615	26.00	92.00	22.70	---	114.70	**139.00**
	Inst	Ea	Sm	UB	.762	21.00	104.00	28.20	---	132.20	**161.00**
10" diameter	Inst	Ea	Lg	UB	.667	24.00	120.00	24.70	---	144.70	**174.00**
	Inst	Ea	Sm	UB	.842	19.00	135.00	31.10	---	166.10	**202.00**
12" diameter	Inst	Ea	Lg	UB	.727	22.00	146.00	26.90	---	172.90	**208.00**
	Inst	Ea	Sm	UB	0.89	18.00	165.00	32.90	---	197.90	**239.00**
14" diameter	Inst	Ea	Lg	UB	.800	20.00	175.00	29.60	---	204.60	**245.00**
	Inst	Ea	Sm	UB	1.00	16.00	198.00	37.00	---	235.00	**282.00**
Insulated tee with insulated tee cap											
6" diameter	Inst	Ea	Lg	UB	.533	30.00	136.00	19.70	---	155.70	**185.00**
	Inst	Ea	Sm	UB	.667	24.00	153.00	24.70	---	177.70	**213.00**
7" diameter	Inst	Ea	Lg	UB	.571	28.00	177.00	21.10	---	198.10	**235.00**
	Inst	Ea	Sm	UB	.727	22.00	200.00	26.90	---	226.90	**270.00**
8" diameter	Inst	Ea	Lg	UB	.615	26.00	200.00	22.70	---	222.70	**264.00**
	Inst	Ea	Sm	UB	.762	21.00	226.00	28.20	---	254.20	**302.00**
10" diameter	Inst	Ea	Lg	UB	.667	24.00	282.00	24.70	---	306.70	**360.00**
	Inst	Ea	Sm	UB	.842	19.00	319.00	31.10	---	350.10	**412.00**
12" diameter	Inst	Ea	Lg	UB	.727	22.00	397.00	26.90	---	423.90	**496.00**
	Inst	Ea	Sm	UB	0.89	18.00	449.00	32.90	---	481.90	**564.00**
14" diameter	Inst	Ea	Lg	UB	.800	20.00	518.00	29.60	---	547.60	**639.00**
	Inst	Ea	Sm	UB	1.00	16.00	585.00	37.00	---	622.00	**727.00**

Description	Oper	Unit	Vol	Crew Size	Man-hours per Unit	Crew Output per Day	Avg Mat'l Unit Cost	Avg Labor Unit Cost	Avg Equip Unit Cost	Avg Total Unit Cost	Avg Price Incl O&P
Joist shield											
6" diameter	Inst	Ea	Lg	UB	.533	30.00	41.40	19.70	---	61.10	**76.80**
	Inst	Ea	Sm	UB	.667	24.00	46.80	24.70	---	71.50	**90.30**
7" diameter	Inst	Ea	Lg	UB	.571	28.00	44.90	21.10	---	66.00	**82.80**
	Inst	Ea	Sm	UB	.727	22.00	50.70	26.90	---	77.60	**98.10**
8" diameter	Inst	Ea	Lg	UB	.615	26.00	55.20	22.70	---	77.90	**97.10**
	Inst	Ea	Sm	UB	.762	21.00	62.40	28.20	---	90.60	**113.00**
10" diameter	Inst	Ea	Lg	UB	.667	24.00	62.70	24.70	---	87.40	**109.00**
	Inst	Ea	Sm	UB	.842	19.00	70.90	31.10	---	102.00	**128.00**
12" diameter	Inst	Ea	Lg	UB	.727	22.00	92.60	26.90	---	119.50	**146.00**
	Inst	Ea	Sm	UB	0.89	18.00	105.00	32.90	---	137.90	**169.00**
14" diameter	Inst	Ea	Lg	UB	.800	20.00	115.00	29.60	---	144.60	**176.00**
	Inst	Ea	Sm	UB	1.00	16.00	130.00	37.00	---	167.00	**204.00**
Round top											
6" diameter	Inst	Ea	Lg	UB	.533	30.00	46.00	19.70	---	65.70	**82.10**
	Inst	Ea	Sm	UB	.667	24.00	52.00	24.70	---	76.70	**96.30**
7" diameter	Inst	Ea	Lg	UB	.571	28.00	62.70	21.10	---	83.80	**103.00**
	Inst	Ea	Sm	UB	.727	22.00	70.90	26.90	---	97.80	**121.00**
8" diameter	Inst	Ea	Lg	UB	.615	26.00	84.00	22.70	---	106.70	**130.00**
	Inst	Ea	Sm	UB	.762	21.00	94.90	28.20	---	123.10	**151.00**
10" diameter	Inst	Ea	Lg	UB	.667	24.00	151.00	24.70	---	175.70	**210.00**
	Inst	Ea	Sm	UB	.842	19.00	170.00	31.10	---	201.10	**242.00**
12" diameter	Inst	Ea	Lg	UB	.727	22.00	215.00	26.90	---	241.90	**287.00**
	Inst	Ea	Sm	UB	0.89	18.00	243.00	32.90	---	275.90	**328.00**
14" diameter	Inst	Ea	Lg	UB	.800	20.00	283.00	29.60	---	312.60	**369.00**
	Inst	Ea	Sm	UB	1.00	16.00	320.00	37.00	---	357.00	**422.00**
Adjustable roof flashing											
6" diameter	Inst	Ea	Lg	UB	.533	30.00	54.60	19.70	---	74.30	**92.00**
	Inst	Ea	Sm	UB	.667	24.00	61.80	24.70	---	86.50	**107.00**
7" diameter	Inst	Ea	Lg	UB	.571	28.00	62.10	21.10	---	83.20	**103.00**
	Inst	Ea	Sm	UB	.727	22.00	70.20	26.90	---	97.10	**120.00**
8" diameter	Inst	Ea	Lg	UB	.615	26.00	67.90	22.70	---	90.60	**112.00**
	Inst	Ea	Sm	UB	.762	21.00	76.70	28.20	---	104.90	**130.00**
10" diameter	Inst	Ea	Lg	UB	.667	24.00	86.80	24.70	---	111.50	**136.00**
	Inst	Ea	Sm	UB	.842	19.00	98.20	31.10	---	129.30	**159.00**
12" diameter	Inst	Ea	Lg	UB	.727	22.00	112.00	26.90	---	138.90	**169.00**
	Inst	Ea	Sm	UB	0.89	18.00	127.00	32.90	---	159.90	**194.00**
14" diameter	Inst	Ea	Lg	UB	.800	20.00	140.00	29.60	---	169.60	**205.00**
	Inst	Ea	Sm	UB	1.00	16.00	159.00	37.00	---	196.00	**237.00**
Minimum Job Charge											
	Inst	Job	Lg	UB	5.33	3.00	---	197.00	---	197.00	**292.00**
	Inst	Job	Sm	UB	6.67	2.40	---	246.00	---	246.00	**365.00**

Wallpaper

Most wall coverings today are really not wallpapers in the technical sense. Wallpaper is a paper material (which may or not be coated with washable plastic). Most of today's products, though still called wallpaper, are actually composed of a vinyl on a fabric, not a paper, backing. Some vinyl/fabric coverings come prepasted.

Vinyl coverings with non-woven fabric, woven fabric or synthetic fiber backing are fire, mildew and fade resistant, and they can be stripped from plaster walls. Coverings with paper backing must be steamed or scraped from the walls. Vinyl coverings may also be stripped from gypsum wallboard, but unless the vinyl covering has a synthetic fiber backing, it's likely to damage the wallboard paper surface.

One gallon of paste (approximately two-thirds pound of dry paste and water) should adequately cover 12 rolls with paper backing and 6 rolls with fabric backing. The rougher the texture of the surface to be pasted, the greater the quantity of wet paste needed.

1. **Dimensions**

   a. Single roll is 36 SF. The paper is 18" wide x 24'-0" long.

   b. Double roll is 72 SF. The paper is 18" wide x 48'-0" long.

2. **Installation.** New paper may be applied over existing paper if the existing paper has butt joints, a smooth surface, and is tight to the wall. When new paper is applied direct to plaster or drywall, the wall should receive a coat of glue size before the paper is applied.

3. **Estimating Technique.** Determine the gross area, deduct openings and other areas not to be papered, and add 20% to the net area for waste. To find the number of rolls needed, divide the net area plus 20% by the number of SF per roll.

The steps in wallpapering

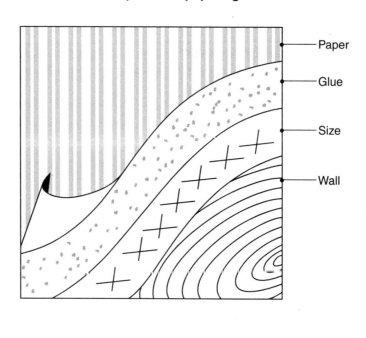

Paper — Glue — Size — Wall

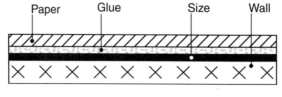

Paper   Glue   Size   Wall

Description	Oper	Unit	Vol	Crew Size	Man-hours per Unit	Crew Output per Day	Avg Mat'l Unit Cost	Avg Labor Unit Cost	Avg Equip Unit Cost	Avg Total Unit Cost	Avg Price Incl O&P

Wallpaper

Butt joint

Includes application of glue sizing. Material prices are not given because they vary radically

Description	Oper	Unit	Vol	Crew Size	Man-hours per Unit	Crew Output per Day	Avg Mat'l Unit Cost	Avg Labor Unit Cost	Avg Equip Unit Cost	Avg Total Unit Cost	Avg Price Incl O&P
Paper, single rolls, 36 SF/roll	Inst	Roll	Lg	QA	.357	22.40	---	12.60	4.14	16.74	**22.70**
	Inst	Roll	Sm	QA	.549	14.56	---	19.30	6.36	25.66	**35.00**
Vinyl, with fabric backing (prepasted), single rolls, 36 SF/roll											
	Inst	Roll	Lg	QA	.400	20.00	---	14.10	4.63	18.73	**25.50**
	Inst	Roll	Sm	QA	.615	13.00	---	21.70	7.13	28.83	**39.20**
Vinyl, with fabric backing (not prepasted), single rolls, 36 SF/roll											
	Inst	Roll	Lg	QA	.435	18.40	---	15.30	5.04	20.34	**27.70**
	Inst	Roll	Sm	QA	.669	11.96	---	23.60	7.75	31.35	**42.60**

Labor adjustments

ADD

Description	Oper	Unit	Vol	Crew Size	Man-hours per Unit	Crew Output per Day	Avg Mat'l Unit Cost	Avg Labor Unit Cost	Avg Equip Unit Cost	Avg Total Unit Cost	Avg Price Incl O&P
For ceiling 9'-0" high or more	Inst	%	Lg	QA	---	---	---	20.0	---	---	---
	Inst	%	Sm	QA	---	---	---	20.0	---	---	---
For kitchens, baths, and other rooms where floor area is less than 50.0 SF											
	Inst	%	Lg	QA	---	---	---	32.0	---	---	---
	Inst	%	Sm	QA	---	---	---	32.0	---	---	---

Water closets. See Toilets, page 428

Water heaters

(A.O. Smith Products)

Detach & reset operations

Description	Oper	Unit	Vol	Crew Size	Man-hours per Unit	Crew Output per Day	Avg Mat'l Unit Cost	Avg Labor Unit Cost	Avg Equip Unit Cost	Avg Total Unit Cost	Avg Price Incl O&P
Water heater	Inst	Ea	Lg	SA	1.33	6.00	---	50.10	---	50.10	**73.60**
	Inst	Ea	Sm	SA	1.78	4.50	---	67.10	---	67.10	**98.60**

Remove operations

Description	Oper	Unit	Vol	Crew Size	Man-hours per Unit	Crew Output per Day	Avg Mat'l Unit Cost	Avg Labor Unit Cost	Avg Equip Unit Cost	Avg Total Unit Cost	Avg Price Incl O&P
Water heater	Demo	Ea	Lg	SA	1.00	8.00	---	37.70	---	37.70	**55.40**
	Demo	Ea	Sm	SA	1.33	6.00	---	50.10	---	50.10	**73.60**

Replace operations

Electric

Standard model, tall

6-year Energy Saver, foam insulated

Description	Oper	Unit	Vol	Crew Size	Man-hours per Unit	Crew Output per Day	Avg Mat'l Unit Cost	Avg Labor Unit Cost	Avg Equip Unit Cost	Avg Total Unit Cost	Avg Price Incl O&P
30 gal round	Inst	Ea	Lg	SA	2.00	4.00	493.00	75.30	---	568.30	**677.00**
	Inst	Ea	Sm	SA	2.67	3.00	598.00	101.00	---	699.00	**836.00**
40 gal round	Inst	Ea	Lg	SA	2.00	4.00	525.00	75.30	---	600.30	**714.00**
	Inst	Ea	Sm	SA	2.67	3.00	638.00	101.00	---	739.00	**881.00**
50 gal round	Inst	Ea	Lg	SA	2.00	4.00	588.00	75.30	---	663.30	**787.00**
	Inst	Ea	Sm	SA	2.67	3.00	714.00	101.00	---	815.00	**969.00**

Description	Oper	Unit	Vol	Crew Size	Man-hours per Unit	Crew Output per Day	Avg Mat'l Unit Cost	Avg Labor Unit Cost	Avg Equip Unit Cost	Avg Total Unit Cost	Avg Price Incl O&P
55 gal round	Inst	Ea	Lg	SA	2.00	4.00	588.00	75.30	---	663.30	**787.00**
	Inst	Ea	Sm	SA	2.67	3.00	714.00	101.00	---	815.00	**969.00**
66 gal round	Inst	Ea	Lg	SA	2.67	3.00	798.00	101.00	---	899.00	**1070.00**
	Inst	Ea	Sm	SA	3.56	2.25	969.00	134.00	---	1103.00	**1310.00**
80 gal round	Inst	Ea	Lg	SA	2.67	3.00	888.00	101.00	---	989.00	**1170.00**
	Inst	Ea	Sm	SA	3.56	2.25	1080.00	134.00	---	1214.00	**1440.00**
119 gal round	Inst	Ea	Lg	SA	2.67	3.00	1260.00	101.00	---	1361.00	**1600.00**
	Inst	Ea	Sm	SA	3.56	2.25	1530.00	134.00	---	1664.00	**1960.00**

Standard model, short

6-year Energy Saver, foam insulated

Description	Oper	Unit	Vol	Crew Size	Man-hours per Unit	Crew Output per Day	Avg Mat'l Unit Cost	Avg Labor Unit Cost	Avg Equip Unit Cost	Avg Total Unit Cost	Avg Price Incl O&P
30 gal round	Inst	Ea	Lg	SA	2.00	4.00	493.00	75.30	---	568.30	**677.00**
	Inst	Ea	Sm	SA	2.67	3.00	598.00	101.00	---	699.00	**836.00**
40 gal round	Inst	Ea	Lg	SA	2.00	4.00	525.00	75.30	---	600.30	**714.00**
	Inst	Ea	Sm	SA	2.67	3.00	638.00	101.00	---	739.00	**881.00**
50 gal round	Inst	Ea	Lg	SA	2.00	4.00	435.00	75.30	---	510.30	**611.00**
	Inst	Ea	Sm	SA	2.67	3.00	529.00	101.00	---	630.00	**756.00**

Standard model, low boy, top connection

6-year Energy Saver, foam insulated

Description	Oper	Unit	Vol	Crew Size	Man-hours per Unit	Crew Output per Day	Avg Mat'l Unit Cost	Avg Labor Unit Cost	Avg Equip Unit Cost	Avg Total Unit Cost	Avg Price Incl O&P
6 gal round	Inst	Ea	Lg	SA	1.60	5.00	337.00	60.30	---	397.30	**477.00**
	Inst	Ea	Sm	SA	2.13	3.75	410.00	80.20	---	490.20	**589.00**
15 gal round	Inst	Ea	Lg	SA	1.60	5.00	360.00	60.30	---	420.30	**502.00**
	Inst	Ea	Sm	SA	2.13	3.75	437.00	80.20	---	517.20	**620.00**
19.9 gal round	Inst	Ea	Lg	SA	1.60	5.00	431.00	60.30	---	491.30	**584.00**
	Inst	Ea	Sm	SA	2.13	3.75	524.00	80.20	---	604.20	**720.00**
29 gal round	Inst	Ea	Lg	SA	2.00	4.00	517.00	75.30	---	592.30	**705.00**
	Inst	Ea	Sm	SA	2.67	3.00	627.00	101.00	---	728.00	**869.00**
40 gal round	Inst	Ea	Lg	SA	2.00	4.00	609.00	75.30	---	684.30	**811.00**
	Inst	Ea	Sm	SA	2.67	3.00	740.00	101.00	---	841.00	**998.00**
50 gal round	Inst	Ea	Lg	SA	2.00	4.00	609.00	75.30	---	684.30	**811.00**
	Inst	Ea	Sm	SA	2.67	3.00	740.00	101.00	---	841.00	**998.00**

Standard model, low boy, side connection

6-year Energy Saver, foam insulated

Description	Oper	Unit	Vol	Crew Size	Man-hours per Unit	Crew Output per Day	Avg Mat'l Unit Cost	Avg Labor Unit Cost	Avg Equip Unit Cost	Avg Total Unit Cost	Avg Price Incl O&P
10 gal round	Inst	Ea	Lg	SA	1.60	5.00	365.00	60.30	---	425.30	**509.00**
	Inst	Ea	Sm	SA	2.13	3.75	444.00	80.20	---	524.20	**628.00**
15 gal round	Inst	Ea	Lg	SA	1.60	5.00	353.00	60.30	---	413.30	**494.00**
	Inst	Ea	Sm	SA	2.13	3.75	428.00	80.20	---	508.20	**611.00**
20 gal round	Inst	Ea	Lg	SA	1.60	5.00	417.00	60.30	---	477.30	**568.00**
	Inst	Ea	Sm	SA	2.13	3.75	507.00	80.20	---	587.20	**701.00**
29 gal round	Inst	Ea	Lg	SA	2.00	4.00	452.00	75.30	---	527.30	**631.00**
	Inst	Ea	Sm	SA	2.67	3.00	549.00	101.00	---	650.00	**779.00**
40 gal round	Inst	Ea	Lg	SA	2.00	4.00	480.00	75.30	---	555.30	**663.00**
	Inst	Ea	Sm	SA	2.67	3.00	583.00	101.00	---	684.00	**818.00**

Description	Oper	Unit	Vol	Crew Size	Man-hours per Unit	Crew Output per Day	Avg Mat'l Unit Cost	Avg Labor Unit Cost	Avg Equip Unit Cost	Avg Total Unit Cost	Avg Price Incl O&P

Deluxe model, tall

10-year Energy Saver, foam insulated

Description	Oper	Unit	Vol	Crew Size	MH	Out	Mat'l	Labor	Equip	Total	O&P
30 gal round	Inst	Ea	Lg	SA	2.00	4.00	619.00	75.30	---	694.30	**822.00**
	Inst	Ea	Sm	SA	2.67	3.00	751.00	101.00	---	852.00	**1010.00**
40 gal round	Inst	Ea	Lg	SA	2.00	4.00	651.00	75.30	---	726.30	**859.00**
	Inst	Ea	Sm	SA	2.67	3.00	791.00	101.00	---	892.00	**1060.00**
50 gal round	Inst	Ea	Lg	SA	2.00	4.00	714.00	75.30	---	789.30	**932.00**
	Inst	Ea	Sm	SA	2.67	3.00	867.00	101.00	---	968.00	**1140.00**
55 gal round	Inst	Ea	Lg	SA	2.00	4.00	651.00	75.30	---	726.30	**859.00**
	Inst	Ea	Sm	SA	2.67	3.00	791.00	101.00	---	892.00	**1060.00**
66 gal round	Inst	Ea	Lg	SA	2.67	3.00	924.00	101.00	---	1025.00	**1210.00**
	Inst	Ea	Sm	SA	3.56	2.25	1120.00	134.00	---	1254.00	**1490.00**
80 gal round	Inst	Ea	Lg	SA	2.67	3.00	1010.00	101.00	---	1111.00	**1310.00**
	Inst	Ea	Sm	SA	3.56	2.25	1230.00	134.00	---	1364.00	**1610.00**
119 gal round	Inst	Ea	Lg	SA	2.67	3.00	1270.00	101.00	---	1371.00	**1610.00**
	Inst	Ea	Sm	SA	3.56	2.25	1540.00	134.00	---	1674.00	**1970.00**

Deluxe model, short

10-year Energy Saver, foam insulated

Description	Oper	Unit	Vol	Crew Size	MH	Out	Mat'l	Labor	Equip	Total	O&P
30 gal round	Inst	Ea	Lg	SA	2.00	4.00	563.00	75.30	---	638.30	**758.00**
	Inst	Ea	Sm	SA	2.67	3.00	683.00	101.00	---	784.00	**934.00**
40 gal round	Inst	Ea	Lg	SA	2.00	4.00	592.00	75.30	---	667.30	**792.00**
	Inst	Ea	Sm	SA	2.67	3.00	719.00	101.00	---	820.00	**975.00**

Deluxe model, low boy, top connection

10-year Energy Saver, foam insulated

Description	Oper	Unit	Vol	Crew Size	MH	Out	Mat'l	Labor	Equip	Total	O&P
6 gal round	Inst	Ea	Lg	SA	1.60	5.00	449.00	60.30	---	509.30	**605.00**
	Inst	Ea	Sm	SA	2.13	3.75	546.00	80.20	---	626.20	**745.00**
15 gal round	Inst	Ea	Lg	SA	1.60	5.00	472.00	60.30	---	532.30	**631.00**
	Inst	Ea	Sm	SA	2.13	3.75	573.00	80.20	---	653.20	**777.00**
19.9 gal round	Inst	Ea	Lg	SA	1.60	5.00	543.00	60.30	---	603.30	**713.00**
	Inst	Ea	Sm	SA	2.13	3.75	660.00	80.20	---	740.20	**876.00**
29 gal round	Inst	Ea	Lg	SA	2.00	4.00	584.00	75.30	---	659.30	**782.00**
	Inst	Ea	Sm	SA	2.67	3.00	709.00	101.00	---	810.00	**963.00**
40 gal round	Inst	Ea	Lg	SA	2.00	4.00	613.00	75.30	---	688.30	**816.00**
	Inst	Ea	Sm	SA	2.67	3.00	745.00	101.00	---	846.00	**1000.00**
50 gal round	Inst	Ea	Lg	SA	2.00	4.00	669.00	75.30	---	744.30	**880.00**
	Inst	Ea	Sm	SA	2.67	3.00	813.00	101.00	---	914.00	**1080.00**

Description	Oper	Unit	Vol	Crew Size	Man-hours per Unit	Crew Output per Day	Avg Mat'l Unit Cost	Avg Labor Unit Cost	Avg Equip Unit Cost	Avg Total Unit Cost	Avg Price Incl O&P

Deluxe model, low boy, side connection

10-year Energy Saver, foam insulated

Description	Oper	Unit	Vol	Crew Size	Man-hours per Unit	Crew Output per Day	Avg Mat'l Unit Cost	Avg Labor Unit Cost	Avg Equip Unit Cost	Avg Total Unit Cost	Avg Price Incl O&P
10 gal round	Inst	Ea	Lg	SA	1.60	5.00	354.00	60.30	---	414.30	**496.00**
	Inst	Ea	Sm	SA	2.13	3.75	430.00	80.20	---	510.20	**613.00**
15 gal round	Inst	Ea	Lg	SA	1.60	5.00	368.00	60.30	---	428.30	**512.00**
	Inst	Ea	Sm	SA	2.13	3.75	447.00	80.20	---	527.20	**632.00**
20 gal round	Inst	Ea	Lg	SA	1.60	5.00	392.00	60.30	---	452.30	**539.00**
	Inst	Ea	Sm	SA	2.13	3.75	476.00	80.20	---	556.20	**665.00**
29 gal round	Inst	Ea	Lg	SA	2.00	4.00	449.00	75.30	---	524.30	**628.00**
	Inst	Ea	Sm	SA	2.67	3.00	546.00	101.00	---	647.00	**775.00**
40 gal round	Inst	Ea	Lg	SA	2.00	4.00	472.00	75.30	---	547.30	**653.00**
	Inst	Ea	Sm	SA	2.67	3.00	573.00	101.00	---	674.00	**807.00**

Gas

Standard model, high recovery, tall

6-year Energy Saver

Description	Oper	Unit	Vol	Crew Size	Man-hours per Unit	Crew Output per Day	Avg Mat'l Unit Cost	Avg Labor Unit Cost	Avg Equip Unit Cost	Avg Total Unit Cost	Avg Price Incl O&P
38 gal round, 50,000 btu	Inst	Ea	Lg	SA	4.44	1.80	675.00	167.00	---	842.00	**1020.00**
	Inst	Ea	Sm	SA	6.15	1.30	819.00	232.00	---	1051.00	**1280.00**
48 gal round, 50,000 btu	Inst	Ea	Lg	SA	4.44	1.80	715.00	167.00	---	882.00	**1070.00**
	Inst	Ea	Sm	SA	6.15	1.30	869.00	232.00	---	1101.00	**1340.00**
50 gal round, 65,000 btu	Inst	Ea	Lg	SA	4.71	1.70	752.00	177.00	---	929.00	**1130.00**
	Inst	Ea	Sm	SA	6.67	1.20	913.00	251.00	---	1164.00	**1420.00**
65 gal round, 65,000 btu	Inst	Ea	Lg	SA	4.71	1.70	843.00	177.00	---	1020.00	**1230.00**
	Inst	Ea	Sm	SA	6.67	1.20	1020.00	251.00	---	1271.00	**1550.00**
74 gal round, 75,100 btu	Inst	Ea	Lg	SA	4.71	1.70	993.00	177.00	---	1170.00	**1400.00**
	Inst	Ea	Sm	SA	6.67	1.20	1210.00	251.00	---	1461.00	**1760.00**
98 gal round, 75,100 btu	Inst	Ea	Lg	SA	4.71	1.70	183.00	177.00	---	360.00	**472.00**
	Inst	Ea	Sm	SA	6.67	1.20	223.00	251.00	---	474.00	**625.00**

Standard model, tall

6-year Energy Saver

Description	Oper	Unit	Vol	Crew Size	Man-hours per Unit	Crew Output per Day	Avg Mat'l Unit Cost	Avg Labor Unit Cost	Avg Equip Unit Cost	Avg Total Unit Cost	Avg Price Incl O&P
28 gal round, 40,000 btu	Inst	Ea	Lg	SA	4.44	1.80	615.00	167.00	---	782.00	**953.00**
	Inst	Ea	Sm	SA	6.15	1.30	746.00	232.00	---	978.00	**1200.00**
38 gal round, 40,000 btu	Inst	Ea	Lg	SA	4.44	1.80	640.00	167.00	---	807.00	**982.00**
	Inst	Ea	Sm	SA	6.15	1.30	777.00	232.00	---	1009.00	**1230.00**
50 gal round, 40,000 btu	Inst	Ea	Lg	SA	4.71	1.70	682.00	177.00	---	859.00	**1040.00**
	Inst	Ea	Sm	SA	6.67	1.20	828.00	251.00	---	1079.00	**1320.00**

Standard model, short

6-year Energy Saver

Description	Oper	Unit	Vol	Crew Size	Man-hours per Unit	Crew Output per Day	Avg Mat'l Unit Cost	Avg Labor Unit Cost	Avg Equip Unit Cost	Avg Total Unit Cost	Avg Price Incl O&P
30 gal round, 40,000 btu	Inst	Ea	Lg	SA	4.44	1.80	615.00	167.00	---	782.00	**953.00**
	Inst	Ea	Sm	SA	6.15	1.30	746.00	232.00	---	978.00	**1200.00**
40 gal round, 40,000 btu	Inst	Ea	Lg	SA	4.44	1.80	640.00	167.00	---	807.00	**982.00**
	Inst	Ea	Sm	SA	6.15	1.30	777.00	232.00	---	1009.00	**1230.00**
50 gal round, 40,000 btu	Inst	Ea	Lg	SA	4.71	1.70	682.00	177.00	---	859.00	**1040.00**
	Inst	Ea	Sm	SA	6.67	1.20	828.00	251.00	---	1079.00	**1320.00**

Description	Oper	Unit	Vol	Crew Size	Man-hours per Unit	Crew Output per Day	Avg Mat'l Unit Cost	Avg Labor Unit Cost	Avg Equip Unit Cost	Avg Total Unit Cost	Avg Price Incl O&P
Deluxe model, high recovery, tall											
10-year Energy Saver											
38 gal round, 50,000 btu	Inst	Ea	Lg	SA	4.44	1.80	787.00	167.00	---	954.00	**1150.00**
	Inst	Ea	Sm	SA	6.15	1.30	955.00	232.00	---	1187.00	**1440.00**
48 gal round, 50,000 btu	Inst	Ea	Lg	SA	4.44	1.80	827.00	167.00	---	994.00	**1200.00**
	Inst	Ea	Sm	SA	6.15	1.30	1000.00	232.00	---	1232.00	**1500.00**
50 gal round, 65,000 btu	Inst	Ea	Lg	SA	4.71	1.70	864.00	177.00	---	1041.00	**1250.00**
	Inst	Ea	Sm	SA	6.67	1.20	1050.00	251.00	---	1301.00	**1580.00**
65 gal round, 65,000 btu	Inst	Ea	Lg	SA	4.71	1.70	885.00	177.00	---	1062.00	**1280.00**
	Inst	Ea	Sm	SA	6.67	1.20	1070.00	251.00	---	1321.00	**1600.00**
74 gal round, 75,100 btu	Inst	Ea	Lg	SA	4.71	1.70	1100.00	177.00	---	1277.00	**1530.00**
	Inst	Ea	Sm	SA	6.67	1.20	1340.00	251.00	---	1591.00	**1910.00**
98 gal round, 75,100 btu	Inst	Ea	Lg	SA	4.71	1.70	1700.00	177.00	---	1877.00	**2210.00**
	Inst	Ea	Sm	SA	6.67	1.20	2060.00	251.00	---	2311.00	**2740.00**
Deluxe model, tall											
10-year Energy Saver											
28 gal round, 40,000 btu	Inst	Ea	Lg	SA	4.44	1.80	795.00	167.00	---	962.00	**1160.00**
	Inst	Ea	Sm	SA	6.15	1.30	966.00	232.00	---	1198.00	**1450.00**
38 gal round, 40,000 btu	Inst	Ea	Lg	SA	4.44	1.80	822.00	167.00	---	989.00	**1190.00**
	Inst	Ea	Sm	SA	6.15	1.30	998.00	232.00	---	1230.00	**1490.00**
50 gal round, 40,000 btu	Inst	Ea	Lg	SA	4.71	1.70	864.00	177.00	---	1041.00	**1250.00**
	Inst	Ea	Sm	SA	6.67	1.20	1050.00	251.00	---	1301.00	**1580.00**
Deluxe model, short											
10-year Energy Saver											
30 gal round, 40,000 btu	Inst	Ea	Lg	SA	4.44	1.80	727.00	167.00	---	894.00	**1080.00**
	Inst	Ea	Sm	SA	6.15	1.30	882.00	232.00	---	1114.00	**1360.00**
40 gal round, 40,000 btu	Inst	Ea	Lg	SA	4.44	1.80	752.00	167.00	---	919.00	**1110.00**
	Inst	Ea	Sm	SA	6.15	1.30	913.00	232.00	---	1145.00	**1390.00**
50 gal round, 40,000 btu	Inst	Ea	Lg	SA	4.71	1.70	794.00	177.00	---	971.00	**1170.00**
	Inst	Ea	Sm	SA	6.67	1.20	964.00	251.00	---	1215.00	**1480.00**
Optional accessories											
Water heater cabinets											
24" x 24" x 72"											
heaters up to 40 gal	---	Ea	Lg	---	---	---	105.00	---	---	105.00	**121.00**
	---	Ea	Sm	---	---	---	128.00	---	---	128.00	**147.00**
30" x 30" x 72"											
heaters up to 75 gal	---	Ea	Lg	---	---	---	130.00	---	---	130.00	**150.00**
	---	Ea	Sm	---	---	---	158.00	---	---	158.00	**182.00**
36" x 36" x 83"											
heaters up to 100 gal	---	Ea	Lg	---	---	---	259.00	---	---	259.00	**298.00**
	---	Ea	Sm	---	---	---	315.00	---	---	315.00	**362.00**

Description	Oper	Unit	Vol	Crew Size	Man-hours per Unit	Crew Output per Day	Avg Mat'l Unit Cost	Avg Labor Unit Cost	Avg Equip Unit Cost	Avg Total Unit Cost	Avg Price Incl O&P
Water heater pans, by diameter											
20" pan (up to 18" unit)	---	Ea	Lg	---	---	---	14.60	---	---	14.60	**16.70**
	---	Ea	Sm	---	---	---	17.70	---	---	17.70	**20.30**
24" pan (up to 22" unit)	---	Ea	Lg	---	---	---	17.40	---	---	17.40	**20.00**
	---	Ea	Sm	---	---	---	21.10	---	---	21.10	**24.20**
26" pan (up to 24" unit)	---	Ea	Lg	---	---	---	21.10	---	---	21.10	**24.30**
	---	Ea	Sm	---	---	---	25.70	---	---	25.70	**29.50**
28" pan (up to 26" unit)	---	Ea	Lg	---	---	---	24.50	---	---	24.50	**28.20**
	---	Ea	Sm	---	---	---	29.80	---	---	29.80	**34.20**
Water heater stands											
22" x 22", heaters up to 52 gal	---	Ea	Lg	---	---	---	50.40	---	---	50.40	**58.00**
	---	Ea	Sm	---	---	---	61.20	---	---	61.20	**70.40**
26" x 26", heaters up to 75 gal	---	Ea	Lg	---	---	---	71.40	---	---	71.40	**82.10**
	---	Ea	Sm	---	---	---	86.70	---	---	86.70	**99.70**
Free-standing restraint system											
75 gal capacity	---	Ea	Lg	---	---	---	225.00	---	---	225.00	**259.00**
	---	Ea	Sm	---	---	---	274.00	---	---	274.00	**315.00**
100 gal capacity	---	Ea	Lg	---	---	---	240.00	---	---	240.00	**276.00**
	---	Ea	Sm	---	---	---	291.00	---	---	291.00	**335.00**
Wall mount platform restraint system											
75 gal capacity	---	Ea	Lg	---	---	---	194.00	---	---	194.00	**223.00**
	---	Ea	Sm	---	---	---	236.00	---	---	236.00	**271.00**
100 gal capacity	---	Ea	Lg	---	---	---	215.00	---	---	215.00	**247.00**
	---	Ea	Sm	---	---	---	261.00	---	---	261.00	**300.00**

Solar water heating systems

Complete closed loop solar system with solar electric water heater with exchanger, differential control, heating element, circulator, collector and tank, and panel sensors. The material costs given include the subcontractor's overhead and profit

82-gallon capacity collector

Description	Oper	Unit	Vol	Crew Size	Man-hours per Unit	Crew Output per Day	Avg Mat'l Unit Cost	Avg Labor Unit Cost	Avg Equip Unit Cost	Avg Total Unit Cost	Avg Price Incl O&P
One deluxe collector	---	LS	Lg	---	---	---	2750.00	---	---	2750.00	---
	---	LS	Sm	---	---	---	3140.00	---	---	3140.00	---
Two deluxe collectors											
Standard collectors	---	LS	Lg	---	---	---	3410.00	---	---	3410.00	---
	---	LS	Sm	---	---	---	3880.00	---	---	3880.00	---
Deluxe collectors	---	LS	Lg	---	---	---	3510.00	---	---	3510.00	---
	---	LS	Sm	---	---	---	4000.00	---	---	4000.00	---
Three collectors											
Economy collectors	---	LS	Lg	---	---	---	4000.00	---	---	4000.00	---
	---	LS	Sm	---	---	---	4560.00	---	---	4560.00	---
Standard collectors	---	LS	Lg	---	---	---	4100.00	---	---	4100.00	---
	---	LS	Sm	---	---	---	4670.00	---	---	4670.00	---
Four collectors											
Economy collectors	---	LS	Lg	---	---	---	4320.00	---	---	4320.00	---
	---	LS	Sm	---	---	---	4920.00	---	---	4920.00	---
Standard collectors	---	LS	Lg	---	---	---	4430.00	---	---	4430.00	---
	---	LS	Sm	---	---	---	5040.00	---	---	5040.00	---

Description	Oper	Unit	Vol	Crew Size	Man-hours per Unit	Crew Output per Day	Avg Mat'l Unit Cost	Avg Labor Unit Cost	Avg Equip Unit Cost	Avg Total Unit Cost	Avg Price Incl O&P
120-gallon capacity system											
Three collectors											
Economy collectors	---	LS	Lg	---	---	---	4960.00	---	---	4960.00	---
	---	LS	Sm	---	---	---	5650.00	---	---	5650.00	---
Standard collectors	---	LS	Lg	---	---	---	5080.00	---	---	5080.00	---
	---	LS	Sm	---	---	---	5780.00	---	---	5780.00	---
Four standard collectors											
	---	LS	Lg	---	---	---	5940.00	---	---	5940.00	---
	---	LS	Sm	---	---	---	6770.00	---	---	6770.00	---
Five collectors											
Economy collectors	---	LS	Lg	---	---	---	6480.00	---	---	6480.00	---
	---	LS	Sm	---	---	---	7380.00	---	---	7380.00	---
Standard collectors	---	LS	Lg	---	---	---	6590.00	---	---	6590.00	---
	---	LS	Sm	---	---	---	7500.00	---	---	7500.00	---
Six collectors											
Economy collectors	---	LS	Lg	---	---	---	7020.00	---	---	7020.00	---
	---	LS	Sm	---	---	---	8000.00	---	---	8000.00	---
Standard collectors	---	LS	Lg	---	---	---	7130.00	---	---	7130.00	---
	---	LS	Sm	---	---	---	8120.00	---	---	8120.00	---
Material adjustments											
Collector mounting kits, one required per collector panel											
Adjustable position hinge	---	Ea	Lg	---	---	---	61.60	---	---	61.60	---
	---	Ea	Sm	---	---	---	70.10	---	---	70.10	---
Integral flange	---	Ea	Lg	---	---	---	5.40	---	---	5.40	---
	---	Ea	Sm	---	---	---	6.15	---	---	6.15	---
Additional panels											
Economy collectors	---	Ea	Lg	---	---	---	540.00	---	---	540.00	---
	---	Ea	Sm	---	---	---	615.00	---	---	615.00	---
Standard collectors	---	Ea	Lg	---	---	---	648.00	---	---	648.00	---
	---	Ea	Sm	---	---	---	738.00	---	---	738.00	---
Deluxe collectors	---	Ea	Lg	---	---	---	756.00	---	---	756.00	---
	---	Ea	Sm	---	---	---	861.00	---	---	861.00	---

Description	Oper	Unit	Vol	Crew Size	Man-hours per Unit	Crew Output per Day	Avg Mat'l Unit Cost	Avg Labor Unit Cost	Avg Equip Unit Cost	Avg Total Unit Cost	Avg Price Incl O&P
Individual components											
Solar storage tanks, glass lined with fiberglass insulation											
66 gallon	---	Ea	Lg	---	---	---	643.00	---	---	643.00	---
	---	Ea	Sm	---	---	---	732.00	---	---	732.00	---
82 gallon	---	Ea	Lg	---	---	---	729.00	---	---	729.00	---
	---	Ea	Sm	---	---	---	830.00	---	---	830.00	---
120 gallon	---	Ea	Lg	---	---	---	1110.00	---	---	1110.00	---
	---	Ea	Sm	---	---	---	1260.00	---	---	1260.00	---
Solar electric water heaters, glass lined with fiberglass insulation, heating element with thermostat											
66 gallon, 4.5 KW	---	Ea	Lg	---	---	---	642.00	---	---	642.00	---
	---	Ea	Sm	---	---	---	731.00	---	---	731.00	---
82 gallon, 4.5 KW	---	Ea	Lg	---	---	---	738.00	---	---	738.00	---
	---	Ea	Sm	---	---	---	840.00	---	---	840.00	---
120 gallon, 4.5 KW	---	Ea	Lg	---	---	---	879.00	---	---	879.00	---
	---	Ea	Sm	---	---	---	1000.00	---	---	1000.00	---
Additional element	---	Ea	Lg	---	---	---	35.60	---	---	35.60	---
	---	Ea	Sm	---	---	---	40.60	---	---	40.60	---
Solar electric water heaters with heat exchangers, glass lined with fiberglass insulation, two copper exchangers, powered circulator, differential control, adjustable thermostat											
82 gallon, 4.5 KW, 1/20 HP	---	Ea	Lg	---	---	---	959.00	---	---	959.00	---
	---	Ea	Sm	---	---	---	1090.00	---	---	1090.00	---
120 gallon, 4.5 KW, 1/20 HP	---	Ea	Lg	---	---	---	1260.00	---	---	1260.00	---
	---	Ea	Sm	---	---	---	1430.00	---	---	1430.00	---

Description	Oper	Unit	Vol	Crew Size	Man-hours per Unit	Crew Output per Day	Avg Mat'l Unit Cost	Avg Labor Unit Cost	Avg Equip Unit Cost	Avg Total Unit Cost	Avg Price Incl O&P

Water softeners
(Bruner/Calgon)

Automatic water softeners, complete with yoke with 3/4" I.P.S. supply

Single-tank units

Description	Oper	Unit	Vol	Crew Size	Man-hours per Unit	Crew Output per Day	Avg Mat'l Unit Cost	Avg Labor Unit Cost	Avg Equip Unit Cost	Avg Total Unit Cost	Avg Price Incl O&P
8,000 grain exchange capacity, 160 lbs salt storage, 6 gpm											
	---	Ea	Lg	---	---	450.0	410.00	---	---	410.00	**471.00**
	---	Ea	Sm	---	---	432.0	522.00	---	---	522.00	**600.00**
15,000 grain exchange capacity, 200 lbs salt storage, 8.8 gpm											
	---	Ea	Lg	---	---	495.0	450.00	---	---	450.00	**518.00**
	---	Ea	Sm	---	---	476.0	574.00	---	---	574.00	**660.00**
25,000 grain exchange capacity, 175 lbs salt storage, 11.3 gpm											
	---	Ea	Lg	---	---	690.0	628.00	---	---	628.00	**722.00**
	---	Ea	Sm	---	---	677.0	800.00	---	---	800.00	**920.00**

Two-tank units, side-by-side

Description	Oper	Unit	Vol	Crew Size	Man-hours per Unit	Crew Output per Day	Avg Mat'l Unit Cost	Avg Labor Unit Cost	Avg Equip Unit Cost	Avg Total Unit Cost	Avg Price Incl O&P
30,000 grain exchange capacity, 1.0 CF											
	---	Ea	Lg	---	---	790.0	719.00	---	---	719.00	**827.00**
	---	Ea	Sm	---	---	590.0	916.00	---	---	916.00	**1050.00**
45,000 grain exchange capacity, 1.5 CF											
	---	Ea	Lg	---	---	890.0	810.00	---	---	810.00	**931.00**
	---	Ea	Sm	---	---	674.0	1030.00	---	---	1030.00	**1190.00**
60,000 grain exchange capacity, 2.0 CF											
	---	Ea	Lg	---	---	1290	1170.00	---	---	1170.00	**1350.00**
	---	Ea	Sm	---	---	---	1500.00	---	---	1500.00	**1720.00**
90,000 grain exchange capacity, 3.0 CF											
	---	Ea	Lg	---	---	1390	1260.00	---	---	1260.00	**1450.00**
	---	Ea	Sm	---	---	---	1610.00	---	---	1610.00	**1850.00**

Automatic water filters, complete with yoke and media with 3/4" I.P.S. supply

Two-tank units, side-by-side

Description	Oper	Unit	Vol	Crew Size	Man-hours per Unit	Crew Output per Day	Avg Mat'l Unit Cost	Avg Labor Unit Cost	Avg Equip Unit Cost	Avg Total Unit Cost	Avg Price Incl O&P
Eliminate rotten egg odor and rust											
	---	Ea	Lg	---	---	547.0	498.00	---	---	498.00	**572.00**
	---	Ea	Sm	---	---	547.0	635.00	---	---	635.00	**730.00**
Eliminate chlorine taste	---	Ea	Lg	---	---	460.0	419.00	---	---	419.00	**481.00**
	---	Ea	Sm	---	---	460.0	534.00	---	---	534.00	**614.00**
Clear up water discoloration	---	Ea	Lg	---	---	412.0	375.00	---	---	375.00	**431.00**
	---	Ea	Sm	---	---	412.0	478.00	---	---	478.00	**550.00**
Clear up corroding pipes	---	Ea	Lg	---	---	428.0	389.00	---	---	389.00	**448.00**
	---	Ea	Sm	---	---	428.0	496.00	---	---	496.00	**571.00**

Description	Oper	Unit	Vol	Crew Size	Man-hours per Unit	Crew Output per Day	Avg Mat'l Unit Cost	Avg Labor Unit Cost	Avg Equip Unit Cost	Avg Total Unit Cost	Avg Price Incl O&P
Manual water filters, complete fiberglass tank with 3/4" pipe supplies with mineral packs											
Eliminate taste and odor	---	Ea	Lg	---	---	408.0	371.00	---	---	371.00	**427.00**
	---	Ea	Sm	---	---	408.0	473.00	---	---	473.00	**544.00**
Eliminate acid water	---	Ea	Lg	---	---	347.0	316.00	---	---	316.00	**363.00**
	---	Ea	Sm	---	---	347.0	403.00	---	---	403.00	**463.00**
Eliminate iron in solution	---	Ea	Lg	---	---	420.0	382.00	---	---	382.00	**440.00**
	---	Ea	Sm	---	---	420.0	487.00	---	---	487.00	**560.00**
Eliminate sediment	---	Ea	Lg	---	---	347.0	316.00	---	---	316.00	**363.00**
	---	Ea	Sm	---	---	347.0	403.00	---	---	403.00	**463.00**
Chemical feed pumps											
115 volt, 9 gal/day	---	Ea	Lg	---	---	242.0	220.00	---	---	220.00	**253.00**
	---	Ea	Sm	---	---	242.0	281.00	---	---	281.00	**323.00**
230 volt, 9 gal/day	---	Ea	Lg	---	---	277.0	252.00	---	---	252.00	**290.00**
	---	Ea	Sm	---	---	277.0	321.00	---	---	321.00	**370.00**

Weathervanes. See Cupolas, page 108

Windows

Windows, with related trim and frame

Description	Oper	Unit	Vol	Crew Size	Man-hours per Unit	Crew Output per Day	Avg Mat'l Unit Cost	Avg Labor Unit Cost	Avg Equip Unit Cost	Avg Total Unit Cost	Avg Price Incl O&P
To 12 SF											
Aluminum	Demo	Ea	Lg	LB	.762	21.00	---	20.90	---	20.90	**31.10**
	Demo	Ea	Sm	LB	1.17	13.70	---	32.10	---	32.10	**47.80**
Wood	Demo	Ea	Lg	LB	1.00	16.00	---	27.40	---	27.40	**40.90**
	Demo	Ea	Sm	LB	1.54	10.40	---	42.20	---	42.20	**62.90**
13 SF to 50 SF											
Aluminum	Demo	Ea	Lg	LB	1.23	13.00	---	33.70	---	33.70	**50.30**
	Demo	Ea	Sm	LB	1.88	8.50	---	51.60	---	51.60	**76.80**
Wood	Demo	Fa	Lg	LB	1.60	10.00	---	43.90	---	43.90	**65.40**
	Demo	Ea	Sm	LB	2.46	6.50	---	67.50		67.50	**101.00**

Aluminum

Vertical slide, satin anodized finish, includes screen and hardware

Description	Oper	Unit	Vol	Crew Size	Man-hours per Unit	Crew Output per Day	Avg Mat'l Unit Cost	Avg Labor Unit Cost	Avg Equip Unit Cost	Avg Total Unit Cost	Avg Price Incl O&P
Single glazed											
1'-6" x 3'-0" H	Inst	Set	Lg	CA	1.14	7.00	83.80	37.90	---	121.70	**153.00**
	Inst	Set	Sm	CA	1.74	4.60	90.40	57.90	---	148.30	**191.00**
2'-0" x 2'-0" H	Inst	Set	Lg	CA	1.14	7.00	76.20	37.90	---	114.10	**145.00**
	Inst	Set	Sm	CA	1.74	4.60	82.20	57.90	---	140.10	**181.00**
2'-0" x 2'-6" H	Inst	Set	Lg	CA	1.14	7.00	86.40	37.90	---	124.30	**156.00**
	Inst	Set	Sm	CA	1.74	4.60	93.20	57.90	---	151.10	**194.00**
2'-0" x 3'-0" H	Inst	Set	Lg	CA	1.23	6.50	92.70	40.90	---	133.60	**168.00**
	Inst	Set	Sm	CA	1.90	4.20	100.00	63.20	---	163.20	**210.00**
2'-0" x 3'-6" H	Inst	Set	Lg	CA	1.23	6.50	99.10	40.90	---	140.00	**175.00**
	Inst	Set	Sm	CA	1.90	4.20	107.00	63.20	---	170.20	**218.00**
2'-0" x 4'-0" H	Inst	Set	Lg	CA	1.23	6.50	104.00	40.90	---	144.90	**181.00**
	Inst	Set	Sm	CA	1.90	4.20	112.00	63.20	---	175.20	**224.00**

Description	Oper	Unit	Vol	Crew Size	Man-hours per Unit	Crew Output per Day	Avg Mat'l Unit Cost	Avg Labor Unit Cost	Avg Equip Unit Cost	Avg Total Unit Cost	Avg Price Incl O&P
2'-0" x 4'-6" H	Inst	Set	Lg	CA	1.23	6.50	112.00	40.90	---	152.90	**190.00**
	Inst	Set	Sm	CA	1.90	4.20	121.00	63.20	---	184.20	**233.00**
2'-0" x 5'-0" H	Inst	Set	Lg	CA	1.33	6.00	117.00	44.30	---	161.30	**201.00**
	Inst	Set	Sm	CA	2.05	3.90	126.00	68.20	---	194.20	**247.00**
2'-0" x 6'-0" H	Inst	Set	Lg	CA	1.33	6.00	128.00	44.30	---	172.30	**214.00**
	Inst	Set	Sm	CA	2.05	3.90	138.00	68.20	---	206.20	**261.00**
2'-6" x 3'-0" H	Inst	Set	Lg	CA	1.23	6.50	102.00	40.90	---	142.90	**178.00**
	Inst	Set	Sm	CA	1.90	4.20	110.00	63.20	---	173.20	**221.00**
2'-6" x 3'-6" H	Inst	Set	Lg	CA	1.23	6.50	108.00	40.90	---	148.90	**186.00**
	Inst	Set	Sm	CA	1.90	4.20	116.00	63.20	---	179.20	**229.00**
2'-6" x 4'-0" H	Inst	Set	Lg	CA	1.23	6.50	114.00	40.90	---	154.90	**193.00**
	Inst	Set	Sm	CA	1.90	4.20	123.00	63.20	---	186.20	**237.00**
2'-6" x 4'-6" H	Inst	Set	Lg	CA	1.23	6.50	123.00	40.90	---	163.90	**203.00**
	Inst	Set	Sm	CA	1.90	4.20	133.00	63.20	---	196.20	**248.00**
2'-6" x 5'-0" H	Inst	Set	Lg	CA	1.33	6.00	128.00	44.30	---	172.30	**214.00**
	Inst	Set	Sm	CA	2.05	3.90	138.00	68.20	---	206.20	**261.00**
2'-6" x 6'-0" H	Inst	Set	Lg	CA	1.33	6.00	141.00	44.30	---	185.30	**229.00**
	Inst	Set	Sm	CA	2.05	3.90	152.00	68.20	---	220.20	**277.00**
3'-0" x 2'-0" H	Inst	Set	Lg	CA	1.23	6.50	97.80	40.90	---	138.70	**174.00**
	Inst	Set	Sm	CA	1.90	4.20	105.00	63.20	---	168.20	**216.00**
3'-0" x 3'-0" H	Inst	Set	Lg	CA	1.23	6.50	109.00	40.90	---	149.90	**187.00**
	Inst	Set	Sm	CA	1.90	4.20	118.00	63.20	---	181.20	**230.00**
3'-0" x 3'-6" H	Inst	Set	Lg	CA	1.23	6.50	117.00	40.90	---	157.90	**196.00**
	Inst	Set	Sm	CA	1.90	4.20	126.00	63.20	---	189.20	**240.00**
3'-0" x 4'-0" H	Inst	Set	Lg	CA	1.23	6.50	124.00	40.90	---	164.90	**205.00**
	Inst	Set	Sm	CA	1.90	4.20	134.00	63.20	---	197.20	**249.00**
3'-0" x 4'-6" H	Inst	Set	Lg	CA	1.23	6.50	135.00	40.90	---	175.90	**216.00**
	Inst	Set	Sm	CA	1.90	4.20	145.00	63.20	---	208.20	**262.00**
3'-0" x 5'-0" H	Inst	Set	Lg	CJ	2.76	5.80	140.00	83.80	---	223.80	**286.00**
	Inst	Set	Sm	CJ	4.21	3.80	151.00	128.00	---	279.00	**365.00**
3'-0" x 6'-0" H	Inst	Set	Lg	CJ	3.08	5.20	154.00	93.50	---	247.50	**317.00**
	Inst	Set	Sm	CJ	4.71	3.40	166.00	143.00	---	309.00	**405.00**
3'-6" x 3'-6" H	Inst	Set	Lg	CA	1.23	6.50	126.00	40.90	---	166.90	**206.00**
	Inst	Set	Sm	CA	1.90	4.20	136.00	63.20	---	199.20	**251.00**
3'-6" x 4'-0" H	Inst	Set	Lg	CA	1.45	5.50	135.00	48.20	---	183.20	**227.00**
	Inst	Set	Sm	CA	2.22	3.60	145.00	73.90	---	218.90	**278.00**
3'-6" x 4'-6" H	Inst	Set	Lg	CA	1.78	4.50	147.00	59.20	---	206.20	**258.00**
	Inst	Set	Sm	CA	2.76	2.90	159.00	91.80	---	250.80	**321.00**
3'-6" x 5'-0" H	Inst	Set	Lg	CJ	3.08	5.20	151.00	93.50	---	244.50	**314.00**
	Inst	Set	Sm	CJ	4.71	3.40	163.00	143.00	---	306.00	**402.00**
4'-0" x 3'-0" H	Inst	Set	Lg	CA	1.23	6.50	126.00	40.90	---	166.90	**206.00**
	Inst	Set	Sm	CA	1.90	4.20	136.00	63.20	---	199.20	**251.00**
4'-0" x 3'-6" H	Inst	Set	Lg	CA	1.45	5.50	135.00	48.20	---	183.20	**227.00**
	Inst	Set	Sm	CA	2.22	3.60	145.00	73.90	---	218.90	**278.00**
4'-0" x 4'-0" H	Inst	Set	Lg	CJ	2.76	5.80	144.00	83.80	---	227.80	**291.00**
	Inst	Set	Sm	CJ	4.21	3.80	155.00	128.00	---	283.00	**370.00**
4'-0" x 4'-6" H	Inst	Set	Lg	CJ	2.76	5.80	152.00	83.80	---	235.80	**301.00**
	Inst	Set	Sm	CJ	4.21	3.80	164.00	128.00	---	292.00	**381.00**
4'-0" x 5'-0" H	Inst	Set	Lg	CJ	3.08	5.20	163.00	93.50	---	256.50	**327.00**
	Inst	Set	Sm	CJ	4.71	3.40	175.00	143.00	---	318.00	**416.00**

Description	Oper	Unit	Vol	Crew Size	Man-hours per Unit	Crew Output per Day	Avg Mat'l Unit Cost	Avg Labor Unit Cost	Avg Equip Unit Cost	Avg Total Unit Cost	Avg Price Incl O&P
Dual glazed											
1'-6" x 3'-0" H	Inst	Set	Lg	CA	1.18	6.80	113.00	39.30	---	152.30	**189.00**
	Inst	Set	Sm	CA	1.82	4.40	122.00	60.60	---	182.60	**231.00**
2'-0" x 2'-6" H	Inst	Set	Lg	CA	1.18	6.80	117.00	39.30	---	156.30	**193.00**
	Inst	Set	Sm	CA	1.82	4.40	126.00	60.60	---	186.60	**236.00**
2'-0" x 3'-0" H	Inst	Set	Lg	CA	1.27	6.30	124.00	42.30	---	166.30	**207.00**
	Inst	Set	Sm	CA	1.95	4.10	134.00	64.90	---	198.90	**252.00**
2'-0" x 3'-6" H	Inst	Set	Lg	CA	1.27	6.30	133.00	42.30	---	175.30	**217.00**
	Inst	Set	Sm	CA	1.95	4.10	144.00	64.90	---	208.90	**263.00**
2'-0" x 4'-0" H	Inst	Set	Lg	CA	1.27	6.30	141.00	42.30	---	183.30	**226.00**
	Inst	Set	Sm	CA	1.95	4.10	152.00	64.90	---	216.90	**272.00**
2'-0" x 4'-6" H	Inst	Set	Lg	CA	1.27	6.30	155.00	42.30	---	197.30	**242.00**
	Inst	Set	Sm	CA	1.95	4.10	167.00	64.90	---	231.90	**290.00**
2'-0" x 5'-0" H	Inst	Set	Lg	CA	1.38	5.80	166.00	45.90	---	211.90	**260.00**
	Inst	Set	Sm	CA	2.11	3.80	179.00	70.20	---	249.20	**312.00**
2'-0" x 6'-0" H	Inst	Set	Lg	CA	1.38	5.80	183.00	45.90	---	228.90	**279.00**
	Inst	Set	Sm	CA	2.11	3.80	197.00	70.20	---	267.20	**332.00**
2'-6" x 3'-0" H	Inst	Set	Lg	CA	1.27	6.30	137.00	42.30	---	179.30	**221.00**
	Inst	Set	Sm	CA	1.95	4.10	148.00	64.90	---	212.90	**267.00**
2'-6" x 3'-6" H	Inst	Set	Lg	CA	1.27	6.30	146.00	42.30	---	188.30	**231.00**
	Inst	Set	Sm	CA	1.95	4.10	158.00	64.90	---	222.90	**279.00**
2'-6" x 4'-0" H	Inst	Set	Lg	CA	1.27	6.30	156.00	42.30	---	198.30	**243.00**
	Inst	Set	Sm	CA	1.95	4.10	169.00	64.90	---	233.90	**291.00**
2'-6" x 4'-6" H	Inst	Set	Lg	CA	1.27	6.30	170.00	42.30	---	212.30	**259.00**
	Inst	Set	Sm	CA	1.95	4.10	184.00	64.90	---	248.90	**308.00**
2'-6" x 5'-0" H	Inst	Set	Lg	CA	1.38	5.80	183.00	45.90	---	228.90	**279.00**
	Inst	Set	Sm	CA	2.11	3.80	197.00	70.20	---	267.20	**332.00**
2'-6" x 6'-0" H	Inst	Set	Lg	CA	1.38	5.80	202.00	45.90	---	247.90	**301.00**
	Inst	Set	Sm	CA	2.11	3.80	218.00	70.20	---	288.20	**356.00**
3'-0" x 3'-0" H	Inst	Set	Lg	CA	1.27	6.30	149.00	42.30	---	191.30	**234.00**
	Inst	Set	Sm	CA	1.95	4.10	160.00	64.90	---	224.90	**282.00**
3'-0" x 3'-6" H	Inst	Set	Lg	CA	1.27	6.30	160.00	42.30	---	202.30	**247.00**
	Inst	Set	Sm	CA	1.95	4.10	173.00	64.90	---	237.90	**296.00**
3'-0" x 4'-0" H	Inst	Set	Lg	CA	1.27	6.30	170.00	42.30	---	212.30	**259.00**
	Inst	Set	Sm	CA	1.95	4.10	184.00	64.90	---	248.90	**308.00**
3'-0" x 4'-6" H	Inst	Set	Lg	CA	1.27	6.30	187.00	42.30	---	229.30	**278.00**
	Inst	Set	Sm	CA	1.95	4.10	201.00	64.90	---	265.90	**329.00**
3'-0" x 5'-0" H	Inst	Set	Lg	CJ	2.86	5.60	199.00	86.80	---	285.80	**360.00**
	Inst	Set	Sm	CJ	4.44	3.60	215.00	135.00	---	350.00	**450.00**
3'-0" x 6'-0" H	Inst	Set	Lg	CJ	3.20	5.00	220.00	97.10	---	317.10	**398.00**
	Inst	Set	Sm	CJ	4.85	3.30	237.00	147.00	---	384.00	**493.00**
3'-6" x 3'-6" H	Inst	Set	Lg	CA	1.27	6.30	179.00	42.30	---	221.30	**269.00**
	Inst	Set	Sm	CA	1.95	4.10	193.00	64.90	---	257.90	**319.00**
3'-6" x 4'-0" H	Inst	Set	Lg	CA	1.51	5.30	185.00	50.20	---	235.20	**289.00**
	Inst	Set	Sm	CA	2.35	3.40	200.00	78.20	---	278.20	**347.00**
3'-6" x 4'-6" H	Inst	Set	Lg	CA	1.86	4.30	203.00	61.90	---	264.90	**327.00**
	Inst	Set	Sm	CA	2.86	2.80	219.00	95.20	---	314.20	**395.00**
3'-6" x 5'-0" H	Inst	Set	Lg	CJ	3.20	5.00	215.00	97.10	---	312.10	**393.00**
	Inst	Set	Sm	CJ	4.85	3.30	232.00	147.00	---	379.00	**487.00**

Description	Oper	Unit	Vol	Crew Size	Man-hours per Unit	Crew Output per Day	Avg Mat'l Unit Cost	Avg Labor Unit Cost	Avg Equip Unit Cost	Avg Total Unit Cost	Avg Price Incl O&P
4'-0" x 3'-0" H	Inst	Set	Lg	CA	1.27	6.30	173.00	42.30	---	215.30	**262.00**
	Inst	Set	Sm	CA	1.95	4.10	186.00	64.90	---	250.90	**312.00**
4'-0" x 3'-6" H	Inst	Set	Lg	CA	1.51	5.30	187.00	50.20	---	237.20	**290.00**
	Inst	Set	Sm	CA	2.35	3.40	201.00	78.20	---	279.20	**349.00**
4'-0" x 4'-0" H	Inst	Set	Lg	CJ	2.86	5.60	199.00	86.80	---	285.80	**360.00**
	Inst	Set	Sm	CJ	4.44	3.60	215.00	135.00	---	350.00	**450.00**
4'-0" x 4'-6" H	Inst	Set	Lg	CJ	2.86	5.60	222.00	86.80	---	308.80	**386.00**
	Inst	Set	Sm	CJ	4.44	3.60	240.00	135.00	---	375.00	**478.00**
4'-0" x 5'-0" H	Inst	Set	Lg	CJ	3.20	5.00	241.00	97.10	---	338.10	**423.00**
	Inst	Set	Sm	CJ	4.85	3.30	260.00	147.00	---	407.00	**520.00**
For bronze finish ADD	Inst	%	Lg	---	---	---	12.0	---	---	---	**---**
	Inst	%	Sm	---	---	---	12.0	---	---	---	**---**

Horizontal slide, satin anodized finish, includes screen and hardware

1 sliding lite, 1 fixed lite
Single glazed

Description	Oper	Unit	Vol	Crew Size	Man-hours per Unit	Crew Output per Day	Avg Mat'l Unit Cost	Avg Labor Unit Cost	Avg Equip Unit Cost	Avg Total Unit Cost	Avg Price Incl O&P
2'-0" x 2'-0" H	Inst	Set	Lg	CA	1.23	6.50	55.90	40.90	---	96.80	**126.00**
	Inst	Set	Sm	CA	1.90	4.20	60.30	63.20	---	123.50	**164.00**
2'-0" x 3'-0" H	Inst	Set	Lg	CA	1.23	6.50	67.30	40.90	---	108.20	**139.00**
	Inst	Set	Sm	CA	1.90	4.20	72.60	63.20	---	135.80	**178.00**
2'-6" x 3'-0" H	Inst	Set	Lg	CA	1.23	6.50	72.40	40.90	---	113.30	**145.00**
	Inst	Set	Sm	CA	1.90	4.20	78.10	63.20	---	141.30	**185.00**
3'-0" x 1'-0" H	Inst	Set	Lg	CA	1.23	6.50	52.10	40.90	---	93.00	**121.00**
	Inst	Set	Sm	CA	1.90	4.20	56.20	63.20	---	119.40	**159.00**
3'-0" x 1'-6" H	Inst	Set	Lg	CA	1.23	6.50	58.40	40.90	---	99.30	**129.00**
	Inst	Set	Sm	CA	1.90	4.20	63.00	63.20	---	126.20	**167.00**
3'-0" x 2'-0" H	Inst	Set	Lg	CA	1.23	6.50	66.00	40.90	---	106.90	**137.00**
	Inst	Set	Sm	CA	1.90	4.20	71.20	63.20	---	134.40	**177.00**
3'-0" x 2'-6" H	Inst	Set	Lg	CA	1.23	6.50	72.40	40.90	---	113.30	**145.00**
	Inst	Set	Sm	CA	1.90	4.20	78.10	63.20	---	141.30	**185.00**
3'-0" x 3'-0" H	Inst	Set	Lg	CA	1.23	6.50	78.70	40.90	---	119.60	**152.00**
	Inst	Set	Sm	CA	1.90	4.20	84.90	63.20	---	148.10	**193.00**
3'-0" x 3'-6" H	Inst	Set	Lg	CA	1.45	5.50	85.10	48.20	---	133.30	**170.00**
	Inst	Set	Sm	CA	2.22	3.60	91.80	73.90	---	165.70	**216.00**
3'-0" x 4'-0" H	Inst	Set	Lg	CA	1.78	4.50	91.40	59.20	---	150.60	**194.00**
	Inst	Set	Sm	CA	2.76	2.90	98.60	91.80	---	190.40	**251.00**
3'-0" x 5'-0" H	Inst	Set	Lg	CJ	4.10	3.90	105.00	124.00	---	229.00	**308.00**
	Inst	Set	Sm	CJ	6.40	2.50	114.00	194.00	---	308.00	**422.00**
3'-6" x 2'-0" H	Inst	Set	Lg	CA	1.23	6.50	69.90	40.90	---	110.80	**142.00**
	Inst	Set	Sm	CA	1.90	4.20	75.40	63.20	---	138.60	**181.00**
3'-6" x 2'-6" H	Inst	Set	Lg	CA	1.23	6.50	77.50	40.90	---	118.40	**150.00**
	Inst	Set	Sm	CA	1.90	4.20	83.60	63.20	---	146.80	**191.00**
3'-6" x 3'-0" H	Inst	Set	Lg	CA	1.45	5.50	83.80	48.20	---	132.00	**169.00**
	Inst	Set	Sm	CA	2.22	3.60	90.40	73.90	---	164.30	**215.00**
3'-6" x 3'-6" H	Inst	Set	Lg	CA	1.45	5.50	91.40	48.20	---	139.60	**178.00**
	Inst	Set	Sm	CA	2.22	3.60	98.60	73.90	---	172.50	**224.00**
3'-6" x 4'-0" H	Inst	Set	Lg	CA	1.78	4.50	97.80	59.20	---	157.00	**201.00**
	Inst	Set	Sm	CA	2.76	2.90	105.00	91.80	---	196.80	**259.00**

Description	Oper	Unit	Vol	Crew Size	Man-hours per Unit	Crew Output per Day	Avg Mat'l Unit Cost	Avg Labor Unit Cost	Avg Equip Unit Cost	Avg Total Unit Cost	Avg Price Incl O&P
4'-0" x 1'-0" H	Inst	Set	Lg	CA	1.45	5.50	59.70	48.20	---	107.90	**141.00**
	Inst	Set	Sm	CA	2.22	3.60	64.40	73.90	---	138.30	**185.00**
4'-0" x 1'-6" H	Inst	Set	Lg	CA	1.45	5.50	66.00	48.20	---	114.20	**148.00**
	Inst	Set	Sm	CA	2.22	3.60	71.20	73.90	---	145.10	**193.00**
4'-0" x 2'-0" H	Inst	Set	Lg	CA	1.78	4.50	74.90	59.20	---	134.10	**175.00**
	Inst	Set	Sm	CA	2.76	2.90	80.80	91.80	---	172.60	**231.00**
4'-0" x 2'-6" H	Inst	Set	Lg	CA	1.78	4.50	82.60	59.20	---	141.80	**184.00**
	Inst	Set	Sm	CA	2.76	2.90	89.10	91.80	---	180.90	**240.00**
4'-0" x 3'-0" H	Inst	Set	Lg	CA	1.78	4.50	90.20	59.20	---	149.40	**193.00**
	Inst	Set	Sm	CA	2.76	2.90	97.30	91.80	---	189.10	**250.00**
4'-0" x 3'-6" H	Inst	Set	Lg	CA	1.90	4.20	97.80	63.20	---	161.00	**207.00**
	Inst	Set	Sm	CA	2.96	2.70	105.00	98.50	---	203.50	**269.00**
4'-0" x 4'-0" H	Inst	Set	Lg	CA	1.90	4.20	105.00	63.20	---	168.20	**216.00**
	Inst	Set	Sm	CA	2.96	2.70	114.00	98.50	---	212.50	**279.00**
4'-0" x 5'-0" H	Inst	Set	Lg	CJ	4.57	3.50	121.00	139.00	---	260.00	**347.00**
	Inst	Set	Sm	CJ	6.96	2.30	130.00	211.00	---	341.00	**467.00**
5'-0" x 2'-0" H	Inst	Set	Lg	CJ	3.81	4.20	83.80	116.00	---	199.80	**270.00**
	Inst	Set	Sm	CJ	5.93	2.70	90.40	180.00	---	270.40	**374.00**
5'-0" x 2'-6" H	Inst	Set	Lg	CJ	3.81	4.20	92.70	116.00	---	208.70	**280.00**
	Inst	Set	Sm	CJ	5.93	2.70	100.00	180.00	---	280.00	**385.00**
5'-0" x 3'-0" H	Inst	Set	Lg	CJ	4.10	3.90	100.00	124.00	---	224.00	**302.00**
	Inst	Set	Sm	CJ	6.40	2.50	108.00	194.00	---	302.00	**416.00**
5'-0" x 3'-6" H	Inst	Set	Lg	CJ	4.10	3.90	109.00	124.00	---	233.00	**312.00**
	Inst	Set	Sm	CJ	6.40	2.50	118.00	194.00	---	312.00	**427.00**
5'-0" x 4'-0" H	Inst	Set	Lg	CJ	4.57	3.50	118.00	139.00	---	257.00	**344.00**
	Inst	Set	Sm	CJ	6.96	2.30	127.00	211.00	---	338.00	**463.00**
5'-0" x 5'-0" H	Inst	Set	Lg	CJ	4.57	3.50	142.00	139.00	---	281.00	**372.00**
	Inst	Set	Sm	CJ	6.96	2.30	153.00	211.00	---	364.00	**493.00**
6'-0" x 2'-0" H	Inst	Set	Lg	CJ	3.90	4.10	92.70	118.00	---	210.70	**284.00**
	Inst	Set	Sm	CJ	5.93	2.70	100.00	180.00	---	280.00	**385.00**
6'-0" x 2'-6" H	Inst	Set	Lg	CJ	3.90	4.10	102.00	118.00	---	220.00	**294.00**
	Inst	Set	Sm	CJ	5.93	2.70	110.00	180.00	---	290.00	**396.00**
6'-0" x 3'-0" H	Inst	Set	Lg	CJ	4.10	3.90	112.00	124.00	---	236.00	**315.00**
	Inst	Set	Sm	CJ	6.40	2.50	121.00	194.00	---	315.00	**430.00**
8'-0" x 2'-0" H	Inst	Set	Lg	CJ	4.10	3.90	116.00	124.00	---	240.00	**320.00**
	Inst	Set	Sm	CJ	6.40	2.50	125.00	194.00	---	319.00	**435.00**
8'-0" x 2'-6" H	Inst	Set	Lg	CJ	4.10	3.90	128.00	124.00	---	252.00	**334.00**
	Inst	Set	Sm	CJ	6.40	2.50	138.00	194.00	---	332.00	**451.00**
8'-0" x 3'-0" H	Inst	Set	Lg	CJ	4.57	3.50	141.00	139.00	---	280.00	**370.00**
	Inst	Set	Sm	CJ	6.96	2.30	152.00	211.00	---	363.00	**492.00**
8'-0" x 4'-0" H	Inst	Set	Lg	CJ	5.00	3.20	178.00	152.00	---	330.00	**432.00**
	Inst	Set	Sm	CJ	7.62	2.10	192.00	231.00	---	423.00	**568.00**
8'-0" x 5'-0" H	Inst	Set	Lg	CJ	5.33	3.00	215.00	162.00	---	377.00	**490.00**
	Inst	Set	Sm	CJ	8.00	2.00	232.00	243.00	---	475.00	**631.00**
For bronze finish, ADD	Inst	%	Lg	---	---	---	20.0	---	---	---	---
	Inst	%	Sm	---	---	---	20.0	---	---	---	---

Description	Oper	Unit	Vol	Crew Size	Man-hours per Unit	Crew Output per Day	Avg Mat'l Unit Cost	Avg Labor Unit Cost	Avg Equip Unit Cost	Avg Total Unit Cost	Avg Price Incl O&P
Dual glazed											
2'-0" x 2'-0" H	Inst	Set	Lg	CA	1.27	6.30	90.20	42.30	---	132.50	**167.00**
	Inst	Set	Sm	CA	1.95	4.10	97.30	64.90	---	162.20	**209.00**
2'-0" x 2'-6" H	Inst	Set	Lg	CA	1.27	6.30	99.10	42.30	---	141.40	**177.00**
	Inst	Set	Sm	CA	1.95	4.10	107.00	64.90	---	171.90	**220.00**
2'-0" x 3'-0" H	Inst	Set	Lg	CA	1.27	6.30	108.00	42.30	---	150.30	**188.00**
	Inst	Set	Sm	CA	1.95	4.10	116.00	64.90	---	180.90	**231.00**
2'-0" x 3'-6" H	Inst	Set	Lg	CA	1.27	6.30	117.00	42.30	---	159.30	**198.00**
	Inst	Set	Sm	CA	1.95	4.10	126.00	64.90	---	190.90	**242.00**
2'-0" x 4'-0" H	Inst	Set	Lg	CA	1.27	6.30	126.00	42.30	---	168.30	**208.00**
	Inst	Set	Sm	CA	1.95	4.10	136.00	64.90	---	200.90	**253.00**
2'-6" x 2'-0" H	Inst	Set	Lg	CA	1.27	6.30	96.50	42.30	---	138.80	**174.00**
	Inst	Set	Sm	CA	1.95	4.10	104.00	64.90	---	168.90	**217.00**
2'-6" x 2'-6" H	Inst	Set	Lg	CA	1.27	6.30	108.00	42.30	---	150.30	**188.00**
	Inst	Set	Sm	CA	1.95	4.10	116.00	64.90	---	180.90	**231.00**
2'-6" x 3'-0" H	Inst	Set	Lg	CA	1.31	6.10	117.00	43.60	---	160.60	**200.00**
	Inst	Set	Sm	CA	2.00	4.00	126.00	66.50	---	192.50	**245.00**
2'-6" x 3'-6" H	Inst	Set	Lg	CA	1.31	6.10	127.00	43.60	---	170.60	**211.00**
	Inst	Sct	Sm	CA	2.00	4.00	137.00	66.50	---	203.50	**257.00**
2'-6" x 4'-0" H	Inst	Set	Lg	CA	1.38	5.80	137.00	45.90	---	182.90	**227.00**
	Inst	Set	Sm	CA	2.11	3.80	148.00	70.20	---	218.20	**275.00**
3'-0" x 2'-0" H	Inst	Set	Lg	CA	1.27	6.30	104.00	42.30	---	146.30	**183.00**
	Inst	Set	Sm	CA	1.95	4.10	112.00	64.90	---	176.90	**227.00**
3'-0" x 2'-6" H	Inst	Set	Lg	CA	1.27	6.30	116.00	42.30	---	158.30	**196.00**
	Inst	Set	Sm	CA	1.95	4.10	125.00	64.90	---	189.90	**241.00**
3'-0" x 3'-0" H	Inst	Set	Lg	CA	1.27	6.30	126.00	42.30	---	168.30	**208.00**
	Inst	Set	Sm	CA	1.95	4.10	136.00	64.90	---	200.90	**253.00**
3'-0" x 3'-6" H	Inst	Set	Lg	CA	1.51	5.30	137.00	50.20	---	187.20	**233.00**
	Inst	Set	Sm	CA	2.35	3.40	148.00	78.20	---	226.20	**287.00**
3'-0" x 4'-0" H	Inst	Set	Lg	CA	1.86	4.30	149.00	61.90	---	210.90	**264.00**
	Inst	Set	Sm	CA	2.86	2.80	160.00	95.20	---	255.20	**327.00**
3'-0" x 5'-0" H	Inst	Set	Lg	CA	2.16	3.70	171.00	71.90	---	242.90	**305.00**
	Inst	Set	Sm	CA	3.33	2.40	185.00	111.00	---	296.00	**379.00**
3'-6" x 2'-0" H	Inst	Set	Lg	CA	1.27	6.30	112.00	42.30		154.30	**192.00**
	Inst	Set	Sm	CA	1.95	4.10	121.00	64.90	---	185.90	**236.00**
3'-6" x 2'-6" H	Inst	Set	Lg	CA	1.27	6.30	123.00	42.30	---	165.30	**205.00**
	Inst	Set	Sm	CA	1.95	4.10	133.00	64.90	---	197.90	**250.00**
3'-6" x 3'-0" H	Inst	Set	Lg	CA	1.51	5.30	136.00	50.20	---	186.20	**232.00**
	Inst	Set	Sm	CA	2.35	3.40	147.00	78.20	---	225.20	**286.00**
3'-6" x 3'-6" H	Inst	Set	Lg	CA	1.51	5.30	147.00	50.20	---	197.20	**245.00**
	Inst	Set	Sm	CA	2.35	3.40	159.00	78.20	---	237.20	**300.00**
3'-6" x 4'-0" H	Inst	Set	Lg	CA	1.86	4.30	160.00	61.90	---	221.90	**277.00**
	Inst	Set	Sm	CA	2.86	2.80	173.00	95.20	---	268.20	**341.00**
3'-6" x 5'-0" H	Inst	Set	Lg	CA	1.86	4.30	187.00	61.90	---	248.90	**308.00**
	Inst	Set	Sm	CA	2.86	2.80	201.00	95.20	---	296.20	**374.00**
4'-0" x 2'-0" H	Inst	Set	Lg	CA	1.86	4.30	118.00	61.90	---	179.90	**229.00**
	Inst	Set	Sm	CA	2.86	2.80	127.00	95.20	---	222.20	**289.00**
4'-0" x 2'-6" H	Inst	Set	Lg	CA	1.86	4.30	132.00	61.90	---	193.90	**245.00**
	Inst	Set	Sm	CA	2.86	2.80	142.00	95.20	---	237.20	**307.00**

Description	Oper	Unit	Vol	Crew Size	Man-hours per Unit	Crew Output per Day	Avg Mat'l Unit Cost	Avg Labor Unit Cost	Avg Equip Unit Cost	Avg Total Unit Cost	Avg Price Incl O&P
4'-0" x 3'-0" H	Inst	Set	Lg	CA	1.86	4.30	145.00	61.90	---	206.90	**259.00**
	Inst	Set	Sm	CA	2.86	2.80	156.00	95.20	---	251.20	**322.00**
4'-0" x 3'-6" H	Inst	Set	Lg	CA	2.00	4.00	157.00	66.50	---	223.50	**281.00**
	Inst	Set	Sm	CA	3.08	2.60	170.00	102.00	---	272.00	**349.00**
4'-0" x 4'-0" H	Inst	Set	Lg	CA	2.00	4.00	171.00	66.50	---	237.50	**297.00**
	Inst	Set	Sm	CA	3.08	2.60	185.00	102.00	---	287.00	**366.00**
4'-0" x 5'-0" H	Inst	Set	Lg	CA	2.42	3.30	199.00	80.50	---	279.50	**350.00**
	Inst	Set	Sm	CA	3.81	2.10	215.00	127.00	---	342.00	**438.00**
5'-0" x 2'-0" H	Inst	Set	Lg	CA	2.00	4.00	133.00	66.50	---	199.50	**253.00**
	Inst	Set	Sm	CA	3.08	2.60	144.00	102.00	---	246.00	**319.00**
5'-0" x 2'-6" H	Inst	Set	Lg	CA	2.00	4.00	149.00	66.50	---	215.50	**271.00**
	Inst	Set	Sm	CA	3.08	2.60	160.00	102.00	---	262.00	**338.00**
5'-0" x 3'-0" H	Inst	Set	Lg	CJ	4.32	3.70	164.00	131.00	---	295.00	**385.00**
	Inst	Set	Sm	CJ	6.67	2.40	177.00	202.00	---	379.00	**507.00**
5'-0" x 3'-6" H	Inst	Set	Lg	CJ	4.32	3.70	179.00	131.00	---	310.00	**403.00**
	Inst	Set	Sm	CJ	6.67	2.40	193.00	202.00	---	395.00	**526.00**
5'-0" x 4'-0" H	Inst	Set	Lg	CJ	4.85	3.30	194.00	147.00	---	341.00	**444.00**
	Inst	Set	Sm	CJ	7.62	2.10	210.00	231.00	---	441.00	**588.00**
5'-0" x 5'-0" H	Inst	Set	Lg	CJ	4.85	3.30	243.00	147.00	---	390.00	**500.00**
	Inst	Set	Sm	CJ	7.62	2.10	262.00	231.00	---	493.00	**648.00**
6'-0" x 2'-0" H	Inst	Set	Lg	CJ	4.10	3.90	147.00	124.00	---	271.00	**356.00**
	Inst	Set	Sm	CJ	6.40	2.50	159.00	194.00	---	353.00	**474.00**
6'-0" x 2'-6" H	Inst	Set	Lg	CJ	4.10	3.90	165.00	124.00	---	289.00	**377.00**
	Inst	Set	Sm	CJ	6.40	2.50	178.00	194.00	---	372.00	**496.00**
6'-0" x 3'-0" H	Inst	Set	Lg	CJ	4.32	3.70	182.00	131.00	---	313.00	**406.00**
	Inst	Set	Sm	CJ	6.67	2.40	196.00	202.00	---	398.00	**529.00**
6'-0" x 3'-6" H	Inst	Set	Lg	CJ	4.85	3.30	197.00	147.00	---	344.00	**447.00**
	Inst	Set	Sm	CJ	7.62	2.10	212.00	231.00	---	443.00	**591.00**
6'-0" x 4'-0" H	Inst	Set	Lg	CJ	4.85	3.30	217.00	147.00	---	364.00	**471.00**
	Inst	Set	Sm	CJ	7.62	2.10	234.00	231.00	---	465.00	**616.00**
6'-0" x 5'-0" H	Inst	Set	Lg	CJ	4.85	3.30	274.00	147.00	---	421.00	**536.00**
	Inst	Set	Sm	CJ	7.62	2.10	296.00	231.00	---	527.00	**687.00**
For bronze finish, ADD	Inst	%	Lg	---	---	---	15.0	---	---	---	**---**
	Inst	%	Sm	---	---	---	15.0	---	---	---	**---**

2 sliding lites, 1 fixed lite

Single glazed

Description	Oper	Unit	Vol	Crew Size	Man-hours per Unit	Crew Output per Day	Avg Mat'l Unit Cost	Avg Labor Unit Cost	Avg Equip Unit Cost	Avg Total Unit Cost	Avg Price Incl O&P
6'-0" x 2'-0" H	Inst	Set	Lg	CJ	4.32	3.70	117.00	131.00	---	248.00	**331.00**
	Inst	Set	Sm	CJ	6.67	2.40	126.00	202.00	---	328.00	**449.00**
6'-0" x 2'-6" H	Inst	Set	Lg	CJ	4.32	3.70	128.00	131.00	---	259.00	**344.00**
	Inst	Set	Sm	CJ	6.67	2.40	138.00	202.00	---	340.00	**463.00**
6'-0" x 3'-0" H	Inst	Set	Lg	CJ	4.57	3.50	140.00	139.00	---	279.00	**369.00**
	Inst	Set	Sm	CJ	6.96	2.30	151.00	211.00	---	362.00	**490.00**
6'-0" x 4'-0" H	Inst	Set	Lg	CJ	4.57	3.50	166.00	139.00	---	305.00	**399.00**
	Inst	Set	Sm	CJ	6.96	2.30	179.00	211.00	---	390.00	**523.00**
6'-0" x 5'-0" H	Inst	Set	Lg	CJ	4.57	3.50	193.00	139.00	---	332.00	**430.00**
	Inst	Set	Sm	CJ	6.96	2.30	208.00	211.00	---	419.00	**556.00**

Description	Oper	Unit	Vol	Crew Size	Man-hours per Unit	Crew Output per Day	Avg Mat'l Unit Cost	Avg Labor Unit Cost	Avg Equip Unit Cost	Avg Total Unit Cost	Avg Price Incl O&P
7'-0" x 2'-0" H	Inst	Set	Lg	CJ	4.32	3.70	123.00	131.00	---	254.00	**338.00**
	Inst	Set	Sm	CJ	6.67	2.40	133.00	202.00	---	335.00	**457.00**
7'-0" x 2'-6" H	Inst	Set	Lg	CJ	4.32	3.70	137.00	131.00	---	268.00	**354.00**
	Inst	Set	Sm	CJ	6.67	2.40	148.00	202.00	---	350.00	**474.00**
7'-0" x 3'-0" H	Inst	Set	Lg	CJ	4.57	3.50	151.00	139.00	---	290.00	**382.00**
	Inst	Set	Sm	CJ	6.96	2.30	163.00	211.00	---	374.00	**504.00**
7'-0" x 4'-0" H	Inst	Set	Lg	CJ	4.57	3.50	192.00	139.00	---	331.00	**429.00**
	Inst	Set	Sm	CJ	6.96	2.30	207.00	211.00	---	418.00	**555.00**
7'-0" x 5'-0" H	Inst	Set	Lg	CJ	5.71	2.80	243.00	173.00	---	416.00	**539.00**
	Inst	Set	Sm	CJ	8.89	1.80	262.00	270.00	---	532.00	**706.00**
8'-0" x 2'-0" H	Inst	Set	Lg	CJ	4.57	3.50	136.00	139.00	---	275.00	**364.00**
	Inst	Set	Sm	CJ	6.96	2.30	147.00	211.00	---	358.00	**485.00**
8'-0" x 2'-6" H	Inst	Set	Lg	CJ	4.57	3.50	150.00	139.00	---	289.00	**380.00**
	Inst	Set	Sm	CJ	6.96	2.30	162.00	211.00	---	373.00	**503.00**
8'-0" x 3'-0" H	Inst	Set	Lg	CJ	4.57	3.50	166.00	139.00	---	305.00	**399.00**
	Inst	Set	Sm	CJ	6.96	2.30	179.00	211.00	---	390.00	**523.00**
8'-0" x 3'-6" H	Inst	Set	Lg	CJ	4.57	3.50	180.00	139.00	---	319.00	**415.00**
	Inst	Set	Sm	CJ	6.96	2.30	195.00	211.00	---	406.00	**541.00**
8'-0" x 4'-0" H	Inst	Set	Lg	CJ	5.71	2.80	196.00	173.00	---	369.00	**485.00**
	Inst	Set	Sm	CJ	8.89	1.80	211.00	270.00	---	481.00	**647.00**
8'-0" x 5'-0" H	Inst	Set	Lg	CJ	5.71	2.80	245.00	173.00	---	418.00	**542.00**
	Inst	Set	Sm	CJ	8.89	1.80	264.00	270.00	---	534.00	**709.00**
10'-0" x 2'-0" H	Inst	Set	Lg	CJ	4.57	3.50	175.00	139.00	---	314.00	**410.00**
	Inst	Set	Sm	CJ	6.96	2.30	189.00	211.00	---	400.00	**534.00**
10'-0" x 3'-0" H	Inst	Set	Lg	CJ	4.57	3.50	201.00	139.00	---	340.00	**439.00**
	Inst	Set	Sm	CJ	6.96	2.30	216.00	211.00	---	427.00	**566.00**
10'-0" x 3'-6" H	Inst	Set	Lg	CJ	5.71	2.80	230.00	173.00	---	403.00	**524.00**
	Inst	Set	Sm	CJ	8.89	1.80	248.00	270.00	---	518.00	**690.00**
10'-0" x 4'-0" H	Inst	Set	Lg	CJ	5.71	2.80	241.00	173.00	---	414.00	**537.00**
	Inst	Set	Sm	CJ	8.89	1.80	260.00	270.00	---	530.00	**704.00**
10'-0" x 5'-0" H	Inst	Set	Lg	CJ	5.71	2.80	339.00	173.00	---	512.00	**650.00**
	Inst	Set	Sm	CJ	8.89	1.80	366.00	270.00	---	636.00	**825.00**
For bronze finish, ADD	Inst	%	Lg	---	---	---	13.5	---	---	---	**---**
	Inst	%	Sm	---	---	---	13.5	---	---	---	**---**

Description	Oper	Unit	Vol	Crew Size	Man-hours per Unit	Crew Output per Day	Avg Mat'l Unit Cost	Avg Labor Unit Cost	Avg Equip Unit Cost	Avg Total Unit Cost	Avg Price Incl O&P
Dual glazed											
6'-0" x 2'-0" H	Inst	Set	Lg	CJ	4.57	3.50	183.00	139.00	---	322.00	**418.00**
	Inst	Set	Sm	CJ	6.96	2.30	197.00	211.00	---	408.00	**544.00**
6'-0" x 2'-6" H	Inst	Set	Lg	CJ	4.57	3.50	203.00	139.00	---	342.00	**442.00**
	Inst	Set	Sm	CJ	6.96	2.30	219.00	211.00	---	430.00	**569.00**
6'-0" x 3'-0" H	Inst	Set	Lg	CJ	4.85	3.30	224.00	147.00	---	371.00	**478.00**
	Inst	Set	Sm	CJ	7.62	2.10	241.00	231.00	---	472.00	**624.00**
6'-0" x 3'-6" H	Inst	Set	Lg	CJ	4.85	3.30	244.00	147.00	---	391.00	**501.00**
	Inst	Set	Sm	CJ	7.62	2.10	263.00	231.00	---	494.00	**649.00**
6'-0" x 4'-0" H	Inst	Set	Lg	CJ	4.85	3.30	276.00	147.00	---	423.00	**538.00**
	Inst	Set	Sm	CJ	7.62	2.10	297.00	231.00	---	528.00	**689.00**
6'-0" x 5'-0" H	Inst	Set	Lg	CJ	4.85	3.30	319.00	147.00	---	466.00	**587.00**
	Inst	Set	Sm	CJ	7.62	2.10	344.00	231.00	---	575.00	**742.00**
7'-0" x 2'-0" H	Inst	Set	Lg	CJ	4.32	3.70	202.00	131.00	---	333.00	**429.00**
	Inst	Set	Sm	CJ	6.67	2.40	218.00	202.00	---	420.00	**554.00**
7'-0" x 2'-6" H	Inst	Set	Lg	CJ	4.32	3.70	222.00	131.00	---	353.00	**452.00**
	Inst	Set	Sm	CJ	6.67	2.40	240.00	202.00	---	442.00	**579.00**
7'-0" x 3'-0" H	Inst	Set	Lg	CJ	4.57	3.50	245.00	139.00	---	384.00	**490.00**
	Inst	Set	Sm	CJ	6.96	2.30	264.00	211.00	---	475.00	**621.00**
7'-0" x 3'-6" H	Inst	Set	Lg	CJ	4.85	3.30	281.00	147.00	---	428.00	**544.00**
	Inst	Set	Sm	CJ	7.62	2.10	303.00	231.00	---	534.00	**695.00**
7'-0" x 4'-0" H	Inst	Set	Lg	CJ	4.85	3.30	306.00	147.00	---	453.00	**573.00**
	Inst	Set	Sm	CJ	7.62	2.10	330.00	231.00	---	561.00	**727.00**
7'-0" x 5'-0" H	Inst	Set	Lg	CJ	6.15	2.60	376.00	187.00	---	563.00	**712.00**
	Inst	Set	Sm	CJ	9.41	1.70	406.00	286.00	---	692.00	**895.00**
8'-0" x 2'-0" H	Inst	Set	Lg	CJ	4.85	3.30	213.00	147.00	---	360.00	**466.00**
	Inst	Set	Sm	CJ	7.62	2.10	230.00	231.00	---	461.00	**612.00**
8'-0" x 2'-6" H	Inst	Set	Lg	CJ	4.85	3.30	237.00	147.00	---	384.00	**494.00**
	Inst	Set	Sm	CJ	7.62	2.10	256.00	231.00	---	487.00	**642.00**
8'-0" x 3'-0" H	Inst	Set	Lg	CJ	4.85	3.30	274.00	147.00	---	421.00	**536.00**
	Inst	Set	Sm	CJ	7.62	2.10	296.00	231.00	---	527.00	**687.00**
8'-0" x 3'-6" H	Inst	Set	Lg	CJ	4.85	3.30	298.00	147.00	---	445.00	**564.00**
	Inst	Set	Sm	CJ	7.62	2.10	322.00	231.00	---	553.00	**717.00**
8'-0" x 4'-0" H	Inst	Set	Lg	CJ	6.15	2.60	324.00	187.00	---	511.00	**652.00**
	Inst	Set	Sm	CJ	9.41	1.70	349.00	286.00	---	635.00	**830.00**
8'-0" x 5'-0" H	Inst	Set	Lg	CJ	6.15	2.60	385.00	187.00	---	572.00	**723.00**
	Inst	Set	Sm	CJ	9.41	1.70	415.00	286.00	---	701.00	**906.00**
10'-0" x 2'-0" H	Inst	Set	Lg	CJ	4.57	3.50	300.00	139.00	---	439.00	**553.00**
	Inst	Set	Sm	CJ	6.96	2.30	323.00	211.00	---	534.00	**689.00**
10'-0" x 3'-0" H	Inst	Set	Lg	CJ	4.57	3.50	368.00	139.00	---	507.00	**632.00**
	Inst	Set	Sm	CJ	6.96	2.30	397.00	211.00	---	608.00	**774.00**
10'-0" x 3'-6" H	Inst	Set	Lg	CJ	5.71	2.80	419.00	173.00	---	592.00	**742.00**
	Inst	Set	Sm	CJ	8.89	1.80	452.00	270.00	---	722.00	**925.00**
10'-0" x 4'-0" H	Inst	Set	Lg	CJ	5.71	2.80	453.00	173.00	---	626.00	**781.00**
	Inst	Set	Sm	CJ	8.89	1.80	489.00	270.00	---	759.00	**967.00**
10'-0" x 5'-0" H	Inst	Set	Lg	CJ	5.71	2.80	550.00	173.00	---	723.00	**892.00**
	Inst	Set	Sm	CJ	8.89	1.80	593.00	270.00	---	863.00	**1090.00**
For bronze finish, ADD	Inst	%	Lg	---	---	---	15.0	---	---	---	---
	Inst	%	Sm	---	---	---	15.0	---	---	---	---

Description	Oper	Unit	Vol	Crew Size	Man-hours per Unit	Crew Output per Day	Avg Mat'l Unit Cost	Avg Labor Unit Cost	Avg Equip Unit Cost	Avg Total Unit Cost	Avg Price Incl O&P

Wood

Awning windows; pine frames, exterior treated and primed, interior natural finish; glazed 1/2" T insulating glass; unit includes weatherstripping and exterior trim

Unit is one ventilating lite wide

20", 24" x 32" W

Description	Oper	Unit	Vol	Crew Size	Man-hours per Unit	Crew Output per Day	Avg Mat'l Unit Cost	Avg Labor Unit Cost	Avg Equip Unit Cost	Avg Total Unit Cost	Avg Price Incl O&P
High quality workmanship	Inst	Set	Lg	CA	1.25	6.40	292.00	41.60	---	333.60	**398.00**
	Inst	Set	Sm	CA	1.90	4.20	315.00	63.20	---	378.20	**457.00**
Good quality workmanship	Inst	Set	Lg	CA	1.00	8.00	260.00	33.30	---	293.30	**349.00**
	Inst	Set	Sm	CA	1.54	5.20	281.00	51.20	---	332.20	**400.00**
Average quality workmanship	Inst	Set	Lg	CA	.833	9.60	216.00	27.70	---	243.70	**290.00**
	Inst	Set	Sm	CA	1.29	6.20	233.00	42.90	---	275.90	**332.00**

32", 36" x 32" W

Description	Oper	Unit	Vol	Crew Size	Man-hours per Unit	Crew Output per Day	Avg Mat'l Unit Cost	Avg Labor Unit Cost	Avg Equip Unit Cost	Avg Total Unit Cost	Avg Price Incl O&P
High quality workmanship	Inst	Set	Lg	CA	1.25	6.40	387.00	41.60	---	428.60	**508.00**
	Inst	Set	Sm	CA	1.90	4.20	418.00	63.20	---	481.20	**575.00**
Good quality workmanship	Inst	Set	Lg	CA	1.00	8.00	343.00	33.30	---	376.30	**444.00**
	Inst	Set	Sm	CA	1.54	5.20	370.00	51.20	---	421.20	**502.00**
Average quality workmanship	Inst	Set	Lg	CA	.833	9.60	286.00	27.70	---	313.70	**370.00**
	Inst	Set	Sm	CA	1.29	6.20	308.00	42.90	---	350.90	**419.00**

20", 24" x 40" W

Description	Oper	Unit	Vol	Crew Size	Man-hours per Unit	Crew Output per Day	Avg Mat'l Unit Cost	Avg Labor Unit Cost	Avg Equip Unit Cost	Avg Total Unit Cost	Avg Price Incl O&P
High quality workmanship	Inst	Set	Lg	CA	1.25	6.40	318.00	41.60	---	359.60	**428.00**
	Inst	Set	Sm	CA	1.90	4.20	343.00	63.20	---	406.20	**489.00**
Good quality workmanship	Inst	Set	Lg	CA	1.00	8.00	305.00	33.30	---	338.30	**400.00**
	Inst	Set	Sm	CA	1.54	5.20	329.00	51.20	---	380.20	**455.00**
Average quality workmanship	Inst	Set	Lg	CA	.833	9.60	254.00	27.70	---	281.70	**334.00**
	Inst	Set	Sm	CA	1.29	6.20	274.00	42.90	---	316.90	**379.00**

32", 36" x 40" W

Description	Oper	Unit	Vol	Crew Size	Man-hours per Unit	Crew Output per Day	Avg Mat'l Unit Cost	Avg Labor Unit Cost	Avg Equip Unit Cost	Avg Total Unit Cost	Avg Price Incl O&P
High quality workmanship	Inst	Set	Lg	CA	1.25	6.40	400.00	41.60	---	441.60	**522.00**
	Inst	Set	Sm	CA	1.90	4.20	432.00	63.20	---	495.20	**591.00**
Good quality workmanship	Inst	Set	Lg	CA	1.00	8.00	337.00	33.30	---	370.30	**437.00**
	Inst	Set	Sm	CA	1.54	5.20	363.00	51.20	---	414.20	**494.00**
Average quality workmanship	Inst	Set	Lg	CA	.833	9.60	279.00	27.70	---	306.70	**363.00**
	Inst	Set	Sm	CA	1.29	6.20	301.00	42.90	---	343.90	**411.00**

20", 24" x 48" W

Description	Oper	Unit	Vol	Crew Size	Man-hours per Unit	Crew Output per Day	Avg Mat'l Unit Cost	Avg Labor Unit Cost	Avg Equip Unit Cost	Avg Total Unit Cost	Avg Price Incl O&P
High quality workmanship	Inst	Set	Lg	CA	1.25	6.40	400.00	41.60	---	441.60	**522.00**
	Inst	Set	Sm	CA	1.90	4.20	432.00	63.20	---	495.20	**591.00**
Good quality workmanship	Inst	Set	Lg	CA	1.00	8.00	330.00	33.30	---	363.30	**430.00**
	Inst	Set	Sm	CA	1.54	5.20	356.00	51.20	---	407.20	**486.00**
Average quality workmanship	Inst	Set	Lg	CA	.833	9.60	274.00	27.70	---	301.70	**357.00**
	Inst	Set	Sm	CA	1.29	6.20	296.00	42.90	---	338.90	**405.00**

32", 36" x 48" W

Description	Oper	Unit	Vol	Crew Size	Man-hours per Unit	Crew Output per Day	Avg Mat'l Unit Cost	Avg Labor Unit Cost	Avg Equip Unit Cost	Avg Total Unit Cost	Avg Price Incl O&P
High quality workmanship	Inst	Set	Lg	CA	1.25	6.40	603.00	41.60	---	644.60	**756.00**
	Inst	Set	Sm	CA	1.90	4.20	651.00	63.20	---	714.20	**843.00**
Good quality workmanship	Inst	Set	Lg	CA	1.00	8.00	375.00	33.30	---	408.30	**481.00**
	Inst	Set	Sm	CA	1.54	5.20	404.00	51.20	---	455.20	**542.00**
Average quality workmanship	Inst	Set	Lg	CA	.833	9.60	311.00	27.70	---	338.70	**399.00**
	Inst	Set	Sm	CA	1.29	6.20	336.00	42.90	---	378.90	**450.00**

Description	Oper	Unit	Vol	Crew Size	Man-hours per Unit	Crew Output per Day	Avg Mat'l Unit Cost	Avg Labor Unit Cost	Avg Equip Unit Cost	Avg Total Unit Cost	Avg Price Incl O&P
Unit is two ventilating lites wide											
20", 24" x 48" W											
High quality workmanship	Inst	Set	Lg	CJ	3.33	4.80	495.00	101.00	---	596.00	**721.00**
	Inst	Set	Sm	CJ	5.16	3.10	534.00	157.00	---	691.00	**849.00**
Good quality workmanship	Inst	Set	Lg	CJ	2.67	6.00	457.00	81.00	---	538.00	**647.00**
	Inst	Set	Sm	CJ	4.10	3.90	493.00	124.00	---	617.00	**754.00**
Average quality workmanship	Inst	Set	Lg	CJ	2.22	7.20	381.00	67.40	---	448.40	**539.00**
	Inst	Set	Sm	CJ	3.40	4.70	411.00	103.00	---	514.00	**627.00**
32", 36" x 48" W											
High quality workmanship	Inst	Set	Lg	CJ	3.33	4.80	546.00	101.00	---	647.00	**780.00**
	Inst	Set	Sm	CJ	5.16	3.10	589.00	157.00	---	746.00	**912.00**
Good quality workmanship	Inst	Set	Lg	CJ	2.67	6.00	521.00	81.00	---	602.00	**720.00**
	Inst	Set	Sm	CJ	4.10	3.90	562.00	124.00	---	686.00	**833.00**
Average quality workmanship	Inst	Set	Lg	CJ	2.22	7.20	433.00	67.40	---	500.40	**599.00**
	Inst	Set	Sm	CJ	3.40	4.70	467.00	103.00	---	570.00	**692.00**
20", 24" x 72" W											
High quality workmanship	Inst	Set	Lg	CJ	3.33	4.80	540.00	101.00	---	641.00	**772.00**
	Inst	Set	Sm	CJ	5.16	3.10	582.00	157.00	---	739.00	**905.00**
Good quality workmanship	Inst	Set	Lg	CJ	2.67	6.00	508.00	81.00	---	589.00	**706.00**
	Inst	Set	Sm	CJ	4.10	3.90	548.00	124.00	---	672.00	**817.00**
Average quality workmanship	Inst	Set	Lg	CJ	2.22	7.20	423.00	67.40	---	490.40	**587.00**
	Inst	Set	Sm	CJ	3.40	4.70	456.00	103.00	---	559.00	**679.00**
32", 36" x 72" W											
High quality workmanship	Inst	Set	Lg	CJ	3.33	4.80	635.00	101.00	---	736.00	**882.00**
	Inst	Set	Sm	CJ	5.16	3.10	685.00	157.00	---	842.00	**1020.00**
Good quality workmanship	Inst	Set	Lg	CJ	2.67	6.00	603.00	81.00	---	684.00	**815.00**
	Inst	Set	Sm	CJ	4.10	3.90	651.00	124.00	---	775.00	**935.00**
Average quality workmanship	Inst	Set	Lg	CJ	2.22	7.20	502.00	67.40	---	569.40	**678.00**
	Inst	Set	Sm	CJ	3.40	4.70	541.00	103.00	---	644.00	**777.00**
20", 24" x 80" W											
High quality workmanship	Inst	Set	Lg	CJ	3.33	4.80	584.00	101.00	---	685.00	**823.00**
	Inst	Set	Sm	CJ	5.16	3.10	630.00	157.00	---	787.00	**960.00**
Good quality workmanship	Inst	Set	Lg	CJ	2.67	6.00	550.00	81.00	---	631.00	**754.00**
	Inst	Set	Sm	CJ	4.10	3.90	593.00	124.00	---	717.00	**869.00**
Average quality workmanship	Inst	Set	Lg	CJ	2.22	7.20	458.00	67.40	---	525.40	**628.00**
	Inst	Set	Sm	CJ	3.40	4.70	495.00	103.00	---	598.00	**724.00**
32", 36" x 80" W											
High quality workmanship	Inst	Set	Lg	CJ	3.33	4.80	673.00	101.00	---	774.00	**926.00**
	Inst	Set	Sm	CJ	5.16	3.10	726.00	157.00	---	883.00	**1070.00**
Good quality workmanship	Inst	Set	Lg	CJ	2.67	6.00	635.00	81.00	---	716.00	**852.00**
	Inst	Set	Sm	CJ	4.10	3.90	685.00	124.00	---	809.00	**974.00**
Average quality workmanship	Inst	Set	Lg	CJ	2.22	7.20	528.00	67.40	---	595.40	**709.00**
	Inst	Set	Sm	CJ	3.40	4.70	570.00	103.00	---	673.00	**810.00**

Description	Oper	Unit	Vol	Crew Size	Man-hours per Unit	Crew Output per Day	Avg Mat'l Unit Cost	Avg Labor Unit Cost	Avg Equip Unit Cost	Avg Total Unit Cost	Avg Price Incl O&P
20", 24" x 96" W											
High quality workmanship	Inst	Set	Lg	CJ	3.33	4.80	635.00	101.00	---	736.00	**882.00**
	Inst	Set	Sm	CJ	5.16	3.10	685.00	157.00	---	842.00	**1020.00**
Good quality workmanship	Inst	Set	Lg	CJ	2.67	6.00	597.00	81.00	---	678.00	**808.00**
	Inst	Set	Sm	CJ	4.10	3.90	644.00	124.00	---	768.00	**927.00**
Average quality workmanship	Inst	Set	Lg	CJ	2.22	7.20	499.00	67.40	---	566.40	**675.00**
	Inst	Set	Sm	CJ	3.40	4.70	538.00	103.00	---	641.00	**774.00**
32", 36" x 96" W											
High quality workmanship	Inst	Set	Lg	CJ	3.33	4.80	749.00	101.00	---	850.00	**1010.00**
	Inst	Set	Sm	CJ	5.16	3.10	808.00	157.00	---	965.00	**1160.00**
Good quality workmanship	Inst	Set	Lg	CJ	2.67	6.00	705.00	81.00	---	786.00	**932.00**
	Inst	Set	Sm	CJ	4.10	3.90	760.00	124.00	---	884.00	**1060.00**
Average quality workmanship	Inst	Set	Lg	CJ	2.22	7.20	588.00	67.40	---	655.40	**777.00**
	Inst	Set	Sm	CJ	3.40	4.70	634.00	103.00	---	737.00	**884.00**

Unit is three ventilating lites wide

Description	Oper	Unit	Vol	Crew Size	Man-hours per Unit	Crew Output per Day	Avg Mat'l Unit Cost	Avg Labor Unit Cost	Avg Equip Unit Cost	Avg Total Unit Cost	Avg Price Incl O&P
20", 24" x 72" W											
High quality workmanship	Inst	Set	Lg	CJ	4.44	3.60	749.00	135.00	---	884.00	**1060.00**
	Inst	Set	Sm	CJ	6.96	2.30	808.00	211.00	---	1019.00	**1250.00**
Good quality workmanship	Inst	Set	Lg	CJ	3.56	4.50	699.00	108.00	---	807.00	**965.00**
	Inst	Set	Sm	CJ	5.52	2.90	754.00	168.00	---	922.00	**1120.00**
Average quality workmanship	Inst	Set	Lg	CJ	2.96	5.40	582.00	89.80	---	671.80	**804.00**
	Inst	Set	Sm	CJ	4.57	3.50	627.00	139.00	---	766.00	**930.00**
32", 36" x 72" W											
High quality workmanship	Inst	Set	Lg	CJ	4.44	3.60	826.00	135.00	---	961.00	**1150.00**
	Inst	Set	Sm	CJ	6.96	2.30	891.00	211.00	---	1102.00	**1340.00**
Good quality workmanship	Inst	Set	Lg	CJ	3.56	4.50	775.00	108.00	---	883.00	**1050.00**
	Inst	Set	Sm	CJ	5.52	2.90	836.00	168.00	---	1004.00	**1210.00**
Average quality workmanship	Inst	Set	Lg	CJ	2.96	5.40	648.00	89.80	---	737.80	**880.00**
	Inst	Set	Sm	CJ	4.57	3.50	699.00	139.00	---	838.00	**1010.00**
20", 24" x 108" W											
High quality workmanship	Inst	Set	Lg	CJ	4.44	3.60	813.00	135.00	---	948.00	**1140.00**
	Inst	Set	Sm	CJ	6.96	2.30	877.00	211.00	---	1088.00	**1330.00**
Good quality workmanship	Inst	Set	Lg	CJ	3.56	4.50	762.00	108.00	---	870.00	**1040.00**
	Inst	Set	Sm	CJ	5.52	2.90	822.00	168.00	---	990.00	**1200.00**
Average quality workmanship	Inst	Set	Lg	CJ	2.96	5.40	635.00	89.80	---	724.80	**865.00**
	Inst	Set	Sm	CJ	4.57	3.50	685.00	139.00	---	824.00	**996.00**
32", 36" x 108" W											
High quality workmanship	Inst	Set	Lg	CJ	4.44	3.60	965.00	135.00	---	1100.00	**1310.00**
	Inst	Set	Sm	CJ	6.96	2.30	1040.00	211.00	---	1251.00	**1510.00**
Good quality workmanship	Inst	Set	Lg	CJ	3.56	4.50	908.00	108.00	---	1016.00	**1210.00**
	Inst	Set	Sm	CJ	5.52	2.90	980.00	168.00	---	1148.00	**1380.00**
Average quality workmanship	Inst	Set	Lg	CJ	2.96	5.40	762.00	89.80	---	851.80	**1010.00**
	Inst	Set	Sm	CJ	4.57	3.50	822.00	139.00	---	961.00	**1150.00**

Description	Oper	Unit	Vol	Crew Size	Man-hours per Unit	Crew Output per Day	Avg Mat'l Unit Cost	Avg Labor Unit Cost	Avg Equip Unit Cost	Avg Total Unit Cost	Avg Price Incl O&P
20", 24" x 120" W											
High quality workmanship	Inst	Set	Lg	CJ	4.44	3.60	889.00	135.00	---	1024.00	**1220.00**
	Inst	Set	Sm	CJ	6.96	2.30	959.00	211.00	---	1170.00	**1420.00**
Good quality workmanship	Inst	Set	Lg	CJ	3.56	4.50	838.00	108.00	---	946.00	**1130.00**
	Inst	Set	Sm	CJ	5.52	2.90	904.00	168.00	---	1072.00	**1290.00**
Average quality workmanship	Inst	Set	Lg	CJ	2.96	5.40	699.00	89.80	---	788.80	**938.00**
	Inst	Set	Sm	CJ	4.57	3.50	754.00	139.00	---	893.00	**1070.00**
32", 36" x 120" W											
High quality workmanship	Inst	Set	Lg	CJ	4.44	3.60	1020.00	135.00	---	1155.00	**1380.00**
	Inst	Set	Sm	CJ	6.96	2.30	1100.00	211.00	---	1311.00	**1590.00**
Good quality workmanship	Inst	Set	Lg	CJ	3.56	4.50	959.00	108.00	---	1067.00	**1260.00**
	Inst	Set	Sm	CJ	5.52	2.90	1030.00	168.00	---	1198.00	**1440.00**
Average quality workmanship	Inst	Set	Lg	CJ	2.96	5.40	800.00	89.80	---	889.80	**1050.00**
	Inst	Set	Sm	CJ	4.57	3.50	863.00	139.00	---	1002.00	**1200.00**
20", 24" x 144" W											
High quality workmanship	Inst	Set	Lg	CJ	4.44	3.60	965.00	135.00	---	1100.00	**1310.00**
	Inst	Set	Sm	CJ	6.96	2.30	1040.00	211.00	---	1251.00	**1510.00**
Good quality workmanship	Inst	Set	Lg	CJ	3.56	4.50	908.00	108.00	---	1016.00	**1210.00**
	Inst	Set	Sm	CJ	5.52	2.90	980.00	168.00	---	1148.00	**1380.00**
Average quality workmanship	Inst	Set	Lg	CJ	2.96	5.40	756.00	89.80	---	845.80	**1000.00**
	Inst	Set	Sm	CJ	4.57	3.50	815.00	139.00	---	954.00	**1150.00**
32", 36" x 144" W											
High quality workmanship	Inst	Set	Lg	CJ	4.44	3.60	1130.00	135.00	---	1265.00	**1500.00**
	Inst	Set	Sm	CJ	6.96	2.30	1220.00	211.00	---	1431.00	**1720.00**
Good quality workmanship	Inst	Set	Lg	CJ	3.56	4.50	1060.00	108.00	---	1168.00	**1380.00**
	Inst	Set	Sm	CJ	5.52	2.90	1140.00	168.00	---	1308.00	**1570.00**
Average quality workmanship	Inst	Set	Lg	CJ	2.96	5.40	889.00	89.80	---	978.80	**1160.00**
	Inst	Set	Sm	CJ	4.57	3.50	959.00	139.00	---	1098.00	**1310.00**
For aluminum clad (baked white enamel exterior)											
ADD	Inst	%	Lg	---	---	---	12.0	---	---	---	---
	Inst	%	Sm	---	---	---	12.0	---	---	---	---

Casement windows; glazed 1/2" T insulating glass; unit includes: hardware, drip cap and weatherstripping; does not include screens; rough opening sizes

1 ventilating lite, no fixed lites; 20", 24", 28" W

Description	Oper	Unit	Vol	Crew Size	Man-hours per Unit	Crew Output per Day	Avg Mat'l Unit Cost	Avg Labor Unit Cost	Avg Equip Unit Cost	Avg Total Unit Cost	Avg Price Incl O&P
36" H											
High quality workmanship	Inst	Set	Lg	CJ	2.00	8.00	292.00	60.70	---	352.70	**427.00**
	Inst	Set	Sm	CJ	3.08	5.20	315.00	93.50	---	408.50	**503.00**
Good quality workmanship	Inst	Set	Lg	CJ	1.60	10.00	269.00	48.60	---	317.60	**382.00**
	Inst	Set	Sm	CJ	2.46	6.50	290.00	74.70	---	304.70	**446.00**
Average quality workmanship	Inst	Set	Lg	CJ	1.33	12.00	225.00	40.40	---	265.40	**319.00**
	Inst	Set	Sm	CJ	2.05	7.80	242.00	62.20	---	304.20	**372.00**
41" H											
High quality workmanship	Inst	Set	Lg	CJ	2.11	7.60	311.00	64.00	---	375.00	**454.00**
	Inst	Set	Sm	CJ	3.27	4.90	336.00	99.20	---	435.20	**535.00**
Good quality workmanship	Inst	Set	Lg	CJ	1.68	9.50	288.00	51.00	---	339.00	**408.00**
	Inst	Set	Sm	CJ	2.58	6.20	311.00	78.30	---	389.30	**475.00**
Average quality workmanship	Inst	Set	Lg	CJ	1.40	11.40	240.00	42.50	---	282.50	**340.00**
	Inst	Set	Sm	CJ	2.16	7.40	259.00	65.60	---	324.60	**396.00**

Description	Oper	Unit	Vol	Crew Size	Man-hours per Unit	Crew Output per Day	Avg Mat'l Unit Cost	Avg Labor Unit Cost	Avg Equip Unit Cost	Avg Total Unit Cost	Avg Price Incl O&P
48" H											
High quality workmanship	Inst	Set	Lg	CJ	2.22	7.20	356.00	67.40	---	423.40	**510.00**
	Inst	Set	Sm	CJ	3.40	4.70	384.00	103.00	---	487.00	**596.00**
Good quality workmanship	Inst	Set	Lg	CJ	1.78	9.00	329.00	54.00	---	383.00	**459.00**
	Inst	Set	Sm	CJ	2.71	5.90	355.00	82.30	---	437.30	**531.00**
Average quality workmanship	Inst	Set	Lg	CJ	1.48	10.80	274.00	44.90	---	318.90	**383.00**
	Inst	Set	Sm	CJ	2.29	7.00	296.00	69.50	---	365.50	**445.00**
53" H											
High quality workmanship	Inst	Set	Lg	CJ	2.35	6.80	387.00	71.30	---	458.30	**552.00**
	Inst	Set	Sm	CJ	3.64	4.40	418.00	110.00	---	528.00	**646.00**
Good quality workmanship	Inst	Set	Lg	CJ	1.88	8.50	358.00	57.10	---	415.10	**497.00**
	Inst	Set	Sm	CJ	2.91	5.50	386.00	88.30	---	474.30	**577.00**
Average quality workmanship	Inst	Set	Lg	CJ	1.57	10.20	298.00	47.70	---	345.70	**415.00**
	Inst	Set	Sm	CJ	2.42	6.60	322.00	73.50	---	395.50	**480.00**
60" H											
High quality workmanship	Inst	Set	Lg	CJ	2.50	6.40	425.00	75.90	---	500.90	**603.00**
	Inst	Set	Sm	CJ	3.81	4.20	459.00	116.00	---	575.00	**701.00**
Good quality workmanship	Inst	Set	Lg	CJ	2.00	8.00	394.00	60.70	---	454.70	**544.00**
	Inst	Set	Sm	CJ	3.08	5.20	425.00	93.50	---	518.50	**629.00**
Average quality workmanship	Inst	Set	Lg	CJ	1.67	9.60	328.00	50.70	---	378.70	**453.00**
	Inst	Set	Sm	CJ	2.58	6.20	353.00	78.30	---	431.30	**524.00**
66" H											
High quality workmanship	Inst	Set	Lg	CJ	2.50	6.40	457.00	75.90	---	532.90	**640.00**
	Inst	Set	Sm	CJ	3.81	4.20	493.00	116.00	---	609.00	**741.00**
Good quality workmanship	Inst	Set	Lg	CJ	2.00	8.00	423.00	60.70	---	483.70	**577.00**
	Inst	Set	Sm	CJ	3.08	5.20	456.00	93.50	---	549.50	**665.00**
Average quality workmanship	Inst	Set	Lg	CJ	1.67	9.60	352.00	50.70	---	402.70	**481.00**
	Inst	Set	Sm	CJ	2.58	6.20	379.00	78.30	---	457.30	**554.00**
72" H											
High quality workmanship	Inst	Set	Lg	CJ	2.67	6.00	489.00	81.00	---	570.00	**684.00**
	Inst	Set	Sm	CJ	4.10	3.90	527.00	124.00	---	651.00	**793.00**
Good quality workmanship	Inst	Set	Lg	CJ	2.13	7.50	452.00	64.70	---	516.70	**617.00**
	Inst	Set	Sm	CJ	3.27	4.90	488.00	99.20	---	587.20	**710.00**
Average quality workmanship	Inst	Set	Lg	CJ	1.78	9.00	377.00	54.00	---	431.00	**515.00**
	Inst	Set	Sm	CJ	2.71	5.90	407.00	82.30	---	489.30	**591.00**

2 ventilating lites, no fixed lites; 40" W

Description	Oper	Unit	Vol	Crew Size	Man-hours per Unit	Crew Output per Day	Avg Mat'l Unit Cost	Avg Labor Unit Cost	Avg Equip Unit Cost	Avg Total Unit Cost	Avg Price Incl O&P
36" H											
High quality workmanship	Inst	Set	Lg	CJ	2.50	6.40	483.00	75.90	---	558.90	**669.00**
	Inst	Set	Sm	CJ	3.81	4.20	521.00	116.00	---	637.00	**772.00**
Good quality workmanship	Inst	Set	Lg	CJ	2.00	8.00	446.00	60.70	---	506.70	**604.00**
	Inst	Set	Sm	CJ	3.08	5.20	481.00	93.50	---	574.50	**693.00**
Average quality workmanship	Inst	Set	Lg	CJ	1.67	9.60	372.00	50.70	---	422.70	**504.00**
	Inst	Set	Sm	CJ	2.58	6.20	401.00	78.30	---	479.30	**579.00**

Description	Oper	Unit	Vol	Crew Size	Man-hours per Unit	Crew Output per Day	Avg Mat'l Unit Cost	Avg Labor Unit Cost	Avg Equip Unit Cost	Avg Total Unit Cost	Avg Price Incl O&P
41" H											
High quality workmanship	Inst	Set	Lg	CJ	2.67	6.00	508.00	81.00	---	589.00	**706.00**
	Inst	Set	Sm	CJ	4.10	3.90	548.00	124.00	---	672.00	**817.00**
Good quality workmanship	Inst	Set	Lg	CJ	2.13	7.50	470.00	64.70	---	534.70	**637.00**
	Inst	Set	Sm	CJ	3.27	4.90	507.00	99.20	---	606.20	**732.00**
Average quality workmanship	Inst	Set	Lg	CJ	1.78	9.00	391.00	54.00	---	445.00	**531.00**
	Inst	Set	Sm	CJ	2.71	5.90	422.00	82.30	---	504.30	**609.00**
48" H											
High quality workmanship	Inst	Set	Lg	CJ	2.86	5.60	584.00	86.80	---	670.80	**802.00**
	Inst	Set	Sm	CJ	4.44	3.60	630.00	135.00	---	765.00	**927.00**
Good quality workmanship	Inst	Set	Lg	CJ	2.29	7.00	540.00	69.50	---	609.50	**725.00**
	Inst	Set	Sm	CJ	3.48	4.60	582.00	106.00	---	688.00	**828.00**
Average quality workmanship	Inst	Set	Lg	CJ	1.90	8.40	450.00	57.70	---	507.70	**604.00**
	Inst	Set	Sm	CJ	2.91	5.50	485.00	88.30	---	573.30	**690.00**
53" H											
High quality workmanship	Inst	Set	Lg	CJ	3.08	5.20	635.00	93.50	---	728.50	**870.00**
	Inst	Set	Sm	CJ	4.71	3.40	685.00	143.00	---	828.00	**1000.00**
Good quality workmanship	Inst	Set	Lg	CJ	2.46	6.50	588.00	74.70	---	662.70	**788.00**
	Inst	Set	Sm	CJ	3.81	4.20	634.00	116.00	---	750.00	**903.00**
Average quality workmanship	Inst	Set	Lg	CJ	2.05	7.80	489.00	62.20	---	551.20	**656.00**
	Inst	Set	Sm	CJ	3.14	5.10	527.00	95.30	---	622.30	**750.00**
60" H											
High quality workmanship	Inst	Set	Lg	CJ	3.33	4.80	705.00	101.00	---	806.00	**962.00**
	Inst	Set	Sm	CJ	5.16	3.10	760.00	157.00	---	917.00	**1110.00**
Good quality workmanship	Inst	Set	Lg	CJ	2.67	6.00	652.00	81.00	---	733.00	**871.00**
	Inst	Set	Sm	CJ	4.10	3.90	703.00	124.00	---	827.00	**995.00**
Average quality workmanship	Inst	Set	Lg	CJ	2.22	7.20	544.00	67.40	---	611.40	**726.00**
	Inst	Set	Sm	CJ	3.40	4.70	586.00	103.00	---	689.00	**829.00**
66" H											
High quality workmanship	Inst	Set	Lg	CJ	3.33	4.80	737.00	101.00	---	838.00	**999.00**
	Inst	Set	Sm	CJ	5.16	3.10	795.00	157.00	---	952.00	**1150.00**
Good quality workmanship	Inst	Set	Lg	CJ	2.67	6.00	682.00	81.00	---	763.00	**906.00**
	Inst	Set	Sm	CJ	4.10	3.90	736.00	124.00	---	860.00	**1030.00**
Average quality workmanship	Inst	Set	Lg	CJ	2.22	7.20	568.00	67.40	---	635.40	**754.00**
	Inst	Set	Sm	CJ	3.40	4.70	612.00	103.00	---	715.00	**859.00**
72" H											
High quality workmanship	Inst	Set	Lg	CJ	3.64	4.40	768.00	110.00	---	078.00	**1050.00**
	Inst	Set	Sm	CJ	5.52	2.90	829.00	168.00	---	997.00	**1200.00**
Good quality workmanship	Inst	Set	Lg	CJ	2.91	5.50	711.00	88.30	---	799.30	**950.00**
	Inst	Set	Sm	CJ	4.44	3.60	767.00	135.00	---	902.00	**1080.00**
Average quality workmanship	Inst	Set	Lg	CJ	2.42	6.60	592.00	73.50	---	665.50	**791.00**
	Inst	Set	Sm	CJ	3.72	4.30	638.00	113.00	---	751.00	**904.00**

Description	Oper	Unit	Vol	Crew Size	Man-hours per Unit	Crew Output per Day	Avg Mat'l Unit Cost	Avg Labor Unit Cost	Avg Equip Unit Cost	Avg Total Unit Cost	Avg Price Incl O&P
2 ventilating lites, no fixed lites; 48" W											
36" H											
High quality workmanship	Inst	Set	Lg	CJ	2.50	6.40	521.00	75.90	---	596.90	**713.00**
	Inst	Set	Sm	CJ	3.81	4.20	562.00	116.00	---	678.00	**819.00**
Good quality workmanship	Inst	Set	Lg	CJ	2.00	8.00	483.00	60.70	---	543.70	**646.00**
	Inst	Set	Sm	CJ	3.08	5.20	521.00	93.50	---	614.50	**739.00**
Average quality workmanship	Inst	Set	Lg	CJ	1.67	9.60	401.00	50.70	---	451.70	**538.00**
	Inst	Set	Sm	CJ	2.58	6.20	433.00	78.30	---	511.30	**615.00**
41" H											
High quality workmanship	Inst	Set	Lg	CJ	2.67	6.00	565.00	81.00	---	646.00	**771.00**
	Inst	Set	Sm	CJ	4.10	3.90	610.00	124.00	---	734.00	**888.00**
Good quality workmanship	Inst	Set	Lg	CJ	2.13	7.50	523.00	64.70	---	587.70	**699.00**
	Inst	Set	Sm	CJ	3.27	4.90	564.00	99.20	---	663.20	**798.00**
Average quality workmanship	Inst	Set	Lg	CJ	1.78	9.00	436.00	54.00	---	490.00	**582.00**
	Inst	Set	Sm	CJ	2.71	5.90	470.00	82.30	---	552.30	**664.00**
48" H											
High quality workmanship	Inst	Set	Lg	CJ	3.33	4.80	622.00	101.00	---	723.00	**867.00**
	Inst	Set	Sm	CJ	5.16	3.10	671.00	157.00	---	828.00	**1010.00**
Good quality workmanship	Inst	Set	Lg	CJ	2.67	6.00	575.00	81.00	---	656.00	**783.00**
	Inst	Set	Sm	CJ	4.10	3.90	621.00	124.00	---	745.00	**900.00**
Average quality workmanship	Inst	Set	Lg	CJ	2.22	7.20	480.00	67.40	---	547.40	**653.00**
	Inst	Set	Sm	CJ	3.40	4.70	518.00	103.00	---	621.00	**750.00**
53" H											
High quality workmanship	Inst	Set	Lg	CJ	3.08	5.20	679.00	93.50	---	772.50	**922.00**
	Inst	Set	Sm	CJ	4.71	3.40	733.00	143.00	---	876.00	**1060.00**
Good quality workmanship	Inst	Set	Lg	CJ	2.46	6.50	629.00	74.70	---	703.70	**835.00**
	Inst	Set	Sm	CJ	3.81	4.20	678.00	116.00	---	794.00	**953.00**
Average quality workmanship	Inst	Set	Lg	CJ	2.05	7.80	523.00	62.20	---	585.20	**695.00**
	Inst	Set	Sm	CJ	3.14	5.10	564.00	95.30	---	659.30	**792.00**
60" H											
High quality workmanship	Inst	Set	Lg	CJ	3.33	4.80	718.00	101.00	---	819.00	**977.00**
	Inst	Set	Sm	CJ	5.16	3.10	774.00	157.00	---	931.00	**1130.00**
Good quality workmanship	Inst	Set	Lg	CJ	2.67	6.00	664.00	81.00	---	745.00	**885.00**
	Inst	Set	Sm	CJ	4.10	3.90	717.00	124.00	---	841.00	**1010.00**
Average quality workmanship	Inst	Set	Lg	CJ	2.22	7.20	552.00	67.40	---	619.40	**736.00**
	Inst	Set	Sm	CJ	3.40	4.70	596.00	103.00	---	699.00	**840.00**
66" H											
High quality workmanship	Inst	Set	Lg	CJ	3.33	4.80	768.00	101.00	---	869.00	**1040.00**
	Inst	Set	Sm	CJ	5.16	3.10	829.00	157.00	---	986.00	**1190.00**
Good quality workmanship	Inst	Set	Lg	CJ	2.67	6.00	711.00	81.00	---	792.00	**939.00**
	Inst	Set	Sm	CJ	4.10	3.90	767.00	124.00	---	891.00	**1070.00**
Average quality workmanship	Inst	Set	Lg	CJ	2.22	7.20	592.00	67.40	---	659.40	**782.00**
	Inst	Set	Sm	CJ	3.40	4.70	638.00	103.00	---	741.00	**889.00**
72" H											
High quality workmanship	Inst	Set	Lg	CJ	3.64	4.40	819.00	110.00	---	929.00	**1110.00**
	Inst	Set	Sm	CJ	5.52	2.90	884.00	168.00	---	1052.00	**1270.00**
Good quality workmanship	Inst	Set	Lg	CJ	2.91	5.50	758.00	88.30	---	846.30	**1000.00**
	Inst	Set	Sm	CJ	4.44	3.60	818.00	135.00	---	953.00	**1140.00**
Average quality workmanship	Inst	Set	Lg	CJ	2.42	6.60	631.00	73.50	---	704.50	**836.00**
	Inst	Set	Sm	CJ	3.72	4.30	681.00	113.00	---	794.00	**952.00**

Description	Oper	Unit	Vol	Crew Size	Man-hours per Unit	Crew Output per Day	Avg Mat'l Unit Cost	Avg Labor Unit Cost	Avg Equip Unit Cost	Avg Total Unit Cost	Avg Price Incl O&P
2 ventilating lites, no fixed lites; 56" W											
36" H											
High quality workmanship	Inst	Set	Lg	CJ	2.50	6.40	565.00	75.90	---	640.90	**764.00**
	Inst	Set	Sm	CJ	3.81	4.20	610.00	116.00	---	726.00	**875.00**
Good quality workmanship	Inst	Set	Lg	CJ	2.00	8.00	523.00	60.70	---	583.70	**693.00**
	Inst	Set	Sm	CJ	3.08	5.20	564.00	93.50	---	657.50	**789.00**
Average quality workmanship	Inst	Set	Lg	CJ	1.67	9.60	436.00	50.70	---	486.70	**577.00**
	Inst	Set	Sm	CJ	2.58	6.20	470.00	78.30	---	548.30	**658.00**
41" H											
High quality workmanship	Inst	Set	Lg	CJ	2.67	6.00	622.00	81.00	---	703.00	**837.00**
	Inst	Set	Sm	CJ	4.10	3.90	671.00	124.00	---	795.00	**959.00**
Good quality workmanship	Inst	Set	Lg	CJ	2.13	7.50	575.00	64.70	---	639.70	**759.00**
	Inst	Set	Sm	CJ	3.27	4.90	621.00	99.20	---	720.20	**863.00**
Average quality workmanship	Inst	Set	Lg	CJ	1.78	9.00	480.00	54.00	---	534.00	**633.00**
	Inst	Set	Sm	CJ	2.71	5.90	518.00	82.30	---	600.30	**719.00**
48" H											
High quality workmanship	Inst	Set	Lg	CJ	2.86	5.60	692.00	86.80	---	778.80	**926.00**
	Inst	Set	Sm	CJ	4.44	3.60	747.00	135.00	---	882.00	**1060.00**
Good quality workmanship	Inst	Set	Lg	CJ	2.29	7.00	640.00	69.50	---	709.50	**840.00**
	Inst	Set	Sm	CJ	3.48	4.60	690.00	106.00	---	796.00	**953.00**
Average quality workmanship	Inst	Set	Lg	CJ	1.90	8.40	533.00	57.70	---	590.70	**700.00**
	Inst	Set	Sm	CJ	2.91	5.50	575.00	88.30	---	663.30	**794.00**
53" H											
High quality workmanship	Inst	Set	Lg	CJ	3.08	5.20	759.00	93.50	---	852.50	**1010.00**
	Inst	Set	Sm	CJ	4.71	3.40	819.00	143.00	---	962.00	**1160.00**
Good quality workmanship	Inst	Set	Lg	CJ	2.46	6.50	702.00	74.70	---	776.70	**920.00**
	Inst	Set	Sm	CJ	3.81	4.20	758.00	116.00	---	874.00	**1040.00**
Average quality workmanship	Inst	Set	Lg	CJ	2.05	7.80	585.00	62.20	---	647.20	**767.00**
	Inst	Set	Sm	CJ	3.14	5.10	632.00	95.30	---	727.30	**869.00**
60" H											
High quality workmanship	Inst	Set	Lg	CJ	3.33	4.80	826.00	101.00	---	927.00	**1100.00**
	Inst	Set	Sm	CJ	5.16	3.10	891.00	157.00	---	1048.00	**1260.00**
Good quality workmanship	Inst	Set	Lg	CJ	2.67	6.00	765.00	81.00	---	846.00	**1000.00**
	Inst	Set	Sm	CJ	4.10	3.90	825.00	124.00	---	949.00	**1140.00**
Average quality workmanship	Inst	Set	Lg	CJ	2.22	7.20	636.00	67.40	---	703.40	**833.00**
	Inst	Set	Sm	CJ	3.40	4.70	686.00	103.00	---	789.00	**944.00**
66" H											
High quality workmanship	Inst	Set	Lg	CJ	3.33	4.80	883.00	101.00	---	984.00	**1170.00**
	Inst	Set	Sm	CJ	5.16	3.10	952.00	157.00	---	1109.00	**1330.00**
Good quality workmanship	Inst	Set	Lg	CJ	2.67	6.00	818.00	81.00	---	899.00	**1060.00**
	Inst	Set	Sm	CJ	4.10	3.90	882.00	124.00	---	1006.00	**1200.00**
Average quality workmanship	Inst	Set	Lg	CJ	2.22	7.20	681.00	67.40	---	748.40	**884.00**
	Inst	Set	Sm	CJ	3.40	4.70	734.00	103.00	---	837.00	**999.00**
72" H											
High quality workmanship	Inst	Set	Lg	CJ	3.64	4.40	940.00	110.00	---	1050.00	**1250.00**
	Inst	Set	Sm	CJ	5.52	2.90	1010.00	168.00	---	1178.00	**1420.00**
Good quality workmanship	Inst	Set	Lg	CJ	2.91	5.50	870.00	88.30	---	958.30	**1130.00**
	Inst	Set	Sm	CJ	4.44	3.60	938.00	135.00	---	1073.00	**1280.00**
Average quality workmanship	Inst	Set	Lg	CJ	2.42	6.60	724.00	73.50	---	797.50	**943.00**
	Inst	Set	Sm	CJ	3.72	4.30	781.00	113.00	---	894.00	**1070.00**

Description	Oper	Unit	Vol	Crew Size	Man-hours per Unit	Crew Output per Day	Avg Mat'l Unit Cost	Avg Labor Unit Cost	Avg Equip Unit Cost	Avg Total Unit Cost	Avg Price Incl O&P

2 ventilating lites, no fixed lites; 72" W

36" H

Description	Oper	Unit	Vol	Crew Size	Man-hours per Unit	Crew Output per Day	Avg Mat'l Unit Cost	Avg Labor Unit Cost	Avg Equip Unit Cost	Avg Total Unit Cost	Avg Price Incl O&P
High quality workmanship	Inst	Set	Lg	CJ	2.86	5.60	737.00	86.80	---	823.80	**977.00**
	Inst	Set	Sm	CJ	4.44	3.60	795.00	135.00	---	930.00	**1120.00**
Good quality workmanship	Inst	Set	Lg	CJ	2.29	7.00	682.00	69.50	---	751.50	**889.00**
	Inst	Set	Sm	CJ	3.48	4.60	736.00	106.00	---	842.00	**1000.00**
Average quality workmanship	Inst	Set	Lg	CJ	1.90	8.40	568.00	57.70	---	625.70	**739.00**
	Inst	Set	Sm	CJ	2.91	5.50	612.00	88.30	---	700.30	**837.00**

41" H

Description	Oper	Unit	Vol	Crew Size	Man-hours per Unit	Crew Output per Day	Avg Mat'l Unit Cost	Avg Labor Unit Cost	Avg Equip Unit Cost	Avg Total Unit Cost	Avg Price Incl O&P
High quality workmanship	Inst	Set	Lg	CJ	3.08	5.20	813.00	93.50	---	906.50	**1070.00**
	Inst	Set	Sm	CJ	4.71	3.40	877.00	143.00	---	1020.00	**1220.00**
Good quality workmanship	Inst	Set	Lg	CJ	2.46	6.50	752.00	74.70	---	826.70	**977.00**
	Inst	Set	Sm	CJ	3.81	4.20	811.00	116.00	---	927.00	**1110.00**
Average quality workmanship	Inst	Set	Lg	CJ	2.05	7.80	626.00	62.20	---	688.20	**813.00**
	Inst	Set	Sm	CJ	3.14	5.10	675.00	95.30	---	770.30	**920.00**

48" H

Description	Oper	Unit	Vol	Crew Size	Man-hours per Unit	Crew Output per Day	Avg Mat'l Unit Cost	Avg Labor Unit Cost	Avg Equip Unit Cost	Avg Total Unit Cost	Avg Price Incl O&P
High quality workmanship	Inst	Set	Lg	CJ	3.33	4.80	883.00	101.00	---	984.00	**1170.00**
	Inst	Set	Sm	CJ	5.16	3.10	952.00	157.00	---	1109.00	**1330.00**
Good quality workmanship	Inst	Set	Lg	CJ	2.67	6.00	817.00	81.00	---	898.00	**1060.00**
	Inst	Set	Sm	CJ	4.10	3.90	881.00	124.00	---	1005.00	**1200.00**
Average quality workmanship	Inst	Set	Lg	CJ	2.22	7.20	681.00	67.40	---	748.40	**884.00**
	Inst	Set	Sm	CJ	3.40	4.70	734.00	103.00	---	837.00	**999.00**

60" H

Description	Oper	Unit	Vol	Crew Size	Man-hours per Unit	Crew Output per Day	Avg Mat'l Unit Cost	Avg Labor Unit Cost	Avg Equip Unit Cost	Avg Total Unit Cost	Avg Price Incl O&P
High quality workmanship	Inst	Set	Lg	CJ	3.33	4.80	1000.00	101.00	---	1101.00	**1310.00**
	Inst	Set	Sm	CJ	5.16	3.10	1080.00	157.00	---	1237.00	**1480.00**
Good quality workmanship	Inst	Set	Lg	CJ	2.67	6.00	928.00	81.00	---	1009.00	**1190.00**
	Inst	Set	Sm	CJ	4.10	3.90	1000.00	124.00	---	1124.00	**1340.00**
Average quality workmanship	Inst	Set	Lg	CJ	2.22	7.20	773.00	67.40	---	840.40	**991.00**
	Inst	Set	Sm	CJ	3.40	4.70	834.00	103.00	---	937.00	**1110.00**

4 ventilating lites, no fixed lites; 68" W

36" H

Description	Oper	Unit	Vol	Crew Size	Man-hours per Unit	Crew Output per Day	Avg Mat'l Unit Cost	Avg Labor Unit Cost	Avg Equip Unit Cost	Avg Total Unit Cost	Avg Price Incl O&P
High quality workmanship	Inst	Set	Lg	CJ	3.33	4.80	940.00	101.00	---	1041.00	**1230.00**
	Inst	Set	Sm	CJ	5.16	3.10	1010.00	157.00	---	1167.00	**1400.00**
Good quality workmanship	Inst	Set	Lg	CJ	2.67	6.00	870.00	81.00	---	951.00	**1120.00**
	Inst	Set	Sm	CJ	4.10	3.90	938.00	124.00	---	1062.00	**1270.00**
Average quality workmanship	Inst	Set	Lg	CJ	2.22	7.20	724.00	67.40	---	791.40	**934.00**
	Inst	Set	Sm	CJ	3.40	4.70	781.00	103.00	---	884.00	**1050.00**

41" H

Description	Oper	Unit	Vol	Crew Size	Man-hours per Unit	Crew Output per Day	Avg Mat'l Unit Cost	Avg Labor Unit Cost	Avg Equip Unit Cost	Avg Total Unit Cost	Avg Price Incl O&P
High quality workmanship	Inst	Set	Lg	CJ	3.64	4.40	1050.00	110.00	---	1160.00	**1380.00**
	Inst	Set	Sm	CJ	5.52	2.90	1140.00	168.00	---	1308.00	**1560.00**
Good quality workmanship	Inst	Set	Lg	CJ	2.91	5.50	975.00	88.30	---	1063.30	**1250.00**
	Inst	Set	Sm	CJ	4.44	3.60	1050.00	135.00	---	1185.00	**1410.00**
Average quality workmanship	Inst	Set	Lg	CJ	2.42	6.60	813.00	73.50	---	886.50	**1040.00**
	Inst	Set	Sm	CJ	3.72	4.30	877.00	113.00	---	990.00	**1180.00**

Description	Oper	Unit	Vol	Crew Size	Man-hours per Unit	Crew Output per Day	Avg Mat'l Unit Cost	Avg Labor Unit Cost	Avg Equip Unit Cost	Avg Total Unit Cost	Avg Price Incl O&P
48" H											
High quality workmanship	Inst	Set	Lg	CJ	4.00	4.00	1210.00	121.00	---	1331.00	**1570.00**
	Inst	Set	Sm	CJ	6.15	2.60	1300.00	187.00	---	1487.00	**1780.00**
Good quality workmanship	Inst	Set	Lg	CJ	3.20	5.00	1120.00	97.10	---	1217.10	**1430.00**
	Inst	Set	Sm	CJ	4.85	3.30	1210.00	147.00	---	1357.00	**1610.00**
Average quality workmanship	Inst	Set	Lg	CJ	2.67	6.00	931.00	81.00	---	1012.00	**1190.00**
	Inst	Set	Sm	CJ	4.10	3.90	1000.00	124.00	---	1124.00	**1340.00**
60" H											
High quality workmanship	Inst	Set	Lg	CJ	4.44	3.60	1490.00	135.00	---	1625.00	**1920.00**
	Inst	Set	Sm	CJ	6.96	2.30	1610.00	211.00	---	1821.00	**2170.00**
Good quality workmanship	Inst	Set	Lg	CJ	3.56	4.50	1380.00	108.00	---	1488.00	**1750.00**
	Inst	Set	Sm	CJ	5.52	2.90	1490.00	168.00	---	1658.00	**1960.00**
Average quality workmanship	Inst	Set	Lg	CJ	2.96	5.40	1150.00	89.80	---	1239.80	**1460.00**
	Inst	Set	Sm	CJ	4.57	3.50	1240.00	139.00	---	1379.00	**1630.00**
72" H											
High quality workmanship	Inst	Set	Lg	CJ	5.00	3.20	1640.00	152.00	---	1792.00	**2110.00**
	Inst	Set	Sm	CJ	7.62	2.10	1770.00	231.00	---	2001.00	**2380.00**
Good quality workmanship	Inst	Set	Lg	CJ	4.00	4.00	1520.00	121.00	---	1641.00	**1930.00**
	Inst	Set	Sm	CJ	6.15	2.60	1640.00	187.00	---	1827.00	**2160.00**
Average quality workmanship	Inst	Set	Lg	CJ	3.33	4.80	1260.00	101.00	---	1361.00	**1600.00**
	Inst	Set	Sm	CJ	5.16	3.10	1360.00	157.00	---	1517.00	**1800.00**

4 ventilating lites, no fixed lites; 82" W

Description	Oper	Unit	Vol	Crew Size	Man-hours per Unit	Crew Output per Day	Avg Mat'l Unit Cost	Avg Labor Unit Cost	Avg Equip Unit Cost	Avg Total Unit Cost	Avg Price Incl O&P
36" H											
High quality workmanship	Inst	Set	Lg	CJ	3.33	4.80	1020.00	101.00	---	1121.00	**1330.00**
	Inst	Set	Sm	CJ	5.16	3.10	1100.00	157.00	---	1257.00	**1500.00**
Good quality workmanship	Inst	Set	Lg	CJ	2.67	6.00	946.00	81.00	---	1027.00	**1210.00**
	Inst	Set	Sm	CJ	4.10	3.90	1020.00	124.00	---	1144.00	**1360.00**
Average quality workmanship	Inst	Set	Lg	CJ	2.22	7.20	789.00	67.40	---	856.40	**1010.00**
	Inst	Set	Sm	CJ	3.40	4.70	851.00	103.00	---	954.00	**1130.00**
41" H											
High quality workmanship	Inst	Set	Lg	CJ	3.64	4.40	1080.00	110.00	---	1190.00	**1410.00**
	Inst	Set	Sm	CJ	5.52	2.90	1160.00	168.00	---	1328.00	**1590.00**
Good quality workmanship	Inst	Set	Lg	CJ	2.91	5.50	999.00	88.30	---	1087.30	**1280.00**
	Inst	Set	Sm	CJ	4.44	3.60	1080.00	135.00	---	1215.00	**1440.00**
Average quality workmanship	Inst	Set	Lg	CJ	2.42	6.60	832.00	73.50	---	905.50	**1070.00**
	Inst	Set	Sm	CJ	3.72	4.30	897.00	113.00	---	1010.00	**1200.00**
48" H											
High quality workmanship	Inst	Set	Lg	CJ	4.00	4.00	1240.00	121.00	---	1361.00	**1610.00**
	Inst	Set	Sm	CJ	6.15	2.60	1340.00	187.00	---	1527.00	**1820.00**
Good quality workmanship	Inst	Set	Lg	CJ	3.20	5.00	1150.00	97.10	---	1247.10	**1460.00**
	Inst	Set	Sm	CJ	4.85	3.30	1240.00	147.00	---	1387.00	**1640.00**
Average quality workmanship	Inst	Set	Lg	CJ	2.67	6.00	955.00	81.00	---	1036.00	**1220.00**
	Inst	Set	Sm	CJ	4.10	3.90	1030.00	124.00	---	1154.00	**1370.00**

Description	Oper	Unit	Vol	Crew Size	Man-hours per Unit	Crew Output per Day	Avg Mat'l Unit Cost	Avg Labor Unit Cost	Avg Equip Unit Cost	Avg Total Unit Cost	Avg Price Incl O&P
60" H											
High quality workmanship	Inst	Set	Lg	CJ	4.44	3.60	1510.00	135.00	---	1645.00	**1940.00**
	Inst	Set	Sm	CJ	6.96	2.30	1630.00	211.00	---	1841.00	**2190.00**
Good quality workmanship	Inst	Set	Lg	CJ	3.56	4.50	1400.00	108.00	---	1508.00	**1770.00**
	Inst	Set	Sm	CJ	5.52	2.90	1510.00	168.00	---	1678.00	**1990.00**
Average quality workmanship	Inst	Set	Lg	CJ	2.96	5.40	1170.00	89.80	---	1259.80	**1480.00**
	Inst	Set	Sm	CJ	4.57	3.50	1260.00	139.00	---	1399.00	**1650.00**
72" H											
High quality workmanship	Inst	Set	Lg	CJ	5.00	3.20	1730.00	152.00	---	1882.00	**2220.00**
	Inst	Set	Sm	CJ	7.62	2.10	1870.00	231.00	---	2101.00	**2500.00**
Good quality workmanship	Inst	Set	Lg	CJ	4.00	4.00	1600.00	121.00	---	1721.00	**2030.00**
	Inst	Set	Sm	CJ	6.15	2.60	1730.00	187.00	---	1917.00	**2270.00**
Average quality workmanship	Inst	Set	Lg	CJ	3.33	4.80	1340.00	101.00	---	1441.00	**1690.00**
	Inst	Set	Sm	CJ	5.16	3.10	1440.00	157.00	---	1597.00	**1890.00**

4 ventilating lites, no fixed lites; 96" W

Description	Oper	Unit	Vol	Crew Size	Man-hours per Unit	Crew Output per Day	Avg Mat'l Unit Cost	Avg Labor Unit Cost	Avg Equip Unit Cost	Avg Total Unit Cost	Avg Price Incl O&P
36" H											
High quality workmanship	Inst	Set	Lg	CJ	3.33	4.80	1090.00	101.00	---	1191.00	**1410.00**
	Inst	Set	Sm	CJ	5.16	3.10	1180.00	157.00	---	1337.00	**1590.00**
Good quality workmanship	Inst	Set	Lg	CJ	2.67	6.00	1010.00	81.00	---	1091.00	**1280.00**
	Inst	Set	Sm	CJ	4.10	3.90	1090.00	124.00	---	1214.00	**1440.00**
Average quality workmanship	Inst	Set	Lg	CJ	2.22	7.20	842.00	67.40	---	909.40	**1070.00**
	Inst	Set	Sm	CJ	3.40	4.70	908.00	103.00	---	1011.00	**1200.00**
41" H											
High quality workmanship	Inst	Set	Lg	CJ	3.64	4.40	1190.00	110.00	---	1300.00	**1540.00**
	Inst	Set	Sm	CJ	5.52	2.90	1290.00	168.00	---	1458.00	**1730.00**
Good quality workmanship	Inst	Set	Lg	CJ	2.91	5.50	1100.00	88.30	---	1188.30	**1400.00**
	Inst	Set	Sm	CJ	4.44	3.60	1190.00	135.00	---	1325.00	**1570.00**
Average quality workmanship	Inst	Set	Lg	CJ	2.42	6.60	921.00	73.50	---	994.50	**1170.00**
	Inst	Set	Sm	CJ	3.72	4.30	993.00	113.00	---	1106.00	**1310.00**
48" H											
High quality workmanship	Inst	Set	Lg	CJ	4.00	4.00	1310.00	121.00	---	1431.00	**1690.00**
	Inst	Set	Sm	CJ	6.15	2.60	1410.00	187.00	---	1597.00	**1900.00**
Good quality workmanship	Inst	Set	Lg	CJ	3.20	5.00	1210.00	97.10	---	1307.10	**1540.00**
	Inst	Set	Sm	CJ	4.85	3.30	1310.00	147.00	---	1457.00	**1720.00**
Average quality workmanship	Inst	Set	Lg	CJ	2.67	6.00	1010.00	81.00	---	1091.00	**1280.00**
	Inst	Set	Sm	CJ	4.10	3.90	1090.00	124.00	---	1214.00	**1440.00**
60" H											
High quality workmanship	Inst	Set	Lg	CJ	4.44	3.60	1640.00	135.00	---	1775.00	**2090.00**
	Inst	Set	Sm	CJ	6.96	2.30	1770.00	211.00	---	1981.00	**2360.00**
Good quality workmanship	Inst	Set	Lg	CJ	3.56	4.50	1520.00	108.00	---	1628.00	**1910.00**
	Inst	Set	Sm	CJ	5.52	2.90	1640.00	168.00	---	1808.00	**2140.00**
Average quality workmanship	Inst	Set	Lg	CJ	2.96	5.40	1270.00	89.80	---	1359.80	**1590.00**
	Inst	Set	Sm	CJ	4.57	3.50	1370.00	139.00	---	1509.00	**1780.00**
72" H											
High quality workmanship	Inst	Set	Lg	CJ	5.00	3.20	1810.00	152.00	---	1962.00	**2310.00**
	Inst	Set	Sm	CJ	7.62	2.10	1950.00	231.00	---	2181.00	**2590.00**
Good quality workmanship	Inst	Set	Lg	CJ	4.00	4.00	1680.00	121.00	---	1801.00	**2110.00**
	Inst	Set	Sm	CJ	6.15	2.60	1810.00	187.00	---	1997.00	**2360.00**
Average quality workmanship	Inst	Set	Lg	CJ	3.33	4.80	1400.00	101.00	---	1501.00	**1760.00**
	Inst	Set	Sm	CJ	5.16	3.10	1510.00	157.00	---	1667.00	**1970.00**

Description	Oper	Unit	Vol	Crew Size	Man-hours per Unit	Crew Output per Day	Avg Mat'l Unit Cost	Avg Labor Unit Cost	Avg Equip Unit Cost	Avg Total Unit Cost	Avg Price Incl O&P
4 ventilating lites, no fixed lites; 114" W											
36" H											
High quality workmanship	Inst	Set	Lg	CJ	3.33	4.80	1190.00	101.00	---	1291.00	**1520.00**
	Inst	Set	Sm	CJ	5.16	3.10	1280.00	157.00	---	1437.00	**1710.00**
Good quality workmanship	Inst	Set	Lg	CJ	2.67	6.00	1100.00	81.00	---	1181.00	**1390.00**
	Inst	Set	Sm	CJ	4.10	3.90	1190.00	124.00	---	1314.00	**1550.00**
Average quality workmanship	Inst	Set	Lg	CJ	2.22	7.20	916.00	67.40	---	983.40	**1150.00**
	Inst	Set	Sm	CJ	3.40	4.70	988.00	103.00	---	1091.00	**1290.00**
41" H											
High quality workmanship	Inst	Set	Lg	CJ	3.64	4.40	1280.00	110.00	---	1390.00	**1630.00**
	Inst	Set	Sm	CJ	5.52	2.90	1380.00	168.00	---	1548.00	**1830.00**
Good quality workmanship	Inst	Set	Lg	CJ	2.91	5.50	1180.00	88.30	---	1268.30	**1490.00**
	Inst	Set	Sm	CJ	4.44	3.60	1270.00	135.00	---	1405.00	**1670.00**
Average quality workmanship	Inst	Set	Lg	CJ	2.42	6.60	984.00	73.50	---	1057.50	**1240.00**
	Inst	Set	Sm	CJ	3.72	4.30	1060.00	113.00	---	1173.00	**1390.00**
48" H											
High quality workmanship	Inst	Set	Lg	CJ	4.00	4.00	1450.00	121.00	---	1571.00	**1850.00**
	Inst	Set	Sm	CJ	6.15	2.60	1570.00	187.00	---	1757.00	**2080.00**
Good quality workmanship	Inst	Set	Lg	CJ	3.20	5.00	1350.00	97.10	---	1447.10	**1690.00**
	Inst	Set	Sm	CJ	4.85	3.30	1450.00	147.00	---	1597.00	**1890.00**
Average quality workmanship	Inst	Set	Lg	CJ	2.67	6.00	1120.00	81.00	---	1201.00	**1410.00**
	Inst	Set	Sm	CJ	4.10	3.90	1210.00	124.00	---	1334.00	**1580.00**
60" H											
High quality workmanship	Inst	Set	Lg	CJ	4.44	3.60	1730.00	135.00	---	1865.00	**2200.00**
	Inst	Set	Sm	CJ	6.96	2.30	1870.00	211.00	---	2081.00	**2470.00**
Good quality workmanship	Inst	Set	Lg	CJ	3.56	4.50	1600.00	108.00	---	1708.00	**2010.00**
	Inst	Set	Sm	CJ	5.52	2.90	1730.00	168.00	---	1898.00	**2240.00**
Average quality workmanship	Inst	Set	Lg	CJ	2.96	5.40	1340.00	89.80	---	1429.80	**1670.00**
	Inst	Set	Sm	CJ	4.57	3.50	1440.00	139.00	---	1579.00	**1870.00**
72" H											
High quality workmanship	Inst	Set	Lg	CJ	5.00	3.20	1970.00	152.00	---	2122.00	**2500.00**
	Inst	Set	Sm	CJ	7.62	2.10	2130.00	231.00	---	2361.00	**2800.00**
Good quality workmanship	Inst	Set	Lg	CJ	4.00	4.00	1830.00	121.00	---	1951.00	**2280.00**
	Inst	Set	Sm	CJ	6.15	2.60	1970.00	187.00	---	2157.00	**2550.00**
Average quality workmanship	Inst	Set	Lg	CJ	3.33	4.80	1520.00	101.00	---	1621.00	**1900.00**
	Inst	Set	Sm	CJ	5.16	3.10	1640.00	157.00	---	1797.00	**2120.00**
4 ventilating lites, 1 fixed lite; 86" W											
36" H											
High quality workmanship	Inst	Set	Lg	CJ	4.00	4.00	1190.00	121.00	---	1311.00	**1550.00**
	Inst	Set	Sm	CJ	6.15	2.60	1280.00	187.00	---	1467.00	**1750.00**
Good quality workmanship	Inst	Set	Lg	CJ	3.20	5.00	1100.00	97.10	---	1197.10	**1410.00**
	Inst	Set	Sm	CJ	4.85	3.30	1190.00	147.00	---	1337.00	**1580.00**
Average quality workmanship	Inst	Set	Lg	CJ	2.67	6.00	916.00	81.00	---	997.00	**1170.00**
	Inst	Set	Sm	CJ	4.10	3.90	988.00	124.00	---	1112.00	**1320.00**
41" H											
High quality workmanship	Inst	Set	Lg	CJ	4.44	3.60	1330.00	135.00	---	1465.00	**1740.00**
	Inst	Set	Sm	CJ	6.96	2.30	1440.00	211.00	---	1651.00	**1970.00**
Good quality workmanship	Inst	Set	Lg	CJ	3.56	4.50	1230.00	108.00	---	1338.00	**1580.00**
	Inst	Set	Sm	CJ	5.52	2.90	1330.00	168.00	---	1498.00	**1780.00**
Average quality workmanship	Inst	Set	Lg	CJ	2.96	5.40	1030.00	89.80	---	1119.80	**1320.00**
	Inst	Set	Sm	CJ	4.57	3.50	1110.00	139.00	---	1249.00	**1480.00**

Description	Oper	Unit	Vol	Crew Size	Man-hours per Unit	Crew Output per Day	Avg Mat'l Unit Cost	Avg Labor Unit Cost	Avg Equip Unit Cost	Avg Total Unit Cost	Avg Price Incl O&P
48" H											
High quality workmanship	Inst	Set	Lg	CJ	5.00	3.20	1510.00	152.00	---	1662.00	**1970.00**
	Inst	Set	Sm	CJ	7.62	2.10	1630.00	231.00	---	1861.00	**2220.00**
Good quality workmanship	Inst	Set	Lg	CJ	4.00	4.00	1170.00	121.00	---	1291.00	**1520.00**
	Inst	Set	Sm	CJ	6.15	2.60	1260.00	187.00	---	1447.00	**1730.00**
Average quality workmanship	Inst	Set	Lg	CJ	3.33	4.80	972.00	101.00	---	1073.00	**1270.00**
	Inst	Set	Sm	CJ	5.16	3.10	1050.00	157.00	---	1207.00	**1440.00**
64" H											
High quality workmanship	Inst	Set	Lg	CJ	5.71	2.80	2040.00	173.00	---	2213.00	**2600.00**
	Inst	Set	Sm	CJ	8.89	1.80	2200.00	270.00	---	2470.00	**2930.00**
Good quality workmanship	Inst	Set	Lg	CJ	4.57	3.50	1890.00	139.00	---	2029.00	**2380.00**
	Inst	Set	Sm	CJ	6.96	2.30	2040.00	211.00	---	2251.00	**2660.00**
Average quality workmanship	Inst	Set	Lg	CJ	3.81	4.20	1570.00	116.00	---	1686.00	**1980.00**
	Inst	Set	Sm	CJ	5.93	2.70	1700.00	180.00	---	1880.00	**2220.00**
72" H											
High quality workmanship	Inst	Set	Lg	CJ	6.67	2.40	2240.00	202.00	---	2442.00	**2880.00**
	Inst	Set	Sm	CJ	10.0	1.60	2420.00	304.00	---	2724.00	**3240.00**
Good quality workmanship	Inst	Set	Lg	CJ	5.33	3.00	2080.00	162.00	---	2242.00	**2630.00**
	Inst	Set	Sm	CJ	8.00	2.00	2240.00	243.00	---	2483.00	**2940.00**
Average quality workmanship	Inst	Set	Lg	CJ	4.44	3.60	1730.00	135.00	---	1865.00	**2190.00**
	Inst	Set	Sm	CJ	6.96	2.30	1860.00	211.00	---	2071.00	**2460.00**

4 ventilating lites, 1 fixed lite; 112" W

Description	Oper	Unit	Vol	Crew Size	Man-hours per Unit	Crew Output per Day	Avg Mat'l Unit Cost	Avg Labor Unit Cost	Avg Equip Unit Cost	Avg Total Unit Cost	Avg Price Incl O&P
36" H											
High quality workmanship	Inst	Set	Lg	CJ	4.00	4.00	1280.00	121.00	---	1401.00	**1650.00**
	Inst	Set	Sm	CJ	6.15	2.60	1380.00	187.00	---	1567.00	**1860.00**
Good quality workmanship	Inst	Set	Lg	CJ	3.20	5.00	1180.00	97.10	---	1277.10	**1500.00**
	Inst	Set	Sm	CJ	4.85	3.30	1270.00	147.00	---	1417.00	**1690.00**
Average quality workmanship	Inst	Set	Lg	CJ	2.67	6.00	984.00	81.00	---	1065.00	**1250.00**
	Inst	Set	Sm	CJ	4.10	3.90	1060.00	124.00	---	1184.00	**1410.00**
41" H											
High quality workmanship	Inst	Set	Lg	CJ	4.44	3.60	1360.00	135.00	---	1495.00	**1760.00**
	Inst	Set	Sm	CJ	6.96	2.30	1470.00	211.00	---	1681.00	**2000.00**
Good quality workmanship	Inst	Set	Lg	CJ	3.56	4.50	1260.00	108.00	---	1368.00	**1610.00**
	Inst	Set	Sm	CJ	5.52	2.90	1360.00	168.00	---	1528.00	**1810.00**
Average quality workmanship	Inst	Set	Lg	CJ	2.96	5.40	1050.00	89.80	---	1139.80	**1340.00**
	Inst	Set	Sm	CJ	4.57	3.50	1130.00	139.00	---	1269.00	**1510.00**
48" H											
High quality workmanship	Inst	Set	Lg	CJ	5.00	3.20	1540.00	152.00	---	1692.00	**2000.00**
	Inst	Set	Sm	CJ	7.62	2.10	1660.00	231.00	---	1891.00	**2260.00**
Good quality workmanship	Inst	Set	Lg	CJ	4.00	4.00	1430.00	121.00	---	1551.00	**1830.00**
	Inst	Set	Sm	CJ	6.15	2.60	1540.00	187.00	---	1727.00	**2050.00**
Average quality workmanship	Inst	Set	Lg	CJ	3.33	4.80	1190.00	101.00	---	1291.00	**1520.00**
	Inst	Set	Sm	CJ	5.16	3.10	1280.00	157.00	---	1437.00	**1710.00**

Description	Oper	Unit	Vol	Crew Size	Man-hours per Unit	Crew Output per Day	Avg Mat'l Unit Cost	Avg Labor Unit Cost	Avg Equip Unit Cost	Avg Total Unit Cost	Avg Price Incl O&P
60" H											
High quality workmanship	Inst	Set	Lg	CJ	5.71	2.80	1870.00	173.00	---	2043.00	**2410.00**
	Inst	Set	Sm	CJ	8.89	1.80	2020.00	270.00	---	2290.00	**2730.00**
Good quality workmanship	Inst	Set	Lg	CJ	4.57	3.50	1730.00	139.00	---	1869.00	**2200.00**
	Inst	Set	Sm	CJ	6.96	2.30	1870.00	211.00	---	2081.00	**2470.00**
Average quality workmanship	Inst	Set	Lg	CJ	3.81	4.20	1450.00	116.00	---	1566.00	**1840.00**
	Inst	Set	Sm	CJ	5.93	2.70	1560.00	180.00	---	1740.00	**2060.00**
72" H											
High quality workmanship	Inst	Set	Lg	CJ	6.67	2.40	2060.00	202.00	---	2262.00	**2680.00**
	Inst	Set	Sm	CJ	10.0	1.60	2230.00	304.00	---	2534.00	**3020.00**
Good quality workmanship	Inst	Set	Lg	CJ	5.33	3.00	1910.00	162.00	---	2072.00	**2440.00**
	Inst	Set	Sm	CJ	8.00	2.00	2060.00	243.00	---	2303.00	**2730.00**
Average quality workmanship	Inst	Set	Lg	CJ	4.44	3.60	1590.00	135.00	---	1725.00	**2030.00**
	Inst	Set	Sm	CJ	6.96	2.30	1720.00	211.00	---	1931.00	**2290.00**

4 ventilating lites, 1 fixed lite; 120" W

Description	Oper	Unit	Vol	Crew Size	Man-hours per Unit	Crew Output per Day	Avg Mat'l Unit Cost	Avg Labor Unit Cost	Avg Equip Unit Cost	Avg Total Unit Cost	Avg Price Incl O&P
36" H											
High quality workmanship	Inst	Set	Lg	CJ	4.00	4.00	1360.00	121.00	---	1481.00	**1740.00**
	Inst	Set	Sm	CJ	6.15	2.60	1470.00	187.00	---	1657.00	**1970.00**
Good quality workmanship	Inst	Set	Lg	CJ	3.20	5.00	1260.00	97.10	---	1357.10	**1590.00**
	Inst	Set	Sm	CJ	4.85	3.30	1360.00	147.00	---	1507.00	**1780.00**
Average quality workmanship	Inst	Set	Lg	CJ	2.67	6.00	1050.00	81.00	---	1131.00	**1330.00**
	Inst	Set	Sm	CJ	4.10	3.90	1130.00	124.00	---	1254.00	**1490.00**
41" H											
High quality workmanship	Inst	Set	Lg	CJ	4.44	3.60	1500.00	135.00	---	1635.00	**1930.00**
	Inst	Set	Sm	CJ	6.96	2.30	1620.00	211.00	---	1831.00	**2180.00**
Good quality workmanship	Inst	Set	Lg	CJ	3.56	4.50	1390.00	108.00	---	1498.00	**1760.00**
	Inst	Set	Sm	CJ	5.52	2.90	1500.00	168.00	---	1668.00	**1970.00**
Average quality workmanship	Inst	Set	Lg	CJ	2.96	5.40	1160.00	89.80	---	1249.80	**1460.00**
	Inst	Set	Sm	CJ	4.57	3.50	1250.00	139.00	---	1389.00	**1640.00**
48" H											
High quality workmanship	Inst	Set	Lg	CJ	5.00	3.20	1640.00	152.00	---	1792.00	**2110.00**
	Inst	Set	Sm	CJ	7.62	2.10	1770.00	231.00	---	2001.00	**2380.00**
Good quality workmanship	Inst	Set	Lg	CJ	4.00	4.00	1520.00	121.00	---	1641.00	**1930.00**
	Inst	Set	Sm	CJ	6.15	2.60	1640.00	187.00	---	1827.00	**2160.00**
Average quality workmanship	Inst	Set	Lg	CJ	3.33	4.80	1260.00	101.00	---	1361.00	**1600.00**
	Inst	Set	Sm	CJ	5.16	3.10	1360.00	157.00	---	1517.00	**1800.00**
60" H											
High quality workmanship	Inst	Set	Lg	CJ	5.71	2.80	1890.00	173.00	---	2063.00	**2430.00**
	Inst	Set	Sm	CJ	8.89	1.80	2030.00	270.00	---	2300.00	**2740.00**
Good quality workmanship	Inst	Set	Lg	CJ	4.57	3.50	1750.00	139.00	---	1889.00	**2220.00**
	Inst	Set	Sm	CJ	6.96	2.30	1880.00	211.00	---	2091.00	**2480.00**
Average quality workmanship	Inst	Set	Lg	CJ	3.81	4.20	1450.00	116.00	---	1566.00	**1850.00**
	Inst	Set	Sm	CJ	5.93	2.70	1570.00	180.00	---	1750.00	**2070.00**
72" H											
High quality workmanship	Inst	Set	Lg	CJ	6.67	2.40	2200.00	202.00	---	2402.00	**2840.00**
	Inst	Set	Sm	CJ	10.00	1.60	2380.00	304.00	---	2684.00	**3190.00**
Good quality workmanship	Inst	Set	Lg	CJ	5.33	3.00	2040.00	162.00	---	2202.00	**2590.00**
	Inst	Set	Sm	CJ	8.00	2.00	2200.00	243.00	---	2443.00	**2890.00**
Average quality workmanship	Inst	Set	Lg	CJ	4.44	3.60	1700.00	135.00	---	1835.00	**2160.00**
	Inst	Set	Sm	CJ	6.96	2.30	1830.00	211.00	---	2041.00	**2420.00**

Description	Oper	Unit	Vol	Crew Size	Man-hours per Unit	Crew Output per Day	Avg Mat'l Unit Cost	Avg Labor Unit Cost	Avg Equip Unit Cost	Avg Total Unit Cost	Avg Price Incl O&P
4 ventilating lites, 1 fixed lite; 142" W											
36" H											
High quality workmanship	Inst	Set	Lg	CJ	4.00	4.00	1490.00	121.00	---	1611.00	**1890.00**
	Inst	Set	Sm	CJ	6.15	2.60	1600.00	187.00	---	1787.00	**2120.00**
Good quality workmanship	Inst	Set	Lg	CJ	3.20	5.00	1380.00	97.10	---	1477.10	**1730.00**
	Inst	Set	Sm	CJ	4.85	3.30	1480.00	147.00	---	1627.00	**1930.00**
Average quality workmanship	Inst	Set	Lg	CJ	2.67	6.00	1150.00	81.00	---	1231.00	**1440.00**
	Inst	Set	Sm	CJ	4.10	3.90	1240.00	124.00	---	1364.00	**1610.00**
41" H											
High quality workmanship	Inst	Set	Lg	CJ	4.00	4.00	1600.00	121.00	---	1721.00	**2020.00**
	Inst	Set	Sm	CJ	6.15	2.60	1730.00	187.00	---	1917.00	**2270.00**
Good quality workmanship	Inst	Set	Lg	CJ	3.20	5.00	1480.00	97.10	---	1577.10	**1850.00**
	Inst	Set	Sm	CJ	4.85	3.30	1600.00	147.00	---	1747.00	**2060.00**
Average quality workmanship	Inst	Set	Lg	CJ	2.67	6.00	1230.00	81.00	---	1311.00	**1540.00**
	Inst	Set	Sm	CJ	4.10	3.90	1330.00	124.00	---	1454.00	**1720.00**
48" H											
High quality workmanship	Inst	Set	Lg	CJ	5.00	3.20	1820.00	152.00	---	1972.00	**2320.00**
	Inst	Set	Sm	CJ	7.62	2.10	1970.00	231.00	---	2201.00	**2610.00**
Good quality workmanship	Inst	Set	Lg	CJ	4.00	4.00	1690.00	121.00	---	1811.00	**2120.00**
	Inst	Set	Sm	CJ	6.15	2.60	1820.00	187.00	---	2007.00	**2370.00**
Average quality workmanship	Inst	Set	Lg	CJ	3.33	4.80	1410.00	101.00	---	1511.00	**1770.00**
	Inst	Set	Sm	CJ	5.16	3.10	1520.00	157.00	---	1677.00	**1980.00**
60" H											
High quality workmanship	Inst	Set	Lg	CJ	5.71	2.80	2180.00	173.00	---	2353.00	**2770.00**
	Inst	Set	Sm	CJ	8.89	1.80	2360.00	270.00	---	2630.00	**3110.00**
Good quality workmanship	Inst	Set	Lg	CJ	4.57	3.50	2020.00	139.00	---	2159.00	**2530.00**
	Inst	Set	Sm	CJ	6.96	2.30	2180.00	211.00	---	2391.00	**2830.00**
Average quality workmanship	Inst	Set	Lg	CJ	3.81	4.20	1690.00	116.00	---	1806.00	**2110.00**
	Inst	Set	Sm	CJ	5.93	2.70	1820.00	180.00	---	2000.00	**2360.00**
72" H											
High quality workmanship	Inst	Set	Lg	CJ	6.67	2.40	2500.00	202.00	---	2702.00	**3180.00**
	Inst	Set	Sm	CJ	10.0	1.60	2700.00	304.00	---	3004.00	**3560.00**
Good quality workmanship	Inst	Set	Lg	CJ	5.33	3.00	2320.00	162.00	---	2482.00	**2910.00**
	Inst	Set	Sm	CJ	8.00	2.00	2500.00	243.00	---	2743.00	**3240.00**
Average quality workmanship	Inst	Set	Lg	CJ	4.44	3.60	1930.00	135.00	---	2065.00	**2420.00**
	Inst	Set	Sm	CJ	6.96	2.30	2080.00	211.00	---	2291.00	**2710.00**
For aluminum clad (baked white enamel) exterior											
ADD	Inst	%	Lg	---	---	---	20.0	---	---	---	**---**
	Inst	%	Sm	---	---	---	20.0	---	---	---	**---**

Description	Oper	Unit	Vol	Crew Size	Man-hours per Unit	Crew Output per Day	Avg Mat'l Unit Cost	Avg Labor Unit Cost	Avg Equip Unit Cost	Avg Total Unit Cost	Avg Price Incl O&P

Double hung windows; glazed 7/16" T insulating glass; two lites per unit; unit includes screens and weatherstripping; rough opening sizes

1'-6" x 2'-0", 2'-6", 3'-0", 3'-6"

Description	Oper	Unit	Vol	Crew Size	Man-hours per Unit	Crew Output per Day	Avg Mat'l Unit Cost	Avg Labor Unit Cost	Avg Equip Unit Cost	Avg Total Unit Cost	Avg Price Incl O&P
High quality workmanship	Inst	Set	Lg	CA	2.50	3.20	305.00	83.20	---	388.20	**475.00**
	Inst	Set	Sm	CA	3.81	2.10	329.00	127.00	---	456.00	**568.00**
Good quality workmanship	Inst	Set	Lg	CA	2.00	4.00	281.00	66.50	---	347.50	**423.00**
	Inst	Set	Sm	CA	3.08	2.60	303.00	102.00	---	405.00	**502.00**
Average quality workmanship	Inst	Set	Lg	CA	1.67	4.80	249.00	55.60	---	304.60	**370.00**
	Inst	Set	Sm	CA	2.58	3.10	269.00	85.80	---	354.80	**438.00**

1'-6" x 4'-0", 4'-6", 5'-0", 5'-6"

Description	Oper	Unit	Vol	Crew Size	Man-hours per Unit	Crew Output per Day	Avg Mat'l Unit Cost	Avg Labor Unit Cost	Avg Equip Unit Cost	Avg Total Unit Cost	Avg Price Incl O&P
High quality workmanship	Inst	Set	Lg	CA	2.50	3.20	437.00	83.20	---	520.20	**627.00**
	Inst	Set	Sm	CA	3.81	2.10	471.00	127.00	---	598.00	**732.00**
Good quality workmanship	Inst	Set	Lg	CA	2.00	4.00	398.00	66.50	---	464.50	**557.00**
	Inst	Set	Sm	CA	3.08	2.60	429.00	102.00	---	531.00	**647.00**
Average quality workmanship	Inst	Set	Lg	CA	1.67	4.80	345.00	55.60	---	400.60	**481.00**
	Inst	Set	Sm	CA	2.58	3.10	373.00	85.80	---	458.80	**557.00**

2'-0" x 2'-0", 2'-6", 3'-0", 3'-6"

Description	Oper	Unit	Vol	Crew Size	Man-hours per Unit	Crew Output per Day	Avg Mat'l Unit Cost	Avg Labor Unit Cost	Avg Equip Unit Cost	Avg Total Unit Cost	Avg Price Incl O&P
High quality workmanship	Inst	Set	Lg	CA	2.50	3.20	315.00	83.20	---	398.20	**487.00**
	Inst	Set	Sm	CA	3.81	2.10	340.00	127.00	---	467.00	**581.00**
Good quality workmanship	Inst	Set	Lg	CA	2.00	4.00	278.00	66.50	---	344.50	**420.00**
	Inst	Set	Sm	CA	3.08	2.60	300.00	102.00	---	402.00	**499.00**
Average quality workmanship	Inst	Set	Lg	CA	1.67	4.80	260.00	55.60	---	315.60	**383.00**
	Inst	Set	Sm	CA	2.58	3.10	281.00	85.80	---	366.80	**452.00**

2'-0" x 4'-0", 4'-6", 5'-0", 6'-0"

Description	Oper	Unit	Vol	Crew Size	Man-hours per Unit	Crew Output per Day	Avg Mat'l Unit Cost	Avg Labor Unit Cost	Avg Equip Unit Cost	Avg Total Unit Cost	Avg Price Incl O&P
High quality workmanship	Inst	Set	Lg	CA	2.50	3.20	414.00	83.20	---	497.20	**601.00**
	Inst	Set	Sm	CA	3.81	2.10	447.00	127.00	---	574.00	**704.00**
Good quality workmanship	Inst	Set	Lg	CA	2.00	4.00	384.00	66.50	---	450.50	**541.00**
	Inst	Set	Sm	CA	3.08	2.60	414.00	102.00	---	516.00	**630.00**
Average quality workmanship	Inst	Set	Lg	CA	1.67	4.80	356.00	55.60	---	411.60	**492.00**
	Inst	Set	Sm	CA	2.58	3.10	384.00	85.80	---	469.80	**570.00**

2'-6" x 2'-0", 2'-6", 3'-0", 3'-6"

Description	Oper	Unit	Vol	Crew Size	Man-hours per Unit	Crew Output per Day	Avg Mat'l Unit Cost	Avg Labor Unit Cost	Avg Equip Unit Cost	Avg Total Unit Cost	Avg Price Incl O&P
High quality workmanship	Inst	Set	Lg	CA	2.50	3.20	328.00	83.20	---	411.20	**502.00**
	Inst	Set	Sm	CA	3.81	2.10	353.00	127.00	---	480.00	**597.00**
Good quality workmanship	Inst	Set	Lg	CA	2.00	4.00	304.00	66.50	---	370.50	**449.00**
	Inst	Set	Sm	CA	3.08	2.60	327.00	102.00	---	429.00	**530.00**
Average quality workmanship	Inst	Set	Lg	CA	1.67	4.80	269.00	55.60	---	324.60	**393.00**
	Inst	Set	Sm	CA	2.58	3.10	290.00	85.80	---	375.00	**463.00**

2'-6" x 4'-0", 4'-6", 5'-0", 6'-0"

Description	Oper	Unit	Vol	Crew Size	Man-hours per Unit	Crew Output per Day	Avg Mat'l Unit Cost	Avg Labor Unit Cost	Avg Equip Unit Cost	Avg Total Unit Cost	Avg Price Incl O&P
High quality workmanship	Inst	Set	Lg	CA	2.50	3.20	462.00	83.20	---	545.20	**656.00**
	Inst	Set	Sm	CA	3.81	2.10	499.00	127.00	---	626.00	**764.00**
Good quality workmanship	Inst	Set	Lg	CA	2.00	4.00	414.00	66.50	---	480.50	**576.00**
	Inst	Set	Sm	CA	3.08	2.60	447.00	102.00	---	549.00	**667.00**
Average quality workmanship	Inst	Set	Lg	CA	1.67	4.80	387.00	55.60	---	442.60	**529.00**
	Inst	Set	Sm	CA	2.58	3.10	418.00	85.80	---	503.80	**609.00**

Description	Oper	Unit	Vol	Crew Size	Man-hours per Unit	Crew Output per Day	Avg Mat'l Unit Cost	Avg Labor Unit Cost	Avg Equip Unit Cost	Avg Total Unit Cost	Avg Price Incl O&P
3'-0" x 2'-0", 2'-6", 3'-0", 3'-6"											
High quality workmanship	Inst	Set	Lg	CA	2.50	3.20	380.00	83.20	---	463.20	**561.00**
	Inst	Set	Sm	CA	3.81	2.10	410.00	127.00	---	537.00	**661.00**
Good quality workmanship	Inst	Set	Lg	CA	2.00	4.00	320.00	66.50	---	386.50	**468.00**
	Inst	Set	Sm	CA	3.08	2.60	345.00	102.00	---	447.00	**551.00**
Average quality workmanship	Inst	Set	Lg	CA	1.67	4.80	292.00	55.60	---	347.60	**419.00**
	Inst	Set	Sm	CA	2.58	3.10	315.00	85.80	---	400.80	**491.00**
3'-0" x 4'-0", 4'-6", 5'-0", 6'-0"											
High quality workmanship	Inst	Set	Lg	CA	2.50	3.20	509.00	83.20	---	592.20	**710.00**
	Inst	Set	Sm	CA	3.81	2.10	549.00	127.00	---	676.00	**822.00**
Good quality workmanship	Inst	Set	Lg	CA	2.00	4.00	478.00	66.50	---	544.50	**649.00**
	Inst	Set	Sm	CA	3.08	2.60	515.00	102.00	---	617.00	**746.00**
Average quality workmanship	Inst	Set	Lg	CA	1.67	4.80	422.00	55.60	---	477.60	**568.00**
	Inst	Set	Sm	CA	2.58	3.10	455.00	85.80	---	540.80	**652.00**
3'-6" x 3'-0", 3'-6", 4'-0"											
High quality workmanship	Inst	Set	Lg	CA	2.50	3.20	455.00	83.20	---	538.20	**648.00**
	Inst	Set	Sm	CA	3.81	2.10	490.00	127.00	---	617.00	**754.00**
Good quality workmanship	Inst	Set	Lg	CA	2.00	4.00	386.00	66.50	---	452.50	**544.00**
	Inst	Set	Sm	CA	3.08	2.60	416.00	102.00	---	518.00	**633.00**
Average quality workmanship	Inst	Set	Lg	CA	1.67	4.80	342.00	55.60	---	397.60	**476.00**
	Inst	Set	Sm	CA	2.58	3.10	369.00	85.80	---	454.80	**553.00**
3'-6" x 4'-6", 5'-0", 6'-0"											
High quality workmanship	Inst	Set	Lg	CA	2.50	3.20	625.00	83.20	---	708.20	**843.00**
	Inst	Set	Sm	CA	3.81	2.10	674.00	127.00	---	801.00	**965.00**
Good quality workmanship	Inst	Set	Lg	CA	2.00	4.00	573.00	66.50	---	639.50	**759.00**
	Inst	Set	Sm	CA	3.08	2.60	618.00	102.00	---	720.00	**864.00**
Average quality workmanship	Inst	Set	Lg	CA	1.67	4.80	512.00	55.60	---	567.60	**672.00**
	Inst	Set	Sm	CA	2.58	3.10	552.00	85.80	---	637.80	**764.00**
4'-0" x 3'-0", 3'-6", 4'-0"											
High quality workmanship	Inst	Set	Lg	CA	2.50	3.20	663.00	83.20	---	746.20	**887.00**
	Inst	Set	Sm	CA	3.81	2.10	715.00	127.00	---	842.00	**1010.00**
Good quality workmanship	Inst	Set	Lg	CA	2.00	4.00	533.00	66.50	---	599.50	**713.00**
	Inst	Set	Sm	CA	3.08	2.60	575.00	102.00	---	677.00	**815.00**
Average quality workmanship	Inst	Set	Lg	CA	1.67	4.80	399.00	55.60	---	454.60	**542.00**
	Inst	Set	Sm	CA	2.58	3.10	430.00	85.80	---	515.80	**623.00**
4'-0" x 4'-6", 5'-0", 6'-0"											
High quality workmanship	Inst	Set	Lg	CA	2.50	3.20	699.00	83.20	---	782.20	**928.00**
	Inst	Set	Sm	CA	3.81	2.10	754.00	127.00	---	881.00	**1060.00**
Good quality workmanship	Inst	Set	Lg	CA	2.00	4.00	635.00	66.50	---	701.50	**830.00**
	Inst	Set	Sm	CA	3.08	2.60	685.00	102.00	---	787.00	**941.00**
Average quality workmanship	Inst	Set	Lg	CA	1.67	4.80	558.00	55.60	---	613.60	**725.00**
	Inst	Set	Sm	CA	2.58	3.10	601.00	85.80	---	686.80	**820.00**

Description	Oper	Unit	Vol	Crew Size	Man-hours per Unit	Crew Output per Day	Avg Mat'l Unit Cost	Avg Labor Unit Cost	Avg Equip Unit Cost	Avg Total Unit Cost	Avg Price Incl O&P
4'-6" x 4'-0", 4'-6											
High quality workmanship	Inst	Set	Lg	CA	2.50	3.20	737.00	83.20	---	820.20	**972.00**
	Inst	Set	Sm	CA	3.81	2.10	795.00	127.00	---	922.00	**1100.00**
Good quality workmanship	Inst	Set	Lg	CA	2.00	4.00	572.00	66.50	---	638.50	**757.00**
	Inst	Set	Sm	CA	3.08	2.60	617.00	102.00	---	719.00	**863.00**
Average quality workmanship	Inst	Set	Lg	CA	1.67	4.80	480.00	55.60	---	535.60	**635.00**
	Inst	Set	Sm	CA	2.58	3.10	518.00	85.80	---	603.80	**724.00**
4'-6" x 5'-0", 6'-0"											
High quality workmanship	Inst	Set	Lg	CA	2.50	3.20	800.00	83.20	---	883.20	**1040.00**
	Inst	Set	Sm	CA	3.81	2.10	863.00	127.00	---	990.00	**1180.00**
Good quality workmanship	Inst	Set	Lg	CA	2.00	4.00	660.00	66.50	---	726.50	**859.00**
	Inst	Set	Sm	CA	3.08	2.60	712.00	102.00	---	814.00	**973.00**
Average quality workmanship	Inst	Set	Lg	CA	1.67	4.80	593.00	55.60	---	648.60	**765.00**
	Inst	Set	Sm	CA	2.58	3.10	640.00	85.80	---	725.80	**865.00**
5'-0" x 4'-0", 4'-6											
High quality workmanship	Inst	Set	Lg	CA	2.50	3.20	864.00	83.20	---	947.20	**1120.00**
	Inst	Set	Sm	CA	3.81	2.10	932.00	127.00	---	1059.00	**1260.00**
Good quality workmanship	Inst	Set	Lg	CA	2.00	4.00	749.00	66.50	---	815.50	**962.00**
	Inst	Set	Sm	CA	3.08	2.60	808.00	102.00	---	910.00	**1080.00**
Average quality workmanship	Inst	Set	Lg	CA	1.67	4.80	518.00	55.60	---	573.60	**679.00**
	Inst	Set	Sm	CA	2.58	3.10	559.00	85.80	---	644.80	**772.00**
5'-0" x 5'-0", 6'-0"											
High quality workmanship	Inst	Set	Lg	CA	2.50	3.20	927.00	83.20	---	1010.20	**1190.00**
	Inst	Set	Sm	CA	3.81	2.10	1000.00	127.00	---	1127.00	**1340.00**
Good quality workmanship	Inst	Set	Lg	CA	2.00	4.00	800.00	66.50	---	866.50	**1020.00**
	Inst	Set	Sm	CA	3.08	2.60	863.00	102.00	---	965.00	**1150.00**
Average quality workmanship	Inst	Set	Lg	CA	1.67	4.80	631.00	55.60	---	686.60	**809.00**
	Inst	Set	Sm	CA	2.58	3.10	681.00	85.80	---	766.80	**912.00**

Picture windows

Unit with one fixed lite, 3/4" T insulating glass, frame sizes

4'-0" W x 3'-6" H; 3'-6" W x 4'-0" H

Description	Oper	Unit	Vol	Crew Size	Man-hours per Unit	Crew Output per Day	Avg Mat'l Unit Cost	Avg Labor Unit Cost	Avg Equip Unit Cost	Avg Total Unit Cost	Avg Price Incl O&P
High quality workmanship	Inst	Ea	Lg	CJ	1.82	8.80	386.00	55.20	---	441.20	**527.00**
	Inst	Ea	Sm	CJ	2.81	5.70	416.00	85.30	---	501.30	**607.00**
Good quality workmanship	Inst	Ea	Lg	CJ	1.45	11.00	356.00	44.00	---	400.00	**475.00**
	Inst	Ea	Sm	CJ	2.22	7.20	384.00	67.40	---	451.40	**542.00**
Average quality workmanship	Inst	Ea	Lg	CJ	1.21	13.20	297.00	36.70	---	333.70	**397.00**
	Inst	Ea	Sm	CJ	1.86	8.60	321.00	56.50	---	377.50	**453.00**
4'-8" W x 3'-6" H; 4'-0" W x 4'-0" H; 3'-6" W x 4'-8"H											
High quality workmanship	Inst	Ea	Lg	CJ	1.82	8.80	445.00	55.20	---	500.20	**594.00**
	Inst	Ea	Sm	CJ	2.81	5.70	480.00	85.30	---	565.30	**679.00**
Good quality workmanship	Inst	Ea	Lg	CJ	1.45	11.00	411.00	44.00	---	455.00	**539.00**
	Inst	Ea	Sm	CJ	2.22	7.20	444.00	67.40	---	511.40	**612.00**
Average quality workmanship	Inst	Ea	Lg	CJ	1.21	13.20	343.00	36.70	---	379.70	**449.00**
	Inst	Ea	Sm	CJ	1.86	8.60	370.00	56.50	---	426.50	**510.00**

Description	Oper	Unit	Vol	Crew Size	Man-hours per Unit	Crew Output per Day	Avg Mat'l Unit Cost	Avg Labor Unit Cost	Avg Equip Unit Cost	Avg Total Unit Cost	Avg Price Incl O&P
4'-8" W x 4'-0" H; 4'-0" W x 4'-10" H											
High quality workmanship	Inst	Ea	Lg	CJ	1.82	8.80	489.00	55.20	---	544.20	**645.00**
	Inst	Ea	Sm	CJ	2.81	5.70	527.00	85.30	---	612.30	**735.00**
Good quality workmanship	Inst	Ea	Lg	CJ	1.45	11.00	452.00	44.00	---	496.00	**586.00**
	Inst	Ea	Sm	CJ	2.22	7.20	488.00	67.40	---	555.40	**662.00**
Average quality workmanship	Inst	Ea	Lg	CJ	1.21	13.20	377.00	36.70	---	413.70	**489.00**
	Inst	Ea	Sm	CJ	1.86	8.60	407.00	56.50	---	463.50	**553.00**
4'-8" W x 4'-8" H											
High quality workmanship	Inst	Ea	Lg	CJ	2.00	8.00	533.00	60.70	---	593.70	**704.00**
	Inst	Ea	Sm	CJ	3.08	5.20	575.00	93.50	---	668.50	**802.00**
Good quality workmanship	Inst	Ea	Lg	CJ	1.60	10.00	493.00	48.60	---	541.60	**640.00**
	Inst	Ea	Sm	CJ	2.46	6.50	532.00	74.70	---	606.70	**723.00**
Average quality workmanship	Inst	Ea	Lg	CJ	1.33	12.00	411.00	40.40	---	451.40	**534.00**
	Inst	Ea	Sm	CJ	2.05	7.80	444.00	62.20	---	506.20	**604.00**
4'-8" W x 5'-6" H; 5'-4" W x 4'-8" H											
High quality workmanship	Inst	Ea	Lg	CJ	2.00	8.00	673.00	60.70	---	733.70	**865.00**
	Inst	Ea	Sm	CJ	3.08	5.20	726.00	93.50	---	819.50	**975.00**
Good quality workmanship	Inst	Ea	Lg	CJ	1.60	10.00	622.00	48.60	---	670.60	**788.00**
	Inst	Ea	Sm	CJ	2.46	6.50	671.00	74.70	---	745.70	**884.00**
Average quality workmanship	Inst	Ea	Lg	CJ	1.33	12.00	518.00	40.40	---	558.40	**656.00**
	Inst	Ea	Sm	CJ	2.05	7.80	559.00	62.20	---	621.20	**736.00**
5'-4" W x 5'-6" H											
High quality workmanship	Inst	Ea	Lg	CJ	2.00	8.00	762.00	60.70	---	822.70	**967.00**
	Inst	Ea	Sm	CJ	3.08	5.20	822.00	93.50	---	915.50	**1090.00**
Good quality workmanship	Inst	Ea	Lg	CJ	1.60	10.00	705.00	48.60	---	753.60	**883.00**
	Inst	Ea	Sm	CJ	2.46	6.50	760.00	74.70	---	834.70	**986.00**
Average quality workmanship	Inst	Ea	Lg	CJ	1.33	12.00	587.00	40.40	---	627.40	**735.00**
	Inst	Ea	Sm	CJ	2.05	7.80	633.00	62.20	---	695.20	**821.00**

Unit with one fixed lite, glazed 3/4" T insulating glass and two glazed S.S.B. casement flankers, 15", 19" or 23" W; unit includes: hardware, casing and molding; rough opening sizes

40" H

88" W

Description	Oper	Unit	Vol	Crew Size	Man-hours per Unit	Crew Output per Day	Avg Mat'l Unit Cost	Avg Labor Unit Cost	Avg Equip Unit Cost	Avg Total Unit Cost	Avg Price Incl O&P
High quality workmanship	Inst	Set	Lg	CJ	2.50	6.40	1040.00	75.90	---	1115.90	**1300.00**
	Inst	Set	Sm	CJ	3.81	4.20	1120.00	116.00	---	1236.00	**1460.00**
Good quality workmanship	Inst	Set	Lg	CJ	2.00	8.00	958.00	60.70	---	1018.70	**1190.00**
	Inst	Set	Sm	CJ	3.08	5.20	1030.00	93.50	---	1123.50	**1330.00**
Average quality workmanship	Inst	Set	Lg	CJ	1.67	9.60	798.00	50.70	---	848.70	**993.00**
	Inst	Set	Sm	CJ	2.58	6.20	860.00	78.30	---	938.30	**1110.00**

96" W

Description	Oper	Unit	Vol	Crew Size	Man-hours per Unit	Crew Output per Day	Avg Mat'l Unit Cost	Avg Labor Unit Cost	Avg Equip Unit Cost	Avg Total Unit Cost	Avg Price Incl O&P
High quality workmanship	Inst	Set	Lg	CJ	2.50	6.40	1050.00	75.90	---	1125.90	**1320.00**
	Inst	Set	Sm	CJ	3.81	4.20	1130.00	116.00	---	1246.00	**1470.00**
Good quality workmanship	Inst	Set	Lg	CJ	2.00	8.00	969.00	60.70	---	1029.70	**1210.00**
	Inst	Set	Sm	CJ	3.08	5.20	1050.00	93.50	---	1143.50	**1340.00**
Average quality workmanship	Inst	Set	Lg	CJ	1.67	9.60	808.00	50.70	---	858.70	**1000.00**
	Inst	Set	Sm	CJ	2.58	6.20	871.00	78.30	---	949.30	**1120.00**

Description	Oper	Unit	Vol	Crew Size	Man-hours per Unit	Crew Output per Day	Avg Mat'l Unit Cost	Avg Labor Unit Cost	Avg Equip Unit Cost	Avg Total Unit Cost	Avg Price Incl O&P
104" W											
High quality workmanship	Inst	Set	Lg	CJ	2.50	6.40	1190.00	75.90	---	1265.90	**1490.00**
	Inst	Set	Sm	CJ	3.81	4.20	1290.00	116.00	---	1406.00	**1650.00**
Good quality workmanship	Inst	Set	Lg	CJ	2.00	8.00	1100.00	60.70	---	1160.70	**1360.00**
	Inst	Set	Sm	CJ	3.08	5.20	1190.00	93.50	---	1283.50	**1510.00**
Average quality workmanship	Inst	Set	Lg	CJ	1.67	9.60	921.00	50.70	---	971.70	**1130.00**
	Inst	Set	Sm	CJ	2.58	6.20	993.00	78.30	---	1071.30	**1260.00**
112" W											
High quality workmanship	Inst	Set	Lg	CJ	2.50	6.40	1210.00	75.90	---	1285.90	**1500.00**
	Inst	Set	Sm	CJ	3.81	4.20	1300.00	116.00	---	1416.00	**1670.00**
Good quality workmanship	Inst	Set	Lg	CJ	2.00	8.00	1120.00	60.70	---	1180.70	**1370.00**
	Inst	Set	Sm	CJ	3.08	5.20	1200.00	93.50	---	1293.50	**1530.00**
Average quality workmanship	Inst	Set	Lg	CJ	1.67	9.60	927.00	50.70	---	977.70	**1140.00**
	Inst	Set	Sm	CJ	2.58	6.20	1000.00	78.30	---	1078.30	**1270.00**
48" H											
88" W											
High quality workmanship	Inst	Set	Lg	CJ	2.50	6.40	1150.00	75.90	---	1225.90	**1440.00**
	Inst	Set	Sm	CJ	3.81	4.20	1240.00	116.00	---	1356.00	**1600.00**
Good quality workmanship	Inst	Set	Lg	CJ	2.00	8.00	1060.00	60.70	---	1120.70	**1310.00**
	Inst	Set	Sm	CJ	3.08	5.20	1150.00	93.50	---	1243.50	**1460.00**
Average quality workmanship	Inst	Set	Lg	CJ	1.67	9.60	886.00	50.70	---	936.70	**1100.00**
	Inst	Set	Sm	CJ	2.58	6.20	956.00	78.30	---	1034.30	**1220.00**
96" W											
High quality workmanship	Inst	Set	Lg	CJ	2.86	5.60	1160.00	86.80	---	1246.80	**1470.00**
	Inst	Set	Sm	CJ	4.44	3.60	1250.00	135.00	---	1385.00	**1640.00**
Good quality workmanship	Inst	Set	Lg	CJ	2.29	7.00	1080.00	69.50	---	1149.50	**1340.00**
	Inst	Set	Sm	CJ	3.48	4.60	1160.00	106.00	---	1266.00	**1490.00**
Average quality workmanship	Inst	Set	Lg	CJ	1.90	8.40	897.00	57.70	---	954.70	**1120.00**
	Inst	Set	Sm	CJ	2.91	5.50	967.00	88.30	---	1055.30	**1240.00**
104" W											
High quality workmanship	Inst	Set	Lg	CJ	2.86	5.60	1350.00	86.80	---	1436.80	**1690.00**
	Inst	Set	Sm	CJ	4.44	3.60	1460.00	135.00	---	1595.00	**1880.00**
Good quality workmanship	Inst	Set	Lg	CJ	2.29	7.00	1250.00	69.50	---	1319.50	**1540.00**
	Inst	Set	Sm	CJ	3.48	4.60	1350.00	106.00	---	1456.00	**1710.00**
Average quality workmanship	Inst	Set	Lg	CJ	1.90	8.40	1040.00	57.70	---	1097.70	**1290.00**
	Inst	Set	Sm	CJ	2.91	5.50	1120.00	88.30	---	1208.30	**1430.00**
112" W											
High quality workmanship	Inst	Set	Lg	CJ	2.86	5.60	1370.00	86.80	---	1456.80	**1700.00**
	Inst	Set	Sm	CJ	4.44	3.60	1470.00	135.00	---	1605.00	**1900.00**
Good quality workmanship	Inst	Set	Lg	CJ	2.29	7.00	1260.00	69.50	---	1329.50	**1560.00**
	Inst	Set	Sm	CJ	3.48	4.60	1360.00	106.00	---	1466.00	**1730.00**
Average quality workmanship	Inst	Set	Lg	CJ	1.90	8.40	1050.00	57.70	---	1107.70	**1300.00**
	Inst	Set	Sm	CJ	2.91	5.50	1140.00	88.30	---	1228.30	**1440.00**

Description	Oper	Unit	Vol	Crew Size	Man-hours per Unit	Crew Output per Day	Avg Mat'l Unit Cost	Avg Labor Unit Cost	Avg Equip Unit Cost	Avg Total Unit Cost	Avg Price Incl O&P
60" H											
88" W											
High quality workmanship	Inst	Set	Lg	CJ	2.86	5.60	1360.00	86.80	---	1446.80	**1690.00**
	Inst	Set	Sm	CJ	4.44	3.60	1470.00	135.00	---	1605.00	**1890.00**
Good quality workmanship	Inst	Set	Lg	CJ	2.29	7.00	1260.00	69.50	---	1329.50	**1550.00**
	Inst	Set	Sm	CJ	3.48	4.60	1360.00	106.00	---	1466.00	**1720.00**
Average quality workmanship	Inst	Set	Lg	CJ	1.90	8.40	1050.00	57.70	---	1107.70	**1290.00**
	Inst	Set	Sm	CJ	2.91	5.50	1130.00	88.30	---	1218.30	**1430.00**
96" W											
High quality workmanship	Inst	Set	Lg	CJ	2.86	5.60	1370.00	86.80	---	1456.80	**1710.00**
	Inst	Set	Sm	CJ	4.44	3.60	1480.00	135.00	---	1615.00	**1900.00**
Good quality workmanship	Inst	Set	Lg	CJ	2.29	7.00	1270.00	69.50	---	1339.50	**1560.00**
	Inst	Set	Sm	CJ	3.48	4.60	1370.00	106.00	---	1476.00	**1730.00**
Average quality workmanship	Inst	Set	Lg	CJ	1.90	8.40	1060.00	57.70	---	1117.70	**1300.00**
	Inst	Set	Sm	CJ	2.91	5.50	1140.00	88.30	---	1228.30	**1440.00**
104" W											
High quality workmanship	Inst	Set	Lg	CJ	3.33	4.80	1720.00	101.00	---	1821.00	**2130.00**
	Inst	Set	Sm	CJ	5.16	3.10	1860.00	157.00	---	2017.00	**2370.00**
Good quality workmanship	Inst	Set	Lg	CJ	2.67	6.00	1590.00	81.00	---	1671.00	**1950.00**
	Inst	Set	Sm	CJ	4.10	3.90	1720.00	124.00	---	1844.00	**2160.00**
Average quality workmanship	Inst	Set	Lg	CJ	2.22	7.20	1330.00	67.40	---	1397.40	**1630.00**
	Inst	Set	Sm	CJ	3.40	4.70	1430.00	103.00	---	1533.00	**1800.00**
112" W											
High quality workmanship	Inst	Set	Lg	CJ	3.33	4.80	1730.00	101.00	---	1831.00	**2150.00**
	Inst	Set	Sm	CJ	5.16	3.10	1870.00	157.00	---	2027.00	**2390.00**
Good quality workmanship	Inst	Set	Lg	CJ	2.67	6.00	1600.00	81.00	---	1681.00	**1970.00**
	Inst	Set	Sm	CJ	4.10	3.90	1730.00	124.00	---	1854.00	**2180.00**
Average quality workmanship	Inst	Set	Lg	CJ	2.22	7.20	1340.00	67.40	---	1407.40	**1640.00**
	Inst	Set	Sm	CJ	3.40	4.70	1440.00	103.00	---	1543.00	**1810.00**
72" H											
88" W											
High quality workmanship	Inst	Set	Lg	CJ	3.33	4.80	1600.00	101.00	---	1701.00	**1990.00**
	Inst	Set	Sm	CJ	5.16	3.10	1730.00	157.00	---	1887.00	**2220.00**
Good quality workmanship	Inst	Set	Lg	CJ	2.67	6.00	1480.00	81.00	---	1561.00	**1820.00**
	Inst	Set	Sm	CJ	4.10	3.90	1600.00	124.00	---	1724.00	**2020.00**
Average quality workmanship	Inst	Set	Lg	CJ	2.22	7.20	1230.00	67.40	---	1297.40	**1520.00**
	Inst	Set	Sm	CJ	3.40	4.70	1330.00	103.00	---	1433.00	**1690.00**
96" W											
High quality workmanship	Inst	Set	Lg	CJ	3.33	4.80	1610.00	101.00	---	1711.00	**2010.00**
	Inst	Set	Sm	CJ	5.16	3.10	1740.00	157.00	---	1897.00	**2240.00**
Good quality workmanship	Inst	Set	Lg	CJ	2.67	6.00	1490.00	81.00	---	1571.00	**1840.00**
	Inst	Set	Sm	CJ	4.10	3.90	1610.00	124.00	---	1734.00	**2040.00**
Average quality workmanship	Inst	Set	Lg	CJ	2.22	7.20	1240.00	67.40	---	1307.40	**1530.00**
	Inst	Set	Sm	CJ	3.40	4.70	1340.00	103.00	---	1443.00	**1700.00**
Material adjustments											
For glazed 1/2" insulated flankers,											
ADD	Inst	%	Lg	---	---	---	15.0	---	---	---	---
	Inst	%	Sm	---	---	---	15.0	---	---	---	---
For screen per flanker,											
ADD	Inst	Ea	Lg	---	---	---	6.35	---	---	6.35	**7.30**
	Inst	Ea	Sm	---	---	---	6.85	---	---	6.85	**7.88**

Description	Oper	Unit	Vol	Crew Size	Man-hours per Unit	Crew Output per Day	Avg Mat'l Unit Cost	Avg Labor Unit Cost	Avg Equip Unit Cost	Avg Total Unit Cost	Avg Price Incl O&P

Garden windows, unit extends out 13-1/4" from wall, bronze tempered glass in top panels of all units, wood shelves included

1 fixed lite
Single glazed

3'-0" x 3'-0" H

Description	Oper	Unit	Vol	Crew Size	Man-hours per Unit	Crew Output per Day	Avg Mat'l Unit Cost	Avg Labor Unit Cost	Avg Equip Unit Cost	Avg Total Unit Cost	Avg Price Incl O&P
High quality workmanship	Inst	Ea	Lg	CJ	4.44	3.60	530.00	135.00	---	665.00	**811.00**
	Inst	Ea	Sm	CJ	6.96	2.30	571.00	211.00	---	782.00	**974.00**
Good quality workmanship	Inst	Ea	Lg	CJ	3.56	4.50	485.00	108.00	---	593.00	**720.00**
	Inst	Ea	Sm	CJ	5.52	2.90	523.00	168.00	---	691.00	**853.00**
Average quality workmanship	Inst	Ea	Lg	CJ	2.96	5.40	405.00	89.80	---	494.80	**601.00**
	Inst	Ea	Sm	CJ	4.57	3.50	437.00	139.00	---	576.00	**711.00**

3'-0" x 4'-0" H

Description	Oper	Unit	Vol	Crew Size	Man-hours per Unit	Crew Output per Day	Avg Mat'l Unit Cost	Avg Labor Unit Cost	Avg Equip Unit Cost	Avg Total Unit Cost	Avg Price Incl O&P
High quality workmanship	Inst	Ea	Lg	CJ	4.44	3.60	574.00	135.00	---	709.00	**862.00**
	Inst	Ea	Sm	CJ	6.96	2.30	619.00	211.00	---	830.00	**1030.00**
Good quality workmanship	Inst	Ea	Lg	CJ	3.56	4.50	527.00	108.00	---	635.00	**768.00**
	Inst	Ea	Sm	CJ	5.52	2.90	569.00	168.00	---	737.00	**905.00**
Average quality workmanship	Inst	Ea	Lg	CJ	2.96	5.40	441.00	89.80	---	530.80	**642.00**
	Inst	Ea	Sm	CJ	4.57	3.50	475.00	139.00	---	614.00	**755.00**

4'-0" x 3'-0" H

Description	Oper	Unit	Vol	Crew Size	Man-hours per Unit	Crew Output per Day	Avg Mat'l Unit Cost	Avg Labor Unit Cost	Avg Equip Unit Cost	Avg Total Unit Cost	Avg Price Incl O&P
High quality workmanship	Inst	Ea	Lg	CJ	4.44	3.60	599.00	135.00	---	734.00	**892.00**
	Inst	Ea	Sm	CJ	6.96	2.30	647.00	211.00	---	858.00	**1060.00**
Good quality workmanship	Inst	Ea	Lg	CJ	3.56	4.50	550.00	108.00	---	658.00	**794.00**
	Inst	Ea	Sm	CJ	5.52	2.90	593.00	168.00	---	761.00	**934.00**
Average quality workmanship	Inst	Ea	Lg	CJ	2.96	5.40	458.00	89.80	---	547.80	**662.00**
	Inst	Ea	Sm	CJ	4.57	3.50	495.00	139.00	---	634.00	**777.00**

4'-0" x 4'-0" H

Description	Oper	Unit	Vol	Crew Size	Man-hours per Unit	Crew Output per Day	Avg Mat'l Unit Cost	Avg Labor Unit Cost	Avg Equip Unit Cost	Avg Total Unit Cost	Avg Price Incl O&P
High quality workmanship	Inst	Ea	Lg	CJ	5.00	3.20	615.00	152.00	---	767.00	**935.00**
	Inst	Ea	Sm	CJ	7.62	2.10	663.00	231.00	---	894.00	**1110.00**
Good quality workmanship	Inst	Ea	Lg	CJ	4.00	4.00	564.00	121.00	---	685.00	**831.00**
	Inst	Ea	Sm	CJ	6.15	2.60	608.00	187.00	---	795.00	**980.00**
Average quality workmanship	Inst	Ea	Lg	CJ	3.33	4.80	470.00	101.00	---	571.00	**692.00**
	Inst	Ea	Sm	CJ	5.16	3.10	507.00	157.00	---	664.00	**818.00**

4'-0" x 5'-0" H

Description	Oper	Unit	Vol	Crew Size	Man-hours per Unit	Crew Output per Day	Avg Mat'l Unit Cost	Avg Labor Unit Cost	Avg Equip Unit Cost	Avg Total Unit Cost	Avg Price Incl O&P
High quality workmanship	Inst	Ea	Lg	CJ	5.71	2.80	659.00	173.00	---	832.00	**1020.00**
	Inst	Ea	Sm	CJ	8.89	1.80	711.00	270.00	---	981.00	**1220.00**
Good quality workmanship	Inst	Ea	Lg	CJ	4.57	3.50	605.00	139.00	---	744.00	**903.00**
	Inst	Ea	Sm	CJ	6.96	2.30	652.00	211.00	---	863.00	**1070.00**
Average quality workmanship	Inst	Ea	Lg	CJ	3.81	4.20	504.00	116.00	---	620.00	**753.00**
	Inst	Ea	Sm	CJ	5.93	2.70	544.00	180.00	---	724.00	**895.00**

5'-0" x 3'-0" H

Description	Oper	Unit	Vol	Crew Size	Man-hours per Unit	Crew Output per Day	Avg Mat'l Unit Cost	Avg Labor Unit Cost	Avg Equip Unit Cost	Avg Total Unit Cost	Avg Price Incl O&P
High quality workmanship	Inst	Ea	Lg	CJ	5.00	3.20	612.00	152.00	---	764.00	**932.00**
	Inst	Ea	Sm	CJ	7.62	2.10	660.00	231.00	---	891.00	**1110.00**
Good quality workmanship	Inst	Ea	Lg	CJ	4.00	4.00	561.00	121.00	---	682.00	**828.00**
	Inst	Ea	Sm	CJ	6.15	2.60	606.00	187.00	---	793.00	**976.00**
Average quality workmanship	Inst	Ea	Lg	CJ	3.33	4.80	469.00	101.00	---	570.00	**691.00**
	Inst	Ea	Sm	CJ	5.16	3.10	506.00	157.00	---	663.00	**816.00**

Description	Oper	Unit	Vol	Crew Size	Man-hours per Unit	Crew Output per Day	Avg Mat'l Unit Cost	Avg Labor Unit Cost	Avg Equip Unit Cost	Avg Total Unit Cost	Avg Price Incl O&P
5'-0" x 3'-8" H											
High quality workmanship	Inst	Ea	Lg	CJ	5.71	2.80	649.00	173.00	---	822.00	**1010.00**
	Inst	Ea	Sm	CJ	8.89	1.80	700.00	270.00	---	970.00	**1210.00**
Good quality workmanship	Inst	Ea	Lg	CJ	4.57	3.50	596.00	139.00	---	735.00	**893.00**
	Inst	Ea	Sm	CJ	6.96	2.30	643.00	211.00	---	854.00	**1060.00**
Average quality workmanship	Inst	Ea	Lg	CJ	3.81	4.20	497.00	116.00	---	613.00	**745.00**
	Inst	Ea	Sm	CJ	5.93	2.70	536.00	180.00	---	716.00	**886.00**
5'-0" x 4'-0" H											
High quality workmanship	Inst	Ea	Lg	CJ	5.71	2.80	659.00	173.00	---	832.00	**1020.00**
	Inst	Ea	Sm	CJ	8.89	1.80	711.00	270.00	---	981.00	**1220.00**
Good quality workmanship	Inst	Ea	Lg	CJ	4.57	3.50	605.00	139.00	---	744.00	**903.00**
	Inst	Ea	Sm	CJ	6.96	2.30	652.00	211.00	---	863.00	**1070.00**
Average quality workmanship	Inst	Ea	Lg	CJ	3.81	4.20	504.00	116.00	---	620.00	**753.00**
	Inst	Ea	Sm	CJ	5.93	2.70	544.00	180.00	---	724.00	**895.00**
5'-0" x 5'-0" H											
High quality workmanship	Inst	Ea	Lg	CJ	6.67	2.40	718.00	202.00	---	920.00	**1130.00**
	Inst	Ea	Sm	CJ	10.0	1.60	774.00	304.00	---	1078.00	**1350.00**
Good quality workmanship	Inst	Ea	Lg	CJ	5.33	3.00	658.00	162.00	---	820.00	**999.00**
	Inst	Ea	Sm	CJ	8.00	2.00	710.00	243.00	---	953.00	**1180.00**
Average quality workmanship	Inst	Ea	Lg	CJ	4.44	3.60	549.00	135.00	---	684.00	**833.00**
	Inst	Ea	Sm	CJ	6.96	2.30	592.00	211.00	---	803.00	**998.00**
6'-0" x 3'-0" H											
High quality workmanship	Inst	Ea	Lg	CJ	5.71	2.80	654.00	173.00	---	827.00	**1010.00**
	Inst	Ea	Sm	CJ	8.89	1.80	706.00	270.00	---	976.00	**1220.00**
Good quality workmanship	Inst	Ea	Lg	CJ	4.57	3.50	599.00	139.00	---	738.00	**897.00**
	Inst	Ea	Sm	CJ	6.96	2.30	647.00	211.00	---	858.00	**1060.00**
Average quality workmanship	Inst	Ea	Lg	CJ	3.81	4.20	500.00	116.00	---	616.00	**749.00**
	Inst	Ea	Sm	CJ	5.93	2.70	540.00	180.00	---	720.00	**891.00**
6'-0" x 3'-8" H											
High quality workmanship	Inst	Ea	Lg	CJ	6.15	2.60	683.00	187.00	---	870.00	**1070.00**
	Inst	Ea	Sm	CJ	9.41	1.70	737.00	286.00	---	1023.00	**1280.00**
Good quality workmanship	Inst	Ea	Lg	CJ	5.00	3.20	627.00	152.00	---	779.00	**949.00**
	Inst	Ea	Sm	CJ	7.62	2.10	677.00	231.00	---	908.00	**1130.00**
Average quality workmanship	Inst	Ea	Lg	CJ	4.21	3.80	523.00	128.00	---	651.00	**793.00**
	Inst	Ea	Sm	CJ	6.40	2.50	564.00	194.00	---	758.00	**941.00**
6'-0" x 4'-0" H											
High quality workmanship	Inst	Ea	Lg	CJ	6.67	2.40	712.00	202.00	---	914.00	**1120.00**
	Inst	Ea	Sm	CJ	10.0	1.60	769.00	304.00	---	1073.00	**1340.00**
Good quality workmanship	Inst	Ea	Lg	CJ	5.33	3.00	653.00	162.00	---	815.00	**993.00**
	Inst	Ea	Sm	CJ	8.00	2.00	704.00	243.00	---	947.00	**1170.00**
Average quality workmanship	Inst	Ea	Lg	CJ	4.44	3.60	545.00	135.00	---	680.00	**829.00**
	Inst	Ea	Sm	CJ	6.96	2.30	588.00	211.00	---	799.00	**993.00**

Description	Oper	Unit	Vol	Crew Size	Man-hours per Unit	Crew Output per Day	Avg Mat'l Unit Cost	Avg Labor Unit Cost	Avg Equip Unit Cost	Avg Total Unit Cost	Avg Price Incl O&P
Dual glazed											
3'-0" x 3'-0" H											
High quality workmanship	Inst	Ea	Lg	CJ	4.44	3.60	725.00	135.00	---	860.00	**1040.00**
	Inst	Ea	Sm	CJ	6.96	2.30	782.00	211.00	---	993.00	**1220.00**
Good quality workmanship	Inst	Ea	Lg	CJ	3.56	4.50	665.00	108.00	---	773.00	**927.00**
	Inst	Ea	Sm	CJ	5.52	2.90	718.00	168.00	---	886.00	**1080.00**
Average quality workmanship	Inst	Ea	Lg	CJ	2.96	5.40	555.00	89.80	---	644.80	**773.00**
	Inst	Ea	Sm	CJ	4.57	3.50	599.00	139.00	---	738.00	**897.00**
3'-0" x 4'-0" H											
High quality workmanship	Inst	Ea	Lg	CJ	4.44	3.60	799.00	135.00	---	934.00	**1120.00**
	Inst	Ea	Sm	CJ	6.96	2.30	862.00	211.00	---	1073.00	**1310.00**
Good quality workmanship	Inst	Ea	Lg	CJ	3.56	4.50	733.00	108.00	---	841.00	**1000.00**
	Inst	Ea	Sm	CJ	5.52	2.90	790.00	168.00	---	958.00	**1160.00**
Average quality workmanship	Inst	Ea	Lg	CJ	2.96	5.40	611.00	89.80	---	700.80	**837.00**
	Inst	Ea	Sm	CJ	4.57	3.50	659.00	139.00	---	798.00	**966.00**
4'-0" x 3'-0" H											
High quality workmanship	Inst	Ea	Lg	CJ	4.44	3.60	791.00	135.00	---	926.00	**1110.00**
	Inst	Ea	Sm	CJ	6.96	2.30	854.00	211.00	---	1065.00	**1300.00**
Good quality workmanship	Inst	Ea	Lg	CJ	3.56	4.50	726.00	108.00	---	834.00	**997.00**
	Inst	Ea	Sm	CJ	5.52	2.90	784.00	168.00	---	952.00	**1150.00**
Average quality workmanship	Inst	Ea	Lg	CJ	2.96	5.40	606.00	89.80	---	695.80	**831.00**
	Inst	Ea	Sm	CJ	4.57	3.50	653.00	139.00	---	792.00	**960.00**
4'-0" x 4'-0" H											
High quality workmanship	Inst	Ea	Lg	CJ	5.00	3.20	859.00	152.00	---	1011.00	**1210.00**
	Inst	Ea	Sm	CJ	7.62	2.10	926.00	231.00	---	1157.00	**1410.00**
Good quality workmanship	Inst	Ea	Lg	CJ	4.00	4.00	787.00	121.00	---	908.00	**1090.00**
	Inst	Ea	Sm	CJ	6.15	2.60	849.00	187.00	---	1036.00	**1260.00**
Average quality workmanship	Inst	Ea	Lg	CJ	3.33	4.80	657.00	101.00	---	758.00	**907.00**
	Inst	Ea	Sm	CJ	5.16	3.10	708.00	157.00	---	865.00	**1050.00**
4'-0" x 5'-0" H											
High quality workmanship	Inst	Ea	Lg	CJ	5.71	2.80	942.00	173.00	---	1115.00	**1340.00**
	Inst	Ea	Sm	CJ	8.89	1.80	1020.00	270.00	---	1290.00	**1570.00**
Good quality workmanship	Inst	Ea	Lg	CJ	4.57	3.50	865.00	139.00	---	1004.00	**1200.00**
	Inst	Ea	Sm	CJ	6.96	2.30	933.00	211.00	---	1144.00	**1390.00**
Average quality workmanship	Inst	Ea	Lg	CJ	3.81	4.20	721.00	116.00	---	837.00	**1000.00**
	Inst	Ea	Sm	CJ	5.93	2.70	778.00	180.00	---	958.00	**1160.00**
5'-0" x 3'-0" H											
High quality workmanship	Inst	Ea	Lg	CJ	5.00	3.20	851.00	152.00	---	1003.00	**1210.00**
	Inst	Ea	Sm	CJ	7.62	2.10	918.00	231.00	---	1149.00	**1400.00**
Good quality workmanship	Inst	Ea	Lg	CJ	4.00	4.00	780.00	121.00	---	901.00	**1080.00**
	Inst	Ea	Sm	CJ	6.15	2.60	841.00	187.00	---	1028.00	**1250.00**
Average quality workmanship	Inst	Ea	Lg	CJ	3.33	4.80	650.00	101.00	---	751.00	**899.00**
	Inst	Ea	Sm	CJ	5.16	3.10	701.00	157.00	---	858.00	**1040.00**

Description	Oper	Unit	Vol	Crew Size	Man-hours per Unit	Crew Output per Day	Avg Mat'l Unit Cost	Avg Labor Unit Cost	Avg Equip Unit Cost	Avg Total Unit Cost	Avg Price Incl O&P
5'-0" x 3'-8" H											
High quality workmanship	Inst	Ea	Lg	CJ	5.71	2.80	908.00	173.00	---	1081.00	**1300.00**
	Inst	Ea	Sm	CJ	8.89	1.80	980.00	270.00	---	1250.00	**1530.00**
Good quality workmanship	Inst	Ea	Lg	CJ	4.57	3.50	833.00	139.00	---	972.00	**1170.00**
	Inst	Ea	Sm	CJ	6.96	2.30	899.00	211.00	---	1110.00	**1350.00**
Average quality workmanship	Inst	Ea	Lg	CJ	3.81	4.20	695.00	116.00	---	811.00	**972.00**
	Inst	Ea	Sm	CJ	5.93	2.70	749.00	180.00	---	929.00	**1130.00**
5'-0" x 4'-0" H											
High quality workmanship	Inst	Ea	Lg	CJ	5.71	2.80	927.00	173.00	---	1100.00	**1330.00**
	Inst	Ea	Sm	CJ	8.89	1.80	1000.00	270.00	---	1270.00	**1550.00**
Good quality workmanship	Inst	Ea	Lg	CJ	4.57	3.50	851.00	139.00	---	990.00	**1190.00**
	Inst	Ea	Sm	CJ	6.96	2.30	918.00	211.00	---	1129.00	**1370.00**
Average quality workmanship	Inst	Ea	Lg	CJ	3.81	4.20	709.00	116.00	---	825.00	**988.00**
	Inst	Ea	Sm	CJ	5.93	2.70	764.00	180.00	---	944.00	**1150.00**
5'-0" x 5'-0" H											
High quality workmanship	Inst	Ea	Lg	CJ	6.67	2.40	1030.00	202.00	---	1232.00	**1490.00**
	Inst	Ea	Sm	CJ	10.0	1.60	1110.00	304.00	---	1414.00	**1730.00**
Good quality workmanship	Inst	Fa	Lg	CJ	5.33	3.00	946.00	162.00	---	1108.00	**1330.00**
	Inst	Ea	Sm	CJ	8.00	2.00	1020.00	243.00	---	1263.00	**1540.00**
Average quality workmanship	Inst	Ea	Lg	CJ	4.44	3.60	789.00	135.00	---	924.00	**1110.00**
	Inst	Ea	Sm	CJ	6.96	2.30	851.00	211.00	---	1062.00	**1300.00**
6'-0" x 3'-0" H											
High quality workmanship	Inst	Ea	Lg	CJ	5.71	2.80	927.00	173.00	---	1100.00	**1330.00**
	Inst	Ea	Sm	CJ	8.89	1.80	1000.00	270.00	---	1270.00	**1550.00**
Good quality workmanship	Inst	Ea	Lg	CJ	4.57	3.50	851.00	139.00	---	990.00	**1190.00**
	Inst	Ea	Sm	CJ	6.96	2.30	918.00	211.00	---	1129.00	**1370.00**
Average quality workmanship	Inst	Ea	Lg	CJ	3.81	4.20	709.00	116.00	---	825.00	**988.00**
	Inst	Ea	Sm	CJ	5.93	2.70	764.00	180.00	---	944.00	**1150.00**
6'-0" x 3'-8" H											
High quality workmanship	Inst	Ea	Lg	CJ	6.15	2.60	978.00	187.00	---	1165.00	**1400.00**
	Inst	Ea	Sm	CJ	9.41	1.70	1050.00	286.00	---	1336.00	**1640.00**
Good quality workmanship	Inst	Ea	Lg	CJ	5.00	3.20	897.00	152.00	---	1049.00	**1260.00**
	Inst	Ea	Sm	CJ	7.62	2.10	967.00	231.00	---	1198.00	**1460.00**
Average quality workmanship	Inst	Ea	Lg	CJ	4.21	3.80	748.00	128.00	---	876.00	**1050.00**
	Inst	Ea	Sm	CJ	6.40	2.50	807.00	194.00	---	1001.00	**1220.00**
6'-0" x 4'-0" H											
High quality workmanship	Inst	Ea	Lg	CJ	6.67	2.40	1030.00	202.00	---	1232.00	**1490.00**
	Inst	Ea	Sm	CJ	10.0	1.60	1110.00	304.00	---	1414.00	**1730.00**
Good quality workmanship	Inst	Ea	Lg	CJ	5.33	3.00	944.00	162.00	---	1106.00	**1330.00**
	Inst	Ea	Sm	CJ	8.00	2.00	1020.00	243.00	---	1263.00	**1530.00**
Average quality workmanship	Inst	Ea	Lg	CJ	4.44	3.60	786.00	135.00	---	921.00	**1110.00**
	Inst	Ea	Sm	CJ	6.96	2.30	848.00	211.00	---	1059.00	**1290.00**

Description	Oper	Unit	Vol	Crew Size	Man-hours per Unit	Crew Output per Day	Avg Mat'l Unit Cost	Avg Labor Unit Cost	Avg Equip Unit Cost	Avg Total Unit Cost	Avg Price Incl O&P
1 fixed lite, 1 sliding lite											
Single glazed											
3'-0" x 3'-0" H											
High quality workmanship	Inst	Ea	Lg	CJ	4.44	3.60	676.00	135.00	---	811.00	**979.00**
	Inst	Ea	Sm	CJ	6.96	2.30	729.00	211.00	---	940.00	**1160.00**
Good quality workmanship	Inst	Ea	Lg	CJ	3.56	4.50	620.00	108.00	---	728.00	**875.00**
	Inst	Ea	Sm	CJ	5.52	2.90	669.00	168.00	---	837.00	**1020.00**
Average quality workmanship	Inst	Ea	Lg	CJ	2.96	5.40	517.00	89.80	---	606.80	**729.00**
	Inst	Ea	Sm	CJ	4.57	3.50	558.00	139.00	---	697.00	**849.00**
3'-0" x 4'-0" H											
High quality workmanship	Inst	Ea	Lg	CJ	4.44	3.60	743.00	135.00	---	878.00	**1060.00**
	Inst	Ea	Sm	CJ	6.96	2.30	801.00	211.00	---	1012.00	**1240.00**
Good quality workmanship	Inst	Ea	Lg	CJ	3.56	4.50	681.00	108.00	---	789.00	**945.00**
	Inst	Ea	Sm	CJ	5.52	2.90	734.00	168.00	---	902.00	**1100.00**
Average quality workmanship	Inst	Ea	Lg	CJ	2.96	5.40	568.00	89.80	---	657.80	**788.00**
	Inst	Ea	Sm	CJ	4.57	3.50	612.00	139.00	---	751.00	**912.00**
4'-0" x 3'-0" H											
High quality workmanship	Inst	Ea	Lg	CJ	4.44	3.60	726.00	135.00	---	861.00	**1040.00**
	Inst	Ea	Sm	CJ	6.96	2.30	784.00	211.00	---	995.00	**1220.00**
Good quality workmanship	Inst	Ea	Lg	CJ	3.56	4.50	665.00	108.00	---	773.00	**927.00**
	Inst	Ea	Sm	CJ	5.52	2.90	718.00	168.00	---	886.00	**1080.00**
Average quality workmanship	Inst	Ea	Lg	CJ	2.96	5.40	555.00	89.80	---	644.80	**773.00**
	Inst	Ea	Sm	CJ	4.57	3.50	599.00	139.00	---	738.00	**897.00**
4'-0" x 4'-0" H											
High quality workmanship	Inst	Ea	Lg	CJ	5.00	3.20	786.00	152.00	---	938.00	**1130.00**
	Inst	Ea	Sm	CJ	7.62	2.10	848.00	231.00	---	1079.00	**1320.00**
Good quality workmanship	Inst	Ea	Lg	CJ	4.00	4.00	723.00	121.00	---	844.00	**1010.00**
	Inst	Ea	Sm	CJ	6.15	2.60	780.00	187.00	---	967.00	**1180.00**
Average quality workmanship	Inst	Ea	Lg	CJ	3.33	4.80	602.00	101.00	---	703.00	**844.00**
	Inst	Ea	Sm	CJ	5.16	3.10	649.00	157.00	---	806.00	**982.00**
4'-0" x 5'-0" H											
High quality workmanship	Inst	Ea	Lg	CJ	5.71	2.80	865.00	173.00	---	1038.00	**1250.00**
	Inst	Ea	Sm	CJ	8.89	1.80	933.00	270.00	---	1203.00	**1480.00**
Good quality workmanship	Inst	Ea	Lg	CJ	4.57	3.50	794.00	139.00	---	933.00	**1120.00**
	Inst	Ea	Sm	CJ	6.96	2.30	856.00	211.00	---	1067.00	**1300.00**
Average quality workmanship	Inst	Ea	Lg	CJ	3.81	4.20	662.00	116.00	---	778.00	**934.00**
	Inst	Ea	Sm	CJ	5.93	2.70	714.00	180.00	---	894.00	**1090.00**
5'-0" x 3'-0" H											
High quality workmanship	Inst	Ea	Lg	CJ	5.00	3.20	776.00	152.00	---	928.00	**1120.00**
	Inst	Ea	Sm	CJ	7.62	2.10	837.00	231.00	---	1068.00	**1310.00**
Good quality workmanship	Inst	Ea	Lg	CJ	4.00	4.00	711.00	121.00	---	832.00	**1000.00**
	Inst	Ea	Sm	CJ	6.15	2.60	767.00	187.00	---	954.00	**1160.00**
Average quality workmanship	Inst	Ea	Lg	CJ	3.33	4.80	593.00	101.00	---	694.00	**834.00**
	Inst	Ea	Sm	CJ	5.16	3.10	640.00	157.00	---	797.00	**971.00**

Description	Oper	Unit	Vol	Crew Size	Man-hours per Unit	Crew Output per Day	Avg Mat'l Unit Cost	Avg Labor Unit Cost	Avg Equip Unit Cost	Avg Total Unit Cost	Avg Price Incl O&P
5'-0" x 3'-8" H											
High quality workmanship	Inst	Ea	Lg	CJ	5.71	2.80	832.00	173.00	---	1005.00	**1220.00**
	Inst	Ea	Sm	CJ	8.89	1.80	897.00	270.00	---	1167.00	**1440.00**
Good quality workmanship	Inst	Ea	Lg	CJ	4.57	3.50	763.00	139.00	---	902.00	**1090.00**
	Inst	Ea	Sm	CJ	6.96	2.30	823.00	211.00	---	1034.00	**1260.00**
Average quality workmanship	Inst	Ea	Lg	CJ	3.81	4.20	636.00	116.00	---	752.00	**905.00**
	Inst	Ea	Sm	CJ	5.93	2.70	686.00	180.00	---	866.00	**1060.00**
5'-0" x 4'-0" H											
High quality workmanship	Inst	Ea	Lg	CJ	5.71	2.80	841.00	173.00	---	1014.00	**1230.00**
	Inst	Ea	Sm	CJ	8.89	1.80	907.00	270.00	---	1177.00	**1450.00**
Good quality workmanship	Inst	Ea	Lg	CJ	4.57	3.50	771.00	139.00	---	910.00	**1090.00**
	Inst	Ea	Sm	CJ	6.96	2.30	832.00	211.00	---	1043.00	**1270.00**
Average quality workmanship	Inst	Ea	Lg	CJ	3.81	4.20	643.00	116.00	---	759.00	**912.00**
	Inst	Ea	Sm	CJ	5.93	2.70	693.00	180.00	---	873.00	**1070.00**
5'-0" x 5'-0" H											
High quality workmanship	Inst	Ea	Lg	CJ	6.67	2.40	945.00	202.00	---	1147.00	**1390.00**
	Inst	Ea	Sm	CJ	10.0	1.60	1020.00	304.00	---	1324.00	**1630.00**
Good quality workmanship	Inst	Ea	Lg	CJ	5.33	3.00	867.00	162.00	---	1029.00	**1240.00**
	Inst	Ea	Sm	CJ	8.00	2.00	936.00	243.00	---	1179.00	**1440.00**
Average quality workmanship	Inst	Ea	Lg	CJ	4.44	3.60	701.00	135.00	---	836.00	**1010.00**
	Inst	Ea	Sm	CJ	6.96	2.30	756.00	211.00	---	967.00	**1190.00**
6'-0" x 3'-0" H											
High quality workmanship	Inst	Ea	Lg	CJ	5.71	2.80	815.00	173.00	---	988.00	**1200.00**
	Inst	Ea	Sm	CJ	8.89	1.80	880.00	270.00	---	1150.00	**1420.00**
Good quality workmanship	Inst	Ea	Lg	CJ	4.57	3.50	748.00	139.00	---	887.00	**1070.00**
	Inst	Ea	Sm	CJ	6.96	2.30	807.00	211.00	---	1018.00	**1240.00**
Average quality workmanship	Inst	Ea	Lg	CJ	3.81	4.20	624.00	116.00	---	740.00	**891.00**
	Inst	Ea	Sm	CJ	5.93	2.70	673.00	180.00	---	853.00	**1040.00**
6'-0" x 3'-8" H											
High quality workmanship	Inst	Ea	Lg	CJ	6.67	2.40	865.00	202.00	---	1067.00	**1300.00**
	Inst	Ea	Sm	CJ	10.0	1.60	933.00	304.00	---	1237.00	**1530.00**
Good quality workmanship	Inst	Ea	Lg	CJ	5.33	3.00	794.00	162.00	---	956.00	**1160.00**
	Inst	Ea	Sm	CJ	8.00	2.00	856.00	243.00	---	1099.00	**1350.00**
Average quality workmanship	Inst	Ea	Lg	CJ	4.44	3.60	662.00	135.00	---	797.00	**963.00**
	Inst	Ea	Sm	CJ	6.96	2.30	714.00	211.00	---	925.00	**1140.00**
6'-0" x 4'-0" H											
High quality workmanship	Inst	Ea	Lg	CJ	6.67	2.40	886.00	202.00	---	1088.00	**1320.00**
	Inst	Ea	Sm	CJ	10.0	1.60	956.00	304.00	---	1260.00	**1560.00**
Good quality workmanship	Inst	Ea	Lg	CJ	5.33	3.00	813.00	162.00	---	975.00	**1180.00**
	Inst	Ea	Sm	CJ	8.00	2.00	877.00	243.00	---	1120.00	**1370.00**
Average quality workmanship	Inst	Ea	Lg	CJ	4.44	3.60	678.00	135.00	---	813.00	**982.00**
	Inst	Ea	Sm	CJ	6.96	2.30	732.00	211.00	---	943.00	**1160.00**

Wiring. See Electrical, page 151

Index

493

Practical References for Builders

Renovating & Restyling Older Homes

Any builder can turn a run-down old house into a showcase of perfection — if the customer has unlimited funds to spend. Unfortunately, most customers are on a tight budget. They usually want more improvements than they can afford — and they expect you to deliver. This book shows how to add economical improvements that can increase the property value by two, five or even ten times the cost of the remodel. Sound impossible? Here you'll find the secrets of a builder who has been putting these techniques to work on Victorian and Craftsman-style houses for twenty years. You'll see what to repair, what to replace and what to leave, so you can remodel or restyle older homes for the least amount of money and the greatest increase in value. **416 pages, 8½ x 11, $33.50**

CD Estimator

CD Estimator (for *Windows*™) puts at your fingertips over 135,000 construction costs for new construction, remodeling, renovation & insurance repair, home improvement, framing & finish carpentry, electrical, concrete & masonry, painting, and plumbing & HVAC. Monthly cost updates are available at no charge on the Internet. You'll also have the National Estimator program — a stand-alone estimating program for *Windows*™ that Remodeling magazine called a "computer wiz," and *Job Cost Wizard*, a program that lets you export your estimates to QuickBooks Pro for actual job costing. A 60-minute interactive video teaches you how to use this CD-ROM to estimate construction costs. And to top it off, to help you create professional-looking estimates, the disk includes over 40 construction estimating and bidding forms. **CD Estimator is $78.50**

National Framing & Finish Carpentry Estimator

Easy-to-follow instructions and cost-saving tips for estimating and installing everything from the basement stairway to the ridge board, including hundreds of pictures, tables and illustrations. Complete material and labor costs, including manhours, for estimating the rough and finish carpentry work you're likely to face. Includes estimates for floors — girders and beams, foundation plates, joists and plywood; walls — plates, studs, wind bracing, sheathing, gable studs; ceilings — 2 x 4 to 2 x 12 joists; roofs

— rafters and roof sheathing. Even includes nail quantity and demolition labor estimates. For finish carpentry, you'll find costs and manhours for ceiling sheathing, interior trim, cabinets, hardwood flooring and wood stairways. Includes a CD with an electronic version of the book with *National Estimator*, a stand-alone *Windows*™ estimating program, plus an interactive multimedia video that shows how to use the disk to compile construction cost estimates. **408 pages, 8½ x 11, $57.25. Revised annually**

National Renovation & Insurance Repair Estimator

Current prices in dollars and cents for hard-to-find items needed on most insurance, repair, remodeling, and renovation jobs. All price items include labor, material, and equipment breakouts, plus special charts that tell you exactly how these costs are calculated. Includes a CD with an electronic version of the book with *National Estimator*, a stand-alone *Windows*™ estimating program, plus an interactive multimedia video that shows how to use the disk to compile construction cost estimates. **488 pages, 8½ x 11, $59.50. Revised annually**

Construction Forms & Contracts

125 forms you can copy and use — or load into your computer (from the FREE disk enclosed). Then you can customize the forms to fit your company, fill them out, and print. Loads into Word for Windows, Lotus 1-2-3, WordPerfect, Works, or Excel programs. You'll find forms covering accounting, estimating, fieldwork, contracts, and general office. Each form comes with complete instructions on when to use it and how to fill it out. These forms were designed, tested and used by contractors, and will help keep your business organized, profitable and out of legal, accounting and collection troubles. Includes a CD for Windows™ and Mac™. **432 pages, 8½ x 11, $41.75**

Contractor's Plain-English Legal Guide

For today's contractors, legal problems are like snakes in the swamp - you might not see them, but you know they're there. This book tells you where the snakes are hiding and directs you to the safe path. With the directions in this easy-to-read handbook you're less likely to need a $200-an-hour lawyer. Includes simple directions for starting your business, writing contracts that cover just about any eventuality, collecting what's owed you, filing liens, protecting yourself from unethical subcontractors, and more. For about the price of 15 minutes in a lawyer's office, you'll have a guide that will make many of those visits unnecessary. Includes a CD with blank copies of all the forms and contracts in the book. **272 pages, 8½ x 11, $49.50**

Plumber's Handbook Revised

Explains how to install plumbing systems that will pass inspection — the first time. Clearly illustrated, with diagrams, charts and tables that make it easy to select the right material and install it correctly. Covers vents, waste piping, drainage, septic tanks, hot and cold water supply systems, wells, fire protection piping, fixtures, solar energy systems, gas piping and more. Completely updated to comply with the current editions of the *International Plumbing Code (IPC)* and the *Uniform Plumbing Code (UPC)* that are standards for most cities and code jurisdictions. New tables, illustrations and chapters bring this book current with recent amendments to the plumbing codes. **304 pages, 8½ x 11, $36.50**

Basic Lumber Engineering for Builders

Beam and lumber requirements for many jobs aren't always clear, especially with changing building codes and lumber products. Most of the time you rely on your own "rules of thumb" when figuring spans or lumber engineering. This book can help you fill the gap between what you can find in the building code span tables and what you need to pay a certified engineer to do. With its large, clear illustrations and examples, this book shows you how to figure stresses for pre-engineered wood or wood structural members, how to calculate loads, and how to design your own girders, joists and beams. Included FREE with the book — an easy-to-use limited version of NorthBridge Software's Wood Beam Sizing program. **272 pages, 8½ x 11, $38.00**

National Home Improvement Estimator

Current labor and material prices for home improvement projects. Provides manhours for each job, recommended crew size, and the labor cost for the removal and installation work. Material prices are current, with location adjustment factors and free monthly updates on the Web. Gives step-by-step instructions for the work, with helpful diagrams, and home improvement shortcuts and tips from an expert. Includes a CD with an electronic version of the book, and *National Estimator*, a stand-alone *Windows*™ estimating program, plus an interactive multimedia tutorial that shows how to use the disk to compile home improvement cost estimates. **520 pages, 8½ x 11, $58.75. Revised annually**

Contractor's Guide to QuickBooks Pro 2007

This user-friendly manual walks you through QuickBooks Pro's detailed setup procedure and explains step-by-step how to create a first-rate accounting system. You'll learn in days, rather than weeks, how to use *QuickBooks Pro* to get your contracting business organized, with simple, fast accounting procedures. On the CD included with the book you'll find a *QuickBooks Pro* file for a construction company (open it, enter your own company's data, and add info on your suppliers and subs). You also get a complete estimating program, including a database, and a job costing program that lets you export your estimates to *QuickBooks Pro*. It even includes many useful construction forms to use in your business. **344 pages, 8½ x 11, $53.00**

Previous versions also available; Prices listed on order form.

National Construction Estimator

Current building costs for residential, commercial, and industrial construction. Estimated prices for every common building material. Provides man-hours, recommended crew, and gives the labor cost for installation. Includes a CD with an electronic version of the book with National Estimator, a stand-alone Windows™ estimating program, plus an interactive multimedia video that shows how to use the disk to compile construction cost estimates. **592 pages, 8½ x 11, $57.50. Revised annually**

Professional Kitchen Design

Remodeling kitchens requires a "special" touch — one that blends artistic flair with function to create a kitchen with charm and personality as well as one that is easy to work in. Here you'll find how to make the best use of the space available in any kitchen design job, as well as tips and lessons on how to design one-wall, two-wall, L-shaped, U-shaped, peninsula and island kitchens. Also includes what you need to know to run a profitable kitchen design business. **176 pages, 8½ x 11, $24.50**

Roofing Construction & Estimating

Installation, repair and estimating for nearly every type of roof covering available today in residential and commercial structures: asphalt shingles, roll roofing, wood shingles and shakes, clay tile, slate, metal, built-up, and elastomeric. Covers sheathing and underlayment techniques, as well as secrets for installing leakproof valleys. Many estimating tips help you minimize waste, as well as insure a profit on every job. Troubleshooting techniques help you identify the true source of most leaks. Over 300 large, clear illustrations help you find the answer to just about all your roofing questions. **432 pages, 8½ x 11, $38.00**

Profits in Buying & Renovating Homes

Step-by-step instructions for selecting, repairing, improving, and selling highly profitable "fixer-uppers." Shows which price ranges offer the highest profit-to-investment ratios, which neighborhoods offer the best return, practical directions for repairs, and tips on dealing with buyers, sellers, and real estate agents. Shows you how to determine your profit before you buy, what "bargains" to avoid, and how to make simple, profitable, inexpensive upgrades. **304 pages, 8½ x 11, $24.75**

Craftsman's Construction Installation Encyclopedia

Step-by-step installation instructions for just about any residential construction, remodeling or repair task, arranged alphabetically, from *Acoustic tile* to *Wood flooring*. Includes hundreds of illustrations that show how to build, install, or remodel each part of the job, as well as manhour tables for each work item so you can estimate and bid with confidence. Also includes a CD with all the material in the book, handy look-up features, and the ability to capture and print out for your crew the instructions and diagrams for any job. **792 pages, 8½ x 11, $65.00**

Builder's Guide to Room Additions

How to tackle problems that are unique to additions, such as requirements for basement conversions, reinforcing ceiling joists for second-story conversions, handling problems in attic conversions, what's required for footings, foundations, and slabs, how to design the best bathroom for the space, and much more. Besides actual construction, you'll even find help in designing, planning, and estimating your room addition jobs. **352 pages, 8½ x 11, $34.95**

2006 International Residential Code

Replacing the *CABO One-* and *Two-Family Dwelling Code*, this book has the latest technological advances in building design and construction. Among the changes are provisions for steel framing and energy savings. Also contains mechanical, fuel gas and plumbing provisions that coordinate with the *International Mechanical Code* and *International Plumbing Code*. **604 pages, 8½ x 11, $76.50**
Also available: **2003 International Residential Code, $72.50**
2000 International Residential Code, $59.00
2000 International Residential Code on interactive CD, $48.00

Home Remodeler's Costbook

Over 5,000 remodeling costs in CSI format, with a separate section on square-foot costs to help with quick estimates based on project type and size. Includes detailed descriptions of the work estimated, unit of measurement — whether square foot, lineal foot, or each. This book was developed in conjunction with Home Builder Press (a division of NAHB). **380 pages, 8½ x 11, 75.95**

Download all of Craftsman's most popular costbooks for one low price with the Craftsman Site License. http://www.craftsmansitelicense.com